Spon's Online Set up and

Use of *Spon's Online* is subject to the single user licence at the back of this book.

Electronic access to your price book is now provided as an ebook on the VitalSource® Bookshelf platform. You can access it online or offline on your PC/Mac, smartphone or tablet. You can browse and search the content across all the books you've got, make notes and highlights and share these notes with other users.

Setting up

1. Create a VitalSource Bookshelf account at https://online.vitalsource.com/user/new (or log into your existing account if you already have one).

2. Retrieve the code by scratching off the security-protected label inside the front cover of this book. The code in the scratch-off panel can only be used once. Log in to Bookshelf and click the **Account** menu at the top right of the screen. Select **Redeem** and enter the code in the **Code To Redeem** box. Press **Redeem**. Once the code has been redeemed your Spon's price book will download and appear in your **library**.

You will now be able to access the ebook online or offline on your smartphone, tablet or PC/Mac. Your notes and highlights will automatically stay in sync no matter where you make them.

Use ONLINE

1. Log in to your Bookshelf account at https://online.vitalsource.com).
2. Double-click on the title in your **library** to open the ebook.

Use OFFLINE

Download BookShelf to your PC, Mac, iOS device, Android device or Kindle Fire, and log in to your Bookshelf account to access your ebook, as follows:

On your PC
Go to http:/ … and follow the instructions to download the free VitalSource Bookshelf app to your PC or Mac. Double-click the VitalSource Bookshelf icon that appears on your desktop and log into your Bookshelf account. Select **All Titles** from the menu on the left - you should see your price book on your Bookshelf. Double-click the price book to open it.

On your iPhone/iPod Touch/iPad
Download the free VitalSource Bookshelf App available via the iTunes App Store. Open the Bookshelf app and log into your Bookshelf account. Select **All Titles** - you should see your price book on your Bookshelf. Select the price book to open it. You can find more information at http://support.vitalsource.com /kb/bookshelf-touch/2004

On your Android™ smartphone or tablet
Download the free VitalSource Bookshelf App available via Google Play. Open the Bookshelf app and log into your Bookshelf account. You should see your price book on your Bookshelf. Select the price book to open it. You can find more information at http://support.vitalsource.com/kb/android/welcome

On your Kindle Fire
Go to http://support.vitalsource.com/kb/Kindle-Fire/app-installation-guide and follow the instructions to download the free VitalSource Bookshelf App. Open the Bookshelf app and log into your Bookshelf account. Select All Titles - you should see your price book on your Bookshelf. Select the price book to open it. You can find more information at http://support.vitalsource.com/kb/kindle-fire/welcome

Support

If you have any questions about downloading Bookshelf, creating your account, or accessing and using your ebook edition, please visit http://support.vitalsource.com/

For questions or comments on content, please contact us on sponsonline@tandf.co.uk

---

Free Updates

with three easy steps…

1. Register today on www.pricebooks.co.uk/updates

2. We'll alert you by email when new updates are posted on our website

3. Then go to www.pricebooks.co.uk/updates
   and download the update.

All four Spon Price Books – *Architects' and Builders'*, *Civil Engineering and Highway Works*, *External Works and Landscape* and *Mechanical and Electrical Services* – are supported by an updating service. Two or three updates are loaded on our website during the year, typically in November, February and May. Each gives details of changes in prices of materials, wage rates and other significant items, with regional price level adjustments for Northern Ireland, Scotland and Wales and regions of England. The updates terminate with the publication of the next annual edition.

As a purchaser of a Spon Price Book you are entitled to this updating service for this 2014 edition – free of charge. Simply register via the website www.pricebooks.co.uk/updates and we will send you an email when each update becomes available.

If you haven't got internet access or if you've some questions about the updates please write to us at Spon Price Book Updates, Spon Press Marketing Department, 3 Park Square, Milton Park, Abingdon, Oxfordshire, OX14 4RN.

Find out more about Spon books
Visit www.pricebooks.co.uk for more details.

Spon's Mechanical and Electrical Services Price Book

2014

Spon's Mechanical and Electrical Services Price Book

Edited by

Davis Langdon, An AECOM Company

2014

Forty-fifth edition

CRC Press
Taylor & Francis Group

First edition 1968
Forty-fifth edition published 2014
by CRC Press
2 Park Square, Milton Park, Abingdon, Oxon, OX14 4RN

and by CRC Press
Taylor & Francis, Broken Sound Parkway, NW, Suite 300, Boca Raton, FL 33487

CRC Press is an imprint of the Taylor & Francis Group, an informa business

British Library Cataloguing in Publication Data
A catalogue record for this book is available from the British Library

ISBN13: 978-1-4822-0414-8
Ebook: 978-1-4822-0415-5

ISSN: 0305-4543

Typeset in Arial by Taylor & Francis Books

MIX
Paper from responsible sources
FSC www.fsc.org FSC® C013056

Printed and bound in Great Britain by
TJ International Ltd, Padstow, Cornwall

Contents

Preface vii
Special Acknowledgements ix
Acknowledgements xi

PART 1: ENGINEERING FEATURES

Building Information Modelling (BIM) 3
Building Management Systems (BMS) 7
Renewable Energy Options 11
Grey Water Recycling and Rainwater Harvesting 15
Ground Water Cooling 19
Fuel Cells 25
Biomass Energy 29
Capital Allowances 33
Enhanced Capital Allowances (ECAs), The Energy Technology List (ETL) and The Water Technology List (WTL) 39
LED Lighting 43
Getting the Connection 47
Feed-In Tariffs (FITs) 51
Carbon Trading 55
Renewable Obligation Certificates (ROCs) 59
Typical Engineering Details 63

PART 2: APPROXIMATE ESTIMATING

Directions 89
Cost Indices 90
RIBA Stage A/B Feasibility Costs 93
RIBA Stage C Elemental Rates 98
All-in Rates 106
Building Models – Elemental Cost Summaries 128

PART 3: MATERIAL COSTS/MEASURED WORK PRICES

Mechanical Installations
Directions 157
R: Disposal Systems 163
R10: Rainwater Pipework/Gutters 163
R11: Above Ground Drainage 188
S: Piped Supply Systems 218
S10: Cold Water 218
S11: Hot Water 281
S32: Natural Gas 286
S41: Fuel Oil Storage/Distribution 291
S60: Fire Hose Reels 293
S61: Dry Risers 294
S63: Sprinklers 295
S65: Fire Hydrants 302
T: Mechanical/Cooling/Heating Systems 304
T10: Gas/Oil Fired Boilers 304
T13: Packaged Steam Generators 321
T31: Low Temperature Hot Water Heating 322
T33: Steam Heating 414
T42: Local Heating Units 418
T60: Central Refrigeration Plant 419
T61: Chilled Water 428
T70: Local Cooling Units 435
U: Ventilation/Air Conditioning Systems 437
U10: General Ventilation 437
U14: Ductwork: Fire-Rated 533
U30: Low Velocity Air Conditioning 554
U31: VAV Air Conditioning 557

U41: Fan Coil Air Conditioning 558
U70: Air Curtains 563
Electrical Installations
Directions 567
V: Electrical Supply/Power/Lighting 571
V10: Electrical Generation Plant 571
V11: MV Supply 574
V20: LV Distribution 579
V21: General Lighting 657
V22: General LV Power 667
V32: Uninterruptible Power Supply 676
V40: Emergency Lighting 678
W: Communications/Security/Control 683
W10: Telecommunications 683
W20: Radio/Television 686
W23: Clocks 688
W30: Data Transmission 690
W40: Access Control 697
W41: Security Detection and Alarm 698
W50: Fire Detection and Alarm 700
W51: Earthing and Bonding 703
W52: Lightning Protection 704
W60: Central Control/Building Management 709

PART 4: RATES OF WAGES

Mechanical Installations 715
Electrical Installations 723

PART 5: DAYWORK

Heating and Ventilating Industry 731
Electrical Industry 734
Building Industry Plant Hire Costs 737

PART 6: TABLES AND MEMORANDA **757**

INDEX **851**

Preface

The forty-fifth edition of *SPON'S Mechanical and Electrical Services Price Book* continues to cover the widest range and depth of services, reflecting the many alternative systems and products that are commonly used in the industry as well as current industry trends.

Prices in the first quarter of 2013 saw a small uptick as contractors have been faced with some material price increases that they find impossible to fully absorb. There have been two quarters of increase in construction activity, whether this is the start of a sustained upturn is yet to be determined and there is as yet no indication of this growth influencing tender prices.

Key influences

- Some indications of increased construction activity
- Construction inflation measures are marginal (as at 1Q 2013)
- Consumer price inflation higher than construction price inflation and forecast fall being stubbornly resisted
- Lower metal costs owing to a slowing demand from China

2013 Wage increase with both Mechanical and Electrical trades have been agreed but lower than inflation in the period since the last agreements.

Before referring to prices or other information in the book, readers are advised to study the 'Directions' which precede each section of the Materials Costs/Measured Work Prices. As before, no allowance has been made in any of the sections for Value Added Tax.

The order of the book reflects the order of the estimating process, from broad outline costs through to detailed unit rate items.

The approximate estimating section has been thoroughly reviewed to provide up to date key data in terms of square metre rates, all-in rates for key elements and selected specialist activities and elemental analyses on a comprehensive range of building types.

The prime purpose of the Materials Costs/Measured Work Prices part is to provide industry average prices for mechanical and electrical services, giving a reasonably accurate indication of their likely cost. Supplementary information is included which will enable readers to make adjustments to suit their own requirements. It cannot be emphasized too strongly that it is not intended that these prices should be used in the preparation of an actual tender without adjustment for the circumstances of the particular project in terms of productivity, locality, project size and current market conditions. Adjustments should be made to standard rates for time, location, local conditions, site constraints and any other factor likely to affect the costs of a specific scheme. Readers are referred to the build up of the gang rates, where allowances are included for supervision, labour related insurances, and where the percentage allowances for overhead, profit and preliminaries are defined.

Readers are reminded of the service available on the Spon's website detailing significant changes to the published information. www.pricebooks.co.uk/updates.

As with previous editions the Editors invite the views of readers, critical or otherwise, which might usefully be considered when preparing future editions of this work.

Whilst every effort is made to ensure the accuracy of the information given in this publication, neither the Editors nor Publishers in any way accept liability for loss of any kind resulting from the use made by any person of such information.

In conclusion, the Editors record their appreciation of the indispensable assistance received from the many individuals and organizations in compiling this book.

DAVIS LANGDON, An AECOM Company
MidCity Place
71 High Holborn
London WC1V 6QS

Telephone: 020 7061 7000
Facsimile: 020 7061 7061

Special Acknowledgements

The Editors wish to record their appreciation of the special assistance given by the following organizations in the compilation of this edition.

redstone

160 Centennial Business Park
Centennial Avenue
Elstree
Hertfordshire
WD6 3SH
Tel: 0845 201 0000
Email: salesenquiries@redstone.com
www.redstone.com

HOTCHKISS

Hampden Park Industrial Estate
Eastbourne
East Sussex
BN22 9AX
Tel: 01323 501234
Fax: 01323 508752
Email: info@Hotchkiss.co.uk
www.Hotchkiss.co.uk

Abbey

Abbey Thermal Insulation Ltd
23–24 Riverside House
Lower Southend Road
Wickford
Essex
SS11 8BB
Tel: 01268 572116
Fax: 01268 572117
Email: general@abbeythermal.com

MJL

Michael J Lonsdale
22–24 Cowper Street
London
EC2A 4AP
Tel: 0845 241 6622
Fax: 0845 241 633

T.Clarke
BUILDING SERVICES GROUP

Stanhope House
116–118 Walworth Road
London
SE17 1JY
Tel: 020 7358 5000
Email: info@tclarke.co.uk
www.tclarke.co.uk

Gardner&Co

Gardner & Co Ltd
1–5 Bermondsey Street
London
SE1 2ER
Tel: 020 7403 5894
Fax: 020 7378 0316
www.gardnerco.net

Acknowledgements

The editors wish to record their appreciation of the assistance given by many individuals and organizations in the compilation of this edition.

Manufacturers, Distributors and Subcontractors who have contributed this year include: -

A C Plastics Industries Ltd
Armstrong Road
Daneshill East
Basingstoke RG24 8NU
Tel: 01256 329334
Fax: 01256 817862
www.acplastiques.com
GRP Water Storage Tanks

Actionair
Joseph Wilson Ind. Estate
South Street
Whitstable
Kent CT5 3DU
Tel: 01227 276100
Fax: 01227 264262
Email: sales@actionair.co.uk
www.actionair.co.uk
Dampers

Alfa Laval Limited
Unit 1, 6 Wellheads Road
Farburn Industrial Estate
Dyce
Aberdeen AB21 7HG
Tel: 01224 424300
Fax: 01224 725213
www.alfalaval.com
Heat Exchangers

Aquilar Limited
Dial Post Court
Horsham Road
Rusper
West Sussex RH12 4QX
Tel: 08707 940310
Fax: 08707 940320
www.aquilar.co.uk
Leak Detection

Balmoral Tanks
Balmoral Park
Loirston
Aberdeen AB12 3GY
Tel: 01224 859000
Fax: 01224 859123
www.balmoral-group.com
GRP Water Storage Tanks

Biddle Air Systems Ltd
St. Mary's Road, Nuneaton
Warwickshire CV11 5AU
Tel: 02476 384233
Fax: 02476 373621
Email: info@biddle-air.co.uk
Air Curtains

Braithwaite Engineers Ltd
Neptune Works
Uskway
Newport
South Wales NP9 2UY
Tel: 01633 262141
Fax: 01633 250631
www.braithwaite.co.uk
Sectional Steel Water Storage Tanks

Broadcrown Limited
Alliance Works
Airfield Industrial Estate
Hixon
Staffs ST18 0PF
Tel: 01889 272200
Fax: 01889 272220
www.broadcrown.co.uk
Generators

Caradon Stelrad Ideal Boilers
PO Box 103
National Avenue
Kingston-upon-Hall
North Humberside HU5 4JN
Tel: 08708 400030
Fax: 08708 400059
www.rycroft.com
Boilers/Heating Products

Carrier Air Conditioning
United Technologies House
Guildford Road
Leatherhead
Surrey KT22 9UT
Tel: 0870 6001100
Fax: 01372 220221
www.carrier.uk.com
Chilled Water Plant

Chloride Power Protection
Unit C, George Curl Way
Southampton SO18 2RY
Tel: 023 8061 0311
Fax: 023 8061 0852
www.chloridepower.com
Static UPS Systems

Cooper Lighting and Security
London Project Office
Suite 8, King Harold Court
Sun Street
Waltham Abbey
Essex EN9 1ER
Tel: 01302 303303
Fax: 01392 367155
www.cooper-ls.com
Emergency Lighting and Luminaires

Danfoss Flowmetering Ltd
Magflo House
Ebley Road
Stonehouse
Glos GL10 2LU
Tel: 01453 828891
Fax: 01453 853860
www.danfoss-randall.co.uk
Energy Meters

Dewey Waters Limited
Cox's Green
Wrington
Bristol BS40 5QS
Tel: 01934 862601
Fax: 01934 862604
www.deweywaters.co.uk
Tanks

Dunham-Bush Limited
8 Downley Road
Havant
Hampshire PO9 2JD
Tel: 02392 477700
Fax: 02392 450396
www.dunham-bush.com
Convectors and Heaters

EMS Radio Fire & Security Systems Limited
Technology House
Sea Street
Herne Bay
Kent CT6 8JZ
Tel: 01227 369570
Fax: 01227 369679
www.emsgroup.co.uk
Security

Engineering Appliances Ltd
Unit 11
Sunbury Cross Ind Est
Brooklands Close
Sunbury On Thames TW16 7DX
Tel: 01932 788888
Fax: 01932 761263
Email: info@engineering-appliances.co.uk
www.engineeringappliances.com
Expansion Joints, Air and Dirt Separators

FKI Hawker Siddeley
Falcon Works
P O Box 7713
Meadow Lane
Loughborough
Leicestershire LE11 1ZF
Tel: 01495 331024
Fax: 01495 331019
www.fkiswitchgear.com
MV Supply, Cables and MV Switchgear and Transformers

Flakt Woods Limited
Axial Way
Colchester CO4 5ZD
Tel: 01206 222555
Fax: 01206 222777
Fans

Hall Fire Protection Limited
186 Moorside Road
Swinton
Manchester M27 9HA
Tel: 0161 793 4822
Fax: 0161 794 4950
www.hallfire.co.uk
Fire Protection Equipment

Halton
5 Waterside Business Park
Witham
Essex CM8 3YQ
Tel: 01376 503040
Fax: 01376 503060
www.haltongroup.com
Chilled Beams

Hattersley, Newman, Hender Ltd
Burscough Road
Ormskirk
Lancashire L39 2XG
Tel: 01695 577199
Fax: 01695 578775
Email: uksales@hattersley-valves.co.uk
www.hattersley.com
Valves

Hitec Power Protection Limited
Unit B21a
Holly Farm Business Park
Honiley
Kenilworth
Warwickshire CV8 1NP
Tel: 01926 484535
Fax: 01926 484336
www.hitecups.co.uk
Uninterruptible Power Supply (Rotary/Diesel)

Honeywell CS Limited
Honeywell House
Anchor Boulevard
Crossways Business Park
Dartford
Kent DA2 6QH
Tel: 01322 484800
Fax: 01322 484898
www.honeywell.com
Control Components

Hoval Limited
Northgate
Newark
Notts NG24 1JN
Tel: 01636 672711
Fax: 01636 673532
www.hoval.co.uk
Boilers

HRS Hevac Ltd
10–12 Caxton Way
Watford Business Park
Watford
Herts WD18 8JY
Tel: 01923 232335
Fax: 01923 230266
www.hrshevac.co.uk
Heat Exchangers

Hudevad
Bridge House
Bridge Street
Walton on Thames
Tel: 01932 247835
Fax: 01932 247694
www.hudevad.co.uk
Radiators

Hydrotec (UK) Limited
Hydrotec House
5 Mannor Courtyard
Hughenden Avenue
High Wycombe HP13 5RE
Tel: 01494 796040
Fax: 01494 796049
www.hydrotec.com
Chemical Treatment

IAC
IEC House
Moorside Road
Winchester
Hampshire SO23 7US
Tel: 01962 873000
Fax: 01962 873102
www.industrialacoustics.com
Attenuators

IC Service & Maintenance Ltd
Unit K3 Temple Court
Knights Place
Knight Road
Strood
Kent ME2 2LT
Tel: 01634 290300
Fax: 01634 290700
www.icservice.biz
Fire Detection & Alarm

Ideal Boilers
P O Box 103
National Avenue
Kingston Upon Hull
East Yorkshire HU5 4JN
Tel: 01482 492251
Fax: 01482 448858
www.idealboilers.com
Boilers

Kampmann
Benson Environmental Limited
47 Central Avenue
West Molesey
Surrey KT8 2QZ
Tel: 020 8783 0033
Fax: 020 8783 0140
www.diffusionenv.com
Trench Heating

Kiddie Fire Protection Services
Enterprise House
Jasmine Grove
London SE20 8JW
Tel: (020) 8659 7235
Fax: (020) 8659 7237
www.kfp.co.uk
Fire Protection Equipment

Metc.raft Ltd
Harwood Industrial Estate
Littlehampton
West Sussex BN17 7BB
Tel: 01903 714226
Fax: 01903 723206
www.metcraft.co.uk
Oil Storage Tanks

Mitsubishi Electric Europe BV
Unit 8, Electra Park
Bidder Street
Canning Town
London E16 4ES
Tel (switchboard): 020 7511 5664
www.mitsubishi-lifts.co.uk
Lifts and Escalators

Osma Underfloor Heating
18 Apple Lane
Sowton Trade City
Exeter
Devon EX2 5GL
Tel: 01392 444122
Fax: 01392 444135
www.osmaufh.co.uk
Underfloor Heating

Pullen Pumps Limited
158 Beddington Lane
Croydon CR9 4PT
Tel: 020 8684 9521
Fax: 020 8689 8892
www.pullenpumps.co.uk
Pumps, Booster Sets

Reliance Hi-tech
Boundary House
Cricketfield Road
Uxbridge
Middlesex UB8 1QG
Tel: 01895 205000
Fax: 01895 205100
www.reliancesecurity.co.uk
Access Control and Security Detection and Alarm

Rycroft
Duncombe Road
Bradford BD8 9TB
Tel: 01274 490911
Fax: 01274 498580
www.rycroft.com
Storage Cylinders

Saint-Gobain PAM UK Ltd
Lows Lane
Stanton-By-Dale
Ilkeston
Derbyshire DE7 4RU
Tel: 0115930 5000
Fax: 0115932 9513
www.saint-gobain-pam.co.uk
Cast Iron Pipework

Senior Hargreaves
Lord Street
Bury
Lancashire
BL9 0RG
Tel: 0161 764 5082
Fax: 0161 762 2333
Email: postbox@senior-hargreaves.co.uk

Schneider Electric Limited
120 New Cavendish Street
London W1W 6XX
Tel: 0870 608 8608
www.schneider-electric.com

SF Limited
Pottington Business Park
Barnstaple
Devon EX31 1LZ
Tel: 01271 326633
Fax: 01271 334303
Flues

Simmtronic Limited
Waterside
Charlton Mead Lane
Hoddesdon
Hertfordshire EN11 0QR
Tel: 01992 456869
Fax: 01992 445132
www.simmtronic.com
Lighting Controls

Socomec Limited
Knowl Piece
Wilbury Way
Hitchin
Hertfordshire SG4 0TY
Tel: 01462 440033
Fax: 01462 431143
www.socomec.com
Automatic Transfer Switches

Spirax-Sarco Ltd
Charlton House
Cheltenham
Gloucestershire GL53 8ER
Tel: 01242 521361
Fax: 01242 573342
www.spiraxsarco.com
Traps and Valves

Tyco Limited
Unit 6 West Point Enterprize Park
Clarence Avenue
Trafford Park
Manchester M17 1QS
Tel: 0161 875 0400
Fax: 0161 875 0491
www.tyco.com
Fire Protection

Utile Engineering Company Ltd
Irthlingborough
Northants NN9 5UG
Tel: 01933 650216
Fax: 01933 652738
www.utileengineering.com
Gas Boosters

Woods of Colchester
Tufnell Way
Colchester
Essex CO4 5AR
Tel: 01206 544122
Fax: 01206 574434
Air Distribution, Fans, Anti-Vibration Mountings

At the heart of AECOM, we join the dots that create, enhance and sustain built, natural and social environments.

PART 1

Engineering Features

This section on Engineering Features, deals with current issues and/or technical advancements within the industry. These shall be complimented by cost models and/or itemized prices for items that form part of such.

The intention is that the book shall develop to provide more than just a schedule of prices to assist the user in the preparation and evaluation of costs.

	page
Building Information Modelling (BIM)	3
Building Management Systems (BMS)	7
Renewable Energy Options	11
Grey Water Recycling and Rainwater Harvesting	15
Ground Water Cooling	19
Fuel Cells	25
Biomass Energy	29
Capital Allowances	33
Enhanced Capital Allowances (ECAs), The Energy Technology List (ETL) and The Water Technology List (WTL)	39
LED Lighting	43
Getting the Connection	47
Feed-In Tariffs (FITs)	51
Carbon Trading	55
Renewable Obligation Certificates (ROCs)	59
Typical Engineering Details	63

Building Information Modelling (BIM)

With the UK Government mandating the use of Building Information Modelling (BIM) on all its construction projects by 2016, BIM is currently the hot topic of the construction industry. Some say it will revolutionize building design and construction where many processes will become highly-integrated, efficient and effective. Others say they have 'seen it all before' and argue that BIM is not much more than a new set of design tools that construction must adapt to and work with. However the majority agree that BIM is introducing new ways of working for design and construction teams bringing considerable benefits for all involved in construction – and most of all for clients.

So what is BIM?

BIM exploits the potential in computer-based modelling technologies to provide a new way of designing buildings and managing design and construction processes. BIM goes well beyond the effective application of 3D CAD. It provides a major step-change in the ability of design and construction teams to structure and exchange information around shared, computer-based models of a building project. This of course can bring great benefits, including better design coordination, reductions in design costs and improved communications throughout the design and construction process. But the real benefits of BIM go further. By working together to develop sophisticated, information-rich and coordinated computer models, the different design and construction disciplines can 'prototype' projects before they are built. Designs can be developed and tested 'virtually' so that the performance and cost is optimized. They can be coordinated so that many potential problems are either designed-out or avoided altogether. Ultimately, the benefits not only lie in more efficient and effective design and construction processes, but better and more certain project outcomes – better buildings that are more fit-for-purpose and meet their brief requirements and design intent.

How does BIM work?

BIM is based on digital models of a building that store information about the project (relating to architecture, engineering, construction and so on) in a way that enables it to be shared across and between different design and construction disciplines. Not only that, 3D dynamic modelling software can be used to develop and manipulate these digital models to refine the design, and also to test and validate its potential performance across a range of criteria, including buildability, energy performance-in-use, whole life costing etc. The potential for all key project information to be stored and manipulated on a computer is what sets BIM apart from more conventional approaches, and BIM-based design solutions differ from their traditional counterparts in that they:

- Are created and developed on digital databases which enable collaboration and effective data exchange between different disciplines;
- Allow change to be managed through these databases, so that changes in one part of the database are reflected in (and coordinated through) changes in other parts; and
- Capture and preserve information for reuse by all members of the design and construction team, including facilities management (FM), and user operation and management.

Conventionally, a good deal of design and construction work is document based. Information is communicated and stored via a variety of drawings and reports that, despite being stored and distributed in digital form, are essentially 'unstructured' and thus of limited use. Not only is this information unstructured, it is also held in a variety of forms and locations that are not formally coordinated (information on individual building components, for example, are contained on drawings, specifications, bills of quantity descriptions, etc.). Such an approach has considerable potential for data conflicts and redundancy as well as risks to data integrity and security. Conversely, by providing an intelligent, digital structure for project information – and ultimately a means by which the information can all be held centrally as a 'single' model – BIM opens up a wide range of possibilities for improvement. These include better ways of generating, exchanging, storing and reusing project information that greatly improve communications between different design and construction disciplines through the life of the asset.

While the idea of a single, central project model is something of a 'holy grail' for BIM, in practice digital information is held in a number of project models that are used for different purposes – architectural 'rendering', structural analysis, costing, and so on. But the key feature of BIM working is that these models are co-ordinated and work from shared databases. Thus BIM is as much a process of generating, sharing and managing project information throughout the lifecycle of the asset, as it is the digital model itself. That is why BIM is sometimes referred to as BIM (M) – Building Information Modelling and Management.

Of course, digitally structuring information can take many forms, from simple temporary models created for a specific purpose (i.e. a room schedule with key room dimensions in spreadsheet form) through to shared 3D whole-building models containing architectural, structural, servicing and other data all in the same place ('Full BIM').

And the greater the degree of information-sharing and collaboration in the development of the models, the more accurate and complete the models will be and the greater are the benefits of using BIM. As shared models are developed, problems – such as clashes between structure and services – tend to get ironed out, and designs become more consistent and coordinated. As all project information comes to be based on a single model, there is considerably less scope for misinterpretation and consequential change, disruption and re-work. And all of this helps greatly to improve the predictability and certainty of project outcomes.

BIM benefits

Generally, BIM is recognized as providing a wide range of valuable benefits including:

- **Design:** improved coordination of design and deliverables between disciplines; improved project understanding through visualization; improved design management and control, including change control; and improved understanding of design changes and implications through parametric modelling.
- **Compliance:** ability to perform simulation and analysis for regulatory compliance; and ability to simulate and optimize energy and wider sustainability performance.
- **Costing/economics:** ability to perform cost analysis as the design develops, and to check for adherence to budget/cost targets; ability to understand cost impacts of design changes; and improved accuracy of cost estimates.
- **Construction:** reduction of construction risks through identification of constructability issues early in the design process; early detection and avoidance of clashes; ability to model impact of design changes on schedule and programme; and ability to integrate contractor/subcontractor design input directly to the model.
- **Operation and management:** creation of an FM database directly from the project (as built) model; ability to perform FM costing and procurement from the model; and ability to update the model with real-time information on actual performance through the life of the building.

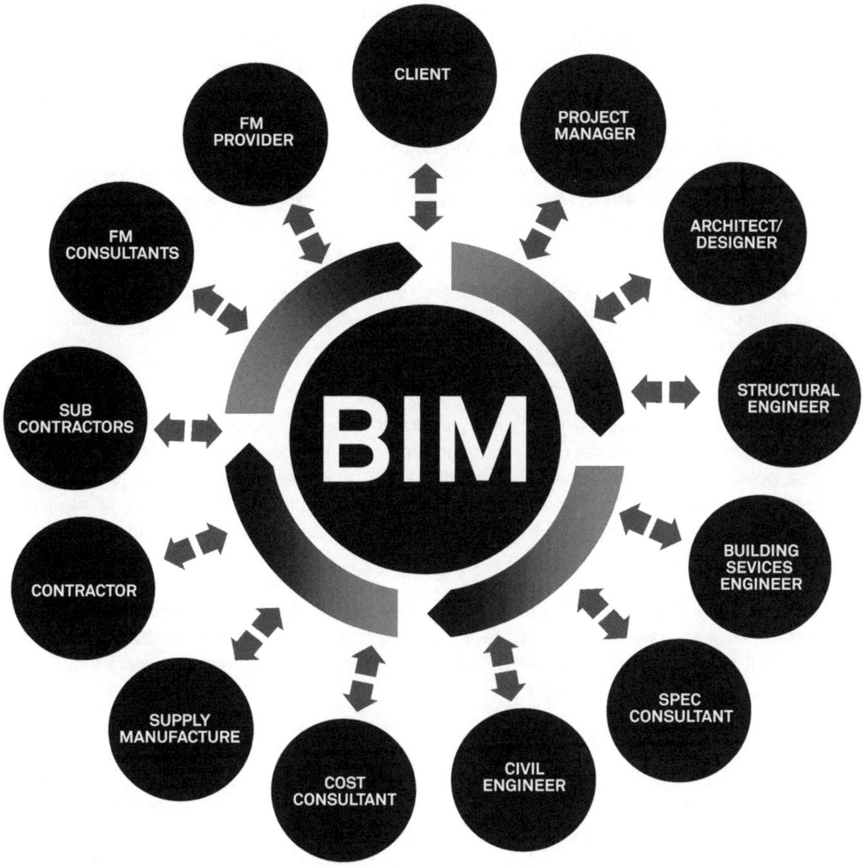

Figure 1: Information sharing (BIM)
Source: BuildingSMART

Ongoing BIM issues and risks

Primarily because BIM is a relatively new approach that has not yet achieved widespread adoption in the construction sector, there are a number of uncertainties and potential risks in its adoption, including:

- **Legal, contractual and insurance issues:** Fundamentally, there is a lack of precedent to provide clear guidance on issues such as ownership of intellectual property rights (IPR) in project models and information. Standard forms of building contract – including design and consultant appointments – do not currently make specific provision for BIM. The implications for design liability and associated insurance arrangements of the greater integration of design, construction and operation information envisaged under BIM have not yet been fully worked through.
- **Current practice:** Related to the issues of liability and insurance is the question of the standards of professional practice (and associated duties of care) that are expected across the sector. New modelling capabilities are blurring the boundaries between who does what in design and construction teams; 'best practice' in BIM is not yet clearly defined and different disciplines are developing their own methods and standards; and training requirements are not yet clear.

- **IT and software:** While a critical requirement of BIM is interoperability of systems and data, not all software and associated IT platforms currently used for BIM deploy IFC–compliant information standards, for example, and manufacturers have not yet agreed on a single standard. The variety of systems in use can impose heavy training burdens on firms who need to operate with some or all of them.
- **Awareness and expectations:** Paradoxically, BIM suffers from relatively low levels of client and sector awareness combined with very high expectations typical of new information technology developments. Because the ultimate potential of BIM is so considerable there is a tendency for some proponents to believe that all the possible benefits are available today, and to over-promise on what can currently be achieved.

A balanced view

The benefits outweigh the risks to a very significant degree and the sheer transformative affect of BIM should not be underestimated. More efficient and effective design processes; greatly improved information quality and co-ordination between design and construction; and better prototyping prior to construction are all made possible through BIM.

Potentially these bring enormous benefits, not only by improving the efficiency of design and construction, but by improving its effectiveness also, ultimately providing greater certainty of project outcomes and better buildings. More than that, the potential for project models to support the management of facilities in the post-construction phase is considerable and could lead to more effective operation of buildings through the whole life cycle. One of these benefits is to provide new learning about what really works in design and construction which can be fed back to inform and improve these processes.

Conversely, the downside risks, though real, are generally resolvable and do not create insurmountable barriers to BIM adoption. Indeed, as BIM use becomes more widespread, it may be expected that, amongst other things, issues associated with IPR, insurances and best practice will all become clearer and more easily managed through improved contracts and management arrangements.

In conclusion

The most important decision clients will face is the selection and appointment of the design and construction team. Finding the right team of people with a positive approach to using BIM, to sharing information and above all to learning from the experience is key. In this regard, clients should expect all their consultants, contractors and specialists to be familiar with BIM and its requirements; to be positively engaged in its adoption; and to be actively developing ways in which processes can be made more value-adding and effective. Many projects have seen the adoption of BIM for clash detection and coordination purposes with a BIM consultant being appointed to manage the process. Greater understanding from the design teams and demand from Clients will see more traditionally based tasks being achieved through BIM in the subsequent years to come.

Building Management Systems (BMS)

A **Building Management System (BMS)** is a computer based control system installed in buildings that controls and monitors the building's mechanical and electrical equipment in the most efficient way.

Benefits

- Switches plant on and off automatically according to time, time of day and environmental conditions (e.g. senses conditions and activates devices to correct settings)
- Optimizes plant operation and services
- Monitors plant status and environmental conditions and thus improves standards of operation and maintenance
- Allows remote access, control and monitoring (e.g. using graphical information and alarms)

Types of BMS points

- Is the physical connection of a BMS point to a controller or outstation I/O terminals? The points schedule is critical when designing the system
- Input – gathers data/monitor; Output – provide command/control
- Analogue points – has a variable value 0% to 100% (e.g. temperature sensor, modulating control valves)
- Digital points – binary I/O (0 or 1) (e.g. on/off control, run/off/fault status)
- Network integration points (NI) – allows to communicate or integrate to stand alone controller such as chiller, meters, boilers, generators

BMS components – hardware and software/programming

Hardware

- **Primary network/BMS backbone**
  - Head end equipment (desktop computer, laptop, monitors, keyboard, dot matrix and colour printer, UPS supply, etc.)
  - Ethernet Network Cabling (normally Cat 5e or Cat 6 cables)
  - Ethernet switch (located inside the control panels)
- **Secondary network**
  - Control enclosure (CE) or Control panel (CP) or BMS outstation
  - Motor control centre (MCC)
  - Unitary controller for FCU/VAV/Chilled beams
  - Power cables and isolators (from MCC panel to mechanical equipments)
  - Control cables (from MCC/CE panel to field devices)
  - Field devices (control valves, meter, sensors, VSD/Inverter, DOL starter, star-delta starter, actuators)

Software/programming

- **BMS programme protocol** – Refers to the communication of BMS from PC software to the different types of controller/sensors *(e.g. Modbus, LonWorks, BACnet)*
- **Graphics** – Provide true representation of the installed plant. Colour graphics shall incorporate automatic updating of real time field data.
- **Software System Integration** (*e.g. energy management system, chillers, boiler, CHP, generator, water treatment plant, closed control units, blind control, lighting control, security and access controls)*

BMS Pricing – *Quick Estimate for a typical medium size office building:*

Shell and Core:

1.	Head end equipment (Workstation) *Extra cost:* • *Additional remote workstation*	£10,000 – £15,000 *£5,000 – £10,000*
2.	Network connection (BMS backbone – Copper)	£10,000 – £20,000
3.	Main plant – cost per BMS points *Factors that affect the cost:* • *UPS and touch pad display for each control panel* • *Panels enclosure – fire-rated or GRP enclosure*	£500 – £800 per pts
4.	Control enclosure (CE) for future tenants. Additional I/O modules will be part of fitting out cost *Factors that affect the cost:* • *UPS and touch pad display for each control panel* • *Panels enclosure – fire-rated or GRP enclosure*	£5,000 – £10,000 per panel
5.	Landlord FCUs – depending on points per/ FCU controller *Factors that affect the cost:* • *Local controller for cellular office, meeting rooms, etc.*	£400 – £700 per unit
6.	Software, graphics, commissioning, engineering	Included in BMS pts rate
7.	Trade contract preliminaries	20% to 25%

Category A, B or C Office Fit-out:

1.	Tenant plants – cost per points utilizing existing MCC/CE panel but require additional I/O modules	£300 – £500 per pts
2.	Energy meters a. LTHW/CHW meters b. Electricity meters (meters by Electrical TC) *Factors that affect the cost:* • *Size of heat meter or type of meter such as ultrasonic flow meter*	£750 – £1,500 £200 – £300
3.	Intelligent unitary controller (IUC) – depends on points per/ IUC controller and the air conditioning solution such as 4-pipe FCU, 2-pipe w/electric heating, VAV, chilled beams, chilled ceiling *Factors that affect the cost:* • *Local controller for cellular office/meeting room* • *For chilled beam and chilled ceiling, controls depends on the zoning* • *Depend also on zone controls*	£400 – £700 per unit
4.	Software, graphics, commissioning *Extra over cost:* • *Integration to other standalone system (e.g. lighting control, AV, fire alarm, security, central battery)*	Included in pts rate
5.	Trade contract preliminaries	20% to 25%

Renewable Energy Options

This article focuses on building-integrated options rather than large-scale utility solutions such as wind farms, which are addressed separately, and provides an analysis of where they may be best installed.

The legislative background, imperatives and incentives

In recognition of the causes and effects of global climate change, the Kyoto protocol was signed by the UK and other nations in 1992, with a commitment to reduce the emission of greenhouse gases relative to 1990 as the base year.

The first phase of European Union Emission Trading Scheme (EU ETS) covered the power sector and high-energy users such as oil refineries, metal processing, mineral and paper pulp industries. From 1 January 2005, all such companies in the EU had to limit their CO_2 emissions to allocated levels in line with Kyoto. The EU ETS is now in its third phase, running from 2013 to 2020 and covers the 27 EU Member States, as well as Iceland, Liechtenstein, Norway and Croatia. The main change from previous phases being that additional sectors and gases are now included. Key principles of the EU ETS are that participating organizations can:

- Meet the targets by reducing their own emissions (e.g. by implementing energy efficiency measures, using renewable energy sources), or
- Exceed the targets and sell or bank their excess emission allowances, or
- Fail to meet the targets and buy emission allowances from other participants

The EU ETS is designated as a 'cap and trade' system, where participation is mandatory for the sectors covered and it accepts credits from emission-saving projects carried out under the Kyoto Protocol's Clean Development Mechanism (CDM) and Joint Implementation instrument (JI).

In the UK, the Utilities Act (2000) requires power suppliers to provide some electricity from renewables, starting at 3% in 2003 and rising to15% by 2015. In a similar way to the EU ETS, generating companies receive and can trade Renewables Obligation Certificates (ROCs) for the qualifying electricity they generate.

The focus is on Greenhouse Gases (GHG) and Carbon Dioxide (CO_2) in particular as the main direct contributor to the greenhouse effect. The goals set by the UK government are:

- 34% GHG emissions reduction by 2020 (below 1990 baseline)
- Around 30% of electricity from renewables by 2020
- 80% GHG emissions reduction by 2050 (below 1990 baseline)

Four years after the introduction of ROCs, it was estimated that less than 3% of UK electricity was being generated from renewable sources. A 'step change' in policy was required, and the Office of the Deputy Prime Minister (ODPM) published 'Planning Policy Statement 22 (PPS 22): Renewable Energy' in order to promote renewable energy through the UK's regional and local planning authorities. PPS 22 was replaced by the National Planning Policy Framework, published in March 2012 by the Department for Communities and Local Government (DCLG), which encourages the use of renewable resources.

Local planning & building regulations

More than 100 local authorities embraced PPS 22 by adopting pro-renewables planning policies typically requiring a percentage (e.g. 10%) of a development's electricity or thermal energy needs to be derived from renewable sources. Government is developing further guidance to support local authorities addressing the sustainability of planned developments and to ensure a level playing field across the country.

Government is also committed to successive improvements in national new-build standards through changes to the Building Regulations, Part L, Conservation of Fuel and Power. In October 2010, new regulations introduced a 25% improvement on 2006 standards. The next review, due in 2013, is expected to further strengthen standards, in line with developing zero carbon policies.

Energy performance certificates (EPCs)

A Recast of the 2002 EU Directive on the Energy Performance of Buildings (EPBD) was published in 2010. It requires that EPCs be produced for buildings constructed, sold, or rented out to new tenants. For buildings occupied by public authorities and frequently visited by the public, the EPC must be displayed in a prominent place clearly visible to the public. These provisions enable prospective buyers and tenants to be informed of a building's energy performance, ensure the public sector leads by example, and raise public awareness. The Recast was transposed in the UK by national Governments, for example in England & Wales through the Energy Performance of Buildings (England & Wales) Regulations 2012 and other regulations. Other requirements of the Recast include: adopting a method to calculate the energy performance of buildings, setting minimum energy performance requirements and for new buildings to be Nearly Zero Energy Buildings by 2020.

Assessing the regulated carbon emissions associated with new buildings is now an important part of the design and building permitting process with the regulatory approach set out in Part L of the Building Regulations (Conservation of Fuel & Power), the associated National Calculation Methodology (NCM) Modelling Guide and the Standard Assessment Procedure (SAP). On-site renewable energy sources are taken into account and there are limits on design flexibility to discourage inappropriate trade-offs such as buildings with poor insulation standards offset by renewable energy systems.

Technology options and applications

- **Wind generators** – In a suitable location, wind energy can be an effective source of renewable power. Without grant, an installed cost range of £3000 to £5000 per kW of generator capacity may be achieved for small building-mounted turbines. A common arrangement was for a turbine with three blades on a horizontal axis, all mounted on a tower or, for small generators in inner city areas, on a building. Such arrangements typically compare poorly against other renewable options, as they are highly dependent on wind speed at the turbine, obstructions (e.g. nearby buildings and trees), turbulences, the elevation of the turbine above ground, and mitigating other impacts such as aesthetics consideration for planning permission, noise, and vibrations. With suitable conditions, average site wind speeds of 4 m/s can produce useful amounts of energy from a small generator up to about 3 kW, but larger generators require at least 7 m/s. A small increase in average site wind speed will typically result a large increase in output. There will be a need for inverters, synchronizing equipment, and metering for a grid connection.

  Larger, stand-alone turbines typically compare more favourably than smaller building-mounted turbines. Third party provision through an Energy Service Company (ESCo) can be successful for larger (stand-alone) installations located within or close to the host building's site, especially in industrial settings where there may be less aesthetic or noise issues than inner city locations. The ESCo provides funding, installs and operates the plant and the client signs up for the renewable electrical energy at an agreed price for a period of time.

- **Building integrated photovoltaics (BIPV)** – Photovoltaic materials, commonly known as solar cells, generate direct current electrical power when exposed to light. Solar cells are constructed from semi-conducting materials that absorb solar radiation; electrons are displaced within the material, thus starting a flow of current through an external connected circuit. PVs are available in a number of forms including monocrystalline, polycrystalline, amorphous silicon (thin film) or hybrid panels that are mounted on or integrated into the roofs or facades of buildings. Conversion efficiency of solar energy to electrical power is improving with advances in technology and ranges from 10% to 20%. In practice, allowing for UK weather conditions, an installation of 7 m^2 of monocrystalline modules (south facing at 30° from horizontal) typically produces 1000 watts peak (1 kWp), yielding about 800 kWh in a year. Installed costs range from £1200 to £2500/kWp.

- **Ground source heat pumps (heating & cooling)** – At a particular depth (about 10 m in the UK), the ground temperature remains substantially constant throughout the year. Heat (and coolth) may be extracted through either an 'open' system – discharging ground water to river or sewer after passing it through a heat exchanger, or a 'closed' system – circulating a fluid (often water) through a heat exchanger and (typically) vertical pipes extending below the ground water table. An electrically driven heat pump is then used to raise the fluid temperature via the refrigeration cycle, and low temperature hot water is delivered to the building.

  Most inner city ground heating and cooling systems consist of a cluster of pipes inserted into vertical holes typically 50 to 100 metres deep depending on space and ground conditions. Horizontal systems can be used where site circumstances allow. Costs for the drilling operation vary according to location, site accessibility, and ground conditions. Geological investigations are recommended to confirm ground conditions, reduce risks, and improve design and cost certainty.

  Such systems may achieve a Coefficient of Performance (COP = heat output/ electrical energy input) of between 3 and 4, achieving good savings of energy compared with conventional fossil fuels based systems. Installed costs are in the range £600 to £1700/kW depending on system type (vertical or horizontal), its size and complexity.

  Note that there is some debate on the status of ground source heat pumps as a renewable source of energy as it requires an external source of power which may not be renewable, typically electricity from the grid.

- **Solar water heating** – The basic principle is to collect heat from the sun via a fluid which is circulated in a roof solar panel or 'collector'. The heated fluid is then used to preheat hot water for space heating or domestic hot water, either in a separate tank or a twin coil hot water cylinder. Purpose-designed 'evacuated tube collectors' were developed to increase performance against the typical 'flat plate collectors'. A typical residential 'evacuated tube collectors' system has a cost ranging from £700 to £1000/m^2 depending on pipe runs and complexity. Such system may produce approximately 500 to 800 kWh/m^2 per year. Commercial systems are larger and more complex, and may achieve similar performance, providing there is sufficient hot water demand. Low-density residential, retail and leisure developments with washrooms and showers may also be suitable applications providing adequate demand for hot water.

- **Biomass boilers** – Wood chips or pellets derived from waste or farmed coppices or forests are available commercially and are considered carbon neutral, having absorbed carbon dioxide during growth. With a suitable fuel storage hopper and automatic screw drive and controls, biomass boilers can replace conventional boilers with little technical or aesthetic impact. However, they do depend on a viable source of fuel, and there are requirements for fuel deliveries access (in particular for inner city or remote locations), fuel storage, ash removal/disposal, as well as periodic de-coking. In individual dwellings, space may be a problem because a biomass boiler does not integrate readily into a typical modern kitchen. However, communal systems (serving multiple dwellings/flats) may be a viable domestic application. Biomass boilers are available in a wide range of domestic and commercial sizes. For a large installation, biomass boilers are more likely to form part of a modular system rather than to displace conventional boilers entirely. There is a cost premium for the biomass storage and feed system, and the cost of the fuel is currently comparable with other solid fuels. Installed costs of a biomass boiler range from £200 to £350/kW.

- **Biomass combined heat & power (CHP)** – Conventional CHP installations consist of either an internal combustion engine or a gas turbine driving an alternator, with maximum recovery of heat, particularly from the exhaust system. For best efficiency, there needs to be a convenient and constant requirement for the heat energy output and the generated electricity should also be utilized locally, with any excess exported to the grid.

  Considering the cost implications for biomass storage and handling as described for biomass boilers above, biomass CHP would only be viable in specific circumstances, with installed system costs in the order of £2500 to £3500/kW (electrical). Note that, at the time of writing, the authors are not aware of any small-scale biomass CHP system successfully operated in the UK over any significant period.

Grey Water Recycling and Rainwater Harvesting

The potential for grey water recycling and rainwater harvesting for both domestic residential and for various types of commercial building, considering the circumstances in which the systems offer benefits, both as stand-alone installations and combined.

Water usage trends

Water usage in the UK has increased dramatically over last century or so and it is still accelerating. The current average per capita usage is estimated to be at least 150 litres per day, and the population is predicted to rise from 60 million now to 65 million by 2017 and to 75 million by 2031, with an attendant increase in loading on water supply and drainage infrastructures.

Even at the current levels of consumption, it is clear from recent experience that long, dry summers can expose the drier regions of the UK to water shortages and restrictions. The predicted effects of climate change include reduced summer rainfall, more extreme weather patterns, and an increase in the frequency of exceptionally warm dry summers. This is likely to result in a corresponding increase in demand to satisfy more irrigation of gardens, parks, additional usage of sports facilities and other open spaces, together with additional needs for agriculture. The net effect, therefore, at least in the drier regions of the UK, is for increased demand coincident with a reduction in water resource, thereby increasing the risk of shortages.

Water applications and reuse opportunities

Average domestic water utilization can be summarized as follows, as a percentage of total usage (Source: Three Valleys Water)

- Wash hand basin 8%
- Toilet 35%
- Dishwasher 4%
- Washing machine 12%
- Shower 5%
- Kitchen sink 15%
- Bath 15%
- External use 6%

Water for drinking and cooking makes up less than 20% of the total, and more than a third of the total is used for toilet flushing. The demand for garden watering, although still relatively small, is increasing year by year and coincides with summer shortages, thereby exacerbating the problem.

In many types of building it is feasible to collect rainwater from the roof area and to store it, after suitable filtration, in order to meet the demand for toilet flushing, cleaning, washing machines and outdoor use – thereby saving in many cases a third of the water demand. The other, often complementary recycling approach, is to collect and disinfect 'greywater' – the waste water from baths, showers and washbasins. Hotels, leisure centres, care homes and apartment blocks generate large volumes of waste water and therefore present a greater opportunity for recycling. With intelligent design, even offices can make worthwhile water savings by recycling greywater, not necessarily to flush all of the toilets in the building but perhaps just those in one or two primary cores, with the greywater plant and distribution pipework dimensioned accordingly.

Intuitively, rainwater harvesting and greywater recycling seem like 'the right things to do' and rainwater harvesting is already common practice in many counties in Northern continental Europe. In Germany, for example, some 60,000 to 80,000 systems are being installed every year – compared with perhaps 2,000 systems in the UK. Greywater recycling systems, which are less widespread than rainwater harvesting, have been developed over the last 20 years and the 'state of the art' is to use biological and UV (ultra-violet) disinfection rather than chemicals, and to reduce the associated energy use through advanced technology membrane micro filters.

In assessing the environmental credentials of new developments, the sustainable benefits are recognized by the Building Research Establishment Environmental Assessment Method (BREEAM) whereby additional points can be gained for efficient systems – those designed to achieve enough water savings to satisfy at least 50% of the relevant demand. Furthermore, rainwater harvesting systems from several manufacturers are included in the 'Energy Technology Product List' and thereby qualify for Enhanced Capital Allowances (ECAs). Claims are allowed not only for the equipment, but also to directly associated project costs including:

- Transportation – the cost of getting equipment to the site.
- Installation – cranage (to lift heavy equipment into place), project management costs and labour, plus any necessary modifications to the site or existing equipment.
- Professional Fees – if they are directly related to the acquisition and installation of the equipment.

Rainwater harvesting

A typical domestic rainwater harvesting system can be installed at reasonable cost if properly designed and installed at the same time as building the house. The collection tank can either be buried or installed in a basement area. Rainwater enters the drainage system through sealed gullies and passes through a pre-filter to remove leaves and other debris before passing into the collection tank. A submersible pump, under the control of the monitoring and sensing panel, delivers recycled rainwater on demand. The non-potable distribution pipework to the washing machine, cleaner's tap, outside tap and toilets etc. could be either a boosted system or configured for a header tank in the loft, with mains supply back-up, monitors and sensors located there instead of at the control panel.

Calculating the collection tank size brings into play the concept of system efficiency – relating the water volume saved to the annual demand. In favourable conditions – ample rainfall and large roof collection area – it would be possible in theory to achieve almost 100%. In practice, systems commonly achieve 50 to 70% efficiency, with enough storage to meet demand for typically one week, though this is subject to several variables. As well as reducing the demand for drinking quality mains supply water, rainwater harvesting tanks act as an effective storm water attenuator, thereby reducing the drainage burden and the risk of local flooding which is a benefit to the wider community. Many urban buildings are located where conditions are unfavourable for rainwater harvesting – low rainfall and small roof collection area. In these circumstances it may still be worth considering water savings through greywater recycling.

Greywater recycling

In a greywater recycling system, waste water from baths, showers and washbasins is collected by conventional fittings and pipework, to enter a pretreatment sedimentation tank which removes the larger dirt particles. This is followed by the aerobic treatment tank in which cleaning bacteria ensure that all biodegradable substances are broken down. The water then passes onto a third tank, where an ultra-filtration membrane removes all particles larger than 0.00005 mm, (this includes viruses and bacteria) effectively disinfecting the recycled greywater. The clean water is then stored in the fourth tank from where it is pumped on demand under the control of monitors and sensors in the control panel. Recycled greywater may then be used for toilet and urinal flushing, for laundry and general cleaning, and for outdoor use such as vehicle washing and garden irrigation – a substantial water saving for premises such as hotels. If the tank becomes depleted, the distribution is switched automatically to the mains water back-up supply. If there is insufficient plant room space, then the tanks may be buried but with adequate arrangements for maintenance access. Overflow soakaways are recommended where feasible but are not an inherent part of the rainwater harvesting and greywater recycling systems.

Combined rainwater/greywater systems

Rainwater can be integrated into a greywater scheme with very little added complication other than increased tank size. In situations where adequate rainwater can be readily collected and diverted to pretreatment, then heavy demands such as garden irrigation can be met more easily than with greywater alone. An additional benefit is that they reduce the risk of flooding by keeping collected storm water on site instead of passing it immediately into the drains.

Indicative system cost and payback considerations

Rainwater harvesting scenario: Office building in Leeds having a roof area of 2000 m² and accommodating 435 people over 3 floors. Local annual rainfall is 875 mm and the application is for toilet and urinal flushing. An underground collection tank of 25,000 litres has been specified to give 4.5 storage days.

Rainwater harvesting cost breakdown		**Cost £**
System tanks and filters and controls		21,000
Mains water back-up and distribution pump arrangement		5,000
Non-potable distribution pipework		1,000
Connections to drainage		1,000
Civil works and tank installation (assumption normal ground conditions)		8,000
System installation & commissioning		2,000
	TOTAL COST	38,000

Rainwater harvesting payback considerations

Annual water saving: 1100 m³ @ average cost £2.30/m³ = £2,530

Annual maintenance and system energy cost = £800

Indicative payback period = 38,000/1,730 = 22 years

Greywater recycling scenario: Urban leisure hotel building with 200 bedrooms offers little opportunity for rainwater harvesting but has a greywater demand of up to 12000 litres per day for toilet and urinal flushing, plus a laundry. The greywater recycling plant is located in a basement plant room.

Greywater recycling cost breakdown		**Cost £**
System tanks and controls		37,000
Mains water back-up and distribution pump arrangement		4,000
Non-potable distribution pipework		4,000
Greywater waste collection pipework		5,000
Connections to drainage		1,000
System installation & commissioning		6,000
	TOTAL COST	57,000

Greywater recycling payback considerations

Annual water saving: 4260 m³ @ average cost £2.30 / m³ = £9,798

Annual maintenance and system energy cost = £2,200

Indicative payback period = 57,000/7,598 = 7.5 years

Exclusions

- Site organization and management costs other than specialist contractor's allowances
- Contingency/design reserve
- Main contractor's overhead and profit or management fee
- Professional fees
- Tax allowances
- Value Added Tax

Conclusions

There is a justified and growing interest in saving and recycling water by way of both greywater recycling and rainwater harvesting. The financial incentive at today's water cost is not great for small or inefficient systems but water costs are predicted to rise and demand to increase – not least as a result of population growth.

The payback periods for the above scenarios are not intended to compare to potential payback periods of rainwater harvesting against greywater recycling but rather to illustrate the importance of choosing 'horses for courses'. The office building with its relatively small roof area and limited demand for toilet flushing results in a fairly inefficient system. Burying the collection tank also adds a cost so that payback exceeds 20 years. Payback periods of less than 10 years are feasible for buildings with large roofs and a large demand for toilet flushing or for other uses. Therefore sports stadia, exhibition halls, supermarkets, schools and similar structures are likely to be suitable.

The hotel scenario is good application for greywater recycling. Many hotels and residential developments will generate more than enough greywater to meet the demand for toilet flushing etc. and in these circumstances large quantities of water can be saved and recycled with attractive payback periods.

Ground Water Cooling

The use of ground water cooling systems, considering the technical and cost implications of this renewable energy technology

The application of ground water cooling systems is quickly becoming an established technology in the UK with numerous installations having been completed for a wide range of building types, both new build and existing (refurbished).

Buildings in the UK are significant users of energy, accounting for 45% of UK carbon emissions in relation to their construction and occupation. The drivers for considering renewable technologies such as groundwater cooling are well documented and can briefly be summarized as follows:

- Government set targets – The Energy White Paper, published in 2003, setting a target of producing 10% of UK electricity from renewable sources by 2010 and the aspiration of doubling this by 2020.
- The proposed revision to the Building Regulations Part L 2006, in raising the overall energy efficiency of non domestic buildings, through the reduction in carbon emissions. This was subsequently updated by 2010 to require a 25% reduction of carbon emissions from the 2006 requirement and in 2013 it is expected that the requirement will be a 44% reduction from 2006.
- Local Government policy for sustainable development. In the case of London, major new developments (i.e. City of London schemes over 30,000 m^2) are required to demonstrate how they will generate a proportion of the site's delivered energy requirements from on-site renewable sources where feasible. The GLA's expectation is that, overall, large developments will contribute 10% of their energy requirement using renewables, although the actual requirement will vary from site to site. Local authorities are also likely to set lower targets for buildings which fall below the GLA's renewables threshold.
- Company policies of building developers and end users to minimize detrimental impact to the environment.

The ground as a heat source/sink

The thermal capacity of the ground can provide an efficient means of tempering the internal climate of buildings. Whereas the annual swing in mean air temperature in the UK is around 20 K, the temperature of the ground is far more stable. At the modest depth of 2 m, the swing in temperature reduces to 8 K, while at a depth of 50 m the temperature of the ground is stable at 11–13°C. This stability and ambient temperature therefore makes groundwater a useful source of renewable energy for heating and cooling systems in buildings.

Furthermore, former industrial cities like Nottingham, Birmingham, Liverpool and London have a particular problem with rising ground water as they no longer need to abstract water from below ground for use in manufacturing. The use of groundwater for cooling is therefore encouraged by the Environment Agency in areas with rising groundwater as a means of combating this problem.

System types

Ground water cooling systems may be defined as either open or closed loop.

Open loop systems

Open loop systems generally involve the direct abstraction and use of ground water, typically from aquifers (porous water bearing rock). Water is abstracted via one or more boreholes and passed through a heat exchanger and is returned via a separate borehole or boreholes, discharged to foul water drainage or released into a suitable available source such as a river. Typical ground water supply temperatures are in the range 10–12°C and typical re-injection temperatures around 3°C warmer than the extraction temperature (subject to the requirements of the extraction licence).

Open loop systems fed by groundwater at 8°C, can typically cool water to 12°C on the secondary side of the heat exchanger to serve conventional cooling systems.

Open loop systems are thermally efficient but overtime can suffer from blockages caused by silt, and corrosion due to dissolved salts. As a result, additional cost may be incurred in having to provide filtration or water treatment, before the water can be used in the building.

Abstraction licence and discharge consent needs to be obtained for each installation, and this together with the maintenance and durability issues can significantly affect whole life operating costs, making this system less attractive.

Closed loop systems

Closed loop systems do not rely on the direct abstraction of water, but instead comprise a continuous pipework loop buried in the ground. Water circulates in the pipework and provides the means of heat transfer with the ground. Since ground water is not being directly used, closed loop systems therefore suffer fewer of the operational problems of open loop systems, being designed to be virtually maintenance free, but do not contribute to the control of groundwater levels.

There are two types of closed loop system:

Vertical Boreholes – Vertical loops are inserted as U tubes into pre-drilled boreholes, typically less than 150 mm in diameter. These are backfilled with a high conductivity grout to seal the bore, prevent any cross contamination and to ensure good thermal conductivity between the pipe wall and surrounding ground. Vertical boreholes have the highest performance and means of heat rejection, but also have the highest cost due to associated drilling and excavation requirements.

As an alternative to having a separate borehole housing the pipe loop, it can also be integrated with the piling, where the loop is encased within the structural piles. This obviously saves on the costs of drilling and excavation since these would be carried out as part of the piling installation. The feasibility of this option would depend on marrying up the piling layout with the load requirement, and hence the number of loops, for the building.

Horizontal Loops – These are single (or pairs) of pipes laid in 2 m deep trenches, which are backfilled with fine aggregate. These obviously require a greater physical area than vertical loops but are cheaper to install. As they are located closer to the surface where ground temperatures are less stable, efficiency is lower compared to open systems. Alternatively, coiled pipework can also be used where excavation is more straightforward and a large amount of land is available. Although performance may be reduced with this system as the pipe overlaps itself, it does represent a cost effective way of maximizing the length of pipe installed and hence overall system capacity.

The case for heat pumps

Instead of using the groundwater source directly in the building, referred to as passive cooling, when coupled to a reverse cycle heat pump, substantially increased cooling loads can be achieved.

Heat is extracted from the building and transferred by the heat pump into the water circulating through the loop. As it circulates, it gives up heat to the cooler earth, with the cooler water returning to the heat pump to pick up more heat. In heating mode the cycle is reversed, with the heat being extracted from the earth and being delivered to the HVAC system.

The use of heat pumps provides greater flexibility for heating and cooling applications within the building than passive systems. Ground source heat pumps are inherently more efficient than air source heat pumps, their energy requirement is therefore lower and their associated CO_2 emissions are also reduced, so they are well suited for connection to a groundwater source.

Closed loop systems can typically achieve outputs of 50W/m (of bore length), although this will vary with geology and borehole construction. When coupled to a reverse cycle heat pump, 1 m of vertical borehole will typically deliver 140 kWh of useful heating and 110 kWh of cooling per annum, although this will depend on hours run and length of heating and cooling seasons.

Key factors affecting cost

- The cost is obviously dependent on the type of system used. Deciding on what system is best suited to a particular project is dependent on the peak cooling and heating loads of the building and its likely load profile. This in turn determines the performance required from the ground loop, in terms of area of coverage in the case of the horizontal looped system, and in the case of vertical boreholes, the depth and number or bores. The cost of the system is therefore a function of the building load.
- In the case of vertical boreholes, drilling costs are significant factor, as specific ground conditions can be variable, and there are potential problems in drilling through sand layers, pebble beds, gravels and clay, which may mean additional costs through having to drill additional holes or the provision of sleeving etc. The costs of excavation obviously make the vertical borehole solution significantly more expensive than the equivalent horizontal loop.
- The thermal efficiency of the building is also a factor. The higher load associated with a thermally inefficient building obviously results in the requirement for a greater number of boreholes or greater area of horizontal loop coverage, however in the case of boreholes the associated cost differential between a thermally inefficient building and a thermally efficient one is substantially greater than the equivalent increase in the cost of conventional plant. Reducing the energy consumption of the building is cheaper than producing the energy from renewables and the use of renewable energy only becomes cost effective, and indeed should only be considered, when a building is energy efficient.
- With open loop systems, the principal risk in terms of operation is that the user is not in control of the quantity or quality of the water being taken out of the ground, this being dependent on the local ground conditions. Reduced performance due to blockage (silting etc.) may lead to the system not delivering the design duties whilst bacteriological contamination may lead to the expensive water treatment or the system being taken temporarily out of operation. In order to mitigate the above risk, it may be decided to provide additional means of heat rejection and heating by mechanical means as a back up to the borehole system, in the event of operational problems. This obviously carries a significant cost. If this additional plant were not provided, then there are space savings to be had over conventional systems due to the absence of heating, heat rejection and possibly refrigeration plant.
- Open loop systems may lend themselves particularly well to certain applications increasing their cost effectiveness, i.e. in the case of a leisure centre, the removal of heat from the air-conditioned parts of the centre and the supply of fresh water to the swimming pool.
- In terms of the requirements for abstraction and disposal of the water for open loop systems, there are risks associated with the future availability and cost of the necessary licenses; particularly in areas of high forecast energy consumption, such as the South East of England, which needs to be borne in mind when selecting a suitable system.

Whilst open loop systems would suit certain applications or end user clients, for commercial buildings the risks associated with this system tend to mean that closed loop applications are the system of choice. When coupled to a reversible heat pump, the borehole acts simply as a heat sink or heat source so the problems associated with open loop systems do not arise.

Typical costs

Table 1 gives details of the typical borehole cost to an existing site in Central London, using one 140 m deep borehole working on the open loop principle, providing heat rejection for the 600 kW of cooling provided to the building. The borehole passes through rubble, river gravel terraces, clay and finally chalk, and is lined above the chalk level to prevent the hole collapsing. The breakdown includes all costs associated with the provision of a working borehole up to the well head, including the manhole chamber and manhole. The costs of any plant or equipment from the well head are not included.

Heat is drawn out of the cooling circuit and the water is discharged into the Thames at an elevated temperature. In this instance, although the boreholes are more expensive than the dry air cooler alternative, the operating cost is significantly reduced as the system can operate at around three times the efficiency of conventional dry air coolers, so the payback period is a reasonable one. Additionally, the borehole system does not generate any noise, does not require rooftop space and does not require as much maintenance.

This is representative of a typical cost of providing a borehole for an open loop scheme within the London basin. There are obviously economies of scale to be had in drilling more than one well at the same time, with two wells saving approximately 10% of the comparative cost of two separate wells and four wells typically saving 15%.

Table 2 provides a summary of the typical range of costs that could expected for the different types of system based on current prices.

Table 1: Breakdown of the Cost of a Typical Open Loop Borehole System

Description	**Cost £**
<u>General Items</u>	
• Mobilization, Insurances, demobilization on completion	21,000
• Fencing around working area for the duration of drilling and testing	2,000
• Modifications to existing LV panel and installation of new power supplies for borehole installation	15,000
<u>Trial Hole</u>	
• Allowance for breakout access to nearest walkway (Existing borehole on site used for trial purposes, hence no drilling costs included)	3,000
<u>Construct Borehole</u>	
• Drilling, using temporary casing where required, permanent casing and grouting	32,000
<u>Borehole Cap and Chamber</u>	
• Cap borehole with PN16 flange, construct manhole chamber in roadway, rising main, header pipework, valves, flow meter	12,000
• Permanent pump	14,000
<u>Samples</u>	
• Water samples	1,000
<u>Acidization</u>	
• Mobilization, set up and removal of equipment for acidization of borehole, carry out acidization	12,000
<u>Development and Test Pumping</u>	
• Mobilize pumping equipment and materials and remove on completion of testing	4,000
• Calibration test, pretest monitoring, step testing	4,000
• Constant rate testing and monitoring	20,000
• Waste removal and disposal	3,000
<u>Reinstatement</u>	
• Reinstatement and making good	2,000
Total	**145,000**

Table 2: Summary of the Range of Costs for Different Systems

		Range		
System	**Small–4 kWth**	**Medium–50 kWth**	**Large–400 kWth**	**Notes**
Heat pump (per unit)	£3,000–5,000	£30,000–42,000	£145,000–175,000	
Slinky pipe (per installation) including excavation	£3,000–4,000	£42,000–52,000	£360,000–390,000$^{(1)}$	$^{(1)}$ Based on 90 nr 50 m lengths
Vertical, closed (per installation) using structural piles	N/A	£42,000–63,000	N/A	Based on 50 nr piles. Includes borehole cap and header pipework but excludes connection to pump room and heat pumps
Vertical, closed (per installation) including excavation	£2,000–3,000	£63,000–85,000	£370,000–400,000	Includes borehole cap and header pipework but excludes connection to pump room and heat pumps
Vertical, open (per installation) including excavation	£2,000–3,000	£45,000–65,000	£335,000–370,000	Excludes connection to pump room and heat exchangers

Fuel Cells

The application of fuel cell technology within buildings

Fuel cells are electrochemical devices that convert the chemical energy in fuel into electrical energy directly, without combustion, with high electrical efficiency and low pollutant emissions. They represent a new type of power generation technology that offers modularity, efficient operation across a wide range of load conditions, and opportunities for integration into co-generation systems. With the publication of the energy white paper in February this year, the Government confirmed its commitment to the development of fuel cells as a key technology in the UK's future energy system, as the move is made away from a carbon based economy.

There are currently very few fuel cells available commercially, and those that are available are not financially viable. Demand has therefore been limited to niche applications, where the end user is willing to pay the premium for what they consider to be the associated key benefits. Indeed, the UK currently has only one fuel cell in regular commercial operation. However, fuel cell technology has made significant progress in recent years, with prices predicted to approach those of the principal competition in the near future.

Fuel cell technology

A fuel cell is composed of an anode (a negative electrode that repels electrons), an electrolyte membrane in the centre, and a cathode (a positive electrode that attracts electrons). As hydrogen flows into the cell on the anode side, a platinum coating on the anode facilitates the separation of the hydrogen gas into electrons and protons. The electrolyte membrane only allows the protons to pass through to the cathode side of the fuel cell. The electrons cannot pass through this membrane and flow through an external circuit to form an electric current.

As oxygen flows into the fuel cell cathode, another platinum coating helps the oxygen, protons, and electrons combine to produce pure water and heat.

The voltage from a single cell is about 0.7 volts, just enough for a light bulb. However by stacking the cells, higher outputs are achieved, with the number of cells in the stack determining the total voltage, and the surface area of each cell determining the total current. Multiplying the two together yields the total electrical power generated.

In a fuel cell the conversion process from chemical energy to electricity is direct. In contrast, conventional energy conversion processes first transform chemical energy to heat through combustion and then convert heat to electricity through some form of power cycle (e.g. gas turbine or internal combustion engine) together with a generator. The fuel cell is therefore not limited by the Carnot efficiency limits of an internal combustion engine in converting fuel to power, resulting in efficiencies 2 to 3 times greater.

Fuel cell systems

In addition to the fuel cell itself, the system comprises the following subsystems:

- A fuel processor – This allows the cell to operate with available hydrocarbon fuels, by cleaning the fuel and converting (or reforming) it as required.
- A power conditioner – This regulates the dc electricity output of the cell to meet the application, and to power the fuel cell auxiliary systems.
- An air management system – This delivers air at the required temperature, pressure and humidity to the fuel stack and fuel processor.
- A thermal management system – This heats or cools the various process streams entering and leaving the fuel cell and fuel processor, as required.
- A water management system – Pure water is required for fuel processing in all fuel cell systems, and for dehumidification in the PEMFC.

The overall electrical conversion efficiency of a fuel cell system (defined as the electrical power out divided by the chemical energy into the system, taking into account the individual efficiencies of the subsystems) ranges from 35–55%. Taking into account the thermal energy available from the system, the overall or cogeneration efficiency is 75–90%.

Also, unlike most conventional generating systems (which operate most efficiently near full load, and then suffer declining efficiency as load decreases), fuel cell systems can maintain high efficiency at loads as low as 20% of full load.

Fuel cell systems also offer the following potential benefits:

- At operating temperature, they respond quickly to load changes, the limiting factor usually being the response time of the auxiliary systems.
- They are modular and can be built in a wide range of outputs. This also allows them to be located close to the point of electricity use, facilitating cogeneration systems.
- Noise levels are comparable with residential or light commercial air conditioning systems.
- Commercially available systems are designed to operate unattended and manufactured as packaged units.
- Since the fuel cell stack has no moving parts, other than the replacement of the stack at 3–5 year intervals there is little on-site maintenance. The maintenance requirements are well established for the auxiliary system plant.
- Fuel cell stacks fuelled by hydrogen produce only water, therefore the fuel processor is the primary source of emissions, and these are significantly lower than emissions from conventional combustion systems.
- Since fuel cell technology generates 50% more electricity than the conventional equivalent without directly burning any fuel, CO_2 emissions are significantly reduced in the production of the source fuel.
- Potentially zero carbon emissions when using hydrogen produced from renewable energy sources.
- The facilitation of embedded generation, where electricity is generated close to the point of use, minimizing transmission losses.
- The fast response times of fuel cells offer potential for use in UPS systems, replacing batteries and standby generators.

Types of fuel cell

There are four main types of fuel cell technology that are applicable for building systems, classed in terms of the electrolyte they use. The chemical reactions involved in each cell are very different.

Phosphoric Acid Fuel Cells (PAFCs) are the dominant current technology for large stationary applications and have been available commercially for some time. There is less potential for PAFC unit cost reduction than for some other fuel cell systems, and this technology may be superseded in time by the other technologies.

The Solid Oxide Fuel Cell (SOFC) offers significant flexibility due to its large power range and wide fuel compatibility. SOFCs represent one of the most promising technologies for stationary applications. There are difficulties when operating at high temperatures with the stability of the materials, however, significant further development and cost reduction is anticipated with this type.

The relative complexity of Molten Carbonate Fuel Cells (MCFCs) has tended to limit developments to large-scale stationary applications, although the technology is still very much in the development stages.

The quick start-up times and size range make Polymer Electrolyte Membrane Fuel Cells (PEMFCs) suitable for small to medium sized stationary applications. They have a high power density and can vary output quickly, making them well suited for transport applications as well as UPS systems. The development efforts in the transport sector suggest there will continue to be substantial cost reductions over both the short and long term.

All four technologies remain the subject of extensive research and development programmes to reduce initial costs and improve reliability through improvements in materials, optimization of operating conditions and advances in manufacturing.

The market for fuel cells

The stationary applications market for fuel cells can be sectaries as follows:

- Distributed generation/CHP – For large-scale applications, there are no drivers specifically advantageous to fuel cells, with economics (and specifically initial cost) therefore being the main consideration. So, until cost competitive and thoroughly proven and reliable fuel cells are available, their use is likely to be limited to niche applications such as environmentally sensitive areas.
- Domestic and small-scale CHP – The drivers for the use of fuel cells in this emerging market are better value for customers than separate gas and electricity purchase, reduction in domestic CO_2 emissions, and potential reduction in electricity transmissioncosts. However, the barriers of resistance to distributed generation, high capital costs and competition from Stirling engines needs to be overcome.
- Small generator sets and remote power – The drivers for the use of fuel cells are high reliability, low noise and low refuelling frequencies, which cannot be met by existing technologies. Since cost is often not the primary consideration, fuel cells will find early markets in this sector. Existing PEMFC systems are close to meeting the requirements in terms of cost, size and performance. Small SOFCs have potential in this market, but require further development.

Costs

Fuel cell technologies are still significantly more expensive than the existing technologies typically and, dependent on the fuel cell type, between £3,000 – £5,000/kW. To extend fuel cell application beyond niche markets, their cost needs to reduce significantly. The successful and widespread commercial application of fuel cells is dependent on the projected cost reductions indicated, with electricity generated from fuel cells being competitive with current centralized and distributed power generation.

Conclusions

Despite significant growth in recent years, fuel cells are still at a relatively early stage of commercial development, with prohibitively high capital costs preventing them from competing with the incumbent technology in the market place. However, costs are forecast to reduce significantly over the next five years as the technology moves from niche applications, and into mass production.

However, in order for these projected cost reductions to be achieved, customers need to be convinced that the end product is not only cost competitive but also thoroughly proven, and Government support represents a key part in achieving this.

The Governments of Canada, USA, Japan and Germany have all been active in supporting development of the fuel cell sector through integrated strategies, however the UK has been slow in this respect, and support has to date been small in comparison. It is clear that without Government intervention, fuel cell applications may struggle to reach the cost and performance requirements of the emerging fuel cell market.

Biomass Energy

The potential for biomass energy systems, with regards to the adequacy of the fuel supply and the viability of various system types at different scales

Biomass heating and combined heat and power (CHP) systems have become a major component of the low-carbon strategy for many projects, as they can provide a large renewable energy component at a relatively low initial cost. Work by the Carbon Trust has demonstrated that both large and small biomass systems were viable even before recent increases in gas and fuel oil prices, so it is no surprise that recent research by South Bank University into the renewables strategies to large London projects has found that 25% feature biomass or biofuel systems.

These proposals are not without risk, however. Although the technology is well established, few schemes are in operation in the UK and the long-term success depends more on the effectiveness of the local supply chain than the quality of the design and installation.

How the biomass market works

Biomass is defined as living or recently dead biological material that can be used as an energy source. Biomass is generally used to provide heat, generate electricity or drive CHP engines. The biomass family includes biofuels, which are being specified in city centre schemes, but which provide lower energy outputs and could transfer farmland away from food production.

In the UK, much of the focus in biomass development is on the better utilization of waste materials such as timber and the use of set-aside land for low-intensity energy crops such as willow, rather than expansion of the biofuels sector. There are a variety of drivers behind the development of a biomass strategy. In addition to carbon neutrality, another policy goal is the promotion of the UK's energy security through the development of independent energy sources. A third objective is to address energy poverty, particularly for off-grid energy users, who are most vulnerable to the effects of high long-term costs of fuel oil and bottled gas.

Biomass' position in the zero-carbon hierarchy is a little ambiguous in that its production, transport and combustion all produce carbon emissions, albeit most is offset during a plant's growth cycle. The key to neutrality is that the growing and combustion cycles need to occur over a short period, so that combustion emissions are genuinely offset. Biomass strategy is also concerned about minimizing waste and use of landfill, and the ash produced by combustion can be used as a fertilizer.

Dramatic increases in fossil fuel prices have swung considerations decisively in favour of technologies such as biomass. Research by the Carbon Trust has demonstrated that, with oil at $50 (£25) a barrel, rates of return of more than 10% could be achieved with both small and large heating installations. CHP and electricity-only schemes have more complex viability issues linked to renewable incentives, but with oil currently trading at over ($100) £50 a barrel and a plentiful supply of source material, it is argued that biomass input prices will not rise and so the sector should become increasingly competitive.

The main sources of biomass in the UK include:

- Forestry crops, including the waste products of tree surgery industry
- Industrial waste, particularly timber, paper and card: timber pallets account for 30% of this waste stream by weight
- Woody energy crops, particularly those grown through 'short rotation' methods such as willow coppicing
- Wastes and residues taken from food, agriculture and manufacturing

Biomass is an emerging UK energy sector. Most suppliers are small and there remains a high level of commercial risk associated with finding appropriate, reliable sources of biomass. This is particularly the case for larger-scale schemes such as those proposed for Greater London, which will have sourced biomass either from multiple UK suppliers or from overseas. Many have adopted biofuels as an alternative.

The UK's only large-scale biomass CHP in Slough has a throughput of 180,000 tonnes of biomass per year requiring the total production of more than 20 individual suppliers – not a recipe for easy management or product consistency. However, Carbon Trust research has identified significant potential capacity in waste wood (5 million to 6 million tonnes) and short-rotation coppicing, which could create the conditions for wider adoption of small and large-scale biomass.

Biomass technologies

A wide range of technologies have been developed for processing various forms of biomass, including anaerobic digesters and gasifiers. However, the main biomass technology is solid fuel combustion, as a heat source, CHP unit or energy source for electricity generation.

Solid fuel units use either wood chippings or wood pellets. Wood chippings are largely unprocessed and need few material inputs, other than seasoning, chipping and transport. Wood pellets are formed from compressed sawdust. As a result they have a lower moisture content than wood chippings and consistent dimensions, so are easier to handle but are about twice as expensive.

Solid fuel burners operate in the same way as other fossil fuel-based heat sources, with the following key differences:

- Biomass heat output can be controlled but not instantaneously, so systems cannot respond to rapid load changes. Solutions to provide more flexibility include provision of peak capacity from gas-fired systems, or the use of thermal stores that capture excess heat energy during off-peak periods, enabling extended operation of the biomass system itself
- Heat output cannot be throttled back by as much as gas-fired systems, so for heat-only installations it may be necessary to have an alternative summer system for water heating, such as a solar collector
- Biomass feedstock is bulky and needs a mechanized feed system as well as extensive storage
- Biomass systems are large, and the combustion unit, feed hopper and fuel store take up substantial floor area. A large unit with an output of 500 kW has a footprint of 7.5 m × 2 m
- Biomass systems need maintenance related to fuel deliveries, combustion efficiency, ash removal, adding to the lifetime cost
- Fuel stores need to be physically isolated from the boiler and the rest of the building in order to minimize fire risk. The fuel store needs to be sized to provide for at least 100 hours of operation, which is approximately 100 m^2 for a 500 kW boiler. The space taken up by storage and delivery access may compromise other aspects of site planning
- Fire-protection measures include anti-blowback arrangements on conveyors and fire dampers, together with the specification of elements such as flues for higher operating temperatures
- Collocation of the fuel source and burner at ground level require larger, free-standing flues

As a result of these issues, which drive up initial costs, affect development efficiency and add to management overheads, take-up of biomass has initially been mostly at the small-scale, heat-only end of the market, based on locally sourced feedstock. In such systems the initial cost premium of the biomass boiler can be offset against long-term savings in fuel costs.

Sourcing biomass

Compared with solar or wind power installations, the initial costs of biomass systems are low, the technology is well established and energy output is dependable. As a result, the real challenge for successful operation of a biomass system is associated with the reliable sourcing of feedstock.

Heat-only systems themselves cost between £150 and £750 per kW (excluding costs of storage), depending on scale and technology adopted. This compares with a typical cost of £50 to £300 per kW for a gas-fired boiler – which does not require further investment in fuel or thermal storage bunkers.

As a high proportion of lifetime cost is associated with the operation of a system, availability of good-quality, locally sourced feedstock is essential for long-term viability – particularly in areas where incentivization through policies like the Merton rule is driving up demand.

Research funded by BioRegional in connection with medium-scale biomass systems in the South-east shows that considerable feedstock is already in the system but far more is required to respond to emerging requirements.

Fortunately, the scale of the UK's untapped resource is considerable, with 5 million to 6 million tonnes of waste wood going to landfill annually, and 680,000 ha of set-aside land that could be used for energy crops without affecting agricultural output.

Based on these figures, it is estimated that 15% of the UK's building-related energy load could be supported without recourse to imported material. However, the supply chain is fragmented in terms of producers, processors and distributors – presenting potential biomass users with a range of complexities that gas users simply do not need to worry about. These include:

- Ensuring quality. Guaranteeing biomass quality is important for the assurance of performance and reliability. Variation in moisture content affects combustion, while inconsistent woodchip size or differences in sawdust content can result in malfunction. The presence of contaminants in waste wood causes problems too. High-profile schemes including the 180,000-tonne generator in Slough have had to shutdown because of variations in fuel quality. Use of pellets reduces to risk, but they are more expensive and require more energy for processing and transport
- Functioning markets. The scale of trade in biomass compares unfavourably with gas or oil, in that there are no standard contracts, fixed-price deals or opportunities for hedging which enable major users to manage their energy cost risk
- Security of supply. The potential for competing uses could lead to price inflation. Biofuels carry the greatest such risk, but many biomass streams have alternative uses. Lack of capacity in the marketplace is another security issue, with no mechanism to encourage strategic stockpiling for improved response to crop failure or fluctuations in demand
- Installation and maintenance infrastructure. The different technologies used in biomass systems creates maintenance requirements not yet met by a readily available pool of skilled system engineers

Optimum uses of biomass technology

Biomass is a high-grade, locally available source of energy that can be used at a range of scales to support domestic and commercial use. Following increases in fossil fuel prices, one of the main barriers to adoption is, now, the capability of the supply chain.

The Carbon Trust's biomass sector review, completed before the large energy price rises in recent years, drew the following key conclusions about the most effective application of the technologies:

- Returns on CHP and electricity-generating systems depend heavily on government incentives such as renewable obligations certificates. Under the present arrangements, large CHP systems provide the best returns
- Heat-only systems are very responsive to changes in fuel prices
- Small-scale heat-only plants produce the best returns, because the cost of the displaced fuel (typically fuel oil) is more expensive
- Small-scale electricity and large-scale heat-only installations produce very poor returns
- There is little difference in the impact of fuel type in the returns generated by projects

The study also concluded that heat installations at all scales had the greater potential for carbon saving, based on a finite supply of biomass. This is because heat-generating processes have the greatest efficiency and, in the case of small-scale systems in isolated, off-grid dwellings, displace fuels such as oil that have the greatest carbon intensity. Ninety percent of the UK's existing biomass resource of 5.6 m tonnes per annum could be used in displacing carbon-intensive off-grid heating, saving 2.5 m tonnes of carbon emissions.

Small-scale systems are well established in Europe and the existing local supply chain suits the demand pattern. In addition, since the target market is in rural, off-grid locations, affected dwellings are less likely to suffer space constraints related to storage. As fuel costs continue to rise, the benefits of avoiding fuel poverty, combined with the effective reduction of carbon emissions from existing buildings, mean smaller systems are likely to offer the best mid-term use of the existing biomass supply base, with large-scale systems being developed as the supply chain matures and expands.

Large-scale systems also offer the opportunity to generate significant returns, but the barriers that developers or operators face are significant, particularly if there is an electricity supply component, which requires a supply agreement. However, while developers are required to delivery renewable energy on site, biomass in the form of biofuels, has the great attraction of being able to provide a scale of renewable energy generation that other systems such as ground source heating or photovoltaics simply cannot compete with.

Whether biomass plant should be used on commercial schemes in urban locations is potentially a policy issue. Sizing of both CHP and heat-only systems should be determined by the heat load, which for city-centre schemes may not be that large – affecting the potential for the CHP component. The costs of a district heating element on these schemes may also be prohibitively high, and considerations of biomass transport and storage also make it harder to get city-centre schemes to stack up.

It may be a more appropriate policy to encourage industrial users or large-scale regenerators to take first call on the expanding biomass resource, rather than commercial schemes. The launch of a 45 MW biomass power station in Scotland illustrates this trend. Data shows that 50% of the market potential for industrial applications of CHP could utilize 100% of the UK's available biomass resource. The issues that city-centre biomass schemes face in connection with storage, transport, emissions and supply chain management might be better addressed by industrial users or their energy suppliers in low-cost locations rather than by developers in prime city-centre sites.

Indicative costs

System	**Indicative load kW**	**Capital cost (£/kWh)**
Gas-fired boiler	50	90
	400	50
Biomass-fired boiler	50	530
	500	265
Biomass-fired CHP	1,000	480

Allowance for stand-alone boiler house and fuel store £30,000–60,000 for 50 kWh system indicative costs exclude flues and plant room installation

Fuel costs (typical bulk prices at small commercial scale – February 2012)

Wood chip: 2.9 p/kWh

Wood pellet: 4.2p/kWh

Fuel oil: 6.0 p/kWh

Natural gas: 4.8 p/kWh

Bottled LPG: 7.6 p/kWh

Capital Allowances

Introduction

Capital Allowances provide tax relief by prescribing a statutory rate of depreciation for tax purposes in place of that used for accounting purposes. They are utilized by government to provide an incentive to invest in capital equipment, including assets within commercial property, by allowing the majority of taxpayers a deduction from taxable profits for certain types of capital expenditure, thereby reducing or deferring tax liabilities.

The capital allowances most commonly applicable to real estate are those given for capital expenditure on existing commercial buildings in disadvantaged areas, and plant and machinery in all buildings other than residential dwellings. Relief for certain expenditure on industrial buildings and hotels was withdrawn from April 2011, although the ability to claim plant and machinery remains.

Enterprise Zone Allowances are also available for capital expenditure within designated areas only where there is a focus on high value manufacturing. Enhanced rates of allowances are available on certain types of energy and water saving plant and machinery assets, whilst reduced rates apply to 'integral features' and items with an expected economic life of more than 25 years.

The Act

The primary legislation is contained in the Capital Allowances Act 2001. Major changes to the system were announced by the Government in 2007 and there have been further changes in subsequent Finance Acts.

The Act is arranged in 12 Parts (plus two addenda) and was published with an accompanying set of Explanatory Notes.

Plant and machinery

The Finance Act 1994 introduced major changes to the availability of Capital Allowances on real estate. A definition was introduced which precludes expenditure on the provision of a building from qualifying for plant and machinery, with prescribed exceptions.

List A in Section 21 of the 2001 Act sets out those assets treated as parts of buildings:

- *Walls, floors, ceilings, doors, gates, shutters, windows and stairs.*
- *Mains services, and systems, for water, electricity and gas.*
- *Waste disposal systems.*
- *Sewerage and drainage systems.*
- *Shafts or other structures in which lifts, hoists, escalators and moving walkways are installed.*
- *Fire safety systems.*

Similarly, List B in Section 22 identifies excluded structures and other assets.

Both sections are, however, subject to Section 23. This section sets out expenditure, which although being part of a building, may still be expenditure on the provision of Plant and Machinery.

List C in Section 23 is reproduced below:

Sections 21 and 22 do not affect the question whether expenditure on any item in List C is expenditure on the provision of plant or machinery

1. Machinery (including devices for providing motive power) not within any other item in this list.
2. Gas and sewerage systems provided mainly –
   a. to meet the particular requirements of the qualifying activity, or
   b. to serve particular plant or machinery used for the purposes of the qualifying activity.
3. Omitted.
4. Manufacturing or processing equipment; storage equipment (including cold rooms); display equipment; and counters, checkouts and similar equipment.
5. Cookers, washing machines, dishwashers, refrigerators and similar equipment; washbasins, sinks, baths, showers, sanitaryware and similar equipment; and furniture and furnishings.
6. Hoists.
7. Sound insulation provided mainly to meet the particular requirements of the qualifying activity.
8. Computer, telecommunication and surveillance systems (including their wiring or other links).
9. Refrigeration or cooling equipment.
10. Fire alarm systems; sprinkler and other equipment for extinguishing or containing fires.
11. Burglar alarm systems.
12. Strong rooms in bank or building society premises; safes.
13. Partition walls, where moveable and intended to be moved in the course of the qualifying activity.
14. Decorative assets provided for the enjoyment of the public in hotel, restaurant or similar trades.
15. Advertising hoardings; signs, displays and similar assets.
16. Swimming pools (including diving boards, slides & structures on which such boards or slides are mounted).
17. Any glasshouse constructed so that the required environment (namely, air, heat, light, irrigation and temperature) for the growing of plants is provided automatically by means of devices forming an integral part of its structure.
18. Cold stores.
19. Caravans provided mainly for holiday lettings.
20. Buildings provided for testing aircraft engines run within the buildings.
21. Moveable buildings intended to be moved in the course of the qualifying activity.
22. The alteration of land for the purpose only of installing Plant or Machinery.
23. The provision of dry docks.
24. The provision of any jetty or similar structure provided mainly to carry Plant or Machinery.
25. The provision of pipelines or underground ducts or tunnels with a primary purpose of carrying utility conduits.
26. The provision of towers to support floodlights.
27. The provision of –
   a. any reservoir incorporated into a water treatment works, or
   b. any service reservoir of treated water for supply within any housing estate or other particular locality.
28. The provision of –
   a. silos provided for temporary storage, or
   b. storage tanks.
29. The provision of slurry pits or silage clamps.
30. The provision of fish tanks or fish ponds.
31. The provision of rails, sleepers and ballast for a railway or tramway.
32. The provision of structures and other assets for providing the setting for any ride at an amusement park or exhibition.
33. The provision of fixed zoo cages.

Capital allowances on plant and machinery are given in the form of writing down allowances at the rate of 18% per annum on a reducing balance basis. For every £100 of qualifying expenditure £18 is claimable in year 1, £14.76 in year 2 and so on until either the all the allowances have been claimed or the asset is sold.

Integral features

The category of qualifying expenditure on 'integral features' was introduced with effect from April 2008. The following items are integral features:

- An electrical system (including a lighting system)
- A cold water system
- A space or water heating system, a powered system of ventilation, air cooling or air purification, and any floor or ceiling comprised in such a system
- A lift, an escalator or a moving walkway
- External solar shading

A reduced writing down allowance of 8% per annum is available on integral features.

Thermal insulation

For many years the addition of thermal insulation to an existing industrial building has been treated as qualifying for plant and machinery allowances. From April 2008 this has been extended to include all commercial buildings but not residential buildings.

A reduced writing down allowance of 8% per annum is available on thermal insulation.

Long-life assets

A reduced writing down allowance of 8% per annum is available on long-life assets. Allowances were given at the rate of 6% before April 2008.

A long-life asset is defined as plant and machinery that can reasonably be expected to have a useful economic life of at least 25 years. The useful economic life is taken as the period from first use until it is likely to cease to be used as a fixed asset of any business. It is important to note that this likely to be a shorter period than an item's physical life.

Plant and machinery provided for use in a building used wholly or mainly as dwelling house, showroom, hotel, office or retail shop or similar premises, or for purposes ancillary to such use, cannot be long-life assets.

In contrast plant and machinery assets in buildings such as factories, cinemas, hospitals and so on are all potentially long-life assets.

Case law

The fact that an item appears in List C does not automatically mean that it will qualify for capital allowances. It only means that it may potentially qualify.

Guidance about the meaning of plant has to be found in case law. The cases go back a long way, beginning in 1887. The current state of the law on the meaning of plant derives from the decision in the case of *Wimpy International Ltd and Associated Restaurants Ltd v Warland* in the late 1980s.

The Judge in that case said that there were three tests to be applied when considering whether or not an item is plant.

1. Is the item stock in trade? If the answer yes, then the item is not plant.
2. Is the item used for carrying on the business? In order to pass the business use test the item must be employed in carrying on the business; it is not enough for the asset to be simply used in the business. For example, product display lighting in a retail store may be plant but general lighting in a warehouse would fail the test.

3. Is the item the business premises or part of the business premises? An item cannot be plant if it fails the premises test, i.e. if the business use is as the premises (or part of the premises) or place on which the business is conducted. The meaning of part of the premises in this context should not be confused with the law of real property. The Inland Revenue's internal manuals suggest there are four general factors to be considered, each of which is a question of fact and degree:
   - Does the item appear visually to retain a separate identity
   - With what degree of permanence has it been attached to the building
   - To what extent is the structure complete without it
   - To what extent is it intended to be permanent or alternatively is it likely to be replaced within a short period

There is obviously a core list of items that will usually qualify in the majority of cases. However, many other still need to be looked at on a case-by-case basis. For example, decorative assets in a hotel restaurant may be plant but similar assets in an office reception area would almost certainly not be.

One of the benefits of the integral features rules, apart from simplification, is that items that did not qualify by applying these rules, such as general lighting in a warehouse or an office building, will now qualify albeit at a reduced rate.

Refurbishment schemes

Building refurbishment projects will typically be a mixture of capital costs and revenue expenses, unless the works are so extensive that they are more appropriately classified a redevelopment. A straightforward repair or a 'like for like' replacement of part of an asset would be a revenue expense, meaning that the entire amount can be deducted from taxable profits in the same year.

Where capital expenditure is incurred that is incidental to the installation of plant or machinery then Section 25 of the 2001 Act allows it to be treated as part of the expenditure on the qualifying item. Incidental expenditure will often include parts of the building that would be otherwise disallowed, as shown in the Lists reproduced above. For example, the cost of forming a lift shaft inside an existing building would be deemed to be part of the expenditure on the provision of the lift.

The extent of the application of section 25 was reviewed for the first time by the Special Commissioners in December 2007 and by the First Tier Tribunal (Tax Chamber) in December 2009, in the case of JD Wetherspoon. The key areas of expenditure considered were overheads and preliminaries where it was held that such costs could be allocated on a pro-rata basis; decorative timber panelling which was found to be part of the premises and so ineligible for allowances; toilet lighting which was considered to provide an attractive ambience and qualified for allowances; and incidental building alterations of which enclosing walls to toilets and kitchens and floor finishes did not qualify but tiled splash backs, toilet cubicles and drainage did qualify along with the related sanitary fittings and kitchen equipment.

Annual investment allowance

The annual investment allowance is available to all businesses of any size and allows a deduction for the whole of the first £250,000 from January 2013 (£25,000 before January 2013) of qualifying expenditure on plant and machinery, including integral features and long life assets.

The enhanced capital allowances scheme

The scheme is one of a series of measures introduced to ensure that the UK meets its target for reducing greenhouse gases under the Kyoto Protocol. 100% first year allowances are available on products included on the Energy Technology List published on the website at www.eca.gov.uk and other technologies supported by the scheme. All businesses will be able to claim the enhanced allowances, but only investments in new and unused machinery and plant can qualify.

There are currently 17 technologies and 53 sub-technologies currently covered by the scheme:

- Air-to-air energy recovery
- Automatic monitoring and targeting (AMT)
- Boiler equipment
- Combined heat and power (CHP)
- Compact heat exchangers
- Compressed air equipment
- Heat pumps
- Heating ventilation and air conditioning equipment
- High speed hand air dryers
- Lighting
- Motors and drives
- Pipework insulation
- Radiant and warm air heaters
- Refrigeration equipment
- Solar thermal systems
- Thermal screens
- Uninterruptible power supplies

The Finance Act 2003 introduced a new category of environmentally beneficial plant and machinery qualifying for 100% first-year allowances. The Water Technology List includes 14 technologies:

- Cleaning in place equipment
- Efficient showers
- Efficient taps
- Efficient toilets
- Efficient washing machines
- Flow controllers
- Leakage detection equipment
- Meters and monitoring equipment
- Rainwater harvesting equipment.
- Small-scale slurry and sludge dewatering equipment.
- Vehicle wash water reclaim units
- Water efficient industrial cleaning equipment
- Water management equipment for mechanical seals
- Water reuse

Buildings and structures and long life assets as defined above cannot qualify under the scheme. However, following the introduction of the integral features rules, lighting in any non residential building may potentially qualify for enhanced capital allowances if it meets the relevant criteria.

A limited payable ECA tax credit equal to 19% of the loss surrendered was also introduced for UK companies in April 2008.

From April 2012 expenditure on plant and machinery for which tariff payments are received under the renewable energy schemes introduced by the Department of Energy and Climate Change (Feed-in Tariffs or Renewable Heat Incentives) will not be entitled to enhanced capital allowances.

Enterprise zones

The creation of 11 new enterprise zones was announced in the 2011 Budget. Additional zones have since been added bringing the number to 24 in total. Originally introduced in the early 1980s as a stimulus to commercial development and investment, they had virtually faded from the real estate psyche.

The original zones benefited from a 100% first year allowance on capital expenditure incurred on the construction (or the purchase within two years of first use) of any commercial building within a designated enterprise zone, within ten years of the site being so designated. Like other allowances given under the industrial buildings code the building has a life of twenty-five years for tax purposes.

The majority of these enterprise zones had reached the end of their ten-year life by 1993. However, in certain very limited circumstances it may still be possible to claim these allowances up to twenty years after the site was first designated.

Enterprise zones benefit from a number of reliefs, including a 100% first year allowance for new and unused non-leased plant and machinery assets, where there is a focus on high-value manufacturing.

Flat conversion allowances

Tax relief is available on capital expenditure incurred on or after 11 May 2001 on the renovation or conversion of vacant or underused space above shops and other commercial premises to provide flats for rent.

In order to qualify the property must have been built before 1980 and the expenditure incurred on, or in connection with:

- Converting part of a qualifying building into a qualifying flat.
- Renovating an existing flat in a qualifying building if the flat is, or will be a qualifying flat.
- Repairs incidental to conversion or renovation of a qualifying flat, and
- The cost of providing access to the flat(s).

The property must not have more than four storeys above the ground floor and it must appear that, when the property was constructed, the floors above the ground floor were primarily for residential use. The ground floor must be authorized for business use at the time of the conversion work and for the period during which the flat is held for letting. Each new flat must be a self-contained dwelling, with external access separate from the ground-floor premises. It must have no more than 4 rooms, excluding kitchen and bathroom. None of the flats can be 'high value' flats, as defined in the legislation. The new flats must be available for letting as a dwelling for a period of not more than 5 years.

An initial allowance of 100% is available or, alternatively, a lower amount may be claimed, in which case the balance may be claimed at a rate of 25% per annum in subsequent a years. The allowances may be recovered if the flat is sold or ceases to be let within 7 years.

At Budget 2011 the Government announced that the relief would be abolished from April 2013.

Business premises renovation allowance

The Business premises renovation allowance (BPRA) was first announced in December 2003. The idea behind the scheme is to bring long-term vacant properties back into productive use by providing 100% capital allowances for the cost of renovating and converting unused premises in disadvantaged areas. The legislation was included in Finance Act 2005 and was finally implemented on 11 April 2007 following EU state aid approval.

The legislation is identical in many respects to that for flat conversion allowances. The scheme will apply to properties within one of the areas specified in the Assisted Areas Order 2007 and Northern Ireland.

BPRA will be available to both individuals and companies who own or lease business property that has been unused for 12 months or more. Allowances will be available to a person who incurs qualifying capital expenditure on the renovation of business premises.

An announcement to extend the scheme by a further five years to 2017 was made within the 2011 Budget.

Other capital allowances

Other types of allowances include those available for capital expenditure on Mineral Extraction, Research and Development, Know-How, Patents, Dredging and Assured Tenancy.

Enhanced Capital Allowances (ECAs), The Energy Technology List (ETL) and The Water Technology List (WTL)

Background

This bulletin is intended to raise awareness of the existence, location and relevance of the ETL and the WTL which are often referred to for simplicity as 'The Technology Lists'. The lists comprise the technologies and products which qualify for the UK Government's Enhanced Capital Allowances (ECA) scheme. Also included in the lists are the energy-saving and water-saving performance criteria for each product or technology.

What are ECAs?

Enhanced Capital Allowances are a form of tax relief enabling a business to claim 100% first-year capital allowances on their spending on qualifying plant and machinery instead of the normal reducing balance writing down allowances of 8% for integral features and 18% for main pool plant. There are two building services-related schemes for ECAs:

- Energy-saving plant and machinery
- Water conservation plant and machinery

Businesses can write off the whole of the capital cost of their investment in these technologies against their taxable profits of the period during which they make the investment. This can deliver a helpful cash flow boost and a shortened payback period via both the energy, or water, saved and the allowances claimed. Alternatively a loss making business can surrender the ECA value of the loss to obtain a 19% tax credit.

ECAs are only available to investors of qualifying plant and machinery and not developers. Developers are 'traders' and they incur revenue expenditure and not capital expenditure and so cannot claim capital allowances. However the purchaser of a newly constructed development for investment can still make use of ECAs provided they are sold unused and the purchaser is a taxpayer. This can be a valuable selling point for developers if highlighted in the marketing literature.

Interest in ECAs has been highlighted by the changes to the existing capital allowances regime which have reduced the value of allowances for many qualifying items. The introduction of Energy Performance Certificates and the Carbon Reduction Commitment Energy Saving Scheme will also promote the types of technologies that could qualify for ECAs.

Who manages the technology lists and where are they?

The lists are part of the ECA scheme which was developed by the Treasury, HM Revenue & and the Department of Energy & Climate Change (DECC). The ETL is managed and maintained by the Carbon Trust and can be found at http://etl.decc.gov.uk/. The WTL is managed by Business Link and the Department for Environment Food & Rural Affairs (DEFRA) and can be found at http://wtl.defra.gov.uk.

What technologies and products are included in the ETL? (as at 2013)

- Air-to-air energy recovery
- Automatic monitoring and targeting
- Boiler equipment
- Combined heat and power (CHP)
- Compact heat exchangers
- Compressed air equipment
- Heat pumps
- Heating ventilation and air conditioning (HVAC) equipment
- High speed hand air dryers
- Lighting (high efficiency lighting units, lighting controls and LEDs)
- Motors and drives
- Pipework insulation
- Radiant and warm air heaters
- Refrigeration equipment
- Solar thermal systems
- Thermal screens
- Uninterruptible power supplies

And the WTL?

- Cleaning in place equipment
- Efficient showers
- Efficient taps
- Efficient toilets
- Efficient washing machines
- Flow controllers
- Leakage detection equipment
- Meters and monitoring equipment
- Rainwater harvesting equipment
- Small-scale slurry and sludge dewatering equipment
- Vehicle wash water reclaim units
- Water efficient industrial cleaning equipment
- Water management equipment for mechanical seals
- Water reuse

Eligibility criteria

It is not possible to summarize the eligibility criteria here because each technology has different definitions and considerations. Most of the above technologies have specific products listed that are eligible, but others are defined by detailed performance criteria. The key technologies worthy of note which operate on a Performance criteria are Lighting, CHP, Pipework Insulation, Automatic Monitoring & Targeting equipment and Water Reuse. Qualification is generally extremely strict. To give a brief example; Lighting doesn't merely concentrate on the lamp type used but also has very strict requirements for the light fitting itself, having to meet high Light Output Ratios, colour rendering requirements and use high frequency control equipment. CHP and AMT equipment is also heavily guided.

Scope of ECA claims

For products on the Technology Lists, claims will be considered for the cost of the equipment itself, and other costs directly involved in installing it. These include:

- Transportation – the cost of getting equipment to the site.
- Installation – cranage (to lift heavy equipment into place), project management costs and labour, plus any necessary modifications to the site or existing equipment.
- Preliminary costs and oncosts – if they are directly related to the acquisition and installation of the equipment.
- Professional Fees – if they are directly related to the acquisition and installation of the equipment.

Making a claim for ECAs

If investing in eligible technology claimants should submit their ECA claims as part of their normal Income or Corporation Tax return. The best advice we can give clients on relevant projects is to highlight the fact that their investment may be eligible, and introduce Davis Langdon Banking Tax & Finance to the client as early as possible in the design process so that they can offer a view on the potential for worthwhile tax relief before, during and after the investment programme.

Follow-up

For further general advice on any aspect of this Bulletin please contact Andy White of Davis Langdon Banking Tax & Finance.

LED Lighting

LED Lighting – is it the 'best thing since sliced bread'?

Background

Current LED technology is becoming a viable alternative to other energy efficient lighting solutions in a variety of applications which include external lighting, industrial and office environments. With careful selection, LED systems can deliver the qualitative design requirements without compromise to the overall visual comfort and ambiance of the space. LED characteristics of high light output, good light colour quality and rendering and engineered product efficiency are beginning to become comparable to fluorescent lighting systems, however, there is an indeterminate risk regarding the longevity and stability of LED operating characteristics simply because the latest LEDs have not been around long enough. They are relatively untested on the marketplace when compared to a proven technology of fluorescent.

Currently available types of LED luminaries

The initial concept of LED lighting was developed for electronic devices, signal lighting, and decorative external lighting. In continued development, high output 'white' LED have been designed and engineered to replace traditional lighting equipment for more energy efficient systems.

Current market LEDs are available with lumen efficacy up to 120 lumens, provide high lumen maintenance, long life, and provide instant dimmable light. LEDs are also noted for their benefits of no infrared and ultra-violet radiation, no mercury content, and a high degree of recyclable content for a sustainable lighting solution.

Recent LED installations have included office environments, restaurants, hotels, roadways and car parks, and feature lighting. With improvements in high colour rendering white LEDs, street lighting installations have the side benefit of dramatically improving the resolution of CCTV coverage to areas such as car parks and building elevations.

- Colour temperature range (2,500–10,000°K)
- Colour rendering (65–95r CIE Ra)
- RGB vivid colours (millions)
- Dimmable
- Instant light
- Compact & robust (can integrate into fabric of built environment)
- Retrofit & new build applications
- Most of content recyclable & no mercury
- No IR or UV radiation

Warning though!

It should be noted the marketplace is inundated with varying qualities of LEDs. There are no agreed standardized formats leading to a myriad of different styles, types and colours available. As LED technology is pioneered and developed by semi-conductor electronic manufacturers, the industry can no longer rely on the traditional lamp manufacturers to set relatively clear guidance to qualitative aspects of their products. With this lack of clarity, some suppliers are taking advantage by flogging inferior quality LEDs with misleading and unreliable data.

The LED industry is also hindered by its own high speed of development. Month by month new developments are introduced. This is leading to hesitation in commitment to the technology from lighting manufacturers, designers, and clients leading to questions like, 'will a replacement LED module be available in the event of failure?'

Currently, a great deal of trust and faith is put in manufacturer's ability to provide proper documentation and support for their products. It is important to use the manufacturers that can provide sufficient data and that they have a reliable reputation in the marketplace. A manufacturer should be able to advise on issues of thermal management, colour consistency and appearance (binning), colour rendering, dimming capability, true lumen output and efficacy, operational lumen depreciation, and ideally provide independent tested photometric reports, evidence of electrical safety and thermal testing.

Finally, it is important to get the manufacturer to prove their products and design. For a true LED replacement lighting system, the design scheme should not compromise on the qualitative demands defined in good practice design guidances. For example, the market place has various 'replacement tubes' available where on the face of it you simply install a LED 'tube' – this is not the case. The 'tube' is often simply a linear strip of LEDs within a glass tube. The retrofitted luminaire may require modifications, and the photometric performance may not be directly comparable. It should also be remembered that all diffusers are specially designed for a fluorescent tube and will probably adversely affect the light distribution and will be completely different to that of a normal tube. So caution, when dealing with these, if requested by a client.

Cost implications

Manufacturers and suppliers make great hay regarding the energy savings that can be made against more traditional light sources.

The manufacturers' cost calculations should be carefully examined as in some instances they base their calculations on several incorrect assumptions such as they may state 365 twelve hour days but, even after correcting these incorrect assumptions, the payback and subsequent energy cost savings are dramatic.

Payback

As the LED luminaires are currently more expensive than more traditional sources of illumination the payback period is of particular importance as actual energy savings will only be realized after the LED luminaires if you like have paid for itself.

Therefore based on the following material costs only as it is assumed the labour content will be similar:

Power cost (£)/kWh	0.12
Days in use/year	260
Hours in use/day	12

	Typical LED Downlight	Compact Fluorescent Downlight	LV Halogen (IRC) Downlight
Input Power (Watts)	16	28	35
Lifetime (hours)	50,000	12,000	4,000
Replacement Lamp Cost including labour (£)	–	15.00	15.00
Annual Energy Cost per lamp (£)	6.04	10.57	13.21
Total Energy Cost for 25 luminaires per year (£)	151.01	264.28	330.28
Luminaire Unit Cost (£)	130.00	70.00	80.00

Supply only cost of 25 luminaires (£)	3,250.00	1,750.00	2,000.00
Allowance for 4nr emergency battery packs (£)	320.00	320.00	320.00
Lamps (£)	Included	375.00	375.00
Total cost of luminaries/emergencies/ lamps (£)	3,570.00	2,445.00	2,695.00

Extra over cost of LED luminaires and emergency luminaires compared to traditional luminaires is (£)		1,125.00	875.00
Yearly cost of energy and yearly re-lamping allowance (compact fluorescent @ every 4 years; halogen @ every 1.5 year) (£)	151.01	358.03	580.28
The calculated yearly energy and re-lamping saving of using LED luminaires when compared to traditional luminaires is (£)		207.02	429.27
Therefore time taken in years to 'pay' for the additional cost of the LED luminaires based on the energy and re-lamping costs of 'traditional' luminaires in **years** is		**5.43**	**2.04**

Power cost (£)/kWh	0.12
Days in use/year	260
Hours in use/day	12

	Recessed LED Office Luminaire	Recessed Fluorescent Office Luminaire
Input Power (Watts)	37.5	42
Typical Lifetime (hours)	50,000	12,000
Replacement Lamp Cost including labour (£)	–	20.00
Annual Energy Cost per lamp (£)	14.15	15.85
Total Energy Cost for 25 luminaires per year (£)	353.93	396.28
Luminaire Unit Cost including lamps (£)	225.00	165.00
Supply only cost of 25 luminaires including lamps (£)	5,625.00	4,125.00
Allowance for 4nr emergency battery packs (£)	520.00	400.00
Total cost of luminaries/emergencies/lamps (£)	6,145.00	4,525.00

Extra over cost of LED luminaires and emergency luminaires compared to fluorescent luminaires is (£)		1,620.00
Yearly cost of energy and yearly re-lamping allowance (fluorescent @ every 4 years) (£)	353.93	521.28
The calculated yearly energy and re-lamping saving of using LED luminaires when compared to fluorescent luminaires is (£)		167.35
Therefore time taken in years to 'pay' for the additional cost of the LED luminaires based on the energy and re-lamping costs of fluorescent luminaires in **years** is		**9.68**

Notes and conclusions

1. LED luminaires do save energy.
2. LED luminaires are generally twice the price of a 'traditional' downlight but LED luminaire prices are becoming more competitive as the technology develops.
3. Some LED luminaires may not be converted to emergency so 'additional' luminaires may be required.
4. LED replacement lamps may have completely different light emitting characteristics. And, the luminaire may require modification to accept the retrofit LED lamp.

Getting the Connection

Details the process involved in the provision of a new electrical supply to a site

The provision of an electricity connection has both a physical and contractual element. Physically, it involves the design, planning and construction of electrical infrastructure (cables, switchgear, civil works), whilst contractually, it requires legal agreements to be drawn up and agreed (construction, connection, adoption).

Planning

The Planning Stage, typically RIBA Stage C, is where the site's developer (the Developer) should be formulating their plans for the scheme and, in doing so, consult the local distribution network operator's (the Host)[1] long-term development statement, which will identify potential connection point opportunities. This information is normally readily available from the Host and should provide an early indication of whether the Host's network may need to be reinforced before the development can be connected.

As the Planning Stage progresses, the Developer should discuss its proposals with the Host. Relatively simple connections for single building supplies are straight forward, however schemes of a more complex nature may require some form of feasibility study to be carried out to assess connection options and provide indicative costs for the contestable[2] and non-contestable[3] work elements.

Design

The Design Stage, typically RIBA Stage D, is the point at which the Developer submits its formal connection request to the Host. It is important that this is completed in accordance with the specific procedures of the Host, since if it does not include all supporting information required by the Host's application process, there is likely to be a delay in the processing of a firm offer. The Host is expected to provide a documented application process to assist the applicant make a complete application.

The convention is to request a Section 16/16 A connection[4] where the terms are standard and non-negotiable. The alternative is a Section 22 connection[2] offer, where the terms are fully negotiable. If the Section 22 route is chosen then caution is required, as disputes over Section 22 agreements cannot be referred to the regulator for determination after the connection agreement has been signed; a post contract dispute will have to be pursued as a civil action. It is always advised that a Section 16/16 A offer is requested before considering the Section 22 option.

On receipt of the Host's firm offer[5], the Developer generally has 90 days to accept its terms and to undertake its own review of the offer to ensure it meets its requirements. If for any reason the Developer and Host are unable to reach agreement of the terms, it is recommended that the Developer seeks specialist advice. It should be noted that in extreme circumstances, it may be necessary to refer an issue to regulator for determination but, be warned, this can be a lengthy process, taking upto 16 weeks to conclude.

1. A distribution network operator (DNO) is a company that is responsible for the design, construction operation and maintenance of a public electricity distribution network. The host DNO is the electricity distribution network to which the development site will directly connect.
2. Contestable is work in providing the connection that can be carried out by an accredited independent party.
3. Non contestable is work that can only be carried by the Host.
4. The Electricity Act 1989.
5. Unless the offer becomes interactive, i.e. another connection scheme is vying for network capacity. Where the connection request becomes interactive the Developer has 30 days to accept the offer. The Host will inform the Developer if the connection request is interactive.

Competition

One of the key decisions affecting the way in which the connection process proceeds is whether the Developer wishes to introduce competition into the procurement process by appointing a third party to design and construct the contestable connection works. In these circumstances, the Developer can requisition a non contestable quotation from the Host or ask the third party to do so on its behalf. The Host is obliged to provide information within standardized time frames against which its performance is monitored by the industry regulator (Ofgem).

Contestable works are those that may be carried out either by the Host or by an approved contractor, on the Developer's behalf. This contrasts with non contestable works, which can only be carried out by the Host. Broadly speaking, the Host will make the connection to its network for the new supply and undertake any upstream network reinforcement works. All works downstream from the point of connection to the Host network into the site and to each building are contestable works. Therefore, the extent of the contestable works can vary significantly depending on where the point of connection is designated by the Host. Alternatively, the Developer can request an independent licensed network operator (IDNO) to adopt assets constructed by the third party.

If the Developer decides to contract with a third party to construct the contestable works, it is the Developer's responsibility to ensure that the construction works meet the Host's network adoption requirements.

Network reinforcement

Reinforcement works may be required to increase the capacity of the network to enable the connection to meet a site's projected demand. In terms of the capacity made available by reinforcement, the following possible scenarios arise:

- Where reinforcement works are necessary for sole use by the Developer, the Developer is charged the full cost of the works. In some circumstances the most economic method of reinforcement may introduce spare network capacity in excess of the Developer's requirements. In such circumstances the Developer can receive a rebate where this spare capacity is absorbed by subsequent developments. However, the possibility of a rebate is time constrained and the original development will only qualify for a rebate for up to 5 years after the connection is completed.
- Where reinforcement works are necessary but also cater for the Host's future network requirements, the Developer is charged for a proportion of the cost of the works in the form of a capital contribution. This is calculated using cost apportionment factors for security (the ratio of capacity requested to that which is made available) and fault level (the ratio of fault level contribution of connection to that which is made available).

The charges levied for reinforcement works are attributable to those reinforcement works undertaken at one voltage level above the connection voltage only. That is, if connection voltage is 400 V then costs are chargeable for reinforcement works undertaken at the next highest voltage, which is usually 11 kV, and not for works conducted at the next highest voltage, which is usually 33 kV[6]. These deeper network reinforcement costs are recovered as part of the system charges built into the electricity supply tariffs.

Costs

The cost of the Developer's connection depends on the nature and extent of the works to be undertaken. The distance between the site and the Host's network, the size of the customer demand in relation to available capacity (and hence the potential need for reinforcement), customer-specific timescales and to an extent market value of raw material and labour, are all significant factors that will affect the cost.

As part of the firm offer received from the Host, the Developer is provided with a charging statement, which includes the charges to be levied for the following items:

6. Some DNOs use voltages of 6.6 kV and 22 kV

- Assessment and Design – to identify and design the most appropriate point on the existing network for the connection
- Design Approval – to ensure design of a connection meets the safety and operation requirements of the Host
- Non-contestable Works/Reinforcement – to include circuits and plant forming part of the connection that can be undertaken by the Host only, including land rights issues and consents
- Contestable Connection Works – to include circuits and plant forming part of the connection that can be undertaken by approved contractors or the Host
- Inspection of Works – to ensure that works are being constructed in accordance with the design requirements of the Host
- Commissioning of Works – to include circuit outages and testing that will ensure that connection is safe to be energized.

In some circumstances, in addition to the cost of the physical connection works there may be chargeable costs associated with operation, maintenance repair and replacement of the new or modified connection. These are known as O&M costs and are chargeable as capitalized up-front costs where the Developer requests a solution that is in excess of the minimum necessary to provide the connection. For example, where extra resilience is requested for a connection, over and above that which the Host is obligated to provide, then O&M costs can be levied but only for the extra resilience element of the connection, not the total cost of the connection.

Feed-In Tariffs (FITs)

To ensure that everyone plays their part towards achieving the Renewable Energy Strategy's (RES) target of 15% for renewable energy by 2020, the need to develop new methods of renewable energy in all sectors is apparent.

On 1 February 2010, the Department for Energy and Climate Change (DECC) announced their intention to implement tariff levels following the Energy Act 2008 introduction of Feed-in Tariffs (FITs). The FIT scheme is intended to encourage the adoption of small-scale low carbon technologies up to a maximum of 5MW through tariff payments made on both generation and export of produced renewable energy.

Feed-in Tariffs (FITs) are effectively payments which are made for every kilowatt per hour of renewable electricity (kWhr) that is generated. Homeowners/businesses will have the opportunity to benefit financially from the scheme either by consuming it on-site, earning a generation tariff, or by selling spare capacity to the grid, generating an export tariff.

Amendment 2 to the Feed-in Tariffs Order 2010 came into effect on 1 August 2012. The schemes policy and tariff rates are set by the Government with the scheme being administered by energy suppliers and Ofgem.

Feed-in Tariffs will be paid by electricity suppliers. Generators between 50 kW and 5MW should apply to Ofgem for accreditation. Micro-generators can obtain accreditation via the Microgeneration Certification Scheme (MCS), before then applying to a supplier for a FIT agreement. All generation will be metered and FIT payments will be made in accordance with the Electricity Act 1989. The consumer will be responsible for the capital of the plant; the necessary access/connections to the electricity and transmission systems and organization of the payment receipts.

Since the scheme was introduced on 1 April 2010 and up to December 2012 a total of 358,658 renewable installations have been registered, with 1,657 MW of total installed capacity also being registered.

FITs review

The first review of the Feed-in Tariff scheme was scheduled to be carried out in 2012 however it was brought forward by a year due to the surge in uptake which placed a huge strain on the FITs budget. The surge was due to the sharp fall in the cost of solar PV installations and also the creation of large-scale solar farms, which went against the ethos of the scheme which was to promote small scale installations.

August 2011 saw the introduction of lower rates for large-scale solar installations. October 2011 the Department of Energy and Climate Control (DECC) launched a consultation into cuts for smaller-scale solar installations. The consultation resulted in the decision to reduce FITs to approximately half the original tariff, however this was brought to the High Court and after appeals were heard the ruling found the cuts to tariffs were illegal.

February 2012 saw the DECC announce wide ranging reforms to the FIT scheme which they hope will provide a better framework for consumers and communities. There are five key areas that the DECC has identified as needing refinements;

- The tariff for micro-CHP installations will be increased to recognize the benefits this technology could bring and to encourage its development
- A tariff of 21p/kWh to take effect from 1 April 2012 for domestic-size solar panels with an eligibility date on or after 3 March 2012
- Installations of solar PV panels on or after 1 April 2012 will be required to produce and Energy Performance Certificate rating of 'D' or above to qualify for a full FIT.

- 'multi-installation' tariff rates set at 80% of the standard tariffs will be introduced for solar PV installations where a single individual or organization is already receiving FITs for other solar PV installations.
- Individuals or organizations with 25 or fewer installations will still be eligible for the individual rate. DECC is consulting on a proposal that social housing, community projects and distributed energy schemes be exempt from multi-installation tariff rates.

Following the FITs review the Department of Energy and Climate Change (DECC) announced the first scheduled reduction in tariff rates for all PV installations with an eligibility date on or after 1 November 2012. The new tariff rates will be published by Ofgem by 31 August 2012. Tariff rates will be continuously reviewed as set out in the degression mechanism as follows:

- Degression will take place on a quarterly basis, with tariffs changing on the first day of the first month of the quarter for new installations with an eligibility date from that date.
- Tariffs will be published at least two months before the start of the new quarter, and will be based on deployment of new PV generating capacity in the previous quarter.
- The baseline degression rate will be 3.5% per quarter.
- Degression will be skipped if deployment is below a floor threshold.
- Deployment statistics will be published on a monthly basis by DECC.

Eligibility

Small-scale (up to a maximum capacity of 5MW) low-carbon electricity technologies eligible for FITs are:

- Wind
- Solar Photovoltaics (PV)
- Hydro
- Anaerobic Digestion
- Domestic scale micro Combined Heat & Power (CHP), eligible up to 2 kW.

As of 1 April 2010, micro generators (<50 kW) in Anaerobic Digestion, Hydro, Solar PV and Wind will be ineligible for support under the Renewable Obligation (RO) due to FIT inclusion. Refurbished/renovated installations will be ineligible. Small generators that applied for accreditation on or after the 15 July 2009 and before 1 April 2010 will be eligible to transfer to FITs.

Generating stations whose electricity is sold under a Non-Fossil Fuel Obligation (NFFO) arrangement will be ineligible for FITs; however will remain eligible to receive RO support.

Any existing systems installed before 15 July 2009 will only qualify for the FIT scheme if they are under 50 kW and registered with the RO.

From 1 December 2012, new provisions have been introduced for schools and some community organizations.

Tariffs

The Feed-in Tariffs (FITs) provide three main financial benefits to the consumer:

- A 'Generation' tariff (up to 21.65p/kWh depending on type and size) based on the total generation and the energy type.
- An 'export' tariff (4.50p/kWh) for any surplus energy produced exported into the grid and due to producing a percentage of the energy used.
- On-site usage of generated electricity, therefore reducing the amount purchased from an electricity supplier.

Energy Source	Scale	Effective Dates	Applicable Rates	Tariff (p/k/Wh)
Anaerobic Digestion	≤250 kW	to 31 March 2014		15.16
Anaerobic Digestion	>250 kW–500 kW	to 31 March 2014		14.02
Anaerobic Digestion	>500 kW–5 MW	to 31 March 2014		9.24
Hydro generating station	≤15 kW	to 31 March 2014		21.65
Hydro generating station	>15 kW–100 kW	to 31 March 2014		20.21
Hydro generating station	>100 kW–500 kW	from 15 March 2013		15.98
Hydro generating station	>500 kW–2 MW	to 31 March 2014		12.48
Hydro generating station	>2 MW–5 MW	to 31 March 2014		3.23
Combined Heat & Power (CHP)	≤2 kW	from 15 March 2013	Limited to 30,000 units	12.89
Solar photovoltaic	≤4 kW	to 30 September 2013	Higher	14.90
Solar photovoltaic	≤4 kW	to 30 September 2013	Middle	13.41
Solar photovoltaic	≤4 kW	to 30 September 2013	Lower	6.85
Solar photovoltaic	>4 kW–10 kW	to 30 September 2013	Higher	13.50
Solar photovoltaic	>4 kW–10 kW	to 30 September 2013	Middle	12.15
Solar photovoltaic	>4 kW–10 kW	to 30 September 2013	Lower	6.85
Solar photovoltaic	>10 kW–50 kW	to 30 September 2013	Higher	12.57
Solar photovoltaic	>10 kW–50 kW	to 30 September 2013	Middle	11.31
Solar photovoltaic	>10 kW–50 kW	to 30 September 2013	Lower	6.85
Solar photovoltaic	>50 kW–100 kW	to 30 September 2013	Higher	11.10
Solar photovoltaic	>50 kW −100 kW	to 30 September 2013	Middle	9.99
Solar photovoltaic	>50 kW–100 kW	to 30 September 2013	Lower	6.85
Solar photovoltaic	>100 kW–150 kW	to 30 September 2013	Higher	11.10
Solar photovoltaic	>100 kW–150 kW	to 30 September 2013	Middle	9.99
Solar photovoltaic	>100 kW–150 kW	to 30 September 2013	Lower	6.85
Solar photovoltaic	>150 kW–250 kW	to 30 September 2013	Higher	10.62
Solar photovoltaic	>150 kW–250 kW	to 30 September 2013	Middle	9.56
Solar photovoltaic	>150 kW–250 kW	to 30 September 2013	Lower	6.85
Solar photovoltaic	>250 kW–5 MW	to 30 September 2013		6.85
Wind	≤1.5 kW	to 31 March 2014		21.65
Wind	>1.5 kW–15 kW	to 31 March 2014		21.65
Wind	>15 kW–100 kW	to 31 March 2014		21.65
Wind	>100 kW–500 kW	to 31 March 2014		18.04
Wind	>500 kW–1.5 MW	to 31 March 2014		9.79
Wind	>1.5MW–5MW	to 31 March 2014		4.15

Carbon Trading

What is carbon trading?

Carbon trading is currently the central pillar of the Kyoto Protocol and other international agreements aimed at slowing climate change.

Emissions' trading (also known as cap and trade) is a market-based approach used to control pollution by providing economic incentives for achieving reductions in the emissions of pollutants. Carbon emissions' trading is a form of emissions trading that specifically targets carbon dioxide.

A central authority (usually a governmental body) sets a limit or cap on the amount of a pollutant that can be emitted. The limit or cap is allocated or sold to firms in the form of emissions permits which represent the right to emit or discharge a specific volume of the specified pollutant. Firms are required to hold a number of permits (or credits) equivalent to their emission levels. The total number of permits cannot exceed the cap, limiting total emissions to that level. Firms that need to increase their emission permits must buy permits from those who require fewer permits. The transfer of permits is referred to as a trade. In effect, the buyer is paying a charge for polluting, while the seller is being rewarded for having reduced emissions. Thus, in theory, those who can reduce emissions most cheaply do so, achieving the pollution reduction at the lowest cost to society.

Carbon trading in the UK

New rules set out by the government will pitch some of the UK's largest organizations against each other in a drive to cut carbon dioxide (CO_2) emissions.

For the first time large non-energy intensive organizations, which account for about 10% of UK CO_2 emissions, will be legally bound to closely monitor and report their emissions from energy use in preparation for carbon trading.

The Government scheme, known as the CRC Energy Efficiency Scheme, will include businesses will be ranked according to reductions in energy use and improvements in energy efficiency alongside public sector organizations such as NHS trusts, local authorities and government departments.

Analysis for the Environment Agency suggests that the scheme could reduce CO_2 emissions by up to 11.6 million tonnes per year by 2020, the equivalent of taking four million cars off the road. It is also expected to save organizations money through reduced energy bills, benefiting the economy by at least £1billion by 2020.

During the introductory phase in 2012, allowances are being sold at a fixed price of £12 per tonne of CO_2. A further 15,000 organizations that use less than 6,000MWh, but still have at least one half hourly electricity meter, will be obliged to register and declare their electricity use.

How does the EU Emission Trading System (EU ETS) work?

Each EU member state must develop a National Allocation Plan (NAP) approved by the European Commission. This sets an overall cap on the total emissions allowed from all the installations covered by the system. This is converted into allowances (one allowance equals one tonne of CO_2), which are then distributed by EU member states to installations covered by the system.

At the end of each year, installations are required to surrender allowances to account for their actual emissions. They may use all or part of their allocation. Installations can emit more than their allocation by buying allowances from the market. Similarly, an installation that emits less than its allocation can sell its surplus allowances. The environmental outcome is not affected because the amount of allowances allocated is fixed.

CRC scheme

The Carbon Reduction Commitment Energy Efficiency Scheme (CRC) was introduced in 2010 following years of design and many rounds of consultation. Registration and the making of Information Disclosures took considerable effort for many organizations and most hoped that things would settle down from 30 September 2010. But before a month was out the Comprehensive Spending Review heralded further change, which was consulted on over a 4 week period in the run up to Christmas 2010. And in January 2011 a further five Government discussions papers were issued in an attempt to engage business in further simplification measures.

Qualification for the scheme is based on electricity supply across organizations and groups of undertakings, rather than at an individual site basis. Organizations that are a single entity, that is, they are not part of a group, will need to establish qualification for CRC across the sites that are part of that organization.

Organizations that are part of a group will be grouped under the highest parent undertaking.

Organizations qualified as participants for Phase 1 if, during the 2008 calendar year, they had at least one half-hourly electricity meter (HHM), settled on the half hourly market and if they consumed at least 6,000 MWh (megawatt hours) through all half-hourly meters.

If an organization consumed less than 6,000 MWh of electricity, they did not need to register for Phase 1.

A megawatt hour (MWh) is an amount of electricity. It does not mean watts per hour. A megawatt hour is the equivalent of a million-watt device being used for a period of one hour.

To put the CRC energy use into perspective, 100 computers running constantly 24 hours a day, for 273 years would equal 6,000 MWh. There are a 1,000 kilowatt hours (KWh) in a megawatt hour.

Organizations that qualified for Phase 1 may not for Phase 2. This will be especially important for organizations that were only just above the Introductory Phase threshold of 6000MWh electricity usage in 2008, and so qualify, but have made progress in energy reduction projects since then. If an organization consumes more than 6,000 MWh during the Phase 2 qualification year (1 April 2012 to 31 March 2013) then they will qualify for the scheme. This same principal applies to all planned future phases.

Phase 1 registration and information disclosures are closed. The Environment Agency is currently contacting those organizations that it thinks are responsible for half-hourly electricity meters that were not declared in those registrations and information disclosures.

Participants in Phase 1 (those who made full registrations) should be tracking and recording ALL their energy consumptions from 1 April 2010.

Participants should also be conducting internal audits, keeping records of them and be prepared for an Environment Agency audit. There is planned 20–25% coverage of random audits each year.

Year 1 (Qualification year) full energy reports needed to be submitted summer 2011 – Participants in Phase 1 needed to report their full year 1 energy consumptions (1 April 2010 to 31 March 2011) in the period 1 April 2011 to 31 July 2011. The report had to include ALL energy used, not just the electricity through the half hourly meters that were declared during registration. CO_2 credits are not required and do not need to be paid for in the qualification period. At the end of the 2nd year (registration period), CO_2 credits will be purchased based on the second year's carbon emissions.

Participants in Phase 1 have to buy their first tranche of emissions allowances in summer 2012 and they must buy enough to cover their actual emissions in year 2 (1 April 2011 to 31 March 2012). In other words they will be buying in arrears. Prices will be fixed at £12/tonne, equivalent to approximately 10% of the total cost of the energy itself.

There is no revenue recycling so CRC is now a flat tax at £12/tonne of CO_2 or roughly 10% of the energy bill itself.

The league table will still be published, based on the same performance metrics as before – absolute emissions, relative emissions and early action metrics – with the early action metrics tapering away over the first three years.

CRC Phase 1 has been extended and now runs until 31 March 2014, in other words it has four reporting years rather than the previous three.

The Phase 2 qualification year is 1 April 2012 to 31 March 2013. Phase 2 registration occurs in the period 2013/2014. The first reporting year in Phase 2 will be 1 April 2013 to 31 March 2014 and the fifth reporting year of Phase 2 will end 31 March 2018. The third phase will begin on 1 April 2018.

Simplifying the CRC energy efficiency acheme

In 2011, the Government set out its proposals for the future of the scheme. Responding to feedback received, the government proposed a number of simplifications to the CRC scheme which, subject to further analysis and formal consultation, will be implemented from Phase 2 onwards (where registration for Phase 2 begins in April 2013).

Amongst other simplifications the Government is proposing to:

- **Reduce the number of fuels covered by the scheme**
  Under the current scheme, businesses have to report on the emissions from 29 different fuels and we propose reducing this to four: electricity, gas, kerosene and diesel for heating. This will cover over approximately 95% of emissions captured under the CRC scheme while significantly reducing the administration burden of the scheme.
- **Move to fixed price allowance sales**
  Instead of establishing an emissions cap and holding annual auctions as proposed by the previous Administration we propose from the start of phase 2 in 2014 there should be two sales per year where the price of allowances is fixed. This would remove the need for businesses to come up with auctioning strategies and give price certainty to help investment decisions. A lower price in the first sale will incentivize good energy management and reward those who successfully forecast energy use.
- **Simplify the organizational rules**
  Abolish the need for large businesses to participate in groups which do not reflect their natural structure.
- **Make qualification processes easier**
  To make qualification a one step process instead of two. Previously businesses had to firstly prove they had a qualifying electricity meter and then declare they used a particular amount of electricity. This would be abolished in favor of participants just having to prove they use a certain amount of electricity from the qualifying meter.
- **Reducing overlap with other schemes**
  Any CCA or EU ETS site would be automatically exempt from the CRC scheme.

Carbon floor

What is a carbon floor price and why is it needed?

Creating a carbon floor price in the UK essentially requires our industries to pay a top up if the market price for carbon falls below a certain level.

A carbon floor price is a regulatory/taxation policy that states that polluters must pay a minimum amount of money for the right to pollute. This is likely to take the form of a tax that requires those who qualify to make a payment to the Treasury. It is expected to replace the existing Climate Change Levy, which is a downstream tax on energy use rather than a direct upstream tax on greenhouse gas pollution.

Roughly half of Europe's emissions are covered by a European regulation that caps emissions (the EU Emissions Trading Scheme) requiring them to submit sufficient permits to cover their emissions. Permits, known as allowances, can be freely traded and the price someone is willing to pay to acquire them determines the price of pollution. At present because there are too many allowances available in the market compared to the demand prices are relatively low – at around €15 per tonne of carbon dioxide equivalent. This low price is not necessarily enough to dissuade polluters from continuing to emit and does not provide an attractive enough return for would-be investors in low carbon solutions. There is also the risk that it could fall even lower. This lack of price certainty is seen as a potentially important barrier to investment.

A carbon floor price is therefore primarily designed to attract low carbon investment into a country by making the price of pollution higher and increasing the rewards for low carbon projects. As explained below it is not in and of itself an environmental policy and in terms of value for money it must be assessed as an industrial policy.

The UK could decide to impose an immediately effective floor price, which takes the price of pollution all the way up to this projected level of cost. This would however have a very significant impact on the competitiveness of UK industry relative to competitors in the rest of Europe. It is much more likely that the floor price will initially be set either at today's market levels (around £13–14 per tonne) or slightly above, incorporating the cost of the Climate Change Levy which is currently equivalent to around £4–6 per tonne. A price escalator may also be built-in to steadily increase the floor over time.

Renewable Obligation Certificates (ROCs)

The Renewable Obligation (RO) is the main support scheme for renewable electricity projects in the UK, placing an obligation on licensed UK suppliers of electricity to increase their proportion of electricity production from renewable sources, or result in a penalty.

Since its introduction in 2002, it has succeeded in tripling the level of renewable electricity in the UK from 1.8% to 5.4% and is currently worth around £1 billion/year in support to the renewable electricity industry. The RO applies to all powered plant with a power capacity greater than 5 MW. The target started at 3% and is presently at 11.4% rising incrementally to 15.4% by 2015. It is likely to be extended to 20% by 2020.

In April 2009, the introduction of banding under the Renewables Obligation Order 2009 meant different technologies receive different levels of support, providing a greater incentive to those that are further from the market.

The RO was extended from its current end date of 2027 to 2037, in April 2010, for new projects with a view to providing greater long-term certainty for investors and an increase in support for offshore wind projects.

The RO is administered by the Office of the Gas and Electricity Markets (Ofgem) and suppliers of electricity have to prove they have met this obligation, producing Renewable Obligation Certificates (ROCs) to renewable electricity generators at the end of each year.

A Renewable Obligation Certificate (ROC) is a green certificate that is issued by Ofgem to an accredited generator for eligible renewable electricity generated within the UK and supplied to customers in the UK by a licensed supplier. A ROC is issued for each megawatt hour (MWh) of eligible renewable output generated.

Failing to meet the obligation results in 'buy-out' fines being paid to Ofgem on the shortfall of every MWh sold that was not renewable. Ofgem then distributes the funds to all electricity supply companies possessing ROCs, the amount received being in proportion to the number of ROCs held (at the end of the year). If a supplier meets part or all of its RO, but other companies do not, the supplier who has ROCs will be rewarded with a share of the fines.

Previously, 1 ROC was issued for each megawatt hour (MWh) of eligible generation, regardless of technology. Since April 2009, the reforms introduced means that new generators joining the RO now receive different numbers of ROCs, depending on their costs and potential for large-scale deployment. For example, onshore wind continues to receive 1 ROC/MWh, whereas offshore wind and energy crops currently receive 2 ROCs/MWh.

Obligation periods are valid for a year, beginning on the 1 April to 31 March. Supply companies have until the 31 September following the period to submit sufficient ROCs to cover their obligation, or submit sufficient payment to Ofgem to cover their shortfall.

Buy-out price

Suppliers can meet all, or part of their obligations by making a buy-out payment. The buy-out price set by Ofgem for the compliance period of 2011–2012 is £38.69 per Renewables Obligation Certificate (ROC). The buy-out price sets the rate which suppliers must pay if they fail to meet their obligations under the scheme and is adjusted annually in accordance with the Retail Prices Index (RPI).

Buy-out fund redistribution

At the end of the year, the funds made to Ofgem are distributed to all the electricity suppliers possessing ROCs, with the amount received in proportion to the number of ROCs held. If a supplier meets all or part of its RO, it will be rewarded with a share of the buy-out fines.

Pricing

Due to ROCs having the potential to save the supplier from having to commit to a buy-out payment, it increases the price of the electricity. When the renewable generator sells the electricity to a supplier it is not uncommon for the ROC to be sold in addition ultimately forcing the cost of electricity upwards. Also, due to the fact that ROCs entitles suppliers to a share of the 'buy-out' fund at the end of year, increases its value. Electrical suppliers can benefit financially by participating in the RO system due to the renewable targets set by the Government likely to be under-fulfilled and the fact that the RO is not over-subscribed will result in the ROCs and their recycled values being worth more than the £38.69 per MWh.

e-ROC

The most efficient method of buying and selling Renewable Obligation Certificates (ROCs) is through the e-ROC on-line auctions. They offer renewable generators access to the whole supplier market in the UK, delivering high ROC prices for low fees. The average price of ROCs sold through the auctions 25 May 2011 was £51.24 and with the fees set at only 50p per ROC (subject to a minimum fee of £50), indicates a profitable return for those in participation. Auctions are operated by NFPAS, a subsidiary of the Non-Fossil Purchasing Agency Limited (NFPA) and are usually held four times a year. NFPAS runs regular e-ROC on-line auctions for the sale of Renewable Obligation Certificates (ROCs).

Eligibility

The reforms stated in Renewable Obligations Order 2009 introduced the concept of 'banding' for the Renewable Obligation Certificates (ROCs). The aim of ROC banding is to establish the number of ROCs per MWh that can be obtained according to the type of technology that is used to generate the renewable electricity.

There are 28 renewable technologies covered by ROCs Banding, resulting in an increasingly complex regulatory environment for technology providers, project developers and finance providers to navigate.

Band	**Renewable technology**	**Level of banding (ROCs/MWh)**
Established 1	Landfill Gas	0.25
Established 2	Sewage Gas Co-Firing of Non-Energy Crops (Regular) Biomass	0.5
Reference	Onshore Wind Hydro-Electric Co-Firing of Energy Crops Co-Firing of Biomass with CHP Energy from Waste with CHP Geo Pressure Pre-Banded Gasification Pre-Banded Pyrolysis Standard Gasification Standard Pyrolysis	1
Post-Demonstration	Offshore Wind Dedicated Regular Biomass Co-Firing of Energy Crops with CHP	1.5

Band	**Renewable technology**	**Level of banding (ROCs/ MWh)**
Emerging	Wave Tidal Steam Advanced Gasification Advanced Pyrolysis Anaerobic Digestion Dedicated Energy Crops Dedicated Energy Crops with CHP Dedicated Regular Biomass with HP Solar Photovoltaic Geothermal Tidal Lagoons Tidal Barrages	2

Typical Engineering Details

In addition to the Engineering Features, Typical Engineering Details are included. These are indicative schematics to assist in the compilation of costing exercises. The user should note that these are only examples and cannot be construed to reflect the design for each and every situation. They are merely provided to assist the user with gaining an understanding of the Engineering concepts and elements making up such.

ELECTRICAL

- Urban Network Mainly Underground
- Urban Network Mainly Underground with Reinforcement
- Urban Network Mainly Underground with Substation Reinforcement
- Typical Simple 11 kV Network Connection For LV Intakes Up To 1000 kVA
- Typical 11 kV Network Connections For MV Intakes 1000 kVA To 6000 kVA
- Static UPS System – Simplified Single Line Schematic For a Single Module
- Typical Data Transmission (Structured Cabling)
- Typical Networked Lighting Control System
- Typical Standby Power System, Single Line Schematic
- Typical Fire Detection and Alarm Schematic
- Typical Block Diagram – Access Control System (ACS)
- Typical Block Diagram – Intruder Detection System (IDS)
- Typical Block Diagram – Digital CCTV

MECHANICAL

- Fan Coil Unit System
- Displacement System
- Chilled Ceiling System (Passive System)
- Chilled Beam System (Passive or Active)
- Variable Air Volume (VAV)
- Variable Refrigerant Volume System (VRV)
- Reverse cycle heat pump

Urban network mainly underground

Details: Connection to small housing development 10 houses, 60 m of LV cable from local 11/LV substation route in footpath and verge, 10 m of service cable to each plot in verge

Supply Capacity: 200 kVA

Connection Voltage: LV

Nr of Phases: 1Φ

Breakdown of detailed cost information

	Labour	Plant	Materials	Overheads	Total
Cable	£250	£120	£ 1,200	£450	£2,020
Jointing	£1,600	£400	£600	£750	£3,350
Switchgear	–	–	–	–	–
Termination	£400	£100	£100	£170	£770
Transformer	–	–	–	–	–
Trench/Reinstate	£1,000	£600	£1,150	£1,100	£3,850
OHL LV	–	–	–	–	–
OHL MV	–	–	–	–	–
Other	–	–	–	–	–
Special/One-offs					–
Total Calculated Price	£3,250	£1,220	£3,050	£2,470	£9,990
Non-contestable Elements and Associated Charges ranges between 15–20%					
Total Non-contestable Elements and Associated Charges					£1,500–£2,000
Grand Total Calculated Price excl. VAT					£11,490–£11,990

Urban network mainly underground with reinforcement

Details: Connection to small housing development 10 houses, 60 m of LV cable from local 11/LV substation route in footpath and verge, 10 m of service cable to each plot in verge. Scheme includes reinforcement of LV distribution board

Supply Capacity: 200 kVA

Connection Voltage: LV

Nr of Phases: 1Φ

Breakdown of detailed cost information

	Labour	Plant	Materials	Overheads	Total
Cable	£250	£120	£1,200	£450	£2,020
Jointing	£1,600	£400	£600	£750	£3,350
Switchgear	£1,850	£600	£2,700	£1,450	£6,600
Termination	£400	£100	£100	£170	£770
Transformer	–	–	–	–	–
Trench/Reinstate	£1,000	£600	£1,150	£1,100	£3,850
OHL LV	–	–	–	–	–
OHL MV	–	–	–	–	–
Other	–	–	–	–	–
Special/One-offs					–
Total Calculated Price	£5,100	£1,820	£5,750	£3,920	£16,590
Non-contestable Elements and Associated Charges ranges between 15–20%					
Total Non-contestable Elements and Associated Charges					£2,500–£3,500
Grand Total Calculated Price excl. VAT					£19,090–£20,090

Urban network mainly underground with substation reinforcement

Details: Connection to small housing development 10 houses, 60 m of LV cable from local 11/LV substation route in footpath and verge, 10 m of service cable to each plot in verge. Scheme includes reinforcement of LV distribution board and new substation and 20 m of MV cable

Supply Capacity: 200 kVA

Connection Voltage: LV

Nr of Phases: 1Φ

Breakdown of detailed cost information

	Labour	Plant	Materials	Overheads	Total
Cable	£300	£150	£1,650	£600	£2,700
Jointing	£1,900	£500	£800	£900	£4,100
Switchgear	£5,600	£1,800	£16,000	£6,700	£30,100
Termination	£650	£160	£310	£320	£1,440
Transformer	£2,900	£950	£8,000	£3,400	£15,250
Trench/Reinstate	£1,600	£1,000	£2,000	£1,840	£6,440
OHL LV	–	–	–	–	–
OHL MV	–	–	–	–	–
Other	–	–	–	–	–
Special/One-offs					–
Total Calculated Price	£12,950	£4,560	£28,760	£13,760	£60,030
Non-contestable Elements and Associated Charges ranges between 15–20%					
Total Non-contestable Elements and Associated Charges					£9,000–£12,000
Grand Total Calculated Price excl. VAT					£69,030–£72,030

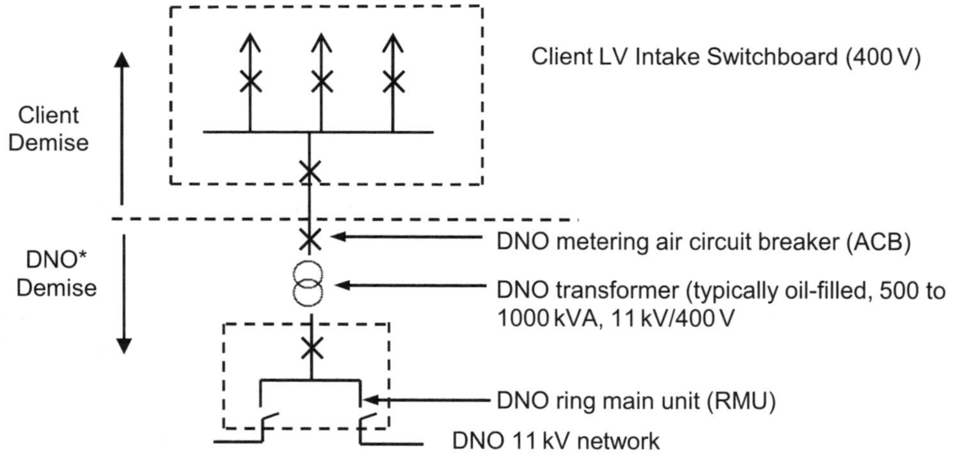

Note: * DNO – Distribution Network Operator

<u>Typical Simple 11 kV Network Connection For LV Intakes Up To 1000 kVA</u>

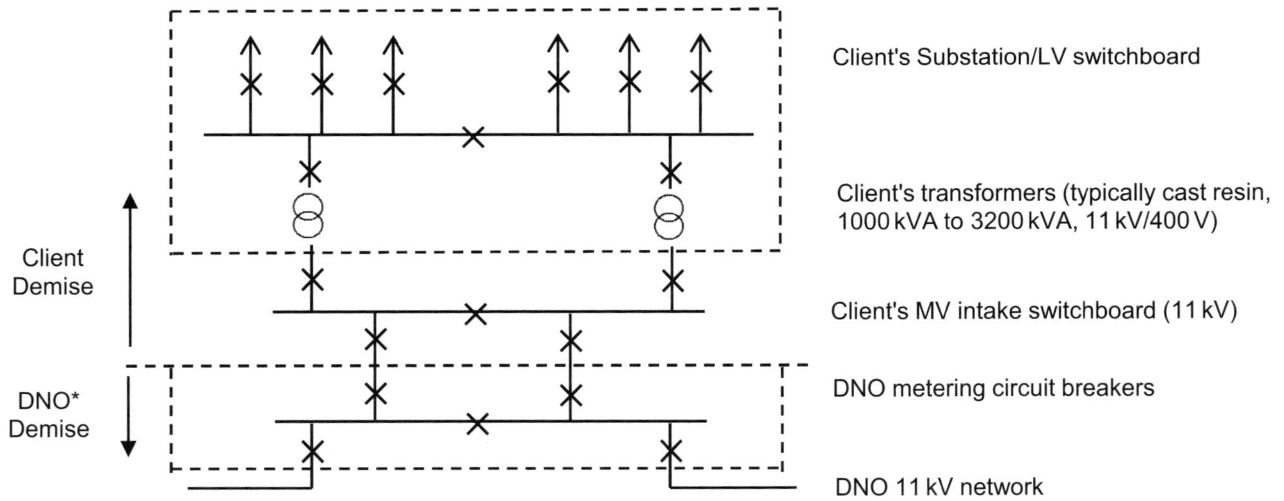

Note: * DNO – Distribution Network Operator

Typical 11 kV Network Connections For MV Intakes 1000 kVA To 6000 kVA

STATIC UPS SYSTEM - SIMPLIFIED SCHEMATIC FOR SINGLE MODLE

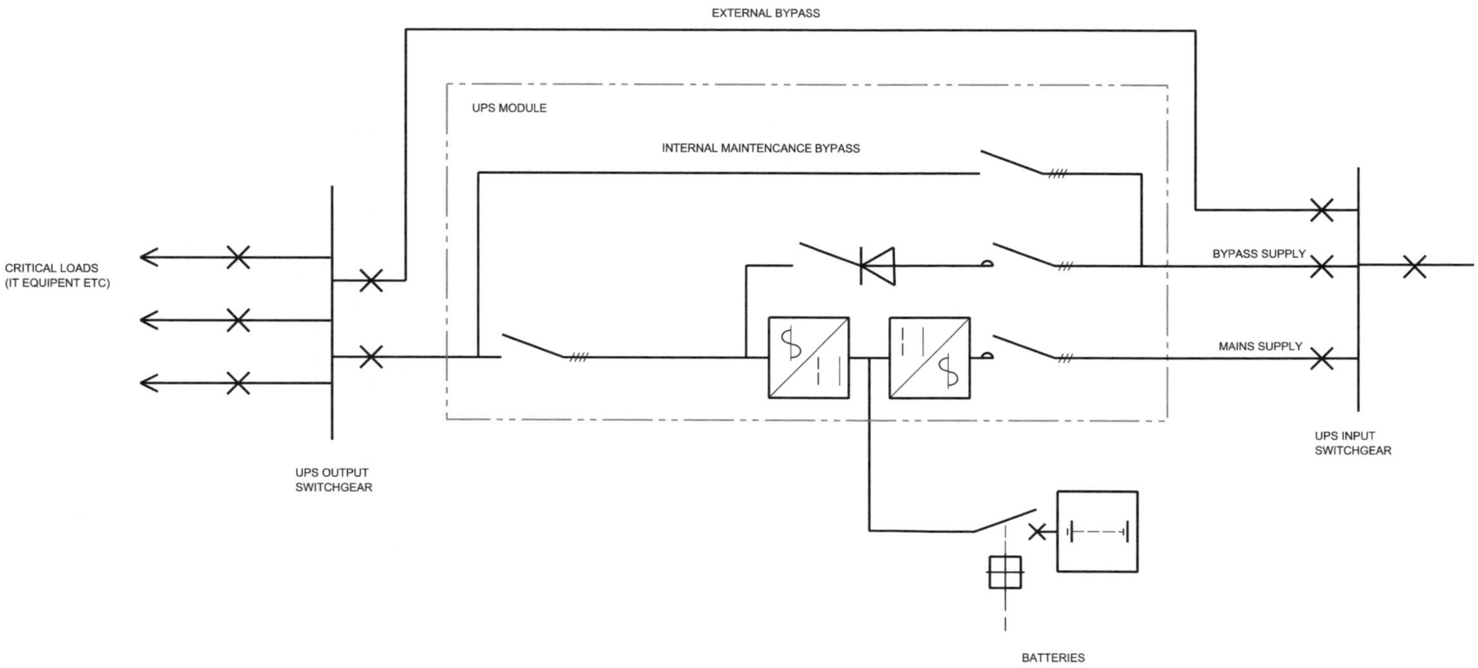

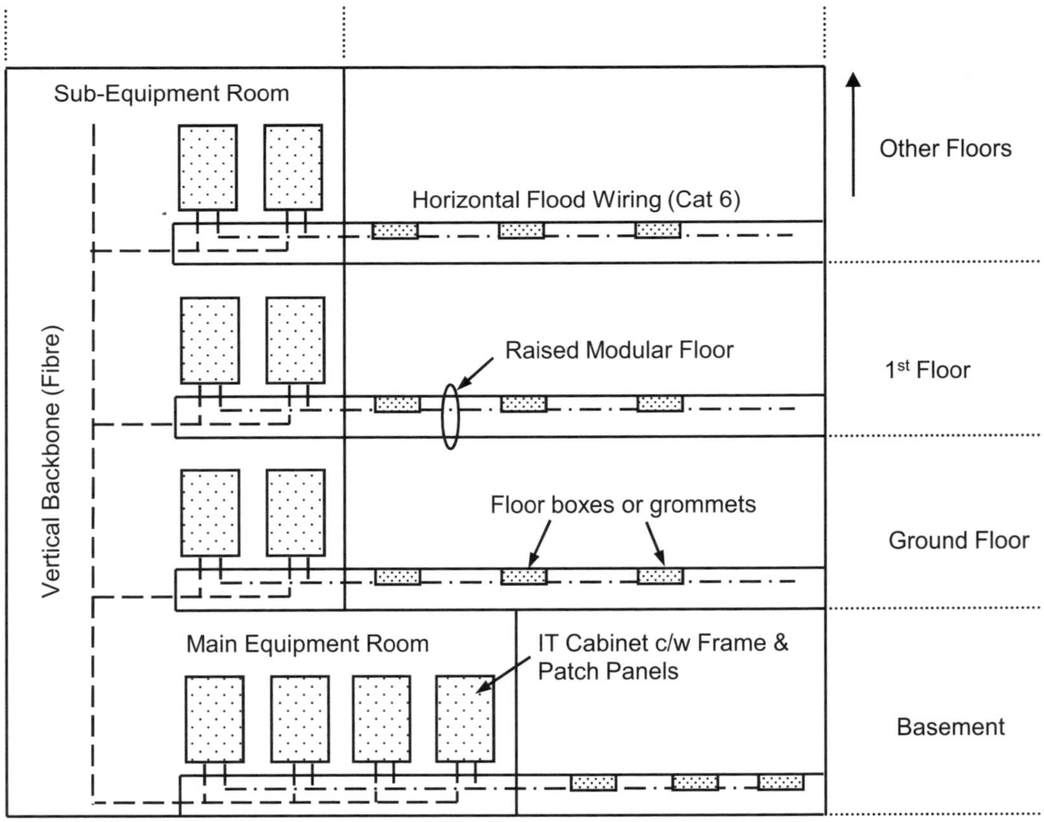

Typical Data Transmission (Structured Cabling)

TYPICAL NETWORKED LIGHITNG CONTROL SYSTEM

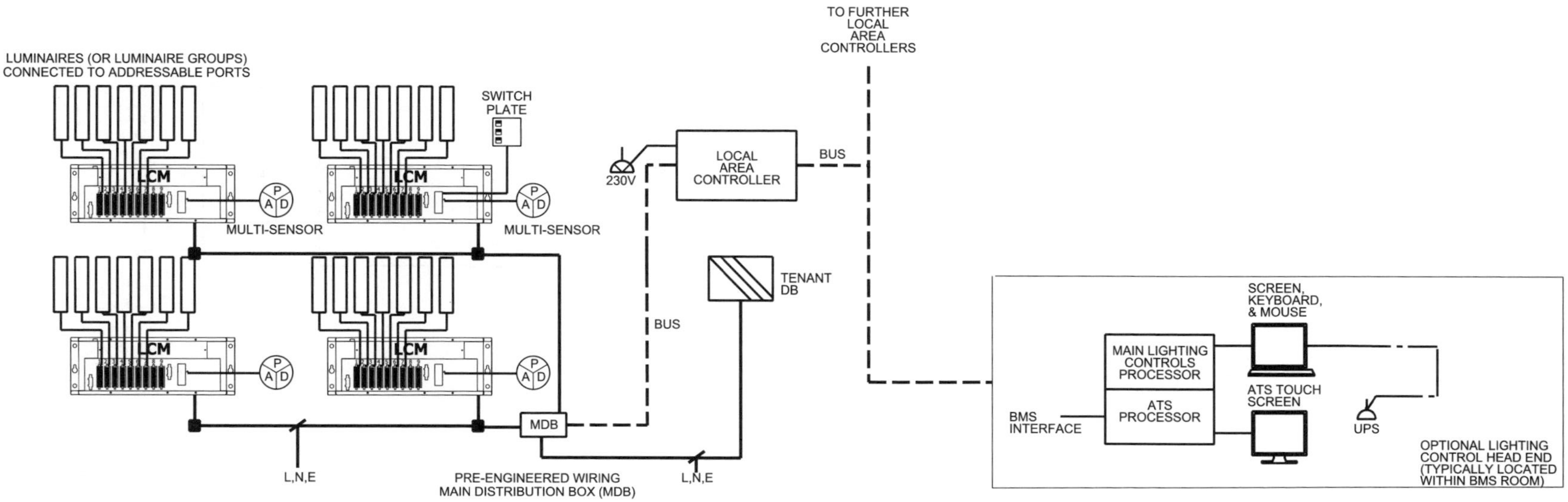

TYPICAL STANDBY POWER SYSTEM

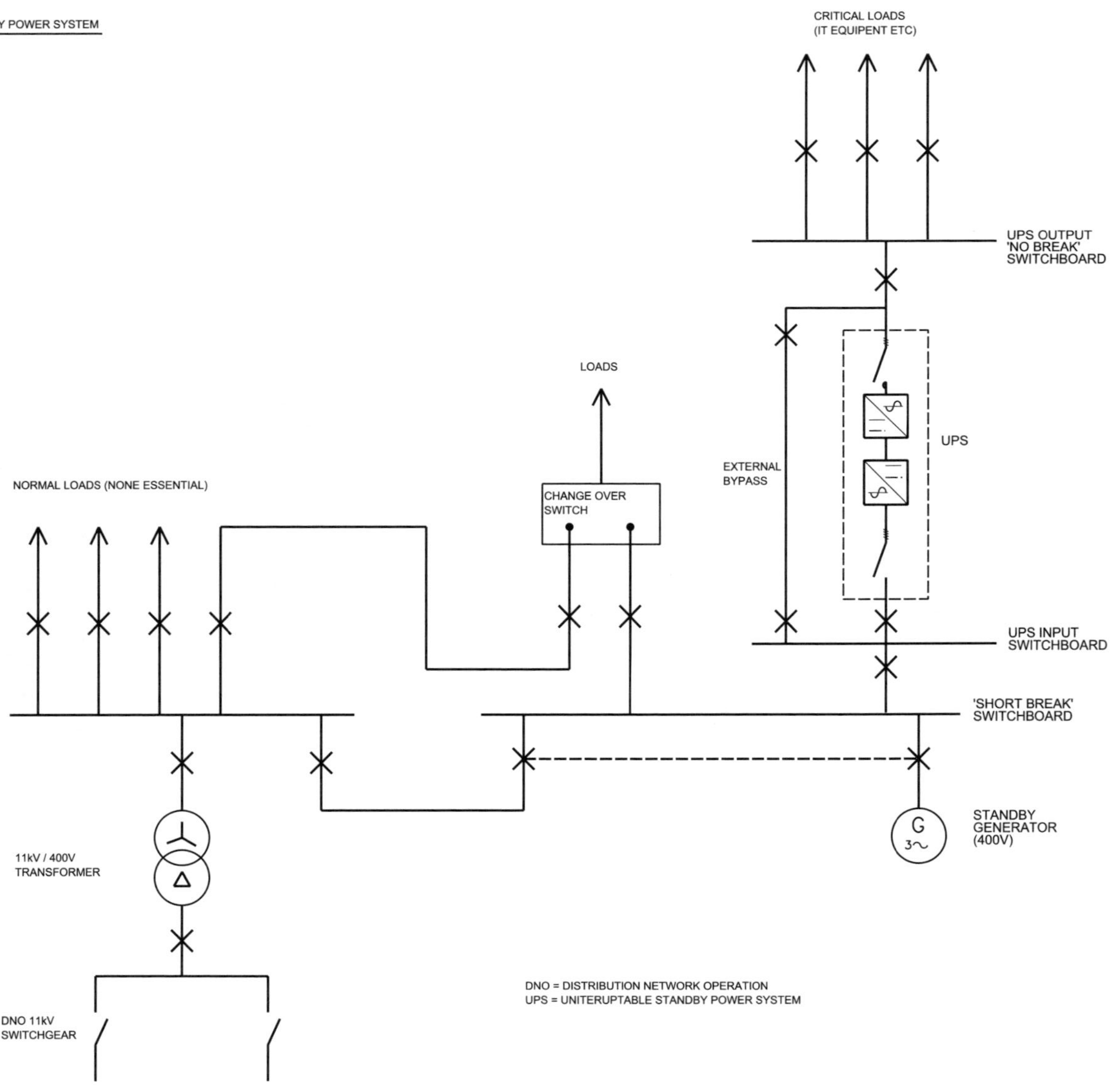

DNO = DISTRIBUTION NETWORK OPERATION
UPS = UNITERUPTABLE STANDBY POWER SYSTEM

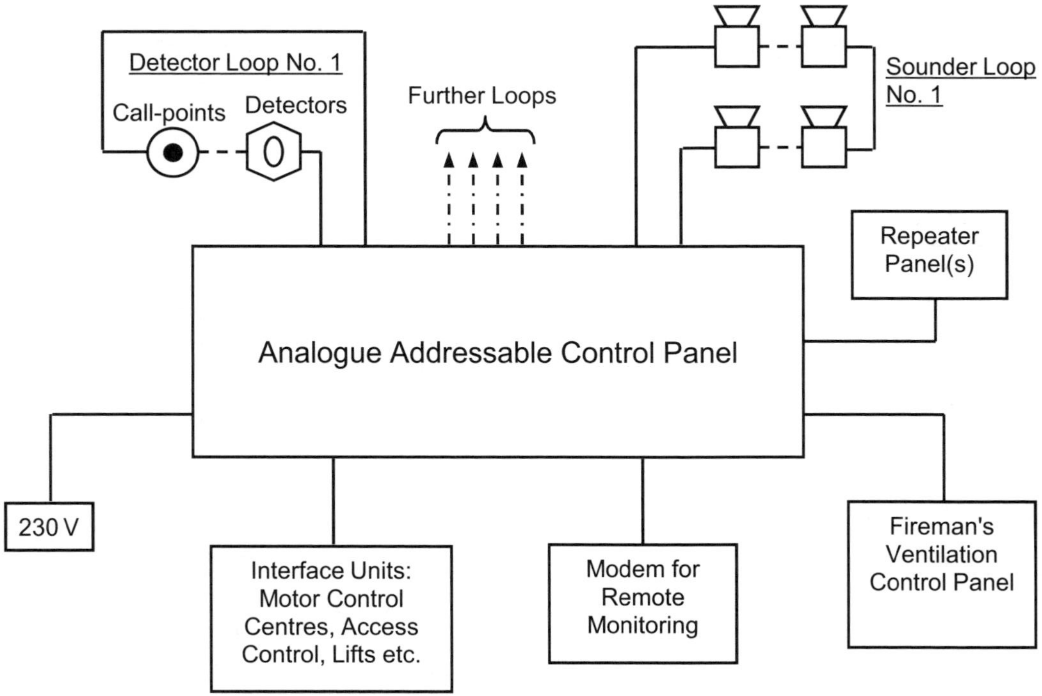

Typical Fire Detection and Alarm Schematic

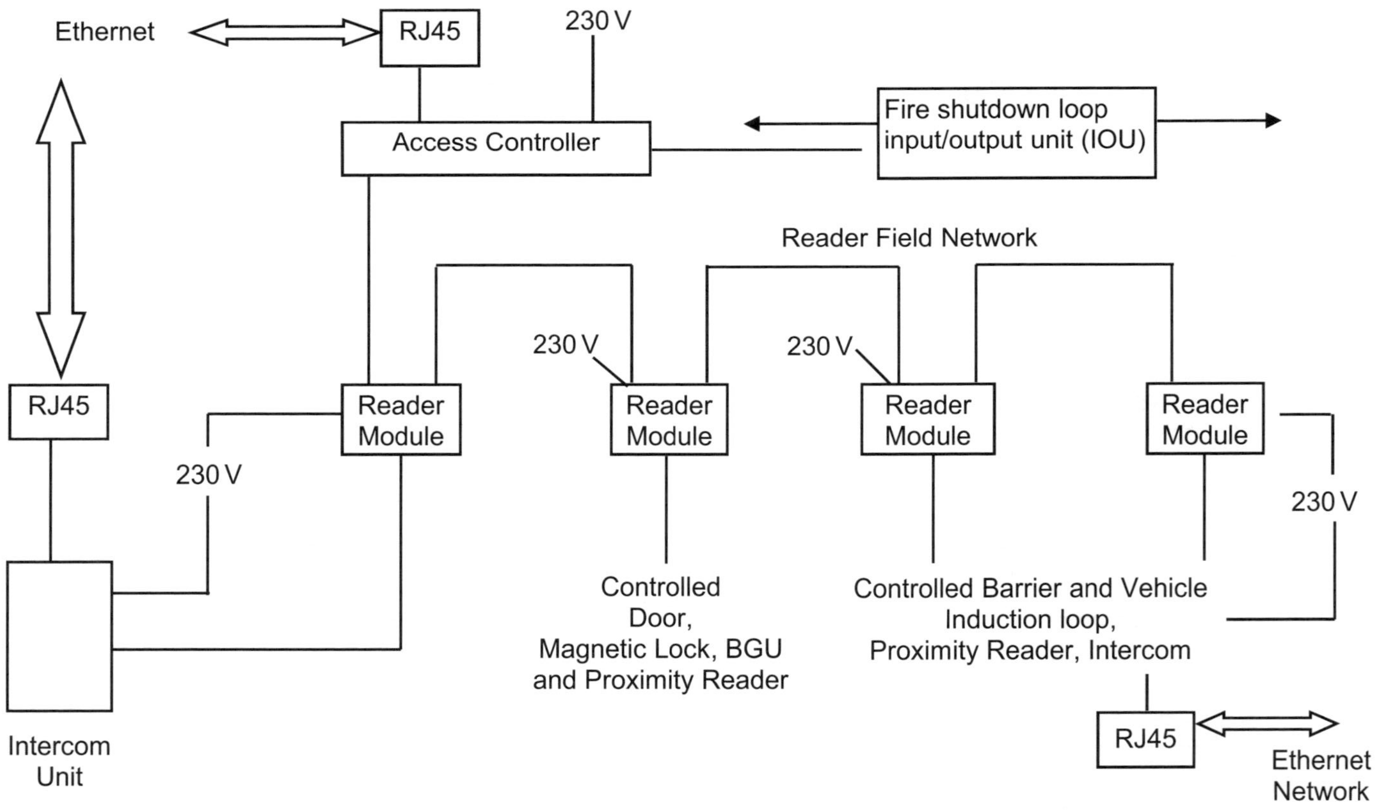

Typical Block Diagram – Access Control System (ACS)

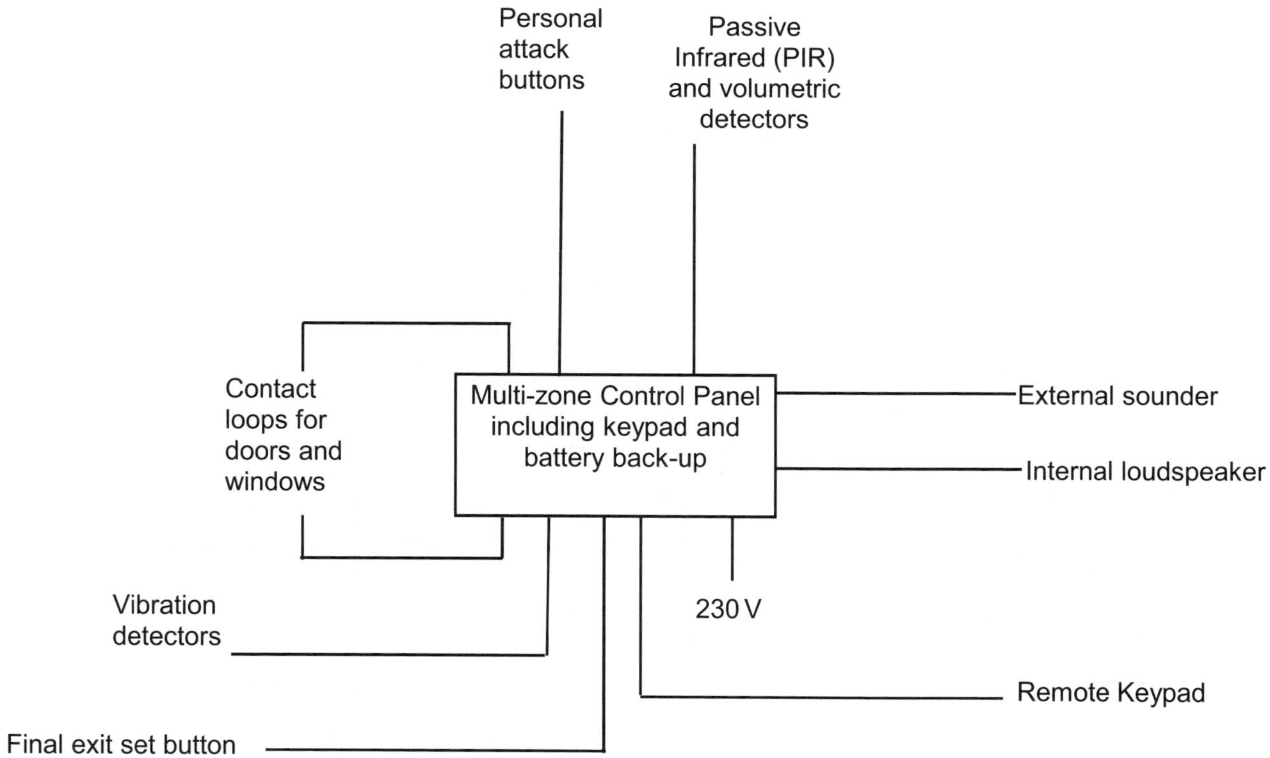

Typical Block Diagram – Intruder Detection System (IDS)

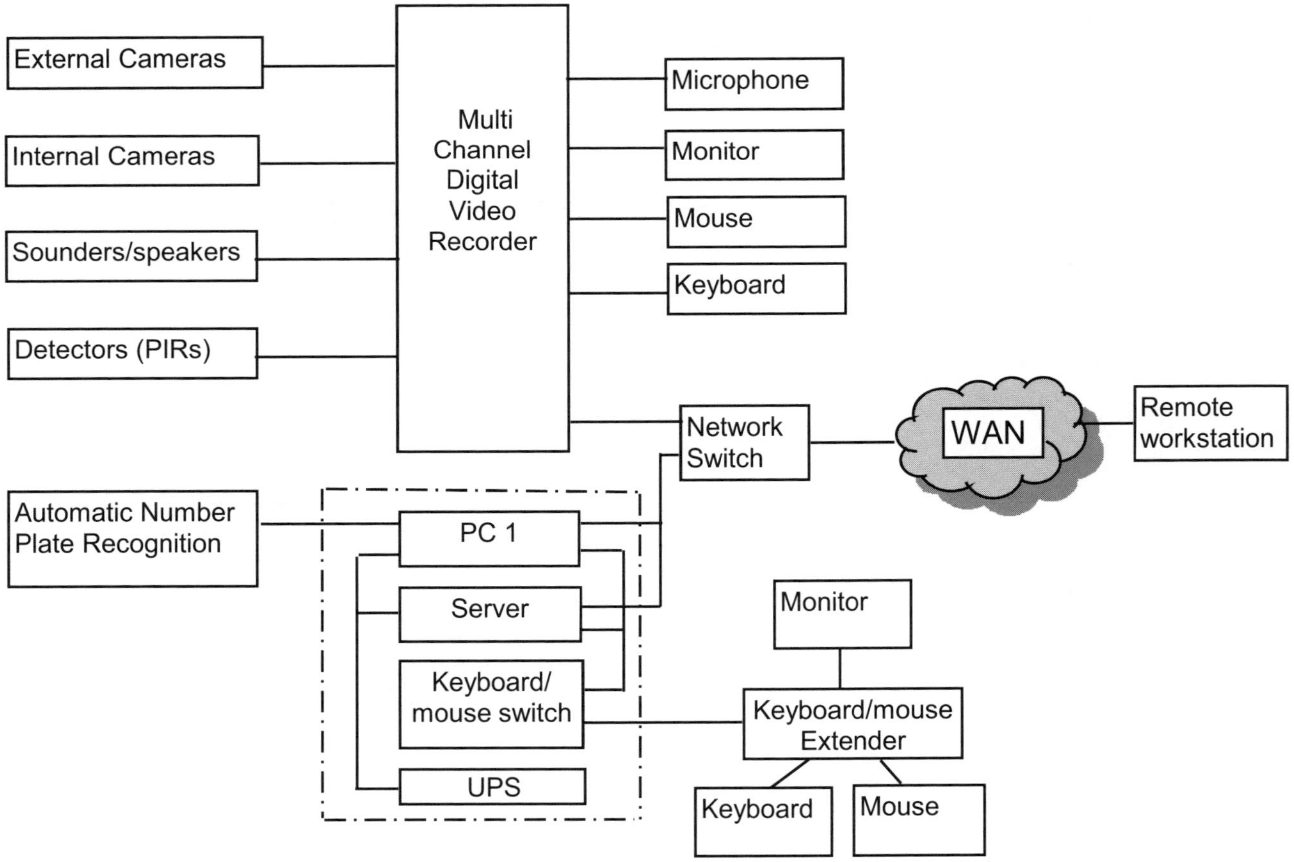

Typical Block Diagram – Digital CCTV

FAN COIL UNIT SYSTEM

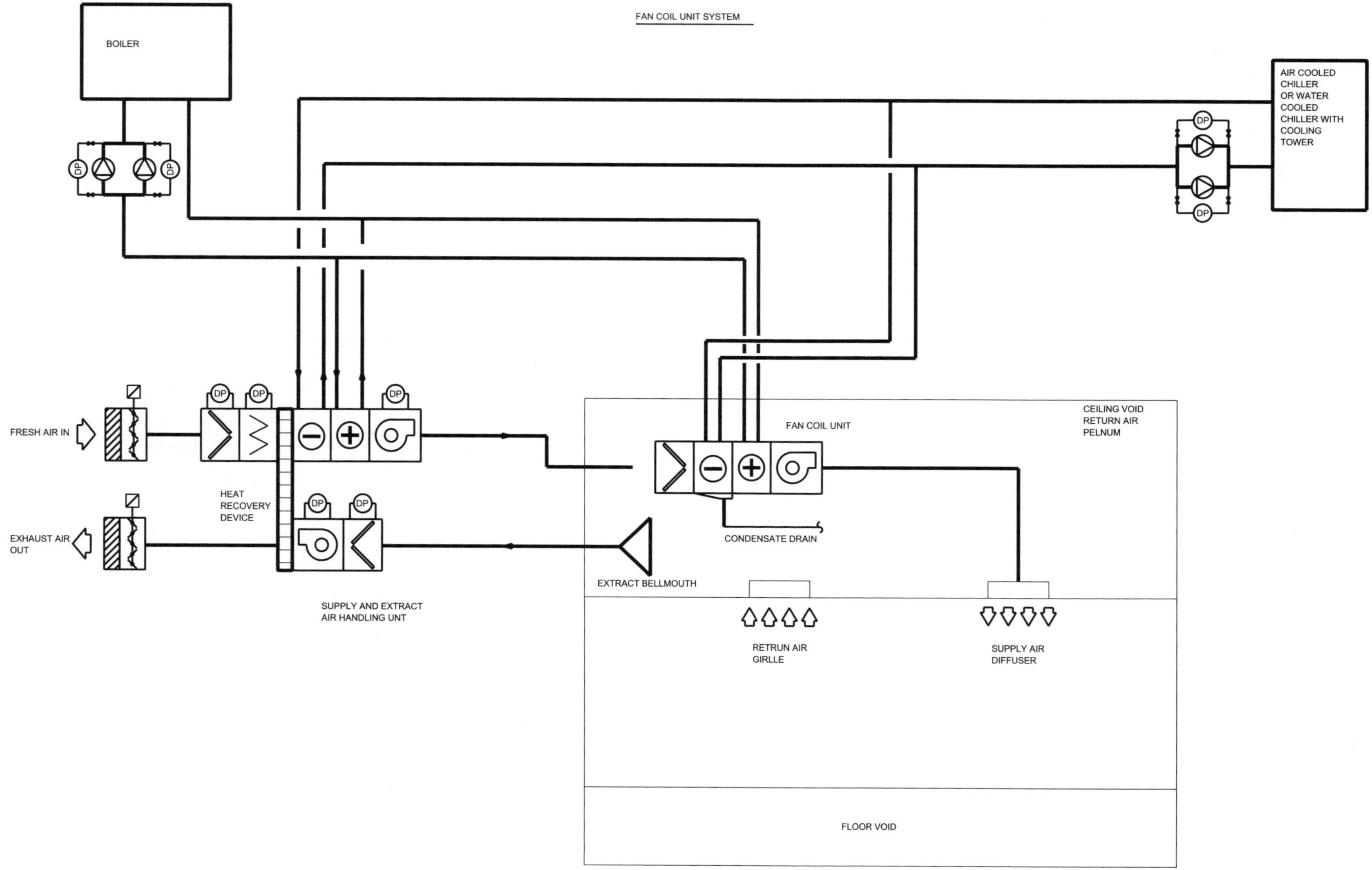

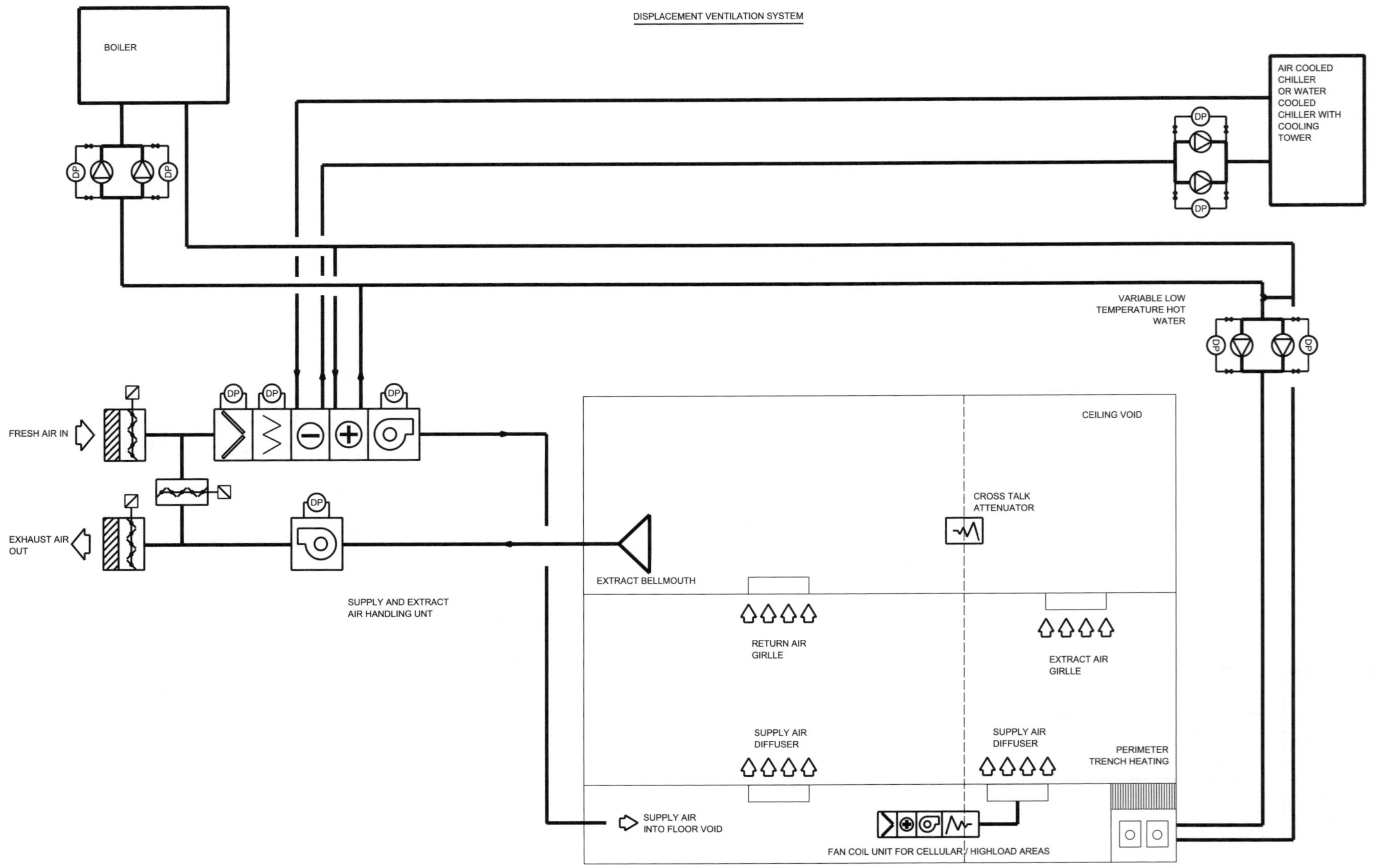
DISPLACEMENT VENTILATION SYSTEM
BOILER
DP
AIR COOLED CHILLER OR WATER COOLED CHILLER WITH COOLING TOWER
VARIABLE LOW TEMPERATURE HOT WATER
FRESH AIR IN
EXHAUST AIR OUT
SUPPLY AND EXTRACT AIR HANDLING UNT
CEILING VOID
CROSS TALK ATTENUATOR
EXTRACT BELLMOUTH
RETURN AIR GIRLLE
EXTRACT AIR GIRLLE
SUPPLY AIR DIFFUSER
SUPPLY AIR DIFFUSER
PERIMETER TRENCH HEATING
SUPPLY AIR INTO FLOOR VOID
FAN COIL UNIT FOR CELLULAR / HIGHLOAD AREAS

CHILLED CEILING SYSTEM

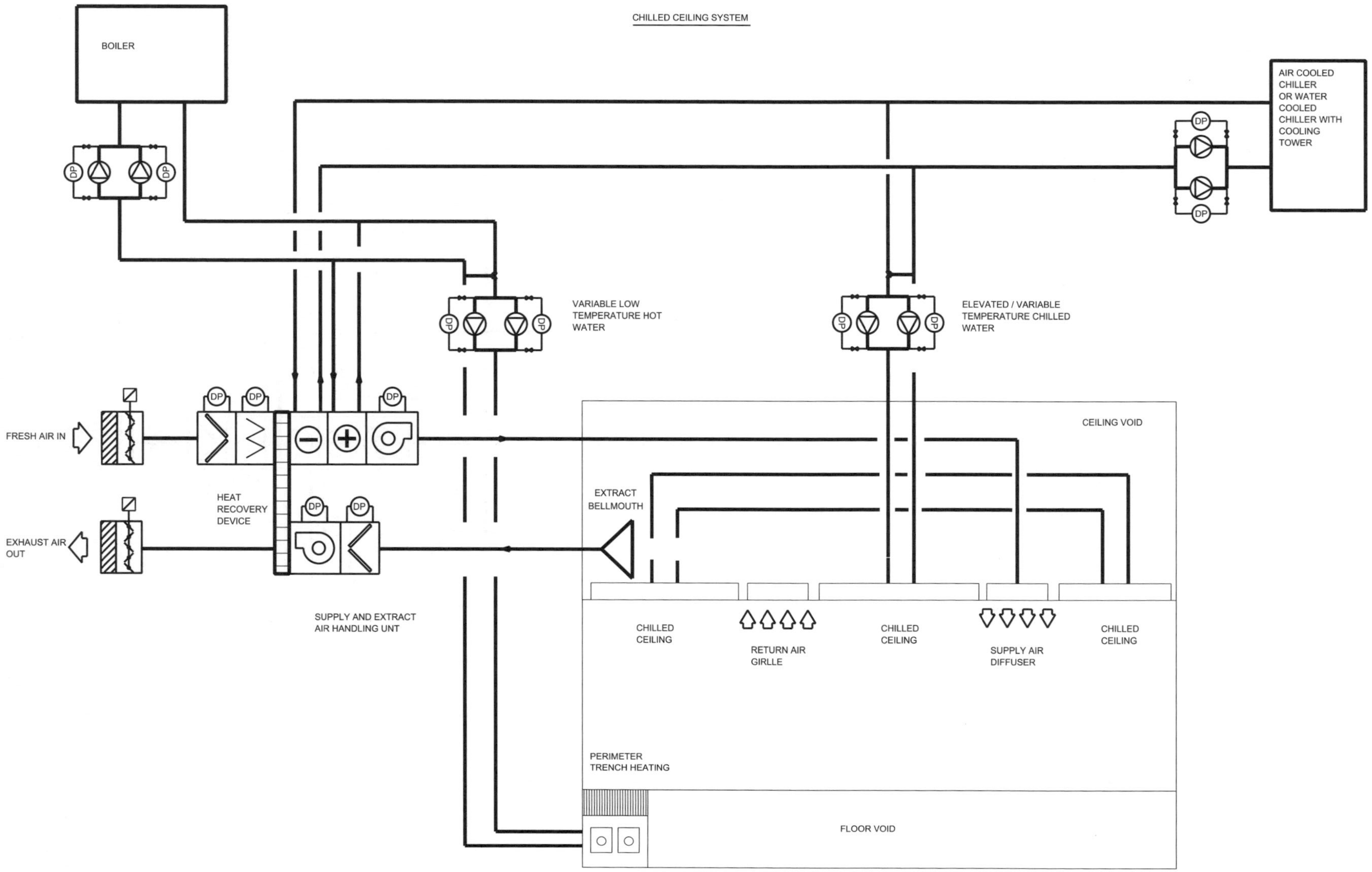

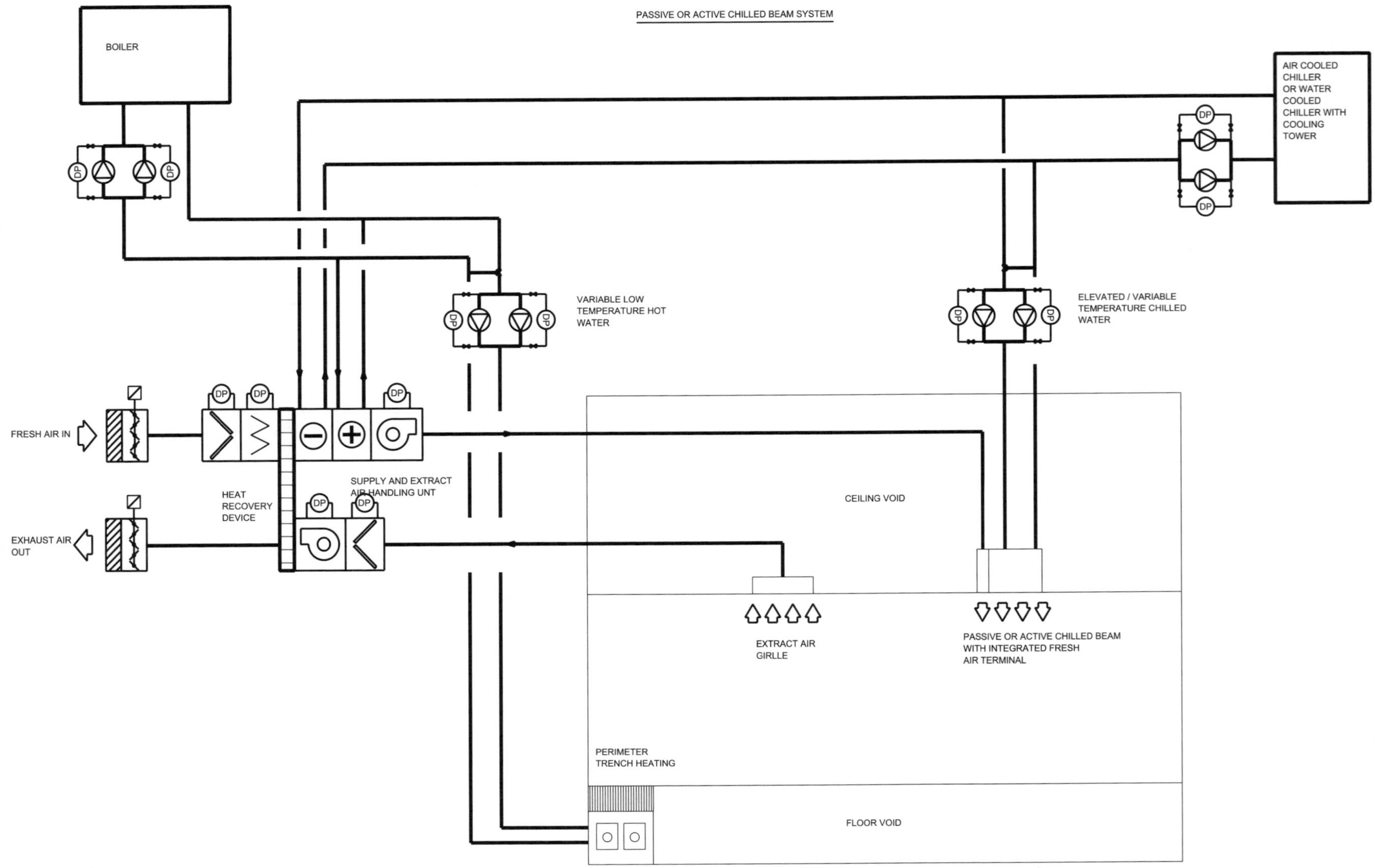
PASSIVE OR ACTIVE CHILLED BEAM SYSTEM
BOILER
AIR COOLED CHILLER OR WATER COOLED CHILLER WITH COOLING TOWER
DP
VARIABLE LOW TEMPERATURE HOT WATER
ELEVATED / VARIABLE TEMPERATURE CHILLED WATER
FRESH AIR IN
EXHAUST AIR OUT
HEAT RECOVERY DEVICE
SUPPLY AND EXTRACT AIR HANDLING UNT
CEILING VOID
EXTRACT AIR GIRLLE
PASSIVE OR ACTIVE CHILLED BEAM WITH INTEGRATED FRESH AIR TERMINAL
PERIMETER TRENCH HEATING
FLOOR VOID

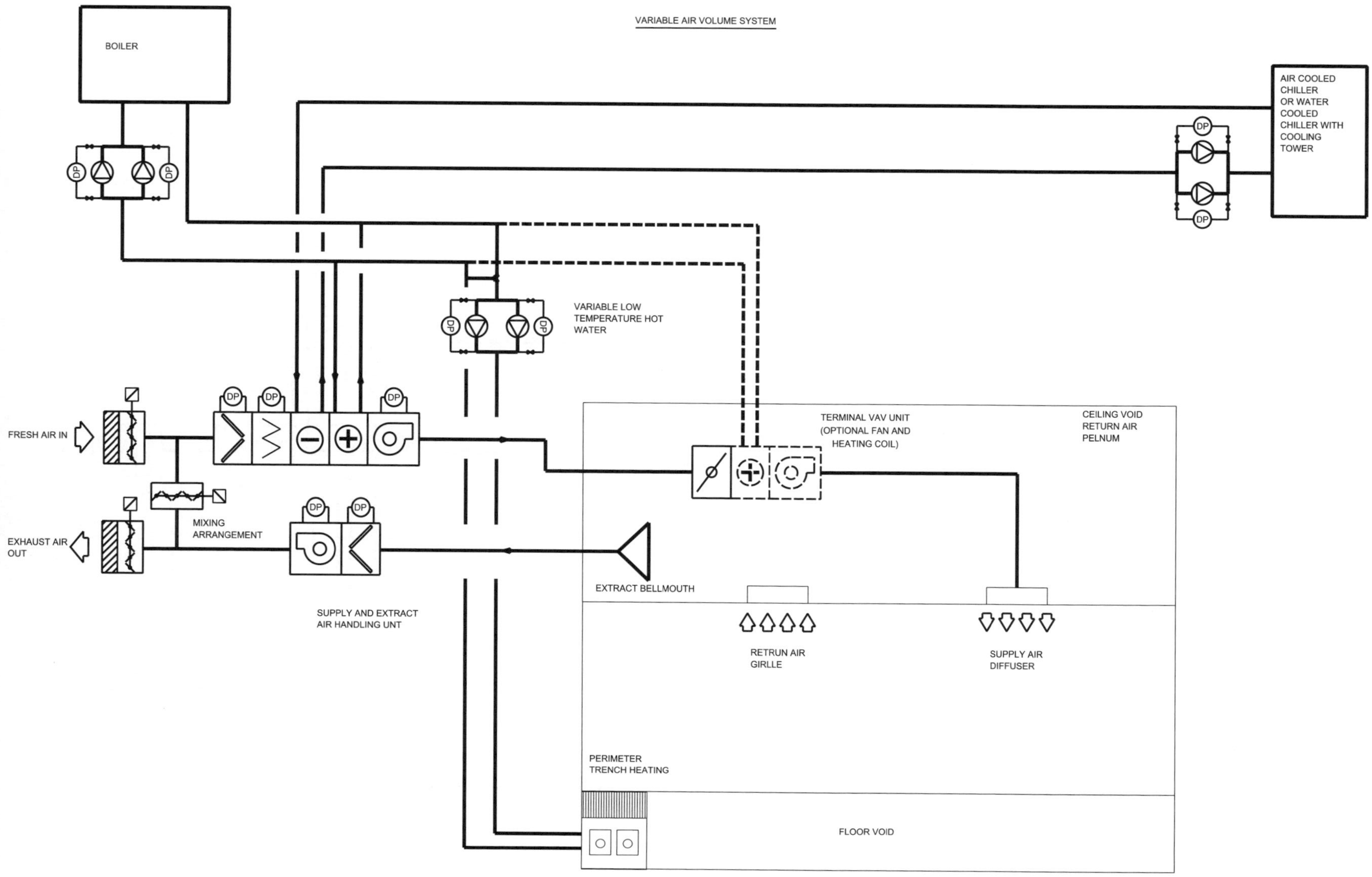
VARIABLE AIR VOLUME SYSTEM
BOILER
AIR COOLED CHILLER OR WATER COOLED CHILLER WITH COOLING TOWER
DP
VARIABLE LOW TEMPERATURE HOT WATER
FRESH AIR IN
EXHAUST AIR OUT
MIXING ARRANGEMENT
SUPPLY AND EXTRACT AIR HANDLING UNT
TERMINAL VAV UNIT (OPTIONAL FAN AND HEATING COIL)
CEILING VOID RETURN AIR PELNUM
EXTRACT BELLMOUTH
RETRUN AIR GIRLLE
SUPPLY AIR DIFFUSER
PERIMETER TRENCH HEATING
FLOOR VOID

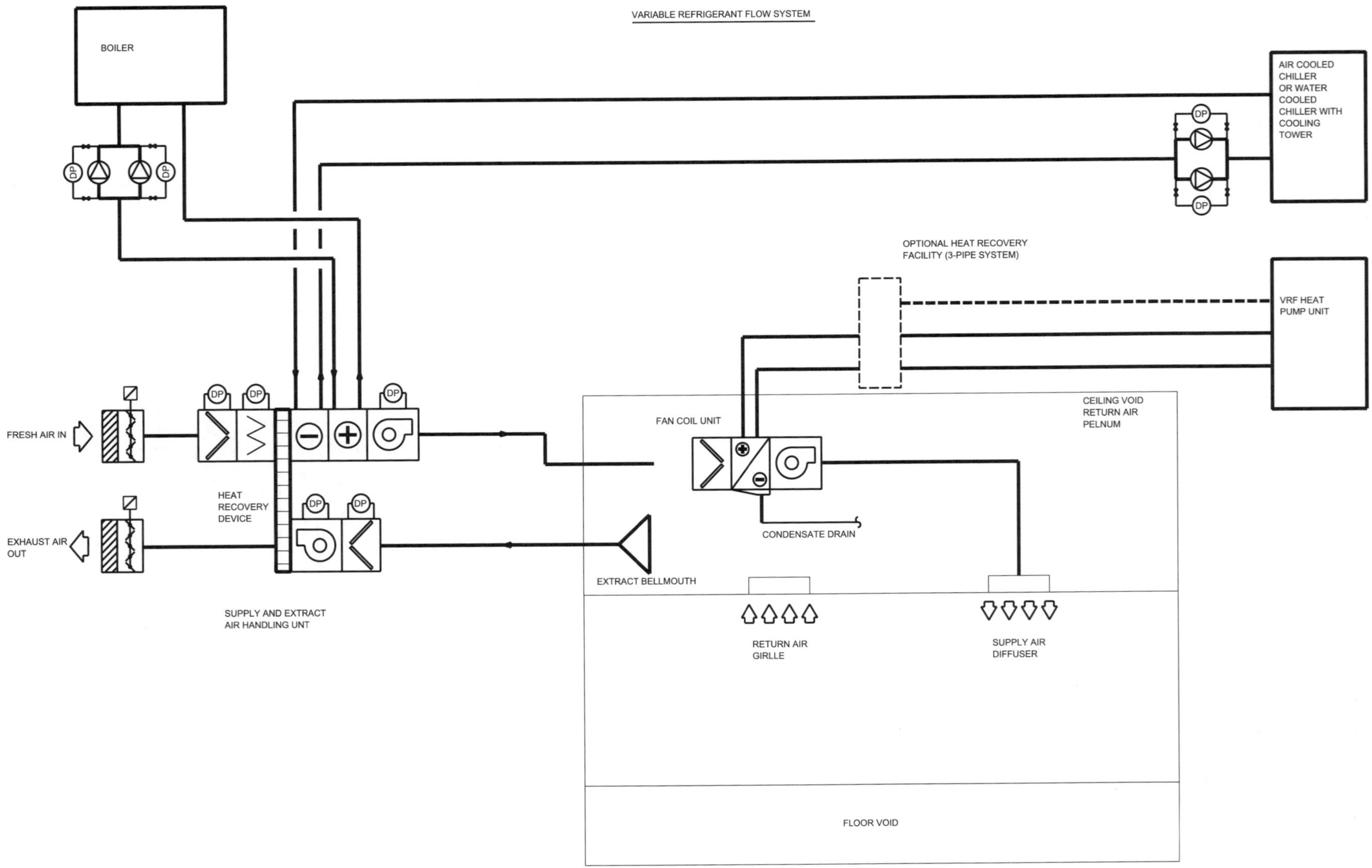
VARIABLE REFRIGERANT FLOW SYSTEM
BOILER
DP
AIR COOLED CHILLER OR WATER COOLED CHILLER WITH COOLING TOWER
OPTIONAL HEAT RECOVERY FACILITY (3-PIPE SYSTEM)
VRF HEAT PUMP UNIT
FRESH AIR IN
EXHAUST AIR OUT
HEAT RECOVERY DEVICE
SUPPLY AND EXTRACT AIR HANDLING UNT
FAN COIL UNIT
CEILING VOID RETURN AIR PELNUM
CONDENSATE DRAIN
EXTRACT BELLMOUTH
RETURN AIR GIRLLE
SUPPLY AIR DIFFUSER
FLOOR VOID

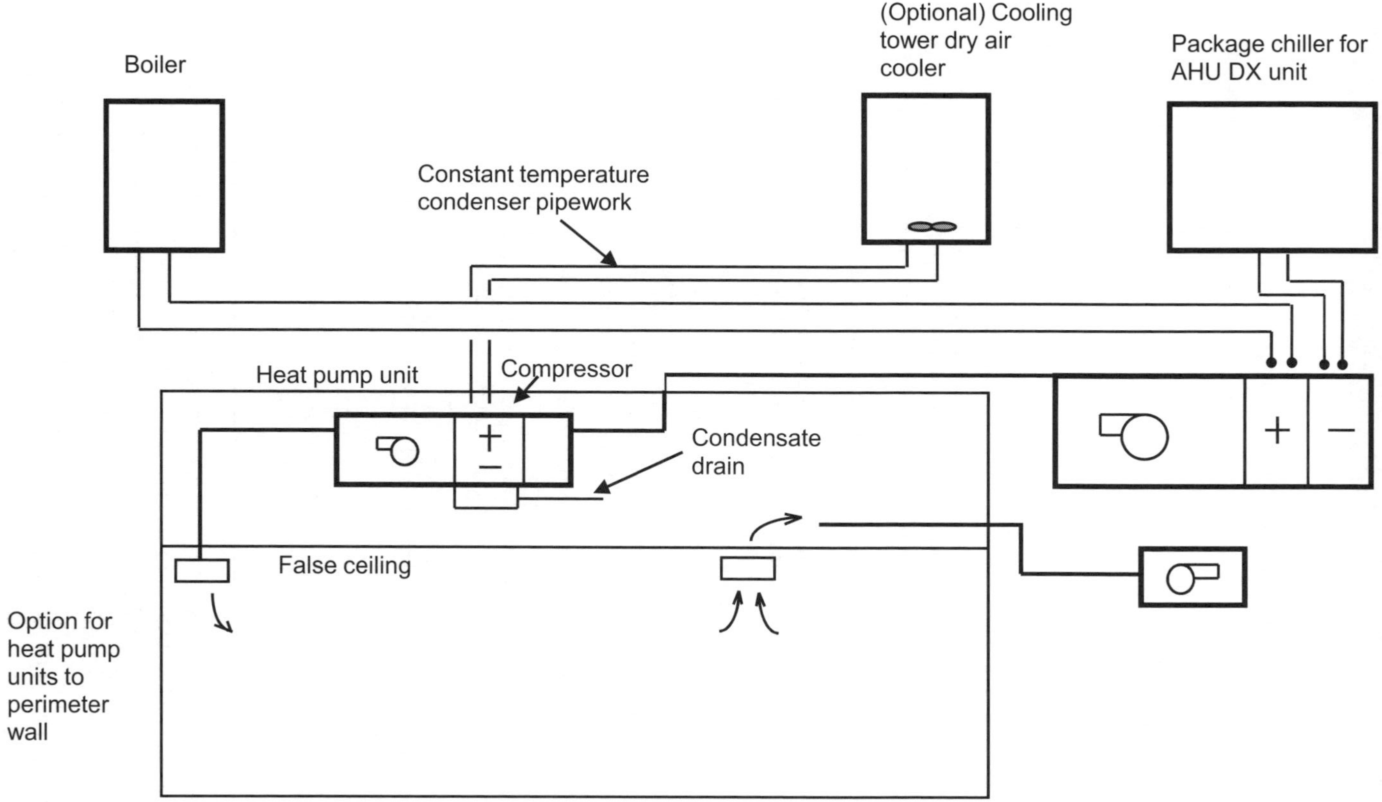

Reverse Cycle Heat Pump

PART 2

Approximate Estimating

DIRECTIONS

The prices shown in this section of the book are average prices on a fixed price basis for typical buildings tendered during the second quarter of 2013. Unless otherwise noted, they exclude external services and professional fees.

The information in this section has been arranged to follow more closely the order in which estimates may be developed, in accordance with RIBA work stages.

a) Cost Indices and Regional Variations – These provide information regarding the adjustments to be made to estimates taking into account current pricing levels for different locations in the UK.

b) Feasibility Costs – These provide a range of data (based on a rate per square metre) for all-in engineering costs, excluding lifts, associated with a wide variety of building types. These would typically be used at work stage A/B (feasibility) of a project.

c) Elemental Rates – The outline costs for offices have been developed further to provide rates for the alternative solutions for each of the services elements. These would typically be used at work stage C, outline proposal.

   Where applicable, costs have been identified as Shell and Core and Fit Out to reflect projects where the choice of procurement has dictated that the project is divided into two distinctive contractual parts.

   Such detail would typically be required at work stage D, detailed proposals.

d) All-in Rates – These are provided for a number of items and complete parts of a system i.e. boiler plant, ductwork, pipework, electrical switchgear and small power distribution, together with lifts and escalators. Refer to the relevant section for further guidance notes.

e) Elemental Costs – These are provided for a diverse range of building types; offices, laboratory, shopping mall, airport terminal building, supermarket, performing arts centre, sports hall, luxury hotel, hospital and secondary school. Also included is a separate analysis of a building management system for an office block. In each case, a full analysis of engineering services costs is given to show the division between all elements and their relative costs to the total building area. A regional variation factor has been applied to bring these analyses to a common London base.

Prices should be applied to the total floor area of all storeys of the building under consideration. The area should be measured between the external walls without deduction for internal walls and staircases/lift shafts i.e. GIA (Gross Internal Area).

Although prices are reviewed in the light of recent tenders it has only been possible to provide a range of prices for each building type. This should serve to emphasize that these can only be average prices for typical requirements and that such prices can vary widely depending on variations in size, location, phasing, specification, site conditions, procurement route, programme, market conditions and net to gross area efficiencies. Rates per square metre should not therefore be used indiscriminately and each case needs to be assessed on its own merits.

The prices do not include for incidental builder's work nor for profit and attendance by a Main Contractor where the work is executed as a subcontract: they do however include for preliminaries, profit and overheads for the services contractor. Capital contributions to statutory authorities and public undertakings and the cost of work carried out by them have been excluded.

Where services works are procured indirectly, i.e. ductwork via a mechanical Subcontractor, the reader should make due allowance for the addition of a further level of profit etc.

COST INDICES

The following tables reflect the major changes in cost to contractors but do not necessarily reflect changes in tender levels. In addition to changes in labour and materials costs, tenders are affected by other factors such as the degree of competition in the particular industry, the area where the work is to be carried out, the availability of labour and the prevailing economic conditions. This has meant in recent years that, when there has been an abundance of work, tender levels have tended to increase at a greater rate than can be accounted for solely by increases in basic labour and material costs and, conversely, when there is a shortage of work this has tended to result in keener tenders. Allowances for these factors are impossible to assess on a general basis and can only be based on experience and knowledge of the particular circumstances.

In compiling the tables the cost of labour has been calculated on the basis of a notional gang as set out elsewhere in the book. The proportion of labour to materials has been assumed as follows:

Mechanical Services – 30:70, Electrical Services – 50:50, (1976 = 100)

Mechanical Services				
Year	*First Quarter*	*Second Quarter*	*Third Quarter*	*Fourth Quarter*
2000	384	386	388	400
2001	401	401	405	411
2002	411	411	410	442
2003	443	446	447	456
2004	458	464	467	482
2005	486	488	492	508
2006	513	522	527	533
2007	535	541	546	555
2008	555	560	567	584
2009	582	578	577	577
2010	576	583	585	592
2011	600	606	607	605
2012	607	606	605	608
2013	P 613	P 617	F 619	F 620
2014	F 622	F 628	F 629	F 630

Electrical Services				
Year	*First Quarter*	*Second Quarter*	*Third Quarter*	*Fourth Quarter*
2000	458	464	465	468
2001	485	484	484	487
2002	508	508	508	513
2003	530	533	533	541
2004	571	574	576	589
2005	607	608	607	615
2006	631	636	641	649
2007	665	666	668	676
2008	693	697	700	710
2009	720	721	724	737
2010	760	766	766	768

COST INDICES

Electrical Services				
Year	*First Quarter*	*Second Quarter*	*Third Quarter*	*Fourth Quarter*
2011	777	780	778	783
2012	777	776	775	792
2013	P 789	P 789	F 789	F 801
2014	F 803	F 803	F 804	F 811

(P = Provisional)

(F = Forecast)

COST INDICES

Regional variations

Prices throughout this Book apply to work in the London area (see Directions at the beginning of the Mechanical Installations and Electrical Installations sections). However, prices for mechanical and electrical services installations will of course vary from region to region, largely as a result of differing labour costs but also depending on the degree of accessibility, urbanization and local market conditions.

The following table of regional factors is intended to provide readers with indicative adjustments that may be made to the prices in the Book for locations outside of London. The figures are of necessity averages for regions and further adjustments should be considered for city centre or very isolated locations, or other known local factors.

Greater London	1.03	Yorkshire & Humberside	0.95
South East	1.02 (Excl GL)	North West	0.90
South West	0.97	Northern	0.90
East Midlands	0.94	Scotland	0.93
West Midlands	0.95	Wales	0.93
East Anglia	1.00	Northern Ireland	0.69

RIBA STAGE A/B FEASIBILITY COSTS

Item	Unit	Range £		
Typical square metre rates for Engineering Services				
The following examples indicate the range of rates within each building type for engineering services, excluding lifts etc., utilities services and professional fees. Based on Gross Internal Area (GIA).				
Industrial Buildings				
Factories				
Owner occupation: Includes for rainwater, soil/waste, LTHW heating via HL radiant heaters, BMS, LV installations, lighting, fire alarms, security, earthing	m²	110.00	to	130.00
Owner occupation: Includes for rainwater, soil/waste, sprinklers, LTHW heating via HL gas fired heaters, local air conditioning, BMS, MV/LV installations, lighting, fire alarms, security, earthing	m²	160.00	to	190.00
Warehouses				
High bay for owner occupation: Includes for rainwater, soil/waste, LTHW heating via HL gas fired heaters, BMS, MV/LV installations, lighting, fire alarms, security, earthing	m²	86.00	to	105.00
High bay for owner occupation: Includes for rainwater, soil/waste, sprinklers, LTHW heating via HL radiant heaters, local air condtioning, BMS, MV/LV installations, lighting, fire alarms, security, earthing	m²	170.00	to	210.00
Distribution Centres				
High bay for letting: Includes for rainwater, soil/waste, LTHW heating via HL gas fired heaters, BMS, MV/LV installations, lighting, fire alarms, security, earthing	m²	110.00	to	130.00
High bay for owner occupation: Includes for rainwater, soil/waste, sprinklers, LTHW heating via HL radiant heaters, local air conditioning, BMS, MV/LV installations, lighting, fire alarms, security	m²	180.00	to	220.00
Office Buildings 5,000 m² to 15,000 m²				
Offices for Letting				
Shell & Core and Cat A non air-conditioned: Includes for rainwater, soil/waste, cold water, hot water via local electrical heaters, LTHW heating via radiator heaters, toilet extract, controls, LV installations, lighting, small power (landlords), fire alarms, earthing, security & IT wireways	m²	240.00	to	300.00
Shell & Core and Cat A non air-conditioned: Includes for rainwater, soil/waste, cold water, hot water, LTHW heating via perimeter heaters, toilet extract, controls, LV installations, lighting, small power (landlords), fire alarms, earthing, security wireways, IT wireways	m²	255.00	to	310.00
Shell & Core and Cat A air-conditioned: Includes for rainwater, soil/waste, VRV 3 pipe heat pumps, toilet extract, BMS, LV installations, lighting, small power (landlords), fire alarms, earthing, security & IT wireways	m²	355.00	to	435.00
Shell & Core and Cat A air-conditioned: Includes for rainwater, soil/waste, cold water, hot water via local electrical heaters, LTHW heating via perimeter heaters, 2 pipe, toilet extract, BMS, LV installations, lighting, small power (landlords), fire alarms, earthing, security & IT wireways	m²	390.00	to	480.00

RIBA STAGE A/B FEASIBILITY COSTS

Item	Unit	Range £
Typical square metre rates for Engineering Services – cont		
Office Buildings 5,000 m² to 15,000 m² – cont		
Offices for Owner Occupation		
Non air-conditioned: Includes for rainwater, soil/waste, cold water, hot water via local electrical heaters, dry risers, LTHW heating via radiator heaters, toilet extract, controls, LV installations, lighting, small power (landlords), fire alarms, earthing, security, IT wireways	m²	255.00 to 310.00
Non air-conditioned: Includes for rainwater, soil/waste, cold water, hot water, dry risers, LTHW heating via perimeter heaters, toilet extract, BMS, LV installations, lighting, small power (landlords), fire alarms, earthing, security, IT wireways	m²	275.00 to 335.00
Air-conditioned: Includes for rainwater, soil/waste, cold water, hot water via electrical heaters, dry risers, LTHW heating via perimeter heating, 4 pipe air conditioning, toilet extract, kitchen extract, sprinkler protection, BMS, LV installations, life safety standby generators, lighting, small power, fire alarms, earthing, security, IT wireways small power, fire alarms L1/P1, earthing, security and IT wireways	m²	445.00 to 540.00
Health and Welfare Facilities		
District General Hospitals		
Natural ventilation: Includes for rainwater, soil/waste, cold water, hot water, dry risers, medical gases, LTHW heating via perimeter heating, toilet extract, kitchen extract, BMS, LV installations, standby generation, lighting, small power, fire alarms, earthing/lightning protection, nurse call systems, security, IT wireways	m²	390.00 to 480.00
Natural ventilation: Includes for rainwater, soil/waste, cold water, hot water, dry risers, LTHW heating via perimeter heaters, localized VAV air conditioning, kitchen/toilet extract, BMS, LV installations, standby generation, lighting, small power, fire alarms, earthing/lightning protection, nurse call systems, security, IT wireways	m²	550.00 to 680.00
Private Hospitals		
Air-conditioned: Includes for rainwater, soil/waste, cold water, hot water, dry risers, medical gases, LTHW heating, 2 pipe air conditioning, toilet extract, kitchen extract, BMS, LV installations, standby generation, lighting, small power, fire alarms, earthing, nurse call systems, security, IT wireways	m²	620.00 to 750.00
Air-conditioned: Includes for rainwater, soil/waste, cold water, hot water, dry risers, medical gases, LTHW heating, 4 pipe air conditioning, kitchen/toilet extract, BMS, LV installations, standby generation, lighting, small power, fire alarms, earthing, nurse call system, security, IT wireways	m²	660.00 to 810.00

RIBA STAGE A/B FEASIBILITY COSTS

Item	Unit	Range £		
Day Care Unit				
Natural ventilation: Includes for rainwater, soil/waste, cold water, hot water via local electrical heaters, medical gases, LTHW heating, toilet extract, kitchen extract, BMS, LV installations, lighting, small power, fire alarms, earthing, security, IT wireways	m²	390.00	to	480.00
Comfort cooled: Includes for rainwater, soil/waste, cold water, hot water, medical gases, LTHW heating, DX air conditioning, kitchen/toilet extract, BMS, LV installations, lighting, small power, fire alarms, earthing, security, IT wireways	m²	420.00	to	510.00
Entertainment and Recreation Buildings				
Non Performing				
Natural ventilation: Includes for rainwater, soil/waste, cold water, central hot water, dry risers, LTHW heating, toilet extract, kitchen extract, controls, LV installations, lighting, small power, fire alarms, earthing, security, IT wireways	m²	280.00	to	345.00
Comfort cooled: Includes for rainwater, soil/waste, cold water, hot water via electrical heaters, sprinklers/dry risers, LTHW heating, DX air conditioning, kitchen/toilet extract, BMS, LV installations, lighting, small power, fire alarms, earthing, security, IT wireways	m²	485.00	to	590.00
Performing Arts (with Theatre)				
Natural ventilation: Includes for rainwater, soil/waste, cold water, central hot water, sprinklers/dry risers, LTHW heating, toilet extract, kitchen extract, controls, LV installations, lighting, small power, fire alarms, earthing, security, IT wireways	m²	450.00	to	550.00
Comfort cooled: Includes for rainwater, soil/waste, cold water, central hot water, sprinklers/dry risers, LTHW heating, DX air conditioning, kitchen/toilet extract, BMS, LV installations, lighting including enhanced dimming/scene setting, small power, fire alarms, earthing, security, IT wireways including production, audio and video recording, EPOS system	m²	610.00	to	740.00
Sports Halls				
Natural ventilation: Includes for rainwater, soil/waste, cold water, hot water gas fired heaters, LTHW heating, toilet extract, BMS, LV installations, lighting, small power, fire alarms, earthing, security	m²	200.00	to	250.00
Comfort cooled: Includes for rainwater, soil/waste, cold water, hot water via LTHW heat exchangers, LTHW heating, air conditioning via AHU's, toilet extract, BMS, LV installations, lighting, small power, fire alarms, earthing, security	m²	315.00	to	385.00
Multi-Purpose Leisure Centre				
Natural ventilation: Includes for rainwater, soil/waste, cold water, hot water gas fired heaters, LTHW heating, toilet extract, BMS, LV installations, lighting, small power, fire alarms, earthing, security, IT wireways	m²	290.00	to	350.00
Comfort cooled: Includes for rainwater, soil/waste, cold water, hot water LTHW heat exchangers, LTHW heating, air conditioning via AHU, pool hall supply/extract, kitchen/toilet extract, BMS, LV installations, lighting, small power, fire alarms, earthing, security, IT wireways	m²	495.00	to	610.00

RIBA STAGE A/B FEASIBILITY COSTS

Item	Unit	Range £
Typical square metre rates for Engineering Services – cont		
Retail Buildings		
Open Arcade		
Natural ventilation: Includes for rainwater, soil/waste, cold water, hot water, sprinklers/dry risers, LTHW heating, toilet extract, smoke extract, BMS, LV installations, life safety standby generators, lighting, small power, fire alarms, public address, earthing/lightning protection, security, IT wireways	m²	290.00 to 360.00
Enclosed Shopping Mall		
Air-conditioned: Includes for rainwater, soil/waste, cold water, hot water, sprinklers/dry risers, LTHW heating, air conditioning via AHU's, toilet extract, smoke extract, BMS, LV installations, life safety standby generators, lighting, small power, fire alarms, public address, earthing/lightning protection, CCTV/security, IT wireways people counting systems	m²	430.00 to 520.00
Department Stores		
Air-conditioned: Includes for sanitaryware, soil/waste, cold water, hot water, sprinklers/dry risers, LTHW heating, air conditioning via AHU's, toilet extract, smoke extract, BMS, LV installations, life safety standby generators, lighting, small power, fire alarms, public address, earthing, lightning protection, CCTV/security, IT installation wireways	m²	340.00 to 410.00
Supermarkets		
Air-conditioned: Includes for rainwater, soil/waste, cold water, hot water, sprinklers, LTHW heating, air conditioning via AHU's, toilet extract, BMS, LV installations, lighting, small power, fire alarms, earthing, lightning protection, security, IT wireways, refrigeration	m²	490.00 to 600.00
Educational Buildings		
Secondary Schools (Academy)		
Natural Ventilation: Includes for rainwater, soil/waste, cold water, central hot water, LTHW heating, toilet extract, BMS, LV installations, lighting, small power, fire alarms, earthing, security, IT wireways	m²	280.00 to 345.00
Natural vent with comfort cooling to selected areas (BB93 compliant): Includes for rainwater, soil/waste, cold water, central hot water, LTHW heating, DX air conditioning, general supply/extract, toilet extract, BMS, LV installations, lighting, small power, fire alarms, earthing, security, IT wireways	m²	340.00 to 410.00

RIBA STAGE A/B FEASIBILITY COSTS

Item	Unit	Range £
Scientific Buildings		
Educational Research		
Comfort cooled: Includes for rainwater, soil/waste, cold water, central hot water, dry risers, compressed air, medical gasses, LTHW heating, 4 pipe air conditioning, toilet extract, fume, BMS, LV installations, lighting, small power, fire alarms, earthing, lightning protection, security, IT wireways	m²	630.00 to 770.00
Air-conditioned: Includes for rainwater, soil/waste, laboratory waste, cold water, central hot water, specialist water, dry risers, compressed air, medical gasses, steam, LTHW heating, VAV air conditioning, Comm's room cooling, toilet extract, fume extract, BMS, LV installations, UPS, standby generators, lighting, small power, fire alarms, earthing, lightning protection, security, IT wireways	m²	1125.00 to 1375.00
Commercial Research		
Air-conditioned; Includes for rainwater, soil/waste, laboratory waste, cold water, central hot water, specialist water, dry risers, compressed air, medical gasses, steam, LTHW heating, VAV air conditioning, Comm room cooling, toilet extract, fume extract, BMS, LV installations, UPS, standby generators, lighting, small power, fire alarms, earthing, lightning protection, security, IT wireways	m²	1175.00 to 1425.00
Residential Facilities		
Hotels		
1 to 3 Star: Includes for rainwater, soil/waste, cold water, hot water, dry risers, LTHW heating via radiators, toilet/bathroom extract, kitchen extract, BMS, LV installations, lighting, small power, fire alarms, earthing, lightning protection, security, IT wireways	m²	380.00 to 470.00
4 to 5 Star: Includes for rainwater, soil/waste, cold water, hot water, sprinklers, dry risers, 4 pipe air conditioning, kitchen extract, toilet/bathroom extract, BMS, LV installations, life safety standby generators, lighting, small power, fire alarms, earthing, security, IT wireways	m²	680.00 to 830.00

RIBA STAGE C ELEMENTAL RATES

Item	Unit	Range £	
Elemental Rates for Alternative Engineering Services Solutions – Offices			
The following examples of building types indicate the range of rates for alternative design solutions for each of the engineering services elements based on Gross Internal Area for the Shell and Core and Net Internal Area for the Fit Out. Fit Out is assumed to be to Cat A Standard.			
Consideration should be made for the size of the building, which may affect the economies of scale for rates, i.e. the larger the building the lower the rates.			
OFFICES			
Mechanical and Electrical Services			
Shell & Core			
5.1 Sanitary Installations			
Building up to 3,000 m²	m² GIA	8.10 to	9.90
Building over 3,000 m² up to 15,000 m² (low rise)	m² GIA	6.75 to	8.25
5.3 Disposal Installations			
Building up to 3,000 m²	m² GIA	18.00 to	22.00
Building over 3,000 m² up to 15,000 m²	m² GIA	16.20 to	19.80
5.4 Water Installations			
Building up to 3,000 m²	m² GIA	18.00 to	22.00
Building over 3,000 m² up to 15,000 m²	m² GIA	18.00 to	22.00
5.6 Space Heating and Air Conditioning			
LPHW Heating Installation; including gas installations			
Building up to 3,000 m²	m² GIA	36.00 to	44.00
Building over 3,000 m² up to 15,000 m²	m² GIA	38.00 to	47.00
Comfort cooling;			
2 pipe fan coil for building up to 3,000 m²	m² GIA	56.00 to	69.00
2 pipe fan coil for building over 3,000 m² up to 15,000 m²	m² GIA	52.00 to	63.00
2 pipe variable refrigerant volume (VRV) for building up to 3,000 m²	m² GIA	40.50 to	49.50
Full air conditioning;			
4 pipe fan coil for building up to 3,000 m²	m² GIA	81.00 to	99.00
4 pipe fan coil for building over 3,000 m² up to 15,000 m²	m² GIA	81.00 to	99.00
4 pipe variable refrigerant volume for building up to 3,000 m²	m² GIA	70.00 to	85.00
Ventilated (active) chilled beams for building over 3,000 m² up to 15,000 m²	m² GIA	81.00 to	99.00
Chilled beam exposed services for building over 3,000 m² to 15,000 m²	m² GIA	81.00 to	99.00
Concealed passive chilled beams for building over 3,000 m² up to 15,000 m²	m² GIA	81.00 to	99.00
Chilled ceiling for building over 3,000 m² to 15,000 m²	m² GIA	81.00 to	99.00
Chilled ceiling/perimeter beams for building over 3,000 m² up to 15,000 m²	m² GIA	81.00 to	99.00
Displacement for building over 3,000 m² up to 15,000 m²	m² GIA	81.00 to	99.00
5.7 Ventilation Systems (excluding smoke extract)			
Building up to 3,000 m²	m² GIA	31.50 to	38.50
Building over 3,000 m² up to 15,000 m²	m² GIA	29.00 to	36.00

RIBA STAGE C ELEMENTAL RATES

Item	Unit	Range £
5.8 Electrical Installations		
LV Installations		
Standby generators (life safety only)		
Buildings over 3,000 m² up to 15,000 m²	m² GIA	16.20 to 19.80
LV Distribution		
Buildings up to 3,000 m²	m² GIA	25.00 to 30.00
Buildings over 3,000 m² up to 15,000 m²	m² GIA	31.50 to 38.50
Lighting Installations (including lighting controls and luminaires)		
Buildings up to 3,000 m²	m² GIA	16.20 to 19.80
Buildings over 3,000 m² up to 15,000 m²	m² GIA	16.20 to 19.80
Small Power		
Buildings up to 3,000 m²	m² GIA	5.40 to 6.60
Buildings over 3,000 m² to up to 15,000 m²	m² GIA	4.95 to 6.05
Electrical Installations for Mechanical Plant		
Buildings up to 3,000 m²	m² GIA	6.30 to 7.70
Buildings over 3,000 m² to up to 15,000 m²	m² GIA	7.20 to 8.80
5.11 Fire and Lightning Protection		
Fire Protection over 3,000 m² to 15,000 m²		
Dry risers	m² GIA	3.60 to 4.40
Sprinkler installation	m² GIA	16.20 to 19.80
Protective Installations		
Earthing		
Buildings up to 3,000 m²	m² GIA	2.25 to 2.75
Buildings over 3,000 m² to 15,000 m²	m² GIA	2.25 to 2.75
Lightning Protection		
Buildings up to 3,000 m²	m² GIA	3.15 to 3.85
Buildings over 3,000 m² to 15,000 m²	m² GIA	2.25 to 2.75
5.12 Communication and Security Installations		
Fire Alarms (single stage)		
Buildings up to 3,000 m²	m² GIA	8.10 to 9.90
Buildings over 3,000 m² to 15,000 m²	m² GIA	10.30 to 12.70
Fire Alarms (phased evacuation)		
Buildings over 3,000 m² to 15,000 m²	m² GIA	12.20 to 14.80
IT (Wireways only)		
Buildings up to 3,000 m²	m² GIA	2.25 to 2.75
Buildings over 3,000 m² to 15,000 m²	m² GIA	2.25 to 2.75
Security		
Buildings up to 3,000 m²	m² GIA	6.30 to 7.70
Buildings over 3,000 m² to 15,000 m²	m² GIA	7.20 to 8.80
BMS Controls; including MCC panels and control cabling		
Full air conditioning	m² GIA	16.20 to 19.80

RIBA STAGE C ELEMENTAL RATES

Item	Unit	Range £
Elemental Rates for Alternative Engineering Services Solutions – Offices – cont		
Fit Out		
5.6 Space Heating and Air Conditioning		
LPHW Heating Installation; including gas installations		
Building up to 3,000 m²	m² NIA	56.00 to 69.00
Building over 3,000 m² to 15,000 m²	m² NIA	45.00 to 55.00
Comfort cooling;		
2 pipe fan coil for building up to 3,000 m²	m² NIA	81.00 to 99.00
2 pipe fan coil for building over 3,000 m² to 15,000 m²	m² NIA	79.00 to 96.00
2 pipe variable refrigerant volume (VRV) for building up to 3,000 m²	m² NIA	61.00 to 74.00
Full air conditioning;		
4 pipe fan coil for building up to 3,000 m²	m² NIA	120.00 to 150.00
4 pipe fan coil for building over 3,000 m² to 15,000 m²	m² NIA	110.00 to 130.00
4 pipe variable refrigerant volume for building up to 3,000 m²	m² NIA	100.00 to 120.00
Ventilated (active) chilled beams for building over 3,000 m² to 15,000 m²	m² NIA	120.00 to 150.00
Chilled beam exposed services for building over 3,000 m² to 15,000 m²	m² NIA	185.00 to 230.00
Concealed passive chilled beams for building over 3,000 m² to 15,000 m²	m² NIA	110.00 to 130.00
Chilled ceiling for building over 3,000 m² to 15,000 m²	m² NIA	190.00 to 230.00
Chilled ceiling/perimeter beams for building over 3,000 m² to 15,000 m²	m² NIA	200.00 to 245.00
Displacement for building over 3,000 m² to 15,000 m²	m² NIA	72.00 to 88.00
5.8 Electrical Installations		
Lighting Installations (including lighting controls and luminaires)		
Buildings up to 3,000 m²	m² NIA	45.00 to 55.00
Buildings over 3,000 m² to 15,000 m²	m² NIA	56.00 to 69.00
Electrical Installations for Mechanical Plant		
Buildings up to 3,000 m²	m² NIA	6.75 to 8.25
Buildings over 3,000 m² to 15,000 m²	m² NIA	6.75 to 8.25
5.11 Fire and Lightning Protection		
Fire Protection over 3,000 m² to 15,000 m²		
Sprinkler installation	m² NIA	25.00 to 30.00
Earthing		
Buildings up to 3,000 m²	m² NIA	2.25 to 2.75
Buildings over 3,000 m² to 15,000 m²	m² NIA	2.25 to 2.75
5.12 Communications and Security Installations		
Fire Alarms (single stage)		
Buildings up to 3,000 m²	m² NIA	10.30 to 12.70
Buildings over 3,000 m² to 15,000 m²	m² NIA	10.30 to 12.70
Fire Alarms (phased evacuation)		
Buildings over 3,000 m² to 15,000 m²	m² NIA	14.00 to 17.00
BMS Controls; including MCC panels and control cabling		
Full air conditioning	m² NIA	16.20 to 19.80

RIBA STAGE C ELEMENTAL RATES

Item	Unit	Range £		
Elemental Rates for Alternative Engineering Services Solutions – Hotels				
The following examples of building types indicate the range of rates for alternative design solutions for each of the engineering services elements based on Gross Internal Area for the Shell and Core and Net Internal Area for the Fit Out. Fit Out is assumed to be to Cat A Standard.				
Consideration should be made for the size of the building, which may affect the economies of scale for rates, i.e. the larger the building the lower the rates.				
HOTELS				
5 Services				
Shell & Core				
5.1 Sanitary Installations				
2 to 3 star	m² GIA	23.00	to	28.00
4 to 5 star	m² GIA	33.00	to	40.00
5.3 Disposal Installations				
2 to 3 star	m² GIA	25.50	to	31.00
4 to 5 star	m² GIA	26.00	to	32.00
5.4 Water Installations				
2 to 3 star	m² GIA	30.50	to	37.50
4 to 5 star	m² GIA	42.00	to	52.00
5.6 Space Heating and Air Conditioning				
LPHW Heating Installation; including gas installations				
2 to 3 star	m² GIA	34.50	to	42.00
4 to 5 star	m² GIA	35.00	to	43.00
Air Conditioning; including ventilation				
2 to 3 star – 4 pipe Fan Coil	m² GIA	200.00	to	250.00
4 to 5 star – 4 pipe Fan Coil	m² GIA	210.00	to	255.00
2 to 3 star – 3 pipe variable refrigerant volume	m² GIA	130.00	to	160.00
4 to 5 star – 3 pipe variable refrigerant volume	m² GIA	130.00	to	160.00
5.8 Electrical Services				
LV Installations				
Standby generators (life safety only)				
2 to 3 star	m² GIA	13.90	to	17.10
4 to 5 star	m² GIA	14.40	to	17.60
LV distribution				
2 to 3 star	m² GIA	30.00	to	37.00
4 to 5 star	m² GIA	40.00	to	49.00

RIBA STAGE C ELEMENTAL RATES

Item	Unit	Range £		
Elemental Rates for Alternative Engineering Services Solutions – Hotels – cont				
Shell & Core – cont				
Lighting Installations				
2 to 3 star	m² GIA	21.00	to	26.00
4 to 5 star	m² GIA	30.50	to	37.50
Small Power				
2 to 3 star	m² GIA	7.20	to	8.80
4 to 5 star	m² GIA	12.20	to	14.80
Electrical installations for mechanical plant				
2 to 3 star	m² GIA	6.30	to	7.70
4 to 5 star	m² GIA	6.30	to	7.70
5.11 Fire and Lightning Protection				
Fire Protection				
2 to 3 star – Dry risers	m² GIA	9.45	to	11.60
4 to 5 star – Dry risers	m² GIA	9.45	to	11.60
2 to 3 star – Sprinkler installation	m² GIA	21.00	to	25.00
4 to 5 star – Sprinkler installation	m² GIA	23.00	to	28.50
Earthing				
2 to 3 star	m² GIA	1.40	to	1.70
4 to 5 star	m² GIA	1.40	to	1.70
Lightning Protection				
2 to 3 star	m² GIA	1.40	to	1.70
4 to 5 star	m² GIA	1.40	to	1.70
5.12 Communications and Security Installations				
Fire Alarms				
2 to 3 star	m² GIA	13.90	to	17.10
4 to 5 star	m² GIA	14.40	to	17.60
IT				
2 to 3 star	m² GIA	11.70	to	14.30
4 to 5 star	m² GIA	11.70	to	14.30
Security				
2 to 3 star	m² GIA	18.40	to	22.50
4 to 5 star	m² GIA	18.90	to	23.00
BMS Controls; including MCC panels and control cabling				
2 to 3 star	m² GIA	10.30	to	12.70
4 to 5 star	m² GIA	25.50	to	31.00

RIBA STAGE C ELEMENTAL RATES

Item	Unit	Range £		
Elemental Rates for Alternative Engineering Services Solutions – Residential				
The following examples of building types indicate the range of rates for alternative design solutions for each of the engineering services elements based on Gross Internal Area for the Shell and Core and Net Internal Area for the Fit Out. Fit Out is assumed to be to Cat A Standard.				
Consideration should be made for the size of the building, which may affect the economies of scale for rates, i.e. the larger the building the lower the rates.				
RESIDENTIAL APARTMENTS				
5 Services				
Shell & Core				
5.1 Sanitary Installations				
Affordable	m² GIA	1.35	to	1.65
Private	m² GIA	3.70	to	4.50
5.3 Disposal Installations				
Affordable	m² GIA	11.70	to	14.30
Private	m² GIA	15.80	to	19.30
5.4 Water Installations				
Affordable	m² GIA	25.00	to	30.00
Private	m² GIA	41.00	to	50.00
5.5 Heat Source				
Affordable	m² GIA	9.00	to	11.00
Private	m² GIA	17.10	to	21.00
5.6 Space Heating & Air Conditioning				
Affordable	m² GIA	8.10	to	9.90
Private	m² GIA	32.00	to	39.00
5.7 Ventilation Systems				
Affordable	m² GIA	15.80	to	19.30
Private	m² GIA	28.00	to	34.00
5.8 Electrical Installations				
Affordable	m² GIA	34.00	to	41.00
Private	m² GIA	47.00	to	58.00
5.9 Fuel Installations (Gas)				
Affordable	m² GIA	2.25	to	2.75
Private	m² GIA	3.60	to	4.40

RIBA STAGE C ELEMENTAL RATES

Item	Unit	Range £		
Elemental Rates for Alternative Engineering Services Solutions – Residential – cont				
Shell & Core – cont				
5.11 Fire and Lightning Protection				
Affordable	m² GIA	9.00	to	11.00
Private	m² GIA	11.70	to	14.30
5.12 Communications and Security Installations				
Affordable	m² GIA	23.00	to	28.00
Private	m² GIA	28.00	to	34.00
5.13 Special Installations				
Affordable	m² GIA	6.75	to	8.25
Private	m² GIA	13.90	to	17.10
Fit Out				
5.1 Sanitary Installations				
Affordable	m² NIA	29.00	to	36.00
Private	m² NIA	54.00	to	66.00
5.3 Disposal Installations				
Affordable	m² NIA	9.00	to	11.00
Private	m² NIA	11.30	to	13.80
5.4 Water Installation				
Affordable	m² NIA	29.00	to	36.00
Private	m² NIA	45.00	to	55.00
5.5 Heat Source				
Affordable	m² NIA	13.50	to	16.50
Private	m² NIA	18.40	to	22.50
5.6 Space Heating and Air Conditioning				
Affordable	m² NIA	45.00	to	55.00
Private	m² NIA	170.00	to	210.00
5.7 Ventilating Systems				
Affordable (whole house vent)	m² NIA	22.50	to	27.50
Private (whole house vent)	m² NIA	37.00	to	45.00
5.8 Electrical Installations				
Affordable	m² NIA	45.00	to	55.00
Private	m² NIA	92.00	to	110.00

RIBA STAGE C ELEMENTAL RATES

Item	Unit	Range £		
5.9 Fuel Installations (Gas)				
Affordable	m² NIA	9.00	to	11.00
Private	m² NIA	10.80	to	13.20
5.11 Fire and Lightning Protection				
Affordable	m² NIA	18.00	to	22.00
Private	m² NIA	20.00	to	25.00
5.12 Communication and Security Installations				
Affordable	m² NIA	34.00	to	41.00
Private	m² NIA	68.00	to	83.00
5.13 Special Installations				
Affordable	m² NIA	6.75	to	8.25
Private	m² NIA	20.00	to	25.00
Note: The range in cost differs due to the vast diversity in services strategies available. The lower end of the scale reflects local plant within apartment schemes, such as combi boilers, local ventilation, etc. The high end of the scale is based on central plant installations with good quality apartments, which includes comfort cooling, sprinklers, home network installations, video entry, higher quality of sanitaryware and good quality LED lighting.				

ALL-IN RATES

Item	Unit	Range £
5 Services		
5.3 Disposal Installations		
ABOVE GROUND DRAINAGE		
Soil and waste	Point	320.00 to 390.00
5.4 Water Installations		
WATER INSTALLATIONS		
Cold Water	Point	320.00 to 390.00
Hot Water	Point	365.00 to 445.00
PIPEWORK		
HOT AND COLD WATER		
Excludes insulation, valves and ancillaries, etc.		
Light gauge copper tube to EN1057 R250 (TX) formerly BS 2871 Part 1 Table X with joints as described including allowance for waste, fittings and supports assuming average runs with capillary joints up to 54 mm and bronze welded thereafter		
Horizontal High Level Distribution		
15 mm	m	29.00 to 35.00
22 mm	m	30.00 to 36.00
28 mm	m	38.00 to 46.00
35 mm	m	42.00 to 52.00
42 mm	m	49.50 to 61.00
54 mm	m	63.00 to 77.00
67 mm	m	88.00 to 110.00
76 mm	m	92.00 to 110.00
108 mm	m	140.00 to 170.00
Risers		
15 mm	m	16.20 to 19.80
22 mm	m	18.90 to 23.00
28 mm	m	29.00 to 35.00
35 mm	m	32.50 to 39.50
42 mm	m	39.50 to 48.50
54 mm	m	46.00 to 56.00
67 mm	m	79.00 to 97.00
76 mm	m	88.00 to 110.00
108 mm	m	115.00 to 140.00
Toilet Areas, etc., at Low Level		
15 mm	m	49.50 to 61.00
22 mm	m	60.00 to 74.00
28 mm	m	83.00 to 100.00

ALL-IN RATES

Item	Unit	Range £		
LTHW AND CHILLED WATER				
Excludes insulation, valves and ancillaries, etc.				
Black heavy weight mild steel tube to BS 1387 with joints in the running length, allowance for waste, fittings and supports assuming average runs				
Horizontal Distribution – Basements, etc.				
15 mm	m	35.00	to	43.00
20 mm	m	39.00	to	47.00
25 mm	m	42.00	to	52.00
32 mm	m	49.50	to	61.00
40 mm	m	57.00	to	69.00
50 mm	m	69.00	to	85.00
65 mm	m	75.00	to	91.00
80 mm	m	100.00	to	120.00
100 mm	m	130.00	to	160.00
125 mm	m	180.00	to	225.00
150 mm	m	240.00	to	300.00
200 mm	m	330.00	to	400.00
250 mm	m	405.00	to	490.00
300 mm	m	480.00	to	580.00
Risers				
15 mm	m	21.50	to	26.00
20 mm	m	25.00	to	31.00
25 mm	m	28.00	to	34.00
32 mm	m	31.50	to	38.50
40 mm	m	35.00	to	43.00
50 mm	m	44.00	to	54.00
65 mm	m	55.00	to	67.00
80 mm	m	75.00	to	91.00
100 mm	m	92.00	to	110.00
125 mm	m	115.00	to	140.00
150 mm	m	150.00	to	185.00
200 mm	m	240.00	to	290.00
250 mm	m	310.00	to	375.00
300 mm	m	350.00	to	425.00
On Floor Distribution				
15 mm	m	33.00	to	41.00
20 mm	m	39.00	to	47.00
25 mm	m	41.50	to	51.00
32 mm	m	48.50	to	59.00
40 mm	m	55.00	to	67.00
50 mm	m	65.00	to	79.00

ALL-IN RATES

Item	Unit	Range £
5.4 Water Installations – cont		
LTHW AND CHILLED WATER – cont		
Plantroom Areas, etc.		
15 mm	m	35.00 to 43.00
20 mm	m	39.50 to 48.50
25 mm	m	45.00 to 55.00
32 mm	m	51.00 to 63.00
40 mm	m	59.00 to 72.00
50 mm	m	70.00 to 86.00
65 mm	m	77.00 to 94.00
80 mm	m	110.00 to 135.00
100 mm	m	130.00 to 160.00
125 mm	m	190.00 to 235.00
150 mm	m	255.00 to 310.00
200 mm	m	340.00 to 415.00
250 mm	m	420.00 to 520.00
300 mm	m	495.00 to 610.00
5.5 Heat Source		
HEAT SOURCE		
Gas fired boilers including gas train and controls	kW	29.00 to 36.00
Gas fired boilers including gas train, controls, flue, plantroom pipework, valves and insulation, pumps and pressurization unit	kW	81.00 to 99.00
5.6 Space Heating and Air Conditioning		
CHILLED WATER		
Air cooled R134a refrigerant chiller including control panel, anti-vibration mountings	kW	110.00 to 140.00
Air cooled R134a refrigerant chiller including control panel, anti-vibration mountings, plantroom pipework, valves, insulation, pumps and pressurization units	kW	190.00 to 235.00
Water cooled R134a refrigerant chiller including control panel, anti-vibration mountings	kW	68.00 to 83.00
Water cooled R134a refrigerant chiller including control panel, anti-vibration mountings, plantroom pipework, valves, insulation, pumps and pressurization units	kW	140.00 to 175.00
Absorption steam medium chiller including control panel, anti-vibration mountings, plantroom pipework, valves, insulation, pumps and pressurization units	kW	260.00 to 320.00
HEAT REJECTION		
Open circuit, forced draft cooling tower	kW	52.00 to 63.00
Closed circuit, forced draft cooling tower	kW	63.00 to 77.00
Dry air	kW	56.00 to 69.00

ALL-IN RATES

Item	Unit	Range £
PUMPS		
Pumps including flexible connections, anti-vibration mountings	kPa	31.50 to 38.50
Pumps including flexible connections, anti-vibration mountings, plantroom pipework, valves, insulation and accessories	kPa	79.00 to 96.00
DUCTWORK		
The rates below allow for ductwork and for all other labour and material in fabrication, fittings, supports and jointing to equipment, stop and capped ends, elbows, bends, diminishing and transition pieces, regular and reducing couplings, volume control dampers, branch diffuser and 'snap on' grille connections, ties, 'Ys', crossover spigots, etc., turning vanes, regulating dampers, access doors and openings, hand-holes, test holes and covers, blanking plates, flanges, stiffeners, tie rods and all supports and brackets fixed to structure.		
Rectangular galvanized mild steel ductwork as HVCA DW 144 up to 1000 mm longest side	m²	52.00 to 63.00
Rectangular galvanized mild steel ductwork as HVCA DW 144 up to 2500 mm longest side	m²	56.00 to 69.00
Rectangular galvanized mild steel ductwork as HVCA DW 144 up to 3000 mm longest side and above	m²	77.00 to 94.00
Circular galvanized mild steel ductwork as HVCA DW 144	m²	56.00 to 69.00
Flat oval galvanized mild steel ductwork as HVCA DW 144 up to 545 mm wide	m²	56.00 to 69.00
Flat oval galvanized mild steel ductwork as HVCA DW 144 up to 880 mm wide	m²	63.00 to 77.00
Flat oval galvanized mild steel ductwork as HVCA DW 144 up to 1785 mm wide	m²	72.00 to 88.00
PACKAGED AIR HANDLING UNITS		
Air handling unit including LPHW preheater coil, pre-filter panel, LPHW heater coils, chilled water coil, filter panels, inverter drive, motorized volume control dampers, sound attenuation, flexible connections to ductwork and all anti-vibration mountings	m³/s	6300.00 to 7700.00
5.7 Ventilating Systems		
EXTRACT FANS		
Extract fan including inverter drive, sound attenuation, flexible connections to ductwork and all anti-vibration mountings	m³/s	1250.00 to 1550.00

ALL-IN RATES

Item	Unit	Range £		
5.8 Electrical Installations				
MV/LV INSTALLATIONS				
The cost of MV/LV equipment will vary according to the electricity supplier's requirements, the duty required and the actual location of the site. For estimating purposes, the items indicated below are typical of the equipment required in a MV substation incorporated into a building.				
RING MAIN UNIT				
Ring Main Unit, 11 kV including electrical terminations	Point	13500.00	to	16500.00
TRANSFORMERS				
Oil filled transformers, 11 kV to 415 kV including electrical terminations	KVA	14.80	to	18.10
Cast resin transformers, 11 kV to 415 kV including electrical terminations	KVA	15.80	to	19.30
Midal filled transformers, 11 kV to 415 kV including electrical terminations	KVA	17.10	to	21.00
MV SWITCHGEAR				
Cubicle section MV switchpanel, Form 4 type 6 including air circuit breakers, meters and electrical terminations	Section	20000.00	to	25000.00
LV SWITCHGEAR				
LV switchpanel, Form 3 including all isolators, fuses, meters and electrical terminations	Isolat	2025.00	to	2475.00
LV switchpanel, Form 4 type 5 including all isolators, fuses, meters and electrical terminations	Isolat	3200.00	to	3900.00
EXTERNAL PACKAGED SUBSTATION				
Extra over cost for prefabricated packaged substation housing, excludes base and protective security fencing	Item	20000.00	to	25000.00
STANDBY GENERATING SETS				
Diesel powered including control panel, flue, oil day tank and attenuation				
Approximate installed cost, LV	KVA	250.00	to	310.00
Approximate installed cost, MV	KVA	290.00	to	350.00
UNINTERRUPTIBLE POWER SUPPLY				
Rotary UPS including control panel and choke transformer (excludes distribution)				
Approximate installed cost (range 1000 kVA to 2500 kVA)	KVA	405.00	to	495.00
Static UPS including control panel, automatic bypass, DC isolator and batteries for 30 minutes standby (excludes distribution)				
Approximate installed cost (range 500 kVA to 1000 kVA)	KVA	220.00	to	270.00

ALL-IN RATES

Item	Unit	Range £		
SMALL POWER				
Approximate prices for wiring of power points of length not exceeding 20 m, including accessories, wireways but excluding distribution boards				
13 amp accessories				
Wired in PVC insulated twin and earth cable in ring main circuit				
Domestic properties	Point	49.50	to	61.00
Commercial properties	Point	68.00	to	83.00
Industrial properties	Point	68.00	to	83.00
Wired in PVC insulated twin and earth cable in radial circuit				
Domestic properties	Point	68.00	to	83.00
Commercial properties	Point	81.00	to	99.00
Industrial properties	Point	81.00	to	99.00
Wired in LSF insulated single cable in ring main circuit				
Commercial properties	Point	81.00	to	99.00
Industrial properties	Point	81.00	to	99.00
Wired in LSF insulated single cable in radial circuit				
Commercial properties	Point	99.00	to	120.00
Industrial properties	Point	99.00	to	120.00
45 amp wired in PVC insulated twin and earth cable				
Domestic properties	Point	99.00	to	120.00
Low voltage power circuits				
Three phase four wire radial circuit feeding an individual load, wired in LSF insulated single cable including all wireways, isolator, not exceeding 10 metres; in commercial properties				
1.5	Point	160.00	to	200.00
2.5	Point	180.00	to	220.00
4	Point	190.00	to	230.00
6	Point	210.00	to	250.00
10	Point	240.00	to	300.00
16	Point	270.00	to	330.00
Three phase four core radial circuit feeding an individual load item, wired in LSF/SWA/XLPE insulated cable including terimations, isolator, clipped to surface, not exceeding 10 metres; in commercial properties				
1.5	Point	120.00	to	150.00
2.5	Point	135.00	to	165.00
4	Point	150.00	to	180.00
6	Point	160.00	to	200.00
10	Point	250.00	to	300.00
16	Point	315.00	to	385.00

ALL-IN RATES

Item	Unit	Range £		
5.8 Electrical Installations – cont				
LIGHTING				
Approximate prices for wiring of lighting points including rose, wireways but excluding distribution boards, luminaires and switches				
Final circuits				
Wired in PVC insulated twin and earth cable				
Domestic properties	Point	36.00	to	44.00
Commercial properties	Point	45.00	to	55.00
Industrial properties	Point	45.00	to	55.00
Wired in LSF insulated single cable				
Commercial properties	Point	59.00	to	72.00
Industrial properties	Point	59.00	to	72.00
ELECTRICAL WORKS IN CONNECTION WITH MECHANICAL SERVICES				
The cost of electrical connections to mechanical services equipment will vary depending on the type of building and complexity of the equipment.				
Typical rate for power wiring, isolators and associated wireways.	m^2	6.30	to	7.70
EXTERNAL LIGHTING				
Estate road lighting				
Post type road lighting lantern 70 watt CDM-T 3000k complete with 5 m high column with hinged lockable door, control gear and cut-out including 2.5 mm two core butyl cable internal wiring, interconnectionns and earthing fed by 16 mm² four core XLPE/SWA/LSF cable and terminations. Approximate installed price per metre road length (based on 300 metres run) including time switch but excluding builder's work in connection				
Columns erected at 30 m intervals along road (cost per m of road)	m	63.00	to	77.00
Bollard lighting				
Bollard lighting fitting 26 watt TC-D 3500k including control gear, all internal wiring, interconnectionns, earthing and 25 metres of 2.5 mm² three core XLPE/SWA/LSF cable				
Approximate installed price excluding builder's work in connection	EA	900.00	to	1100.00
Outdoor flood lighting				
Wall mounted outdoor flood light fitting complete with tungsten halogen lamp, mounting bracket, wire guard and all internal wiring and containment, fixed to brickwork or concrete and connected				
Installed price 500 watt	Point	180.00	to	220.00
Installed price 1000 watt	Point	220.00	to	270.00

ALL-IN RATES

Item	Unit	Range £
Pedestal mounted outdoor flood light fitting complete with 1000 watt MBF/U lamp, control gear, contained in weatherproof steel box, all internal wiring and containment, interconnections and earthing, fixed to brickwork or concrete and connected		
Approximate installed price excluding builder's work in connection	EA	900.00 to 1100.00
5.10 Lift and Conveyor Systems		
LIFT INSTALLATIONS		
The cost of lift installations will vary depending upon a variety of circumstances. The following prices assume a car height of 2.2 metres, manufacturer's standard car finish, brushed stainless steel 2 panel centre opening doors to BSEN81 Part 1 & 2 and Lift Regulations 1997.		
Passenger lifts, machine room above		
Electrically operated AC drive serving 2 levels with directional collective controls and a speed of 1.0 m/s		
8 Person	Item	52000.00 to 64000.00
10 Person	Item	57000.00 to 69000.00
13 Person	Item	59000.00 to 73000.00
17 Person	Item	68000.00 to 83000.00
21 Person	Item	77000.00 to 93000.00
26 Person	Item	87000.00 to 105000.00
Electrically operated AC drive serving 4 levels and a speed of 1.0 m/s		
8 Person	Item	61000.00 to 75000.00
10 Person	Item	65000.00 to 80000.00
13 Person	Item	68000.00 to 84000.00
17 Person	Item	77000.00 to 95000.00
21 Person	Item	87000.00 to 105000.00
26 Person	Item	99000.00 to 120000.00
Electrically operated AC drive serving 6 levels and a speed of 1.0 m/s		
8 Person	Item	69000.00 to 85000.00
10 Person	Item	74000.00 to 90000.00
13 Person	Item	77000.00 to 95000.00
17 Person	Item	86000.00 to 105000.00
21 Person	Item	97000.00 to 120000.00
26 Person	Item	110000.00 to 135000.00
Electrically operated AC drive serving 8 levels and a speed of 1.0 m/s		
8 Person	Item	78000.00 to 95000.00
10 Person	Item	82000.00 to 100000.00
13 Person	Item	86000.00 to 105000.00
17 Person	Item	96000.00 to 120000.00
21 Person	Item	110000.00 to 130000.00
26 Person	Item	120000.00 to 150000.00

ALL-IN RATES

Item	Unit	Range £
5.10 Lift and Conveyor Systems – cont		
Passenger lifts, machine room above – cont		
Electrically operated AC drive serving 10 levels and a speed of 1.0 m/s		
8 Person	Item	86000.00 to 105000.00
10 Person	Item	91000.00 to 110000.00
13 Person	Item	94000.00 to 115000.00
17 Person	Item	105000.00 to 130000.00
21 Person	Item	120000.00 to 145000.00
26 Person	Item	135000.00 to 165000.00
Electrically operated AC drive serving 12 levels and a speed of 1.0 m/s		
8 Person	Item	94000.00 to 115000.00
10 Person	Item	99000.00 to 120000.00
13 Person	Item	100000.00 to 125000.00
17 Person	Item	115000.00 to 140000.00
21 Person	Item	130000.00 to 160000.00
26 Person	Item	145000.00 to 180000.00
Electrically operated AC drive serving 14 levels and a speed of 1.0 m/s		
8 Person	Item	105000.00 to 130000.00
10 Person	Item	110000.00 to 130000.00
13 Person	Item	110000.00 to 140000.00
17 Person	Item	125000.00 to 155000.00
21 Person	Item	140000.00 to 170000.00
26 Person	Item	160000.00 to 195000.00
Add to above for:		
Increase speed from 1.0 m/s to 1.6 m/s		
8 Person	Item	3600.00 to 4400.00
10 Person	Item	3700.00 to 4500.00
13 Person	Item	3800.00 to 4600.00
17 Person	Item	3800.00 to 4600.00
21 Person	Item	3800.00 to 4600.00
26 Person	Item	3800.00 to 4600.00
Increase speed from 1.6 m/s to 2.0 m/s		
8 Person	Item	4950.00 to 6100.00
10 Person	Item	5400.00 to 6600.00
13 Person	Item	5900.00 to 7200.00
17 Person	Item	2600.00 to 3200.00
21 Person	Item	2700.00 to 3300.00
26 Person	Item	2700.00 to 3300.00
Increase speed from 2.0 m/s to 2.5 m/s		
8 Person	Item	1825.00 to 2250.00
10 Person	Item	1825.00 to 2250.00
13 Person	Item	2250.00 to 2700.00

ALL-IN RATES

Item	Unit	Range £
17 Person	Item	2250.00 to 2700.00
21 Person	Item	2600.00 to 3200.00
26 Person	Item	2600.00 to 3200.00
Enhanced finish to car – Centre mirror, flat ceiling, carpet		
8 Person	Item	2500.00 to 3100.00
10 Person	Item	2700.00 to 3300.00
13 Person	Item	2700.00 to 3300.00
17 Person	Item	3050.00 to 3750.00
21 Person	Item	3500.00 to 4300.00
26 Person	Item	4150.00 to 5100.00
Bottom motor room		
8 Person	Item	6300.00 to 7700.00
10 Person	Item	6300.00 to 7700.00
13 Person	Item	6300.00 to 7700.00
17 Person	Item	7700.00 to 9400.00
21 Person	Item	7700.00 to 9400.00
26 Person	Item	7700.00 to 9500.00
Fire fighting control		
8 Person	Item	4950.00 to 6100.00
10 Person	Item	4950.00 to 6100.00
13 Person	Item	4950.00 to 6100.00
17 Person	Item	4950.00 to 6100.00
21 Person	Item	4950.00 to 6100.00
26 Person	Item	4950.00 to 6100.00
Glass back		
8 Person	Item	2250.00 to 2750.00
10 Person	Item	2500.00 to 3100.00
13 Person	Item	3000.00 to 3600.00
17 Person	Item	3600.00 to 4400.00
21 Person	Item	3600.00 to 4400.00
26 Person	Item	3600.00 to 4400.00
Glass doors		
8 Person	Item	17000.00 to 21000.00
10 Person	Item	17000.00 to 21000.00
13 Person	Item	19000.00 to 23000.00
17 Person	Item	20000.00 to 25000.00
21 Person	Item	19000.00 to 23500.00
26 Person	Item	19000.00 to 23500.00
Painting to entire pit		
8 Person	Item	1800.00 to 2200.00
10 Person	Item	1800.00 to 2200.00
13 Person	Item	1800.00 to 2200.00
17 Person	Item	1800.00 to 2200.00
21 Person	Item	1800.00 to 2200.00
26 Person	Item	1800.00 to 2200.00

ALL-IN RATES

Item	Unit	Range £
5.10 Lift and Conveyor Systems – cont		
Add to above for: – cont		
Dual seal shaft		
8 Person	Item	3600.00 to 4400.00
10 Person	Item	3600.00 to 4400.00
13 Person	Item	3600.00 to 4400.00
17 Person	Item	4300.00 to 5300.00
21 Person	Item	4300.00 to 5300.00
26 Person	Item	4300.00 to 5300.00
Dust sealing machine room		
8 Person	Item	720.00 to 880.00
10 Person	Item	720.00 to 880.00
13 Person	Item	1175.00 to 1425.00
17 Person	Item	1175.00 to 1425.00
21 Person	Item	1175.00 to 1425.00
26 Person	Item	1175.00 to 1425.00
Intercom to reception desk and security room		
8 Person	Item	360.00 to 440.00
10 Person	Item	360.00 to 440.00
13 Person	Item	360.00 to 440.00
17 Person	Item	360.00 to 440.00
21 Person	Item	360.00 to 440.00
26 Person	Item	360.00 to 440.00
Heating, cooling and ventilation to machine room		
8 Person	Item	900.00 to 1100.00
10 Person	Item	900.00 to 1100.00
13 Person	Item	900.00 to 1100.00
17 Person	Item	900.00 to 1100.00
21 Person	Item	900.00 to 1100.00
26 Person	Item	900.00 to 1100.00
Shaft lighting/small power		
8 Person	Item	3150.00 to 3850.00
10 Person	Item	3150.00 to 3850.00
13 Person	Item	3150.00 to 3850.00
17 Person	Item	3150.00 to 3850.00
21 Person	Item	3150.00 to 3850.00
26 Person	Item	3150.00 to 3850.00
Motor room lighting/small power		
8 Person	Item	1175.00 to 1425.00
10 Person	Item	1175.00 to 1425.00
13 Person	Item	1350.00 to 1650.00
17 Person	Item	1450.00 to 1750.00
21 Person	Item	1450.00 to 1750.00
26 Person	Item	1450.00 to 1750.00

ALL-IN RATES

Item	Unit	Range £
Lifting beams		
8 Person	Item	1350.00 to 1650.00
10 Person	Item	1350.00 to 1650.00
13 Person	Item	1350.00 to 1650.00
17 Person	Item	1525.00 to 1875.00
21 Person	Item	1525.00 to 1875.00
26 Person	Item	1525.00 to 1875.00
10 mm equipotential bonding of all entrance metalwork		
8 Person	Item	720.00 to 880.00
10 Person	Item	720.00 to 880.00
13 Person	Item	720.00 to 880.00
17 Person	Item	720.00 to 880.00
21 Person	Item	720.00 to 880.00
26 Person	Item	720.00 to 880.00
Shaft secondary steelwork		
8 Person	Item	5200.00 to 6400.00
10 Person	Item	5400.00 to 6600.00
13 Person	Item	5600.00 to 6800.00
17 Person	Item	5800.00 to 7000.00
21 Person	Item	5800.00 to 7000.00
26 Person	Item	5800.00 to 7000.00
Independent insurance inspection		
8 Person	Item	1700.00 to 2075.00
10 Person	Item	1700.00 to 2075.00
13 Person	Item	1700.00 to 2075.00
17 Person	Item	1700.00 to 2075.00
21 Person	Item	1700.00 to 2075.00
26 Person	Item	1700.00 to 2075.00
12 month warranty service		
8 Person	Item	1800.00 to 2200.00
10 Person	Item	1800.00 to 2200.00
13 Person	Item	1800.00 to 2200.00
17 Person	Item	1800.00 to 2200.00
21 Person	Item	1800.00 to 2200.00
26 Person	Item	1800.00 to 2200.00
Passenger lifts, machine room less		
Electrically operated AC drive serving 2 levels with directional collective controls and a speed of 1.0 m/s		
8 Person	Item	47000.00 to 58000.00
10 Person	Item	52000.00 to 64000.00
13 Person	Item	55000.00 to 68000.00
17 Person	Item	67000.00 to 82000.00
21 Person	Item	73000.00 to 90000.00
26 Person	Item	81000.00 to 99000.00

ALL-IN RATES

Item	Unit	Range £
5.10 Lift and Conveyor Systems – cont		
Passenger lifts, machine room less – cont		
Electrically operated AC drive serving 4 levels and a speed of 1.0 m/s		
8 Person	Item	55000.00 to 67000.00
10 Person	Item	59000.00 to 73000.00
13 Person	Item	64000.00 to 78000.00
17 Person	Item	76000.00 to 93000.00
21 Person	Item	83000.00 to 100000.00
26 Person	Item	91000.00 to 110000.00
Electrically operated AC drive serving 6 levels and a speed of 1.0 m/s		
8 Person	Item	63000.00 to 78000.00
10 Person	Item	68000.00 to 83000.00
13 Person	Item	72000.00 to 88000.00
17 Person	Item	85000.00 to 100000.00
21 Person	Item	92000.00 to 110000.00
26 Person	Item	100000.00 to 120000.00
Electrically operated AC drive serving 8 levels and a speed of 1.0 m/s		
8 Person	Item	72000.00 to 87000.00
10 Person	Item	76000.00 to 92000.00
13 Person	Item	80000.00 to 98000.00
17 Person	Item	94000.00 to 115000.00
21 Person	Item	100000.00 to 120000.00
26 Person	Item	110000.00 to 135000.00
Electrically operated AC drive serving 10 levels and a speed of 1.0 m/s		
8 Person	Item	80000.00 to 97000.00
10 Person	Item	84000.00 to 100000.00
13 Person	Item	89000.00 to 110000.00
17 Person	Item	100000.00 to 125000.00
21 Person	Item	110000.00 to 135000.00
26 Person	Item	120000.00 to 150000.00
Electrically operated AC drive serving 12 levels and a speed of 1.0 m/s		
8 Person	Item	88000.00 to 110000.00
10 Person	Item	92000.00 to 110000.00
13 Person	Item	97000.00 to 120000.00
17 Person	Item	110000.00 to 135000.00
21 Person	Item	120000.00 to 145000.00
26 Person	Item	130000.00 to 160000.00
Electrically operated AC drive serving 14 levels and a speed of 1.0 m/s		
8 Person	Item	98000.00 to 120000.00
10 Person	Item	100000.00 to 120000.00
13 Person	Item	110000.00 to 130000.00
17 Person	Item	120000.00 to 150000.00
21 Person	Item	130000.00 to 160000.00
26 Person	Item	140000.00 to 175000.00

ALL-IN RATES

Item	Unit	Range £
Add to above for:		
Increase speed from 1.0 m/s to 1.6 m/s		
8 Person	Item	2350.00 to 2850.00
10 Person	Item	2350.00 to 2850.00
13 Person	Item	2425.00 to 3000.00
17 Person	Item	3800.00 to 4600.00
21 Person	Item	4300.00 to 5300.00
26 Person	Item	5600.00 to 6800.00
Enhanced finish to car – Centre mirror, flat ceiling, carpet		
8 Person	Item	2500.00 to 3100.00
10 Person	Item	2700.00 to 3300.00
13 Person	Item	2700.00 to 3300.00
17 Person	Item	3050.00 to 3750.00
21 Person	Item	3500.00 to 4300.00
26 Person	Item	4150.00 to 5100.00
Fire fighting control		
8 Person	Item	4950.00 to 6100.00
10 Person	Item	4950.00 to 6100.00
13 Person	Item	4950.00 to 6100.00
17 Person	Item	4950.00 to 6100.00
21 Person	Item	4950.00 to 6100.00
26 Person	Item	4950.00 to 6100.00
Painting to entire pit		
8 Person	Item	1800.00 to 2200.00
10 Person	Item	1800.00 to 2200.00
13 Person	Item	1800.00 to 2200.00
17 Person	Item	1800.00 to 2200.00
21 Person	Item	1800.00 to 2200.00
26 Person	Item	1800.00 to 2200.00
Dual seal shaft		
8 Person	Item	3600.00 to 4400.00
10 Person	Item	3600.00 to 4400.00
13 Person	Item	3600.00 to 4400.00
17 Person	Item	4300.00 to 5300.00
21 Person	Item	4300.00 to 5300.00
26 Person	Item	4300.00 to 5300.00
Shaft lighting/small power		
8 Person	Item	3150.00 to 3850.00
10 Person	Item	3150.00 to 3850.00
13 Person	Item	3150.00 to 3850.00
17 Person	Item	3150.00 to 3850.00
21 Person	Item	3150.00 to 3850.00
26 Person	Item	3150.00 to 3850.00

ALL-IN RATES

Item	Unit	Range £
5.10 Lift and Conveyor Systems – cont		
Add to above for: – cont		
Intercom to reception desk and security room		
8 Person	Item	360.00 to 440.00
10 Person	Item	360.00 to 440.00
13 Person	Item	360.00 to 440.00
17 Person	Item	360.00 to 440.00
21 Person	Item	360.00 to 440.00
26 Person	Item	360.00 to 440.00
Lifting beams		
8 Person	Item	1350.00 to 1650.00
10 Person	Item	1350.00 to 1650.00
13 Person	Item	1350.00 to 1650.00
17 Person	Item	1525.00 to 1875.00
21 Person	Item	1525.00 to 1875.00
26 Person	Item	1525.00 to 1875.00
10 mm Equipotential bonding of all entrance metalwork		
8 Person	Item	720.00 to 880.00
10 Person	Item	720.00 to 880.00
13 Person	Item	720.00 to 880.00
17 Person	Item	720.00 to 880.00
21 Person	Item	720.00 to 880.00
26 Person	Item	720.00 to 880.00
Shaft secondary steelwork		
8 Person	Item	5200.00 to 6400.00
10 Person	Item	5400.00 to 6600.00
13 Person	Item	5600.00 to 6800.00
17 Person	Item	5800.00 to 7000.00
21 Person	Item	5800.00 to 7000.00
26 Person	Item	5800.00 to 7000.00
Independent insurance inspection		
8 Person	Item	1700.00 to 2075.00
10 Person	Item	1700.00 to 2075.00
13 Person	Item	1700.00 to 2075.00
17 Person	Item	1700.00 to 2075.00
21 Person	Item	1700.00 to 2075.00
26 Person	Item	1700.00 to 2075.00
12 month warranty service		
8 Person	Item	900.00 to 1100.00
10 Person	Item	900.00 to 1100.00
13 Person	Item	900.00 to 1100.00
17 Person	Item	900.00 to 1100.00
21 Person	Item	900.00 to 1100.00
26 Person	Item	900.00 to 1100.00

ALL-IN RATES

Item	Unit	Range £
Goods lifts, machine room above		
Electrically operated two speed serving 2 levels to take 1000 kg load, prime coated internal finish and a speed of 1.0 m/s		
2000 kg	Item	88000.00 to 110000.00
2250 kg	Item	97000.00 to 120000.00
2500 kg	Item	98000.00 to 120000.00
3000 kg	Item	105000.00 to 130000.00
Electrically operated two speed serving 4 levels and a speed of 1.0 m/s		
2000 kg	Item	100000.00 to 120000.00
2250 kg	Item	110000.00 to 130000.00
2500 kg	Item	110000.00 to 135000.00
3000 kg	Item	125000.00 to 150000.00
Electrically operated two speed serving 6 levels and a speed of 1.0 m/s		
2000 kg	Item	110000.00 to 135000.00
2250 kg	Item	120000.00 to 150000.00
2500 kg	Item	120000.00 to 150000.00
3000 kg	Item	140000.00 to 170000.00
Electrically operated two speed serving 8 levels and a speed of 1.0 m/s		
2000 kg	Item	120000.00 to 150000.00
2250 kg	Item	130000.00 to 160000.00
2500 kg	Item	135000.00 to 165000.00
3000 kg	Item	150000.00 to 190000.00
Electrically operated two speed serving 10 levels and a speed of 1.0 m/s		
2000 kg	Item	135000.00 to 165000.00
2250 kg	Item	145000.00 to 180000.00
2500 kg	Item	150000.00 to 180000.00
3000 kg	Item	170000.00 to 205000.00
Electrically operated two speed serving 12 levels and a speed of 1.0 m/s		
2000 kg	Item	145000.00 to 180000.00
2250 kg	Item	160000.00 to 190000.00
2500 kg	Item	155000.00 to 190000.00
3000 kg	Item	180000.00 to 225000.00
Electrically operated two speed serving 14 levels and a speed of 1.0 m/s		
2000 kg	Item	160000.00 to 195000.00
2250 kg	Item	170000.00 to 210000.00
2500 kg	Item	170000.00 to 210000.00
3000 kg	Item	200000.00 to 240000.00

ALL-IN RATES

Item	Unit	Range £
5.10 Lift and Conveyor Systems – cont		
Add to above for:		
Increase speed of travel from 1.0 m/s to 1.6 m/s		
2000 kg	Item	1175.00 to 1425.00
Enhanced finish to car – centre mirror, flat ceiling, carpet		
2000 kg	Item	3150.00 to 3850.00
2250 kg	Item	3150.00 to 3850.00
2500 kg	Item	3150.00 to 3850.00
Bottom motor room		
2000 kg	Item	7700.00 to 9400.00
Painting to entire pit		
2000 kg	Item	1800.00 to 2200.00
2250 kg	Item	1800.00 to 2200.00
2500 kg	Item	1800.00 to 2200.00
3000 kg	Item	1800.00 to 2200.00
Dual seal shaft		
2000 kg	Item	3600.00 to 4400.00
2250 kg	Item	3600.00 to 4400.00
2500 kg	Item	3600.00 to 4400.00
3000 kg	Item	3600.00 to 4400.00
Intercom to reception desk and security room		
2000 kg	Item	360.00 to 440.00
2250 kg	Item	360.00 to 440.00
2500 kg	Item	360.00 to 440.00
3000 kg	Item	360.00 to 440.00
Heating, cooling and ventilation to machine room		
2000 kg	Item	900.00 to 1100.00
2250 kg	Item	900.00 to 1100.00
2500 kg	Item	900.00 to 1100.00
3000 kg	Item	900.00 to 1100.00
Lifting beams		
2000 kg	Item	1350.00 to 1650.00
2250 kg	Item	1350.00 to 1650.00
2500 kg	Item	1350.00 to 1650.00
3000 kg	Item	1350.00 to 1650.00
10 mm equipotential bonding of all entrance metalwork		
2000 kg	Item	720.00 to 880.00
2250 kg	Item	720.00 to 880.00
2500 kg	Item	720.00 to 880.00
3000 kg	Item	720.00 to 880.00

ALL-IN RATES

Item	Unit	Range £
Independent insurance inspection		
2000 kg	Item	1700.00 to 2075.00
2250 kg	Item	1700.00 to 2075.00
2500 kg	Item	1700.00 to 2075.00
3000 kg	Item	1700.00 to 2075.00
12 month warranty service		
2000 kg	Item	900.00 to 1100.00
2250 kg	Item	1075.00 to 1325.00
2500 kg	Item	1075.00 to 1325.00
3000 kg	Item	1075.00 to 1325.00
Goods lift, machine room less		
Electrically operated two speed serving 2 levels to take 1000 kg load, prime coated internal finish and a speed of 1.0 m/s		
2000 kg	Item	79000.00 to 97000.00
2250 kg	Item	85000.00 to 100000.00
2500 kg	Item	90000.00 to 110000.00
Electrically operated two speed serving 4 levels and a speed of 1.0 m/s		
2000 kg	Item	89000.00 to 110000.00
2250 kg	Item	95000.00 to 115000.00
2500 kg	Item	100000.00 to 120000.00
Electrically operated two speed serving 6 levels and a speed of 1.0 m/s		
2000 kg	Item	99000.00 to 120000.00
2250 kg	Item	99000.00 to 120000.00
2500 kg	Item	110000.00 to 140000.00
Electrically operated two speed serving 8 levels and a speed of 1.0 m/s		
2000 kg	Item	110000.00 to 130000.00
2250 kg	Item	115000.00 to 140000.00
2500 kg	Item	120000.00 to 150000.00
Electrically operated two speed serving 10 levels and a speed of 1.0 m/s		
2000 kg	Item	120000.00 to 145000.00
2250 kg	Item	125000.00 to 150000.00
2500 kg	Item	130000.00 to 160000.00
Electrically operated two speed serving 12 levels and a speed of 1.0 m/s		
2000 kg	Item	130000.00 to 160000.00
2250 kg	Item	135000.00 to 170000.00
2500 kg	Item	140000.00 to 175000.00
Electrically operated two speed serving 14 levels and a speed of 1.0 m/s		
2000 kg	Item	140000.00 to 170000.00
2250 kg	Item	150000.00 to 180000.00
2500 kg	Item	150000.00 to 190000.00

ALL-IN RATES

Item	Unit	Range £
5.10 Lift and Conveyor Systems – cont		
Add to above for:		
Increase speed of travel from 1.0 m/s to 1.6 m/s		
2000 kg	Item	4300.00 to 5300.00
2250 kg	Item	6900.00 to 8500.00
Enhanced finish to car – centre mirror, flat ceiling, carpet		
2000 kg	Item	3700.00 to 4500.00
2250 kg	Item	4850.00 to 5900.00
2500 kg	Item	7200.00 to 8800.00
Painting to entire pit		
2000 kg	Item	1800.00 to 2200.00
2250 kg	Item	1800.00 to 2200.00
2500 kg	Item	1800.00 to 2200.00
Dual seal shaft		
2000 kg	Item	3600.00 to 4400.00
2250 kg	Item	3600.00 to 4400.00
2500 kg	Item	3600.00 to 4400.00
Intercom to reception desk and security room		
2000 kg	Item	360.00 to 440.00
2250 kg	Item	360.00 to 440.00
2500 kg	Item	360.00 to 440.00
Lifting beams		
2000 kg	Item	1350.00 to 1650.00
2250 kg	Item	1350.00 to 1650.00
2500 kg	Item	1350.00 to 1650.00
10 mm equipotential bonding of all entrance metalwork		
2000 kg	Item	720.00 to 880.00
2250 kg	Item	720.00 to 880.00
2500 kg	Item	720.00 to 880.00
Independent insurance inspection		
2000 kg	Item	1700.00 to 2075.00
2250 kg	Item	1700.00 to 2075.00
2500 kg	Item	1700.00 to 2075.00
12 month warranty service		
2000 kg	Item	900.00 to 1100.00
2250 kg	Item	900.00 to 1100.00
2500 kg	Item	900.00 to 1100.00

ALL-IN RATES

Item	Unit	Range £
ESCALATOR INSTALLATIONS		
30Ø pitch escalator with a rise of 3 to 6 metres with standard balustrades		
1000 mm step width	Item	72000.00 to 88000.00
Add to above for:		
Balustrade lighting	Item	2250.00 to 2750.00
Skirting lighting	Item	8800.00 to 11000.00
Emergency stop button pedestals	Item	4050.00 to 4950.00
Truss cladding – stainless steel	Item	22500.00 to 27500.00
Truss cladding – spray painted steel	Item	20000.00 to 24000.00
5.11 Fire and Lightning Protection		
SPRINKLER INSTALLATION		
Recommended maximum area coverage per sprinkler head:		
Extra light hazard, 21 m² of floor area		
Ordinary hazard, 12 m² of floor area		
Extra high hazard, 9 m² of floor area		
Equipment		
Sprinkler equipment installation, pipework, valve sets, booster pumps and water storage	Item	63000.00 to 77000.00
Price per sprinkler head; including pipework, valves and supports	Point	160.00 to 200.00
HOSE REELS AND DRY RISERS		
Wall mounted concealed hose reel with 36 metre hose including approximately 15 metres of pipework and isolating valve:		
Price per hose reel	Point	1625.00 to 1975.00
100 mm dry riser main including 2 way breeching valve and box, 65 mm landing valve, complete with padlock and leather strap and automatic air vent and drain valve:		
Price per landing	Point	1350.00 to 1650.00

ALL-IN RATES

Item	Unit	Range £
5.12 Communications and Security Installations		
FIRE ALARMS		
Cost per point for two core MICC insulated wired system including all terminations, supports and wireways		
Call point	Point	225.00 to 275.00
Smoke detector	Point	200.00 to 240.00
Smoke/heat detector	Point	225.00 to 275.00
Heat detector	Point	215.00 to 260.00
Heat detector and sounder	Point	190.00 to 230.00
Input/output/relay units	Point	270.00 to 330.00
Alarm sounder	Point	215.00 to 260.00
Alarm sounder/beacon	Point	250.00 to 310.00
Speakers/voice sounders	Point	250.00 to 310.00
Speakers/voice sounders (weatherproof)	Point	270.00 to 330.00
Beacon/strobe	Point	200.00 to 240.00
Beacon/strobe (weatherproof)	Point	290.00 to 350.00
Door release units	Point	290.00 to 350.00
Beam detector	Point	810.00 to 990.00
For costs for zone control panel, battery chargers and batteries, see 'Prices for Measured Work' section.		
ACCESS CONTROL SYSTEMS		
Door mounted access control unit inclusive of door furniture, lock plus software; including up to 50 metres of cable and termination; including documentation testing and commissioning		
Internal single leaf door	Point	990.00 to 1200.00
Internal double door	Point	1075.00 to 1325.00
External single leaf door	Point	1175.00 to 1425.00
External double leaf door	Point	1350.00 to 1650.00
Management control PC with printer software and commissioning up to 1000 users	Point	13500.00 to 16500.00
CCTV INSTALLATIONS		
CCTV equipment inclusive of 50 m of cable including testing and commissioning		
Internal camera with bracket	Point	810.00 to 990.00
Internal camera with housing	Point	990.00 to 1200.00
Internal PTZ camera with bracket	Point	1525.00 to 1875.00
External fixed camera with housing	Point	1075.00 to 1325.00
External PTZ camera dome	Point	2150.00 to 2600.00
External PTZ camera dome with power	Point	2600.00 to 3200.00

ALL-IN RATES

Item	Unit	Range £		
IT INSTALLATIONS				
DATA CABLING				
Complete channel link including patch leads, cable, panels, testing and documentation (excludes cabinets and/or frames, patch cords, backbone/ harness connectivity as well as containment)				
Low Level				
Cat 5e (up to 5,000 outlets)	Point	45.00	to	55.00
Cat 5e (5,000 to 15,000 outlets)	Point	36.00	to	44.00
Cat 6 (up to 5,000 outlets)	Point	54.00	to	66.00
Cat 6 (5,000 to 15,000 outlets)	Point	49.50	to	61.00
Cat 6a (up to 5,000 outlets)	Point	68.00	to	83.00
Cat 6a (5,000 to 15,000 outlets)	Point	63.00	to	77.00
Cat 7 (up to 5,000 outlets)	Point	81.00	to	99.00
Cat 7 (5,000 to 15,000 outlets)	Point	77.00	to	94.00
Note: LSZH cable based on average of 50 metres false floor low level installation assuming 1 workstation in 2.5 m × 2.5 m (to 3.2 m × 3.2 m) density, with 4 data points per workstation. Not applicable for installations with less than 250 No. outlets.				
High Level				
Cat 5e (up to 500 outlets)	Point	54.00	to	66.00
Cat 5e (over 500 outlets)	Point	45.00	to	55.00
Cat 6 (up to 500 outlets)	Point	63.00	to	77.00
Cat 6 (over 500 outlets)	Point	59.00	to	72.00
Cat 6a (up to 500 outlets)	Point	77.00	to	94.00
Cat 6a (over 500 outlets)	Point	72.00	to	88.00
Cat 7 (up to 500 outlets)	Point	90.00	to	110.00
Cat 7 (over 500 outlets)	Point	86.00	to	105.00
Note: High level at 10 × 10 m grid.				

BUILDING MODELS – ELEMENTAL COST SUMMARIES

Item	Unit	£
AIRPORT TERMINAL BUILDING		
New build airport terminal building, premium quality, located in the South East, handling both domestic and international flights with a gross internal floor area (GIA) of 25,000 m². These costs exclude baggage handling, check-in systems, pre-check in and boarding security systems, vertical transportation and services to aircraft stands. Heat source is via district mains (excluded). Costs assume the works are executed under landside access/logistics environment. Total M&E services value approximately £21 million		
Cost Summary		
5 Services		
5.1 Sanitary Installations	m²	3.00
5.3 Disposal Installations		
rainwater	m²	6.00
soil and waste	m²	7.00
condensate	m²	1.00
5.4 Water Installations	m²	17.00
5.6 Space Heating and Air Treatment		
LTHW heating system	m²	73.00
chilled water system	m²	69.00
supply and extract air conditioning system	m²	99.00
allowance for service to communications room	m²	9.00
5.7 Ventilating Services		
mechanical ventilation to baggage handling and plantrooms	m²	23.00
toilet extract ventilation	m²	8.00
smoke extract installation	m²	9.00
kitchen extract system	m²	3.00
5.8 Electrical Installation		
MV/LV Switchgear	m²	61.00
standby generator	m²	37.00
mains and sub-mains installation	m²	42.00
small power installation	m²	14.00
lighting and luminaries	m²	91.00
emergency llighting installation	m²	11.00
power to mechanical services	m²	6.00
5.9 Fuel Installations (gas)	m²	2.00

BUILDING MODELS – ELEMENTAL COST SUMMARIES

Item	Unit	£
5.11 Fire and Lightning Protection		
lightning protection	m²	3.00
earthing and bonding	m²	4.00
sprinkler installations	m²	32.00
dry riser and hosereels installation	m²	6.00
fire suppression installation to comunications room	m²	3.00
5.12 Communications and Security Installations		
fire and smoke detection and alarm system	m²	28.00
voice/public address system	m²	19.00
intruder detection	m²	9.00
security, CCTV and access control	m²	32.00
wireways for telephones, data and structure cable	m²	14.00
structured cable installation	m²	32.00
flight information display system	m²	23.00
BMS installation	m²	50.00
Total Cost/m² (based on GIFA of 25,000 m²)	£/m²	850.00

BUILDING MODELS – ELEMENTAL COST SUMMARIES

Item	Unit	£
SHOPPING MALL (TENANT'S FIT OUT EXCLUDED)		
Natural ventilation shopping mall with approximately 33,000 m² two storey retail area and a 13,000 m² above ground, mechanically ventilated, covered car park, situated in a town centre in South East England. Total M&E services value approximately £10,500,000		
Cost Summary		
RETAIL BUILDING		
5 Services		
5.1 Sanitary Installations	m²	1.00
5.3 Sanitary Installation		
rainwater	m²	6.00
soil, waste and vent	m²	6.00
5.4 Water Installations		
cold water installation	m²	6.00
hot water installation	m²	5.00
5.5 Heat Source		
5.6 Space Heating and Air Conditioning		
condenser water system	m²	30.00
LTHW installation	m²	4.00
air conditioning system	m²	30.00
over door heaters at entrances	m²	1.00
5.7 Ventilation Systems		
public toilet ventilation	m²	1.00
plant room ventilation	m²	4.00
supply and extract systems to shop units	m²	13.00
toilet extract systems to shop units	m²	2.00
smoke ventilation to Mall area	m²	10.00
service corridor ventilation	m²	2.00
other miscellaneous ventilation	m²	17.00
5.8 Electrical Installations		
LV distribution	m²	22.00
standby power	m²	4.00
general lighting	m²	60.00
external lighting	m²	5.00
emergency lighting	m²	12.00
small power	m²	10.00
mechanical services power supplies	m²	3.00
general earthing	m²	1.00
UPS for security and CCTV equipment	m²	1.00
5.9 Fuel Installations		
gas supply and boilers	m²	2.00
gas supplies to anchr (major) stores	m²	1.00

BUILDING MODELS – ELEMENTAL COST SUMMARIES

Item	Unit	£
5.11 Fire and Lightning Protection		
lightning protection	m²	1.00
sprinkler installations	m²	14.00
dry risers	Excl	–
hosereel installation	Excl	–
5.12 Communications, Security and Control Installations		
fire alarm installation	m²	8.00
public address/voice alarm	m²	5.00
security installation	m²	11.00
general containment	m²	12.00
5.13 Special Installations		
BMS/Controls	m²	17.00
Total Cost/m² (based on GIFA of 33,000 m²)	£/m²	327.00
CAR PARK – 13,000m²		
5 Services		
5.3 Disposal Intallations		
car park drainage	m²	5.00
5.7 Ventilation Systems		
car park ventilation (impulse fans)	m²	30.00
5.8 Electrical Installations		
LV distribution	m²	10.00
standby power	Incl	–
general lighting	m²	20.00
external lighting	Excl	–
emergency lighting	m²	5.00
small power	m²	3.00
mechanical services power supplies	m²	5.00
general earthing	m²	1.00
ramp frost protection	m²	2.00
5.11 Fire and Lightning Protection		
sprinkler installation	m²	20.00
dry riser and hosereel installation	Excl	–
fire alarm installtions	m²	20.00
5.12 Communication, Security and Control Installations		
security installations	m²	9.00
BMS/Controls	m²	6.00
5.13 Special Installations		
entry/exit barriers, pay stations	m²	6.00
Total Cost/m² (based on GIFA of 13,000 m²)	£/m²	142.00

BUILDING MODELS – ELEMENTAL COST SUMMARIES

Item	Unit	£
SUPERMARKET		
Supermarket located in the South East with a total gross floor area of 4,000 m², including a sales area of 2,350 m². The building is on one level and incorporates a main sales, coffee shop, bakery, offices and amenities areas and warehouse. Total M&E services value is approximately £1,000,000		
Cost Summary		
5 Services		
5.1 Sanitary Installations	m²	1.50
5.3 Disposal Installations	m²	3.50
5.4 Water Installations		
hot and cold water services	m²	25.00
5.6 Space Heating and Air Conditioning		
heating & ventilation with cooling via DX units	m²	10.00
5.7 Ventilating Systems		
supply and extract sytems	m²	2.50
5.8 Electrical Installations		
panels/boards	m²	29.00
containment	m²	1.50
general lighting	m²	10.00
small power	m²	10.00
mechanical serviceds wiring	m²	2.00
5.9 Fuel Installations	N/A	–
5.11 Fire and Lightning Protection		
lightning protection	m²	1.00
5.12 Communications and Security Installations		
fire alarms, detection and public address	m²	2.50
CCTV	m²	2.50
intruder alarm, detection and store security	m²	2.50
telecom and structured cabling	m²	0.50
BMS	m²	7.00
data cabinet	m²	2.50
controls wiring	m²	3.50
5.13 Special Installations		
Refrigeration		
installation	m²	28.00
plant	m²	28.00
cold store	m²	10.00
cabinets	m²	76.00
Total Cost/m² (based on GIFA of 4,000 m²)	£	260.00

BUILDING MODELS – ELEMENTAL COST SUMMARIES

Item	Unit	£
OFFICE BUILDING		
Speculative 15 storey office in Central London for multiple tenant occupancy with a gross floor area of 19,300 m². A four pipe fan coil system, with roof mounted cooling towers, gas fired boilers and basement mounted water cooled chillers. Total M&E services value approximately £6,800,000 (Shell & Core) and £3,500,000 (Cat A fit-out)		
Cost Summary		
SHELL & CORE – 19,300m²		
5 Services		
5.1 Sanitary Installations	m²	7.00
5.3 Disposal Installations		
rainwater/soil and waste	m²	15.00
condensate	m²	2.00
5.4 Water Installations		
hot and cold water service	m²	16.00
5.5 Heat Source	m²	9.00
5.6 Space Heating and Air Conditioning		
LTHW heating	m²	15.00
chilled water	m²	43.00
ductwork	m²	41.00
5.7 Ventilating Systems		
toilet extract ventilation	m²	6.00
basement extract	m²	15.00
miscellaneous ventilation systems	m²	18.00
5.8 Electrical Installations		
generator	m²	6.00
MV/LV supply/distribution	m²	57.00
general lighting	m²	21.00
general power	m²	4.00
electrical services for mechanical equipment	m²	5.00
5.9 Fuel Installatiuons (gas)	m²	1.00
5.11 Fire and Lighting Protection		
wet risers	m²	10.00
sprinklers	m²	17.00
earting and bonding	m²	2.00
lightning protection	m²	2.00

BUILDING MODELS – ELEMENTAL COST SUMMARIES

Item	Unit	£
OFFICE BUILDING – cont		
5 Services – cont		
5.12 Communication and Security Installations		
fire and voice alarms	m²	14.00
voice and data (wireways)	m²	1.00
security (wireways)	m²	1.00
disbled/refuge alarms	m²	3.00
CCTV/Access control/intruder detection	m²	4.00
BMS	m²	21.00
Total Cost/m² (based on GIFA of 19,300 m²)	£	355.00
CATEGORY 'A' FIT OUT – 12,500m² NIA		
5 Services		
5.6 Space Heating and Air Conditioning		
LTHW heating	m²	27.00
chilled water	m²	33.00
ductwork and grilles	m²	81.00
condensate	m²	8.00
5.8 Electrical Installations		
lighting installation	m²	68.00
electrical services in connection	m²	3.00
tenant distribution board	m²	6.00
5.11 Fire and Lightning Protection		
sprinkler installation	m²	21.00
5.12 Communications and Security Installations		
fire and voice alarms	m²	13.00
BMS	m²	19.00
Total Cost/m² (based on NIFA of 12,500 m²)	£	280.00

BUILDING MODELS – ELEMENTAL COST SUMMARIES

Item	Unit	£
BUSINESS PARK		
New build office in South East within M25 part of a speculative business park with a gross floor area of 10,000 m². A full air displacement system with roof mounted air cooled chillers, gas fired boilers and air handling plant. Total M&E services value approximately £2,500,000 (shell & core) and £1,000,000 (Cat A Fit-out)		
Cost Summary		
SHELL & CORE – 10,000m² GIA		
5 Services		
5.1 Sanitary Installations	m²	7.00
5.3 Disposal Installations		
rainwater	m²	3.00
soil and waste	m²	8.00
5.4 Water Installations		
cold water services	m²	9.00
hot water services	m²	3.00
5.5 Heat Source	m²	7.00
5.6 Space Heating and Air Conditioning		
LTHW heating; plantroom and risers	m²	11.00
chilled water; plantroom and risers	m²	26.00
ductwork; plantroom and risers	m²	77.00
5.7 Ventilating Systems		
toilet and miscellaneous ventilation	m²	7.00
5.8 Electrical Installations		
LV supply/distribution	m²	21.00
general lighting	m²	18.00
general power	m²	5.00
electrical services in connection with mechanical services	m²	3.00
5.9 Fuel Installations (gas)	m²	2.00
5.11 Fire and Lightning Protection		
earthing and bonding	m²	2.00
lightning protection	m²	2.00
dry risers	m²	1.00

BUILDING MODELS – ELEMENTAL COST SUMMARIES

Item	Unit	£
BUSINESS PARK – cont		
5 Services – cont		
5.12 Communication and Security Installations		
fire alarms	m²	9.00
security (wireways)	m²	2.00
data and voice (wireways)	m²	2.00
BMS	m²	19.00
Total Cost/m² (based on GIFA of 10,000 m²)	£	244.00
CATEGORY 'A' FIT OUT – 8,000m² NIA		
5 Services		
5.6 Space Heating and Air Conditioning		
LTHW heating and perimeter heaters	m²	40.00
floor swirl diffusers and supply ductwork	m²	22.00
5.8 Electrical Installations		
distribution boards	m²	2.00
general lightining, recessed including lighting controls	m²	47.00
5.11 Fire and Lightning Protection		
earthing and bonding	m²	1.00
5.12 Communication and Security Installations		
fire alarms	m²	5.00
BMS	m²	6.00
Total Cost/m² (based on NIFA of 8,000 m²)	£	123.00

BUILDING MODELS – ELEMENTAL COST SUMMARIES

Item	Unit	£
PERFORMING ARTS CENTRE (MEDIUM SPECIFICATION)		
Performing Arts centre with a Gross Internal Area (GIA) of approximately 8,000 m², based on a medium specification with cooling to the Auditorium. Total M&E service value is approximately £4,900,00		
The development comprises dance studios and a theatre auditorium in the outer London area. The theatre would require all the necessary stage lighting, machinery and equipment installed in a modern professional theatre (these are excluded from the model, as assumed to be FF&E, but the containment and power wiring is included).		
Cost Summary		
5 Services		
5.1 Sanitary Installations	m²	12.00
5.3 Disposal Installations		
soil, waste and rainwater	m²	11.00
5.4 Water Installations		
cold water services	m²	10.00
hot water services	m²	9.00
5.5 Heat Source	m²	16.00
5.6 Space Heating and Air Conditioning		
heating	m²	70.00
chilled water system	m²	24.00
supply and extract air systems	m²	110.00
5.7 Ventilation Services		
ventilation and extract systems to toilets, kitchen and workshop	m²	14.00
5.8 Electrical Installations		
LV supply/distribution	m²	48.00
general lighting	m²	110.00
small power	m²	28.00
5.9 Fuel Installations (gas)	m²	4.00
5.11 Fire and Lightning Protection		
lightning protection	m²	1.00
5.12 Communication and Security Installations		
fire alarms and detection	m²	23.00
voice and data complete installation (excluding active equipment)	m²	28.00
security; access; control; disabled alarms; staff paging	m²	26.00
BMS	m²	42.00
5.13 Special Installations		
theatre systems includes for containment and power wiring	m²	20.00
Total Cost/m² (based on GIFA of 8,000 m²)	£	606.00

BUILDING MODELS – ELEMENTAL COST SUMMARIES

Item	Unit	£
SPORTS HALL		
Single storey sports hall, located in the South East, with a gross internal area of 1,200 m² (40 m × 30 m). Total M&E services value approximately £220,000		
Cost Summary		
5 Services		
5.1 Sanitary Installations	m²	10.00
5.3 Disposal Installations		
rainwater	m²	3.00
soil and waste	m²	6.00
5.4 Water Installations		
hot and cold water services	m²	14.00
5.1 Heat Source		
boilers, flues, pumps and controls	m²	12.00
5.6 Space Heating and Air Conditioning		
warm air heating to sports hall area	m²	17.00
radiator heating to ancillary areas	m²	20.00
5.7 Ventilating Systems		
ventilation to changing, fitness and sports hall areas	m²	13.00
5.8 Electrical Installations		
main swithcgear and sub-mains	m²	11.00
small power	m²	10.00
lighting and luminaires to sports areas	m²	17.00
lightging and luminaires to ancillary areas	m²	19.00
5.9 Fuel Installations (included in 5.5)	Incl	–
5.11 Protective and Lightning Protection		
lightning protection	m²	3.00
5.12 Communications and Security Installations		
fire, smoke detection and alarm system, intruder detection	m²	11.00
CCTV installation	m²	13.00
public address and music systems	m²	6.00
wireways for voice and data	m²	2.00
Total Cost/m² (based on GIFA of 1,200 m²)	£	187.00

BUILDING MODELS – ELEMENTAL COST SUMMARIES

Item	Unit	£
STADIUM – NEW		
A three storey stadium, located in Greater London with a gross internal area of 85,000 m² and incorporating 60,000 spectator seats. Total M&E services value approximately £28,000,000		
Cost Summary		
5 Services		
5.1 Sanitary Installations	m²	10.00
5.3 Disposal Installations		
rainwater	m²	4.00
above ground drainage	m²	11.00
5.4 Water Installations		
hot and cold water	m²	20.00
5.5 Heat Source	m²	9.00
5.6 Space Heating and Air Conditioning		
heating	m²	6.00
cooling	m²	13.00
5.7 Ventilating Systems	m²	48.00
5.8 Electrical Installation		
MV/LV Supply	m²	10.00
LV distribution	m²	23.00
general lighting	m²	57.00
small power	m²	14.00
earting and bonding	m²	1.00
power supply to mechanical equipmemt	m²	1.00
pitch lighting	m²	10.00
5.9 Fuel Installations (gas)	m²	1.00
5.11 Fire and Lightning Protection		
lightning protection	m²	1.00
hydrants	m²	2.00
5.12 Communications and Security Installations		
wireways for data, TV, telecom and PA	m²	8.00
public address	m²	15.00
security	m²	14.00
data voice installations	m²	30.00
fire alarms	m²	8.00
disabled/refuse alarm/call systems	m²	3.00
BMS	m²	14.00
Total Cost/m² (based on GIFA of 85,000 m²)	£	333.00
Total Cost/seat (based on 60,000 seats)	£	470.00

BUILDING MODELS – ELEMENTAL COST SUMMARIES

Item	Unit	£
HOTELS		
200 bedroom, four star hotel, situated in Central London, with a gross internal floor area of 16,500 m². Total M&E services value approximately £10,000,000		
The development comprises a ten storey building with large suites on each guest floor, together with banqueting, meeting rooms and leisure facilities.		
Cost Summary		
5 Services		
5.1 Sanitary Installations	m²	35.00
5.3 Disposal Installations rainwater, soil and waste	m²	28.00
5.4 Water Installations hot and cold water services	m²	45.00
5.5 Heat Source condensing boiler and pumps, etc.	m²	15.00
5.6 Space Heating and Air Conditioning Air conditioning system; chillers, pumps, air handling units, ductwork, fan coil units, etc., to guest rooms, public areas, meeting and banquet rooms	m²	135.00
5.7 Ventilating Systems general toilet extract and ventilation to kitchens and bathrooms, etc.	m²	50.00
5.8 Electrical Installation MV/LV Installation, standby power, lighting, emergency lighting and small power to guest floors and public areas including earthing and lightning protection	m²	150.00
5.9 Fuel Installations (gas)	m²	2.50
5.11 Fire and Lightning Protection dry risers and sprinkler installation	m²	29.00
5.12 Communications and Security Installations		
fire, smoke detection and alarm system.security CCTV	m²	42.00
background music, AV wireways	m²	14.00
telecommunications, data and TV wiring (no hotel management and head end equipment)	m²	24.00
BMS	m²	26.00
Total Cost/m² (basec on GIFA of 16,500 m²)	£	600.00

BUILDING MODELS – ELEMENTAL COST SUMMARIES

Item	Unit	£
PRIVATE HOSPITAL		
New build project building. The works consist of a new 80 bed hospital of approximately 15,000 m², eight storey with a plant room. Total M&E services value £10,000,000		
All heat is provided from existing steam boiler plant, medical gases are also served from existing plant. The project includes the provision of additional standby electrical generation to serve the wider site requirements.		
This hospital has six operating theatres, ITU/HDU department, pathology facilities, diagnostic imaging, out patient facilities and physiotherapy.		
Cost Summary		
5 Services		
5.1 Sanitary Installations	m²	19.00
5.3 Disposal Installations		
rainwater	m²	2.00
soil and waste	m²	22.00
specialist drainage (above ground)	m²	1.00
5.4 Water Installations		
hot and cold water services	m²	48.00
5.5 Heat Source (included in 5.6)	Incl	–
5.6 Space Heating and Air Conditioning		
LPHW heating	m²	38.00
chilled water	m²	36.00
steam and condensate	m²	22.00
ventilation, comfort cooling and air conditioning	m²	120.00
5.7 Ventilating Systems (included in 5.6)	Incl	–
5.8 Electrical Installations		
MV distribution	m²	2.00
LV supply/distribution	m²	26.00
standby power	m²	31.00
UPS	m²	22.00
general lighting	m²	38.00
general power	m²	37.00
emergency lighting	m²	15.00
theatre lighting	m²	13.00
specialist lighting	m²	12.00
external lighting	m²	2.00
electrical supplies for mechanical equipment	m²	11.00
oil installations	m²	1.00

BUILDING MODELS – ELEMENTAL COST SUMMARIES

Item	Unit	£
PRIVATE HOSPITAL – cont		
5 Services – cont		
5.9 Fuel Installations		
gas installations	m²	3.00
oil installations	m²	6.00
5.11 Fire and Lightning Protection		
dry risers	m²	1.00
lightning protection	m²	0.50
5.12 Communications and Security Installations		
fire alarms and detection	m²	18.00
voice and data	m²	11.00
security and CCTV	m²	3.00
nurse call and cardiac alarm system	m²	16.00
personnel paging	m²	4.00
hospital radio (entertainment)	m²	22.00
BMS	m²	37.00
5.13 Special Installations		
pneumatic tube conveying system	m²	3.00
group 1 equipment	m²	36.00
Total Cost/m² (based on GIFA of 15,000 m²)	£	680.00

BUILDING MODELS – ELEMENTAL COST SUMMARIES

Item	Unit	£
SCHOOL		
New build secondary school (Academy) located in Southern England, with a gross internal floor area of 10,000 m². Total M&E services value approximately £4,000,000		
The building comprises a three storey teaching block, including provision for music, drama, catering, sports hall, science laboratories, food technology, workshops and reception/admin (BB93 compliant). Excludes IT cabling and sprinkler protection.		
Cost Summary		
5 Services		
5.1 Sanitary Installations		
toilet cores and changing facilities only	m²	9.20
5.3 Disposal Installations		
rainwater installations	m²	4.00
soil and waste	m²	11.00
5.4 Water Installations		
potable hot and cold water services	m²	27.50
non potable hot and cold water services to labs and art rooms	m²	10.00
5.5 Heat Source		
gas fired boiler installation	m²	14.00
5.6 Space Heating and Air Conditioning		
LTHW heating system (primary)	m²	34.00
LTHW heating system (secondary)	m²	7.00
DX cooling system to ICT server rooms	m²	4.65
mechanical supply and extract ventilation including DX type cooling to Music, Drama, Kitchen/Dining and Sports Hall	m²	40.00
5.7 Ventilating Systems		
toilet extract systems	m²	7.00
changing area extract systems	m²	4.00
extract ventilation from design/food technology and science labs	m²	8.00
5.8 Electrical Installations		
mains and sub-mains distribution	m²	35.00
lighting and luminaires including emergency fittings	m²	65.00
small power installation	m²	28.00
earthig and bonding	m²	2.00
5.9 Fuel Installations (gas)	m²	9.00

BUILDING MODELS – ELEMENTAL COST SUMMARIES

Item	Unit	£
SCHOOL – cont		
5 Services – cont		
5.11 Fire and Lightning Protection		
lightning protection	m²	1.50
5.12 Communications and Security Installations		
containment for telephone, IT data, AV and security systems	m²	3.00
fire, smoke detection and alarm system	m²	14.20
security installations including CCTV, access control and intruder alarm	m²	18.80
disabled toilet, refuge and induction loop systems	m²	4.00
BMS – to plant	m²	25.00
BMS – to opening vents/windows	m²	10.00
Total Cost/m² (based on GIFA of 10,000 m²)	£	396.00

BUILDING MODELS – ELEMENTAL COST SUMMARIES

Item	Unit	£
AFFORDABLE RESIDENTIAL DEVELOPMENT		
A 12 storey, 50 apartment affordable residential development with a gross internal area of 3,400 m² and a net internal area of 2,400 m², situated within the London area. The development does not include a car park and is based on 71% efficiency. Total M&E services value approximately £550,000.		
Based on an individual radiator LTHW system within each apartment, with local gas combi boilers exhausting to building facade. Kitchens and bathrooms are also ventilated to the building facade, there are pendant light fittings, an audio entry system, telephone and satellite installation. Sanitaryware is of lower quality with plastic pipework runs. Full sprinkler installation installed throughout. Excludes remote metering, with local controls only, no BMS.		
Cost Summary		
SHELL & CORE		
5 Services		
5.1 Sanitary Installations	m²	1.50
5.3 Disposal Installations	m²	13.00
5.4 Water Installations	m²	27.00
5.6 Space Heating and Air Treatment	m²	19.00
5.7 Ventilating Services	m²	18.00
5.8 Electrical Installation	m²	38.00
5.9 Fuel Installations – Gas	m²	3.00
5.11 Fire and Lightning Protection	m²	10.00
5.12 Communication, Security and Control Installations	m²	28.00
5.13 Specialist Installations	m²	8.00
Total Cost/m² (based on GIFA of 3,400 m²)	£/m²	165.00

BUILDING MODELS – ELEMENTAL COST SUMMARIES

Item	Unit	£
AFFORDABLE RESIDENTIAL DEVELOPMENT – cont		
FITTING OUT		
5 Services		
5.1 Sanitaryware	m²	33.00
5.3 Disposal Installations	m²	10.00
5.4 Water Installations	m²	33.00
5.5 Heat source	m²	15.00
5.6 Space Heating and Air Treatment	m²	50.00
5.7 Ventilating Services	m²	25.00
5.8 Electrical Installation	m²	50.00
5.9 Fuel Installations – Gas	m²	10.00
5.11 Fire and Lightning Protection	m²	20.00
5.12 Communication, Security and Control Installations	m²	38.00
5.13 Specialist Installations	m²	7.00
Total Cost/m² (based on a NIA of 2,400 m²)	£/m²	290.00
Total Cost per apartment – Shall & Core and Fit Out	£	25,400.00

BUILDING MODELS – ELEMENTAL COST SUMMARIES

Item	Unit	£
PRIVATE RESIDENTIAL DEVELOPMENT		
A 20 storey, 250 apartment private residential developments with a gross internal area of 22,750 m² and a net internal area of 20,415 m², situated within the London area. The development does not include a car park and is based on 90% efficiency. Total M&E service value is £15,000,000.		
Included is a central boiler and hot water installation with apartment heat exchanger. Good quality sanitary appliances. Perimeter trench heating to each apartment and 4 pipe fan coil unit installation. Central air cooled chiller systems. No gas to apartments. Whole house ventilation system discharging to local facade, wet riser with full sprinkler installation. Video entry and TV satellite installation. Flood wiring for apartment home automation and sound system only. Public Health apartment runs in plastic. Local apartment controls only. Lutron or Creston type system excluded.		
Cost Summary		
SHELL & CORE		
5 Services		
5.1 Sanitary Installations	m²	4.00
5.3 Disposal Installations	m²	18.00
5.4 Water Installations	m²	47.00
5.5 Heat Source	m²	19.00
5.6 Space Heating and Air Conditioning	m²	36.00
5.7 Ventilating Systems	m²	31.00
5.8 Electrical Installations	m²	52.00
5.9 Fuel Installations (gas)	m²	4.00
5.11 Fire and Lightning Protection	m²	13.00
5.12 Communications and Security Installations	m²	31.00
5.13 Special Installations	m²	16.00
Total Cost/m² (based on GIFA of 22,750 m²)	£	270.00

BUILDING MODELS – ELEMENTAL COST SUMMARIES

Item	Unit	£
PRIVATE RESIDENTIAL DEVELOPMENT – cont		
FITTING OUT		
5 Services		
5.1 Sanitary Installations	m²	60.00
5.3 Disposal Installations	m²	13.00
5.4 Water Installations	m²	50.00
5.5 Heat Source	m²	20.00
5.6 Space Heating and Air Conditioning	m²	190.00
5.7 Ventilating Systems	m²	40.00
5.8 Electrical Installations	m²	100.00
5.11 Fire and Lightning Protection	m²	15.00
5.12 Communications and Security Installations	m²	75.00
5.13 Special Installations	m²	25.00
Total Cost/m² (based on NIFA of 20,415 m²)	£	590.00
Total cost per apartment – Shell & Core and Fit Out	£	73000.00

BUILDING MODELS – ELEMENTAL COST SUMMARIES

Item	Unit	£
DISTRIBUTION CENTRE		
Distribution centre located in London with a total gross floor area of 75,000 m², including a refrigerated cold box of 17,500 m². Total M&E services value approximately £15,000,000		
The building is on one level and also incorporates an office area, vehicle recovery unit, gate house and plantrooms.		
Cost Summary		
5 Services		
5.3 Disposal Installations		
soil and waste	m²	2.00
rainwater	m²	6.00
5.4 Water Installations		
hot and cold water services	m²	2.00
5.5 Space Heating and Air Conditioning		
heating with ventilation to offices, displacement system to main warehouse	m²	21.00
5.7 Ventilating Systems		
smoke extract system	m²	5.00
5.8 Electrical Installations		
generator	m²	15.00
main MV installation	m²	19.00
MV distribution	m²	12.00
lighting installation	m²	10.00
small power installation	m²	12.00
5.9 Fuel Installations		
gas mains services to plantroom	m²	0.50
5.11 Fire and Lightning Protection		
sprinklers including racking protection	m²	42.00
earthing/lightning protection	m²	0.50
5.12 Communications ans Security Installations		
fire alarms, detection and public address	m²	11.00
CCTV	m²	7.00
BMS	m²	5.00
5.13 Special Installations		
refrigeration	m²	35.00
Total Cost/m² (based on GIFA of 75,000 m²)	£	205.00

BUILDING MODELS – ELEMENTAL COST SUMMARIES

Item	Unit	£
DATA CENTRE		
New build data centre located in the London area/proximity to M25. Net Technical Area (NTA) provided at 2,000 m² with typically other areas of 250 m² office space, 250 m² ancillary space and 1,000 m² internal plant. Total GIA 3,500 m². Total M&E services value approximately £16,500,000		
Power and cooling to Technical Space @ 1,500w/m². Cost/m² against Nett Technical Area – NTA.		
Cost Summary		
5 Services		
5.3 Disposal Installations		
soil and waste	m²	14.00
rainwater	m²	10.00
condensate	m²	10.00
5.4 Water Installations		
hot and cold water services	m²	24.00
5.6 Space Heating and Air Conditioning		
chilled water plant with redundancy of N+1 to provide 1,000w/m² net technical space cooling	m²	370.00
chilled water distribution to data centre to free-standing cooling units and distribution to ancillary office and workshop/build areas	m²	680.00
free-standing cooling units with redundancy of N+20% to technical space and switchrooms. Based on single coil cooling units 30% of units with humidification	m²	345.00
floor grilles	m²	71.00
5.7 Ventilating Systems		
supply and extract ventilation systems to data centre, switchrooms and ancillary spaces including dedicated gas extract and hot aisle extract system	m²	425.00
5.8 Electrical Installations		
main MV installations including transformers with redundant capacity of N+1	m²	650.00
LV distribution including cabling and busbar installations to provide fulll system – System dual supplies/redundancy including supplies to mechanical services, office and ancillary areas	m²	780.00
Generator Installation		
standby rated containerized generators with redundant capacity of N+1 and including synchronization panel, 72 hours bulk fuel store and controls	m²	910.00
Uninterruptible Power Supplies		
static UPS to provide 2 × (N+1) system redundancy with 10 minute battery autonomy	m²	1000.00

BUILDING MODELS – ELEMENTAL COST SUMMARIES

Item	Unit	£
LV Switchgear		
Incoming LV switchgear, UPS input and output boards for electrical and mechanical services systems	m²	850.00
Power Distribution Units (PDUs)		
PDUs to provide a redundant capacity of 2 Nr PDU to include static transfer switch and isolating transformer	m²	780.00
Lighting Installation		
lighting to technical, workshop and plant areas including office/ancillary spaces and external areas	m²	100.00
5.9 Fuel Installation (Assumed not required)		
5.11 Protective Installations		
gaseous suppression to technical areas and switchrooms	m²	245.00
lightning protection	m²	14.00
earthing and clean earth	m²	29.00
leak detection	m²	14.00
5.12 Communications and Security Installations		
very early smoke detection alarm (VESDA) to technical areas	m²	135.00
CCTV installations	m²	160.00
access control installations	m²	140.00
BMS	m²	265.00
PLC/Electrical monitoring system	m²	240.00
Total Cost/m² (based on NTA of 2,000 m²)	£	8200.00

BUILDING MODELS – ELEMENTAL COST SUMMARIES

Item	Unit	£
BUILDING MANAGEMENT INSTALLATIONS		
Category A Fit-Out		
Option 1 – 185 Nr four pipe fan coil – 740 points		
1.0 Field Equipment Network devices; Valves/actuators; Sensing devices	point	67.00
2.0 Cabling Power – from local isolator to DDC controller; Control – from DDC controller to field equipment	point	38.00
3.0 Programming Software – central facility; Software – network devices; Graphics	point	21.00
4.0 On site testing and commissioning Equipment; Programming/graphics; Power and control cabling	point	22.00
Total Option 1 – Four pipefan coil	point	150.00
Cost/FCU	£	600.00
Option 2 – 185 Nr two pipe fan coil system with electric heating – 740 points		
1.0 Field Equipment Network devices; Valves/actuators/thyristors; Sensing devices	point	87.00
2.0 Cabling Power – from local isolator to DDC controller; Control – from DDC controller to field equipment	point	40.00
3.0 Programming Software – central facility; Software – network devices; Graphics	point	21.00
4.0 On site testing and commissioning Equipment; Programming/graphics; Power and control cabling	point	22.50
Total Option 2 – Two pipe fan coil with electric heating	point	170.00
Cost/FCU	£	680.00
Option 3 – 180 Nr chilled beams with perimeter heating – 567 points		
1.0 Field Equipment Network devices; Valves/actuators; Sensing devices	point	83.00

BUILDING MODELS – ELEMENTAL COST SUMMARIES

Item	Unit	£
2.0 Cabling Power – from local isolator to DDC controller; Control – from DDC controller to field equipment	point	56.00
3.0 Programming Cost/Point	point	28.00
4.0 On site testing and commissioning Equipment; Programming/Graphics; Power and control cabling	point	30.00
Total Option 3 – Chilled beams with perimeter heating	point	200.00
Cost/Chilled Beam	£	630.00
Shell & Core Only		
Main Plant – Cost/BMS Points	£	650.00
Landlord FCUs – Cost/Terminal Units	£	650.00
Trade Contract Preliminaries	%	25.00
Notes: The following are included in points, rates		
– DDC Controllers/Control Enclosures/Control Panels		
– Motor Control Centre (MCC)		
– Field Devices		
– Control and Power Cabling from DDC Controllers/MCC		
– Programming		
– On Site Testing and Commissioning		

PART 3

Material Costs/Measured Work Prices

Mechanical Installations 157

R: Disposal Systems 163
R10: Rainwater Pipework/Gutter 163
R11: Above Ground Drainage 188

S: Piped Supply Systems 218
S10: Cold Water 218
S11: Hot Water 281
S32: Natural Gas 286
S41: Fuel Oil Storage/Distribution 291
S60: Fire Hose Reels 293
S61: Dry Risers 294
S63: Sprinklers 295
S65: Fire Hydrants 302

T: Mechanical/Cooling/Heating Systems 304
T10: Gas/Oil Fired Boilers 304
T13: Packaged Steam Generators 321
T31: Low Temperature Hot Water Heating 322
T33: Steam Heating 414
T42: Local Heating Units 418
T60: Central Refrigeration Plant 419
T61: Chilled Water 428
T70: Local Cooling Units 435

U: Ventilation/Air Conditioning Systems 437
U10: General Ventilation 437
U14: Ductwork: Fire-Rated 533
U30: Low Velocity Air Conditioning 554
U31: VAV Air Conditioning 557
U41: Fan Coil Air Conditioning 558
U70: Air Curtains 563

Electrical Installations 567

V: Electrical Supply/Power/Lighting 571
V10: Electrical Generation Plant 571
V11: MV Supply 574
V20: LV Distribution 579
V21: General Lighting 657
V22: General LV Power 667
V32: Uninterruptible Power Supply 676
V40: Emergency Lighting 678

W: Communications/Security/Control 683
W10: Telecommunications 683
W20: Radio/Television 686
W23: Clocks 688
W30: Data Transmission 690
W40: Access Control 697
W41: Security Detection and Alarm 698
W50: Fire Detection and Alarm 700
W51: Earthing and Bonding 703
W52: Lightning Protection 704
W60: Central Control/Building Management 709

Mechanical Installations

Material Costs/Measured Work Prices

DIRECTIONS

The following explanations are given for each of the column headings and letter codes.

Unit	Prices for each unit are given as singular (i.e. 1 metre, 1 nr) unless stated otherwise.
Net price	Industry tender prices, plus nominal allowance for fixings (unless measured separately), waste and applicable trade discounts.
Material cost	Net price plus percentage allowance for overheads, profit and preliminaries.
Labour norms	In man-hours for each operation.
Labour cost	Labour constant multiplied by the appropriate all-in man-hour cost based on gang rate (See also relevant Rates of Wages Section) plus percentage allowance for overheads, profit and preliminaries.
Measured work price (total rate)	Material cost plus Labour cost.

MATERIAL COSTS

The Material Costs given are based at Second Quarter 2013 but exclude any charges in respect of VAT. The average rate of copper during this quarter is US$6,800/UK£4,400 per tonne. Users of the book are advised to register on the SPON's website www.pricebooks.co.uk/updates to receive the free quarterly updates – alerts will then be provided by email as changes arise.

MEASURED WORK PRICES

These prices are intended to apply to new work in the London area. The prices are for reasonable quantities of work and the user should make suitable adjustments if the quantities are especially small or especially large. Adjustments may also be required for locality (e.g. outside London – refer to cost indices in approximate estimating section for details of adjustment factors) and for the market conditions e.g. volume of work secured or being tendered) at the time of use.

MECHANICAL INSTALLATIONS

The labour rate has been based on average gang rates per man hour effective from 1 April 2013. To this rate has been added 7.5% and 3% to cover preliminary items, site and head office overheads together with 2% for profit, resulting in an inclusive rate of £24.71 per man hour. The rate has been calculated on a working year of 2,016 hours; a detailed build-up of the rate is given at the end of these directions.

The rates and allowances will be subject to increases as covered by the National Agreements. Future changes will be published in the free Spon's quarterly update by registering on their website.

DIRECTIONS

DUCTWORK INSTALLATIONS

The labour rate basis is as per Mechanical above and to this rate has been added 5.5% plus 12.5% to cover site and head office overheads only (factory overhead is included in the material rate) and preliminary items together with 3% for profit, resulting in an inclusive rate of £25.87 per man hour. The rate has been calculated on a working year of 2,016 hours; a detailed build-up of the rate is given at the end of these directions.

In calculating the 'Measured Work Prices' the following assumptions have been made:

(a) That the work is carried out as a subcontract under the Standard Form of Building Contract.
(b) That, unless otherwise stated, the work is being carried out in open areas at a height which would not require more than simple scaffolding.
(c) That the building in which the work is being carried out is no more than six storey's high.

Where these assumptions are not valid, as for example where work is carried out in ducts and similar confined spaces or in multi-storey structures when additional time is needed to get to and from upper floors, then an appropriate adjustment must be made to the prices. Such adjustment will normally be to the labour element only.

Note: The rates do not include for any uplift applied if the ductwork package is procured via the Mechanical Sucontractor

DIRECTIONS

LABOUR RATE – MECHANICAL & PUBLIC HEALTH

The annual cost of a notional twelve man gang of 12 men

	FOREMAN	SENIOR CRAFTSMAN (+2 Welding skill)	SENIOR CRAFTSMAN	CRAFTSMAN	INSTALLER	MATE (Over 18)	SUB TOTALS
	1 NR	1 NR	2 NR	4 NR	2 NR	2 NR	
Hourly Rate from 1 April 2013	**15.23**	**13.11**	**12.59**	**12.07**	**10.46**	**8.82**	
Working hours per annum per man	1,702.40	1,702.40	1,702.40	1,702.40	1,702.40	1,702.40	
x Hourly rate × nr of men = £ per annum	**25,927.55**	**22,318.46**	**42,866.43**	**82,191.87**	**35,614.21**	**30,030.34**	**238,948.86**
Overtime Rate	21.46	18.47	17.75	16.28	14.75	12.42	
Overtime hours per annum per man	313.60	313.60	313.60	313.60	313.60	313.60	
x Hourly rate × nr of men = £ per annum	**6,729.86**	**5,792.19**	**11,132.80**	**20,421.63**	**9,251.20**	**7,789.82**	**61,117.50**
Total	**32,657.41**	**28,110.66**	**53,999.23**	**102,613.50**	**44,865.41**	**37,820.16**	**300,066.37**
Incentive schemes 0.00%	**0.00**	**0.00**	**0.00**	**0.00**	**0.00**	**0.00**	**0.00**
Daily Travel Time Allowance (15–20 miles each way)	9.45	9.45	9.45	9.45	9.45	9.09	
Days per annum per man	224.00	224.00	224.00	224.00	224.00	224.00	
x nr of men = £ per annum	**2,116.80**	**2,116.80**	**4,233.60**	**8,467.20**	**4,233.60**	**4,072.32**	**25,240.32**
Daily Travel Fare (15–20 miles each way)	11.00	11.00	11.00	11.00	11.00	11.00	
Days per annum per man	224.00	224.00	224.00	224.00	224.00	224.00	
x nr of men = £ per annum	**2,464.00**	**2,464.00**	**4,928.00**	**9,856.00**	**4,928.00**	**4,928.00**	**29,568.00**
Employer's Pension contributions at 01/04/13:							
£ Contributions/annum	**487.76**	**424.32**	**766.48**	**1,532.96**	**598.00**	**469.04**	**4,278.56**
National Insurance Contributions:							
Gross pay – subject to NI	39,943.12	34,767.13	69,244.66	134,599.12	58,846.78	50,665.24	
% of NI Contributions	13.80	13.80	13.80	13.80	13.80	13.80	
£ Contributions/annum	**4,140.98**	**3,426.70**	**6,813.43**	**13,090.01**	**5,378.52**	**4,249.47**	**37,099.10**
Holiday Credit and Welfare contributions:							
Number of weeks	52	52	52	52	52	52	
Total weekly £ contribution each	81.95	71.64	69.13	64.14	58.86	50.89	
X nr of men = £ Contributions/annum	**4,261.40**	**3,725.28**	**7,189.52**	**13,341.12**	**6,121.44**	**5,292.56**	**39,931.32**
Holiday Top-up Funding including overtime	13.14	11.35	10.91	12.31	9.02	7.62	
Cost	**683.51**	**590.39**	**1,134.31**	**2,561.30**	**938.33**	**792.20**	**6,700.04**

	SUBTOTAL		**442,883.71**
	TRAINING (INCLUDING ANY TRADE REGISTRATIONS) – SAY	1.00%	**4,428.84**
	SEVERANCE PAY AND SUNDRY COSTS – SAY	1.50%	**6,709.69**
	EMPLOYER'S LIABILITY AND THIRD PARTY INSURANCE – SAY	2.00%	**9,080.44**
	ANNUAL COST OF NOTIONAL GANG		**463,102.68**
MEN ACTUALLY WORKING = 10.5	THEREFORE ANNUAL COST PER PRODUCTIVE MAN		**44,105.02**
AVERAGE NR OF HOURS WORKED PER MAN = 2016	THEREFORE ALL IN MAN HOURS		**21.88**
	PRELIMINARY ITEMS – SAY	7.50%	**1.64**
	SITE AND HEAD OFFICE OVERHEADS – SAY	3.00%	**0.71**
	PROFIT – SAY	2.00%	**0.48**
	THEREFORE INCLUSIVE MAN HOUR RATE		**24.71**

DIRECTIONS

Notes:

1) The following assumptions have been made in the above calculations:
   a) The working week of 38 hours i.e. the normal working week as defined by the National Agreement.
   b) The actual hours worked are five days of 9 hours each.
   c) A working year of 2016 hours.
   d) Five days in the year are lost through sickness or similar reason.
2) The incentive scheme addition of 5% is intended to reflect bonus schemes typically in use.
3) National insurance contributions are those effective from 6 April 2013. The National Insurance concession will be withdrawn for the construction industry on 30 October 2012 therefore NI will be paid for 52 weeks.
4) Weekly Holiday Credit/Welfare Stamp values are those effective from 1 April 2013.
5) Rates are based from 1 April 2013.
6) Overtime rates are based on Premium Rate 1.
7) Fares with Oyster Card (New Malden to Waterloo + Zone 1 – Anytime fare) current at June 2013 (TfL).

DIRECTIONS

LABOUR RATE – DUCTWORK

The annual cost of notional eight man gang of 8 men

		FOREMAN	SENIOR CRAFTSMAN	CRAFTSMAN	INSTALLER	SUB TOTALS
		1 NR	1 NR	4 NR	2 NR	
Hourly Rate from 1 April 2013		**15.23**	**12.59**	**12.07**	**10.46**	
Working hours per annum per man		1,702.40	1,702.40	1,702.40	1,702.40	
x Hourly rate × nr of men = £ per annum		**25,927.55**	**21,433.22**	**82,191.87**	**35,614.21**	**165,166.85**
Overtime Rate		21.46	17.75	16.28	14.75	
Overtime hours per annum per man		313.60	313.60	313.60	313.60	
x hourly rate × nr of men = £ per annum		**6,729.86**	**5,566.40**	**20,421.63**	**9,251.20**	**41,969.09**
Total		**32,657.41**	**26,999.62**	**102,613.50**	**44,865.41**	**207,135.94**
Incentive schemes	**0.00%**	**0.00**	**0.00**	**0.00**	**0.00**	**0.00**
Daily Travel Time Allowance (15–20 miles each way)		9.45	9.45	9.45	9.45	
Days per annum per man		224.00	224.00	224.00	224.00	
x nr of men = £ per annum		**2,116.80**	**2,116.80**	**8,467.20**	**4,233.60**	**16,934.40**
Daily Travel Fare (15–20 miles each way)		11.00	11.00	11.00	11.00	
Days per annum per man		224.00	224.00	224.00	224.00	
x nr of men = £ per annum		**2,464.00**	**2,464.00**	**9,856.00**	**4,928.00**	**19,712.00**
Employer's Pension Contributions at 3% Basic Pay from 7 October 2013						
£ Contributions/annum		**487.76**	**424.32**	**1,532.96**	**598.00**	**3,043.04**
National Insurance Contributions:						
Gross pay – subject to NI		42,183.12	35,742.33	136,839.12	61,086.78	
% of NI Contributions		13.8	13.8	13.8	13.8	
£ Contributions/annum		**4,450.10**	**3,561.27**	**13,399.13**	**5,687.64**	**27,098.14**
Holiday Credit and Welfare contributions:						
Number of weeks		52	52	52	52	
Total weekly £ contribution each		81.95	69.13	64.14	58.86	
x nr of men = £ Contributions/annum		**4,261.40**	**3,594.76**	**13,341.12**	**6,121.44**	**27,318.72**
Holiday Top-up Funding including overtime		**13.14**	**10.91**	**12.31**	**9.02**	
Cost		**683.51**	**567.15**	**2,561.30**	**938.33**	**4,750.29**

	SUBTOTAL		**305,992.53**
	TRAINING (INCLUDING ANY TRADE REGISTRATIONS) – SAY	**1.00%**	**3,059.93**
	SEVERANCE PAY AND SUNDRY COSTS – SAY	**1.50%**	**4,635.79**
	EMPLOYER'S LIABILITY AND THIRD PARTY INSURANCE – SAY	**2.00%**	**6,273.76**
	ANNUAL COST OF NOTIONAL GANG		**319,962.01**
MEN ACTUALLY WORKING = >7.5	THEREFORE ANNUAL COST PER PRODUCTIVE MAN		**42,661.60**
AVERAGE NR OF HOURS WORKED PER MAN = 2016	THEREFORE ALL IN MAN HOURS		**21.16**
	PRELIMINARY ITEMS – SAY	**12.50%**	**2.65**
	SITE AND HEAD OFFICE OVERHEADS – SAY	**5.50%**	**1.31**
	PROFIT – SAY	**3.00%**	**0.75**
	THEREFORE INCLUSIVE MAN HOUR RATE		**25.87**

DIRECTIONS

Notes:

1) The following assumptions have been made in the above calculations:
   a) The working week of 38 hours i.e. the normal working week as defined by the National Agreement.
   b) The actual hours worked are five days of 9 hours each.
   c) A working year of 2016 hours.
   d) Five days in the year are lost through sickness or similar reason.
2) The incentive scheme addition of 5% is intended to reflect bonus schemes typically in use.
3) National insurance contributions are those effective from 6 April 2013. The National Insurance concession will be withdrawn for the construction industry on 30 October 2012 therefore NI will be paid for 52 weeks.
4) Weekly Holiday Credit/Welfare Stamp values are those effective from 1 April 2013.
5) Rates are based from 1 April 2013.
6) Fares with Oyster Card (New Malden to Waterloo + Zone 1 – Anytime fare) current at June 2013 (TfL).
7) Overtime rates are based on Premium Rate 1.

R: DISPOSAL SYSTEMS

Item	Net Price £	Material £	Labour hours	Labour £	Unit	Total rate £
R10: RAINWATER PIPEWORK/GUTTERS						
PVC-U gutters: push fit joints; fixed with brackets to backgrounds; BS 4576 BS EN 607						
Half round gutter, with brackets measured separately						
75 mm	3.47	3.92	0.69	17.05	m	**20.97**
100 mm	6.15	6.95	0.64	15.81	m	**22.76**
150 mm	8.04	9.08	0.82	20.26	m	**29.34**
Brackets: including fixing to backgrounds. For minimum fixing distances, refer to the Tables and Memoranda at the rear of the book						
75 mm; Fascia	1.36	1.54	0.15	3.70	nr	**5.24**
100 mm; Jointing	2.16	2.44	0.16	3.96	nr	**6.40**
100 mm; Support	0.88	0.99	0.16	3.96	nr	**4.95**
150 mm; Fascia	2.63	2.97	0.16	3.96	nr	**6.93**
Bracket supports: including fixing to backgrounds. For minimum fixing distances, refer to the Tables and Memoranda at the rear of the book						
Side rafter	3.45	3.90	0.16	3.96	nr	**7.86**
Top rafter	3.45	3.90	0.16	3.96	nr	**7.86**
Rise and fall	3.73	4.21	0.16	3.96	nr	**8.17**
Extra over fittings half round PVC-U gutter						
Union						
75 mm	2.02	2.28	0.19	4.70	nr	**6.98**
100 mm	2.16	2.44	0.24	5.93	nr	**8.37**
150 mm	8.03	9.07	0.28	6.93	nr	**16.00**
Rainwater pipe outlets						
Running: 75 × 53 mm dia.	4.65	5.25	0.12	2.97	nr	**8.22**
Running: 100 × 68 mm dia.	3.18	3.59	0.12	2.97	nr	**6.56**
Running: 150 × 110 mm dia.	14.86	16.78	0.12	2.97	nr	**19.75**
Stop end: 100 × 68 mm dia.	3.56	4.02	0.12	2.97	nr	**6.99**
Internal stop ends: short						
75 mm	2.02	3.42	0.09	2.22	nr	**5.64**
100 mm	1.02	2.30	0.09	2.22	nr	**4.52**
150 mm	5.36	6.05	0.09	2.22	nr	**8.27**

R: DISPOSAL SYSTEMS

Item	Net Price £	Material £	Labour hours	Labour £	Unit	Total rate £
R10: RAINWATER PIPEWORK/GUTTERS – cont						
Extra over fittings half round PVC-U gutter – cont						
External stop ends: short						
75 mm	4.02	4.54	0.09	2.22	nr	**6.76**
100 mm	4.79	5.41	0.09	2.22	nr	**7.63**
150 mm	5.36	6.05	0.09	2.22	nr	**8.27**
Angles						
75 mm; 45°	5.49	6.20	0.20	4.95	nr	**11.15**
75 mm; 90°	5.49	6.20	0.20	4.95	nr	**11.15**
100 mm; 90°	3.79	4.28	0.20	4.95	nr	**9.23**
100 mm; 120°	4.19	4.73	0.20	4.95	nr	**9.68**
100 mm; 135°	4.19	4.73	0.20	4.95	nr	**9.68**
100 mm; Prefabricated to special angle	23.66	26.72	0.23	5.68	nr	**32.40**
100 mm; Prefabricated to raked angle	25.60	28.91	0.23	5.68	nr	**34.59**
150 mm; 90°	13.58	15.34	0.20	4.95	nr	**20.29**
Gutter adaptors						
100 mm; Stainless steel clip	1.95	2.20	0.16	3.96	nr	**6.16**
100 mm; Cast iron spigot	5.35	6.04	0.23	5.68	nr	**11.72**
100 mm; Cast iron socket	5.35	6.04	0.23	5.68	nr	**11.72**
100 mm; Cast iron 'ogee' spigot	5.42	6.12	0.23	5.68	nr	**11.80**
100 mm; Cast iron 'ogee' socket	5.42	6.12	0.23	5.68	nr	**11.80**
100 mm; Half round to Square PVC-U	9.70	10.95	0.23	5.68	nr	**16.63**
100 mm; Gutter overshoot guard	12.00	13.56	0.58	14.33	nr	**27.89**
Square gutter, with brackets measured separately						
120 mm	3.28	3.70	0.82	20.26	m	**23.96**
Brackets: including fixing to backgrounds. For minimum fixing distances, refer to the Tables and Memoranda at the rear of the book						
Jointing	2.26	2.55	0.16	3.96	nr	**6.51**
Support	0.99	1.12	0.16	3.96	nr	**5.08**
Bracket support: including fixing to backgrounds. For minimum fixing distances, refer to the Tables and Memoranda at the rear of the book						
Side rafter	3.45	3.90	0.16	3.96	nr	**7.86**
Top rafter	3.45	3.90	0.16	3.96	nr	**7.86**
Rise and fall	3.73	4.21	0.16	3.96	nr	**8.17**

R: DISPOSAL SYSTEMS

Item	Net Price £	Material £	Labour hours	Labour £	Unit	Total rate £
Extra over fittings square PVC-U gutter						
Rainwater pipe outlets						
Running: 62 mm²	3.19	3.60	0.12	2.97	nr	**6.57**
Stop end: 62 mm²	3.61	4.08	0.12	2.97	nr	**7.05**
Stop ends: short						
External	1.66	1.88	0.09	2.22	nr	**4.10**
Angles						
90°	4.10	4.63	0.20	4.95	nr	**9.58**
120°	10.26	11.59	0.20	4.95	nr	**16.54**
135°	4.80	5.42	0.20	4.95	nr	**10.37**
Prefabricated to special angle	24.26	27.40	0.23	5.68	nr	**33.08**
Prefabricated to raked angle	29.91	33.78	0.23	5.68	nr	**39.46**
Gutter adaptors						
Cast iron	15.57	17.58	0.23	5.68	nr	**23.26**
High capacity square gutter, with brackets measured separately						
137 mm	7.48	8.45	0.82	20.26	m	**28.71**
Brackets: including fixing to backgrounds. For minimum fixing distances, refer to the Tables and Memoranda at the rear of the book						
Jointing	6.98	7.88	0.16	3.96	nr	**11.84**
Support	2.91	3.28	0.16	3.96	nr	**7.24**
Overslung	2.71	3.06	0.16	3.96	nr	**7.02**
Bracket supports: including fixing to backgrounds. For minimum fixing distances, refer to the Tables and Memoranda at the rear of the book						
Side rafter	4.16	4.70	0.16	3.96	nr	**8.66**
Top rafter	4.16	4.70	0.16	3.96	nr	**8.66**
Rise and fall	6.68	7.55	0.16	3.96	nr	**11.51**
Extra over fittings high capacity square UPV-C						
Rainwater pipe outlets						
Running: 75 mm²	11.65	13.16	0.12	2.97	nr	**16.13**
Running: 82 mm dia.	11.65	13.16	0.12	2.97	nr	**16.13**
Running: 110 mm dia.	10.19	11.51	0.12	2.97	nr	**14.48**
Screwed outlet adaptor						
75 mm² pipe	7.41	8.36	0.23	5.68	nr	**14.04**

R: DISPOSAL SYSTEMS

Item	Net Price £	Material £	Labour hours	Labour £	Unit	Total rate £
R10: RAINWATER PIPEWORK/GUTTERS – cont						
Extra over fittings high capacity square UPV-C – cont						
Stop ends: short						
External	4.02	4.54	0.09	2.22	nr	**6.76**
Angles						
90°	10.01	11.30	0.20	4.95	nr	**16.25**
135°	11.78	13.30	0.20	4.95	nr	**18.25**
Prefabricated to special angle	34.03	38.43	0.23	5.68	nr	**44.11**
Prefabricated to raked internal angle	59.25	66.91	0.23	5.68	nr	**72.59**
Prefabricated to raked external angle	59.25	66.91	0.23	5.68	nr	**72.59**
Deep eliptical gutter, with brackets measured separately						
137 mm	3.98	4.50	0.82	20.26	m	**24.76**
Brackets: including fixing to backgrounds. For minimum fixing distances, refer to the Tables and Memoranda at the rear of the book						
Jointing	3.15	3.56	0.16	3.96	nr	**7.52**
Support	1.34	1.51	0.16	3.96	nr	**5.47**
Bracket support: including fixing to backgrounds. For minimum fixing distances, refer to the Tables and Memoranda at the rear of the book						
Side rafter	4.36	4.93	0.16	3.96	nr	**8.89**
Top rafter	4.36	4.93	0.16	3.96	nr	**8.89**
Rise and fall	7.01	7.92	0.16	3.96	nr	**11.88**
Extra over fittings deep eliptical PVC-U gutter						
Rainwater pipe outlets						
Running: 68 mm dia.	5.01	5.66	0.12	2.97	nr	**8.63**
Running: 82 mm dia.	4.76	5.38	0.12	2.97	nr	**8.35**
Stop end: 68 mm dia.	5.09	5.75	0.12	2.97	nr	**8.72**
Stop ends: short						
External	2.45	2.76	0.09	2.22	nr	**4.98**
Angles						
90°	5.07	5.72	0.20	4.95	nr	**10.67**
135°	6.01	6.78	0.20	4.95	nr	**11.73**
Prefabricated to special angle	17.58	19.85	0.23	5.68	nr	**25.53**

R: DISPOSAL SYSTEMS

Item	Net Price £	Material £	Labour hours	Labour £	Unit	**Total rate £**
Gutter adaptors						
Stainless steel clip	4.31	4.87	0.16	3.96	nr	**8.83**
Marley deepflow	3.48	3.93	0.23	5.68	nr	**9.61**
Ogee profile PVC-U gutter, with brackets measured separately						
122 mm	4.19	4.73	0.82	20.26	m	**24.99**
Brackets: including fixing to backgrounds. For minimum fixing distances, refer to the Tables and Memoranda at the rear of the book						
Jointing	3.33	3.76	0.16	3.96	nr	**7.72**
Support	1.28	1.45	0.16	3.96	nr	**5.41**
Overslung	1.28	1.45	0.16	3.96	nr	**5.41**
Extra over fittings Ogee profile PVC-U gutter						
Rainwater pipe outlets						
Running: 68 mm dia.	4.76	5.38	0.12	2.97	nr	**8.35**
Stop ends: short						
Internal/External: left or right hand	2.41	2.72	0.09	2.22	nr	**4.94**
Angles						
90°: internal or external	4.82	5.45	0.20	4.95	nr	**10.40**
135°: internal or external	4.82	5.45	0.20	4.95	nr	**10.40**
PVC-U rainwater pipe: dry push fit joints; fixed with brackets to backgrounds; BS 4576/ BS EN 607						
Pipe: circular, with brackets measured separately						
53 mm	5.41	6.11	0.61	15.08	m	**21.19**
68 mm	5.64	6.36	0.61	15.08	m	**21.44**
Pipe clip: including fixing to backgrounds. For minimum fixing distances, refer to the Tables and Memoranda at the rear of the book						
68 mm	1.26	1.43	0.16	3.96	nr	**5.39**
Pipe clip adjustable: including fixing to backgrounds. For minimum fixing distances, refer to the Tables and Memoranda at the rear of the book						
53 mm	1.93	2.18	0.16	3.96	nr	**6.14**
68 mm	2.70	3.05	0.16	3.96	nr	**7.01**

R: DISPOSAL SYSTEMS

Item	Net Price £	Material £	Labour hours	Labour £	Unit	**Total rate £**
R10: RAINWATER PIPEWORK/GUTTERS – cont						
Pipe: circular, with brackets measured separately – cont						
Pipe clip drive in: including fixing to backgrounds. For minimum fixing distances, refer to the Tables and Memoranda at the rear of the book						
68 mm	3.03	3.42	0.16	3.96	nr	**7.38**
Extra over fittings circular pipework PVC-U						
Pipe coupler: PVC-U to PVC-U						
68 mm	1.57	1.77	0.12	2.97	nr	**4.74**
Pipe coupler: PVC-U to Cast Iron						
68 mm: to 3' cast iron	6.17	6.97	0.17	4.20	nr	**11.17**
68 mm: to 3.¾' cast iron	21.20	23.94	0.17	4.20	nr	**28.14**
Access pipe: single socket						
68 mm	9.85	11.13	0.15	3.70	nr	**14.83**
Bend: short radius						
53 mm: 67.5°	2.29	2.59	0.20	4.95	nr	**7.54**
68 mm: 92.5°	3.37	3.80	0.20	4.95	nr	**8.75**
68 mm: 112.5°	3.37	3.80	0.20	4.95	nr	**8.75**
Bend: long radius						
68 mm: 112°	3.19	3.60	0.20	4.95	nr	**8.55**
Branch						
68 mm: 92°	20.70	23.38	0.23	5.68	nr	**29.06**
68 mm: 112°	20.77	23.46	0.23	5.68	nr	**29.14**
Double branch						
68 mm: 112°	40.98	46.28	0.24	5.93	nr	**52.21**
Shoe						
53 mm	3.40	3.84	0.12	2.97	nr	**6.81**
68 mm	5.01	5.66	0.12	2.97	nr	**8.63**
Rainwater head: including fixing to backgrounds						
68 mm	9.85	11.13	0.29	7.17	nr	**18.30**

R: DISPOSAL SYSTEMS

Item	Net Price £	Material £	Labour hours	Labour £	Unit	Total rate £
Pipe: square, with brackets measured separately						
62 mm	3.45	3.90	0.45	11.13	m	**15.03**
75 mm	6.56	7.41	0.45	11.13	m	**18.54**
Pipe clip: including fixing to backgrounds. For minimum fixing distances, refer to the Tables and Memoranda at the rear of the book						
62 mm	1.16	1.31	0.16	3.96	nr	**5.27**
75 mm	2.56	2.89	0.16	3.96	nr	**6.85**
Pipe clip adjustable: including fixing to backgrounds. For minimum fixing distances, refer to the Tables and Memoranda at the rear of the book						
62 mm	3.57	4.03	0.16	3.96	nr	**7.99**
Extra over fittings square pipework PVC-U						
Pipe coupler: PVC-U to PVC-U						
62 mm	1.69	1.91	0.20	4.95	nr	**6.86**
75 mm	2.63	2.97	0.20	4.95	nr	**7.92**
Square to circular adaptor: single socket						
62 mm to 68 mm	3.30	3.72	0.20	4.95	nr	**8.67**
Square to circular adaptor: single socket						
75 mm to 62 mm	4.26	4.81	0.20	4.95	nr	**9.76**
Access pipe						
62 mm	19.71	22.26	0.16	3.96	nr	**26.22**
75 mm	20.80	23.49	0.16	3.96	nr	**27.45**
Bends						
62 mm: 92.5°	3.05	3.45	0.20	4.95	nr	**8.40**
62 mm: 112.5°	2.22	2.51	0.20	4.95	nr	**7.46**
75 mm: 112.5°	5.56	6.28	0.20	4.95	nr	**11.23**
Bends: prefabricated special angle						
62 mm	19.80	22.36	0.23	5.68	nr	**28.04**
75 mm	27.51	31.07	0.23	5.68	nr	**36.75**
Offset						
62 mm	4.77	5.39	0.20	4.95	nr	**10.34**
75 mm	13.14	14.84	0.20	4.95	nr	**19.79**
Offset: prefabricated special angle						
62 mm	19.80	22.36	0.23	5.68	nr	**28.04**

R: DISPOSAL SYSTEMS

Item	Net Price £	Material £	Labour hours	Labour £	Unit	Total rate £
R10: RAINWATER PIPEWORK/GUTTERS – cont						
Extra over fittings square pipework PVC-U – cont						
Shoe						
62 mm	2.80	3.16	0.12	2.97	nr	**6.13**
75 mm	4.82	5.45	0.12	2.97	nr	**8.42**
Branch						
62 mm	6.92	7.81	0.23	5.68	nr	**13.49**
75 mm	33.76	38.13	0.23	5.68	nr	**43.81**
Double branch						
62 mm	37.15	41.95	0.24	5.93	nr	**47.88**
Rainwater head						
62 mm	31.15	35.18	0.29	7.17	nr	**42.35**
75 mm	32.77	37.01	3.45	85.21	nr	**122.22**
PVC-U rainwater pipe: solvent welded joints; fixed with brackets to backgrounds; BS 4576/ BS EN 607						
Pipe: circular, with brackets measured separately						
82 mm	10.63	12.01	0.35	8.65	m	**20.66**
Pipe clip: galvanized; including fixing to backgrounds. For minimum fixing distances, refer to the Tables and Memoranda at the rear of the book						
82 mm	4.32	4.88	0.58	14.33	nr	**19.21**
Pipe clip: galvanized plastic coated; including fixing to backgrounds. For minimum fixing distances, refer to the Tables and Memoranda at the rear of the book						
82 mm	5.98	6.75	0.58	14.33	nr	**21.08**
Pipe clip: PVC-U including fixing to backgrounds. For minimum fixing distances, refer to the Tables and Memoranda at the rear of the book						
82 mm	3.21	3.62	0.58	14.33	nr	**17.95**

R: DISPOSAL SYSTEMS

Item	Net Price £	Material £	Labour hours	Labour £	Unit	Total rate £
Pipe clip: PVC-U adjustable: including fixing to backgrounds. For minimum fixing distances, refer to the Tables and Memoranda at the rear of the book						
82 mm	5.88	6.64	0.58	14.33	nr	**20.97**
Extra over fittings circular pipework PVC-U						
Pipe coupler: PVC-U to PVC-U						
82 mm	5.39	6.09	0.21	5.18	nr	**11.27**
Access pipe						
82 mm	33.18	37.47	0.23	5.68	nr	**43.15**
Bend						
82 mm: 92, 112.5 and 135°	13.55	15.30	0.29	7.17	nr	**22.47**
Shoe						
82 mm	8.36	9.45	0.29	7.17	nr	**16.62**
110 mm	10.49	11.84	0.32	7.91	nr	**19.75**
Branch						
82 mm: 92, 112.5 and 135°	20.18	22.79	0.35	8.65	nr	**31.44**
Rainwater head						
82 mm	17.71	20.00	0.58	14.33	nr	**34.33**
110 mm	16.13	18.22	0.58	14.33	nr	**32.55**
Roof outlets: 178 dia.; Flat						
50 mm	17.70	19.99	1.15	28.42	nr	**48.41**
82 mm	17.70	19.99	1.15	28.42	nr	**48.41**
Roof outlets: 178 mm dia.; Domed						
50 mm	17.70	19.99	1.15	28.42	nr	**48.41**
82 mm	17.70	19.99	1.15	28.42	nr	**48.41**
Roof outlets: 406 mm dia.; Flat						
82 mm	34.59	39.07	1.15	28.42	nr	**67.49**
110 mm	34.59	39.07	1.15	28.42	nr	**67.49**
Roof outlets: 406 mm dia.; Domed						
82 mm	34.59	39.07	1.15	28.42	nr	**67.49**
110 mm	34.59	39.07	1.15	28.42	nr	**67.49**
Roof outlets: 406 mm dia.; Inverted						
82 mm	76.14	85.99	1.15	28.42	nr	**114.41**
110 mm	76.14	85.99	1.15	28.42	nr	**114.41**

R: DISPOSAL SYSTEMS

Item	Net Price £	Material £	Labour hours	Labour £	Unit	Total rate £
R10: RAINWATER PIPEWORK/GUTTERS – cont						
Extra over fittings circular pipework PVC-U – cont						
Roof outlets: 406 mm dia.; Vent Pipe						
82 mm	50.01	56.48	1.15	28.42	nr	**84.90**
110 mm	50.01	56.48	1.15	28.42	nr	**84.90**
Balcony outlets: screed						
82 mm	29.60	33.43	1.15	28.42	nr	**61.85**
Balcony outlets: asphalt						
82 mm	29.44	33.25	1.15	28.42	nr	**61.67**
Adaptors						
82 mm × 62 mm^2 pipe	3.96	4.47	0.21	5.18	nr	**9.65**
82 mm × 68 mm circular pipe	3.96	4.47	0.21	5.18	nr	**9.65**
For 110 mm dia. pipework and fittings refer to R11: Above Ground Drainage						
Cast iron gutters: mastic and bolted joints; BS 460; fixed with brackets to backgrounds						
Half round gutter, with brackets measured separately						
100 mm	13.92	15.72	0.85	21.00	m	**36.72**
115 mm	14.51	16.39	0.97	23.96	m	**40.35**
125 mm	16.97	19.17	0.97	23.96	m	**43.13**
150 mm	29.01	32.76	1.12	27.68	m	**60.44**
Brackets; fixed to backgrounds. For minimum fixing distances, refer to the Tables and Memoranda at the rear of the book						
Fascia						
100 mm	3.26	3.68	0.16	3.96	nr	**7.64**
115 mm	3.26	3.68	0.16	3.96	nr	**7.64**
125 mm	3.26	3.68	0.16	3.96	nr	**7.64**
150 mm	4.13	4.66	0.16	3.96	nr	**8.62**
Rise and fall						
100 mm	6.48	7.31	0.39	9.63	nr	**16.94**
115 mm	6.48	7.31	0.39	9.63	nr	**16.94**
125 mm	6.66	7.52	0.39	9.63	nr	**17.15**
150 mm	6.78	7.66	0.39	9.63	nr	**17.29**

R: DISPOSAL SYSTEMS

Item	Net Price £	Material £	Labour hours	Labour £	Unit	**Total rate £**
Top rafter						
100 mm	3.99	4.51	0.16	3.96	nr	**8.47**
115 mm	3.99	4.51	0.16	3.96	nr	**8.47**
125 mm	5.40	6.10	0.16	3.96	nr	**10.06**
150 mm	7.34	8.29	0.16	3.96	nr	**12.25**
Side rafter						
100 mm	3.99	4.51	0.16	3.96	nr	**8.47**
115 mm	3.99	4.51	0.16	3.96	nr	**8.47**
125 mm	5.40	6.10	0.16	3.96	nr	**10.06**
150 mm	7.34	8.29	0.16	3.96	nr	**12.25**
Extra over fittings half round gutter cast iron BS 460						
Union						
100 mm	7.65	8.64	0.39	9.63	nr	**18.27**
115 mm	9.55	10.78	0.48	11.86	nr	**22.64**
125 mm	10.79	12.19	0.48	11.86	nr	**24.05**
150 mm	12.10	13.67	0.55	13.59	nr	**27.26**
Stop end; internal						
100 mm	3.90	4.41	0.12	2.97	nr	**7.38**
115 mm	5.05	5.70	0.15	3.70	nr	**9.40**
125 mm	5.05	5.70	0.15	3.70	nr	**9.40**
150 mm	7.01	7.92	0.20	4.95	nr	**12.87**
Stop end; external						
100 mm	3.90	4.41	0.12	2.97	nr	**7.38**
115 mm	4.95	5.59	0.15	3.70	nr	**9.29**
125 mm	5.05	5.70	0.15	3.70	nr	**9.40**
150 mm	7.01	7.92	0.20	4.95	nr	**12.87**
90° angle; single socket						
100 mm	11.60	13.10	0.39	9.63	nr	**22.73**
115 mm	11.94	13.48	0.43	10.63	nr	**24.11**
125 mm	14.07	15.89	0.43	10.63	nr	**26.52**
150 mm	25.72	29.05	0.50	12.35	nr	**41.40**
90° angle; double socket						
100 mm	14.06	15.88	0.39	9.63	nr	**25.51**
115 mm	14.92	16.85	0.43	10.63	nr	**27.48**
125 mm	19.31	21.81	0.43	10.63	nr	**32.44**
135° angle; single socket						
100 mm	11.85	13.38	0.39	9.63	nr	**23.01**
115 mm	11.96	13.50	0.43	10.63	nr	**24.13**
125 mm	17.67	19.95	0.43	10.63	nr	**30.58**
150 mm	26.22	29.61	0.50	12.35	nr	**41.96**

R: DISPOSAL SYSTEMS

Item	Net Price £	Material £	Labour hours	Labour £	Unit	Total rate £
R10: RAINWATER PIPEWORK/GUTTERS – cont						
Extra over fittings half round gutter cast iron BS 460 – cont						
Running outlet						
65 mm outlet						
100 mm	11.32	12.78	0.39	9.63	nr	**22.41**
115 mm	12.32	13.91	0.43	10.63	nr	**24.54**
125 mm	14.07	15.89	0.43	10.63	nr	**26.52**
75 mm outlet						
100 mm	11.32	12.78	0.39	9.63	nr	**22.41**
115 mm	12.32	13.91	0.43	10.63	nr	**24.54**
125 mm	14.07	15.89	0.43	10.63	nr	**26.52**
150 mm	24.38	27.53	0.50	12.35	nr	**39.88**
100 mm outlet						
150 mm	24.38	27.53	0.50	12.35	nr	**39.88**
Stop end outlet; socket						
65 mm outlet						
100 mm	13.27	14.98	0.39	9.63	nr	**24.61**
115 mm	14.88	16.81	0.43	10.63	nr	**27.44**
75 mm outlet						
125 mm	13.27	14.98	0.43	10.63	nr	**25.61**
150 mm	27.89	59.02	0.50	12.35	nr	**71.37**
100 mm outlet						
150 mm	24.38	27.53	0.50	12.35	nr	**39.88**
Stop end outlet; spigot						
65 mm outlet						
100 mm	13.27	14.98	0.39	9.63	nr	**24.61**
115 mm	14.88	16.81	0.43	10.63	nr	**27.44**
75 mm outlet						
125 mm	13.27	14.98	0.43	10.63	nr	**25.61**
150 mm	27.89	31.50	0.50	12.35	nr	**43.85**
100 mm outlet						
150 mm	27.89	31.50	0.50	12.35	nr	**43.85**

R: DISPOSAL SYSTEMS

Item	Net Price £	Material £	Labour hours	Labour £	Unit	**Total rate £**
Half round; 3 mm thick double beaded gutter, with brackets measured separately						
100 mm	14.17	16.00	0.85	21.00	m	**37.00**
115 mm	14.79	16.71	0.85	21.00	m	**37.71**
125 mm	16.96	19.16	0.97	23.96	m	**43.12**
Brackets; fixed to backgrounds. For minimum fixing distances, refer to the Tables and Memoranda at the rear of the book						
Fascia						
100 mm	3.26	3.68	0.16	3.96	nr	**7.64**
115 mm	3.26	3.68	0.16	3.96	nr	**7.64**
125 mm	3.26	3.68	0.16	3.96	nr	**7.64**
Extra over fittings half round 3 mm thick gutter BS 460						
Union						
100 mm	7.65	8.64	0.38	9.38	nr	**18.02**
115 mm	9.33	10.54	0.38	9.38	nr	**19.92**
125 mm	10.79	12.19	0.43	10.63	nr	**22.82**
Stop end; internal						
100 mm	3.90	4.41	0.12	2.97	nr	**7.38**
115 mm	5.05	5.70	0.12	2.97	nr	**8.67**
125 mm	5.07	5.72	0.15	3.70	nr	**9.42**
Stop end; external						
100 mm	3.90	4.41	0.12	2.97	nr	**7.38**
115 mm	5.05	5.70	0.12	2.97	nr	**8.67**
125 mm	5.07	5.72	0.15	3.70	nr	**9.42**
90° angle; single socket						
100 mm	12.05	13.61	0.38	9.38	nr	**22.99**
115 mm	12.22	13.80	0.38	9.38	nr	**23.18**
125 mm	14.07	15.89	0.43	10.63	nr	**26.52**
135° angle; single socket						
100 mm	11.85	13.38	0.38	9.38	nr	**22.76**
115 mm	11.94	13.48	0.38	9.38	nr	**22.86**
125 mm	14.89	16.82	0.43	10.63	nr	**27.45**
Running outlet						
65 mm outlet						
100 mm	12.15	13.72	0.38	9.38	nr	**23.10**
115 mm	12.56	14.19	0.38	9.38	nr	**23.57**
125 mm	14.58	16.46	0.43	10.63	nr	**27.09**

R: DISPOSAL SYSTEMS

Item	Net Price £	Material £	Labour hours	Labour £	Unit	Total rate £
R10: RAINWATER PIPEWORK/GUTTERS – cont						
Extra over fittings half round 3 mm thick gutter BS 460 – cont						
75 mm outlet						
115 mm	12.56	14.19	0.38	9.38	nr	**23.57**
125 mm	14.58	16.46	0.43	10.63	nr	**27.09**
Stop end outlet; socket						
65 mm outlet						
100 mm	13.27	14.98	0.38	9.38	nr	**24.36**
115 mm	14.88	16.81	0.38	9.38	nr	**26.19**
125 mm	16.63	18.78	0.43	10.63	nr	**29.41**
75 mm outlet						
125 mm	16.93	19.11	0.43	10.63	nr	**29.74**
Stop end outlet; spigot						
65 mm outlet						
100 mm	13.27	14.98	0.38	9.38	nr	**24.36**
115 mm	14.88	16.81	0.38	9.38	nr	**26.19**
125 mm	16.63	18.78	0.43	10.63	nr	**29.41**
Deep half round gutter, with brackets measured separately						
100 × 75 mm	23.30	26.32	0.85	21.00	m	**47.32**
125 × 75 mm	30.11	34.01	0.97	23.96	m	**57.97**
Brackets; fixed to backgrounds. For minimum fixing distances, refer to the Tables and Memoranda at the rear of the book						
Fascia						
100 × 75 mm	11.00	12.42	0.16	3.96	nr	**16.38**
125 × 75 mm	13.56	15.31	0.16	3.96	nr	**19.27**
Extra over fittings deep half round gutter BS 460						
Union						
100 × 75 mm	12.80	14.45	0.38	9.38	nr	**23.83**
125 × 75 mm	13.56	15.31	0.43	10.63	nr	**25.94**

R: DISPOSAL SYSTEMS

Item	Net Price £	Material £	Labour hours	Labour £	Unit	Total rate £
Stop end; internal						
100 × 75 mm	11.20	12.65	0.12	2.97	nr	**15.62**
125 × 75 mm	13.82	15.61	0.15	3.70	nr	**19.31**
Stop end; external						
100 × 75 mm	11.20	12.65	0.12	2.97	nr	**15.62**
125 × 75 mm	13.82	15.61	0.15	3.70	nr	**19.31**
90° angle; single socket						
100 × 75 mm	31.94	36.07	0.38	9.38	nr	**45.45**
125 × 75 mm	40.55	45.80	0.43	10.63	nr	**56.43**
135° angle; single socket						
100 × 75 mm	31.94	36.07	0.38	9.38	nr	**45.45**
125 × 75 mm	40.55	45.80	0.43	10.63	nr	**56.43**
Running outlet						
65 mm outlet						
100 × 75 mm	31.94	36.07	0.38	9.38	nr	**45.45**
125 × 75 mm	40.55	45.80	0.43	10.63	nr	**56.43**
75 mm outlet						
100 × 75 mm	31.94	36.07	0.38	9.38	nr	**45.45**
125 × 75 mm	29.43	33.23	0.43	10.63	nr	**43.86**
Stop end outlet; socket						
65 mm outlet						
100 × 75 mm	43.15	48.74	0.38	9.38	nr	**58.12**
75 mm outlet						
100 × 75 mm	43.15	48.74	0.38	9.38	nr	**58.12**
125 × 75 mm	43.15	48.74	0.43	10.63	nr	**59.37**
Stop end outlet; spigot						
65 mm outlet						
100 × 75 mm	43.15	48.74	0.38	9.38	nr	**58.12**
75 mm outlet						
100 × 75 mm	43.15	48.74	0.38	9.38	nr	**58.12**
125 × 75 mm	43.15	48.74	0.43	10.63	nr	**59.37**
Ogee gutter, with brackets measured separately						
100 mm	15.52	17.52	0.85	21.00	m	**38.52**
115 mm	17.07	19.28	0.97	23.96	m	**43.24**
125 mm	17.91	20.23	0.97	23.96	m	**44.19**

R: DISPOSAL SYSTEMS

Item	Net Price £	Material £	Labour hours	Labour £	Unit	Total rate £
R10: RAINWATER PIPEWORK/GUTTERS – cont						
Ogee gutter, with brackets measured separately – cont						
Brackets; fixed to backgrounds.For minimum fixing distances, refer to the Tables and Memoranda at the rear of the book						
Fascia						
100 mm	3.55	4.01	0.16	3.96	nr	**7.97**
115 mm	3.55	4.01	0.16	3.96	nr	**7.97**
125 mm	4.00	4.52	0.16	3.96	nr	**8.48**
Extra over fittings ogee cast iron gutter BS 460						
Union						
100 mm	7.66	8.65	0.38	9.38	nr	**18.03**
115 mm	9.33	10.54	0.43	10.63	nr	**21.17**
125 mm	10.79	12.19	0.43	10.63	nr	**22.82**
Stop end; internal						
100 mm	3.99	4.51	0.12	2.97	nr	**7.48**
115 mm	5.16	5.82	0.15	3.70	nr	**9.52**
125 mm	5.16	5.82	0.15	3.70	nr	**9.52**
Stop end; external						
100 mm	3.99	4.51	0.12	2.97	nr	**7.48**
115 mm	5.16	5.82	0.15	3.70	nr	**9.52**
125 mm	5.16	5.82	0.15	3.70	nr	**9.52**
90° angle; internal						
100 mm	12.10	13.67	0.38	9.38	nr	**23.05**
115 mm	13.12	14.82	0.43	10.63	nr	**25.45**
125 mm	14.32	16.18	0.43	10.63	nr	**26.81**
90° angle; external						
100 mm	12.10	13.67	0.38	9.38	nr	**23.05**
115 mm	13.12	14.82	0.43	10.63	nr	**25.45**
125 mm	14.32	16.18	0.43	10.63	nr	**26.81**
135° angle; internal						
100 mm	12.57	14.20	0.38	9.38	nr	**23.58**
115 mm	13.39	15.13	0.43	10.63	nr	**25.76**
125 mm	17.64	19.92	0.43	10.63	nr	**30.55**

R: DISPOSAL SYSTEMS

Item	Net Price £	Material £	Labour hours	Labour £	Unit	**Total rate £**
135° angle; external						
100 mm	12.57	14.20	0.38	9.38	nr	**23.58**
115 mm	13.39	15.13	0.43	10.63	nr	**25.76**
125 mm	17.64	19.92	0.43	10.63	nr	**30.55**
Running outlet						
65 mm outlet						
100 mm	12.33	13.92	0.38	9.38	nr	**23.30**
115 mm	13.13	14.83	0.43	10.63	nr	**25.46**
125 mm	14.32	16.18	0.43	10.63	nr	**26.81**
75 mm outlet						
125 mm	14.32	16.18	0.43	10.63	nr	**26.81**
Stop end outlet; socket						
65 mm outlet						
100 mm	19.47	21.99	0.38	9.38	nr	**31.37**
115 mm	19.47	21.99	0.43	10.63	nr	**32.62**
125 mm	19.47	21.99	0.43	10.63	nr	**32.62**
75 mm outlet						
125 mm	19.47	21.99	0.43	10.63	nr	**32.62**
Stop end outlet; spigot						
65 mm outlet						
100 mm	19.47	21.99	0.38	9.38	nr	**31.37**
115 mm	19.47	21.99	0.43	10.63	nr	**32.62**
125 mm	19.47	21.99	0.43	10.63	nr	**32.62**
75 mm outlet						
125 mm	19.47	21.99	0.43	10.63	nr	**32.62**
Notts ogee gutter, with brackets measured separately						
115 mm	27.57	31.14	0.85	21.00	m	**52.14**
Brackets; fixed to backgrounds. For minimum fixing distances, refer to the Tables and Memoranda at the rear of the book						
Fascia						
115 mm	10.75	12.14	0.16	3.96	nr	**16.10**

R: DISPOSAL SYSTEMS

Item	Net Price £	Material £	Labour hours	Labour £	Unit	Total rate £
R10: RAINWATER PIPEWORK/GUTTERS – cont						
Extra over fittings Notts ogee cast iron gutter BS 460						
Union						
115 mm	13.06	14.75	0.38	9.38	nr	**24.13**
Stop end; internal						
115 mm	10.75	12.14	0.16	3.96	nr	**16.10**
Stop end; external						
115 mm	10.75	12.14	0.16	3.96	nr	**16.10**
90° angle; internal						
115 mm	31.19	35.22	0.43	10.63	nr	**45.85**
90° angle; external						
115 mm	31.19	35.22	0.43	10.63	nr	**45.85**
135° angle; internal						
115 mm	31.78	35.89	0.43	10.63	nr	**46.52**
135° angle; external						
115 mm	31.78	35.89	0.43	10.63	nr	**46.52**
Running outlet						
65 mm outlet						
115 mm	37.39	42.23	0.43	10.63	nr	**52.86**
75 mm outlet						
115 mm	37.39	42.23	0.43	10.63	nr	**52.86**
Stop end outlet; socket						
65 mm outlet						
115 mm	48.14	54.37	0.43	10.63	nr	**65.00**
Stop end outlet; spigot						
65 mm outlet						
115 mm	48.14	54.37	0.43	10.63	nr	**65.00**
No 46 moulded gutter, with brackets measured separately						
100 × 75 mm	26.66	30.11	0.85	21.00	m	**51.11**
125 × 100 mm	39.09	44.15	0.97	23.96	m	**68.11**

R: DISPOSAL SYSTEMS

Item	Net Price £	Material £	Labour hours	Labour £	Unit	**Total rate £**
Brackets; fixed to backgrounds. For minimum fixing distances, refer to the Tables and Memoranda at the rear of the book						
Fascia						
100 × 75 mm	5.95	6.72	0.16	3.96	nr	**10.68**
125 × 100 mm	5.95	6.72	0.16	3.96	nr	**10.68**
Extra over fittings						
Union						
100 × 75 mm	12.56	14.19	0.38	9.38	nr	**23.57**
125 × 100 mm	14.56	16.44	0.43	10.63	nr	**27.07**
Stop end; internal						
100 × 75 mm	11.25	12.71	0.12	2.97	nr	**15.68**
125 × 100 mm	14.56	16.44	0.15	3.70	nr	**20.14**
Stop end; external						
100 × 75 mm	11.25	12.71	0.12	2.97	nr	**15.68**
125 × 100 mm	14.56	16.44	0.15	3.70	nr	**20.14**
90° angle; internal						
100 × 75 mm	29.45	33.26	0.38	9.38	nr	**42.64**
125 × 100 mm	42.33	47.81	0.43	10.63	nr	**58.44**
90° angle; external						
100 × 75 mm	29.45	33.26	0.38	9.38	nr	**42.64**
125 × 100 mm	42.33	47.81	0.43	10.63	nr	**58.44**
135° angle; internal						
100 × 75 mm	30.02	33.90	0.38	9.38	nr	**43.28**
125 × 100 mm	42.33	47.81	0.43	10.63	nr	**58.44**
135° angle; external						
100 × 75 mm	30.02	33.90	0.38	9.38	nr	**43.28**
125 × 100 mm	42.33	47.81	0.43	10.63	nr	**58.44**
Running outlet						
65 mm outlet						
100 × 75 mm	30.02	33.90	0.38	9.38	nr	**43.28**
125 × 100 mm	42.33	47.81	0.43	10.63	nr	**58.44**
75 mm outlet						
100 × 75 mm	30.02	33.90	0.38	9.38	nr	**43.28**
125 × 100 mm	42.33	47.81	0.43	10.63	nr	**58.44**

R: DISPOSAL SYSTEMS

Item	Net Price £	Material £	Labour hours	Labour £	Unit	Total rate £
R10: RAINWATER PIPEWORK/GUTTERS – cont						
Extra over fittings – cont						
100 mm outlet						
100 × 75 mm	30.02	33.90	0.38	9.38	nr	**43.28**
125 × 100 mm	42.33	47.81	0.43	10.63	nr	**58.44**
100 × 75 mm outlet						
125 × 100 mm	30.02	33.90	0.43	10.63	nr	**44.53**
Stop end outlet; socket						
65 mm outlet						
100 × 75 mm	56.89	64.25	0.38	9.38	nr	**73.63**
75 mm outlet						
125 × 100 mm	56.89	64.25	0.43	10.63	nr	**74.88**
Stop end outlet; spigot						
65 mm outlet						
100 × 75 mm	56.89	64.25	0.38	9.38	nr	**73.63**
75 mm outlet						
125 × 100 mm	56.89	64.25	0.43	10.63	nr	**74.88**
Box gutter, with brackets measured separately						
100 × 75 mm	43.23	48.82	0.85	21.00	m	**69.82**
Brackets; fixed to backgrounds. For minimum fixing distances, refer to the Tables and Memoranda at the rear of the book						
Fascia						
100 × 75 mm	6.85	7.73	0.16	3.96	nr	**11.69**
Extra over fittings box cast iron gutter BS 460						
Union						
100 × 75 mm	9.09	10.26	0.38	9.38	nr	**19.64**
Stop end; external						
100 × 75 mm	6.85	7.73	0.12	2.97	nr	**10.70**

R: DISPOSAL SYSTEMS

Item	Net Price £	Material £	Labour hours	Labour £	Unit	**Total rate £**
90° angle						
100 × 75 mm	34.55	39.01	0.38	9.38	nr	**48.39**
135° angle						
100 × 75 mm	34.55	39.01	0.38	9.38	nr	**48.39**
Running outlet						
65 mm outlet						
100 × 75 mm	34.55	39.01	0.38	9.38	nr	**48.39**
75 mm outlet						
100 × 75 mm	34.55	39.01	0.38	9.38	nr	**48.39**
100 × 75 mm outlet						
100 × 75 mm	34.55	39.01	0.38	9.38	nr	**48.39**
Cast iron rainwater pipe; dry joints; BS 460; fixed to backgrounds						
Circular						
Plain socket pipe, with brackets measured separately						
65 mm	25.36	28.64	0.69	17.05	m	**45.69**
75 mm	25.36	28.64	0.69	17.05	m	**45.69**
100 mm	34.62	39.10	0.69	17.05	m	**56.15**
Bracket; fixed to backgrounds. For minimum fixing distances, refer to the Tables and Memoranda at the rear of the book						
65 mm	9.27	10.47	0.29	7.17	nr	**17.64**
75 mm	9.30	10.51	0.29	7.17	nr	**17.68**
100 mm	9.43	10.65	0.29	7.17	nr	**17.82**
Eared socket pipe, with wall spacers measured separately						
65 mm	27.11	30.62	0.62	15.32	m	**45.94**
75 mm	27.11	30.62	0.62	15.32	m	**45.94**
100 mm	36.39	41.10	0.62	15.32	m	**56.42**
Wall spacer plate; eared pipework						
65 mm	5.98	6.75	0.16	3.96	nr	**10.71**
75 mm	6.09	6.87	0.16	3.96	nr	**10.83**
100 mm	9.43	10.65	0.16	3.96	nr	**14.61**

R: DISPOSAL SYSTEMS

Item	Net Price £	Material £	Labour hours	Labour £	Unit	Total rate £
R10: RAINWATER PIPEWORK/GUTTERS – cont						
Extra over fittings circular cast iron pipework BS 460						
Loose sockets						
Plain socket						
65 mm	7.32	8.26	0.23	5.68	nr	**13.94**
75 mm	7.32	8.26	0.23	5.68	nr	**13.94**
100 mm	11.34	12.81	0.23	5.68	nr	**18.49**
Eared socket						
65 mm	10.60	11.97	0.29	7.17	nr	**19.14**
75 mm	10.60	11.97	0.29	7.17	nr	**19.14**
100 mm	14.32	16.18	0.29	7.17	nr	**23.35**
Shoe; front projection						
Plain socket						
65 mm	22.81	25.77	0.23	5.68	nr	**31.45**
75 mm	22.81	25.77	0.23	5.68	nr	**31.45**
100 mm	30.74	34.72	0.23	5.68	nr	**40.40**
Eared socket						
65 mm	26.43	29.85	0.29	7.17	nr	**37.02**
75 mm	26.43	29.85	0.29	7.17	nr	**37.02**
100 mm	35.08	39.62	0.29	7.17	nr	**46.79**
Access pipe						
65 mm	41.17	46.49	0.23	5.68	nr	**52.17**
75 mm	43.23	48.82	0.23	5.68	nr	**54.50**
100 mm	76.67	86.59	0.23	5.68	nr	**92.27**
100 mm; eared	86.56	97.76	0.29	7.17	nr	**104.93**
Bends; any degree						
65 mm	16.47	18.60	0.23	5.68	nr	**24.28**
75 mm	19.63	22.16	0.23	5.68	nr	**27.84**
100 mm	27.73	31.31	0.23	5.68	nr	**36.99**
Branch						
92.5°						
65 mm	31.78	35.89	0.29	7.17	nr	**43.06**
75 mm	35.05	39.59	0.29	7.17	nr	**46.76**
100 mm	40.85	46.13	0.29	7.17	nr	**53.30**

R: DISPOSAL SYSTEMS

Item	Net Price £	Material £	Labour hours	Labour £	Unit	**Total rate £**
112.5°						
65 mm	31.78	35.89	0.29	7.17	nr	**43.06**
75 mm	35.05	39.59	0.29	7.17	nr	**46.76**
135°						
65 mm	31.78	35.89	0.29	7.17	nr	**43.06**
75 mm	35.05	39.59	0.29	7.17	nr	**46.76**
Offsets						
75 to 150 mm projection						
65 mm	24.75	27.95	0.25	6.18	nr	**34.13**
75 mm	24.75	27.95	0.25	6.18	nr	**34.13**
100 mm	46.69	52.73	0.25	6.18	nr	**58.91**
225 mm projection						
65 mm	25.23	28.49	0.25	6.18	nr	**34.67**
75 mm	25.23	28.49	0.25	6.18	nr	**34.67**
100 mm	46.69	52.73	0.25	6.18	nr	**58.91**
305 mm projection						
65 mm	33.76	38.13	0.25	6.18	nr	**44.31**
75 mm	35.43	40.01	0.25	6.18	nr	**46.19**
100 mm	57.66	65.12	0.25	6.18	nr	**71.30**
380 mm projection						
65 mm	67.36	76.07	0.25	6.18	nr	**82.25**
75 mm	67.36	76.07	0.25	6.18	nr	**82.25**
100 mm	91.94	103.84	0.25	6.18	nr	**110.02**
455 mm projection						
65 mm	78.85	89.05	0.25	6.18	nr	**95.23**
75 mm	78.85	89.05	0.25	6.18	nr	**95.23**
100 mm	114.35	129.14	0.25	6.18	nr	**135.32**
Rectangular						
Plain socket						
100 × 75 mm	71.68	80.95	1.04	25.70	m	**106.65**
Bracket; fixed to backgrounds. For minimum fixing distances, refer to the Tables and Memoranda at the rear of the book						
100 × 75 mm; build in holdabat	40.79	46.06	0.35	8.65	nr	**54.71**
100 × 75 mm; trefoil earband	32.50	36.70	0.29	7.17	nr	**43.87**
100 × 75 mm; plain earband	31.43	35.50	0.29	7.17	nr	**42.67**

R: DISPOSAL SYSTEMS

Item	Net Price £	Material £	Labour hours	Labour £	Unit	Total rate £
R10: RAINWATER PIPEWORK/GUTTERS – cont						
Rectangular – cont						
Eared Socket, with wall spacers measured separately						
100 × 75 mm	72.86	82.28	1.16	28.66	m	**110.94**
Wall spacer plate; eared pipework						
100 × 75	9.43	10.65	0.16	3.96	nr	**14.61**
Extra over fittings rectangular cast iron pipework BS 460						
Loose socket						
100 × 75 mm; plain	30.40	34.33	0.23	5.68	nr	**40.01**
100 × 75 mm; eared	50.37	56.89	0.29	7.17	nr	**64.06**
Shoe; front						
100 × 75 mm; plain	80.64	91.07	0.23	5.68	nr	**96.75**
100 × 75 mm; eared	98.53	111.27	0.29	7.17	nr	**118.44**
Shoe; side						
100 × 75 mm; plain	97.83	110.49	0.23	5.68	nr	**116.17**
100 × 75 mm; eared	122.32	138.14	0.29	7.17	nr	**145.31**
Bends; side; any degree						
100 × 75 mm; plain	74.20	83.79	0.25	6.18	nr	**89.97**
100 × 75 mm; 135°; plain	75.79	85.59	0.25	6.18	nr	**91.77**
Bends; side; any degree						
100 × 75 mm; eared	93.82	105.96	0.25	6.18	nr	**112.14**
Bends; front; any degree						
100 × 75 mm; plain	70.27	79.36	0.25	6.18	nr	**85.54**
100 × 75 mm; eared	79.63	89.93	0.25	6.18	nr	**96.11**
Offset; side;						
Plain socket						
75 mm projection	97.76	110.40	0.25	6.18	nr	**116.58**
115 mm projection	101.68	114.83	0.25	6.18	nr	**121.01**
225 mm projection	126.68	143.07	0.25	6.18	nr	**149.25**
305 mm projection	146.00	164.88	0.25	6.18	nr	**171.06**

R: DISPOSAL SYSTEMS

Item	Net Price £	Material £	Labour hours	Labour £	Unit	**Total rate £**
Offset; front						
Plain socket						
75 mm projection	74.38	84.00	0.25	6.18	nr	**90.18**
150 mm projection	82.21	92.84	0.25	6.18	nr	**99.02**
225 mm projection	101.76	114.92	0.25	6.18	nr	**121.10**
305 mm projection	121.06	136.72	0.25	6.18	nr	**142.90**
Eared socket						
75 mm projection	95.14	107.45	0.25	6.18	nr	**113.63**
150 mm projection	102.60	115.87	0.25	6.18	nr	**122.05**
225 mm projection	121.36	137.06	0.25	6.18	nr	**143.24**
305 mm projection	141.21	159.48	0.25	6.18	nr	**165.66**
Offset; plinth						
115 mm projection; plain	76.77	86.70	0.25	6.18	nr	**92.88**
115 mm projection; eared	98.87	111.66	0.25	6.18	nr	**117.84**
Rainwater heads						
Flat hopper						
210 × 160 × 185 mm; 65 mm outlet	58.41	65.96	0.40	9.88	nr	**75.84**
210 × 160 × 185 mm; 75 mm outlet	58.41	65.96	0.40	9.88	nr	**75.84**
250 × 215 × 215 mm; 100 mm outlet	69.22	78.17	0.40	9.88	nr	**88.05**
Flat rectangular						
225 × 125 × 125 mm; 65 mm outlet	81.21	91.72	0.40	9.88	nr	**101.60**
225 × 125 × 125 mm; 75 mm outlet	81.21	91.72	0.40	9.88	nr	**101.60**
280 × 150 × 130 mm; 100 mm outlet	112.12	126.62	0.40	9.88	nr	**136.50**
Rectangular						
250 × 180 × 175 mm; 75 mm outlet	75.72	85.52	0.40	9.88	nr	**95.40**
250 × 180 × 175 mm; 100 mm outlet	75.72	85.52	0.40	9.88	nr	**95.40**
300 × 250 × 200 mm; 65 mm outlet	105.46	119.11	0.40	9.88	nr	**128.99**
300 × 250 × 200 mm; 75 mm outlet	105.46	119.11	0.40	9.88	nr	**128.99**
300 × 250 × 200 mm; 100 mm outlet	105.46	119.11	0.40	9.88	nr	**128.99**
300 × 250 × 200 mm; 100 × 75 mm outlet	105.46	119.11	0.40	9.88	nr	**128.99**
Castellated rectangular						
250 × 180 × 175 mm; 65 mm outlet	75.72	85.52	0.40	9.88	nr	**95.40**

R: DISPOSAL SYSTEMS

Item	Net Price £	Material £	Labour hours	Labour £	Unit	Total rate £
R11: ABOVE GROUND DRAINAGE						
Pricing note: degree angles are only indicated where material prices differ						
PVC-U overflow pipe; solvent welded joints; fixed with clips to backgrounds						
Pipe, with brackets measured separately						
19 mm	1.14	1.29	0.21	5.18	m	**6.47**
Fixings						
Pipe clip: including fixing to backgrounds. For minimum fixing distances, refer to the Tables and Memoranda at the rear of the book						
19 mm	0.45	0.51	0.18	4.45	nr	**4.96**
Extra over fittings overflow pipework PVC-U						
Straight coupler						
19 mm	1.21	1.37	0.17	4.20	nr	**5.57**
Bend						
19 mm: 91.25°	1.43	1.61	0.17	4.20	nr	**5.81**
19 mm: 135°	1.44	1.62	0.17	4.20	nr	**5.82**
Tee						
19 mm	1.56	1.76	0.18	4.45	nr	**6.21**
Reverse nut connector						
19 mm	0.57	0.64	0.15	3.70	nr	**4.34**
BSP adaptor: solvent welded socket to threaded socket						
19 mm × ¾'	2.08	2.35	0.14	3.46	nr	**5.81**
Straight tank connector						
19 mm	1.89	2.13	0.21	5.18	nr	**7.31**
32 mm	0.93	1.05	0.28	6.93	nr	**7.98**
40 mm	0.93	1.05	0.30	7.41	nr	**8.46**
Bent tank connector						
19 mm	2.24	2.53	0.21	5.18	nr	**7.71**
Tundish						
19 mm	29.57	33.39	0.38	9.38	nr	**42.77**

R: DISPOSAL SYSTEMS

Item	Net Price £	Material £	Labour hours	Labour £	Unit	Total rate £
MuPVC waste pipe; solvent welded joints; fixed with clips to backgrounds; BS 5255						
Pipe, with brackets measured separately						
32 mm	1.80	2.03	0.23	5.68	m	**7.71**
40 mm	2.24	2.53	0.23	5.68	m	**8.21**
50 mm	3.38	3.81	0.26	6.43	m	**10.24**
Fixings						
Pipe clip: including fixing to backgrounds. For minimum fixing distances, refer to the Tables and Memoranda at the rear of the book						
32 mm	0.37	0.42	0.13	3.20	nr	**3.62**
40 mm	0.43	0.49	0.13	3.20	nr	**3.69**
50 mm	0.57	0.64	0.13	3.20	nr	**3.84**
Pipe clip: expansion: including fixing to backgrounds. For minimum fixing distances, refer to the Tables and Memoranda at the rear of the book						
32 mm	0.45	0.51	0.13	3.20	nr	**3.71**
40 mm	0.47	0.53	0.13	3.20	nr	**3.73**
50 mm	1.08	1.22	0.13	3.20	nr	**4.42**
Pipe clip: metal; including fixing to backgrounds. For minimum fixing distances, refer to the Tables and Memoranda at the rear of the book						
32 mm	1.62	1.83	0.13	3.20	nr	**5.03**
40 mm	1.92	2.17	0.13	3.20	nr	**5.37**
50 mm	2.45	2.76	0.13	3.20	nr	**5.96**
Extra over fittings waste pipework MuPVC						
Screwed access plug						
32 mm	1.05	1.18	0.18	4.45	nr	**5.63**
40 mm	1.05	1.18	0.18	4.45	nr	**5.63**
50 mm	1.51	1.70	0.25	6.18	nr	**7.88**
Straight coupling						
32 mm	1.13	1.28	0.27	6.67	nr	**7.95**
40 mm	1.13	1.28	0.27	6.67	nr	**7.95**
50 mm	2.05	2.32	0.27	6.67	nr	**8.99**
Expansion coupling						
32 mm	1.99	2.24	0.27	6.67	nr	**8.91**
40 mm	2.39	2.70	0.27	6.67	nr	**9.37**
50 mm	3.24	3.66	0.27	6.67	nr	**10.33**

R: DISPOSAL SYSTEMS

Item	Net Price £	Material £	Labour hours	Labour £	Unit	**Total rate £**
R11: ABOVE GROUND DRAINAGE – cont						
Extra over fittings waste pipework MuPVC – cont						
MuPVC to copper coupling						
32 mm	1.99	2.24	0.27	6.67	nr	**8.91**
40 mm	2.39	2.70	0.27	6.67	nr	**9.37**
50 mm	3.24	3.66	0.27	6.67	nr	**10.33**
Spigot and socket coupling						
32 mm	1.99	2.24	0.27	6.67	nr	**8.91**
40 mm	2.39	2.70	0.27	6.67	nr	**9.37**
50 mm	3.24	3.66	0.27	6.67	nr	**10.33**
Union						
32 mm	4.72	5.33	0.28	6.93	nr	**12.26**
40 mm	6.18	6.98	0.28	6.93	nr	**13.91**
50 mm	7.05	7.97	0.28	6.93	nr	**14.90**
Reducer: socket						
32 × 19 mm	1.70	1.92	0.27	6.67	nr	**8.59**
40 × 32 mm	1.13	1.28	0.27	6.67	nr	**7.95**
50 × 32 mm	1.63	1.84	0.27	6.67	nr	**8.51**
50 × 40 mm	1.99	2.24	0.27	6.67	nr	**8.91**
Reducer: level invert						
40 × 32 mm	1.41	1.59	0.27	6.67	nr	**8.26**
50 × 32 mm	1.74	1.97	0.27	6.67	nr	**8.64**
50 × 40 mm	1.74	1.97	0.27	6.67	nr	**8.64**
Swept bend						
32 mm	1.15	1.30	0.27	6.67	nr	**7.97**
32 mm: 165°	1.19	1.35	0.27	6.67	nr	**8.02**
40 mm	1.28	1.45	0.27	6.67	nr	**8.12**
40 mm: 165°	2.25	2.54	0.27	6.67	nr	**9.21**
50 mm	2.24	2.53	0.30	7.41	nr	**9.94**
50 mm: 165°	2.97	3.36	0.30	7.41	nr	**10.77**
Knuckle bend						
32 mm	1.05	1.18	0.27	6.67	nr	**7.85**
40 mm	1.16	1.31	0.27	6.67	nr	**7.98**
Spigot and socket bend						
32 mm	1.88	2.12	0.27	6.67	nr	**8.79**
32 mm: 150°	1.94	2.19	0.27	6.67	nr	**8.86**
40 mm	2.15	2.43	0.27	6.67	nr	**9.10**
50 mm	3.06	3.46	0.30	7.41	nr	**10.87**

R: DISPOSAL SYSTEMS

Item	Net Price £	Material £	Labour hours	Labour £	Unit	**Total rate £**
Swept tee						
32 mm: 91.25°	1.54	1.74	0.31	7.66	nr	**9.40**
32 mm: 135°	1.85	2.09	0.31	7.66	nr	**9.75**
40 mm: 91.25°	1.96	2.21	0.31	7.66	nr	**9.87**
40 mm: 135°	2.44	2.75	0.31	7.66	nr	**10.41**
50 mm	3.81	4.30	0.31	7.66	nr	**11.96**
Swept cross						
40 mm: 91.25°	4.74	5.36	0.31	7.66	nr	**13.02**
50 mm: 91.25°	4.97	5.61	0.43	10.63	nr	**16.24**
50 mm: 135°	6.27	7.08	0.31	7.66	nr	**14.74**
Male iron adaptor						
32 mm	1.72	1.94	0.28	6.93	nr	**8.87**
40 mm	2.01	2.27	0.28	6.93	nr	**9.20**
Female iron adaptor						
32 mm	2.01	2.27	0.28	6.93	nr	**9.20**
40 mm	2.01	2.27	0.28	6.93	nr	**9.20**
50 mm	2.90	3.27	0.31	7.66	nr	**10.93**
Reverse nut adaptor						
32 mm	2.57	2.91	0.20	4.95	nr	**7.86**
40 mm	2.57	2.91	0.20	4.95	nr	**7.86**
Automatic air admittance valve						
32 mm	14.39	16.25	0.27	6.67	nr	**22.92**
40 mm	14.39	16.25	0.28	6.93	nr	**23.18**
50 mm	14.39	16.25	0.31	7.66	nr	**23.91**
MuPVC to metal adpator: including heat shrunk joint to metal						
50 mm	6.19	6.99	0.38	9.38	nr	**16.37**
Caulking bush: including joint to metal						
32 mm	2.70	3.05	0.31	7.66	nr	**10.71**
40 mm	2.70	3.05	0.31	7.66	nr	**10.71**
50 mm	2.70	3.05	0.32	7.91	nr	**10.96**
Weathering apron						
50 mm	2.38	2.69	0.65	16.07	nr	**18.76**
Vent cowl						
50 mm	2.71	3.06	0.19	4.70	nr	**7.76**

R: DISPOSAL SYSTEMS

Item	Net Price £	Material £	Labour hours	Labour £	Unit	Total rate £
R11: ABOVE GROUND DRAINAGE – cont						
ABS waste pipe; solvent welded joints; fixed with clips to backgrounds; BS 5255						
Pipe, with brackets measured separately						
32 mm	1.77	2.00	0.23	5.68	m	**7.68**
40 mm	2.21	2.50	0.23	5.68	m	**8.18**
50 mm	2.79	3.15	0.26	6.43	m	**9.58**
Fixings						
Pipe clip: including fixing to backgrounds. For minimum fixing distances, refer to the Tables and Memoranda at the rear of the book						
32 mm	0.28	0.32	0.17	4.20	nr	**4.52**
40 mm	0.36	0.41	0.17	4.20	nr	**4.61**
50 mm	1.07	1.20	0.17	4.20	nr	**5.40**
Pipe clip: expansion: including fixing to backgrounds. For minimum fixing distances, refer to the Tables and Memoranda at the rear of the book						
32 mm	0.28	0.32	0.17	4.20	nr	**4.52**
40 mm	0.36	0.41	0.17	4.20	nr	**4.61**
50 mm	1.07	1.20	0.17	4.20	nr	**5.40**
Pipe clip: metal; including fixing to backgrounds. For minimum fixing distances, refer to the Tables and Memoranda at the rear of the book						
32 mm	1.62	1.83	0.17	4.20	nr	**6.03**
40 mm	1.92	2.17	0.17	4.20	nr	**6.37**
50 mm	2.45	2.76	0.17	4.20	nr	**6.96**
Extra over fittings waste pipework ABS						
Screwed access plug						
32 mm	1.10	1.24	0.18	4.45	nr	**5.69**
40 mm	1.10	1.24	0.18	4.45	nr	**5.69**
50 mm	2.28	2.57	0.25	6.18	nr	**8.75**
Straight coupling						
32 mm	1.10	1.24	0.27	6.67	nr	**7.91**
40 mm	1.10	1.24	0.27	6.67	nr	**7.91**
50 mm	2.28	2.57	0.27	6.67	nr	**9.24**

R: DISPOSAL SYSTEMS

Item	Net Price £	Material £	Labour hours	Labour £	Unit	**Total rate £**
Expansion coupling						
32 mm	2.28	2.57	0.27	6.67	nr	**9.24**
40 mm	2.28	2.57	0.27	6.67	nr	**9.24**
50 mm	3.50	3.96	0.27	6.67	nr	**10.63**
ABS to copper coupling						
32 mm	2.28	2.57	0.27	6.67	nr	**9.24**
40 mm	2.28	2.57	0.27	6.67	nr	**9.24**
50 mm	3.50	3.96	0.27	6.67	nr	**10.63**
Reducer: socket						
40 × 32 mm	1.10	1.24	0.27	6.67	nr	**7.91**
50 × 32 mm	2.51	2.84	0.27	6.67	nr	**9.51**
50 × 40 mm	2.51	2.84	0.27	6.67	nr	**9.51**
Swept bend						
32 mm	1.10	1.24	0.27	6.67	nr	**7.91**
40 mm	1.10	1.24	0.27	6.67	nr	**7.91**
50 mm	2.28	2.57	0.30	7.41	nr	**9.98**
Knuckle bend						
32 mm	1.10	1.24	0.27	6.67	nr	**7.91**
40 mm	1.10	1.24	0.27	6.67	nr	**7.91**
Swept tee						
32 mm	1.60	1.81	0.31	7.66	nr	**9.47**
40 mm	1.60	1.81	0.31	7.66	nr	**9.47**
50 mm	4.18	4.72	0.31	7.66	nr	**12.38**
Swept cross						
40 mm	5.33	6.02	0.23	5.68	nr	**11.70**
50 mm	6.09	6.87	0.43	10.63	nr	**17.50**
Male iron adaptor						
32 mm	2.28	2.57	0.28	6.93	nr	**9.50**
40 mm	2.28	2.57	0.28	6.93	nr	**9.50**
Female iron adapator						
32 mm	2.28	2.57	0.28	6.93	nr	**9.50**
40 mm	2.28	2.57	0.28	6.93	nr	**9.50**
50 mm	3.42	3.87	0.31	7.66	nr	**11.53**
Tank connectors						
32 mm	1.58	1.79	0.29	7.17	nr	**8.96**
40 mm	1.74	1.97	0.29	7.17	nr	**9.14**
Caulking bush: including joint to pipework						
50 mm	2.70	3.05	0.50	12.35	nr	**15.40**

R: DISPOSAL SYSTEMS

Item	Net Price £	Material £	Labour hours	Labour £	Unit	Total rate £
R11: ABOVE GROUND DRAINAGE – cont						
Fixings						
Polypropylene waste pipe; push fit joints; fixed with clips to backgrounds; BS 5254						
Pipe, with brackets measured separately						
32 mm	0.97	1.09	0.21	5.18	m	**6.27**
40 mm	1.20	1.36	0.21	5.18	m	**6.54**
50 mm	2.00	2.25	0.38	9.38	m	**11.63**
Pipe clip: saddle; including fixing to backgrounds. For minimum fixing distances, refer to the Tables and Memoranda at the rear of the book						
32 mm	0.33	0.38	0.17	4.20	nr	**4.58**
40 mm	0.33	0.38	0.17	4.20	nr	**4.58**
Pipe clip: including fixing to backgrounds. For minimum fixing distances, refer to the Tables and Memoranda at the rear of the book						
50 mm	0.79	0.89	0.17	4.20	nr	**5.09**
Extra over fittings waste pipework polypropylene						
Screwed access plug						
32 mm	0.93	1.05	0.16	3.96	nr	**5.01**
40 mm	0.93	1.05	0.16	3.96	nr	**5.01**
50 mm	1.60	1.81	0.20	4.95	nr	**6.76**
Straight coupling						
32 mm	0.93	1.05	0.19	4.70	nr	**5.75**
40 mm	0.93	1.05	0.19	4.70	nr	**5.75**
50 mm	1.60	1.81	0.20	4.95	nr	**6.76**
Universal waste pipe coupler						
32 mm dia.	1.57	1.77	0.20	4.95	nr	**6.72**
40 mm dia.	1.78	2.01	0.20	4.95	nr	**6.96**
Reducer						
40 × 32 mm	1.60	1.81	0.19	4.70	nr	**6.51**
50 × 32 mm	1.66	1.88	0.19	4.70	nr	**6.58**
50 × 40 mm	1.72	1.94	0.20	4.95	nr	**6.89**
Swept bend						
32 mm	0.93	1.05	0.19	4.70	nr	**5.75**
40 mm	0.93	1.05	0.19	4.70	nr	**5.75**
50 mm	1.60	1.81	0.20	4.95	nr	**6.76**

R: DISPOSAL SYSTEMS

Item	Net Price £	Material £	Labour hours	Labour £	Unit	Total rate £
Knuckle bend						
32 mm	0.93	1.05	0.19	4.70	nr	**5.75**
40 mm	0.93	1.05	0.19	4.70	nr	**5.75**
50 mm	1.60	1.81	0.20	4.95	nr	**6.76**
Spigot and socket bend						
32 mm	0.93	1.05	0.19	4.70	nr	**5.75**
40 mm	0.93	1.05	0.19	4.70	nr	**5.75**
Swept tee						
32 mm	1.02	1.15	0.22	5.44	nr	**6.59**
40 mm	1.02	1.15	0.22	5.44	nr	**6.59**
50 mm	1.72	1.94	0.23	5.68	nr	**7.62**
Male iron adaptor						
32 mm	0.93	1.05	0.13	3.20	nr	**4.25**
40 mm	0.93	1.05	0.19	4.70	nr	**5.75**
50 mm	1.60	1.81	0.15	3.70	nr	**5.51**
Tank connector						
32 mm	0.93	1.05	0.24	5.93	nr	**6.98**
40 mm	0.93	1.05	0.24	5.93	nr	**6.98**
50 mm	1.60	1.81	0.35	8.65	nr	**10.46**
Polypropylene traps; including fixing to appliance and connection to pipework; BS 3943						
Tubular P trap; 75 mm seal						
32 mm dia.	3.77	4.25	0.20	4.95	nr	**9.20**
40 mm dia.	4.35	4.92	0.20	4.95	nr	**9.87**
Tubular S trap; 75 mm seal						
32 mm dia.	4.75	5.37	0.20	4.95	nr	**10.32**
40 mm dia.	5.59	6.31	0.20	4.95	nr	**11.26**
Running tubular P trap; 75 mm seal						
32 mm dia.	5.78	6.53	0.20	4.95	nr	**11.48**
40 mm dia.	6.31	7.13	0.20	4.95	nr	**12.08**
Running tubular S trap; 75 mm seal						
32 mm dia.	6.95	7.85	0.20	4.95	nr	**12.80**
40 mm dia.	7.49	8.46	0.20	4.95	nr	**13.41**
Spigot and socket bend; converter from P to S Trap						
32 mm	1.47	1.66	0.20	4.95	nr	**6.61**
40 mm	1.58	1.79	0.21	5.18	nr	**6.97**

R: DISPOSAL SYSTEMS

Item	Net Price £	Material £	Labour hours	Labour £	Unit	Total rate £
R11: ABOVE GROUND DRAINAGE – cont						
Polypropylene traps – cont						
Bottle P trap; 75 mm seal						
32 mm dia.	4.20	4.74	0.20	4.95	nr	**9.69**
40 mm dia.	5.01	5.66	0.20	4.95	nr	**10.61**
Bottle S trap; 75 mm seal						
32 mm dia.	5.06	5.71	0.20	4.95	nr	**10.66**
40 mm dia.	6.15	6.95	0.25	6.18	nr	**13.13**
Bottle P trap; resealing; 75 mm seal						
32 mm dia.	5.22	5.90	0.20	4.95	nr	**10.85**
40 mm dia.	6.11	6.91	0.25	6.18	nr	**13.09**
Bottle S trap; resealing; 75 mm seal						
32 mm dia.	5.98	6.75	0.20	4.95	nr	**11.70**
40 mm dia.	6.93	7.82	0.25	6.18	nr	**14.00**
Bath trap, low level; 38 mm seal						
40 mm dia.	5.23	5.91	0.25	6.18	nr	**12.09**
Bath trap, low level; 38 mm seal complete with overflow hose						
40 mm dia.	8.10	9.15	0.25	6.18	nr	**15.33**
Bath trap; 75 mm seal complete with overlow hose						
40 mm dia.	8.06	9.10	0.25	6.18	nr	**15.28**
Bath trap; 75 mm seal complete with overflow hose and overflow outlet						
40 mm dia.	13.75	15.52	0.20	4.95	nr	**20.47**
Bath trap; 75 mm seal complete with overflow hose, overflow outlet and ABS chrome waste						
40 mm dia.	19.02	21.48	0.20	4.95	nr	**26.43**
Washing machine trap; 75 mm seal including stand pipe						
40 mm dia.	10.95	12.36	0.25	6.18	nr	**18.54**
Washing machine standpipe						
40 mm dia.	5.71	6.45	0.25	6.18	nr	**12.63**

R: DISPOSAL SYSTEMS

Item	Net Price £	Material £	Labour hours	Labour £	Unit	Total rate £
Plastic unslotted chrome plated basin/sink waste including plug						
32 mm	6.88	7.77	0.34	8.40	nr	**16.17**
40 mm	9.20	10.39	0.34	8.40	nr	**18.79**
Plastic slotted chrome plated basin/sink waste including plug						
32 mm	5.45	6.15	0.34	8.40	nr	**14.55**
40 mm	9.13	10.31	0.34	8.40	nr	**18.71**
Bath overflow outlet; plastic; white						
42 mm	4.79	5.41	0.37	9.15	nr	**14.56**
Bath overlow outlet; plastic; chrome plated						
42 mm	6.34	7.16	0.37	9.15	nr	**16.31**
Combined cistern and bath overflow outlet; plastic; white						
42 mm	9.84	11.12	0.39	9.63	nr	**20.75**
Combined cistern and bath overlow outlet; plastic; chrome plated						
42 mm	9.83	11.10	0.39	9.63	nr	**20.73**
Cistern overflow outlet; plastic; white						
42 mm	7.92	8.95	0.15	3.70	nr	**12.65**
Cistern overlow outlet; plastic; chrome plated						
42 mm	7.14	8.07	0.15	3.70	nr	**11.77**
PVC-U soil and waste pipe; solvent welded joints; fixed with clips to backgrounds; BS 4514/ BS EN 607						
Pipe, with brackets measured separately						
82 mm	8.19	9.25	0.35	8.65	m	**17.90**
110 mm	8.35	9.44	0.41	10.13	m	**19.57**
160 mm	21.64	24.44	0.51	12.61	m	**37.05**
Fixings						
Galvanized steel pipe clip: including fixing to backgrounds. For minimum fixing distances, refer to the Tables and Memoranda at the rear of the book						
82 mm	3.40	3.84	0.18	4.45	nr	**8.29**
110 mm	3.53	3.99	0.18	4.45	nr	**8.44**
160 mm	8.49	9.59	0.18	4.45	nr	**14.04**

R: DISPOSAL SYSTEMS

Item	Net Price £	Material £	Labour hours	Labour £	Unit	Total rate £
R11: ABOVE GROUND DRAINAGE – cont						
Fixings – cont						
Plastic coated steel pipe clip: including fixing to backgrounds. For minimum fixing distances, refer to the Tables and Memoranda at the rear of the book						
82 mm	4.76	5.38	0.18	4.45	nr	**9.83**
110 mm	6.48	7.31	0.18	4.45	nr	**11.76**
160 mm	8.18	9.24	0.18	4.45	nr	**13.69**
Plastic pipe clip: including fixing to backgrounds. For minimum fixing distances, refer to the Tables and Memoranda at the rear of the book						
82 mm	2.51	2.84	0.18	4.45	nr	**7.29**
110 mm	4.84	5.47	0.18	4.45	nr	**9.92**
Plastic coated steel pipe clip: adjustable; including fixing to backgrounds. For minimum fixing distances, refer to the Tables and Memoranda at the rear of the book						
82 mm	3.85	4.35	0.20	4.95	nr	**9.30**
110 mm	4.80	5.42	0.20	4.95	nr	**10.37**
Galvanized steel pipe clip: drive in; including fixing to backgrounds. For minimum fixing distances, refer to the Tables and Memoranda at the rear of the book						
110 mm	7.33	8.28	0.22	5.44	nr	**13.72**
Extra over fittings solvent welded pipework PVC-U						
Straight coupling						
82 mm	4.35	4.92	0.21	5.18	nr	**10.10**
110 mm	5.43	6.13	0.22	5.44	nr	**11.57**
160 mm	15.67	17.70	0.24	5.93	nr	**23.63**
Expansion coupling						
82 mm	6.50	7.34	0.21	5.18	nr	**12.52**
110 mm	6.65	7.51	0.22	5.44	nr	**12.95**
160 mm	20.00	22.59	0.24	5.93	nr	**28.52**
Slip coupling; double ring socket						
82 mm	13.50	15.25	0.21	5.18	nr	**20.43**
110 mm	16.91	19.09	0.22	5.44	nr	**24.53**
160 mm	26.85	30.32	0.24	5.93	nr	**36.25**

R: DISPOSAL SYSTEMS

Item	Net Price £	Material £	Labour hours	Labour £	Unit	**Total rate £**
Puddle flanges						
110 mm	131.97	149.04	0.45	11.13	nr	**160.17**
160 mm	213.33	240.93	0.55	13.59	nr	**254.52**
Socket reducer						
82 to 50 mm	6.61	7.47	0.18	4.45	nr	**11.92**
110 to 50 mm	8.24	9.30	0.18	4.45	nr	**13.75**
110 to 82 mm	8.49	9.59	0.22	5.44	nr	**15.03**
160 to 110 mm	17.21	19.44	0.26	6.43	nr	**25.87**
Socket plugs						
82 mm	5.12	5.78	0.15	3.70	nr	**9.48**
110 mm	7.49	8.46	0.20	4.95	nr	**13.41**
160 mm	13.78	15.57	0.27	6.67	nr	**22.24**
Access door; including cutting into pipe						
82 mm	14.52	16.40	0.28	6.93	nr	**23.33**
110 mm	14.52	16.40	0.34	8.40	nr	**24.80**
160 mm	25.95	29.30	0.46	11.36	nr	**40.66**
Screwed access cap						
82 mm	10.28	11.61	0.15	3.70	nr	**15.31**
110 mm	12.12	13.69	0.20	4.95	nr	**18.64**
160 mm	22.81	25.77	0.27	6.67	nr	**32.44**
Access pipe: spigot and socket						
110 mm	17.92	20.24	0.22	5.44	nr	**25.68**
Access pipe: double socket						
110 mm	17.92	20.24	0.22	5.44	nr	**25.68**
Swept bend						
82 mm	10.90	12.31	0.29	7.17	nr	**19.48**
110 mm	12.77	14.42	0.32	7.91	nr	**22.33**
160 mm	31.80	35.91	0.49	12.11	nr	**48.02**
Bend; special angle						
82 mm	21.01	23.73	0.29	7.17	nr	**30.90**
110 mm	25.05	28.29	0.32	7.91	nr	**36.20**
160 mm	42.29	47.77	0.49	12.11	nr	**59.88**
Spigot and socket bend						
82 mm	10.55	11.91	0.26	6.43	nr	**18.34**
110 mm	12.38	13.98	0.32	7.91	nr	**21.89**
110 mm: 135°	13.89	15.69	0.32	7.91	nr	**23.60**
160 mm: 135°	30.75	34.73	0.44	10.87	nr	**45.60**
Variable bend: single socket						
110 mm	23.94	27.04	0.33	8.15	nr	**35.19**

R: DISPOSAL SYSTEMS

Item	Net Price £	Material £	Labour hours	Labour £	Unit	**Total rate £**
R11: ABOVE GROUND DRAINAGE – cont						
Extra over fittings solvent welded pipework PVC-U – cont						
Variable bend: double socket						
110 mm	23.88	26.97	0.33	8.15	nr	**35.12**
Access bend						
110 mm	35.42	40.00	0.33	8.15	nr	**48.15**
Single branch: two bosses						
82 mm	15.24	17.21	0.35	8.65	nr	**25.86**
82 mm: 104°	16.26	18.36	0.35	8.65	nr	**27.01**
110 mm	16.89	19.07	0.42	10.38	nr	**29.45**
110 mm: 135°	17.61	19.89	0.42	10.38	nr	**30.27**
160 mm	35.86	40.50	0.50	12.35	nr	**52.85**
160 mm: 135°	36.68	41.42	0.50	12.35	nr	**53.77**
Single branch; four bosses						
110 mm	20.56	23.23	0.42	10.38	nr	**33.61**
Single access branch						
82 mm	49.41	55.80	0.35	8.65	nr	**64.45**
110 mm	28.92	32.66	0.42	10.38	nr	**43.04**
Unequal single branch						
160 × 160 × 110 mm	40.49	45.73	0.50	12.35	nr	**58.08**
160 × 160 × 110 mm: 135°	43.50	49.13	0.50	12.35	nr	**61.48**
Double branch						
110 mm	43.63	49.28	0.42	10.38	nr	**59.66**
110 mm: 135°	41.73	47.13	0.42	10.38	nr	**57.51**
Corner branch						
110 mm	73.72	83.26	0.42	10.38	nr	**93.64**
Unequal double branch						
160 × 160 × 110 mm	75.40	85.16	0.50	12.35	nr	**97.51**
Single boss pipe; single socket						
110 × 110 × 32 mm	4.57	5.16	0.24	5.93	nr	**11.09**
110 × 110 × 40 mm	4.57	5.16	0.24	5.93	nr	**11.09**
110 × 110 × 50 mm	4.81	5.44	0.24	5.93	nr	**11.37**
Single boss pipe; triple socket						
110 × 110 × 40 mm	7.41	8.36	0.24	5.93	nr	**14.29**

R: DISPOSAL SYSTEMS

Item	Net Price £	Material £	Labour hours	Labour £	Unit	**Total rate £**
Waste boss; including cutting into pipe						
82 to 32 mm	5.97	6.74	0.29	7.17	nr	**13.91**
82 to 40 mm	5.97	6.74	0.29	7.17	nr	**13.91**
110 to 32 mm	5.97	6.74	0.29	7.17	nr	**13.91**
110 to 40 mm	5.97	6.74	0.29	7.17	nr	**13.91**
110 to 50 mm	6.18	6.98	0.29	7.17	nr	**14.15**
160 to 32 mm	8.45	9.55	0.30	7.41	nr	**16.96**
160 to 40 mm	8.45	9.55	0.35	8.65	nr	**18.20**
160 to 50 mm	8.45	9.55	0.40	9.88	nr	**19.43**
Self locking waste boss; including cutting into pipe						
110 to 32 mm	7.98	9.02	0.30	7.41	nr	**16.43**
110 to 40 mm	8.32	9.39	0.30	7.41	nr	**16.80**
110 to 50 mm	10.76	12.15	0.30	7.41	nr	**19.56**
Adaptor saddle; including cutting to pipe						
82 to 32 mm	3.98	4.50	0.29	7.17	nr	**11.67**
110 to 40 mm	4.91	5.55	0.29	7.17	nr	**12.72**
160 to 50 mm	8.88	10.03	0.29	7.17	nr	**17.20**
Branch boss adaptor						
32 mm	2.37	2.67	0.26	6.43	nr	**9.10**
40 mm	2.37	2.67	0.26	6.43	nr	**9.10**
50 mm	3.39	3.83	0.26	6.43	nr	**10.26**
Branch boss adaptor bend						
32 mm	3.24	3.66	0.26	6.43	nr	**10.09**
40 mm	3.54	4.00	0.26	6.43	nr	**10.43**
50 mm	4.24	4.78	0.26	6.43	nr	**11.21**
Automatic air admittance valve						
82 to 110 mm	37.22	42.03	0.19	4.70	nr	**46.73**
PVC-U to metal adpator: including heat shrunk joint to metal						
110 mm	10.36	11.70	0.57	14.09	nr	**25.79**
Caulking bush: including joint to pipework						
82 mm	10.20	11.52	0.46	11.36	nr	**22.88**
110 mm	10.20	11.52	0.46	11.36	nr	**22.88**
Vent cowl						
82 mm	3.07	3.47	0.13	3.20	nr	**6.67**
110 mm	3.10	3.50	0.13	3.20	nr	**6.70**
160 mm	8.13	9.18	0.13	3.20	nr	**12.38**

R: DISPOSAL SYSTEMS

Item	Net Price £	Material £	Labour hours	Labour £	Unit	Total rate £
R11: ABOVE GROUND DRAINAGE – cont						
Extra over fittings solvent welded pipework PVC-U – cont						
Weathering apron; to lead slates						
82 mm	3.07	3.47	1.15	28.42	nr	**31.89**
110 mm	3.53	3.99	1.15	28.42	nr	**32.41**
160 mm	10.61	11.98	1.15	28.42	nr	**40.40**
Weathering apron; to asphalt						
82 mm	12.97	14.65	1.10	27.18	nr	**41.83**
110 mm	12.97	14.65	1.10	27.18	nr	**41.83**
Weathering slate; flat; 406 × 406 mm						
82 mm	36.54	41.27	1.04	25.70	nr	**66.97**
110 mm	36.54	41.27	1.04	25.70	nr	**66.97**
Weathering slate; flat; 457 × 457 mm						
82 mm	37.48	42.33	1.04	25.70	nr	**68.03**
110 mm	37.48	42.33	1.04	25.70	nr	**68.03**
Weathering slate; angled; 610 × 610 mm						
82 mm	50.66	57.21	1.04	25.70	nr	**82.91**
110 mm	50.66	57.21	1.04	25.70	nr	**82.91**
PVC-U soil and waste pipe; ring seal joints; fixed with clips to backgrounds; BS 4514/ BS EN 607						
Pipe, with brackets measured separately						
82 mm dia.	7.22	8.15	0.35	8.65	m	**16.80**
110 mm dia.	7.28	8.22	0.41	10.13	m	**18.35**
160 mm dia.	26.09	29.47	0.51	12.61	m	**42.08**
Fixings						
Galvanized steel pipe clip: including fixing to backgrounds. For minimum fixing distances, refer to the Tables and Memoranda at the rear of the book						
82 mm	3.40	3.84	0.18	4.45	nr	**8.29**
110 mm	3.53	3.99	0.18	4.45	nr	**8.44**
160 mm	8.49	9.59	0.18	4.45	nr	**14.04**

R: DISPOSAL SYSTEMS

Item	Net Price £	Material £	Labour hours	Labour £	Unit	**Total rate £**
Plastic coated steel pipe clip: including fixing to backgrounds. For minimum fixing distances, refer to the Tables and Memoranda at the rear of the book						
82 mm	4.76	5.38	0.18	4.45	nr	**9.83**
110 mm	6.48	7.31	0.18	4.45	nr	**11.76**
160 mm	8.18	9.24	0.18	4.45	nr	**13.69**
Plastic pipe clip: including fixing to backgrounds. For minimum fixing distances, refer to the Tables and Memoranda at the rear of the book						
82 mm	2.51	2.84	0.18	4.45	nr	**7.29**
110 mm	4.84	5.47	0.18	4.45	nr	**9.92**
Plastic coated steel pipe clip: adjustable; including fixing to backgrounds. For minimum fixing distances, refer to the Tables and Memoranda at the rear of the book						
82 mm	3.85	4.35	0.20	4.95	nr	**9.30**
110 mm	4.80	5.42	0.20	4.95	nr	**10.37**
Galvanized steel pipe clip: drive in; including fixing to backgrounds. For minimum fixing distances, refer to the Tables and Memoranda at the rear of the book						
110 mm	7.33	8.28	0.22	5.44	nr	**13.72**
Extra over fittings ring seal pipework PVC-U						
Straight coupling						
82 mm	4.73	5.34	0.21	5.18	nr	**10.52**
110 mm	5.31	6.00	0.22	5.44	nr	**11.44**
160 mm	17.92	20.24	0.24	5.93	nr	**26.17**
Straight coupling; double socket						
82 mm	8.35	9.44	0.21	5.18	nr	**14.62**
110 mm	7.58	8.56	0.22	5.44	nr	**14.00**
160 mm	17.92	20.24	0.24	5.93	nr	**26.17**
Reducer; socket						
82 to 50 mm	7.79	8.80	0.15	3.70	nr	**12.50**
110 to 50 mm	10.86	12.26	0.15	3.70	nr	**15.96**
110 to 82 mm	10.86	12.26	0.19	4.70	nr	**16.96**
160 to 110 mm	15.64	17.67	0.31	7.66	nr	**25.33**
Access cap						
82 mm	8.48	9.58	0.15	3.70	nr	**13.28**
110 mm	9.99	11.28	0.17	4.20	nr	**15.48**

R: DISPOSAL SYSTEMS

Item	Net Price £	Material £	Labour hours	Labour £	Unit	Total rate £
R11: ABOVE GROUND DRAINAGE – cont						
Extra over fittings ring seal pipework PVC-U – cont						
Access cap; pressure plug						
160 mm	23.48	26.52	0.33	8.15	nr	**34.67**
Access pipe						
82 mm	20.80	23.49	0.22	5.44	nr	**28.93**
110 mm	20.76	23.45	0.22	5.44	nr	**28.89**
160 mm	43.60	49.25	0.24	5.93	nr	**55.18**
Bend						
82 mm	10.58	11.94	0.29	7.17	nr	**19.11**
82 mm; adjustable radius	23.86	26.95	0.29	7.17	nr	**34.12**
110 mm	11.38	12.85	0.32	7.91	nr	**20.76**
110 mm; adjustable radius	21.24	23.99	0.32	7.91	nr	**31.90**
160 mm	34.17	38.59	0.49	12.11	nr	**50.70**
160 mm; adjustable radius	44.83	50.63	0.49	12.11	nr	**62.74**
Bend; spigot and socket						
110 mm	11.49	12.97	0.32	7.91	nr	**20.88**
Bend; offset						
82 mm	9.31	10.52	0.21	5.18	nr	**15.70**
110 mm	14.93	16.86	0.32	7.91	nr	**24.77**
160 mm	32.96	37.22	0.32	7.91	nr	**45.13**
Bend; access						
110 mm	31.34	35.39	0.33	8.15	nr	**43.54**
Single branch						
82 mm	16.61	18.76	0.35	8.65	nr	**27.41**
110 mm	15.35	17.34	0.42	10.38	nr	**27.72**
110 mm; 45°	15.80	17.84	0.31	7.66	nr	**25.50**
160 mm	39.59	44.72	0.50	12.35	nr	**57.07**
Single branch; access						
82 mm	24.89	28.11	0.35	8.65	nr	**36.76**
110 mm	36.03	40.69	0.42	10.38	nr	**51.07**
Unequal single branch						
160 × 160 × 110 mm	40.48	45.72	0.50	12.35	nr	**58.07**
160 × 160 × 110 mm; 45°	43.50	49.13	0.50	12.35	nr	**61.48**
Double branch; 4 bosses						
110 mm	40.20	45.40	0.49	12.11	nr	**57.51**

R: DISPOSAL SYSTEMS

Item	Net Price £	Material £	Labour hours	Labour £	Unit	**Total rate £**
Corner branch; 2 bosses						
110 mm	73.71	83.25	0.49	12.11	nr	**95.36**
Multi-branch; 4 bosses						
110 mm	22.00	24.85	0.52	12.85	nr	**37.70**
Boss branch						
110 × 32 mm	4.56	5.15	0.34	8.40	nr	**13.55**
110 × 40 mm	4.56	5.15	0.34	8.40	nr	**13.55**
Strap on boss						
110 × 32 mm	5.07	5.72	0.30	7.41	nr	**13.13**
110 × 40 mm	5.07	5.72	0.30	7.41	nr	**13.13**
110 × 50 mm	5.11	5.77	0.30	7.41	nr	**13.18**
Patch boss						
82 × 32 mm	5.97	6.74	0.31	7.66	nr	**14.40**
82 × 40 mm	5.97	6.74	0.31	7.66	nr	**14.40**
82 × 50 mm	5.97	6.74	0.31	7.66	nr	**14.40**
Boss pipe; collar 4 boss						
110 mm	7.41	8.36	0.35	8.65	nr	**17.01**
Boss adaptor; rubber; push fit						
32 mm	2.77	3.13	0.26	6.43	nr	**9.56**
40 mm	2.77	3.13	0.26	6.43	nr	**9.56**
50 mm	3.14	3.55	0.26	6.43	nr	**9.98**
WC connector; cap and seal; solvent socket						
110 mm	7.71	8.71	0.23	5.68	nr	**14.39**
110 mm; 90°	12.40	14.00	0.27	6.67	nr	**20.67**
Vent terminal						
82 mm	3.07	3.47	0.13	3.20	nr	**6.67**
110 mm	3.10	3.50	0.13	3.20	nr	**6.70**
160 mm	8.14	9.19	0.13	3.20	nr	**12.39**
Weathering slate; inclined; 610 × 610 mm						
82 mm	50.66	57.21	1.04	25.70	nr	**82.91**
110 mm	50.66	57.21	1.04	25.70	nr	**82.91**
Weathering slate; inclined; 450 × 450 mm						
82 mm	37.48	42.33	1.04	25.70	nr	**68.03**
110 mm	37.48	42.33	1.04	25.70	nr	**68.03**
Weathering slate; flat; 400 × 400 mm						
82 mm	36.57	41.30	1.04	25.70	nr	**67.00**
110 mm	36.57	41.30	1.04	25.70	nr	**67.00**

R: DISPOSAL SYSTEMS

Item	Net Price £	Material £	Labour hours	Labour £	Unit	Total rate £
R11: ABOVE GROUND DRAINAGE – cont						
Extra over fittings ring seal pipework PVC-U – cont						
Air admittance valve						
82 mm	37.22	42.03	0.19	4.70	nr	**46.73**
110 mm	37.22	42.03	0.19	4.70	nr	**46.73**
Cast iron pipe; nitrile rubber gasket joint with continuity clip BSEN 877; fixed vertically to backgrounds						
Pipe, with brackets and jointing couplings measured separately						
50 mm	18.42	20.81	0.31	7.66	m	**28.47**
75 mm	21.43	24.20	0.34	8.40	m	**32.60**
100 mm	23.88	26.97	0.37	9.15	m	**36.12**
150 mm	49.86	56.31	0.60	14.83	m	**71.14**
Fixings						
Brackets; fixed to backgrounds. For minimum fixing distances, refer to the Tables and Memoranda at the rear of the book						
50 mm	6.64	7.50	0.15	3.70	nr	**11.20**
75 mm	7.27	8.21	0.18	4.45	nr	**12.66**
100 mm	7.45	8.42	0.18	4.45	nr	**12.87**
150 mm	13.81	15.60	0.20	4.95	nr	**20.55**
Extra over fittings nitrile gasket cast iron pipework BS 416/6087, with jointing couplings measured separately						
Standard coupling						
50 mm	8.91	10.07	0.17	4.20	nr	**14.27**
75 mm	9.83	11.10	0.17	4.20	nr	**15.30**
100 mm	12.85	14.51	0.17	4.20	nr	**18.71**
150 mm	25.65	28.97	0.17	4.20	nr	**33.17**
Conversion coupling						
65 × 75 mm	10.42	11.77	0.60	14.83	nr	**26.60**
70 × 75 mm	10.42	11.77	0.60	14.83	nr	**26.60**
90 × 100 mm	13.43	15.17	0.67	16.55	nr	**31.72**
Access pipe; round door						
50 mm	26.29	29.69	0.41	10.13	nr	**39.82**
75 mm	37.78	42.67	0.46	11.36	nr	**54.03**
100 mm	39.71	44.85	0.67	16.58	nr	**61.43**
150 mm	66.05	74.59	0.83	20.52	nr	**95.11**

R: DISPOSAL SYSTEMS

Item	Net Price £	Material £	Labour hours	Labour £	Unit	**Total rate £**
Access pipe; square door						
100 mm	78.24	88.36	0.67	16.58	nr	**104.94**
150 mm	119.77	135.27	0.83	20.52	nr	**155.79**
Taper reducer						
75 mm	21.60	24.40	0.60	14.83	nr	**39.23**
100 mm	27.41	30.96	0.67	16.55	nr	**47.51**
150 mm	53.37	60.27	0.83	20.52	nr	**80.79**
Blank cap						
50 mm	5.45	6.15	0.24	5.93	nr	**12.08**
75 mm	6.81	7.69	0.26	6.43	nr	**14.12**
100 mm	6.61	7.47	0.32	7.91	nr	**15.38**
150 mm	9.55	10.78	0.40	9.88	nr	**20.66**
Blank cap; 50 mm screwed tapping						
75 mm	13.42	15.16	0.26	6.43	nr	**21.59**
100 mm	15.57	17.58	0.32	7.91	nr	**25.49**
150 mm	17.29	19.52	0.40	9.88	nr	**29.40**
Universal connector						
50 × 56/48/40 mm	8.11	9.16	0.33	8.15	nr	**17.31**
Change piece; BS416						
100 mm	17.29	19.52	0.47	11.61	nr	**31.13**
WC connector						
100 mm	27.83	31.43	0.49	12.11	nr	**43.54**
Boss pipe; 2" BSPT socket						
50 mm	22.22	25.09	0.58	14.33	nr	**39.42**
75 mm	32.51	36.72	0.65	16.07	nr	**52.79**
100 mm	38.83	43.85	0.79	19.52	nr	**63.37**
150 mm	60.50	68.33	0.86	21.26	nr	**89.59**
Boss pipe; 2" BSPT socket; 135°						
100 mm	46.78	52.84	0.79	19.52	nr	**72.36**
Boss pipe; 2 × 2" BSPT socket; opposed						
75 mm	43.06	48.63	0.65	16.07	nr	**64.70**
100 mm	50.21	56.71	0.79	19.52	nr	**76.23**
Boss pipe; 2 × 2" BSPT socket; in line						
100 mm	50.71	57.27	0.79	19.52	nr	**76.79**
Boss pipe; 2 × 2" BSPT socket; 90°						
100 mm	50.21	56.71	0.79	19.52	nr	**76.23**

R: DISPOSAL SYSTEMS

Item	Net Price £	Material £	Labour hours	Labour £	Unit	Total rate £
R11: ABOVE GROUND DRAINAGE – cont						
Fixings – cont						
Bend; short radius						
50 mm	15.70	17.73	0.50	12.35	nr	**30.08**
75 mm	17.77	20.07	0.60	14.83	nr	**34.90**
100 mm	21.73	24.54	0.67	16.55	nr	**41.09**
100 mm; 11°	18.74	21.16	0.67	16.55	nr	**37.71**
100 mm; 67°	21.73	24.54	0.67	16.55	nr	**41.09**
150 mm	38.83	43.85	0.83	20.52	nr	**64.37**
Access bend; short radius						
50 mm	38.70	43.71	0.50	12.35	nr	**56.06**
75 mm	41.98	47.41	0.60	14.83	nr	**62.24**
100 mm	40.79	46.06	0.67	16.55	nr	**62.61**
100 mm; 45°	45.98	51.93	0.67	16.55	nr	**68.48**
150 mm	65.29	73.74	0.83	20.52	nr	**94.26**
150 mm; 45°	65.29	73.74	0.83	20.52	nr	**94.26**
Long radius bend						
75 mm	29.65	33.49	0.60	14.83	nr	**48.32**
100 mm	35.20	39.76	0.67	16.55	nr	**56.31**
100 mm; 5°	21.73	24.54	0.67	16.55	nr	**41.09**
150 mm	76.75	86.68	0.83	20.52	nr	**107.20**
150 mm; 22.5°	80.45	90.86	0.83	20.52	nr	**111.38**
Access bend; long radius						
75 mm	52.63	59.44	0.60	14.83	nr	**74.27**
100 mm	59.46	67.16	0.67	16.55	nr	**83.71**
150 mm	104.74	118.29	0.83	20.52	nr	**138.81**
Long tail bend						
100 × 250 mm long	28.08	31.71	0.70	17.30	nr	**49.01**
100 × 815 mm long	89.27	100.82	0.70	17.30	nr	**118.12**
Offset						
75 mm projection						
75 mm	23.18	26.18	0.53	13.10	nr	**39.28**
100 mm	22.87	25.83	0.66	16.31	nr	**42.14**
115 mm projection						
75 mm	25.48	28.77	0.53	13.10	nr	**41.87**
100 mm	28.57	32.26	0.66	16.31	nr	**48.57**
150 mm projection						
75 mm	29.93	33.80	0.53	13.10	nr	**46.90**
100 mm	29.93	33.80	0.66	16.31	nr	**50.11**

R: DISPOSAL SYSTEMS

Item	Net Price £	Material £	Labour hours	Labour £	Unit	**Total rate £**
225 mm projection						
100 mm	32.72	36.95	0.66	16.31	nr	**53.26**
300 mm projection						
100 mm	35.20	39.76	0.66	16.31	nr	**56.07**
Branch; equal and unequal						
50 mm	23.62	26.67	0.78	19.28	nr	**45.95**
75 mm	26.74	30.20	0.85	21.00	nr	**51.20**
100 mm	33.59	37.93	1.00	24.71	nr	**62.64**
150 mm	83.29	94.06	1.20	29.67	nr	**123.73**
150 × 100 mm; 87.5°	63.74	71.99	1.21	29.78	nr	**101.77**
150 × 100 mm; 45°	93.49	105.59	1.21	29.78	nr	**135.37**
Branch; 2’ BSPT screwed socket						
100 mm	45.01	50.84	1.00	24.71	nr	**75.55**
Branch; long tail						
100 × 915 mm long	93.07	105.11	1.00	24.71	nr	**129.82**
Access branch; equal and unequal						
50 mm	51.46	58.12	0.78	19.28	nr	**77.40**
75 mm	51.46	58.12	0.85	21.00	nr	**79.12**
100 mm	57.85	65.33	1.02	25.21	nr	**90.54**
150 mm	123.16	139.10	1.20	29.67	nr	**168.77**
150 × 100 mm; 87.5°	103.51	116.90	1.20	29.67	nr	**146.57**
150 × 100 mm; 45°	112.20	126.71	1.20	29.67	nr	**156.38**
Parallel branch						
100 mm	35.20	39.76	1.00	24.71	nr	**64.47**
Double branch						
75 mm	39.71	44.85	0.95	23.48	nr	**68.33**
100 mm	41.56	46.94	1.30	32.13	nr	**79.07**
150 × 100 mm	117.03	132.17	1.56	38.55	nr	**170.72**
Double access branch						
100 mm	65.80	74.32	1.43	35.35	nr	**109.67**
Corner branch						
100 mm	59.08	66.73	1.30	32.13	nr	**98.86**
Puddle flange; grey epoxy coated						
100 mm	39.50	44.61	1.00	24.71	nr	**69.32**
Roof vent connector; asphalt						
75 mm	48.08	54.30	0.90	22.24	nr	**76.54**
100 mm	37.54	42.40	0.97	23.97	nr	**66.37**

R: DISPOSAL SYSTEMS

Item	Net Price £	Material £	Labour hours	Labour £	Unit	Total rate £
R11: ABOVE GROUND DRAINAGE – cont						
Fixings – cont						
P trap						
100 mm	34.84	39.35	1.00	24.71	nr	**64.06**
P trap with access						
50 mm	53.31	60.21	0.77	19.03	nr	**79.24**
75 mm	53.31	60.21	0.90	22.24	nr	**82.45**
100 mm	59.09	66.74	1.16	28.66	nr	**95.40**
150 mm	103.19	116.55	1.77	43.74	nr	**160.29**
Bellmouth gully inlet						
100 mm	50.43	56.96	1.08	26.68	nr	**83.64**
Balcony gully inlet						
100 mm	152.25	171.95	1.08	26.68	nr	**198.63**
Roof outlet						
Flat grate						
75 mm	93.99	106.15	0.83	20.51	nr	**126.66**
100 mm	132.23	149.34	1.08	26.68	nr	**176.02**
Dome grate						
75 mm	93.99	106.15	0.83	20.51	nr	**126.66**
100 mm	148.31	167.50	1.08	26.68	nr	**194.18**
Top hat						
100 mm	200.79	226.77	1.08	26.68	nr	**253.45**
Cast iron pipe; EPDM rubber gasket joint with continuity clip; BS EN877; fixed to backgrounds						
Pipe, with brackets and jointing couplings measured separately						
50 mm	19.23	21.72	0.31	7.66	m	**29.38**
70 mm	22.49	25.40	0.34	8.40	m	**33.80**
100 mm	26.76	30.22	0.37	9.15	m	**39.37**
125 mm	42.94	48.50	0.65	16.07	m	**64.57**
150 mm	53.02	59.88	0.70	17.30	m	**77.18**
200 mm	88.59	100.05	1.14	28.16	m	**128.21**
250 mm	123.79	139.81	1.25	30.89	m	**170.70**
300 mm	154.15	174.09	1.53	37.81	m	**211.90**

R: DISPOSAL SYSTEMS

Item	Net Price £	Material £	Labour hours	Labour £	Unit	**Total rate £**
Fixings						
Brackets; fixed to backgrounds. For minimum fixing distances, refer to the Tables and Memoranda at the rear of the book						
Ductile iron						
50 mm	8.17	9.23	0.10	2.47	nr	**11.70**
70 mm	8.17	9.23	0.10	2.47	nr	**11.70**
100 mm	9.44	10.66	0.15	3.70	nr	**14.36**
150 mm	17.49	19.76	0.20	4.95	nr	**24.71**
200 mm	64.86	73.26	0.25	6.18	nr	**79.44**
Mild steel; vertical						
125 mm	15.91	17.97	0.15	3.70	nr	**21.67**
Mild steel; stand off						
250 mm	34.66	39.15	0.25	6.18	nr	**45.33**
300 mm	38.18	43.12	0.25	6.18	nr	**49.30**
Stack support; rubber seal						
70 mm	32.42	36.62	0.55	13.59	nr	**50.21**
100 mm	36.06	40.73	0.65	16.07	nr	**56.80**
125 mm	39.99	45.17	0.74	18.29	nr	**63.46**
150 mm	57.08	64.46	0.86	21.26	nr	**85.72**
Wall spacer plate; cast iron (eared sockets)						
50 mm	6.95	7.85	0.10	2.47	nr	**10.32**
70 mm	6.95	7.85	0.10	2.47	nr	**10.32**
100 mm	6.95	7.85	0.10	2.47	nr	**10.32**
Extra over fittings EPDM rubber jointed cast iron pipework BS EN 877, with jointing couplings measured separately						
Coupling						
50 mm	8.73	9.86	0.10	2.47	nr	**12.33**
70 mm	9.62	10.86	0.10	2.47	nr	**13.33**
100 mm	12.52	14.14	0.10	2.47	nr	**16.61**
125 mm	15.55	17.56	0.15	3.70	nr	**21.26**
150 mm	25.09	28.34	0.30	7.41	nr	**35.75**
200 mm	56.12	63.38	0.35	8.65	nr	**72.03**
250 mm	80.38	90.78	0.40	9.88	nr	**100.66**
300 mm	93.01	105.05	0.50	12.35	nr	**117.40**
Plain socket						
50 mm	21.94	24.78	0.10	2.47	nr	**27.25**
70 mm	21.94	24.78	0.10	2.47	nr	**27.25**
100 mm	25.20	28.46	0.10	2.47	nr	**30.93**
150 mm	42.51	48.01	0.10	2.47	nr	**50.48**

R: DISPOSAL SYSTEMS

Item	Net Price £	Material £	Labour hours	Labour £	Unit	**Total rate £**
R11: ABOVE GROUND DRAINAGE – cont						
Extra over fittings EPDM rubber jointed cast iron pipework BS EN 877, with jointing couplings measured separately – cont						
Eared socket						
50 mm	22.60	25.52	0.25	6.18	nr	**31.70**
70 mm	22.60	25.52	0.25	6.18	nr	**31.70**
100 mm	27.39	30.94	0.25	6.18	nr	**37.12**
150 mm	44.91	50.71	0.25	6.18	nr	**56.89**
Slip socket						
50 mm	28.43	32.11	0.25	6.18	nr	**38.29**
70 mm	28.43	32.11	0.25	6.18	nr	**38.29**
100 mm	33.20	37.50	0.25	6.18	nr	**43.68**
150 mm	49.98	56.45	0.25	6.18	nr	**62.63**
Stack support pipe						
70 mm	20.98	23.69	0.74	18.29	nr	**41.98**
100 mm	28.82	32.55	0.88	21.74	nr	**54.29**
125 mm	30.82	34.81	1.00	24.71	nr	**59.52**
150 mm	41.76	47.16	1.19	29.41	nr	**76.57**
Access pipe; round door						
50 mm	40.28	45.49	0.27	6.67	nr	**52.16**
70 mm	42.62	48.13	0.30	7.41	nr	**55.54**
100 mm	46.85	52.91	0.32	7.91	nr	**60.82**
150 mm	84.80	95.77	0.71	17.54	nr	**113.31**
Access pipe; square door						
100 mm	90.59	102.32	0.32	7.91	nr	**110.23**
125 mm	94.28	106.48	0.67	16.55	nr	**123.03**
150 mm	141.84	160.19	0.71	17.54	nr	**177.73**
200 mm	281.75	318.21	1.21	29.90	nr	**348.11**
250 mm	443.32	500.69	1.31	32.36	nr	**533.05**
300 mm	553.01	624.57	1.41	34.84	nr	**659.41**
Taper reducer						
70 mm	23.26	26.27	0.30	7.41	nr	**33.68**
100 mm	27.35	30.89	0.32	7.91	nr	**38.80**
125 mm	27.49	31.05	0.64	15.81	nr	**46.86**
150 mm	52.50	59.29	0.67	16.55	nr	**75.84**
200 mm	85.26	96.29	1.15	28.42	nr	**124.71**
250 mm	176.10	198.89	1.25	30.89	nr	**229.78**
300 mm	242.09	273.41	1.37	33.86	nr	**307.27**

R: DISPOSAL SYSTEMS

Item	Net Price £	Material £	Labour hours	Labour £	Unit	**Total rate £**
Blank cap						
50 mm	6.09	6.87	0.24	5.93	nr	**12.80**
70 mm	6.43	7.26	0.26	6.43	nr	**13.69**
100 mm	7.47	8.44	0.32	7.91	nr	**16.35**
125 mm	10.57	11.93	0.35	8.65	nr	**20.58**
150 mm	10.80	12.20	0.40	9.88	nr	**22.08**
200 mm	47.70	53.88	0.60	14.83	nr	**68.71**
250 mm	97.50	110.12	0.65	16.07	nr	**126.19**
300 mm	138.94	156.92	0.72	17.79	nr	**174.71**
Blank cap; 50 mm screwed tapping						
70 mm	14.33	16.19	0.26	6.43	nr	**22.62**
100 mm	15.47	17.47	0.32	7.91	nr	**25.38**
150 mm	18.60	21.00	0.40	9.88	nr	**30.88**
Universal connector; EPDM rubber						
50 × 56/48/40 mm	12.43	14.04	0.30	7.41	nr	**21.45**
Blank end; push fit						
100 × 38/32 mm	13.88	15.68	0.32	7.91	nr	**23.59**
Boss pipe; 2" BSPT socket						
50 mm	34.06	38.46	0.27	6.67	nr	**45.13**
75 mm	34.06	38.46	0.30	7.41	nr	**45.87**
100 mm	41.63	47.01	0.32	7.91	nr	**54.92**
150 mm	67.88	76.66	0.71	17.54	nr	**94.20**
Boss pipe; 2 × 2" BSPT socket; opposed						
100 mm	53.78	60.74	0.32	7.91	nr	**68.65**
Boss pipe; 2 × 2" BSPT socket; 90°						
100 mm	53.78	60.74	0.32	7.91	nr	**68.65**
Manifold connector						
100 mm	83.04	93.79	0.64	15.81	nr	**109.60**
150 mm	115.69	130.65	1.00	24.71	nr	**155.36**
Bend; short radius						
50 mm	15.12	17.07	0.27	6.67	nr	**23.74**
70 mm	17.02	19.23	0.30	7.41	nr	**26.64**
100 mm	20.15	22.76	0.32	7.91	nr	**30.67**
125 mm	35.73	40.35	0.62	15.33	nr	**55.68**
150 mm	36.18	40.86	0.67	16.55	nr	**57.41**
200 mm; 45°	107.72	121.66	1.21	29.90	nr	**151.56**
250 mm; 45°	210.19	237.38	1.31	32.36	nr	**269.74**
300 mm; 45°	295.68	333.94	1.43	35.34	nr	**369.28**

R: DISPOSAL SYSTEMS

Item	Net Price £	Material £	Labour hours	Labour £	Unit	**Total rate £**
R11: ABOVE GROUND DRAINAGE – cont						
Extra over fittings EPDM rubber jointed cast iron pipework BS EN 877, with jointing couplings measured separately – cont						
Access bend; short radius						
70 mm	32.98	37.25	0.30	7.41	nr	**44.66**
100 mm	48.18	54.42	0.32	7.91	nr	**62.33**
150 mm	84.78	95.75	0.67	16.55	nr	**112.30**
Bend; long radius bend						
100 mm; 88°	51.21	57.83	0.32	7.91	nr	**65.74**
100 mm; 22°	37.76	42.65	0.32	7.91	nr	**50.56**
150 mm; 88°	146.79	165.78	0.67	16.55	nr	**182.33**
Access bend; long radius						
100 mm	62.33	70.39	0.32	7.91	nr	**78.30**
150 mm	151.42	171.00	0.32	7.91	nr	**178.91**
Bend; long tail						
100 mm	35.30	39.87	0.32	7.91	nr	**47.78**
Bend; long tail double						
70 mm	54.71	61.79	0.30	7.41	nr	**69.20**
100 mm	60.35	68.16	0.32	7.91	nr	**76.07**
Offset; 75 mm projection						
100 mm	30.97	34.98	0.32	7.91	nr	**42.89**
Offset; 130 mm projection						
50 mm	25.62	28.94	0.27	6.67	nr	**35.61**
70 mm	38.87	43.90	0.30	7.41	nr	**51.31**
100 mm	51.05	57.66	0.32	7.91	nr	**65.57**
125 mm	64.70	73.07	0.67	16.55	nr	**89.62**
Branch; equal and unequal						
50 mm	24.25	27.39	0.37	9.15	nr	**36.54**
70 mm	25.61	28.93	0.40	9.88	nr	**38.81**
100 mm	35.13	39.68	0.42	10.38	nr	**50.06**
125 mm	70.41	79.52	0.76	18.78	nr	**98.30**
150 mm	76.61	86.53	0.97	23.97	nr	**110.50**
200 mm	196.52	221.95	1.51	37.31	nr	**259.26**
250 mm	251.15	283.65	1.63	40.27	nr	**323.92**
300 mm	414.42	468.05	1.77	43.74	nr	**511.79**

R: DISPOSAL SYSTEMS

Item	Net Price £	Material £	Labour hours	Labour £	Unit	**Total rate £**
Branch; radius; equal and unequal						
70 mm	31.20	35.24	0.40	9.88	nr	**45.12**
100 mm	41.06	46.37	0.42	10.38	nr	**56.75**
150 mm	88.82	100.32	1.37	33.86	nr	**134.18**
200 mm	249.94	282.29	1.51	37.31	nr	**319.60**
Branch; long tail						
100 mm	114.57	129.40	0.52	12.85	nr	**142.25**
Access branch; radius; equal and unequal						
70 mm	45.74	51.66	0.40	9.88	nr	**61.54**
100 mm	62.01	70.03	0.42	10.38	nr	**80.41**
150 mm	148.06	167.22	0.97	23.97	nr	**191.19**
Double branch; equal and unequal						
100 mm	34.59	39.07	0.52	12.85	nr	**51.92**
100 mm; 70°	51.60	58.27	0.52	12.85	nr	**71.12**
150 mm	148.06	167.22	1.37	33.86	nr	**201.08**
200 mm	253.21	285.98	1.51	37.31	nr	**323.29**
Double branch; radius; equal and unequal						
100 mm	44.61	50.38	0.52	12.85	nr	**63.23**
150 mm	182.63	206.26	1.37	33.86	nr	**240.12**
Corner branch						
100 mm	90.24	101.92	0.52	12.85	nr	**114.77**
150 mm	99.77	112.68	0.52	12.85	nr	**125.53**
Corner branch; long tail						
100 mm	134.63	152.05	0.52	12.85	nr	**164.90**
Roof vent connector; asphalt						
100 mm	56.77	64.12	0.32	7.91	nr	**72.03**
Movement connector						
100 mm	72.12	81.45	0.32	7.91	nr	**89.36**
150 mm	133.49	150.77	0.67	16.55	nr	**167.32**
Expansion plugs						
70 mm	16.05	18.13	0.32	7.91	nr	**26.04**
100 mm	19.99	22.57	0.39	9.63	nr	**32.20**
150 mm	35.89	40.53	0.55	13.59	nr	**54.12**
P trap						
100 mm dia.	37.40	42.24	0.32	7.91	nr	**50.15**
P trap with access						
50 mm	57.11	64.50	0.27	6.67	nr	**71.17**
70 mm	57.11	64.50	0.30	7.41	nr	**71.91**
100 mm	61.86	69.86	0.32	7.91	nr	**77.77**
150 mm	110.59	124.90	0.67	16.55	nr	**141.45**

R: DISPOSAL SYSTEMS

Item	Net Price £	Material £	Labour hours	Labour £	Unit	Total rate £
R11: ABOVE GROUND DRAINAGE – cont						
Extra over fittings EPDM rubber jointed cast iron pipework BS EN 877, with jointing couplings measured separately – cont						
Branch trap						
100 mm	135.54	153.08	0.42	10.38	nr	**163.46**
Stench trap						
100 mm	263.60	297.70	0.42	10.38	nr	**308.08**
Balcony gully inlet						
100 mm	152.25	171.95	1.00	24.71	nr	**196.66**
Roof outlet						
Flat grate						
70 mm	93.99	106.15	1.00	24.71	nr	**130.86**
100 mm	132.23	149.34	1.00	24.71	nr	**174.05**
Dome grate						
70 mm	93.99	106.15	1.00	24.71	nr	**130.86**
100 mm	148.31	167.50	1.00	24.71	nr	**192.21**
Top hat						
100 mm	200.79	226.77	1.00	24.71	nr	**251.48**
Floor drains; for cast iron pipework BS 416 and BS EN877						
Adjustable clamp plate body						
100 mm; 165 mm nickel bronze grate and frame	89.51	101.09	0.50	12.35	nr	**113.44**
100 mm; 165 mm nickel bronze rodding eye	104.16	117.64	0.50	12.35	nr	**129.99**
100 mm; 150 × 150 mm nickel bronze grate and frame	100.43	113.42	0.50	12.35	nr	**125.77**
100 mm; 150 × 150 nickel bronze rodding eye	104.16	117.64	0.50	12.35	nr	**129.99**
Deck plate body						
100 mm; 165 mm nickel bronze grate and frame	89.51	101.09	0.50	12.35	nr	**113.44**
100 mm; 165 mm nickel bronze rodding eye	104.16	117.64	0.50	12.35	nr	**129.99**
100 mm; 150 × 150 mm nickel bronze grate and frame	100.43	113.42	0.50	12.35	nr	**125.77**

R: DISPOSAL SYSTEMS

Item	Net Price £	Material £	Labour hours	Labour £	Unit	**Total rate £**
100 mm; 150 × 150 mm nickel bronze rodding eye	104.16	117.64	0.50	12.35	nr	**129.99**
Extra for						
100 mm; Srewed extension piece	34.91	39.42	0.30	7.41	nr	**46.83**
100 mm; Grating extension piece; screwed or spigot	26.48	29.91	0.30	7.41	nr	**37.32**
100 mm; Brewary trap	963.28	1087.92	2.00	49.42	nr	**1137.34**

S: PIPED SUPPLY SYSTEMS

Item	Net Price £	Material £	Labour hours	Labour £	Unit	Total rate £
S10: COLD WATER						
Y10 – PIPELINES						
COPPER PIPEWORK						
Copper pipe; capillary or compression joints in the running length; EN1057 R250 (TX) formerly BS 2871 Table X						
Fixed vertically or at low level, with brackets measured separately						
12 mm dia.	1.46	1.65	0.39	9.63	m	**11.28**
15 mm dia.	1.65	1.87	0.40	9.88	m	**11.75**
22 mm dia.	3.30	3.72	0.47	11.61	m	**15.33**
28 mm dia.	4.18	4.72	0.51	12.61	m	**17.33**
35 mm dia.	9.90	11.18	0.58	14.33	m	**25.51**
42 mm dia.	12.06	13.62	0.66	16.31	m	**29.93**
54 mm dia.	15.51	17.51	0.72	17.79	m	**35.30**
67 mm dia.	20.28	22.91	0.75	18.53	m	**41.44**
76 mm dia.	28.69	32.41	0.76	18.78	m	**51.19**
108 mm dia.	41.27	46.61	0.78	19.28	m	**65.89**
133 mm dia.	53.55	60.48	1.05	25.94	m	**86.42**
159 mm dia.	84.86	95.84	1.15	28.42	m	**124.26**
Fixed horizontally at high level or suspended, with brackets measured separately						
12 mm dia.	1.46	1.65	0.45	11.13	m	**12.78**
15 mm dia.	1.65	1.87	0.46	11.36	m	**13.23**
22 mm dia.	3.30	3.72	0.54	13.35	m	**17.07**
28 mm dia.	4.18	9.45	0.59	14.58	m	**24.03**
35 mm dia.	9.90	11.18	0.67	16.55	m	**27.73**
42 mm dia.	12.06	13.62	0.76	18.78	m	**32.40**
54 mm dia.	15.51	17.51	0.83	20.51	m	**38.02**
67 mm dia.	20.28	22.91	0.86	21.26	m	**44.17**
76 mm dia.	28.69	32.41	0.87	21.50	m	**53.91**
108 mm dia.	41.27	46.61	0.90	22.24	m	**68.85**
133 mm dia.	53.55	60.48	1.21	29.90	m	**90.38**
159 mm dia.	84.86	95.84	1.32	32.62	m	**128.46**
Copper pipe; capillary or compression joints in the running length; EN1057 R250 (TY) formerly BS 2871 Table Y						
Fixed vertically or at low level with brackets measured separately (Refer to Copper Pipe Table X Section)						
12 mm dia.	1.89	2.13	0.41	10.13	m	**12.26**
15 mm dia.	2.70	3.05	0.43	10.63	m	**13.68**
22 mm dia.	4.75	5.37	0.50	12.35	m	**17.72**
28 mm dia.	6.14	6.94	0.54	13.35	m	**20.29**
35 mm dia.	8.95	10.11	0.62	15.33	m	**25.44**

S: PIPED SUPPLY SYSTEMS

Item	Net Price £	Material £	Labour hours	Labour £	Unit	**Total rate £**
42 mm dia.	10.85	12.25	0.71	17.54	m	**29.79**
54 mm dia.	18.50	20.89	0.78	19.28	m	**40.17**
67 mm dia.	24.56	27.73	0.82	20.26	m	**47.99**
76 mm dia.	35.89	40.53	0.60	14.83	m	**55.36**
108 mm dia.	49.85	56.30	0.88	21.74	m	**78.04**
Copper pipe; capillary or compression joints in the running length; EN1057 R250 (TX) formerly BS 2871 Table X						
Plastic coated gas and cold water service pipe for corrosive environments, fixed vertically or at low level with brackets measured separtely						
15 mm dia. (white)	3.78	4.27	0.59	14.58	m	**18.85**
22 mm dia. (white)	6.68	7.55	0.68	16.81	m	**24.36**
28 mm dia. (white)	6.89	7.78	0.74	18.30	m	**26.08**
FIXINGS						
For copper pipework						
Saddle band						
6 mm dia.	0.06	0.08	0.05	1.26	nr	**1.34**
8 mm dia.	0.06	0.08	0.07	1.69	nr	**1.77**
10 mm dia.	0.08	0.10	0.09	2.11	nr	**2.21**
12 mm dia.	0.08	0.10	0.11	2.64	nr	**2.74**
15 mm dia.	0.08	0.10	0.13	3.20	nr	**3.30**
22 mm dia.	0.14	0.17	0.13	3.20	nr	**3.37**
28 mm dia.	0.09	0.11	0.16	3.96	nr	**4.07**
35 mm dia.	0.27	0.31	0.18	4.45	nr	**4.76**
42 mm dia.	0.57	0.64	0.21	5.18	nr	**5.82**
54 mm dia.	0.76	0.86	0.21	5.18	nr	**6.04**
Single spacing clip						
15 mm dia.	0.16	0.19	0.14	3.46	nr	**3.65**
22 mm dia.	0.16	0.19	0.15	3.70	nr	**3.89**
28 mm dia.	0.39	0.44	0.17	4.20	nr	**4.64**
Two piece spacing clip						
8 mm dia. bottom	0.13	0.15	0.11	2.72	nr	**2.87**
8 mm dia. top	0.13	0.15	0.11	2.72	nr	**2.87**
12 mm dia. bottom	0.13	0.15	0.13	3.20	nr	**3.35**
12 mm dia. top	0.13	0.15	0.13	3.20	nr	**3.35**
15 mm dia. bottom	0.14	0.17	0.13	3.20	nr	**3.37**
15 mm dia. top	0.14	0.17	0.13	3.20	nr	**3.37**
22 mm dia. bottom	0.14	0.17	0.14	3.46	nr	**3.63**
22 mm dia. top	0.16	0.19	0.14	3.46	nr	**3.65**
28 mm dia. bottom	0.14	0.17	0.16	3.96	nr	**4.13**
28 mm dia. top	0.26	0.30	0.16	3.96	nr	**4.26**
35 mm dia. bottom	0.16	0.19	0.21	5.18	nr	**5.37**
35 mm dia. top	0.38	0.43	0.21	5.18	nr	**5.61**

S: PIPED SUPPLY SYSTEMS

Item	Net Price £	Material £	Labour hours	Labour £	Unit	Total rate £
S10: COLD WATER – cont						
For copper pipework – cont						
Single pipe bracket						
15 mm dia.	1.31	1.48	0.14	3.46	nr	**4.94**
22 mm dia.	1.33	1.50	0.14	3.46	nr	**4.96**
28 mm dia.	1.83	2.07	0.17	4.20	nr	**6.27**
Single pipe ring						
15 mm dia.	2.30	2.60	0.26	6.43	nr	**9.03**
22 mm dia.	2.50	2.83	0.26	6.43	nr	**9.26**
28 mm dia.	3.13	3.54	0.31	7.66	nr	**11.20**
35 mm dia.	4.41	4.98	0.32	7.91	nr	**12.89**
42 mm dia.	5.12	5.78	0.32	7.91	nr	**13.69**
54 mm dia.	6.56	7.41	0.34	8.40	nr	**15.81**
67 mm dia.	11.00	12.42	0.35	8.65	nr	**21.07**
76 mm dia.	22.61	25.53	0.42	10.38	nr	**35.91**
108 mm dia.	41.60	46.98	0.42	10.38	nr	**57.36**
Double pipe ring						
15 mm dia.	3.16	3.57	0.26	6.43	nr	**10.00**
22 mm dia.	3.95	4.46	0.26	6.43	nr	**10.89**
28 mm dia.	6.22	7.03	0.31	7.66	nr	**14.69**
35 mm dia.	8.30	9.37	0.32	7.91	nr	**17.28**
42 mm dia.	9.34	10.55	0.32	7.91	nr	**18.46**
54 mm dia.	9.63	10.87	0.34	8.40	nr	**19.27**
67 mm dia.	12.13	13.70	0.35	8.65	nr	**22.35**
76 mm dia.	18.40	20.78	0.42	10.38	nr	**31.16**
108 mm dia.	25.13	28.39	0.42	10.38	nr	**38.77**
Wall bracket						
15 mm dia.	2.61	2.95	0.05	1.23	nr	**4.18**
22 mm dia.	3.42	3.87	0.05	1.23	nr	**5.10**
28 mm dia.	4.09	4.62	0.05	1.23	nr	**5.85**
35 mm dia.	6.14	6.94	0.05	1.23	nr	**8.17**
42 mm dia.	8.10	9.15	0.05	1.23	nr	**10.38**
54 mm dia.	12.64	14.28	0.05	1.23	nr	**15.51**
Hospital bracket						
15 mm dia.	4.49	5.07	0.26	6.43	nr	**11.50**
22 mm dia.	5.31	6.00	0.26	6.43	nr	**12.43**
28 mm dia.	6.22	7.03	0.31	7.66	nr	**14.69**
35 mm dia.	6.59	7.45	0.32	7.91	nr	**15.36**
42 mm dia.	9.43	10.65	0.32	7.91	nr	**18.56**
54 mm dia.	12.62	14.25	0.34	8.40	nr	**22.65**
Screw on backplate, female						
All sizes 15 mm to 54 mm × 10 mm	1.45	1.64	0.10	2.47	nr	**4.11**

S: PIPED SUPPLY SYSTEMS

Item	Net Price £	Material £	Labour hours	Labour £	Unit	**Total rate £**
Screw on backplate, male						
All sizes 15 mm to 54 mm × 10 mm	1.91	2.15	0.10	2.47	nr	**4.62**
Pipe joist clips, single						
15 mm dia.	0.91	1.03	0.08	1.98	nr	**3.01**
22 mm dia.	0.91	1.03	0.08	1.98	nr	**3.01**
Pipe joist clips, double						
15 mm dia.	1.27	1.44	0.08	1.98	nr	**3.42**
22 mm dia.	0.69	0.78	0.08	1.98	nr	**2.76**
Extra over channel sections for fabricated hangers and brackets						
Galvanized steel; including inserts, bolts, nuts, washers; fixed to backgrounds						
41 × 21 mm	5.65	6.39	0.29	7.17	m	**13.56**
41 × 41 mm	6.77	7.65	0.29	7.17	m	**14.82**
Threaded rods; metric thread; including nuts, washers etc.						
10 mm dia. × 600 mm long for ring clips up to 54 mm	1.86	2.10	0.18	4.45	nr	**6.55**
12 mm dia. × 600 mm long for ring clips from 54 mm	2.86	3.23	0.18	4.45	nr	**7.68**
Extra over copper pipes; capillary fittings; BS 864						
Stop end						
15 mm dia.	1.32	1.49	0.13	3.20	nr	**4.69**
22 mm dia.	2.47	2.78	0.14	3.46	nr	**6.24**
28 mm dia.	4.41	4.98	0.17	4.20	nr	**9.18**
35 mm dia.	9.74	11.00	0.19	4.70	nr	**15.70**
42 mm dia.	16.77	18.94	0.22	5.44	nr	**24.38**
54 mm dia.	23.41	26.44	0.23	5.68	nr	**32.12**
Straight coupling; copper to copper						
6 mm dia.	1.39	1.57	0.23	5.68	nr	**7.25**
8 mm dia.	1.42	1.60	0.23	5.68	nr	**7.28**
10 mm dia.	0.72	0.82	0.23	5.68	nr	**6.50**
15 mm dia.	0.16	0.19	0.23	5.68	nr	**5.87**
22 mm dia.	0.42	0.48	0.26	6.43	nr	**6.91**
28 mm dia.	1.22	1.38	0.30	7.41	nr	**8.79**
35 mm dia.	3.98	4.50	0.34	8.40	nr	**12.90**
42 mm dia.	6.65	7.51	0.38	9.38	nr	**16.89**
54 mm dia.	12.26	13.84	0.42	10.38	nr	**24.22**
67 mm dia.	36.40	41.11	0.53	13.10	nr	**54.21**

S: PIPED SUPPLY SYSTEMS

Item	Net Price £	Material £	Labour hours	Labour £	Unit	Total rate £
S10: COLD WATER – cont						
Extra over copper pipes – cont						
Adaptor coupling; imperial to metric						
½" × 15 mm dia.	2.88	3.25	0.27	6.67	nr	**9.92**
¾" × 22 mm dia.	2.52	2.85	0.31	7.66	nr	**10.51**
1" × 28 mm dia.	4.96	5.60	0.36	8.90	nr	**14.50**
1 ¼" × 35 mm dia.	8.29	9.36	0.41	10.13	nr	**19.49**
1 ½" × 42 mm dia.	10.55	11.91	0.46	11.36	nr	**23.27**
Reducing coupling						
15 × 10 mm dia.	3.03	3.42	0.23	5.68	nr	**9.10**
22 × 10 mm dia.	4.45	5.03	0.26	6.43	nr	**11.46**
22 × 15 mm dia.	4.75	5.37	0.27	6.67	nr	**12.04**
28 × 15 mm dia.	5.47	6.18	0.28	6.93	nr	**13.11**
28 × 22 mm dia.	5.51	6.22	0.30	7.41	nr	**13.63**
35 × 28 mm dia.	7.89	8.91	0.34	8.40	nr	**17.31**
42 × 35 mm dia.	11.57	13.07	0.38	9.38	nr	**22.45**
54 × 35 mm dia.	20.31	22.94	0.42	10.38	nr	**33.32**
54 × 42 mm dia.	22.16	25.03	0.42	10.38	nr	**35.41**
Straight female connector						
15 mm × ½" dia.	3.01	3.40	0.27	6.67	nr	**10.07**
22 mm × ¾" dia.	4.36	4.93	0.31	7.66	nr	**12.59**
28 mm × 1" dia.	8.24	9.30	0.36	8.90	nr	**18.20**
35 mm × 1 ¼" dia.	14.24	16.09	0.41	10.13	nr	**26.22**
42 mm × 1 ½" dia.	18.48	20.87	0.46	11.36	nr	**32.23**
54 mm × 2" dia.	29.31	33.10	0.52	12.85	nr	**45.95**
Straight male connector						
15 mm × ½" dia.	2.56	2.89	0.27	6.67	nr	**9.56**
22 mm × ¾" dia.	4.58	5.17	0.31	7.66	nr	**12.83**
28 mm × 1" dia.	7.39	8.34	0.36	8.90	nr	**17.24**
35 mm × 1 ¼" dia.	12.98	14.66	0.41	10.13	nr	**24.79**
42 mm × 1 ½" dia.	16.72	18.88	0.46	11.36	nr	**30.24**
54 mm × 2" dia.	25.38	28.66	0.52	12.85	nr	**41.51**
67 mm × 2 ½" dia.	40.54	45.79	0.63	15.57	nr	**61.36**
Female reducing connector						
15 mm × ¾" dia.	7.48	8.45	0.27	6.67	nr	**15.12**
Male reducing connector						
15 mm × ¾" dia.	6.68	7.55	0.27	6.67	nr	**14.22**
22 mm × 1" dia.	10.18	11.50	0.31	7.66	nr	**19.16**
Flanged connector						
28 mm dia.	47.78	53.96	0.36	8.90	nr	**62.86**
35 mm dia.	60.50	68.33	0.41	10.13	nr	**78.46**
42 mm dia.	72.31	81.67	0.46	11.36	nr	**93.03**

S: PIPED SUPPLY SYSTEMS

Item	Net Price £	Material £	Labour hours	Labour £	Unit	**Total rate £**
54 mm dia.	109.32	123.46	0.52	12.85	nr	**136.31**
67 mm dia.	128.54	145.18	0.61	15.08	nr	**160.26**
Tank connector						
15 mm × ½" dia.	6.40	7.23	0.25	6.18	nr	**13.41**
22 mm × ¾" dia.	9.76	11.03	0.28	6.93	nr	**17.96**
28 mm × 1" dia.	12.83	14.49	0.32	7.91	nr	**22.40**
35 mm × 1 ¼" dia.	16.45	18.57	0.37	9.15	nr	**27.72**
42 mm × 1 ½" dia.	21.56	24.35	0.43	10.63	nr	**34.98**
54 mm × 2" dia.	32.96	37.22	0.46	11.36	nr	**48.58**
Tank connector with long thread						
15 mm × ½" dia.	8.29	9.36	0.30	7.41	nr	**16.77**
22 mm × ¾" dia.	11.80	13.33	0.33	8.15	nr	**21.48**
28 mm × 1" dia.	14.61	16.50	0.39	9.63	nr	**26.13**
Reducer						
15 × 10 mm dia.	1.02	1.15	0.23	5.68	nr	**6.83**
22 × 15 mm dia.	0.71	0.81	0.26	6.43	nr	**7.24**
28 × 15 mm dia.	2.70	3.05	0.28	6.93	nr	**9.98**
28 × 22 mm dia.	2.07	2.34	0.30	7.41	nr	**9.75**
35 × 22 mm dia.	7.66	8.65	0.34	8.40	nr	**17.05**
42 × 22 mm dia.	13.84	15.63	0.36	8.90	nr	**24.53**
42 × 35 mm dia.	10.70	12.09	0.38	9.38	nr	**21.47**
54 × 35 mm dia.	22.46	25.37	0.40	9.88	nr	**35.25**
54 × 42 mm dia.	19.37	21.88	0.42	10.38	nr	**32.26**
67 × 54 mm dia.	26.33	29.73	0.53	13.10	nr	**42.83**
Adaptor; copper to female iron						
15 mm × ½" dia.	5.18	5.85	0.27	6.67	nr	**12.52**
22 mm × ¾" dia.	7.90	8.93	0.31	7.66	nr	**16.59**
28 mm × 1" dia.	11.14	12.58	0.36	8.90	nr	**21.48**
35 mm × 1 ¼" dia.	20.15	22.76	0.41	10.13	nr	**32.89**
42 mm × 1 ½" dia.	25.38	28.66	0.46	11.36	nr	**40.02**
54 mm × 2" dia.	30.54	34.50	0.52	12.85	nr	**47.35**
Adaptor; copper to male iron						
15 mm × ½" dia.	5.29	5.98	0.27	6.67	nr	**12.65**
22 mm × ¾" dia.	6.77	7.65	0.31	7.66	nr	**15.31**
28 mm × 1" dia.	11.30	12.76	0.36	8.90	nr	**21.66**
35 mm × 1 ¼" dia.	16.46	18.59	0.41	10.13	nr	**28.72**
42 mm × 1 ½" dia.	22.74	25.68	0.46	11.36	nr	**37.04**
54 mm × 2" dia.	30.54	34.50	0.52	12.85	nr	**47.35**
Union coupling						
15 mm dia.	7.17	8.10	0.41	10.13	nr	**18.23**
22 mm dia.	11.47	12.95	0.45	11.13	nr	**24.08**
28 mm dia.	16.72	18.88	0.51	12.61	nr	**31.49**

S: PIPED SUPPLY SYSTEMS

Item	Net Price £	Material £	Labour hours	Labour £	Unit	Total rate £
S10: COLD WATER – cont						
Extra over copper pipes – cont						
Union coupling – cont						
35 mm dia.	21.94	24.78	0.64	15.81	nr	**40.59**
42 mm dia.	32.06	36.21	0.68	16.81	nr	**53.02**
54 mm dia.	61.00	68.89	0.78	19.28	nr	**88.17**
67 mm dia.	103.29	116.66	0.96	23.72	nr	**140.38**
Elbow						
15 mm dia.	0.28	0.32	0.23	5.68	nr	**6.00**
22 mm dia.	0.73	0.83	0.26	6.43	nr	**7.26**
28 mm dia.	1.96	2.21	0.31	7.66	nr	**9.87**
35 mm dia.	8.53	9.63	0.35	8.65	nr	**18.28**
42 mm dia.	14.08	15.90	0.41	10.13	nr	**26.03**
54 mm dia.	29.09	32.85	0.44	10.87	nr	**43.72**
67 mm dia.	75.50	85.27	0.54	13.35	nr	**98.62**
Backplate elbow						
15 mm dia.	5.38	6.08	0.51	12.61	nr	**18.69**
22 mm dia.	11.55	13.05	0.54	13.35	nr	**26.40**
Overflow bend						
22 mm dia.	16.27	18.37	0.26	6.43	nr	**24.80**
Return bend						
15 mm dia.	8.06	9.10	0.23	5.68	nr	**14.78**
22 mm dia.	15.83	17.88	0.26	6.43	nr	**24.31**
28 mm dia.	20.23	22.85	0.31	7.66	nr	**30.51**
Obtuse elbow						
15 mm dia.	1.03	1.16	0.23	5.68	nr	**6.84**
22 mm dia.	2.16	2.44	0.26	6.43	nr	**8.87**
28 mm dia.	4.12	4.65	0.31	7.66	nr	**12.31**
35 mm dia.	12.86	14.52	0.36	8.90	nr	**23.42**
42 mm dia.	22.90	25.87	0.41	10.13	nr	**36.00**
54 mm dia.	41.41	46.77	0.44	10.87	nr	**57.64**
67 mm dia.	75.14	84.86	0.54	13.35	nr	**98.21**
Straight tap connector						
15 mm × ½" dia.	1.38	1.56	0.13	3.20	nr	**4.76**
22 mm × ¾" dia.	2.06	2.33	0.14	3.46	nr	**5.79**
Bent tap connector						
15 mm × ½" dia.	2.05	2.32	0.13	3.20	nr	**5.52**
22 mm × ¾" dia.	6.27	7.08	0.14	3.46	nr	**10.54**

S: PIPED SUPPLY SYSTEMS

Item	Net Price £	Material £	Labour hours	Labour £	Unit	**Total rate £**
Bent male union connector						
15 mm × ½" dia.	10.48	11.83	0.41	10.13	nr	**21.96**
22 mm × ¾" dia.	13.62	15.38	0.45	11.13	nr	**26.51**
28 mm × 1" dia.	19.48	22.00	0.51	12.61	nr	**34.61**
35 mm × 1 ¼" dia.	31.79	35.90	0.64	15.81	nr	**51.71**
42 mm × 1 ½" dia.	51.67	58.35	0.68	16.81	nr	**75.16**
54 mm × 2" dia.	81.64	92.21	0.78	19.28	nr	**111.49**
Bent female union connector						
15 mm dia.	10.48	11.83	0.41	10.13	nr	**21.96**
22 mm × ¾" dia.	13.62	15.38	0.45	11.13	nr	**26.51**
28 mm × 1" dia.	19.48	22.00	0.51	12.61	nr	**34.61**
35 mm × 1 ¼" dia.	31.79	35.90	0.64	15.81	nr	**51.71**
42 mm × 1 ½" dia.	51.67	58.35	0.68	16.81	nr	**75.16**
54 mm × 2" dia.	81.64	92.21	0.78	19.28	nr	**111.49**
Straight union adaptor						
15 mm × ¾" dia.	4.48	5.06	0.41	10.13	nr	**15.19**
22 mm × 1" dia.	6.36	7.18	0.45	11.13	nr	**18.31**
28 mm × 1 ¼" dia.	10.28	11.61	0.51	12.61	nr	**24.22**
35 mm × 1 ½" dia.	15.83	17.88	0.64	15.81	nr	**33.69**
42 mm × 2" dia.	19.98	22.56	0.68	16.81	nr	**39.37**
54 mm × 2 ½" dia.	30.86	34.85	0.78	19.28	nr	**54.13**
Straight male union connector						
15 mm × ½" dia.	8.93	10.09	0.41	10.13	nr	**20.22**
22 mm × ¾" dia.	11.57	13.07	0.45	11.13	nr	**24.20**
28 mm × 1" dia.	17.25	19.48	0.51	12.61	nr	**32.09**
35 mm × 1 ¼" dia.	24.85	28.07	0.64	15.81	nr	**43.88**
42 mm × 1 ½" dia.	39.05	44.10	0.68	16.81	nr	**60.91**
54 mm × 2" dia.	56.11	63.37	0.78	19.28	nr	**82.65**
Straight female union connector						
15 mm × ½" dia.	8.93	10.09	0.41	10.13	nr	**20.22**
22 mm × ¾" dia.	11.57	13.07	0.45	11.13	nr	**24.20**
28 mm × 1" dia.	17.25	19.48	0.51	12.61	nr	**32.09**
35 mm × 1 ¼" dia.	24.85	28.07	0.64	15.81	nr	**43.88**
42 mm × 1 ½" dia.	39.05	44.10	0.68	16.81	nr	**60.91**
54 mm × 2" dia.	56.11	63.37	0.78	19.28	nr	**82.65**
Male nipple						
¾ × ½" dia.	2.11	2.39	0.24	5.87	nr	**8.26**
1 × ¾" dia.	2.52	2.85	0.32	7.91	nr	**10.76**
1 ¼ × 1" dia.	2.78	3.14	0.37	9.15	nr	**12.29**
1 ½ × 1 ¼" dia.	10.31	11.65	0.42	10.38	nr	**22.03**
2 × 1 ½" dia.	21.11	23.84	0.46	11.36	nr	**35.20**
2 ½ × 2" dia.	27.50	31.06	0.56	13.83	nr	**44.89**

S: PIPED SUPPLY SYSTEMS

Item	Net Price £	Material £	Labour hours	Labour £	Unit	Total rate £
S10: COLD WATER – cont						
Extra over copper pipes – cont						
Female nipple						
¾ × ½" dia.	3.27	3.69	0.19	4.70	nr	**8.39**
1 × ¾" dia.	5.11	5.77	0.32	7.91	nr	**13.68**
1 ¼ × 1" dia.	6.97	7.87	0.37	9.15	nr	**17.02**
1 ½ × 1 ¼" dia.	10.31	11.65	0.42	10.38	nr	**22.03**
2 × 1 ½" dia.	21.11	23.84	0.46	11.36	nr	**35.20**
2 ½ × 2" dia.	27.50	31.06	0.56	13.83	nr	**44.89**
Equal tee						
10 mm dia.	2.79	3.15	0.25	6.18	nr	**9.33**
15 mm dia.	0.28	0.32	0.36	8.90	nr	**9.22**
22 mm dia.	0.73	0.83	0.39	9.63	nr	**10.46**
28 mm dia.	5.45	6.15	0.43	10.63	nr	**16.78**
35 mm dia.	13.88	15.68	0.57	14.09	nr	**29.77**
42 mm dia.	22.26	25.14	0.60	14.83	nr	**39.97**
54 mm dia.	44.89	50.69	0.65	16.07	nr	**66.76**
67 mm dia.	61.08	68.98	0.78	19.28	nr	**88.26**
Female tee, reducing branch FI						
15 × 15 mm × ¼" dia.	6.71	7.58	0.36	8.90	nr	**16.48**
22 × 22 mm × ½" dia.	8.23	9.29	0.39	9.65	nr	**18.94**
28 × 28 mm × ¾" dia.	16.11	18.20	0.43	10.63	nr	**28.83**
35 × 35 mm × ¾" dia.	23.24	26.24	0.47	11.61	nr	**37.85**
42 × 42 mm × ½" dia.	27.92	31.53	0.60	14.83	nr	**46.36**
Backplate tee						
15 × 15 mm × ½" dia.	12.72	14.36	0.62	15.33	nr	**29.69**
Heater tee						
½ × ½" × 15 mm dia.	11.44	12.92	0.36	8.90	nr	**21.82**
Union heater tee						
½ × ½" × 15 mm dia.	11.44	12.92	0.36	8.90	nr	**21.82**
Sweep tee – equal						
15 mm dia.	9.09	10.26	0.36	8.90	nr	**19.16**
22 mm dia.	11.69	13.20	0.39	9.63	nr	**22.83**
28 mm dia.	19.68	22.23	0.43	10.63	nr	**32.86**
35 mm dia.	27.91	31.52	0.57	14.09	nr	**45.61**
42 mm dia.	41.38	46.74	0.60	14.83	nr	**61.57**
54 mm dia.	45.83	51.76	0.65	16.07	nr	**67.83**
67 mm dia.	62.48	70.56	0.78	19.28	nr	**89.84**

S: PIPED SUPPLY SYSTEMS

Item	Net Price £	Material £	Labour hours	Labour £	Unit	**Total rate £**
Sweep tee – reducing						
22 × 22 × 15 mm dia.	9.80	11.07	0.39	9.63	nr	**20.70**
28 × 28 × 22 mm dia.	16.65	18.81	0.43	10.63	nr	**29.44**
35 × 35 × 22 mm dia.	27.91	31.52	0.57	14.09	nr	**45.61**
Sweep tee – double						
15 mm dia.	10.28	11.61	0.36	8.90	nr	**20.51**
22 mm dia.	13.99	15.80	0.39	9.63	nr	**25.43**
28 mm dia.	21.26	24.01	0.43	10.63	nr	**34.64**
Cross						
15 mm dia.	13.60	15.36	0.48	11.86	nr	**27.22**
22 mm dia.	15.19	17.16	0.53	13.10	nr	**30.26**
28 mm dia.	21.81	24.63	0.61	15.08	nr	**39.71**
Extra over copper pipes; high duty capillary fittings; BS 864						
Stop end						
15 mm dia.	6.37	7.19	0.16	3.96	nr	**11.15**
Straight coupling; copper to copper						
15 mm dia.	2.91	3.28	0.27	6.67	nr	**9.95**
22 mm dia.	4.67	5.27	0.32	7.91	nr	**13.18**
28 mm dia.	6.24	7.05	0.37	9.15	nr	**16.20**
35 mm dia.	11.66	13.17	0.43	10.63	nr	**23.80**
42 mm dia.	12.75	14.40	0.50	12.35	nr	**26.75**
54 mm dia.	18.75	21.18	0.54	13.35	nr	**34.53**
Reducing coupling						
15 × 12 mm dia.	5.49	6.20	0.27	6.67	nr	**12.87**
22 × 15 mm dia.	6.37	7.19	0.32	7.91	nr	**15.10**
28 × 22 mm dia.	8.77	9.90	0.37	9.15	nr	**19.05**
Straight female connector						
15 mm × ½" dia.	7.17	8.10	0.32	7.91	nr	**16.01**
22 mm × ¾" dia.	8.08	9.13	0.36	8.90	nr	**18.03**
28 mm × 1" dia.	11.92	13.46	0.42	10.38	nr	**23.84**
Straight male connector						
15 mm × ½" dia.	6.97	7.87	0.32	7.91	nr	**15.78**
22 mm × ¾" dia.	8.08	9.13	0.36	8.90	nr	**18.03**
28 mm × 1" dia.	11.92	13.46	0.42	10.38	nr	**23.84**
42 mm × 1 ½" dia.	23.25	26.25	0.53	13.10	nr	**39.35**
54 mm × 2' dia.	37.79	42.68	0.62	15.33	nr	**58.01**
Reducer						
15 × 12 mm dia.	3.60	4.07	0.27	6.67	nr	**10.74**
22 × 15 mm dia.	3.52	3.98	0.32	7.91	nr	**11.89**
28 × 22 mm dia.	6.37	7.19	0.37	9.15	nr	**16.34**

S: PIPED SUPPLY SYSTEMS

Item	Net Price £	Material £	Labour hours	Labour £	Unit	Total rate £
S10: COLD WATER – cont						
Extra over copper pipes – cont						
Reducer – cont						
35 × 28 mm dia.	8.08	9.13	0.43	10.63	nr	**19.76**
42 × 35 mm dia.	10.41	11.76	0.50	12.35	nr	**24.11**
54 × 42 mm dia.	16.79	18.96	0.39	9.63	nr	**28.59**
Straight union adaptor						
15 mm × ¾" dia.	5.82	6.57	0.27	6.67	nr	**13.24**
22 mm × 1" dia.	7.88	8.90	0.32	7.91	nr	**16.81**
28 mm × 1 ¼" dia.	10.41	11.76	0.37	9.15	nr	**20.91**
35 mm × 1 ½" dia.	18.87	21.31	0.43	10.63	nr	**31.94**
42 mm × 2" dia.	23.91	27.00	0.50	12.35	nr	**39.35**
Bent union adaptor						
15 mm × ¾" dia.	15.14	17.10	0.27	6.67	nr	**23.77**
22 mm × 1" dia.	20.43	23.07	0.32	7.91	nr	**30.98**
28 mm × 1 ¼" dia.	27.50	31.06	0.37	9.15	nr	**40.21**
Adaptor; male copper to FI						
15 mm × ½" dia.	6.81	7.69	0.27	6.67	nr	**14.36**
22 mm × ¾" dia.	11.69	13.20	0.32	7.91	nr	**21.11**
Union coupling						
15 mm dia.	13.12	14.82	0.54	13.35	nr	**28.17**
22 mm dia.	16.79	18.96	0.60	14.83	nr	**33.79**
28 mm dia.	23.31	26.33	0.68	16.81	nr	**43.14**
35 mm dia.	40.69	45.95	0.83	20.51	nr	**66.46**
42 mm dia.	47.93	54.13	0.89	21.99	nr	**76.12**
Elbow						
15 mm dia.	8.45	9.55	0.27	6.67	nr	**16.22**
22 mm dia.	9.03	10.20	0.32	7.91	nr	**18.11**
28 mm dia.	13.40	15.14	0.37	9.15	nr	**24.29**
35 mm dia.	20.96	23.67	0.43	10.63	nr	**34.30**
42 mm dia.	26.11	29.49	0.50	12.35	nr	**41.84**
54 mm dia.	45.42	51.30	0.52	12.85	nr	**64.15**
Return bend						
28 mm dia.	20.23	22.85	0.37	9.15	nr	**32.00**
35 mm dia.	24.92	28.14	0.43	10.63	nr	**38.77**
Bent male union connector						
15 mm × ½" dia.	19.58	22.11	0.54	13.35	nr	**35.46**
22 mm × ¾" dia.	26.38	29.79	0.60	14.83	nr	**44.62**
28 mm × 1" dia.	47.93	54.13	0.68	16.81	nr	**70.94**

S: PIPED SUPPLY SYSTEMS

Item	Net Price £	Material £	Labour hours	Labour £	Unit	**Total rate £**
Composite flange						
35 mm dia.	3.96	4.47	0.38	9.39	nr	**13.86**
42 mm dia.	5.14	5.80	0.41	10.13	nr	**15.93**
54 mm dia.	6.55	7.39	0.43	10.63	nr	**18.02**
Equal tee						
15 mm dia.	9.70	10.95	0.44	10.87	nr	**21.82**
22 mm dia.	12.20	13.78	0.47	11.61	nr	**25.39**
28 mm dia.	16.08	18.16	0.53	13.10	nr	**31.26**
35 mm dia.	27.50	31.06	0.70	17.30	nr	**48.36**
42 mm dia.	35.01	39.54	0.84	20.76	nr	**60.30**
54 mm dia.	55.13	62.26	0.79	19.52	nr	**81.78**
Reducing tee						
15 × 12 mm dia.	13.30	15.02	0.44	10.87	nr	**25.89**
22 × 15 mm dia.	15.72	17.76	0.47	11.61	nr	**29.37**
28 × 22 mm dia.	22.44	25.35	0.53	13.10	nr	**38.45**
35 × 28 mm dia.	35.59	40.20	0.73	18.03	nr	**58.23**
42 × 28 mm dia.	45.56	51.46	0.84	20.76	nr	**72.22**
54 × 28 mm dia.	71.94	81.25	1.01	24.96	nr	**106.21**
Extra over copper pipes; compression fittings; BS 864						
Stop end						
15 mm dia.	1.49	1.68	0.10	2.47	nr	**4.15**
22 mm dia.	1.77	2.00	0.12	2.97	nr	**4.97**
28 mm dia.	2.03	2.30	0.15	3.70	nr	**6.00**
Straight connector; copper to copper						
15 mm dia.	3.09	3.49	0.18	4.45	nr	**7.94**
22 mm dia.	4.10	4.63	0.21	5.18	nr	**9.81**
28 mm dia.	5.44	6.14	0.24	5.93	nr	**12.07**
Straight connector; copper to imperial copper						
22 mm dia.	5.10	5.76	0.21	5.18	nr	**10.94**
Male coupling; copper to MI (BSP)						
15 mm dia.	0.88	0.99	0.19	4.70	nr	**5.69**
22 mm dia.	1.38	1.56	0.23	5.68	nr	**7.24**
28 mm dia.	3.21	3.62	0.26	6.43	nr	**10.05**
Male coupling with long thread and backnut						
15 mm dia.	5.66	6.40	0.19	4.70	nr	**11.10**
22 mm dia.	7.18	8.11	0.23	5.68	nr	**13.79**
Female coupling; copper to FI (BSP)						
15 mm dia.	1.08	1.22	0.19	4.70	nr	**5.92**
22 mm dia.	1.54	1.74	0.23	5.68	nr	**7.42**
28 mm dia.	4.49	5.07	0.27	6.67	nr	**11.74**

S: PIPED SUPPLY SYSTEMS

Item	Net Price £	Material £	Labour hours	Labour £	Unit	Total rate £
S10: COLD WATER – cont						
Extra over copper pipes – cont						
Elbow						
15 mm dia.	1.16	1.31	0.18	4.45	nr	**5.76**
22 mm dia.	1.97	2.22	0.21	5.18	nr	**7.40**
28 mm dia.	6.64	7.50	0.24	5.93	nr	**13.43**
Male elbow; copper to FI (BSP)						
15 mm × ½" dia.	2.21	2.50	0.19	4.70	nr	**7.20**
22 mm × ¾" dia.	2.87	3.24	0.23	5.68	nr	**8.92**
28 mm × 1" dia.	6.94	7.83	0.27	6.67	nr	**14.50**
Female elbow; copper to FI (BSP)						
15 mm × ½" dia.	3.39	3.83	0.19	4.70	nr	**8.53**
22 mm × ¾" dia.	4.90	5.54	0.23	5.68	nr	**11.22**
28 mm × 1" dia.	8.65	9.77	0.27	6.67	nr	**16.44**
Backplate elbow						
15 mm × ½" dia.	4.90	5.54	0.50	12.35	nr	**17.89**
Tank coupling; long thread						
22 mm dia.	7.18	8.11	0.46	11.36	nr	**19.47**
Tee equal						
15 mm dia.	1.63	1.84	0.28	6.93	nr	**8.77**
22 mm dia.	2.75	3.10	0.30	7.41	nr	**10.51**
28 mm dia.	12.50	14.12	0.34	8.40	nr	**22.52**
Tee reducing						
22 mm dia.	7.07	7.99	0.30	7.41	nr	**15.40**
Backplate tee						
15 mm dia.	17.26	19.49	0.62	15.33	nr	**34.82**
Extra over fittings; silver brazed welded joints						
Reducer						
76 × 67 mm dia.	28.50	32.19	1.40	34.60	nr	**66.79**
108 × 76 mm dia.	59.72	67.44	1.80	44.47	nr	**111.91**
133 × 108 mm dia.	119.09	134.50	2.20	54.37	nr	**188.87**
159 × 133 mm dia.	152.31	172.02	2.60	64.25	nr	**236.27**
90° elbow						
76 mm dia.	68.85	77.75	1.60	39.54	nr	**117.29**
108 mm dia.	126.73	143.13	2.00	49.42	nr	**192.55**
133 mm dia.	267.90	302.56	2.40	59.30	nr	**361.86**
159 mm dia.	332.58	375.61	2.80	69.19	nr	**444.80**

S: PIPED SUPPLY SYSTEMS

Item	Net Price £	Material £	Labour hours	Labour £	Unit	**Total rate £**
45° elbow						
76 mm dia.	64.50	72.85	1.60	39.54	nr	**112.39**
108 mm dia.	101.16	114.25	2.00	49.42	nr	**163.67**
133 mm dia.	258.68	292.15	2.40	59.30	nr	**351.45**
159 mm dia.	372.33	420.51	2.80	69.19	nr	**489.70**
Equal tee						
76 mm dia.	83.84	94.69	2.40	59.30	nr	**153.99**
108 mm dia.	130.22	147.07	3.00	74.13	nr	**221.20**
133 mm dia.	306.70	346.38	3.60	88.96	nr	**435.34**
159 mm dia.	365.78	413.11	4.20	103.80	nr	**516.91**
Extra over copper pipes; dezincification resistant compression fittings; BS 864						
Stop end						
15 mm dia.	2.13	2.41	0.10	2.47	nr	**4.88**
22 mm dia.	3.09	3.49	0.13	3.20	nr	**6.69**
28 mm dia.	6.61	7.47	0.15	3.70	nr	**11.17**
35 mm dia.	10.37	11.71	0.18	4.45	nr	**16.16**
42 mm dia.	17.26	19.49	0.20	4.95	nr	**24.44**
Straight coupling; copper to copper						
15 mm dia.	1.71	1.93	0.18	4.45	nr	**6.38**
22 mm dia.	2.79	3.15	0.21	5.18	nr	**8.33**
28 mm dia.	6.32	7.14	0.24	5.93	nr	**13.07**
35 mm dia.	13.38	15.12	0.29	7.17	nr	**22.29**
42 mm dia.	17.59	19.87	0.33	8.15	nr	**28.02**
54 mm dia.	26.31	29.71	0.38	9.38	nr	**39.09**
Straight swivel connector; copper to imperial copper						
22 mm dia.	6.29	7.10	0.20	4.95	nr	**12.05**
Male coupling; copper to MI (BSP)						
15 mm × ½" dia.	1.52	1.71	0.19	4.70	nr	**6.41**
22 mm × ¾" dia.	2.31	2.61	0.23	5.68	nr	**8.29**
28 mm × 1" dia.	4.48	5.06	0.26	6.43	nr	**11.49**
35 mm × 1 ¼" dia.	10.17	11.49	0.32	7.91	nr	**19.40**
42 mm × 1 ½" dia.	15.25	17.23	0.37	9.15	nr	**26.38**
54 mm × 2" dia.	22.53	25.45	0.57	14.09	nr	**39.54**
Male coupling with long thread and backnuts						
22 mm dia.	8.65	9.77	0.23	5.68	nr	**15.45**
28 mm dia.	9.58	10.82	0.24	5.93	nr	**16.75**
Female coupling; copper to FI (BSP)						
15 mm × ½" dia.	1.83	2.07	0.19	4.70	nr	**6.77**
22 mm × ¾" dia.	2.68	3.03	0.23	5.68	nr	**8.71**
28 mm × 1" dia.	5.80	6.55	0.27	6.67	nr	**13.22**

S: PIPED SUPPLY SYSTEMS

Item	Net Price £	Material £	Labour hours	Labour £	Unit	Total rate £
S10: COLD WATER – cont						
Extra over copper pipes – cont						
Female coupling – cont						
35 mm × 1 ¼" dia.	12.22	13.80	0.32	7.91	nr	**21.71**
42 mm × 1 ½" dia.	16.42	18.54	0.37	9.15	nr	**27.69**
54 mm × 2" dia.	24.09	27.20	0.42	10.38	nr	**37.58**
Elbow						
15 mm dia.	2.06	2.33	0.18	4.45	nr	**6.78**
22 mm dia.	3.28	3.70	0.21	5.18	nr	**8.88**
28 mm dia.	8.16	9.22	0.24	5.93	nr	**15.15**
35 mm dia.	18.06	20.40	0.29	7.17	nr	**27.57**
42 mm dia.	24.46	27.62	0.33	8.15	nr	**35.77**
54 mm dia.	42.09	47.53	0.38	9.38	nr	**56.91**
Male elbow; copper to MI (BSP)						
15 mm × ½" dia.	3.57	4.03	0.19	4.70	nr	**8.73**
22 mm × ¾" dia.	3.99	4.51	0.23	5.68	nr	**10.19**
28 mm × 1" dia.	7.48	8.45	0.27	6.67	nr	**15.12**
Female elbow; copper to FI (BSP)						
15 mm × ½" dia.	3.82	4.31	0.19	4.70	nr	**9.01**
22 mm × ¾" dia.	5.50	6.21	0.23	5.68	nr	**11.89**
28 mm × 1" dia.	9.13	10.31	0.27	6.67	nr	**16.98**
Backplate elbow						
15 mm × ½" dia.	5.56	6.28	0.50	12.35	nr	**18.63**
Straight tap connector						
15 mm dia.	3.16	3.57	0.13	3.20	nr	**6.77**
22 mm dia.	6.85	7.73	0.15	3.70	nr	**11.43**
Tank coupling						
15 mm dia.	4.35	4.92	0.19	4.70	nr	**9.62**
22 mm dia.	4.81	5.44	0.23	5.68	nr	**11.12**
28 mm dia.	10.16	11.48	0.27	6.67	nr	**18.15**
35 mm dia.	17.94	20.26	0.32	7.91	nr	**28.17**
42 mm dia.	29.17	32.95	0.37	9.15	nr	**42.10**
54 mm dia.	37.57	42.43	0.31	7.66	nr	**50.09**
Tee equal						
15 mm dia.	2.89	3.26	0.28	6.93	nr	**10.19**
22 mm dia.	4.77	5.39	0.30	7.41	nr	**12.80**
28 mm dia.	13.02	14.71	0.34	8.40	nr	**23.11**
35 mm dia.	23.49	26.53	0.43	10.63	nr	**37.16**
42 mm dia.	36.94	41.72	0.46	11.36	nr	**53.08**
54 mm dia.	59.33	67.00	0.54	13.35	nr	**80.35**

S: PIPED SUPPLY SYSTEMS

Item	Net Price £	Material £	Labour hours	Labour £	Unit	**Total rate £**
Tee reducing						
22 mm dia.	7.62	8.61	0.30	7.41	nr	**16.02**
28 mm dia.	12.57	14.20	0.34	8.40	nr	**22.60**
35 mm dia.	22.95	25.92	0.43	10.63	nr	**36.55**
42 mm dia.	35.49	40.09	0.46	11.36	nr	**51.45**
54 mm dia.	59.33	67.00	0.54	13.35	nr	**80.35**
Extra over copper pipes; bronze one piece brazing flanges; metric, including jointing ring and bolts						
Bronze flange; PN6						
15 mm dia.	26.96	30.45	0.27	6.67	nr	**37.12**
22 mm dia.	32.14	36.30	0.32	7.91	nr	**44.21**
28 mm dia.	35.30	39.87	0.36	8.90	nr	**48.77**
35 mm dia.	42.18	47.63	0.47	11.61	nr	**59.24**
42 mm dia.	46.01	51.96	0.54	13.35	nr	**65.31**
54 mm dia.	53.37	60.27	0.63	15.57	nr	**75.84**
67 mm dia.	62.92	71.06	0.77	19.03	nr	**90.09**
76 mm dia.	75.83	85.64	0.93	22.98	nr	**108.62**
108 mm dia.	100.19	113.16	1.14	28.16	nr	**141.32**
133 mm dia.	137.28	155.04	1.41	34.84	nr	**189.88**
159 mm dia.	198.05	223.68	1.74	42.99	nr	**266.67**
Bronze flange; PN10						
15 mm dia.	30.88	34.87	0.27	6.67	nr	**41.54**
22 mm dia.	32.87	37.13	0.32	7.91	nr	**45.04**
28 mm dia.	36.15	40.83	0.38	9.38	nr	**50.21**
35 mm dia.	43.69	49.35	0.47	11.61	nr	**60.96**
42 mm dia.	48.98	55.31	0.54	13.35	nr	**68.66**
54 mm dia.	53.60	60.54	0.63	15.57	nr	**76.11**
67 mm dia.	62.97	71.11	0.77	19.03	nr	**90.14**
76 mm dia.	75.83	85.64	0.93	22.98	nr	**108.62**
108 mm dia.	95.84	108.24	1.14	28.16	nr	**136.40**
133 mm dia.	127.97	144.52	1.41	34.84	nr	**179.36**
159 mm dia.	162.19	183.17	1.74	42.99	nr	**226.16**
Bronze flange; PN16						
15 mm dia.	33.97	38.36	0.27	6.67	nr	**45.03**
22 mm dia.	36.16	40.84	0.32	7.91	nr	**48.75**
28 mm dia.	39.76	44.90	0.38	9.38	nr	**54.28**
35 mm dia.	48.06	54.27	0.47	11.61	nr	**65.88**
42 mm dia.	53.89	60.86	0.54	13.35	nr	**74.21**
54 mm dia.	58.95	66.58	0.63	15.57	nr	**82.15**
67 mm dia.	69.26	78.22	0.77	19.03	nr	**97.25**
76 mm dia.	83.41	94.21	0.93	22.98	nr	**117.19**
108 mm dia.	105.43	119.07	1.14	28.16	nr	**147.23**
133 mm dia.	140.76	158.98	1.41	34.84	nr	**193.82**
159 mm dia.	178.41	201.49	1.74	42.99	nr	**244.48**

S: PIPED SUPPLY SYSTEMS

Item	Net Price £	Material £	Labour hours	Labour £	Unit	Total rate £
S10: COLD WATER – cont						
Extra over copper pipes; bronze blank flanges; metric, including jointing ring and bolts						
Gunmetal blank flange; PN6						
15 mm dia.	22.63	25.56	0.27	6.67	nr	**32.23**
22 mm dia.	28.89	32.63	0.27	6.67	nr	**39.30**
28 mm dia.	29.69	33.53	0.27	6.67	nr	**40.20**
35 mm dia.	48.58	54.87	0.32	7.91	nr	**62.78**
42 mm dia.	65.53	74.01	0.32	7.91	nr	**81.92**
54 mm dia.	72.27	81.62	0.34	8.40	nr	**90.02**
67 mm dia.	88.92	100.43	0.36	8.90	nr	**109.33**
76 mm dia.	114.54	129.36	0.37	9.15	nr	**138.51**
108 mm dia.	181.99	205.54	0.41	10.13	nr	**215.67**
133 mm dia.	214.70	242.48	0.58	14.33	nr	**256.81**
159 mm dia.	268.35	303.07	0.61	15.08	nr	**318.15**
Gunmetal blank flange; PN10						
15 mm dia.	27.38	30.93	0.27	6.67	nr	**37.60**
22 mm dia.	35.42	40.00	0.27	6.67	nr	**46.67**
28 mm dia.	39.22	44.30	0.27	6.67	nr	**50.97**
35 mm dia.	48.58	54.87	0.32	7.91	nr	**62.78**
42 mm dia.	91.05	102.84	0.32	7.91	nr	**110.75**
54 mm dia.	103.81	117.24	0.34	8.40	nr	**125.64**
67 mm dia.	111.36	125.77	0.46	11.36	nr	**137.13**
76 mm dia.	147.99	167.14	0.47	11.61	nr	**178.75**
108 mm dia.	173.68	196.16	0.51	12.61	nr	**208.77**
133 mm dia.	230.71	260.56	0.58	14.33	nr	**274.89**
159 mm dia.	314.22	354.88	0.71	17.54	nr	**372.42**
Gunmetal blank flange; PN16						
15 mm dia.	27.38	30.93	0.27	6.67	nr	**37.60**
22 mm dia.	35.96	40.62	0.27	6.67	nr	**47.29**
28 mm dia.	39.22	44.30	0.27	6.67	nr	**50.97**
35 mm dia.	48.58	54.87	0.32	7.91	nr	**62.78**
42 mm dia.	91.05	102.84	0.32	7.91	nr	**110.75**
54 mm dia.	103.81	117.24	0.34	8.40	nr	**125.64**
67 mm dia.	129.35	146.08	0.46	11.36	nr	**157.44**
76 mm dia.	147.99	167.14	0.47	11.61	nr	**178.75**
108 mm dia.	173.68	196.16	0.51	12.61	nr	**208.77**
133 mm dia.	398.90	450.51	0.58	14.33	nr	**464.84**
159 mm dia.	476.47	538.12	0.71	17.54	nr	**555.66**

S: PIPED SUPPLY SYSTEMS

Item	Net Price £	Material £	Labour hours	Labour £	Unit	**Total rate £**
Extra over copper pipes; bronze screwed flanges; metric, including jointing ring and bolts						
Gunmetal screwed flange; 6 BSP						
15 mm dia.	22.63	25.56	0.35	8.65	nr	**34.21**
22 mm dia.	26.18	29.57	0.47	11.61	nr	**41.18**
28 mm dia.	27.28	30.81	0.52	12.85	nr	**43.66**
35 mm dia.	36.77	41.52	0.62	15.33	nr	**56.85**
42 mm dia.	44.14	49.85	0.70	17.30	nr	**67.15**
54 mm dia.	59.98	67.74	0.84	20.76	nr	**88.50**
67 mm dia.	75.28	85.02	1.03	25.46	nr	**110.48**
76 mm dia.	90.87	102.63	1.22	30.14	nr	**132.77**
108 mm dia.	143.89	162.51	1.41	34.84	nr	**197.35**
133 mm dia.	170.57	192.64	1.75	43.25	nr	**235.89**
159 mm dia.	217.14	245.24	2.21	54.61	nr	**299.85**
Gunmetal screwed flange; 10 BSP						
15 mm dia.	27.41	30.96	0.35	8.65	nr	**39.61**
22 mm dia.	31.92	36.05	0.47	11.61	nr	**47.66**
28 mm dia.	35.26	39.82	0.52	12.85	nr	**52.67**
35 mm dia.	50.24	56.74	0.62	15.33	nr	**72.07**
42 mm dia.	61.34	69.28	0.70	17.30	nr	**86.58**
54 mm dia.	87.23	98.52	0.84	20.76	nr	**119.28**
67 mm dia.	102.12	115.33	1.03	25.46	nr	**140.79**
76 mm dia.	115.28	130.19	1.22	30.14	nr	**160.33**
108 mm dia.	153.04	172.84	1.41	34.84	nr	**207.68**
133 mm dia.	185.46	209.46	1.75	43.25	nr	**252.71**
159 mm dia.	329.14	371.73	2.21	54.61	nr	**426.34**
Gunmetal screwed flange; 16 BSP						
15 mm dia.	21.54	24.33	0.35	8.65	nr	**32.98**
22 mm dia.	27.23	30.75	0.47	11.61	nr	**42.36**
28 mm dia.	31.52	35.60	0.52	12.85	nr	**48.45**
35 mm dia.	48.17	54.41	0.62	15.33	nr	**69.74**
42 mm dia.	57.98	65.48	0.70	17.30	nr	**82.78**
54 mm dia.	74.57	84.22	0.84	20.76	nr	**104.98**
67 mm dia.	107.79	121.74	1.03	25.46	nr	**147.20**
76 mm dia.	123.61	139.61	1.22	30.14	nr	**169.75**
108 mm dia.	144.63	163.34	1.41	34.84	nr	**198.18**
133 mm dia.	234.35	264.67	1.75	43.25	nr	**307.92**
159 mm dia.	289.90	327.41	2.21	54.61	nr	**382.02**
Extra over copper pipes; labour						
Made bend						
15 mm dia.	–	–	0.26	6.43	nr	**6.43**
22 mm dia.	–	–	0.28	6.93	nr	**6.93**
28 mm dia.	–	–	0.31	7.66	nr	**7.66**
35 mm dia.	–	–	0.42	10.38	nr	**10.38**
42 mm dia.	–	–	0.51	12.61	nr	**12.61**
54 mm dia.	–	–	0.58	14.33	nr	**14.33**
67 mm dia.	–	–	0.69	17.05	nr	**17.05**
76 mm dia.	–	–	0.80	19.77	nr	**19.77**

S: PIPED SUPPLY SYSTEMS

Item	Net Price £	Material £	Labour hours	Labour £	Unit	Total rate £
S10: COLD WATER – cont						
Extra over copper pipes – cont						
Bronze butt weld						
15 mm dia.	–	–	0.25	6.18	nr	**6.18**
22 mm dia.	–	–	0.31	7.66	nr	**7.66**
28 mm dia.	–	–	0.37	9.15	nr	**9.15**
35 mm dia.	–	–	0.49	12.11	nr	**12.11**
42 mm dia.	–	–	0.58	14.33	nr	**14.33**
54 mm dia.	–	–	0.72	17.79	nr	**17.79**
67 mm dia.	–	–	0.88	21.74	nr	**21.74**
76 mm dia.	–	–	1.08	26.68	nr	**26.68**
108 mm dia.	–	–	1.37	33.86	nr	**33.86**
133 mm dia.	–	–	1.73	42.75	nr	**42.75**
159 mm dia.	–	–	2.03	50.16	nr	**50.16**
PRESS FIT (copper fittings)						
Mechanical press fit joints; butyl rubber O ring						
Coupler						
15 mm dia.	0.97	1.09	0.36	8.90	nr	**9.99**
22 mm dia.	1.52	1.71	0.36	8.90	nr	**10.61**
28 mm dia.	3.12	3.52	0.44	10.87	nr	**14.39**
35 mm dia.	3.93	4.44	0.44	10.87	nr	**15.31**
42 mm dia.	7.08	8.00	0.52	12.85	nr	**20.85**
54 mm dia.	9.04	10.21	0.60	14.83	nr	**25.04**
Stop end						
22 mm dia.	2.45	2.76	0.18	4.45	nr	**7.21**
28 mm dia.	3.86	4.36	0.22	5.44	nr	**9.80**
35 mm dia.	6.62	7.48	0.22	5.44	nr	**12.92**
42 mm dia.	9.96	11.25	0.26	6.43	nr	**17.68**
54 mm dia.	11.99	13.55	0.30	7.41	nr	**20.96**
Reducer						
22 × 15 mm dia.	1.12	1.26	0.36	8.90	nr	**10.16**
28 × 15 mm dia.	2.98	3.37	0.40	9.88	nr	**13.25**
28 × 22 mm dia.	3.08	3.48	0.40	9.88	nr	**13.36**
35 × 22 mm dia.	3.69	4.17	0.40	9.88	nr	**14.05**
35 × 28 mm dia.	4.09	4.62	0.44	10.87	nr	**15.49**
42 × 22 mm dia.	6.39	7.22	0.44	10.87	nr	**18.09**
42 × 28 mm dia.	6.08	6.86	0.48	11.86	nr	**18.72**
42 × 35 mm dia.	6.08	6.86	0.48	11.86	nr	**18.72**
54 × 35 mm dia.	8.24	9.30	0.52	12.85	nr	**22.15**
54 × 42 mm dia.	8.24	9.30	0.56	13.83	nr	**23.13**

S: PIPED SUPPLY SYSTEMS

Item	Net Price £	Material £	Labour hours	Labour £	Unit	**Total rate £**
90° elbow						
15 mm dia.	1.04	1.17	0.36	8.90	nr	**10.07**
22 mm dia.	1.76	1.99	0.36	8.90	nr	**10.89**
28 mm dia.	3.78	4.27	0.44	10.87	nr	**15.14**
35 mm dia.	7.79	8.80	0.44	10.87	nr	**19.67**
42 mm dia.	14.63	16.52	0.52	12.85	nr	**29.37**
54 mm dia.	20.30	22.93	0.60	14.83	nr	**37.76**
45° elbow						
15 mm dia.	1.29	1.46	0.36	8.90	nr	**10.36**
22 mm dia.	1.78	2.01	0.36	8.90	nr	**10.91**
28 mm dia.	5.35	6.04	0.44	10.87	nr	**16.91**
35 mm dia.	7.62	8.61	0.44	10.87	nr	**19.48**
42 mm dia.	12.69	14.33	0.52	12.85	nr	**27.18**
54 mm dia.	18.05	20.39	0.60	14.83	nr	**35.22**
Equal tee						
15 mm dia.	1.67	1.89	0.54	13.35	nr	**15.24**
22 mm dia.	3.03	3.42	0.54	13.35	nr	**16.77**
28 mm dia.	5.45	6.15	0.66	16.31	nr	**22.46**
35 mm dia.	9.41	10.63	0.66	16.31	nr	**26.94**
42 mm dia.	18.60	21.00	0.78	19.28	nr	**40.28**
54 mm dia.	23.21	26.21	0.90	22.24	nr	**48.45**
Reducing tee						
22 × 15 mm dia.	2.47	2.78	0.54	13.35	nr	**16.13**
28 × 15 mm dia.	4.79	5.41	0.62	15.33	nr	**20.74**
28 × 22 mm dia.	6.44	7.27	0.62	15.33	nr	**22.60**
35 × 22 mm dia.	8.38	9.47	0.62	15.33	nr	**24.80**
35 × 28 mm dia.	9.34	10.55	0.62	15.33	nr	**25.88**
42 × 28 mm dia.	16.88	19.06	0.70	17.30	nr	**36.36**
42 × 35 mm dia.	16.88	19.06	0.70	17.30	nr	**36.36**
54 × 35 mm dia.	28.53	32.22	0.82	20.26	nr	**52.48**
54 × 42 mm dia.	28.53	32.22	0.82	20.26	nr	**52.48**
Male iron connector; BSP thread						
15 mm dia.	3.52	3.98	0.18	4.45	nr	**8.43**
22 mm dia.	5.22	5.90	0.18	4.45	nr	**10.35**
28 mm dia.	6.99	7.89	0.22	5.44	nr	**13.33**
35 mm dia.	12.64	14.28	0.22	5.44	nr	**19.72**
42 mm dia.	16.94	19.14	0.26	6.43	nr	**25.57**
54 mm dia.	32.69	36.92	0.30	7.41	nr	**44.33**
90° elbow; male iron BSP thread						
15 mm dia.	5.70	6.44	0.36	8.90	nr	**15.34**
22 mm dia.	8.93	10.09	0.36	8.90	nr	**18.99**
28 mm dia.	13.67	15.44	0.44	10.87	nr	**26.31**
35 mm dia.	17.78	20.08	0.44	10.87	nr	**30.95**
42 mm dia.	23.19	26.19	0.52	12.85	nr	**39.04**
54 mm dia.	33.88	38.26	0.60	14.83	nr	**53.09**

S: PIPED SUPPLY SYSTEMS

Item	Net Price £	Material £	Labour hours	Labour £	Unit	Total rate £
S10: COLD WATER – cont						
Mechanical press fit joints – cont						
Female iron connector; BSP thread						
15 mm dia.	4.02	4.54	0.18	4.45	nr	**8.99**
22 mm dia.	5.31	6.00	0.18	4.45	nr	**10.45**
28 mm dia.	7.16	8.09	0.22	5.44	nr	**13.53**
35 mm dia.	13.99	15.80	0.22	5.44	nr	**21.24**
42 mm dia.	20.00	22.59	0.26	6.43	nr	**29.02**
54 mm dia.	34.29	38.73	0.30	7.41	nr	**46.14**
90° elbow; female iron BSP thread						
15 mm dia.	4.81	5.44	0.36	8.90	nr	**14.34**
22 mm dia.	7.08	8.00	0.36	8.90	nr	**16.90**
28 mm dia.	11.69	13.20	0.44	10.87	nr	**24.07**
35 mm dia.	15.12	17.07	0.44	10.87	nr	**27.94**
42 mm dia.	20.56	23.23	0.52	12.85	nr	**36.08**
54 mm dia.	30.25	34.16	0.60	14.83	nr	**48.99**
STAINLESS STEEL PIPEWORK						
Stainless steel pipes; capillary or compression joints; BS 4127, vertical or at low level, with brackets measured separately						
Grade 304; satin finish						
15 mm dia.	3.42	3.87	0.41	10.13	m	**14.00**
22 mm dia.	4.80	5.42	0.51	12.61	m	**18.03**
28 mm dia.	6.55	7.39	0.58	14.33	m	**21.72**
35 mm dia.	9.88	11.16	0.65	16.08	m	**27.24**
42 mm dia.	12.54	14.16	0.71	17.55	m	**31.71**
54 mm dia.	17.47	19.73	0.80	19.77	m	**39.50**
Grade 316 satin finish						
15 mm dia.	4.40	4.97	0.61	15.08	m	**20.05**
22 mm dia.	8.23	9.29	0.76	18.79	m	**28.08**
28 mm dia.	9.76	11.03	0.87	21.50	m	**32.53**
35 mm dia.	17.71	20.00	0.98	24.22	m	**44.22**
42 mm dia.	22.91	25.88	1.06	26.20	m	**52.08**
54 mm dia.	26.70	30.15	1.16	28.66	m	**58.81**

S: PIPED SUPPLY SYSTEMS

Item	Net Price £	Material £	Labour hours	Labour £	Unit	Total rate £
FIXINGS						
For stainless steel pipework						
Single pipe ring						
15 mm dia.	11.39	12.86	0.26	6.43	nr	**19.29**
22 mm dia.	13.27	14.98	0.26	6.43	nr	**21.41**
28 mm dia.	13.90	15.70	0.31	7.66	nr	**23.36**
35 mm dia.	15.79	17.83	0.32	7.91	nr	**25.74**
42 mm dia.	17.94	20.26	0.32	7.91	nr	**28.17**
54 mm dia.	20.48	23.13	0.34	8.40	nr	**31.53**
Screw on backplate, female						
All sizes 15 mm to 54 mm dia.	10.27	11.60	0.10	2.47	nr	**14.07**
Screw on backplate, male						
All sizes 15 mm to 54 mm dia.	11.70	13.21	0.10	2.47	nr	**15.68**
Stainless steel threaded rods; metric thread; including nuts, washers etc.						
10 mm dia. × 600 mm long	12.24	13.82	0.18	4.45	nr	**18.27**
Extra over stainless steel pipes; capillary fittings						
Straight coupling						
15 mm dia.	1.26	1.43	0.25	6.18	nr	**7.61**
22 mm dia.	2.03	2.30	0.28	6.93	nr	**9.23**
28 mm dia.	2.69	3.04	0.33	8.15	nr	**11.19**
35 mm dia.	6.19	6.99	0.37	9.15	nr	**16.14**
42 mm dia.	7.14	8.07	0.42	10.38	nr	**18.45**
54 mm dia.	10.74	12.13	0.45	11.13	nr	**23.26**
45° bend						
15 mm dia.	6.76	7.64	0.25	6.18	nr	**13.82**
22 mm dia.	8.88	10.03	0.30	7.33	nr	**17.36**
28 mm dia.	10.90	12.31	0.33	8.15	nr	**20.46**
35 mm dia.	12.92	14.60	0.37	9.15	nr	**23.75**
42 mm dia.	16.63	18.78	0.42	10.38	nr	**29.16**
54 mm dia.	20.55	23.21	0.45	11.13	nr	**34.34**
90° bend						
15 mm dia.	3.49	3.94	0.28	6.93	nr	**10.87**
22 mm dia.	4.71	5.32	0.28	6.93	nr	**12.25**
28 mm dia.	6.65	7.51	0.33	8.15	nr	**15.66**
35 mm dia.	16.21	18.31	0.37	9.15	nr	**27.46**
42 mm dia.	22.32	25.20	0.42	10.38	nr	**35.58**
54 mm dia.	30.25	34.16	0.45	11.13	nr	**45.29**

S: PIPED SUPPLY SYSTEMS

Item	Net Price £	Material £	Labour hours	Labour £	Unit	**Total rate £**
S10: COLD WATER – cont						
STAINLESS STEEL PIPEWORK – cont						
Extra over stainless steel pipes – cont						
Reducer						
22 × 15 mm dia.	8.06	9.10	0.28	6.93	nr	**16.03**
28 × 22 mm dia.	8.98	10.14	0.33	8.15	nr	**18.29**
35 × 28 mm dia.	10.99	12.41	0.37	9.15	nr	**21.56**
42 × 35 mm dia.	11.85	13.38	0.42	10.38	nr	**23.76**
54 × 42 mm dia.	35.13	39.68	0.48	11.88	nr	**51.56**
Tap connector						
15 mm dia.	16.99	19.19	0.13	3.20	nr	**22.39**
22 mm dia.	22.46	25.37	0.14	3.46	nr	**28.83**
28 mm dia.	31.18	35.21	0.17	4.20	nr	**39.41**
Tank connector						
15 mm dia.	21.95	24.79	0.13	3.20	nr	**27.99**
22 mm dia.	32.67	36.89	0.13	3.20	nr	**40.09**
28 mm dia.	40.23	45.43	0.15	3.70	nr	**49.13**
35 mm dia.	58.16	65.69	0.18	4.45	nr	**70.14**
42 mm dia.	76.80	86.74	0.21	5.18	nr	**91.92**
54 mm dia.	116.29	131.34	0.24	5.93	nr	**137.27**
Tee equal						
15 mm dia.	6.26	7.07	0.37	9.15	nr	**16.22**
22 mm dia.	7.79	8.80	0.40	9.88	nr	**18.68**
28 mm dia.	9.41	10.63	0.45	11.13	nr	**21.76**
35 mm dia.	22.61	25.53	0.59	14.60	nr	**40.13**
42 mm dia.	27.90	31.51	0.62	15.33	nr	**46.84**
54 mm dia.	56.34	63.63	0.67	16.55	nr	**80.18**
Unequal tee						
22 × 15 mm dia.	12.69	14.33	0.37	9.15	nr	**23.48**
28 × 15 mm dia.	14.29	16.14	0.45	11.13	nr	**27.27**
28 × 22 mm dia.	14.29	16.14	0.45	11.13	nr	**27.27**
35 × 22 mm dia.	24.98	28.21	0.59	14.60	nr	**42.81**
35 × 28 mm dia.	24.98	28.21	0.59	14.60	nr	**42.81**
42 × 28 mm dia.	30.70	34.67	0.62	15.33	nr	**50.00**
42 × 35 mm dia.	30.70	34.67	0.62	15.33	nr	**50.00**
54 × 35 mm dia.	63.57	71.80	0.67	16.55	nr	**88.35**
54 × 42 mm dia.	63.57	71.80	0.67	16.55	nr	**88.35**

S: PIPED SUPPLY SYSTEMS

Item	Net Price £	Material £	Labour hours	Labour £	Unit	**Total rate £**
Union, conical seat						
15 mm dia.	28.14	31.78	0.25	6.18	nr	**37.96**
22 mm dia.	44.33	50.06	0.28	6.93	nr	**56.99**
28 mm dia.	57.30	64.72	0.33	8.15	nr	**72.87**
35 mm dia.	75.23	84.97	0.37	9.15	nr	**94.12**
42 mm dia.	94.88	107.16	0.42	10.38	nr	**117.54**
54 mm dia.	125.52	141.76	0.45	11.13	nr	**152.89**
Union, flat seat						
15 mm dia.	29.39	33.19	0.25	6.18	nr	**39.37**
22 mm dia.	45.77	51.69	0.28	6.93	nr	**58.62**
28 mm dia.	59.17	66.83	0.33	8.15	nr	**74.98**
35 mm dia.	77.30	87.30	0.37	9.15	nr	**96.45**
42 mm dia.	97.39	110.00	0.42	10.38	nr	**120.38**
54 mm dia.	130.59	147.49	0.45	11.13	nr	**158.62**
Extra over stainless steel pipes; compression fittings						
Straight coupling						
15 mm dia.	24.46	27.62	0.18	4.45	nr	**32.07**
22 mm dia.	46.62	52.65	0.22	5.44	nr	**58.09**
28 mm dia.	62.74	70.86	0.25	6.18	nr	**77.04**
35 mm dia.	96.87	109.41	0.30	7.41	nr	**116.82**
42 mm dia.	113.07	127.70	0.40	9.88	nr	**137.58**
90° bend						
15 mm dia.	30.84	34.83	0.18	4.45	nr	**39.28**
22 mm dia.	61.27	69.20	0.22	5.44	nr	**74.64**
28 mm dia.	83.56	94.37	0.25	6.18	nr	**100.55**
35 mm dia.	169.19	191.09	0.33	8.15	nr	**199.24**
42 mm dia.	247.25	279.25	0.35	8.65	nr	**287.90**
Reducer						
22 × 15 mm dia.	44.39	50.13	0.28	6.93	nr	**57.06**
28 × 22 mm dia.	60.80	68.67	0.28	6.93	nr	**75.60**
35 × 28 mm dia.	88.84	100.34	0.30	7.41	nr	**107.75**
42 × 35 mm dia.	118.20	133.50	0.37	9.15	nr	**142.65**

S: PIPED SUPPLY SYSTEMS

Item	Net Price £	Material £	Labour hours	Labour £	Unit	Total rate £
S10: COLD WATER – cont						
STAINLESS STEEL PIPEWORK – cont						
Extra over stainless steel pipes – cont						
Stud coupling						
15 mm dia.	25.43	28.72	0.42	10.38	nr	**39.10**
22 mm dia.	43.02	48.58	0.25	6.18	nr	**54.76**
28 mm dia.	59.72	67.44	0.25	6.18	nr	**73.62**
35 mm dia.	95.66	108.04	0.37	9.15	nr	**117.19**
42 mm dia.	113.07	127.70	0.42	10.38	nr	**138.08**
Equal tee						
15 mm dia.	43.43	49.05	0.37	9.15	nr	**58.20**
22 mm dia.	89.72	101.33	0.40	9.88	nr	**111.21**
28 mm dia.	122.71	138.59	0.45	11.13	nr	**149.72**
35 mm dia.	244.01	275.58	0.59	14.60	nr	**290.18**
42 mm dia.	338.34	382.12	0.62	15.33	nr	**397.45**
Running tee						
15 mm dia.	53.46	60.37	0.37	9.15	nr	**69.52**
22 mm dia.	96.31	108.77	0.40	9.88	nr	**118.65**
28 mm dia.	163.07	184.17	0.59	14.60	nr	**198.77**
PRESS FIT (stainless steel)						
Press fit jointing system; butyl rubber O ring mechanical joint						
Pipework						
15 mm dia.	4.86	5.49	0.46	11.36	m	**16.85**
22 mm dia.	7.76	8.76	0.48	11.86	m	**20.62**
28 mm dia.	9.57	10.81	0.52	12.85	m	**23.66**
35 mm dia.	14.09	15.91	0.56	13.83	m	**29.74**
42 mm dia.	17.34	19.58	0.58	14.33	m	**33.91**
54 mm dia.	22.07	24.93	0.66	16.31	m	**41.24**

S: PIPED SUPPLY SYSTEMS

Item	Net Price £	Material £	Labour hours	Labour £	Unit	**Total rate £**
FIXINGS						
For stainless steel pipes						
Refer to fixings for stainless steel pipes; capillary or compression joints; BS 4127						
Extra over stainless steel pipes; press fit jointing system;						
Coupling						
15 mm dia.	4.95	5.59	0.36	8.90	nr	**14.49**
22 mm dia.	6.22	7.03	0.36	8.90	nr	**15.93**
28 mm dia.	7.00	7.91	0.44	10.87	nr	**18.78**
35 mm dia.	8.70	9.82	0.44	10.87	nr	**20.69**
42 mm dia.	11.88	13.41	0.52	12.85	nr	**26.26**
54 mm dia.	14.29	16.14	0.60	14.83	nr	**30.97**
Stop end						
22 mm dia.	4.70	5.30	0.18	4.45	nr	**9.75**
28 mm dia.	5.47	6.18	0.22	5.44	nr	**11.62**
35 mm dia.	8.95	10.11	0.22	5.44	nr	**15.55**
42 mm dia.	12.56	14.19	0.26	6.43	nr	**20.62**
54 mm dia.	14.54	16.42	0.30	7.41	nr	**23.83**
Reducer						
22 × 15 mm dia.	5.89	6.65	0.36	8.90	nr	**15.55**
28 × 15 mm dia.	6.67	7.54	0.40	9.88	nr	**17.42**
28 × 22 mm dia.	6.90	7.79	0.40	9.88	nr	**17.67**
35 × 22 mm dia.	8.43	9.52	0.40	9.88	nr	**19.40**
35 × 28 mm dia.	10.44	11.79	0.44	10.87	nr	**22.66**
42 × 35 mm dia.	11.00	12.42	0.48	11.86	nr	**24.28**
54 × 42 mm dia.	12.58	14.21	0.56	13.83	nr	**28.04**
90° bend						
15 mm dia.	7.07	7.99	0.36	8.90	nr	**16.89**
22 mm dia.	9.88	11.16	0.36	8.90	nr	**20.06**
28 mm dia.	12.47	14.09	0.44	10.87	nr	**24.96**
35 mm dia.	19.62	22.15	0.44	10.87	nr	**33.02**
42 mm dia.	32.76	37.00	0.52	12.85	nr	**49.85**
54 mm dia.	45.25	51.10	0.60	14.83	nr	**65.93**
45° bend						
15 mm dia.	9.59	10.83	0.36	8.90	nr	**19.73**
22 mm dia.	11.93	13.47	0.36	8.90	nr	**22.37**
28 mm dia.	13.87	15.67	0.44	10.87	nr	**26.54**
35 mm dia.	16.28	18.39	0.44	10.87	nr	**29.26**
42 mm dia.	26.21	29.60	0.52	12.85	nr	**42.45**
54 mm dia.	34.05	38.45	0.60	14.83	nr	**53.28**

S: PIPED SUPPLY SYSTEMS

Item	Net Price £	Material £	Labour hours	Labour £	Unit	Total rate £
S10: COLD WATER – cont						
STAINLESS STEEL PIPEWORK – cont						
Extra over stainless steel pipes – cont						
Equal tee						
15 mm dia.	11.58	13.08	0.54	13.35	nr	**26.43**
22 mm dia.	14.23	16.08	0.54	13.35	nr	**29.43**
28 mm dia.	16.62	18.77	0.66	16.31	nr	**35.08**
35 mm dia.	21.05	23.78	0.66	16.31	nr	**40.09**
42 mm dia.	29.87	33.73	0.78	19.28	nr	**53.01**
54 mm dia.	35.76	40.39	0.90	22.24	nr	**62.63**
Reducing tee						
22 × 15 mm dia.	12.16	13.73	0.54	13.35	nr	**27.08**
28 × 15 mm dia.	14.74	16.65	0.62	15.33	nr	**31.98**
28 × 22 mm dia.	15.95	18.01	0.62	15.33	nr	**33.34**
35 × 22 mm dia.	18.92	21.37	0.62	15.33	nr	**36.70**
35 × 28 mm dia.	19.74	22.30	0.62	15.33	nr	**37.63**
42 × 28 mm dia.	28.07	31.70	0.70	17.30	nr	**49.00**
42 × 35 mm dia.	28.91	32.65	0.70	17.30	nr	**49.95**
54 × 35 mm dia.	32.66	36.88	0.82	20.26	nr	**57.14**
54 × 42 mm dia.	33.58	37.92	0.82	20.26	nr	**58.18**
FIXINGS						
For stainless steel pipes						
Refer to fixings for stainless steel pipes; capillary or compression joints; BS 4127						
MEDIUM DENSITY POLYETHYLENE – BLUE						
Note: MDPE is sized on Outside Dia. ie OD not ID						
Pipes for water distribution; laid underground; electrofusion joints in the running length; BS 6572						
Coiled service pipe						
20 mm dia.	0.77	0.87	0.37	9.15	m	**10.02**
25 mm dia.	1.00	1.13	0.41	10.13	m	**11.26**
32 mm dia.	1.66	1.88	0.47	11.62	m	**13.50**
50 mm dia.	3.97	4.49	0.53	13.10	m	**17.59**
63 mm dia.	6.23	7.04	0.60	14.83	m	**21.87**

S: PIPED SUPPLY SYSTEMS

Item	Net Price £	Material £	Labour hours	Labour £	Unit	**Total rate £**
Mains service pipe						
90 mm dia.	9.40	10.62	0.90	22.24	m	**32.86**
110 mm dia.	14.07	15.89	1.10	27.18	m	**43.07**
125 mm dia.	17.80	20.10	1.20	29.67	m	**49.77**
160 mm dia.	28.37	32.04	1.48	36.57	m	**68.61**
180 mm dia.	36.96	41.74	1.50	37.10	m	**78.84**
225 mm dia.	56.15	63.41	1.77	43.74	m	**107.15**
250 mm dia.	70.88	80.05	1.75	43.28	m	**123.33**
315 mm dia.	109.45	123.61	1.90	46.95	m	**170.56**
Extra over fittings; MDPE blue; electrofusion joints						
Coupler						
20 mm dia.	6.39	7.22	0.36	8.90	nr	**16.12**
25 mm dia.	6.39	7.22	0.40	9.88	nr	**17.10**
32 mm dia.	6.39	7.22	0.44	10.87	nr	**18.09**
40 mm dia.	9.44	10.66	0.48	11.86	nr	**22.52**
50 mm dia.	9.07	10.24	0.52	12.85	nr	**23.09**
63 mm dia.	11.85	13.38	0.58	14.33	nr	**27.71**
90 mm dia.	17.42	19.68	0.67	16.55	nr	**36.23**
110 mm dia.	28.01	31.63	0.74	18.29	nr	**49.92**
125 mm dia.	31.60	35.69	0.83	20.51	nr	**56.20**
160 mm dia.	50.26	56.76	1.00	24.71	nr	**81.47**
180 mm dia.	59.10	66.75	1.25	30.89	nr	**97.64**
225 mm dia.	94.46	106.68	1.35	33.36	nr	**140.04**
250 mm dia.	138.30	156.19	1.50	37.07	nr	**193.26**
315 mm dia.	228.08	257.59	1.80	44.47	nr	**302.06**
Extra over fittings; MDPE blue; butt fused joints						
Cap						
25 mm dia.	12.13	13.70	0.20	4.95	nr	**18.65**
32 mm dia.	12.13	13.70	0.22	5.44	nr	**19.14**
40 mm dia.	12.86	14.52	0.24	5.93	nr	**20.45**
50 mm dia.	18.75	21.18	0.26	6.43	nr	**27.61**
63 mm dia.	21.51	24.30	0.32	7.91	nr	**32.21**
90 mm dia.	35.47	40.06	0.37	9.15	nr	**49.21**
110 mm dia.	71.42	80.66	0.40	9.88	nr	**90.54**
125 mm dia.	57.02	64.40	0.46	11.36	nr	**75.76**
160 mm dia.	65.19	73.62	0.50	12.35	nr	**85.97**
180 mm dia.	109.04	123.14	0.60	14.83	nr	**137.97**
225 mm dia.	128.79	145.45	0.68	16.81	nr	**162.26**
250 mm dia.	189.04	213.50	0.75	18.53	nr	**232.03**
315 mm dia.	243.75	275.29	0.90	22.24	nr	**297.53**

S: PIPED SUPPLY SYSTEMS

Item	Net Price £	Material £	Labour hours	Labour £	Unit	Total rate £
S10: COLD WATER – cont						
MEDIUM DENSITY POLYETHYLENE – BLUE – cont						
Extra over fittings – cont						
Reducer						
63 × 32 mm dia.	16.38	18.50	0.54	13.35	nr	**31.85**
63 × 50 mm dia.	19.17	21.65	0.60	14.83	nr	**36.48**
90 × 63 mm dia.	24.26	27.40	0.67	16.55	nr	**43.95**
110 × 90 mm dia.	33.42	37.74	0.74	18.29	nr	**56.03**
125 × 90 mm dia.	48.61	54.90	0.83	20.51	nr	**75.41**
125 × 110 mm dia.	53.55	60.48	1.00	24.71	nr	**85.19**
160 × 110 mm dia.	82.28	92.92	1.10	27.18	nr	**120.10**
180 × 125 mm dia.	89.36	100.92	1.25	30.89	nr	**131.81**
225 × 160 mm dia.	145.72	164.58	1.40	34.60	nr	**199.18**
250 × 180 mm dia.	111.96	126.45	1.80	44.47	nr	**170.92**
315 × 250 mm dia.	128.80	145.46	2.40	59.30	nr	**204.76**
Bend; 45°						
50 mm dia.	25.01	28.24	0.50	12.35	nr	**40.59**
63 mm dia.	30.25	34.16	0.58	14.33	nr	**48.49**
90 mm dia.	46.85	52.91	0.67	16.55	nr	**69.46**
110 mm dia.	68.69	77.58	0.74	18.29	nr	**95.87**
125 mm dia.	76.78	86.71	0.83	20.51	nr	**107.22**
160 mm dia.	142.29	160.70	1.00	24.71	nr	**185.41**
180 mm dia.	161.98	182.94	1.25	30.89	nr	**213.83**
225 mm dia.	206.46	233.17	1.40	34.58	nr	**267.75**
250 mm dia.	223.41	252.32	1.80	44.47	nr	**296.79**
315 mm dia.	278.35	314.36	2.40	59.30	nr	**373.66**
Bend; 90°						
50 mm dia.	25.01	28.24	0.50	12.35	nr	**40.59**
63 mm dia.	30.25	34.16	0.58	14.33	nr	**48.49**
90 mm dia.	46.85	52.91	0.67	16.55	nr	**69.46**
110 mm dia.	68.69	77.58	0.74	18.29	nr	**95.87**
125 mm dia.	76.78	86.71	0.83	20.51	nr	**107.22**
160 mm dia.	142.29	160.70	1.00	24.71	nr	**185.41**
180 mm dia.	276.36	312.12	1.25	30.89	nr	**343.01**
225 mm dia.	349.43	394.65	1.40	34.60	nr	**429.25**
250 mm dia.	382.22	431.67	1.80	44.47	nr	**476.14**
315 mm dia.	478.50	540.42	2.40	59.30	nr	**599.72**
Equal tee						
50 mm dia.	27.34	30.88	0.70	17.30	nr	**48.18**
63 mm dia.	29.84	33.70	0.75	18.53	nr	**52.23**
90 mm dia.	54.52	61.58	0.87	21.50	nr	**83.08**

S: PIPED SUPPLY SYSTEMS

Item	Net Price £	Material £	Labour hours	Labour £	Unit	**Total rate £**
110 mm dia.	81.28	91.80	1.00	24.71	nr	**116.51**
125 mm dia.	104.95	118.53	1.08	26.68	nr	**145.21**
160 mm dia.	172.56	194.89	1.35	33.36	nr	**228.25**
180 mm dia.	176.57	199.42	1.63	40.27	nr	**239.69**
225 mm dia.	214.00	241.69	1.90	46.95	nr	**288.64**
250 mm dia.	295.27	333.48	2.70	66.73	nr	**400.21**
315 mm dia.	736.09	831.34	3.60	88.96	nr	**920.30**
Extra over plastic fittings, compression joints						
Straight connector						
20 mm dia.	2.82	3.18	0.38	9.39	nr	**12.57**
25 mm dia.	2.97	3.36	0.45	11.13	nr	**14.49**
32 mm dia.	7.04	7.96	0.50	12.35	nr	**20.31**
50 mm dia.	16.21	18.31	0.68	16.81	nr	**35.12**
63 mm dia.	24.40	27.56	0.85	21.00	nr	**48.56**
Reducing connector						
25 mm dia.	5.81	6.56	0.38	9.39	nr	**15.95**
32 mm dia.	9.38	10.60	0.45	11.13	nr	**21.73**
50 mm dia.	25.99	29.36	0.50	12.35	nr	**41.71**
63 mm dia.	36.27	40.96	0.62	15.33	nr	**56.29**
Straight connector; polyethylene to MI						
20 mm dia.	2.56	2.89	0.31	7.67	nr	**10.56**
25 mm dia.	4.36	4.93	0.35	8.65	nr	**13.58**
32 mm dia.	4.72	5.33	0.40	9.88	nr	**15.21**
50 mm dia.	12.06	13.62	0.55	13.60	nr	**27.22**
63 mm dia.	16.99	19.19	0.65	16.08	nr	**35.27**
Straight connector; polyethylene to FI						
20 mm dia.	3.45	3.90	0.31	7.67	nr	**11.57**
25 mm dia.	3.72	4.20	0.35	8.65	nr	**12.85**
32 mm dia.	4.46	5.04	0.40	9.88	nr	**14.92**
50 mm dia.	14.16	15.99	0.55	13.60	nr	**29.59**
63 mm dia.	19.84	22.41	0.75	18.53	nr	**40.94**
Elbow						
20 mm dia.	3.77	4.25	0.38	9.39	nr	**13.64**
25 mm dia.	5.57	6.29	0.45	11.13	nr	**17.42**
32 mm dia.	8.11	9.16	0.50	12.35	nr	**21.51**
50 mm dia.	18.83	21.27	0.68	16.81	nr	**38.08**
63 mm dia.	25.60	28.92	0.80	19.77	nr	**48.69**
Elbow; polyethylene to MI						
25 mm dia.	4.79	5.41	0.35	8.65	nr	**14.06**

S: PIPED SUPPLY SYSTEMS

Item	Net Price £	Material £	Labour hours	Labour £	Unit	Total rate £
S10: COLD WATER – cont						
MEDIUM DENSITY POLYETHYLENE – BLUE – cont						
Extra over plastic fittings, compression joints – cont						
Elbow; polyethylene to FI						
20 mm dia.	3.42	3.87	0.31	7.67	nr	**11.54**
25 mm dia.	4.66	5.26	0.35	8.65	nr	**13.91**
32 mm dia.	6.98	7.88	0.42	10.38	nr	**18.26**
50 mm dia.	16.57	18.72	0.50	12.35	nr	**31.07**
63 mm dia.	21.73	24.54	0.55	13.60	nr	**38.14**
Tank coupling						
25 mm dia.	7.19	8.12	0.42	10.38	nr	**18.50**
Equal tee						
20 mm dia.	5.07	5.72	0.53	13.10	nr	**18.82**
25 mm dia.	7.92	8.95	0.55	13.60	nr	**22.55**
32 mm dia.	9.93	11.21	0.64	15.82	nr	**27.03**
50 mm dia.	23.16	26.15	0.75	18.53	nr	**44.68**
63 mm dia.	35.89	40.53	0.87	21.50	nr	**62.03**
Equal tee; FI branch						
20 mm dia.	4.91	5.55	0.45	11.13	nr	**16.68**
25 mm dia.	7.83	8.84	0.50	12.35	nr	**21.19**
32 mm dia.	9.53	10.76	0.60	14.83	nr	**25.59**
50 mm dia.	21.94	24.78	0.68	16.81	nr	**41.59**
63 mm dia.	30.80	34.78	0.81	20.02	nr	**54.80**
Equal tee; MI branch						
25 mm dia.	7.68	8.67	0.50	12.35	nr	**21.02**
ABS PIPEWORK						
Pipes; solvent welded joints in the running length, brackets measured separately						
Class C (9 bar pressure)						
1" dia.	5.12	5.78	0.30	7.41	m	**13.19**
1 ¼" dia.	8.60	9.71	0.33	8.15	m	**17.86**
1 ½" dia.	10.94	12.35	0.36	8.90	m	**21.25**
2" dia.	14.73	16.64	0.39	9.63	m	**26.27**
3" dia.	30.34	34.26	0.46	11.36	m	**45.62**
4" dia.	50.02	56.49	0.53	13.10	m	**69.59**
6" dia.	98.90	111.70	0.76	18.78	m	**130.48**
8" dia.	169.84	191.82	0.97	23.97	m	**215.79**

S: PIPED SUPPLY SYSTEMS

Item	Net Price £	Material £	Labour hours	Labour £	Unit	**Total rate £**
Class E (15 bar pressure)						
½" dia.	3.90	4.41	0.24	5.93	m	**10.34**
¾" dia.	6.01	6.78	0.27	6.67	m	**13.45**
1" dia.	7.92	8.95	0.30	7.41	m	**16.36**
1 ¼" dia.	11.83	13.36	0.33	8.15	m	**21.51**
1 ½" dia.	15.59	17.61	0.36	8.90	m	**26.51**
2" dia.	19.54	22.07	0.39	9.63	m	**31.70**
3" dia.	39.28	44.36	0.49	12.11	m	**56.47**
4" dia.	63.18	71.36	0.57	14.09	m	**85.45**
FIXINGS						
Refer to steel pipes; galvanized iron. For minimum fixing dimensions, refer to the Tables and Memoranda at the rear of the book						
Extra over fittings; solvent welded joints						
Cap						
½" dia.	1.68	1.90	0.16	3.96	nr	**5.86**
¾" dia.	1.93	2.18	0.19	4.70	nr	**6.88**
1" dia.	2.22	2.51	0.22	5.44	nr	**7.95**
1 ¼" dia.	3.70	4.18	0.25	6.18	nr	**10.36**
1 ½" dia.	5.71	6.45	0.28	6.93	nr	**13.38**
2" dia.	7.24	8.18	0.31	7.66	nr	**15.84**
3" dia.	21.74	24.55	0.36	8.90	nr	**33.45**
4" dia.	33.24	37.54	0.44	10.87	nr	**48.41**
Elbow 90°						
½" dia.	2.33	2.63	0.29	7.17	nr	**9.80**
¾" dia.	2.79	3.15	0.34	8.40	nr	**11.55**
1" dia.	3.90	4.41	0.40	9.88	nr	**14.29**
1 ¼" dia.	6.60	7.46	0.45	11.13	nr	**18.59**
1 ½" dia.	8.58	9.69	0.51	12.61	nr	**22.30**
2" dia.	13.05	14.74	0.56	13.83	nr	**28.57**
3" dia.	37.49	42.34	0.65	16.07	nr	**58.41**
4" dia.	55.99	63.23	0.80	19.77	nr	**83.00**
6" dia.	225.39	254.55	1.21	29.90	nr	**284.45**
8" dia.	343.99	388.50	1.45	35.83	nr	**424.33**
Elbow 45°						
½" dia.	4.51	5.09	0.29	7.17	nr	**12.26**
¾" dia.	4.58	5.17	0.34	8.40	nr	**13.57**
1" dia.	5.71	6.45	0.40	9.88	nr	**16.33**
1 ¼" dia.	8.36	9.45	0.45	11.13	nr	**20.58**
1 ½" dia.	10.37	11.71	0.51	12.61	nr	**24.32**
2" dia.	14.40	16.26	0.56	13.83	nr	**30.09**
3" dia.	33.90	38.29	0.65	16.07	nr	**54.36**
4" dia.	70.28	79.38	0.80	19.77	nr	**99.15**
6" dia.	145.70	164.56	1.21	29.90	nr	**194.46**
8" dia.	313.53	354.10	1.45	35.83	nr	**389.93**

S: PIPED SUPPLY SYSTEMS

Item	Net Price £	Material £	Labour hours	Labour £	Unit	Total rate £
S10: COLD WATER – cont						
ABS PIPEWORK – cont						
Extra over fittings – cont						
Reducing bush						
¾" × ½" dia.	1.72	1.94	0.42	10.38	nr	**12.32**
1" × ½" dia.	2.22	2.51	0.45	11.13	nr	**13.64**
1" × ¾" dia.	2.22	2.51	0.45	11.13	nr	**13.64**
1 ¼" × 1" dia.	2.99	3.38	0.48	11.86	nr	**15.24**
1 ½" × ¾" dia.	3.90	4.41	0.51	12.61	nr	**17.02**
1 ½" × 1" dia.	3.90	4.41	0.51	12.61	nr	**17.02**
1 ½" × 1 ¼" dia.	3.90	4.41	0.51	12.61	nr	**17.02**
2" × 1" dia.	5.12	5.78	0.56	13.83	nr	**19.61**
2" × 1 ¼" dia.	5.12	5.78	0.56	13.83	nr	**19.61**
2" × 1 ½" dia.	5.12	5.78	0.56	13.83	nr	**19.61**
3" × 1 ½" dia.	14.40	16.26	0.65	16.07	nr	**32.33**
3" × 2" dia.	14.40	16.26	0.65	16.07	nr	**32.33**
4" × 3" dia.	19.85	22.42	0.80	19.77	nr	**42.19**
6" × 4" dia.	61.15	69.06	1.21	29.90	nr	**98.96**
Union						
½" dia.	9.28	10.49	0.34	8.40	nr	**18.89**
¾" dia.	10.00	11.29	0.39	9.63	nr	**20.92**
1" dia.	13.48	15.23	0.43	10.63	nr	**25.86**
1 ¼" dia.	16.54	18.68	0.50	12.35	nr	**31.03**
1 ½" dia.	22.76	25.70	0.57	14.09	nr	**39.79**
2" dia.	29.69	33.53	0.62	15.33	nr	**48.86**
Sockets						
½" dia.	1.72	1.94	0.34	8.40	nr	**10.34**
¾" dia.	1.93	2.18	0.39	9.63	nr	**11.81**
1" dia.	2.22	2.51	0.43	10.63	nr	**13.14**
1 ¼" dia.	3.90	4.41	0.50	12.35	nr	**16.76**
1 ½" dia.	4.70	5.30	0.57	14.09	nr	**19.39**
2" dia.	6.60	7.46	0.62	15.33	nr	**22.79**
3" dia.	26.55	29.99	0.70	17.30	nr	**47.29**
4" dia.	37.69	42.56	0.70	17.30	nr	**59.86**
6" dia.	94.16	106.35	1.26	31.14	nr	**137.49**
8" dia.	188.08	212.42	1.55	38.30	nr	**250.72**
Barrel nipple						
½" dia.	3.25	3.67	0.34	8.40	nr	**12.07**
¾" dia.	4.22	4.76	0.39	9.63	nr	**14.39**
1" dia.	5.47	6.18	0.43	10.63	nr	**16.81**
1 ¼" dia.	7.58	8.56	0.50	12.35	nr	**20.91**
1 ½" dia.	8.94	10.10	0.57	14.09	nr	**24.19**
2" dia.	10.83	12.23	0.62	15.33	nr	**27.56**
3" dia.	28.78	32.51	0.70	17.30	nr	**49.81**

S: PIPED SUPPLY SYSTEMS

Item	Net Price £	Material £	Labour hours	Labour £	Unit	**Total rate £**
Tee, 90°						
½" dia.	2.66	3.01	0.41	10.13	nr	**13.14**
¾" dia.	3.70	4.18	0.47	11.61	nr	**15.79**
1" dia.	5.12	5.78	0.55	13.59	nr	**19.37**
1 ¼" dia.	7.36	8.31	0.64	15.81	nr	**24.12**
1 ½" dia.	10.83	12.23	0.71	17.54	nr	**29.77**
2" dia.	16.54	18.68	0.78	19.28	nr	**37.96**
3" dia.	48.21	54.45	0.91	22.49	nr	**76.94**
4" dia.	70.76	79.92	1.12	27.68	nr	**107.60**
6" dia.	247.26	279.26	1.69	41.77	nr	**321.03**
8" dia.	385.53	435.42	2.03	50.16	nr	**485.58**
Full face flange						
½" dia.	28.80	32.53	0.10	2.47	nr	**35.00**
¾" dia.	29.48	33.29	0.13	3.20	nr	**36.49**
1" dia.	31.93	36.06	0.15	3.70	nr	**39.76**
1 ¼" dia.	35.49	40.09	0.18	4.45	nr	**44.54**
1 ½" dia.	42.68	48.21	0.21	5.18	nr	**53.39**
2" dia.	57.78	65.26	0.29	7.17	nr	**72.43**
3" dia.	99.06	111.87	0.37	9.15	nr	**121.02**
4" dia.	129.82	146.61	0.41	10.13	nr	**156.74**
PVC-U PIPEWORK						
Pipes; solvent welded joints in the running length, brackets measured separately						
Class C (9 bar pressure)						
2" dia.	11.99	13.55	0.41	10.13	m	**23.68**
3" dia.	22.96	25.93	0.47	11.61	m	**37.54**
4" dia.	40.75	46.02	0.50	12.35	m	**58.37**
6" dia.	88.18	99.59	1.76	43.49	m	**143.08**
Class D (12 bar pressure)						
1 ¼" dia.	6.98	7.88	0.41	10.13	m	**18.01**
1 ½" dia.	9.60	10.84	0.42	10.38	m	**21.22**
2" dia.	14.88	16.81	0.45	11.13	m	**27.94**
3" dia.	31.88	36.01	0.48	11.86	m	**47.87**
4" dia.	53.37	60.27	0.53	13.10	m	**73.37**
6" dia.	99.05	111.86	0.58	14.33	m	**126.19**
Class E (15 bar pressure)						
½" dia.	3.41	3.86	0.38	9.38	m	**13.24**
¾" dia.	4.88	5.51	0.40	9.88	m	**15.39**
1" dia.	5.68	6.42	0.41	10.13	m	**16.55**
1 ¼" dia.	8.34	9.41	0.41	10.13	m	**19.54**
1 ½" dia.	10.85	12.25	0.42	10.38	m	**22.63**
2" dia.	16.95	19.15	0.45	11.13	m	**30.28**
3" dia.	36.71	41.46	0.47	11.61	m	**53.07**
4" dia.	60.28	68.08	0.50	12.35	m	**80.43**
6" dia.	130.58	147.47	0.53	13.10	m	**160.57**

S: PIPED SUPPLY SYSTEMS

Item	Net Price £	Material £	Labour hours	Labour £	Unit	Total rate £
S10: COLD WATER – cont						
PVC-U PIPEWORK – cont						
Pipes – cont						
Class 7						
½" dia.	6.05	6.83	0.32	7.91	m	**14.74**
¾" dia.	8.47	9.57	0.33	8.15	m	**17.72**
1" dia.	12.92	14.60	0.40	9.88	m	**24.48**
1 ¼" dia.	17.75	20.04	0.40	9.88	m	**29.92**
1 ½" dia.	21.97	24.82	0.41	10.13	m	**34.95**
2" dia.	36.52	41.25	0.43	10.63	m	**51.88**
FIXINGS						
Refer to steel pipes; galvanized iron. For minimum fixing dimensions, refer to the Tables and Memoranda at the rear of the book						
Extra over fittings; solvent welded joints						
End cap						
½" dia.	1.17	1.33	0.17	4.20	nr	**5.53**
¾" dia.	1.36	1.54	0.19	4.70	nr	**6.24**
1" dia.	1.52	1.71	0.22	5.44	nr	**7.15**
1 ¼" dia.	2.39	2.70	0.25	6.18	nr	**8.88**
1 ½" dia.	4.01	4.53	0.28	6.93	nr	**11.46**
2" dia.	4.91	5.55	0.31	7.66	nr	**13.21**
3" dia.	15.07	17.02	0.36	8.90	nr	**25.92**
4" dia.	23.27	26.29	0.44	10.87	nr	**37.16**
6" dia.	56.24	63.52	0.67	16.55	nr	**80.07**
Socket						
½" dia.	1.23	1.39	0.31	7.66	nr	**9.05**
¾" dia.	1.36	1.54	0.35	8.65	nr	**10.19**
1" dia.	1.59	1.80	0.42	10.38	nr	**12.18**
1 ¼" dia.	2.87	3.24	0.45	11.13	nr	**14.37**
1 ½" dia.	3.36	3.79	0.51	12.61	nr	**16.40**
2" dia.	4.77	5.39	0.56	13.83	nr	**19.22**
3" dia.	18.22	20.57	0.65	16.07	nr	**36.64**
4" dia.	26.41	29.82	0.80	19.77	nr	**49.59**
6" dia.	66.23	74.80	1.21	29.90	nr	**104.70**
Reducing socket						
¾ × ½" dia.	1.45	1.64	0.31	7.66	nr	**9.30**
1 × ¾" dia.	1.81	2.04	0.35	8.65	nr	**10.69**
1 ¼ × 1" dia.	3.46	3.91	0.42	10.38	nr	**14.29**

S: PIPED SUPPLY SYSTEMS

Item	Net Price £	Material £	Labour hours	Labour £	Unit	**Total rate £**
1 ½ × 1 ¼" dia.	3.86	4.36	0.45	11.13	nr	**15.49**
2 × 1 ½" dia.	5.82	6.57	0.51	12.61	nr	**19.18**
3 × 2" dia.	17.72	20.01	0.56	13.83	nr	**33.84**
4 × 3" dia.	26.23	29.62	0.65	16.07	nr	**45.69**
6 × 4" dia.	95.59	107.96	0.80	19.77	nr	**127.73**
8 × 6" dia.	148.05	167.21	1.21	29.90	nr	**197.11**
Elbow, 90°						
½" dia.	1.63	1.84	0.31	7.66	nr	**9.50**
¾" dia.	1.96	2.21	0.35	8.65	nr	**10.86**
1" dia.	2.72	3.07	0.42	10.38	nr	**13.45**
1 ¼" dia.	4.77	5.39	0.45	11.13	nr	**16.52**
1 ½" dia.	6.14	6.94	0.45	11.13	nr	**18.07**
2" dia.	9.09	10.26	0.56	13.83	nr	**24.09**
3" dia.	26.23	29.62	0.65	16.07	nr	**45.69**
4" dia.	39.50	44.61	0.80	19.77	nr	**64.38**
6" dia.	156.41	176.64	1.21	29.90	nr	**206.54**
Elbow 45°						
½" dia.	3.10	3.50	0.31	7.66	nr	**11.16**
¾" dia.	3.31	3.73	0.35	8.65	nr	**12.38**
1" dia.	4.01	4.53	0.45	11.13	nr	**15.66**
1 ¼" dia.	5.75	6.50	0.45	11.13	nr	**17.63**
1 ½" dia.	7.21	8.14	0.51	12.61	nr	**20.75**
2" dia.	10.16	11.48	0.56	13.83	nr	**25.31**
3" dia.	23.93	27.03	0.65	16.07	nr	**43.10**
4" dia.	49.18	55.54	0.80	19.77	nr	**75.31**
6" dia.	101.48	114.61	1.21	29.90	nr	**144.51**
Bend 90° (long radius)						
3" dia.	73.13	82.59	0.65	16.07	nr	**98.66**
4" dia.	147.72	166.83	0.80	19.77	nr	**186.60**
6" dia.	324.61	366.61	1.21	29.90	nr	**396.51**
Bend 45° (long radius)						
1 ½" dia.	17.36	19.60	0.51	12.61	nr	**32.21**
2" dia.	28.38	32.05	0.56	13.83	nr	**45.88**
3" dia.	60.64	68.48	0.65	16.07	nr	**84.55**
4" dia.	118.03	133.30	0.80	19.77	nr	**153.07**
Socket union						
½" dia.	6.30	7.12	0.34	8.40	nr	**15.52**
¾" dia.	7.21	8.14	0.39	9.63	nr	**17.77**
1" dia.	9.36	10.57	0.45	11.13	nr	**21.70**
1 ¼" dia.	11.65	13.16	0.50	12.35	nr	**25.51**
1 ½" dia.	15.97	18.03	0.57	14.09	nr	**32.12**
2" dia.	20.65	23.32	0.62	15.33	nr	**38.65**
3" dia.	76.89	86.84	0.70	17.30	nr	**104.14**
4" dia.	104.10	117.57	0.89	21.99	nr	**139.56**

S: PIPED SUPPLY SYSTEMS

Item	Net Price £	Material £	Labour hours	Labour £	Unit	Total rate £
S10: COLD WATER – cont						
PVC-U PIPEWORK – cont						
Extra over fittings – cont						
Saddle plain						
2" × 1 ¼" dia.	16.23	18.33	0.42	10.38	nr	**28.71**
3" × 1 ½" dia.	22.80	25.76	0.48	11.86	nr	**37.62**
4" × 2" dia.	25.69	29.02	0.68	16.81	nr	**45.83**
6" × 2" dia.	30.16	34.06	0.91	22.49	nr	**56.55**
Straight tank connector						
½" dia.	4.19	4.73	0.13	3.20	nr	**7.93**
¾" dia.	4.74	5.36	0.14	3.46	nr	**8.82**
1" dia.	10.07	11.37	0.14	3.46	nr	**14.83**
1 ¼" dia.	25.58	28.89	0.16	3.96	nr	**32.85**
1 ½" dia.	28.04	31.67	0.18	4.45	nr	**36.12**
2" dia.	33.60	37.94	0.24	5.93	nr	**43.87**
3" dia.	34.45	38.90	0.29	7.17	nr	**46.07**
Equal tee						
½" dia.	1.89	2.13	0.44	10.87	nr	**13.00**
¾" dia.	2.39	2.70	0.48	11.86	nr	**14.56**
1" dia.	3.60	4.07	0.54	13.35	nr	**17.42**
1 ¼" dia.	5.09	5.75	0.70	17.30	nr	**23.05**
1 ½" dia.	7.35	8.30	0.74	18.29	nr	**26.59**
2" dia.	11.65	13.16	0.80	19.77	nr	**32.93**
3" dia.	33.77	38.14	1.04	25.70	nr	**63.84**
4" dia.	49.52	55.93	1.28	31.63	nr	**87.56**
6" dia.	172.47	194.79	1.93	47.70	nr	**242.49**
PVC-C PIPEWORK						
Pipes; solvent welded in the running length, brackets measured separately						
Pipe; 3 m long; PN25						
16 × 2.0 mm	3.85	4.35	0.20	4.95	m	**9.30**
20 × 2.3 mm	5.82	6.57	0.20	4.95	m	**11.52**
25 × 2.8 mm	7.54	8.52	0.20	4.95	m	**13.47**
32 × 3.6 mm	11.20	12.65	0.20	4.95	m	**17.60**
Pipe; 5 m long; PN25						
40 × 4.5 mm	13.65	15.41	0.20	4.95	m	**20.36**
50 × 5.6 mm	20.56	23.23	0.20	4.95	m	**28.18**
63 × 7.0 mm	31.74	35.84	0.20	4.95	m	**40.79**

S: PIPED SUPPLY SYSTEMS

Item	Net Price £	Material £	Labour hours	Labour £	Unit	**Total rate £**
FIXINGS						
Refer to steel pipes; galvanized iron. For minimum fixing dimensions, refer to the Tables and Memoranda at the rear of the book						
Extra over fittings; solvent welded joints						
Straight coupling; PN25						
16 mm	0.51	0.57	0.20	4.95	nr	**5.52**
20 mm	0.72	0.82	0.20	4.95	nr	**5.77**
25 mm	0.90	1.02	0.20	4.95	nr	**5.97**
32 mm	2.77	3.13	0.20	4.95	nr	**8.08**
40 mm	3.56	4.02	0.20	4.95	nr	**8.97**
50 mm	4.77	5.39	0.20	4.95	nr	**10.34**
63 mm	8.41	9.50	0.20	4.95	nr	**14.45**
Elbow; 90°; PN25						
16 mm	0.83	0.94	0.20	4.95	nr	**5.89**
20 mm	1.28	1.45	0.20	4.95	nr	**6.40**
25 mm	1.59	1.80	0.20	4.95	nr	**6.75**
32 mm	3.31	3.73	0.20	4.95	nr	**8.68**
40 mm	5.10	5.76	0.20	4.95	nr	**10.71**
50 mm	7.07	7.99	0.20	4.95	nr	**12.94**
63 mm	12.08	13.65	0.20	4.95	nr	**18.60**
Elbow; 45°; PN25						
20 mm	1.28	1.45	0.20	4.95	nr	**6.40**
25 mm	1.59	1.80	0.20	4.95	nr	**6.75**
32 mm	3.31	3.73	0.20	4.95	nr	**8.68**
40 mm	5.10	5.76	0.20	4.95	nr	**10.71**
50 mm	7.07	7.99	0.20	4.95	nr	**12.94**
63 mm	12.08	13.65	0.20	4.95	nr	**18.60**
Reducer fitting; single stage reduction						
20/16 mm	0.90	1.02	0.20	4.95	nr	**5.97**
25/20 mm	1.09	1.23	0.20	4.95	nr	**6.18**
32/25 mm	2.18	2.46	0.20	4.95	nr	**7.41**
40/32 mm	2.87	3.24	0.20	4.95	nr	**8.19**
50/40 mm	3.31	3.73	0.20	4.95	nr	**8.68**
63/50 mm	5.02	5.67	0.20	4.95	nr	**10.62**
Equal tee; 90°; PN25						
16 mm	1.39	1.57	0.20	4.95	nr	**6.52**
20 mm	1.90	2.14	0.20	4.95	nr	**7.09**
25 mm	2.41	2.72	0.20	4.95	nr	**7.67**
32 mm	3.93	4.44	0.20	4.95	nr	**9.39**
40 mm	6.80	7.68	0.20	4.95	nr	**12.63**
50 mm	10.18	11.50	0.20	4.95	nr	**16.45**
63 mm	17.17	19.39	0.20	4.95	nr	**24.34**

S: PIPED SUPPLY SYSTEMS

Item	Net Price £	Material £	Labour hours	Labour £	Unit	Total rate £
S10: COLD WATER – cont						
PVC-C PIPEWORK – cont						
Extra over fittings – cont						
Cap; PN25						
20 mm	0.96	1.08	0.20	4.95	nr	**6.03**
25 mm	1.28	1.45	0.20	4.95	nr	**6.40**
32 mm	1.86	2.10	0.20	4.95	nr	**7.05**
40 mm	2.55	2.88	0.20	4.95	nr	**7.83**
50 mm	3.56	4.02	0.20	4.95	nr	**8.97**
63 mm	5.67	6.41	0.20	4.95	nr	**11.36**
SCREWED STEEL PIPEWORK						
Galvanized steel pipes; screwed and socketed joints; BS 1387: 1985						
Galvanized; medium, fixed vertically, with brackets measured separately, screwed joints are within the running length, but any flanges are additional						
10 mm dia.	4.13	4.66	0.51	12.61	m	**17.27**
15 mm dia.	3.72	4.20	0.52	12.85	m	**17.05**
20 mm dia.	4.20	4.74	0.55	13.59	m	**18.33**
25 mm dia.	5.88	6.64	0.60	14.83	m	**21.47**
32 mm dia.	7.27	8.21	0.67	16.55	m	**24.76**
40 mm dia.	8.45	9.55	0.75	18.53	m	**28.08**
50 mm dia.	11.86	13.39	0.85	21.00	m	**34.39**
65 mm dia.	16.07	18.15	0.93	22.98	m	**41.13**
80 mm dia.	20.82	23.51	1.07	26.44	m	**49.95**
100 mm dia.	29.44	33.25	1.46	36.08	m	**69.33**
125 mm dia.	46.83	52.89	1.72	42.50	m	**95.39**
150 mm dia.	54.37	61.40	1.96	48.43	m	**109.83**
Galvanized; heavy, fixed vertically, with brackets measured separately, screwed joints are within the running length, but any flanges are additional						
15 mm dia.	4.41	4.98	0.52	12.85	m	**17.83**
20 mm dia.	5.00	5.65	0.55	13.59	m	**19.24**
25 mm dia.	7.14	8.07	0.60	14.83	m	**22.90**
32 mm dia.	8.85	10.00	0.67	16.55	m	**26.55**
40 mm dia.	10.33	11.67	0.75	18.53	m	**30.20**
50 mm dia.	14.30	16.15	0.85	21.00	m	**37.15**
65 mm dia.	19.43	21.94	0.93	22.98	m	**44.92**

S: PIPED SUPPLY SYSTEMS

Item	Net Price £	Material £	Labour hours	Labour £	Unit	**Total rate £**
80 mm dia.	24.69	27.89	1.07	26.44	m	**54.33**
100 mm dia.	34.40	38.85	1.46	36.08	m	**74.93**
125 mm dia.	49.85	56.30	1.72	42.50	m	**98.80**
150 mm dia.	58.30	65.84	1.96	48.43	m	**114.27**
Galvanized; medium, fixed horizontaly or suspended at high level, with brackets measured separately, screwed joints are within the running length, but any flanges are additional						
10 mm dia.	4.13	4.66	0.51	12.61	m	**17.27**
15 mm dia.	3.72	4.20	0.52	12.85	m	**17.05**
20 mm dia.	4.20	4.74	0.55	13.59	m	**18.33**
25 mm dia.	5.88	6.64	0.60	14.83	m	**21.47**
32 mm dia.	7.27	8.21	0.67	16.55	m	**24.76**
40 mm dia.	8.45	9.55	0.75	18.53	m	**28.08**
50 mm dia.	11.86	13.39	0.85	21.00	m	**34.39**
65 mm dia.	16.07	18.15	0.93	22.98	m	**41.13**
80 mm dia.	20.82	23.51	1.07	26.44	m	**49.95**
100 mm dia.	29.44	33.25	1.46	36.08	m	**69.33**
125 mm dia.	46.83	52.89	1.72	42.50	m	**95.39**
150 mm dia.	54.37	61.40	1.96	48.43	m	**109.83**
Galvanized; heavy, fixed horizontaly or suspended at high level, with brackets measured separately, screwed joints are within the running length, but any flanges are additional						
15 mm dia.	4.41	4.98	0.52	12.85	m	**17.83**
20 mm dia.	5.00	5.65	0.55	13.59	m	**19.24**
25 mm dia.	7.14	8.07	0.60	14.83	m	**22.90**
32 mm dia.	8.85	10.00	0.67	16.55	m	**26.55**
40 mm dia.	10.33	11.67	0.75	18.53	m	**30.20**
50 mm dia.	14.30	16.15	0.85	21.00	m	**37.15**
65 mm dia.	19.43	21.94	0.93	22.98	m	**44.92**
80 mm dia.	24.69	27.89	1.07	26.44	m	**54.33**
100 mm dia.	34.40	38.85	1.46	36.08	m	**74.93**
125 mm dia.	49.85	56.30	1.72	42.50	m	**98.80**
150 mm dia.	58.30	65.84	1.96	48.43	m	**114.27**

S: PIPED SUPPLY SYSTEMS

Item	Net Price £	Material £	Labour hours	Labour £	Unit	Total rate £
S10: COLD WATER – cont						
FIXINGS						
For steel pipes; galvanized iron. For minimum fixing dimensions, refer to the Tables and Memoranda at the rear of the book						
Single pipe bracket, screw on, galvanized iron; screwed to wood						
15 mm dia.	1.08	1.22	0.14	3.46	nr	**4.68**
20 mm dia.	1.20	1.36	0.14	3.46	nr	**4.82**
25 mm dia.	1.40	1.58	0.17	4.20	nr	**5.78**
32 mm dia.	1.91	2.15	0.19	4.70	nr	**6.85**
40 mm dia.	2.85	3.22	0.22	5.44	nr	**8.66**
50 mm dia.	3.77	4.25	0.22	5.44	nr	**9.69**
65 mm dia.	4.45	5.03	0.28	6.93	nr	**11.96**
80 mm dia.	6.97	7.87	0.32	7.91	nr	**15.78**
100 mm dia.	10.08	11.38	0.35	8.65	nr	**20.03**
Single pipe bracket, screw on, galvanized iron; plugged and screwed						
15 mm dia.	1.08	1.22	0.25	6.18	nr	**7.40**
20 mm dia.	1.20	1.36	0.25	6.18	nr	**7.54**
25 mm dia.	1.40	1.58	0.30	7.41	nr	**8.99**
32 mm dia.	1.91	2.15	0.32	7.91	nr	**10.06**
40 mm dia.	2.85	3.22	0.32	7.91	nr	**11.13**
50 mm dia.	3.77	4.25	0.32	7.91	nr	**12.16**
65 mm dia.	4.45	5.03	0.35	8.65	nr	**13.68**
80 mm dia.	6.97	7.87	0.42	10.38	nr	**18.25**
100 mm dia.	10.08	11.38	0.42	10.38	nr	**21.76**
Single pipe bracket for building in, galvanized iron						
15 mm dia.	1.14	1.29	0.10	2.47	nr	**3.76**
20 mm dia.	1.28	1.45	0.11	2.72	nr	**4.17**
25 mm dia.	1.40	1.58	0.12	2.97	nr	**4.55**
32 mm dia.	1.48	1.67	0.14	3.46	nr	**5.13**
40 mm dia.	1.91	2.15	0.15	3.70	nr	**5.85**
50 mm dia.	2.41	2.72	0.16	3.96	nr	**6.68**
Pipe ring, single socket, galvanized iron						
15 mm dia.	1.14	1.29	0.10	2.47	nr	**3.76**
20 mm dia.	1.28	1.45	0.11	2.72	nr	**4.17**
25 mm dia.	1.40	1.58	0.12	2.97	nr	**4.55**
32 mm dia.	1.48	1.67	0.15	3.70	nr	**5.37**
40 mm dia.	1.91	2.15	0.15	3.70	nr	**5.85**
50 mm dia.	2.41	2.72	0.16	3.96	nr	**6.68**
65 mm dia.	3.51	3.97	0.30	7.41	nr	**11.38**

S: PIPED SUPPLY SYSTEMS

Item	Net Price £	Material £	Labour hours	Labour £	Unit	Total rate £
80 mm dia.	4.20	4.74	0.35	8.65	nr	**13.39**
100 mm dia.	6.36	7.18	0.40	9.88	nr	**17.06**
125 mm dia.	12.86	14.52	0.60	14.83	nr	**29.35**
150 mm dia.	15.53	17.54	0.77	19.03	nr	**36.57**
Pipe ring, double socket, galvanized iron						
15 mm dia.	10.13	11.44	0.10	2.47	nr	**13.91**
20 mm dia.	11.65	13.16	0.11	2.72	nr	**15.88**
25 mm dia.	12.84	14.50	0.12	2.97	nr	**17.47**
32 mm dia.	14.15	15.98	0.14	3.46	nr	**19.44**
40 mm dia.	17.87	20.19	0.15	3.70	nr	**23.89**
50 mm dia.	20.33	22.96	0.16	3.96	nr	**26.92**
Screw on backplate (Male), galvanized iron; plugged and screwed						
All sizes 15 mm to 50 mm × M12	0.90	1.02	0.10	2.47	nr	**3.49**
Screw on backplate (Female), galvanized iron; plugged and screwed						
All sizes 15 mm to 50 mm × M12	0.90	1.02	0.10	2.47	nr	**3.49**
Extra over channel sections for fabricated hangers and brackets						
Galvanized steel; including inserts, bolts, nuts, washers; fixed to backgrounds						
41 × 21 mm	5.65	6.39	0.29	7.17	m	**13.56**
41 × 41 mm	6.77	7.65	0.29	7.17	m	**14.82**
Threaded rods; metric thread; including nuts, washers etc.						
10 mm dia. × 600 mm long	1.86	2.10	0.18	4.45	nr	**6.55**
12 mm dia. × 600 mm long	2.86	3.23	0.18	4.45	nr	**7.68**
Extra over steel flanges, screwed and drilled; metric; BS 4504						
Screwed flanges; PN6						
15 mm dia.	14.16	15.99	0.35	8.65	nr	**24.64**
20 mm dia.	14.16	15.99	0.47	11.61	nr	**27.60**
25 mm dia.	14.16	15.99	0.53	13.10	nr	**29.09**
32 mm dia.	14.16	15.99	0.62	15.33	nr	**31.32**
40 mm dia.	14.16	15.99	0.70	17.30	nr	**33.29**
50 mm dia.	15.09	17.04	0.84	20.76	nr	**37.80**
65 mm dia.	21.00	23.72	1.03	25.46	nr	**49.18**
80 mm dia.	29.64	33.48	1.23	30.40	nr	**63.88**
100 mm dia.	35.06	39.60	1.41	34.84	nr	**74.44**
125 mm dia.	64.31	72.63	1.77	43.74	nr	**116.37**
150 mm dia.	64.31	72.63	2.21	54.61	nr	**127.24**

S: PIPED SUPPLY SYSTEMS

Item	Net Price £	Material £	Labour hours	Labour £	Unit	**Total rate £**
S10: COLD WATER – cont						
FIXINGS – cont						
Extra over steel flanges, screwed and drilled – cont						
Screwed flanges; PN16						
15 mm dia.	18.01	20.34	0.35	8.65	nr	**28.99**
20 mm dia.	18.01	20.34	0.47	11.61	nr	**31.95**
25 mm dia.	18.01	20.34	0.53	13.10	nr	**33.44**
32 mm dia.	19.23	21.72	0.62	15.33	nr	**37.05**
40 mm dia.	19.23	21.72	0.70	17.30	nr	**39.02**
50 mm dia.	23.45	26.49	0.84	20.76	nr	**47.25**
65 mm dia.	29.30	33.09	1.03	25.46	nr	**58.55**
80 mm dia.	35.68	40.30	1.23	30.40	nr	**70.70**
100 mm dia.	40.01	45.19	1.41	34.84	nr	**80.03**
125 mm dia.	70.02	79.08	1.77	43.74	nr	**122.82**
150 mm dia.	68.91	77.83	2.21	54.61	nr	**132.44**
Extra over steel flanges, screwed and drilled; imperial; BS 10						
Screwed flanges; table E						
½" dia.	24.20	27.34	0.35	8.65	nr	**35.99**
¾" dia.	24.20	27.34	0.47	11.61	nr	**38.95**
1" dia.	24.20	27.34	0.53	13.10	nr	**40.44**
1 ¼" dia.	24.20	27.34	0.62	15.33	nr	**42.67**
1 ½" dia.	24.20	27.34	0.70	17.30	nr	**44.64**
2" dia.	24.20	27.34	0.84	20.76	nr	**48.10**
2 ½" dia.	28.79	32.52	1.03	25.46	nr	**57.98**
3" dia.	34.69	39.18	1.23	30.40	nr	**69.58**
4" dia.	44.14	49.85	1.41	34.84	nr	**84.69**
5" dia.	93.54	105.64	1.77	43.74	nr	**149.38**
Extra over steel flange connections						
Bolted connection between pair of flanges; including gasket, bolts, nuts and washers						
50 mm dia.	41.75	47.15	0.53	13.10	nr	**60.25**
65 mm dia.	52.59	59.39	0.53	13.10	nr	**72.49**
80 mm dia.	60.32	68.13	0.53	13.10	nr	**81.23**
100 mm dia.	71.99	81.30	0.53	13.10	nr	**94.40**
125 mm dia.	135.98	153.57	0.61	15.08	nr	**168.65**
150 mm dia.	139.86	157.96	0.90	22.24	nr	**180.20**

S: PIPED SUPPLY SYSTEMS

Item	Net Price £	Material £	Labour hours	Labour £	Unit	Total rate £
Extra over heavy steel tubular fittings; BS 1387						
Long screw connection with socket and backnut						
15 mm dia.	5.58	6.30	0.63	15.57	nr	**21.87**
20 mm dia.	7.00	7.91	0.84	20.76	nr	**28.67**
25 mm dia.	9.16	10.34	0.95	23.48	nr	**33.82**
32 mm dia.	12.07	13.63	1.11	27.44	nr	**41.07**
40 mm dia.	14.69	16.60	1.28	31.63	nr	**48.23**
50 mm dia.	21.67	24.47	1.53	37.81	nr	**62.28**
65 mm dia.	49.78	56.22	1.87	46.22	nr	**102.44**
80 mm dia.	64.39	72.73	2.21	54.61	nr	**127.34**
100 mm dia.	73.04	82.49	3.05	75.37	nr	**157.86**
Running nipple						
15 mm dia.	1.40	1.58	0.50	12.35	nr	**13.93**
20 mm dia.	1.75	1.98	0.68	16.81	nr	**18.79**
25 mm dia.	1.87	2.11	0.77	19.03	nr	**21.14**
32 mm dia.	3.01	3.40	0.90	22.24	nr	**25.64**
40 mm dia.	4.07	4.60	1.03	25.46	nr	**30.06**
50 mm dia.	6.19	6.99	1.23	30.40	nr	**37.39**
65 mm dia.	13.32	15.04	1.50	37.07	nr	**52.11**
80 mm dia.	20.76	23.45	1.78	43.99	nr	**67.44**
100 mm dia.	32.53	36.74	2.38	58.80	nr	**95.54**
Barrel nipple						
15 mm dia.	1.17	1.33	0.50	12.35	nr	**13.68**
20 mm dia.	1.77	2.00	0.68	16.81	nr	**18.81**
25 mm dia.	1.98	2.23	0.77	19.03	nr	**21.26**
32 mm dia.	3.28	3.70	0.90	22.24	nr	**25.94**
40 mm dia.	3.66	4.13	1.03	25.46	nr	**29.59**
50 mm dia.	5.24	5.92	1.23	30.40	nr	**36.32**
65 mm dia.	11.16	12.61	1.50	37.07	nr	**49.68**
80 mm dia.	15.58	17.59	1.78	43.99	nr	**61.58**
100 mm dia.	28.17	31.81	2.38	58.80	nr	**90.61**
125 mm dia.	52.34	59.11	2.87	70.93	nr	**130.04**
150 mm dia.	82.45	93.12	3.39	83.76	nr	**176.88**
Close taper nipple						
15 mm dia.	1.66	1.88	0.50	12.35	nr	**14.23**
20 mm dia.	2.15	2.43	0.68	16.81	nr	**19.24**
25 mm dia.	2.81	3.17	0.77	19.03	nr	**22.20**
32 mm dia.	4.20	4.74	0.90	22.24	nr	**26.98**
40 mm dia.	5.21	5.89	1.03	25.46	nr	**31.35**
50 mm dia.	8.00	9.04	1.23	30.40	nr	**39.44**
65 mm dia.	12.64	14.28	1.50	37.07	nr	**51.35**
80 mm dia.	20.71	23.39	1.78	43.99	nr	**67.38**
100 mm dia.	39.37	44.46	2.38	58.80	nr	**103.26**

S: PIPED SUPPLY SYSTEMS

Item	Net Price £	Material £	Labour hours	Labour £	Unit	Total rate £
S10: COLD WATER – cont						
FIXINGS – cont						
Extra over heavy steel tubular fittings – cont						
90° bend with socket						
15 mm dia.	4.58	5.17	0.64	15.81	nr	**20.98**
20 mm dia.	6.15	6.95	0.85	21.00	nr	**27.95**
25 mm dia.	9.41	10.63	0.97	23.97	nr	**34.60**
32 mm dia.	13.49	15.24	1.12	27.68	nr	**42.92**
40 mm dia.	16.48	18.61	1.29	31.89	nr	**50.50**
50 mm dia.	25.63	28.95	1.55	38.30	nr	**67.25**
65 mm dia.	51.68	58.36	1.89	46.70	nr	**105.06**
80 mm dia.	76.72	86.65	2.24	55.36	nr	**142.01**
100 mm dia.	136.04	153.64	3.09	76.36	nr	**230.00**
125 mm dia.	333.19	376.30	3.92	96.87	nr	**473.17**
150 mm dia.	500.24	564.97	4.74	117.13	nr	**682.10**
Extra over heavy steel fittings; BS 1740						
Plug						
15 mm dia.	1.27	1.44	0.28	6.93	nr	**8.37**
20 mm dia.	1.97	2.22	0.38	9.38	nr	**11.60**
25 mm dia.	3.46	3.91	0.44	10.87	nr	**14.78**
32 mm dia.	5.37	6.07	0.51	12.61	nr	**18.68**
40 mm dia.	5.92	6.68	0.59	14.58	nr	**21.26**
50 mm dia.	8.47	9.57	0.70	17.30	nr	**26.87**
65 mm dia.	20.26	22.88	0.85	21.00	nr	**43.88**
80 mm dia.	37.92	42.83	1.00	24.71	nr	**67.54**
100 mm dia.	72.81	82.23	1.44	35.59	nr	**117.82**
Socket						
15 mm dia.	1.40	1.58	0.64	15.81	nr	**17.39**
20 mm dia.	1.59	1.80	0.85	21.00	nr	**22.80**
25 mm dia.	2.25	2.54	0.97	23.97	nr	**26.51**
32 mm dia.	3.25	3.67	1.12	27.68	nr	**31.35**
40 mm dia.	3.95	4.46	1.29	31.89	nr	**36.35**
50 mm dia.	6.08	6.86	1.55	38.30	nr	**45.16**
65 mm dia.	12.06	13.62	1.89	46.70	nr	**60.32**
80 mm dia.	15.59	17.61	2.24	55.36	nr	**72.97**
100 mm dia.	29.35	33.15	3.09	76.36	nr	**109.51**
150 mm dia.	70.07	79.14	4.74	117.13	nr	**196.27**
Elbow, female/female						
15 mm dia.	8.06	9.10	0.64	15.81	nr	**24.91**
20 mm dia.	10.50	11.86	0.85	21.00	nr	**32.86**
25 mm dia.	14.26	16.11	0.97	23.97	nr	**40.08**

S: PIPED SUPPLY SYSTEMS

Item	Net Price £	Material £	Labour hours	Labour £	Unit	**Total rate £**
32 mm dia.	26.55	29.99	1.12	27.68	nr	**57.67**
40 mm dia.	31.66	35.76	1.29	31.89	nr	**67.65**
50 mm dia.	51.90	58.62	1.55	38.30	nr	**96.92**
65 mm dia.	126.80	143.21	1.89	46.70	nr	**189.91**
80 mm dia.	151.20	170.77	2.24	55.36	nr	**226.13**
100 mm dia.	261.82	295.70	3.09	76.36	nr	**372.06**
Equal tee						
15 mm dia.	10.01	11.30	0.91	22.49	nr	**33.79**
20 mm dia.	11.64	13.15	1.22	30.14	nr	**43.29**
25 mm dia.	17.15	19.37	1.40	34.60	nr	**53.97**
32 mm dia.	35.42	40.00	1.62	40.04	nr	**80.04**
40 mm dia.	38.56	43.55	1.86	45.97	nr	**89.52**
50 mm dia.	62.70	70.81	2.21	54.61	nr	**125.42**
65 mm dia.	151.62	171.24	2.72	67.21	nr	**238.45**
80 mm dia.	162.71	183.76	3.21	79.32	nr	**263.08**
100 mm dia.	261.86	295.74	4.44	109.72	nr	**405.46**
Extra over malleable iron fittings; BS 143						
Cap						
15 mm dia.	3.13	3.54	0.32	7.91	nr	**11.45**
20 mm dia.	3.29	3.71	0.43	10.63	nr	**14.34**
25 mm dia.	5.94	6.71	0.49	12.11	nr	**18.82**
32 mm dia.	11.26	12.72	0.58	14.33	nr	**27.05**
40 mm dia.	12.09	13.66	0.66	16.31	nr	**29.97**
50 mm dia.	19.23	21.72	0.78	19.28	nr	**41.00**
65 mm dia.	33.04	37.31	0.96	23.72	nr	**61.03**
80 mm dia.	54.52	61.58	1.13	27.92	nr	**89.50**
100 mm dia.	92.94	104.97	1.70	42.01	nr	**146.98**
Plain plug, hollow						
15 mm dia.	1.52	1.71	0.28	6.93	nr	**8.64**
20 mm dia.	2.37	2.67	0.38	9.38	nr	**12.05**
25 mm dia.	4.16	4.70	0.44	10.87	nr	**15.57**
32 mm dia.	6.44	7.27	0.51	12.61	nr	**19.88**
40 mm dia.	7.11	8.03	0.59	14.58	nr	**22.61**
50 mm dia.	10.17	11.49	0.70	17.30	nr	**28.79**
65 mm dia.	24.32	27.47	0.85	21.00	nr	**48.47**
80 mm dia.	45.50	51.39	1.00	24.71	nr	**76.10**
100 mm dia.	87.39	98.70	1.44	35.59	nr	**134.29**
Plain plug, solid						
15 mm dia.	1.89	2.13	0.29	7.17	nr	**9.30**
20 mm dia.	2.02	2.28	0.38	9.38	nr	**11.66**
25 mm dia.	2.68	3.03	0.44	10.87	nr	**13.90**
32 mm dia.	4.11	4.64	0.51	12.61	nr	**17.25**
40 mm dia.	5.55	6.27	0.59	14.58	nr	**20.85**
50 mm dia.	7.28	8.22	0.70	17.30	nr	**25.52**

S: PIPED SUPPLY SYSTEMS

Item	Net Price £	Material £	Labour hours	Labour £	Unit	**Total rate £**
S10: COLD WATER – cont						
FIXINGS – cont						
Extra over malleable iron fittings – cont						
Elbow, male/female						
15 mm dia.	0.99	1.12	0.64	15.81	nr	**16.93**
20 mm dia.	1.32	1.49	0.85	21.00	nr	**22.49**
25 mm dia.	2.20	2.49	0.97	23.97	nr	**26.46**
32 mm dia.	5.00	5.65	1.12	27.68	nr	**33.33**
40 mm dia.	6.90	7.79	1.29	31.89	nr	**39.68**
50 mm dia.	8.88	10.03	1.55	38.30	nr	**48.33**
65 mm dia.	19.81	22.37	1.89	46.70	nr	**69.07**
80 mm dia.	27.08	30.58	2.24	55.36	nr	**85.94**
100 mm dia.	47.35	53.48	3.09	76.36	nr	**129.84**
Elbow						
15 mm dia.	0.88	0.99	0.64	15.81	nr	**16.80**
20 mm dia.	1.21	1.37	0.85	21.00	nr	**22.37**
25 mm dia.	1.88	2.12	0.97	23.97	nr	**26.09**
32 mm dia.	3.91	4.42	1.12	27.68	nr	**32.10**
40 mm dia.	5.86	6.62	1.29	31.89	nr	**38.51**
50 mm dia.	6.86	7.75	1.55	38.30	nr	**46.05**
65 mm dia.	15.29	17.27	1.89	46.70	nr	**63.97**
80 mm dia.	22.46	25.37	2.24	55.36	nr	**80.73**
100 mm dia.	38.58	43.57	3.09	76.36	nr	**119.93**
125 mm dia.	92.55	104.53	4.44	109.72	nr	**214.25**
150 mm dia.	172.33	194.63	5.79	143.09	nr	**337.72**
45° elbow						
15 mm dia.	2.28	2.57	0.64	15.81	nr	**18.38**
20 mm dia.	2.81	3.17	0.85	21.00	nr	**24.17**
25 mm dia.	3.87	4.38	0.97	23.97	nr	**28.35**
32 mm dia.	9.01	10.18	1.12	27.68	nr	**37.86**
40 mm dia.	10.60	11.97	1.29	31.89	nr	**43.86**
50 mm dia.	14.54	16.42	1.55	38.30	nr	**54.72**
65 mm dia.	20.43	23.07	1.89	46.70	nr	**69.77**
80 mm dia.	30.72	34.69	2.24	55.36	nr	**90.05**
100 mm dia.	59.24	66.90	3.09	76.36	nr	**143.26**
150 mm dia.	180.30	203.63	5.79	143.09	nr	**346.72**
Bend, male/female						
15 mm dia.	1.75	1.98	0.64	15.81	nr	**17.79**
20 mm dia.	2.87	3.24	0.85	21.00	nr	**24.24**
25 mm dia.	4.02	4.54	0.97	23.97	nr	**28.51**
32 mm dia.	6.80	7.68	1.12	27.68	nr	**35.36**
40 mm dia.	9.97	11.26	1.29	31.89	nr	**43.15**
50 mm dia.	18.75	21.18	1.55	38.30	nr	**59.48**
65 mm dia.	28.70	32.42	1.89	46.70	nr	**79.12**
80 mm dia.	44.18	49.90	2.24	55.36	nr	**105.26**
100 mm dia.	109.43	123.59	3.09	76.36	nr	**199.95**

S: PIPED SUPPLY SYSTEMS

Item	Net Price £	Material £	Labour hours	Labour £	Unit	**Total rate £**
Bend, male						
15 mm dia.	3.99	4.51	0.64	15.81	nr	**20.32**
20 mm dia.	4.48	5.06	0.85	21.00	nr	**26.06**
25 mm dia.	6.58	7.44	0.97	23.97	nr	**31.41**
32 mm dia.	14.56	16.44	1.12	27.68	nr	**44.12**
40 mm dia.	20.42	23.06	1.29	31.89	nr	**54.95**
50 mm dia.	30.98	34.99	1.55	38.30	nr	**73.29**
Bend, female						
15 mm dia.	1.79	2.02	0.64	15.81	nr	**17.83**
20 mm dia.	2.56	2.89	0.85	21.00	nr	**23.89**
25 mm dia.	3.58	4.04	0.97	23.97	nr	**28.01**
32 mm dia.	6.98	7.88	1.12	27.68	nr	**35.56**
40 mm dia.	8.31	9.38	1.29	31.89	nr	**41.27**
50 mm dia.	13.10	14.79	1.55	38.30	nr	**53.09**
65 mm dia.	28.70	32.42	1.89	46.70	nr	**79.12**
80 mm dia.	42.55	48.05	2.24	55.36	nr	**103.41**
100 mm dia.	89.28	100.84	3.09	76.36	nr	**177.20**
125 mm dia.	180.12	203.43	4.44	109.72	nr	**313.15**
150 mm dia.	395.73	446.93	5.79	143.09	nr	**590.02**
Return bend						
15 mm dia.	9.14	10.32	0.64	15.81	nr	**26.13**
20 mm dia.	14.77	16.68	0.85	21.00	nr	**37.68**
25 mm dia.	18.43	20.82	0.97	23.97	nr	**44.79**
32 mm dia.	25.59	28.90	1.12	27.68	nr	**56.58**
40 mm dia.	34.67	39.16	1.29	31.89	nr	**71.05**
50 mm dia.	52.91	59.75	1.55	38.30	nr	**98.05**
Equal socket, parallel thread						
15 mm dia.	0.86	0.97	0.64	15.81	nr	**16.78**
20 mm dia.	1.12	1.26	0.85	21.00	nr	**22.26**
25 mm dia.	1.42	1.60	0.97	23.97	nr	**25.57**
32 mm dia.	3.16	3.57	1.12	27.68	nr	**31.25**
40 mm dia.	4.28	4.83	1.29	31.89	nr	**36.72**
50 mm dia.	6.19	6.99	1.55	38.30	nr	**45.29**
65 mm dia.	9.83	11.10	1.89	46.70	nr	**57.80**
80 mm dia.	13.93	15.73	2.24	55.36	nr	**71.09**
100 mm dia.	22.99	25.97	3.09	76.36	nr	**102.33**
Concentric reducing socket						
20 × 15 mm dia.	1.37	1.55	0.76	18.78	nr	**20.33**
25 × 15 mm dia.	1.79	2.02	0.86	21.26	nr	**23.28**
25 × 20 mm dia.	1.69	1.91	0.86	21.26	nr	**23.17**
32 × 25 mm dia.	3.25	3.67	1.01	24.96	nr	**28.63**
40 × 25 mm dia.	4.28	4.83	1.16	28.66	nr	**33.49**
40 × 32 mm dia.	4.75	5.37	1.16	28.66	nr	**34.03**
50 × 25 mm dia.	8.23	9.29	1.38	34.10	nr	**43.39**

S: PIPED SUPPLY SYSTEMS

Item	Net Price £	Material £	Labour hours	Labour £	Unit	Total rate £
S10: COLD WATER – cont						
FIXINGS – cont						
Extra over malleable iron fittings – cont						
Concentric reducing socket – cont						
50 × 40 mm dia.	6.65	7.51	1.38	34.10	nr	**41.61**
65 × 50 mm dia.	11.60	13.10	1.69	41.77	nr	**54.87**
80 × 50 mm dia.	14.46	16.33	2.00	49.42	nr	**65.75**
100 × 50 mm dia.	28.85	32.58	2.75	67.95	nr	**100.53**
100 × 80 mm dia.	86.29	97.45	2.75	67.95	nr	**165.40**
150 × 100 mm dia.	80.18	90.56	4.10	101.32	nr	**191.88**
Eccentric reducing socket						
20 × 15 mm dia.	3.10	3.50	0.76	18.78	nr	**22.28**
25 × 15 mm dia.	8.81	9.95	0.86	21.26	nr	**31.21**
25 × 20 mm dia.	10.00	11.29	0.86	21.26	nr	**32.55**
32 × 25 mm dia.	12.93	14.61	1.01	24.96	nr	**39.57**
40 × 25 mm dia.	14.82	16.74	1.16	28.66	nr	**45.40**
40 × 32 mm dia.	16.13	18.22	1.16	28.66	nr	**46.88**
50 × 25 mm dia.	16.17	18.26	1.18	29.16	nr	**47.42**
50 × 40 mm dia.	16.22	18.32	1.28	31.63	nr	**49.95**
65 × 50 mm dia.	16.43	18.55	1.69	41.77	nr	**60.32**
80 × 50 mm dia.	17.50	19.77	2.00	49.42	nr	**69.19**
Hexagon bush						
20 × 15 mm dia.	0.77	0.87	0.37	9.15	nr	**10.02**
25 × 15 mm dia.	1.06	1.19	0.43	10.63	nr	**11.82**
25 × 20 mm dia.	0.99	1.12	0.43	10.63	nr	**11.75**
32 × 25 mm dia.	1.33	1.50	0.51	12.61	nr	**14.11**
40 × 25 mm dia.	2.00	2.25	0.58	14.33	nr	**16.58**
40 × 32 mm dia.	2.00	2.25	0.58	14.33	nr	**16.58**
50 × 25 mm dia.	4.22	4.76	0.71	17.54	nr	**22.30**
50 × 40 mm dia.	3.95	4.46	0.71	17.54	nr	**22.00**
65 × 50 mm dia.	7.26	8.20	0.84	20.76	nr	**28.96**
80 × 50 mm dia.	10.96	12.38	1.00	24.71	nr	**37.09**
100 × 50 mm dia.	25.38	28.66	1.52	37.57	nr	**66.23**
100 × 80 mm dia.	21.12	23.86	1.52	37.57	nr	**61.43**
150 × 100 mm dia.	75.88	85.70	2.48	61.28	nr	**146.98**
Hexagon nipple						
15 mm dia.	0.83	0.94	0.28	6.93	nr	**7.87**
20 mm dia.	0.94	1.06	0.38	9.38	nr	**10.44**
25 mm dia.	1.32	1.49	0.44	10.87	nr	**12.36**
32 mm dia.	2.82	3.18	0.51	12.61	nr	**15.79**
40 mm dia.	3.25	3.67	0.59	14.58	nr	**18.25**
50 mm dia.	5.91	6.67	0.70	17.30	nr	**23.97**
65 mm dia.	9.90	11.18	0.85	21.00	nr	**32.18**

S: PIPED SUPPLY SYSTEMS

Item	Net Price £	Material £	Labour hours	Labour £	Unit	**Total rate £**
80 mm dia.	13.69	15.46	1.00	24.71	nr	**40.17**
100 mm dia.	24.30	27.45	1.44	35.59	nr	**63.04**
150 mm dia.	67.26	75.96	2.32	57.32	nr	**133.28**
Union, male/female						
15 mm dia.	4.16	4.70	0.64	15.81	nr	**20.51**
20 mm dia.	5.09	5.75	0.85	21.00	nr	**26.75**
25 mm dia.	5.91	6.67	0.97	23.97	nr	**30.64**
32 mm dia.	10.45	11.80	1.12	27.68	nr	**39.48**
40 mm dia.	13.37	15.10	1.29	31.89	nr	**46.99**
50 mm dia.	21.04	23.77	1.55	38.30	nr	**62.07**
65 mm dia.	47.07	53.16	1.89	46.70	nr	**99.86**
Union, female						
15 mm dia.	3.95	4.46	0.64	15.81	nr	**20.27**
20 mm dia.	4.32	4.88	0.85	21.00	nr	**25.88**
25 mm dia.	5.07	5.72	0.97	23.97	nr	**29.69**
32 mm dia.	8.74	9.87	1.12	27.68	nr	**37.55**
40 mm dia.	9.87	11.15	1.29	31.89	nr	**43.04**
50 mm dia.	14.71	16.62	1.55	38.30	nr	**54.92**
65 mm dia.	37.68	42.55	1.89	46.70	nr	**89.25**
80 mm dia.	49.82	56.26	2.24	55.36	nr	**111.62**
100 mm dia.	94.86	107.13	3.09	76.36	nr	**183.49**
Union elbow, male/female						
15 mm dia.	6.28	7.09	0.64	15.81	nr	**22.90**
20 mm dia.	7.86	8.87	0.85	21.00	nr	**29.87**
25 mm dia.	11.04	12.46	0.97	23.97	nr	**36.43**
Twin elbow						
15 mm dia.	6.04	6.82	0.91	22.49	nr	**29.31**
20 mm dia.	6.67	7.54	1.22	30.14	nr	**37.68**
25 mm dia.	10.80	12.20	1.39	34.34	nr	**46.54**
32 mm dia.	21.64	24.44	1.62	40.04	nr	**64.48**
40 mm dia.	27.40	30.95	1.86	45.97	nr	**76.92**
50 mm dia.	35.21	39.77	2.21	54.61	nr	**94.38**
65 mm dia.	56.91	64.27	2.72	67.21	nr	**131.48**
80 mm dia.	64.59	72.95	3.21	79.32	nr	**152.27**
Equal tee						
15 mm dia.	1.21	1.37	0.91	22.49	nr	**23.86**
20 mm dia.	1.77	2.00	1.22	30.14	nr	**32.14**
25 mm dia.	2.53	2.86	1.39	34.34	nr	**37.20**
32 mm dia.	5.36	6.05	1.62	40.04	nr	**46.09**
40 mm dia.	7.34	8.29	1.86	45.97	nr	**54.26**
50 mm dia.	10.56	11.92	2.21	54.61	nr	**66.53**
65 mm dia.	24.75	27.95	2.72	67.21	nr	**95.16**
80 mm dia.	28.85	32.58	3.21	79.32	nr	**111.90**
100 mm dia.	52.31	59.08	4.44	109.72	nr	**168.80**
125 mm dia.	97.02	109.58	5.38	132.94	nr	**242.52**
150 mm dia.	232.03	262.06	6.31	155.93	nr	**417.99**

S: PIPED SUPPLY SYSTEMS

Item	Net Price £	Material £	Labour hours	Labour £	Unit	Total rate £
S10: COLD WATER – cont						
FIXINGS – cont						
Extra over malleable iron fittings – cont						
Tee reducing on branch						
20 × 15 mm dia.	1.81	2.04	1.22	30.14	nr	**32.18**
25 × 15 mm dia.	2.47	2.78	1.39	34.34	nr	**37.12**
25 × 20 mm dia.	2.81	3.17	1.39	34.34	nr	**37.51**
32 × 25 mm dia.	5.45	6.15	1.62	40.04	nr	**46.19**
40 × 25 mm dia.	6.90	7.79	1.86	45.97	nr	**53.76**
40 × 32 mm dia.	10.13	11.44	1.86	45.97	nr	**57.41**
50 × 25 mm dia.	9.16	10.34	2.21	54.61	nr	**64.95**
50 × 40 mm dia.	14.24	16.09	2.21	54.61	nr	**70.70**
65 × 50 mm dia.	21.97	24.82	2.72	67.21	nr	**92.03**
80 × 50 mm dia.	29.71	33.56	3.21	79.32	nr	**112.88**
100 × 50 mm dia.	49.18	55.54	4.44	109.72	nr	**165.26**
100 × 80 mm dia.	75.88	85.70	4.44	109.72	nr	**195.42**
150 × 100 mm dia.	123.09	139.02	6.28	155.19	nr	**294.21**
Equal pitcher tee						
15 mm dia.	4.20	4.74	0.91	22.49	nr	**27.23**
20 mm dia.	5.18	5.85	1.22	30.14	nr	**35.99**
25 mm dia.	7.76	8.76	1.39	34.34	nr	**43.10**
32 mm dia.	12.03	13.59	1.62	40.04	nr	**53.63**
40 mm dia.	18.60	21.00	1.86	45.97	nr	**66.97**
50 mm dia.	26.11	29.49	2.21	54.61	nr	**84.10**
65 mm dia.	37.15	41.95	2.72	67.21	nr	**109.16**
80 mm dia.	57.96	65.46	3.21	79.32	nr	**144.78**
100 mm dia.	130.41	147.29	4.44	109.72	nr	**257.01**
Cross						
15 mm dia.	3.49	3.94	1.00	24.71	nr	**28.65**
20 mm dia.	5.46	6.17	1.33	32.86	nr	**39.03**
25 mm dia.	6.92	7.81	1.51	37.31	nr	**45.12**
32 mm dia.	10.29	11.62	1.77	43.74	nr	**55.36**
40 mm dia.	13.85	15.65	2.02	49.92	nr	**65.57**
50 mm dia.	21.52	24.31	2.42	59.80	nr	**84.11**
65 mm dia.	30.72	34.69	2.97	73.39	nr	**108.08**
80 mm dia.	40.86	46.14	3.50	86.49	nr	**132.63**
100 mm dia.	84.32	95.23	4.84	119.61	nr	**214.84**

S: PIPED SUPPLY SYSTEMS

Item	Net Price £	Material £	Labour hours	Labour £	Unit	**Total rate £**
Y11 – PIPELINE ANCILLARIES						
VALVES						
Regulators						
Gunmetal; self-acting two port thermostat; single seat; screwed; normally closed; with adjustable or fixed bleed device						
25 mm dia.	616.00	695.71	1.46	36.08	nr	**731.79**
32 mm dia.	633.77	715.77	1.45	35.83	nr	**751.60**
40 mm dia.	677.43	765.08	1.55	38.31	nr	**803.39**
50 mm dia.	815.48	921.00	1.68	41.51	nr	**962.51**
Self acting temperature regulator for storage calorifier; integral sensing element and pocket; screwed ends						
15 mm dia.	606.07	684.49	1.32	32.62	nr	**717.11**
25 mm dia.	665.26	751.34	1.52	37.57	nr	**788.91**
32 mm dia.	859.09	970.25	1.79	44.24	nr	**1014.49**
40 mm dia.	1050.92	1186.90	1.99	49.17	nr	**1236.07**
50 mm dia.	1228.47	1387.42	2.26	55.84	nr	**1443.26**
Self acting temperature regulator for storage calorifier; integral sensing element and pocket; flanged ends; bolted connection						
15 mm dia.	889.71	1004.83	0.61	15.08	nr	**1019.91**
25 mm dia.	1018.27	1150.03	0.72	17.79	nr	**1167.82**
32 mm dia.	1283.55	1449.63	0.94	23.24	nr	**1472.87**
40 mm dia.	1520.28	1717.00	1.03	25.46	nr	**1742.46**
50 mm dia.	1765.15	1993.55	1.18	29.16	nr	**2022.71**
Chrome plated thermostatic mixing valves including non-return valves and inlet swivel connections with strainers; copper compression fittings						
15 mm dia.	98.20	110.90	0.69	17.05	nr	**127.95**
Chrome plated thermostatic mixing valves including non-return valves and inlet swivel connections with angle pattern combined isolating valves and strainers; copper compression fittings						
15 mm dia.	161.15	182.00	0.69	17.05	nr	**199.05**
Gunmetal thermostatic mixing valves including non-return valves and inlet swivel connections with strainers; copper compression fittings						
15 mm dia.	203.79	230.16	0.69	17.05	nr	**247.21**

S: PIPED SUPPLY SYSTEMS

Item	Net Price £	Material £	Labour hours	Labour £	Unit	Total rate £
S10: COLD WATER – cont						
Y11 – PIPELINE ANCILLARIES – cont						
Gunmetal thermostatic mixing valves including non-return valves and inlet swivel connections with angle pattern combined isolating valves and strainers; copper compression fittings						
15 mm dia.	214.26	241.98	0.69	17.05	nr	**259.03**
Ball float valves						
Bronze, equilibrium; copper float; working pressure cold services up to 16 bar; flanged ends; BS 4504 Table 16/21; bolted connections						
25 mm dia.	132.99	150.20	1.04	25.70	nr	**175.90**
32 mm dia.	184.50	208.38	1.22	30.14	nr	**238.52**
40 mm dia.	250.23	282.61	1.38	34.10	nr	**316.71**
50 mm dia.	393.23	444.11	1.66	41.02	nr	**485.13**
65 mm dia.	421.69	476.26	1.93	47.70	nr	**523.96**
80 mm dia.	470.19	531.03	2.16	53.38	nr	**584.41**
Heavy, equilibrium; with long tail and backnut; copper float; screwed for iron						
25 mm dia.	122.51	138.36	1.58	39.05	nr	**177.41**
32 mm dia.	168.19	189.95	1.78	43.99	nr	**233.94**
40 mm dia.	237.11	267.79	1.90	46.95	nr	**314.74**
50 mm dia.	311.36	351.65	2.65	65.48	nr	**417.13**
Brass, ball valve; BS 1212; copper float; screwed						
15 mm dia.	6.62	7.48	0.25	6.18	nr	**13.66**
22 mm dia.	12.03	13.59	0.29	7.17	nr	**20.76**
28 mm dia.	52.43	59.21	0.35	8.65	nr	**67.86**
Gate valves						
DZR copper alloy wedge non-rising stem; capillary joint to copper						
15 mm dia.	13.79	15.58	0.84	20.76	nr	**36.34**
22 mm dia.	16.92	19.10	1.01	24.96	nr	**44.06**
28 mm dia.	22.88	25.84	1.19	29.41	nr	**55.25**
35 mm dia.	41.05	46.36	1.38	34.10	nr	**80.46**
42 mm dia.	69.75	78.77	1.62	40.04	nr	**118.81**
54 mm dia.	97.71	110.35	1.94	47.94	nr	**158.29**

S: PIPED SUPPLY SYSTEMS

Item	Net Price £	Material £	Labour hours	Labour £	Unit	**Total rate £**
Cocks; capillary joints to copper						
Stopcock; brass head with gun metal body						
15 mm dia.	6.08	6.86	0.45	11.13	nr	**17.99**
22 mm dia.	11.36	12.83	0.46	11.36	nr	**24.19**
28 mm dia.	32.30	36.48	0.54	13.35	nr	**49.83**
Lockshield stop cocks; brass head with gun metal body						
15 mm dia.	7.03	7.94	0.45	11.13	nr	**19.07**
22 mm dia.	7.03	7.94	0.46	11.36	nr	**19.30**
28 mm dia.	7.03	7.94	0.54	13.35	nr	**21.29**
DZR stopcock; brass head with gun metal body						
15 mm dia.	16.67	18.83	0.45	11.13	nr	**29.96**
22 mm dia.	28.87	32.61	0.46	11.36	nr	**43.97**
28 mm dia.	48.11	54.34	0.54	13.35	nr	**67.69**
Gunmetal stopcock						
35 mm dia.	62.89	71.02	0.69	17.05	nr	**88.07**
42 mm dia.	83.52	94.33	0.71	17.54	nr	**111.87**
54 mm dia.	124.76	140.90	0.81	20.01	nr	**160.91**
Double union stopcock						
15 mm dia.	17.92	20.24	0.60	14.83	nr	**35.07**
22 mm dia.	25.19	28.45	0.60	14.83	nr	**43.28**
28 mm dia.	44.82	50.62	0.69	17.05	nr	**67.67**
Double union DZR stopcock						
15 mm dia.	29.25	33.04	0.60	14.83	nr	**47.87**
22 mm dia.	35.95	40.61	0.61	15.08	nr	**55.69**
28 mm dia.	66.46	75.06	0.69	17.05	nr	**92.11**
Double union gun metal stopcock						
35 mm dia.	110.73	125.06	0.63	15.57	nr	**140.63**
42 mm dia.	151.97	171.64	0.67	16.55	nr	**188.19**
54 mm dia.	239.07	270.00	0.85	21.00	nr	**291.00**
Double union stopcock with easy clean cover						
15 mm dia.	31.04	35.06	0.60	14.83	nr	**49.89**
22 mm dia.	38.72	43.73	0.61	15.08	nr	**58.81**
28 mm dia.	72.00	81.31	0.69	17.05	nr	**98.36**
Combined stopcock and drain						
15 mm dia.	30.79	34.77	0.67	16.55	nr	**51.32**
22 mm dia.	37.58	42.44	0.68	16.81	nr	**59.25**
Combined DZR stopcock and drain						
15 mm dia.	40.70	45.97	0.67	16.55	nr	**62.52**

S: PIPED SUPPLY SYSTEMS

Item	Net Price £	Material £	Labour hours	Labour £	Unit	Total rate £
S10: COLD WATER – cont						
Y11 – PIPELINE ANCILLARIES – cont						
Gate valve						
DZR copper alloy wedge non-rising stem; compression joint to copper						
15 mm dia.	13.79	15.58	0.84	20.76	nr	**36.34**
22 mm dia.	16.92	19.10	1.01	24.96	nr	**44.06**
28 mm dia.	22.88	25.84	1.19	29.41	nr	**55.25**
35 mm dia.	41.05	46.36	1.38	34.10	nr	**80.46**
42 mm dia.	69.75	78.77	1.62	40.04	nr	**118.81**
54 mm dia.	97.71	110.35	1.94	47.94	nr	**158.29**
Cocks; compression joints to copper						
Stopcock; brass head gun metal body						
15 mm dia.	5.53	6.24	0.42	10.38	nr	**16.62**
22 mm dia.	10.33	11.67	0.42	10.38	nr	**22.05**
28 mm dia.	29.37	33.17	0.45	11.13	nr	**44.30**
Lockshield stopcock; brass head gun metal body						
15 mm dia.	5.86	6.62	0.42	10.38	nr	**17.00**
22 mm dia.	8.39	9.48	0.42	10.38	nr	**19.86**
28 mm dia.	14.93	16.86	0.45	11.13	nr	**27.99**
DZR Stopcock						
15 mm dia.	13.90	15.70	0.38	9.38	nr	**25.08**
22 mm dia.	24.07	27.18	0.39	9.63	nr	**36.81**
28 mm dia.	40.09	45.28	0.40	9.88	nr	**55.16**
35 mm dia.	62.89	71.02	0.52	12.85	nr	**83.87**
42 mm dia.	83.52	94.33	0.54	13.35	nr	**107.68**
54 mm dia.	124.76	140.90	0.63	15.57	nr	**156.47**
DZR Lockshield stopcock						
15 mm dia.	5.86	6.62	0.38	9.38	nr	**16.00**
22 mm dia.	9.25	10.44	0.39	9.63	nr	**20.07**
Combined stop/draincock						
15 mm dia.	30.79	34.77	0.22	5.44	nr	**40.21**
22 mm dia.	39.62	44.75	0.45	11.13	nr	**55.88**
DZR combined stop/draincock						
15 mm dia.	40.70	45.97	0.41	10.13	nr	**56.10**
22 mm dia.	52.73	59.56	0.42	10.38	nr	**69.94**
Stopcock to polyethylene						
15 mm dia.	21.27	24.02	0.38	9.38	nr	**33.40**
20 mm dia.	26.80	30.26	0.39	9.63	nr	**39.89**
25 mm dia.	35.64	40.25	0.40	9.88	nr	**50.13**

S: PIPED SUPPLY SYSTEMS

Item	Net Price £	Material £	Labour hours	Labour £	Unit	**Total rate £**
Draw off coupling						
15 mm dia.	12.92	14.60	0.38	9.38	nr	**23.98**
DZR draw off coupling						
15 mm dia.	12.92	14.60	0.38	9.38	nr	**23.98**
22 mm dia.	14.95	16.88	0.39	9.63	nr	**26.51**
Draw off elbow						
15 mm dia.	14.18	16.01	0.38	9.38	nr	**25.39**
22 mm dia.	17.48	19.74	0.39	9.63	nr	**29.37**
Lockshield drain cock						
15 mm dia.	14.83	16.75	0.41	10.13	nr	**26.88**
Check valves						
DZR copper alloy and bronze, WRC approved cartridge double check valve; BS 6282; working pressure cold services up to 10 bar at 65°C; screwed ends						
32 mm dia.	88.55	100.01	1.38	34.10	nr	**134.11**
40 mm dia.	100.89	113.94	1.62	40.04	nr	**153.98**
50 mm dia.	165.45	186.85	1.94	47.94	nr	**234.79**
Y20 – PUMPS						
Packaged cold water pressure booster set; fully automatic; 3 phase supply; includes fixing in position; electrical work elsewhere.						
Pressure booster set						
0.75 l/s @ 30 m head	3962.17	4474.85	9.38	231.78	nr	**4706.63**
1.5 l/s @ 30 m head	4705.08	5313.89	9.38	231.78	nr	**5545.67**
3 l/s @ 30 m head	5748.68	6492.53	10.38	256.50	nr	**6749.03**
6 l/s @ 30 m head	12983.19	14663.15	10.38	256.50	nr	**14919.65**
12 l/s @ 30 m head	16220.13	18318.93	12.38	305.92	nr	**18624.85**
0.75 l/s @ 50 m head	4510.52	5094.16	9.38	231.78	nr	**5325.94**
1.5 l/s @ 50 m head	5748.68	6492.53	9.38	231.78	nr	**6724.31**
3 l/s @ 50 m head	6350.10	7171.77	10.38	256.50	nr	**7428.27**
6 l/s @ 50 m head	14500.84	16377.18	10.38	256.50	nr	**16633.68**
12 l/s @ 50 m head	17812.07	20116.86	12.38	305.92	nr	**20422.78**
0.75 l/s @ 70 m head	4935.02	5573.59	9.38	231.78	nr	**5805.37**
1.5 l/s @ 70 m head	6346.55	7167.76	9.38	231.78	nr	**7399.54**
3 l/s @ 70 m head	6756.92	7631.23	10.38	256.50	nr	**7887.73**
6 l/s @ 70 m head	15830.99	17879.44	10.38	256.50	nr	**18135.94**
12 l/s @ 70 m head	19244.81	21735.00	12.38	305.92	nr	**22040.92**

S: PIPED SUPPLY SYSTEMS

Item	Net Price £	Material £	Labour hours	Labour £	Unit	Total rate £
S10: COLD WATER – cont						
Y20 – PUMPS – cont						
Packaged cold water pressure booster set – cont						
Automatic sump pump for clear and drainage water; single stage centrifugal pump, presure tight electric motor; single phase supply; includes fixing in position; electrical work elsewhere						
Single pump						
1 l/s @ 2.68 m total head	209.41	236.51	3.50	86.49	nr	**323.00**
1 l/s @ 4.68 m total head	230.50	260.32	3.50	86.49	nr	**346.81**
1 l/s @ 6.68 m total head	304.32	343.70	3.50	86.49	nr	**430.19**
2 l/s @ 4.38 m total head	304.32	343.70	4.00	98.85	nr	**442.55**
2 l/s @ 6.38 m total head	304.32	343.70	4.00	98.85	nr	**442.55**
2 l/s @ 8.38 m total head	391.69	442.37	4.00	98.85	nr	**541.22**
3 l/s @ 3.7 m total head	304.32	343.70	4.50	111.20	nr	**454.90**
3 l/s @ 5.7 m total head	391.69	442.37	4.50	111.20	nr	**553.57**
4 l/s @ 2.9 m total head	304.32	343.70	5.00	123.55	nr	**467.25**
4 l/s @ 4.9 m total head	391.69	442.37	5.00	123.55	nr	**565.92**
4 l/s @ 6.9 m total head	1086.20	1226.74	5.00	123.55	nr	**1350.29**
Extra for high level alarm box with single float switch, local alarm and volt free contacts for remote alarm.	318.74	359.98	–	–	nr	**359.98**
Duty/standby pump unit						
1 l/s @ 2.68 m total head	397.71	449.17	5.00	123.55	nr	**572.72**
1 l/s @ 4.68 m total head	436.89	493.42	5.00	123.55	nr	**616.97**
1 l/s @ 6.68 m total head	590.56	666.98	5.00	123.55	nr	**790.53**
2 l/s @ 4.38 m total head	590.56	666.98	5.50	135.92	nr	**802.90**
2 l/s @ 6.38 m total head	590.56	666.98	5.50	135.92	nr	**802.90**
2 l/s @ 8.38 m total head	756.27	854.13	5.50	135.92	nr	**990.05**
3 l/s @ 3.7 m total head	590.56	666.98	6.00	148.27	nr	**815.25**
3 l/s @ 5.7 m total head	756.27	854.13	6.00	148.27	nr	**1002.40**
4 l/s @ 2.9 m total head	590.56	666.98	6.50	160.62	nr	**827.60**
4 l/s @ 4.9 m total head	756.27	854.13	6.50	160.62	nr	**1014.75**
4 /s @ 6.9 m total head	2009.69	2269.73	7.00	172.98	nr	**2442.71**
Extra for 4nr float switches to give pump on, off and high level alarm	333.02	376.11	–	–	nr	**376.11**
Extra for dual pump control panel, internal wall mounted IP54, including volt free contacts	1585.81	1791.01	4.00	98.85	nr	**1889.86**

S: PIPED SUPPLY SYSTEMS

Item	Net Price £	Material £	Labour hours	Labour £	Unit	Total rate £
Y21 – TANKS						
Cisterns; fibreglass; complete with ball valve, fixing plate and fitted covers						
Rectangular						
60 litres capacity	187.20	211.43	1.33	32.86	nr	**244.29**
100 litres capacity	206.96	233.74	1.40	34.60	nr	**268.34**
150 litres capacity	235.04	265.46	1.61	39.79	nr	**305.25**
250 litres capacity	266.24	300.69	1.61	39.79	nr	**340.48**
420 litres capacity	371.28	419.32	1.99	49.17	nr	**468.49**
730 litres capacity	626.08	707.09	3.31	81.79	nr	**788.88**
800 litres capacity	718.64	811.62	3.60	88.96	nr	**900.58**
1700 litres capacity	1101.36	1243.87	13.32	329.15	nr	**1573.02**
2250 litres capacity	1078.48	1218.03	20.18	498.68	nr	**1716.71**
3400 litres capacity	1660.88	1875.79	24.50	605.42	nr	**2481.21**
4500 litres capacity	1977.04	2232.86	29.91	739.11	nr	**2971.97**
Cisterns; polypropylene; complete with ball valve, fixing plate and cover; includes placing in position						
Rectangular						
18 litres capacity	65.87	74.39	1.00	24.71	nr	**99.10**
68 litres capacity	88.76	100.25	1.00	24.71	nr	**124.96**
91 litres capacity	89.34	100.90	1.00	24.71	nr	**125.61**
114 litres capacity	99.40	112.26	1.00	24.71	nr	**136.97**
182 litres capacity	131.28	148.27	1.00	24.71	nr	**172.98**
227 litres capacity	167.21	188.84	1.00	24.71	nr	**213.55**
Circular						
114 litres capacity	92.99	105.02	1.00	24.71	nr	**129.73**
227 litres capacity	94.82	107.09	1.00	24.71	nr	**131.80**
318 litres capacity	157.71	178.11	1.00	24.71	nr	**202.82**
455 litres capacity	181.27	204.72	1.00	24.71	nr	**229.43**
Steel sectional water storage tank; hot pressed steel tank to BS 1564 TYPE 1; 5 mm plate; pre-insulated and complete with all connections and fittings to comply with BSEN 13280; 2001 and WRAS water supply (water fittings) regulations 1999; externally flanged base and sides; cost of erection (on prepared base) is included within the net price, labour cost allows for offloading and positioning materials						
Note: Prices are based on the most economical tank size for each volume, and the cost will vary with differing tank dimensions, for the same volume						

S: PIPED SUPPLY SYSTEMS

Item	Net Price £	Material £	Labour hours	Labour £	Unit	Total rate £
S10: COLD WATER – cont						
Y21 – TANKS – cont						
Volume, size						
4,900 litres, 3.66 m × 1.22 m × 1,22 m (h)	6810.88	7692.18	6.00	148.27	nr	**7840.45**
20,300 litres, 3.66 m × 2.4 m × 2.4 m (h)	14508.26	16385.56	12.00	296.53	nr	**16682.09**
52,000 litres, 6.1 m × 3.6 m × 2.4 m (h)	25868.93	29216.24	19.00	469.52	nr	**29685.76**
94,000 litres, 7.3 m × 3.6 m × 3.6 m (h)	38918.08	43953.88	28.00	691.92	nr	**44645.80**
140,000 litres, 9.7 m × 6.1 m × 2.44 m (h)	50827.20	57403.99	28.00	691.92	nr	**58095.91**
GRP sectional water storage tank; pre-insulated and complete with all connections and fittings to comply with BSEN 13280; 2001 and WRAS water supply (water fittings) regulations 1999; externally flanged base and sides; cost of erection (on prepared base) is included within the net price, labour cost allows for offloading and positioning materials						
Note: Prices are based on the most economical tank size for each volume, and the cost will vary with differing tank dimensions, for the same volume						
Volume, size						
4,500 litres, 3 m × 1 m × 1.5 m (h)	3060.13	3456.10	5.00	123.55	nr	**3579.65**
10,000 litres, 2.5 m × 2 m × 2 m (h)	4261.11	4812.47	7.00	172.98	nr	**4985.45**
20,000 litres, 4 m × 2.5 m × 2 m (h)	6263.43	7073.88	10.00	247.12	nr	**7321.00**
30,000 litres 5 m × 3 m × 2 m (h)	8036.06	9075.89	12.00	296.53	nr	**9372.42**
40,000 litres, 5 m × 4 m × 2 m (h)	9530.59	10763.81	12.00	296.53	nr	**11060.34**
50,000 litres, 5 m × 4 m × 2.5 m (h)	11984.05	13534.73	14.00	345.95	nr	**13880.68**
60,000 litres, 6 m × 4 m × 2.5 m (h)	13675.31	15444.83	16.00	395.38	nr	**15840.21**
70,000 litres, 7 m × 4 m × 2.5 m (h)	15373.78	17363.07	16.00	395.38	nr	**17758.45**
80,000 litres, 8 m × 4 m × 2.5 m (h)	16886.85	19071.92	16.00	395.38	nr	**19467.30**
90,000 litres, 6 m × 5 m × 3 m (h)	17656.26	19940.89	16.00	395.38	nr	**20336.27**
105,000 litres, 7 m × 5 m × 3 m (h)	19454.64	21971.97	24.00	593.07	nr	**22565.04**
120,000 litres, 8 m × 5 m × 3 m (h)	21285.98	24040.28	24.00	593.07	nr	**24633.35**
135,000 litres, 9 m × 6 m × 2.5 m (h)	23470.61	26507.59	24.00	593.07	nr	**27100.66**
144,000 litres, 8 m × 6 m × 3 m (h)	23779.61	26856.57	24.00	593.07	nr	**27449.64**

S: PIPED SUPPLY SYSTEMS

Item	Net Price £	Material £	Labour hours	Labour £	Unit	Total rate £
Y25 – CLEANING AND CHEMICAL TREATMENT						
Electromagnetic water conditioner, complete with control box; maximum inlet pressure 16 bar; electrical work elsewhere						
Connection size, nominal flow rate at 50 mbar						
20 mm dia., 0.3 l/s	1659.20	1873.89	1.25	30.89	nr	**1904.78**
25 mm dia., 0.6 l/s	2179.32	2461.31	1.45	35.83	nr	**2497.14**
32 mm dia., 1.2 l/s	3105.15	3506.94	1.55	38.30	nr	**3545.24**
40 mm dia., 1.7 l/s	3718.89	4200.09	1.65	40.77	nr	**4240.86**
50 mm dia., 3.4 l/s	4863.17	5492.43	1.75	43.25	nr	**5535.68**
65 mm dia., 5.2 l/s	5331.28	6021.12	1.90	46.95	nr	**6068.07**
100 mm dia., 30.5 l/s	10138.82	11450.73	3.00	74.13	nr	**11524.86**
Ultraviolet water sterillising unit, complete with control unit; UV lamp housed in quartz tube; unit complete with UV intensity sensor, flushing and discharge valve and facilities for remote alarm; electrical work elsewhere						
Maximum flow rate (@ 250 J/m^2 exposure), connection size						
0.82 l/s, 40 mm dia.	2508.82	2833.45	1.98	48.93	nr	**2882.38**
1.28 l/s, 40 mm dia.	2880.71	3253.46	1.98	48.93	nr	**3302.39**
2.00 l/s, 40 mm dia.	3767.53	4255.03	1.98	48.93	nr	**4303.96**
4.14 l/s, 50 mm dia.	4139.41	4675.03	2.10	51.90	nr	**4726.93**
1.28 l/s, 40 mm dia.	4110.81	4642.72	1.98	48.93	nr	**4691.65**
2.00 l/s, 40 mm dia.	5712.79	6452.00	1.98	48.93	nr	**6500.93**
4.14 l/s, 50 mm dia.	6542.39	7388.94	2.10	51.90	nr	**7440.84**
7.4 l/s, 80 mm dia.	10967.88	12387.07	3.60	88.96	nr	**12476.03**
16.8 l/s 100 mm dia.	13679.81	15449.91	3.60	88.96	nr	**15538.87**
32.3 l/s 100 mm dia.	15367.61	17356.11	3.60	88.96	nr	**17445.07**
Base exchange water softener complete with resin tank, brine tank and consumption data monitoring facilities						
Capacities of softeners are based on 300ppm hardness and quoted in m^3 of softened water produced. Design flow rates are recommended for continuous use						
Simplex configuration						
Design flow rate, min-max softenend water produced						
1 l/s, 5.8 m^3–11.2 m^3	1807.95	2041.89	8.00	197.69	nr	**2239.58**
1.3 l/s, 11.7 m^3–21.4 m^3	2471.63	2791.44	8.00	197.69	nr	**2989.13**

S: PIPED SUPPLY SYSTEMS

Item	Net Price £	Material £	Labour hours	Labour £	Unit	Total rate £
S10: COLD WATER – cont						
Y25 – CLEANING AND CHEMICAL TREATMENT – cont						
Capacities of softeners are based on 300ppm hardness and quoted in m³ of softened water produced. – cont						
Design flow rate, min-max softenend water produced – cont						
1.3 l/s, 15.5 m³–28.5 m³	2814.92	3179.16	10.00	247.12	nr	**3426.28**
1.6 l/s, 23.3 m³–42.7 m³	3575.86	4038.56	10.00	247.12	nr	**4285.68**
1.6 l/s, 38.8 m³–71.2 m³	4027.85	4549.04	12.00	296.53	nr	**4845.57**
1.9 l/s, 11.7 m³–21.4 m³	4062.18	4587.81	12.00	296.53	nr	**4884.34**
3.2 l/s, 19.4 m³–35.6 m³	4359.69	4923.82	12.00	296.53	nr	**5220.35**
4.4 l/s, 31 m³–57 m³	5137.79	5802.60	15.00	370.67	nr	**6173.27**
5.1 l/s, 46.6 m³–85.4 m³	7655.20	8645.74	15.00	370.67	nr	**9016.41**
5.1 l/s, 77.7 m³–142.4 m³	9360.17	10571.33	18.00	444.80	nr	**11016.13**
Duplex configuration						
Design flow rate, min-max softenend water produced						
2 l/s, 5.8 m³–22.4 m³	3009.44	3398.84	12.00	296.53	nr	**3695.37**
2.6 l/s, 11.7 m³–42.8 m³	4085.06	4613.64	12.00	296.53	nr	**4910.17**
2.6 l/s, 15.5 m³–57 m³	4279.59	4833.35	15.00	370.67	nr	**5204.02**
3.2 l/s, 23.3 m³–85.4 m³	5526.85	6241.99	15.00	370.67	nr	**6612.66**
3.2 l/s, 38.8 m³–142.4 m³	6602.47	7456.79	18.00	444.80	nr	**7901.59**
3.8 l/s, 11.7 m³–42.8 m³	7117.39	8038.34	18.00	444.80	nr	**8483.14**
6.4 l/s, 19.4 m³–71.2 m³	7655.20	8645.74	18.00	444.80	nr	**9090.54**
8.8 l/s, 31.1 m³–114 m³	8776.59	9912.24	23.00	568.35	nr	**10480.59**
10.2 l/s, 46.6 m³–170.8 m³	14063.14	15882.84	23.00	568.35	nr	**16451.19**
10.2 l/s, 77.7 m³–284.8 m³	17438.75	19695.24	27.00	667.20	nr	**20362.44**
Triplex configuration						
Design flow rate, min-max softenend water produced						
3 l/s, 5.8 m³–33.6 m³	4182.33	4723.50	15.00	370.67	nr	**5094.17**
3.9 l/s, 11.7 m³–64.2 m³	5927.34	6694.31	15.00	370.67	nr	**7064.98**
3.9 l/s, 15.5 m³–85.5 m³	6259.18	7069.09	18.00	444.80	nr	**7513.89**
4.8 l/s, 23.3 m³–128.1 m³	8147.24	9201.45	18.00	444.80	nr	**9646.25**
4.8 l/s, 38.8 m³–213.6 m³	9360.17	10571.33	22.00	543.65	nr	**11114.98**
5.7 l/s, 11.7 m³–64.2 m³	10481.56	11837.82	22.00	543.65	nr	**12381.47**
9.6 l/s, 19.4 m³–106.8 m³	11362.65	12832.92	22.00	543.65	nr	**13376.57**
13.2 l/s, 31.1 m³–171.0 m³	13033.29	14719.73	27.00	667.20	nr	**15386.93**
15.3 l/s, 46.6 m³–256.2 m³	20963.12	23675.64	27.00	667.20	nr	**24342.84**
15.3 l/s, 77.7 m³–427.2 m³	26192.45	29581.62	32.00	790.75	nr	**30372.37**

S: PIPED SUPPLY SYSTEMS

Item	Net Price £	Material £	Labour hours	Labour £	Unit	**Total rate £**
Y50 – THERMAL INSULATION						
Flexible closed cell walled insulation; Class 1/Class O; adhesive joints; including around fittings						
6 mm wall thickness						
15 mm dia.	1.15	1.30	0.15	3.70	m	**5.00**
22 mm dia.	1.37	1.55	0.15	3.70	m	**5.25**
28 mm dia.	1.73	1.96	0.15	3.70	m	**5.66**
9 mm wall thickness						
15 mm dia.	1.32	1.49	0.15	3.70	m	**5.19**
22 mm dia.	1.67	1.89	0.15	3.70	m	**5.59**
28 mm dia.	1.77	2.00	0.15	3.70	m	**5.70**
35 mm dia.	2.01	2.27	0.15	3.70	m	**5.97**
42 mm dia.	2.30	2.60	0.15	3.70	m	**6.30**
54 mm dia.	2.38	2.69	0.15	3.70	m	**6.39**
13 mm wall thickness						
15 mm dia.	1.69	1.91	0.15	3.70	m	**5.61**
22 mm dia.	2.09	2.36	0.15	3.70	m	**6.06**
28 mm dia.	2.48	2.81	0.15	3.70	m	**6.51**
35 mm dia.	2.67	3.02	0.15	3.70	m	**6.72**
42 mm dia.	3.14	3.55	0.15	3.70	m	**7.25**
54 mm dia.	4.37	4.94	0.15	3.70	m	**8.64**
67 mm dia.	6.46	7.29	0.15	3.70	m	**10.99**
76 mm dia.	7.60	8.59	0.15	3.70	m	**12.29**
108 mm dia.	7.96	8.99	0.15	3.70	m	**12.69**
19 mm wall thickness						
15 mm dia.	2.76	3.12	0.15	3.70	m	**6.82**
22 mm dia.	3.35	3.78	0.15	3.70	m	**7.48**
28 mm dia.	4.55	5.14	0.15	3.70	m	**8.84**
35 mm dia.	5.27	5.96	0.15	3.70	m	**9.66**
42 mm dia.	6.22	7.03	0.15	3.70	m	**10.73**
54 mm dia.	7.87	8.88	0.15	3.70	m	**12.58**
67 mm dia.	9.44	10.66	0.15	3.70	m	**14.36**
76 mm dia.	10.90	12.31	0.22	5.44	m	**17.75**
108 mm dia.	14.88	16.81	0.22	5.44	m	**22.25**
25 mm wall thickness						
15 mm dia.	5.40	6.10	0.15	3.70	m	**9.80**
22 mm dia.	5.92	6.68	0.15	3.70	m	**10.38**
28 mm dia.	6.69	7.56	0.15	3.70	m	**11.26**
35 mm dia.	7.42	8.38	0.15	3.70	m	**12.08**
42 mm dia.	7.91	8.94	0.15	3.70	m	**12.64**
54 mm dia.	9.30	10.51	0.15	3.70	m	**14.21**
67 mm dia.	12.53	14.15	0.15	3.70	m	**17.85**
76 mm dia.	14.12	15.94	0.22	5.44	m	**21.38**

S: PIPED SUPPLY SYSTEMS

Item	Net Price £	Material £	Labour hours	Labour £	Unit	**Total rate £**
S10: COLD WATER – cont						
Y50 – THERMAL INSULATION – cont						
32 mm wall thickness						
15 mm dia.	6.77	7.65	0.15	3.70	m	**11.35**
22 mm dia.	7.42	8.38	0.15	3.70	m	**12.08**
28 mm dia.	8.61	9.72	0.15	3.70	m	**13.42**
35 mm dia.	8.89	10.04	0.15	3.70	m	**13.74**
42 mm dia.	10.34	11.68	0.15	3.70	m	**15.38**
54 mm dia.	13.01	14.70	0.15	3.70	m	**18.40**
76 mm dia.	19.46	21.98	0.22	5.44	m	**27.42**
Note: For mineral fibre sectional insulation; bright class O foil faced; bright class O foil taped joints; 19 mm aluminium bands rates, refer to section T31 – Low Temperature Hot Water Heating, Y50 Thermal Insluation						
Note: For mineral fibre sectional insulation; bright class O foil faced; bright class O foil taped joints; 22 swg plain/embossed aluminium cladding; pop riveted rates, refer to section T31 – Low Temperature Hot Water Heating, Y50 Thermal Insluation						
Note: For mineral fibre sectional insulation; bright class O foil faced; bright class O foil taped joints; 0.8 mm polyisobutylene sheeting; welded joints rates, refer to section T31 – Low Temperature Hot Water Heating, Y50 Thermal Insluation						

S: PIPED SUPPLY SYSTEMS

Item	Net Price £	Material £	Labour hours	Labour £	Unit	Total rate £
S11: HOT WATER						
Y10 – PIPELINES						
Note: For pipework prices refer to section S10 – Cold Water						
Y11 – PIPELINE ANCILLARIES						
Note: For prices for ancillaries refer to section S10 – Cold Water						
Y23 – STORAGE CYLINDERS/ CALORIFIERS						
CYLINDERS						
Insulated copper storage cylinders; BS 699; includes placing in position						
Grade 3 (maximum 10 m working head)						
BS size 6; 115 litres capacity; 400 mm dia.; 1050 mm high	177.86	200.88	1.50	37.10	nr	**237.98**
BS size 7; 120 litres capacity; 450 mm dia.; 900 mm high	210.28	237.49	2.00	49.42	nr	**286.91**
BS size 8; 144 litres capacity; 450 mm dia.; 1050 mm high	223.54	252.46	2.80	69.22	nr	**321.68**
Grade 4 (maximum 6 m working head)						
BS size 2; 96 litres capacity; 400 mm dia.; 900 mm high	137.71	155.53	1.50	37.10	nr	**192.63**
BS size 7; 120 litres capacity; 450 mm dia.; 900 mm high	144.72	163.44	1.50	37.10	nr	**200.54**
BS size 8; 144 litres capacity; 450 mm dia.; 1050 mm high	172.75	195.11	1.50	37.10	nr	**232.21**
BS size 9; 166 litres capacity; 450 mm dia.; 1200 mm high	252.60	285.28	1.50	37.10	nr	**322.38**
Storage cylinders; brazed copper construction; to BS 699; screwed bosses; includes placing in position						
Tested to 2.2 bar, 15 m maximum head						
144 litres	782.88	884.18	3.00	74.22	nr	**958.40**
160 litres	885.02	999.54	3.00	74.22	nr	**1073.76**
200 litres	912.23	1030.27	3.76	92.90	nr	**1123.17**
255 litres	1038.16	1172.49	3.76	92.90	nr	**1265.39**
290 litres	1402.36	1583.82	3.76	92.90	nr	**1676.72**
370 litres	1606.62	1814.51	4.50	111.31	nr	**1925.82**
450 litres	2188.54	2471.73	5.00	123.55	nr	**2595.28**

S: PIPED SUPPLY SYSTEMS

Item	Net Price £	Material £	Labour hours	Labour £	Unit	Total rate £
S11: HOT WATER – cont						
Y23 – STORAGE CYLINDERS/ CALORIFIERS – cont						
Storage cylinders – cont						
Tested to 2.55 bar, 17 m maximum head						
550 litres	2369.76	2676.40	5.00	123.55	nr	**2799.95**
700 litres	2770.47	3128.95	6.02	148.87	nr	**3277.82**
800 litres	3204.39	3619.02	6.54	161.52	nr	**3780.54**
900 litres	3467.48	3916.16	8.00	197.69	nr	**4113.85**
1000 litres	3659.17	4132.65	8.00	197.69	nr	**4330.34**
1250 litres	4007.62	4526.19	13.16	325.15	nr	**4851.34**
1500 litres	6138.41	6932.69	15.15	374.40	nr	**7307.09**
2000 litres	7366.64	8319.84	17.24	426.05	nr	**8745.89**
3000 litres	10347.59	11686.52	24.39	602.72	nr	**12289.24**
Indirect cylinders; copper; bolted top; up to 5 tappings for connections; BS 1586; includes placing in position						
Grade 3, tested to 1.45 bar, 10 m maximum head						
74 litres capacity	358.95	405.40	1.50	37.10	nr	**442.50**
96 litres capacity	365.44	412.72	1.50	37.10	nr	**449.82**
114 litres capacity	375.23	423.78	1.50	37.10	nr	**460.88**
117 litres capacity	389.46	439.85	2.00	49.42	nr	**489.27**
140 litres capacity	401.32	453.25	2.50	61.78	nr	**515.03**
162 litres capacity	561.22	633.84	3.00	74.22	nr	**708.06**
190 litres capacity	613.46	692.84	3.51	86.70	nr	**779.54**
245 litres capacity	717.86	810.75	3.80	93.95	nr	**904.70**
280 litres capacity	1272.54	1437.20	4.00	98.85	nr	**1536.05**
360 litres capacity	1376.96	1555.13	4.50	111.31	nr	**1666.44**
440 litres capacity	1598.84	1805.73	4.50	111.31	nr	**1917.04**
Grade 2, tested to 2.2 bar, 15 m maximum head						
117 litres capacity	518.79	585.92	2.00	49.42	nr	**635.34**
140 litres capacity	564.50	637.54	2.50	61.78	nr	**699.32**
162 litres capacity	646.06	729.66	2.80	69.22	nr	**798.88**
190 litres capacity	750.52	847.63	3.00	74.22	nr	**921.85**
245 litres capacity	907.12	1024.50	4.00	98.85	nr	**1123.35**
280 litres capacity	1448.74	1636.20	4.00	98.85	nr	**1735.05**
360 litres capacity	1598.84	1805.73	4.50	111.31	nr	**1917.04**
440 litres capacity	1892.49	2137.37	4.50	111.31	nr	**2248.68**
Grade 1, tested 3.65 bar, 25 m maximum head						
190 litres capacity	1115.95	1260.35	3.00	74.22	nr	**1334.57**
245 litres capacity	1269.25	1433.49	3.00	74.22	nr	**1507.71**
280 litres capacity	1794.63	2026.84	4.00	98.85	nr	**2125.69**
360 litres capacity	2271.05	2564.91	4.50	111.31	nr	**2676.22**
440 litres capacity	2757.17	3113.94	4.50	111.31	nr	**3225.25**

S: PIPED SUPPLY SYSTEMS

Item	Net Price £	Material £	Labour hours	Labour £	Unit	**Total rate £**
Indirect cylinders, including manhole; BS 853						
Grade 3, tested to 1.5 bar, 10 m maximum head						
550 litres capacity	2320.87	2621.18	5.21	128.70	nr	**2749.88**
700 litres capacity	2569.56	2902.05	6.02	148.87	nr	**3050.92**
800 litres capacity	2984.00	3370.11	6.54	161.52	nr	**3531.63**
1000 litres capacity	3730.03	4212.68	7.04	174.03	nr	**4386.71**
1500 litres capacity	4310.23	4867.95	10.00	247.12	nr	**5115.07**
2000 litres capacity	5968.04	6740.27	16.13	398.56	nr	**7138.83**
Grade 2, tested to 2.55 bar, 15 m maximum head						
550 litres capacity	2573.13	2906.08	5.21	128.70	nr	**3034.78**
700 litres capacity	3220.44	3637.15	6.02	148.87	nr	**3786.02**
800 litres capacity	3398.45	3838.19	6.54	161.52	nr	**3999.71**
1000 litres capacity	4207.62	4752.07	7.04	174.03	nr	**4926.10**
1500 litres capacity	5178.61	5848.70	10.00	247.12	nr	**6095.82**
2000 litres capacity	6473.29	7310.90	16.13	398.56	nr	**7709.46**
Grade 1, tested to 4 bar, 25 m maximum head						
550 litres capacity	2993.88	3381.27	5.21	128.70	nr	**3509.97**
700 litres capacity	3398.45	3838.19	6.02	148.87	nr	**3987.06**
800 litres capacity	3641.22	4112.37	6.54	161.52	nr	**4273.89**
1000 litres capacity	4854.98	5483.19	7.04	174.03	nr	**5657.22**
1500 litres capacity	5825.91	6579.75	10.00	247.12	nr	**6826.87**
2000 litres capacity	7120.58	8041.95	16.13	398.56	nr	**8440.51**
Storage calorifiers; copper; heater battery capable of raising temperature of contents from 10°C to 65°C in one hour; static head not exceeding 1.35 bar; BS 853; includes fixing in position on cradles or legs						
Horizontal; primary LPHW at 82°C/71°C						
400 litres capacity	3322.44	3752.35	7.04	174.03	nr	**3926.38**
1000 litres capacity	5315.88	6003.73	8.00	197.69	nr	**6201.42**
2000 litres capacity	10631.83	12007.53	14.08	348.04	nr	**12355.57**
3000 litres capacity	13123.64	14821.77	25.00	617.78	nr	**15439.55**
4000 litres capacity	15947.70	18011.25	40.00	988.45	nr	**18999.70**
4500 litres capacity	17972.35	20297.88	50.00	1235.56	nr	**21533.44**
Vertical; primary LPHW at 82°C/71°C						
400 litres capacity	3744.40	4228.91	7.04	174.03	nr	**4402.94**
1000 litres capacity	6017.81	6796.48	8.00	197.69	nr	**6994.17**
2000 litres capacity	11462.43	12945.62	14.08	348.04	nr	**13293.66**
3000 litres capacity	14328.05	16182.02	25.00	617.78	nr	**16799.80**
4000 litres capacity	17575.67	19849.87	40.00	988.45	nr	**20838.32**
4500 litres capacity	19868.18	22439.02	50.00	1235.56	nr	**23674.58**

S: PIPED SUPPLY SYSTEMS

Item	Net Price £	Material £	Labour hours	Labour £	Unit	Total rate £
S11: HOT WATER – cont						
Y23 – STORAGE CYLINDERS/ CALORIFIERS – cont						
Storage calorifiers; galvanized mild steel; heater battery capable of raising temperature of contents from 10°C to 65°C in one hour; static head not exceeding 1.35 bar; BS 853; includes fixing in position on cradles or legs						
Horizontal; primary LPHW at 82°C/71°C						
400 litres capacity	3322.44	3752.35	7.04	174.03	nr	**3926.38**
1000 litres capacity	5315.88	6003.73	8.00	197.69	nr	**6201.42**
2000 litres capacity	10631.83	12007.53	14.08	348.04	nr	**12355.57**
3000 litres capacity	13123.64	14821.77	25.00	617.78	nr	**15439.55**
4000 litres capacity	15947.70	18011.25	40.00	988.45	nr	**18999.70**
4500 litres capacity	17972.35	20297.88	50.00	1235.56	nr	**21533.44**
Vertical; primary LPHW at 82°C/71°C						
400 litres capacity	3744.40	4228.91	7.04	174.03	nr	**4402.94**
1000 litres capacity	6017.81	6796.48	8.00	197.69	nr	**6994.17**
2000 litres capacity	11462.43	12945.62	14.08	348.04	nr	**13293.66**
3000 litres capacity	14328.05	16182.02	25.00	617.78	nr	**16799.80**
4000 litres capacity	17575.67	19849.87	40.00	988.45	nr	**20838.32**
4500 litres capacity	19868.18	22439.02	50.00	1235.56	nr	**23674.58**
Indirect cylinders; mild steel, welded throughout, galvanized; with bolted connections; includes placing in position						
3.2 mm plate						
136 litres capacity	1282.80	1448.79	2.50	61.78	nr	**1510.57**
159 litres capacity	1381.50	1560.26	2.80	69.22	nr	**1629.48**
182 litres capacity	1697.31	1916.94	3.00	74.22	nr	**1991.16**
227 litres capacity	2091.99	2362.69	3.00	74.22	nr	**2436.91**
273 litres capacity	2440.30	2756.06	4.00	98.85	nr	**2854.91**
364 litres capacity	2841.94	3209.67	4.50	111.31	nr	**3320.98**
455 litres capacity	2935.38	3315.20	5.00	123.55	nr	**3438.75**
683 litres capacity	4750.33	5365.00	6.02	148.87	nr	**5513.87**
910 litres capacity	5517.14	6231.03	7.04	174.03	nr	**6405.06**

S: PIPED SUPPLY SYSTEMS

Item	Net Price £	Material £	Labour hours	Labour £	Unit	Total rate £
Y50 – INSULATION						
Refer to sections S10 – Cold Water and T31 – Low Temperature Hot Water Heating for details						
Local electric hot water heaters						
Unvented multi-point water heater; providing hot water for one or more outlets: Used with conventional taps or mixers: Factory fitted temperature and pressure relief valve: Externally adjustable thermostat: Elemental 'on' indicator: Fitted with 1 metre of 3 core cable: Electrical supply and connection excluded						
5 litre capacity, 2.2 kW rating	181.73	205.24	1.50	37.07	nr	**242.31**
10 litre capacity, 2.2 kW rating	202.68	228.91	1.50	37.07	nr	**265.98**
15 litre capacity, 2.2 kW rating	225.22	254.36	1.50	37.07	nr	**291.43**
30 litre capacity, 3 kW rating	401.40	453.34	2.00	49.42	nr	**502.76**
50 litre capacity, 3 kW rating	422.60	477.28	2.00	49.42	nr	**526.70**
80 litre capacity, 3 kW rating	831.55	939.14	2.00	49.42	nr	**988.56**
100 litre capacity, 3 kW rating	892.72	1008.23	2.00	49.42	nr	**1057.65**
Accessories						
Pressure reducing valve and expansion kit	147.77	166.89	2.00	49.42	nr	**216.31**
Thermostatic blending valve	78.64	88.81	1.00	24.71	nr	**113.52**

S: PIPED SUPPLY SYSTEMS

Item	Net Price £	Material £	Labour hours	Labour £	Unit	**Total rate £**
S32: NATURAL GAS						
Y10 – PIPELINES						
MEDIUM DENSITY POLYETHYLENE – YELLOW						
Pipe; laid underground; electrofusion joints in the running length; BS 6572; BGT PL2 standards						
Coiled service pipe						
20 mm dia.	1.39	1.57	0.37	9.15	m	**10.72**
25 mm dia.	1.81	2.04	0.41	10.13	m	**12.17**
32 mm dia.	2.99	3.38	0.47	11.62	m	**15.00**
63 mm dia.	11.38	12.85	0.60	14.83	m	**27.68**
90 mm dia.	14.94	16.87	0.90	22.24	m	**39.11**
Mains service pipe						
63 mm dia.	11.05	12.48	0.60	14.83	m	**27.31**
90 mm dia.	14.51	16.39	0.90	22.24	m	**38.63**
125 mm dia.	28.01	31.63	1.20	29.66	m	**61.29**
180 mm dia.	57.83	65.31	1.50	37.07	m	**102.38**
250 mm dia.	106.45	120.23	1.75	43.25	m	**163.48**
Extra over fittings, electrofusion joints						
Straight connector						
32 mm dia.	8.66	9.78	0.47	11.61	nr	**21.39**
63 mm dia.	16.27	18.37	0.58	14.33	nr	**32.70**
90 mm dia.	24.01	27.12	0.67	16.55	nr	**43.67**
125 mm dia.	44.95	50.77	0.83	20.51	nr	**71.28**
180 mm dia.	80.94	91.41	1.25	30.89	nr	**122.30**
Reducing connector						
90 × 63 mm dia.	33.51	37.84	0.67	16.55	nr	**54.39**
125 × 90 mm dia.	67.21	75.91	0.83	20.51	nr	**96.42**
180 × 125 mm dia.	123.32	139.28	1.25	30.89	nr	**170.17**
Bend; 45°						
90 mm dia.	64.78	73.16	0.67	16.55	nr	**89.71**
125 mm dia.	105.96	119.67	0.83	20.51	nr	**140.18**
180 mm dia.	240.37	271.47	1.25	30.89	nr	**302.36**
Bend; 90°						
63 mm dia.	41.97	47.40	0.58	14.33	nr	**61.73**
90 mm dia.	64.78	73.16	0.67	16.55	nr	**89.71**
125 mm dia.	105.96	119.67	0.83	20.51	nr	**140.18**
180 mm dia.	240.37	271.47	1.25	30.89	nr	**302.36**

S: PIPED SUPPLY SYSTEMS

Item	Net Price £	Material £	Labour hours	Labour £	Unit	**Total rate £**
Extra over malleable iron fittings, compression joints						
Straight connector						
20 mm dia.	12.36	13.96	0.38	9.39	nr	**23.35**
25 mm dia.	13.47	15.21	0.45	11.13	nr	**26.34**
32 mm dia.	15.11	17.06	0.50	12.35	nr	**29.41**
63 mm dia.	30.34	34.26	0.85	21.00	nr	**55.26**
Straight connector; polyethylene to MI						
20 mm dia.	10.50	11.86	0.31	7.67	nr	**19.53**
25 mm dia.	11.45	12.93	0.35	8.65	nr	**21.58**
32 mm dia.	12.79	14.44	0.40	9.88	nr	**24.32**
63 mm dia.	21.43	24.20	0.65	16.08	nr	**40.28**
Straight connector; polyethylene to FI						
20 mm dia.	10.50	11.86	0.31	7.67	nr	**19.53**
25 mm dia.	12.79	14.44	0.35	8.65	nr	**23.09**
32 mm dia.	12.79	14.44	0.40	9.88	nr	**24.32**
63 mm dia.	21.43	24.20	0.75	18.53	nr	**42.73**
Elbow						
20 mm dia.	16.07	18.15	0.38	9.39	nr	**27.54**
25 mm dia.	17.53	19.80	0.45	11.13	nr	**30.93**
32 mm dia.	19.64	22.19	0.50	12.35	nr	**34.54**
63 mm dia.	39.46	44.56	0.80	19.77	nr	**64.33**
Equal tee						
20 mm dia.	21.43	24.20	0.53	13.10	nr	**37.30**
25 mm dia.	25.05	28.29	0.55	13.60	nr	**41.89**
32 mm dia.	31.54	35.62	0.64	15.82	nr	**51.44**
SCREWED STEEL						
For prices for steel pipework refer to section T31 – Low Temperature Hot Water Heating						

S: PIPED SUPPLY SYSTEMS

Item	Net Price £	Material £	Labour hours	Labour £	Unit	Total rate £
S32: NATURAL GAS – cont						
PIPE IN PIPE						
Note: For pipe in pipe, a sleeve size two pipe sizes bigger than actual pipe size has been allowed. All rates refer to actual pipe size						
Black steel pipes – screwed and socketed joints; BS 1387: 1985 upto 50 mm pipe size. Butt welded joints; BS 1387: 1985 65 mm pipe size and above.						
Pipe						
25 mm	13.36	15.09	1.73	42.75	m	**57.84**
32 mm	17.75	20.04	1.95	48.20	m	**68.24**
40 mm	22.83	25.79	2.16	53.38	m	**79.17**
50 mm	29.97	33.84	2.44	60.30	m	**94.14**
65 mm	41.10	46.42	2.95	72.90	m	**119.32**
80 mm	49.67	56.10	3.42	84.52	m	**140.62**
100 mm	62.04	70.06	4.00	98.85	m	**168.91**
Extra over black steel pipes – screwed pipework; black malleable iron fittings; BS 143. Welded pipework; butt welded steel fittings; BS 1965						
Bend, 90°						
25 mm	7.70	8.70	2.91	71.91	m	**80.61**
32 mm	11.75	13.27	3.45	85.26	m	**98.53**
40 mm	16.45	18.57	5.34	131.96	m	**150.53**
50 mm	18.54	20.94	6.53	161.36	m	**182.30**
65 mm	25.05	28.29	8.84	218.44	m	**246.73**
80 mm	42.69	48.22	10.73	265.15	m	**313.37**
100 mm	56.70	64.04	12.76	315.31	m	**379.35**
Bend, 45°						
25 mm	9.75	11.02	2.91	71.91	m	**82.93**
32 mm	12.64	14.28	3.45	85.26	m	**99.54**
40 mm	14.12	15.94	5.34	131.96	m	**147.90**
50 mm	16.84	19.02	6.53	161.36	m	**180.38**
65 mm	21.99	24.84	8.84	218.44	m	**243.28**
80 mm	37.62	42.48	10.73	265.15	m	**307.63**
100 mm	48.19	54.43	12.76	315.31	m	**369.74**
Equal tee						
25 mm	64.91	73.31	4.18	103.30	m	**176.61**
32 mm	66.63	75.26	4.94	122.07	m	**197.33**
40 mm	81.98	92.59	7.28	179.90	m	**272.49**

S: PIPED SUPPLY SYSTEMS

Item	Net Price £	Material £	Labour hours	Labour £	Unit	**Total rate £**
50 mm	84.12	95.00	8.48	209.55	m	**304.55**
65 mm	111.99	126.48	11.47	283.44	m	**409.92**
80 mm	192.16	217.03	14.23	351.63	m	**568.66**
100 mm	217.48	245.62	17.92	442.82	m	**688.44**
Copper pipe; capillary or compression joints in the running length; EN1057 R250 (TX) formerly BS 2871 Table X						
Plastic coated gas service pipe for corrosive environments, fixed vertically or at low level with brackets measured separtely						
15 mm dia. (yellow)	7.81	8.82	0.85	21.00	m	**29.82**
22 mm dia. (yellow)	15.42	17.41	0.96	23.74	m	**41.15**
28 mm dia. (yellow)	19.57	22.10	1.06	26.19	m	**48.29**
Copper pipe; capillary or compression joints in the running length; EN1057 R250 (TY) formerly BS 2871 Table Y						
Plastic coated gas and cold water service pipe for corrosive environments, fixed vertically or at low level with brackets measured separately (Refer to Copper Pipe Table X Section)						
15 mm dia. (yellow)	9.04	10.21	0.61	15.08	m	**25.29**
22 mm dia. (yellow)	16.43	18.55	0.69	17.05	m	**35.60**
FIXINGS						
Refer to Section S10 Cold Water						
Extra over copper pipes; capillary fittings; BS 864						
Refer to Section S10 Cold Water						

S: PIPED SUPPLY SYSTEMS

Item	Net Price £	Material £	Labour hours	Labour £	Unit	Total rate £
S32: NATURAL GAS – cont						
GAS BOOSTERS						
Complete skid mounted gas booster set, including AV mounts, flexible connections, low pressure switch,control panel and NRV (for run/standby unit); 3 phase supply; in accordance with IGE/UP/2; includes delivery, offloading and positioning						
Single unit						
Flow, pressure range						
0–200 m³/hour, 0.1–2.6 kPa	3362.71	3797.83	10.00	247.12	nr	**4044.95**
0–200 m³/hour, 0.1–4.0 kPa	3836.94	4333.42	10.00	247.12	nr	**4580.54**
0–200 m³/hour, 0.1–7 kPa	4465.11	5042.87	10.00	247.12	nr	**5289.99**
0–200 m³/hour, 0.1–9.5 kPa	4649.88	5251.55	10.00	247.12	nr	**5498.67**
0–200 m³/hour 0.1–11.0 kPa	5194.95	5867.15	10.00	247.12	nr	**6114.27**
0–400 m³/hour, 0.1–4.0 kPa	4237.24	4785.51	10.00	247.12	nr	**5032.63**
0–1000 m³/hour, 0.1–7.4 kPa	5869.31	6628.77	10.00	247.12	nr	**6875.89**
50–1000 m³/hour, 0.1–16.0 kPa	10965.73	12384.64	20.00	494.22	nr	**12878.86**
50–1000 m³/hour, 0.1–24.5 kPa	12397.62	14001.81	20.00	494.22	nr	**14496.03**
50–1000 m³/hour, 0.1–31.0 kPa	14100.54	15925.08	20.00	494.22	nr	**16419.30**
50–1000 m³/hour, 0.1–41.0 kPa	15563.27	17577.08	20.00	494.22	nr	**18071.30**
50–1000 m³/hour, 0.1–51.0 kPa	16148.31	18237.82	20.00	494.22	nr	**18732.04**
100–1800 m³/hour, 3.5–23.5 kPa	16570.23	18714.34	20.00	494.22	nr	**19208.56**
100–1800 m³/hour, 4.5–27.0 kPa	17872.80	20185.45	20.00	494.22	nr	**20679.67**
100–1800 m³/hour, 6.0–32.5 kPa	19794.35	22355.64	20.00	494.22	nr	**22849.86**
100–1800 m³/hour, 7.2–39.0 kPa	22772.11	25718.71	20.00	494.22	nr	**26212.93**
100–1800 m³/hour, 9.0–42.0 kPa	23936.13	27033.35	20.00	494.22	nr	**27527.57**
Run/Standby unit						
Flow, pressure range						
0–200 m³/hour, 0.1–2.6 kPa	17295.73	19533.71	16.00	395.38	nr	**19929.09**
0–200 m³/hour, 0.1–4.0 kPa	17822.49	20128.63	16.00	395.38	nr	**20524.01**
0–200 m³/hour, 0.1–7 kPa	18235.27	20594.82	16.00	395.38	nr	**20990.20**
0–200 m³/hour, 0.1–9.5 kPa	18687.13	21105.15	16.00	395.38	nr	**21500.53**
0–200 m³/hour 0.1–11.0 kPa	18996.71	21454.79	16.00	395.38	nr	**21850.17**
0–400 m³/hour, 0.1–4.0 kPa	20949.10	23659.81	16.00	395.38	nr	**24055.19**
0–1000 m³/hour, 0.1–7.4 kPa	26703.05	30158.29	25.00	617.78	nr	**30776.07**
50–1000 m³/hour, 0.1–16.0 kPa	36598.84	41334.55	25.00	617.78	nr	**41952.33**
50–1000 m³/hour, 0.1–24.5 kPa	41376.62	46730.55	25.00	617.78	nr	**47348.33**
50–1000 m³/hour, 0.1–31.0 kPa	46143.21	52113.91	25.00	617.78	nr	**52731.69**
50–1000 m³/hour, 0.1–41.0 kPa	50923.77	57513.05	25.00	617.78	nr	**58130.83**
50–1000 m³/hour, 0.1–51.0 kPa	51152.47	57771.34	25.00	617.78	nr	**58389.12**
100–1800 m³/hour, 3.5–23.5 kPa	53220.37	60106.82	25.00	617.78	nr	**60724.60**
100–1800 m³/hour, 4.5–27.0 kPa	54679.49	61754.75	25.00	617.78	nr	**62372.53**
100–1800 m³/hour, 6.0–32.5 kPa	58091.78	65608.56	25.00	617.78	nr	**66226.34**
100–1800 m³/hour, 7.2–39.0 kPa	61745.55	69735.12	25.00	617.78	nr	**70352.90**
100–1800 m³/hour, 9.0–42.0 kPa	64902.85	73300.95	25.00	617.78	nr	**73918.73**

S: PIPED SUPPLY SYSTEMS

Item	Net Price £	Material £	Labour hours	Labour £	Unit	**Total rate £**
S41: FUEL OIL STORAGE/DISTRIBUTION						
Y10 – PIPELINES						
For pipework prices refer to Section T31 – Low Temperature Hot Water Heating						
Y21 – TANKS						
Fuel storage tanks; mild steel; with all necessary screwed bosses; oil resistant joint rings; includes placing in position						
Rectangular						
1360 litres (300 gallon) capacity; 2 mm plate	358.13	404.47	12.03	297.28	nr	**701.75**
2730 litres (600 gallon) capacity; 2.5 mm plate	478.56	540.49	18.60	459.63	nr	**1000.12**
4550 litres (1000 gallon) capacity; 3 mm plate	1003.33	1133.16	25.00	617.78	nr	**1750.94**
Fuel storage tanks; 5 mm plate mild steel to BS 799 type J; complete with raised neck manhole with bolted cover, screwed connections, vent and fill connections, drain valve, gauge and overfill alarm; includes placing in position; excludes pumps and control panel						
Nominal capacity, size						
5,600 litres, 2.5 m × 1.5 m × 1.5 m high	2341.83	2644.85	20.00	494.22	nr	**3139.07**
Extra for bund unit (internal use)	1346.73	1520.99	30.00	741.34	nr	**2262.33**
Extra for external use with bund (watertight)	852.93	963.30	2.00	49.42	nr	**1012.72**
10,200 litres, 3.05 m × 1.83 m × 1.83 m high	2910.44	3287.03	30.00	741.34	nr	**4028.37**
Extra for bund unit (internal use)	1960.25	2213.90	40.00	988.45	nr	**3202.35**
Extra for external use with bund (watertight)	1047.46	1183.00	2.00	49.42	nr	**1232.42**
15,000 litres, 3.75 m × 2 m × 2 m high	3681.08	4157.40	40.00	988.45	nr	**5145.85**
Extra for bund unit (internal use)	2633.62	2974.40	55.00	1359.11	nr	**4333.51**
Extra for external use with bund (watertight)	1227.02	1385.79	2.00	49.42	nr	**1435.21**
20,000 litres, 4 m × 2.5 m × 2 m high	4855.74	5484.05	50.00	1235.56	nr	**6719.61**
Extra for bund unit (internal use)	3194.78	3608.17	65.00	1606.22	nr	**5214.39**
Extra for external use with bund (watertight)	1451.48	1639.29	2.00	49.42	nr	**1688.71**
Extra for BMS output (all tank sizes)	508.76	574.59	–	–	nr	**574.59**

S: PIPED SUPPLY SYSTEMS

Item	Net Price £	Material £	Labour hours	Labour £	Unit	**Total rate £**
S41: FUEL OIL STORAGE/DISTRIBUTION – cont						
Fuel storage tanks; plastic; with all necessary screwed bosses; oil resistant joint rings; includes placing in position						
Cylindrical; horizontal						
1250 litres (285 gallon) capacity	289.76	327.26	3.73	92.17	nr	**419.43**
1350 litres (300 gallon) capacity	296.03	334.34	4.30	106.25	nr	**440.59**
2500 litres (550 gallon) capacity	455.87	514.86	4.88	120.58	nr	**635.44**
Cylindrical; vertical						
1365 litres (300 gallon) capacity	183.79	207.57	3.73	92.17	nr	**299.74**
2600 litres (570 gallon) capacity	279.73	315.92	4.88	120.58	nr	**436.50**
3635 litres (800 gallon) capacity	436.10	492.53	4.88	120.58	nr	**613.11**
5455 litres (1200 gallon) capacity	635.29	717.49	5.95	147.03	nr	**864.52**
Bunded tanks						
1135 litres (250 gallon) capacity	589.20	665.44	4.30	106.25	nr	**771.69**
1590 litres (350 gallon) capacity	700.94	791.64	4.88	120.58	nr	**912.22**
2500 litres (550 gallon) capacity	833.16	940.97	5.95	147.03	nr	**1088.00**
5000 litres (1100 gallon) capacity	1681.25	1898.79	6.53	161.36	nr	**2060.15**

S: PIPED SUPPLY SYSTEMS

Item	Net Price £	Material £	Labour hours	Labour £	Unit	**Total rate £**
S60: FIRE HOSE REELS						
Y10 – PIPELINES						
For pipework prices refer to section S10 – Cold Water						
Y11 – PIPELINE ANCILLARIES						
For prices for ancillaries refer to section S10 – Cold Water						
Hose reels; automatic; connection to 25 mm screwed joint; reel with 30.5 metres, 19 mm rubber hose; suitable for working pressure up to 7 bar						
Reels						
Non-swing pattern	279.28	315.40	3.75	92.67	nr	**408.07**
Recessed non-swing pattern	312.52	352.94	3.75	92.67	nr	**445.61**
Swinging pattern	329.02	371.58	3.75	92.67	nr	**464.25**
Recessed swinging pattern	329.02	371.58	3.75	92.67	nr	**464.25**
Hose reels; manual; connection to 25 mm screwed joint; reel with 30.5 metres, 19 mm rubber hose; suitable for working pressure up to 7 bar						
Reels						
Non-swing pattern	279.28	315.40	3.25	80.30	nr	**395.70**
Recessed non-swing pattern	312.52	352.94	3.25	80.30	nr	**433.24**
Swinging pattern	329.02	371.58	3.25	80.30	nr	**451.88**
Recessed swinging pattern	329.02	371.58	3.25	80.30	nr	**451.88**

S: PIPED SUPPLY SYSTEMS

Item	Net Price £	Material £	Labour hours	Labour £	Unit	Total rate £
S61: DRY RISERS						
Y10 – PIPELINES						
For pipework prices refer to section S10 – Cold Water						
Y11 – PIPELINE ANCILLARIES						
VALVES (BS 5041, parts 2 and 3)						
Bronze/gunmetal inlet breeching for pumping in with 65 mm dia. instantaneous male coupling; with cap, chain and 25 mm drain valve						
Double inlet with back pressure valve, flanged to steel	250.40	282.81	1.75	43.25	nr	**326.06**
Quadruple inlet with back pressure valve, flanged to steel	560.65	633.20	1.75	43.25	nr	**676.45**
Bronze/gunmetal gate type outlet valve with 65 mm dia. instantaneous female coupling; cap and chain; wheel head secured by padlock and leather strap						
Flanged to BS 4504 PN6 (bolted connection to counter flanges measured separately)	200.81	226.80	1.75	43.25	nr	**270.05**
Bronze/gunmetal landing type outlet valve, with 65 mm dia. instantaneous female coupling; cap and chain; wheelhead secured by padlock and leather strap; bolted connections to counter flanges measured separately						
Horizontal, flanged to BS 4504 PN6	218.70	247.00	1.50	37.07	nr	**284.07**
Oblique, flanged to BS 4504 PN6	218.70	247.00	1.50	37.07	nr	**284.07**
Air valve, screwed joints to steel						
25 mm dia.	43.47	49.09	0.55	13.59	nr	**62.68**
INLET BOXES (BS 5041, part 5)						
Steel dry riser inlet box with hinged wire glazed door suitably lettered (fixing by others)						
610 × 460 × 325 mm; double inlet	266.49	300.97	3.00	74.13	nr	**375.10**
610 × 610 × 356 mm; quadruple inlet	497.37	561.72	3.00	74.13	nr	**635.85**
OUTLET BOXES (BS 5041, part 5)						
Steel dry riser outlet box with hinged wire glazed door suitably lettered (fixing by others)						
610 × 460 × 325 mm; single outlet	260.24	293.91	3.00	74.13	nr	**368.04**

S: PIPED SUPPLY SYSTEMS

Item	Net Price £	Material £	Labour hours	Labour £	Unit	**Total rate £**
S63: SPRINKLERS						
Y10 – PIPELINES						
Prefabricated black steel pipework; screwed joints, including all coupliings, unions and the like to BS 1387:1985; includes fixing to backgrounds, with brackets measured separately						
Heavy weight						
25 mm dia.	5.46	6.17	0.47	11.61	m	**17.78**
32 mm dia.	6.78	7.66	0.53	13.10	m	**20.76**
40 mm dia.	7.90	8.93	0.58	14.33	m	**23.26**
50 mm dia.	10.97	12.39	0.63	15.57	m	**27.96**
FIXINGS						
For steel pipes; black malleable iron. For minimum fixing distances, refer to the Tables and Memoranda at the rear of the book						
Pipe ring, single socket, black malleable iron						
25 mm dia.	0.92	1.04	0.12	2.97	nr	**4.01**
32 mm dia.	0.98	1.11	0.14	3.46	nr	**4.57**
40 mm dia.	1.27	1.44	0.15	3.70	nr	**5.14**
50 mm dia.	1.60	1.81	0.16	3.96	nr	**5.77**
Extra over channel sections for fabricated hangers and brackets						
Galvanized steel; including inserts, bolts, nuts, washers; fixed to backgrounds						
41 × 21 mm	5.65	6.39	0.29	7.17	m	**13.56**
41 × 41 mm	6.77	7.65	0.29	7.17	m	**14.82**
Threaded rods; metric thread; including nuts, washers etc.						
12 mm dia. × 600 mm long	2.86	3.23	0.18	4.45	nr	**7.68**
Extra over for black malleable iron fittings; BS 143						
Plain plug, solid						
25 mm dia.	1.95	2.20	0.40	9.88	nr	**12.08**
32 mm dia.	2.80	3.16	0.44	10.87	nr	**14.03**
40 mm dia.	3.95	4.46	0.48	11.86	nr	**16.32**
50 mm dia.	5.18	5.85	0.56	13.83	nr	**19.68**

S: PIPED SUPPLY SYSTEMS

Item	Net Price £	Material £	Labour hours	Labour £	Unit	**Total rate £**
S63: SPRINKLERS – cont						
Extra over for black malleable iron fittings – cont						
Concentric reducing socket						
32 mm dia.	2.31	2.61	0.48	11.86	nr	**14.47**
40 mm dia.	3.01	3.40	0.55	13.59	nr	**16.99**
50 mm dia.	4.21	4.75	0.60	14.83	nr	**19.58**
Elbow; 90° female/female						
25 mm dia.	1.36	1.54	0.44	10.87	nr	**12.41**
32 mm dia.	2.48	2.81	0.53	13.10	nr	**15.91**
40 mm dia.	4.17	4.71	0.60	14.83	nr	**19.54**
50 mm dia.	4.88	5.51	0.65	16.07	nr	**21.58**
Tee						
25 mm dia. equal	1.85	2.09	0.51	12.61	nr	**14.70**
32 mm dia. reducing to 25 mm dia.	3.37	3.80	0.54	13.35	nr	**17.15**
40 mm dia.	5.18	5.85	0.65	16.07	nr	**21.92**
50 mm dia.	7.45	8.42	0.78	19.28	nr	**27.70**
Cross tee						
25 mm dia. equal	5.02	5.67	1.16	28.66	nr	**34.33**
32 mm dia.	7.32	8.27	1.40	34.60	nr	**42.87**
40 mm dia.	9.43	10.65	1.60	39.54	nr	**50.19**
50 mm dia.	15.31	17.29	1.68	41.51	nr	**58.80**
Prefabricated black steel pipework; welded joints, including all coupliings, unions and the like to BS 1387:1985; fixing to backgrounds						
Heavy weight						
65 mm dia.	14.93	16.86	0.65	16.07	m	**32.93**
80 mm dia.	19.00	21.46	0.70	17.30	m	**38.76**
100 mm dia.	26.51	29.94	0.85	21.00	m	**50.94**
150 mm dia.	35.87	40.51	1.15	28.42	m	**68.93**
FIXINGS						
For steel pipes; black malleable iron. For minimum fixing distances, refer to the Tables and Memoranda at the rear of the book						
Pipe ring, single socket, black malleable iron						
65 mm dia.	2.31	2.61	0.30	7.41	nr	**10.02**
80 mm dia.	2.77	3.13	0.35	8.65	nr	**11.78**
100 mm dia.	4.19	4.73	0.40	9.88	nr	**14.61**
150 mm dia.	9.50	10.73	0.77	19.03	nr	**29.76**

S: PIPED SUPPLY SYSTEMS

Item	Net Price £	Material £	Labour hours	Labour £	Unit	**Total rate £**
Extra over fittings						
Reducer (one size down)						
65 mm dia.	7.66	8.65	2.70	66.73	nr	**75.38**
80 mm dia.	9.97	11.26	2.86	70.68	nr	**81.94**
100 mm dia.	19.88	22.45	3.22	79.57	nr	**102.02**
150 mm dia.	55.24	62.38	4.20	103.80	nr	**166.18**
Elbow; 90°						
65 mm dia.	10.54	11.90	3.06	75.61	nr	**87.51**
80 mm dia.	15.48	17.48	3.40	84.02	nr	**101.50**
100 mm dia.	19.35	21.86	3.70	91.43	nr	**113.29**
150 mm dia.	29.02	32.77	5.20	128.50	nr	**161.27**
Branch bend						
65 mm dia.	10.73	12.12	3.60	88.96	nr	**101.08**
80 mm dia.	12.56	14.19	3.80	93.90	nr	**108.09**
100 mm dia.	19.33	21.83	5.10	126.03	nr	**147.86**
150 mm dia.	50.00	56.47	7.50	185.33	nr	**241.80**
Prefabricated black steel pipe; victaulic joints; including all couplings and the like to BS 1387: 1985; fixing to backgrounds						
Heavy weight						
65 mm dia.	12.88	14.55	0.70	17.30	m	**31.85**
80 mm dia.	16.73	18.89	0.78	19.28	m	**38.17**
100 mm dia.	24.39	27.55	0.93	22.98	m	**50.53**
150 mm dia.	48.56	54.85	1.25	30.89	m	**85.74**
FIXINGS						
For fixings refer to For steel pipes; black malleable iron.						
Extra over fittings						
Coupling						
65 mm dia.	9.43	10.65	0.26	6.43	nr	**17.08**
80 mm dia.	10.10	11.40	0.26	6.43	nr	**17.83**
100 mm dia.	13.24	14.95	0.32	7.91	nr	**22.86**
150 mm dia.	23.31	26.33	0.35	8.65	nr	**34.98**
Reducer						
65 mm dia.	7.66	8.65	0.48	11.86	nr	**20.51**
80 mm dia.	9.97	11.26	0.43	10.63	nr	**21.89**
100 mm dia.	19.88	22.45	0.46	11.36	nr	**33.81**
150 mm dia.	29.82	33.68	0.45	11.13	nr	**44.81**

S: PIPED SUPPLY SYSTEMS

Item	Net Price £	Material £	Labour hours	Labour £	Unit	Total rate £
S63: SPRINKLERS – cont						
Extra over fittings – cont						
Elbow; any degree						
65 mm dia.	10.54	11.90	0.56	13.83	nr	**25.73**
80 mm dia.	15.48	17.48	0.63	15.57	nr	**33.05**
100 mm dia.	29.85	33.71	0.71	17.54	nr	**51.25**
150 mm dia.	44.78	50.57	0.80	19.77	nr	**70.34**
Equal tee						
65 mm dia.	16.33	18.44	0.74	18.29	nr	**36.73**
80 mm dia.	19.66	22.21	0.83	20.51	nr	**42.72**
100 mm dia.	36.03	40.69	0.94	23.24	nr	**63.93**
150 mm dia.	54.05	61.05	1.05	25.94	nr	**86.99**
Y11 – PIPELINE ANCILLARIES						
SPRINKLER HEADS						
Sprinkler heads; brass body; frangible glass bulb; manufactured to standard operating temperature of 57–141°C; quick response; RTI<50						
Conventional pattern; 15 mm dia.	4.31	4.87	0.15	3.70	nr	**8.57**
Sidewall pattern; 15 mm dia.	6.34	7.16	0.15	3.70	nr	**10.86**
Conventional pattern; 15 mm dia.; satin chrome plated	5.06	5.71	0.15	3.70	nr	**9.41**
Sidewall pattern; 15 mm dia.; satin chrome plated	6.58	7.44	0.15	3.70	nr	**11.14**
Fully concealed; fusible link; 15 mm dia.	15.82	17.87	0.15	3.70	nr	**21.57**
VALVES						
Wet system alarm valves; including internal non-return valve; working pressure up to 12.5 bar; BS4504 PN16 flanged ends; bolted connections						
100 mm dia.	1368.71	1545.81	25.00	617.78	nr	**2163.59**
150 mm dia.	1671.91	1888.24	25.00	617.78	nr	**2506.02**
Wet system bypass alarm valves; including internal non-return valve; working pressure up to 12.5 bar; BS4504 PN16 flanged ends; bolted connections						
100 mm dia.	2412.66	2724.85	25.00	617.78	nr	**3342.63**
150 mm dia.	3050.41	3445.12	25.00	617.78	nr	**4062.90**

S: PIPED SUPPLY SYSTEMS

Item	Net Price £	Material £	Labour hours	Labour £	Unit	**Total rate £**
Alternate system wet/dry alarm station; including butterfly valve, wet alarm valve, dry pipe differential pressure valve and pressure gauges; working pressure up to 10.5 bar; BS4505 PN16 flanged ends; bolted connections						
100 mm dia.	2990.12	3377.03	40.00	988.45	nr	**4365.48**
150 mm dia.	3484.65	3935.55	40.00	988.45	nr	**4924.00**
Alternate system wet/dry alarm station; including electrically operated butterfly valve, water supply accelerator set, wet alarm valve, dry pipe differential pressure valve and pressure gauges; working pressure up to 10.5 bar; BS4505 PN16 flanged ends; bolted connections						
100 mm dia.	3408.37	3849.40	45.00	1112.00	nr	**4961.40**
150 mm dia.	3907.60	4413.22	45.00	1112.00	nr	**5525.22**
ALARM/GONGS						
Water operated motor alarm and gong; stainless steel and aluminum body and gong; screwed connections						
Connection to sprinkler system and drain pipework	417.20	471.18	6.00	148.27	nr	**619.45**
Y21 – WATER TANKS						
Note: Prices are based on the most economical tank size for each volume, and the cost will vary with differing tank dimensions, for the same volume						
Steel sectional sprinkler tank; ordinary hazard, life safety classification; two compartment tank, complete with all fittings and accessories to comply with LPCB type A requirements; cost of erection (on prepared supports) is included within net price, labour cost allows for offloading and positioning of materials						
Volume, size						
70 m³, 6.1 m × 4.88 m × 2.44 m (h)	25008.87	28244.89	24.00	593.07	nr	**28837.96**
105 m³, 7.3 m × 6.1 m × 2.44 m (h)	29346.48	33143.77	28.00	691.92	nr	**33835.69**
168 m³, 9.76 m × 4.88 m × 3.66 m (h)	36258.35	40950.00	28.00	691.92	nr	**41641.92**
211 m³, 9.76 m × 6.1 m × 3.66 m (h)	40029.63	45209.27	32.00	790.75	nr	**46000.02**

S: PIPED SUPPLY SYSTEMS

Item	Net Price £	Material £	Labour hours	Labour £	Unit	Total rate £
S63: SPRINKLERS – cont						
Y21 – WATER TANKS – cont						
Steel sectional sprinkler tank; ordinary hazard, property protection classification; single compartment tank, complete with all fittings and accessories to comply with LPCB type A requirements; cost of erection (on prepared supports) is included within net price, labour cost allows for offloading and positioning of materials						
Volume, size						
70 m³, 6.1 m × 4.88 m × 2.44 m (h)	17633.63	19915.34	24.00	593.07	nr	**20508.41**
105 m³, 7.3 m × 6.1 m × 2.44 m (h)	21784.62	24603.44	28.00	691.92	nr	**25295.36**
168 m³, 9.76 m × 4.88 m × 3.66 m (h)	28149.46	31791.86	28.00	691.92	nr	**32483.78**
211 m³, 9.76 m × 6.1 m × 3.66 m (h)	31586.08	35673.16	32.00	790.75	nr	**36463.91**
GRP sectional sprinkler tank; ordinary hazard, life safety classification; two compartment tank, complete with all fittings and accessories to comply with LPCB type A requirements; cost of erection (on prepared supports) is included within net price, labour cost allows for offloading and positioning of materials						
Volume, size						
55 m³, 6 m × 4 m × 3 m (h)	27907.74	31518.87	14.00	345.95	nr	**31864.82**
70 m³, 6 m × 5 m × 3 m (h)	31202.88	35240.38	16.00	395.38	nr	**35635.76**
80 m³, 8 m × 4 m × 3 m (h)	32337.34	36521.63	16.00	395.38	nr	**36917.01**
105 m³, 10 m × 4 m × 3 m (h)	36592.22	41327.07	24.00	593.07	nr	**41920.14**
125 m³, 10 m × 5 m × 3 m (h)	41267.92	46607.78	24.00	593.07	nr	**47200.85**
140 m³, 8 m × 7 m × 3 m (h)	42977.01	48538.02	24.00	593.07	nr	**49131.09**
135 m³, 9 m × 6 m × 3 m (h)	44276.36	50005.50	24.00	593.07	nr	**50598.57**
160 m³, 13 m × 5 m × 3 m (h)	48611.21	54901.26	24.00	593.07	nr	**55494.33**
185 m³, 12 m × 6 m × 3 m (h)	51421.55	58075.24	24.00	593.07	nr	**58668.31**
GRP sectional sprinkler tank; ordinary hazard, property protection classification; single compartment tank, complete with all fittings and accessories to comply with LPCB type A requirements; cost of erection (on prepared supports) is within net price, labour cost allows for offloading and positioning of materials						
Volume, size						
55 m³, 6 m × 4 m × 3 m (h)	21470.05	24248.16	14.00	345.95	nr	**24594.11**
70 m³, 6 m × 5 m × 3 m (h)	24255.79	27394.36	16.00	395.38	nr	**27789.74**
80 m³, 8 m × 4 m × 3 m (h)	25889.80	29239.81	16.00	395.38	nr	**29635.19**

S: PIPED SUPPLY SYSTEMS

Item	Net Price £	Material £	Labour hours	Labour £	Unit	**Total rate £**
105 m³, 10 m × 4 m × 3 m (h)	30330.49	34255.11	24.00	593.07	nr	**34848.18**
125 m³, 10 m × 5 m × 3 m (h)	34297.44	38735.36	24.00	593.07	nr	**39328.43**
140 m³, 8 m × 7 m × 3 m (h)	35505.74	40100.00	24.00	593.07	nr	**40693.07**
135 m³, 9 m × 6 m × 3 m (h)	37304.83	42131.89	24.00	593.07	nr	**42724.96**
160 m³, 13 m × 5 m × 3 m (h)	41821.62	47233.13	24.00	593.07	nr	**47826.20**
185 m³, 12 m × 6 m × 3 m (h)	43934.30	49619.18	24.00	593.07	nr	**50212.25**

S: PIPED SUPPLY SYSTEMS

Item	Net Price £	Material £	Labour hours	Labour £	Unit	Total rate £
S65: FIRE HYDRANTS						
EXTINGUISHERS						
Fire extinguishers; hand held; BS 5423; placed in position						
Water type; cartridge operated; for Class A fires						
Water type, 9 litres capacity; 55 gm CO_2 cartridge; Class A fires (fire-rating 13 A)	79.60	89.90	1.00	24.71	nr	**114.61**
Foam type, 9 litres capacity; 75 gm CO_2 cartridge; Class A & B fires (fire-rating 13 A:183B)	94.40	106.61	1.00	24.71	nr	**131.32**
Dry powder type; cartridge operated; for Class A, B & C fires and electrical equipment fires						
Dry powder type, 1 kg capacity; 12 gm CO_2 cartridge; Class A, B & C fires (fire-rating 5 A:34B)	33.66	38.02	1.00	24.71	nr	**62.73**
Dry powder type, 2 kg capacity; 28 gm CO_2 cartridge; Class A, B & C fires (fire-rating 13 A:55B)	45.33	51.19	1.00	24.71	nr	**75.90**
Dry powder type, 4 kg capacity; 90 gm CO_2 cartridge; Class A, B & C fires (fire-rating 21 A:183B)	80.75	91.20	1.00	24.71	nr	**115.91**
Dry powder type, 9 kg capacity; 190 gm CO_2 cartridge; Class A, B & C fires (fire-rating 43 A:233B)	108.19	122.19	1.00	24.71	nr	**146.90**
Dry powder type; stored pressure type; for Class A, B & C fires and electrical equipment fires						
Dry powder type, 1 kg capacity; Class A, B & C fires (fire-rating 5 A:34B)	31.01	35.03	1.00	24.71	nr	**59.74**
Dry powder type, 2 kg capacity; Class A, B & C fires (fire-rating 13 A:55B)	36.81	41.58	1.00	24.71	nr	**66.29**
Dry powder type, 4 kg capacity; Class A, B & C fires (fire-rating 21 A:183B)	68.76	77.65	1.00	24.71	nr	**102.36**
Dry powder type, 9 kg capacity; Class A, B & C fires (fire-rating 43 A:233B)	89.09	100.61	1.00	24.71	nr	**125.32**
Carbon dioxide type; for Class B fires and electrical equipment fires						
CO_2 type with hose and horn, 2 kg capacity, Class B fires (fire-rating 34B)	85.17	96.19	1.00	24.71	nr	**120.90**
CO_2 type with hose and horn, 5 kg capacity, Class B fires (fire-rating 55B)	133.45	150.72	1.00	24.71	nr	**175.43**

S: PIPED SUPPLY SYSTEMS

Item	Net Price £	Material £	Labour hours	Labour £	Unit	**Total rate £**
Glass fibre blanket, in GRP container						
1100 × 1100 mm	27.03	30.53	0.50	12.35	nr	**42.88**
1200 × 1200 mm	29.62	33.46	0.50	12.35	nr	**45.81**
1800 × 1200 mm	39.51	44.63	0.50	12.35	nr	**56.98**
HYDRANTS						
Fire hydrants; bolted connections						
Underground hydrants, complete with frost plug to BS 750						
sluice valve pattern type 1	222.26	251.02	4.50	111.20	nr	**362.22**
screw down pattern type 2	161.44	182.32	4.50	111.20	nr	**293.52**
Stand pipe for underground hydrant; screwed base; light alloy						
Single outlet	127.83	144.37	1.00	24.71	nr	**169.08**
Double outlet	187.04	211.24	1.00	24.71	nr	**235.95**
64 mm dia. bronze/gunmetal outlet valves						
Oblique flanged landing valve	139.91	158.02	1.00	24.71	nr	**182.73**
Oblique screwed landing valve	139.91	158.02	1.00	24.71	nr	**182.73**
Cast iron surface box; fixing by others						
400 × 200 × 100 mm	122.51	138.36	1.00	24.71	nr	**163.07**
500 × 200 × 150 mm	165.07	186.43	1.00	24.71	nr	**211.14**
Frost plug	29.73	33.58	0.25	6.18	nr	**39.76**

T: MECHANICAL/COOLING/HEATING SYSTEMS

Item	Net Price £	Material £	Labour hours	Labour £	Unit	Total rate £
T10: GAS/OIL FIRED BOILERS						
DOMESTIC						
Domestic water boilers; stove enamelled casing; electric controls; placing in position; assembling and connecting; electrical work elsewhere						
Gas fired; floor standing; connected to conventional flue						
9 to 12 kW	581.64	656.90	8.59	212.27	nr	**869.17**
12 to 15 kW	611.89	691.07	8.59	212.27	nr	**903.34**
15 to 18 kW	651.72	736.05	8.88	219.43	nr	**955.48**
18 to 21 kW	757.16	855.14	9.92	245.14	nr	**1100.28**
21 to 23 kW	839.39	948.00	10.66	263.42	nr	**1211.42**
23 to 29 kW	1088.67	1229.54	11.81	291.83	nr	**1521.37**
29 to 37 kW	1287.63	1454.24	11.81	291.83	nr	**1746.07**
37 to 41 kW	1339.09	1512.36	12.68	313.34	nr	**1825.70**
Gas fired; wall hung; connected to conventional flue						
9 to 12 kW	511.21	577.36	8.59	212.27	nr	**789.63**
12 to 15 kW	660.00	745.40	8.59	212.27	nr	**957.67**
13 to 18 kW	837.60	945.98	8.59	212.27	nr	**1158.25**
Gas fired; floor standing; connected to balanced flue						
9 to 12 kW	726.57	820.58	9.16	226.36	nr	**1046.94**
12 to 15 kW	755.96	853.78	10.95	270.60	nr	**1124.38**
15 to 18 kW	813.73	919.02	11.98	296.03	nr	**1215.05**
18 to 21 kW	958.90	1082.97	12.78	315.81	nr	**1398.78**
21 to 23 kW	1105.97	1249.08	12.78	315.81	nr	**1564.89**
23 to 29 kW	1409.68	1592.09	15.45	381.80	nr	**1973.89**
29 to 37 kW	2709.97	3060.62	17.65	436.15	nr	**3496.77**
Gas fired; wall hung; connected to balanced flue						
6 to 9 kW	548.59	619.58	9.16	226.36	nr	**845.94**
9 to 12 kW	627.76	708.99	9.16	226.36	nr	**935.35**
12 to 15 kW	710.85	802.83	9.45	233.53	nr	**1036.36**
15 to 18 kW	852.35	962.64	9.74	240.69	nr	**1203.33**
18 to 22 kW	904.74	1021.81	9.74	240.69	nr	**1262.50**
Gas fired; wall hung; connected to fan flue (including flue kit)						
6 to 9 kW	627.42	708.60	9.16	226.36	nr	**934.96**
9 to 12 kW	699.20	789.67	9.16	226.36	nr	**1016.03**
12 to 15 kW	759.01	857.22	10.95	270.60	nr	**1127.82**

T: MECHANICAL/COOLING/HEATING SYSTEMS

Item	Net Price £	Material £	Labour hours	Labour £	Unit	Total rate £
15 to 18 kW	817.58	923.38	11.98	296.03	nr	**1219.41**
18 to 23 kW	1065.27	1203.11	12.78	315.81	nr	**1518.92**
23 to 29 kW	1379.50	1558.00	15.45	381.80	nr	**1939.80**
29 to 35 kW	1693.63	1912.78	17.65	436.15	nr	**2348.93**
Oil fired; floor standing; connected to conventional flue						
12 to 15	1089.39	1230.35	10.38	256.50	nr	**1486.85**
15 to 19	1136.40	1283.45	12.20	301.48	nr	**1584.93**
21 to 25	1296.95	1464.77	14.30	353.37	nr	**1818.14**
26 to 32	1421.94	1605.93	15.80	390.44	nr	**1996.37**
35 to 50	1605.42	1813.15	20.46	505.58	nr	**2318.73**
Fire place mounted natural gas fire and back boiler; cast iron water boiler; electric control box; fire output 3 kW with wood surround						
10.50 kW	317.82	358.95	8.88	219.43	nr	**578.38**
FORCED DRAFT						
Commercial cast iron sectional floor standing boilers; pressure jet burner; including controls, enamelled jacket, insulation, assembly and commissioning; electrical work elsewhere						
Gas fired (on/off type), connected to conventional flue						
16–26 kW; 3 sections; 125 mm dia. flue	2311.99	2611.15	8.00	197.69	nr	**2808.84**
26–33 kW; 4 sections; 125 mm dia. flue	2465.00	2783.96	8.00	197.69	nr	**2981.65**
33–40 kW; 5 sections; 125 mm dia. flue	2788.00	3148.75	8.00	197.69	nr	**3346.44**
35–50 kW; 3 sections; 153 mm dia. flue	3060.00	3455.95	8.00	197.69	nr	**3653.64**
50–65 kW; 4 sections; 153 mm dia. flue	3297.99	3724.73	8.00	197.69	nr	**3922.42**
65–80 kW; 5 sections; 153 mm dia. flue	3586.99	4051.12	8.00	197.69	nr	**4248.81**
80–100 kW; 6 sections; 180 mm dia. flue	4168.39	4707.76	8.00	197.69	nr	**4905.45**
100–120 kW; 7 sections; 180 mm dia. flue	5269.99	5951.90	8.00	197.69	nr	**6149.59**
105–140 kW; 5 sections; 180 mm dia. flue	6119.99	6911.89	8.00	197.69	nr	**7109.58**
140–180 kW; 6 sections; 180 mm dia. flue	6969.99	7871.87	8.00	197.69	nr	**8069.56**
180–230 kW; 7 sections; 200 mm dia. flue	7819.98	8831.84	8.00	197.69	nr	**9029.53**
230–280 kW; 8 sections; 200 mm dia. flue	8330.00	9407.86	8.00	197.69	nr	**9605.55**
280–330 kW; 9 sections; 200 mm dia. flue	9349.99	10559.84	8.00	197.69	nr	**10757.53**
Gas fired (high/low type), connected to conventional flue						
105–140 kW; 5 sections; 180 mm dia. flue	7309.99	8255.87	8.00	197.69	nr	**8453.56**
140–180 kW; 6 sections; 180 mm dia. flue	7854.00	8870.27	8.00	197.69	nr	**9067.96**
180–230 kW; 7 sections; 200 mm dia. flue	8907.99	10060.64	8.00	197.69	nr	**10258.33**
230–280 kW; 8 sections; 200 mm dia. flue	9519.98	10751.82	8.00	197.69	nr	**10949.51**
280–330 kW; 9 sections; 200 mm dia. flue	10199.99	11519.82	8.00	197.69	nr	**11717.51**
300–390 kW; 8 sections; 250 mm dia. flue	12239.98	13823.77	12.00	296.53	nr	**14120.30**
390–450 kW; 9 sections; 250 mm dia. flue	13480.98	15225.36	12.00	296.53	nr	**15521.89**

T: MECHANICAL/COOLING/HEATING SYSTEMS

Item	Net Price £	Material £	Labour hours	Labour £	Unit	Total rate £
T10: GAS/OIL FIRED BOILERS – cont						
FORCED DRAFT – cont						
Commercial cast iron sectional floor standing boilers – cont						
Gas fired (high/low type), connected to conventional flue – cont						
450–540 kW; 10 sections; 250 mm dia. flue	14789.98	16703.73	12.00	296.53	nr	**17000.26**
540–600 kW; 11 sections; 300 mm dia. flue	15129.99	17087.73	12.00	296.53	nr	**17384.26**
600–670 kW; 12 sections; 300 mm dia. flue	18573.09	22099.01	12.00	296.53	nr	**22395.54**
670–720 kW; 13 sections; 300 mm dia. flue	18913.10	22483.02	12.00	296.53	nr	**22779.55**
720–780 kW; 14 sections; 300 mm dia. flue	19168.09	22771.01	12.00	296.53	nr	**23067.54**
754–812 kW; 14 sections; 400 mm dia. flue	19338.09	22963.00	12.00	296.53	nr	**23259.53**
812–870 kW; 15 sections; 400 mm dia. flue	24438.08	28722.90	12.00	296.53	nr	**29019.43**
870–928 kW; 16 sections; 400 mm dia. flue	26478.09	31026.89	12.00	296.53	nr	**31323.42**
928–986 kW; 17 sections; 400 mm dia. flue	27158.09	31794.88	12.00	296.53	nr	**32091.41**
986–1044 kW; 18 sections; 400 mm dia. flue	28348.09	33138.84	12.00	296.53	nr	**33435.37**
1044–1102 kW; 19 sections; 400 mm dia. flue	29028.08	33906.83	12.00	296.53	nr	**34203.36**
1102–1160 kW; 20 sections; 400 mm dia. flue	29776.08	34751.62	12.00	296.53	nr	**35048.15**
1160–1218 kW; 21 sections; exceeding 400 mm dia. flue	31408.08	36594.79	12.00	296.53	nr	**36891.32**
1218–1276 kW; 22 sections; exceeding 400 mm dia. flue	33018.15	38574.62	12.00	296.53	nr	**38871.15**
1276–1334 kW; 23 sections; exceeding 400 mm dia. flue	34191.15	39899.39	12.00	296.53	nr	**40195.92**
1334–1392 kW; 24 sections; exceeding 400 mm dia. flue	35721.15	41627.35	12.00	296.53	nr	**41923.88**
1392–1450 kW; 25 sections; exceeding 400 mm dia. flue	36741.16	42779.35	12.00	296.53	nr	**43075.88**
Oil fired (on/off type), connected to conventional flue						
16–26 kW; 3 sections; 125 mm dia. flue	1700.00	1919.97	8.00	197.69	nr	**2117.66**
26–33 kW; 4 sections; 125 mm dia. flue	1869.99	2111.96	8.00	197.69	nr	**2309.65**
33–40 kW; 5 sections; 125 mm dia. flue	2040.00	2303.97	8.00	197.69	nr	**2501.66**
35–50 kW; 3 sections; 153 mm dia. flue	2397.00	2707.16	8.00	197.69	nr	**2904.85**
50–65 kW; 4 sections; 153 mm dia. flue	2516.00	2841.56	8.00	197.69	nr	**3039.25**
65–80 kW; 5 sections; 153 mm dia. flue	2618.00	2956.76	8.00	197.69	nr	**3154.45**
80–100 kW; 6 sections; 180 mm dia. flue	3910.00	4415.94	8.00	197.69	nr	**4613.63**
100–120 kW; 7 sections; 180 mm dia. flue	4419.99	4991.91	8.00	197.69	nr	**5189.60**
105–140 kW; 5 sections; 180 mm dia. flue	5099.99	5759.90	8.00	197.69	nr	**5957.59**
140–180 kW; 6 sections; 180 mm dia. flue	5780.00	6527.90	8.00	197.69	nr	**6725.59**
180–230 kW; 7 sections; 200 mm dia. flue	7139.99	8063.87	8.00	197.69	nr	**8261.56**
230–280 kW; 8 sections; 200 mm dia. flue	7819.98	8831.84	8.00	197.69	nr	**9029.53**
280–330 kW; 9 sections; 200 mm dia. flue	8669.98	9791.84	8.00	197.69	nr	**9989.53**

T: MECHANICAL/COOLING/HEATING SYSTEMS

Item	Net Price £	Material £	Labour hours	Labour £	Unit	Total rate £
Oil fired (high/low type), connected to conventional flue						
105–140 kW; 5 sections; 180 mm dia. flue	4097.00	4627.13	8.00	197.69	nr	**4824.82**
140–180 kW; 6 sections; 180 mm dia. flue	4419.99	4991.91	8.00	197.69	nr	**5189.60**
180–230 kW; 7 sections; 200 mm dia. flue	4929.99	5567.90	8.00	197.69	nr	**5765.59**
230–280 kW; 8 sections; 200 mm dia. flue	5814.00	6566.30	8.00	197.69	nr	**6763.99**
280–330 kW; 9 sections; 200 mm dia. flue	9180.00	10367.84	8.00	197.69	nr	**10565.53**
300–390 kW; 8 sections; 250 mm dia. flue	10199.99	11519.82	12.00	296.53	nr	**11816.35**
390–450 kW; 9 sections; 250 mm dia. flue	10964.98	12383.79	12.00	296.53	nr	**12680.32**
450–540 kW; 10 sections; 250 mm dia. flue	12409.98	14015.77	12.00	296.53	nr	**14312.30**
540–600 kW; 11 sections; 300 mm dia. flue	13038.98	14726.16	12.00	296.53	nr	**15022.69**
600–670 kW; 12 sections; 300 mm dia. flue	13939.98	15743.74	12.00	296.53	nr	**16040.27**
670–720 kW; 13 sections; 300 mm dia. flue	15809.98	17855.71	12.00	296.53	nr	**18152.24**
720–780 kW; 14 sections; 300 mm dia. flue	16064.97	18143.70	12.00	296.53	nr	**18440.23**
754–812 kW; 14 sections; 400 mm dia. flue	16285.98	18393.30	12.00	296.53	nr	**18689.83**
812–870 kW; 15 sections; 400 mm dia. flue	21759.98	24575.61	12.00	296.53	nr	**24872.14**
870–928 kW; 16 sections; 400 mm dia. flue	22592.97	25516.39	12.00	296.53	nr	**25812.92**
928–986 kW; 17 sections; 400 mm dia. flue	23374.96	26399.56	12.00	296.53	nr	**26696.09**
986–1044 kW; 18 sections; 400 mm dia. flue	24309.97	27455.55	12.00	296.53	nr	**27752.08**
1044–1102 kW; 19 sections; 400 mm dia. flue	25499.97	28799.54	12.00	296.53	nr	**29096.07**
1102–1160 kW; 20 sections; 400 mm dia. flue	26927.97	30412.31	12.00	296.53	nr	**30708.84**
1160–1218 kW; 21 sections; exceeding 400 mm dia. flue	28423.97	32101.89	12.00	296.53	nr	**32398.42**
1218–1276 kW; 22 sections; exceeding 400 mm dia. flue	29834.97	33695.47	12.00	296.53	nr	**33992.00**
1276–1334 kW; 23 sections; exceeding 400 mm dia. flue	30888.97	34885.85	12.00	296.53	nr	**35182.38**
1334–1392 kW; 24 sections; exceeding 400 mm dia. flue	32699.45	36930.60	12.00	296.53	nr	**37227.13**
1392–1450 kW; 25 sections; exceeding 400 mm dia. flue	33897.96	38284.19	12.00	296.53	nr	**38580.72**
Commercial steel shell floor standing boilers; pressure jet burner; including controls, enamelled jacket, insulation, placing in position and commissioning; electrical work elsewhere						
Gas fired (on/off type), connected to conventional flue						
130–190 kW	6095.23	6883.92	8.00	197.69	nr	**7081.61**
200–250 kW	6903.97	7797.31	8.00	197.69	nr	**7995.00**
280–360 kW	8268.92	9338.88	8.00	197.69	nr	**9536.57**
375–500 kW	10077.86	11381.88	8.00	197.69	nr	**11579.57**

T: MECHANICAL/COOLING/HEATING SYSTEMS

Item	Net Price £	Material £	Labour hours	Labour £	Unit	Total rate £
T10: GAS/OIL FIRED BOILERS – cont						
FORCED DRAFT – cont						
Commercial steel shell floor standing boilers – cont						
Gas fired (high/low type), connected to conventional flue						
130–190 kW	7423.87	8384.48	8.00	197.69	nr	**8582.17**
200–250 kW	7928.92	8954.89	8.00	197.69	nr	**9152.58**
280–360 kW	9477.08	10703.37	8.00	197.69	nr	**10901.06**
375–500 kW	10573.00	11941.09	8.00	197.69	nr	**12138.78**
580–730 kW	13345.81	15072.69	10.00	247.12	nr	**15319.81**
655–820 kW	13618.14	15380.26	10.00	247.12	nr	**15627.38**
830–1040 kW	13666.00	15434.31	12.00	296.53	nr	**15730.84**
1070–1400 kW	18074.45	20413.19	12.00	296.53	nr	**20709.72**
1420–1850 kW	22507.65	25420.03	12.00	296.53	nr	**25716.56**
1850–2350 kW	25564.34	28872.24	14.00	345.95	nr	**29218.19**
2300–3000 kW	30164.24	34067.34	14.00	345.95	nr	**34413.29**
2800–3500 kW	39119.76	44181.66	14.00	345.95	nr	**44527.61**
Oil fired (on/off type), connected to conventional flue						
130–190 kW	5580.29	6302.36	8.00	197.69	nr	**6500.05**
200–250 kW	6093.59	6882.07	8.00	197.69	nr	**7079.76**
Oil fired (high/low type), connected to conventional flue						
130–190 kW	6073.78	6859.69	8.00	197.69	nr	**7057.38**
200–250 kW	6587.08	7439.41	8.00	197.69	nr	**7637.10**
280–360 kW	7593.88	8576.49	8.00	197.69	nr	**8774.18**
375–500 kW	9079.32	10254.14	8.00	197.69	nr	**10451.83**
580–730 kW	10099.30	11406.10	10.00	247.12	nr	**11653.22**
655–820 kW	10369.98	11711.80	10.00	247.12	nr	**11958.92**
830–1040 kW	11997.37	13549.77	12.00	296.53	nr	**13846.30**
1070–1400 kW	14659.59	16556.47	12.00	296.53	nr	**16853.00**
1420–1850 kW	17750.94	20047.82	12.00	296.53	nr	**20344.35**
1850–2350 kW	22732.10	25673.52	14.00	345.95	nr	**26019.47**
2300–3000 kW	26552.97	29988.80	14.00	345.95	nr	**30334.75**
2800–3500 kW	35508.50	40103.13	14.00	345.95	nr	**40449.08**

T: MECHANICAL/COOLING/HEATING SYSTEMS

Item	Net Price £	Material £	Labour hours	Labour £	Unit	**Total rate £**
ATMOSPHERIC						
Commercial cast iron sectional floor standing boilers; atmospheric; including controls, enamelled jacket, insulation, assembly and commissioning; electrical work elsewhere						
Gas (on/off type), connected to conventional flue						
30–40 kW	2177.97	2459.79	8.00	197.69	nr	**2657.48**
40–50 kW	2387.80	2696.77	8.00	197.69	nr	**2894.46**
50–60 kW	2537.95	2866.35	8.00	197.69	nr	**3064.04**
60–70 kW	2770.03	3128.46	8.00	197.69	nr	**3326.15**
70–80 kW	2935.94	3315.84	8.00	197.69	nr	**3513.53**
80–90 kW	3430.41	3874.29	8.00	197.69	nr	**4071.98**
90–100 kW	3568.67	4030.44	8.00	197.69	nr	**4228.13**
100–110 kW	3781.76	4271.10	8.00	197.69	nr	**4468.79**
110–120 kW	3999.71	4517.25	8.00	197.69	nr	**4714.94**
Gas (high/low type), connected to conventional flue						
30–40 kW	2472.38	2792.29	8.00	197.69	nr	**2989.98**
40–50 kW	2622.02	2961.29	8.00	197.69	nr	**3158.98**
50–60 kW	2818.84	3183.58	8.00	197.69	nr	**3381.27**
60–70 kW	3074.20	3471.99	8.00	197.69	nr	**3669.68**
70–80 kW	3254.76	3675.91	8.00	197.69	nr	**3873.60**
80–90 kW	3617.48	4085.56	8.00	197.69	nr	**4283.25**
90–100 kW	3814.30	4307.85	8.00	197.69	nr	**4505.54**
100–110 kW	4038.76	4561.36	8.00	197.69	nr	**4759.05**
110–120 kW	4290.87	4846.09	8.00	197.69	nr	**5043.78**
120–140 kW	6436.31	7269.13	8.00	197.69	nr	**7466.82**
140–160 kW	6598.96	7452.83	8.00	197.69	nr	**7650.52**
160–180 kW	6711.21	7579.61	8.00	197.69	nr	**7777.30**
180–200 kW	7339.05	8288.68	8.00	197.69	nr	**8486.37**
200–220 kW	7656.23	8646.91	10.00	247.12	nr	**8894.03**
220–260 kW	8319.87	9396.42	10.00	247.12	nr	**9643.54**
260–300 kW	9072.97	10246.97	10.00	247.12	nr	**10494.09**
300–340 kW	9676.43	10928.51	12.00	296.53	nr	**11225.04**

T: MECHANICAL/COOLING/HEATING SYSTEMS

Item	Net Price £	Material £	Labour hours	Labour £	Unit	Total rate £
T10: GAS/OIL FIRED BOILERS – cont						
CONDENSING						
Low Nox wall mounted condensing boiler with high efficiency modulating premix burner: Aluminium heat exchanger: Placing in position						
Maximum output						
35 kW	1817.92	2053.15	11.00	271.82	nr	**2324.97**
45 kW	1878.93	2122.06	11.00	271.82	nr	**2393.88**
60 kW	2122.91	2397.60	11.00	271.82	nr	**2669.42**
80 kW	2554.66	2885.22	11.00	271.82	nr	**3157.04**
Low Nox floor standing condensing boiler with high efficiency modulating premix burner: Stainless steel heat exchanger: Including controls: Placing in position.						
Maximum output						
50 kW	2987.61	3374.19	8.00	197.69	nr	**3571.88**
60 kW	3193.33	3606.53	8.00	197.69	nr	**3804.22**
80 kW	3486.35	3937.47	8.00	197.69	nr	**4135.16**
100 kW	4009.00	4527.75	8.00	197.69	nr	**4725.44**
125 kW	4982.55	5627.27	10.00	247.12	nr	**5874.39**
150 kW	5513.57	6227.00	10.00	247.12	nr	**6474.12**
200 kW	7622.13	8608.39	10.00	247.12	nr	**8855.51**
250 kW	8344.51	9424.25	10.00	247.12	nr	**9671.37**
300 kW	10564.30	11931.27	10.00	247.12	nr	**12178.39**
350 kW	11510.33	12999.71	10.00	247.12	nr	**13246.83**
400 kW	12382.22	13984.41	10.00	247.12	nr	**14231.53**
450 kW	12589.12	14218.09	10.00	247.12	nr	**14465.21**
500 kW	12796.04	14451.79	10.00	247.12	nr	**14698.91**
650 kW	13455.03	15196.04	10.00	247.12	nr	**15443.16**
700 kW	21494.56	24275.85	10.00	247.12	nr	**24522.97**
800 kW	22985.98	25960.25	12.00	296.53	nr	**26256.78**
900 kW	23777.73	26854.45	12.00	296.53	nr	**27150.98**
1000 kW	24571.88	27751.35	12.00	296.53	nr	**28047.88**
1300 kW	26066.88	29439.80	12.00	296.53	nr	**29736.33**

T: MECHANICAL/COOLING/HEATING SYSTEMS

Item	Net Price £	Material £	Labour hours	Labour £	Unit	Total rate £
FLUE SYSTEMS						
Flues; suitable for domestic, medium sized industrial and commercial oil and gas appliances; stainless steel, twin wall, insulated; for use internally or externally						
Straight length; 120 mm long; including one locking band						
127 mm dia.	40.62	45.88	0.49	12.11	nr	**57.99**
152 mm dia.	45.52	51.41	0.51	12.61	nr	**64.02**
175 mm dia.	52.84	59.68	0.54	13.35	nr	**73.03**
203 mm dia.	60.12	67.90	0.58	14.33	nr	**82.23**
254 mm dia.	71.10	80.30	0.70	17.30	nr	**97.60**
304 mm dia.	89.07	100.59	0.74	18.29	nr	**118.88**
355 mm dia.	127.55	144.05	0.80	19.77	nr	**163.82**
Straight length; 300 mm long; including one locking band						
127 mm dia.	62.67	70.78	0.52	12.85	nr	**83.63**
152 mm dia.	70.84	80.01	0.52	12.85	nr	**92.86**
178 mm dia.	81.40	91.93	0.55	13.59	nr	**105.52**
203 mm dia.	92.12	104.04	0.64	15.81	nr	**119.85**
254 mm dia.	102.16	115.38	0.79	19.52	nr	**134.90**
304 mm dia.	122.56	138.41	0.86	21.26	nr	**159.67**
355 mm dia.	134.49	151.89	0.94	23.24	nr	**175.13**
400 mm dia.	143.92	162.55	1.03	25.46	nr	**188.01**
450 mm dia.	164.61	185.91	1.03	25.46	nr	**211.37**
500 mm dia.	176.56	199.41	1.10	27.18	nr	**226.59**
550 mm dia.	194.85	220.06	1.10	27.18	nr	**247.24**
600 mm dia.	215.00	242.82	1.10	27.18	nr	**270.00**
Straight length; 500 mm long; including one locking band						
127 mm dia.	73.65	83.18	0.55	13.59	nr	**96.77**
152 mm dia.	82.11	92.74	0.55	13.59	nr	**106.33**
178 mm dia.	92.47	104.44	0.63	15.57	nr	**120.01**
203 mm dia.	108.09	122.07	0.63	15.57	nr	**137.64**
254 mm dia.	125.54	141.78	0.86	21.26	nr	**163.04**
304 mm dia.	150.41	169.87	0.95	23.48	nr	**193.35**
355 mm dia.	168.91	190.77	1.03	25.46	nr	**216.23**
400 mm dia.	184.44	208.30	1.12	27.68	nr	**235.98**
450 mm dia.	213.01	240.58	1.12	27.68	nr	**268.26**
500 mm dia.	229.52	259.22	1.19	29.41	nr	**288.63**
550 mm dia.	253.00	285.73	1.19	29.41	nr	**315.14**
600 mm dia.	262.08	295.99	1.19	29.41	nr	**325.40**

T: MECHANICAL/COOLING/HEATING SYSTEMS

Item	Net Price £	Material £	Labour hours	Labour £	Unit	Total rate £
T10: GAS/OIL FIRED BOILERS – cont						
FLUE SYSTEMS – cont						
Straight length; 1000 mm long; including one locking band						
127 mm dia.	131.66	148.70	0.62	15.33	nr	**164.03**
152 mm dia.	146.74	165.73	0.68	16.81	nr	**182.54**
178 mm dia.	165.10	186.47	0.74	18.29	nr	**204.76**
203 mm dia.	194.47	219.64	0.80	19.77	nr	**239.41**
254 mm dia.	220.39	248.91	0.87	21.50	nr	**270.41**
304 mm dia.	254.39	287.30	1.06	26.19	nr	**313.49**
355 mm dia.	291.92	329.69	1.16	28.66	nr	**358.35**
400 mm dia.	312.62	353.07	1.26	31.14	nr	**384.21**
450 mm dia.	329.87	372.55	1.26	31.14	nr	**403.69**
500 mm dia.	357.98	404.30	1.33	32.86	nr	**437.16**
550 mm dia.	393.70	444.64	1.33	32.86	nr	**477.50**
600 mm dia.	413.61	467.13	1.33	32.86	nr	**499.99**
Adjustable length; boiler removal; internal use only; including one locking band						
127 mm dia.	60.35	68.16	0.52	12.85	nr	**81.01**
152 mm dia.	68.20	77.02	0.55	13.59	nr	**90.61**
178 mm dia.	77.29	87.29	0.59	14.58	nr	**101.87**
203 mm dia.	88.63	100.10	0.64	15.81	nr	**115.91**
254 mm dia.	131.42	148.42	0.79	19.52	nr	**167.94**
304 mm dia.	157.40	177.77	0.86	21.26	nr	**199.03**
355 mm dia.	176.51	199.35	0.99	24.46	nr	**223.81**
400 mm dia.	308.88	348.85	0.91	22.49	nr	**371.34**
450 mm dia.	330.63	373.41	0.91	22.49	nr	**395.90**
500 mm dia.	360.60	407.26	0.99	24.46	nr	**431.72**
550 mm dia.	393.39	444.29	0.99	24.46	nr	**468.75**
600 mm dia.	411.82	465.11	0.99	24.46	nr	**489.57**
Inspection length; 500 mm long; including one locking band						
127 mm dia.	155.12	175.20	0.55	13.59	nr	**188.79**
152 mm dia.	160.62	181.41	0.55	13.59	nr	**195.00**
178 mm dia.	169.18	191.07	0.63	15.57	nr	**206.64**
203 mm dia.	179.15	202.33	0.63	15.57	nr	**217.90**
254 mm dia.	231.63	261.60	0.86	21.26	nr	**282.86**
304 mm dia.	251.04	283.52	0.95	23.48	nr	**307.00**
355 mm dia.	283.45	320.13	1.03	25.46	nr	**345.59**
400 mm dia.	454.76	513.60	1.12	27.68	nr	**541.28**
450 mm dia.	467.14	527.58	1.12	27.68	nr	**555.26**
500 mm dia.	512.58	578.90	1.19	29.41	nr	**608.31**
550 mm dia.	536.00	605.36	1.19	29.41	nr	**634.77**
600 mm dia.	545.11	615.64	1.19	29.41	nr	**645.05**

T: MECHANICAL/COOLING/HEATING SYSTEMS

Item	Net Price £	Material £	Labour hours	Labour £	Unit	**Total rate £**
Adapters						
127 mm dia.	12.97	14.65	0.49	12.11	nr	**26.76**
152 mm dia.	14.20	16.03	0.51	12.61	nr	**28.64**
178 mm dia.	15.18	17.15	0.54	13.35	nr	**30.50**
203 mm dia.	17.28	19.51	0.58	14.33	nr	**33.84**
254 mm dia.	19.01	21.47	0.70	17.30	nr	**38.77**
304 mm dia.	23.72	26.79	0.74	18.29	nr	**45.08**
355 mm dia.	28.81	32.54	0.80	19.77	nr	**52.31**
400 mm dia.	32.32	36.51	0.89	21.99	nr	**58.50**
450 mm dia.	34.52	38.98	0.89	21.99	nr	**60.97**
500 mm dia.	36.63	41.37	0.96	23.72	nr	**65.09**
550 mm dia.	42.60	48.11	0.96	23.72	nr	**71.83**
600 mm dia.	51.18	57.80	0.96	23.72	nr	**81.52**
Fittings for flue system						
90° insulated tee; including two locking bands						
127 mm dia.	150.82	170.34	1.89	46.70	nr	**217.04**
152 mm dia.	174.06	196.58	2.04	50.42	nr	**247.00**
178 mm dia.	190.24	214.85	2.39	59.06	nr	**273.91**
203 mm dia.	222.75	251.57	2.56	63.26	nr	**314.83**
254 mm dia.	225.66	254.86	2.95	72.90	nr	**327.76**
304 mm dia.	280.62	316.93	3.41	84.26	nr	**401.19**
355 mm dia.	357.76	404.05	3.77	93.17	nr	**497.22**
400 mm dia.	471.43	532.43	4.25	105.02	nr	**637.45**
450 mm dia.	491.36	554.94	4.76	117.63	nr	**672.57**
500 mm dia.	556.01	627.95	5.12	126.53	nr	**754.48**
550 mm dia.	597.67	675.01	5.61	138.63	nr	**813.64**
600 mm dia.	621.88	702.35	5.98	147.77	nr	**850.12**
135° insulated tee; including two locking bands						
127 mm dia.	195.60	220.91	1.89	46.70	nr	**267.61**
152 mm dia.	211.49	238.85	2.04	50.42	nr	**289.27**
178 mm dia.	231.23	261.15	2.39	59.06	nr	**320.21**
203 mm dia.	293.46	331.43	2.56	63.26	nr	**394.69**
254 mm dia.	333.45	376.59	2.95	72.90	nr	**449.49**
304 mm dia.	251.24	283.75	3.41	84.26	nr	**368.01**
355 mm dia.	494.35	558.32	3.77	93.17	nr	**651.49**
400 mm dia.	650.76	734.96	4.25	105.02	nr	**839.98**
450 mm dia.	704.48	795.64	4.76	117.63	nr	**913.27**
500 mm dia.	820.71	926.90	5.12	126.53	nr	**1053.43**
550 mm dia.	842.05	951.01	5.61	138.63	nr	**1089.64**
600 mm dia.	888.70	1003.69	5.98	147.77	nr	**1151.46**
Wall sleeve; for 135° tee through wall						
127 mm dia.	20.02	22.61	1.89	46.70	nr	**69.31**
152 mm dia.	26.86	30.33	2.04	50.42	nr	**80.75**
178 mm dia.	28.13	31.77	2.39	59.06	nr	**90.83**

T: MECHANICAL/COOLING/HEATING SYSTEMS

Item	Net Price £	Material £	Labour hours	Labour £	Unit	Total rate £
T10: GAS/OIL FIRED BOILERS – cont						
FLUE SYSTEMS – cont						
Fittings for flue system – cont						
Wall sleeve – cont						
203 mm dia.	31.65	35.74	2.56	63.26	nr	**99.00**
254 mm dia.	35.29	39.85	2.95	72.90	nr	**112.75**
304 mm dia.	41.13	46.45	3.41	84.26	nr	**130.71**
355 mm dia.	45.70	51.61	3.77	93.17	nr	**144.78**
15° insulated elbow; including two locking bands						
127 mm dia.	104.66	118.20	1.57	38.79	nr	**156.99**
152 mm dia.	116.32	131.38	1.79	44.24	nr	**175.62**
178 mm dia.	124.40	140.49	2.05	50.65	nr	**191.14**
203 mm dia.	131.88	148.94	2.33	57.58	nr	**206.52**
254 mm dia.	135.83	153.41	2.45	60.55	nr	**213.96**
304 mm dia.	171.73	193.95	3.43	84.76	nr	**278.71**
355 mm dia.	229.74	259.47	4.71	116.38	nr	**375.85**
30° insulated elbow; including two locking bands						
127 mm dia.	104.66	118.20	1.44	35.59	nr	**153.79**
152 mm dia.	116.32	131.38	1.62	40.04	nr	**171.42**
178 mm dia.	124.40	140.49	1.89	46.70	nr	**187.19**
203 mm dia.	131.88	148.94	2.17	53.62	nr	**202.56**
254 mm dia.	135.83	153.41	2.16	53.38	nr	**206.79**
304 mm dia.	171.73	193.95	2.74	67.71	nr	**261.66**
355 mm dia.	229.30	258.97	3.17	78.34	nr	**337.31**
400 mm dia.	229.74	259.47	3.53	87.23	nr	**346.70**
450 mm dia.	266.25	300.71	3.88	95.87	nr	**396.58**
500 mm dia.	279.14	315.26	4.24	104.77	nr	**420.03**
550 mm dia.	299.58	338.34	4.61	113.92	nr	**452.26**
600 mm dia.	326.66	368.92	4.96	122.56	nr	**491.48**
45° insulated elbow; including two locking bands						
127 mm dia.	104.66	118.20	1.44	35.59	nr	**153.79**
152 mm dia.	116.32	131.38	1.51	37.31	nr	**168.69**
178 mm dia.	124.40	140.49	1.58	39.05	nr	**179.54**
203 mm dia.	131.88	148.94	1.66	41.02	nr	**189.96**
254 mm dia.	135.83	153.41	1.72	42.50	nr	**195.91**
304 mm dia.	171.73	193.95	1.80	44.47	nr	**238.42**
355 mm dia.	229.74	259.47	1.94	47.94	nr	**307.41**
400 mm dia.	292.97	330.88	2.01	49.67	nr	**380.55**
450 mm dia.	306.87	346.58	2.09	51.64	nr	**398.22**
500 mm dia.	329.17	371.76	2.16	53.38	nr	**425.14**
550 mm dia.	359.00	405.45	2.23	55.10	nr	**460.55**
600 mm dia.	368.10	415.73	2.30	56.83	nr	**472.56**

T: MECHANICAL/COOLING/HEATING SYSTEMS

Item	Net Price £	Material £	Labour hours	Labour £	Unit	Total rate £
Flue supports						
Wall support, galvanized; including plate and brackets						
127 mm dia.	63.54	71.76	2.24	55.36	nr	**127.12**
152 mm dia.	69.86	78.90	2.44	60.30	nr	**139.20**
178 mm dia.	77.12	87.10	2.52	62.27	nr	**149.37**
203 mm dia.	80.31	90.70	2.77	68.45	nr	**159.15**
254 mm dia.	95.78	108.17	2.98	73.63	nr	**181.80**
304 mm dia.	108.29	122.30	3.46	85.50	nr	**207.80**
355 mm dia.	144.94	163.69	4.08	100.82	nr	**264.51**
400 mm dia.; with 300 mm support length and collar	372.32	420.49	4.80	118.61	nr	**539.10**
450 mm dia.; with 300 mm support length and collar	396.95	448.31	5.62	138.88	nr	**587.19**
500 mm dia.; with 300 mm support length and collar	433.66	489.77	6.24	154.19	nr	**643.96**
550 mm dia.; with 300 mm support length and collar	474.45	535.84	6.97	172.24	nr	**708.08**
600 mm dia.; with 300 mm support length and collar	500.86	565.67	7.49	185.09	nr	**750.76**
Ceiling/floor support						
127 mm dia.	19.67	22.22	1.86	45.97	nr	**68.19**
152 mm dia.	21.86	24.68	2.14	52.88	nr	**77.56**
178 mm dia.	26.18	29.57	1.93	47.70	nr	**77.27**
203 mm dia.	39.42	44.52	2.74	67.71	nr	**112.23**
254 mm dia.	44.97	50.79	3.21	79.32	nr	**130.11**
304 mm dia.	51.40	58.05	3.68	90.94	nr	**148.99**
355 mm dia.	61.25	69.18	4.28	105.76	nr	**174.94**
400 mm dia.	78.24	88.36	4.86	120.09	nr	**208.45**
450 mm dia.	84.96	95.95	5.46	134.92	nr	**230.87**
500 mm dia.	103.15	116.49	6.04	149.26	nr	**265.75**
550 mm dia.	126.19	142.51	6.65	164.32	nr	**306.83**
600 mm dia.	181.10	204.53	7.24	178.91	nr	**383.44**
Ceiling/floor firestop spacer						
127 mm dia.	3.83	4.32	0.66	16.31	nr	**20.63**
152 mm dia.	4.26	4.81	0.69	17.05	nr	**21.86**
178 mm dia.	4.82	5.45	0.70	17.30	nr	**22.75**
203 mm dia.	5.71	6.45	0.87	21.50	nr	**27.95**
254 mm dia.	5.89	6.65	0.91	22.49	nr	**29.14**
304 mm dia.	7.08	8.00	0.95	23.48	nr	**31.48**
355 mm dia.	12.90	14.57	0.99	24.46	nr	**39.03**
Wall band; internal or external use						
127 mm dia.	23.17	26.16	1.03	25.46	nr	**51.62**
152 mm dia.	24.19	27.32	1.07	26.44	nr	**53.76**
178 mm dia.	25.15	28.41	1.11	27.44	nr	**55.85**

T: MECHANICAL/COOLING/HEATING SYSTEMS

Item	Net Price £	Material £	Labour hours	Labour £	Unit	Total rate £
T10: GAS/OIL FIRED BOILERS – cont						
FLUE SYSTEMS – cont						
Flue supports – cont						
Wall band – cont						
203 mm dia.	26.54	29.98	1.18	29.16	nr	**59.14**
254 mm dia.	27.63	31.20	1.30	32.12	nr	**63.32**
304 mm dia.	29.85	33.71	1.45	35.83	nr	**69.54**
355 mm dia.	31.76	35.87	1.65	40.77	nr	**76.64**
400 mm dia.	39.39	44.48	1.85	45.72	nr	**90.20**
450 mm dia.	41.97	47.40	2.39	59.06	nr	**106.46**
500 mm dia.	50.45	56.98	2.25	55.60	nr	**112.58**
550 mm dia.	52.84	59.68	2.45	60.55	nr	**120.23**
600 mm dia.	55.86	63.09	2.66	65.73	nr	**128.82**
Flashings and terminals						
Insulated top stub; including one locking band						
127 mm dia.	57.06	64.44	1.49	36.82	nr	**101.26**
152 mm dia.	64.53	72.88	1.90	46.95	nr	**119.83**
178 mm dia.	69.47	78.46	1.92	47.45	nr	**125.91**
203 mm dia.	73.92	83.49	2.20	54.37	nr	**137.86**
254 mm dia.	78.49	88.65	2.49	61.53	nr	**150.18**
304 mm dia.	107.30	121.19	2.79	68.95	nr	**190.14**
355 mm dia.	141.65	159.98	3.19	78.84	nr	**238.82**
400 mm dia.	136.57	154.24	3.59	88.71	nr	**242.95**
450 mm dia.	148.37	167.57	3.97	98.10	nr	**265.67**
500 mm dia.	171.65	193.86	4.38	108.23	nr	**302.09**
550 mm dia.	179.60	202.84	4.78	118.13	nr	**320.97**
600 mm dia.	185.87	209.92	5.17	127.75	nr	**337.67**
Rain cap; including one locking band						
127 mm dia.	30.50	34.45	1.49	36.82	nr	**71.27**
152 mm dia.	31.88	36.01	1.54	38.06	nr	**74.07**
178 mm dia.	35.11	39.66	1.72	42.50	nr	**82.16**
203 mm dia.	41.98	47.41	2.00	49.42	nr	**96.83**
254 mm dia.	55.19	62.33	2.49	61.53	nr	**123.86**
304 mm dia.	37.79	42.68	2.80	69.19	nr	**111.87**
355 mm dia.	99.70	112.60	3.19	78.84	nr	**191.44**
400 mm dia.	99.47	112.34	3.45	85.26	nr	**197.60**
450 mm dia.	108.18	122.18	3.97	98.10	nr	**220.28**
500 mm dia.	116.97	132.11	4.38	108.23	nr	**240.34**
550 mm dia.	125.50	141.74	4.78	118.13	nr	**259.87**
600 mm dia.	134.18	151.54	5.17	127.75	nr	**279.29**
Round top; including one locking band						
127 mm dia.	58.32	65.86	1.49	36.82	nr	**102.68**
152 mm dia.	63.60	71.83	1.65	40.77	nr	**112.60**
178 mm dia.	72.50	81.89	1.92	47.45	nr	**129.34**

T: MECHANICAL/COOLING/HEATING SYSTEMS

Item	Net Price £	Material £	Labour hours	Labour £	Unit	**Total rate £**
203 mm dia.	85.32	96.36	2.20	54.37	nr	**150.73**
254 mm dia.	100.87	113.92	2.49	61.53	nr	**175.45**
304 mm dia.	132.49	149.63	2.80	69.19	nr	**218.82**
355 mm dia.	176.76	199.63	3.19	78.84	nr	**278.47**
Coping cap; including one locking band						
127 mm dia.	32.66	36.88	1.49	36.82	nr	**73.70**
152 mm dia.	34.22	38.65	1.65	40.77	nr	**79.42**
178 mm dia.	37.65	42.52	1.92	47.45	nr	**89.97**
203 mm dia.	45.17	51.01	2.20	54.37	nr	**105.38**
254 mm dia.	55.19	62.33	2.49	61.53	nr	**123.86**
304 mm dia.	74.42	84.05	2.79	68.95	nr	**153.00**
355 mm dia.	99.70	112.60	3.19	78.84	nr	**191.44**
Storm collar						
127 mm dia.	6.19	6.99	0.52	12.85	nr	**19.84**
152 mm dia.	6.63	7.49	0.55	13.59	nr	**21.08**
178 mm dia.	7.34	8.29	0.57	14.09	nr	**22.38**
203 mm dia.	7.69	8.68	0.66	16.31	nr	**24.99**
254 mm dia.	9.63	10.87	0.66	16.31	nr	**27.18**
304 mm dia.	10.04	11.34	0.72	17.79	nr	**29.13**
355 mm dia.	10.73	12.12	0.77	19.03	nr	**31.15**
400 mm dia.	28.45	32.13	0.82	20.26	nr	**52.39**
450 mm dia.	31.32	35.37	0.87	21.50	nr	**56.87**
500 mm dia.	34.15	38.57	0.92	22.74	nr	**61.31**
550 mm dia.	37.01	41.80	0.98	24.21	nr	**66.01**
600 mm dia.	39.85	45.00	1.03	25.46	nr	**70.46**
Flat flashing; including storm collar and sealant						
127 mm dia.	34.78	39.28	1.49	36.82	nr	**76.10**
152 mm dia.	35.94	40.59	1.65	40.77	nr	**81.36**
178 mm dia.	37.70	42.57	1.92	47.45	nr	**90.02**
203 mm dia.	41.29	46.63	2.20	54.37	nr	**101.00**
254 mm dia.	56.42	63.72	2.49	61.53	nr	**125.25**
304 mm dia.	67.41	76.13	2.80	69.19	nr	**145.32**
355 mm dia.	106.24	119.98	3.20	79.08	nr	**199.06**
400 mm dia.	148.21	167.39	3.59	88.71	nr	**256.10**
450 mm dia.	170.26	192.29	3.97	98.10	nr	**290.39**
500 mm dia.	184.45	208.31	4.38	108.23	nr	**316.54**
550 mm dia.	195.84	221.18	4.78	118.13	nr	**339.31**
600 mm dia.	202.92	229.17	5.17	127.75	nr	**356.92**
5°–30° rigid adjustable flashing; including storm collar and sealant						
127 mm dia.	59.23	66.89	1.49	36.82	nr	**103.71**
152 mm dia.	62.41	70.48	1.65	40.77	nr	**111.25**
178 mm dia.	66.37	74.96	1.92	47.45	nr	**122.41**

T: MECHANICAL/COOLING/HEATING SYSTEMS

Item	Net Price £	Material £	Labour hours	Labour £	Unit	Total rate £
T10: GAS/OIL FIRED BOILERS – cont						
FLUE SYSTEMS – cont						
Flashings and terminals – cont						
5°−30° rigid adjustable flashing – cont						
203 mm dia.	69.79	78.82	2.20	54.37	nr	**133.19**
254 mm dia.	73.46	82.97	2.49	61.53	nr	**144.50**
304 mm dia.	90.91	102.67	2.80	69.19	nr	**171.86**
355 mm dia.	103.39	116.77	3.19	78.84	nr	**195.61**
400 mm dia.	333.34	376.47	3.59	88.71	nr	**465.18**
450 mm dia.	389.01	439.34	3.97	98.10	nr	**537.44**
500 mm dia.	415.30	469.04	4.38	108.23	nr	**577.27**
550 mm dia.	434.59	490.82	4.77	117.87	nr	**608.69**
600 mm dia.	472.83	534.01	5.17	127.75	nr	**661.76**
Domestic and small commercial; twin walled gas vent system suitable for gas fired appliances; domestic gas boilers; small commercial boilers with internal or external flues						
152 mm long						
100 mm dia.	6.34	7.16	0.52	12.85	nr	**20.01**
125 mm dia.	7.79	8.80	0.52	12.85	nr	**21.65**
150 mm dia.	8.43	9.52	0.52	12.85	nr	**22.37**
305 mm long						
100 mm dia.	9.61	10.85	0.52	12.85	nr	**23.70**
125 mm dia.	11.28	12.74	0.52	12.85	nr	**25.59**
150 mm dia.	13.38	15.12	0.52	12.85	nr	**27.97**
457 mm long						
100 mm dia.	10.64	12.02	0.55	13.59	nr	**25.61**
125 mm dia.	11.96	13.50	0.55	13.59	nr	**27.09**
150 mm dia.	14.80	16.72	0.55	13.59	nr	**30.31**
914 mm long						
100 mm dia.	19.00	21.46	0.62	15.33	nr	**36.79**
125 mm dia.	22.15	25.02	0.62	15.33	nr	**40.35**
150 mm dia.	25.40	28.68	0.62	15.33	nr	**44.01**
1524 mm long						
100 mm dia.	27.45	31.00	0.82	20.26	nr	**51.26**
125 mm dia.	33.78	38.15	0.84	20.76	nr	**58.91**
150 mm dia.	36.27	40.96	0.84	20.76	nr	**61.72**
Adjustable length 305 mm long						
100 mm dia.	12.17	13.75	0.56	13.83	nr	**27.58**
125 mm dia.	13.66	15.43	0.56	13.83	nr	**29.26**
150 mm dia.	17.23	19.46	0.56	13.83	nr	**33.29**

T: MECHANICAL/COOLING/HEATING SYSTEMS

Item	Net Price £	Material £	Labour hours	Labour £	Unit	Total rate £
Adjustable length 457 mm long						
100 mm dia.	16.41	18.53	0.56	13.83	nr	**32.36**
125 mm dia.	19.89	22.46	0.56	13.83	nr	**36.29**
150 mm dia.	22.13	24.99	0.56	13.83	nr	**38.82**
Adjustable elbow 0°–90°						
100 mm dia.	13.88	15.68	0.48	11.86	nr	**27.54**
125 mm dia.	16.41	18.53	0.48	11.86	nr	**30.39**
150 mm dia.	20.55	23.21	0.48	11.86	nr	**35.07**
Draughthood connector						
100 mm dia.	4.28	4.83	0.48	11.86	nr	**16.69**
125 mm dia.	4.83	5.46	0.48	11.86	nr	**17.32**
150 mm dia.	5.24	5.92	0.48	11.86	nr	**17.78**
Adaptor						
100 mm dia.	10.39	11.73	0.48	11.86	nr	**23.59**
125 mm dia.	10.61	11.98	0.48	11.86	nr	**23.84**
150 mm dia.	10.83	12.23	0.48	11.86	nr	**24.09**
Support plate						
100 mm dia.	7.50	8.47	0.48	11.86	nr	**20.33**
125 mm dia.	7.97	9.00	0.48	11.86	nr	**20.86**
150 mm dia.	8.54	9.65	0.48	11.86	nr	**21.51**
Wall band						
100 mm dia.	6.77	7.65	0.48	11.86	nr	**19.51**
125 mm dia.	7.23	8.17	0.48	11.86	nr	**20.03**
150 mm dia.	9.16	10.34	0.48	11.86	nr	**22.20**
Firestop						
100 mm dia.	2.94	3.33	0.48	11.86	nr	**15.19**
125 mm dia.	2.94	3.33	0.48	11.86	nr	**15.19**
150 mm dia.	3.37	3.80	0.48	11.86	nr	**15.66**
Flat flashing						
125 mm dia.	19.98	22.56	0.55	13.59	nr	**36.15**
150 mm dia.	27.79	31.39	0.55	13.59	nr	**44.98**
Adjustable flashing 5°–30°						
100 mm dia.	52.06	58.79	0.55	13.59	nr	**72.38**
125 mm dia.	81.28	91.80	0.55	13.59	nr	**105.39**
Storm collar						
100 mm dia.	4.15	4.69	0.55	13.59	nr	**18.28**
125 mm dia.	4.24	4.78	0.55	13.59	nr	**18.37**
150 mm dia.	4.35	4.92	0.55	13.59	nr	**18.51**

T: MECHANICAL/COOLING/HEATING SYSTEMS

Item	Net Price £	Material £	Labour hours	Labour £	Unit	Total rate £
T10: GAS/OIL FIRED BOILERS – cont						
FLUE SYSTEMS – cont						
Domestic and small commercial – cont						
Gas vent terminal						
100 mm dia.	15.84	17.89	0.55	13.59	nr	**31.48**
125 mm dia.	17.41	19.67	0.55	13.59	nr	**33.26**
150 mm dia.	22.34	25.23	0.55	13.59	nr	**38.82**
Twin wall galvanized steel flue box, 125 mm dia.; fitted for gas fire, where no chimney exists						
Free-standing	100.68	113.71	2.15	53.13	nr	**166.84**
Recess	100.68	113.71	2.15	53.13	nr	**166.84**
Back boiler	74.18	83.78	2.40	59.30	nr	**143.08**

T: MECHANICAL/COOLING/HEATING SYSTEMS

Item	Net Price £	Material £	Labour hours	Labour £	Unit	**Total rate £**
T13: PACKAGED STEAM GENERATORS						
Packaged steam boilers; boiler mountings centrifugal water feed pump; insulation; and sheet steel wrap around casing; plastic coated						
Gas fired						
293 kW rating	18736.24	21160.61	86.45	2136.29	nr	**23296.90**
1465 kW rating	39963.92	45135.05	148.22	3662.69	nr	**48797.74**
2930 kW rating	57389.65	64815.58	207.50	5127.57	nr	**69943.15**
Oil fired						
293 kW rating	16993.67	19192.56	86.45	2136.29	nr	**21328.85**
1465 kW rating	36968.44	41751.98	148.22	3662.69	nr	**45414.67**
2930 kW rating	55949.51	63189.09	207.50	5127.57	nr	**68316.66**

T: MECHANICAL/COOLING/HEATING SYSTEMS

Item	Net Price £	Material £	Labour hours	Labour £	Unit	**Total rate £**
T31: LOW TEMPERATURE HOT WATER HEATING						
Y10 – PIPELINES						
SCREWED STEEL						
Black steel pipes; screwed and socketed joints; BS 1387: 1985. Fixed vertically, brackets measured separately. Screwed joints are within the running length, but any flanges are additional						
Medium weight						
10 mm dia.	4.19	4.73	0.37	9.15	m	**13.88**
15 mm dia.	2.65	2.99	0.37	9.15	m	**12.14**
20 mm dia.	3.11	3.51	0.37	9.15	m	**12.66**
25 mm dia.	4.47	5.05	0.41	10.13	m	**15.18**
32 mm dia.	5.54	6.25	0.48	11.86	m	**18.11**
40 mm dia.	6.44	7.27	0.52	12.85	m	**20.12**
50 mm dia.	9.05	10.22	0.62	15.33	m	**25.55**
65 mm dia.	12.28	13.87	0.65	16.07	m	**29.94**
80 mm dia.	15.96	18.02	1.10	27.18	m	**45.20**
100 mm dia.	22.61	25.53	1.31	32.36	m	**57.89**
125 mm dia.	28.77	32.50	1.66	41.02	m	**73.52**
150 mm dia.	33.41	37.73	1.88	46.45	m	**84.18**
Heavy weight						
15 mm dia.	3.15	3.56	0.37	9.15	m	**12.71**
20 mm dia.	3.73	4.21	0.37	9.15	m	**13.36**
25 mm dia.	5.46	6.17	0.41	10.13	m	**16.30**
32 mm dia.	6.78	7.66	0.48	11.86	m	**19.52**
40 mm dia.	7.90	8.93	0.52	12.85	m	**21.78**
50 mm dia.	10.97	12.39	0.62	15.33	m	**27.72**
65 mm dia.	14.93	16.86	0.64	15.81	m	**32.67**
80 mm dia.	19.00	21.46	1.10	27.18	m	**48.64**
100 mm dia.	26.51	29.94	1.31	32.36	m	**62.30**
125 mm dia.	30.68	34.65	1.66	41.02	m	**75.67**
150 mm dia.	35.87	40.51	1.88	46.45	m	**86.96**
200 mm dia.	56.53	63.84	2.99	73.89	m	**137.73**
250 mm dia.	70.87	80.04	3.49	86.24	m	**166.28**
300 mm dia.	84.39	95.31	3.91	96.62	m	**191.93**

T: MECHANICAL/COOLING/HEATING SYSTEMS

Item	Net Price £	Material £	Labour hours	Labour £	Unit	Total rate £
Black steel pipes; screwed and socketed joints; BS 1387: 1985. Fixed at high level or suspended, brackets measured separately. Screwed joints are within the running length, but any flanges are additional						
Medium weight						
10 mm dia.	4.19	4.73	0.58	14.33	m	**19.06**
15 mm dia.	2.65	2.99	0.58	14.33	m	**17.32**
20 mm dia.	3.11	3.51	0.58	14.33	m	**17.84**
25 mm dia.	4.47	5.05	0.60	14.83	m	**19.88**
32 mm dia.	5.54	6.25	0.68	16.81	m	**23.06**
40 mm dia.	6.44	7.27	0.73	18.03	m	**25.30**
50 mm dia.	9.05	10.22	0.85	21.00	m	**31.22**
65 mm dia.	12.28	13.87	0.88	21.74	m	**35.61**
80 mm dia.	15.96	18.02	1.45	35.83	m	**53.85**
100 mm dia.	22.61	25.53	1.74	42.99	m	**68.52**
125 mm dia.	28.77	32.50	2.21	54.61	m	**87.11**
150 mm dia.	33.41	37.73	2.50	61.78	m	**99.51**
Heavy weight						
15 mm dia.	3.15	3.56	0.58	14.33	m	**17.89**
20 mm dia.	3.73	4.21	0.58	14.33	m	**18.54**
25 mm dia.	5.46	6.17	0.60	14.83	m	**21.00**
32 mm dia.	6.78	7.66	0.68	16.81	m	**24.47**
40 mm dia.	7.90	8.93	0.73	18.03	m	**26.96**
50 mm dia.	10.97	12.39	0.85	21.00	m	**33.39**
65 mm dia.	14.93	16.86	0.88	21.74	m	**38.60**
80 mm dia.	19.00	21.46	1.45	35.83	m	**57.29**
100 mm dia.	26.51	29.94	1.74	42.99	m	**72.93**
125 mm dia.	30.68	34.65	2.21	54.61	m	**89.26**
150 mm dia.	35.87	40.51	2.50	61.78	m	**102.29**
200 mm dia.	56.53	63.84	2.99	73.89	m	**137.73**
250 mm dia.	70.87	80.04	3.49	86.24	m	**166.28**
300 mm dia.	84.39	95.31	3.91	96.62	m	**191.93**
FIXINGS						
For steel pipes; black malleable iron. For minimum fixing distances, refer to the Tables and Memoranda to the rear of the book						
Single pipe bracket, screw on, black malleable iron; screwed to wood						
15 mm dia.	0.66	0.74	0.14	3.46	nr	**4.20**
20 mm dia.	0.73	0.83	0.14	3.46	nr	**4.29**
25 mm dia.	0.86	0.97	0.17	4.20	nr	**5.17**
32 mm dia.	1.56	1.76	0.19	4.70	nr	**6.46**
40 mm dia.	1.48	1.67	0.22	5.44	nr	**7.11**
50 mm dia.	2.07	2.34	0.22	5.44	nr	**7.78**
65 mm dia.	2.72	3.07	0.28	6.93	nr	**10.00**
80 mm dia.	3.73	4.21	0.32	7.91	nr	**12.12**
100 mm dia.	5.45	6.15	0.35	8.65	nr	**14.80**

T: MECHANICAL/COOLING/HEATING SYSTEMS

Item	Net Price £	Material £	Labour hours	Labour £	Unit	**Total rate £**
T31: LOW TEMPERATURE HOT WATER HEATING – cont						
FIXINGS – cont						
For steel pipes – cont						
Single pipe bracket, screw on, black malleable iron; plugged and screwed						
15 mm dia.	0.66	0.74	0.25	6.18	nr	**6.92**
20 mm dia.	0.73	0.83	0.25	6.18	nr	**7.01**
25 mm dia.	0.86	0.97	0.30	7.41	nr	**8.38**
32 mm dia.	1.56	1.76	0.32	7.91	nr	**9.67**
40 mm dia.	1.48	1.67	0.32	7.91	nr	**9.58**
50 mm dia.	2.07	2.34	0.32	7.91	nr	**10.25**
65 mm dia.	2.72	3.07	0.35	8.65	nr	**11.72**
80 mm dia.	3.73	4.21	0.42	10.38	nr	**14.59**
100 mm dia.	5.45	6.15	0.42	10.38	nr	**16.53**
Single pipe bracket for building in, black malleable iron						
15 mm dia.	1.92	2.17	0.10	2.47	nr	**4.64**
20 mm dia.	1.92	2.17	0.11	2.72	nr	**4.89**
25 mm dia.	1.92	2.17	0.12	2.97	nr	**5.14**
32 mm dia.	2.18	2.46	0.14	3.46	nr	**5.92**
40 mm dia.	2.20	2.49	0.15	3.70	nr	**6.19**
50 mm dia.	2.29	2.59	0.16	3.96	nr	**6.55**
Pipe ring, single socket, black malleable iron						
15 mm dia.	0.76	0.86	0.10	2.47	nr	**3.33**
20 mm dia.	0.84	0.95	0.11	2.72	nr	**3.67**
25 mm dia.	0.92	1.04	0.12	2.97	nr	**4.01**
32 mm dia.	0.98	1.11	0.14	3.46	nr	**4.57**
40 mm dia.	1.27	1.44	0.15	3.70	nr	**5.14**
50 mm dia.	1.60	1.81	0.16	3.96	nr	**5.77**
65 mm dia.	2.31	2.61	0.30	7.41	nr	**10.02**
80 mm dia.	2.77	3.13	0.35	8.65	nr	**11.78**
100 mm dia.	4.19	4.73	0.40	9.88	nr	**14.61**
125 mm dia.	8.48	9.58	0.60	14.83	nr	**24.41**
150 mm dia.	9.50	10.73	0.77	19.03	nr	**29.76**
200 mm dia.	26.28	29.68	0.90	22.24	nr	**51.92**
250 mm dia.	32.84	37.09	1.10	27.18	nr	**64.27**
300 mm dia.	39.42	44.52	1.25	30.89	nr	**75.41**
350 mm dia.	46.01	51.96	1.50	37.07	nr	**89.03**
400 mm dia.	54.44	61.49	1.75	43.25	nr	**104.74**
Pipe ring, double socket, black malleable iron						
15 mm dia.	0.88	0.99	0.10	2.47	nr	**3.46**
20 mm dia.	1.01	1.14	0.11	2.72	nr	**3.86**
25 mm dia.	1.13	1.28	0.12	2.97	nr	**4.25**

T: MECHANICAL/COOLING/HEATING SYSTEMS

Item	Net Price £	Material £	Labour hours	Labour £	Unit	Total rate £
32 mm dia.	1.34	1.51	0.14	3.46	nr	**4.97**
40 mm dia.	1.45	1.64	0.15	3.70	nr	**5.34**
50 mm dia.	2.27	2.56	0.16	3.96	nr	**6.52**
Screw on backplate (Male), black malleable iron; plugged and screwed						
M12	0.48	0.54	0.10	2.47	nr	**3.01**
Screw on backplate (Female), black malleable iron; plugged and screwed						
M12	0.48	0.54	0.10	2.47	nr	**3.01**
Extra over channel sections for fabricated hangers and brackets						
Galvanized steel; including inserts, bolts, nuts, washers; fixed to backgrounds						
41 × 21 mm	5.65	6.39	0.29	7.17	m	**13.56**
41 × 41 mm	6.77	7.65	0.29	7.17	m	**14.82**
Threaded rods; metric thread; including nuts, washers etc.						
12 mm dia. × 600 mm long	2.86	3.23	0.18	4.45	nr	**7.68**
Pipe roller and chair						
Roller and chair; black malleable						
Up to 50 mm dia.	14.57	16.45	0.20	4.95	nr	**21.40**
65 mm dia.	15.01	16.95	0.20	4.95	nr	**21.90**
80 mm dia.	16.29	18.40	0.20	4.95	nr	**23.35**
100 mm dia.	16.57	18.72	0.20	4.95	nr	**23.67**
125 mm dia.	23.22	26.22	0.20	4.95	nr	**31.17**
150 mm dia.	23.22	26.22	0.30	7.41	nr	**33.63**
175 mm dia.	39.19	44.26	0.30	7.41	nr	**51.67**
200 mm dia.	60.02	67.79	0.30	7.41	nr	**75.20**
250 mm dia.	78.31	88.44	0.30	7.41	nr	**95.85**
300 mm dia.	98.60	111.35	0.30	7.41	nr	**118.76**
Roller bracket; black malleable						
25 mm dia.	3.06	3.46	0.20	4.95	nr	**8.41**
32 mm dia.	3.21	3.62	0.20	4.95	nr	**8.57**
40 mm dia.	3.43	3.88	0.20	4.95	nr	**8.83**
50 mm dia.	3.62	4.09	0.20	4.95	nr	**9.04**
65 mm dia.	4.77	5.39	0.20	4.95	nr	**10.34**
80 mm dia.	6.87	7.76	0.20	4.95	nr	**12.71**
100 mm dia.	7.64	8.63	0.20	4.95	nr	**13.58**
125 mm dia.	12.60	14.23	0.20	4.95	nr	**19.18**
150 mm dia.	12.60	14.23	0.30	7.41	nr	**21.64**
175 mm dia.	28.09	31.72	0.30	7.41	nr	**39.13**
200 mm dia.	28.09	31.72	0.30	7.41	nr	**39.13**

T: MECHANICAL/COOLING/HEATING SYSTEMS

Item	Net Price £	Material £	Labour hours	Labour £	Unit	Total rate £
T31: LOW TEMPERATURE HOT WATER HEATING – cont						
FIXINGS – cont						
Pipe roller and chair – cont						
Roller bracket – cont						
250 mm dia.	37.34	42.17	0.30	7.41	nr	**49.58**
300 mm dia.	45.87	51.81	0.30	7.41	nr	**59.22**
350 mm dia.	73.75	83.29	0.30	7.41	nr	**90.70**
400 mm dia.	84.30	95.21	0.30	7.41	nr	**102.62**
Extra over black steel screwed pipes; black steel flanges, screwed and drilled; metric; BS 4504						
Screwed flanges; PN6						
15 mm dia.	7.79	8.80	0.35	8.65	nr	**17.45**
20 mm dia.	7.79	8.80	0.47	11.61	nr	**20.41**
25 mm dia.	7.79	8.80	0.53	13.10	nr	**21.90**
32 mm dia.	7.79	8.80	0.62	15.33	nr	**24.13**
40 mm dia.	7.79	8.80	0.70	17.30	nr	**26.10**
50 mm dia.	8.30	9.37	0.84	20.76	nr	**30.13**
65 mm dia.	11.55	13.05	1.03	25.46	nr	**38.51**
80 mm dia.	16.31	18.42	1.23	30.40	nr	**48.82**
100 mm dia.	19.28	21.78	1.41	34.84	nr	**56.62**
125 mm dia.	35.37	39.94	1.77	43.74	nr	**83.68**
150 mm dia.	35.37	39.94	2.21	54.61	nr	**94.55**
Screwed flanges; PN16						
15 mm dia.	9.90	11.18	0.35	8.65	nr	**19.83**
20 mm dia.	9.90	11.18	0.47	11.61	nr	**22.79**
25 mm dia.	9.90	11.18	0.53	13.10	nr	**24.28**
32 mm dia.	10.58	11.94	0.62	15.33	nr	**27.27**
40 mm dia.	10.58	11.94	0.70	17.30	nr	**29.24**
50 mm dia.	12.90	14.57	0.84	20.76	nr	**35.33**
65 mm dia.	16.12	18.21	1.03	25.46	nr	**43.67**
80 mm dia.	19.63	22.17	1.23	30.40	nr	**52.57**
100 mm dia.	22.01	24.86	1.41	34.84	nr	**59.70**
125 mm dia.	38.51	43.49	1.77	43.74	nr	**87.23**
150 mm dia.	37.90	42.80	2.21	54.61	nr	**97.41**

T: MECHANICAL/COOLING/HEATING SYSTEMS

Item	Net Price £	Material £	Labour hours	Labour £	Unit	Total rate £
Extra over black steel screwed pipes; black steel flanges, screwed and drilled; imperial; BS 10						
Screwed flanges; Table E						
½" dia.	14.79	16.71	0.35	8.65	nr	**25.36**
¾" dia.	14.79	16.71	0.47	11.61	nr	**28.32**
1" dia.	14.77	16.68	0.53	13.10	nr	**29.78**
1 ¼" dia.	14.79	16.71	0.62	15.33	nr	**32.04**
1 ½" dia.	14.79	16.71	0.70	17.30	nr	**34.01**
2" dia.	14.79	16.71	0.84	20.76	nr	**37.47**
2 ½" dia.	17.59	19.87	1.03	25.46	nr	**45.33**
3" dia.	19.94	22.52	1.23	30.40	nr	**52.92**
4" dia.	26.98	30.47	1.41	34.84	nr	**65.31**
5" dia.	57.16	64.56	1.77	43.74	nr	**108.30**
6" dia.	57.16	64.56	2.21	54.61	nr	**119.17**
Extra over black steel screwed pipes; black steel flange connections						
Bolted connection between pair of flanges; including gasket, bolts, nuts and washers						
50 mm dia.	7.17	8.10	0.53	13.10	nr	**21.20**
65 mm dia.	8.48	9.58	0.53	13.10	nr	**22.68**
80 mm dia.	12.74	14.39	0.53	13.10	nr	**27.49**
100 mm dia.	13.83	15.62	0.61	15.08	nr	**30.70**
125 mm dia.	16.26	18.36	0.61	15.08	nr	**33.44**
150 mm dia.	23.24	26.24	0.90	22.24	nr	**48.48**
Extra over black steel screwed pipes; black heavy steel tubular fittings; BS 1387						
Long screw connection with socket and backnut						
15 mm dia.	5.26	5.94	0.63	15.57	nr	**21.51**
20 mm dia.	6.61	7.47	0.84	20.76	nr	**28.23**
25 mm dia.	8.67	9.79	0.95	23.48	nr	**33.27**
32 mm dia.	11.63	13.14	1.11	27.44	nr	**40.58**
40 mm dia.	13.98	15.79	1.28	31.63	nr	**47.42**
50 mm dia.	20.60	23.27	1.53	37.81	nr	**61.08**
65 mm dia.	28.81	32.54	1.87	46.22	nr	**78.76**
80 mm dia.	65.87	74.39	2.21	54.61	nr	**129.00**
100 mm dia.	106.72	120.53	3.05	75.37	nr	**195.90**
Running nipple						
15 mm dia.	1.29	1.46	0.50	12.35	nr	**13.81**
20 mm dia.	1.63	1.84	0.68	16.81	nr	**18.65**
25 mm dia.	2.01	2.27	0.77	19.03	nr	**21.30**
32 mm dia.	2.81	3.17	0.90	22.24	nr	**25.41**
40 mm dia.	3.79	4.28	1.03	25.46	nr	**29.74**
50 mm dia.	5.77	6.52	1.23	30.40	nr	**36.92**

T: MECHANICAL/COOLING/HEATING SYSTEMS

Item	Net Price £	Material £	Labour hours	Labour £	Unit	**Total rate £**
T31: LOW TEMPERATURE HOT WATER HEATING – cont						
FIXINGS – cont						
Extra over black steel screwed pipes – cont						
Running nipple – cont						
65 mm dia.	12.43	14.04	1.50	37.07	nr	**51.11**
80 mm dia.	19.37	21.88	1.78	43.99	nr	**65.87**
100 mm dia.	30.36	34.29	2.38	58.80	nr	**93.09**
Barrel nipple						
15 mm dia.	0.73	0.83	0.50	12.35	nr	**13.18**
20 mm dia.	1.17	1.33	0.68	16.81	nr	**18.14**
25 mm dia.	1.54	1.74	0.77	19.03	nr	**20.77**
32 mm dia.	2.31	2.61	0.90	22.24	nr	**24.85**
40 mm dia.	2.85	3.22	1.03	25.46	nr	**28.68**
50 mm dia.	4.06	4.59	1.23	30.40	nr	**34.99**
65 mm dia.	8.68	9.80	1.50	37.07	nr	**46.87**
80 mm dia.	12.10	13.67	1.78	43.99	nr	**57.66**
100 mm dia.	21.93	24.77	2.38	58.80	nr	**83.57**
125 mm dia.	40.70	45.97	2.87	70.93	nr	**116.90**
150 mm dia.	64.14	72.44	3.39	83.76	nr	**156.20**
Close taper nipple						
15 mm dia.	1.54	1.74	0.50	12.35	nr	**14.09**
20 mm dia.	2.01	2.27	0.68	16.81	nr	**19.08**
25 mm dia.	2.61	2.95	0.77	19.03	nr	**21.98**
32 mm dia.	3.92	4.43	0.90	22.24	nr	**26.67**
40 mm dia.	4.86	5.49	1.03	25.46	nr	**30.95**
50 mm dia.	7.48	8.45	1.23	30.40	nr	**38.85**
65 mm dia.	14.52	16.40	1.50	37.07	nr	**53.47**
80 mm dia.	17.90	20.22	1.78	43.99	nr	**64.21**
100 mm dia.	36.75	41.50	2.38	58.80	nr	**100.30**
Extra over black steel screwed pipes; black malleable iron fittings; BS 143						
Cap						
15 mm dia.	0.62	0.70	0.32	7.91	nr	**8.61**
20 mm dia.	0.71	0.81	0.43	10.63	nr	**11.44**
25 mm dia.	0.90	1.02	0.49	12.11	nr	**13.13**
32 mm dia.	1.57	1.77	0.58	14.33	nr	**16.10**
40 mm dia.	1.86	2.10	0.66	16.31	nr	**18.41**
50 mm dia.	3.87	4.38	0.78	19.28	nr	**23.66**
65 mm dia.	6.46	7.29	0.96	23.72	nr	**31.01**
80 mm dia.	7.30	8.24	1.13	27.92	nr	**36.16**
100 mm dia.	15.99	18.05	1.70	42.01	nr	**60.06**

T: MECHANICAL/COOLING/HEATING SYSTEMS

Item	Net Price £	Material £	Labour hours	Labour £	Unit	**Total rate £**
Plain plug, hollow						
15 mm dia.	0.43	0.49	0.28	6.93	nr	**7.42**
20 mm dia.	0.56	0.63	0.38	9.38	nr	**10.01**
25 mm dia.	0.76	0.86	0.44	10.87	nr	**11.73**
32 mm dia.	1.06	1.19	0.51	12.61	nr	**13.80**
40 mm dia.	1.99	2.24	0.59	14.58	nr	**16.82**
50 mm dia.	2.78	3.14	0.70	17.30	nr	**20.44**
65 mm dia.	4.55	5.14	0.85	21.00	nr	**26.14**
80 mm dia.	7.09	8.01	1.00	24.71	nr	**32.72**
100 mm dia.	13.06	14.75	1.44	35.59	nr	**50.34**
Plain plug, solid						
15 mm dia.	1.36	1.54	0.28	6.93	nr	**8.47**
20 mm dia.	1.31	1.48	0.38	9.38	nr	**10.86**
25 mm dia.	1.95	2.20	0.44	10.87	nr	**13.07**
32 mm dia.	2.80	3.16	0.51	12.61	nr	**15.77**
40 mm dia.	3.95	4.46	0.59	14.58	nr	**19.04**
50 mm dia.	5.18	5.85	0.70	17.30	nr	**23.15**
90° elbow, male/female						
15 mm dia.	0.71	0.81	0.64	15.81	nr	**16.62**
20 mm dia.	0.97	1.09	0.85	21.00	nr	**22.09**
25 mm dia.	1.60	1.81	0.97	23.97	nr	**25.78**
32 mm dia.	2.92	3.29	1.12	27.68	nr	**30.97**
40 mm dia.	4.91	5.55	1.29	31.89	nr	**37.44**
50 mm dia.	6.31	7.13	1.55	38.30	nr	**45.43**
65 mm dia.	13.65	15.41	1.89	46.70	nr	**62.11**
80 mm dia.	18.65	21.06	2.24	55.36	nr	**76.42**
100 mm dia.	32.61	36.83	3.09	76.36	nr	**113.19**
90° elbow						
15 mm dia.	0.64	0.72	0.64	15.81	nr	**16.53**
20 mm dia.	0.88	0.99	0.85	21.00	nr	**21.99**
25 mm dia.	1.36	1.54	0.97	23.97	nr	**25.51**
32 mm dia.	2.48	2.81	1.12	27.68	nr	**30.49**
40 mm dia.	4.17	4.71	1.29	31.89	nr	**36.60**
50 mm dia.	4.88	5.51	1.55	38.30	nr	**43.81**
65 mm dia.	10.54	11.90	1.89	46.70	nr	**58.60**
80 mm dia.	15.48	17.48	2.24	55.36	nr	**72.84**
100 mm dia.	29.85	33.71	3.09	76.36	nr	**110.07**
125 mm dia.	72.61	82.01	4.44	109.72	nr	**191.73**
150 mm dia.	135.18	152.67	5.79	143.09	nr	**295.76**
45° elbow						
15 mm dia.	1.53	1.72	0.64	15.81	nr	**17.53**
20 mm dia.	1.88	2.12	0.85	21.00	nr	**23.12**
25 mm dia.	2.80	3.16	0.97	23.97	nr	**27.13**
32 mm dia.	5.72	6.46	1.12	27.68	nr	**34.14**
40 mm dia.	7.02	7.93	1.29	31.89	nr	**39.82**
50 mm dia.	9.63	10.87	1.55	38.30	nr	**49.17**
65 mm dia.	14.08	15.90	1.89	46.70	nr	**62.60**

T: MECHANICAL/COOLING/HEATING SYSTEMS

Item	Net Price £	Material £	Labour hours	Labour £	Unit	**Total rate £**
T31: LOW TEMPERATURE HOT WATER HEATING – cont						
FIXINGS – cont						
Extra over black steel screwed pipes – cont						
45° elbow – cont						
80 mm dia.	21.16	23.90	2.24	55.36	nr	**79.26**
100 mm dia.	46.42	52.43	3.09	76.36	nr	**128.79**
150 mm dia.	129.62	146.39	5.79	143.09	nr	**289.48**
90° bend, male/female						
15 mm dia.	1.26	1.43	0.64	15.81	nr	**17.24**
20 mm dia.	1.83	2.07	0.85	21.00	nr	**23.07**
25 mm dia.	2.71	3.06	0.97	23.97	nr	**27.03**
32 mm dia.	4.31	4.87	1.12	27.68	nr	**32.55**
40 mm dia.	6.80	7.68	1.29	31.89	nr	**39.57**
50 mm dia.	11.88	13.41	1.55	38.30	nr	**51.71**
65 mm dia.	19.77	22.33	1.89	46.70	nr	**69.03**
80 mm dia.	30.43	34.36	2.24	55.36	nr	**89.72**
100 mm dia.	75.40	85.16	3.09	76.36	nr	**161.52**
90° bend, male						
15 mm dia.	2.90	3.27	0.64	15.81	nr	**19.08**
20 mm dia.	3.25	3.67	0.85	21.00	nr	**24.67**
25 mm dia.	4.77	5.39	0.97	23.97	nr	**29.36**
32 mm dia.	10.35	11.69	1.12	27.68	nr	**39.37**
40 mm dia.	14.53	16.41	1.29	31.89	nr	**48.30**
50 mm dia.	22.05	24.90	1.55	38.30	nr	**63.20**
90° bend, female						
15 mm dia.	1.16	1.31	0.64	15.81	nr	**17.12**
20 mm dia.	1.65	1.87	0.85	21.00	nr	**22.87**
25 mm dia.	2.32	2.62	0.97	23.97	nr	**26.59**
32 mm dia.	4.39	4.96	1.12	27.68	nr	**32.64**
40 mm dia.	5.86	6.62	1.29	31.89	nr	**38.51**
50 mm dia.	8.25	9.31	1.55	38.30	nr	**47.61**
65 mm dia.	17.62	19.90	1.89	46.70	nr	**66.60**
80 mm dia.	28.07	31.70	2.24	55.36	nr	**87.06**
100 mm dia.	61.52	69.48	3.09	76.36	nr	**145.84**
125 mm dia.	145.12	163.89	4.44	109.72	nr	**273.61**
150 mm dia.	263.93	298.08	5.79	143.09	nr	**441.17**
Return bend						
15 mm dia.	6.62	7.48	0.64	15.81	nr	**23.29**
20 mm dia.	10.71	12.10	0.85	21.00	nr	**33.10**
25 mm dia.	13.36	15.09	0.97	23.97	nr	**39.06**

T: MECHANICAL/COOLING/HEATING SYSTEMS

Item	Net Price £	Material £	Labour hours	Labour £	Unit	**Total rate £**
32 mm dia.	20.71	23.39	1.12	27.68	nr	**51.07**
40 mm dia.	24.66	27.85	1.29	31.89	nr	**59.74**
50 mm dia.	37.64	42.51	1.55	38.30	nr	**80.81**
Equal socket, parallel thread						
15 mm dia.	0.62	0.70	0.64	15.81	nr	**16.51**
20 mm dia.	0.76	0.86	0.85	21.00	nr	**21.86**
25 mm dia.	1.04	1.17	0.97	23.97	nr	**25.14**
32 mm dia.	2.09	2.36	1.12	27.68	nr	**30.04**
40 mm dia.	3.05	3.45	1.29	31.89	nr	**35.34**
50 mm dia.	4.40	4.97	1.55	38.30	nr	**43.27**
65 mm dia.	6.77	7.65	1.89	46.70	nr	**54.35**
80 mm dia.	9.59	10.83	2.24	55.36	nr	**66.19**
100 mm dia.	15.83	17.88	3.09	76.36	nr	**94.24**
Concentric reducing socket						
20 × 15 mm dia.	0.98	1.11	0.76	18.78	nr	**19.89**
25 × 15 mm dia.	1.16	1.31	0.85	21.00	nr	**22.31**
25 × 20 mm dia.	1.22	1.38	0.86	21.26	nr	**22.64**
32 × 25 mm dia.	2.31	2.61	1.01	24.96	nr	**27.57**
40 × 25 mm dia.	2.90	3.27	1.16	28.66	nr	**31.93**
40 × 32 mm dia.	3.01	3.40	1.16	28.66	nr	**32.06**
50 × 25 mm dia.	4.33	4.89	1.38	34.10	nr	**38.99**
50 × 40 mm dia.	4.21	4.75	1.38	34.10	nr	**38.85**
65 × 50 mm dia.	7.66	8.65	1.69	41.77	nr	**50.42**
80 × 50 mm dia.	9.97	11.26	2.00	49.42	nr	**60.68**
100 × 50 mm dia.	19.88	22.45	2.75	67.95	nr	**90.40**
100 × 80 mm dia.	27.42	30.97	2.75	67.95	nr	**98.92**
150 × 100 mm dia.	55.24	62.38	4.10	101.32	nr	**163.70**
Eccentric reducing socket						
20 × 15 mm dia.	1.76	1.99	0.73	18.03	nr	**20.02**
25 × 15 mm dia.	5.02	5.67	0.85	21.00	nr	**26.67**
25 × 20 mm dia.	5.69	6.43	0.85	21.00	nr	**27.43**
32 × 25 mm dia.	7.78	8.78	1.01	24.96	nr	**33.74**
40 × 25 mm dia.	8.89	10.04	1.16	28.66	nr	**38.70**
40 × 32 mm dia.	4.47	5.05	1.16	28.66	nr	**33.71**
50 × 25 mm dia.	5.77	6.52	1.38	34.10	nr	**40.62**
50 × 40 mm dia.	7.96	8.99	1.38	34.10	nr	**43.09**
65 × 50 mm dia.	9.26	10.45	1.69	41.77	nr	**52.22**
80 × 50 mm dia.	16.21	18.31	2.00	49.42	nr	**67.73**
Hexagon bush						
20 × 15 mm dia.	0.56	0.63	0.37	9.15	nr	**9.78**
25 × 15 mm dia.	0.68	0.77	0.43	10.63	nr	**11.40**
25 × 20 mm dia.	0.71	0.81	0.43	10.63	nr	**11.44**
32 × 25 mm dia.	0.92	1.04	0.51	12.61	nr	**13.65**
40 × 25 mm dia.	1.46	1.65	0.58	14.33	nr	**15.98**
40 × 32 mm dia.	1.33	1.50	0.58	14.33	nr	**15.83**
50 × 25 mm dia.	3.01	3.40	0.71	17.54	nr	**20.94**

T: MECHANICAL/COOLING/HEATING SYSTEMS

Item	Net Price £	Material £	Labour hours	Labour £	Unit	Total rate £
T31: LOW TEMPERATURE HOT WATER HEATING – cont						
FIXINGS – cont						
Extra over black steel screwed pipes – cont						
Hexagon bush – cont						
50 × 40 mm dia.	2.78	3.14	0.71	17.54	nr	**20.68**
65 × 50 mm dia.	4.66	5.26	0.85	21.00	nr	**26.26**
80 × 50 mm dia.	7.56	8.54	1.00	24.71	nr	**33.25**
100 × 50 mm dia.	17.47	19.73	1.52	37.57	nr	**57.30**
100 × 80 mm dia.	24.10	27.21	1.52	37.57	nr	**64.78**
150 × 100 mm dia.	52.27	59.04	2.57	63.51	nr	**122.55**
Hexagon nipple						
15 mm dia.	0.60	0.67	0.28	6.93	nr	**7.60**
20 mm dia.	0.68	0.77	0.38	9.38	nr	**10.15**
25 mm dia.	0.97	1.09	0.44	10.87	nr	**11.96**
32 mm dia.	1.99	2.24	0.51	12.61	nr	**14.85**
40 mm dia.	2.31	2.61	0.59	14.58	nr	**17.19**
50 mm dia.	4.18	4.72	0.70	17.30	nr	**22.02**
65 mm dia.	6.82	7.70	0.85	21.00	nr	**28.70**
80 mm dia.	9.86	11.14	1.00	24.71	nr	**35.85**
100 mm dia.	16.74	18.91	1.44	35.59	nr	**54.50**
150 mm dia.	37.35	42.19	2.32	57.32	nr	**99.51**
Union, male/female						
15 mm dia.	3.15	3.56	0.64	15.81	nr	**19.37**
20 mm dia.	3.85	4.35	0.85	21.00	nr	**25.35**
25 mm dia.	4.81	5.44	0.97	23.97	nr	**29.41**
32 mm dia.	7.41	8.36	1.12	27.68	nr	**36.04**
40 mm dia.	9.49	10.72	1.29	31.89	nr	**42.61**
50 mm dia.	14.95	16.88	1.55	38.30	nr	**55.18**
65 mm dia.	28.53	32.22	1.89	46.70	nr	**78.92**
80 mm dia.	39.22	44.30	2.24	55.36	nr	**99.66**
Union, female						
15 mm dia.	2.57	2.91	0.64	15.81	nr	**18.72**
20 mm dia.	2.80	3.16	0.85	21.00	nr	**24.16**
25 mm dia.	3.27	3.69	0.97	23.97	nr	**27.66**
32 mm dia.	6.16	6.96	1.12	27.68	nr	**34.64**
40 mm dia.	6.96	7.86	1.29	31.89	nr	**39.75**
50 mm dia.	11.53	13.03	1.55	38.30	nr	**51.33**
65 mm dia.	24.85	28.07	1.89	46.70	nr	**74.77**
80 mm dia.	34.32	38.76	2.24	55.36	nr	**94.12**
100 mm dia.	65.35	73.81	3.09	76.36	nr	**150.17**

T: MECHANICAL/COOLING/HEATING SYSTEMS

Item	Net Price £	Material £	Labour hours	Labour £	Unit	**Total rate £**
Union elbow, male/female						
15 mm dia.	4.56	5.15	0.55	13.59	nr	**18.74**
20 mm dia.	5.70	6.44	0.85	21.00	nr	**27.44**
25 mm dia.	8.00	9.04	0.97	23.97	nr	**33.01**
Twin elbow						
15 mm dia.	3.69	4.17	0.91	22.49	nr	**26.66**
20 mm dia.	4.08	4.61	1.22	30.14	nr	**34.75**
25 mm dia.	6.60	7.46	1.39	34.34	nr	**41.80**
32 mm dia.	13.56	15.31	1.62	40.04	nr	**55.35**
40 mm dia.	17.17	19.39	1.86	45.97	nr	**65.36**
50 mm dia.	22.07	24.93	2.21	54.61	nr	**79.54**
65 mm dia.	39.21	44.29	2.72	67.21	nr	**111.50**
80 mm dia.	66.80	75.44	3.21	79.32	nr	**154.76**
Equal tee						
15 mm dia.	0.88	0.99	0.91	22.49	nr	**23.48**
20 mm dia.	1.28	1.45	1.22	30.14	nr	**31.59**
25 mm dia.	1.85	2.09	1.39	34.34	nr	**36.43**
32 mm dia.	3.37	3.80	1.62	40.04	nr	**43.84**
40 mm dia.	5.18	5.85	1.86	45.97	nr	**51.82**
50 mm dia.	7.45	8.42	2.21	54.61	nr	**63.03**
65 mm dia.	16.33	18.44	2.72	67.21	nr	**85.65**
80 mm dia.	19.88	22.45	3.21	79.32	nr	**101.77**
100 mm dia.	36.03	40.69	4.44	109.72	nr	**150.41**
125 mm dia.	104.69	118.24	5.38	132.94	nr	**251.18**
150 mm dia.	166.81	188.39	6.31	155.93	nr	**344.32**
Tee reducing on branch						
20 × 15 mm dia.	1.16	1.31	1.22	30.14	nr	**31.45**
25 × 15 mm dia.	1.60	1.81	1.39	34.34	nr	**36.15**
25 × 20 mm dia.	1.90	2.14	1.39	34.34	nr	**36.48**
32 × 25 mm dia.	3.31	3.73	1.62	40.04	nr	**43.77**
40 × 25 mm dia.	4.38	4.95	1.86	45.97	nr	**50.92**
40 × 32 mm dia.	5.74	6.49	1.86	45.97	nr	**52.46**
50 × 25 mm dia.	6.51	7.35	2.21	54.61	nr	**61.96**
50 × 40 mm dia.	9.70	10.95	2.21	54.61	nr	**65.56**
65 × 50 mm dia.	15.14	17.10	2.72	67.21	nr	**84.31**
80 × 50 mm dia.	20.47	23.12	3.21	79.32	nr	**102.44**
100 × 50 mm dia.	33.89	38.27	4.44	109.72	nr	**147.99**
100 × 80 mm dia.	52.27	59.04	4.44	109.72	nr	**168.76**
150 × 100 mm dia.	122.88	138.78	6.31	155.93	nr	**294.71**
Equal pitcher tee						
15 mm dia.	2.69	3.04	0.91	22.49	nr	**25.53**
20 mm dia.	3.35	3.78	1.22	30.14	nr	**33.92**
25 mm dia.	5.02	5.67	1.39	34.34	nr	**40.01**
32 mm dia.	7.62	8.61	1.62	40.04	nr	**48.65**
40 mm dia.	11.78	13.30	1.86	45.97	nr	**59.27**
50 mm dia.	16.55	18.69	2.21	54.61	nr	**73.30**
65 mm dia.	25.59	28.90	2.72	67.21	nr	**96.11**

T: MECHANICAL/COOLING/HEATING SYSTEMS

Item	Net Price £	Material £	Labour hours	Labour £	Unit	Total rate £
T31: LOW TEMPERATURE HOT WATER HEATING – cont						
FIXINGS – cont						
Extra over black steel screwed pipes – cont						
Equal pitcher tee – cont						
80 mm dia.	39.94	45.10	3.21	79.32	nr	**124.42**
100 mm dia.	89.85	101.48	4.44	109.72	nr	**211.20**
Cross						
15 mm dia.	2.63	2.97	1.00	24.71	nr	**27.68**
20 mm dia.	3.96	4.47	1.33	32.86	nr	**37.33**
25 mm dia.	5.02	5.67	1.51	37.31	nr	**42.98**
32 mm dia.	7.32	8.27	1.77	43.74	nr	**52.01**
40 mm dia.	9.43	10.65	2.02	49.92	nr	**60.57**
50 mm dia.	15.31	17.29	2.42	59.80	nr	**77.09**
65 mm dia.	21.16	23.90	2.97	73.39	nr	**97.29**
80 mm dia.	28.14	31.78	3.50	86.49	nr	**118.27**
100 mm dia.	58.09	65.61	4.84	119.61	nr	**185.22**
BLACK WELDED STEEL						
Black steel pipes; butt welded joints; BS 1387: 1985; including protective painting. Fixed Vertical with brackets measured separately (Refer to Screwed Steel Section). Welded butt joints are within the running length, but any flanges are additional						
Medium weight						
10 mm dia.	2.48	2.81	0.37	9.15	m	**11.96**
15 mm dia.	2.65	2.99	0.37	9.15	m	**12.14**
20 mm dia.	3.11	3.51	0.37	9.15	m	**12.66**
25 mm dia.	4.47	5.05	0.41	10.13	m	**15.18**
32 mm dia.	5.54	6.25	0.48	11.86	m	**18.11**
40 mm dia.	6.44	7.27	0.52	12.85	m	**20.12**
50 mm dia.	9.05	10.22	0.62	15.33	m	**25.55**
65 mm dia.	12.28	13.87	0.64	15.81	m	**29.68**
80 mm dia.	15.96	18.02	1.10	27.18	m	**45.20**
100 mm dia.	22.61	25.53	1.31	32.36	m	**57.89**
125 mm dia.	28.77	32.50	1.66	41.02	m	**73.52**
150 mm dia.	33.41	37.73	1.88	46.45	m	**84.18**

T: MECHANICAL/COOLING/HEATING SYSTEMS

Item	Net Price £	Material £	Labour hours	Labour £	Unit	**Total rate £**
Heavy weight						
15 mm dia.	3.15	3.56	0.37	9.15	m	**12.71**
20 mm dia.	3.73	4.21	0.37	9.15	m	**13.36**
25 mm dia.	5.46	6.17	0.41	10.13	m	**16.30**
32 mm dia.	6.78	7.66	0.48	11.86	m	**19.52**
40 mm dia.	7.90	8.93	0.52	12.85	m	**21.78**
50 mm dia.	10.97	12.39	0.62	15.33	m	**27.72**
65 mm dia.	14.93	16.86	0.64	15.81	m	**32.67**
80 mm dia.	19.00	21.46	1.10	27.18	m	**48.64**
100 mm dia.	26.51	29.94	1.31	32.36	m	**62.30**
125 mm dia.	30.68	34.65	1.66	41.02	m	**75.67**
150 mm dia.	35.87	40.51	1.88	46.45	m	**86.96**
Black steel pipes; butt welded joints; BS 1387: 1985; including protective painting. Fixed at High Level or Suspended with brackets measured separately (Refer to Screwed Steel Section). Welded butt joints are within the running length, but any flanges are additional						
Medium weight						
10 mm dia.	2.48	2.81	0.58	14.33	m	**17.14**
15 mm dia.	2.65	2.99	0.58	14.33	m	**17.32**
20 mm dia.	3.11	3.51	0.58	14.33	m	**17.84**
25 mm dia.	4.47	5.05	0.60	14.83	m	**19.88**
32 mm dia.	5.54	6.25	0.68	16.81	m	**23.06**
40 mm dia.	6.44	7.27	0.73	18.03	m	**25.30**
50 mm dia.	9.05	10.22	0.85	21.00	m	**31.22**
65 mm dia.	12.28	13.87	0.88	21.74	m	**35.61**
80 mm dia.	15.96	18.02	1.45	35.83	m	**53.85**
100 mm dia.	22.61	25.53	1.74	42.99	m	**68.52**
125 mm dia.	28.77	32.50	2.21	54.61	m	**87.11**
150 mm dia.	33.41	37.73	2.50	61.78	m	**99.51**
Heavy weight						
15 mm dia.	3.15	3.56	0.58	14.33	m	**17.89**
20 mm dia.	3.73	4.21	0.58	14.33	m	**18.54**
25 mm dia.	5.46	6.17	0.60	14.83	m	**21.00**
32 mm dia.	6.78	7.66	0.68	16.81	m	**24.47**
40 mm dia.	7.90	8.93	0.73	18.03	m	**26.96**
50 mm dia.	10.97	12.39	0.85	21.00	m	**33.39**
65 mm dia.	14.93	16.86	0.88	21.74	m	**38.60**
80 mm dia.	19.00	21.46	1.45	35.83	m	**57.29**
100 mm dia.	26.51	29.94	1.74	42.99	m	**72.93**
125 mm dia.	30.68	34.65	2.21	54.61	m	**89.26**
150 mm dia.	35.87	40.51	2.50	61.78	m	**102.29**

T: MECHANICAL/COOLING/HEATING SYSTEMS

Item	Net Price £	Material £	Labour hours	Labour £	Unit	Total rate £
T31: LOW TEMPERATURE HOT WATER HEATING – cont						
FIXINGS						
Refer to steel pipes; black malleable iron. For minimum fixing distances, refer to the Tables and Memoranda to the rear of the book						
Extra over black steel butt welded pipes; black steel flanges, welded and drilled; metric; BS 4504						
Welded flanges; PN6						
15 mm dia.	3.28	3.70	0.59	14.58	nr	**18.28**
20 mm dia.	3.28	3.70	0.69	17.05	nr	**20.75**
25 mm dia.	3.28	3.70	0.84	20.76	nr	**24.46**
32 mm dia.	5.45	6.15	1.00	24.71	nr	**30.86**
40 mm dia.	5.80	6.55	1.11	27.44	nr	**33.99**
50 mm dia.	5.97	6.74	1.37	33.86	nr	**40.60**
65 mm dia.	8.78	9.91	1.54	38.06	nr	**47.97**
80 mm dia.	13.18	14.88	1.67	41.27	nr	**56.15**
100 mm dia.	14.40	16.26	2.22	54.86	nr	**71.12**
125 mm dia.	18.88	21.32	2.61	64.50	nr	**85.82**
150 mm dia.	19.07	21.54	2.99	73.89	nr	**95.43**
Welded flanges; PN16						
15 mm dia.	6.57	7.42	0.59	14.58	nr	**22.00**
20 mm dia.	6.57	7.42	0.69	17.05	nr	**24.47**
25 mm dia.	6.57	7.42	0.84	20.76	nr	**28.18**
32 mm dia.	8.87	10.02	1.00	24.71	nr	**34.73**
40 mm dia.	8.87	10.02	1.11	27.44	nr	**37.46**
50 mm dia.	11.65	13.16	1.37	33.86	nr	**47.02**
65 mm dia.	13.44	15.18	1.54	38.06	nr	**53.24**
80 mm dia.	17.08	19.29	1.67	41.27	nr	**60.56**
100 mm dia.	17.43	19.69	2.22	54.86	nr	**74.55**
125 mm dia.	27.68	31.26	2.61	64.50	nr	**95.76**
150 mm dia.	33.03	37.30	2.99	73.89	nr	**111.19**
Blank flanges, slip on for welding; PN6						
15 mm dia.	2.06	2.33	0.48	11.86	nr	**14.19**
20 mm dia.	2.06	2.33	0.55	13.59	nr	**15.92**
25 mm dia.	2.06	2.33	0.64	15.81	nr	**18.14**
32 mm dia.	4.22	4.76	0.76	18.78	nr	**23.54**
40 mm dia.	4.22	4.76	0.84	20.76	nr	**25.52**
50 mm dia.	3.83	4.32	1.01	24.96	nr	**29.28**
65 mm dia.	7.60	8.59	1.30	32.12	nr	**40.71**

T: MECHANICAL/COOLING/HEATING SYSTEMS

Item	Net Price £	Material £	Labour hours	Labour £	Unit	**Total rate £**
80 mm dia.	7.94	8.97	1.41	34.84	nr	**43.81**
100 mm dia.	8.17	9.23	1.78	43.99	nr	**53.22**
125 mm dia.	16.29	18.40	2.06	50.90	nr	**69.30**
150 mm dia.	15.42	17.41	2.35	58.07	nr	**75.48**
Blank flanges, slip on for welding; PN16						
15 mm dia.	1.87	2.11	0.48	11.86	nr	**13.97**
20 mm dia.	2.87	3.24	0.55	13.59	nr	**16.83**
25 mm dia.	3.49	3.94	0.64	15.81	nr	**19.75**
32 mm dia.	3.57	4.03	0.76	18.78	nr	**22.81**
40 mm dia.	4.61	5.20	0.84	20.76	nr	**25.96**
50 mm dia.	7.16	8.09	1.01	24.96	nr	**33.05**
65 mm dia.	8.78	9.91	1.30	32.12	nr	**42.03**
80 mm dia.	11.39	12.86	1.41	34.84	nr	**47.70**
100 mm dia.	14.12	15.94	1.78	43.99	nr	**59.93**
125 mm dia.	20.72	23.40	2.06	50.90	nr	**74.30**
150 mm dia.	26.73	30.19	2.35	58.07	nr	**88.26**
Extra over black steel butt welded pipes; black steel flanges, welding and drilled; imperial; BS 10						
Welded flanges; Table E						
15 mm dia.	6.74	7.61	0.59	14.58	nr	**22.19**
20 mm dia.	6.74	7.61	0.69	17.05	nr	**24.66**
25 mm dia.	6.74	7.61	0.84	20.76	nr	**28.37**
32 mm dia.	6.74	7.61	1.00	24.71	nr	**32.32**
40 mm dia.	6.94	7.83	1.11	27.44	nr	**35.27**
50 mm dia.	8.86	10.01	1.37	33.86	nr	**43.87**
65 mm dia.	10.78	12.18	1.54	38.06	nr	**50.24**
80 mm dia.	14.16	15.99	1.67	41.27	nr	**57.26**
100 mm dia.	20.91	23.61	2.22	54.86	nr	**78.47**
125 mm dia.	40.67	45.93	2.61	64.50	nr	**110.43**
150 mm dia.	40.67	45.93	2.99	73.89	nr	**119.82**
Blank flanges, slip on for welding; Table E						
15 mm dia.	4.63	5.23	0.48	11.86	nr	**17.09**
20 mm dia.	4.63	5.23	0.55	13.59	nr	**18.82**
25 mm dia.	4.63	5.23	0.64	15.81	nr	**21.04**
32 mm dia.	5.84	6.60	0.76	18.78	nr	**25.38**
40 mm dia.	6.34	7.16	0.84	20.76	nr	**27.92**
50 mm dia.	7.00	7.91	1.01	24.96	nr	**32.87**
65 mm dia.	8.07	9.12	1.30	32.12	nr	**41.24**
80 mm dia.	17.62	19.90	1.41	34.84	nr	**54.74**
100 mm dia.	20.01	22.60	1.78	43.99	nr	**66.59**
125 mm dia.	32.91	37.17	2.06	50.90	nr	**88.07**
150 mm dia.	42.31	47.79	2.35	58.07	nr	**105.86**

T: MECHANICAL/COOLING/HEATING SYSTEMS

Item	Net Price £	Material £	Labour hours	Labour £	Unit	Total rate £
T31: LOW TEMPERATURE HOT WATER HEATING – cont						
FIXINGS – cont						
Extra over black steel butt welded pipes; black steel flange connections						
Bolted connection between pair of flanges; including gasket, bolts, nuts and washers						
50 mm dia.	7.17	8.10	0.50	12.35	nr	**20.45**
65 mm dia.	8.48	9.58	0.50	12.35	nr	**21.93**
80 mm dia.	12.74	14.39	0.50	12.35	nr	**26.74**
100 mm dia.	13.83	15.62	0.50	12.35	nr	**27.97**
125 mm dia.	16.26	18.36	0.50	12.35	nr	**30.71**
150 mm dia.	23.24	26.24	0.88	21.74	nr	**47.98**
Extra over fittings; BS 1965; butt welded						
Cap						
25 mm dia.	15.91	17.97	0.47	11.61	nr	**29.58**
32 mm dia.	15.91	17.97	0.59	14.58	nr	**32.55**
40 mm dia.	15.99	18.05	0.70	17.30	nr	**35.35**
50 mm dia.	18.70	21.12	0.99	24.46	nr	**45.58**
65 mm dia.	21.96	24.81	1.35	33.36	nr	**58.17**
80 mm dia.	22.35	25.25	1.66	41.02	nr	**66.27**
100 mm dia.	29.22	33.00	2.23	55.10	nr	**88.10**
125 mm dia.	41.11	46.43	3.03	74.88	nr	**121.31**
150 mm dia.	47.55	53.70	3.79	93.66	nr	**147.36**
Concentric reducer						
20 × 15 mm dia.	8.86	10.01	0.69	17.05	nr	**27.06**
25 × 15 mm dia.	11.05	12.48	0.87	21.50	nr	**33.98**
25 × 20 mm dia.	8.55	9.66	0.87	21.50	nr	**31.16**
32 × 25 mm dia.	11.78	13.30	1.08	26.68	nr	**39.98**
40 × 25 mm dia.	15.35	17.34	1.38	34.10	nr	**51.44**
40 × 32 mm dia.	10.51	11.87	1.38	34.10	nr	**45.97**
50 × 25 mm dia.	14.61	16.50	1.82	44.97	nr	**61.47**
50 × 40 mm dia.	10.35	11.69	1.82	44.97	nr	**56.66**
65 × 50 mm dia.	14.28	16.13	2.52	62.27	nr	**78.40**
80 × 50 mm dia.	14.53	16.41	3.24	80.06	nr	**96.47**
100 × 50 mm dia.	24.05	27.16	4.08	100.82	nr	**127.98**
100 × 80 mm dia.	16.55	18.69	4.08	100.82	nr	**119.51**
125 × 80 mm dia.	36.15	40.83	4.71	116.38	nr	**157.21**
150 × 100 mm dia.	39.09	44.15	5.33	131.71	nr	**175.86**
Eccentric reducer						
20 × 15 mm dia.	13.16	14.86	0.69	17.05	nr	**31.91**
25 × 15 mm dia.	17.24	19.47	0.87	21.50	nr	**40.97**
25 × 20 mm dia.	14.46	16.33	0.87	21.50	nr	**37.83**

T: MECHANICAL/COOLING/HEATING SYSTEMS

Item	Net Price £	Material £	Labour hours	Labour £	Unit	**Total rate £**
32 × 25 mm dia.	15.99	18.05	1.08	26.68	nr	**44.73**
40 × 25 mm dia.	19.75	22.31	1.38	34.10	nr	**56.41**
40 × 32 mm dia.	18.86	21.30	1.38	34.10	nr	**55.40**
50 × 25 mm dia.	22.50	25.41	1.82	44.97	nr	**70.38**
50 × 40 mm dia.	16.22	18.32	1.82	44.97	nr	**63.29**
65 × 50 mm dia.	19.13	21.60	2.52	62.27	nr	**83.87**
80 × 50 mm dia.	23.42	26.45	3.24	80.06	nr	**106.51**
100 × 50 mm dia.	39.67	44.80	4.08	100.82	nr	**145.62**
100 × 80 mm dia.	29.27	33.06	4.08	100.82	nr	**133.88**
125 × 80 mm dia.	78.18	88.29	4.71	116.38	nr	**204.67**
150 × 100 mm dia.	59.21	66.87	5.33	131.71	nr	**198.58**
45° elbow, long radius						
15 mm dia.	3.32	3.75	0.56	13.83	nr	**17.58**
20 mm dia.	3.37	3.80	0.75	18.53	nr	**22.33**
25 mm dia.	4.38	4.95	0.93	22.98	nr	**27.93**
32 mm dia.	5.31	6.00	1.17	28.92	nr	**34.92**
40 mm dia.	5.37	6.07	1.46	36.08	nr	**42.15**
50 mm dia.	7.36	8.31	1.97	48.67	nr	**56.98**
65 mm dia.	9.48	10.71	2.70	66.73	nr	**77.44**
80 mm dia.	8.75	9.88	3.32	82.04	nr	**91.92**
100 mm dia.	13.24	14.95	4.09	101.07	nr	**116.02**
125 mm dia.	28.14	31.78	4.94	122.07	nr	**153.85**
150 mm dia.	34.96	39.48	5.78	142.83	nr	**182.31**
90° elbow, long radius						
15 mm dia.	3.43	3.88	0.56	13.83	nr	**17.71**
20 mm dia.	3.36	3.79	0.75	18.53	nr	**22.32**
25 mm dia.	4.46	5.04	0.93	22.98	nr	**28.02**
32 mm dia.	5.32	6.01	1.17	28.92	nr	**34.93**
40 mm dia.	5.38	6.08	1.46	36.08	nr	**42.16**
50 mm dia.	7.36	8.31	1.97	48.67	nr	**56.98**
65 mm dia.	9.48	10.71	2.70	66.73	nr	**77.44**
80 mm dia.	10.29	11.62	3.32	82.04	nr	**93.66**
100 mm dia.	15.57	17.58	4.09	101.07	nr	**118.65**
125 mm dia.	32.40	36.59	4.94	122.07	nr	**158.66**
150 mm dia.	41.13	46.45	5.78	142.83	nr	**189.28**
Branch bend (based on branch and pipe sizes being the same)						
15 mm dia.	13.69	15.46	0.85	21.00	nr	**36.46**
20 mm dia.	13.57	15.33	0.85	21.00	nr	**36.33**
25 mm dia.	13.72	15.49	1.02	25.20	nr	**40.69**
32 mm dia.	12.95	14.63	1.11	27.44	nr	**42.07**
40 mm dia.	12.78	14.43	1.36	33.61	nr	**48.04**
50 mm dia.	12.43	14.04	1.70	42.01	nr	**56.05**
65 mm dia.	18.36	20.74	1.78	43.99	nr	**64.73**
80 mm dia.	28.69	32.41	1.82	44.97	nr	**77.38**
100 mm dia.	37.41	42.25	1.87	46.22	nr	**88.47**
125 mm dia.	65.93	74.46	2.21	54.61	nr	**129.07**
150 mm dia.	101.38	114.50	2.65	65.48	nr	**179.98**

T: MECHANICAL/COOLING/HEATING SYSTEMS

Item	Net Price £	Material £	Labour hours	Labour £	Unit	**Total rate £**
T31: LOW TEMPERATURE HOT WATER HEATING – cont						
FIXINGS – cont						
Extra over fittings – cont						
Equal tee						
15 mm dia.	32.46	36.66	0.82	20.26	nr	**56.92**
20 mm dia.	32.46	36.66	1.10	27.18	nr	**63.84**
25 mm dia.	32.46	36.66	1.35	33.36	nr	**70.02**
32 mm dia.	32.46	36.66	1.63	40.27	nr	**76.93**
40 mm dia.	32.46	36.66	2.14	52.88	nr	**89.54**
50 mm dia.	34.18	38.61	3.02	74.63	nr	**113.24**
65 mm dia.	49.86	56.31	3.61	89.21	nr	**145.52**
80 mm dia.	49.94	56.41	4.18	103.30	nr	**159.71**
100 mm dia.	62.13	70.17	5.24	129.49	nr	**199.66**
125 mm dia.	142.22	160.62	6.70	165.57	nr	**326.19**
150 mm dia.	155.34	175.44	8.45	208.81	nr	**384.25**
Extra over black steel butt welded pipes; labour						
Made bend						
15 mm dia.	–	–	0.42	10.38	nr	**10.38**
20 mm dia.	–	–	0.42	10.38	nr	**10.38**
25 mm dia.	–	–	0.50	12.35	nr	**12.35**
32 mm dia.	–	–	0.62	15.33	nr	**15.33**
40 mm dia.	–	–	0.74	18.29	nr	**18.29**
50 mm dia.	–	–	0.89	21.99	nr	**21.99**
65 mm dia.	–	–	1.05	25.94	nr	**25.94**
80 mm dia.	–	–	1.13	27.92	nr	**27.92**
100 mm dia.	–	–	2.90	71.67	nr	**71.67**
125 mm dia.	–	–	3.56	87.96	nr	**87.96**
150 mm dia.	–	–	4.18	103.30	nr	**103.30**
Splay cut end						
15 mm dia.	–	–	0.14	3.46	nr	**3.46**
20 mm dia.	–	–	0.16	3.96	nr	**3.96**
25 mm dia.	–	–	0.18	4.45	nr	**4.45**
32 mm dia.	–	–	0.25	6.18	nr	**6.18**
40 mm dia.	–	–	0.27	6.67	nr	**6.67**
50 mm dia.	–	–	0.31	7.66	nr	**7.66**
65 mm dia.	–	–	0.35	8.65	nr	**8.65**
80 mm dia.	–	–	0.40	9.88	nr	**9.88**
100 mm dia.	–	–	0.48	11.86	nr	**11.86**
125 mm dia.	–	–	0.56	13.83	nr	**13.83**
150 mm dia.	–	–	0.64	15.81	nr	**15.81**

T: MECHANICAL/COOLING/HEATING SYSTEMS

Item	Net Price £	Material £	Labour hours	Labour £	Unit	**Total rate £**
Screwed joint to fitting						
15 mm dia.	–	–	0.30	7.41	nr	**7.41**
20 mm dia.	–	–	0.40	9.88	nr	**9.88**
25 mm dia.	–	–	0.46	11.36	nr	**11.36**
32 mm dia.	–	–	0.53	13.10	nr	**13.10**
40 mm dia.	–	–	0.61	15.08	nr	**15.08**
50 mm dia.	–	–	0.73	18.03	nr	**18.03**
65 mm dia.	–	–	0.89	21.99	nr	**21.99**
80 mm dia.	–	–	1.05	25.94	nr	**25.94**
100 mm dia.	–	–	1.46	36.08	nr	**36.08**
125 mm dia.	–	–	2.10	51.90	nr	**51.90**
150 mm dia.	–	–	2.73	67.46	nr	**67.46**
Straight butt weld						
15 mm dia.	–	–	0.31	7.66	nr	**7.66**
20 mm dia.	–	–	0.42	10.38	nr	**10.38**
25 mm dia.	–	–	0.52	12.85	nr	**12.85**
32 mm dia.	–	–	0.69	17.05	nr	**17.05**
40 mm dia.	–	–	0.83	20.51	nr	**20.51**
50 mm dia.	–	–	1.22	30.14	nr	**30.14**
65 mm dia.	–	–	1.57	38.79	nr	**38.79**
80 mm dia.	–	–	1.95	48.20	nr	**48.20**
100 mm dia.	–	–	2.38	58.80	nr	**58.80**
125 mm dia.	–	–	2.83	69.93	nr	**69.93**
150 mm dia.	–	–	3.27	80.80	nr	**80.80**
CARBON WELDED STEEL						
Hot finished seamless carbon steel pipe; BS 806 and BS 3601; wall thickness to BS 3600; butt welded joints; including protective painting, fixed vertically or at low level, brackets measured separately (Refer to Screwed Pipework Section). Welded butt joints are within the running length, but any flanges are additional						
Pipework						
200 mm dia.	82.02	92.64	2.04	50.42	m	**143.06**
250 mm dia.	73.85	83.41	2.59	64.01	m	**147.42**
300 mm dia.	173.63	196.09	2.99	73.89	m	**269.98**
350 mm dia.	183.24	206.95	3.52	86.99	m	**293.94**
400 mm dia.	319.48	360.81	4.08	100.82	m	**461.63**

T: MECHANICAL/COOLING/HEATING SYSTEMS

Item	Net Price £	Material £	Labour hours	Labour £	Unit	**Total rate £**
T31: LOW TEMPERATURE HOT WATER HEATING – cont						
CARBON WELDED STEEL – cont						
Hot finished seamless carbon steel pipe; BS 806 and BS 3601; wall thickness to BS 3600; butt welded joints; including protective painting, fixed at high level or suspended, brackets measured separately (Refer to Screwed Pipework Section). Welded butt joints are within the running length, but any flanges are additional						
Pipework						
200 mm dia.	82.02	92.64	3.70	91.43	m	**184.07**
250 mm dia.	73.85	83.41	4.73	116.88	m	**200.29**
300 mm dia.	173.63	196.09	5.65	139.62	m	**335.71**
350 mm dia.	183.24	206.95	6.68	165.08	m	**372.03**
400 mm dia.	319.48	360.81	7.70	190.28	m	**551.09**
FIXINGS						
Refer to steel pipes; black malleable iron. For minimum fixing distances, refer to the Tables and Memoranda to the rear of the book						
Extra over fittings; BS 1965 part 1; butt welded						
Cap						
200 mm dia.	67.16	75.85	3.70	91.43	nr	**167.28**
250 mm dia.	128.87	145.54	4.73	116.88	nr	**262.42**
300 mm dia.	140.96	159.20	5.65	139.62	nr	**298.82**
350 mm dia.	217.18	245.28	6.68	165.08	nr	**410.36**
400 mm dia.	251.20	283.70	7.70	190.28	nr	**473.98**
Concentric reducer						
200 mm × 150 mm dia.	73.37	82.86	7.27	179.65	nr	**262.51**
250 mm × 150 mm dia.	106.39	120.16	9.05	223.63	nr	**343.79**
250 mm × 200 mm dia.	65.11	73.53	9.10	224.87	nr	**298.40**
300 mm × 150 mm dia.	122.89	138.79	10.75	265.65	nr	**404.44**
300 mm × 200 mm dia.	127.27	143.74	10.75	265.65	nr	**409.39**
300 mm × 250 mm dia.	113.14	127.78	11.15	275.52	nr	**403.30**
350 mm × 200 mm dia.	192.99	217.96	12.50	308.89	nr	**526.85**
350 mm × 250 mm dia.	161.81	182.74	12.70	313.83	nr	**496.57**
350 mm × 300 mm dia.	161.81	182.74	13.00	321.25	nr	**503.99**
400 mm × 250 mm dia.	236.59	267.20	14.46	357.32	nr	**624.52**
400 mm × 300 mm dia.	293.00	330.91	14.51	358.56	nr	**689.47**
400 mm × 350 mm dia.	255.39	288.44	15.16	374.62	nr	**663.06**

T: MECHANICAL/COOLING/HEATING SYSTEMS

Item	Net Price £	Material £	Labour hours	Labour £	Unit	Total rate £
Eccentric reducer						
200 mm × 150 mm dia.	132.50	149.64	7.27	179.65	nr	**329.29**
250 mm × 150 mm dia.	176.56	199.41	9.05	223.63	nr	**423.04**
250 mm × 200 mm dia.	115.53	130.48	9.10	224.87	nr	**355.35**
300 mm × 150 mm dia.	211.81	239.22	10.75	265.65	nr	**504.87**
300 mm × 200 mm dia.	244.30	275.91	10.75	265.65	nr	**541.56**
300 mm × 250 mm dia.	196.65	222.09	11.15	275.52	nr	**497.61**
350 mm × 200 mm dia.	224.38	253.41	12.50	308.89	nr	**562.30**
350 mm × 250 mm dia.	186.54	210.68	12.70	313.83	nr	**524.51**
350 mm × 300 mm dia.	186.54	210.68	13.00	321.25	nr	**531.93**
400 mm × 250 mm dia.	299.59	338.35	14.46	357.32	nr	**695.67**
400 mm × 300 mm dia.	280.55	316.85	14.51	358.56	nr	**675.41**
400 mm × 350 mm dia.	345.68	390.41	15.16	374.62	nr	**765.03**
45° elbow						
200 mm dia.	70.17	79.25	7.75	191.52	nr	**270.77**
250 mm dia.	131.79	148.84	10.05	248.34	nr	**397.18**
300 mm dia.	192.55	217.46	12.20	301.48	nr	**518.94**
350 mm dia.	155.82	175.98	14.65	362.02	nr	**538.00**
400 mm dia.	258.98	292.50	17.12	423.06	nr	**715.56**
90° elbow						
200 mm dia.	82.55	93.23	7.75	191.52	nr	**284.75**
250 mm dia.	155.07	175.13	10.05	248.34	nr	**423.47**
300 mm dia.	226.53	255.85	12.20	301.48	nr	**557.33**
350 mm dia.	311.43	351.73	14.65	362.02	nr	**713.75**
400 mm dia.	397.78	449.25	17.12	423.06	nr	**872.31**
Equal tee						
200 mm dia.	218.10	246.32	11.25	278.00	nr	**524.32**
250 mm dia.	373.96	422.35	14.53	359.06	nr	**781.41**
300 mm dia.	464.11	524.17	17.55	433.67	nr	**957.84**
350 mm dia.	622.66	703.23	20.98	518.44	nr	**1221.67**
400 mm dia.	711.62	803.70	24.38	602.45	nr	**1406.15**
Extra over black steel butt welded pipes; labour						
Straight butt weld						
200 mm dia.	–	–	4.08	100.82	nr	**100.82**
250 mm dia.	–	–	5.20	128.50	nr	**128.50**
300 mm dia.	–	–	6.22	153.70	nr	**153.70**
350 mm dia.	–	–	7.33	181.13	nr	**181.13**
400 mm dia.	–	–	8.41	207.83	nr	**207.83**
Branch weld						
100 mm dia.	–	–	3.46	85.50	nr	**85.50**
125 mm dia.	–	–	4.23	104.53	nr	**104.53**
150 mm dia.	–	–	5.00	123.55	nr	**123.55**

T: MECHANICAL/COOLING/HEATING SYSTEMS

Item	Net Price £	Material £	Labour hours	Labour £	Unit	Total rate £
T31: LOW TEMPERATURE HOT WATER HEATING – cont						
FIXINGS – cont						
Extra over black steel butt welded pipes; black steel flanges, welding and drilled; metric; BS 4504						
Welded flanges; PN16						
200 mm dia.	44.90	50.71	4.10	101.32	nr	**152.03**
250 mm dia.	64.50	72.85	5.33	131.71	nr	**204.56**
300 mm dia.	90.35	102.04	6.40	158.15	nr	**260.19**
350 mm dia.	174.75	197.36	7.43	183.61	nr	**380.97**
400 mm dia.	230.71	260.56	8.45	208.81	nr	**469.37**
Welded flanges; PN25						
200 mm dia.	149.10	168.39	4.10	101.32	nr	**269.71**
250 mm dia.	178.72	201.85	5.33	131.71	nr	**333.56**
300 mm dia.	241.68	272.95	6.40	158.15	nr	**431.10**
Blank flanges, slip on for welding; PN16						
200 mm dia.	103.83	117.27	2.70	66.73	nr	**184.00**
250 mm dia.	157.79	178.20	3.48	86.00	nr	**264.20**
300 mm dia.	168.59	190.40	4.20	103.80	nr	**294.20**
350 mm dia.	263.85	297.99	4.78	118.13	nr	**416.12**
400 mm dia.	348.88	394.03	5.35	132.20	nr	**526.23**
Blank flanges, slip on for welding; PN25						
200 mm dia.	103.83	117.27	2.70	66.73	nr	**184.00**
250 mm dia.	157.79	178.20	3.48	86.00	nr	**264.20**
300 mm dia.	168.59	190.40	4.20	103.80	nr	**294.20**
Extra over black steel butt welded pipes; black steel flange connections						
Bolted connection between pair of flanges; including gasket, bolts, nuts and washers						
200 mm dia.	87.13	98.40	3.83	94.65	nr	**193.05**
250 mm dia.	134.85	152.30	4.93	121.83	nr	**274.13**
300 mm dia.	183.85	207.64	5.90	145.79	nr	**353.43**

T: MECHANICAL/COOLING/HEATING SYSTEMS

Item	Net Price £	Material £	Labour hours	Labour £	Unit	**Total rate £**
PRESS FIT						
Press fit jointing system; operating temperature −20°C to +120°C; operating pressure 16 bar ; butyl rubber 'O' ring mechanical joint. With brackets measured separately (Refer to Screwed Steel Section)						
Carbon steel						
Pipework						
15 mm dia.	1.42	1.60	0.46	11.36	m	**12.96**
20 mm dia.	2.28	2.57	0.48	11.86	m	**14.43**
25 mm dia.	3.20	3.61	0.52	12.85	m	**16.46**
32 mm dia.	4.13	4.66	0.56	13.83	m	**18.49**
40 mm dia.	5.66	6.40	0.58	14.33	m	**20.73**
50 mm dia.	7.34	8.29	0.66	16.31	m	**24.60**
Extra over for carbon steel press fit fittings						
Coupling						
15 mm dia.	0.97	1.09	0.36	8.90	nr	**9.99**
20 mm dia.	1.17	1.33	0.36	8.90	nr	**10.23**
25 mm dia.	1.48	1.67	0.44	10.87	nr	**12.54**
32 mm dia.	2.48	2.81	0.44	10.87	nr	**13.68**
40 mm dia.	3.31	3.73	0.52	12.85	nr	**16.58**
50 mm dia.	3.91	4.42	0.60	14.83	nr	**19.25**
Reducer						
20 × 15 mm dia.	0.89	1.01	0.36	8.90	nr	**9.91**
25 × 15 mm dia.	1.17	1.33	0.40	9.88	nr	**11.21**
25 × 20 mm dia.	1.23	1.39	0.40	9.88	nr	**11.27**
32 × 20 mm dia.	1.37	1.55	0.40	9.88	nr	**11.43**
32 × 25 mm dia.	1.47	1.66	0.44	10.87	nr	**12.53**
40 × 32 mm dia.	3.18	3.59	0.48	11.86	nr	**15.45**
50 × 20 mm dia.	9.17	10.35	0.48	11.86	nr	**22.21**
50 × 25 mm dia.	9.24	10.43	0.52	12.85	nr	**23.28**
50 × 40 mm dia.	9.71	10.97	0.56	13.83	nr	**24.80**
90° elbow						
15 mm dia.	1.40	1.58	0.36	8.90	nr	**10.48**
20 mm dia.	1.83	2.07	0.36	8.90	nr	**10.97**
25 mm dia.	2.51	2.84	0.44	10.87	nr	**13.71**
32 mm dia.	6.27	7.08	0.44	10.87	nr	**17.95**
40 mm dia.	10.03	11.33	0.52	12.85	nr	**24.18**
50 mm dia.	11.98	13.53	0.60	14.83	nr	**28.36**

T: MECHANICAL/COOLING/HEATING SYSTEMS

Item	Net Price £	Material £	Labour hours	Labour £	Unit	Total rate £
T31: LOW TEMPERATURE HOT WATER HEATING – cont						
PRESS FIT – cont						
Extra over for carbon steel press fit fittings – cont						
45° elbow						
15 mm dia.	1.67	1.89	0.36	8.90	nr	**10.79**
20 mm dia.	1.86	2.10	0.36	8.90	nr	**11.00**
25 mm dia.	2.52	2.85	0.44	10.87	nr	**13.72**
32 mm dia.	4.97	5.61	0.44	10.87	nr	**16.48**
40 mm dia.	6.24	7.05	0.52	12.85	nr	**19.90**
50 mm dia.	7.05	7.97	0.60	14.83	nr	**22.80**
Equal tee						
15 mm dia.	2.67	3.02	0.54	13.35	nr	**16.37**
20 mm dia.	3.09	3.49	0.54	13.35	nr	**16.84**
25 mm dia.	4.13	4.66	0.66	16.31	nr	**20.97**
32 mm dia.	6.45	7.28	0.66	16.31	nr	**23.59**
40 mm dia.	9.51	10.74	0.78	19.28	nr	**30.02**
50 mm dia.	11.42	12.89	0.90	22.24	nr	**35.13**
Reducing tee						
20 × 15 mm dia.	3.03	3.42	0.54	13.35	nr	**16.77**
25 × 15 mm dia.	4.09	4.62	0.62	15.33	nr	**19.95**
25 × 20 mm dia.	4.45	5.03	0.62	15.33	nr	**20.36**
32 × 15 mm dia.	6.00	6.77	0.62	15.33	nr	**22.10**
32 × 20 mm dia.	6.49	7.33	0.62	15.33	nr	**22.66**
32 × 25 mm dia.	6.59	7.45	0.62	15.33	nr	**22.78**
40 × 20 mm dia.	8.69	9.81	0.70	17.30	nr	**27.11**
40 × 25 mm dia.	9.01	10.18	0.70	17.30	nr	**27.48**
40 × 32 mm dia.	8.80	9.93	0.70	17.30	nr	**27.23**
50 × 20 mm dia.	10.36	11.70	0.82	20.26	nr	**31.96**
50 × 25 mm dia.	10.56	11.92	0.82	20.26	nr	**32.18**
50 × 32 mm dia.	10.87	12.28	0.82	20.26	nr	**32.54**
50 × 40 mm dia.	11.40	12.87	0.82	20.26	nr	**33.13**

T: MECHANICAL/COOLING/HEATING SYSTEMS

Item	Net Price £	Material £	Labour hours	Labour £	Unit	**Total rate £**
MECHANICAL GROOVED						
Mechanical grooved jointing system; working temperature not exceeding 82° C BS 5750; pipework complete with grooved joints; painted finish. With brackets measured separately (Refer to Screwed Steel Section)						
Grooved joints						
65 mm	12.97	14.65	0.58	14.33	m	**28.98**
80 mm	13.59	15.35	0.68	16.81	m	**32.16**
100 mm	17.64	19.92	0.79	19.52	m	**39.44**
125 mm	27.69	31.27	1.02	25.20	m	**56.47**
150 mm	28.90	32.64	1.15	28.42	m	**61.06**
Extra over mechanical grooved system fittings						
Couplings						
65 mm	12.97	14.65	0.41	10.13	nr	**24.78**
80 mm	13.59	15.35	0.41	10.13	nr	**25.48**
100 mm	17.64	19.92	0.66	16.31	nr	**36.23**
125 mm	27.69	31.27	0.68	16.81	nr	**48.08**
150 mm	28.90	32.64	0.80	19.77	nr	**52.41**
Concentric reducers (one size down)						
80 mm	20.82	23.51	0.59	14.58	nr	**38.09**
100 mm	20.86	23.56	0.71	17.54	nr	**41.10**
125 mm	35.97	40.63	0.85	21.00	nr	**61.63**
150 mm	46.63	52.66	0.98	24.21	nr	**76.87**
Short radius elbow; 90°						
65 mm	23.83	26.92	0.53	13.10	nr	**40.02**
80 mm	24.33	27.48	0.61	15.08	nr	**42.56**
100 mm	32.57	36.78	0.80	19.77	nr	**56.55**
125 mm	53.71	60.66	0.90	22.24	nr	**82.90**
150 mm	69.71	78.73	0.94	23.24	nr	**101.97**
Short radius elbow; 45°						
65 mm	20.41	23.05	0.53	13.10	nr	**36.15**
80 mm	22.93	25.90	0.61	15.08	nr	**40.98**
100 mm	28.50	32.19	0.80	19.77	nr	**51.96**
125 mm	48.21	54.45	0.90	22.24	nr	**76.69**
150 mm	53.07	59.94	0.94	23.24	nr	**83.18**
Equal tee						
65 mm	42.91	48.46	0.83	20.51	nr	**68.97**
80 mm	45.47	51.36	0.93	22.98	nr	**74.34**
100 mm	50.84	57.42	1.18	29.16	nr	**86.58**
125 mm	134.84	152.29	1.37	33.86	nr	**186.15**
150 mm	125.01	141.19	1.43	35.34	nr	**176.53**

T: MECHANICAL/COOLING/HEATING SYSTEMS

Item	Net Price £	Material £	Labour hours	Labour £	Unit	Total rate £
T31: LOW TEMPERATURE HOT WATER HEATING – cont						
PLASTIC PIPEWORK						
Polypropylene PP-R 80 pipe, mechanically stabilized by fibre compound mixture in middle layer; suitable for continuous working temperatures of 0–90 degress C; thermally fused joints in the running length						
Pipe; 4 m long; PN 20						
20 mm dia.	2.52	2.85	0.35	8.65	m	**11.50**
25 mm dia.	3.80	4.29	0.39	9.63	m	**13.92**
32 mm dia.	4.33	4.89	0.43	10.63	m	**15.52**
40 mm dia.	5.79	6.54	0.47	11.61	m	**18.15**
50 mm dia.	8.43	9.52	0.51	12.61	m	**22.13**
63 mm dia.	13.90	15.70	0.52	12.85	m	**28.55**
75 mm dia.	17.98	20.31	0.60	14.83	m	**35.14**
90 mm dia.	27.73	31.31	0.69	17.05	m	**48.36**
110 mm dia.	41.70	47.09	0.69	17.05	m	**64.14**
125 mm dia.	44.68	50.46	0.85	21.00	m	**71.46**
FIXINGS						
Refer to steel pipes; black malleable iron. For minimum fixing distances, refer to the Tables and Memoranda to the rear of the book						
Extra over fittings; thermally fused joints						
Overbridge bow						
20 mm dia.	1.68	1.90	0.51	12.61	nr	**14.51**
25 mm dia.	3.08	3.48	0.56	13.83	nr	**17.31**
32 mm dia.	6.16	6.96	0.65	16.07	nr	**23.03**
Elbow 90°						
20 mm dia.	0.57	0.64	0.44	10.87	nr	**11.51**
25 mm dia.	0.74	0.84	0.52	12.85	nr	**13.69**
32 mm dia.	1.08	1.22	0.59	14.58	nr	**15.80**
40 mm dia.	1.69	1.91	0.66	16.31	nr	**18.22**
50 mm dia.	3.65	4.12	0.73	18.03	nr	**22.15**
63 mm dia.	5.58	6.30	0.85	21.00	nr	**27.30**
75 mm dia.	12.35	13.94	0.85	21.00	nr	**34.94**
90 mm dia.	22.81	25.77	1.04	25.70	nr	**51.47**
110 mm dia.	32.43	36.63	1.04	25.70	nr	**62.33**
125 mm dia.	49.96	56.43	1.30	32.12	nr	**88.55**

T: MECHANICAL/COOLING/HEATING SYSTEMS

Item	Net Price £	Material £	Labour hours	Labour £	Unit	Total rate £
Long bend 90°						
20 mm dia.	3.00	3.39	0.48	11.86	nr	**15.25**
25 mm dia.	3.13	3.54	0.57	14.09	nr	**17.63**
32 mm dia.	3.68	4.15	0.65	16.07	nr	**20.22**
40 mm dia.	6.74	7.61	0.73	18.03	nr	**25.64**
Elbow 90°, female/male						
20 mm dia.	0.58	0.65	0.44	10.87	nr	**11.52**
25 mm dia.	0.76	0.86	0.52	12.85	nr	**13.71**
32 mm dia.	1.09	1.23	0.59	14.58	nr	**15.81**
Elbow 45°						
20 mm dia.	0.58	0.65	0.44	10.87	nr	**11.52**
25 mm dia.	0.76	0.86	0.52	12.85	nr	**13.71**
32 mm dia.	1.09	1.23	0.59	14.58	nr	**15.81**
40 mm dia.	1.70	1.92	0.66	16.31	nr	**18.23**
50 mm dia.	3.65	4.12	0.73	18.03	nr	**22.15**
63 mm dia.	5.59	6.31	0.85	21.00	nr	**27.31**
75 mm dia.	12.36	13.96	0.85	21.00	nr	**34.96**
90 mm dia.	22.81	25.77	1.04	25.70	nr	**51.47**
110 mm dia.	32.44	36.64	1.04	25.70	nr	**62.34**
125 mm dia.	49.97	56.44	1.30	32.12	nr	**88.56**
Elbow 45°, female/male						
20 mm dia.	0.58	0.65	0.44	10.87	nr	**11.52**
25 mm dia.	0.76	0.86	0.52	12.85	nr	**13.71**
32 mm dia.	1.09	1.23	0.59	14.58	nr	**15.81**
T-Piece 90°						
20 mm dia.	0.79	0.89	0.61	15.08	nr	**15.97**
25 mm dia.	1.07	1.20	0.72	17.79	nr	**18.99**
32 mm dia.	1.39	1.57	0.83	20.51	nr	**22.08**
40 mm dia.	2.13	2.41	0.92	22.74	nr	**25.15**
50 mm dia.	6.07	6.85	1.01	24.96	nr	**31.81**
63 mm dia.	8.70	9.82	1.11	27.44	nr	**37.26**
75 mm dia.	14.51	16.39	1.18	29.16	nr	**45.55**
90 mm dia.	26.70	30.15	1.46	36.08	nr	**66.23**
110 mm dia.	41.66	47.05	1.46	36.08	nr	**83.13**
125 mm dia.	55.35	62.52	1.82	44.97	nr	**107.49**
T-Piece 90° reducing						
25 × 20 × 20 mm	1.09	1.23	0.72	17.79	nr	**19.02**
25 × 20 × 25 mm	1.09	1.23	0.72	17.79	nr	**19.02**
32 × 20 × 32 mm	1.39	1.57	0.83	20.51	nr	**22.08**
32 × 25 × 32 mm	1.39	1.57	0.83	20.51	nr	**22.08**
40 × 20 × 40 mm	2.13	2.41	0.92	22.74	nr	**25.15**
40 × 25 × 40 mm	2.13	2.41	0.92	22.74	nr	**25.15**
40 × 32 × 40 mm	2.13	2.41	0.92	22.74	nr	**25.15**

T: MECHANICAL/COOLING/HEATING SYSTEMS

Item	Net Price £	Material £	Labour hours	Labour £	Unit	Total rate £
T31: LOW TEMPERATURE HOT WATER HEATING – cont						
FIXINGS – cont						
Extra over fittings – cont						
T-Piece 90° reducing – cont						
50 × 25 × 50 mm	6.07	6.85	1.01	24.96	nr	**31.81**
50 × 32 × 50 mm	6.07	6.85	1.01	24.96	nr	**31.81**
50 × 40 × 50 mm	6.07	6.85	1.01	24.96	nr	**31.81**
63 × 20 × 63 mm	8.19	9.25	1.11	27.44	nr	**36.69**
63 × 25 × 63 mm	8.19	9.25	1.11	27.44	nr	**36.69**
63 × 32 × 63 mm	8.19	9.25	1.11	27.44	nr	**36.69**
63 × 40 × 63 mm	8.19	9.25	1.01	24.96	nr	**34.21**
63 × 50 × 63 mm	8.19	9.25	1.01	24.96	nr	**34.21**
75 × 20 × 75 mm	13.32	15.04	1.18	29.16	nr	**44.20**
75 × 25 × 75 mm	13.32	15.04	1.18	29.16	nr	**44.20**
75 × 32 × 75 mm	13.32	15.04	1.18	29.16	nr	**44.20**
75 × 40 × 75 mm	13.32	15.04	1.18	29.16	nr	**44.20**
75 × 50 × 75 mm	13.32	15.04	1.18	29.16	nr	**44.20**
75 × 63 × 75 mm	13.32	15.04	1.18	29.16	nr	**44.20**
90 × 63 × 90 mm	26.70	30.15	1.46	36.08	nr	**66.23**
110 × 75 × 110 mm	41.66	47.05	1.46	36.08	nr	**83.13**
110 × 90 × 110 mm	41.66	47.05	1.46	36.08	nr	**83.13**
125 × 90 × 125 mm	49.55	55.96	1.82	44.97	nr	**100.93**
125 × 110 × 125 mm	50.56	57.10	1.82	44.97	nr	**102.07**
Reducer						
25 × 20 mm	0.62	0.70	0.59	14.58	nr	**15.28**
32 × 20 mm	0.80	0.91	0.62	15.33	nr	**16.24**
32 × 25 mm	0.80	0.91	0.62	15.33	nr	**16.24**
40 × 20 mm	1.28	1.45	0.66	16.31	nr	**17.76**
40 × 25 mm	1.28	1.45	0.66	16.31	nr	**17.76**
40 × 32 mm	1.28	1.45	0.66	16.31	nr	**17.76**
50 × 20 mm	2.05	2.32	0.73	18.03	nr	**20.35**
50 × 25 mm	2.05	2.32	0.73	18.03	nr	**20.35**
50 × 32 mm	2.05	2.32	0.73	18.03	nr	**20.35**
50 × 40 mm	2.05	2.32	0.73	18.03	nr	**20.35**
63 × 40 mm	4.13	4.66	0.78	19.28	nr	**23.94**
63 × 25 mm	4.13	4.66	0.78	19.28	nr	**23.94**
63 × 32 mm	4.13	4.66	0.78	19.28	nr	**23.94**
63 × 50 mm	4.13	4.66	0.78	19.28	nr	**23.94**
75 × 50 mm	4.61	5.20	0.85	21.00	nr	**26.20**
75 × 63 mm	4.61	5.20	0.85	21.00	nr	**26.20**
90 × 63 mm	10.30	11.63	1.04	25.70	nr	**37.33**
90 × 75 mm	10.30	11.63	1.04	25.70	nr	**37.33**
110 × 90 mm	16.65	18.81	1.17	28.92	nr	**47.73**
125 × 110 mm	26.02	29.39	1.43	35.34	nr	**64.73**

T: MECHANICAL/COOLING/HEATING SYSTEMS

Item	Net Price £	Material £	Labour hours	Labour £	Unit	**Total rate £**
Socket						
20 mm dia.	0.56	0.63	0.51	12.61	nr	**13.24**
25 mm dia.	0.62	0.70	0.56	13.83	nr	**14.53**
32 mm dia.	0.80	0.91	0.65	16.07	nr	**16.98**
40 mm dia.	0.99	1.12	0.74	18.29	nr	**19.41**
50 mm dia.	2.05	2.32	0.81	20.01	nr	**22.33**
63 mm dia.	4.13	4.66	0.86	21.26	nr	**25.92**
75 mm dia.	4.61	5.20	0.91	22.49	nr	**27.69**
90 mm dia.	11.90	13.44	0.91	22.49	nr	**35.93**
110 mm dia.	20.23	22.85	0.91	22.49	nr	**45.34**
125 mm dia.	28.19	31.83	1.30	32.12	nr	**63.95**
End cap						
20 mm dia.	0.87	0.98	0.25	6.18	nr	**7.16**
25 mm dia.	1.07	1.20	0.29	7.17	nr	**8.37**
32 mm dia.	1.31	1.48	0.33	8.15	nr	**9.63**
40 mm dia.	2.10	2.38	0.36	8.90	nr	**11.28**
50 mm dia.	2.91	3.28	0.40	9.88	nr	**13.16**
63 mm dia.	4.90	5.54	0.44	10.87	nr	**16.41**
75 mm dia.	7.06	7.98	0.47	11.61	nr	**19.59**
90 mm dia.	15.99	18.05	0.57	14.09	nr	**32.14**
110 mm dia.	19.22	21.71	0.57	14.09	nr	**35.80**
125 mm dia.	29.29	33.08	0.85	21.00	nr	**54.08**
Stub flange with gasket						
32 mm dia.	19.43	21.94	0.23	5.68	nr	**27.62**
40 mm dia.	24.44	27.60	0.27	6.67	nr	**34.27**
50 mm dia.	29.57	33.39	0.38	9.38	nr	**42.77**
63 mm dia.	35.50	40.10	0.43	10.63	nr	**50.73**
75 mm dia.	41.66	47.05	0.48	11.86	nr	**58.91**
90 mm dia.	56.38	63.68	0.53	13.10	nr	**76.78**
110 mm dia.	78.94	89.16	0.53	13.10	nr	**102.26**
125 mm dia.	113.14	127.78	0.75	18.53	nr	**146.31**
Weld in saddle with female thread						
40–½"	1.41	1.59	0.36	8.90	nr	**10.49**
50–½"	1.41	1.59	0.36	8.90	nr	**10.49**
63–½"	1.41	1.59	0.40	9.88	nr	**11.47**
75–½"	1.41	1.59	0.40	9.88	nr	**11.47**
90–½"	1.41	1.59	0.46	11.36	nr	**12.95**
110–½"	1.41	1.59	0.46	11.36	nr	**12.95**
Weld in saddle with male thread						
50–½"	1.41	1.59	0.36	8.90	nr	**10.49**
63–½"	1.41	1.59	0.40	9.88	nr	**11.47**
75–½"	1.41	1.59	0.40	9.88	nr	**11.47**
90–½"	1.41	1.59	0.46	11.36	nr	**12.95**
110–½"	1.41	1.59	0.46	11.36	nr	**12.95**

T: MECHANICAL/COOLING/HEATING SYSTEMS

Item	Net Price £	Material £	Labour hours	Labour £	Unit	Total rate £
T31: LOW TEMPERATURE HOT WATER HEATING – cont						
FIXINGS – cont						
Extra over fittings – cont						
Transition piece, round with female thread						
20 × ½"	3.27	3.69	0.29	7.17	nr	**10.86**
20 × ¾"	4.30	4.86	0.29	7.17	nr	**12.03**
25 × ½"	3.27	3.69	0.33	8.15	nr	**11.84**
25 × ¾"	4.30	4.86	0.33	8.15	nr	**13.01**
Transition piece, hexagon with female thread						
32 × 1"	12.16	13.73	0.36	8.90	nr	**22.63**
40 × 1 ¼"	19.24	21.73	0.36	8.90	nr	**30.63**
50 × 1 ½"	22.33	25.21	0.36	8.90	nr	**34.11**
63 × 2"	34.60	39.08	0.40	9.88	nr	**48.96**
75 × 2"	36.09	40.76	0.40	9.88	nr	**50.64**
125 × 5"	69.16	78.11	0.51	12.61	nr	**90.72**
Stop valve for surface assembly						
20 mm dia.	13.43	15.17	0.25	6.18	nr	**21.35**
25 mm dia.	13.43	15.17	0.29	7.17	nr	**22.34**
32 mm dia.	25.29	28.56	0.33	8.15	nr	**36.71**
Ball valve						
20 mm dia.	47.28	53.40	0.25	6.18	nr	**59.58**
25 mm dia.	50.66	57.21	0.29	7.17	nr	**64.38**
32 mm dia.	60.87	68.75	0.33	8.15	nr	**76.90**
40 mm dia.	77.69	87.74	0.36	8.90	nr	**96.64**
50 mm dia.	106.66	120.46	0.40	9.88	nr	**130.34**
63 mm dia.	120.25	135.81	0.44	10.87	nr	**146.68**
Floor or ceiling cover plates						
Plastic						
15 mm dia.	0.46	0.52	0.16	3.96	nr	**4.48**
20 mm dia.	0.51	0.57	0.22	5.44	nr	**6.01**
25 mm dia.	0.54	0.61	0.22	5.44	nr	**6.05**
32 mm dia.	0.61	0.69	0.24	5.93	nr	**6.62**
40 mm dia.	1.34	1.51	0.26	6.43	nr	**7.94**
50 mm dia.	1.49	1.68	0.26	6.43	nr	**8.11**
Chromium plated						
15 mm dia.	3.12	3.52	0.16	3.96	nr	**7.48**
20 mm dia.	3.33	3.76	0.17	4.20	nr	**7.96**
25 mm dia.	3.46	3.91	0.21	5.18	nr	**9.09**
32 mm dia.	3.53	3.99	0.22	5.44	nr	**9.43**
40 mm dia.	3.99	4.51	0.26	6.43	nr	**10.94**
50 mm dia.	4.77	5.39	0.26	6.43	nr	**11.82**

T: MECHANICAL/COOLING/HEATING SYSTEMS

Item	Net Price £	Material £	Labour hours	Labour £	Unit	**Total rate £**
Y11 – PIPELINE ANCILLARIES						
EXPANSION JOINTS						
Axial movement bellows expansion joints; stainless steel						
Screwed ends for steel pipework; up to 6 bar G at 100°C						
15 mm dia.	75.56	85.33	0.68	16.81	nr	**102.14**
20 mm dia.	80.71	91.16	0.81	20.01	nr	**111.17**
25 mm dia.	83.84	94.69	0.93	22.98	nr	**117.67**
32 mm dia.	88.77	100.26	1.06	26.19	nr	**126.45**
40 mm dia.	94.02	106.18	1.16	28.66	nr	**134.84**
50 mm dia.	93.67	105.79	1.19	29.41	nr	**135.20**
Screwed ends for steel pipework; aluminium and steel outer sleeves; up to 16 bar G at 120°C						
20 mm dia.	82.32	92.97	1.32	32.62	nr	**125.59**
25 mm dia.	83.84	94.69	1.52	37.57	nr	**132.26**
32 mm dia.	88.77	100.26	1.80	44.47	nr	**144.73**
40 mm dia.	94.02	106.18	2.03	50.16	nr	**156.34**
50 mm dia.	95.54	107.91	2.26	55.84	nr	**163.75**
Flanged ends for steel pipework; aluminium and steel outer sleeves; up to 16 bar G at 120°C						
20 mm dia.	72.83	82.25	0.53	13.10	nr	**95.35**
25 mm dia.	78.03	88.13	0.64	15.81	nr	**103.94**
32 mm dia.	107.50	121.41	0.74	18.29	nr	**139.70**
40 mm dia.	116.17	131.20	0.82	20.26	nr	**151.46**
50 mm dia.	137.85	155.68	0.89	21.99	nr	**177.67**
Flanged ends for steel pipework; up to 16 bar G at 120°C						
65 mm dia.	127.44	143.93	1.10	27.18	nr	**171.11**
80 mm dia.	136.98	154.70	1.31	32.36	nr	**187.06**
100 mm dia.	156.92	177.22	1.78	43.99	nr	**221.21**
150 mm dia.	240.15	271.23	3.08	76.11	nr	**347.34**
Screwed ends for non-ferrous pipework; up to 6 bar G at 100°C						
20 mm dia.	82.32	92.97	0.72	17.79	nr	**110.76**
25 mm dia.	83.84	94.69	0.84	20.76	nr	**115.45**
32 mm dia.	88.77	100.26	1.02	25.20	nr	**125.46**
40 mm dia.	94.02	106.18	1.11	27.44	nr	**133.62**
50 mm dia.	95.54	107.91	1.18	29.16	nr	**137.07**

T: MECHANICAL/COOLING/HEATING SYSTEMS

Item	Net Price £	Material £	Labour hours	Labour £	Unit	Total rate £
T31: LOW TEMPERATURE HOT WATER HEATING – cont						
EXPANSION JOINTS – cont						
Flanged ends for steel, copper or non-ferrous pipework; up to 16 bar G at 120°C						
65 mm dia.	163.86	185.06	0.87	21.50	nr	**206.56**
80 mm dia.	169.06	190.93	0.95	23.48	nr	**214.41**
100 mm dia.	195.07	220.31	1.15	28.42	nr	**248.73**
150 mm dia.	319.91	361.30	1.36	33.61	nr	**394.91**
Angular movement bellows expansion joints; stainless steel						
Flanged ends for steel pipework; up to 16 bar G at 120°C						
50 mm dia.	188.89	213.33	0.71	17.54	nr	**230.87**
65 mm dia.	229.24	258.91	0.83	20.51	nr	**279.42**
80 mm dia.	267.87	302.53	0.91	22.49	nr	**325.02**
100 mm dia.	345.83	390.58	0.97	23.97	nr	**414.55**
125 mm dia.	456.79	515.90	1.16	28.66	nr	**544.56**
150 mm dia.	543.17	613.45	1.18	29.16	nr	**642.61**
VALVES						
Isolating valves						
Bronze gate valve; non-rising stem; BS 5154, series B, PN 32; working pressure up to 14 bar for saturated steam, 32 bar from −10°C to 100°C; screwed ends to steel						
15 mm dia.	39.37	44.46	1.11	27.44	nr	**71.90**
20 mm dia.	51.07	57.68	1.28	31.63	nr	**89.31**
25 mm dia.	69.18	78.13	1.49	36.82	nr	**114.95**
32 mm dia.	95.54	107.91	1.88	46.45	nr	**154.36**
40 mm dia.	128.30	144.90	2.31	57.08	nr	**201.98**
50 mm dia.	180.27	203.59	2.80	69.19	nr	**272.78**
Bronze gate valve; non-rising stem; BS 5154, series B, PN 20; working pressure up to 9 bar for saturated steam, 20 bar from −10°C to 100°C; screwed ends to steel						
15 mm dia.	26.85	30.32	0.84	20.76	nr	**51.08**
20 mm dia.	38.16	43.09	1.01	24.96	nr	**68.05**
25 mm dia.	49.37	55.75	1.19	29.41	nr	**85.16**
32 mm dia.	70.45	79.57	1.38	34.10	nr	**113.67**
40 mm dia.	96.42	108.90	1.62	40.04	nr	**148.94**
50 mm dia.	137.78	155.61	1.94	47.94	nr	**203.55**

T: MECHANICAL/COOLING/HEATING SYSTEMS

Item	Net Price £	Material £	Labour hours	Labour £	Unit	**Total rate £**
Bronze gate valve; non-rising stem; BS 5154, series B, PN 16; working pressure up to 7 bar for saturated steam, 16 bar from −10°C to 100°C; BS4504 flanged ends; bolted connections						
15 mm dia.	125.88	142.17	1.18	29.16	nr	**171.33**
20 mm dia.	133.34	150.59	1.24	30.64	nr	**181.23**
25 mm dia.	175.59	198.31	1.31	32.36	nr	**230.67**
32 mm dia.	220.31	248.82	1.43	35.34	nr	**284.16**
40 mm dia.	266.36	300.83	1.53	37.81	nr	**338.64**
50 mm dia.	368.23	415.87	1.63	40.27	nr	**456.14**
65 mm dia.	571.88	645.87	1.71	42.25	nr	**688.12**
80 mm dia.	809.41	914.14	1.88	46.45	nr	**960.59**
100 mm dia.	1470.93	1661.26	2.03	50.16	nr	**1711.42**
Cast iron gate valve; bronze trim; non rising stem; BS 5150, PN6; working pressure 6 bar from −10°C to 120°C; BS4504 flanged ends; bolted connections						
50 mm dia.	239.49	270.48	1.85	45.72	nr	**316.20**
65 mm dia.	250.32	282.71	2.00	49.42	nr	**332.13**
80 mm dia.	289.29	326.73	2.27	56.10	nr	**382.83**
100 mm dia.	365.93	413.28	2.76	68.21	nr	**481.49**
125 mm dia.	523.81	591.59	6.05	149.50	nr	**741.09**
150 mm dia.	602.73	680.72	8.03	198.43	nr	**879.15**
200 mm dia.	1148.06	1296.61	9.17	226.60	nr	**1523.21**
250 mm dia.	1766.52	1995.10	10.72	264.90	nr	**2260.00**
300 mm dia.	2095.22	2366.33	11.75	290.35	nr	**2656.68**
Cast iron gate valve; bronze trim; non rising stem; BS 5150, PN10; working pressure up to 8.4 bar for saturated steam, 10 bar from −10° C to 120°C; BS4504 flanged ends; bolted connections						
50 mm dia.	251.21	283.71	1.85	45.72	nr	**329.43**
65 mm dia.	312.19	352.58	2.00	49.42	nr	**402.00**
80 mm dia.	347.57	392.55	2.27	56.10	nr	**448.65**
100 mm dia.	462.67	522.54	2.76	68.21	nr	**590.75**
125 mm dia.	646.88	730.59	6.05	149.50	nr	**880.09**
150 mm dia.	763.20	861.95	8.03	198.43	nr	**1060.38**
200 mm dia.	1402.22	1583.66	9.17	226.60	nr	**1810.26**
250 mm dia.	2108.84	2381.71	10.72	264.90	nr	**2646.61**
300 mm dia.	2373.06	2680.12	11.75	290.35	nr	**2970.47**
350 mm dia.	2760.13	3117.27	12.67	313.09	nr	**3430.36**

T: MECHANICAL/COOLING/HEATING SYSTEMS

Item	Net Price £	Material £	Labour hours	Labour £	Unit	Total rate £
T31: LOW TEMPERATURE HOT WATER HEATING – cont						
VALVES – cont						
Isolating valves – cont						
Cast iron gate valve; bronze trim; non rising stem; BS 5163 series A, PN16; working pressure for cold water services up to 16 bar; BS4504 flanged ends; bolted connections						
50 mm dia.	251.21	283.71	1.85	45.72	nr	**329.43**
65 mm dia.	312.19	352.58	2.00	49.42	nr	**402.00**
80 mm dia.	347.57	392.55	2.27	56.10	nr	**448.65**
100 mm dia.	462.67	522.54	2.76	68.21	nr	**590.75**
125 mm dia.	646.88	730.59	6.05	149.50	nr	**880.09**
150 mm dia.	763.20	861.95	8.03	198.43	nr	**1060.38**
Ball valves						
Malleable iron body; lever operated stainless steel ball and stem; working pressure up to 12 bar; flanged ends to BS 4504 16/11; bolted connections						
40 mm dia.	333.87	377.07	1.54	38.06	nr	**415.13**
50 mm dia.	420.71	475.15	1.64	40.52	nr	**515.67**
80 mm dia.	680.55	768.61	1.92	47.45	nr	**816.06**
100 mm dia.	703.64	794.69	2.80	69.19	nr	**863.88**
150 mm dia.	1300.61	1468.90	12.05	297.77	nr	**1766.67**
Malleable iron body; lever operated stainless steel ball and stem; working pressure up to 16 bar; screwed ends to steel						
20 mm dia.	39.66	44.79	1.34	33.13	nr	**77.92**
25 mm dia.	40.78	46.05	1.40	34.62	nr	**80.67**
32 mm dia.	56.18	63.45	1.46	36.13	nr	**99.58**
40 mm dia.	56.18	63.45	1.54	38.08	nr	**101.53**
50 mm dia.	67.18	75.88	1.64	40.58	nr	**116.46**
Carbon steel body; lever operated stainless steel ball and stem; Class 150; working pressure up to 19 bar; screwed ends to steel						
15 mm dia.	73.37	82.86	0.84	20.77	nr	**103.63**
20 mm dia.	82.47	93.14	1.14	28.18	nr	**121.32**
25 mm dia.	89.30	100.86	1.30	32.13	nr	**132.99**

T: MECHANICAL/COOLING/HEATING SYSTEMS

Item	Net Price £	Material £	Labour hours	Labour £	Unit	**Total rate £**
Globe valves						
Bronze; rising stem; renewable disc; BS 5154 series B, PN32; working pressure up to 14 bar for saturated steam, 32 bar from −10°C to 100°C; screwed ends to steel						
15 mm dia.	40.40	45.62	0.77	19.03	nr	**64.65**
20 mm dia.	55.24	62.38	1.03	25.46	nr	**87.84**
25 mm dia.	79.25	89.50	1.19	29.41	nr	**118.91**
32 mm dia.	111.58	126.02	1.38	34.10	nr	**160.12**
40 mm dia.	155.39	175.50	1.62	40.04	nr	**215.54**
50 mm dia.	238.51	269.37	1.61	39.79	nr	**309.16**
Bronze; needle valve; rising stem; BS 5154, series B, PN32; working pressure up to 14 bar for saturated steam, 32 bar from −10°C to 100°C; screwed ends to steel						
15 mm dia.	30.71	34.68	1.07	26.44	nr	**61.12**
20 mm dia.	51.98	58.70	1.18	29.16	nr	**87.86**
25 mm dia.	73.30	82.78	1.27	31.39	nr	**114.17**
32 mm dia.	152.59	172.34	1.35	33.36	nr	**205.70**
40 mm dia.	240.40	271.50	1.47	36.32	nr	**307.82**
50 mm dia.	304.47	343.86	1.61	39.79	nr	**383.65**
Bronze; rising stem; renewable disc; BS 5154, series B, PN16; working pressure upto 7 bar for saturated steam, 16 bar from −10°C to 100°C; BS4504 flanged ends; bolted connections						
15 mm dia.	81.63	92.19	1.16	28.66	nr	**120.85**
20 mm dia.	94.41	106.63	1.26	31.14	nr	**137.77**
25 mm dia.	165.56	186.99	1.38	34.10	nr	**221.09**
32 mm dia.	209.00	236.05	1.47	36.32	nr	**272.37**
40 mm dia.	237.56	268.30	1.56	38.55	nr	**306.85**
50 mm dia.	340.84	384.95	1.71	42.25	nr	**427.20**
Bronze; rising stem; renewable disc; BS 2060, class 250; working pressure up to 24 bar for saturated steam, 38 bar from −10°C to 100°C; flanged ends (BS 10 table H); bolted connections						
15 mm dia.	207.87	234.76	1.16	28.66	nr	**263.42**
20 mm dia.	241.58	272.84	1.26	31.14	nr	**303.98**
25 mm dia.	331.98	374.93	1.38	34.10	nr	**409.03**
32 mm dia.	433.69	489.80	1.47	36.32	nr	**526.12**
40 mm dia.	519.34	586.54	1.56	38.55	nr	**625.09**
50 mm dia.	807.55	912.04	1.71	42.25	nr	**954.29**
65 mm dia.	965.76	1090.73	1.88	46.45	nr	**1137.18**
80 mm dia.	1227.23	1386.03	2.03	50.16	nr	**1436.19**

T: MECHANICAL/COOLING/HEATING SYSTEMS

Item	Net Price £	Material £	Labour hours	Labour £	Unit	Total rate £
T31: LOW TEMPERATURE HOT WATER HEATING – cont						
VALVES – cont						
Check valves						
Bronze; swing pattern; BS 5154 series B, PN 25; working pressure up to 10.5 bar for saturated steam, 25 bar from −10°C to 100°C; screwed ends to steel						
15 mm dia.	18.60	21.00	0.77	19.03	nr	**40.03**
20 mm dia.	22.09	24.95	1.03	25.46	nr	**50.41**
25 mm dia.	30.62	34.58	1.19	29.41	nr	**63.99**
32 mm dia.	51.87	58.58	1.38	34.10	nr	**92.68**
40 mm dia.	64.54	72.89	1.62	40.04	nr	**112.93**
50 mm dia.	98.98	111.79	1.94	47.94	nr	**159.73**
65 mm dia.	202.65	228.87	2.45	60.55	nr	**289.42**
80 mm dia.	286.54	323.62	2.83	69.93	nr	**393.55**
Bronze; vertical lift pattern; BS 5154 series B, PN32; working pressure up to 14 bar for saturated steam, 32 bar from −10°C to 100°C; screwed ends to steel						
15 mm dia.	23.60	26.65	0.96	23.72	nr	**50.37**
20 mm dia.	33.56	37.90	1.07	26.44	nr	**64.34**
25 mm dia.	49.27	55.64	1.17	28.92	nr	**84.56**
32 mm dia.	75.05	84.76	1.33	32.86	nr	**117.62**
40 mm dia.	97.79	110.45	1.41	34.84	nr	**145.29**
50 mm dia.	146.54	165.51	1.55	38.30	nr	**203.81**
65 mm dia.	442.45	499.70	1.80	44.47	nr	**544.17**
80 mm dia.	661.47	747.06	1.99	49.17	nr	**796.23**
Bronze; oblique swing pattern; BS 5154 series A, PN32; working pressure up to 14 bar for saturated steam, 32 bar from −10°C to 120°C; screwed connections to steel						
15 mm dia.	29.97	33.84	0.96	23.72	nr	**57.56**
20 mm dia.	35.61	40.22	1.07	26.44	nr	**66.66**
25 mm dia.	49.36	55.74	1.17	28.92	nr	**84.66**
32 mm dia.	73.81	83.36	1.33	32.86	nr	**116.22**
40 mm dia.	95.14	107.45	1.41	34.84	nr	**142.29**
50 mm dia.	157.61	178.00	1.55	38.30	nr	**216.30**

T: MECHANICAL/COOLING/HEATING SYSTEMS

Item	Net Price £	Material £	Labour hours	Labour £	Unit	Total rate £
Cast iron; swing pattern; BS 5153 PN6; working pressure up to 6 bar from −10°C to 120°C; BS 4504 flanged ends; bolted connections						
50 mm dia.	523.18	590.88	1.86	45.97	nr	**636.85**
65 mm dia.	572.83	646.96	2.00	49.42	nr	**696.38**
80 mm dia.	644.77	728.20	2.56	63.26	nr	**791.46**
100 mm dia.	825.97	932.85	2.76	68.21	nr	**1001.06**
125 mm dia.	1232.23	1391.68	6.05	149.50	nr	**1541.18**
150 mm dia.	1385.48	1564.75	8.11	200.41	nr	**1765.16**
200 mm dia.	3063.98	3460.44	9.26	228.83	nr	**3689.27**
250 mm dia.	3829.87	4325.43	10.72	264.90	nr	**4590.33**
300 mm dia.	4595.86	5190.55	11.75	290.35	nr	**5480.90**
Cast iron; horizontal lift pattern; BS 5153 PN16; working pressure up to 13 bar for saturated steam, 16 bar from −10°C to 120°C; BS 4504 flanged ends; bolted connections						
50 mm dia.	472.45	533.58	1.86	45.97	nr	**579.55**
65 mm dia.	472.45	533.58	2.00	49.42	nr	**583.00**
80 mm dia.	584.94	660.62	2.56	63.26	nr	**723.88**
100 mm dia.	681.18	769.32	2.96	73.14	nr	**842.46**
125 mm dia.	1016.30	1147.81	7.76	191.76	nr	**1339.57**
150 mm dia.	1142.63	1290.48	10.50	259.47	nr	**1549.95**
Cast iron; semi lugged butterfly valve; BS5155 PN16; working pressure 16 bar from −10°C to 120°C; BS 4504 flanged ends; bolted connections						
50 mm dia.	151.68	171.31	2.20	54.37	nr	**225.68**
65 mm dia.	156.53	176.79	2.31	57.08	nr	**233.87**
80 mm dia.	184.39	208.25	2.88	71.17	nr	**279.42**
100 mm dia.	256.33	289.50	3.11	76.86	nr	**366.36**
125 mm dia.	374.12	422.52	5.02	124.05	nr	**546.57**
150 mm dia.	429.43	485.00	6.98	172.48	nr	**657.48**
200 mm dia.	685.50	774.20	8.25	203.87	nr	**978.07**
250 mm dia.	1011.78	1142.70	10.47	258.72	nr	**1401.42**
300 mm dia.	1517.69	1714.07	11.48	283.68	nr	**1997.75**
Commissioning valves						
Bronze commissioning set; metering station; double regulating valve; BS5154 PN20 Series B; working pressure 20 bar from −10°C to 100°C; screwed ends to steel						
15 mm dia.	79.15	89.39	1.08	26.68	nr	**116.07**
20 mm dia.	127.18	143.64	1.46	36.08	nr	**179.72**
25 mm dia.	151.23	170.80	1.68	41.51	nr	**212.31**
32 mm dia.	202.77	229.01	1.95	48.20	nr	**277.21**
40 mm dia.	294.06	332.11	2.27	56.10	nr	**388.21**
50 mm dia.	443.78	501.21	2.73	67.46	nr	**568.67**

T: MECHANICAL/COOLING/HEATING SYSTEMS

Item	Net Price £	Material £	Labour hours	Labour £	Unit	Total rate £
T31: LOW TEMPERATURE HOT WATER HEATING – cont						
VALVES – cont						
Commissioning valves – cont						
Cast iron commissioning set; metering station; double regulating valve; BS5152 PN16; working pressure 16 bar from −10°C to 90°C; flanged ends (BS 4504, Part 1, Table 16); bolted connections						
65 mm dia.	672.32	759.32	1.80	44.47	nr	**803.79**
80 mm dia.	805.04	909.21	2.56	63.26	nr	**972.47**
100 mm dia.	1120.74	1265.76	2.30	56.83	nr	**1322.59**
125 mm dia.	1641.61	1854.02	2.44	60.30	nr	**1914.32**
150 mm dia.	2162.50	2442.32	2.90	71.67	nr	**2513.99**
200 mm dia.	5535.15	6251.37	8.26	204.11	nr	**6455.48**
250 mm dia.	6924.39	7820.37	10.49	259.22	nr	**8079.59**
300 mm dia.	7520.04	8493.09	11.49	283.93	nr	**8777.02**
Cast iron variable orifice double regulating valve; orifice valve; BS5152 PN16; working pressure 16 bar from −10° to 90°C; flanged ends (BS 4504, Part 1, Table 16); bolted connections						
65 mm dia.	492.24	555.93	2.00	49.42	nr	**605.35**
80 mm dia.	602.50	680.46	2.56	63.26	nr	**743.72**
100 mm dia.	826.83	933.82	2.96	73.14	nr	**1006.96**
125 mm dia.	1238.18	1398.39	7.76	191.76	nr	**1590.15**
150 mm dia.	1590.41	1796.20	10.50	259.47	nr	**2055.67**
200 mm dia.	4168.28	4707.64	8.26	204.11	nr	**4911.75**
250 mm dia.	6337.85	7157.93	10.49	259.22	nr	**7417.15**
300 mm dia.	11368.83	12839.90	11.49	283.93	nr	**13123.83**
Cast iron globe valve with double regulating feature; BS5152 PN16; working pressure 16 bar from −10°C to 120°C; flanged ends (BS 4504, Part 1, Table 16); bolted connections						
65 mm dia.	586.38	662.26	2.00	49.42	nr	**711.68**
80 mm dia.	709.13	800.88	2.56	63.26	nr	**864.14**
100 mm dia.	934.13	1055.01	2.96	73.14	nr	**1128.15**
125 mm dia.	1302.42	1470.94	7.76	191.76	nr	**1662.70**
150 mm dia.	1648.39	1861.68	10.50	259.47	nr	**2121.15**
200 mm dia.	4267.55	4819.74	8.26	204.11	nr	**5023.85**
250 mm dia.	6456.70	7292.16	10.49	259.22	nr	**7551.38**
300 mm dia.	10144.40	11457.04	11.49	283.93	nr	**11740.97**

T: MECHANICAL/COOLING/HEATING SYSTEMS

Item	Net Price £	Material £	Labour hours	Labour £	Unit	**Total rate £**
Bronze autoflow commissioning valve; PN25 ; working pressure 25 bar up to 100°C; screwed ends to steel						
15 mm dia.	89.74	101.35	0.82	20.26	nr	**121.61**
20 mm dia.	95.03	107.32	1.08	26.68	nr	**134.00**
25 mm dia.	118.33	133.64	1.27	31.39	nr	**165.03**
32 mm dia.	184.80	208.71	1.50	37.07	nr	**245.78**
40 mm dia.	202.36	228.54	1.76	43.49	nr	**272.03**
50 mm dia.	299.19	337.91	2.13	52.63	nr	**390.54**
Ductile iron autoflow commissioning valves; PN16; working pressure 16 bar from −10°C to 120°C; for ANSI 150 flanged ends						
65 mm dia.	787.91	889.86	2.31	57.08	nr	**946.94**
80 mm dia.	871.14	983.86	2.88	71.17	nr	**1055.03**
100 mm dia.	1112.84	1256.83	3.11	76.86	nr	**1333.69**
150 mm dia.	2292.47	2589.11	6.98	172.48	nr	**2761.59**
200 mm dia.	3339.20	3771.28	8.26	204.11	nr	**3975.39**
250 mm dia.	4637.66	5237.75	10.49	259.22	nr	**5496.97**
300 mm dia.	5923.01	6689.41	11.49	283.93	nr	**6973.34**
Strainers						
Bronze strainer; Y type; PN32 ; working pressure 32 bar from −10°C to 100°C; screwed ends to steel						
15 mm dia.	35.97	40.63	0.82	20.26	nr	**60.89**
20 mm dia.	45.69	51.60	1.08	26.68	nr	**78.28**
25 mm dia.	60.61	68.45	1.27	31.39	nr	**99.84**
32 mm dia.	97.89	110.56	1.50	37.07	nr	**147.63**
40 mm dia.	126.61	142.99	1.76	43.49	nr	**186.48**
50 mm dia.	212.06	239.50	2.13	52.63	nr	**292.13**
Cast iron strainer; Y type; PN16; working pressure 16 bar from −10°C to 120°C; BS 4504 flanged ends						
65 mm dia.	223.73	252.68	2.31	57.08	nr	**309.76**
80 mm dia.	264.83	299.09	2.88	71.17	nr	**370.26**
100 mm dia.	387.47	437.61	3.11	76.86	nr	**514.47**
125 mm dia.	714.29	806.72	5.02	124.05	nr	**930.77**
150 mm dia.	916.44	1035.02	6.98	172.48	nr	**1207.50**
200 mm dia.	1538.20	1737.23	8.26	204.11	nr	**1941.34**
250 mm dia.	2224.33	2512.15	10.49	259.22	nr	**2771.37**
300 mm dia.	3622.29	4091.00	11.49	283.93	nr	**4374.93**

T: MECHANICAL/COOLING/HEATING SYSTEMS

Item	Net Price £	Material £	Labour hours	Labour £	Unit	Total rate £
T31: LOW TEMPERATURE HOT WATER HEATING – cont						
VALVES – cont						
Regulators						
Gunmetal; self-acting two port thermostatic regulator; single seat; screwed ends; complete with sensing element, 2 m long capillary tube						
15 mm dia.	627.42	708.60	1.37	33.86	nr	**742.46**
20 mm dia.	647.59	731.38	1.24	30.64	nr	**762.02**
25 mm dia.	664.60	750.60	1.34	33.11	nr	**783.71**
Gunmetal; self-acting two port thermostatic regulator; double seat; flanged ends (BS 4504 PN25); with sensing element, 2 m long capillary tube; steel body						
65 mm dia.	2326.81	2627.89	1.23	33.60	nr	**2661.49**
80 mm dia.	2746.29	3101.65	1.62	44.26	nr	**3145.91**
Control valves; electrically operated (electrical work elsewhere)						
Cast iron; butterfly type; two position electrically controlled 240 V motor and linkage mechanism; for low pressure hot water; maximum pressure 6 bar at 120°C; flanged ends						
25 mm dia.	779.93	880.85	1.47	36.32	nr	**917.17**
32 mm dia.	805.88	910.16	1.52	37.57	nr	**947.73**
40 mm dia.	1126.08	1271.79	1.61	39.79	nr	**1311.58**
50 mm dia.	1158.09	1307.95	1.71	42.25	nr	**1350.20**
65 mm dia.	1187.09	1340.70	2.51	62.03	nr	**1402.73**
80 mm dia.	1233.51	1393.12	2.69	66.47	nr	**1459.59**
100 mm dia.	1294.46	1461.96	2.81	69.43	nr	**1531.39**
125 mm dia.	1433.78	1619.30	2.94	72.65	nr	**1691.95**
150 mm dia.	1553.73	1754.78	3.33	82.28	nr	**1837.06**
200 mm dia.	1927.15	2176.52	3.67	90.69	nr	**2267.21**
Cast iron; three way 240V motorized; for low pressure hot water; maximum pressure 6 bar 120°C; flanged ends, drilled (BS 10, Table F)						
25 mm dia.	871.68	984.47	1.99	49.17	nr	**1033.64**
40 mm dia.	914.26	1032.56	2.13	52.63	nr	**1085.19**
50 mm dia.	902.65	1019.45	3.21	79.32	nr	**1098.77**
65 mm dia.	1008.55	1139.05	3.23	79.81	nr	**1218.86**
80 mm dia.	1130.10	1276.33	3.50	86.49	nr	**1362.82**

T: MECHANICAL/COOLING/HEATING SYSTEMS

Item	Net Price £	Material £	Labour hours	Labour £	Unit	**Total rate £**
Two port normally closed motorized valve; electric actuator; spring return; domestic usage						
22 mm dia.	123.59	139.59	1.18	29.16	nr	**168.75**
28 mm dia.	167.51	189.19	1.35	33.36	nr	**222.55**
Two port on/off motorized valve; electric actuator; spring return; domestic usage						
22 mm dia.	123.59	139.59	1.18	29.16	nr	**168.75**
Three port motorized valve; electric actuator; spring return; domestic usage						
22 mm dia.	180.33	203.66	1.18	29.16	nr	**232.82**
Safety and relief valves						
Bronze relief valve; spring type; side outlet; working pressure up to 20.7 bar at 120°C; screwed ends to steel						
15 mm dia.	137.83	155.66	0.26	6.43	nr	**162.09**
20 mm dia.	151.16	170.72	0.36	8.90	nr	**179.62**
Bronze relief valve; spring type; side outlet; working pressure up to 17.2 bar at 120°C; screwed ends to steel						
25 mm dia.	206.74	233.49	0.38	9.38	nr	**242.87**
32 mm dia.	280.10	316.34	0.48	11.86	nr	**328.20**
Bronze relief valve; spring type; side outlet; working pressure up to 13.8 bar at 120°C; screwed ends to steel						
40 mm dia.	365.68	413.00	0.64	15.81	nr	**428.81**
50 mm dia.	476.83	538.53	0.76	18.78	nr	**557.31**
65 mm dia.	882.86	997.10	0.94	23.24	nr	**1020.34**
80 mm dia.	1158.31	1308.19	1.10	27.18	nr	**1335.37**
Cocks; screwed joints to steel						
Bronze gland cock; complete with malleable iron lever; working pressure up to 10 bar at 100°C; screwed ends to steel						
15 mm dia.	48.43	54.69	0.77	19.03	nr	**73.72**
20 mm dia.	68.90	77.82	1.03	25.46	nr	**103.28**
25 mm dia.	92.85	104.87	1.19	29.41	nr	**134.28**
32 mm dia.	140.30	158.46	1.38	34.10	nr	**192.56**
40 mm dia.	195.90	221.25	1.62	40.04	nr	**261.29**
50 mm dia.	298.99	337.68	1.94	47.94	nr	**385.62**

T: MECHANICAL/COOLING/HEATING SYSTEMS

Item	Net Price £	Material £	Labour hours	Labour £	Unit	**Total rate £**
T31: LOW TEMPERATURE HOT WATER HEATING – cont						
VALVES – cont						
Cocks – cont						
Bronze three-way plug cock; complete with malleable iron lever; working pressure up to 10 bar at 100°C; screwed ends to steel						
15 mm dia.	98.19	110.89	0.77	19.03	nr	**129.92**
20 mm dia.	113.65	128.36	1.03	25.46	nr	**153.82**
25 mm dia.	158.88	179.44	1.19	29.41	nr	**208.85**
32 mm dia.	225.33	254.49	1.38	34.10	nr	**288.59**
40 mm dia.	271.68	306.84	1.62	40.04	nr	**346.88**
Air vents; including regulating, adjusting and testing						
Automatic air vent; maximum pressure up to 7 bar at 93°C; screwed ends to steel						
15 mm dia.	12.36	13.96	0.80	19.77	nr	**33.73**
Automatic air vent; maximum pressure up to 7 bar at 93°C; lockhead isolating valve; screwed ends to steel						
15 mm dia.	12.36	13.96	0.83	20.51	nr	**34.47**
Automatic air vent; maximum pressure up to 17 bar at 200°C; flanged ends (BS10, Table H); bolted connections to counter flange (measured separately)						
15 mm dia.	444.49	502.00	0.83	20.51	nr	**522.51**
Radiator valves						
Bronze; wheelhead or lockshield; chromium plated finish; screwed joints to steel						
Straight						
15 mm dia.	15.77	17.81	0.59	14.58	nr	**32.39**
20 mm dia.	24.54	27.71	0.73	18.03	nr	**45.74**
25 mm dia.	30.81	34.79	0.85	21.00	nr	**55.79**
Angled						
15 mm dia.	53.36	60.26	0.59	14.58	nr	**74.84**
20 mm dia.	70.36	79.47	0.73	18.03	nr	**97.50**
25 mm dia.	90.45	102.15	0.85	21.00	nr	**123.15**

T: MECHANICAL/COOLING/HEATING SYSTEMS

Item	Net Price £	Material £	Labour hours	Labour £	Unit	**Total rate £**
Bronze; wheelhead or lockshield; chromium plated finish; compression joints to copper						
Straight						
15 mm dia.	28.24	31.90	0.59	14.58	nr	**46.48**
20 mm dia.	35.59	40.20	0.73	18.03	nr	**58.23**
25 mm dia.	44.73	50.52	0.85	21.00	nr	**71.52**
Angled						
15 mm dia.	21.32	24.08	0.59	14.58	nr	**38.66**
20 mm dia.	22.36	25.26	0.73	18.03	nr	**43.29**
25 mm dia.	27.72	31.30	0.85	21.00	nr	**52.30**
Twin entry						
8 mm dia.	29.63	33.47	0.23	5.68	nr	**39.15**
10 mm dia.	33.62	37.97	0.23	5.68	nr	**43.65**
Bronze; thermostatic head; chromium plated finish; compression joints to copper						
Straight						
15 mm dia.	11.67	13.18	0.59	14.58	nr	**27.76**
20 mm dia.	17.13	19.35	0.73	18.03	nr	**37.38**
Angled						
15 mm dia.	10.87	12.28	0.59	14.58	nr	**26.86**
20 mm dia.	15.93	17.99	0.73	18.03	nr	**36.02**
GAUGES						
Thermometers and pressure gauges						
Dial thermometer; coated steel case and dial; glass window; brass pocket; BS 5235; pocket length 100 mm; screwed end						
Back/bottom entry						
100 mm dia. face	59.44	67.13	0.81	20.02	nr	**87.15**
150 mm dia. face	97.38	109.98	0.81	20.02	nr	**130.00**
Dial pressure/altitude gauge; bronze bourdon tube type; coated steel case and dial; glass window BS 1780; screwed end						
100 mm dia. face	88.11	99.51	0.81	20.02	nr	**119.53**
150 mm dia. face	98.68	111.45	0.81	20.02	nr	**131.47**

T: MECHANICAL/COOLING/HEATING SYSTEMS

Item	Net Price £	Material £	Labour hours	Labour £	Unit	Total rate £
T31: LOW TEMPERATURE HOT WATER HEATING – cont						
EQUIPMENT						
PRESSURIZATION UNITS						
LTHW pressurization unit complete with expansion vessel(s), interconnecting pipework and all necessary isolating and drain valves; includes placing in position; electrical work elsewhere. Selection based on a final working pressure of 4 bar, a 3 m static head and system operating temperatures of 82/71° C						
System volume						
2,400 litres	1851.30	2090.85	15.00	370.67	nr	**2461.52**
6,000–20,000 litres	4233.00	4780.73	22.00	543.65	nr	**5324.38**
25,000 litres	6519.84	7363.47	22.00	543.65	nr	**7907.12**
DIRT SEPARATORS						
Dirt seperator; maximum operating temperature and pressure of 110°C and 10 bar; fitted with drain valve						
Bore size, flow rate (at 1.0 m/s velocity); threaded connections						
32 mm dia., 3.7 m³/h	83.33	94.12	2.29	56.59	nr	**150.71**
40 mm dia., 5.0 m³/h	99.20	112.04	2.45	60.55	nr	**172.59**
Bore size, flow rate (at 1.5 m/s velocity); flanged connections to PN16						
50 mm dia., 13.0 m³/h	560.82	633.39	3.00	74.13	nr	**707.52**
65 mm dia., 21.0 m³/h	584.63	660.28	3.00	74.13	nr	**734.41**
80 mm dia., 29.0 m³/h	822.71	929.17	3.84	94.89	nr	**1024.06**
100 mm dia., 49.0 m³/h	870.33	982.94	4.44	109.72	nr	**1092.66**
125 mm dia., 74.0 m³/h	1667.91	1883.73	11.64	287.63	nr	**2171.36**
150 mm dia., 109.0 m³/h	1739.33	1964.39	15.75	389.20	nr	**2353.59**
200 mm dia., 181.0 m³/h	2621.56	2960.77	15.75	389.20	nr	**3349.97**
250 mm dia., 288.0 m³/h	4646.59	5247.84	15.75	389.20	nr	**5637.04**
300 mm dia., 407.0 m³/h	6791.99	7670.84	17.24	426.02	nr	**8096.86**

T: MECHANICAL/COOLING/HEATING SYSTEMS

Item	Net Price £	Material £	Labour hours	Labour £	Unit	Total rate £
MICROBUBBLE DEAERATORS						
Microbubble deaerator; maximum operating temperature and pressure of 110°C and 10 bar; fitted with drain valve						
Bore size, flow rate (at 1.0 m/s velocity); threaded connections						
32 mm dia., 3.7 m³/h	76.72	86.65	2.29	56.59	nr	**143.24**
40 mm dia., 5.0 m³/h	91.27	103.08	2.45	60.55	nr	**163.63**
Bore size, flow rate (at 1.5 m/s velocity); flanged connections to PN16						
50 mm dia., 13.0 m³/h	560.82	633.39	3.00	74.13	nr	**707.52**
65 mm dia., 21.0 m³/h	583.52	659.02	3.00	74.13	nr	**733.15**
80 mm dia., 29.0 m³/h	822.71	929.17	3.84	94.89	nr	**1024.06**
100 mm dia., 49.0 m³/h	870.33	982.94	4.44	109.72	nr	**1092.66**
125 mm dia., 74.0 m³/h	1667.91	1883.73	11.64	287.63	nr	**2171.36**
150 mm dia., 109.0 m³/h	1739.33	1964.39	15.75	389.20	nr	**2353.59**
200 mm dia., 181.0 m³/h	2621.56	2960.77	15.75	389.20	nr	**3349.97**
250 mm dia., 288.0 m³/h	4646.59	5247.84	15.75	389.20	nr	**5637.04**
300 mm dia., 407.0 m³/h	6815.80	7697.73	17.24	426.02	nr	**8123.75**
COMBINED MICROBUBBLE DEAERATORS AND DIRT SEPARATORS						
Combined deaerator and dirt separators; maximum operating temperature and pressure of 110°C and 10 bar; fitted with drain valve						
Bore size, flow rate (at 1.5 m/s velocity); threaded connections						
25 mm dia., 2.0 m³/h	105.81	119.50	2.75	67.95	nr	**187.45**
Bore size, flow rate (at 1.5 m/s velocity); flanged connections to PN16						
50 mm dia., 13.0 m³/h	715.57	808.16	3.60	88.96	nr	**897.12**
65 mm dia., 21.0 m³/h	739.38	835.05	3.60	88.96	nr	**924.01**
80 mm dia., 29.0 m³/h	1001.27	1130.83	4.61	113.92	nr	**1244.75**
100 mm dia., 49.0 m³/h	1048.89	1184.61	5.33	131.71	nr	**1316.32**
125 mm dia., 74.0 m³/h	2001.22	2260.17	13.97	345.21	nr	**2605.38**
150 mm dia., 109.0 m³/h	2073.97	2342.33	18.90	467.04	nr	**2809.37**
200 mm dia., 181.0 m³/h	3216.77	3633.01	18.90	467.04	nr	**4100.05**
250 mm dia., 288.0 m³/h	5671.67	6405.56	18.90	467.04	nr	**6872.60**
300 mm dia., 407.0 m³/h	8697.98	9823.46	20.69	511.28	nr	**10334.74**

T: MECHANICAL/COOLING/HEATING SYSTEMS

Item	Net Price £	Material £	Labour hours	Labour £	Unit	Total rate £
T31: LOW TEMPERATURE HOT WATER HEATING – cont						
Y20 – PUMPS						
Centrifugal heating and chilled water pump; belt drive; 3 phase, 1450 rpm motor; max. pressure 1000 kN/m²; max. temperature 125°C; bed plate; coupling guard; bolted connections; supply only mating flanges;includes fixing on prepared concrete base; electrical work elsewhere						
40 mm pump size; 4.0 l/s at 70 kPa max head; 0.25 kW max motor rating	1344.98	1519.01	7.59	187.56	nr	**1706.57**
40 mm pump size; 4.0 l/s at 130 kPa max head; 1.5 kW max motor rating	1683.23	1901.04	8.09	199.91	nr	**2100.95**
50 mm pump size; 8.5 l/s at 90 kPa max head; 2.2 kW max motor rating	1735.87	1960.48	8.67	214.25	nr	**2174.73**
50 mm pump size; 8.5 l/s at 190 kPa max head; 3 kW max motor rating	3017.03	3407.42	11.20	276.77	nr	**3684.19**
50 mm pump size; 8.5 l/s at 215 kPa max head; 4 kW max motor rating	1972.01	2227.18	11.70	289.13	nr	**2516.31**
65 mm pump size; 14.0 l/s at 90 kPa max head; 3 kW max motor rating	1729.50	1953.29	11.70	289.13	nr	**2242.42**
65 mm pump size; 14.0 l/s at 160 kPa max head; 4 kW max motor rating	1917.76	2165.91	11.70	289.13	nr	**2455.04**
80 mm pump size; 14.5 l/s at 210 kPa max head; 5.5 kW max motor rating	2942.06	3322.75	11.70	289.13	nr	**3611.88**
80 mm pump size; 22.0 l/s at 130 kPa max head; 5.5 kW max motor rating	2942.06	3322.75	13.64	337.06	nr	**3659.81**
80 mm pump size; 22.0 l/s at 200 kPa max head; 7.5 kW max motor rating	3080.85	3479.50	13.64	337.06	nr	**3816.56**
100 mm pump size; 22.0 l/s at 250 kPa max head; 11 kW max motor rating	4323.73	4883.20	13.64	337.06	nr	**5220.26**
100 mm pump size; 30.0 l/s at 100 kPa max head; 4.0 kW max motor rating	2420.33	2733.51	19.15	473.22	nr	**3206.73**
100 mm pump size; 36.0 l/s at 250 kPa max head; 15.0 kW max motor rating	4579.00	5171.50	19.15	473.22	nr	**5644.72**
100 mm pump size; 36.0 l/s at 550 kPa max head; 30.0 kW max motor rating	6214.35	7018.46	19.15	473.22	nr	**7491.68**

T: MECHANICAL/COOLING/HEATING SYSTEMS

Item	Net Price £	Material £	Labour hours	Labour £	Unit	**Total rate £**
Centrifugal heating and chilled water pump; twin head belt drive; 3 phase, 1450 rpm motor; max. pressure 1000 kN/m²; max. temperature 125°C; bed plate; coupling guard; bolted connections; supply only mating flanges; includes fixing on prepared concrete base; electrical work elsewhere						
40 mm pump size; 4.0 l/s at 70 kPa max head; 0.75 kW max motor rating	2685.24	3032.69	7.59	187.56	nr	**3220.25**
40 mm pump size; 4.0 l/s at 130 kPa max head; 1.5 kW max motor rating	3232.39	3650.64	8.09	199.91	nr	**3850.55**
50 mm pump size; 8.5 l/s at 90 kPa max head; 2.2 kW max motor rating	3302.28	3729.58	8.67	214.25	nr	**3943.83**
50 mm pump size; 8.5 l/s at 190 kPa max head; 4 kW max motor rating	3806.32	4298.84	11.20	276.77	nr	**4575.61**
65 mm pump size; 8.5 l/s at 215 kPa max head; 4 kW max motor rating	5230.68	5907.50	11.70	289.13	nr	**6196.63**
65 mm pump size; 14.0 l/s at 90 kPa max head; 3 kW max motor rating	5492.21	6202.88	11.70	289.13	nr	**6492.01**
65 mm pump size; 14.0 l/s at 160 kPa max head; 4 kW max motor rating	5766.83	6513.03	11.70	289.13	nr	**6802.16**
80 mm pump size; 14.5 l/s at 210 kPa max head; 7.5 kW max motor rating	5963.72	6735.40	13.64	337.06	nr	**7072.46**
Centrifugal heating and chilled water pump; close coupled; 3 phase, 1450 rpm motor; max. pressure 1000 kN/m²; max. temperature 110°C; bed plate; coupling guard; bolted connections; supply only mating flanges; includes fixing on prepared concrete base; electrical work elsewhere						
40 mm pump size; 4.0 l/s at 23 kPa max head; 0.55 kW max motor rating	753.83	851.37	7.31	180.63	nr	**1032.00**
50 mm pump size; 4.0 l/s at 75 kPa max head; 0.75 kW max motor rating	835.61	943.73	7.31	180.63	nr	**1124.36**
50 mm pump size; 7.0 l/s at 65 kPa max head; 0.75 kW max motor rating	835.61	943.73	8.01	197.94	nr	**1141.67**
65 mm pump size; 10.0 l/s at 33 kPa max head; 0.75 kW max motor rating	947.13	1069.68	8.01	197.94	nr	**1267.62**
50 mm pump size; 4.0 l/s at 120 kPa max head; 1.5 kW max motor rating	1140.41	1287.97	8.01	197.94	nr	**1485.91**
80 mm pump size; 16.0 l/s at 80 kPa max head; 2.2 kW max motor rating	1476.44	1667.49	12.35	305.18	nr	**1972.67**
80 mm pump size; 16.0 l/s at 120 kPa max head; 4.0 kW max motor rating	1439.26	1625.49	12.35	305.18	nr	**1930.67**
100 mm pump size; 28.0 l/s at 40 kPa max head; 2.2 kW max motor rating	1458.60	1647.33	17.86	441.34	nr	**2088.67**
100 mm pump size; 28.0 l/s at 90 kPa max head; 4.0 kW max motor rating	1610.24	1818.60	17.86	441.34	nr	**2259.94**
125 mm pump size; 40.0 l/s at 50 kPa max head; 3.0 kW max motor rating	1535.91	1734.65	25.85	638.79	nr	**2373.44**

T: MECHANICAL/COOLING/HEATING SYSTEMS

Item	Net Price £	Material £	Labour hours	Labour £	Unit	Total rate £
T31: LOW TEMPERATURE HOT WATER HEATING – cont						
Y20 – PUMPS – cont						
Centrifugal heating and chilled water pump – cont						
125 mm pump size; 40.0 l/s at 120 kPa max head; 7.5 kW max motor rating	1895.72	2141.02	25.85	638.79	nr	**2779.81**
150 mm pump size; 70.0 l/s at 75 kPa max head; 11 kW max motor rating	2786.34	3146.87	30.43	751.96	nr	**3898.83**
150 mm pump size; 70.0 l/s at 120 kPa max head; 15.0 kW max motor rating	2978.14	3363.50	30.43	751.96	nr	**4115.46**
150 mm pump size; 70.0 l/s at 150 kPa max head; 15.0 kW max motor rating	2978.14	3363.50	30.43	751.96	nr	**4115.46**
Centrifugal heating & chilled water pump; close coupled; 3 phase, variable speed motor; max. system pressure 1000 kN/m²; max. temperature 110°C; bed plate; coupling guard; bolted connections; supply only mating flanges; includes fixing on prepared concrete base; electrical work elsewhere.						
40 mm pump size; 4.0 l/s at 23 kPa max head; 0.55 kW max motor rating	1360.46	1536.50	7.31	180.63	nr	**1717.13**
40 mm pump size; 4.0 l/s at 75 kPa max head; 0.75 kW max motor rating	1501.72	1696.04	7.31	180.63	nr	**1876.67**
50 mm pump size; 7.0 l/s at 65 kPa max head; 1.5 kW max motor rating	1842.20	2080.58	8.01	197.94	nr	**2278.52**
50 mm pump size; 10.0 l/s at 33 kPa max head; 1.5 kW max motor rating	1842.20	2080.58	8.01	197.94	nr	**2278.52**
50 mm pump size; 4.0 l/s at 120 kPa max head; 1.5 kW max motor rating	1842.20	2080.58	8.01	197.94	nr	**2278.52**
80 mm pump size; 16.0 l/s at 80 kPa max head; 2.2 kW max motor rating	2401.25	2711.96	12.35	305.18	nr	**3017.14**
80 mm pump size; 16.0 l/s at 120 kPa max head; 3.0 kW max motor rating	2590.09	2925.24	12.35	305.18	nr	**3230.42**
100 mm pump size; 28.0 l/s at 40 kPa max head; 2.2 kW max motor rating	2460.72	2779.12	17.86	441.34	nr	**3220.46**
100 mm pump size; 28.0 l/s at 90 kPa max head; 4.0.kW max motor rating	2935.03	3314.81	17.86	441.34	nr	**3756.15**
125 mm pump size; 40.0 l/s at 50 kPa max head; 3.0 kW max motor rating	2817.57	3182.15	25.85	638.79	nr	**3820.94**
125 mm pump size; 40.0 l/s at 120 kPa max head; 7.5 kW max motor rating	3965.40	4478.50	25.85	638.79	nr	**5117.29**
150 mm pump size; 70.0 l/s at 75 kPa max head; 7.5 kW max motor rating	4583.93	5177.07	30.43	751.96	nr	**5929.03**

T: MECHANICAL/COOLING/HEATING SYSTEMS

Item	Net Price £	Material £	Labour hours	Labour £	Unit	Total rate £
Glandless domestic heating pump; for low pressure domestic hot water heating systems; 240 volt; 50 Hz electric motor; max working pressure 1000 N/m² and max temperature of 130°C; includes fixing in position; electrical work elsewhere						
1' BSP unions – 2 speed	135.61	153.15	1.58	39.05	nr	**192.20**
1.25' BSP unions – 3 speed	199.83	225.69	1.58	39.05	nr	**264.74**
Glandless pumps; for hot water secondary supply; silent running; 3 phase; max pressure 1000kN/m²; max temperature 130° C ; bolted connections; supply only mating flanges; including fixing in position; electrical elsewhere						
1' BSP unions – 3 speed	263.17	297.22	1.58	39.05	nr	**336.27**
Pipeline mounted circulator; for heating and chilled water; silent running; 3 phase; 1450 rpm motor; max pressure 1000 kN/m²; max temperature 120°C; bolted connections; supply only mating flanges; includes fixing in position; electrical elsewhere						
32 mm pump size; 2.0 l/s at 17 kPa max head; 0.2 kW max motor rating	526.35	594.46	6.44	159.14	nr	**753.60**
50 mm pump size; 3.0 l/s at 20 kPa max head; 0.2 kW max motor rating	523.37	591.09	6.86	169.52	nr	**760.61**
65 mm pump size; 5.0 l/s at 30 kPa max head; 0.37 kW max motor rating	832.64	940.38	7.48	184.83	nr	**1125.21**
65 mm pump size; 8.0 l/s at 37 kPa max head; 0.75 kW max motor rating	932.25	1052.87	7.48	184.83	nr	**1237.70**
80 mm pump size; 12.0 l/s at 42 kPa max head; 1.1 kW max motor rating	1083.91	1224.16	8.01	197.94	nr	**1422.10**
100 mm pump size; 25.0 l/s at 37 kPa max head; 2.2 kW max motor rating	1544.83	1744.72	9.11	225.12	nr	**1969.84**
Dual pipeline mounted circulator; for heating & chilled water; silent running; 3 phase; 1450 rpm motor; max pressure 1000 kN/m²; max temperature 120°C; bolted connections; supply only mating flanges; includes fixing in position; electrical work elsewhere						
40 mm pump size; 2.0 l/s at 17 kPa max head; 0.8 kW max motor rating	970.92	1096.55	7.88	194.72	nr	**1291.27**
50 mm pump size; 3.0 l/s at 20 kPa max head; 0.2 kW max motor rating	973.88	1099.90	8.01	197.94	nr	**1297.84**
65 mm pump size; 5.0 l/s at 30 kPa max head; 0.37 kW max motor rating	1580.51	1785.02	9.20	227.35	nr	**2012.37**
65 mm pump size; 8.0 l/s at 37 kPa max head; 0.75 kW max motor rating	1827.33	2063.78	9.20	227.35	nr	**2291.13**
100 mm pump size; 12.0 l/s at 42 kPa max head; 1.1 kW max motor rating	2066.72	2334.15	9.45	233.53	nr	**2567.68**

T: MECHANICAL/COOLING/HEATING SYSTEMS

Item	Net Price £	Material £	Labour hours	Labour £	Unit	Total rate £
T31: LOW TEMPERATURE HOT WATER HEATING – cont						
Y20 – PUMPS – cont						
Glandless accelerator pumps; for low and medium pressure heating services; silent running; 3 phase; 1450 rpm motor; max pressure 1000 kN/m²; max temperature 130°C; bolted connections; supply only mating flanges; includes fixing in position; electrical work elsewhere						
40 mm pump size; 4.0 l/s at 15 kPa max head; 0.35 kW max motor rating	430.08	485.73	6.94	171.50	nr	**657.23**
50 mm pump size; 6.0 l/s at 20 kPa max head; 0.45 kW max motor rating	458.34	517.65	7.35	181.63	nr	**699.28**
80 mm pump size; 13.0 l/s at 28 kPa max head; 0.58 kW max motor rating	903.17	1020.03	7.76	191.76	nr	**1211.79**
Y22 – HEAT EXCHANGERS						
Plate heat exchanger; for use in LTHW systems; painted carbon steel frame; stainless steel plates, nitrile rubber gaskets; design pressure of 10 bar and operating temperature of 110/135°C						
Primary side; 80°C in, 69°C out; secondary side; 82°C in, 71°C out						
107 KW, 2.38 l/s	1995.95	2254.22	10.00	247.12	nr	**2501.34**
245 kW, 5.46 l/s	3067.07	3463.93	10.00	247.12	nr	**3711.05**
287 kW, 6.38 l/s	3404.61	3845.14	10.00	247.12	nr	**4092.26**
328 kW, 7.31 l/s	3714.97	4195.67	10.00	247.12	nr	**4442.79**
364 kW, 8.11 l/s	4408.85	4979.33	10.00	247.12	nr	**5226.45**
403 kW, 8.96 l/s	4628.31	5227.19	10.00	247.12	nr	**5474.31**
453 kW, 10.09 l/s	4815.36	5438.45	10.00	247.12	nr	**5685.57**
490 kW, 10.89 l/s	4991.97	5637.91	10.00	247.12	nr	**5885.03**
1000 kW, 21.7 l/s	3713.93	4194.49	12.00	296.53	nr	**4491.02**
1500 kW, 32.6 l/s	4731.76	5344.02	12.00	296.53	nr	**5640.55**
2000 kW, 43.4 l/s	5676.44	6410.94	15.00	370.67	nr	**6781.61**
2500 kW, 54.3 l/s	6977.47	7880.32	15.00	370.67	nr	**8250.99**
Note: For temperature conditions different to those above, the cost of the units can vary significantly, and so manufacturers advice should be sought.						

T: MECHANICAL/COOLING/HEATING SYSTEMS

Item	Net Price £	Material £	Labour hours	Labour £	Unit	Total rate £
Y23 – CALORIFIERS						
Non-storage calorifiers; mild steel; heater battery duty 116°C/90°C to BS 853, maximum test on shell 11.55 bar, tubes 26.25 bar						
Horizontal or vertical; primary water at 116°C on, 90°C off						
40 kW capacity	827.45	934.51	3.00	74.13	nr	**1008.64**
88 kW capacity	963.99	1088.73	5.00	123.55	nr	**1212.28**
176 kW capacity	1104.68	1247.62	7.04	174.03	nr	**1421.65**
293 kW capacity	1531.96	1730.19	9.01	222.63	nr	**1952.82**
586 kW capacity	2282.31	2577.63	22.22	549.14	nr	**3126.77**
879 kW capacity	3178.56	3589.85	28.57	706.03	nr	**4295.88**
1465 kW capacity	5106.53	5767.29	50.00	1235.56	nr	**7002.85**
2000 kW capacity	9294.57	10497.24	60.00	1482.67	nr	**11979.91**
HEAT EMITTERS						
Perimeter convector heating; metal casing with standard finish; aluminium extruded grille; including backplates						
Top/sloping/flat front outlet						
60 × 200 mm	32.38	36.57	2.00	49.42	m	**85.99**
60 × 300 mm	35.21	39.77	2.00	49.42	m	**89.19**
60 × 450 mm	43.65	49.30	2.00	49.42	m	**98.72**
60 × 525 mm	46.46	52.47	2.00	49.42	m	**101.89**
60 × 600 mm	50.68	57.24	2.00	49.42	m	**106.66**
90 × 260 mm	35.21	39.77	2.00	49.42	m	**89.19**
90 × 300 mm	36.60	41.34	2.00	49.42	m	**90.76**
90 × 450 mm	45.07	50.90	2.00	49.42	m	**100.32**
90 × 525 mm	49.29	55.67	2.00	49.42	m	**105.09**
90 × 600 mm	52.11	58.85	2.00	49.42	m	**108.27**
Extra over for dampers						
Damper	15.48	17.48	0.25	6.18	nr	**23.66**
Extra over for fittings						
60 mm end caps	12.39	13.99	0.25	6.18	nr	**20.17**
90 mm end caps	19.50	22.02	0.25	6.18	nr	**28.20**
60 mm corners	26.34	29.74	0.25	6.18	nr	**35.92**
90 mm corners	38.73	43.74	0.25	6.18	nr	**49.92**

T: MECHANICAL/COOLING/HEATING SYSTEMS

Item	Net Price £	Material £	Labour hours	Labour £	Unit	Total rate £
T31: LOW TEMPERATURE HOT WATER HEATING – cont						
HEAT EMITTERS – cont						
Radiant strip heaters						
Suitable for connection to hot water system; aluminium sheet panels with steel pipe clamped to upper surface; including insulation, sliding brackets, cover plates, end closures; weld or screwed BSP ends						
One pipe						
1500 mm long	65.47	73.94	3.11	76.86	nr	**150.80**
3000 mm long	103.56	116.96	3.11	76.86	nr	**193.82**
4500 mm long	140.47	158.65	3.11	76.86	nr	**235.51**
6000 mm long	193.64	218.70	3.11	76.86	nr	**295.56**
Two pipe						
1500 mm long	122.23	138.05	4.15	102.55	nr	**240.60**
3000 mm long	193.07	218.06	4.15	102.55	nr	**320.61**
4500 mm long	263.66	297.78	4.15	102.55	nr	**400.33**
6000 mm long	355.22	401.19	4.15	102.55	nr	**503.74**
Pressed steel panel type radiators; fixed with and including brackets; taking down once for decoration; refixing						
300 mm high; single panel						
500 mm length	17.87	20.19	2.03	50.16	nr	**70.35**
1000 mm length	35.72	40.34	2.03	50.16	nr	**90.50**
1500 mm length	46.80	52.86	2.03	50.16	nr	**103.02**
2000 mm length	52.50	59.29	2.47	61.04	nr	**120.33**
2500 mm length	58.19	65.72	2.97	73.39	nr	**139.11**
3000 mm length	69.64	78.65	3.22	79.57	nr	**158.22**
300 mm high; double panel; convector						
500 mm length	34.36	38.81	2.13	52.63	nr	**91.44**
1000 mm length	68.77	77.67	2.13	52.63	nr	**130.30**
1500 mm length	103.13	116.47	2.13	52.63	nr	**169.10**
2000 mm length	137.52	155.32	2.57	63.51	nr	**218.83**
2500 mm length	171.88	194.12	3.07	75.86	nr	**269.98**
3000 mm length	206.28	232.97	3.31	81.79	nr	**314.76**

T: MECHANICAL/COOLING/HEATING SYSTEMS

Item	Net Price £	Material £	Labour hours	Labour £	Unit	**Total rate £**
450 mm high; single panel						
500 mm length	16.67	18.83	2.08	51.40	nr	**70.23**
1000 mm length	33.37	37.69	2.08	51.40	nr	**89.09**
1600 mm length	53.39	60.30	2.53	62.53	nr	**122.83**
2000 mm length	66.74	75.38	2.97	73.39	nr	**148.77**
2400 mm length	80.07	90.43	3.47	85.74	nr	**176.17**
3000 mm length	100.09	113.04	3.82	94.39	nr	**207.43**
450 mm high; double panel; convector						
500 mm length	30.56	34.52	2.18	53.88	nr	**88.40**
1000 mm length	61.10	69.00	2.18	53.88	nr	**122.88**
1600 mm length	111.91	126.39	2.63	64.98	nr	**191.37**
2000 mm length	188.13	212.48	3.06	75.61	nr	**288.09**
2400 mm length	225.77	254.98	3.37	83.28	nr	**338.26**
3000 mm length	282.21	318.73	3.92	96.87	nr	**415.60**
600 mm high; single panel						
500 mm length	22.34	25.23	2.18	53.88	nr	**79.11**
1000 mm length	44.70	50.48	2.43	60.05	nr	**110.53**
1600 mm length	71.50	80.75	3.13	77.34	nr	**158.09**
2000 mm length	89.37	100.93	3.77	93.17	nr	**194.10**
2400 mm length	107.26	121.14	4.07	100.57	nr	**221.71**
3000 mm length	134.08	151.43	5.11	126.28	nr	**277.71**
600 mm high; double panel; convector						
500 mm length	38.47	43.45	2.28	56.34	nr	**99.79**
1000 mm length	76.94	86.89	2.28	56.34	nr	**143.23**
1600 mm length	140.93	159.16	3.23	79.81	nr	**238.97**
2000 mm length	236.92	267.58	3.87	95.64	nr	**363.22**
2400 mm length	284.29	321.08	4.17	103.05	nr	**424.13**
3000 mm length	355.36	401.34	5.24	129.49	nr	**530.83**
700 mm high; single panel						
500 mm length	26.14	29.52	2.23	55.10	nr	**84.62**
1000 mm length	52.27	59.04	2.83	69.93	nr	**128.97**
1600 mm length	83.64	94.46	3.73	92.17	nr	**186.63**
2000 mm length	104.56	118.09	4.46	110.21	nr	**228.30**
2400 mm length	125.49	141.73	4.48	110.70	nr	**252.43**
3000 mm length	156.83	177.12	5.24	129.49	nr	**306.61**
700 mm high; double panel; convector						
500 mm length	50.00	56.47	2.33	57.58	nr	**114.05**
1000 mm length	134.52	151.93	3.08	76.11	nr	**228.04**
1600 mm length	215.22	243.07	3.83	94.65	nr	**337.72**
2000 mm length	269.03	303.84	4.17	103.05	nr	**406.89**
2400 mm length	322.85	364.63	4.37	108.00	nr	**472.63**
3000 mm length	403.55	455.77	4.82	119.11	nr	**574.88**

T: MECHANICAL/COOLING/HEATING SYSTEMS

Item	Net Price £	Material £	Labour hours	Labour £	Unit	Total rate £
T31: LOW TEMPERATURE HOT WATER HEATING – cont						
HEAT EMITTERS – cont						
Flat panel type steel radiators; fixed with and including brackets; taking down once for decoration; refixing						
300 mm high; single panel (44 mm deep)						
500 mm length	150.26	169.71	2.03	50.16	nr	**219.87**
1000 mm length	225.58	254.77	2.03	50.16	nr	**304.93**
1500 mm length	300.90	339.83	2.03	50.16	nr	**389.99**
2000 mm length	376.23	424.91	2.47	61.04	nr	**485.95**
2400 mm length	436.49	492.97	2.97	73.39	nr	**566.36**
3000 mm length	650.68	734.88	3.22	79.57	nr	**814.45**
300 mm high; double panel (100 mm deep)						
500 mm length	346.87	391.75	2.03	50.16	nr	**441.91**
1000 mm length	543.87	614.24	2.03	50.16	nr	**664.40**
1500 mm length	740.87	836.74	2.03	50.16	nr	**886.90**
2000 mm length	937.86	1059.22	2.47	61.04	nr	**1120.26**
2400 mm length	1095.46	1237.21	2.97	73.39	nr	**1310.60**
3000 mm length	1455.67	1644.03	3.22	79.57	nr	**1723.60**
500 mm high; single panel (44 mm deep)						
500 mm length	164.74	186.06	2.13	52.63	nr	**238.69**
1000 mm length	254.55	287.49	2.13	52.63	nr	**340.12**
1500 mm length	344.36	388.92	2.13	52.63	nr	**441.55**
2000 mm length	434.17	490.34	2.57	63.51	nr	**553.85**
2400 mm length	506.01	571.49	3.07	75.86	nr	**647.35**
3000 mm length	737.59	833.03	3.31	81.79	nr	**914.82**
500 mm high; double panel (100 mm deep)						
500 mm length	401.91	453.91	2.08	51.40	nr	**505.31**
1000 mm length	653.96	738.58	2.08	51.40	nr	**789.98**
1500 mm length	906.00	1023.23	2.53	62.53	nr	**1085.76**
2000 mm length	1158.04	1307.88	2.97	73.39	nr	**1381.27**
2400 mm length	1359.67	1535.60	3.47	85.74	nr	**1621.34**
3000 mm length	1785.93	2017.02	3.82	94.39	nr	**2111.41**
600 mm high; single panel (44 mm deep)						
500 mm length	194.49	219.66	2.18	53.88	nr	**273.54**
1000 mm length	304.57	343.98	2.18	53.88	nr	**397.86**
1500 mm length	414.66	812.30	2.63	64.98	nr	**877.28**
2000 mm length	524.75	592.65	3.06	75.61	nr	**668.26**
2400 mm length	612.82	692.11	3.37	83.28	nr	**775.39**
3000 mm length	868.73	981.14	3.92	96.87	nr	**1078.01**

T: MECHANICAL/COOLING/HEATING SYSTEMS

Item	Net Price £	Material £	Labour hours	Labour £	Unit	**Total rate £**
600 mm high; double panel (100 mm deep)						
500 mm length	458.50	517.82	2.18	53.88	nr	**571.70**
1000 mm length	748.21	845.03	2.43	60.05	nr	**905.08**
1500 mm length	1037.91	1172.21	3.13	77.34	nr	**1249.55**
2000 mm length	1327.61	1499.40	3.77	93.17	nr	**1592.57**
2400 mm length	1559.37	1761.14	4.07	100.57	nr	**1861.71**
3000 mm length	2030.82	2293.60	5.11	126.28	nr	**2419.88**
700 mm high; single panel (44 mm deep)						
500 mm length	206.07	232.73	2.28	56.34	nr	**289.07**
1000 mm length	327.75	370.16	2.28	56.34	nr	**426.50**
1500 mm length	449.43	507.58	3.23	79.81	nr	**587.39**
2000 mm length	571.10	645.00	3.87	95.64	nr	**740.64**
2400 mm length	668.44	754.93	4.17	103.05	nr	**857.98**
3000 mm length	938.26	1059.67	5.24	129.49	nr	**1189.16**
700 mm high; double panel (100 mm deep)						
500 mm length	496.16	560.36	2.23	55.10	nr	**615.46**
1000 mm length	823.53	930.09	2.83	69.93	nr	**1000.02**
1500 mm length	1150.89	1299.81	3.73	92.17	nr	**1391.98**
2000 mm length	1478.26	1669.54	4.46	110.21	nr	**1779.75**
2400 mm length	1740.15	1965.32	4.48	110.70	nr	**2076.02**
3000 mm length	1958.81	2212.27	5.24	129.49	nr	**2341.76**
Fan convector; sheet metal casing with lockable access panel; centrifugal fan; air filter; LPHW heating coil; extruded aluminium grilles; 3 speed; includes fixing in position; electrical work elsewhere						
Free-standing flat top, 695 mm high, medium speed rating						
Entering air temperature, 18°C						
695 mm long, 1 row 1.94 kW, 75 l/s	643.53	726.80	2.73	67.46	nr	**794.26**
695 mm long, 2 row 2.64 kW, 75 l/s	643.53	726.80	2.73	67.46	nr	**794.26**
895 mm long, 1 row 4.02 kW, 150 l/s	725.15	818.98	2.73	67.46	nr	**886.44**
895 mm long, 2 row 5.62 kW, 150 l/s	725.15	818.98	2.73	67.46	nr	**886.44**
1195 mm long, 1 row 6.58 kW, 250 l/s	825.59	932.41	3.00	74.13	nr	**1006.54**
1195 mm long, 2 row 9.27 kW, 250 l/s	825.59	932.41	3.00	74.13	nr	**1006.54**
1495 mm long, 1 row 9.04 kW, 340 l/s	921.35	1040.56	3.26	80.56	nr	**1121.12**
1495 mm long, 2 row 12.73 kW, 340 l/s	921.35	1040.56	3.26	80.56	nr	**1121.12**
Free-standing flat top, 695 mm high, medium speed rating, c/w floor plinth						
695 mm long, 1 row 1.94 kW, 75 l/s	675.70	763.13	2.73	67.46	nr	**830.59**
695 mm long, 2 row 2.64 kW, 75 l/s	675.70	763.13	2.73	67.46	nr	**830.59**
895 mm long, 1 row 4.02 kW, 150 l/s	761.40	859.92	2.73	67.46	nr	**927.38**

T: MECHANICAL/COOLING/HEATING SYSTEMS

Item	Net Price £	Material £	Labour hours	Labour £	Unit	Total rate £
T31: LOW TEMPERATURE HOT WATER HEATING – cont						
HEAT EMITTERS – cont						
Fan convector – cont						
Free-standing flat top, 695 mm high, medium speed rating, c/w floor plinth – cont						
895 mm long, 2 row 5.62 kW, 150 l/s	761.40	859.92	2.73	67.46	nr	**927.38**
1195 mm long, 1 row 6.58 kW, 250 l/s	866.88	979.05	3.00	74.13	nr	**1053.18**
1195 mm long, 2 row 9.27 kW, 250 l/s	866.88	979.05	3.00	74.13	nr	**1053.18**
1495 mm long, 1 row 9.04 kW, 340 l/s	967.40	1092.57	3.26	80.56	nr	**1173.13**
1495 mm long, 2 row 12.73 kW, 340 l/s	967.40	1092.57	3.26	80.56	nr	**1173.13**
Free-standing sloping top, 695 mm high, medium speed rating, c/w floor plinth						
695 mm long, 1 row 1.94 kW, 75 l/s	699.24	789.71	2.73	67.46	nr	**857.17**
695 mm long, 2 row 2.64 kW, 75 l/s	699.24	789.71	2.73	67.46	nr	**857.17**
895 mm long, 1 row 4.02 kW, 150 l/s	784.94	886.50	2.73	67.46	nr	**953.96**
895 mm long, 2 row 5.62 kW, 150 l/s	784.94	886.50	2.73	67.46	nr	**953.96**
1195 mm long, 1 row 6.58 kW, 250 l/s	890.42	1005.64	3.00	74.13	nr	**1079.77**
1195 mm long, 2 row 9.27 kW, 250 l/s	890.42	1005.64	3.00	74.13	nr	**1079.77**
1495 mm long, 1 row 9.04 kW, 340 l/s	990.95	1119.17	3.26	80.56	nr	**1199.73**
1495 mm long, 2 row 12.73 kW, 340 l/s	990.95	1119.17	3.26	80.56	nr	**1199.73**
Wall mounted high level sloping discharge						
695 mm long, 1 row 1.94 kW, 75 l/s	706.30	797.69	2.73	67.46	nr	**865.15**
695 mm long, 2 row 2.64 kW, 75 l/s	706.30	797.69	2.73	67.46	nr	**865.15**
895 mm long, 1 row 4.02 kW, 150 l/s	726.73	820.76	2.73	67.46	nr	**888.22**
895 mm long, 2 row 5.62 kW, 150 l/s	726.73	820.76	2.73	67.46	nr	**888.22**
1195 mm long, 1 row 6.58 kW, 250 l/s	884.44	998.89	3.00	74.13	nr	**1073.02**
1195 mm long, 2 row 9.27 kW, 250 l/s	884.44	998.89	3.00	74.13	nr	**1073.02**
1495 mm long, 1 row 9.04 kW, 340 l/s	959.00	1083.09	3.26	80.56	nr	**1163.65**
1495 mm long, 2 row 12.73 kW, 340 l/s	959.00	1083.09	3.26	80.56	nr	**1163.65**
Ceiling mounted sloping inlet/outlet 665 mm wide						
895 mm long, 1 row 4.02 kW, 150 l/s	798.91	902.28	4.15	102.55	nr	**1004.83**
895 mm long, 2 row 5.62 kW, 150 l/s	798.91	902.28	4.15	102.55	nr	**1004.83**
1195 mm long, 1 row 6.58 kW, 250 l/s	897.80	1013.97	4.15	102.55	nr	**1116.52**
1195 mm long, 2 row 9.27 kW, 250 l/s	897.80	1013.97	4.15	102.55	nr	**1116.52**
1495 mm long, 1 row 9.04 kW, 340 l/s	985.69	1113.24	4.15	102.55	nr	**1215.79**
1495 mm long, 2 row 12.73 kW, 340 l/s	985.69	1113.24	4.15	102.55	nr	**1215.79**
Free-standing unit, extended height 1700/1900/2100 mm						
895 mm long, 1 row 4.02 kW, 150 l/s	924.79	1044.45	3.11	76.86	nr	**1121.31**
895 mm long, 2 row 5.62 kW, 150 l/s	924.49	1044.11	3.11	76.86	nr	**1120.97**
1195 mm long, 1 row 6.58 kW, 250 l/s	1079.87	1219.60	3.11	76.86	nr	**1296.46**

T: MECHANICAL/COOLING/HEATING SYSTEMS

Item	Net Price £	Material £	Labour hours	Labour £	Unit	Total rate £
1195 mm long, 2 row 9.27 kW, 250 l/s	1079.87	1219.60	3.11	76.86	nr	**1296.46**
1495 mm long, 1 row 9.04 kW, 340 l/s	1188.18	1341.92	3.11	76.86	nr	**1418.78**
1495 mm long, 2 row 12.73 kW, 340 l/s	1188.18	1341.92	3.11	76.86	nr	**1418.78**
LTHW trench heating; water temperatures 90°C/70°C; room air temperature 20°C; convector with copper tubes and aluminium fins within steel duct; Includes fixing within floor screed; electrical work elsewhere						
Natural convection type						
Normal capacity, 182 mm width, complete with linear, natural anodized aluminium grille (grille also costed separately below)						
92 mm deep						
1250 mm long, 234 W output	226.01	255.25	2.00	49.42	nr	**304.67**
2250 mm long, 471 W output	361.61	408.40	4.00	98.85	nr	**507.25**
3250 mm long, 709 W output	494.64	558.64	5.00	123.55	nr	**682.19**
4250 mm long, 946 W output	631.53	713.25	7.00	172.98	nr	**886.23**
5000 mm long, 1124 W output	862.69	974.31	8.00	197.69	nr	**1172.00**
120 mm deep						
1250 mm long, 294 W output	263.46	297.55	2.00	49.42	nr	**346.97**
2250 mm long, 471 W output	414.57	468.21	4.00	98.85	nr	**567.06**
3250 mm long, 891 W output	564.37	637.40	5.00	123.55	nr	**760.95**
4250 mm long, 1190 W output	714.18	806.60	7.00	172.98	nr	**979.58**
5000 mm long, 1414 W output	962.15	1086.65	8.00	197.69	nr	**1284.34**
150 mm deep						
1250 mm long, 329 W output	269.92	304.85	2.00	49.42	nr	**354.27**
2250 mm long, 664 W output	426.19	481.34	4.00	98.85	nr	**580.19**
3250 mm long, 998 W output	579.87	654.90	5.00	123.55	nr	**778.45**
4250 mm long, 1333 W output	737.43	832.85	7.00	172.98	nr	**1005.83**
5000 mm long, 1584 W output	989.26	1117.27	8.00	197.69	nr	**1314.96**
200 mm deep						
1250 mm long, 396 W output	281.54	317.97	2.00	49.42	nr	**367.39**
2250 mm long, 799 W output	515.30	581.98	4.00	98.85	nr	**680.83**
3250 mm long, 1201 W output	613.45	692.82	5.00	123.55	nr	**816.37**
4250 mm long, 1603 W output	785.21	886.81	7.00	172.98	nr	**1059.79**
5000 mm long, 1905 W output	1043.50	1178.53	8.00	197.69	nr	**1376.22**

T: MECHANICAL/COOLING/HEATING SYSTEMS

Item	Net Price £	Material £	Labour hours	Labour £	Unit	Total rate £
T31: LOW TEMPERATURE HOT WATER HEATING – cont						
HEAT EMITTERS – cont						
Fan assisted type (outputs assume fan at 50%)						
Normal capacity, 182 mm width, complete with Natural anodized aluminium grille						
112 mm deep						
1250 mm long, 437 W output	596.65	673.85	2.00	49.42	nr	**723.27**
2250 mm long, 1019 W output	721.94	815.36	4.00	98.85	nr	**914.21**
3250 mm long, 1488 W output	848.49	958.28	5.00	123.55	nr	**1081.83**
4250 mm long, 1845 W output	973.77	1099.77	7.00	172.98	nr	**1272.75**
5000 mm long, 2038 W output	1099.04	1241.25	8.00	197.69	nr	**1438.94**
Linear grille anodized aluminium, 170 mm width (if supplied as a separate item)	140.78	159.00	–	–	m	**159.00**
Roll up grille, natural anodized aluminium	140.78	159.00	–	–	m	**159.00**
Thermostatic valve with remote regulator (c/w valve body)	58.12	65.64	4.00	98.85	nr	**164.49**
Fan speed controller	28.41	32.09	2.00	49.42	nr	**81.51**
Note: As an alternative to thermostatic control, the system can be controlled via two port valves. Refer to valve section in T31 for valve prices						
LTHW underfloor heating; water flow and return temperatures of 60°C and 70°C; pipework at 300 mm centres; pipe fixings; flow and return manifolds and zone actuators; wiring block; insulation; includes fixing in position; excludes secondary pump, mixing valve, zone thermostats and floor finishes; electrical work elsewhere						
Note: All rates are expressed on a m² basis, for the following example areas						
Screeded floor with 15–25 mm stone/marble finish (producing 80–100 W/m²)						
250 m² area (single zone)	24.68	27.88	0.14	3.46	m²	**31.34**
1000 m² area (single zone)	24.04	27.15	0.12	2.97	m²	**30.12**
5000 m² area (multi-zone)	23.99	27.09	0.10	2.47	m²	**29.56**

T: MECHANICAL/COOLING/HEATING SYSTEMS

Item	Net Price £	Material £	Labour hours	Labour £	Unit	Total rate £
Screeded floor with 10 mm carpet tile (producing 80–100 W/m²)						
250 m² area (single zone)	24.68	27.88	0.14	3.46	m²	**31.34**
1000 m² area (single zone)	24.04	27.15	0.12	2.97	m²	**30.12**
5000 m² area (multi-zone)	23.99	27.09	0.10	2.47	m²	**29.56**
Floating timber floor with 20 mm timber finish (producing 70–80 Wm²)						
250 m² are (single zone)	31.10	35.13	0.14	3.46	m²	**38.59**
1000 m² area(single zone)	33.34	37.66	0.12	2.97	m²	**40.63**
5000 m² area (multi-zone)	31.85	35.98	0.10	2.47	m²	**38.45**
Floating timber floor with 10 mm carpet tile (producing 70–80 W/m²)						
250 m² are (single zone)	40.48	45.72	0.16	3.96	m²	**49.68**
1000 m² area(single zone)	37.54	42.40	0.12	2.97	m²	**45.37**
5000 m² area (multi-zone)	35.91	40.56	0.10	2.47	m²	**43.03**
PIPE FREEZING						
Freeze isolation of carbon steel or copper pipelines containing static water, either side of work location, freeze duration not exceeding 4 hours assuming that flow and return circuits are treated concurrently and activities undertaken during normal working hours						
Up to 4 freezes						
50 mm dia.	404.60	456.95	1.00	24.71	nr	**481.66**
65 mm dia.	404.60	456.95	1.00	24.71	nr	**481.66**
80 mm dia.	462.54	522.39	1.00	24.71	nr	**547.10**
100 mm dia.	520.50	587.85	1.00	24.71	nr	**612.56**
150 mm dia.	866.07	978.14	1.00	24.71	nr	**1002.85**
200 mm dia.	1327.54	1499.32	1.00	24.71	nr	**1524.03**

T: MECHANICAL/COOLING/HEATING SYSTEMS

Item	Net Price £	Material £	Labour hours	Labour £	Unit	Total rate £
T31: LOW TEMPERATURE HOT WATER HEATING – cont						
ENERGY METERS						
Ultrasonic						
Energy meter for measuring energy use in LTHW systems; includes ultrasonic flow meter (with sensor and signal converter), energy calculator, pair of temperature sensors with brass pockets, and 3 m of interconnecting cable; includes fixing in position; electrical work elsewhere						
Pipe size (flanged connections to PN16); maximum flow rate						
50 mm, 36 m³/hr	1172.30	1323.99	1.80	44.47	nr	**1368.46**
65 mm, 60 m³/hr	1292.23	1459.44	2.32	57.32	nr	**1516.76**
80 mm, 100 m³/hr	1448.49	1635.92	2.56	63.26	nr	**1699.18**
125 mm, 250 m³/hr	1678.22	1895.37	3.60	88.96	nr	**1984.33**
150 mm, 360 m³/hr	1821.79	2057.52	4.80	118.61	nr	**2176.13**
200 mm, 600 m³/hr	2035.48	2298.87	6.24	154.19	nr	**2453.06**
250 mm, 1000 m³/hr	2348.82	2652.74	9.60	237.23	nr	**2889.97**
300 mm, 1500 m³/hr	2760.99	3118.25	10.80	266.87	nr	**3385.12**
350 mm, 2000 m³/hr	3326.87	3757.35	13.20	326.20	nr	**4083.55**
400 mm, 2500 m³/hr	3808.29	4301.06	15.60	385.50	nr	**4686.56**
500 mm, 3000 m³/hr	4321.79	4881.01	24.00	593.07	nr	**5474.08**
600 mm, 3500 m³/hr	4855.58	5483.87	28.00	691.92	nr	**6175.79**
Y53 – CONTROL COMPONENTS – MECHANICAL						
Room thermostats; light and medium duty; installed and connected						
Range 3°C to 27°C; 240 volt						
1 amp; on/off type	25.80	29.14	0.30	7.41	nr	**36.55**
Range 0°C to +15°C; 240 volt						
6 amp; frost thermostat	17.40	19.66	0.30	7.41	nr	**27.07**
Range 3°C to 27°C; 250 volt						
2 amp; changeover type; dead zone	43.15	48.74	0.30	7.41	nr	**56.15**
2 amp; changeover type	20.02	22.61	0.30	7.41	nr	**30.02**
2 amp; changeover type; concealed setting	25.25	28.52	0.30	7.41	nr	**35.93**
6 amp; on/off type	15.58	17.59	0.30	7.41	nr	**25.00**
6 amp; temperature set-back	32.25	36.42	0.30	7.41	nr	**43.83**
16 amp; on/off type	24.00	27.10	0.30	7.41	nr	**34.51**
16 amp; on/off type; concealed setting	26.16	29.55	0.30	7.41	nr	**36.96**

T: MECHANICAL/COOLING/HEATING SYSTEMS

Item	Net Price £	Material £	Labour hours	Labour £	Unit	**Total rate £**
20 amp; on/off type; concealed setting	22.28	25.16	0.30	7.41	nr	**32.57**
20 amp; indicated 'off' position	28.10	31.73	0.30	7.41	nr	**39.14**
20 amp; manual; double pole on/off and neon indicator	51.15	57.77	0.30	7.41	nr	**65.18**
20 amp; indicated 'off' position	33.48	37.81	0.30	7.41	nr	**45.22**
Range 10°C to 40°C; 240 volt						
20 amp; changeover contacts	30.50	34.45	0.30	7.41	nr	**41.86**
2 amp; 'heating-cooling' switch	65.97	74.51	0.30	7.41	nr	**81.92**
Surface thermostats						
Cylinder thermostat						
6 amp; changeover type; with cable	18.01	20.34	0.25	6.18	nr	**26.52**
Electrical thermostats; installed and connected						
Range 5°C to 30°C; 230 volt standard port single time						
10 amp with sensor	25.01	28.24	0.30	7.41	nr	**35.65**
Range 5°C to 30°C; 230 volt standard port double time						
10 amp with sensor	28.34	32.01	0.30	7.41	nr	**39.42**
10 amp with sensor and on/off switch	43.23	48.83	0.30	7.41	nr	**56.24**
Radiator thermostats						
Angled valve body; thermostatic head; built in sensor						
15 mm; liquid filled	15.59	17.61	0.84	20.77	nr	**38.38**
15 mm; wax filled	15.59	17.61	0.84	20.77	nr	**38.38**
Immersion thermostats; stem type; domestic water boilers; fitted; electrical work elsewhere						
Temperature range 0°C to 40°C						
Non standard; 280 mm stem	9.49	10.72	0.25	6.18	nr	**16.90**
Temperature range 18°C to 88°C						
13 amp; 178 mm stem	6.30	7.12	0.25	6.18	nr	**13.30**
20 amp; 178 mm stem	9.80	11.07	0.25	6.18	nr	**17.25**
Non standard; pocket clip; 280 mm stem	9.09	10.26	0.25	6.18	nr	**16.44**

T: MECHANICAL/COOLING/HEATING SYSTEMS

Item	Net Price £	Material £	Labour hours	Labour £	Unit	**Total rate £**
T31: LOW TEMPERATURE HOT WATER HEATING – cont						
Y53 – CONTROL COMPONENTS – MECHANICAL – cont						
Immersion thermostats – cont						
Temperature range 40°C to 80°C						
13 amp; 178 mm stem	3.46	3.91	0.25	6.18	nr	**10.09**
20 amp; 178 mm stem	6.54	7.38	0.25	6.18	nr	**13.56**
Non standard; pocket clip; 280 mm stem	10.18	11.50	0.25	6.18	nr	**17.68**
13 amp; 457 mm stem	4.10	4.63	0.25	6.18	nr	**10.81**
20 amp; 457 mm stem	7.16	8.09	0.25	6.18	nr	**14.27**
Temperature range 50°C to 100°C						
Non standard; 1780 mm stem	8.57	9.68	0.25	6.18	nr	**15.86**
Non standard; 280 mm stem	8.94	10.10	0.25	6.18	nr	**16.28**
Pockets for thermostats						
For 178 mm stem	11.21	12.66	0.25	6.18	nr	**18.84**
For 280 mm stem	11.40	12.87	0.25	6.18	nr	**19.05**
Immersion thermostats; stem type; industrial installations; fitted; electrical work elsewhere						
Temperature range 5°C to 105°C						
For 305 mm stem	140.83	159.05	0.50	12.35	nr	**171.40**

T: MECHANICAL/COOLING/HEATING SYSTEMS

Item	Net Price £	Material £	Labour hours	Labour £	Unit	**Total rate £**
Y50 – THERMAL INSULATION						
For flexible closed cell insulation see Section S10 – Cold Water						
Mineral fibre sectional insulation; bright class O foil faced; bright class O foil taped joints; 19 mm aluminium bands						
Concealed pipework						
20 mm thick						
15 mm dia.	3.11	3.51	0.15	3.70	m	**7.21**
20 mm dia.	3.31	3.73	0.15	3.70	m	**7.43**
25 mm dia.	3.55	4.01	0.15	3.70	m	**7.71**
32 mm dia.	3.97	4.49	0.15	3.70	m	**8.19**
40 mm dia.	4.25	4.80	0.15	3.70	m	**8.50**
50 mm dia.	4.85	5.48	0.15	3.70	m	**9.18**
Extra over for fittings concealed insulation						
Flange/union						
15 mm dia.	1.55	1.75	0.13	3.20	nr	**4.95**
20 mm dia.	1.66	1.88	0.13	3.20	nr	**5.08**
25 mm dia.	1.78	2.01	0.13	3.20	nr	**5.21**
32 mm dia.	1.98	2.23	0.13	3.20	nr	**5.43**
40 mm dia.	2.12	2.40	0.13	3.20	nr	**5.60**
50 mm dia.	2.42	2.73	0.13	3.20	nr	**5.93**
Valves						
15 mm dia.	3.11	3.51	0.15	3.70	nr	**7.21**
20 mm dia.	3.55	4.01	0.15	3.70	nr	**7.71**
25 mm dia.	3.55	4.01	0.15	3.70	nr	**7.71**
32 mm dia.	3.97	4.49	0.15	3.70	nr	**8.19**
40 mm dia.	4.25	4.80	0.15	3.70	nr	**8.50**
50 mm dia.	4.85	5.48	0.15	3.70	nr	**9.18**
Expansion bellows						
15 mm dia.	6.20	7.00	0.22	5.44	nr	**12.44**
20 mm dia.	6.63	7.49	0.22	5.44	nr	**12.93**
25 mm dia.	7.12	8.04	0.22	5.44	nr	**13.48**
32 mm dia.	7.93	8.96	0.22	5.44	nr	**14.40**
40 mm dia.	8.49	9.59	0.22	5.44	nr	**15.03**
50 mm dia.	9.70	10.95	0.22	5.44	nr	**16.39**
25 mm thick						
15 mm dia.	3.42	3.87	0.15	3.70	m	**7.57**
20 mm dia.	3.68	4.15	0.15	3.70	m	**7.85**
25 mm dia.	4.13	4.66	0.15	3.70	m	**8.36**
32 mm dia.	4.48	5.06	0.15	3.70	m	**8.76**
40 mm dia.	4.80	5.42	0.15	3.70	m	**9.12**
50 mm dia.	5.49	6.20	0.15	3.70	m	**9.90**
65 mm dia.	6.28	7.09	0.15	3.70	m	**10.79**

T: MECHANICAL/COOLING/HEATING SYSTEMS

Item	Net Price £	Material £	Labour hours	Labour £	Unit	Total rate £
T31: LOW TEMPERATURE HOT WATER HEATING – cont						
Y50 – THERMAL INSULATION – cont						
Extra over for fittings concealed insulation – cont						
25 mm thick – cont						
80 mm dia.	6.88	7.77	0.22	5.44	m	**13.21**
100 mm dia.	9.08	10.25	0.22	5.44	m	**15.69**
125 mm dia.	10.49	11.85	0.22	5.44	m	**17.29**
150 mm dia.	12.50	14.12	0.22	5.44	m	**19.56**
200 mm dia.	17.67	19.96	0.25	6.18	m	**26.14**
250 mm dia.	21.15	23.89	0.25	6.18	m	**30.07**
300 mm dia.	22.61	25.53	0.25	6.18	m	**31.71**
Extra over for fittings concealed insulation						
Flange/union						
15 mm dia.	1.70	1.92	0.13	3.20	nr	**5.12**
20 mm dia.	1.84	2.08	0.13	3.20	nr	**5.28**
25 mm dia.	2.06	2.33	0.13	3.20	nr	**5.53**
32 mm dia.	2.25	2.54	0.13	3.20	nr	**5.74**
40 mm dia.	2.39	2.70	0.13	3.20	nr	**5.90**
50 mm dia.	2.75	3.10	0.13	3.20	nr	**6.30**
65 mm dia.	3.14	3.55	0.13	3.20	nr	**6.75**
80 mm dia.	3.45	3.90	0.18	4.45	nr	**8.35**
100 mm dia.	4.54	5.13	0.18	4.45	nr	**9.58**
125 mm dia.	5.25	5.93	0.18	4.45	nr	**10.38**
150 mm dia.	6.26	7.07	0.18	4.45	nr	**11.52**
200 mm dia.	8.83	9.98	0.22	5.44	nr	**15.42**
250 mm dia.	10.58	11.94	0.22	5.44	nr	**17.38**
300 mm dia.	11.31	12.77	0.22	5.44	nr	**18.21**
Valves						
15 mm dia.	3.42	3.87	0.15	3.70	nr	**7.57**
20 mm dia.	3.68	4.15	0.15	3.70	nr	**7.85**
25 mm dia.	4.13	4.66	0.15	3.70	nr	**8.36**
32 mm dia.	4.48	5.06	0.15	3.70	nr	**8.76**
40 mm dia.	4.80	5.42	0.15	3.70	nr	**9.12**
50 mm dia.	5.49	6.20	0.15	3.70	nr	**9.90**
65 mm dia.	6.28	7.09	0.15	3.70	nr	**10.79**
80 mm dia.	6.88	7.77	0.20	4.95	nr	**12.72**
100 mm dia.	9.08	10.25	0.20	4.95	nr	**15.20**
125 mm dia.	10.49	11.85	0.20	4.95	nr	**16.80**
150 mm dia.	12.50	14.12	0.20	4.95	nr	**19.07**
200 mm dia.	17.67	19.96	0.25	6.18	nr	**26.14**
250 mm dia.	21.15	23.89	0.25	6.18	nr	**30.07**
300 mm dia.	22.61	25.53	0.25	6.18	nr	**31.71**

T: MECHANICAL/COOLING/HEATING SYSTEMS

Item	Net Price £	Material £	Labour hours	Labour £	Unit	**Total rate £**
Expansion bellows						
15 mm dia.	6.83	7.71	0.22	5.44	nr	**13.15**
20 mm dia.	7.37	8.32	0.22	5.44	nr	**13.76**
25 mm dia.	8.22	9.28	0.22	5.44	nr	**14.72**
32 mm dia.	8.96	10.12	0.22	5.44	nr	**15.56**
40 mm dia.	9.61	10.85	0.22	5.44	nr	**16.29**
50 mm dia.	11.01	12.43	0.22	5.44	nr	**17.87**
65 mm dia.	12.56	14.19	0.22	5.44	nr	**19.63**
80 mm dia.	13.75	15.52	0.29	7.17	nr	**22.69**
100 mm dia.	18.14	20.49	0.29	7.17	nr	**27.66**
125 mm dia.	20.99	23.70	0.29	7.17	nr	**30.87**
150 mm dia.	25.01	28.24	0.29	7.17	nr	**35.41**
200 mm dia.	35.32	39.89	0.36	8.90	nr	**48.79**
250 mm dia.	42.31	47.79	0.36	8.90	nr	**56.69**
300 mm dia.	45.23	51.08	0.36	8.90	nr	**59.98**
30 mm thick						
15 mm dia.	4.44	5.02	0.15	3.70	m	**8.72**
20 mm dia.	4.75	5.37	0.15	3.70	m	**9.07**
25 mm dia.	5.03	5.68	0.15	3.70	m	**9.38**
32 mm dia.	5.48	6.19	0.15	3.70	m	**9.89**
40 mm dia.	5.81	6.56	0.15	3.70	m	**10.26**
50 mm dia.	6.64	7.50	0.15	3.70	m	**11.20**
65 mm dia.	7.51	8.49	0.15	3.70	m	**12.19**
80 mm dia.	8.21	9.27	0.22	5.44	m	**14.71**
100 mm dia.	10.60	11.97	0.22	5.44	m	**17.41**
125 mm dia.	12.22	13.80	0.22	5.44	m	**19.24**
150 mm dia.	14.35	16.21	0.22	5.44	m	**21.65**
200 mm dia.	20.05	22.64	0.25	6.18	m	**28.82**
250 mm dia.	23.85	26.94	0.25	6.18	m	**33.12**
300 mm dia.	25.28	28.55	0.25	6.18	m	**34.73**
350 mm dia.	27.78	31.38	0.25	6.18	m	**37.56**
Extra over for fittings concealed insulation						
Flange/union						
15 mm dia.	2.21	2.50	0.13	3.20	nr	**5.70**
20 mm dia.	2.38	2.69	0.13	3.20	nr	**5.89**
25 mm dia.	2.51	2.84	0.13	3.20	nr	**6.04**
32 mm dia.	2.75	3.10	0.13	3.20	nr	**6.30**
40 mm dia.	2.90	3.27	0.13	3.20	nr	**6.47**
50 mm dia.	3.33	3.76	0.13	3.20	nr	**6.96**
65 mm dia.	3.77	4.25	0.13	3.20	nr	**7.45**
80 mm dia.	4.09	4.62	0.18	4.45	nr	**9.07**
100 mm dia.	5.31	6.00	0.18	4.45	nr	**10.45**
125 mm dia.	6.10	6.88	0.18	4.45	nr	**11.33**
150 mm dia.	7.19	8.12	0.18	4.45	nr	**12.57**
200 mm dia.	10.03	11.33	0.22	5.44	nr	**16.77**
250 mm dia.	11.94	13.48	0.22	5.44	nr	**18.92**
300 mm dia.	12.66	14.30	0.22	5.44	nr	**19.74**
350 mm dia.	13.89	15.69	0.22	5.44	nr	**21.13**

T: MECHANICAL/COOLING/HEATING SYSTEMS

Item	Net Price £	Material £	Labour hours	Labour £	Unit	Total rate £
T31: LOW TEMPERATURE HOT WATER HEATING – cont						
Y50 – THERMAL INSULATION – cont						
Extra over for fittings concealed insulation – cont						
Valves						
15 mm dia.	4.75	5.37	0.15	3.70	nr	**9.07**
20 mm dia.	4.75	5.37	0.15	3.70	nr	**9.07**
25 mm dia.	5.03	5.68	0.15	3.70	nr	**9.38**
32 mm dia.	5.48	6.19	0.15	3.70	nr	**9.89**
40 mm dia.	5.81	6.56	0.15	3.70	nr	**10.26**
50 mm dia.	6.64	7.50	0.15	3.70	nr	**11.20**
65 mm dia.	7.51	8.49	0.15	3.70	nr	**12.19**
80 mm dia.	8.21	9.27	0.20	4.95	nr	**14.22**
100 mm dia.	10.60	11.97	0.20	4.95	nr	**16.92**
125 mm dia.	12.22	13.80	0.20	4.95	nr	**18.75**
150 mm dia.	14.35	16.21	0.20	4.95	nr	**21.16**
200 mm dia.	20.05	22.64	0.25	6.18	nr	**28.82**
250 mm dia.	23.85	26.94	0.25	6.18	nr	**33.12**
300 mm dia.	25.28	28.55	0.25	6.18	nr	**34.73**
350 mm dia.	27.78	31.38	0.25	6.18	nr	**37.56**
Expansion bellows						
15 mm dia.	8.89	10.04	0.22	5.44	nr	**15.48**
20 mm dia.	9.49	10.72	0.22	5.44	nr	**16.16**
25 mm dia.	10.06	11.36	0.22	5.44	nr	**16.80**
32 mm dia.	10.99	12.41	0.22	5.44	nr	**17.85**
40 mm dia.	11.61	13.12	0.22	5.44	nr	**18.56**
50 mm dia.	13.28	14.99	0.22	5.44	nr	**20.43**
65 mm dia.	15.03	16.97	0.22	5.44	nr	**22.41**
80 mm dia.	16.40	18.52	0.29	7.17	nr	**25.69**
100 mm dia.	21.23	23.98	0.29	7.17	nr	**31.15**
125 mm dia.	24.42	27.58	0.29	7.17	nr	**34.75**
150 mm dia.	28.69	32.41	0.29	7.17	nr	**39.58**
200 mm dia.	40.11	45.30	0.36	8.90	nr	**54.20**
250 mm dia.	47.71	53.89	0.36	8.90	nr	**62.79**
300 mm dia.	50.59	57.14	0.36	8.90	nr	**66.04**
350 mm dia.	55.55	62.74	0.36	8.90	nr	**71.64**
40 mm thick						
15 mm dia.	5.68	6.42	0.15	3.70	m	**10.12**
20 mm dia.	5.86	6.62	0.15	3.70	m	**10.32**
25 mm dia.	6.31	7.13	0.15	3.70	m	**10.83**
32 mm dia.	6.73	7.60	0.15	3.70	m	**11.30**
40 mm dia.	7.06	7.98	0.15	3.70	m	**11.68**
50 mm dia.	7.99	9.03	0.15	3.70	m	**12.73**
65 mm dia.	8.97	10.13	0.15	3.70	m	**13.83**
80 mm dia.	9.75	11.02	0.22	5.44	m	**16.46**
100 mm dia.	12.67	14.31	0.22	5.44	m	**19.75**

T: MECHANICAL/COOLING/HEATING SYSTEMS

Item	Net Price £	Material £	Labour hours	Labour £	Unit	**Total rate £**
125 mm dia.	14.32	16.18	0.22	5.44	m	**21.62**
150 mm dia.	16.70	18.86	0.22	5.44	m	**24.30**
200 mm dia.	23.08	26.07	0.25	6.18	m	**32.25**
250 mm dia.	27.05	30.55	0.25	6.18	m	**36.73**
300 mm dia.	28.92	32.66	0.25	6.18	m	**38.84**
350 mm dia.	31.98	36.12	0.25	6.18	m	**42.30**
400 mm dia.	35.64	40.25	0.25	6.18	m	**46.43**
Extra over for fittings concealed insulation						
Flange/union						
15 mm dia.	2.85	3.22	0.13	3.20	nr	**6.42**
20 mm dia.	2.93	3.30	0.13	3.20	nr	**6.50**
25 mm dia.	3.15	3.56	0.13	3.20	nr	**6.76**
32 mm dia.	3.35	3.78	0.13	3.20	nr	**6.98**
40 mm dia.	3.53	3.99	0.13	3.20	nr	**7.19**
50 mm dia.	4.00	4.52	0.13	3.20	nr	**7.72**
65 mm dia.	4.48	5.06	0.13	3.20	nr	**8.26**
80 mm dia.	4.89	5.52	0.18	4.45	nr	**9.97**
100 mm dia.	6.33	7.15	0.18	4.45	nr	**11.60**
125 mm dia.	7.14	8.07	0.18	4.45	nr	**12.52**
150 mm dia.	8.36	9.45	0.18	4.45	nr	**13.90**
200 mm dia.	11.55	13.05	0.22	5.44	nr	**18.49**
250 mm dia.	13.52	15.27	0.22	5.44	nr	**20.71**
300 mm dia.	14.45	16.32	0.22	5.44	nr	**21.76**
350 mm dia.	16.00	18.07	0.22	5.44	nr	**23.51**
400 mm dia.	17.84	20.15	0.22	5.44	nr	**25.59**
Valves						
15 mm dia.	5.68	6.42	0.15	3.70	nr	**10.12**
20 mm dia.	5.86	6.62	0.15	3.70	nr	**10.32**
25 mm dia.	6.31	7.13	0.15	3.70	nr	**10.83**
32 mm dia.	6.73	7.60	0.15	3.70	nr	**11.30**
40 mm dia.	7.06	7.98	0.15	3.70	nr	**11.68**
50 mm dia.	7.99	9.03	0.15	3.70	nr	**12.73**
65 mm dia.	8.97	10.13	0.15	3.70	nr	**13.83**
80 mm dia.	9.75	11.02	0.20	4.95	nr	**15.97**
100 mm dia.	12.67	14.31	0.20	4.95	nr	**19.26**
125 mm dia.	14.32	16.18	0.20	4.95	nr	**21.13**
150 mm dia.	16.70	18.86	0.20	4.95	nr	**23.81**
200 mm dia.	23.08	26.07	0.25	6.18	nr	**32.25**
250 mm dia.	27.05	30.55	0.25	6.18	nr	**36.73**
300 mm dia.	28.92	32.66	0.25	6.18	nr	**38.84**
350 mm dia.	31.98	36.12	0.25	6.18	nr	**42.30**
400 mm dia.	35.64	40.25	0.25	6.18	nr	**46.43**
Expansion bellows						
15 mm dia.	11.39	12.86	0.22	5.44	nr	**18.30**
20 mm dia.	11.71	13.23	0.22	5.44	nr	**18.67**
25 mm dia.	12.62	14.25	0.22	5.44	nr	**19.69**

T: MECHANICAL/COOLING/HEATING SYSTEMS

Item	Net Price £	Material £	Labour hours	Labour £	Unit	**Total rate £**
T31: LOW TEMPERATURE HOT WATER HEATING – cont						
Y50 – THERMAL INSULATION – cont						
Extra over for fittings concealed insulation – cont						
Expansion bellows – cont						
32 mm dia.	13.44	15.18	0.22	5.44	nr	**20.62**
40 mm dia.	14.14	15.97	0.22	5.44	nr	**21.41**
50 mm dia.	16.00	18.07	0.22	5.44	nr	**23.51**
65 mm dia.	17.92	20.24	0.22	5.44	nr	**25.68**
80 mm dia.	19.53	22.05	0.29	7.17	nr	**29.22**
100 mm dia.	25.32	28.60	0.29	7.17	nr	**35.77**
125 mm dia.	28.62	32.32	0.29	7.17	nr	**39.49**
150 mm dia.	33.42	37.74	0.29	7.17	nr	**44.91**
200 mm dia.	46.16	52.13	0.36	8.90	nr	**61.03**
250 mm dia.	54.10	61.10	0.36	8.90	nr	**70.00**
300 mm dia.	57.84	65.32	0.36	8.90	nr	**74.22**
350 mm dia.	31.98	36.12	0.36	8.90	nr	**45.02**
400 mm dia.	71.30	80.53	0.36	8.90	nr	**89.43**
50 mm thick						
15 mm dia.	7.91	8.94	0.15	3.70	m	**12.64**
20 mm dia.	8.32	9.39	0.15	3.70	m	**13.09**
25 mm dia.	8.84	9.99	0.15	3.70	m	**13.69**
32 mm dia.	9.24	10.43	0.15	3.70	m	**14.13**
40 mm dia.	9.72	10.98	0.15	3.70	m	**14.68**
50 mm dia.	10.90	12.31	0.15	3.70	m	**16.01**
65 mm dia.	11.92	13.46	0.15	3.70	m	**17.16**
80 mm dia.	12.75	14.40	0.22	5.44	m	**19.84**
100 mm dia.	16.30	18.41	0.22	5.44	m	**23.85**
125 mm dia.	18.29	20.65	0.22	5.44	m	**26.09**
150 mm dia.	21.12	23.86	0.22	5.44	m	**29.30**
200 mm dia.	28.82	32.55	0.25	6.18	m	**38.73**
250 mm dia.	33.29	37.60	0.25	6.18	m	**43.78**
300 mm dia.	35.28	39.84	0.25	6.18	m	**46.02**
350 mm dia.	38.93	43.97	0.25	6.18	m	**50.15**
400 mm dia.	43.18	48.77	0.25	6.18	m	**54.95**
Extra over for fittings concealed insulation						
Flange/union						
15 mm dia.	3.96	4.47	0.13	3.20	nr	**7.67**
20 mm dia.	4.16	4.70	0.13	3.20	nr	**7.90**
25 mm dia.	4.43	5.01	0.13	3.20	nr	**8.21**
32 mm dia.	4.63	5.23	0.13	3.20	nr	**8.43**
40 mm dia.	4.87	5.50	0.13	3.20	nr	**8.70**
50 mm dia.	5.44	6.14	0.13	3.20	nr	**9.34**
65 mm dia.	5.95	6.72	0.13	3.20	nr	**9.92**

T: MECHANICAL/COOLING/HEATING SYSTEMS

Item	Net Price £	Material £	Labour hours	Labour £	Unit	**Total rate £**
80 mm dia.	6.38	7.20	0.18	4.45	nr	**11.65**
100 mm dia.	8.16	9.22	0.18	4.45	nr	**13.67**
125 mm dia.	9.15	10.33	0.18	4.45	nr	**14.78**
150 mm dia.	10.56	11.92	0.18	4.45	nr	**16.37**
200 mm dia.	14.42	16.29	0.22	5.44	nr	**21.73**
250 mm dia.	16.65	18.81	0.22	5.44	nr	**24.25**
300 mm dia.	17.64	19.92	0.22	5.44	nr	**25.36**
350 mm dia.	19.48	22.00	0.22	5.44	nr	**27.44**
400 mm dia.	21.60	24.40	0.22	5.44	nr	**29.84**
Valves						
15 mm dia.	7.91	8.94	0.15	3.70	nr	**12.64**
20 mm dia.	8.32	9.39	0.15	3.70	nr	**13.09**
25 mm dia.	8.84	9.99	0.15	3.70	nr	**13.69**
32 mm dia.	9.24	10.43	0.15	3.70	nr	**14.13**
40 mm dia.	9.72	10.98	0.15	3.70	nr	**14.68**
50 mm dia.	10.90	12.31	0.15	3.70	nr	**16.01**
65 mm dia.	11.92	13.46	0.15	3.70	nr	**17.16**
80 mm dia.	12.75	14.40	0.20	4.95	nr	**19.35**
100 mm dia.	16.30	18.41	0.20	4.95	nr	**23.36**
125 mm dia.	18.29	20.65	0.20	4.95	nr	**25.60**
150 mm dia.	21.12	23.86	0.20	4.95	nr	**28.81**
200 mm dia.	28.82	32.55	0.25	6.18	nr	**38.73**
250 mm dia.	33.29	37.60	0.25	6.18	nr	**43.78**
300 mm dia.	35.28	39.84	0.25	6.18	nr	**46.02**
350 mm dia.	38.93	43.97	0.25	6.18	nr	**50.15**
400 mm dia.	43.18	48.77	0.25	6.18	nr	**54.95**
Expansion bellows						
15 mm dia.	15.79	17.83	0.22	5.44	nr	**23.27**
20 mm dia.	16.64	18.79	0.22	5.44	nr	**24.23**
25 mm dia.	17.68	19.97	0.22	5.44	nr	**25.41**
32 mm dia.	18.50	20.89	0.22	5.44	nr	**26.33**
40 mm dia.	19.44	21.95	0.22	5.44	nr	**27.39**
50 mm dia.	21.79	24.61	0.22	5.44	nr	**30.05**
65 mm dia.	23.81	26.89	0.22	5.44	nr	**32.33**
80 mm dia.	25.50	28.79	0.29	7.17	nr	**35.96**
100 mm dia.	32.59	36.81	0.29	7.17	nr	**43.98**
125 mm dia.	36.58	41.31	0.29	7.17	nr	**48.48**
150 mm dia.	42.24	47.71	0.29	7.17	nr	**54.88**
200 mm dia.	57.68	65.15	0.36	8.90	nr	**74.05**
250 mm dia.	66.58	75.19	0.36	8.90	nr	**84.09**
300 mm dia.	70.58	79.71	0.36	8.90	nr	**88.61**
350 mm dia.	77.89	87.96	0.36	8.90	nr	**96.86**
400 mm dia.	86.36	97.53	0.36	8.90	nr	**106.43**

T: MECHANICAL/COOLING/HEATING SYSTEMS

Item	Net Price £	Material £	Labour hours	Labour £	Unit	Total rate £
T31: LOW TEMPERATURE HOT WATER HEATING – cont						
Y50 – THERMAL INSULATION – cont						
Mineral fibre sectional insulation; bright class O foil faced; bright class O foil taped joints; 22 swg plain/embossed aluminium cladding; pop riveted						
Plantroom pipework						
20 mm thick						
15 mm dia.	5.07	5.72	0.44	10.87	m	**16.59**
20 mm dia.	5.36	6.05	0.44	10.87	m	**16.92**
25 mm dia.	5.73	6.47	0.44	10.87	m	**17.34**
32 mm dia.	6.21	7.02	0.44	10.87	m	**17.89**
40 mm dia.	6.55	7.39	0.44	10.87	m	**18.26**
50 mm dia.	7.29	8.23	0.44	10.87	m	**19.10**
Extra over for fittings plantroom insulation						
Flange/union						
15 mm dia.	5.66	6.40	0.58	14.33	nr	**20.73**
20 mm dia.	6.01	6.78	0.58	14.33	nr	**21.11**
25 mm dia.	6.43	7.26	0.58	14.33	nr	**21.59**
32 mm dia.	7.04	7.96	0.58	14.33	nr	**22.29**
40 mm dia.	7.43	8.39	0.58	14.33	nr	**22.72**
50 mm dia.	8.34	9.41	0.58	14.33	nr	**23.74**
Bends						
15 mm dia.	2.78	3.14	0.44	10.87	nr	**14.01**
20 mm dia.	2.96	3.35	0.44	10.87	nr	**14.22**
25 mm dia.	3.16	3.57	0.44	10.87	nr	**14.44**
32 mm dia.	3.42	3.87	0.44	10.87	nr	**14.74**
40 mm dia.	3.59	4.06	0.44	10.87	nr	**14.93**
50 mm dia.	4.01	4.53	0.44	10.87	nr	**15.40**
Tees						
15 mm dia.	1.67	1.89	0.44	10.87	nr	**12.76**
20 mm dia.	1.77	2.00	0.44	10.87	nr	**12.87**
25 mm dia.	1.89	2.13	0.44	10.87	nr	**13.00**
32 mm dia.	2.05	2.32	0.44	10.87	nr	**13.19**
40 mm dia.	2.16	2.44	0.44	10.87	nr	**13.31**
50 mm dia.	2.40	2.71	0.44	10.87	nr	**13.58**
Valves						
15 mm dia.	3.11	3.51	0.78	19.28	nr	**22.79**
20 mm dia.	3.55	4.01	0.78	19.28	nr	**23.29**
25 mm dia.	3.55	4.01	0.78	19.28	nr	**23.29**
32 mm dia.	3.97	4.49	0.78	19.28	nr	**23.77**
40 mm dia.	4.25	4.80	0.78	19.28	nr	**24.08**
50 mm dia.	4.85	5.48	0.78	19.28	nr	**24.76**

T: MECHANICAL/COOLING/HEATING SYSTEMS

Item	Net Price £	Material £	Labour hours	Labour £	Unit	**Total rate £**
Pumps						
15 mm dia.	16.65	18.81	2.34	57.82	nr	**76.63**
20 mm dia.	17.66	19.94	2.34	57.82	nr	**77.76**
25 mm dia.	18.96	21.41	2.34	57.82	nr	**79.23**
32 mm dia.	20.68	23.36	2.34	57.82	nr	**81.18**
40 mm dia.	21.85	24.67	2.34	57.82	nr	**82.49**
50 mm dia.	24.56	27.73	2.34	57.82	nr	**85.55**
Expansion bellows						
15 mm dia.	13.31	15.03	1.05	25.94	nr	**40.97**
20 mm dia.	14.12	15.94	1.05	25.94	nr	**41.88**
25 mm dia.	15.17	17.14	1.05	25.94	nr	**43.08**
32 mm dia.	16.53	18.67	1.05	25.94	nr	**44.61**
40 mm dia.	17.47	19.73	1.05	25.94	nr	**45.67**
50 mm dia.	19.64	22.19	1.05	25.94	nr	**48.13**
25 mm thick						
15 mm dia.	5.56	6.28	0.44	10.87	m	**17.15**
20 mm dia.	6.01	6.78	0.44	10.87	m	**17.65**
25 mm dia.	6.59	7.45	0.44	10.87	m	**18.32**
32 mm dia.	7.02	7.93	0.44	10.87	m	**18.80**
40 mm dia.	7.54	8.52	0.44	10.87	m	**19.39**
50 mm dia.	8.45	9.55	0.44	10.87	m	**20.42**
65 mm dia.	9.55	10.78	0.44	10.87	m	**21.65**
80 mm dia.	10.36	11.70	0.52	12.85	m	**24.55**
100 mm dia.	12.86	14.52	0.52	12.85	m	**27.37**
125 mm dia.	14.92	16.85	0.52	12.85	m	**29.70**
150 mm dia.	17.38	19.62	0.52	12.85	m	**32.47**
200 mm dia.	23.73	26.81	0.60	14.83	m	**41.64**
250 mm dia.	28.19	31.83	0.60	14.83	m	**46.66**
300 mm dia.	31.57	35.66	0.60	14.83	m	**50.49**
Extra over for fittings plantroom insulation						
Flange/union						
15 mm dia.	6.21	7.02	0.58	14.33	nr	**21.35**
20 mm dia.	6.73	7.60	0.58	14.33	nr	**21.93**
25 mm dia.	7.42	8.38	0.58	14.33	nr	**22.71**
32 mm dia.	7.96	8.99	0.58	14.33	nr	**23.32**
40 mm dia.	8.54	9.65	0.58	14.33	nr	**23.98**
50 mm dia.	9.63	10.87	0.58	14.33	nr	**25.20**
65 mm dia.	10.92	12.33	0.58	14.33	nr	**26.66**
80 mm dia.	11.87	13.40	0.67	16.55	nr	**29.95**
100 mm dia.	15.03	16.97	0.67	16.55	nr	**33.52**
125 mm dia.	17.45	19.71	0.67	16.55	nr	**36.26**
150 mm dia.	20.46	23.10	0.67	16.55	nr	**39.65**
200 mm dia.	28.26	31.92	0.87	21.50	nr	**53.42**
250 mm dia.	33.67	38.03	0.87	21.50	nr	**59.53**
300 mm dia.	37.21	42.02	0.87	21.50	nr	**63.52**

T: MECHANICAL/COOLING/HEATING SYSTEMS

Item	Net Price £	Material £	Labour hours	Labour £	Unit	**Total rate £**
T31: LOW TEMPERATURE HOT WATER HEATING – cont						
Y50 – THERMAL INSULATION – cont						
Extra over for fittings plantroom insulation – cont						
Bends						
15 mm dia.	3.05	3.45	0.44	10.87	nr	**14.32**
20 mm dia.	3.31	3.73	0.44	10.87	nr	**14.60**
25 mm dia.	3.62	4.09	0.44	10.87	nr	**14.96**
32 mm dia.	3.87	4.38	0.44	10.87	nr	**15.25**
40 mm dia.	4.16	4.70	0.44	10.87	nr	**15.57**
50 mm dia.	4.65	5.25	0.44	10.87	nr	**16.12**
65 mm dia.	5.25	5.93	0.44	10.87	nr	**16.80**
80 mm dia.	5.69	6.43	0.52	12.85	nr	**19.28**
100 mm dia.	7.06	7.98	0.52	12.85	nr	**20.83**
125 mm dia.	8.22	9.28	0.52	12.85	nr	**22.13**
150 mm dia.	9.55	10.78	0.52	12.85	nr	**23.63**
200 mm dia.	13.05	14.74	0.60	14.83	nr	**29.57**
250 mm dia.	15.51	17.51	0.60	14.83	nr	**32.34**
300 mm dia.	17.36	19.60	0.60	14.83	nr	**34.43**
Tees						
15 mm dia.	1.84	2.08	0.44	10.87	nr	**12.95**
20 mm dia.	1.97	2.22	0.44	10.87	nr	**13.09**
25 mm dia.	2.17	2.45	0.44	10.87	nr	**13.32**
32 mm dia.	2.32	2.62	0.44	10.87	nr	**13.49**
40 mm dia.	2.49	2.82	0.44	10.87	nr	**13.69**
50 mm dia.	2.78	3.14	0.44	10.87	nr	**14.01**
65 mm dia.	3.15	3.56	0.44	10.87	nr	**14.43**
80 mm dia.	3.42	3.87	0.52	12.85	nr	**16.72**
100 mm dia.	4.24	4.78	0.52	12.85	nr	**17.63**
125 mm dia.	4.93	5.57	0.52	12.85	nr	**18.42**
150 mm dia.	5.73	6.47	0.52	12.85	nr	**19.32**
200 mm dia.	7.82	8.83	0.60	14.83	nr	**23.66**
250 mm dia.	9.32	10.53	0.60	14.83	nr	**25.36**
300 mm dia.	10.43	11.78	0.60	14.83	nr	**26.61**
Valves						
15 mm dia.	9.88	11.16	0.78	19.28	nr	**30.44**
20 mm dia.	10.67	12.05	0.78	19.28	nr	**31.33**
25 mm dia.	11.78	13.30	0.78	19.28	nr	**32.58**
32 mm dia.	12.64	14.28	0.78	19.28	nr	**33.56**
40 mm dia.	13.56	15.31	0.78	19.28	nr	**34.59**
50 mm dia.	15.28	17.26	0.78	19.28	nr	**36.54**
65 mm dia.	17.35	19.59	0.78	19.28	nr	**38.87**
80 mm dia.	18.87	21.31	0.92	22.74	nr	**44.05**
100 mm dia.	23.87	26.96	0.92	22.74	nr	**49.70**
125 mm dia.	27.73	31.31	0.92	22.74	nr	**54.05**
150 mm dia.	32.49	36.69	0.92	22.74	nr	**59.43**

T: MECHANICAL/COOLING/HEATING SYSTEMS

Item	Net Price £	Material £	Labour hours	Labour £	Unit	**Total rate £**
200 mm dia.	44.90	50.71	1.12	27.68	nr	**78.39**
250 mm dia.	53.48	60.40	1.12	27.68	nr	**88.08**
300 mm dia.	59.09	66.74	1.12	27.68	nr	**94.42**
Pumps						
15 mm dia.	18.28	20.64	2.34	57.82	nr	**78.46**
20 mm dia.	19.76	22.32	2.34	57.82	nr	**80.14**
25 mm dia.	21.83	24.65	2.34	57.82	nr	**82.47**
32 mm dia.	23.40	26.43	2.34	57.82	nr	**84.25**
40 mm dia.	25.12	28.37	2.34	57.82	nr	**86.19**
50 mm dia.	28.32	31.99	2.34	57.82	nr	**89.81**
65 mm dia.	32.12	36.27	2.34	57.82	nr	**94.09**
80 mm dia.	34.94	39.46	2.76	68.21	nr	**107.67**
100 mm dia.	44.20	49.92	2.76	68.21	nr	**118.13**
125 mm dia.	51.33	57.98	2.76	68.21	nr	**126.19**
150 mm dia.	60.16	67.94	2.76	68.21	nr	**136.15**
200 mm dia.	83.13	93.89	3.36	83.04	nr	**176.93**
250 mm dia.	99.04	111.85	3.36	83.04	nr	**194.89**
300 mm dia.	109.41	123.56	3.36	83.04	nr	**206.60**
Expansion bellows						
15 mm dia.	14.62	16.51	1.05	25.94	nr	**42.45**
20 mm dia.	15.81	17.86	1.05	25.94	nr	**43.80**
25 mm dia.	17.45	19.71	1.05	25.94	nr	**45.65**
32 mm dia.	18.73	21.15	1.05	25.94	nr	**47.09**
40 mm dia.	20.10	22.71	1.05	25.94	nr	**48.65**
50 mm dia.	22.66	25.59	1.05	25.94	nr	**51.53**
65 mm dia.	25.69	29.02	1.05	25.94	nr	**54.96**
80 mm dia.	27.96	31.58	1.26	31.14	nr	**62.72**
100 mm dia.	35.36	39.93	1.26	31.14	nr	**71.07**
125 mm dia.	41.05	46.36	1.26	31.14	nr	**77.50**
150 mm dia.	48.11	54.34	1.26	31.14	nr	**85.48**
200 mm dia.	66.52	75.12	1.53	37.81	nr	**112.93**
250 mm dia.	79.24	89.49	1.53	37.81	nr	**127.30**
300 mm dia.	87.53	98.86	1.53	37.81	nr	**136.67**
30 mm thick						
15 mm dia.	6.85	7.73	0.44	10.87	m	**18.60**
20 mm dia.	7.25	8.19	0.44	10.87	m	**19.06**
25 mm dia.	7.68	8.67	0.44	10.87	m	**19.54**
32 mm dia.	8.41	9.50	0.44	10.87	m	**20.37**
40 mm dia.	8.76	9.89	0.44	10.87	m	**20.76**
50 mm dia.	9.72	10.98	0.44	10.87	m	**21.85**
65 mm dia.	10.96	12.38	0.44	10.87	m	**23.25**
80 mm dia.	11.98	13.53	0.52	12.85	m	**26.38**
100 mm dia.	14.72	16.63	0.52	12.85	m	**29.48**
125 mm dia.	16.82	18.99	0.52	12.85	m	**31.84**
150 mm dia.	19.61	22.14	0.52	12.85	m	**34.99**
200 mm dia.	26.34	29.74	0.60	14.83	m	**44.57**
250 mm dia.	31.16	35.19	0.60	14.83	m	**50.02**
300 mm dia.	34.31	38.75	0.60	14.83	m	**53.58**
350 mm dia.	38.20	43.15	0.60	14.83	m	**57.98**

T: MECHANICAL/COOLING/HEATING SYSTEMS

Item	Net Price £	Material £	Labour hours	Labour £	Unit	Total rate £
T31: LOW TEMPERATURE HOT WATER HEATING – cont						
Y50 – THERMAL INSULATION – cont						
Extra over for fittings plantroom insulation						
Flange/union						
15 mm dia.	7.79	8.80	0.58	14.33	nr	**23.13**
20 mm dia.	8.27	9.34	0.58	14.33	nr	**23.67**
25 mm dia.	8.77	9.90	0.58	14.33	nr	**24.23**
32 mm dia.	9.61	10.85	0.58	14.33	nr	**25.18**
40 mm dia.	10.04	11.34	0.58	14.33	nr	**25.67**
50 mm dia.	11.25	12.71	0.58	14.33	nr	**27.04**
65 mm dia.	12.71	14.35	0.58	14.33	nr	**28.68**
80 mm dia.	13.89	15.69	0.67	16.55	nr	**32.24**
100 mm dia.	17.35	19.59	0.67	16.55	nr	**36.14**
125 mm dia.	19.85	22.42	0.67	16.55	nr	**38.97**
150 mm dia.	23.23	26.23	0.67	16.55	nr	**42.78**
200 mm dia.	31.62	35.71	0.87	21.50	nr	**57.21**
250 mm dia.	37.50	42.35	0.87	21.50	nr	**63.85**
300 mm dia.	40.79	46.06	0.87	21.50	nr	**67.56**
350 mm dia.	45.25	51.10	0.87	21.50	nr	**72.60**
Bends						
15 mm dia.	3.77	4.25	0.44	10.87	nr	**15.12**
20 mm dia.	3.98	4.50	0.44	10.87	nr	**15.37**
25 mm dia.	4.23	4.77	0.44	10.87	nr	**15.64**
32 mm dia.	4.62	5.22	0.44	10.87	nr	**16.09**
40 mm dia.	4.81	5.44	0.44	10.87	nr	**16.31**
50 mm dia.	5.36	6.05	0.44	10.87	nr	**16.92**
65 mm dia.	6.03	6.81	0.44	10.87	nr	**17.68**
80 mm dia.	6.58	7.44	0.52	12.85	nr	**20.29**
100 mm dia.	8.09	9.14	0.52	12.85	nr	**21.99**
125 mm dia.	9.25	10.44	0.52	12.85	nr	**23.29**
150 mm dia.	10.80	12.20	0.52	12.85	nr	**25.05**
200 mm dia.	14.49	16.36	0.60	14.83	nr	**31.19**
250 mm dia.	17.15	19.37	0.60	14.83	nr	**34.20**
300 mm dia.	18.87	21.31	0.60	14.83	nr	**36.14**
350 mm dia.	21.02	23.74	0.60	14.83	nr	**38.57**
Tees						
15 mm dia.	2.26	2.55	0.44	10.87	nr	**13.42**
20 mm dia.	2.39	2.70	0.44	10.87	nr	**13.57**
25 mm dia.	2.53	2.86	0.44	10.87	nr	**13.73**
32 mm dia.	2.77	3.13	0.44	10.87	nr	**14.00**
40 mm dia.	2.89	3.26	0.44	10.87	nr	**14.13**
50 mm dia.	3.21	3.62	0.44	10.87	nr	**14.49**
65 mm dia.	3.61	4.08	0.44	10.87	nr	**14.95**
80 mm dia.	3.96	4.47	0.52	12.85	nr	**17.32**
100 mm dia.	4.87	5.50	0.52	12.85	nr	**18.35**
125 mm dia.	5.56	6.28	0.52	12.85	nr	**19.13**
150 mm dia.	6.48	7.31	0.52	12.85	nr	**20.16**

T: MECHANICAL/COOLING/HEATING SYSTEMS

Item	Net Price £	Material £	Labour hours	Labour £	Unit	**Total rate £**
200 mm dia.	8.70	9.82	0.60	14.83	nr	**24.65**
250 mm dia.	10.29	11.62	0.60	14.83	nr	**26.45**
300 mm dia.	11.32	12.78	0.60	14.83	nr	**27.61**
350 mm dia.	12.62	14.25	0.60	14.83	nr	**29.08**
Valves						
15 mm dia.	12.39	13.99	0.78	19.28	nr	**33.27**
20 mm dia.	13.14	14.84	0.78	19.28	nr	**34.12**
25 mm dia.	13.92	15.72	0.78	19.28	nr	**35.00**
32 mm dia.	15.26	17.24	0.78	19.28	nr	**36.52**
40 mm dia.	15.96	18.02	0.78	19.28	nr	**37.30**
50 mm dia.	17.87	20.19	0.78	19.28	nr	**39.47**
65 mm dia.	20.20	22.82	0.78	19.28	nr	**42.10**
80 mm dia.	22.05	24.90	0.92	22.74	nr	**47.64**
100 mm dia.	27.54	31.10	0.92	22.74	nr	**53.84**
125 mm dia.	31.54	35.62	0.92	22.74	nr	**58.36**
150 mm dia.	36.88	41.66	0.92	22.74	nr	**64.40**
200 mm dia.	50.22	56.72	1.12	27.68	nr	**84.40**
250 mm dia.	59.55	67.26	1.12	27.68	nr	**94.94**
300 mm dia.	64.78	73.16	1.12	27.68	nr	**100.84**
350 mm dia.	71.88	81.18	1.12	27.68	nr	**108.86**
Pumps						
15 mm dia.	22.94	25.91	2.34	57.82	nr	**83.73**
20 mm dia.	24.34	27.49	2.34	57.82	nr	**85.31**
25 mm dia.	25.78	29.11	2.34	57.82	nr	**86.93**
32 mm dia.	28.25	31.91	2.34	57.82	nr	**89.73**
40 mm dia.	29.55	33.37	2.34	57.82	nr	**91.19**
50 mm dia.	33.10	37.38	2.34	57.82	nr	**95.20**
65 mm dia.	37.39	42.23	2.34	57.82	nr	**100.05**
80 mm dia.	40.83	46.11	2.76	68.21	nr	**114.32**
100 mm dia.	51.01	57.61	2.76	68.21	nr	**125.82**
125 mm dia.	58.42	65.98	2.76	68.21	nr	**134.19**
150 mm dia.	68.30	77.14	2.76	68.21	nr	**145.35**
200 mm dia.	93.01	105.05	3.36	83.04	nr	**188.09**
250 mm dia.	110.30	124.57	3.36	83.04	nr	**207.61**
300 mm dia.	119.99	135.52	3.36	83.04	nr	**218.56**
350 mm dia.	133.10	150.32	3.36	83.04	nr	**233.36**
Expansion bellows						
15 mm dia.	18.35	20.73	1.05	25.94	nr	**46.67**
20 mm dia.	19.47	21.99	1.05	25.94	nr	**47.93**
25 mm dia.	20.63	23.30	1.05	25.94	nr	**49.24**
32 mm dia.	22.60	25.52	1.05	25.94	nr	**51.46**
40 mm dia.	23.64	26.70	1.05	25.94	nr	**52.64**
50 mm dia.	26.47	29.90	1.05	25.94	nr	**55.84**
65 mm dia.	29.92	33.79	1.05	25.94	nr	**59.73**
80 mm dia.	32.68	36.90	1.26	31.14	nr	**68.04**
100 mm dia.	40.80	46.08	1.26	31.14	nr	**77.22**
125 mm dia.	46.75	52.80	1.26	31.14	nr	**83.94**
150 mm dia.	54.65	61.72	1.26	31.14	nr	**92.86**

T: MECHANICAL/COOLING/HEATING SYSTEMS

Item	Net Price £	Material £	Labour hours	Labour £	Unit	Total rate £
T31: LOW TEMPERATURE HOT WATER HEATING – cont						
Y50 – THERMAL INSULATION – cont						
Extra over for fittings plantroom insulation – cont						
Expansion bellows – cont						
200 mm dia.	74.41	84.04	1.53	37.81	nr	**121.85**
250 mm dia.	88.22	99.63	1.53	37.81	nr	**137.44**
300 mm dia.	95.99	108.41	1.53	37.81	nr	**146.22**
350 mm dia.	106.48	120.26	1.53	37.81	nr	**158.07**
40 mm thick						
15 mm dia.	8.39	9.48	0.44	10.87	m	**20.35**
20 mm dia.	8.80	9.93	0.44	10.87	m	**20.80**
25 mm dia.	9.37	10.58	0.44	10.87	m	**21.45**
32 mm dia.	9.82	11.09	0.44	10.87	m	**21.96**
40 mm dia.	10.44	11.79	0.44	10.87	m	**22.66**
50 mm dia.	11.56	13.06	0.44	10.87	m	**23.93**
65 mm dia.	12.88	14.55	0.44	10.87	m	**25.42**
80 mm dia.	13.86	15.66	0.52	12.85	m	**28.51**
100 mm dia.	20.29	22.92	0.52	12.85	m	**35.77**
125 mm dia.	19.01	21.47	0.52	12.85	m	**34.32**
150 mm dia.	22.17	25.04	0.52	12.85	m	**37.89**
200 mm dia.	29.31	33.10	0.60	14.83	m	**47.93**
250 mm dia.	34.35	38.79	0.60	14.83	m	**53.62**
300 mm dia.	37.98	42.89	0.60	14.83	m	**57.72**
350 mm dia.	42.44	47.93	0.60	14.83	m	**62.76**
400 mm dia.	47.72	53.90	0.60	14.83	m	**68.73**
Extra over for fittings plantroom insulation						
Flange/union						
15 mm dia.	9.67	10.92	0.58	14.33	nr	**25.25**
20 mm dia.	10.10	11.40	0.58	14.33	nr	**25.73**
25 mm dia.	10.81	12.21	0.58	14.33	nr	**26.54**
32 mm dia.	11.37	12.84	0.58	14.33	nr	**27.17**
40 mm dia.	12.06	13.62	0.58	14.33	nr	**27.95**
50 mm dia.	13.44	15.18	0.58	14.33	nr	**29.51**
65 mm dia.	15.02	16.96	0.58	14.33	nr	**31.29**
80 mm dia.	16.20	18.30	0.67	16.55	nr	**34.85**
100 mm dia.	20.29	22.92	0.67	16.55	nr	**39.47**
125 mm dia.	22.72	25.66	0.67	16.55	nr	**42.21**
150 mm dia.	26.51	29.94	0.67	16.55	nr	**46.49**
200 mm dia.	35.57	40.17	0.87	21.50	nr	**61.67**
250 mm dia.	41.69	47.08	0.87	21.50	nr	**68.58**
300 mm dia.	45.62	51.52	0.87	21.50	nr	**73.02**
350 mm dia.	50.85	57.43	0.87	21.50	nr	**78.93**
400 mm dia.	57.01	64.38	0.87	21.50	nr	**85.88**

T: MECHANICAL/COOLING/HEATING SYSTEMS

Item	Net Price £	Material £	Labour hours	Labour £	Unit	**Total rate £**
Bends						
15 mm dia.	4.61	5.20	0.44	10.87	nr	**16.07**
20 mm dia.	4.83	5.46	0.44	10.87	nr	**16.33**
25 mm dia.	5.16	5.82	0.44	10.87	nr	**16.69**
32 mm dia.	5.39	6.09	0.44	10.87	nr	**16.96**
40 mm dia.	5.74	6.49	0.44	10.87	nr	**17.36**
50 mm dia.	6.36	7.18	0.44	10.87	nr	**18.05**
65 mm dia.	7.08	8.00	0.44	10.87	nr	**18.87**
80 mm dia.	7.61	8.60	0.52	12.85	nr	**21.45**
100 mm dia.	9.39	10.61	0.52	12.85	nr	**23.46**
125 mm dia.	10.45	11.80	0.52	12.85	nr	**24.65**
150 mm dia.	12.20	13.78	0.52	12.85	nr	**26.63**
200 mm dia.	16.11	18.20	0.60	14.83	nr	**33.03**
250 mm dia.	18.77	21.20	0.60	14.83	nr	**36.03**
300 mm dia.	20.90	23.60	0.60	14.83	nr	**38.43**
350 mm dia.	23.35	26.37	0.60	14.83	nr	**41.20**
400 mm dia.	26.25	29.65	0.60	14.83	nr	**44.48**
Tees						
15 mm dia.	2.77	3.13	0.44	10.87	nr	**14.00**
20 mm dia.	2.89	3.26	0.44	10.87	nr	**14.13**
25 mm dia.	3.09	3.49	0.44	10.87	nr	**14.36**
32 mm dia.	3.24	3.66	0.44	10.87	nr	**14.53**
40 mm dia.	3.45	3.90	0.44	10.87	nr	**14.77**
50 mm dia.	3.81	4.30	0.44	10.87	nr	**15.17**
65 mm dia.	4.25	4.80	0.44	10.87	nr	**15.67**
80 mm dia.	4.57	5.16	0.52	12.85	nr	**18.01**
100 mm dia.	5.64	6.36	0.52	12.85	nr	**19.21**
125 mm dia.	6.28	7.09	0.52	12.85	nr	**19.94**
150 mm dia.	7.32	8.27	0.52	12.85	nr	**21.12**
200 mm dia.	9.67	10.92	0.60	14.83	nr	**25.75**
250 mm dia.	11.34	12.81	0.60	14.83	nr	**27.64**
300 mm dia.	12.52	14.14	0.60	14.83	nr	**28.97**
350 mm dia.	14.01	15.82	0.60	14.83	nr	**30.65**
400 mm dia.	15.75	17.79	0.60	14.83	nr	**32.62**
Valves						
15 mm dia.	15.37	17.36	0.78	19.28	nr	**36.64**
20 mm dia.	16.04	18.12	0.78	19.28	nr	**37.40**
25 mm dia.	17.15	19.37	0.78	19.28	nr	**38.65**
32 mm dia.	18.06	20.40	0.78	19.28	nr	**39.68**
40 mm dia.	19.15	21.62	0.78	19.28	nr	**40.90**
50 mm dia.	21.34	24.10	0.78	19.28	nr	**43.38**
65 mm dia.	23.84	26.93	0.78	19.28	nr	**46.21**
80 mm dia.	25.73	29.06	0.92	22.74	nr	**51.80**
100 mm dia.	32.25	36.42	0.92	22.74	nr	**59.16**
125 mm dia.	36.06	40.73	0.92	22.74	nr	**63.47**
150 mm dia.	42.10	47.55	0.92	22.74	nr	**70.29**
200 mm dia.	56.49	63.80	1.12	27.68	nr	**91.48**
250 mm dia.	66.22	74.79	1.12	27.68	nr	**102.47**
300 mm dia.	72.46	81.83	1.12	27.68	nr	**109.51**
350 mm dia.	80.77	91.22	1.12	27.68	nr	**118.90**

T: MECHANICAL/COOLING/HEATING SYSTEMS

Item	Net Price £	Material £	Labour hours	Labour £	Unit	**Total rate £**
T31: LOW TEMPERATURE HOT WATER HEATING – cont						
Y50 – THERMAL INSULATION – cont						
Extra over for fittings plantroom insulation – cont						
Valves – cont						
400 mm dia.	90.57	102.29	1.12	27.68	nr	**129.97**
Pumps						
15 mm dia.	28.45	32.13	2.34	57.82	nr	**89.95**
20 mm dia.	29.71	33.56	2.34	57.82	nr	**91.38**
25 mm dia.	31.77	35.88	2.34	57.82	nr	**93.70**
32 mm dia.	33.43	37.76	2.34	57.82	nr	**95.58**
40 mm dia.	35.47	40.06	2.34	57.82	nr	**97.88**
50 mm dia.	39.54	44.66	2.34	57.82	nr	**102.48**
65 mm dia.	44.16	49.88	2.34	57.82	nr	**107.70**
80 mm dia.	47.65	53.82	2.76	68.21	nr	**122.03**
100 mm dia.	59.70	67.42	2.76	68.21	nr	**135.63**
125 mm dia.	66.79	75.43	2.76	68.21	nr	**143.64**
150 mm dia.	77.98	88.07	2.76	68.21	nr	**156.28**
200 mm dia.	104.60	118.14	3.36	83.04	nr	**201.18**
250 mm dia.	122.63	138.50	3.36	83.04	nr	**221.54**
300 mm dia.	175.45	198.16	3.36	83.04	nr	**281.20**
350 mm dia.	149.55	168.90	3.36	83.04	nr	**251.94**
400 mm dia.	167.71	189.41	3.36	83.04	nr	**272.45**
Expansion bellows						
15 mm dia.	22.76	25.70	1.05	25.94	nr	**51.64**
20 mm dia.	23.78	26.86	1.05	25.94	nr	**52.80**
25 mm dia.	25.41	28.70	1.05	25.94	nr	**54.64**
32 mm dia.	26.75	30.21	1.05	25.94	nr	**56.15**
40 mm dia.	28.37	32.04	1.05	25.94	nr	**57.98**
50 mm dia.	31.62	35.71	1.05	25.94	nr	**61.65**
65 mm dia.	35.32	39.89	1.05	25.94	nr	**65.83**
80 mm dia.	38.13	43.06	1.26	31.14	nr	**74.20**
100 mm dia.	47.77	53.95	1.26	31.14	nr	**85.09**
125 mm dia.	53.42	60.33	1.26	31.14	nr	**91.47**
150 mm dia.	62.37	70.44	1.26	31.14	nr	**101.58**
200 mm dia.	83.67	94.49	1.53	37.81	nr	**132.30**
250 mm dia.	98.11	110.80	1.53	37.81	nr	**148.61**
300 mm dia.	107.34	121.23	1.53	37.81	nr	**159.04**
350 mm dia.	119.63	135.11	1.53	37.81	nr	**172.92**
400 mm dia.	134.17	151.53	1.53	37.81	nr	**189.34**
50 mm thick						
15 mm dia.	10.93	12.34	0.44	10.87	m	**23.21**
20 mm dia.	11.52	13.02	0.44	10.87	m	**23.89**
25 mm dia.	12.21	13.79	0.44	10.87	m	**24.66**

T: MECHANICAL/COOLING/HEATING SYSTEMS

Item	Net Price £	Material £	Labour hours	Labour £	Unit	**Total rate £**
32 mm dia.	12.84	14.50	0.44	10.87	m	**25.37**
40 mm dia.	13.42	15.16	0.44	10.87	m	**26.03**
50 mm dia.	14.81	16.73	0.44	10.87	m	**27.60**
65 mm dia.	15.92	17.98	0.44	10.87	m	**28.85**
80 mm dia.	17.10	19.31	0.52	12.85	m	**32.16**
100 mm dia.	21.12	23.86	0.52	12.85	m	**36.71**
125 mm dia.	23.37	26.40	0.52	12.85	m	**39.25**
150 mm dia.	26.85	30.32	0.52	12.85	m	**43.17**
200 mm dia.	35.20	39.76	0.60	14.83	m	**54.59**
250 mm dia.	40.68	45.94	0.60	14.83	m	**60.77**
300 mm dia.	44.28	50.01	0.60	14.83	m	**64.84**
350 mm dia.	49.36	55.74	0.60	14.83	m	**70.57**
400 mm dia.	55.21	62.35	0.60	14.83	m	**77.18**
Extra over for fittings plantroom insulation						
Flange/union						
15 mm dia.	12.88	14.55	0.58	14.33	nr	**28.88**
20 mm dia.	13.59	15.35	0.58	14.33	nr	**29.68**
25 mm dia.	14.39	16.25	0.58	14.33	nr	**30.58**
32 mm dia.	15.11	17.06	0.58	14.33	nr	**31.39**
40 mm dia.	15.83	17.88	0.58	14.33	nr	**32.21**
50 mm dia.	17.58	19.86	0.58	14.33	nr	**34.19**
65 mm dia.	18.99	21.45	0.58	14.33	nr	**35.78**
80 mm dia.	20.35	22.98	0.67	16.55	nr	**39.53**
100 mm dia.	25.45	28.74	0.67	16.55	nr	**45.29**
125 mm dia.	28.31	31.98	0.67	16.55	nr	**48.53**
150 mm dia.	32.56	36.77	0.67	16.55	nr	**53.32**
200 mm dia.	43.29	48.89	0.87	21.50	nr	**70.39**
250 mm dia.	50.03	56.51	0.87	21.50	nr	**78.01**
300 mm dia.	53.98	60.97	0.87	21.50	nr	**82.47**
350 mm dia.	60.02	67.79	0.87	21.50	nr	**89.29**
400 mm dia.	66.96	75.62	0.87	21.50	nr	**97.12**
Bend						
15 mm dia.	6.02	6.80	0.44	10.87	nr	**17.67**
20 mm dia.	6.34	7.16	0.44	10.87	nr	**18.03**
25 mm dia.	6.71	7.58	0.44	10.87	nr	**18.45**
32 mm dia.	7.06	7.98	0.44	10.87	nr	**18.85**
40 mm dia.	7.37	8.32	0.44	10.87	nr	**19.19**
50 mm dia.	8.15	9.20	0.44	10.87	nr	**20.07**
65 mm dia.	8.76	9.89	0.44	10.87	nr	**20.76**
80 mm dia.	9.39	10.61	0.52	12.85	nr	**23.46**
100 mm dia.	11.60	13.10	0.52	12.85	nr	**25.95**
125 mm dia.	12.86	14.52	0.52	12.85	nr	**27.37**
150 mm dia.	14.77	16.68	0.52	12.85	nr	**29.53**
200 mm dia.	19.36	21.87	0.60	14.83	nr	**36.70**
250 mm dia.	22.38	25.28	0.60	14.83	nr	**40.11**
300 mm dia.	24.35	27.50	0.60	14.83	nr	**42.33**
350 mm dia.	27.14	30.65	0.60	14.83	nr	**45.48**
400 mm dia.	30.36	34.29	0.60	14.83	nr	**49.12**

T: MECHANICAL/COOLING/HEATING SYSTEMS

Item	Net Price £	Material £	Labour hours	Labour £	Unit	**Total rate £**
T31: LOW TEMPERATURE HOT WATER HEATING – cont						
Y50 – THERMAL INSULATION – cont						
Extra over for fittings plantroom insulation – cont						
Tee						
15 mm dia.	3.61	4.08	0.44	10.87	nr	**14.95**
20 mm dia.	3.80	4.29	0.44	10.87	nr	**15.16**
25 mm dia.	4.02	4.54	0.44	10.87	nr	**15.41**
32 mm dia.	4.24	4.78	0.44	10.87	nr	**15.65**
40 mm dia.	4.43	5.01	0.44	10.87	nr	**15.88**
50 mm dia.	4.89	5.52	0.44	10.87	nr	**16.39**
65 mm dia.	5.25	5.93	0.44	10.87	nr	**16.80**
80 mm dia.	5.64	6.36	0.52	12.85	nr	**19.21**
100 mm dia.	6.97	7.87	0.52	12.85	nr	**20.72**
125 mm dia.	7.71	8.71	0.52	12.85	nr	**21.56**
150 mm dia.	8.86	10.01	0.52	12.85	nr	**22.86**
200 mm dia.	11.60	13.10	0.60	14.83	nr	**27.93**
250 mm dia.	13.43	15.17	0.60	14.83	nr	**30.00**
300 mm dia.	14.62	16.51	0.60	14.83	nr	**31.34**
350 mm dia.	16.29	18.40	0.60	14.83	nr	**33.23**
400 mm dia.	18.22	20.57	0.60	14.83	nr	**35.40**
Valves						
15 mm dia.	20.47	23.12	0.78	19.28	nr	**42.40**
20 mm dia.	21.57	24.36	0.78	19.28	nr	**43.64**
25 mm dia.	22.86	25.82	0.78	19.28	nr	**45.10**
32 mm dia.	24.00	27.10	0.78	19.28	nr	**46.38**
40 mm dia.	25.14	28.40	0.78	19.28	nr	**47.68**
50 mm dia.	27.90	31.51	0.78	19.28	nr	**50.79**
65 mm dia.	30.14	34.04	0.78	19.28	nr	**53.32**
80 mm dia.	32.33	36.52	0.92	22.74	nr	**59.26**
100 mm dia.	40.42	45.66	0.92	22.74	nr	**68.40**
125 mm dia.	44.94	50.76	0.92	22.74	nr	**73.50**
150 mm dia.	51.72	58.42	0.92	22.74	nr	**81.16**
200 mm dia.	68.77	77.67	1.12	27.68	nr	**105.35**
250 mm dia.	79.47	89.75	1.12	27.68	nr	**117.43**
300 mm dia.	86.08	97.22	1.12	27.68	nr	**124.90**
350 mm dia.	95.32	107.65	1.12	27.68	nr	**135.33**
400 mm dia.	106.35	120.12	1.12	27.68	nr	**147.80**
Pumps						
15 mm dia.	37.90	42.80	2.34	57.82	nr	**100.62**
20 mm dia.	39.95	45.11	2.34	57.82	nr	**102.93**
25 mm dia.	42.32	47.80	2.34	57.82	nr	**105.62**
32 mm dia.	44.45	50.20	2.34	57.82	nr	**108.02**
40 mm dia.	46.57	52.59	2.34	57.82	nr	**110.41**
50 mm dia.	51.64	58.32	2.34	57.82	nr	**116.14**
65 mm dia.	73.84	83.40	2.34	57.82	nr	**141.22**

T: MECHANICAL/COOLING/HEATING SYSTEMS

Item	Net Price £	Material £	Labour hours	Labour £	Unit	**Total rate £**
80 mm dia.	59.88	67.63	2.76	68.21	nr	**135.84**
100 mm dia.	74.85	84.54	2.76	68.21	nr	**152.75**
125 mm dia.	83.25	94.02	2.76	68.21	nr	**162.23**
150 mm dia.	95.79	108.18	2.76	68.21	nr	**176.39**
200 mm dia.	127.35	143.83	3.36	83.04	nr	**226.87**
250 mm dia.	147.15	166.19	3.36	83.04	nr	**249.23**
300 mm dia.	158.79	179.34	3.36	83.04	nr	**262.38**
350 mm dia.	176.51	199.35	3.36	83.04	nr	**282.39**
400 mm dia.	196.95	222.43	3.36	83.04	nr	**305.47**
Expansion bellows						
15 mm dia.	30.31	34.23	1.05	25.94	nr	**60.17**
20 mm dia.	31.95	36.09	1.05	25.94	nr	**62.03**
25 mm dia.	33.85	38.23	1.05	25.94	nr	**64.17**
32 mm dia.	35.55	40.15	1.05	25.94	nr	**66.09**
40 mm dia.	37.24	42.05	1.05	25.94	nr	**67.99**
50 mm dia.	41.33	46.68	1.05	25.94	nr	**72.62**
65 mm dia.	44.65	50.43	1.05	25.94	nr	**76.37**
80 mm dia.	47.91	54.11	1.26	31.14	nr	**85.25**
100 mm dia.	59.87	67.62	1.26	31.14	nr	**98.76**
125 mm dia.	66.59	75.20	1.26	31.14	nr	**106.34**
150 mm dia.	76.64	86.56	1.26	31.14	nr	**117.70**
200 mm dia.	101.89	115.08	1.53	37.81	nr	**152.89**
250 mm dia.	117.72	132.96	1.53	37.81	nr	**170.77**
300 mm dia.	127.04	143.48	1.53	37.81	nr	**181.29**
350 mm dia.	141.20	159.47	1.53	37.81	nr	**197.28**
400 mm dia.	157.55	177.94	1.53	37.81	nr	**215.75**

T: MECHANICAL/COOLING/HEATING SYSTEMS

Item	Net Price £	Material £	Labour hours	Labour £	Unit	**Total rate £**
T31: LOW TEMPERATURE HOT WATER HEATING – cont						
Y50 – THERMAL INSULATION – cont						
Mineral fibre sectional insulation; bright class O foil faced; bright class O foil taped joints; 0.8 mm polyisobutylene sheeting; welded joints						
External pipework						
20 mm thick						
15 mm dia.	4.66	5.26	0.30	7.41	m	**12.67**
20 mm dia.	4.97	5.61	0.30	7.41	m	**13.02**
25 mm dia.	5.36	6.05	0.30	7.41	m	**13.46**
32 mm dia.	5.89	6.65	0.30	7.41	m	**14.06**
40 mm dia.	6.28	7.09	0.30	7.41	m	**14.50**
50 mm dia.	7.08	8.00	0.30	7.41	m	**15.41**
Extra over for fittings external insulation						
Flange/union						
15 mm dia.	7.14	8.07	0.75	18.53	nr	**26.60**
20 mm dia.	7.58	8.56	0.75	18.53	nr	**27.09**
25 mm dia.	8.15	9.20	0.75	18.53	nr	**27.73**
32 mm dia.	8.88	10.03	0.75	18.53	nr	**28.56**
40 mm dia.	9.37	10.58	0.75	18.53	nr	**29.11**
50 mm dia.	10.50	11.86	0.75	18.53	nr	**30.39**
Bends						
15 mm dia.	1.18	1.34	0.30	7.41	nr	**8.75**
20 mm dia.	1.24	1.40	0.30	7.41	nr	**8.81**
25 mm dia.	1.35	1.52	0.30	7.41	nr	**8.93**
32 mm dia.	1.47	1.66	0.30	7.41	nr	**9.07**
40 mm dia.	1.57	1.77	0.30	7.41	nr	**9.18**
50 mm dia.	1.77	2.00	0.30	7.41	nr	**9.41**
Tees						
15 mm dia.	1.18	1.34	0.30	7.41	nr	**8.75**
20 mm dia.	1.24	1.40	0.30	7.41	nr	**8.81**
25 mm dia.	1.35	1.52	0.30	7.41	nr	**8.93**
32 mm dia.	1.47	1.66	0.30	7.41	nr	**9.07**
40 mm dia.	1.57	1.77	0.30	7.41	nr	**9.18**
50 mm dia.	1.77	2.00	0.30	7.41	nr	**9.41**
Valves						
15 mm dia.	11.34	12.81	1.03	25.46	nr	**38.27**
20 mm dia.	12.05	13.61	1.03	25.46	nr	**39.07**
25 mm dia.	12.94	14.62	1.03	25.46	nr	**40.08**
32 mm dia.	14.08	15.90	1.03	25.46	nr	**41.36**
40 mm dia.	14.88	16.81	1.03	25.46	nr	**42.27**
50 mm dia.	16.67	18.83	1.03	25.46	nr	**44.29**

T: MECHANICAL/COOLING/HEATING SYSTEMS

Item	Net Price £	Material £	Labour hours	Labour £	Unit	**Total rate £**
Expansion bellows						
15 mm dia.	16.80	18.97	1.42	35.09	nr	**54.06**
20 mm dia.	17.86	20.18	1.42	35.09	nr	**55.27**
25 mm dia.	19.19	21.68	1.42	35.09	nr	**56.77**
32 mm dia.	20.88	23.58	1.42	35.09	nr	**58.67**
40 mm dia.	22.04	24.89	1.42	35.09	nr	**59.98**
50 mm dia.	24.70	27.90	1.42	35.09	nr	**62.99**
25 mm thick						
15 mm dia.	5.15	5.81	0.30	7.41	m	**13.22**
20 mm dia.	5.52	6.23	0.30	7.41	m	**13.64**
25 mm dia.	6.08	6.86	0.30	7.41	m	**14.27**
32 mm dia.	6.58	7.44	0.30	7.41	m	**14.85**
40 mm dia.	7.00	7.91	0.30	7.41	m	**15.32**
50 mm dia.	7.90	8.93	0.30	7.41	m	**16.34**
65 mm dia.	8.96	10.12	0.30	7.41	m	**17.53**
80 mm dia.	9.80	11.07	0.40	9.88	m	**20.95**
100 mm dia.	12.38	13.98	0.40	9.88	m	**23.86**
125 mm dia.	14.25	16.10	0.40	9.88	m	**25.98**
150 mm dia.	16.71	18.87	0.40	9.88	m	**28.75**
200 mm dia.	22.67	25.60	0.50	12.35	m	**37.95**
250 mm dia.	27.06	30.56	0.50	12.35	m	**42.91**
300 mm dia.	29.44	33.25	0.50	12.35	m	**45.60**
Extra over for fittings external insulation						
Flange/union						
15 mm dia.	7.86	8.87	0.75	18.53	nr	**27.40**
20 mm dia.	8.47	9.57	0.75	18.53	nr	**28.10**
25 mm dia.	9.30	10.51	0.75	18.53	nr	**29.04**
32 mm dia.	9.95	11.24	0.75	18.53	nr	**29.77**
40 mm dia.	10.65	12.03	0.75	18.53	nr	**30.56**
50 mm dia.	11.94	13.48	0.75	18.53	nr	**32.01**
65 mm dia.	13.50	15.25	0.75	18.53	nr	**33.78**
80 mm dia.	14.69	16.60	0.89	21.99	nr	**38.59**
100 mm dia.	18.25	20.61	0.89	21.99	nr	**42.60**
125 mm dia.	21.13	23.87	0.89	21.99	nr	**45.86**
150 mm dia.	24.58	27.76	0.89	21.99	nr	**49.75**
200 mm dia.	33.27	37.58	1.15	28.42	nr	**66.00**
250 mm dia.	39.60	44.73	1.15	28.42	nr	**73.15**
300 mm dia.	44.00	49.69	1.15	28.42	nr	**78.11**
Bends						
15 mm dia.	1.28	1.45	0.30	7.41	nr	**8.86**
20 mm dia.	1.39	1.57	0.30	7.41	nr	**8.98**
25 mm dia.	1.50	1.69	0.30	7.41	nr	**9.10**
32 mm dia.	1.64	1.86	0.30	7.41	nr	**9.27**
40 mm dia.	1.75	1.98	0.30	7.41	nr	**9.39**
50 mm dia.	1.97	2.22	0.30	7.41	nr	**9.63**
65 mm dia.	2.25	2.54	0.30	7.41	nr	**9.95**

T: MECHANICAL/COOLING/HEATING SYSTEMS

Item	Net Price £	Material £	Labour hours	Labour £	Unit	**Total rate £**
T31: LOW TEMPERATURE HOT WATER HEATING – cont						
Y50 – THERMAL INSULATION – cont						
Extra over for fittings external insulation – cont						
Bends – cont						
80 mm dia.	2.44	2.75	0.40	9.88	nr	**12.63**
100 mm dia.	3.09	3.49	0.40	9.88	nr	**13.37**
125 mm dia.	3.55	4.01	0.40	9.88	nr	**13.89**
150 mm dia.	4.18	4.72	0.40	9.88	nr	**14.60**
200 mm dia.	5.66	6.40	0.50	12.35	nr	**18.75**
250 mm dia.	6.77	7.65	0.50	12.35	nr	**20.00**
300 mm dia.	7.35	8.30	0.50	12.35	nr	**20.65**
Tees						
15 mm dia.	1.28	1.45	0.30	7.41	nr	**8.86**
20 mm dia.	1.39	1.57	0.30	7.41	nr	**8.98**
25 mm dia.	1.50	1.69	0.30	7.41	nr	**9.10**
32 mm dia.	1.64	1.86	0.30	7.41	nr	**9.27**
40 mm dia.	1.75	1.98	0.30	7.41	nr	**9.39**
50 mm dia.	1.97	2.22	0.30	7.41	nr	**9.63**
65 mm dia.	2.25	2.54	0.30	7.41	nr	**9.95**
80 mm dia.	2.44	2.75	0.40	9.88	nr	**12.63**
100 mm dia.	3.09	3.49	0.40	9.88	nr	**13.37**
125 mm dia.	3.55	4.01	0.40	9.88	nr	**13.89**
150 mm dia.	4.18	4.72	0.40	9.88	nr	**14.60**
200 mm dia.	5.66	6.40	0.50	12.35	nr	**18.75**
250 mm dia.	6.77	7.65	0.50	12.35	nr	**20.00**
300 mm dia.	7.35	8.30	0.50	12.35	nr	**20.65**
Valves						
15 mm dia.	12.49	14.11	1.03	25.46	nr	**39.57**
20 mm dia.	13.46	15.20	1.03	25.46	nr	**40.66**
25 mm dia.	14.76	16.67	1.03	25.46	nr	**42.13**
32 mm dia.	15.82	17.87	1.03	25.46	nr	**43.33**
40 mm dia.	16.90	19.08	1.03	25.46	nr	**44.54**
50 mm dia.	18.98	21.44	1.03	25.46	nr	**46.90**
65 mm dia.	21.46	24.24	1.03	25.46	nr	**49.70**
80 mm dia.	23.32	26.34	1.25	30.89	nr	**57.23**
100 mm dia.	29.00	32.75	1.25	30.89	nr	**63.64**
125 mm dia.	33.55	37.89	1.25	30.89	nr	**68.78**
150 mm dia.	39.07	44.13	1.25	30.89	nr	**75.02**
200 mm dia.	52.86	59.70	1.55	38.30	nr	**98.00**
250 mm dia.	62.90	71.04	1.55	38.30	nr	**109.34**
300 mm dia.	69.87	78.91	1.55	38.30	nr	**117.21**

T: MECHANICAL/COOLING/HEATING SYSTEMS

Item	Net Price £	Material £	Labour hours	Labour £	Unit	**Total rate £**
Expansion bellows						
15 mm dia.	18.51	20.91	1.42	35.09	nr	**56.00**
20 mm dia.	19.94	22.52	1.42	35.09	nr	**57.61**
25 mm dia.	21.87	24.70	1.42	35.09	nr	**59.79**
32 mm dia.	23.45	26.49	1.42	35.09	nr	**61.58**
40 mm dia.	25.05	28.29	1.42	35.09	nr	**63.38**
50 mm dia.	28.09	31.72	1.42	35.09	nr	**66.81**
65 mm dia.	31.78	35.89	1.42	35.09	nr	**70.98**
80 mm dia.	34.55	39.03	1.75	43.25	nr	**82.28**
100 mm dia.	42.95	48.51	1.75	43.25	nr	**91.76**
125 mm dia.	49.71	56.14	1.75	43.25	nr	**99.39**
150 mm dia.	57.87	65.36	1.75	43.25	nr	**108.61**
200 mm dia.	78.30	88.43	2.17	53.62	nr	**142.05**
250 mm dia.	93.18	105.23	2.17	53.62	nr	**158.85**
300 mm dia.	103.51	116.90	3.17	78.34	nr	**195.24**
30 mm thick						
15 mm dia.	7.45	8.42	0.30	7.41	m	**15.83**
20 mm dia.	6.74	7.61	0.30	7.41	m	**15.02**
25 mm dia.	7.14	8.07	0.30	7.41	m	**15.48**
32 mm dia.	7.75	8.75	0.30	7.41	m	**16.16**
40 mm dia.	8.17	9.23	0.30	7.41	m	**16.64**
50 mm dia.	9.19	10.38	0.30	7.41	m	**17.79**
65 mm dia.	10.34	11.68	0.30	7.41	m	**19.09**
80 mm dia.	11.24	12.70	0.40	9.88	m	**22.58**
100 mm dia.	14.06	15.88	0.40	9.88	m	**25.76**
125 mm dia.	16.09	18.18	0.40	9.88	m	**28.06**
150 mm dia.	18.69	21.10	0.40	9.88	m	**30.98**
200 mm dia.	25.17	28.43	0.50	12.35	m	**40.78**
250 mm dia.	29.86	33.72	0.50	12.35	m	**46.07**
300 mm dia.	32.21	36.37	0.50	12.35	m	**48.72**
350 mm dia.	35.22	39.78	0.50	12.35	m	**52.13**
Extra over for fittings external insulation						
Flange/union						
15 mm dia.	9.62	10.86	0.75	18.53	nr	**29.39**
20 mm dia.	10.19	11.51	0.75	18.53	nr	**30.04**
25 mm dia.	10.81	12.21	0.75	18.53	nr	**30.74**
32 mm dia.	11.78	13.30	0.75	18.53	nr	**31.83**
40 mm dia.	12.31	13.90	0.75	18.53	nr	**32.43**
50 mm dia.	13.74	15.51	0.75	18.53	nr	**34.04**
65 mm dia.	15.47	17.47	0.75	18.53	nr	**36.00**
80 mm dia.	16.86	19.04	0.89	21.99	nr	**41.03**
100 mm dia.	20.75	23.44	0.89	21.99	nr	**45.43**
125 mm dia.	23.71	26.78	0.89	21.99	nr	**48.77**
150 mm dia.	27.54	31.10	0.89	21.99	nr	**53.09**
200 mm dia.	36.80	41.56	1.15	28.42	nr	**69.98**
250 mm dia.	43.58	49.21	1.15	28.42	nr	**77.63**
300 mm dia.	47.75	53.93	1.15	28.42	nr	**82.35**
350 mm dia.	52.75	59.58	1.15	28.42	nr	**88.00**

T: MECHANICAL/COOLING/HEATING SYSTEMS

Item	Net Price £	Material £	Labour hours	Labour £	Unit	Total rate £
T31: LOW TEMPERATURE HOT WATER HEATING – cont						
Y50 – THERMAL INSULATION – cont						
Extra over for fittings external insulation – cont						
Bends						
15 mm dia.	1.58	1.79	0.30	7.41	nr	**9.20**
20 mm dia.	1.68	1.90	0.30	7.41	nr	**9.31**
25 mm dia.	1.79	2.02	0.30	7.41	nr	**9.43**
32 mm dia.	1.93	2.18	0.30	7.41	nr	**9.59**
40 mm dia.	2.04	2.31	0.30	7.41	nr	**9.72**
50 mm dia.	2.31	2.61	0.30	7.41	nr	**10.02**
65 mm dia.	2.59	2.93	0.30	7.41	nr	**10.34**
80 mm dia.	2.81	3.17	0.40	9.88	nr	**13.05**
100 mm dia.	3.52	3.98	0.40	9.88	nr	**13.86**
125 mm dia.	4.02	4.54	0.40	9.88	nr	**14.42**
150 mm dia.	4.67	5.27	0.40	9.88	nr	**15.15**
200 mm dia.	6.30	7.12	0.50	12.35	nr	**19.47**
250 mm dia.	7.48	8.45	0.50	12.35	nr	**20.80**
300 mm dia.	8.05	9.09	0.50	12.35	nr	**21.44**
350 mm dia.	8.80	9.93	0.50	12.35	nr	**22.28**
Tees						
15 mm dia.	1.58	1.79	0.30	7.41	nr	**9.20**
20 mm dia.	1.68	1.90	0.30	7.41	nr	**9.31**
25 mm dia.	1.79	2.02	0.30	7.41	nr	**9.43**
32 mm dia.	1.93	2.18	0.30	7.41	nr	**9.59**
40 mm dia.	2.04	2.31	0.30	7.41	nr	**9.72**
50 mm dia.	2.31	2.61	0.30	7.41	nr	**10.02**
65 mm dia.	2.59	2.93	0.30	7.41	nr	**10.34**
80 mm dia.	2.81	3.17	0.40	9.88	nr	**13.05**
100 mm dia.	3.52	3.98	0.40	9.88	nr	**13.86**
125 mm dia.	4.02	4.54	0.40	9.88	nr	**14.42**
150 mm dia.	4.67	5.27	0.40	9.88	nr	**15.15**
200 mm dia.	6.30	7.12	0.50	12.35	nr	**19.47**
250 mm dia.	7.48	8.45	0.50	12.35	nr	**20.80**
300 mm dia.	8.05	9.09	0.50	12.35	nr	**21.44**
350 mm dia.	8.80	9.93	0.50	12.35	nr	**22.28**
Valves						
15 mm dia.	15.28	17.26	1.03	25.46	nr	**42.72**
20 mm dia.	16.19	18.29	1.03	25.46	nr	**43.75**
25 mm dia.	17.16	19.38	1.03	25.46	nr	**44.84**
32 mm dia.	18.70	21.12	1.03	25.46	nr	**46.58**
40 mm dia.	19.57	22.10	1.03	25.46	nr	**47.56**
50 mm dia.	21.83	24.65	1.03	25.46	nr	**50.11**
65 mm dia.	24.56	27.73	1.03	25.46	nr	**53.19**
80 mm dia.	26.77	30.23	1.25	30.89	nr	**61.12**
100 mm dia.	32.94	37.20	1.25	30.89	nr	**68.09**
125 mm dia.	37.64	42.51	1.25	30.89	nr	**73.40**

T: MECHANICAL/COOLING/HEATING SYSTEMS

Item	Net Price £	Material £	Labour hours	Labour £	Unit	**Total rate £**
150 mm dia.	43.73	49.39	1.25	30.89	nr	**80.28**
200 mm dia.	58.46	66.02	1.55	38.30	nr	**104.32**
250 mm dia.	69.22	78.17	1.55	38.30	nr	**116.47**
300 mm dia.	75.85	85.66	1.55	38.30	nr	**123.96**
350 mm dia.	83.78	94.63	1.55	38.30	nr	**132.93**
Expansion bellows						
15 mm dia.	22.62	25.55	1.42	35.09	nr	**60.64**
20 mm dia.	23.99	27.09	1.42	35.09	nr	**62.18**
25 mm dia.	25.42	28.71	1.42	35.09	nr	**63.80**
32 mm dia.	27.71	31.29	1.42	35.09	nr	**66.38**
40 mm dia.	29.00	32.75	1.42	35.09	nr	**67.84**
50 mm dia.	32.32	36.51	1.42	35.09	nr	**71.60**
65 mm dia.	36.40	41.11	1.42	35.09	nr	**76.20**
80 mm dia.	39.67	44.80	1.75	43.25	nr	**88.05**
100 mm dia.	48.80	55.11	1.75	43.25	nr	**98.36**
125 mm dia.	55.77	62.99	1.75	43.25	nr	**106.24**
150 mm dia.	64.78	73.16	1.75	43.25	nr	**116.41**
200 mm dia.	86.61	97.82	2.17	53.62	nr	**151.44**
250 mm dia.	102.56	115.83	2.17	53.62	nr	**169.45**
300 mm dia.	112.36	126.90	2.17	53.62	nr	**180.52**
350 mm dia.	124.13	140.19	2.17	53.62	nr	**193.81**
40 mm thick						
15 mm dia.	7.93	8.96	0.30	7.41	m	**16.37**
20 mm dia.	8.19	9.25	0.30	7.41	m	**16.66**
25 mm dia.	8.76	9.89	0.30	7.41	m	**17.30**
32 mm dia.	9.32	10.53	0.30	7.41	m	**17.94**
40 mm dia.	9.75	11.02	0.30	7.41	m	**18.43**
50 mm dia.	10.87	12.28	0.30	7.41	m	**19.69**
65 mm dia.	12.13	13.70	0.30	7.41	m	**21.11**
80 mm dia.	13.14	14.84	0.40	9.88	m	**24.72**
100 mm dia.	16.41	18.53	0.40	9.88	m	**28.41**
125 mm dia.	18.50	20.89	0.40	9.88	m	**30.77**
150 mm dia.	21.34	24.10	0.40	9.88	m	**33.98**
200 mm dia.	28.47	32.15	0.50	12.35	m	**44.50**
250 mm dia.	33.34	37.66	0.50	12.35	m	**50.01**
300 mm dia.	36.10	40.77	0.50	12.35	m	**53.12**
350 mm dia.	39.67	44.80	0.50	12.35	m	**57.15**
400 mm dia.	44.16	49.88	0.50	12.35	m	**62.23**
Extra over for fittings external insulation						
Flange/union						
15 mm dia.	11.84	13.37	0.75	18.53	nr	**31.90**
20 mm dia.	12.36	13.96	0.75	18.53	nr	**32.49**
25 mm dia.	13.18	14.88	0.75	18.53	nr	**33.41**
32 mm dia.	13.89	15.69	0.75	18.53	nr	**34.22**
40 mm dia.	14.67	16.56	0.75	18.53	nr	**35.09**
50 mm dia.	16.25	18.35	0.75	18.53	nr	**36.88**
65 mm dia.	18.10	20.44	0.75	18.53	nr	**38.97**
80 mm dia.	19.52	22.04	0.89	21.99	nr	**44.03**
100 mm dia.	24.03	27.14	0.89	21.99	nr	**49.13**

T: MECHANICAL/COOLING/HEATING SYSTEMS

Item	Net Price £	Material £	Labour hours	Labour £	Unit	**Total rate £**
T31: LOW TEMPERATURE HOT WATER HEATING – cont						
Y50 – THERMAL INSULATION – cont						
Extra over for fittings external insulation – cont						
Flange/union – cont						
125 mm dia.	26.88	30.36	0.89	21.99	nr	**52.35**
150 mm dia.	31.15	35.18	0.89	21.99	nr	**57.17**
200 mm dia.	41.08	46.40	1.15	28.42	nr	**74.82**
250 mm dia.	48.15	54.38	1.15	28.42	nr	**82.80**
300 mm dia.	52.91	59.75	1.15	28.42	nr	**88.17**
350 mm dia.	58.68	66.27	1.15	28.42	nr	**94.69**
400 mm dia.	65.72	74.23	1.15	28.42	nr	**102.65**
Bends						
15 mm dia.	1.97	2.22	0.30	7.41	nr	**9.63**
20 mm dia.	2.05	2.32	0.30	7.41	nr	**9.73**
25 mm dia.	2.19	2.47	0.30	7.41	nr	**9.88**
32 mm dia.	2.33	2.63	0.30	7.41	nr	**10.04**
40 mm dia.	2.44	2.75	0.30	7.41	nr	**10.16**
50 mm dia.	2.74	3.09	0.30	7.41	nr	**10.50**
65 mm dia.	3.04	3.44	0.30	7.41	nr	**10.85**
80 mm dia.	3.30	3.72	0.40	9.88	nr	**13.60**
100 mm dia.	4.10	4.63	0.40	9.88	nr	**14.51**
125 mm dia.	4.62	5.22	0.40	9.88	nr	**15.10**
150 mm dia.	5.35	6.04	0.40	9.88	nr	**15.92**
200 mm dia.	7.12	8.04	0.50	12.35	nr	**20.39**
250 mm dia.	8.33	9.40	0.50	12.35	nr	**21.75**
300 mm dia.	9.02	10.19	0.50	12.35	nr	**22.54**
350 mm dia.	9.91	11.19	0.50	12.35	nr	**23.54**
400 mm dia.	11.04	12.46	0.50	12.35	nr	**24.81**
Tees						
15 mm dia.	1.97	2.22	0.30	7.41	nr	**9.63**
20 mm dia.	2.05	2.32	0.30	7.41	nr	**9.73**
25 mm dia.	2.19	2.47	0.30	7.41	nr	**9.88**
32 mm dia.	2.33	2.63	0.30	7.41	nr	**10.04**
40 mm dia.	2.44	2.75	0.30	7.41	nr	**10.16**
50 mm dia.	2.74	3.09	0.30	7.41	nr	**10.50**
65 mm dia.	3.04	3.44	0.30	7.41	nr	**10.85**
80 mm dia.	3.30	3.72	0.40	9.88	nr	**13.60**
100 mm dia.	4.10	4.63	0.40	9.88	nr	**14.51**
125 mm dia.	4.62	5.22	0.40	9.88	nr	**15.10**
150 mm dia.	5.35	6.04	0.40	9.88	nr	**15.92**
200 mm dia.	7.12	8.04	0.50	12.35	nr	**20.39**
250 mm dia.	8.33	9.40	0.50	12.35	nr	**21.75**
300 mm dia.	9.02	10.19	0.50	12.35	nr	**22.54**
350 mm dia.	9.91	11.19	0.50	12.35	nr	**23.54**
400 mm dia.	11.04	12.46	0.50	12.35	nr	**24.81**

T: MECHANICAL/COOLING/HEATING SYSTEMS

Item	Net Price £	Material £	Labour hours	Labour £	Unit	**Total rate £**
Valves						
15 mm dia.	18.79	21.23	1.03	25.46	nr	**46.69**
20 mm dia.	19.63	22.17	1.03	25.46	nr	**47.63**
25 mm dia.	20.93	23.63	1.03	25.46	nr	**49.09**
32 mm dia.	22.05	24.90	1.03	25.46	nr	**50.36**
40 mm dia.	23.31	26.33	1.03	25.46	nr	**51.79**
50 mm dia.	25.83	29.17	1.03	25.46	nr	**54.63**
65 mm dia.	28.75	32.47	1.03	25.46	nr	**57.93**
80 mm dia.	31.00	35.01	1.25	30.89	nr	**65.90**
100 mm dia.	38.18	43.12	1.25	30.89	nr	**74.01**
125 mm dia.	42.70	48.23	1.25	30.89	nr	**79.12**
150 mm dia.	49.49	55.90	1.25	30.89	nr	**86.79**
200 mm dia.	65.24	73.68	1.55	38.30	nr	**111.98**
250 mm dia.	76.44	86.33	1.55	38.30	nr	**124.63**
300 mm dia.	84.04	94.91	1.55	38.30	nr	**133.21**
350 mm dia.	93.21	105.27	1.55	38.30	nr	**143.57**
400 mm dia.	104.37	117.87	1.55	38.30	nr	**156.17**
Expansion bellows						
15 mm dia.	27.83	31.43	1.42	35.09	nr	**66.52**
20 mm dia.	29.09	32.85	1.42	35.09	nr	**67.94**
25 mm dia.	31.01	35.03	1.42	35.09	nr	**70.12**
32 mm dia.	32.65	36.87	1.42	35.09	nr	**71.96**
40 mm dia.	34.53	38.99	1.42	35.09	nr	**74.08**
50 mm dia.	38.27	43.22	1.42	35.09	nr	**78.31**
65 mm dia.	42.60	48.11	1.42	35.09	nr	**83.20**
80 mm dia.	45.92	51.86	1.75	43.25	nr	**95.11**
100 mm dia.	56.57	63.89	1.75	43.25	nr	**107.14**
125 mm dia.	63.26	71.44	1.75	43.25	nr	**114.69**
150 mm dia.	73.34	82.83	1.75	43.25	nr	**126.08**
200 mm dia.	96.66	109.17	2.17	53.62	nr	**162.79**
250 mm dia.	113.25	127.91	2.17	53.62	nr	**181.53**
300 mm dia.	124.51	140.62	2.17	53.62	nr	**194.24**
350 mm dia.	138.10	155.97	2.17	53.62	nr	**209.59**
400 mm dia.	154.62	174.62	2.17	53.62	nr	**228.24**
50 mm thick						
15 mm dia.	10.44	11.79	0.30	7.41	m	**19.20**
20 mm dia.	10.95	12.36	0.30	7.41	m	**19.77**
25 mm dia.	11.58	13.08	0.30	7.41	m	**20.49**
32 mm dia.	12.50	14.12	0.30	7.41	m	**21.53**
40 mm dia.	12.70	14.34	0.30	7.41	m	**21.75**
50 mm dia.	14.07	15.89	0.30	7.41	m	**23.30**
65 mm dia.	15.34	17.33	0.30	7.41	m	**24.74**
80 mm dia.	16.41	18.53	0.40	9.88	m	**28.41**
100 mm dia.	20.31	22.94	0.40	9.88	m	**32.82**
125 mm dia.	22.73	25.67	0.40	9.88	m	**35.55**
150 mm dia.	25.98	29.35	0.40	9.88	m	**39.23**
200 mm dia.	34.41	38.86	0.50	12.35	m	**51.21**
250 mm dia.	39.75	44.89	0.50	12.35	m	**57.24**
300 mm dia.	42.63	48.14	0.50	12.35	m	**60.49**
350 mm dia.	46.78	52.84	0.50	12.35	m	**65.19**
400 mm dia.	51.82	58.53	0.50	12.35	m	**70.88**

T: MECHANICAL/COOLING/HEATING SYSTEMS

Item	Net Price £	Material £	Labour hours	Labour £	Unit	**Total rate £**
T31: LOW TEMPERATURE HOT WATER HEATING – cont						
Y50 – THERMAL INSULATION – cont						
Extra over for fittings external insulation						
Flange/union						
15 mm dia.	15.38	17.37	0.75	18.53	nr	**35.90**
20 mm dia.	16.19	18.29	0.75	18.53	nr	**36.82**
25 mm dia.	17.11	19.33	0.75	18.53	nr	**37.86**
32 mm dia.	17.96	20.29	0.75	18.53	nr	**38.82**
40 mm dia.	18.79	21.23	0.75	18.53	nr	**39.76**
50 mm dia.	20.72	23.40	0.75	18.53	nr	**41.93**
65 mm dia.	22.40	25.30	0.75	18.53	nr	**43.83**
80 mm dia.	24.01	27.12	0.89	21.99	nr	**49.11**
100 mm dia.	29.52	33.34	0.89	21.99	nr	**55.33**
125 mm dia.	32.81	37.06	0.89	21.99	nr	**59.05**
150 mm dia.	37.58	42.44	0.89	21.99	nr	**64.43**
200 mm dia.	49.16	55.52	1.15	28.42	nr	**83.94**
250 mm dia.	56.80	64.15	1.15	28.42	nr	**92.57**
300 mm dia.	61.63	69.60	1.15	28.42	nr	**98.02**
350 mm dia.	68.20	77.02	1.15	28.42	nr	**105.44**
400 mm dia.	75.99	85.82	1.15	28.42	nr	**114.24**
Bend						
15 mm dia.	2.61	2.95	0.30	7.41	nr	**10.36**
20 mm dia.	2.75	3.10	0.30	7.41	nr	**10.51**
25 mm dia.	2.89	3.26	0.30	7.41	nr	**10.67**
32 mm dia.	3.04	3.44	0.30	7.41	nr	**10.85**
40 mm dia.	3.18	3.59	0.30	7.41	nr	**11.00**
50 mm dia.	3.52	3.98	0.30	7.41	nr	**11.39**
65 mm dia.	3.83	4.32	0.30	7.41	nr	**11.73**
80 mm dia.	4.09	4.62	0.40	9.88	nr	**14.50**
100 mm dia.	5.08	5.73	0.40	9.88	nr	**15.61**
125 mm dia.	5.67	6.41	0.40	9.88	nr	**16.29**
150 mm dia.	6.50	7.34	0.40	9.88	nr	**17.22**
200 mm dia.	8.61	9.72	0.50	12.35	nr	**22.07**
250 mm dia.	9.94	11.23	0.50	12.35	nr	**23.58**
300 mm dia.	10.66	12.04	0.50	12.35	nr	**24.39**
350 mm dia.	11.70	13.21	0.50	12.35	nr	**25.56**
400 mm dia.	12.95	14.63	0.50	12.35	nr	**26.98**
Tee						
15 mm dia.	2.61	2.95	0.30	7.41	nr	**10.36**
20 mm dia.	2.75	3.10	0.30	7.41	nr	**10.51**
25 mm dia.	2.89	3.26	0.30	7.41	nr	**10.67**
32 mm dia.	3.04	3.44	0.30	7.41	nr	**10.85**
40 mm dia.	3.18	3.59	0.30	7.41	nr	**11.00**
50 mm dia.	3.52	3.98	0.30	7.41	nr	**11.39**
65 mm dia.	3.83	4.32	0.30	7.41	nr	**11.73**

T: MECHANICAL/COOLING/HEATING SYSTEMS

Item	Net Price £	Material £	Labour hours	Labour £	Unit	**Total rate £**
80 mm dia.	4.09	4.62	0.40	9.88	nr	**14.50**
100 mm dia.	5.08	5.73	0.40	9.88	nr	**15.61**
125 mm dia.	5.67	6.41	0.40	9.88	nr	**16.29**
150 mm dia.	6.50	7.34	0.40	9.88	nr	**17.22**
200 mm dia.	8.61	9.72	0.50	12.35	nr	**22.07**
250 mm dia.	9.94	11.23	0.50	12.35	nr	**23.58**
300 mm dia.	10.66	12.04	0.50	12.35	nr	**24.39**
350 mm dia.	11.70	13.21	0.50	12.35	nr	**25.56**
400 mm dia.	12.95	14.63	0.50	12.35	nr	**26.98**
Valves						
15 mm dia.	24.44	27.60	1.03	25.46	nr	**53.06**
20 mm dia.	25.70	29.03	1.03	25.46	nr	**54.49**
25 mm dia.	27.17	30.68	1.03	25.46	nr	**56.14**
32 mm dia.	28.53	32.22	1.03	25.46	nr	**57.68**
40 mm dia.	29.85	33.71	1.03	25.46	nr	**59.17**
50 mm dia.	32.91	37.17	1.03	25.46	nr	**62.63**
65 mm dia.	35.59	40.20	1.03	25.46	nr	**65.66**
80 mm dia.	38.14	43.07	1.25	30.89	nr	**73.96**
100 mm dia.	46.89	52.96	1.25	30.89	nr	**83.85**
125 mm dia.	52.12	58.86	1.25	30.89	nr	**89.75**
150 mm dia.	59.66	67.38	1.25	30.89	nr	**98.27**
200 mm dia.	78.07	88.17	1.55	38.30	nr	**126.47**
250 mm dia.	90.21	101.89	1.55	38.30	nr	**140.19**
300 mm dia.	97.87	110.54	1.55	38.30	nr	**148.84**
350 mm dia.	108.31	122.33	1.55	38.30	nr	**160.63**
400 mm dia.	120.69	136.30	1.55	38.30	nr	**174.60**
Expansion bellows						
15 mm dia.	36.21	40.89	1.42	35.09	nr	**75.98**
20 mm dia.	38.08	43.00	1.42	35.09	nr	**78.09**
25 mm dia.	40.26	45.47	1.42	35.09	nr	**80.56**
32 mm dia.	42.28	47.75	1.42	35.09	nr	**82.84**
40 mm dia.	44.21	49.93	1.42	35.09	nr	**85.02**
50 mm dia.	48.75	55.06	1.42	35.09	nr	**90.15**
65 mm dia.	52.73	59.56	1.42	35.09	nr	**94.65**
80 mm dia.	56.51	63.82	1.75	43.25	nr	**107.07**
100 mm dia.	69.47	78.46	1.75	43.25	nr	**121.71**
125 mm dia.	77.21	87.20	1.75	43.25	nr	**130.45**
150 mm dia.	88.40	99.84	1.75	43.25	nr	**143.09**
200 mm dia.	115.65	130.61	2.17	53.62	nr	**184.23**
250 mm dia.	133.65	150.94	2.17	53.62	nr	**204.56**
300 mm dia.	145.00	163.76	2.17	53.62	nr	**217.38**
350 mm dia.	160.46	181.22	2.17	53.62	nr	**234.84**
400 mm dia.	178.80	201.94	2.17	53.62	nr	**255.56**

T: MECHANICAL/COOLING/HEATING SYSTEMS

Item	Net Price £	Material £	Labour hours	Labour £	Unit	Total rate £
T33: STEAM HEATING						
Y10 – PIPELINES						
For pipework prices refer to Section T31 – Low Temperature Hot Water Heating						
Y11 – PIPELINE ANCILLARIES						
Steam traps and accessories						
Cast iron; inverted bucket type; steam trap pressure range up to 17 bar at 210°C; screwed ends						
½" dia.	94.43	106.65	0.85	21.00	nr	**127.65**
¾" dia.	94.43	106.65	1.13	27.92	nr	**134.57**
1" dia.	159.11	179.69	1.35	33.36	nr	**213.05**
1½" dia.	491.23	554.79	1.80	44.47	nr	**599.26**
2" dia.	549.86	621.01	2.18	53.88	nr	**674.89**
Cast iron; inverted bucket type; steam trap pressure range up to 17 bar at 210°C; flanged ends to BS 4504 PN16; bolted connections						
15 mm dia.	158.16	178.62	1.15	28.42	nr	**207.04**
20 mm dia.	191.14	215.87	1.25	30.89	nr	**246.76**
25 mm dia.	237.26	267.96	1.33	32.86	nr	**300.82**
40 mm dia.	503.35	568.48	1.46	36.08	nr	**604.56**
50 mm dia.	561.97	634.68	1.60	39.54	nr	**674.22**
Steam traps and strainers						
Stainless steel; thermodynamic trap with pressure range up to 42 bar; temperature range to 400°C; screwed ends to steel						
15 mm dia.	63.23	71.41	0.84	20.77	nr	**92.18**
20 mm dia.	71.56	80.81	1.14	28.18	nr	**108.99**
Stainless steel; thermodynamic trap with pressure range up to 24 bar; temperature range to 288°C; flanged ends to DIN 2456 PN64; bolted connections						
15 mm dia.	132.38	149.51	1.24	30.66	nr	**180.17**
20 mm dia.	145.11	163.88	1.34	33.13	nr	**197.01**
25 mm dia.	167.73	189.43	1.40	34.62	nr	**224.05**
Malleable iron pipeline strainer; max steam working pressure 14 bar and temperature range to 230°C; screwed ends to steel						
½" dia.	18.41	20.79	0.84	20.77	nr	**41.56**
¾" dia.	23.01	25.99	1.14	28.18	nr	**54.17**
1" dia.	35.90	40.54	1.30	32.13	nr	**72.67**

T: MECHANICAL/COOLING/HEATING SYSTEMS

Item	Net Price £	Material £	Labour hours	Labour £	Unit	**Total rate £**
1½" dia.	73.74	83.28	1.50	37.10	nr	**120.38**
2" dia.	103.54	116.93	1.74	43.05	nr	**159.98**
Bronze pipeline strainer; max steam working pressure 25 bar; flanged ends to BS 4504 PN25; bolted connections						
15 mm dia.	159.59	180.24	1.24	30.66	nr	**210.90**
20 mm dia.	194.65	219.84	1.34	33.13	nr	**252.97**
25 mm dia.	223.19	252.07	1.40	34.62	nr	**286.69**
32 mm dia.	346.47	391.30	1.46	36.13	nr	**427.43**
40 mm dia.	393.19	444.07	1.54	38.08	nr	**482.15**
50 mm dia.	604.70	682.94	1.64	40.58	nr	**723.52**
65 mm dia.	669.58	756.22	2.50	61.78	nr	**818.00**
80 mm dia.	833.08	940.88	2.91	71.84	nr	**1012.72**
100 mm dia.	1442.98	1629.69	3.51	86.70	nr	**1716.39**
Balanced pressure thermostatic steam trap and strainer; max working pressure up to 13 bar; screwed ends to steel						
½" dia.	131.04	147.99	1.26	31.16	nr	**179.15**
¾" dia.	131.39	148.39	1.71	42.32	nr	**190.71**
Bimetallic thermostatic steam trap and strainer; max working pressure up to 21 bar; flanged ends						
15 mm	170.47	192.52	1.24	30.66	nr	**223.18**
20 mm	187.27	211.50	1.34	33.13	nr	**244.63**
Sight glasses						
Pressed brass; straight; single window; screwed ends to steel						
15 mm dia.	40.22	45.42	0.84	20.77	nr	**66.19**
20 mm dia.	44.64	50.42	1.14	28.18	nr	**78.60**
25 mm dia.	55.79	63.01	1.30	32.13	nr	**95.14**
Gunmetal; straight; double window; screwed ends to steel						
15 mm dia.	64.87	73.27	0.84	20.77	nr	**94.04**
20 mm dia.	71.37	80.60	1.14	28.18	nr	**108.78**
25 mm dia.	88.23	99.64	1.30	32.13	nr	**131.77**
32 mm dia.	145.34	164.15	1.35	33.39	nr	**197.54**
40 mm dia.	145.34	164.15	1.74	43.05	nr	**207.20**
50 mm dia.	176.48	199.32	2.08	51.48	nr	**250.80**
SG Iron flanged; BS 4504, PN 25						
15 mm dia.	132.36	149.49	1.00	24.71	nr	**174.20**
20 mm dia.	155.73	175.88	1.25	30.89	nr	**206.77**
25 mm dia.	198.56	224.26	1.50	37.10	nr	**261.36**

T: MECHANICAL/COOLING/HEATING SYSTEMS

Item	Net Price £	Material £	Labour hours	Labour £	Unit	Total rate £
T33: STEAM HEATING – cont						
Sight glasses – cont						
SG Iron flanged – cont						
32 mm dia.	219.29	247.67	1.70	42.02	nr	**289.69**
40 mm dia.	286.79	323.90	2.00	49.42	nr	**373.32**
50 mm dia.	346.47	391.30	2.30	56.94	nr	**448.24**
Check valve and sight glass; gun metal; screwed						
15 mm dia.	65.51	73.99	0.84	20.77	nr	**94.76**
20 mm dia.	69.41	78.39	1.14	28.18	nr	**106.57**
25 mm dia.	116.78	131.89	1.30	32.13	nr	**164.02**
Pressure reducing valves						
Pressure reducing valve for steam; maximum range of 17 bar and 232°C; screwed ends to steel						
15 mm dia.	490.51	553.98	0.87	21.50	nr	**575.48**
20 mm dia.	530.73	599.40	0.91	22.49	nr	**621.89**
25 mm dia.	572.25	646.29	1.35	33.36	nr	**679.65**
Pressure reducing valve for steam; maximum range of 17 bar and 232°C; flanged ends to BS 4504 PN 25						
25 mm dia.	689.05	778.21	1.70	42.01	nr	**820.22**
32 mm dia.	782.49	883.74	1.87	46.22	nr	**929.96**
40 mm dia.	934.31	1055.20	2.12	52.40	nr	**1107.60**
50 mm dia.	1078.35	1217.88	2.57	63.51	nr	**1281.39**
Safety and relief valves						
Bronze safety valve; 'pop' type; side outlet; including easing lever; working pressure saturated steam up to 20.7 bar; screwed ends to steel						
15 mm dia.	168.25	190.02	0.32	7.91	nr	**197.93**
20 mm dia.	190.18	214.79	0.40	9.88	nr	**224.67**
Bronze safety valve; 'pop' type; side outlet; including easing lever; working pressure saturated steam up to 17.2 bar; screwed ends to steel						
25 mm dia.	270.64	305.66	0.47	11.61	nr	**317.27**
32 mm dia.	323.34	365.18	0.56	13.83	nr	**379.01**

T: MECHANICAL/COOLING/HEATING SYSTEMS

Item	Net Price £	Material £	Labour hours	Labour £	Unit	**Total rate £**
Bronze safety valve; 'pop' type; side outlet; including easing lever; working pressure saturated steam up to 13.8 bar; screwed ends to steel						
40 mm dia.	418.43	472.58	0.64	15.81	nr	**488.39**
50 mm dia.	582.29	657.63	0.76	18.78	nr	**676.41**
65 mm dia.	828.09	935.24	0.94	23.24	nr	**958.48**
80 mm dia.	1073.89	1212.84	1.10	27.18	nr	**1240.02**
EQUIPMENT						
Y23 – CALORIFIERS						
Non-storage calorifiers; mild steel shell construction with indirect steam heating for secondary LPHW at 82°C flow and 71°C return to BS 853; maximum test on shell 11 bar, tubes 26 bar						
Horizontal/vertical, for steam at 3 bar–5.5 bar						
88 kW capacity	624.25	705.02	8.00	197.69	nr	**902.71**
176 kW capacity	908.75	1026.33	12.05	297.72	nr	**1324.05**
293 kW capacity	990.04	1118.14	14.08	348.04	nr	**1466.18**
586 kW capacity	1445.46	1632.50	37.04	915.23	nr	**2547.73**
879 kW capacity	1799.79	2032.68	40.00	988.45	nr	**3021.13**
1465 kW capacity	2194.77	2478.76	45.45	1123.22	nr	**3601.98**

T: MECHANICAL/COOLING/HEATING SYSTEMS

Item	Net Price £	Material £	Labour hours	Labour £	Unit	**Total rate £**
T42: LOCAL HEATING UNITS						
Warm air unit heater for connection to LTHW or steam supplies; suitable for heights upto 3 m; recirculating type; mild steel casing; heating coil; adjustable discharge louvre; axial fan; horizontal or vertical discharge; normal speed; entering air temperature 15°C; complete with enclosures; includes fixing in position; includes connections to primary heating supply; electrical work elsewhere						
Low pressure hot water						
7.5 kW, 265 l/s	353.24	398.94	6.53	161.36	nr	**560.30**
15.4 kW, 575 l/s	427.60	482.93	7.54	186.32	nr	**669.25**
26.9 kW, 1040 l/s	579.42	654.39	8.65	213.75	nr	**868.14**
48.0 kW, 1620 l/s	763.78	862.61	9.35	231.05	nr	**1093.66**
Steam, 2 bar						
9.2 kW, 265 l/s	505.05	570.40	6.53	161.36	nr	**731.76**
18.8 kW, 575 l/s	546.89	617.65	6.82	168.52	nr	**786.17**
34.8 kW, 1040 l/s	629.00	710.39	6.82	168.52	nr	**878.91**
51.6 kW, 1625 l/s	852.11	962.37	7.10	175.45	nr	**1137.82**

T: MECHANICAL/COOLING/HEATING SYSTEMS

Item	Net Price £	Material £	Labour hours	Labour £	Unit	**Total rate £**
T60: CENTRAL REFRIGERATION PLANT						
CHILLERS						
Air cooled						
Selection of air cooled chillers based on chilled water flow and return temperatures 6°C and 12°C, and an outdoor temperature of 35°C						
Air cooled liquid chiller; refrigerant 407C; scroll compressors; twin circuit; integral controls; includes placing in position; electrical work elsewhere						
Cooling load						
100 kW	15423.17	17418.85	8.00	197.69	nr	**17616.54**
150 kW	19437.08	21952.14	8.00	197.69	nr	**22149.83**
200 kW	24058.80	27171.89	8.00	197.69	nr	**27369.58**
Air cooled liquid chiller; refrigerant 407C; reciprocating compressors; twin circuit; integral controls; includes placing in position; electrical work elsewhere						
Cooling load						
400 kW	42918.80	48472.28	8.00	197.69	nr	**48669.97**
550 kW	61200.70	69119.77	8.00	197.69	nr	**69317.46**
700 kW	71099.13	80299.00	9.00	222.40	nr	**80521.40**
Air cooled liquid chiller; refrigerant R134a; screw compressors; twin circuit; integral controls; includes placing in position; electrical work elsewhere						
Cooling load						
250 kW	43275.50	48875.14	8.00	197.69	nr	**49072.83**
400 kW	51844.50	58552.92	8.00	197.69	nr	**58750.61**
600 kW	69597.50	78603.07	8.00	197.69	nr	**78800.76**
800 kW	83650.25	94474.17	9.00	222.40	nr	**94696.57**
1000 kW	98082.25	110773.60	9.00	222.40	nr	**110996.00**
1200 kW	124586.70	140707.59	10.00	247.12	nr	**140954.71**

T: MECHANICAL/COOLING/HEATING SYSTEMS

Item	Net Price £	Material £	Labour hours	Labour £	Unit	Total rate £
T60: CENTRAL REFRIGERATION PLANT – cont						
Air cooled – cont						
Air cooled liquid chiller; ductable for indoor installation; refrigerant 407C; scroll compressors; integral controls; includes placing in position; electrical work elsewhere						
Cooling load						
40 kW	9172.73	10359.64	6.00	148.27	nr	**10507.91**
80 kW	14364.35	16223.03	6.00	148.27	nr	**16371.30**
Higher efficiency air cooled						
Selection of air cooled chillers based on chilled water flow and return temperatures of 6°C and 12°C and an outdoor temperature of 25°C						
These machines have significantly higher part load operating efficiencies than conventional air cooled machines						
Air cooled liquid chiller, refrigerant R410 A; scroll compressors; complete with free cooling facility; integral controls; including placing in position; electrical work elsewhere						
Cooling load						
250 kW	39166.28	44234.20	8.00	197.69	nr	**44431.89**
300 kW	43214.00	48805.67	8.00	197.69	nr	**49003.36**
350 kW	49936.97	56398.56	8.00	197.69	nr	**56596.25**
400 kW	56358.60	63651.12	8.00	197.69	nr	**63848.81**
450 kW	58561.32	66138.86	8.00	197.69	nr	**66336.55**
500 kW	63661.72	71899.23	8.00	197.69	nr	**72096.92**
600 kW	65330.43	73783.86	8.00	197.69	nr	**73981.55**
650 kW	76304.07	86177.43	9.00	222.40	nr	**86399.83**
700 kW	80816.13	91273.33	9.00	222.40	nr	**91495.73**

T: MECHANICAL/COOLING/HEATING SYSTEMS

Item	Net Price £	Material £	Labour hours	Labour £	Unit	**Total rate £**
Water cooled						
Selection of water cooled chillers based on chilled water flow and return temperatures of 6°C and 12°C, and condenser entering and leaving temperatures of 27°C and 33°C						
Water cooled liquid chiller; refrigerant 407C; reciprocating compressors; twin circuit; integral controls; includes placing in position; electrical work elsewhere						
Cooling load						
200 kw	19825.55	22390.88	8.00	197.69	nr	**22588.57**
350 kW	35084.72	39624.51	8.00	197.69	nr	**39822.20**
500 kW	39808.95	44960.03	8.00	197.69	nr	**45157.72**
650 kW	46140.38	52110.72	9.00	222.40	nr	**52333.12**
750 kW	56201.78	63474.01	9.00	222.40	nr	**63696.41**
Water cooled condenserless liquid chiller; refrigerant 407C; reciprocating compressors; twin circuit; integral controls; includes placing in position; electrical work elsewhere						
Cooling load						
200 kW	17110.80	19324.85	8.00	197.69	nr	**19522.54**
350 kW	30036.96	33923.59	8.00	197.69	nr	**34121.28**
500 kW	37962.89	42875.10	8.00	197.69	nr	**43072.79**
650 kW	46429.51	52437.25	9.00	222.40	nr	**52659.65**
750 kW	52831.88	59668.06	9.00	222.40	nr	**59890.46**
Water cooled liquid chiller; refrigerant R134a; screw compressors; twin circuit; integral controls; includes placing in position; electrical work elsewhere						
Cooling load						
300 kW	33059.32	37337.03	8.00	197.69	nr	**37534.72**
500 kW	42135.70	47587.84	8.00	197.69	nr	**47785.53**
700 kW	53132.93	60008.07	9.00	222.40	nr	**60230.47**
900 kW	74476.50	84113.38	9.00	222.40	nr	**84335.78**
1100 kW	82270.60	92916.00	10.00	247.12	nr	**93163.12**
1300 kW	95122.05	107430.37	10.00	247.12	nr	**107677.49**
Water cooled liquid chiller; refrigerant R134a; centrifugal compressors; twin circuit; integral controls; includes placing in position; electrical work elsewhere						
Cooling load						
1600 kW	134193.13	151557.05	11.00	271.82	nr	**151828.87**
1900 kW	163434.62	184582.24	11.00	271.82	nr	**184854.06**

T: MECHANICAL/COOLING/HEATING SYSTEMS

Item	Net Price £	Material £	Labour hours	Labour £	Unit	Total rate £
T60: CENTRAL REFRIGERATION PLANT – cont						
Water cooled – cont						
Cooling load – cont						
2200 kW	194716.08	219911.37	13.00	321.25	nr	**220232.62**
2500 kW	201176.44	227207.66	13.00	321.25	nr	**227528.91**
3000 kW	244811.84	276489.27	15.00	370.67	nr	**276859.94**
3500 kW	333216.11	376332.61	15.00	370.67	nr	**376703.28**
4000 kW	353904.96	399698.50	20.00	494.22	nr	**400192.72**
4500 kW	377579.18	426436.04	20.00	494.22	nr	**426930.26**
5000 kW	460492.21	520077.60	25.00	617.78	nr	**520695.38**
Absorption						
Absorption chiller, for operation using low pressure steam; selection based on chilled water flow and return temperatures of 6°C and 12°C, steam at 1 bar gauge and condenser entering and leaving temperatures of 27°C and 33°C; integral controls; includes placing in position; electrical work elsewhere						
Cooling load						
400 kW	57816.40	65297.55	8.00	197.69	nr	**65495.24**
700 kW	73412.95	82912.22	9.00	222.40	nr	**83134.62**
1000 kW	83753.99	94591.34	10.00	247.12	nr	**94838.46**
1300 kW	99951.25	112884.44	12.00	296.53	nr	**113180.97**
1600 kW	118291.74	133598.10	14.00	345.95	nr	**133944.05**
2000 kW	140197.89	158338.79	15.00	370.67	nr	**158709.46**
Absorption chiller, for operation using low pressure hot water; selection based on chilled water flow and return temperatures of 6°C and 12°C, cooling water temperatures of 27°C and 33°C and hot water at 90°C; integral controls; includes placing in position; electrical work elsewhere						
Cooling load						
700 kW	83753.99	94591.34	9.00	222.40	nr	**94813.74**
1000 kW	104695.56	118242.64	10.00	247.12	nr	**118489.76**
1300 kW	121767.80	137523.95	12.00	296.53	nr	**137820.48**
1600 kW	140197.89	158338.79	14.00	345.95	nr	**158684.74**

T: MECHANICAL/COOLING/HEATING SYSTEMS

Item	Net Price £	Material £	Labour hours	Labour £	Unit	**Total rate £**
HEAT REJECTION						
Dry air liquid coolers						
Dry air liquid cooler; selection based on fluid temperatures 45°C on, 40°C off at 32°C dry bulb ambient temperature; includes 20% ethylene glycol; includes placing in postion; electrical work elsewhere						
Flat coil configuration						
500 kW	18706.25	21126.75	15.00	370.67	nr	**21497.42**
Extra for inverter panels (factory wired and mounted on units)	7148.35	8073.31	15.00	370.67	nr	**8443.98**
800 kW	24087.50	27204.30	15.00	370.67	nr	**27574.97**
Extra for inverter panels (factory wired and mounted on units)	13138.45	14838.50	15.00	370.67	nr	**15209.17**
1100 kW	29215.58	32995.93	15.00	370.67	nr	**33366.60**
Extra for inverter panels (factory wired and mounted on units)	14297.73	16147.78	15.00	370.67	nr	**16518.45**
1400 kW	39462.50	44568.75	15.00	370.67	nr	**44939.42**
Extra over for inverter panels (factory wired and mounted on units)	19707.67	22257.75	15.00	370.67	nr	**22628.42**
1700 kW	48179.10	54413.24	15.00	370.67	nr	**54783.91**
Extra for inverter panels (factory wired and mounted on units)	20037.72	22630.51	15.00	370.67	nr	**23001.18**
2000 kW	57400.00	64827.27	15.00	370.67	nr	**65197.94**
Extra for inverter panels (factory wired and mounted on units)	26373.25	29785.82	15.00	370.67	nr	**30156.49**
Note: Heat rejection capacities above 500 kW require multiple units. Prices are therefore for total number of units.						
'Vee' type coil configuration						
500 kW	16605.00	18753.61	15.00	370.67	nr	**19124.28**
Extra for inverter panels (factory wired and mounted on units)	5028.65	5679.33	15.00	370.67	nr	**6050.00**
800 kW	24087.50	27204.30	15.00	370.67	nr	**27574.97**
Extra for inverter panels (factory wired and mounted on units)	11644.00	13150.68	15.00	370.67	nr	**13521.35**
1100 kW	33324.80	37636.86	15.00	370.67	nr	**38007.53**
Extra for inverter panels (factory wired and mounted on units)	14081.45	15903.52	15.00	370.67	nr	**16274.19**
1400 kW	47457.50	53598.27	15.00	370.67	nr	**53968.94**
Extra for inverter panels (factory wired and mounted on units)	17510.08	19775.80	15.00	370.67	nr	**20146.47**
1700 kW	66717.25	75350.13	15.00	370.67	nr	**75720.80**
Extra for inverter panels (factory wired and mounted on units)	23291.08	26304.83	15.00	370.67	nr	**26675.50**

T: MECHANICAL/COOLING/HEATING SYSTEMS

Item	Net Price £	Material £	Labour hours	Labour £	Unit	Total rate £
T60: CENTRAL REFRIGERATION PLANT – cont						
'Vee' type coil configuration – cont						
2000 kW	66625.00	75245.94	15.00	370.67	nr	**75616.61**
Extra for inverter panels (factory wired and mounted on units)	26264.60	29663.11	15.00	370.67	nr	**30033.78**
Note: Heat rejection capacities above 1100 kW require multiple units. Prices are for total number of units.						
Air cooled condensers						
Air cooled condenser; refrigerant 407C; selection based on condensing temperature of 45°C at 32°C dry bulb ambient; includes placing in position; electrical work elsewhere						
Flat coil configuration						
500 kW	15060.66	17009.44	15.00	370.67	nr	**17380.11**
Extra for inverter panels (factory wired and mounted on units)	6723.25	7593.21	15.00	370.67	nr	**7963.88**
800 kW	24708.53	27905.69	15.00	370.67	nr	**28276.36**
Extra for inverter panels (factory wired and mounted on units)	10549.99	11915.11	15.00	370.67	nr	**12285.78**
1100 kW	33078.37	37358.55	15.00	370.67	nr	**37729.22**
Extra for inverter panels (factory wired and mounted on units)	14998.03	16938.70	15.00	370.67	nr	**17309.37**
1400 kW	41297.83	46641.56	15.00	370.67	nr	**47012.23**
Extra for inverter panels (factory wired and mounted on units)	18443.70	20830.23	15.00	370.67	nr	**21200.90**
1700 kW	53777.39	60735.92	15.00	370.67	nr	**61106.59**
Extra for inverter panels (factory wired and mounted on units)	27665.54	31245.32	15.00	370.67	nr	**31615.99**
2000 kW	61946.76	69962.36	15.00	370.67	nr	**70333.03**
Extra for inverter panels (factory wired and mounted on units)	27665.54	31245.32	15.00	370.67	nr	**31615.99**
Note: Heat rejection capacities above 500 kW require multiple units. Prices are for total number of units						

T: MECHANICAL/COOLING/HEATING SYSTEMS

Item	Net Price £	Material £	Labour hours	Labour £	Unit	**Total rate £**
'Vee' type coil configuration						
500 kW	14785.02	16698.12	15.00	370.67	nr	**17068.79**
Extra for inverter panels (factory wired and mounted on units)	5275.01	5957.56	15.00	370.67	nr	**6328.23**
800 kW	22240.18	25117.95	15.00	370.67	nr	**25488.62**
Extra for inverter panels (factory wired and mounted on units)	10800.59	12198.13	15.00	370.67	nr	**12568.80**
1100 kW	32639.83	36863.26	15.00	370.67	nr	**37233.93**
Extra for inverter panels (factory wired and mounted on units)	18368.52	20745.31	15.00	370.67	nr	**21115.98**
1400 kW	38002.53	42919.87	15.00	370.67	nr	**43290.54**
Extra for inverter panels (factory wired and mounted on units)	21601.17	24396.26	15.00	370.67	nr	**24766.93**
1700 kW	48959.74	55294.88	15.00	370.67	nr	**55665.55**
Extra for inverter panels (factory wired and mounted on units)	27565.30	31132.11	15.00	370.67	nr	**31502.78**
2000 kW	57003.81	64379.82	15.00	370.67	nr	**64750.49**
Extra for inverter panels (factory wired and mounted on units)	32401.76	36594.39	15.00	370.67	nr	**36965.06**
Note: Heat rejection capacities above 1100 kW require multiple units. Prices are for total number of units.						
Cooling towers						
Cooling towers; forced draught, centrifugal fan, conterflow design; based on water temperatures of 35°C on and 29°C off at 21°C wet bulb ambient temperature; includes placing in position; electrical work elsewhere						
Open circuit type						
900 kW	10032.03	11330.13	20.00	494.22	nr	**11824.35**
Extra for stainless steel construction	4497.69	5079.67	–	–	nr	**5079.67**
Extra for intake and discharge sound attenuation	4840.69	5467.05	–	–	nr	**5467.05**
Extra for fan dampers for capacity control	1123.14	1268.47	–	–	nr	**1268.47**
1500 kW	15564.96	17578.99	20.00	494.22	nr	**18073.21**
Extra for stainless steel construction	7204.58	8136.82	–	–	nr	**8136.82**
Extra for intake and discharge sound attenuation	7253.90	8192.52	–	–	nr	**8192.52**
Extra for fan dampers for capacity control	1139.97	1287.47	–	–	nr	**1287.47**
2100 kW	22081.33	24938.54	20.00	494.22	nr	**25432.76**
Extra for stainless steel construction	10218.26	11540.45	–	–	nr	**11540.45**
Extra for intake and discharge sound attenuation	10770.88	12164.58	–	–	nr	**12164.58**

T: MECHANICAL/COOLING/HEATING SYSTEMS

Item	Net Price £	Material £	Labour hours	Labour £	Unit	Total rate £
T60: CENTRAL REFRIGERATION PLANT – cont						
Cooling towers – cont						
Open circuit type – cont						
Extra for fan dampers for capacity control	1301.72	1470.16	–	–	nr	**1470.16**
2700 kW	26667.10	30117.69	20.00	494.22	nr	**30611.91**
Extra for stainless steel construction	12465.04	14077.96	–	–	nr	**14077.96**
Extra for intake and discharge sound attenuation	14510.40	16387.97	–	–	nr	**16387.97**
Extra for fan dampers for capacity control	2026.33	2288.52	–	–	nr	**2288.52**
3300 kW	32862.54	37114.79	20.00	494.22	nr	**37609.01**
Extra for stainless steel construction	15213.39	17181.93	–	–	nr	**17181.93**
Extra for intake and discharge sound attenuation	13684.85	15455.60	–	–	nr	**15455.60**
Extra for fan dampers for capacity control	1657.56	1872.04	–	–	nr	**1872.04**
3900 kW	37039.42	41832.14	20.00	494.22	nr	**42326.36**
Extra for stainless steel construction	17147.87	19366.72	–	–	nr	**19366.72**
Extra for intake and discharge sound attenuation	13947.52	15752.26	–	–	nr	**15752.26**
Extra for fan dampers for capacity control	1657.56	1872.04	–	–	nr	**1872.04**
4500 kW	41684.71	47078.51	23.00	568.35	nr	**47646.86**
Extra for stainless steel construction	19438.85	21954.14	–	–	nr	**21954.14**
Extra for intake and discharge sound attenuation	18498.37	20891.97	–	–	nr	**20891.97**
Extra for fan dampers for capacity control	2608.61	2946.15	–	–	nr	**2946.15**
5100 kW	48151.90	54382.51	23.00	568.35	nr	**54950.86**
Extra for stainless steel construction	22185.84	25056.58	–	–	nr	**25056.58**
Extra for intake and discharge sound attenuation	18449.19	20836.43	–	–	nr	**20836.43**
Extra for fan dampers for capacity control	2608.61	2946.15	–	–	nr	**2946.15**
5700 kW	50906.71	57493.78	30.00	741.34	nr	**58235.12**
Extra for stainless steel construction	23770.74	26846.55	–	–	nr	**26846.55**
Extra for intake and discharge sound attenuation	27320.53	30855.67	–	–	nr	**30855.67**
Extra for fan dampers for capacity control	2868.70	3239.90	–	–	nr	**3239.90**
6300 kW	61379.97	69322.23	30.00	741.34	nr	**70063.57**
Extra for stainless steel construction	28416.28	32093.21	–	–	nr	**32093.21**
Extra for intake and discharge sound attenuation	27715.18	31301.38	–	–	nr	**31301.38**
Extra for fan dampers for capacity control	2868.70	3239.90	–	–	nr	**3239.90**
Closed circuit type (includes 20% ethylene glycol)						
900 kW	27966.22	31584.91	20.00	494.22	nr	**32079.13**
Extra for stainless steel construction	22771.70	25718.24	–	–	nr	**25718.24**
Extra for intake and discharge sound attenuation	8415.39	9504.30	–	–	nr	**9504.30**
Extra for fan dampers for capacity control	1139.97	1287.47	–	–	nr	**1287.47**
1500 kW	51993.60	58721.31	20.00	494.22	nr	**59215.53**
Extra for stainless steel construction	41453.35	46817.20	–	–	nr	**46817.20**

T: MECHANICAL/COOLING/HEATING SYSTEMS

Item	Net Price £	Material £	Labour hours	Labour £	Unit	Total rate £
Extra for intake and discharge sound attenuation	13432.53	15170.63	–	–	nr	**15170.63**
Extra for fan dampers for capacity control	1657.56	1872.04	–	–	nr	**1872.04**
2100 kW	63362.29	71561.06	20.00	494.22	nr	**72055.28**
Extra for stainless steel construction	56317.13	63604.28	–	–	nr	**63604.28**
Extra for intake and discharge sound attenuation	16169.24	18261.46	–	–	nr	**18261.46**
Extra for fan dampers for capacity control	1657.56	1872.04	–	–	nr	**1872.04**
2700 kW	68315.55	77155.24	25.00	617.78	nr	**77773.02**
Extra for stainless steel construction	69779.80	78808.95	–	–	nr	**78808.95**
Extra for intake and discharge sound attenuation	24133.54	27256.30	–	–	nr	**27256.30**
Extra for fan dampers for capacity control	2507.68	2832.16	–	–	nr	**2832.16**
3300 kW	105256.62	118876.30	25.00	617.78	nr	**119494.08**
Extra for stainless steel construction	87051.60	98315.64	–	–	nr	**98315.64**
Extra for intake and discharge sound attenuation	32102.98	36256.94	–	–	nr	**36256.94**
Extra for fan dampers for capacity control	2868.70	3239.90	–	–	nr	**3239.90**
3900 kW	115892.91	130888.87	25.00	617.78	nr	**131506.65**
Extra for stainless steel construction	99688.41	112587.59	–	–	nr	**112587.59**
Extra for intake and discharge sound attenuation	32628.32	36850.26	–	–	nr	**36850.26**
Extra for fan dampers for capacity control	2868.70	3239.90	–	–	nr	**3239.90**
4500 kW	138746.69	156699.82	40.00	988.45	nr	**157688.27**
Extra for stainless steel construction	103418.25	116800.06	–	–	nr	**116800.06**
Extra for intake and discharge sound attenuation	46635.37	52669.75	–	–	nr	**52669.75**
Extra for fan dampers for capacity control	5015.36	5664.33	–	–	nr	**5664.33**
5100 kW	154746.49	174769.91	40.00	988.45	nr	**175758.36**
Extra for stainless steel construction	133646.06	150939.19	–	–	nr	**150939.19**
Extra for intake and discharge sound attenuation	46635.37	52669.75	–	–	nr	**52669.75**
Extra for fan dampers for capacity control	5015.36	5664.33	–	–	nr	**5664.33**
5700 kW	190811.55	215501.61	40.00	988.45	nr	**216490.06**
Extra for stainless steel construction	138047.08	155909.68	–	–	nr	**155909.68**
Extra for intake and discharge sound attenuation	52134.67	58880.63	–	–	nr	**58880.63**
Extra for fan dampers for capacity control	5737.40	6479.79	–	–	nr	**6479.79**
6300 kW	203818.34	230191.42	40.00	988.45	nr	**231179.87**
Extra for stainless steel construction	157903.51	178335.43	–	–	nr	**178335.43**
Extra for intake and discharge sound attenuation	52134.67	58880.63	–	–	nr	**58880.63**
Extra for fan dampers for capacity control	5737.40	6479.79	–	–	nr	**6479.79**

T: MECHANICAL/COOLING/HEATING SYSTEMS

Item	Net Price £	Material £	Labour hours	Labour £	Unit	**Total rate £**
T61: CHILLED WATER						
SCREWED STEEL						
Y10 – PIPELINES						
For pipework prices refer to Section T31 – Low Temperature Hot Water Heating, with the exception of chilled water blocks within brackets as detailed hereafter. For minimum fixing distances, refer to the Tables and Memoranda to the rear of the book						
FIXINGS						
For steel pipes; black malleable iron						
Oversized pipe clip, to contain 30 mm insulation block for vapour barrier						
15 mm dia.	3.57	4.03	0.10	2.47	nr	**6.50**
20 mm dia.	3.60	4.07	0.11	2.72	nr	**6.79**
25 mm dia.	3.79	4.28	0.12	2.97	nr	**7.25**
32 mm dia.	3.90	4.41	0.14	3.46	nr	**7.87**
40 mm dia.	4.12	4.65	0.15	3.70	nr	**8.35**
50 mm dia.	5.88	6.64	0.16	3.96	nr	**10.60**
65 mm dia.	6.35	7.17	0.30	7.41	nr	**14.58**
80 mm dia.	6.74	7.61	0.35	8.65	nr	**16.26**
100 mm dia.	10.68	12.07	0.40	9.88	nr	**21.95**
125 mm dia.	12.39	13.99	0.60	14.83	nr	**28.82**
150 mm dia.	19.38	21.89	0.77	19.03	nr	**40.92**
200 mm dia.	24.74	27.94	0.90	22.24	nr	**50.18**
250 mm dia.	30.54	34.50	1.10	27.18	nr	**61.68**
300 mm dia.	36.36	41.07	1.25	30.89	nr	**71.96**
350 mm dia.	42.17	47.62	1.50	37.07	nr	**84.69**
400 mm dia.	47.23	53.35	1.75	43.25	nr	**96.60**
Screw on backplate (Male), black malleable iron; plugged and screwed						
M12	0.48	0.54	0.10	2.47	nr	**3.01**
Screw on backplate (Female), black malleable iron; plugged and screwed						
M12	0.50	1.11	0.10	2.47	nr	**3.58**

T: MECHANICAL/COOLING/HEATING SYSTEMS

Item	Net Price £	Material £	Labour hours	Labour £	Unit	Total rate £
Extra over channel sections for fabricated hangers and brackets						
Galvanized steel; including inserts, bolts, nuts, washers; fixed to backgrounds						
41 × 21 mm	5.65	6.39	0.29	7.17	m	**13.56**
41 × 41 mm	6.77	7.65	0.29	7.17	m	**14.82**
Threaded rods; metric thread; including nuts, washers etc.						
12 mm dia. × 600 mm long	2.86	3.23	0.18	4.45	nr	**7.68**
For plastic pipework suitable for chilled water systems, refer to ABS pipework details in Section S10 – Cold Water with the exception of chilled water blocks within brackets as detailed for the aforementioned steel pipe. For minimum fixing distances, refer to the Tables and Memoranda to the rear of the book						
For copper pipework, refer to Section S10 – Cold Water with the exception of chilled water blocks within brackets as detailed hereafter. For minimum fixing distances, refer to the Tables and Memoranda to the rear of the book						
FIXINGS						
For copper pipework						
Oversized pipe clip, to contain 30 mm insulation block for vapour barrier						
15 mm dia.	3.57	4.03	0.10	2.47	nr	**6.50**
22 mm dia.	3.60	4.07	0.11	2.72	nr	**6.79**
28 mm dia.	3.79	4.28	0.12	2.97	nr	**7.25**
35 mm dia.	3.90	4.41	0.14	3.46	nr	**7.87**
42 mm dia.	4.12	4.65	0.15	3.70	nr	**8.35**
54 mm dia.	5.88	6.64	0.16	3.96	nr	**10.60**
67 mm dia.	6.35	7.17	0.30	7.41	nr	**14.58**
76 mm dia.	6.74	7.61	0.35	8.65	nr	**16.26**
108 mm dia.	2.77	15.19	0.40	9.88	nr	**25.07**
133 mm dia.	12.39	13.99	0.60	14.83	nr	**28.82**
159 mm dia.	19.38	21.89	0.77	19.03	nr	**40.92**
Screw on backplate, female						
All sizes 15 mm to 54 mm × 10 mm	1.45	1.64	0.10	2.47	nr	**4.11**
Screw on backplate, male						
All sizes 15 mm to 54 mm × 10 mm	1.91	2.15	0.10	2.47	nr	**4.62**

T: MECHANICAL/COOLING/HEATING SYSTEMS

Item	Net Price £	Material £	Labour hours	Labour £	Unit	**Total rate £**
T61: CHILLED WATER – cont						
FIXINGS – cont						
Extra over channel sections for fabricated hangers and brackets						
Galvanized steel; including inserts, bolts, nuts, washers; fixed to backgrounds						
41 × 21 mm	5.65	6.39	0.29	7.17	m	**13.56**
41 × 41 mm	6.77	7.65	0.29	7.17	m	**14.82**
Threaded rods; metric thread; including nuts, washers etc.						
10 mm dia. × 600 mm long for ring clips up to 54 mm	1.86	2.10	0.18	4.45	nr	**6.55**
12 mm dia. × 600 mm long for ring clips from 54 mm	2.86	3.23	0.18	4.45	nr	**7.68**
Y11 – PIPELINE ANCILLARIES						
For prices for ancillaries refer to Section T31 – Low Temperature Hot Water Heating						
Y22 – HEAT EXCHANGERS						
Plate heat exchanger; for use in CHW systems; painted carbon steel frame, stainless steel plates, nitrile rubber gaskets, design pressure of 10 bar and operating temperature of 110/135°C						
Primary side; 13°C in, 8°C out; secondary side; 6°C in, 11°C out						
264 kW, 12.60 l/s	4984.03	5628.94	10.00	247.12	nr	**5876.06**
290 kW, 13.85 l/s	5253.63	5933.42	12.00	296.53	nr	**6229.95**
316 kW, 15.11 l/s	5469.07	6176.74	12.00	296.53	nr	**6473.27**
350 kW, 16.69 l/s	5881.53	6642.57	16.00	395.38	nr	**7037.95**
395 kW, 18.88 l/s	6293.98	7108.39	16.00	395.38	nr	**7503.77**
454 kW, 21.69 l/s	6836.63	7721.26	10.00	247.12	nr	**7968.38**
475 kW, 22.66 l/s	6969.13	7870.90	10.00	247.12	nr	**8118.02**
527 kW, 25.17 l/s	7552.09	8529.29	10.00	247.12	nr	**8776.41**
554 kW, 26.43 l/s	7752.56	8755.70	10.00	247.12	nr	**9002.82**
580 kW, 27.68 l/s	8018.19	9055.70	10.00	247.12	nr	**9302.82**
633 kW, 30.19 l/s	8500.28	9600.18	10.00	247.12	nr	**9847.30**
661 kW, 31.52 l/s	8705.36	9831.79	10.00	247.12	nr	**10078.91**
713 kW, 34.04 l/s	9292.93	10495.39	12.00	296.53	nr	**10791.92**
740 kW, 35.28 l/s	9495.71	10724.40	12.00	296.53	nr	**11020.93**
804 kW, 38.33 l/s	10110.94	11419.25	12.00	296.53	nr	**11715.78**
1925 kW, 91.82 l/s	16972.92	19169.14	15.00	370.67	nr	**19539.81**
2710 kW, 129.26 l/s	22544.54	25461.69	15.00	370.67	nr	**25832.36**
3100 kW, 147.87 l/s	25116.05	28365.94	15.00	370.67	nr	**28736.61**

T: MECHANICAL/COOLING/HEATING SYSTEMS

Item	Net Price £	Material £	Labour hours	Labour £	Unit	**Total rate £**
Note: For temperature conditions different to those above, the cost of the units can vary significantly, and therefore the manufacturers advice should be sought.						
Y24 – TRACE HEATING						
Trace heating; for freeze protection or temperature maintainance of pipework; to BS 6351; including fixing to parent structures by plastic pull ties						
Straight laid						
15 mm	19.68	22.23	0.27	6.67	m	**28.90**
25 mm	19.68	22.23	0.27	6.67	m	**28.90**
28 mm	19.68	22.23	0.27	6.67	m	**28.90**
32 mm	19.68	22.23	0.30	7.41	m	**29.64**
35 mm	19.68	22.23	0.31	7.66	m	**29.89**
50 mm	19.68	22.23	0.34	8.40	m	**30.63**
100 mm	19.68	22.23	0.40	9.88	m	**32.11**
150 mm	19.68	22.23	0.40	9.88	m	**32.11**
Helically wound						
15 mm	25.07	28.32	1.00	24.71	m	**53.03**
25 mm	25.07	28.32	1.00	24.71	m	**53.03**
28 mm	25.07	28.32	1.00	24.71	m	**53.03**
32 mm	25.07	28.32	1.00	24.71	m	**53.03**
35 mm	25.07	28.32	1.00	24.71	m	**53.03**
50 mm	25.07	28.32	1.00	24.71	m	**53.03**
100 mm	25.07	28.32	1.00	24.71	m	**53.03**
150 mm	25.07	28.32	1.00	24.71	m	**53.03**
Accessories for trace heating; weatherproof; polycarbonate enclosure to IP standards; fully installed						
Connection junction box						
100 × 100 × 75 mm	60.40	68.22	1.40	34.60	nr	**102.82**
Single air thermostat						
150 × 150 × 75 mm	132.33	149.45	1.42	35.09	nr	**184.54**
Single capillary thermostat						
150 × 150 × 75 mm	190.39	215.03	1.46	36.08	nr	**251.11**
Twin capillary thermostat						
150 × 150 × 75 mm	341.57	385.76	1.46	36.08	nr	**421.84**

T: MECHANICAL/COOLING/HEATING SYSTEMS

Item	Net Price £	Material £	Labour hours	Labour £	Unit	Total rate £
T61: CHILLED WATER – cont						
EQUIPMENT						
PRESSURIZATION UNITS						
Chilled water packaged pressurization unit complete with expansion vessel(s), interconnecting pipework and necessary isolating and drain valves; includes placing in position; electrical work elasewhere						
Selection based on a final working pressure of 4 bar, a 3 m static head and system operating temperatures of 6°/12°C						
System volume						
1800 litres	1439.00	1625.20	8.00	197.69	nr	**1822.89**
4500 litres	1620.00	1829.63	8.00	197.69	nr	**2027.32**
7200 litres	1755.00	1982.08	10.00	247.12	nr	**2229.20**
9900 litres	1755.00	1982.08	10.00	247.12	nr	**2229.20**
15300 litres	2355.00	2659.72	13.00	321.25	nr	**2980.97**
22500 litres	2470.00	2789.61	20.00	494.22	nr	**3283.83**
27000 litres	2470.00	2789.61	20.00	494.22	nr	**3283.83**
CHILLED BEAMS						
Static (passive) beams; based on water at 14°C flow and 16°C return, 24°C room temperature; 600 mm wide coil providing 350–400 W/m output						
Static cooled beam for exposed installation with standard casing	122.38	138.22	4.00	98.85	m	**237.07**
Static cooled beam for installation above open grid or perforated ceiling	83.92	94.78	4.00	98.85	m	**193.63**

T: MECHANICAL/COOLING/HEATING SYSTEMS

Item	Net Price £	Material £	Labour hours	Labour £	Unit	**Total rate £**
Ventilated (active) beams; based on water at 14°C flow and 16°C return, 24°C room temperature; air supply at 10 l/s/linear metre; 300 mm wide beam providing 250–350 W/m output unless stated otherwise; all exposed beams c/w standard casing; electrical work elsewhere						
Ventilated cooled beam flush mounted within a false ceiling; closed type with integrated secondary air circulation; 600 mm wide beam providing 400 W/m output	189.98	214.57	4.50	111.20	m	**325.77**
Ventilated cooled beam flush mounted within a false ceiling; open type	167.83	189.55	4.00	98.85	m	**288.40**
Ventilated cooled beam for exposed mounting with standard casing	167.83	189.55	4.00	98.85	m	**288.40**
Ventilated cooled beam flush mounted within a false ceiling; open type; with recessed integrated flush mounted 28 W or 35 W T5 light fittings	304.20	343.57	4.00	98.85	m	**442.42**
Ventilated cooled beam for exposed mounting with recessed integrated flush mounted 28 W or 35 W T5 light fittings	304.20	343.57	4.00	98.85	m	**442.42**
Ventilated cooled beam for exposed mounting with recessed integrated flush mounted direct and indirect 28 W or 35 W T5 light fittings	332.18	375.17	4.00	98.85	m	**474.02**
LEAK DETECTION						
Leak detection system consisting of a central control module connected by a leader cable to water sensing cables						
Control modules						
Alarm only	526.00	594.06	4.00	109.28	nr	**703.34**
Alarm and location	2279.00	2573.89	8.00	218.57	nr	**2792.46**
Cables						
Sensing – 3 m length	94.00	106.16	4.00	109.28	nr	**215.44**
Sensing – 7.5 m length	138.00	155.86	4.00	109.28	nr	**265.14**
Sensing – 15 m length	239.00	269.92	8.00	218.57	nr	**488.49**
Leader – 3.5 m length	41.00	46.31	2.00	54.64	nr	**100.95**
End terminal						
End terminal	17.00	19.20	0.05	1.37	nr	**20.57**

T: MECHANICAL/COOLING/HEATING SYSTEMS

Item	Net Price £	Material £	Labour hours	Labour £	Unit	Total rate £
T61: CHILLED WATER – cont						
ENERGY METERS						
Ultrasonic						
Energy meter for measuring energy use in chilled water systems; includes ultrasonic flow meter (with sensor and signal converter), energy calculator, pair of temperature sensors with brass pockets, and 3 m of interconnecting cable; includes fixing in position; electrical work elsewhere						
Pipe size (flanged connections to PN16); maximum flow rate						
50 mm, 36 m³/hr	1172.30	1323.99	1.80	44.47	nr	**1368.46**
65 mm, 60 m³/hr	1292.23	1459.44	2.32	57.32	nr	**1516.76**
80 mm, 100 m³/hr	1448.49	1635.92	2.56	63.26	nr	**1699.18**
125 mm, 250 m³/hr	1678.22	1895.37	3.60	88.96	nr	**1984.33**
150 mm, 360 m³/hr	1821.79	2057.52	4.80	118.61	nr	**2176.13**
200 mm, 600 m³/hr	2035.48	2298.87	6.24	154.19	nr	**2453.06**
250 mm, 1000 m³/hr	2348.82	2652.74	9.60	237.23	nr	**2889.97**
300 mm, 1500 m³/hr	2760.99	3118.25	10.80	266.87	nr	**3385.12**
350 mm, 2000 m³/hr	3326.87	3757.35	13.20	326.20	nr	**4083.55**
400 mm, 2500 m³/hr	3808.29	4301.06	15.60	385.50	nr	**4686.56**
500 mm, 3000 m³/hr	4321.79	4881.01	24.00	593.07	nr	**5474.08**
600 mm, 3500 m³/hr	4855.58	5483.87	28.00	691.92	nr	**6175.79**
Electromagnetic						
Energy meter for measuring energy use in chilled water systems; includes electromagnetic flow meter (with sensor and signal converter), energy calculator, pair of temperature sensors with brass pockets, and 3 m of interconnecting cable; includes fixing in position; electrical work elsewhere						
Pipe size (flanged connections to PN40); maximum flow rate						
25 mm, 17.7 m³/hr	840.46	949.21	1.48	36.57	nr	**985.78**
40 mm, 45 m³/hr	848.50	958.29	1.55	38.30	nr	**996.59**
Pipe size (flanged connections to PN16); maximum flow rate						
50 mm, 70 m³/hr	858.00	969.02	1.80	44.47	nr	**1013.49**
65 mm, 120 m³/hr	861.65	973.14	2.32	57.32	nr	**1030.46**
80 mm, 180 m³/hr	866.04	978.10	2.56	63.26	nr	**1041.36**
125 mm, 450 m³/hr	942.78	1064.77	3.60	88.96	nr	**1153.73**
150 mm, 625 m³/hr	1000.51	1129.97	4.80	118.61	nr	**1248.58**
200 mm, 1100 m³/hr	1076.52	1215.82	6.24	154.19	nr	**1370.01**
250 mm, 1750 m³/hr	1208.07	1364.39	9.60	237.23	nr	**1601.62**
300 mm, 2550 m³/hr	1539.87	1739.12	10.80	266.87	nr	**2005.99**
350 mm, 3450 m³/hr	2003.22	2262.43	13.20	326.20	nr	**2588.63**
400 mm, 4500 m³/hr	2285.32	2581.03	15.60	385.50	nr	**2966.53**

T: MECHANICAL/COOLING/HEATING SYSTEMS

Item	Net Price £	Material £	Labour hours	Labour £	Unit	Total rate £
T70: LOCAL COOLING UNITS						
Split system with ceiling void evaporator unit and external condensing unit						
Ceiling mounted 4 way blow cassette heat pump unit with remote fan speed and load control; refrigerant 407C; includes outdoor unit						
Cooling 3.6 kW, heating 4.1 kW	1604.42	1812.02	35.00	864.89	nr	**2676.91**
Cooling 4.9 kW, heating 5.5 kW	1768.06	1996.83	35.00	864.89	nr	**2861.72**
Cooling 7.1 kW, heating 8.2 kW	2140.06	2416.97	35.00	864.89	nr	**3281.86**
Cooling 10 kW, heating 11.2 kW	2531.27	2858.80	35.00	864.89	nr	**3723.69**
Cooling 12.20 kW, heating 14.60 kW	2769.05	3127.35	35.00	864.89	nr	**3992.24**
Ceiling mounted 4 way blow cooling only unit with remote fan speed and load control; refrigerant 407C; includes outdoor unit						
Cooling 3.80 kW	1452.28	1640.20	35.00	864.89	nr	**2505.09**
Cooling 5.20 kW	1622.31	1832.23	35.00	864.89	nr	**2697.12**
Cooling 7.10 kW	2000.73	2259.62	35.00	864.89	nr	**3124.51**
Cooling 10 kW	2324.17	2624.91	35.00	864.89	nr	**3489.80**
Cooling 12.2 kW	2463.52	2782.28	35.00	864.89	nr	**3647.17**
In ceiling, ducted heat pump unit with remote fan speed and load control; refrigerant 407C; includes outdoor unit						
Cooling 3.60 kW, heating 4.10 kW	1176.14	1328.33	35.00	864.89	nr	**2193.22**
Cooling 4.90 kW, heating 5.50 kW	1352.57	1527.58	35.00	864.89	nr	**2392.47**
Cooling 7.10 kW, heating 8.20 kW	1352.57	1527.58	35.00	864.89	nr	**2392.47**
Cooling 10 kW, heating 11.20 kW	1352.57	1527.58	35.00	864.89	nr	**2392.47**
Cooling 12.20 kW, heating 14.50 kW	2726.87	3079.72	35.00	864.89	nr	**3944.61**
In ceiling, ducted cooling only unit with remote fan speed and load control; refrigerant 407C; includes outdoor unit						
Cooling 3.70 kW	1061.09	1198.39	35.00	864.89	nr	**2063.28**
Cooling 4.90 kW	1261.80	1425.07	35.00	864.89	nr	**2289.96**
Cooling 7.10 kW	1904.85	2151.33	35.00	864.89	nr	**3016.22**
Cooling 10 kW	2172.03	2453.08	35.00	864.89	nr	**3317.97**
Cooling 12.3 kW	2421.32	2734.63	35.00	864.89	nr	**3599.52**
Room units						
Ceiling mounted 4 way blow cassette heat pump unit with remote fan speed and load control; refrigerant 407C; excludes outdoor unit						
Cooling 3.6 kW, heating 4.1 kW	1012.51	1143.52	17.00	420.09	nr	**1563.61**
Cooling 4.9 kW, heating 5.5 kW	1036.80	1170.96	17.00	420.09	nr	**1591.05**
Cooling 7.1 kW, heating 8.2 kW	1122.45	1267.69	17.00	420.09	nr	**1687.78**
Cooling 10 kW, heating 11.2 kW	1215.78	1373.09	17.00	420.09	nr	**1793.18**
Cooling 12.20 kW, heating 14.60 kW	1318.05	1488.60	17.00	420.09	nr	**1908.69**

T: MECHANICAL/COOLING/HEATING SYSTEMS

Item	Net Price £	Material £	Labour hours	Labour £	Unit	Total rate £
T70: LOCAL COOLING UNITS – cont						
Room units – cont						
Ceiling mounted 4 way blow cooling unit with remote fan speed and load control; refrigerant 407C; excludes outdoor unit						
Cooling 3.80 kW	956.25	1079.99	17.00	420.09	nr	**1500.08**
Cooling 5.20 kW	963.93	1088.66	17.00	420.09	nr	**1508.75**
Cooling 7.10 kW	1122.45	1267.69	17.00	420.09	nr	**1687.78**
Cooling 10 kW	1215.78	1373.09	17.00	420.09	nr	**1793.18**
Cooling 12.2 kW	1318.05	1488.60	17.00	420.09	nr	**1908.69**
In ceiling, ducted heat pump unit with remote fan speed and load control; refrigerant 407C; excludes outdoor unit						
Cooling 3.60 kW, heating 4.10 kW	584.24	659.84	17.00	420.09	nr	**1079.93**
Cooling 4.90 kW, heating 5.50 kW	621.31	701.71	17.00	420.09	nr	**1121.80**
Cooling 7.10 kW, heating 8.20 kW	1026.58	1159.41	17.00	420.09	nr	**1579.50**
Cooling 10 kW, heating 11.20 kW	1063.64	1201.27	17.00	420.09	nr	**1621.36**
Cooling 12.20 kW, heating 14.50 kW	1275.87	1440.96	17.00	420.09	nr	**1861.05**
In ceiling , ducted cooling unit only with remote fan speed and load control; refrigerant 407C; excludes outdoor unit						
Cooling 3.70 kW	565.06	638.17	17.00	420.09	nr	**1058.26**
Cooling 4.90 kW	603.42	681.50	17.00	420.09	nr	**1101.59**
Cooling 7.10 kW	1026.58	1159.41	17.00	420.09	nr	**1579.50**
Cooling 10 kW	1063.64	1201.27	17.00	420.09	nr	**1621.36**
Cooling 12.3 kW	1275.87	1440.96	17.00	420.09	nr	**1861.05**
External condensing units suitable for connection to multiple indoor units; inverter driven; refrigerant 407C						
Cooling only						
9 kW	2210.39	2496.40	17.00	420.09	nr	**2916.49**
Heat pump						
Cooling 5.20 kW, heating 6.10 kW	1575.01	1778.81	17.00	420.09	nr	**2198.90**
Cooling 6.80 kW, heating 2.50 kW	2008.40	2268.28	17.00	420.09	nr	**2688.37**
Cooling 8 kW, heating 9.60 kW	2325.43	2626.33	17.00	420.09	nr	**3046.42**
Cooling 14.50 kW, heating 16.50 kW	3859.54	4358.95	21.00	518.94	nr	**4877.89**

U: VENTILATION/AIR CONDITIONING SYSTEMS

Item	Net Price £	Material £	Labour hours	Labour £	Unit	Total rate £
U10: DUCTWORK: CIRCULAR						
Y30 – AIR DUCTLINES						
Galvanized sheet metal DW144 class B spirally wound circular section ductwork; including all necessary stiffeners, joints, couplers in the running length and duct supports						
Straight duct						
80 mm dia.	3.12	3.81	0.87	22.51	m	**26.32**
100 mm dia.	4.02	4.91	0.87	22.51	m	**27.42**
160 mm dia.	4.41	5.39	0.87	22.51	m	**27.90**
200 mm dia.	5.62	6.87	0.87	22.51	m	**29.38**
250 mm dia.	6.83	8.35	1.21	31.29	m	**39.64**
315 mm dia.	8.27	10.11	1.21	31.29	m	**41.40**
355 mm dia.	11.30	13.81	1.21	31.29	m	**45.10**
400 mm dia.	12.66	15.48	1.21	31.29	m	**46.77**
450 mm dia.	13.97	17.08	1.21	31.29	m	**48.37**
500 mm dia.	15.20	18.58	1.21	31.29	m	**49.87**
630 mm dia.	29.07	35.53	1.39	35.96	m	**71.49**
710 mm dia.	32.19	39.36	1.39	35.96	m	**75.32**
800 mm dia.	36.98	45.21	1.44	37.24	m	**82.45**
900 mm dia.	45.65	55.81	1.46	37.76	m	**93.57**
1000 mm dia.	57.02	69.71	1.65	42.67	m	**112.38**
1120 mm dia.	67.52	82.54	2.43	62.86	m	**145.40**
1250 mm dia.	73.79	90.21	2.43	62.86	m	**153.07**
1400 mm dia.	83.24	101.76	2.77	71.65	m	**173.41**
1600 mm dia.	107.40	131.29	3.06	79.16	m	**210.45**
Extra over fittings; circular duct class B						
End cap						
80 mm dia.	1.23	1.50	0.15	3.87	nr	**5.37**
100 mm dia.	1.34	1.64	0.15	3.87	nr	**5.51**
160 mm dia.	2.05	2.50	0.15	3.87	nr	**6.37**
200 mm dia.	2.43	2.97	0.20	5.17	nr	**8.14**
250 mm dia.	3.41	4.17	0.29	7.51	nr	**11.68**
315 mm dia.	4.32	5.28	0.29	7.51	nr	**12.79**
355 mm dia.	6.52	7.97	0.44	11.38	nr	**19.35**
400 mm dia.	6.75	8.25	0.44	11.38	nr	**19.63**
450 mm dia.	7.13	8.71	0.44	11.38	nr	**20.09**
500 mm dia.	7.48	9.15	0.44	11.38	nr	**20.53**
630 mm dia.	19.90	24.33	0.58	15.00	nr	**39.33**
710 mm dia.	23.26	28.44	0.69	17.85	nr	**46.29**
800 mm dia.	30.63	37.44	0.81	20.95	nr	**58.39**
900 mm dia.	34.59	42.28	0.92	23.80	nr	**66.08**
1000 mm dia.	47.69	58.30	1.04	26.90	nr	**85.20**

U: VENTILATION/AIR CONDITIONING SYSTEMS

Item	Net Price £	Material £	Labour hours	Labour £	Unit	Total rate £
U10: DUCTWORK: CIRCULAR – cont						
Extra over fittings – cont						
End cap – cont						
1120 mm dia.	52.75	64.49	1.16	30.01	nr	**94.50**
1250 mm dia.	58.11	71.04	1.16	30.01	nr	**101.05**
1400 mm dia.	73.72	90.13	1.16	30.01	nr	**120.14**
1600 mm dia.	83.33	101.87	1.16	30.01	nr	**131.88**
Reducer						
80 mm dia.	3.98	4.86	0.29	7.51	nr	**12.37**
100 mm dia.	4.14	5.06	0.29	7.51	nr	**12.57**
160 mm dia.	5.44	6.65	0.29	7.51	nr	**14.16**
200 mm dia.	6.38	7.80	0.44	11.38	nr	**19.18**
250 mm dia.	7.59	9.28	0.58	15.00	nr	**24.28**
315 mm dia.	9.65	11.79	0.58	15.00	nr	**26.79**
355 mm dia.	12.65	15.46	0.87	22.51	nr	**37.97**
400 mm dia.	14.23	17.40	0.87	22.51	nr	**39.91**
450 mm dia.	15.29	18.69	0.87	22.51	nr	**41.20**
500 mm dia.	16.02	19.58	0.87	22.51	nr	**42.09**
630 mm dia.	44.46	54.35	0.87	22.51	nr	**76.86**
710 mm dia.	48.40	59.17	0.96	24.83	nr	**84.00**
800 mm dia.	62.00	75.80	1.06	27.42	nr	**103.22**
900 mm dia.	65.65	80.26	1.16	30.01	nr	**110.27**
1000 mm dia.	85.09	104.02	1.25	32.33	nr	**136.35**
1120 mm dia.	93.58	114.40	3.47	89.76	nr	**204.16**
1250 mm dia.	104.01	127.15	3.47	89.76	nr	**216.91**
1400 mm dia.	129.53	158.35	4.05	104.77	nr	**263.12**
1600 mm dia.	122.27	149.47	4.62	119.51	nr	**268.98**
90° segmented radius bend						
80 mm dia.	2.11	2.58	0.29	7.51	nr	**10.09**
100 mm dia.	2.42	2.96	0.29	7.51	nr	**10.47**
160 mm dia.	4.09	5.00	0.29	7.51	nr	**12.51**
200 mm dia.	5.49	6.72	0.44	11.38	nr	**18.10**
250 mm dia.	8.10	9.90	0.58	15.00	nr	**24.90**
315 mm dia.	9.42	11.52	0.58	15.00	nr	**26.52**
355 mm dia.	9.80	11.98	0.87	22.51	nr	**34.49**
400 mm dia.	11.55	14.12	0.87	22.51	nr	**36.63**
450 mm dia.	13.20	16.14	0.87	22.51	nr	**38.65**
500 mm dia.	13.82	16.89	0.87	22.51	nr	**39.40**
630 mm dia.	31.86	38.94	0.87	22.51	nr	**61.45**
710 mm dia.	44.36	54.23	0.96	24.83	nr	**79.06**
800 mm dia.	49.07	59.99	1.06	27.42	nr	**87.41**
900 mm dia.	53.45	65.34	1.16	30.01	nr	**95.35**
1000 mm dia.	99.40	121.52	1.25	32.33	nr	**153.85**
1120 mm dia.	110.46	135.03	3.47	89.76	nr	**224.79**
1250 mm dia.	142.22	173.86	3.47	89.76	nr	**263.62**
1400 mm dia.	253.83	310.31	4.05	104.77	nr	**415.08**
1600 mm dia.	262.68	321.12	4.62	119.51	nr	**440.63**

U: VENTILATION/AIR CONDITIONING SYSTEMS

Item	Net Price £	Material £	Labour hours	Labour £	Unit	**Total rate £**
45° radius bend						
80 mm dia.	2.20	2.69	0.29	7.51	nr	**10.20**
100 mm dia.	2.48	3.03	0.29	7.51	nr	**10.54**
160 mm dia.	3.52	4.31	0.29	7.51	nr	**11.82**
200 mm dia.	4.57	5.58	0.40	10.34	nr	**15.92**
250 mm dia.	6.35	7.77	0.58	15.00	nr	**22.77**
315 mm dia.	8.52	10.41	0.58	15.00	nr	**25.41**
355 mm dia.	9.12	11.14	0.87	22.51	nr	**33.65**
400 mm dia.	10.91	13.34	0.87	22.51	nr	**35.85**
450 mm dia.	11.05	13.51	0.87	22.51	nr	**36.02**
500 mm dia.	11.83	14.46	0.87	22.51	nr	**36.97**
630 mm dia.	34.21	41.82	0.87	22.51	nr	**64.33**
710 mm dia.	42.93	52.48	0.96	24.83	nr	**77.31**
800 mm dia.	50.70	61.99	1.06	27.42	nr	**89.41**
900 mm dia.	55.45	67.78	1.16	30.01	nr	**97.79**
1000 mm dia.	91.86	112.30	1.25	32.33	nr	**144.63**
1120 mm dia.	100.24	122.54	3.47	89.76	nr	**212.30**
1250 mm dia.	111.47	136.27	3.47	89.76	nr	**226.03**
1400 mm dia.	153.22	187.32	4.05	104.77	nr	**292.09**
1600 mm dia.	149.58	182.87	4.62	119.51	nr	**302.38**
90° equal twin bend						
80 mm dia.	6.30	7.70	0.58	15.00	nr	**22.70**
100 mm dia.	6.52	7.97	0.58	15.00	nr	**22.97**
160 mm dia.	11.54	14.11	0.58	15.00	nr	**29.11**
200 mm dia.	16.19	19.80	0.87	22.51	nr	**42.31**
250 mm dia.	24.94	30.49	1.16	30.01	nr	**60.50**
315 mm dia.	34.61	42.31	1.16	30.01	nr	**72.32**
355 mm dia.	38.56	47.14	1.73	44.75	nr	**91.89**
400 mm dia.	42.86	52.40	1.73	44.75	nr	**97.15**
450 mm dia.	46.39	56.71	1.73	44.75	nr	**101.46**
500 mm dia.	49.93	61.04	1.73	44.75	nr	**105.79**
630 mm dia.	96.45	117.91	1.73	44.75	nr	**162.66**
710 mm dia.	128.04	156.53	1.82	47.08	nr	**203.61**
800 mm dia.	156.43	191.24	1.93	49.92	nr	**241.16**
900 mm dia.	181.41	221.78	2.02	52.25	nr	**274.03**
1000 mm dia.	249.97	305.59	2.11	54.59	nr	**360.18**
1120 mm dia.	291.11	355.89	4.62	119.51	nr	**475.40**
1250 mm dia.	348.18	425.65	4.62	119.51	nr	**545.16**
1400 mm dia.	556.83	680.73	4.62	119.51	nr	**800.24**
1600 mm dia.	556.80	680.70	4.62	119.51	nr	**800.21**

U: VENTILATION/AIR CONDITIONING SYSTEMS

Item	Net Price £	Material £	Labour hours	Labour £	Unit	Total rate £
U10: DUCTWORK: CIRCULAR – cont						
Extra over fittings – cont						
Conical branch						
80 mm dia.	8.34	10.20	0.58	15.00	nr	**25.20**
100 mm dia.	8.53	10.42	0.58	15.00	nr	**25.42**
160 mm dia.	9.23	11.29	0.58	15.00	nr	**26.29**
200 mm dia.	9.82	12.01	0.87	22.51	nr	**34.52**
250 mm dia.	12.85	15.71	1.16	30.01	nr	**45.72**
315 mm dia.	14.33	17.52	1.16	30.01	nr	**47.53**
355 mm dia.	15.55	19.01	1.73	44.75	nr	**63.76**
400 mm dia.	16.05	19.62	1.73	44.75	nr	**64.37**
450 mm dia.	19.84	24.26	1.73	44.75	nr	**69.01**
500 mm dia.	20.44	24.99	1.73	44.75	nr	**69.74**
630 mm dia.	41.11	50.25	1.73	44.75	nr	**95.00**
710 mm dia.	45.79	55.98	1.82	47.08	nr	**103.06**
800 mm dia.	59.34	72.54	1.93	49.92	nr	**122.46**
900 mm dia.	62.90	76.90	2.02	52.25	nr	**129.15**
1000 mm dia.	79.74	97.48	2.11	54.59	nr	**152.07**
1120 mm dia.	111.12	135.85	4.62	119.51	nr	**255.36**
1250 mm dia.	111.12	135.85	5.20	134.51	nr	**270.36**
1400 mm dia. .	133.87	163.66	5.20	134.51	nr	**298.17**
1600 mm dia.	157.69	192.77	5.20	134.51	nr	**327.28**
45° branch						
80 mm dia.	6.63	8.11	0.58	15.00	nr	**23.11**
100 mm dia.	6.78	8.29	0.58	15.00	nr	**23.29**
160 mm dia.	10.32	12.62	0.58	15.00	nr	**27.62**
200 mm dia.	10.70	13.08	0.87	22.51	nr	**35.59**
250 mm dia.	11.20	13.69	1.16	30.01	nr	**43.70**
315 mm dia.	11.93	14.58	1.16	30.01	nr	**44.59**
355 mm dia.	15.37	18.79	1.73	44.75	nr	**63.54**
400 mm dia.	15.97	19.52	1.73	44.75	nr	**64.27**
450 mm dia.	16.64	20.34	1.73	44.75	nr	**65.09**
500 mm dia.	17.35	21.21	1.73	44.75	nr	**65.96**
630 mm dia.	34.51	42.19	1.73	44.75	nr	**86.94**
710 mm dia.	38.53	47.10	1.82	47.08	nr	**94.18**
800 mm dia.	54.98	67.22	2.13	55.09	nr	**122.31**
900 mm dia.	62.23	76.08	2.31	59.76	nr	**135.84**
1000 mm dia.	76.83	93.93	2.31	59.76	nr	**153.69**
1120 mm dia.	95.20	116.38	4.62	119.51	nr	**235.89**
1250 mm dia.	101.73	124.36	4.62	119.51	nr	**243.87**
1400 mm dia.	117.49	143.63	4.62	119.51	nr	**263.14**
1600 mm dia.	143.63	175.58	4.62	119.51	nr	**295.09**
For galvanized sheet metal DW144 class C rates, refer to galvanized sheet metal DW144 class B						

U: VENTILATION/AIR CONDITIONING SYSTEMS

Item	Net Price £	Material £	Labour hours	Labour £	Unit	Total rate £
U10: DUCTWORK: FLAT OVAL						
Y30 – AIR DUCTLINES						
Galvanized sheet metal DW144 class B spirally wound flat oval section ductwork; including all necessary stiffeners, joints, couplers in the running length and duct supports						
Straight duct						
345 × 102 mm	18.50	22.62	2.71	70.10	m	**92.72**
427 × 102 mm	21.19	25.90	2.99	77.35	m	**103.25**
508 × 102 mm	23.37	28.57	3.14	81.23	m	**109.80**
559 × 152 mm	31.56	38.58	3.43	88.73	m	**127.31**
531 × 203 mm	32.05	39.18	3.43	88.73	m	**127.91**
851 × 203 mm	41.71	51.00	5.72	147.97	m	**198.97**
582 × 254 mm	42.16	51.54	3.62	93.65	m	**145.19**
823 × 254 mm	42.30	51.72	5.80	150.04	m	**201.76**
1303 × 254 mm	100.27	122.58	8.13	210.31	m	**332.89**
632 × 305 mm	38.17	46.66	3.93	101.66	m	**148.32**
1275 × 305 mm	92.51	113.09	8.13	210.31	m	**323.40**
765 × 356 mm	44.24	54.09	5.72	147.97	m	**202.06**
1247 × 356 mm	93.51	114.32	8.13	210.31	m	**324.63**
1727 × 356 mm	125.97	154.00	10.41	269.29	m	**423.29**
737 × 406 mm	45.07	55.09	5.72	147.97	m	**203.06**
818 × 406 mm	47.60	58.20	6.21	160.64	m	**218.84**
978 × 406 mm	52.45	64.12	6.92	179.01	m	**243.13**
1379 × 406 mm	101.87	124.54	8.75	226.34	m	**350.88**
1699 × 406 mm	103.35	126.35	10.41	269.29	m	**395.64**
709 × 457 mm	45.12	55.16	5.72	147.97	m	**203.13**
1189 × 457 mm	86.01	105.15	8.80	227.64	m	**332.79**
1671 × 457 mm	124.16	151.79	10.31	266.70	m	**418.49**
678 × 508 mm	44.57	54.49	5.72	147.97	m	**202.46**
919 × 508 mm	52.45	64.12	7.30	188.84	m	**252.96**
1321 × 508 mm	101.70	124.33	8.75	226.34	m	**350.67**
Extra over fittings; flat oval duct class B						
End cap						
345 × 102 mm	16.23	19.84	0.20	5.17	nr	**25.01**
427 × 102 mm	16.67	20.38	0.20	5.17	nr	**25.55**
508 × 102 mm	17.11	20.92	0.20	5.17	nr	**26.09**
559 × 152 mm	24.77	30.28	0.29	7.51	nr	**37.79**
531 × 203 mm	24.89	30.43	0.29	7.51	nr	**37.94**
851 × 203 mm	27.62	33.76	0.44	11.38	nr	**45.14**
582 × 254 mm	25.56	31.25	0.44	11.38	nr	**42.63**
823 × 254 mm	27.17	33.22	0.44	11.38	nr	**44.60**
1303 × 254 mm	89.76	109.74	0.69	17.85	nr	**127.59**
632 × 305 mm	25.54	31.22	0.69	17.85	nr	**49.07**
1275 × 305 mm	79.18	96.80	0.69	17.85	nr	**114.65**

U: VENTILATION/AIR CONDITIONING SYSTEMS

Item	Net Price £	Material £	Labour hours	Labour £	Unit	Total rate £
U10: DUCTWORK: FLAT OVAL – cont						
Extra over fittings – cont						
End cap – cont						
765 × 356 mm	26.80	32.76	0.69	17.85	nr	**50.61**
1247 × 356 mm	76.90	94.01	0.69	17.85	nr	**111.86**
1727 × 356 mm	114.25	139.67	0.69	17.85	nr	**157.52**
737 × 406 mm	27.57	33.70	1.04	26.90	nr	**60.60**
818 × 406 mm	28.27	34.56	0.69	17.85	nr	**52.41**
978 × 406 mm	29.65	36.25	0.69	17.85	nr	**54.10**
1379 × 406 mm	105.29	128.72	1.04	26.90	nr	**155.62**
1699 × 406 mm	114.93	140.50	1.04	26.90	nr	**167.40**
709 × 457 mm	26.82	32.78	1.04	26.90	nr	**59.68**
1189 × 457 mm	75.08	91.78	1.04	26.90	nr	**118.68**
1671 × 457 mm	111.90	136.79	1.04	26.90	nr	**163.69**
678 × 508 mm	36.19	44.24	1.04	26.90	nr	**71.14**
919 × 508 mm	39.03	47.71	1.04	26.90	nr	**74.61**
1321 × 508 mm	111.87	136.76	1.04	26.90	nr	**163.66**
Reducer						
345 × 102 mm	30.29	37.03	0.95	24.58	nr	**61.61**
427 × 102 mm	30.74	37.58	1.06	27.42	nr	**65.00**
508 × 102 mm	31.32	38.29	1.13	29.23	nr	**67.52**
559 × 152 mm	43.68	53.40	1.26	32.59	nr	**85.99**
531 × 203 mm	43.99	53.78	1.26	32.59	nr	**86.37**
851 × 203 mm	56.66	69.27	1.34	34.66	nr	**103.93**
582 × 254 mm	46.39	56.71	1.34	34.66	nr	**91.37**
823 × 254 mm	58.24	71.20	1.34	34.66	nr	**105.86**
1303 × 254 mm	132.64	162.15	1.34	34.66	nr	**196.81**
632 × 305 mm	51.00	62.35	0.70	18.11	nr	**80.46**
1275 × 305 mm	135.38	165.50	1.16	30.01	nr	**195.51**
765 × 356 mm	54.48	66.60	1.16	30.01	nr	**96.61**
1247 × 356 mm	137.73	168.37	1.16	30.01	nr	**198.38**
1727 × 356 mm	140.08	171.25	1.25	32.33	nr	**203.58**
737 × 406 mm	60.17	73.56	1.16	30.01	nr	**103.57**
818 × 406 mm	60.91	74.46	1.27	32.85	nr	**107.31**
978 × 406 mm	66.10	80.80	1.44	37.24	nr	**118.04**
1379 × 406 mm	139.79	170.90	1.44	37.24	nr	**208.14**
1699 × 406 mm	158.42	193.67	1.44	37.24	nr	**230.91**
709 × 457 mm	60.82	74.36	1.16	30.01	nr	**104.37**
1189 × 457 mm	123.29	150.72	1.34	34.66	nr	**185.38**
1671 × 457 mm	137.07	167.57	1.44	37.24	nr	**204.81**
678 × 508 mm	64.68	79.07	1.16	30.01	nr	**109.08**
919 × 508 mm	68.91	84.24	1.26	32.59	nr	**116.83**
1321 × 508 mm	138.02	168.73	1.44	37.24	nr	**205.97**
90° radius bend						
345 × 102 mm	39.57	48.38	0.29	7.51	nr	**55.89**
427 × 102 mm	37.84	46.26	0.58	15.00	nr	**61.26**
508 × 102 mm	36.47	44.59	0.58	15.00	nr	**59.59**

U: VENTILATION/AIR CONDITIONING SYSTEMS

Item	Net Price £	Material £	Labour hours	Labour £	Unit	**Total rate £**
559 × 152 mm	50.39	61.60	0.58	15.00	nr	**76.60**
531 × 203 mm	50.94	62.27	0.87	22.51	nr	**84.78**
851 × 203 mm	68.84	84.16	0.87	22.51	nr	**106.67**
582 × 254 mm	48.28	59.02	0.87	22.51	nr	**81.53**
823 × 254 mm	71.75	87.71	0.87	22.51	nr	**110.22**
1303 × 254 mm	161.45	197.38	0.96	24.83	nr	**222.21**
632 × 305 mm	45.58	55.72	0.87	22.51	nr	**78.23**
1275 × 305 mm	204.59	250.11	0.96	24.83	nr	**274.94**
765 × 356 mm	53.05	64.86	0.87	22.51	nr	**87.37**
1247 × 356 mm	205.71	251.48	0.96	24.83	nr	**276.31**
1727 × 356 mm	250.57	306.32	1.25	32.33	nr	**338.65**
737 × 406 mm	54.77	66.96	0.96	24.83	nr	**91.79**
818 × 406 mm	78.96	96.53	0.87	22.51	nr	**119.04**
978 × 406 mm	70.10	85.70	0.96	24.83	nr	**110.53**
1379 × 406 mm	184.02	224.96	1.16	30.01	nr	**254.97**
1699 × 406 mm	314.77	384.81	1.25	32.33	nr	**417.14**
709 × 457 mm	63.09	77.13	0.87	22.51	nr	**99.64**
1189 × 457 mm	152.34	186.23	0.96	24.83	nr	**211.06**
1671 × 457 mm	313.94	383.80	1.25	32.33	nr	**416.13**
678 × 508 mm	60.96	74.52	0.87	22.51	nr	**97.03**
919 × 508 mm	68.70	83.99	0.96	24.83	nr	**108.82**
1321 × 508 mm	190.49	232.87	1.16	30.01	nr	**262.88**
45° radius bend						
345 × 102 mm	28.01	34.25	0.79	20.44	nr	**54.69**
427 × 102 mm	27.46	33.57	0.85	21.99	nr	**55.56**
508 × 102 mm	26.91	32.90	0.95	24.58	nr	**57.48**
559 × 152 mm	37.37	45.68	0.79	20.44	nr	**66.12**
531 × 203 mm	37.61	45.98	0.85	21.99	nr	**67.97**
851 × 203 mm	52.54	64.23	0.98	25.36	nr	**89.59**
582 × 254 mm	36.65	44.80	0.76	19.66	nr	**64.46**
823 × 254 mm	54.60	66.74	0.95	24.58	nr	**91.32**
1303 × 254 mm	158.28	193.50	1.16	30.01	nr	**223.51**
632 × 305 mm	39.82	48.68	0.58	15.00	nr	**63.68**
1275 × 305 mm	182.30	222.86	1.16	30.01	nr	**252.87**
765 × 356 mm	40.87	49.97	0.87	22.51	nr	**72.48**
1247 × 356 mm	182.93	223.63	1.16	30.01	nr	**253.64**
1727 × 356 mm	177.92	217.51	1.26	32.59	nr	**250.10**
737 × 406 mm	43.12	52.72	0.69	17.85	nr	**70.57**
818 × 406 mm	59.85	73.17	0.78	20.17	nr	**93.34**
978 × 406 mm	56.08	68.56	0.87	22.51	nr	**91.07**
1379 × 406 mm	174.37	213.17	1.16	30.01	nr	**243.18**
1699 × 406 mm	210.30	257.10	1.27	32.85	nr	**289.95**
709 × 457 mm	48.57	59.38	0.81	20.95	nr	**80.33**
1189 × 457 mm	144.09	176.15	0.95	24.58	nr	**200.73**
1671 × 457 mm	207.19	253.29	1.26	32.59	nr	**285.88**
678 × 508 mm	47.36	57.90	0.92	23.80	nr	**81.70**
919 × 508 mm	55.07	67.32	1.10	28.46	nr	**95.78**
1321 × 508 mm	177.20	216.63	1.25	32.33	nr	**248.96**

U: VENTILATION/AIR CONDITIONING SYSTEMS

Item	Net Price £	Material £	Labour hours	Labour £	Unit	Total rate £
U10: DUCTWORK: FLAT OVAL – cont						
Extra over fittings – cont						
90° hard bend with turning vanes						
345 × 102 mm	40.60	49.64	0.55	14.23	nr	**63.87**
427 × 102 mm	39.62	48.43	1.16	30.01	nr	**78.44**
508 × 102 mm	38.65	47.25	1.16	30.01	nr	**77.26**
559 × 152 mm	40.91	50.02	1.16	30.01	nr	**80.03**
531 × 203 mm	41.28	50.47	1.73	44.75	nr	**95.22**
851 × 203 mm	63.88	78.09	1.73	44.75	nr	**122.84**
582 × 254 mm	46.65	57.03	1.73	44.75	nr	**101.78**
823 × 254 mm	63.48	77.60	1.73	44.75	nr	**122.35**
1303 × 254 mm	110.84	135.51	1.82	47.08	nr	**182.59**
632 × 305 mm	61.41	75.08	1.73	44.75	nr	**119.83**
1275 × 305 mm	112.87	137.99	1.82	47.08	nr	**185.07**
765 × 356 mm	60.27	73.68	1.73	44.75	nr	**118.43**
1247 × 356 mm	113.69	138.99	1.82	47.08	nr	**186.07**
1727 × 356 mm	191.08	233.59	1.82	47.08	nr	**280.67**
737 × 406 mm	66.20	80.93	1.73	44.75	nr	**125.68**
818 × 406 mm	63.91	78.13	1.73	44.75	nr	**122.88**
978 × 406 mm	69.77	85.29	1.73	44.75	nr	**130.04**
1379 × 406 mm	149.72	183.03	1.82	47.08	nr	**230.11**
1699 × 406 mm	234.34	286.48	2.11	54.59	nr	**341.07**
709 × 457 mm	74.00	90.46	1.73	44.75	nr	**135.21**
1189 × 457 mm	123.75	151.29	1.82	47.08	nr	**198.37**
1671 × 457 mm	194.35	237.59	2.11	54.59	nr	**292.18**
678 × 508 mm	71.54	87.46	1.82	47.08	nr	**134.54**
919 × 508 mm	78.50	95.97	1.82	47.08	nr	**143.05**
1321 × 508 mm	156.71	191.58	2.11	54.59	nr	**246.17**
90° branch						
345 × 102 mm	30.79	37.64	0.58	15.00	nr	**52.64**
427 × 102 mm	31.39	38.38	0.58	15.00	nr	**53.38**
508 × 102 mm	31.96	39.07	1.16	30.01	nr	**69.08**
559 × 152 mm	47.30	57.82	1.16	30.01	nr	**87.83**
531 × 203 mm	48.11	58.81	1.16	30.01	nr	**88.82**
851 × 203 mm	62.52	76.43	1.73	44.75	nr	**121.18**
582 × 254 mm	53.79	65.76	1.73	44.75	nr	**110.51**
823 × 254 mm	67.95	83.07	1.73	44.75	nr	**127.82**
1303 × 254 mm	144.65	176.84	1.82	47.08	nr	**223.92**
632 × 305 mm	60.00	73.35	1.73	44.75	nr	**118.10**
1275 × 305 mm	145.26	177.58	1.82	47.08	nr	**224.66**
765 × 356 mm	67.72	82.79	1.73	44.75	nr	**127.54**
1247 × 356 mm	149.43	182.68	1.82	47.08	nr	**229.76**
1727 × 356 mm	189.75	231.97	2.11	54.59	nr	**286.56**
737 × 406 mm	73.03	89.28	1.73	44.75	nr	**134.03**
818 × 406 mm	74.49	91.06	1.73	44.75	nr	**135.81**
978 × 406 mm	82.91	101.36	1.82	47.08	nr	**148.44**
1379 × 406 mm	160.49	196.20	1.93	49.92	nr	**246.12**
1699 × 406 mm	186.96	228.56	2.11	54.59	nr	**283.15**

U: VENTILATION/AIR CONDITIONING SYSTEMS

Item	Net Price £	Material £	Labour hours	Labour £	Unit	**Total rate £**
709 × 457 mm	74.60	91.20	1.73	44.75	nr	**135.95**
1189 × 457 mm	131.76	161.08	1.82	47.08	nr	**208.16**
1671 × 457 mm	187.38	229.07	2.11	54.59	nr	**283.66**
678 × 508 mm	79.65	97.38	1.73	44.75	nr	**142.13**
919 × 508 mm	88.09	107.69	1.82	47.08	nr	**154.77**
1321 × 508 mm	161.15	197.01	2.11	54.59	nr	**251.60**
45° branch						
345 × 102 mm	32.52	39.76	0.58	15.00	nr	**54.76**
427 × 102 mm	34.23	41.85	0.58	15.00	nr	**56.85**
508 × 102 mm	36.13	44.17	1.16	30.01	nr	**74.18**
559 × 152 mm	51.90	63.45	1.73	44.75	nr	**108.20**
531 × 203 mm	51.57	63.05	1.73	44.75	nr	**107.80**
851 × 203 mm	76.34	93.33	1.73	44.75	nr	**138.08**
582 × 254 mm	57.50	70.30	1.73	44.75	nr	**115.05**
823 × 254 mm	77.38	94.60	1.82	47.08	nr	**141.68**
1303 × 254 mm	178.89	218.69	1.92	49.67	nr	**268.36**
632 × 305 mm	63.89	78.10	1.73	44.75	nr	**122.85**
1275 × 305 mm	176.50	215.77	1.82	47.08	nr	**262.85**
765 × 356 mm	73.49	89.85	1.73	44.75	nr	**134.60**
1247 × 356 mm	177.19	216.62	1.82	47.08	nr	**263.70**
1727 × 356 mm	249.24	304.69	1.82	47.08	nr	**351.77**
737 × 406 mm	77.96	95.31	1.73	44.75	nr	**140.06**
818 × 406 mm	81.79	99.99	1.73	44.75	nr	**144.74**
978 × 406 mm	95.95	117.30	1.73	44.75	nr	**162.05**
1379 × 406 mm	193.16	236.14	1.93	49.92	nr	**286.06**
1699 × 406 mm	240.78	294.35	2.19	56.65	nr	**351.00**
709 × 457 mm	77.20	94.38	1.73	44.75	nr	**139.13**
1189 × 457 mm	152.21	186.08	1.82	47.08	nr	**233.16**
1671 × 457 mm	236.89	289.59	2.11	54.59	nr	**344.18**
678 × 508 mm	80.01	97.81	1.73	44.75	nr	**142.56**
919 × 508 mm	95.57	116.83	1.82	47.08	nr	**163.91**
1321 × 508 mm	186.40	227.88	1.93	49.92	nr	**277.80**
For rates for access doors refer to ancillaries in U10: DUCTWORK: RECTANGULAR: CLASS B						

U: VENTILATION/AIR CONDITIONING SYSTEMS

Item	Net Price £	Material £	Labour hours	Labour £	Unit	Total rate £
U10: DUCTWORK: FLEXIBLE						
Y30 – AIR DUCTLINES						
Aluminium foil flexible ductwork, DW 144 class B; multiply aluminium polyester laminate fabric, with high tensile steel wire helix						
Duct						
102 mm dia.	1.23	1.50	0.33	8.53	m	**10.03**
152 mm dia.	1.80	2.20	0.33	8.53	m	**10.73**
203 mm dia.	2.38	2.90	0.33	8.53	m	**11.43**
254 mm dia.	2.98	3.65	0.33	8.53	m	**12.18**
304 mm dia.	3.65	4.46	0.33	8.53	m	**12.99**
355 mm dia.	4.98	6.09	0.33	8.53	m	**14.62**
406 mm dia.	5.53	6.76	0.33	8.53	m	**15.29**
Insulated aluminium foil flexible ductwork, DW144 class B; laminate construction of aluminium and polyester multiply inner core with 25 mm insulation; outer layer of multiply aluminium polyester laminate, with high tensile steel wire helix						
Duct						
102 mm dia.	2.76	3.38	0.50	12.94	m	**16.32**
152 mm dia.	3.60	4.40	0.50	12.94	m	**17.34**
203 mm dia.	4.37	5.35	0.50	12.94	m	**18.29**
254 mm dia.	5.38	6.58	0.50	12.94	m	**19.52**
304 mm dia.	6.92	8.46	0.50	12.94	m	**21.40**
355 mm dia.	8.58	10.49	0.50	12.94	m	**23.43**
406 mm dia.	9.50	11.62	0.50	12.94	m	**24.56**

U: VENTILATION/AIR CONDITIONING SYSTEMS

Item	Net Price £	Material £	Labour hours	Labour £	Unit	Total rate £
U10: DUCTWORK: PLASTIC						
Y30 – AIR DUCTLINES						
Rigid gey PVC DW 154 circular section ductwork; solvent welded or filler rod welded joints; excludes couplers and supports (these are detailed separately); ductwork to conform to curent HSE regulations						
Straight duct (standard length 6 m)						
110 mm	3.41	4.17	1.74	45.01	m	**49.18**
160 mm	6.45	7.89	1.71	44.23	m	**52.12**
200 mm	8.47	10.35	1.79	46.31	m	**56.66**
225 mm	12.07	14.76	1.96	50.70	m	**65.46**
250 mm	11.87	14.51	1.98	51.22	m	**65.73**
315 mm	16.97	20.74	2.23	57.69	m	**78.43**
355 mm	22.73	27.79	2.25	58.21	m	**86.00**
400 mm	28.58	34.94	2.39	61.82	m	**96.76**
450 mm	35.32	43.18	2.95	76.31	m	**119.49**
500 mm	43.49	53.17	2.98	77.10	m	**130.27**
600 mm	67.24	82.20	3.12	80.71	m	**162.91**
Extra for supports (BZP finish)						
Horizontal – maximum 2.4 m centres						
Vertical – maximum 4.0 m centres						
Duct Size						
110 mm	21.08	25.77	0.59	15.25	m	**41.02**
160 mm	21.15	25.85	0.59	15.25	m	**41.10**
200 mm	21.23	25.96	0.59	15.25	m	**41.21**
225 mm	22.53	27.54	0.63	16.29	m	**43.83**
250 mm	22.54	27.55	0.63	16.29	m	**43.84**
315 mm	22.70	27.75	0.63	16.29	m	**44.04**
355 mm	21.71	26.54	0.81	20.95	m	**47.49**
400 mm	29.48	36.04	0.78	20.17	m	**56.21**
450 mm	30.30	37.04	0.80	20.69	m	**57.73**
500 mm	30.46	37.23	0.80	20.69	m	**57.92**
600 mm	32.33	39.52	0.85	21.99	m	**61.51**
Note: These are maximum figures and may be reduced subject to local conditions (i.e. a high number of changes of direction)						

U: VENTILATION/AIR CONDITIONING SYSTEMS

Item	Net Price £	Material £	Labour hours	Labour £	Unit	Total rate £
U10: DUCTWORK: PLASTIC – cont						
Extra over fittings; rigid grey PVC						
90° bend						
110 mm	17.45	21.33	1.11	28.72	m	**50.05**
160 mm	19.25	23.54	1.34	34.66	m	**58.20**
200 mm	16.10	19.68	2.00	51.74	m	**71.42**
225 mm	24.73	30.23	1.79	46.31	m	**76.54**
250 mm	26.32	32.18	2.01	51.99	m	**84.17**
315 mm	45.44	55.55	2.56	66.22	m	**121.77**
355 mm	54.11	66.15	2.71	70.10	m	**136.25**
400 mm	68.73	84.03	3.58	92.61	m	**176.64**
450 mm	207.74	253.97	4.55	117.70	m	**371.67**
500 mm	241.12	294.78	5.01	129.59	m	**424.37**
600 mm	412.73	504.57	5.78	149.51	m	**654.08**
45° bend						
110 mm	7.18	8.78	0.71	18.36	m	**27.14**
160 mm	8.34	10.20	0.93	24.06	m	**34.26**
200 mm	9.15	11.19	1.14	29.49	m	**40.68**
225 mm	10.33	12.63	1.41	36.48	m	**49.11**
250 mm	11.07	13.53	1.62	41.91	m	**55.44**
315 mm	17.67	21.60	1.93	49.92	m	**71.52**
355 mm	22.99	28.11	2.22	57.43	m	**85.54**
400 mm	26.42	32.30	2.85	73.73	m	**106.03**
450 mm	101.76	124.40	3.77	97.52	m	**221.92**
500 mm	113.80	139.12	4.16	107.61	m	**246.73**
600 mm	185.39	226.64	4.81	124.42	m	**351.06**
Tee						
110 mm	15.23	18.62	1.04	26.90	m	**45.52**
160 mm	18.59	22.72	1.38	35.70	m	**58.42**
200 mm	24.73	30.23	1.79	46.31	m	**76.54**
225 mm	28.49	34.82	2.25	58.21	m	**93.03**
250 mm	34.87	42.63	2.62	67.77	m	**110.40**
315 mm	58.08	71.01	3.17	82.01	m	**153.02**
355 mm	81.23	99.30	3.73	96.49	m	**195.79**
400 mm	79.19	96.81	4.71	121.84	m	**218.65**
450 mm	155.73	190.39	5.72	147.97	m	**338.36**
500 mm	176.99	216.37	6.33	163.74	m	**380.11**
Coupler						
110 mm	4.07	4.97	0.70	18.11	m	**23.08**
160 mm	5.05	6.17	0.91	23.55	m	**29.72**
200 mm	6.26	7.65	1.12	28.97	m	**36.62**
225 mm	8.20	10.02	1.41	36.48	m	**46.50**
250 mm	9.12	11.14	1.60	41.40	m	**52.54**
315 mm	10.87	13.29	1.86	48.12	m	**61.41**
355 mm	11.43	13.98	2.43	62.86	m	**76.84**
400 mm	12.18	14.89	2.66	68.81	m	**83.70**
450 mm	29.29	35.80	3.20	82.77	m	**118.57**
500 mm	33.05	40.41	3.52	91.05	m	**131.46**

U: VENTILATION/AIR CONDITIONING SYSTEMS

Item	Net Price £	Material £	Labour hours	Labour £	Unit	**Total rate £**
Damper						
110 mm	34.12	41.71	0.96	24.83	m	**66.54**
160 mm	34.64	42.34	1.26	32.59	m	**74.93**
200 mm	35.39	43.26	1.52	39.32	m	**82.58**
225 mm	36.58	44.72	1.88	48.63	m	**93.35**
250 mm	36.44	44.55	2.16	55.88	m	**100.43**
315 mm	41.80	51.10	2.94	76.06	m	**127.16**
355 mm	44.66	54.60	3.42	88.48	m	**143.08**
400 mm	47.94	58.61	3.79	98.05	m	**156.66**
Reducer						
160 × 110 mm	8.92	10.91	0.85	21.99	m	**32.90**
200 × 110 mm	12.98	15.87	0.98	25.36	m	**41.23**
200 × 160 mm	11.24	13.74	1.07	27.68	m	**41.42**
225 × 200 mm	14.75	18.04	1.38	35.70	m	**53.74**
250 × 160 mm	15.60	19.08	1.38	35.70	m	**54.78**
250 × 200 mm	14.82	18.12	1.48	38.29	m	**56.41**
250 × 225 mm	14.54	17.78	1.58	40.87	m	**58.65**
315 × 200 mm	15.87	19.41	1.63	42.17	m	**61.58**
315 × 250 mm	16.32	19.95	1.84	47.60	m	**67.55**
355 × 200 mm	24.18	29.56	1.99	51.48	m	**81.04**
355 × 250 mm	18.04	22.05	1.97	50.96	m	**73.01**
355 × 315 mm	22.48	27.48	2.16	55.88	m	**83.36**
400 × 225 mm	27.30	33.37	2.47	63.90	m	**97.27**
400 × 315 mm	27.58	33.71	2.75	71.14	m	**104.85**
400 × 355 mm	29.04	35.50	2.73	70.63	m	**106.13**
450 × 315 mm	31.34	38.32	2.91	75.28	m	**113.60**
Flange						
110 mm	5.88	7.19	0.70	18.11	m	**25.30**
160 mm	5.95	7.27	0.91	23.55	m	**30.82**
200 mm	6.42	7.85	1.12	28.97	m	**36.82**
225 mm	6.24	7.63	1.35	34.93	m	**42.56**
250 mm	6.67	8.16	1.55	40.10	m	**48.26**
315 mm	9.49	11.60	1.85	47.86	m	**59.46**
355 mm	10.04	12.28	2.06	53.29	m	**65.57**
400 mm	10.06	12.30	2.62	67.77	m	**80.07**
Polypropylene (PPS) DW154 circular section ductwork; filler rod welded joints; excludes couplers and supports (these are detailed separately); ductwork to conform to current HSE regulations						
Straight duct (standard length 6 m)						
110 mm	15.96	19.51	0.65	16.81	m	**36.32**
160 mm	17.50	21.39	0.75	19.41	m	**40.80**
200 mm	18.58	22.71	0.85	21.99	m	**44.70**
225 mm	23.82	29.12	1.07	27.68	m	**56.80**
250 mm	24.25	29.64	1.18	30.53	m	**60.17**
315 mm	44.65	54.59	1.41	36.48	m	**91.07**
355 mm	47.42	57.97	1.53	39.57	m	**97.54**
400 mm	57.87	70.75	1.73	44.75	m	**115.50**

U: VENTILATION/AIR CONDITIONING SYSTEMS

Item	Net Price £	Material £	Labour hours	Labour £	Unit	**Total rate £**
U10: DUCTWORK: PLASTIC – cont						
Extra for supports (BZP finish)						
Horizontal – maximum 2.4 m centres						
Vertical – maximum 4.0 m centres						
Duct Size						
110 mm	29.22	35.72	0.42	10.87	m	**46.59**
160 mm	29.22	35.72	0.42	10.87	m	**46.59**
200 mm	29.22	35.72	0.42	10.87	m	**46.59**
225 mm	29.22	35.72	0.42	10.87	m	**46.59**
250 mm	29.22	35.72	0.42	10.87	m	**46.59**
315 mm	29.22	35.72	0.42	10.87	m	**46.59**
355 mm	29.22	35.72	0.55	14.23	m	**49.95**
400 mm	41.36	50.56	0.55	14.23	m	**64.79**
Note: These are maximum figures and may be reduced subject to local conditions (i.e. a high number of changes of direction)						
Extra over fittings; polypropylene (DW 154)						
90° bend						
110 mm	11.97	14.64	1.01	26.12	m	**40.76**
160 mm	14.90	18.21	1.30	33.63	m	**51.84**
200 mm	16.37	20.01	1.58	40.87	m	**60.88**
225 mm	22.82	27.90	2.10	54.33	m	**82.23**
250 mm	23.14	28.28	2.36	61.05	m	**89.33**
315 mm	69.01	84.37	3.14	81.23	m	**165.60**
355 mm	77.51	94.76	3.52	91.05	m	**185.81**
400 mm	75.95	92.85	4.39	113.56	m	**206.41**
45° bend						
110 mm	7.47	9.14	0.84	21.72	m	**30.86**
160 mm	10.88	13.30	1.13	29.23	m	**42.53**
200 mm	11.58	14.15	1.39	35.96	m	**50.11**
225 mm	14.20	17.36	1.73	44.75	m	**62.11**
250 mm	15.29	18.69	1.99	51.48	m	**70.17**
315 mm	43.32	52.96	2.57	66.48	m	**119.44**
355 mm	48.29	59.04	2.93	75.80	m	**134.84**
400 mm	46.34	56.65	3.60	93.13	m	**149.78**
Tee						
110 mm	33.38	40.81	1.55	40.10	m	**80.91**
160 mm	41.34	50.54	2.17	56.14	m	**106.68**
200 mm	47.80	58.43	2.68	69.33	m	**127.76**

U: VENTILATION/AIR CONDITIONING SYSTEMS

Item	Net Price £	Material £	Labour hours	Labour £	Unit	**Total rate £**
225 mm	56.91	69.58	3.38	87.44	m	**157.02**
250 mm	66.07	80.77	3.95	102.18	m	**182.95**
315 mm	93.43	114.22	4.69	121.32	m	**235.54**
355 mm	116.81	142.80	5.44	140.72	m	**283.52**
400 mm	136.49	166.86	6.09	157.53	m	**324.39**
Coupler						
110 mm	7.36	9.00	0.84	21.72	m	**30.72**
160 mm	7.93	9.69	1.10	28.46	m	**38.15**
200 mm	8.88	10.86	1.36	35.18	m	**46.04**
225 mm	9.57	11.70	1.67	43.21	m	**54.91**
250 mm	9.70	11.86	1.91	49.41	m	**61.27**
315 mm	13.41	16.40	2.24	57.95	m	**74.35**
355 mm	17.23	21.06	2.58	66.73	m	**87.79**
400 mm	17.60	21.52	3.30	85.37	m	**106.89**
Damper						
110 mm	69.72	85.23	0.79	20.44	m	**105.67**
160 mm	79.20	96.82	1.13	29.23	m	**126.05**
200 mm	85.32	104.31	1.44	37.24	m	**141.55**
225 mm	89.59	109.52	1.81	46.82	m	**156.34**
250 mm	91.64	112.03	2.08	53.81	m	**165.84**
315 mm	104.61	127.88	2.40	62.08	m	**189.96**
355 mm	113.14	138.32	2.75	71.14	m	**209.46**
400 mm	118.15	144.44	3.57	92.35	m	**236.79**
Reducer						
160 × 110 mm	25.12	30.70	0.87	22.51	m	**53.21**
200 × 160 mm	26.00	31.79	1.16	30.01	m	**61.80**
225 × 200 mm	30.32	37.07	1.46	37.76	m	**74.83**
250 × 200 mm	35.73	43.68	1.61	41.65	m	**85.33**
250 × 225 mm	43.79	53.53	1.70	43.97	m	**97.50**
315 × 200 mm	67.35	82.34	1.66	42.95	m	**125.29**
315 × 250 mm	60.91	74.46	2.03	52.51	m	**126.97**
355 × 200 mm	72.72	88.90	2.08	53.81	m	**142.71**
355 × 250 mm	77.40	94.63	2.05	53.03	m	**147.66**
355 × 315 mm	88.52	108.21	2.22	57.43	m	**165.64**
400 × 315 mm	105.22	128.64	2.88	74.50	m	**203.14**
400 × 355 mm	112.60	137.65	2.92	75.54	m	**213.19**
Flange						
110 mm	14.13	17.27	0.76	19.66	m	**36.93**
160 mm	15.88	19.42	1.02	26.38	m	**45.80**
200 mm	17.56	21.47	1.27	32.85	m	**54.32**
225 mm	17.93	21.92	1.55	40.10	m	**62.02**
250 mm	19.37	23.68	1.79	46.31	m	**69.99**
315 mm	21.89	26.76	2.08	53.81	m	**80.57**
355 mm	23.95	29.28	2.36	61.05	m	**90.33**
400 mm	25.44	31.10	3.07	79.41	m	**110.51**

U: VENTILATION/AIR CONDITIONING SYSTEMS

Item	Net Price £	Material £	Labour hours	Labour £	Unit	Total rate £
U10: DUCTWORK: RECTANGULAR – CLASS B						
Y30 – AIR DUCTLINES						
Galvanized sheet metal DW144 class B rectangular section ductwork; including all necessary stiffeners, joints, couplers in the running length and duct supports						
Ductwork up to 400 mm longest side						
Sum of two sides 200 mm	11.35	13.87	1.16	30.01	m	**43.88**
Sum of two sides 300 mm	12.14	14.84	1.16	30.01	m	**44.85**
Sum of two sides 400 mm	12.68	15.50	1.19	30.79	m	**46.29**
Sum of two sides 500 mm	13.42	16.41	1.19	30.79	m	**47.20**
Sum of two sides 600 mm	12.80	15.65	1.27	32.85	m	**48.50**
Sum of two sides 700 mm	13.47	16.47	1.27	32.85	m	**49.32**
Sum of two sides 800 mm	14.18	17.33	1.27	32.85	m	**50.18**
Extra over fittings; rectangular ductwork class B; upto 400 mm longest side						
End cap						
Sum of two sides 200 mm	7.92	9.68	0.38	9.83	nr	**19.51**
Sum of two sides 300 mm	8.81	10.77	0.38	9.83	nr	**20.60**
Sum of two sides 400 mm	9.70	11.86	0.38	9.83	nr	**21.69**
Sum of two sides 500 mm	10.59	12.95	0.38	9.83	nr	**22.78**
Sum of two sides 600 mm	11.47	14.02	0.38	9.83	nr	**23.85**
Sum of two sides 700 mm	12.37	15.12	0.38	9.83	nr	**24.95**
Sum of two sides 800 mm	13.26	16.21	0.38	9.83	nr	**26.04**
Reducer						
Sum of two sides 200 mm	13.00	15.89	1.40	36.21	nr	**52.10**
Sum of two sides 300 mm	14.69	17.96	1.40	36.21	nr	**54.17**
Sum of two sides 400 mm	23.67	28.93	1.42	36.74	nr	**65.67**
Sum of two sides 500 mm	25.61	31.31	1.42	36.74	nr	**68.05**
Sum of two sides 600 mm	27.55	33.68	1.69	43.71	nr	**77.39**
Sum of two sides 700 mm	29.48	36.04	1.69	43.71	nr	**79.75**
Sum of two sides 800 mm	31.40	38.39	1.92	49.67	nr	**88.06**
Offset						
Sum of two sides 200 mm	19.20	23.47	1.63	42.17	nr	**65.64**
Sum of two sides 300 mm	21.70	26.53	1.63	42.17	nr	**68.70**
Sum of two sides 400 mm	31.77	38.84	1.65	42.67	nr	**81.51**
Sum of two sides 500 mm	34.52	42.20	1.65	42.67	nr	**84.87**
Sum of two sides 600 mm	36.80	44.99	1.92	49.67	nr	**94.66**
Sum of two sides 700 mm	39.31	48.06	1.92	49.67	nr	**97.73**
Sum of two sides 800 mm	41.52	50.76	1.92	49.67	nr	**100.43**

U: VENTILATION/AIR CONDITIONING SYSTEMS

Item	Net Price £	Material £	Labour hours	Labour £	Unit	**Total rate £**
Square to round						
Sum of two sides 200 mm	16.90	20.66	1.63	42.17	nr	**62.83**
Sum of two sides 300 mm	18.98	23.21	1.63	42.17	nr	**65.38**
Sum of two sides 400 mm	25.12	30.70	1.65	42.67	nr	**73.37**
Sum of two sides 500 mm	27.33	33.41	1.65	42.67	nr	**76.08**
Sum of two sides 600 mm	29.54	36.11	1.92	49.67	nr	**85.78**
Sum of two sides 700 mm	31.74	38.80	1.92	49.67	nr	**88.47**
Sum of two sides 800 mm	33.94	41.49	1.92	49.67	nr	**91.16**
90° radius bend						
Sum of two sides 200 mm	12.71	15.54	1.22	31.57	nr	**47.11**
Sum of two sides 300 mm	13.70	16.75	1.22	31.57	nr	**48.32**
Sum of two sides 400 mm	22.75	27.81	1.25	32.33	nr	**60.14**
Sum of two sides 500 mm	24.19	29.57	1.25	32.33	nr	**61.90**
Sum of two sides 600 mm	26.12	31.93	1.33	34.40	nr	**66.33**
Sum of two sides 700 mm	27.79	33.97	1.33	34.40	nr	**68.37**
Sum of two sides 800 mm	29.67	36.28	1.40	36.21	nr	**72.49**
45° radius bend						
Sum of two sides 200 mm	13.72	16.77	0.89	23.02	nr	**39.79**
Sum of two sides 300 mm	15.06	18.41	1.12	28.97	nr	**47.38**
Sum of two sides 400 mm	23.82	29.12	1.10	28.46	nr	**57.58**
Sum of two sides 500 mm	25.51	31.19	1.10	28.46	nr	**59.65**
Sum of two sides 600 mm	27.45	33.56	1.16	30.01	nr	**63.57**
Sum of two sides 700 mm	29.25	35.76	1.16	30.01	nr	**65.77**
Sum of two sides 800 mm	31.18	38.12	1.22	31.57	nr	**69.69**
90° mitre bend						
Sum of two sides 200 mm	21.18	25.89	1.29	33.37	nr	**59.26**
Sum of two sides 300 mm	23.19	28.35	1.29	33.37	nr	**61.72**
Sum of two sides 400 mm	33.37	40.80	1.29	33.37	nr	**74.17**
Sum of two sides 500 mm	35.96	43.96	1.29	33.37	nr	**77.33**
Sum of two sides 600 mm	39.28	48.02	1.39	35.96	nr	**83.98**
Sum of two sides 700 mm	42.36	51.79	1.39	35.96	nr	**87.75**
Sum of two sides 800 mm	45.74	55.92	1.46	37.76	nr	**93.68**
Branch						
Sum of two sides 200 mm	20.98	25.65	0.92	23.80	nr	**49.45**
Sum of two sides 300 mm	23.29	28.47	0.92	23.80	nr	**52.27**
Sum of two sides 400 mm	28.94	35.38	0.95	24.58	nr	**59.96**
Sum of two sides 500 mm	31.46	38.46	0.95	24.58	nr	**63.04**
Sum of two sides 600 mm	33.94	41.49	1.03	26.64	nr	**68.13**
Sum of two sides 700 mm	36.42	44.53	1.03	26.64	nr	**71.17**
Sum of two sides 800 mm	38.89	47.54	1.03	26.64	nr	**74.18**
Grille neck						
Sum of two sides 200 mm	23.93	29.25	1.10	28.46	nr	**57.71**
Sum of two sides 300 mm	26.70	32.64	1.10	28.46	nr	**61.10**
Sum of two sides 400 mm	29.48	36.04	1.16	30.01	nr	**66.05**
Sum of two sides 500 mm	32.25	39.43	1.16	30.01	nr	**69.44**
Sum of two sides 600 mm	35.04	42.84	1.18	30.53	nr	**73.37**
Sum of two sides 700 mm	37.81	46.23	1.18	30.53	nr	**76.76**
Sum of two sides 800 mm	40.59	49.63	1.18	30.53	nr	**80.16**

U: VENTILATION/AIR CONDITIONING SYSTEMS

Item	Net Price £	Material £	Labour hours	Labour £	Unit	Total rate £
U10: DUCTWORK: RECTANGULAR – CLASS B – cont						
Ductwork 401 to 600 mm longest side						
Sum of two sides 600 mm	14.42	17.63	1.27	32.85	m	**50.48**
Sum of two sides 700 mm	15.38	18.80	1.27	32.85	m	**51.65**
Sum of two sides 800 mm	16.32	19.95	1.27	32.85	m	**52.80**
Sum of two sides 900 mm	17.20	21.02	1.27	32.85	m	**53.87**
Sum of two sides 1000 mm	18.06	22.08	1.37	35.44	m	**57.52**
Sum of two sides 1100 mm	19.09	23.34	1.37	35.44	m	**58.78**
Sum of two sides 1200 mm	19.96	24.40	1.37	35.44	m	**59.84**
Extra over fittings; ductwork 401 to 600 mm longest side						
End cap						
Sum of two sides 600 mm	11.28	13.79	0.38	9.83	nr	**23.62**
Sum of two sides 700 mm	12.17	14.87	0.38	9.83	nr	**24.70**
Sum of two sides 800 mm	13.05	15.95	0.38	9.83	nr	**25.78**
Sum of two sides 900 mm	13.94	17.05	0.58	15.00	nr	**32.05**
Sum of two sides 1000 mm	14.82	18.12	0.58	15.00	nr	**33.12**
Sum of two sides 1100 mm	15.70	19.19	0.58	15.00	nr	**34.19**
Sum of two sides 1200 mm	16.59	20.28	0.58	15.00	nr	**35.28**
Reducer						
Sum of two sides 600 mm	26.02	31.81	1.69	43.71	nr	**75.52**
Sum of two sides 700 mm	27.89	34.09	1.69	43.71	nr	**77.80**
Sum of two sides 800 mm	29.74	36.36	1.92	49.67	nr	**86.03**
Sum of two sides 900 mm	31.61	38.65	1.92	49.67	nr	**88.32**
Sum of two sides 1000 mm	33.48	40.93	2.18	56.39	nr	**97.32**
Sum of two sides 1100 mm	35.48	43.37	2.18	56.39	nr	**99.76**
Sum of two sides 1200 mm	37.35	45.66	2.18	56.39	nr	**102.05**
Offset						
Sum of two sides 600 mm	36.39	44.49	1.92	49.67	nr	**94.16**
Sum of two sides 700 mm	39.07	47.76	1.92	49.67	nr	**97.43**
Sum of two sides 800 mm	41.21	50.38	1.92	49.67	nr	**100.05**
Sum of two sides 900 mm	43.38	53.03	1.92	49.67	nr	**102.70**
Sum of two sides 1000 mm	45.79	55.98	2.18	56.39	nr	**112.37**
Sum of two sides 1100 mm	48.02	58.70	2.18	56.39	nr	**115.09**
Sum of two sides 1200 mm	50.14	61.30	2.18	56.39	nr	**117.69**
Square to round						
Sum of two sides 600 mm	27.86	34.06	1.33	34.40	nr	**68.46**
Sum of two sides 700 mm	30.00	36.68	1.33	34.40	nr	**71.08**
Sum of two sides 800 mm	32.12	39.26	1.40	36.21	nr	**75.47**
Sum of two sides 900 mm	34.26	41.88	1.40	36.21	nr	**78.09**
Sum of two sides 1000 mm	36.40	44.50	1.82	47.08	nr	**91.58**
Sum of two sides 1100 mm	38.57	47.15	1.82	47.08	nr	**94.23**
Sum of two sides 1200 mm	40.71	49.77	1.82	47.08	nr	**96.85**

U: VENTILATION/AIR CONDITIONING SYSTEMS

Item	Net Price £	Material £	Labour hours	Labour £	Unit	**Total rate £**
90° radius bend						
Sum of two sides 600 mm	25.29	30.92	1.16	30.01	nr	**60.93**
Sum of two sides 700 mm	26.68	32.62	1.16	30.01	nr	**62.63**
Sum of two sides 800 mm	28.70	35.08	1.22	31.57	nr	**66.65**
Sum of two sides 900 mm	30.74	37.58	1.22	31.57	nr	**69.15**
Sum of two sides 1000 mm	32.47	39.70	1.40	36.21	nr	**75.91**
Sum of two sides 1100 mm	34.62	42.32	1.40	36.21	nr	**78.53**
Sum of two sides 1200 mm	36.68	44.85	1.40	36.21	nr	**81.06**
45° bend						
Sum of two sides 600 mm	26.77	32.72	1.16	30.01	nr	**62.73**
Sum of two sides 700 mm	28.42	34.74	1.39	35.96	nr	**70.70**
Sum of two sides 800 mm	30.38	37.14	1.46	37.76	nr	**74.90**
Sum of two sides 900 mm	32.37	39.57	1.46	37.76	nr	**77.33**
Sum of two sides 1000 mm	34.19	41.80	1.88	48.63	nr	**90.43**
Sum of two sides 1100 mm	36.32	44.40	1.88	48.63	nr	**93.03**
Sum of two sides 1200 mm	38.31	46.83	1.88	48.63	nr	**95.46**
90° mitre bend						
Sum of two sides 600 mm	42.61	52.09	1.39	35.96	nr	**88.05**
Sum of two sides 700 mm	45.09	55.13	2.16	55.88	nr	**111.01**
Sum of two sides 800 mm	48.61	59.43	2.26	58.46	nr	**117.89**
Sum of two sides 900 mm	52.18	63.79	2.26	58.46	nr	**122.25**
Sum of two sides 1000 mm	55.38	67.70	3.01	77.86	nr	**145.56**
Sum of two sides 1100 mm	59.10	72.25	3.01	77.86	nr	**150.11**
Sum of two sides 1200 mm	62.74	76.70	3.01	77.86	nr	**154.56**
Branch						
Sum of two sides 600 mm	33.50	40.95	1.03	26.64	nr	**67.59**
Sum of two sides 700 mm	35.96	43.96	1.03	26.64	nr	**70.60**
Sum of two sides 800 mm	38.44	46.99	1.03	26.64	nr	**73.63**
Sum of two sides 900 mm	40.90	50.00	1.03	26.64	nr	**76.64**
Sum of two sides 1000 mm	43.37	53.02	1.29	33.37	nr	**86.39**
Sum of two sides 1100 mm	45.93	56.15	1.29	33.37	nr	**89.52**
Sum of two sides 1200 mm	48.39	59.15	1.29	33.37	nr	**92.52**
Grille neck						
Sum of two sides 600 mm	34.64	42.34	1.18	30.53	nr	**72.87**
Sum of two sides 700 mm	37.44	45.77	1.18	30.53	nr	**76.30**
Sum of two sides 800 mm	40.23	49.18	1.18	30.53	nr	**79.71**
Sum of two sides 900 mm	43.02	52.59	1.18	30.53	nr	**83.12**
Sum of two sides 1000 mm	45.82	56.01	1.44	37.24	nr	**93.25**
Sum of two sides 1100 mm	48.62	59.44	1.44	37.24	nr	**96.68**
Sum of two sides 1200 mm	51.40	62.84	1.44	37.24	nr	**100.08**

U: VENTILATION/AIR CONDITIONING SYSTEMS

Item	Net Price £	Material £	Labour hours	Labour £	Unit	Total rate £
U10: DUCTWORK: RECTANGULAR – CLASS B – cont						
Ductwork 601 to 800 mm longest side						
Sum of two sides 900 mm	19.60	23.96	1.27	32.85	m	**56.81**
Sum of two sides 1000 mm	20.46	25.01	1.37	35.44	m	**60.45**
Sum of two sides 1100 mm	21.34	26.09	1.37	35.44	m	**61.53**
Sum of two sides 1200 mm	22.36	27.34	1.37	35.44	m	**62.78**
Sum of two sides 1300 mm	23.23	28.40	1.40	36.21	m	**64.61**
Sum of two sides 1400 mm	24.10	29.46	1.40	36.21	m	**65.67**
Sum of two sides 1500 mm	24.97	30.53	1.48	38.29	m	**68.82**
Sum of two sides 1600 mm	25.85	31.60	1.55	40.10	m	**71.70**
Extra over fittings; ductwork 601 to 800 mm longest side						
End cap						
Sum of two sides 900 mm	14.02	17.14	0.58	15.00	nr	**32.14**
Sum of two sides 1000 mm	14.90	18.21	0.58	15.00	nr	**33.21**
Sum of two sides 1100 mm	15.78	19.29	0.58	15.00	nr	**34.29**
Sum of two sides 1200 mm	16.68	20.39	0.58	15.00	nr	**35.39**
Sum of two sides 1300 mm	17.56	21.47	0.58	15.00	nr	**36.47**
Sum of two sides 1400 mm	18.32	22.39	0.58	15.00	nr	**37.39**
Sum of two sides 1500 mm	21.33	26.08	0.58	15.00	nr	**41.08**
Sum of two sides 1600 mm	24.33	29.75	0.58	15.00	nr	**44.75**
Reducer						
Sum of two sides 900 mm	32.08	39.22	1.92	49.67	nr	**88.89**
Sum of two sides 1000 mm	33.97	41.53	2.18	56.39	nr	**97.92**
Sum of two sides 1100 mm	35.84	43.82	2.18	56.39	nr	**100.21**
Sum of two sides 1200 mm	37.86	46.29	2.18	56.39	nr	**102.68**
Sum of two sides 1300 mm	39.73	48.57	2.30	59.50	nr	**108.07**
Sum of two sides 1400 mm	41.35	50.55	2.30	59.50	nr	**110.05**
Sum of two sides 1500 mm	47.47	58.03	2.47	63.90	nr	**121.93**
Sum of two sides 1600 mm	53.57	65.49	2.47	63.90	nr	**129.39**
Offset						
Sum of two sides 900 mm	44.79	54.75	1.92	49.67	nr	**104.42**
Sum of two sides 1000 mm	46.82	57.24	2.18	56.39	nr	**113.63**
Sum of two sides 1100 mm	48.83	59.70	2.18	56.39	nr	**116.09**
Sum of two sides 1200 mm	50.92	62.25	2.18	56.39	nr	**118.64**
Sum of two sides 1300 mm	52.91	64.68	2.30	59.50	nr	**124.18**
Sum of two sides 1400 mm	54.83	67.03	2.30	59.50	nr	**126.53**
Sum of two sides 1500 mm	63.15	77.20	2.47	63.90	nr	**141.10**
Sum of two sides 1600 mm	71.45	87.34	2.47	63.90	nr	**151.24**
Square to round						
Sum of two sides 900 mm	34.16	41.76	1.40	36.21	nr	**77.97**
Sum of two sides 1000 mm	36.30	44.37	1.82	47.08	nr	**91.45**
Sum of two sides 1100 mm	38.46	47.02	1.82	47.08	nr	**94.10**

U: VENTILATION/AIR CONDITIONING SYSTEMS

Item	Net Price £	Material £	Labour hours	Labour £	Unit	**Total rate £**
Sum of two sides 1200 mm	40.65	49.70	1.82	47.08	nr	**96.78**
Sum of two sides 1300 mm	42.79	52.31	2.15	55.61	nr	**107.92**
Sum of two sides 1400 mm	44.61	54.54	2.15	55.61	nr	**110.15**
Sum of two sides 1500 mm	52.06	63.64	2.38	61.56	nr	**125.20**
Sum of two sides 1600 mm	59.50	72.74	2.38	61.56	nr	**134.30**
90° radius bend						
Sum of two sides 900 mm	29.09	35.57	1.22	31.57	nr	**67.14**
Sum of two sides 1000 mm	31.18	38.12	1.40	36.21	nr	**74.33**
Sum of two sides 1100 mm	33.27	40.67	1.40	36.21	nr	**76.88**
Sum of two sides 1200 mm	35.42	43.30	1.40	36.21	nr	**79.51**
Sum of two sides 1300 mm	37.51	45.86	1.91	49.41	nr	**95.27**
Sum of two sides 1400 mm	38.93	47.60	1.91	49.41	nr	**97.01**
Sum of two sides 1500 mm	45.26	55.33	2.11	54.59	nr	**109.92**
Sum of two sides 1600 mm	51.58	63.06	2.11	54.59	nr	**117.65**
45° bend						
Sum of two sides 900 mm	32.09	39.23	1.22	31.57	nr	**70.80**
Sum of two sides 1000 mm	34.11	41.70	1.40	36.21	nr	**77.91**
Sum of two sides 1100 mm	36.11	44.15	1.88	48.63	nr	**92.78**
Sum of two sides 1200 mm	38.25	46.76	1.88	48.63	nr	**95.39**
Sum of two sides 1300 mm	40.81	49.89	2.26	58.46	nr	**108.35**
Sum of two sides 1400 mm	41.81	51.11	2.26	58.46	nr	**109.57**
Sum of two sides 1500 mm	48.06	58.75	2.49	64.42	nr	**123.17**
Sum of two sides 1600 mm	54.30	66.38	2.49	64.42	nr	**130.80**
90° mitre bend						
Sum of two sides 900 mm	49.57	60.59	1.22	31.57	nr	**92.16**
Sum of two sides 1000 mm	53.46	65.35	1.40	36.21	nr	**101.56**
Sum of two sides 1100 mm	57.34	70.10	3.01	77.86	nr	**147.96**
Sum of two sides 1200 mm	61.30	74.94	3.01	77.86	nr	**152.80**
Sum of two sides 1300 mm	65.19	79.69	3.67	94.94	nr	**174.63**
Sum of two sides 1400 mm	68.21	83.39	3.67	94.94	nr	**178.33**
Sum of two sides 1500 mm	78.52	96.00	4.07	105.29	nr	**201.29**
Sum of two sides 1600 mm	88.83	108.59	4.07	105.29	nr	**213.88**
Branch						
Sum of two sides 900 mm	42.18	51.56	1.22	31.57	nr	**83.13**
Sum of two sides 1000 mm	44.70	54.64	1.40	36.21	nr	**90.85**
Sum of two sides 1100 mm	47.23	57.74	1.29	33.37	nr	**91.11**
Sum of two sides 1200 mm	49.86	60.96	1.29	33.37	nr	**94.33**
Sum of two sides 1300 mm	52.39	64.05	1.39	35.96	nr	**100.01**
Sum of two sides 1400 mm	54.58	66.72	1.39	35.96	nr	**102.68**
Sum of two sides 1500 mm	62.74	76.70	1.64	42.43	nr	**119.13**
Sum of two sides 1600 mm	70.92	86.70	1.64	42.43	nr	**129.13**

U: VENTILATION/AIR CONDITIONING SYSTEMS

Item	Net Price £	Material £	Labour hours	Labour £	Unit	Total rate £
U10: DUCTWORK: RECTANGULAR – CLASS B – cont						
Extra over fittings; ductwork 601 to 800 mm longest side – cont						
Grille neck						
Sum of two sides 900 mm	43.25	52.87	1.22	31.57	nr	**84.44**
Sum of two sides 1000 mm	46.07	56.32	1.40	36.21	nr	**92.53**
Sum of two sides 1100 mm	48.88	59.76	1.44	37.24	nr	**97.00**
Sum of two sides 1200 mm	51.68	63.18	1.44	37.24	nr	**100.42**
Sum of two sides 1300 mm	54.49	66.61	1.69	43.71	nr	**110.32**
Sum of two sides 1400 mm	56.90	69.56	1.69	43.71	nr	**113.27**
Sum of two sides 1500 mm	66.06	80.76	1.79	46.31	nr	**127.07**
Sum of two sides 1600 mm	75.22	91.96	1.79	46.31	nr	**138.27**
Ductwork 801 to 1000 mm longest side						
Sum of two sides 1100 mm	31.10	38.02	1.37	35.44	m	**73.46**
Sum of two sides 1200 mm	32.82	40.12	1.37	35.44	m	**75.56**
Sum of two sides 1300 mm	34.53	42.21	1.40	36.21	m	**78.42**
Sum of two sides 1400 mm	36.40	44.50	1.40	36.21	m	**80.71**
Sum of two sides 1500 mm	38.10	46.58	1.48	38.29	m	**84.87**
Sum of two sides 1600 mm	39.82	48.68	1.55	40.10	m	**88.78**
Sum of two sides 1700 mm	41.54	50.78	1.55	40.10	m	**90.88**
Sum of two sides 1800 mm	43.40	53.06	1.61	41.65	m	**94.71**
Sum of two sides 1900 mm	45.12	55.16	1.61	41.65	m	**96.81**
Sum of two sides 2000 mm	46.83	57.25	1.61	41.65	m	**98.90**
Extra over fittings; ductwork 801 to 1000 mm longest side						
End cap						
Sum of two sides 1100 mm	15.78	19.29	1.44	37.24	nr	**56.53**
Sum of two sides 1200 mm	16.68	20.39	1.44	37.24	nr	**57.63**
Sum of two sides 1300 mm	17.56	21.47	1.44	37.24	nr	**58.71**
Sum of two sides 1400 mm	18.32	22.39	1.44	37.24	nr	**59.63**
Sum of two sides 1500 mm	21.33	26.08	1.44	37.24	nr	**63.32**
Sum of two sides 1600 mm	24.33	29.75	1.44	37.24	nr	**66.99**
Sum of two sides 1700 mm	27.33	33.41	1.44	37.24	nr	**70.65**
Sum of two sides 1800 mm	30.34	37.09	1.44	37.24	nr	**74.33**
Sum of two sides 1900 mm	33.34	40.76	1.44	37.24	nr	**78.00**
Sum of two sides 2000 mm	36.35	44.43	1.44	37.24	nr	**81.67**
Reducer						
Sum of two sides 1100 mm	29.77	36.39	1.44	37.24	nr	**73.63**
Sum of two sides 1200 mm	31.19	38.13	1.44	37.24	nr	**75.37**
Sum of two sides 1300 mm	32.61	39.86	1.69	43.71	nr	**83.57**
Sum of two sides 1400 mm	33.88	41.42	1.69	43.71	nr	**85.13**
Sum of two sides 1500 mm	39.54	48.34	2.47	63.90	nr	**112.24**
Sum of two sides 1600 mm	45.19	55.25	2.47	63.90	nr	**119.15**
Sum of two sides 1700 mm	50.85	62.16	2.47	63.90	nr	**126.06**

U: VENTILATION/AIR CONDITIONING SYSTEMS

Item	Net Price £	Material £	Labour hours	Labour £	Unit	Total rate £
Sum of two sides 1800 mm	56.61	69.21	2.59	66.99	nr	**136.20**
Sum of two sides 1900 mm	62.27	76.13	2.71	70.10	nr	**146.23**
Sum of two sides 2000 mm	67.92	83.03	2.71	70.10	nr	**153.13**
Offset						
Sum of two sides 1100 mm	46.00	56.24	1.44	37.24	nr	**93.48**
Sum of two sides 1200 mm	46.92	57.36	1.44	37.24	nr	**94.60**
Sum of two sides 1300 mm	47.70	58.32	1.69	43.71	nr	**102.03**
Sum of two sides 1400 mm	48.06	58.75	1.69	43.71	nr	**102.46**
Sum of two sides 1500 mm	54.92	67.14	2.47	63.90	nr	**131.04**
Sum of two sides 1600 mm	61.66	75.38	2.47	63.90	nr	**139.28**
Sum of two sides 1700 mm	68.28	83.47	2.59	66.99	nr	**150.46**
Sum of two sides 1800 mm	74.81	91.45	2.61	67.52	nr	**158.97**
Sum of two sides 1900 mm	82.08	100.34	2.71	70.10	nr	**170.44**
Sum of two sides 2000 mm	88.33	107.99	2.71	70.10	nr	**178.09**
Square to round						
Sum of two sides 1100 mm	32.28	39.46	1.44	37.24	nr	**76.70**
Sum of two sides 1200 mm	33.98	41.54	1.44	37.24	nr	**78.78**
Sum of two sides 1300 mm	35.67	43.61	1.69	43.71	nr	**87.32**
Sum of two sides 1400 mm	37.06	45.31	1.69	43.71	nr	**89.02**
Sum of two sides 1500 mm	44.03	53.83	2.38	61.56	nr	**115.39**
Sum of two sides 1600 mm	51.02	62.38	2.38	61.56	nr	**123.94**
Sum of two sides 1700 mm	58.00	70.91	2.55	65.97	nr	**136.88**
Sum of two sides 1800 mm	64.99	79.45	2.55	65.97	nr	**145.42**
Sum of two sides 1900 mm	71.97	87.98	2.83	73.20	nr	**161.18**
Sum of two sides 2000 mm	78.97	96.54	2.83	73.20	nr	**169.74**
90° radius bend						
Sum of two sides 1100 mm	23.83	29.13	1.44	37.24	nr	**66.37**
Sum of two sides 1200 mm	25.52	31.20	1.44	37.24	nr	**68.44**
Sum of two sides 1300 mm	27.20	33.25	1.69	43.71	nr	**76.96**
Sum of two sides 1400 mm	28.68	35.06	1.69	43.71	nr	**78.77**
Sum of two sides 1500 mm	34.59	42.28	2.11	54.59	nr	**96.87**
Sum of two sides 1600 mm	40.51	49.52	2.11	54.59	nr	**104.11**
Sum of two sides 1700 mm	46.42	56.75	2.26	58.46	nr	**115.21**
Sum of two sides 1800 mm	52.39	64.05	2.26	58.46	nr	**122.51**
Sum of two sides 1900 mm	57.27	70.01	2.48	64.16	nr	**134.17**
Sum of two sides 2000 mm	63.18	77.24	2.48	64.16	nr	**141.40**
45° bend						
Sum of two sides 1100 mm	31.07	37.99	1.44	37.24	nr	**75.23**
Sum of two sides 1200 mm	32.87	40.18	1.44	37.24	nr	**77.42**
Sum of two sides 1300 mm	34.66	42.37	1.69	43.71	nr	**86.08**
Sum of two sides 1400 mm	36.30	44.37	1.69	43.71	nr	**88.08**
Sum of two sides 1500 mm	42.33	51.75	2.49	64.42	nr	**116.17**
Sum of two sides 1600 mm	48.35	59.11	2.49	64.42	nr	**123.53**
Sum of two sides 1700 mm	54.37	66.47	2.67	69.07	nr	**135.54**
Sum of two sides 1800 mm	60.51	73.97	2.67	69.07	nr	**143.04**
Sum of two sides 1900 mm	66.00	80.69	3.06	79.16	nr	**159.85**
Sum of two sides 2000 mm	72.02	88.04	3.06	79.16	nr	**167.20**

U: VENTILATION/AIR CONDITIONING SYSTEMS

Item	Net Price £	Material £	Labour hours	Labour £	Unit	Total rate £
U10: DUCTWORK: RECTANGULAR – CLASS B – cont						
Extra over fittings – cont						
90° mitre bend						
Sum of two sides 1100 mm	49.73	60.79	1.44	37.24	nr	**98.03**
Sum of two sides 1200 mm	53.10	64.91	1.44	37.24	nr	**102.15**
Sum of two sides 1300 mm	56.46	69.02	1.69	43.71	nr	**112.73**
Sum of two sides 1400 mm	59.47	72.70	1.69	43.71	nr	**116.41**
Sum of two sides 1500 mm	69.18	84.57	2.07	53.55	nr	**138.12**
Sum of two sides 1600 mm	78.90	96.46	2.07	53.55	nr	**150.01**
Sum of two sides 1700 mm	88.61	108.33	2.80	72.43	nr	**180.76**
Sum of two sides 1800 mm	98.35	120.23	2.67	69.07	nr	**189.30**
Sum of two sides 1900 mm	106.92	130.71	2.95	76.31	nr	**207.02**
Sum of two sides 2000 mm	116.67	142.63	2.95	76.31	nr	**218.94**
Branch						
Sum of two sides 1100 mm	47.33	57.87	1.44	37.24	nr	**95.11**
Sum of two sides 1200 mm	49.86	60.96	1.44	37.24	nr	**98.20**
Sum of two sides 1300 mm	52.39	64.05	1.64	42.43	nr	**106.48**
Sum of two sides 1400 mm	54.66	66.83	1.64	42.43	nr	**109.26**
Sum of two sides 1500 mm	62.84	76.82	1.64	42.43	nr	**119.25**
Sum of two sides 1600 mm	71.00	86.80	1.64	42.43	nr	**129.23**
Sum of two sides 1700 mm	79.17	96.79	1.69	43.71	nr	**140.50**
Sum of two sides 1800 mm	87.44	106.89	1.69	43.71	nr	**150.60**
Sum of two sides 1900 mm	95.61	116.88	1.85	47.86	nr	**164.74**
Sum of two sides 2000 mm	103.79	126.89	1.85	47.86	nr	**174.75**
Grille neck						
Sum of two sides 1100 mm	48.88	59.76	1.44	37.24	nr	**97.00**
Sum of two sides 1200 mm	51.68	63.18	1.44	37.24	nr	**100.42**
Sum of two sides 1300 mm	54.49	66.61	1.69	43.71	nr	**110.32**
Sum of two sides 1400 mm	56.90	69.56	1.69	43.71	nr	**113.27**
Sum of two sides 1500 mm	66.06	80.76	1.79	46.31	nr	**127.07**
Sum of two sides 1600 mm	75.22	91.96	1.79	46.31	nr	**138.27**
Sum of two sides 1700 mm	84.37	103.14	1.86	48.12	nr	**151.26**
Sum of two sides 1800 mm	93.52	114.33	2.02	52.25	nr	**166.58**
Sum of two sides 1900 mm	102.68	125.53	2.02	52.25	nr	**177.78**
Sum of two sides 2000 mm	111.83	136.71	2.02	52.25	nr	**188.96**
Ductwork 1001 to 1250 mm longest side						
Sum of two sides 1300 mm	40.69	49.74	1.40	36.21	m	**85.95**
Sum of two sides 1400 mm	42.78	52.30	1.40	36.21	m	**88.51**
Sum of two sides 1500 mm	44.70	54.64	1.48	38.29	m	**92.93**
Sum of two sides 1600 mm	46.77	57.18	1.55	40.10	m	**97.28**
Sum of two sides 1700 mm	48.69	59.52	1.55	40.10	m	**99.62**
Sum of two sides 1800 mm	50.61	61.87	1.61	41.65	m	**103.52**
Sum of two sides 1900 mm	52.54	64.23	1.61	41.65	m	**105.88**
Sum of two sides 2000 mm	54.61	66.76	1.61	41.65	m	**108.41**
Sum of two sides 2100 mm	56.52	69.09	2.17	56.14	m	**125.23**

U: VENTILATION/AIR CONDITIONING SYSTEMS

Item	Net Price £	Material £	Labour hours	Labour £	Unit	**Total rate £**
Sum of two sides 2200 mm	58.45	71.45	2.19	56.65	m	**128.10**
Sum of two sides 2300 mm	60.69	74.19	2.19	56.65	m	**130.84**
Sum of two sides 2400 mm	62.61	76.54	2.38	61.56	m	**138.10**
Sum of two sides 2500 mm	64.68	79.07	2.38	61.56	m	**140.63**
Extra over fittings; ductwork 1001 to 1250 mm longest side						
End cap						
Sum of two sides 1300 mm	17.91	21.90	1.69	43.71	nr	**65.61**
Sum of two sides 1400 mm	18.69	22.85	1.69	43.71	nr	**66.56**
Sum of two sides 1500 mm	21.74	26.57	1.69	43.71	nr	**70.28**
Sum of two sides 1600 mm	24.78	30.29	1.69	43.71	nr	**74.00**
Sum of two sides 1700 mm	27.81	34.00	1.69	43.71	nr	**77.71**
Sum of two sides 1800 mm	30.85	37.72	1.69	43.71	nr	**81.43**
Sum of two sides 1900 mm	33.90	41.45	1.69	43.71	nr	**85.16**
Sum of two sides 2000 mm	36.93	45.14	1.69	43.71	nr	**88.85**
Sum of two sides 2100 mm	39.97	48.86	1.69	43.71	nr	**92.57**
Sum of two sides 2200 mm	43.02	52.59	1.69	43.71	nr	**96.30**
Sum of two sides 2300 mm	46.06	56.31	1.69	43.71	nr	**100.02**
Sum of two sides 2400 mm	49.09	60.01	1.69	43.71	nr	**103.72**
Sum of two sides 2500 mm	52.14	63.74	1.69	43.71	nr	**107.45**
Reducer						
Sum of two sides 1300 mm	31.81	38.89	1.69	43.71	nr	**82.60**
Sum of two sides 1400 mm	32.97	40.30	1.69	43.71	nr	**84.01**
Sum of two sides 1500 mm	38.64	47.24	2.47	63.90	nr	**111.14**
Sum of two sides 1600 mm	44.42	54.30	2.47	63.90	nr	**118.20**
Sum of two sides 1700 mm	50.11	61.26	2.47	63.90	nr	**125.16**
Sum of two sides 1800 mm	55.79	68.21	2.59	66.99	nr	**135.20**
Sum of two sides 1900 mm	61.46	75.14	2.71	70.10	nr	**145.24**
Sum of two sides 2000 mm	67.24	82.20	2.59	66.99	nr	**149.19**
Sum of two sides 2100 mm	72.92	89.15	2.92	75.54	nr	**164.69**
Sum of two sides 2200 mm	78.60	96.09	2.92	75.54	nr	**171.63**
Sum of two sides 2300 mm	84.39	103.16	2.92	75.54	nr	**178.70**
Sum of two sides 2400 mm	90.06	110.10	3.12	80.71	nr	**190.81**
Sum of two sides 2500 mm	95.85	117.17	3.12	80.71	nr	**197.88**
Offset						
Sum of two sides 1300 mm	62.86	76.85	1.69	43.71	nr	**120.56**
Sum of two sides 1400 mm	64.93	79.38	1.69	43.71	nr	**123.09**
Sum of two sides 1500 mm	72.80	89.00	2.47	63.90	nr	**152.90**
Sum of two sides 1600 mm	80.63	98.57	2.47	63.90	nr	**162.47**
Sum of two sides 1700 mm	88.22	107.85	2.59	66.99	nr	**174.84**
Sum of two sides 1800 mm	95.69	116.98	2.61	67.52	nr	**184.50**
Sum of two sides 1900 mm	103.04	125.97	2.71	70.10	nr	**196.07**
Sum of two sides 2000 mm	110.29	134.83	2.71	70.10	nr	**204.93**
Sum of two sides 2100 mm	117.36	143.47	2.92	75.54	nr	**219.01**
Sum of two sides 2200 mm	125.43	153.34	3.26	84.33	nr	**237.67**
Sum of two sides 2300 mm	134.65	164.61	3.26	84.33	nr	**248.94**
Sum of two sides 2400 mm	143.82	175.82	3.48	90.02	nr	**265.84**
Sum of two sides 2500 mm	153.04	187.09	3.48	90.02	nr	**277.11**

U: VENTILATION/AIR CONDITIONING SYSTEMS

Item	Net Price £	Material £	Labour hours	Labour £	Unit	Total rate £
U10: DUCTWORK: RECTANGULAR – CLASS B – cont						
Extra over fittings – cont						
Square to round						
Sum of two sides 1300 mm	34.58	42.27	1.69	43.71	nr	**85.98**
Sum of two sides 1400 mm	35.96	43.96	1.69	43.71	nr	**87.67**
Sum of two sides 1500 mm	42.96	52.52	2.38	61.56	nr	**114.08**
Sum of two sides 1600 mm	49.99	61.11	2.38	61.56	nr	**122.67**
Sum of two sides 1700 mm	57.01	69.70	2.55	65.97	nr	**135.67**
Sum of two sides 1800 mm	64.02	78.27	2.55	65.97	nr	**144.24**
Sum of two sides 1900 mm	71.02	86.82	2.83	73.20	nr	**160.02**
Sum of two sides 2000 mm	78.05	95.42	2.83	73.20	nr	**168.62**
Sum of two sides 2100 mm	85.06	103.99	3.85	99.60	nr	**203.59**
Sum of two sides 2200 mm	92.08	112.57	4.18	108.13	nr	**220.70**
Sum of two sides 2300 mm	99.09	121.14	4.22	109.17	nr	**230.31**
Sum of two sides 2400 mm	106.11	129.72	4.68	121.07	nr	**250.79**
Sum of two sides 2500 mm	113.13	138.30	4.70	121.58	nr	**259.88**
90° radius bend						
Sum of two sides 1300 mm	22.43	27.42	1.69	43.71	nr	**71.13**
Sum of two sides 1400 mm	22.78	27.85	1.69	43.71	nr	**71.56**
Sum of two sides 1500 mm	28.85	35.27	2.11	54.59	nr	**89.86**
Sum of two sides 1600 mm	34.96	42.73	2.11	54.59	nr	**97.32**
Sum of two sides 1700 mm	41.03	50.16	2.19	56.65	nr	**106.81**
Sum of two sides 1800 mm	47.11	57.59	2.19	56.65	nr	**114.24**
Sum of two sides 1900 mm	53.18	65.01	2.48	64.16	nr	**129.17**
Sum of two sides 2000 mm	59.29	72.48	2.26	58.46	nr	**130.94**
Sum of two sides 2100 mm	65.36	79.91	2.48	64.16	nr	**144.07**
Sum of two sides 2200 mm	71.43	87.32	2.48	64.16	nr	**151.48**
Sum of two sides 2300 mm	74.90	91.57	2.48	64.16	nr	**155.73**
Sum of two sides 2400 mm	80.97	98.98	3.90	100.88	nr	**199.86**
Sum of two sides 2500 mm	87.05	106.42	3.90	100.88	nr	**207.30**
45° bend						
Sum of two sides 1300 mm	32.51	39.75	1.69	43.71	nr	**83.46**
Sum of two sides 1400 mm	33.50	40.95	1.69	43.71	nr	**84.66**
Sum of two sides 1500 mm	39.63	48.45	2.49	64.42	nr	**112.87**
Sum of two sides 1600 mm	45.85	56.05	2.49	64.42	nr	**120.47**
Sum of two sides 1700 mm	51.98	63.55	2.67	69.07	nr	**132.62**
Sum of two sides 1800 mm	58.11	71.04	2.67	69.07	nr	**140.11**
Sum of two sides 1900 mm	64.24	78.54	3.06	79.16	nr	**157.70**
Sum of two sides 2000 mm	70.47	86.15	3.06	79.16	nr	**165.31**
Sum of two sides 2100 mm	76.60	93.65	4.05	104.77	nr	**198.42**
Sum of two sides 2200 mm	82.74	101.15	4.05	104.77	nr	**205.92**
Sum of two sides 2300 mm	87.60	107.09	4.39	113.56	nr	**220.65**
Sum of two sides 2400 mm	93.72	114.58	4.85	125.46	nr	**240.04**
Sum of two sides 2500 mm	99.95	122.19	4.85	125.46	nr	**247.65**

U: VENTILATION/AIR CONDITIONING SYSTEMS

Item	Net Price £	Material £	Labour hours	Labour £	Unit	**Total rate £**
90° mitre bend						
Sum of two sides 1300 mm	67.96	83.08	1.69	43.71	nr	**126.79**
Sum of two sides 1400 mm	70.75	86.49	1.69	43.71	nr	**130.20**
Sum of two sides 1500 mm	82.07	100.33	2.80	72.43	nr	**172.76**
Sum of two sides 1600 mm	93.39	114.17	2.80	72.43	nr	**186.60**
Sum of two sides 1700 mm	104.70	128.00	2.95	76.31	nr	**204.31**
Sum of two sides 1800 mm	116.01	141.82	2.95	76.31	nr	**218.13**
Sum of two sides 1900 mm	127.34	155.67	4.05	104.77	nr	**260.44**
Sum of two sides 2000 mm	138.66	169.52	4.05	104.77	nr	**274.29**
Sum of two sides 2100 mm	149.96	183.33	4.07	105.29	nr	**288.62**
Sum of two sides 2200 mm	161.28	197.16	4.07	105.29	nr	**302.45**
Sum of two sides 2300 mm	169.61	207.35	4.39	113.56	nr	**320.91**
Sum of two sides 2400 mm	181.00	221.27	4.85	125.46	nr	**346.73**
Sum of two sides 2500 mm	192.38	235.19	4.85	125.46	nr	**360.65**
Branch						
Sum of two sides 1300 mm	53.75	65.71	1.44	37.24	nr	**102.95**
Sum of two sides 1400 mm	56.00	68.46	1.44	37.24	nr	**105.70**
Sum of two sides 1500 mm	64.28	78.58	1.64	42.43	nr	**121.01**
Sum of two sides 1600 mm	72.65	88.82	1.64	42.43	nr	**131.25**
Sum of two sides 1700 mm	80.92	98.92	1.64	42.43	nr	**141.35**
Sum of two sides 1800 mm	89.20	109.05	1.64	42.43	nr	**151.48**
Sum of two sides 1900 mm	97.46	119.15	1.69	43.71	nr	**162.86**
Sum of two sides 2000 mm	105.83	129.38	1.69	43.71	nr	**173.09**
Sum of two sides 2100 mm	114.10	139.49	1.85	47.86	nr	**187.35**
Sum of two sides 2200 mm	122.38	149.61	1.85	47.86	nr	**197.47**
Sum of two sides 2300 mm	130.75	159.85	2.61	67.52	nr	**227.37**
Sum of two sides 2400 mm	139.02	169.95	2.61	67.52	nr	**237.47**
Sum of two sides 2500 mm	147.40	180.20	2.61	67.52	nr	**247.72**
Grille neck						
Sum of two sides 1300 mm	56.08	68.56	1.79	46.31	nr	**114.87**
Sum of two sides 1400 mm	58.61	71.65	1.79	46.31	nr	**117.96**
Sum of two sides 1500 mm	67.91	83.02	1.79	46.31	nr	**129.33**
Sum of two sides 1600 mm	77.22	94.40	1.79	46.31	nr	**140.71**
Sum of two sides 1700 mm	86.52	105.77	1.86	48.12	nr	**153.89**
Sum of two sides 1800 mm	95.82	117.14	2.02	52.25	nr	**169.39**
Sum of two sides 1900 mm	105.14	128.53	2.02	52.25	nr	**180.78**
Sum of two sides 2000 mm	114.43	139.89	2.02	52.25	nr	**192.14**
Sum of two sides 2100 mm	123.74	151.28	2.61	67.52	nr	**218.80**
Sum of two sides 2200 mm	133.04	162.65	2.61	67.52	nr	**230.17**
Sum of two sides 2300 mm	142.35	174.03	2.61	67.52	nr	**241.55**
Sum of two sides 2400 mm	151.64	185.38	2.88	74.50	nr	**259.88**
Sum of two sides 2500 mm	160.96	196.77	2.88	74.50	nr	**271.27**

U: VENTILATION/AIR CONDITIONING SYSTEMS

Item	Net Price £	Material £	Labour hours	Labour £	Unit	**Total rate £**
U10: DUCTWORK: RECTANGULAR – CLASS B – cont						
Ductwork 1251 to 1600 mm longest side						
Sum of two sides 1700 mm	64.17	78.44	1.55	40.10	m	**118.54**
Sum of two sides 1800 mm	66.54	81.35	1.61	41.65	m	**123.00**
Sum of two sides 1900 mm	68.93	84.26	1.61	41.65	m	**125.91**
Sum of two sides 2000 mm	71.17	87.00	1.61	41.65	m	**128.65**
Sum of two sides 2100 mm	73.42	89.75	2.17	56.14	m	**145.89**
Sum of two sides 2200 mm	75.67	92.50	2.19	56.65	m	**149.15**
Sum of two sides 2300 mm	78.06	95.43	2.19	56.65	m	**152.08**
Sum of two sides 2400 mm	80.42	98.31	2.38	61.56	m	**159.87**
Sum of two sides 2500 mm	82.67	101.06	2.38	61.56	m	**162.62**
Sum of two sides 2600 mm	84.92	103.81	2.64	68.29	m	**172.10**
Sum of two sides 2700 mm	87.16	106.55	2.66	68.81	m	**175.36**
Sum of two sides 2800 mm	89.53	109.45	2.95	76.31	m	**185.76**
Sum of two sides 2900 mm	100.46	122.82	2.96	76.57	m	**199.39**
Sum of two sides 3000 mm	102.71	125.57	3.15	81.48	m	**207.05**
Sum of two sides 3100 mm	105.08	128.46	3.15	81.48	m	**209.94**
Sum of two sides 3200 mm	107.32	131.20	3.18	82.27	m	**213.47**
Extra over fittings; ductwork 1251 to 1600 mm longest side						
End cap						
Sum of two sides 1700 mm	27.52	33.64	0.58	15.00	nr	**48.64**
Sum of two sides 1800 mm	30.53	37.33	0.58	15.00	nr	**52.33**
Sum of two sides 1900 mm	33.54	41.00	0.58	15.00	nr	**56.00**
Sum of two sides 2000 mm	36.54	44.67	0.58	15.00	nr	**59.67**
Sum of two sides 2100 mm	39.55	48.35	0.87	22.51	nr	**70.86**
Sum of two sides 2200 mm	42.56	52.03	0.87	22.51	nr	**74.54**
Sum of two sides 2300 mm	45.58	55.72	0.87	22.51	nr	**78.23**
Sum of two sides 2400 mm	48.58	59.39	0.87	22.51	nr	**81.90**
Sum of two sides 2500 mm	51.59	63.07	0.87	22.51	nr	**85.58**
Sum of two sides 2600 mm	54.60	66.74	0.87	22.51	nr	**89.25**
Sum of two sides 2700 mm	57.61	70.43	0.87	22.51	nr	**92.94**
Sum of two sides 2800 mm	60.61	74.10	1.16	30.01	nr	**104.11**
Sum of two sides 2900 mm	63.62	77.78	1.16	30.01	nr	**107.79**
Sum of two sides 3000 mm	70.34	85.99	1.30	33.63	nr	**119.62**
Sum of two sides 3100 mm	69.01	84.37	1.80	46.57	nr	**130.94**
Sum of two sides 3200 mm	71.17	87.00	1.80	46.57	nr	**133.57**
Reducer						
Sum of two sides 1700 mm	33.86	41.40	2.47	63.90	nr	**105.30**
Sum of two sides 1800 mm	39.11	47.81	2.59	66.99	nr	**114.80**
Sum of two sides 1900 mm	44.43	54.31	2.71	70.10	nr	**124.41**
Sum of two sides 2000 mm	49.66	60.71	2.71	70.10	nr	**130.81**
Sum of two sides 2100 mm	54.89	67.10	2.92	75.54	nr	**142.64**
Sum of two sides 2200 mm	60.12	73.50	2.92	75.54	nr	**149.04**
Sum of two sides 2300 mm	65.42	79.98	2.92	75.54	nr	**155.52**

U: VENTILATION/AIR CONDITIONING SYSTEMS

Item	Net Price £	Material £	Labour hours	Labour £	Unit	**Total rate £**
Sum of two sides 2400 mm	70.68	86.41	3.12	80.71	nr	**167.12**
Sum of two sides 2500 mm	75.90	92.79	3.12	80.71	nr	**173.50**
Sum of two sides 2600 mm	81.14	99.20	3.12	80.71	nr	**179.91**
Sum of two sides 2700 mm	86.36	105.58	3.12	80.71	nr	**186.29**
Sum of two sides 2800 mm	91.62	112.00	3.95	102.18	nr	**214.18**
Sum of two sides 2900 mm	91.87	112.31	3.97	102.70	nr	**215.01**
Sum of two sides 3000 mm	96.03	117.40	4.52	116.93	nr	**234.33**
Sum of two sides 3100 mm	99.91	122.14	4.52	116.93	nr	**239.07**
Sum of two sides 3200 mm	103.44	126.45	4.52	116.93	nr	**243.38**
Offset						
Sum of two sides 1700 mm	86.18	105.36	2.59	66.99	nr	**172.35**
Sum of two sides 1800 mm	95.81	117.13	2.61	67.52	nr	**184.65**
Sum of two sides 1900 mm	102.35	125.12	2.71	70.10	nr	**195.22**
Sum of two sides 2000 mm	108.63	132.80	2.71	70.10	nr	**202.90**
Sum of two sides 2100 mm	114.73	140.26	2.92	75.54	nr	**215.80**
Sum of two sides 2200 mm	120.65	147.50	3.26	84.33	nr	**231.83**
Sum of two sides 2300 mm	126.44	154.57	3.26	84.33	nr	**238.90**
Sum of two sides 2400 mm	135.65	165.83	3.47	89.76	nr	**255.59**
Sum of two sides 2500 mm	141.15	172.56	3.48	90.02	nr	**262.58**
Sum of two sides 2600 mm	150.24	183.67	3.49	90.28	nr	**273.95**
Sum of two sides 2700 mm	159.33	194.78	3.50	90.54	nr	**285.32**
Sum of two sides 2800 mm	168.46	205.95	4.34	112.26	nr	**318.21**
Sum of two sides 2900 mm	168.43	205.91	4.76	123.13	nr	**329.04**
Sum of two sides 3000 mm	177.51	217.01	5.32	137.62	nr	**354.63**
Sum of two sides 3100 mm	184.58	225.65	5.35	138.40	nr	**364.05**
Sum of two sides 3200 mm	191.13	233.66	5.35	138.40	nr	**372.06**
Square to round						
Sum of two sides 1700 mm	40.69	49.74	2.55	65.97	nr	**115.71**
Sum of two sides 1800 mm	47.26	57.77	2.55	65.97	nr	**123.74**
Sum of two sides 1900 mm	53.79	65.76	2.83	73.20	nr	**138.96**
Sum of two sides 2000 mm	60.34	73.77	2.83	73.20	nr	**146.97**
Sum of two sides 2100 mm	66.89	81.77	3.85	99.60	nr	**181.37**
Sum of two sides 2200 mm	73.44	89.79	4.18	108.13	nr	**197.92**
Sum of two sides 2300 mm	79.97	97.77	4.22	109.17	nr	**206.94**
Sum of two sides 2400 mm	86.55	105.81	4.68	121.07	nr	**226.88**
Sum of two sides 2500 mm	93.11	113.83	4.70	121.58	nr	**235.41**
Sum of two sides 2600 mm	99.66	121.84	4.70	121.58	nr	**243.42**
Sum of two sides 2700 mm	106.21	129.84	4.71	121.84	nr	**251.68**
Sum of two sides 2800 mm	112.78	137.88	8.19	211.86	nr	**349.74**
Sum of two sides 2900 mm	113.18	138.36	8.62	222.98	nr	**361.34**
Sum of two sides 3000 mm	119.73	146.37	8.75	226.34	nr	**372.71**
Sum of two sides 3100 mm	124.63	152.36	8.75	226.34	nr	**378.70**
Sum of two sides 3200 mm	129.05	157.77	8.75	226.34	nr	**384.11**

U: VENTILATION/AIR CONDITIONING SYSTEMS

Item	Net Price £	Material £	Labour hours	Labour £	Unit	Total rate £
U10: DUCTWORK: RECTANGULAR – CLASS B – cont						
Extra over fittings – cont						
90° radius bend						
Sum of two sides 1700 mm	83.67	102.29	2.19	56.65	nr	**158.94**
Sum of two sides 1800 mm	91.68	112.07	2.19	56.65	nr	**168.72**
Sum of two sides 1900 mm	103.58	126.63	2.26	58.46	nr	**185.09**
Sum of two sides 2000 mm	115.28	140.93	2.26	58.46	nr	**199.39**
Sum of two sides 2100 mm	127.00	155.26	2.48	64.16	nr	**219.42**
Sum of two sides 2200 mm	138.71	169.57	2.48	64.16	nr	**233.73**
Sum of two sides 2300 mm	150.61	184.12	2.48	64.16	nr	**248.28**
Sum of two sides 2400 mm	158.28	193.50	3.90	100.88	nr	**294.38**
Sum of two sides 2500 mm	169.94	207.75	3.90	100.88	nr	**308.63**
Sum of two sides 2600 mm	181.63	222.05	4.26	110.20	nr	**332.25**
Sum of two sides 2700 mm	193.31	236.32	4.55	117.70	nr	**354.02**
Sum of two sides 2800 mm	200.70	245.36	4.55	117.70	nr	**363.06**
Sum of two sides 2900 mm	200.91	245.61	6.87	177.72	nr	**423.33**
Sum of two sides 3000 mm	212.55	259.85	7.00	181.07	nr	**440.92**
Sum of two sides 3100 mm	221.52	270.81	7.00	181.07	nr	**451.88**
Sum of two sides 3200 mm	229.77	280.89	7.00	181.07	nr	**461.96**
45° bend						
Sum of two sides 1700 mm	40.12	49.05	2.67	69.07	nr	**118.12**
Sum of two sides 1800 mm	44.06	53.86	2.67	69.07	nr	**122.93**
Sum of two sides 1900 mm	50.00	61.13	3.06	79.16	nr	**140.29**
Sum of two sides 2000 mm	55.85	68.28	3.06	79.16	nr	**147.44**
Sum of two sides 2100 mm	61.69	75.42	4.05	104.77	nr	**180.19**
Sum of two sides 2200 mm	67.54	82.56	4.05	104.77	nr	**187.33**
Sum of two sides 2300 mm	73.48	89.83	4.39	113.56	nr	**203.39**
Sum of two sides 2400 mm	77.25	94.44	4.85	125.46	nr	**219.90**
Sum of two sides 2500 mm	83.08	101.57	4.85	125.46	nr	**227.03**
Sum of two sides 2600 mm	88.91	108.70	4.87	125.98	nr	**234.68**
Sum of two sides 2700 mm	94.75	115.83	4.87	125.98	nr	**241.81**
Sum of two sides 2800 mm	98.38	120.27	8.81	227.90	nr	**348.17**
Sum of two sides 2900 mm	107.56	131.49	8.81	227.90	nr	**359.39**
Sum of two sides 3000 mm	103.90	127.02	9.31	240.83	nr	**367.85**
Sum of two sides 3100 mm	108.37	132.48	9.31	240.83	nr	**373.31**
Sum of two sides 3200 mm	112.49	137.52	9.39	242.90	nr	**380.42**
90° mitre bend						
Sum of two sides 1700 mm	90.93	111.16	2.67	69.07	nr	**180.23**
Sum of two sides 1800 mm	97.15	118.77	2.80	72.43	nr	**191.20**
Sum of two sides 1900 mm	108.54	132.69	2.95	76.31	nr	**209.00**
Sum of two sides 2000 mm	119.98	146.67	2.95	76.31	nr	**222.98**
Sum of two sides 2100 mm	131.39	160.63	4.05	104.77	nr	**265.40**
Sum of two sides 2200 mm	142.82	174.60	4.05	104.77	nr	**279.37**
Sum of two sides 2300 mm	154.21	188.52	4.39	113.56	nr	**302.08**

U: VENTILATION/AIR CONDITIONING SYSTEMS

Item	Net Price £	Material £	Labour hours	Labour £	Unit	Total rate £
Sum of two sides 2400 mm	160.45	196.15	4.85	125.46	nr	**321.61**
Sum of two sides 2500 mm	171.92	210.17	4.85	125.46	nr	**335.63**
Sum of two sides 2600 mm	183.39	224.20	4.87	125.98	nr	**350.18**
Sum of two sides 2700 mm	194.87	238.23	4.87	125.98	nr	**364.21**
Sum of two sides 2800 mm	201.03	245.76	8.81	227.90	nr	**473.66**
Sum of two sides 2900 mm	201.96	246.90	14.81	383.11	nr	**630.01**
Sum of two sides 3000 mm	213.48	260.98	15.20	393.19	nr	**654.17**
Sum of two sides 3100 mm	222.92	272.52	15.60	403.55	nr	**676.07**
Sum of two sides 3200 mm	231.89	283.49	15.60	403.55	nr	**687.04**
Branch						
Sum of two sides 1700 mm	80.07	97.89	1.69	43.71	nr	**141.60**
Sum of two sides 1800 mm	88.26	107.90	1.69	43.71	nr	**151.61**
Sum of two sides 1900 mm	96.53	118.01	1.85	47.86	nr	**165.87**
Sum of two sides 2000 mm	104.72	128.02	1.85	47.86	nr	**175.88**
Sum of two sides 2100 mm	112.90	138.02	2.61	67.52	nr	**205.54**
Sum of two sides 2200 mm	121.09	148.03	2.61	67.52	nr	**215.55**
Sum of two sides 2300 mm	129.37	158.16	2.61	67.52	nr	**225.68**
Sum of two sides 2400 mm	138.15	168.89	2.88	74.50	nr	**243.39**
Sum of two sides 2500 mm	145.75	178.18	2.88	74.50	nr	**252.68**
Sum of two sides 2600 mm	153.94	188.19	2.88	74.50	nr	**262.69**
Sum of two sides 2700 mm	162.12	198.19	2.88	74.50	nr	**272.69**
Sum of two sides 2800 mm	170.31	208.20	3.94	101.92	nr	**310.12**
Sum of two sides 2900 mm	178.59	218.33	3.94	101.92	nr	**320.25**
Sum of two sides 3000 mm	186.78	228.34	4.83	124.94	nr	**353.28**
Sum of two sides 3100 mm	193.28	236.28	4.83	124.94	nr	**361.22**
Sum of two sides 3200 mm	199.21	243.53	4.83	124.94	nr	**368.47**
Grille neck						
Sum of two sides 1700 mm	85.61	104.66	1.86	48.12	nr	**152.78**
Sum of two sides 1800 mm	94.82	115.92	2.02	52.25	nr	**168.17**
Sum of two sides 1900 mm	104.03	127.17	2.02	52.25	nr	**179.42**
Sum of two sides 2000 mm	113.23	138.42	2.02	52.25	nr	**190.67**
Sum of two sides 2100 mm	122.44	149.68	2.61	67.52	nr	**217.20**
Sum of two sides 2200 mm	131.64	160.93	2.61	67.52	nr	**228.45**
Sum of two sides 2300 mm	140.85	172.19	2.61	67.52	nr	**239.71**
Sum of two sides 2400 mm	150.05	183.43	2.88	74.50	nr	**257.93**
Sum of two sides 2500 mm	159.26	194.70	2.88	74.50	nr	**269.20**
Sum of two sides 2600 mm	168.47	205.96	2.88	74.50	nr	**280.46**
Sum of two sides 2700 mm	177.67	217.21	2.88	74.50	nr	**291.71**
Sum of two sides 2800 mm	186.88	228.46	3.94	101.92	nr	**330.38**
Sum of two sides 2900 mm	196.08	239.71	4.12	106.57	nr	**346.28**
Sum of two sides 3000 mm	205.29	250.97	5.00	129.34	nr	**380.31**
Sum of two sides 3100 mm	212.61	259.92	5.00	129.34	nr	**389.26**
Sum of two sides 3200 mm	219.27	268.06	5.00	129.34	nr	**397.40**

U: VENTILATION/AIR CONDITIONING SYSTEMS

Item	Net Price £	Material £	Labour hours	Labour £	Unit	Total rate £
U10: DUCTWORK: RECTANGULAR – CLASS B – cont						
Ductwork 1601 to 2000 mm longest side						
Sum of two sides 2100 mm	82.80	101.23	2.17	56.14	m	**157.37**
Sum of two sides 2200 mm	85.29	104.27	2.17	56.14	m	**160.41**
Sum of two sides 2300 mm	87.78	107.32	2.19	56.65	m	**163.97**
Sum of two sides 2400 mm	90.23	110.30	2.38	61.56	m	**171.86**
Sum of two sides 2500 mm	92.74	113.37	2.38	61.56	m	**174.93**
Sum of two sides 2600 mm	95.10	116.26	2.64	68.29	m	**184.55**
Sum of two sides 2700 mm	97.45	119.13	2.66	68.81	m	**187.94**
Sum of two sides 2800 mm	99.81	122.01	2.95	76.31	m	**198.32**
Sum of two sides 2900 mm	102.32	125.08	2.96	76.57	m	**201.65**
Sum of two sides 3000 mm	104.67	127.96	2.96	76.57	m	**204.53**
Sum of two sides 3100 mm	113.56	138.82	2.96	76.57	m	**215.39**
Sum of two sides 3200 mm	115.93	141.73	3.15	81.48	m	**223.21**
Sum of two sides 3300 mm	127.14	155.43	3.15	81.48	m	**236.91**
Sum of two sides 3400 mm	129.49	158.30	3.15	81.48	m	**239.78**
Sum of two sides 3500 mm	131.84	161.17	3.15	81.48	m	**242.65**
Sum of two sides 3600 mm	134.20	164.06	3.18	82.27	m	**246.33**
Sum of two sides 3700 mm	136.71	167.13	3.18	82.27	m	**249.40**
Sum of two sides 3800 mm	139.06	170.00	3.18	82.27	m	**252.27**
Sum of two sides 3900 mm	141.42	172.89	3.18	82.27	m	**255.16**
Sum of two sides 4000 mm	143.77	175.76	3.18	82.27	m	**258.03**
Extra over fittings; ductwork 1601 to 2000 mm longest side						
End cap						
Sum of two sides 2100 mm	39.55	48.35	0.87	22.51	nr	**70.86**
Sum of two sides 2200 mm	42.56	52.03	0.87	22.51	nr	**74.54**
Sum of two sides 2300 mm	45.58	55.72	0.87	22.51	nr	**78.23**
Sum of two sides 2400 mm	48.58	59.39	0.87	22.51	nr	**81.90**
Sum of two sides 2500 mm	51.59	63.07	0.87	22.51	nr	**85.58**
Sum of two sides 2600 mm	54.60	66.74	0.87	22.51	nr	**89.25**
Sum of two sides 2700 mm	57.61	70.43	0.87	22.51	nr	**92.94**
Sum of two sides 2800 mm	60.61	74.10	1.16	30.01	nr	**104.11**
Sum of two sides 2900 mm	63.62	77.78	1.16	30.01	nr	**107.79**
Sum of two sides 3000 mm	66.63	81.45	1.73	44.75	nr	**126.20**
Sum of two sides 3100 mm	69.01	84.37	1.80	46.57	nr	**130.94**
Sum of two sides 3200 mm	71.17	87.00	1.80	46.57	nr	**133.57**
Sum of two sides 3300 mm	73.33	89.65	1.80	46.57	nr	**136.22**
Sum of two sides 3400 mm	75.48	92.28	1.80	46.57	nr	**138.85**
Sum of two sides 3500 mm	77.65	94.92	1.80	46.57	nr	**141.49**
Sum of two sides 3600 mm	79.80	97.55	1.80	46.57	nr	**144.12**
Sum of two sides 3700 mm	81.97	100.21	1.80	46.57	nr	**146.78**
Sum of two sides 3800 mm	84.12	102.84	1.80	46.57	nr	**149.41**
Sum of two sides 3900 mm	86.27	105.46	1.80	46.57	nr	**152.03**
Sum of two sides 4000 mm	88.44	108.12	1.80	46.57	nr	**154.69**

U: VENTILATION/AIR CONDITIONING SYSTEMS

Item	Net Price £	Material £	Labour hours	Labour £	Unit	**Total rate £**
Reducer						
Sum of two sides 2100 mm	54.73	66.91	2.61	67.52	nr	**134.43**
Sum of two sides 2200 mm	59.91	73.24	2.61	67.52	nr	**140.76**
Sum of two sides 2300 mm	65.16	79.66	2.61	67.52	nr	**147.18**
Sum of two sides 2400 mm	70.26	85.89	2.88	74.50	nr	**160.39**
Sum of two sides 2500 mm	75.53	92.34	3.12	80.71	nr	**173.05**
Sum of two sides 2600 mm	80.69	98.64	3.12	80.71	nr	**179.35**
Sum of two sides 2700 mm	85.87	104.98	3.12	80.71	nr	**185.69**
Sum of two sides 2800 mm	91.04	111.30	3.95	102.18	nr	**213.48**
Sum of two sides 2900 mm	96.29	117.72	3.97	102.70	nr	**220.42**
Sum of two sides 3000 mm	101.47	124.04	4.52	116.93	nr	**240.97**
Sum of two sides 3100 mm	101.77	124.41	4.52	116.93	nr	**241.34**
Sum of two sides 3200 mm	102.55	125.37	4.52	116.93	nr	**242.30**
Sum of two sides 3300 mm	104.17	127.35	4.52	116.93	nr	**244.28**
Sum of two sides 3400 mm	104.95	128.31	4.52	116.93	nr	**245.24**
Sum of two sides 3500 mm	108.42	132.54	4.52	116.93	nr	**249.47**
Sum of two sides 3600 mm	111.91	136.81	4.52	116.93	nr	**253.74**
Sum of two sides 3700 mm	120.93	147.84	4.52	116.93	nr	**264.77**
Sum of two sides 3800 mm	124.40	152.08	4.52	116.93	nr	**269.01**
Sum of two sides 3900 mm	127.87	156.32	4.52	116.93	nr	**273.25**
Sum of two sides 4000 mm	131.36	160.59	4.52	116.93	nr	**277.52**
Offset						
Sum of two sides 2100 mm	128.18	156.70	2.61	67.52	nr	**224.22**
Sum of two sides 2200 mm	137.88	168.56	2.61	67.52	nr	**236.08**
Sum of two sides 2300 mm	143.52	175.45	2.61	67.52	nr	**242.97**
Sum of two sides 2400 mm	157.29	192.29	2.88	74.50	nr	**266.79**
Sum of two sides 2500 mm	162.74	198.95	3.48	90.02	nr	**288.97**
Sum of two sides 2600 mm	167.92	205.28	3.49	90.28	nr	**295.56**
Sum of two sides 2700 mm	172.90	211.38	3.50	90.54	nr	**301.92**
Sum of two sides 2800 mm	177.70	217.24	4.34	112.26	nr	**329.50**
Sum of two sides 2900 mm	182.35	222.92	4.76	123.13	nr	**346.05**
Sum of two sides 3000 mm	186.76	228.32	5.32	137.62	nr	**365.94**
Sum of two sides 3100 mm	187.03	228.65	5.35	138.40	nr	**367.05**
Sum of two sides 3200 mm	190.91	233.39	5.35	138.40	nr	**371.79**
Sum of two sides 3300 mm	193.61	236.69	5.35	138.40	nr	**375.09**
Sum of two sides 3400 mm	197.50	241.44	5.35	138.40	nr	**379.84**
Sum of two sides 3500 mm	204.09	249.50	5.35	138.40	nr	**387.90**
Sum of two sides 3600 mm	210.68	257.56	5.35	138.40	nr	**395.96**
Sum of two sides 3700 mm	222.76	272.32	5.35	138.40	nr	**410.72**
Sum of two sides 3800 mm	229.35	280.39	5.35	138.40	nr	**418.79**
Sum of two sides 3900 mm	235.94	288.44	5.35	138.40	nr	**426.84**
Sum of two sides 4000 mm	242.52	296.49	5.35	138.40	nr	**434.89**

U: VENTILATION/AIR CONDITIONING SYSTEMS

Item	Net Price £	Material £	Labour hours	Labour £	Unit	Total rate £
U10: DUCTWORK: RECTANGULAR – CLASS B – cont						
Extra over fittings – cont						
Square to round						
Sum of two sides 2100 mm	82.44	100.79	2.61	67.52	nr	**168.31**
Sum of two sides 2200 mm	90.39	110.50	2.61	67.52	nr	**178.02**
Sum of two sides 2300 mm	98.32	120.20	2.61	67.52	nr	**187.72**
Sum of two sides 2400 mm	106.30	129.96	2.88	74.50	nr	**204.46**
Sum of two sides 2500 mm	114.23	139.65	4.70	121.58	nr	**261.23**
Sum of two sides 2600 mm	122.17	149.35	4.70	121.58	nr	**270.93**
Sum of two sides 2700 mm	130.11	159.06	4.71	121.84	nr	**280.90**
Sum of two sides 2800 mm	138.05	168.77	8.19	211.86	nr	**380.63**
Sum of two sides 2900 mm	145.97	178.45	8.19	211.86	nr	**390.31**
Sum of two sides 3000 mm	153.91	188.16	8.19	211.86	nr	**400.02**
Sum of two sides 3100 mm	155.29	189.84	8.19	211.86	nr	**401.70**
Sum of two sides 3200 mm	159.81	195.37	8.19	211.86	nr	**407.23**
Sum of two sides 3300 mm	160.68	196.43	8.19	211.86	nr	**408.29**
Sum of two sides 3400 mm	165.20	201.96	8.62	222.98	nr	**424.94**
Sum of two sides 3500 mm	170.59	208.54	8.62	222.98	nr	**431.52**
Sum of two sides 3600 mm	175.98	215.14	8.62	222.98	nr	**438.12**
Sum of two sides 3700 mm	184.10	225.07	8.62	222.98	nr	**448.05**
Sum of two sides 3800 mm	189.49	231.66	8.75	226.34	nr	**458.00**
Sum of two sides 3900 mm	194.88	238.24	8.75	226.34	nr	**464.58**
Sum of two sides 4000 mm	200.28	244.84	8.75	226.34	nr	**471.18**
90° radius bend						
Sum of two sides 2100 mm	115.92	141.72	2.61	67.52	nr	**209.24**
Sum of two sides 2200 mm	223.69	273.46	2.61	67.52	nr	**340.98**
Sum of two sides 2300 mm	242.00	295.85	2.61	67.52	nr	**363.37**
Sum of two sides 2400 mm	248.53	303.83	2.88	74.50	nr	**378.33**
Sum of two sides 2500 mm	266.77	326.13	3.90	100.88	nr	**427.01**
Sum of two sides 2600 mm	284.71	348.06	4.26	110.20	nr	**458.26**
Sum of two sides 2700 mm	302.64	369.98	4.55	117.70	nr	**487.68**
Sum of two sides 2800 mm	320.57	391.89	4.55	117.70	nr	**509.59**
Sum of two sides 2900 mm	338.83	414.22	6.87	177.72	nr	**591.94**
Sum of two sides 3000 mm	356.77	436.15	6.87	177.72	nr	**613.87**
Sum of two sides 3100 mm	360.21	440.36	6.87	177.72	nr	**618.08**
Sum of two sides 3200 mm	371.92	454.67	6.87	177.72	nr	**632.39**
Sum of two sides 3300 mm	373.05	456.05	6.87	177.72	nr	**633.77**
Sum of two sides 3400 mm	384.76	470.37	7.00	181.07	nr	**651.44**
Sum of two sides 3500 mm	397.60	486.07	7.00	181.07	nr	**667.14**
Sum of two sides 3600 mm	410.43	501.75	7.00	181.07	nr	**682.82**
Sum of two sides 3700 mm	440.00	537.91	7.00	181.07	nr	**718.98**
Sum of two sides 3800 mm	452.84	553.60	7.00	181.07	nr	**734.67**
Sum of two sides 3900 mm	465.67	569.28	7.00	181.07	nr	**750.35**
Sum of two sides 4000 mm	478.51	584.98	7.00	181.07	nr	**766.05**

U: VENTILATION/AIR CONDITIONING SYSTEMS

Item	Net Price £	Material £	Labour hours	Labour £	Unit	**Total rate £**
45° bend						
Sum of two sides 2100 mm	55.71	68.10	2.61	67.52	nr	**135.62**
Sum of two sides 2200 mm	160.13	195.76	2.61	67.52	nr	**263.28**
Sum of two sides 2300 mm	172.45	210.82	2.61	67.52	nr	**278.34**
Sum of two sides 2400 mm	178.54	218.27	2.88	74.50	nr	**292.77**
Sum of two sides 2500 mm	190.82	233.27	4.85	125.46	nr	**358.73**
Sum of two sides 2600 mm	202.86	247.99	4.87	125.98	nr	**373.97**
Sum of two sides 2700 mm	214.91	262.73	4.87	125.98	nr	**388.71**
Sum of two sides 2800 mm	226.95	277.45	8.81	227.90	nr	**505.35**
Sum of two sides 2900 mm	239.25	292.49	8.81	227.90	nr	**520.39**
Sum of two sides 3000 mm	251.28	307.19	9.31	240.83	nr	**548.02**
Sum of two sides 3100 mm	263.81	322.51	9.31	240.83	nr	**563.34**
Sum of two sides 3200 mm	265.17	324.17	9.31	240.83	nr	**565.00**
Sum of two sides 3300 mm	265.87	325.03	9.31	240.83	nr	**565.86**
Sum of two sides 3400 mm	273.80	334.72	9.31	240.83	nr	**575.55**
Sum of two sides 3500 mm	282.45	345.30	9.39	242.90	nr	**588.20**
Sum of two sides 3600 mm	291.10	355.88	9.39	242.90	nr	**598.78**
Sum of two sides 3700 mm	310.92	380.10	9.39	242.90	nr	**623.00**
Sum of two sides 3800 mm	319.57	390.68	9.39	242.90	nr	**633.58**
Sum of two sides 3900 mm	328.22	401.25	9.39	242.90	nr	**644.15**
Sum of two sides 4000 mm	336.86	411.81	9.39	242.90	nr	**654.71**
90° mitre bend						
Sum of two sides 2100 mm	262.35	320.72	2.61	67.52	nr	**388.24**
Sum of two sides 2200 mm	277.31	339.01	2.61	67.52	nr	**406.53**
Sum of two sides 2300 mm	299.31	365.91	2.61	67.52	nr	**433.43**
Sum of two sides 2400 mm	306.53	374.73	2.88	74.50	nr	**449.23**
Sum of two sides 2500 mm	328.58	401.69	4.85	125.46	nr	**527.15**
Sum of two sides 2600 mm	350.75	428.80	4.87	125.98	nr	**554.78**
Sum of two sides 2700 mm	372.91	455.89	4.87	125.98	nr	**581.87**
Sum of two sides 2800 mm	395.07	482.98	8.81	227.90	nr	**710.88**
Sum of two sides 2900 mm	417.12	509.93	14.81	383.11	nr	**893.04**
Sum of two sides 3000 mm	439.28	537.02	15.20	393.19	nr	**930.21**
Sum of two sides 3100 mm	444.77	543.74	15.20	393.19	nr	**936.93**
Sum of two sides 3200 mm	461.84	564.60	15.20	393.19	nr	**957.79**
Sum of two sides 3300 mm	465.69	569.31	15.20	393.19	nr	**962.50**
Sum of two sides 3400 mm	482.76	590.18	15.20	393.19	nr	**983.37**
Sum of two sides 3500 mm	499.82	611.04	15.60	403.55	nr	**1014.59**
Sum of two sides 3600 mm	516.89	631.91	15.60	403.55	nr	**1035.46**
Sum of two sides 3700 mm	539.30	659.30	15.60	403.55	nr	**1062.85**
Sum of two sides 3800 mm	556.37	680.17	15.60	403.55	nr	**1083.72**
Sum of two sides 3900 mm	573.43	701.02	15.60	403.55	nr	**1104.57**
Sum of two sides 4000 mm	590.51	721.91	15.60	403.55	nr	**1125.46**

U: VENTILATION/AIR CONDITIONING SYSTEMS

Item	Net Price £	Material £	Labour hours	Labour £	Unit	Total rate £
U10: DUCTWORK: RECTANGULAR – CLASS B – cont						
Extra over fittings – cont						
Branch						
Sum of two sides 2100 mm	116.02	141.83	2.61	67.52	nr	**209.35**
Sum of two sides 2200 mm	124.21	151.84	2.61	67.52	nr	**219.36**
Sum of two sides 2300 mm	132.50	161.98	2.61	67.52	nr	**229.50**
Sum of two sides 2400 mm	140.58	171.86	2.88	74.50	nr	**246.36**
Sum of two sides 2500 mm	148.88	182.01	2.88	74.50	nr	**256.51**
Sum of two sides 2600 mm	157.08	192.03	2.88	74.50	nr	**266.53**
Sum of two sides 2700 mm	165.27	202.04	2.88	74.50	nr	**276.54**
Sum of two sides 2800 mm	173.47	212.07	3.94	101.92	nr	**313.99**
Sum of two sides 2900 mm	181.76	222.20	3.94	101.92	nr	**324.12**
Sum of two sides 3000 mm	189.97	232.24	3.94	101.92	nr	**334.16**
Sum of two sides 3100 mm	196.49	240.21	3.94	101.92	nr	**342.13**
Sum of two sides 3200 mm	202.44	247.49	3.94	101.92	nr	**349.41**
Sum of two sides 3300 mm	208.46	254.84	3.94	101.92	nr	**356.76**
Sum of two sides 3400 mm	214.40	262.10	4.83	124.94	nr	**387.04**
Sum of two sides 3500 mm	220.33	269.36	4.83	124.94	nr	**394.30**
Sum of two sides 3600 mm	226.27	276.62	4.83	124.94	nr	**401.56**
Sum of two sides 3700 mm	235.03	287.33	4.83	124.94	nr	**412.27**
Sum of two sides 3800 mm	240.96	294.58	4.83	124.94	nr	**419.52**
Sum of two sides 3900 mm	246.90	301.84	4.83	124.94	nr	**426.78**
Sum of two sides 4000 mm	252.83	309.08	4.83	124.94	nr	**434.02**
Grille neck						
Sum of two sides 2100 mm	122.44	149.68	2.61	67.52	nr	**217.20**
Sum of two sides 2200 mm	131.64	160.93	2.61	67.52	nr	**228.45**
Sum of two sides 2300 mm	140.85	172.19	2.61	67.52	nr	**239.71**
Sum of two sides 2400 mm	150.05	183.43	2.88	74.50	nr	**257.93**
Sum of two sides 2500 mm	159.26	194.70	2.88	74.50	nr	**269.20**
Sum of two sides 2600 mm	168.47	205.96	2.88	74.50	nr	**280.46**
Sum of two sides 2700 mm	177.67	217.21	2.88	74.50	nr	**291.71**
Sum of two sides 2800 mm	186.88	228.46	3.94	101.92	nr	**330.38**
Sum of two sides 2900 mm	196.08	239.71	4.12	106.57	nr	**346.28**
Sum of two sides 3000 mm	205.29	250.97	4.12	106.57	nr	**357.54**
Sum of two sides 3100 mm	212.61	259.92	4.12	106.57	nr	**366.49**
Sum of two sides 3200 mm	219.27	268.06	4.12	106.57	nr	**374.63**
Sum of two sides 3300 mm	225.93	276.20	4.12	106.57	nr	**382.77**
Sum of two sides 3400 mm	232.59	284.34	5.00	129.34	nr	**413.68**
Sum of two sides 3500 mm	239.25	292.49	5.00	129.34	nr	**421.83**
Sum of two sides 3600 mm	245.90	300.62	5.00	129.34	nr	**429.96**
Sum of two sides 3700 mm	252.55	308.74	5.00	129.34	nr	**438.08**
Sum of two sides 3800 mm	259.22	316.90	5.00	129.34	nr	**446.24**
Sum of two sides 3900 mm	265.87	325.03	5.00	129.34	nr	**454.37**
Sum of two sides 4000 mm	272.54	333.18	5.00	129.34	nr	**462.52**

U: VENTILATION/AIR CONDITIONING SYSTEMS

Item	Net Price £	Material £	Labour hours	Labour £	Unit	Total rate £
Ductwork 2001 to 2500 mm longest side						
Sum of two sides 2500 mm	117.66	143.84	2.38	61.56	m	**205.40**
Sum of two sides 2600 mm	120.59	147.42	2.64	68.29	m	**215.71**
Sum of two sides 2700 mm	123.00	150.37	2.66	68.81	m	**219.18**
Sum of two sides 2800 mm	126.20	154.28	2.95	76.31	m	**230.59**
Sum of two sides 2900 mm	128.62	157.24	2.96	76.57	m	**233.81**
Sum of two sides 3000 mm	130.88	160.00	3.15	81.48	m	**241.48**
Sum of two sides 3100 mm	133.27	162.93	3.15	81.48	m	**244.41**
Sum of two sides 3200 mm	135.53	165.69	3.15	81.48	m	**247.17**
Sum of two sides 3300 mm	137.94	168.63	3.15	81.48	m	**250.11**
Sum of two sides 3400 mm	150.08	183.47	2.66	68.81	m	**252.28**
Sum of two sides 3500 mm	140.22	171.42	3.15	81.48	m	**252.90**
Sum of two sides 3600 mm	152.34	186.23	3.18	82.27	m	**268.50**
Sum of two sides 3700 mm	163.24	199.56	3.18	82.27	m	**281.83**
Sum of two sides 3800 mm	165.52	202.35	3.18	82.27	m	**284.62**
Sum of two sides 3900 mm	167.79	205.12	3.18	82.27	m	**287.39**
Sum of two sides 4000 mm	170.89	208.91	3.18	82.27	m	**291.18**
Extra over fittings; ductwork 2001 to 2500 mm longest side						
End cap						
Sum of two sides 2500 mm	52.68	64.41	0.87	22.51	nr	**86.92**
Sum of two sides 2600 mm	55.76	68.17	0.87	22.51	nr	**90.68**
Sum of two sides 2700 mm	58.83	71.92	0.87	22.51	nr	**94.43**
Sum of two sides 2800 mm	61.90	75.67	1.16	30.01	nr	**105.68**
Sum of two sides 2900 mm	64.97	79.42	1.16	30.01	nr	**109.43**
Sum of two sides 3000 mm	68.05	83.19	1.73	44.75	nr	**127.94**
Sum of two sides 3100 mm	70.48	86.16	1.73	44.75	nr	**130.91**
Sum of two sides 3200 mm	72.68	88.85	1.73	44.75	nr	**133.60**
Sum of two sides 3300 mm	74.89	91.56	1.73	44.75	nr	**136.31**
Sum of two sides 3400 mm	77.09	94.25	1.80	46.57	nr	**140.82**
Sum of two sides 3500 mm	79.30	96.94	1.80	46.57	nr	**143.51**
Sum of two sides 3600 mm	81.50	99.63	1.80	46.57	nr	**146.20**
Sum of two sides 3700 mm	83.71	102.34	1.80	46.57	nr	**148.91**
Sum of two sides 3800 mm	85.91	105.03	1.80	46.57	nr	**151.60**
Sum of two sides 3900 mm	88.11	107.72	1.80	46.57	nr	**154.29**
Sum of two sides 4000 mm	90.32	110.42	1.80	46.57	nr	**156.99**
Reducer						
Sum of two sides 2500 mm	62.84	76.82	3.12	80.71	nr	**157.53**
Sum of two sides 2600 mm	68.10	83.25	3.12	80.71	nr	**163.96**
Sum of two sides 2700 mm	73.47	89.82	3.12	80.71	nr	**170.53**
Sum of two sides 2800 mm	78.62	96.11	3.95	102.18	nr	**198.29**
Sum of two sides 2900 mm	83.98	102.67	3.97	102.70	nr	**205.37**
Sum of two sides 3000 mm	89.26	109.12	3.97	102.70	nr	**211.82**
Sum of two sides 3100 mm	93.16	113.89	3.97	102.70	nr	**216.59**
Sum of two sides 3200 mm	96.70	118.21	3.97	102.70	nr	**220.91**
Sum of two sides 3300 mm	100.32	122.64	3.97	102.70	nr	**225.34**
Sum of two sides 3400 mm	102.21	124.95	4.52	116.93	nr	**241.88**
Sum of two sides 3500 mm	102.87	125.76	4.52	116.93	nr	**242.69**

U: VENTILATION/AIR CONDITIONING SYSTEMS

Item	Net Price £	Material £	Labour hours	Labour £	Unit	Total rate £
U10: DUCTWORK: RECTANGULAR – CLASS B – cont						
Extra over fittings – cont						
Reducer – cont						
Sum of two sides 3600 mm	103.87	126.98	4.52	116.93	nr	**243.91**
Sum of two sides 3700 mm	105.74	129.26	4.52	116.93	nr	**246.19**
Sum of two sides 3800 mm	106.41	130.09	4.52	116.93	nr	**247.02**
Sum of two sides 3900 mm	109.95	134.42	4.52	116.93	nr	**251.35**
Sum of two sides 4000 mm	113.53	138.79	4.52	116.93	nr	**255.72**
Offset						
Sum of two sides 2500 mm	157.81	192.92	3.48	90.02	nr	**282.94**
Sum of two sides 2600 mm	167.71	205.02	3.49	90.28	nr	**295.30**
Sum of two sides 2700 mm	171.60	209.78	3.50	90.54	nr	**300.32**
Sum of two sides 2800 mm	187.47	229.19	4.34	112.26	nr	**341.45**
Sum of two sides 2900 mm	191.18	233.72	4.76	123.13	nr	**356.85**
Sum of two sides 3000 mm	194.61	237.91	5.32	137.62	nr	**375.53**
Sum of two sides 3100 mm	195.80	239.37	5.35	138.40	nr	**377.77**
Sum of two sides 3200 mm	196.49	240.21	5.32	137.62	nr	**377.83**
Sum of two sides 3300 mm	196.51	240.24	5.32	137.62	nr	**377.86**
Sum of two sides 3400 mm	196.22	239.88	5.35	138.40	nr	**378.28**
Sum of two sides 3500 mm	195.18	238.61	5.32	137.62	nr	**376.23**
Sum of two sides 3600 mm	201.88	246.80	5.35	138.40	nr	**385.20**
Sum of two sides 3700 mm	204.22	249.66	5.35	138.40	nr	**388.06**
Sum of two sides 3800 mm	205.67	251.43	5.35	138.40	nr	**389.83**
Sum of two sides 3900 mm	212.39	259.65	5.35	138.40	nr	**398.05**
Sum of two sides 4000 mm	219.09	267.84	5.35	138.40	nr	**406.24**
Square to round						
Sum of two sides 2500 mm	102.09	124.81	4.70	121.58	nr	**246.39**
Sum of two sides 2600 mm	110.20	134.72	4.70	121.58	nr	**256.30**
Sum of two sides 2700 mm	118.28	144.60	4.71	121.84	nr	**266.44**
Sum of two sides 2800 mm	126.39	154.51	8.19	211.86	nr	**366.37**
Sum of two sides 2900 mm	134.48	164.40	8.19	211.86	nr	**376.26**
Sum of two sides 3000 mm	142.58	174.31	8.19	211.86	nr	**386.17**
Sum of two sides 3100 mm	148.66	181.73	8.19	211.86	nr	**393.59**
Sum of two sides 3200 mm	154.17	188.47	8.62	222.98	nr	**411.45**
Sum of two sides 3300 mm	159.65	195.17	8.62	222.98	nr	**418.15**
Sum of two sides 3400 mm	165.15	201.90	8.62	222.98	nr	**424.88**
Sum of two sides 3500 mm	165.16	201.91	8.62	222.98	nr	**424.89**
Sum of two sides 3600 mm	170.64	208.61	8.75	226.34	nr	**434.95**
Sum of two sides 3700 mm	172.92	211.40	8.75	226.34	nr	**437.74**
Sum of two sides 3800 mm	175.13	214.10	8.75	226.34	nr	**440.44**
Sum of two sides 3900 mm	180.63	220.82	8.75	226.34	nr	**447.16**
Sum of two sides 4000 mm	186.09	227.50	8.75	226.34	nr	**453.84**

U: VENTILATION/AIR CONDITIONING SYSTEMS

Item	Net Price £	Material £	Labour hours	Labour £	Unit	**Total rate £**
90° radius bend						
Sum of two sides 2500 mm	232.83	284.64	3.90	100.88	nr	**385.52**
Sum of two sides 2600 mm	243.51	297.69	4.26	110.20	nr	**407.89**
Sum of two sides 2700 mm	262.08	320.39	4.55	117.70	nr	**438.09**
Sum of two sides 2800 mm	264.31	323.12	4.55	117.70	nr	**440.82**
Sum of two sides 2900 mm	282.79	345.71	6.87	177.72	nr	**523.43**
Sum of two sides 3000 mm	300.98	367.95	6.87	177.72	nr	**545.67**
Sum of two sides 3100 mm	315.07	385.18	6.87	177.72	nr	**562.90**
Sum of two sides 3200 mm	328.05	401.04	6.87	177.72	nr	**578.76**
Sum of two sides 3300 mm	341.33	417.27	6.87	177.72	nr	**594.99**
Sum of two sides 3400 mm	354.31	433.15	6.87	177.72	nr	**610.87**
Sum of two sides 3500 mm	353.55	432.22	7.00	181.07	nr	**613.29**
Sum of two sides 3600 mm	362.40	443.03	7.00	181.07	nr	**624.10**
Sum of two sides 3700 mm	366.53	448.08	7.00	181.07	nr	**629.15**
Sum of two sides 3800 mm	375.38	458.91	7.00	181.07	nr	**639.98**
Sum of two sides 3900 mm	388.36	474.77	7.00	181.07	nr	**655.84**
Sum of two sides 4000 mm	391.93	479.14	7.00	181.07	nr	**660.21**
45° bend						
Sum of two sides 2500 mm	174.37	213.17	4.85	125.46	nr	**338.63**
Sum of two sides 2600 mm	182.76	223.43	4.87	125.98	nr	**349.41**
Sum of two sides 2700 mm	195.27	238.72	4.87	125.98	nr	**364.70**
Sum of two sides 2800 mm	199.28	243.63	8.81	227.90	nr	**471.53**
Sum of two sides 2900 mm	211.74	258.85	8.81	227.90	nr	**486.75**
Sum of two sides 3000 mm	223.98	273.82	9.31	240.83	nr	**514.65**
Sum of two sides 3100 mm	233.51	285.46	9.31	240.83	nr	**526.29**
Sum of two sides 3200 mm	242.28	296.19	9.31	240.83	nr	**537.02**
Sum of two sides 3300 mm	251.29	307.21	9.31	240.83	nr	**548.04**
Sum of two sides 3400 mm	260.05	317.91	9.31	240.83	nr	**558.74**
Sum of two sides 3500 mm	261.87	320.13	9.31	240.83	nr	**560.96**
Sum of two sides 3600 mm	270.52	330.71	9.39	242.90	nr	**573.61**
Sum of two sides 3700 mm	270.63	330.85	9.39	242.90	nr	**573.75**
Sum of two sides 3800 mm	279.29	341.43	9.39	242.90	nr	**584.33**
Sum of two sides 3900 mm	288.06	352.16	9.39	242.90	nr	**595.06**
Sum of two sides 4000 mm	292.12	357.12	9.39	242.90	nr	**600.02**
90° mitre bend						
Sum of two sides 2500 mm	285.30	348.78	4.85	125.46	nr	**474.24**
Sum of two sides 2600 mm	302.84	370.22	4.87	125.98	nr	**496.20**
Sum of two sides 2700 mm	325.44	397.85	4.87	125.98	nr	**523.83**
Sum of two sides 2800 mm	326.88	399.61	8.81	227.90	nr	**627.51**
Sum of two sides 2900 mm	349.52	427.30	14.81	383.11	nr	**810.41**
Sum of two sides 3000 mm	372.29	455.13	14.81	383.11	nr	**838.24**
Sum of two sides 3100 mm	390.93	477.91	15.20	393.19	nr	**871.10**
Sum of two sides 3200 mm	408.52	499.42	15.20	393.19	nr	**892.61**
Sum of two sides 3300 mm	425.96	520.74	15.20	393.19	nr	**913.93**
Sum of two sides 3400 mm	443.54	542.23	15.20	393.19	nr	**935.42**
Sum of two sides 3500 mm	442.86	541.40	15.20	393.19	nr	**934.59**
Sum of two sides 3600 mm	460.44	562.89	15.60	403.55	nr	**966.44**
Sum of two sides 3700 mm	461.76	564.50	15.60	403.55	nr	**968.05**
Sum of two sides 3800 mm	479.34	586.00	15.60	403.55	nr	**989.55**
Sum of two sides 3900 mm	496.92	607.48	15.60	403.55	nr	**1011.03**
Sum of two sides 4000 mm	502.15	613.88	15.60	403.55	nr	**1017.43**

U: VENTILATION/AIR CONDITIONING SYSTEMS

Item	Net Price £	Material £	Labour hours	Labour £	Unit	Total rate £
U10: DUCTWORK: RECTANGULAR – CLASS B – cont						
Extra over fittings – cont						
Branch						
Sum of two sides 2500 mm	152.33	186.22	2.88	74.50	nr	**260.72**
Sum of two sides 2600 mm	160.69	196.44	2.88	74.50	nr	**270.94**
Sum of two sides 2700 mm	169.16	206.80	2.88	74.50	nr	**281.30**
Sum of two sides 2800 mm	177.41	216.89	3.94	101.92	nr	**318.81**
Sum of two sides 2900 mm	185.88	227.24	3.94	101.92	nr	**329.16**
Sum of two sides 3000 mm	194.26	237.49	3.94	101.92	nr	**339.41**
Sum of two sides 3100 mm	200.92	245.62	3.94	101.92	nr	**347.54**
Sum of two sides 3200 mm	206.98	253.03	3.94	101.92	nr	**354.95**
Sum of two sides 3300 mm	213.13	260.55	3.94	101.92	nr	**362.47**
Sum of two sides 3400 mm	219.20	267.98	3.94	101.92	nr	**369.90**
Sum of two sides 3500 mm	225.57	275.76	4.83	124.94	nr	**400.70**
Sum of two sides 3600 mm	231.62	283.16	4.83	124.94	nr	**408.10**
Sum of two sides 3700 mm	237.78	290.69	4.83	124.94	nr	**415.63**
Sum of two sides 3800 mm	243.84	298.09	4.83	124.94	nr	**423.03**
Sum of two sides 3900 mm	249.91	305.52	4.83	124.94	nr	**430.46**
Sum of two sides 4000 mm	256.05	313.03	4.83	124.94	nr	**437.97**
Grille neck						
Sum of two sides 2500 mm	162.65	198.84	2.88	74.50	nr	**273.34**
Sum of two sides 2600 mm	172.05	210.34	2.88	74.50	nr	**284.84**
Sum of two sides 2700 mm	181.45	221.82	2.88	74.50	nr	**296.32**
Sum of two sides 2800 mm	190.85	233.32	3.94	101.92	nr	**335.24**
Sum of two sides 2900 mm	200.25	244.81	3.94	101.92	nr	**346.73**
Sum of two sides 3000 mm	209.65	256.29	3.94	101.92	nr	**358.21**
Sum of two sides 3100 mm	217.13	265.44	4.12	106.57	nr	**372.01**
Sum of two sides 3200 mm	223.94	273.76	4.12	106.57	nr	**380.33**
Sum of two sides 3300 mm	230.73	282.07	4.12	106.57	nr	**388.64**
Sum of two sides 3400 mm	237.54	290.40	4.12	106.57	nr	**396.97**
Sum of two sides 3500 mm	244.34	298.71	5.00	129.34	nr	**428.05**
Sum of two sides 3600 mm	251.13	307.01	5.00	129.34	nr	**436.35**
Sum of two sides 3700 mm	257.93	315.32	5.00	129.34	nr	**444.66**
Sum of two sides 3800 mm	264.74	323.65	5.00	129.34	nr	**452.99**
Sum of two sides 3900 mm	271.53	331.95	5.00	129.34	nr	**461.29**
Sum of two sides 4000 mm	278.34	340.27	5.00	129.34	nr	**469.61**
Ductwork 2501 to 4000 mm longest side						
Sum of two sides 3000 mm	193.25	236.25	2.38	61.56	m	**297.81**
Sum of two sides 3100 mm	197.77	241.77	2.38	61.56	m	**303.33**
Sum of two sides 3200 mm	202.02	246.97	2.38	61.56	m	**308.53**
Sum of two sides 3300 mm	206.28	252.17	2.38	61.56	m	**313.73**
Sum of two sides 3400 mm	211.75	258.87	2.64	68.29	m	**327.16**
Sum of two sides 3500 mm	216.00	264.06	2.66	68.81	m	**332.87**
Sum of two sides 3600 mm	220.26	269.27	2.95	76.31	m	**345.58**

U: VENTILATION/AIR CONDITIONING SYSTEMS

Item	Net Price £	Material £	Labour hours	Labour £	Unit	Total rate £
Sum of two sides 3700 mm	225.86	276.11	2.96	76.57	m	**352.68**
Sum of two sides 3800 mm	230.12	281.32	3.15	81.48	m	**362.80**
Sum of two sides 3900 mm	242.78	296.80	3.15	81.48	m	**378.28**
Sum of two sides 4000 mm	247.03	302.00	3.15	81.48	m	**383.48**
Sum of two sides 4100 mm	251.42	307.36	3.35	86.66	m	**394.02**
Sum of two sides 4200 mm	255.69	312.58	3.35	86.66	m	**399.24**
Sum of two sides 4300 mm	259.93	317.77	3.60	93.13	m	**410.90**
Sum of two sides 4400 mm	266.14	325.36	3.60	93.13	m	**418.49**
Sum of two sides 4500 mm	270.41	330.58	3.60	93.13	m	**423.71**
Extra over fittings; ductwork 2501 to 4000 mm longest side						
End cap						
Sum of two sides 3000 mm	68.33	83.53	1.73	44.75	nr	**128.28**
Sum of two sides 3100 mm	70.77	86.52	1.73	44.75	nr	**131.27**
Sum of two sides 3200 mm	72.98	89.22	1.73	44.75	nr	**133.97**
Sum of two sides 3300 mm	75.19	91.92	1.73	44.75	nr	**136.67**
Sum of two sides 3400 mm	77.40	94.63	1.73	44.75	nr	**139.38**
Sum of two sides 3500 mm	79.62	97.33	1.73	44.75	nr	**142.08**
Sum of two sides 3600 mm	81.83	100.03	1.73	44.75	nr	**144.78**
Sum of two sides 3700 mm	84.05	102.75	1.73	44.75	nr	**147.50**
Sum of two sides 3800 mm	86.26	105.45	1.73	44.75	nr	**150.20**
Sum of two sides 3900 mm	88.48	108.17	1.80	46.57	nr	**154.74**
Sum of two sides 4000 mm	90.69	110.87	1.80	46.57	nr	**157.44**
Sum of two sides 4100 mm	92.90	113.57	1.80	46.57	nr	**160.14**
Sum of two sides 4200 mm	95.12	116.29	1.80	46.57	nr	**162.86**
Sum of two sides 4300 mm	97.33	118.99	1.88	48.63	nr	**167.62**
Sum of two sides 4400 mm	99.54	121.68	1.88	48.63	nr	**170.31**
Sum of two sides 4500 mm	101.76	124.40	1.88	48.63	nr	**173.03**
Reducer						
Sum of two sides 3000 mm	66.50	81.30	3.12	80.71	nr	**162.01**
Sum of two sides 3100 mm	69.70	85.21	3.12	80.71	nr	**165.92**
Sum of two sides 3200 mm	72.45	88.57	3.12	80.71	nr	**169.28**
Sum of two sides 3300 mm	75.21	91.95	3.12	80.71	nr	**172.66**
Sum of two sides 3400 mm	77.94	95.29	3.12	80.71	nr	**176.00**
Sum of two sides 3500 mm	80.68	98.63	3.12	80.71	nr	**179.34**
Sum of two sides 3600 mm	83.44	102.00	3.95	102.18	nr	**204.18**
Sum of two sides 3700 mm	86.26	105.45	3.97	102.70	nr	**208.15**
Sum of two sides 3800 mm	86.55	105.81	4.52	116.93	nr	**222.74**
Sum of two sides 3900 mm	89.01	108.82	4.52	116.93	nr	**225.75**
Sum of two sides 4000 mm	89.31	109.18	4.52	116.93	nr	**226.11**
Sum of two sides 4100 mm	92.15	112.65	4.52	116.93	nr	**229.58**
Sum of two sides 4200 mm	94.90	116.02	4.52	116.93	nr	**232.95**
Sum of two sides 4300 mm	97.66	119.39	4.92	127.28	nr	**246.67**
Sum of two sides 4400 mm	100.47	122.83	4.92	127.28	nr	**250.11**
Sum of two sides 4500 mm	103.21	126.17	5.12	132.45	nr	**258.62**

U: VENTILATION/AIR CONDITIONING SYSTEMS

Item	Net Price £	Material £	Labour hours	Labour £	Unit	Total rate £
U10: DUCTWORK: RECTANGULAR – CLASS B – cont						
Extra over fittings – cont						
Offset						
Sum of two sides 3000 mm	181.74	222.18	3.48	90.02	nr	**312.20**
Sum of two sides 3100 mm	186.43	227.91	3.48	90.02	nr	**317.93**
Sum of two sides 3200 mm	190.69	233.12	3.48	90.02	nr	**323.14**
Sum of two sides 3300 mm	193.78	236.90	3.48	90.02	nr	**326.92**
Sum of two sides 3400 mm	193.98	237.14	3.50	90.54	nr	**327.68**
Sum of two sides 3500 mm	199.13	243.44	3.49	90.28	nr	**333.72**
Sum of two sides 3600 mm	203.66	248.97	4.25	109.94	nr	**358.91**
Sum of two sides 3700 mm	206.30	252.21	4.76	123.13	nr	**375.34**
Sum of two sides 3800 mm	200.23	244.78	5.32	137.62	nr	**382.40**
Sum of two sides 3900 mm	216.67	264.89	5.35	138.40	nr	**403.29**
Sum of two sides 4000 mm	205.14	250.78	5.35	138.40	nr	**389.18**
Sum of two sides 4100 mm	227.62	278.26	5.85	151.34	nr	**429.60**
Sum of two sides 4200 mm	229.43	280.48	5.85	151.34	nr	**431.82**
Sum of two sides 4300 mm	233.51	285.46	6.15	159.08	nr	**444.54**
Sum of two sides 4400 mm	233.89	285.93	6.30	162.98	nr	**448.91**
Sum of two sides 4500 mm	237.59	290.46	6.15	159.08	nr	**449.54**
Square to round						
Sum of two sides 3000 mm	142.73	174.49	4.70	121.58	nr	**296.07**
Sum of two sides 3100 mm	148.89	182.02	4.70	121.58	nr	**303.60**
Sum of two sides 3200 mm	154.36	188.71	4.70	121.58	nr	**310.29**
Sum of two sides 3300 mm	159.82	195.38	4.70	121.58	nr	**316.96**
Sum of two sides 3400 mm	165.30	202.08	4.71	121.84	nr	**323.92**
Sum of two sides 3500 mm	170.76	208.76	4.71	121.84	nr	**330.60**
Sum of two sides 3600 mm	176.24	215.46	8.19	211.86	nr	**427.32**
Sum of two sides 3700 mm	181.70	222.13	8.62	222.98	nr	**445.11**
Sum of two sides 3800 mm	187.17	228.81	8.62	222.98	nr	**451.79**
Sum of two sides 3900 mm	229.18	280.17	8.62	222.98	nr	**503.15**
Sum of two sides 4000 mm	236.02	288.53	8.75	226.34	nr	**514.87**
Sum of two sides 4100 mm	242.20	296.09	11.23	290.50	nr	**586.59**
Sum of two sides 4200 mm	248.72	304.07	11.23	290.50	nr	**594.57**
Sum of two sides 4300 mm	255.25	312.05	11.25	291.02	nr	**603.07**
Sum of two sides 4400 mm	261.77	320.01	11.25	291.02	nr	**611.03**
Sum of two sides 4500 mm	268.28	327.97	11.26	291.27	nr	**619.24**
90° radius bend						
Sum of two sides 3000 mm	395.42	483.40	3.90	100.88	nr	**584.28**
Sum of two sides 3100 mm	413.70	505.75	3.90	100.88	nr	**606.63**
Sum of two sides 3200 mm	429.98	525.65	4.26	110.20	nr	**635.85**
Sum of two sides 3300 mm	435.55	532.46	4.26	110.20	nr	**642.66**
Sum of two sides 3400 mm	446.26	545.56	4.26	110.20	nr	**655.76**
Sum of two sides 3500 mm	451.55	552.02	4.55	117.70	nr	**669.72**
Sum of two sides 3600 mm	467.54	571.57	4.55	117.70	nr	**689.27**

U: VENTILATION/AIR CONDITIONING SYSTEMS

Item	Net Price £	Material £	Labour hours	Labour £	Unit	**Total rate £**
Sum of two sides 3700 mm	455.45	556.79	6.87	177.72	nr	**734.51**
Sum of two sides 3800 mm	471.16	576.00	6.87	177.72	nr	**753.72**
Sum of two sides 3900 mm	471.62	576.56	6.87	177.72	nr	**754.28**
Sum of two sides 4000 mm	472.29	577.38	7.00	181.07	nr	**758.45**
Sum of two sides 4100 mm	470.74	575.48	7.20	186.24	nr	**761.72**
Sum of two sides 4200 mm	486.17	594.35	7.20	186.24	nr	**780.59**
Sum of two sides 4300 mm	501.61	613.22	7.41	191.69	nr	**804.91**
Sum of two sides 4400 mm	503.06	614.99	7.41	191.69	nr	**806.68**
Sum of two sides 4500 mm	502.40	614.19	7.55	195.31	nr	**809.50**
45° bend						
Sum of two sides 3000 mm	191.46	234.06	4.85	125.46	nr	**359.52**
Sum of two sides 3100 mm	200.53	245.15	4.85	125.46	nr	**370.61**
Sum of two sides 3200 mm	208.61	255.03	4.85	125.46	nr	**380.49**
Sum of two sides 3300 mm	216.68	264.90	4.87	125.98	nr	**390.88**
Sum of two sides 3400 mm	234.84	287.09	4.87	125.98	nr	**413.07**
Sum of two sides 3500 mm	219.05	267.79	4.87	125.98	nr	**393.77**
Sum of two sides 3600 mm	237.17	289.95	8.81	227.90	nr	**517.85**
Sum of two sides 3700 mm	220.70	269.81	8.81	227.90	nr	**497.71**
Sum of two sides 3800 mm	228.50	279.35	9.31	240.83	nr	**520.18**
Sum of two sides 3900 mm	249.56	305.09	9.31	240.83	nr	**545.92**
Sum of two sides 4000 mm	257.17	314.40	9.31	240.83	nr	**555.23**
Sum of two sides 4100 mm	227.57	278.20	10.01	258.94	nr	**537.14**
Sum of two sides 4200 mm	242.88	296.92	9.31	240.83	nr	**537.75**
Sum of two sides 4300 mm	235.22	287.56	10.01	258.94	nr	**546.50**
Sum of two sides 4400 mm	225.58	275.77	10.52	272.13	nr	**547.90**
Sum of two sides 4500 mm	233.01	284.86	10.52	272.13	nr	**556.99**
90° mitre bend						
Sum of two sides 3000 mm	492.10	601.59	4.85	125.46	nr	**727.05**
Sum of two sides 3100 mm	515.74	630.49	4.85	125.46	nr	**755.95**
Sum of two sides 3200 mm	537.83	657.50	4.87	125.98	nr	**783.48**
Sum of two sides 3300 mm	546.96	668.67	4.87	125.98	nr	**794.65**
Sum of two sides 3400 mm	559.92	684.51	4.87	125.98	nr	**810.49**
Sum of two sides 3500 mm	568.80	695.36	8.81	227.90	nr	**923.26**
Sum of two sides 3600 mm	590.63	722.05	8.81	227.90	nr	**949.95**
Sum of two sides 3700 mm	580.39	709.53	14.81	383.11	nr	**1092.64**
Sum of two sides 3800 mm	627.34	766.93	14.81	383.11	nr	**1150.04**
Sum of two sides 3900 mm	628.36	768.17	14.81	383.11	nr	**1151.28**
Sum of two sides 4000 mm	640.02	782.43	15.20	393.19	nr	**1175.62**
Sum of two sides 4100 mm	603.15	737.36	16.30	421.65	nr	**1159.01**
Sum of two sides 4200 mm	624.50	763.46	16.50	426.82	nr	**1190.28**
Sum of two sides 4300 mm	625.39	764.55	17.01	440.02	nr	**1204.57**
Sum of two sides 4400 mm	628.22	768.00	17.01	440.02	nr	**1208.02**
Sum of two sides 4500 mm	645.85	789.56	17.01	440.02	nr	**1229.58**

U: VENTILATION/AIR CONDITIONING SYSTEMS

Item	Net Price £	Material £	Labour hours	Labour £	Unit	**Total rate £**
U10: DUCTWORK: RECTANGULAR – CLASS B – cont						
Extra over fittings – cont						
Branch						
Sum of two sides 3000 mm	204.39	249.87	2.88	74.50	nr	**324.37**
Sum of two sides 3100 mm	211.47	258.52	2.88	74.50	nr	**333.02**
Sum of two sides 3200 mm	217.84	266.31	2.88	74.50	nr	**340.81**
Sum of two sides 3300 mm	224.22	274.11	2.88	74.50	nr	**348.61**
Sum of two sides 3400 mm	230.56	281.86	2.88	74.50	nr	**356.36**
Sum of two sides 3500 mm	236.94	289.66	3.94	101.92	nr	**391.58**
Sum of two sides 3600 mm	243.31	297.44	3.94	101.92	nr	**399.36**
Sum of two sides 3700 mm	249.75	305.32	3.94	101.92	nr	**407.24**
Sum of two sides 3800 mm	256.13	313.12	4.83	124.94	nr	**438.06**
Sum of two sides 3900 mm	262.78	321.25	4.83	124.94	nr	**446.19**
Sum of two sides 4000 mm	269.15	329.03	4.83	124.94	nr	**453.97**
Sum of two sides 4100 mm	275.62	336.94	5.44	140.72	nr	**477.66**
Sum of two sides 4200 mm	282.00	344.75	5.44	140.72	nr	**485.47**
Sum of two sides 4300 mm	288.36	352.52	5.85	151.34	nr	**503.86**
Sum of two sides 4400 mm	294.80	360.40	5.85	151.34	nr	**511.74**
Sum of two sides 4500 mm	301.17	368.18	5.85	151.34	nr	**519.52**
Grille neck						
Sum of two sides 3000 mm	211.72	258.83	2.88	74.50	nr	**333.33**
Sum of two sides 3100 mm	219.26	268.05	2.88	74.50	nr	**342.55**
Sum of two sides 3200 mm	226.14	276.46	2.88	74.50	nr	**350.96**
Sum of two sides 3300 mm	233.00	284.85	2.88	74.50	nr	**359.35**
Sum of two sides 3400 mm	239.87	293.24	3.94	101.92	nr	**395.16**
Sum of two sides 3500 mm	246.74	301.65	3.94	101.92	nr	**403.57**
Sum of two sides 3600 mm	253.61	310.04	3.94	101.92	nr	**411.96**
Sum of two sides 3700 mm	260.48	318.43	4.12	106.57	nr	**425.00**
Sum of two sides 3800 mm	267.36	326.85	4.12	106.57	nr	**433.42**
Sum of two sides 3900 mm	274.23	335.24	4.12	106.57	nr	**441.81**
Sum of two sides 4000 mm	281.09	343.64	5.00	129.34	nr	**472.98**
Sum of two sides 4100 mm	287.96	352.03	5.00	129.34	nr	**481.37**
Sum of two sides 4200 mm	294.83	360.43	5.00	129.34	nr	**489.77**
Sum of two sides 4300 mm	301.71	368.84	5.23	135.29	nr	**504.13**
Sum of two sides 4400 mm	308.58	377.24	5.23	135.29	nr	**512.53**
Sum of two sides 4500 mm	315.45	385.64	5.39	139.43	nr	**525.07**

U: VENTILATION/AIR CONDITIONING SYSTEMS

Item	Net Price £	Material £	Labour hours	Labour £	Unit	**Total rate £**
Y30 – ANCILLARIES						
Access doors, hollow steel construction; 25 mm mineral wool insulation; removeable or hinged; fixed with cams; including sub-frame and integral sealing gaskets						
Rectangular duct						
150 × 150 mm	15.59	19.06	1.25	32.33	nr	**51.39**
200 × 200 mm	17.16	20.98	1.25	32.33	nr	**53.31**
300 × 150 mm	17.63	21.56	1.25	32.33	nr	**53.89**
300 × 300 mm	19.84	24.26	1.25	32.33	nr	**56.59**
400 × 400 mm	22.62	27.66	1.35	34.97	nr	**62.63**
450 × 300 mm	22.62	27.66	1.50	38.84	nr	**66.50**
450 × 450 mm	24.94	30.49	1.50	38.84	nr	**69.33**
Access doors, hollow steel construction; 25 mm mineral wool insulation; removeable or hinged; fixed with cams; including sub-frame and integral sealing gaskets						
Flat oval duct						
235 × 90 mm	31.67	38.72	1.25	32.33	nr	**71.05**
235 × 140 mm	33.76	41.27	1.35	34.97	nr	**76.24**
335 × 235 mm	38.57	47.15	1.50	38.84	nr	**85.99**
535 × 235 mm	43.38	53.03	1.50	38.84	nr	**91.87**

U: VENTILATION/AIR CONDITIONING SYSTEMS

Item	Net Price £	Material £	Labour hours	Labour £	Unit	Total rate £
U10: DUCTWORK: RECTANGULAR – CLASS C						
Y30 – AIR DUCTLINES						
Galvanized sheet metal DW144 class C rectangular section ductwork; including all necessary stiffeners, joints, couplers in the running length and duct supports						
Ductwork up to 400 mm longest side						
Sum of two sides 200 mm	12.18	14.89	1.17	30.27	m	**45.16**
Sum of two sides 300 mm	13.17	16.10	1.17	30.27	m	**46.37**
Sum of two sides 400 mm	13.07	15.98	1.19	30.79	m	**46.77**
Sum of two sides 500 mm	14.01	17.13	1.16	30.01	m	**47.14**
Sum of two sides 600 mm	14.04	17.16	1.19	30.79	m	**47.95**
Sum of two sides 700 mm	14.91	18.23	1.19	30.79	m	**49.02**
Sum of two sides 800 mm	15.83	19.35	1.19	30.79	m	**50.14**
Extra over fittings; ductwork up to 400 mm longest side						
End cap						
Sum of two sides 200 mm	7.95	9.72	0.38	9.83	nr	**19.55**
Sum of two sides 300 mm	8.87	10.85	0.38	9.83	nr	**20.68**
Sum of two sides 400 mm	9.77	11.95	0.38	9.83	nr	**21.78**
Sum of two sides 500 mm	10.68	13.06	0.38	9.83	nr	**22.89**
Sum of two sides 600 mm	11.58	14.15	0.38	9.83	nr	**23.98**
Sum of two sides 700 mm	12.50	15.29	0.38	9.83	nr	**25.12**
Sum of two sides 800 mm	13.40	16.38	0.38	9.83	nr	**26.21**
Reducer						
Sum of two sides 200 mm	13.00	15.89	1.40	36.21	nr	**52.10**
Sum of two sides 300 mm	14.69	17.96	1.40	36.21	nr	**54.17**
Sum of two sides 400 mm	23.66	28.92	1.42	36.74	nr	**65.66**
Sum of two sides 500 mm	25.59	31.28	1.42	36.74	nr	**68.02**
Sum of two sides 600 mm	27.52	33.64	1.69	43.71	nr	**77.35**
Sum of two sides 700 mm	29.45	36.00	1.69	43.71	nr	**79.71**
Sum of two sides 800 mm	31.36	38.34	1.92	49.67	nr	**88.01**
Offset						
Sum of two sides 200 mm	19.20	23.47	1.63	42.17	nr	**65.64**
Sum of two sides 300 mm	21.70	26.53	1.63	42.17	nr	**68.70**
Sum of two sides 400 mm	31.74	38.80	1.65	42.67	nr	**81.47**
Sum of two sides 500 mm	34.49	42.17	1.65	42.67	nr	**84.84**
Sum of two sides 600 mm	36.75	44.93	1.92	49.67	nr	**94.60**
Sum of two sides 700 mm	39.24	47.97	1.92	49.67	nr	**97.64**
Sum of two sides 800 mm	41.44	50.67	1.92	49.67	nr	**100.34**

U: VENTILATION/AIR CONDITIONING SYSTEMS

Item	Net Price £	Material £	Labour hours	Labour £	Unit	**Total rate £**
Square to round						
Sum of two sides 200 mm	16.90	20.66	1.22	31.57	nr	**52.23**
Sum of two sides 300 mm	18.98	23.21	1.22	31.57	nr	**54.78**
Sum of two sides 400 mm	25.10	30.68	1.25	32.33	nr	**63.01**
Sum of two sides 500 mm	27.30	33.37	1.25	32.33	nr	**65.70**
Sum of two sides 600 mm	29.50	36.06	1.33	34.40	nr	**70.46**
Sum of two sides 700 mm	31.72	38.78	1.33	34.40	nr	**73.18**
Sum of two sides 800 mm	33.91	41.46	1.40	36.21	nr	**77.67**
90° radius bend						
Sum of two sides 200 mm	12.71	15.54	1.10	28.46	nr	**44.00**
Sum of two sides 300 mm	13.70	16.75	1.10	28.46	nr	**45.21**
Sum of two sides 400 mm	22.73	27.79	1.12	28.97	nr	**56.76**
Sum of two sides 500 mm	24.16	29.54	1.12	28.97	nr	**58.51**
Sum of two sides 600 mm	26.07	31.87	1.16	30.01	nr	**61.88**
Sum of two sides 700 mm	27.74	33.91	1.16	30.01	nr	**63.92**
Sum of two sides 800 mm	29.61	36.19	1.22	31.57	nr	**67.76**
45° radius bend						
Sum of two sides 200 mm	13.72	16.77	1.29	33.37	nr	**50.14**
Sum of two sides 300 mm	15.06	18.41	1.29	33.37	nr	**51.78**
Sum of two sides 400 mm	23.81	29.11	1.29	33.37	nr	**62.48**
Sum of two sides 500 mm	25.49	31.16	1.29	33.37	nr	**64.53**
Sum of two sides 600 mm	27.43	33.54	1.39	35.96	nr	**69.50**
Sum of two sides 700 mm	29.23	35.73	1.39	35.96	nr	**71.69**
Sum of two sides 800 mm	31.13	38.06	1.46	37.76	nr	**75.82**
90° mitre bend						
Sum of two sides 200 mm	21.18	25.89	2.04	52.78	nr	**78.67**
Sum of two sides 300 mm	23.19	28.35	2.04	52.78	nr	**81.13**
Sum of two sides 400 mm	33.33	40.75	2.09	54.05	nr	**94.80**
Sum of two sides 500 mm	35.93	43.93	2.09	54.05	nr	**97.98**
Sum of two sides 600 mm	39.24	47.97	2.15	55.61	nr	**103.58**
Sum of two sides 700 mm	42.31	51.73	2.15	55.61	nr	**107.34**
Sum of two sides 800 mm	45.68	55.85	2.26	58.46	nr	**114.31**
Branch						
Sum of two sides 200 mm	21.03	25.71	0.92	23.80	nr	**49.51**
Sum of two sides 300 mm	23.37	28.57	0.92	23.80	nr	**52.37**
Sum of two sides 400 mm	29.08	35.56	0.95	24.58	nr	**60.14**
Sum of two sides 500 mm	31.64	38.68	0.95	24.58	nr	**63.26**
Sum of two sides 600 mm	34.16	41.76	1.03	26.64	nr	**68.40**
Sum of two sides 700 mm	36.67	44.83	1.03	26.64	nr	**71.47**
Sum of two sides 800 mm	39.17	47.88	1.03	26.64	nr	**74.52**
Grille neck						
Sum of two sides 200 mm	24.10	29.46	1.10	28.46	nr	**57.92**
Sum of two sides 300 mm	26.96	32.96	1.10	28.46	nr	**61.42**
Sum of two sides 400 mm	29.83	36.47	1.16	30.01	nr	**66.48**
Sum of two sides 500 mm	32.70	39.97	1.16	30.01	nr	**69.98**
Sum of two sides 600 mm	35.57	43.49	1.18	30.53	nr	**74.02**
Sum of two sides 700 mm	38.44	46.99	1.18	30.53	nr	**77.52**
Sum of two sides 800 mm	41.30	50.49	1.18	30.53	nr	**81.02**

U: VENTILATION/AIR CONDITIONING SYSTEMS

Item	Net Price £	Material £	Labour hours	Labour £	Unit	**Total rate £**
U10: DUCTWORK: RECTANGULAR – CLASS C – cont						
Ductwork 401 to 600 mm longest side						
Sum of two sides 600 mm	19.39	23.70	1.17	30.27	m	**53.97**
Sum of two sides 700 mm	21.19	25.90	1.17	30.27	m	**56.17**
Sum of two sides 800 mm	22.94	28.05	1.17	30.27	m	**58.32**
Sum of two sides 900 mm	24.65	30.14	1.27	32.85	m	**62.99**
Sum of two sides 1000 mm	26.35	32.21	1.48	38.29	m	**70.50**
Sum of two sides 1100 mm	28.20	34.47	1.49	38.54	m	**73.01**
Sum of two sides 1200 mm	29.89	36.54	1.49	38.54	m	**75.08**
Extra over fittings; ductwork 401 to 600 mm longest side						
End cap						
Sum of two sides 600 mm	11.28	13.79	0.38	9.83	nr	**23.62**
Sum of two sides 700 mm	12.17	14.87	0.38	9.83	nr	**24.70**
Sum of two sides 800 mm	13.05	15.95	0.38	9.83	nr	**25.78**
Sum of two sides 900 mm	13.94	17.05	0.38	9.83	nr	**26.88**
Sum of two sides 1000 mm	14.82	18.12	0.38	9.83	nr	**27.95**
Sum of two sides 1100 mm	15.70	19.19	0.38	9.83	nr	**29.02**
Sum of two sides 1200 mm	16.59	20.28	0.38	9.83	nr	**30.11**
Reducer						
Sum of two sides 600 mm	24.03	29.38	1.69	43.71	nr	**73.09**
Sum of two sides 700 mm	25.57	31.26	1.69	43.71	nr	**74.97**
Sum of two sides 800 mm	27.08	33.10	1.92	49.67	nr	**82.77**
Sum of two sides 900 mm	28.62	34.99	1.92	49.67	nr	**84.66**
Sum of two sides 1000 mm	30.15	36.86	2.18	56.39	nr	**93.25**
Sum of two sides 1100 mm	31.82	38.90	2.18	56.39	nr	**95.29**
Sum of two sides 1200 mm	33.36	40.78	2.18	56.39	nr	**97.17**
Offset						
Sum of two sides 600 mm	34.07	41.65	1.92	49.67	nr	**91.32**
Sum of two sides 700 mm	36.36	44.45	1.92	49.67	nr	**94.12**
Sum of two sides 800 mm	37.65	46.03	1.92	49.67	nr	**95.70**
Sum of two sides 900 mm	38.85	47.49	1.92	49.67	nr	**97.16**
Sum of two sides 1000 mm	40.45	49.45	2.18	56.39	nr	**105.84**
Sum of two sides 1100 mm	41.51	50.75	2.18	56.39	nr	**107.14**
Sum of two sides 1200 mm	42.35	51.78	2.18	56.39	nr	**108.17**
Square to round						
Sum of two sides 600 mm	25.87	31.63	1.33	34.40	nr	**66.03**
Sum of two sides 700 mm	27.67	33.83	1.33	34.40	nr	**68.23**
Sum of two sides 800 mm	29.46	36.02	1.40	36.21	nr	**72.23**
Sum of two sides 900 mm	31.26	38.21	1.40	36.21	nr	**74.42**
Sum of two sides 1000 mm	33.07	40.43	1.82	47.08	nr	**87.51**
Sum of two sides 1100 mm	34.91	42.67	1.82	47.08	nr	**89.75**
Sum of two sides 1200 mm	36.71	44.88	1.82	47.08	nr	**91.96**

U: VENTILATION/AIR CONDITIONING SYSTEMS

Item	Net Price £	Material £	Labour hours	Labour £	Unit	**Total rate £**
90° radius bend						
Sum of two sides 600 mm	22.58	27.60	1.16	30.01	nr	**57.61**
Sum of two sides 700 mm	23.11	28.25	1.16	30.01	nr	**58.26**
Sum of two sides 800 mm	24.61	30.09	1.22	31.57	nr	**61.66**
Sum of two sides 900 mm	26.16	31.98	1.22	31.57	nr	**63.55**
Sum of two sides 1000 mm	27.09	33.11	1.40	36.21	nr	**69.32**
Sum of two sides 1100 mm	28.70	35.08	1.40	36.21	nr	**71.29**
Sum of two sides 1200 mm	30.22	36.95	1.40	36.21	nr	**73.16**
45° bend						
Sum of two sides 600 mm	25.19	30.80	1.39	35.96	nr	**66.76**
Sum of two sides 700 mm	26.37	32.24	1.39	35.96	nr	**68.20**
Sum of two sides 800 mm	28.05	34.29	1.46	37.76	nr	**72.05**
Sum of two sides 900 mm	29.75	36.37	1.46	37.76	nr	**74.13**
Sum of two sides 1000 mm	31.13	38.06	1.88	48.63	nr	**86.69**
Sum of two sides 1100 mm	32.95	40.28	1.88	48.63	nr	**88.91**
Sum of two sides 1200 mm	34.63	42.33	1.88	48.63	nr	**90.96**
90° mitre bend						
Sum of two sides 600 mm	39.72	48.55	2.15	55.61	nr	**104.16**
Sum of two sides 700 mm	41.21	50.38	2.15	55.61	nr	**105.99**
Sum of two sides 800 mm	44.18	54.01	2.26	58.46	nr	**112.47**
Sum of two sides 900 mm	47.18	57.68	2.26	58.46	nr	**116.14**
Sum of two sides 1000 mm	49.47	60.48	3.03	78.37	nr	**138.85**
Sum of two sides 1100 mm	52.59	64.29	3.03	78.37	nr	**142.66**
Sum of two sides 1200 mm	55.64	68.02	3.03	78.37	nr	**146.39**
Branch						
Sum of two sides 600 mm	33.50	40.95	1.03	26.64	nr	**67.59**
Sum of two sides 700 mm	35.96	43.96	1.03	26.64	nr	**70.60**
Sum of two sides 800 mm	38.44	46.99	1.03	26.64	nr	**73.63**
Sum of two sides 900 mm	40.90	50.00	1.03	26.64	nr	**76.64**
Sum of two sides 1000 mm	43.37	53.02	1.29	33.37	nr	**86.39**
Sum of two sides 1100 mm	45.93	56.15	1.29	33.37	nr	**89.52**
Sum of two sides 1200 mm	48.39	59.15	1.29	33.37	nr	**92.52**
Grille neck						
Sum of two sides 600 mm	34.64	42.34	1.18	30.53	nr	**72.87**
Sum of two sides 700 mm	37.44	45.77	1.18	30.53	nr	**76.30**
Sum of two sides 800 mm	40.23	49.18	1.18	30.53	nr	**79.71**
Sum of two sides 900 mm	43.02	52.59	1.18	30.53	nr	**83.12**
Sum of two sides 1000 mm	45.82	56.01	1.44	37.24	nr	**93.25**
Sum of two sides 1100 mm	48.62	59.44	1.44	37.24	nr	**96.68**
Sum of two sides 1200 mm	51.40	62.84	1.44	37.24	nr	**100.08**

U: VENTILATION/AIR CONDITIONING SYSTEMS

Item	Net Price £	Material £	Labour hours	Labour £	Unit	Total rate £
U10: DUCTWORK: RECTANGULAR – CLASS C – cont						
Ductwork 601 to 800 mm longest side						
Sum of two sides 900 mm	27.04	33.05	1.27	32.85	m	**65.90**
Sum of two sides 1000 mm	28.74	35.13	1.48	38.29	m	**73.42**
Sum of two sides 1100 mm	30.44	37.21	1.49	38.54	m	**75.75**
Sum of two sides 1200 mm	31.57	38.59	1.49	38.54	m	**77.13**
Sum of two sides 1300 mm	33.98	41.54	1.51	39.06	m	**80.60**
Sum of two sides 1400 mm	35.69	43.63	1.55	40.10	m	**83.73**
Sum of two sides 1500 mm	37.39	45.71	1.61	41.65	m	**87.36**
Sum of two sides 1600 mm	39.10	47.80	1.62	41.91	m	**89.71**
Extra over fittings; ductwork 601 to 800 mm longest side						
End cap						
Sum of two sides 900 mm	14.02	17.14	0.38	9.83	nr	**26.97**
Sum of two sides 1000 mm	14.90	18.21	0.38	9.83	nr	**28.04**
Sum of two sides 1100 mm	15.78	19.29	0.38	9.83	nr	**29.12**
Sum of two sides 1200 mm	16.68	20.39	0.38	9.83	nr	**30.22**
Sum of two sides 1300 mm	17.56	21.47	0.38	9.83	nr	**31.30**
Sum of two sides 1400 mm	18.32	22.39	0.38	9.83	nr	**32.22**
Sum of two sides 1500 mm	21.33	26.08	0.38	9.83	nr	**35.91**
Sum of two sides 1600 mm	24.33	29.75	0.38	9.83	nr	**39.58**
Reducer						
Sum of two sides 900 mm	29.08	35.56	1.92	49.67	nr	**85.23**
Sum of two sides 1000 mm	30.62	37.43	2.18	56.39	nr	**93.82**
Sum of two sides 1100 mm	32.17	39.33	2.18	56.39	nr	**95.72**
Sum of two sides 1200 mm	33.84	41.36	2.18	56.39	nr	**97.75**
Sum of two sides 1300 mm	35.38	43.25	2.30	59.50	nr	**102.75**
Sum of two sides 1400 mm	36.67	44.83	2.30	59.50	nr	**104.33**
Sum of two sides 1500 mm	42.45	51.89	2.47	63.90	nr	**115.79**
Sum of two sides 1600 mm	48.22	58.95	2.47	63.90	nr	**122.85**
Offset						
Sum of two sides 900 mm	41.28	50.47	1.92	49.67	nr	**100.14**
Sum of two sides 1000 mm	42.33	51.75	2.18	56.39	nr	**108.14**
Sum of two sides 1100 mm	43.25	52.87	2.18	56.39	nr	**109.26**
Sum of two sides 1200 mm	44.14	53.96	2.18	56.39	nr	**110.35**
Sum of two sides 1300 mm	44.79	54.75	2.47	63.90	nr	**118.65**
Sum of two sides 1400 mm	45.70	55.87	2.47	63.90	nr	**119.77**
Sum of two sides 1500 mm	52.48	64.16	2.47	63.90	nr	**128.06**
Sum of two sides 1600 mm	59.13	72.29	2.47	63.90	nr	**136.19**

U: VENTILATION/AIR CONDITIONING SYSTEMS

Item	Net Price £	Material £	Labour hours	Labour £	Unit	**Total rate £**
Square to round						
Sum of two sides 900 mm	31.14	38.07	1.40	36.21	nr	**74.28**
Sum of two sides 1000 mm	32.97	40.30	1.82	47.08	nr	**87.38**
Sum of two sides 1100 mm	34.79	42.53	1.82	47.08	nr	**89.61**
Sum of two sides 1200 mm	36.63	44.78	1.82	47.08	nr	**91.86**
Sum of two sides 1300 mm	38.44	46.99	2.32	60.01	nr	**107.00**
Sum of two sides 1400 mm	39.94	48.82	2.32	60.01	nr	**108.83**
Sum of two sides 1500 mm	47.03	57.49	2.56	66.22	nr	**123.71**
Sum of two sides 1600 mm	54.15	66.20	2.58	66.73	nr	**132.93**
90° radius bend						
Sum of two sides 900 mm	23.43	28.64	1.22	31.57	nr	**60.21**
Sum of two sides 1000 mm	24.89	30.43	1.40	36.21	nr	**66.64**
Sum of two sides 1100 mm	26.35	32.21	1.40	36.21	nr	**68.42**
Sum of two sides 1200 mm	27.88	34.08	1.40	36.21	nr	**70.29**
Sum of two sides 1300 mm	29.33	35.85	1.91	49.41	nr	**85.26**
Sum of two sides 1400 mm	29.72	36.33	1.91	49.41	nr	**85.74**
Sum of two sides 1500 mm	35.38	43.25	2.11	54.59	nr	**97.84**
Sum of two sides 1600 mm	41.06	50.19	2.11	54.59	nr	**104.78**
45° bend						
Sum of two sides 900 mm	28.93	35.37	1.46	37.76	nr	**73.13**
Sum of two sides 1000 mm	30.58	37.39	1.88	48.63	nr	**86.02**
Sum of two sides 1100 mm	32.25	39.43	1.88	48.63	nr	**88.06**
Sum of two sides 1200 mm	34.03	41.60	1.88	48.63	nr	**90.23**
Sum of two sides 1300 mm	35.69	43.63	2.26	58.46	nr	**102.09**
Sum of two sides 1400 mm	36.68	44.85	2.44	63.12	nr	**107.97**
Sum of two sides 1500 mm	42.56	52.03	2.44	63.12	nr	**115.15**
Sum of two sides 1600 mm	48.45	59.24	2.68	69.33	nr	**128.57**
90° mitre bend						
Sum of two sides 900 mm	43.21	52.83	2.26	58.46	nr	**111.29**
Sum of two sides 1000 mm	46.40	56.72	3.03	78.37	nr	**135.09**
Sum of two sides 1100 mm	49.58	60.62	3.03	78.37	nr	**138.99**
Sum of two sides 1200 mm	52.82	64.57	3.03	78.37	nr	**142.94**
Sum of two sides 1300 mm	56.01	68.47	3.85	99.60	nr	**168.07**
Sum of two sides 1400 mm	57.81	70.67	3.85	99.60	nr	**170.27**
Sum of two sides 1500 mm	67.37	82.36	4.25	109.94	nr	**192.30**
Sum of two sides 1600 mm	76.94	94.06	4.26	110.20	nr	**204.26**
Branch						
Sum of two sides 900 mm	42.18	51.56	1.03	26.64	nr	**78.20**
Sum of two sides 1000 mm	44.70	54.64	1.29	33.37	nr	**88.01**
Sum of two sides 1100 mm	47.23	57.74	1.29	33.37	nr	**91.11**
Sum of two sides 1200 mm	49.86	60.96	1.29	33.37	nr	**94.33**
Sum of two sides 1300 mm	52.39	64.05	1.39	35.96	nr	**100.01**
Sum of two sides 1400 mm	54.58	66.72	1.39	35.96	nr	**102.68**
Sum of two sides 1500 mm	62.74	76.70	1.64	42.43	nr	**119.13**
Sum of two sides 1600 mm	70.92	86.70	1.64	42.43	nr	**129.13**

U: VENTILATION/AIR CONDITIONING SYSTEMS

Item	Net Price £	Material £	Labour hours	Labour £	Unit	Total rate £
U10: DUCTWORK: RECTANGULAR – CLASS C – cont						
Extra over fittings – cont						
Grille neck						
Sum of two sides 900 mm	43.25	52.87	1.18	30.53	nr	**83.40**
Sum of two sides 1000 mm	46.07	56.32	1.44	37.24	nr	**93.56**
Sum of two sides 1100 mm	48.88	59.76	1.44	37.24	nr	**97.00**
Sum of two sides 1200 mm	51.68	63.18	1.44	37.24	nr	**100.42**
Sum of two sides 1300 mm	54.49	66.61	1.69	43.71	nr	**110.32**
Sum of two sides 1400 mm	56.90	69.56	1.69	43.71	nr	**113.27**
Sum of two sides 1500 mm	66.06	80.76	1.79	46.31	nr	**127.07**
Sum of two sides 1600 mm	75.22	91.96	1.79	46.31	nr	**138.27**
Ductwork 801 to 1000 mm longest side						
Sum of two sides 1100 mm	31.10	38.02	1.49	38.54	m	**76.56**
Sum of two sides 1200 mm	32.82	40.12	1.49	38.54	m	**78.66**
Sum of two sides 1300 mm	34.53	42.21	1.51	39.06	m	**81.27**
Sum of two sides 1400 mm	36.40	44.50	1.55	40.10	m	**84.60**
Sum of two sides 1500 mm	38.10	46.58	1.61	41.65	m	**88.23**
Sum of two sides 1600 mm	39.82	48.68	1.62	41.91	m	**90.59**
Sum of two sides 1700 mm	41.54	50.78	1.74	45.01	m	**95.79**
Sum of two sides 1800 mm	43.40	53.06	1.76	45.53	m	**98.59**
Sum of two sides 1900 mm	45.12	55.16	1.81	46.82	m	**101.98**
Sum of two sides 2000 mm	46.83	57.25	1.82	47.08	m	**104.33**
Extra over fittings; ductwork 801 to 1000 mm longest side						
End cap						
Sum of two sides 1100 mm	15.78	19.29	0.38	9.83	nr	**29.12**
Sum of two sides 1200 mm	16.68	20.39	0.38	9.83	nr	**30.22**
Sum of two sides 1300 mm	17.56	21.47	0.38	9.83	nr	**31.30**
Sum of two sides 1400 mm	18.32	22.39	0.38	9.83	nr	**32.22**
Sum of two sides 1500 mm	21.33	26.08	0.38	9.83	nr	**35.91**
Sum of two sides 1600 mm	24.33	29.75	0.38	9.83	nr	**39.58**
Sum of two sides 1700 mm	27.33	33.41	0.58	15.00	nr	**48.41**
Sum of two sides 1800 mm	30.34	37.09	0.58	15.00	nr	**52.09**
Sum of two sides 1900 mm	33.34	40.76	0.58	15.00	nr	**55.76**
Sum of two sides 2000 mm	36.35	44.43	0.58	15.00	nr	**59.43**
Reducer						
Sum of two sides 1100 mm	29.77	36.39	2.18	56.39	nr	**92.78**
Sum of two sides 1200 mm	31.19	38.13	2.18	56.39	nr	**94.52**
Sum of two sides 1300 mm	32.61	39.86	2.30	59.50	nr	**99.36**
Sum of two sides 1400 mm	33.88	41.42	2.30	59.50	nr	**100.92**
Sum of two sides 1500 mm	39.54	48.34	2.47	63.90	nr	**112.24**
Sum of two sides 1600 mm	45.19	55.25	2.47	63.90	nr	**119.15**
Sum of two sides 1700 mm	50.85	62.16	2.59	66.99	nr	**129.15**

U: VENTILATION/AIR CONDITIONING SYSTEMS

Item	Net Price £	Material £	Labour hours	Labour £	Unit	**Total rate £**
Sum of two sides 1800 mm	56.61	69.21	2.59	66.99	nr	**136.20**
Sum of two sides 1900 mm	62.27	76.13	2.71	70.10	nr	**146.23**
Sum of two sides 2000 mm	67.92	83.03	2.71	70.10	nr	**153.13**
Offset						
Sum of two sides 1100 mm	46.00	56.24	2.18	56.39	nr	**112.63**
Sum of two sides 1200 mm	46.92	57.36	2.18	56.39	nr	**113.75**
Sum of two sides 1300 mm	47.70	58.32	2.47	63.90	nr	**122.22**
Sum of two sides 1400 mm	48.06	58.75	2.47	63.90	nr	**122.65**
Sum of two sides 1500 mm	54.92	67.14	2.47	63.90	nr	**131.04**
Sum of two sides 1600 mm	61.66	75.38	2.47	63.90	nr	**139.28**
Sum of two sides 1700 mm	68.28	83.47	2.61	67.52	nr	**150.99**
Sum of two sides 1800 mm	74.81	91.45	2.61	67.52	nr	**158.97**
Sum of two sides 1900 mm	82.08	100.34	2.71	70.10	nr	**170.44**
Sum of two sides 2000 mm	88.33	107.99	2.71	70.10	nr	**178.09**
Square to round						
Sum of two sides 1100 mm	32.28	39.46	1.82	47.08	nr	**86.54**
Sum of two sides 1200 mm	33.98	41.54	1.82	47.08	nr	**88.62**
Sum of two sides 1300 mm	35.67	43.61	2.32	60.01	nr	**103.62**
Sum of two sides 1400 mm	37.06	45.31	2.32	60.01	nr	**105.32**
Sum of two sides 1500 mm	44.03	53.83	2.56	66.22	nr	**120.05**
Sum of two sides 1600 mm	51.02	62.38	2.58	66.73	nr	**129.11**
Sum of two sides 1700 mm	58.00	70.91	2.84	73.46	nr	**144.37**
Sum of two sides 1800 mm	64.99	79.45	2.84	73.46	nr	**152.91**
Sum of two sides 1900 mm	71.97	87.98	3.13	80.97	nr	**168.95**
Sum of two sides 2000 mm	78.97	96.54	3.13	80.97	nr	**177.51**
90° radius bend						
Sum of two sides 1100 mm	23.83	29.13	1.40	36.21	nr	**65.34**
Sum of two sides 1200 mm	25.52	31.20	1.40	36.21	nr	**67.41**
Sum of two sides 1300 mm	27.20	33.25	1.91	49.41	nr	**82.66**
Sum of two sides 1400 mm	28.68	35.06	1.91	49.41	nr	**84.47**
Sum of two sides 1500 mm	34.59	42.28	2.11	54.59	nr	**96.87**
Sum of two sides 1600 mm	40.51	49.52	2.11	54.59	nr	**104.11**
Sum of two sides 1700 mm	46.42	56.75	2.55	65.97	nr	**122.72**
Sum of two sides 1800 mm	52.39	64.05	2.55	65.97	nr	**130.02**
Sum of two sides 1900 mm	57.27	70.01	2.80	72.43	nr	**142.44**
Sum of two sides 2000 mm	63.18	77.24	2.80	72.43	nr	**149.67**
45° bend						
Sum of two sides 1100 mm	31.07	37.99	1.88	48.63	nr	**86.62**
Sum of two sides 1200 mm	32.87	40.18	1.88	48.63	nr	**88.81**
Sum of two sides 1300 mm	34.66	42.37	2.26	58.46	nr	**100.83**
Sum of two sides 1400 mm	36.30	44.37	2.44	63.12	nr	**107.49**
Sum of two sides 1500 mm	42.33	51.75	2.68	69.33	nr	**121.08**
Sum of two sides 1600 mm	48.35	59.11	2.69	69.59	nr	**128.70**
Sum of two sides 1700 mm	54.37	66.47	2.96	76.57	nr	**143.04**
Sum of two sides 1800 mm	60.51	73.97	2.96	76.57	nr	**150.54**
Sum of two sides 1900 mm	66.00	80.69	3.26	84.33	nr	**165.02**
Sum of two sides 2000 mm	72.02	88.04	3.26	84.33	nr	**172.37**

U: VENTILATION/AIR CONDITIONING SYSTEMS

Item	Net Price £	Material £	Labour hours	Labour £	Unit	Total rate £
U10: DUCTWORK: RECTANGULAR – CLASS C – cont						
Extra over fittings – cont						
90° mitre bend						
Sum of two sides 1100 mm	49.73	60.79	3.03	78.37	nr	**139.16**
Sum of two sides 1200 mm	53.10	64.91	3.03	78.37	nr	**143.28**
Sum of two sides 1300 mm	56.46	69.02	3.85	99.60	nr	**168.62**
Sum of two sides 1400 mm	59.47	72.70	3.85	99.60	nr	**172.30**
Sum of two sides 1500 mm	69.18	84.57	4.25	109.94	nr	**194.51**
Sum of two sides 1600 mm	78.90	96.46	4.26	110.20	nr	**206.66**
Sum of two sides 1700 mm	88.61	108.33	4.68	121.07	nr	**229.40**
Sum of two sides 1800 mm	98.35	120.23	4.68	121.07	nr	**241.30**
Sum of two sides 1900 mm	106.92	130.71	4.87	125.98	nr	**256.69**
Sum of two sides 2000 mm	116.67	142.63	4.87	125.98	nr	**268.61**
Branch						
Sum of two sides 1100 mm	47.33	57.87	1.29	33.37	nr	**91.24**
Sum of two sides 1200 mm	49.86	60.96	1.29	33.37	nr	**94.33**
Sum of two sides 1300 mm	52.39	64.05	1.39	35.96	nr	**100.01**
Sum of two sides 1400 mm	54.66	66.83	1.39	35.96	nr	**102.79**
Sum of two sides 1500 mm	62.84	76.82	1.64	42.43	nr	**119.25**
Sum of two sides 1600 mm	71.00	86.80	1.64	42.43	nr	**129.23**
Sum of two sides 1700 mm	79.17	96.79	1.69	43.71	nr	**140.50**
Sum of two sides 1800 mm	87.44	106.89	1.69	43.71	nr	**150.60**
Sum of two sides 1900 mm	95.61	116.88	1.85	47.86	nr	**164.74**
Sum of two sides 2000 mm	103.79	126.89	1.85	47.86	nr	**174.75**
Grille neck						
Sum of two sides 1100 mm	48.88	59.76	1.44	37.24	nr	**97.00**
Sum of two sides 1200 mm	51.68	63.18	1.44	37.24	nr	**100.42**
Sum of two sides 1300 mm	54.49	66.61	1.69	43.71	nr	**110.32**
Sum of two sides 1400 mm	56.90	69.56	1.69	43.71	nr	**113.27**
Sum of two sides 1500 mm	66.06	80.76	1.79	46.31	nr	**127.07**
Sum of two sides 1600 mm	75.22	91.96	1.79	46.31	nr	**138.27**
Sum of two sides 1700 mm	84.37	103.14	1.86	48.12	nr	**151.26**
Sum of two sides 1800 mm	93.52	114.33	1.86	48.12	nr	**162.45**
Sum of two sides 1900 mm	102.68	125.53	2.02	52.25	nr	**177.78**
Sum of two sides 2000 mm	111.83	136.71	2.02	52.25	nr	**188.96**
Ductwork 1001 to 1250 mm longest side						
Sum of two sides 1300 mm	52.17	63.78	1.51	39.06	m	**102.84**
Sum of two sides 1400 mm	54.55	66.69	1.55	40.10	m	**106.79**
Sum of two sides 1500 mm	56.79	69.42	1.61	41.65	m	**111.07**
Sum of two sides 1600 mm	59.17	72.34	1.62	41.91	m	**114.25**
Sum of two sides 1700 mm	61.40	75.07	1.74	45.01	m	**120.08**
Sum of two sides 1800 mm	63.64	77.80	1.76	45.53	m	**123.33**
Sum of two sides 1900 mm	65.87	80.53	1.81	46.82	m	**127.35**
Sum of two sides 2000 mm	68.26	83.45	1.82	47.08	m	**130.53**
Sum of two sides 2100 mm	70.50	86.19	2.53	65.44	m	**151.63**

U: VENTILATION/AIR CONDITIONING SYSTEMS

Item	Net Price £	Material £	Labour hours	Labour £	Unit	Total rate £
Sum of two sides 2200 mm	81.06	99.10	2.55	65.97	m	**165.07**
Sum of two sides 2300 mm	83.61	102.22	2.56	66.22	m	**168.44**
Sum of two sides 2400 mm	85.83	104.93	2.76	71.39	m	**176.32**
Sum of two sides 2500 mm	88.23	107.86	2.77	71.65	m	**179.51**
Extra over fittings; ductwork 1001 to 1250 mm longest side						
End cap						
Sum of two sides 1300 mm	17.91	21.90	0.38	9.83	nr	**31.73**
Sum of two sides 1400 mm	18.69	22.85	0.38	9.83	nr	**32.68**
Sum of two sides 1500 mm	21.74	26.57	0.38	9.83	nr	**36.40**
Sum of two sides 1600 mm	24.78	30.29	0.38	9.83	nr	**40.12**
Sum of two sides 1700 mm	27.81	34.00	0.58	15.00	nr	**49.00**
Sum of two sides 1800 mm	30.85	37.72	0.58	15.00	nr	**52.72**
Sum of two sides 1900 mm	33.90	41.45	0.58	15.00	nr	**56.45**
Sum of two sides 2000 mm	36.93	45.14	0.58	15.00	nr	**60.14**
Sum of two sides 2100 mm	39.97	48.86	0.87	22.51	nr	**71.37**
Sum of two sides 2200 mm	43.02	52.59	0.87	22.51	nr	**75.10**
Sum of two sides 2300 mm	46.06	56.31	0.87	22.51	nr	**78.82**
Sum of two sides 2400 mm	49.09	60.01	0.87	22.51	nr	**82.52**
Sum of two sides 2500 mm	52.14	63.74	0.87	22.51	nr	**86.25**
Reducer						
Sum of two sides 1300 mm	25.54	31.22	2.30	59.50	nr	**90.72**
Sum of two sides 1400 mm	26.54	32.44	2.30	59.50	nr	**91.94**
Sum of two sides 1500 mm	32.05	39.18	2.47	63.90	nr	**103.08**
Sum of two sides 1600 mm	37.66	46.04	2.47	63.90	nr	**109.94**
Sum of two sides 1700 mm	43.17	52.78	2.59	66.99	nr	**119.77**
Sum of two sides 1800 mm	48.67	59.50	2.59	66.99	nr	**126.49**
Sum of two sides 1900 mm	54.17	66.22	2.71	70.10	nr	**136.32**
Sum of two sides 2000 mm	59.79	73.09	2.71	70.10	nr	**143.19**
Sum of two sides 2100 mm	65.30	79.82	2.92	75.54	nr	**155.36**
Sum of two sides 2200 mm	66.26	81.00	2.92	75.54	nr	**156.54**
Sum of two sides 2300 mm	71.87	87.86	2.92	75.54	nr	**163.40**
Sum of two sides 2400 mm	77.38	94.60	3.12	80.71	nr	**175.31**
Sum of two sides 2500 mm	82.99	101.45	3.12	80.71	nr	**182.16**
Offset						
Sum of two sides 1300 mm	61.44	75.11	2.47	63.90	nr	**139.01**
Sum of two sides 1400 mm	63.27	77.35	2.47	63.90	nr	**141.25**
Sum of two sides 1500 mm	70.45	86.13	2.47	63.90	nr	**150.03**
Sum of two sides 1600 mm	77.54	94.79	2.47	63.90	nr	**158.69**
Sum of two sides 1700 mm	84.37	103.14	2.61	67.52	nr	**170.66**
Sum of two sides 1800 mm	91.00	111.25	2.61	67.52	nr	**178.77**
Sum of two sides 1900 mm	97.48	119.17	2.71	70.10	nr	**189.27**
Sum of two sides 2000 mm	103.82	126.92	2.71	70.10	nr	**197.02**
Sum of two sides 2100 mm	109.94	134.40	2.92	75.54	nr	**209.94**
Sum of two sides 2200 mm	108.47	132.60	3.26	84.33	nr	**216.93**
Sum of two sides 2300 mm	117.68	143.86	3.26	84.33	nr	**228.19**
Sum of two sides 2400 mm	126.87	155.10	3.47	89.76	nr	**244.86**
Sum of two sides 2500 mm	136.08	166.36	3.48	90.02	nr	**256.38**

U: VENTILATION/AIR CONDITIONING SYSTEMS

Item	Net Price £	Material £	Labour hours	Labour £	Unit	Total rate £
U10: DUCTWORK: RECTANGULAR – CLASS C – cont						
Extra over fittings – cont						
Square to round						
Sum of two sides 1300 mm	28.32	34.62	2.32	60.01	nr	**94.63**
Sum of two sides 1400 mm	29.52	36.09	2.32	60.01	nr	**96.10**
Sum of two sides 1500 mm	36.37	44.47	2.56	66.22	nr	**110.69**
Sum of two sides 1600 mm	43.22	52.84	2.58	66.73	nr	**119.57**
Sum of two sides 1700 mm	50.06	61.20	2.84	73.46	nr	**134.66**
Sum of two sides 1800 mm	56.90	69.56	2.84	73.46	nr	**143.02**
Sum of two sides 1900 mm	71.02	86.82	3.13	80.97	nr	**167.79**
Sum of two sides 2000 mm	70.60	86.31	3.13	80.97	nr	**167.28**
Sum of two sides 2100 mm	77.43	94.66	4.26	110.20	nr	**204.86**
Sum of two sides 2200 mm	79.73	97.47	4.27	110.46	nr	**207.93**
Sum of two sides 2300 mm	86.59	105.85	4.30	111.24	nr	**217.09**
Sum of two sides 2400 mm	93.43	114.22	4.77	123.38	nr	**237.60**
Sum of two sides 2500 mm	100.28	122.59	4.79	123.91	nr	**246.50**
90° radius bend						
Sum of two sides 1300 mm	10.71	13.09	1.91	49.41	nr	**62.50**
Sum of two sides 1400 mm	13.52	16.53	1.91	49.41	nr	**65.94**
Sum of two sides 1500 mm	16.08	19.66	2.11	54.59	nr	**74.25**
Sum of two sides 1600 mm	21.85	26.71	2.11	54.59	nr	**81.30**
Sum of two sides 1700 mm	27.61	33.75	2.55	65.97	nr	**99.72**
Sum of two sides 1800 mm	33.36	40.78	2.55	65.97	nr	**106.75**
Sum of two sides 1900 mm	39.10	47.80	2.80	72.43	nr	**120.23**
Sum of two sides 2000 mm	44.87	54.86	2.80	72.43	nr	**127.29**
Sum of two sides 2100 mm	50.62	61.88	2.80	72.43	nr	**134.31**
Sum of two sides 2200 mm	47.57	58.15	2.80	72.43	nr	**130.58**
Sum of two sides 2300 mm	49.06	59.98	4.35	112.53	nr	**172.51**
Sum of two sides 2400 mm	54.78	66.97	4.35	112.53	nr	**179.50**
Sum of two sides 2500 mm	60.51	73.97	4.35	112.53	nr	**186.50**
45° bend						
Sum of two sides 1300 mm	26.13	31.94	2.26	58.46	nr	**90.40**
Sum of two sides 1400 mm	26.74	32.69	2.44	63.12	nr	**95.81**
Sum of two sides 1500 mm	32.70	39.97	2.68	69.33	nr	**109.30**
Sum of two sides 1600 mm	38.74	47.36	2.69	69.59	nr	**116.95**
Sum of two sides 1700 mm	44.70	54.64	2.96	76.57	nr	**131.21**
Sum of two sides 1800 mm	50.65	61.92	2.96	76.57	nr	**138.49**
Sum of two sides 1900 mm	56.60	69.20	3.26	84.33	nr	**153.53**
Sum of two sides 2000 mm	62.65	76.59	3.26	84.33	nr	**160.92**
Sum of two sides 2100 mm	68.60	83.86	7.50	194.01	nr	**277.87**
Sum of two sides 2200 mm	69.77	85.29	7.50	194.01	nr	**279.30**
Sum of two sides 2300 mm	73.63	90.01	7.55	195.31	nr	**285.32**
Sum of two sides 2400 mm	79.57	97.27	8.13	210.31	nr	**307.58**
Sum of two sides 2500 mm	85.60	104.65	8.30	214.71	nr	**319.36**

U: VENTILATION/AIR CONDITIONING SYSTEMS

Item	Net Price £	Material £	Labour hours	Labour £	Unit	**Total rate £**
90° mitre bend						
Sum of two sides 1300 mm	60.40	73.84	3.85	99.60	nr	**173.44**
Sum of two sides 1400 mm	62.78	76.75	3.85	99.60	nr	**176.35**
Sum of two sides 1500 mm	74.17	90.67	4.25	109.94	nr	**200.61**
Sum of two sides 1600 mm	85.58	104.62	4.26	110.20	nr	**214.82**
Sum of two sides 1700 mm	104.70	128.00	4.68	121.07	nr	**249.07**
Sum of two sides 1800 mm	116.01	141.82	4.68	121.07	nr	**262.89**
Sum of two sides 1900 mm	127.34	155.67	4.87	125.98	nr	**281.65**
Sum of two sides 2000 mm	131.16	160.34	4.87	125.98	nr	**286.32**
Sum of two sides 2100 mm	142.57	174.30	7.50	194.01	nr	**368.31**
Sum of two sides 2200 mm	143.72	175.70	7.50	194.01	nr	**369.71**
Sum of two sides 2300 mm	150.05	183.43	7.55	195.31	nr	**378.74**
Sum of two sides 2400 mm	161.49	197.42	8.13	210.31	nr	**407.73**
Sum of two sides 2500 mm	172.93	211.41	8.30	214.71	nr	**426.12**
Branch						
Sum of two sides 1300 mm	53.75	65.71	1.39	35.96	nr	**101.67**
Sum of two sides 1400 mm	56.00	68.46	1.39	35.96	nr	**104.42**
Sum of two sides 1500 mm	64.28	78.58	1.64	42.43	nr	**121.01**
Sum of two sides 1600 mm	72.65	88.82	1.64	42.43	nr	**131.25**
Sum of two sides 1700 mm	80.92	98.92	1.69	43.71	nr	**142.63**
Sum of two sides 1800 mm	89.20	109.05	1.69	43.71	nr	**152.76**
Sum of two sides 1900 mm	97.46	119.15	1.85	47.86	nr	**167.01**
Sum of two sides 2000 mm	105.83	129.38	1.85	47.86	nr	**177.24**
Sum of two sides 2100 mm	114.10	139.49	2.61	67.52	nr	**207.01**
Sum of two sides 2200 mm	122.38	149.61	2.61	67.52	nr	**217.13**
Sum of two sides 2300 mm	130.75	159.85	2.61	67.52	nr	**227.37**
Sum of two sides 2400 mm	139.02	169.95	2.88	74.50	nr	**244.45**
Sum of two sides 2500 mm	147.40	180.20	2.88	74.50	nr	**254.70**
Grille neck						
Sum of two sides 1300 mm	56.08	68.56	1.69	43.71	nr	**112.27**
Sum of two sides 1400 mm	58.61	71.65	1.69	43.71	nr	**115.36**
Sum of two sides 1500 mm	67.91	83.02	1.79	46.31	nr	**129.33**
Sum of two sides 1600 mm	77.22	94.40	1.79	46.31	nr	**140.71**
Sum of two sides 1700 mm	86.52	105.77	1.86	48.12	nr	**153.89**
Sum of two sides 1800 mm	95.82	117.14	1.86	48.12	nr	**165.26**
Sum of two sides 1900 mm	105.14	128.53	2.02	52.25	nr	**180.78**
Sum of two sides 2000 mm	114.43	139.89	2.02	52.25	nr	**192.14**
Sum of two sides 2100 mm	123.74	151.28	2.61	67.52	nr	**218.80**
Sum of two sides 2200 mm	133.04	162.65	2.80	72.43	nr	**235.08**
Sum of two sides 2300 mm	142.35	174.03	2.80	72.43	nr	**246.46**
Sum of two sides 2400 mm	151.64	185.38	3.06	79.16	nr	**264.54**
Sum of two sides 2500 mm	160.96	196.77	3.06	79.16	nr	**275.93**

U: VENTILATION/AIR CONDITIONING SYSTEMS

Item	Net Price £	Material £	Labour hours	Labour £	Unit	Total rate £
U10: DUCTWORK: RECTANGULAR – CLASS C – cont						
Ductwork 1251 to 1600 mm longest side						
Sum of two sides 1700 mm	66.06	80.76	1.74	45.01	m	**125.77**
Sum of two sides 1800 mm	68.53	83.78	1.76	45.53	m	**129.31**
Sum of two sides 1900 mm	71.04	86.85	1.81	46.82	m	**133.67**
Sum of two sides 2000 mm	73.40	89.73	1.82	47.08	m	**136.81**
Sum of two sides 2100 mm	75.75	92.61	2.53	65.44	m	**158.05**
Sum of two sides 2200 mm	78.11	95.49	2.55	65.97	m	**161.46**
Sum of two sides 2300 mm	80.61	98.55	2.56	66.22	m	**164.77**
Sum of two sides 2400 mm	83.09	101.58	2.76	71.39	m	**172.97**
Sum of two sides 2500 mm	85.44	104.45	2.77	71.65	m	**176.10**
Sum of two sides 2600 mm	96.31	117.74	2.97	76.84	m	**194.58**
Sum of two sides 2700 mm	98.67	120.62	2.99	77.35	m	**197.97**
Sum of two sides 2800 mm	101.18	123.69	3.30	85.37	m	**209.06**
Sum of two sides 2900 mm	103.69	126.76	3.31	85.62	m	**212.38**
Sum of two sides 3000 mm	106.04	129.64	3.53	91.31	m	**220.95**
Sum of two sides 3100 mm	108.52	132.66	3.55	91.83	m	**224.49**
Sum of two sides 3200 mm	110.87	135.54	3.56	92.09	m	**227.63**
Extra over fittings; ductwork 1251 to 1600 mm longest side						
End cap						
Sum of two sides 1700 mm	27.52	33.64	0.58	15.00	nr	**48.64**
Sum of two sides 1800 mm	30.53	37.33	0.58	15.00	nr	**52.33**
Sum of two sides 1900 mm	33.54	41.00	0.58	15.00	nr	**56.00**
Sum of two sides 2000 mm	36.54	44.67	0.58	15.00	nr	**59.67**
Sum of two sides 2100 mm	39.55	48.35	0.87	22.51	nr	**70.86**
Sum of two sides 2200 mm	42.56	52.03	0.87	22.51	nr	**74.54**
Sum of two sides 2300 mm	45.58	55.72	0.87	22.51	nr	**78.23**
Sum of two sides 2400 mm	48.58	59.39	0.87	22.51	nr	**81.90**
Sum of two sides 2500 mm	51.59	63.07	0.87	22.51	nr	**85.58**
Sum of two sides 2600 mm	54.60	66.74	0.87	22.51	nr	**89.25**
Sum of two sides 2700 mm	57.61	70.43	0.87	22.51	nr	**92.94**
Sum of two sides 2800 mm	60.61	74.10	1.16	30.01	nr	**104.11**
Sum of two sides 2900 mm	63.62	77.78	1.16	30.01	nr	**107.79**
Sum of two sides 3000 mm	72.89	89.11	1.73	44.75	nr	**133.86**
Sum of two sides 3100 mm	69.01	84.37	1.73	44.75	nr	**129.12**
Sum of two sides 3200 mm	71.17	87.00	1.73	44.75	nr	**131.75**
Reducer						
Sum of two sides 1700 mm	37.57	45.93	2.59	66.99	nr	**112.92**
Sum of two sides 1800 mm	42.75	52.26	2.59	66.99	nr	**119.25**
Sum of two sides 1900 mm	48.00	58.68	2.71	70.10	nr	**128.78**
Sum of two sides 2000 mm	62.27	76.13	2.71	70.10	nr	**146.23**
Sum of two sides 2100 mm	58.35	71.34	2.92	75.54	nr	**146.88**
Sum of two sides 2200 mm	63.52	77.65	2.92	75.54	nr	**153.19**
Sum of two sides 2300 mm	68.78	84.08	2.92	75.54	nr	**159.62**

U: VENTILATION/AIR CONDITIONING SYSTEMS

Item	Net Price £	Material £	Labour hours	Labour £	Unit	**Total rate £**
Sum of two sides 2400 mm	73.95	90.40	3.12	80.71	nr	**171.11**
Sum of two sides 2500 mm	79.13	96.74	3.12	80.71	nr	**177.45**
Sum of two sides 2600 mm	79.83	97.59	3.16	81.75	nr	**179.34**
Sum of two sides 2700 mm	83.37	101.92	3.16	81.75	nr	**183.67**
Sum of two sides 2800 mm	88.52	108.21	4.00	103.47	nr	**211.68**
Sum of two sides 2900 mm	99.14	121.20	4.01	103.73	nr	**224.93**
Sum of two sides 3000 mm	104.31	127.52	4.56	117.96	nr	**245.48**
Sum of two sides 3100 mm	108.15	132.21	4.56	117.96	nr	**250.17**
Sum of two sides 3200 mm	111.62	136.45	4.56	117.96	nr	**254.41**
Offset						
Sum of two sides 1700 mm	92.44	113.01	2.61	67.52	nr	**180.53**
Sum of two sides 1800 mm	102.13	124.86	2.61	67.52	nr	**192.38**
Sum of two sides 1900 mm	108.63	132.80	2.71	70.10	nr	**202.90**
Sum of two sides 2000 mm	114.82	140.37	2.71	70.10	nr	**210.47**
Sum of two sides 2100 mm	120.84	147.72	2.92	75.54	nr	**223.26**
Sum of two sides 2200 mm	126.67	154.85	3.26	84.33	nr	**239.18**
Sum of two sides 2300 mm	132.34	161.78	3.26	84.33	nr	**246.11**
Sum of two sides 2400 mm	141.59	173.09	3.47	89.76	nr	**262.85**
Sum of two sides 2500 mm	146.94	179.63	3.48	90.02	nr	**269.65**
Sum of two sides 2600 mm	146.98	179.68	3.49	90.28	nr	**269.96**
Sum of two sides 2700 mm	156.11	190.85	3.50	90.54	nr	**281.39**
Sum of two sides 2800 mm	165.23	201.99	4.33	112.00	nr	**313.99**
Sum of two sides 2900 mm	179.76	219.76	4.74	122.62	nr	**342.38**
Sum of two sides 3000 mm	188.89	230.92	5.31	137.36	nr	**368.28**
Sum of two sides 3100 mm	196.02	239.64	5.34	138.13	nr	**377.77**
Sum of two sides 3200 mm	202.61	247.69	5.35	138.40	nr	**386.09**
Square to round						
Sum of two sides 1700 mm	41.86	51.17	2.84	73.46	nr	**124.63**
Sum of two sides 1800 mm	48.36	59.12	2.84	73.46	nr	**132.58**
Sum of two sides 1900 mm	54.82	67.02	3.13	80.97	nr	**147.99**
Sum of two sides 2000 mm	61.31	74.95	3.13	80.97	nr	**155.92**
Sum of two sides 2100 mm	67.79	82.87	4.26	110.20	nr	**193.07**
Sum of two sides 2200 mm	74.27	90.79	4.27	110.46	nr	**201.25**
Sum of two sides 2300 mm	80.75	98.72	4.30	111.24	nr	**209.96**
Sum of two sides 2400 mm	87.24	106.66	4.77	123.38	nr	**230.04**
Sum of two sides 2500 mm	93.72	114.58	4.79	123.91	nr	**238.49**
Sum of two sides 2600 mm	94.10	115.04	4.95	128.05	nr	**243.09**
Sum of two sides 2700 mm	100.58	122.96	4.95	128.05	nr	**251.01**
Sum of two sides 2800 mm	107.04	130.86	8.49	219.63	nr	**350.49**
Sum of two sides 2900 mm	116.20	142.06	8.88	229.71	nr	**371.77**
Sum of two sides 3000 mm	122.68	149.98	9.02	233.33	nr	**383.31**
Sum of two sides 3100 mm	127.50	155.87	9.02	233.33	nr	**389.20**
Sum of two sides 3200 mm	131.86	161.20	9.09	235.14	nr	**396.34**

U: VENTILATION/AIR CONDITIONING SYSTEMS

Item	Net Price £	Material £	Labour hours	Labour £	Unit	Total rate £
U10: DUCTWORK: RECTANGULAR – CLASS C – cont						
Extra over fittings – cont						
90° radius bend						
Sum of two sides 1700 mm	91.62	112.00	2.55	65.97	nr	**177.97**
Sum of two sides 1800 mm	99.36	121.47	2.55	65.97	nr	**187.44**
Sum of two sides 1900 mm	111.18	135.92	2.80	72.43	nr	**208.35**
Sum of two sides 2000 mm	122.81	150.13	2.80	72.43	nr	**222.56**
Sum of two sides 2100 mm	134.44	164.36	2.61	67.52	nr	**231.88**
Sum of two sides 2200 mm	146.06	178.56	2.62	67.77	nr	**246.33**
Sum of two sides 2300 mm	157.88	193.01	2.63	68.03	nr	**261.04**
Sum of two sides 2400 mm	165.22	201.98	4.34	112.26	nr	**314.24**
Sum of two sides 2500 mm	176.80	216.14	4.35	112.53	nr	**328.67**
Sum of two sides 2600 mm	177.44	216.92	4.53	117.17	nr	**334.09**
Sum of two sides 2700 mm	189.02	231.08	4.53	117.17	nr	**348.25**
Sum of two sides 2800 mm	195.40	238.88	7.13	184.44	nr	**423.32**
Sum of two sides 2900 mm	217.85	266.33	7.17	185.48	nr	**451.81**
Sum of two sides 3000 mm	229.40	280.44	7.26	187.80	nr	**468.24**
Sum of two sides 3100 mm	238.26	291.27	7.26	187.80	nr	**479.07**
Sum of two sides 3200 mm	246.40	301.22	7.31	189.10	nr	**490.32**
45° bend						
Sum of two sides 1700 mm	44.01	53.81	2.96	76.57	nr	**130.38**
Sum of two sides 1800 mm	47.83	58.47	2.96	76.57	nr	**135.04**
Sum of two sides 1900 mm	53.72	65.67	3.26	84.33	nr	**150.00**
Sum of two sides 2000 mm	59.51	72.75	3.26	84.33	nr	**157.08**
Sum of two sides 2100 mm	65.30	79.82	7.50	194.01	nr	**273.83**
Sum of two sides 2200 mm	71.11	86.93	7.50	194.01	nr	**280.94**
Sum of two sides 2300 mm	76.99	94.12	7.55	195.31	nr	**289.43**
Sum of two sides 2400 mm	80.61	98.55	8.13	210.31	nr	**308.86**
Sum of two sides 2500 mm	86.38	105.60	8.30	214.71	nr	**320.31**
Sum of two sides 2600 mm	86.31	105.51	8.56	221.43	nr	**326.94**
Sum of two sides 2700 mm	92.08	112.57	8.62	222.98	nr	**335.55**
Sum of two sides 2800 mm	95.21	116.39	9.09	235.14	nr	**351.53**
Sum of two sides 2900 mm	106.42	130.10	9.09	235.14	nr	**365.24**
Sum of two sides 3000 mm	112.18	137.14	9.62	248.86	nr	**386.00**
Sum of two sides 3100 mm	116.59	142.53	9.62	248.86	nr	**391.39**
Sum of two sides 3200 mm	120.65	147.50	9.62	248.86	nr	**396.36**
90° mitre bend						
Sum of two sides 1700 mm	96.71	118.23	4.68	121.07	nr	**239.30**
Sum of two sides 1800 mm	102.80	125.67	4.68	121.07	nr	**246.74**
Sum of two sides 1900 mm	114.25	139.67	4.87	125.98	nr	**265.65**
Sum of two sides 2000 mm	125.73	153.71	4.87	125.98	nr	**279.69**
Sum of two sides 2100 mm	137.21	167.74	7.50	194.01	nr	**361.75**
Sum of two sides 2200 mm	148.70	181.78	7.50	194.01	nr	**375.79**
Sum of two sides 2300 mm	160.15	195.78	7.55	195.31	nr	**391.09**

U: VENTILATION/AIR CONDITIONING SYSTEMS

Item	Net Price £	Material £	Labour hours	Labour £	Unit	**Total rate £**
Sum of two sides 2400 mm	166.17	203.15	8.13	210.31	nr	**413.46**
Sum of two sides 2500 mm	177.69	217.23	8.30	214.71	nr	**431.94**
Sum of two sides 2600 mm	176.24	215.46	8.56	221.43	nr	**436.89**
Sum of two sides 2700 mm	187.77	229.55	8.62	222.98	nr	**452.53**
Sum of two sides 2800 mm	192.85	235.76	15.20	393.19	nr	**628.95**
Sum of two sides 2900 mm	212.98	260.37	15.20	393.19	nr	**653.56**
Sum of two sides 3000 mm	224.53	274.48	15.60	403.55	nr	**678.03**
Sum of two sides 3100 mm	234.01	286.08	16.04	414.94	nr	**701.02**
Sum of two sides 3200 mm	243.02	297.09	16.04	414.94	nr	**712.03**
Branch						
Sum of two sides 1700 mm	82.60	100.98	1.69	43.71	nr	**144.69**
Sum of two sides 1800 mm	90.79	110.99	1.69	43.71	nr	**154.70**
Sum of two sides 1900 mm	99.08	121.13	1.85	47.86	nr	**168.99**
Sum of two sides 2000 mm	107.28	131.15	1.85	47.86	nr	**179.01**
Sum of two sides 2100 mm	115.48	141.17	2.61	67.52	nr	**208.69**
Sum of two sides 2200 mm	123.67	151.18	2.61	67.52	nr	**218.70**
Sum of two sides 2300 mm	131.97	161.34	2.61	67.52	nr	**228.86**
Sum of two sides 2400 mm	140.15	171.33	2.88	74.50	nr	**245.83**
Sum of two sides 2500 mm	148.36	181.37	2.88	74.50	nr	**255.87**
Sum of two sides 2600 mm	156.56	191.39	2.88	74.50	nr	**265.89**
Sum of two sides 2700 mm	164.76	201.42	2.88	74.50	nr	**275.92**
Sum of two sides 2800 mm	172.93	211.41	3.94	101.92	nr	**313.33**
Sum of two sides 2900 mm	183.92	224.84	3.94	101.92	nr	**326.76**
Sum of two sides 3000 mm	192.12	234.87	4.83	124.94	nr	**359.81**
Sum of two sides 3100 mm	198.64	242.84	4.83	124.94	nr	**367.78**
Sum of two sides 3200 mm	204.58	250.10	4.83	124.94	nr	**375.04**
Grille neck						
Sum of two sides 1700 mm	85.61	104.66	1.86	48.12	nr	**152.78**
Sum of two sides 1800 mm	94.82	115.92	1.86	48.12	nr	**164.04**
Sum of two sides 1900 mm	104.03	127.17	2.02	52.25	nr	**179.42**
Sum of two sides 2000 mm	113.23	138.42	2.02	52.25	nr	**190.67**
Sum of two sides 2100 mm	122.44	149.68	2.80	72.43	nr	**222.11**
Sum of two sides 2200 mm	131.64	160.93	2.80	72.43	nr	**233.36**
Sum of two sides 2300 mm	140.85	172.19	2.80	72.43	nr	**244.62**
Sum of two sides 2400 mm	150.05	183.43	3.06	79.16	nr	**262.59**
Sum of two sides 2500 mm	159.26	194.70	3.06	79.16	nr	**273.86**
Sum of two sides 2600 mm	168.47	205.96	3.08	79.67	nr	**285.63**
Sum of two sides 2700 mm	177.67	217.21	3.08	79.67	nr	**296.88**
Sum of two sides 2800 mm	186.88	228.46	4.13	106.83	nr	**335.29**
Sum of two sides 2900 mm	196.08	239.71	4.13	106.83	nr	**346.54**
Sum of two sides 3000 mm	205.29	250.97	5.02	129.85	nr	**380.82**
Sum of two sides 3100 mm	212.61	259.92	5.02	129.85	nr	**389.77**
Sum of two sides 3200 mm	219.27	268.06	5.02	129.85	nr	**397.91**

U: VENTILATION/AIR CONDITIONING SYSTEMS

Item	Net Price £	Material £	Labour hours	Labour £	Unit	**Total rate £**
U10: DUCTWORK: RECTANGULAR – CLASS C – cont						
Ductwork 1601 to 2000 mm longest side						
Sum of two sides 2100 mm	88.00	107.58	2.53	65.44	m	**173.02**
Sum of two sides 2200 mm	90.74	110.93	2.55	65.97	m	**176.90**
Sum of two sides 2300 mm	93.48	114.28	2.55	65.97	m	**180.25**
Sum of two sides 2400 mm	96.19	117.60	2.56	66.22	m	**183.82**
Sum of two sides 2500 mm	98.95	120.96	2.76	71.39	m	**192.35**
Sum of two sides 2600 mm	101.54	124.14	2.77	71.65	m	**195.79**
Sum of two sides 2700 mm	117.43	143.56	2.97	76.84	m	**220.40**
Sum of two sides 2800 mm	106.75	130.50	2.99	77.35	m	**207.85**
Sum of two sides 2900 mm	109.50	133.87	3.30	85.37	m	**219.24**
Sum of two sides 3000 mm	112.10	137.04	3.31	85.62	m	**222.66**
Sum of two sides 3100 mm	129.96	158.88	3.53	91.31	m	**250.19**
Sum of two sides 3200 mm	132.56	162.06	3.53	91.31	m	**253.37**
Sum of two sides 3300 mm	135.31	165.42	3.53	91.31	m	**256.73**
Sum of two sides 3400 mm	137.91	168.60	3.55	91.83	m	**260.43**
Sum of two sides 3500 mm	140.52	171.78	3.55	91.83	m	**263.61**
Sum of two sides 3600 mm	143.11	174.96	3.55	91.83	m	**266.79**
Sum of two sides 3700 mm	145.87	178.32	3.56	92.09	m	**270.41**
Sum of two sides 3800 mm	148.48	181.52	3.56	92.09	m	**273.61**
Sum of two sides 3900 mm	151.08	184.70	3.56	92.09	m	**276.79**
Sum of two sides 4000 mm	153.68	187.87	3.56	92.09	m	**279.96**
Extra over fittings; ductwork 1601 to 2000 mm longest side						
End cap						
Sum of two sides 2100 mm	39.99	48.88	0.87	22.51	nr	**71.39**
Sum of two sides 2200 mm	43.02	52.59	0.87	22.51	nr	**75.10**
Sum of two sides 2300 mm	46.05	56.30	0.87	22.51	nr	**78.81**
Sum of two sides 2400 mm	49.08	60.00	0.87	22.51	nr	**82.51**
Sum of two sides 2500 mm	52.10	63.70	0.87	22.51	nr	**86.21**
Sum of two sides 2600 mm	55.13	67.39	0.87	22.51	nr	**89.90**
Sum of two sides 2700 mm	58.17	71.11	0.87	22.51	nr	**93.62**
Sum of two sides 2800 mm	61.20	74.82	1.16	30.01	nr	**104.83**
Sum of two sides 2900 mm	64.22	78.51	1.16	30.01	nr	**108.52**
Sum of two sides 3000 mm	67.25	82.21	1.73	44.75	nr	**126.96**
Sum of two sides 3100 mm	69.65	85.15	1.73	44.75	nr	**129.90**
Sum of two sides 3200 mm	71.83	87.82	1.73	44.75	nr	**132.57**
Sum of two sides 3300 mm	74.00	90.46	1.73	44.75	nr	**135.21**
Sum of two sides 3400 mm	76.18	93.13	1.73	44.75	nr	**137.88**
Sum of two sides 3500 mm	78.37	95.81	1.73	44.75	nr	**140.56**
Sum of two sides 3600 mm	80.54	98.46	1.73	44.75	nr	**143.21**
Sum of two sides 3700 mm	82.73	101.14	1.73	44.75	nr	**145.89**
Sum of two sides 3800 mm	84.90	103.79	1.73	44.75	nr	**148.54**
Sum of two sides 3900 mm	87.09	106.47	1.73	44.75	nr	**151.22**
Sum of two sides 4000 mm	89.26	109.12	1.73	44.75	nr	**153.87**

U: VENTILATION/AIR CONDITIONING SYSTEMS

Item	Net Price £	Material £	Labour hours	Labour £	Unit	**Total rate £**
Reducer						
Sum of two sides 2100 mm	54.48	66.60	2.92	75.54	nr	**142.14**
Sum of two sides 2200 mm	59.65	72.92	2.92	75.54	nr	**148.46**
Sum of two sides 2300 mm	64.89	79.33	2.92	75.54	nr	**154.87**
Sum of two sides 2400 mm	69.98	85.55	3.12	80.71	nr	**166.26**
Sum of two sides 2500 mm	75.23	91.97	3.12	80.71	nr	**172.68**
Sum of two sides 2600 mm	80.38	98.26	3.16	81.75	nr	**180.01**
Sum of two sides 2700 mm	85.55	104.59	3.16	81.75	nr	**186.34**
Sum of two sides 2800 mm	90.71	110.89	4.00	103.47	nr	**214.36**
Sum of two sides 2900 mm	95.95	117.30	4.01	103.73	nr	**221.03**
Sum of two sides 3000 mm	101.12	123.62	4.01	103.73	nr	**227.35**
Sum of two sides 3100 mm	109.59	133.97	4.01	103.73	nr	**237.70**
Sum of two sides 3200 mm	112.19	137.15	4.56	117.96	nr	**255.11**
Sum of two sides 3300 mm	106.56	130.27	4.56	117.96	nr	**248.23**
Sum of two sides 3400 mm	110.02	134.50	4.56	117.96	nr	**252.46**
Sum of two sides 3500 mm	113.48	138.73	4.56	117.96	nr	**256.69**
Sum of two sides 3600 mm	116.94	142.96	4.56	117.96	nr	**260.92**
Sum of two sides 3700 mm	120.49	147.30	4.56	117.96	nr	**265.26**
Sum of two sides 3800 mm	123.95	151.53	4.56	117.96	nr	**269.49**
Sum of two sides 3900 mm	127.41	155.76	4.56	117.96	nr	**273.72**
Sum of two sides 4000 mm	130.89	160.01	4.56	117.96	nr	**277.97**
Offset						
Sum of two sides 2100 mm	127.96	156.44	2.92	75.54	nr	**231.98**
Sum of two sides 2200 mm	137.65	168.28	3.26	84.33	nr	**252.61**
Sum of two sides 2300 mm	143.25	175.12	3.26	84.33	nr	**259.45**
Sum of two sides 2400 mm	157.03	191.97	3.47	89.76	nr	**281.73**
Sum of two sides 2500 mm	162.45	198.59	3.48	90.02	nr	**288.61**
Sum of two sides 2600 mm	167.59	204.88	3.49	90.28	nr	**295.16**
Sum of two sides 2700 mm	172.52	210.90	3.50	90.54	nr	**301.44**
Sum of two sides 2800 mm	177.27	216.71	4.33	112.00	nr	**328.71**
Sum of two sides 2900 mm	181.87	222.34	4.74	122.62	nr	**344.96**
Sum of two sides 3000 mm	177.16	216.58	5.34	138.13	nr	**354.71**
Sum of two sides 3100 mm	183.73	224.61	5.35	138.40	nr	**363.01**
Sum of two sides 3200 mm	186.23	227.67	5.31	137.36	nr	**365.03**
Sum of two sides 3300 mm	195.79	239.35	5.35	138.40	nr	**377.75**
Sum of two sides 3400 mm	202.36	247.39	5.35	138.40	nr	**385.79**
Sum of two sides 3500 mm	208.94	255.43	5.35	138.40	nr	**393.83**
Sum of two sides 3600 mm	215.51	263.46	5.35	138.40	nr	**401.86**
Sum of two sides 3700 mm	222.11	271.53	5.35	138.40	nr	**409.93**
Sum of two sides 3800 mm	228.68	279.56	5.35	138.40	nr	**417.96**
Sum of two sides 3900 mm	235.25	287.60	5.35	138.40	nr	**426.00**
Sum of two sides 4000 mm	241.83	295.64	5.35	138.40	nr	**434.04**
Square to round						
Sum of two sides 2100 mm	82.19	100.48	4.26	110.20	nr	**210.68**
Sum of two sides 2200 mm	90.13	110.19	4.27	110.46	nr	**220.65**
Sum of two sides 2300 mm	98.05	119.87	4.30	111.24	nr	**231.11**
Sum of two sides 2400 mm	167.52	204.79	4.77	123.38	nr	**328.17**
Sum of two sides 2500 mm	113.93	139.28	4.79	123.91	nr	**263.19**
Sum of two sides 2600 mm	121.86	148.98	4.95	128.05	nr	**277.03**
Sum of two sides 2700 mm	129.79	158.67	4.95	128.05	nr	**286.72**

U: VENTILATION/AIR CONDITIONING SYSTEMS

Item	Net Price £	Material £	Labour hours	Labour £	Unit	Total rate £
U10: DUCTWORK: RECTANGULAR – CLASS C – cont						
Extra over fittings – cont						
Square to round – cont						
Sum of two sides 2800 mm	137.72	168.36	8.49	219.63	nr	**387.99**
Sum of two sides 2900 mm	145.64	178.05	8.88	229.71	nr	**407.76**
Sum of two sides 3000 mm	148.67	181.75	9.02	233.33	nr	**415.08**
Sum of two sides 3100 mm	153.57	187.74	9.02	233.33	nr	**421.07**
Sum of two sides 3200 mm	154.05	188.33	9.09	235.14	nr	**423.47**
Sum of two sides 3300 mm	162.15	198.23	9.09	235.14	nr	**433.37**
Sum of two sides 3400 mm	167.52	204.79	9.09	235.14	nr	**439.93**
Sum of two sides 3500 mm	172.91	211.39	9.09	235.14	nr	**446.53**
Sum of two sides 3600 mm	178.29	217.96	9.09	235.14	nr	**453.10**
Sum of two sides 3700 mm	183.66	224.53	9.09	235.14	nr	**459.67**
Sum of two sides 3800 mm	189.04	231.10	9.09	235.14	nr	**466.24**
Sum of two sides 3900 mm	194.42	237.68	9.09	235.14	nr	**472.82**
Sum of two sides 4000 mm	199.81	244.26	9.09	235.14	nr	**479.40**
90° radius bend						
Sum of two sides 2100 mm	116.80	142.79	2.61	67.52	nr	**210.31**
Sum of two sides 2200 mm	224.59	274.57	2.62	67.77	nr	**342.34**
Sum of two sides 2300 mm	243.24	297.36	2.63	68.03	nr	**365.39**
Sum of two sides 2400 mm	249.45	304.95	4.34	112.26	nr	**417.21**
Sum of two sides 2500 mm	268.06	327.70	4.35	112.53	nr	**440.23**
Sum of two sides 2600 mm	286.36	350.08	4.53	117.17	nr	**467.25**
Sum of two sides 2700 mm	304.66	372.45	4.53	117.17	nr	**489.62**
Sum of two sides 2800 mm	322.96	394.82	7.13	184.44	nr	**579.26**
Sum of two sides 2900 mm	341.57	417.57	7.17	185.48	nr	**603.05**
Sum of two sides 3000 mm	359.88	439.95	7.26	187.80	nr	**627.75**
Sum of two sides 3100 mm	349.41	427.15	7.26	187.80	nr	**614.95**
Sum of two sides 3200 mm	362.61	443.29	7.31	189.10	nr	**632.39**
Sum of two sides 3300 mm	392.53	479.87	7.31	189.10	nr	**668.97**
Sum of two sides 3400 mm	405.73	496.01	7.31	189.10	nr	**685.11**
Sum of two sides 3500 mm	418.93	512.15	7.31	189.10	nr	**701.25**
Sum of two sides 3600 mm	432.14	528.30	7.31	189.10	nr	**717.40**
Sum of two sides 3700 mm	445.66	544.82	7.31	189.10	nr	**733.92**
Sum of two sides 3800 mm	458.86	560.96	7.31	189.10	nr	**750.06**
Sum of two sides 3900 mm	472.06	577.10	7.31	189.10	nr	**766.20**
Sum of two sides 4000 mm	485.26	593.24	7.31	189.10	nr	**782.34**
45° bend						
Sum of two sides 2100 mm	151.21	184.85	7.50	194.01	nr	**378.86**
Sum of two sides 2200 mm	160.62	196.36	7.50	194.01	nr	**390.37**
Sum of two sides 2300 mm	173.13	211.65	7.55	195.31	nr	**406.96**
Sum of two sides 2400 mm	179.04	218.88	8.13	210.31	nr	**429.19**
Sum of two sides 2500 mm	191.53	234.15	8.30	214.71	nr	**448.86**
Sum of two sides 2600 mm	203.76	249.10	8.56	221.43	nr	**470.53**
Sum of two sides 2700 mm	216.00	264.06	8.62	222.98	nr	**487.04**

U: VENTILATION/AIR CONDITIONING SYSTEMS

Item	Net Price £	Material £	Labour hours	Labour £	Unit	Total rate £
Sum of two sides 2800 mm	228.24	279.03	9.09	235.14	nr	**514.17**
Sum of two sides 2900 mm	240.73	294.29	9.09	235.14	nr	**529.43**
Sum of two sides 3000 mm	249.51	305.02	9.62	248.86	nr	**553.88**
Sum of two sides 3100 mm	252.97	309.26	9.62	248.86	nr	**558.12**
Sum of two sides 3200 mm	258.36	315.85	9.62	248.86	nr	**564.71**
Sum of two sides 3300 mm	278.38	340.32	9.62	248.86	nr	**589.18**
Sum of two sides 3400 mm	287.22	351.13	9.62	248.86	nr	**599.99**
Sum of two sides 3500 mm	296.06	361.93	9.62	248.86	nr	**610.79**
Sum of two sides 3600 mm	304.91	372.76	9.62	248.86	nr	**621.62**
Sum of two sides 3700 mm	314.00	383.87	9.62	248.86	nr	**632.73**
Sum of two sides 3800 mm	322.84	394.68	9.62	248.86	nr	**643.54**
Sum of two sides 3900 mm	331.68	405.48	9.62	248.86	nr	**654.34**
Sum of two sides 4000 mm	340.52	416.28	9.62	248.86	nr	**665.14**
90° mitre bend						
Sum of two sides 2100 mm	261.75	319.99	7.50	194.01	nr	**514.00**
Sum of two sides 2200 mm	276.67	338.23	7.50	194.01	nr	**532.24**
Sum of two sides 2300 mm	298.63	365.07	7.55	195.31	nr	**560.38**
Sum of two sides 2400 mm	305.76	373.80	8.13	210.31	nr	**584.11**
Sum of two sides 2500 mm	327.77	400.70	8.30	214.71	nr	**615.41**
Sum of two sides 2600 mm	349.90	427.76	8.56	221.43	nr	**649.19**
Sum of two sides 2700 mm	372.03	454.81	8.62	222.98	nr	**677.79**
Sum of two sides 2800 mm	394.16	481.86	15.20	393.19	nr	**875.05**
Sum of two sides 2900 mm	416.18	508.78	15.20	393.19	nr	**901.97**
Sum of two sides 3000 mm	426.57	521.49	16.04	414.94	nr	**936.43**
Sum of two sides 3100 mm	438.30	535.83	15.60	403.55	nr	**939.38**
Sum of two sides 3200 mm	443.61	542.32	16.04	414.94	nr	**957.26**
Sum of two sides 3300 mm	470.08	574.68	16.04	414.94	nr	**989.62**
Sum of two sides 3400 mm	487.11	595.49	16.04	414.94	nr	**1010.43**
Sum of two sides 3500 mm	504.14	616.31	16.04	414.94	nr	**1031.25**
Sum of two sides 3600 mm	521.17	637.14	16.04	414.94	nr	**1052.08**
Sum of two sides 3700 mm	538.10	657.83	16.04	414.94	nr	**1072.77**
Sum of two sides 3800 mm	555.13	678.65	16.04	414.94	nr	**1093.59**
Sum of two sides 3900 mm	572.16	699.47	16.04	414.94	nr	**1114.41**
Sum of two sides 4000 mm	589.20	720.30	16.04	414.94	nr	**1135.24**
Branch						
Sum of two sides 2100 mm	117.33	143.44	2.61	67.52	nr	**210.96**
Sum of two sides 2200 mm	125.58	153.52	2.61	67.52	nr	**221.04**
Sum of two sides 2300 mm	133.94	163.74	2.61	67.52	nr	**231.26**
Sum of two sides 2400 mm	142.07	173.68	2.88	74.50	nr	**248.18**
Sum of two sides 2500 mm	150.43	183.91	2.88	74.50	nr	**258.41**
Sum of two sides 2600 mm	158.69	194.00	2.88	74.50	nr	**268.50**
Sum of two sides 2700 mm	166.96	204.10	2.88	74.50	nr	**278.60**
Sum of two sides 2800 mm	175.22	214.21	3.94	101.92	nr	**316.13**
Sum of two sides 2900 mm	183.58	224.43	3.94	101.92	nr	**326.35**
Sum of two sides 3000 mm	191.84	234.52	4.83	124.94	nr	**359.46**
Sum of two sides 3100 mm	198.42	242.56	4.83	124.94	nr	**367.50**
Sum of two sides 3200 mm	204.42	249.91	4.83	124.94	nr	**374.85**
Sum of two sides 3300 mm	213.24	260.68	4.83	124.94	nr	**385.62**
Sum of two sides 3400 mm	219.25	268.04	4.83	124.94	nr	**392.98**
Sum of two sides 3500 mm	225.24	275.36	4.83	124.94	nr	**400.30**

U: VENTILATION/AIR CONDITIONING SYSTEMS

Item	Net Price £	Material £	Labour hours	Labour £	Unit	Total rate £
U10: DUCTWORK: RECTANGULAR – CLASS C – cont						
Extra over fittings – cont						
Branch – cont						
Sum of two sides 3600 mm	231.24	282.69	4.83	124.94	nr	**407.63**
Sum of two sides 3700 mm	237.33	290.14	4.83	124.94	nr	**415.08**
Sum of two sides 3800 mm	243.33	297.47	4.83	124.94	nr	**422.41**
Sum of two sides 3900 mm	249.33	304.81	4.83	124.94	nr	**429.75**
Sum of two sides 4000 mm	255.33	312.14	4.83	124.94	nr	**437.08**
Grille neck						
Sum of two sides 2100 mm	124.62	152.35	2.80	72.43	nr	**224.78**
Sum of two sides 2200 mm	133.93	163.73	2.80	72.43	nr	**236.16**
Sum of two sides 2300 mm	143.23	175.10	2.80	72.43	nr	**247.53**
Sum of two sides 2400 mm	152.54	186.48	3.06	79.16	nr	**265.64**
Sum of two sides 2500 mm	161.84	197.85	3.06	79.16	nr	**277.01**
Sum of two sides 2600 mm	171.16	209.24	3.08	79.67	nr	**288.91**
Sum of two sides 2700 mm	166.96	204.10	3.08	79.67	nr	**283.77**
Sum of two sides 2800 mm	175.22	214.21	4.13	106.83	nr	**321.04**
Sum of two sides 2900 mm	183.58	224.43	4.13	106.83	nr	**331.26**
Sum of two sides 3000 mm	208.39	254.76	5.02	129.85	nr	**384.61**
Sum of two sides 3100 mm	215.82	263.84	5.02	129.85	nr	**393.69**
Sum of two sides 3200 mm	222.59	272.12	5.02	129.85	nr	**401.97**
Sum of two sides 3300 mm	229.34	280.37	5.02	129.85	nr	**410.22**
Sum of two sides 3400 mm	236.11	288.65	5.02	129.85	nr	**418.50**
Sum of two sides 3500 mm	242.87	296.91	5.02	129.85	nr	**426.76**
Sum of two sides 3600 mm	249.63	305.18	5.02	129.85	nr	**435.03**
Sum of two sides 3700 mm	256.39	313.44	5.02	129.85	nr	**443.29**
Sum of two sides 3800 mm	263.16	321.71	5.02	129.85	nr	**451.56**
Sum of two sides 3900 mm	269.93	329.99	5.02	129.85	nr	**459.84**
Sum of two sides 4000 mm	276.68	338.24	5.02	129.85	nr	**468.09**
Ductwork 2001 to 2500 mm longest side						
Sum of two sides 2500 mm	168.13	205.54	2.77	71.65	m	**277.19**
Sum of two sides 2600 mm	173.04	211.54	2.97	76.84	m	**288.38**
Sum of two sides 2700 mm	177.42	216.90	2.99	77.35	m	**294.25**
Sum of two sides 2800 mm	182.59	223.22	3.30	85.37	m	**308.59**
Sum of two sides 2900 mm	186.97	228.57	3.31	85.62	m	**314.19**
Sum of two sides 3000 mm	191.22	233.77	3.53	91.31	m	**325.08**
Sum of two sides 3100 mm	195.57	239.08	3.55	91.83	m	**330.91**
Sum of two sides 3200 mm	199.81	244.26	3.56	92.09	m	**336.35**
Sum of two sides 3300 mm	204.19	249.62	3.56	92.09	m	**341.71**
Sum of two sides 3400 mm	208.42	254.79	3.56	92.09	m	**346.88**
Sum of two sides 3500 mm	228.76	279.67	3.56	92.09	m	**371.76**
Sum of two sides 3600 mm	233.00	284.85	3.56	92.09	m	**376.94**
Sum of two sides 3700 mm	237.37	290.18	3.56	92.09	m	**382.27**
Sum of two sides 3800 mm	241.61	295.37	3.56	92.09	m	**387.46**
Sum of two sides 3900 mm	245.85	300.55	3.56	92.09	m	**392.64**
Sum of two sides 4000 mm	250.93	306.76	3.56	92.09	m	**398.85**

U: VENTILATION/AIR CONDITIONING SYSTEMS

Item	Net Price £	Material £	Labour hours	Labour £	Unit	**Total rate £**
Extra over fittings; ductwork 2001 to 2500 mm longest side						
End cap						
Sum of two sides 2500 mm	53.21	65.04	0.87	22.51	nr	**87.55**
Sum of two sides 2600 mm	56.31	68.83	0.87	22.51	nr	**91.34**
Sum of two sides 2700 mm	59.40	72.61	0.87	22.51	nr	**95.12**
Sum of two sides 2800 mm	62.50	76.41	1.16	30.01	nr	**106.42**
Sum of two sides 2900 mm	65.59	80.19	1.16	30.01	nr	**110.20**
Sum of two sides 3000 mm	68.68	83.97	1.73	44.75	nr	**128.72**
Sum of two sides 3100 mm	71.14	86.97	1.73	44.75	nr	**131.72**
Sum of two sides 3200 mm	73.36	89.68	1.73	44.75	nr	**134.43**
Sum of two sides 3300 mm	75.58	92.40	1.73	44.75	nr	**137.15**
Sum of two sides 3400 mm	77.80	95.11	1.73	44.75	nr	**139.86**
Sum of two sides 3500 mm	80.03	97.84	1.73	44.75	nr	**142.59**
Sum of two sides 3600 mm	82.26	100.56	1.73	44.75	nr	**145.31**
Sum of two sides 3700 mm	84.49	103.29	1.73	44.75	nr	**148.04**
Sum of two sides 3800 mm	86.71	106.01	1.73	44.75	nr	**150.76**
Sum of two sides 3900 mm	88.94	108.73	1.73	44.75	nr	**153.48**
Sum of two sides 4000 mm	91.16	111.45	1.73	44.75	nr	**156.20**
Reducer						
Sum of two sides 2500 mm	44.31	54.17	3.12	80.71	nr	**134.88**
Sum of two sides 2600 mm	48.81	59.67	3.16	81.75	nr	**141.42**
Sum of two sides 2700 mm	53.40	65.28	3.16	81.75	nr	**147.03**
Sum of two sides 2800 mm	57.78	70.64	4.00	103.47	nr	**174.11**
Sum of two sides 2900 mm	62.37	76.25	4.01	103.73	nr	**179.98**
Sum of two sides 3000 mm	66.88	81.76	4.56	117.96	nr	**199.72**
Sum of two sides 3100 mm	70.00	85.57	4.56	117.96	nr	**203.53**
Sum of two sides 3200 mm	72.77	88.96	4.56	117.96	nr	**206.92**
Sum of two sides 3300 mm	75.62	92.44	4.56	117.96	nr	**210.40**
Sum of two sides 3400 mm	78.39	95.83	4.56	117.96	nr	**213.79**
Sum of two sides 3500 mm	79.40	97.07	4.56	117.96	nr	**215.03**
Sum of two sides 3600 mm	79.68	97.41	4.56	117.96	nr	**215.37**
Sum of two sides 3700 mm	80.77	98.75	4.56	117.96	nr	**216.71**
Sum of two sides 3800 mm	83.52	102.10	4.56	117.96	nr	**220.06**
Sum of two sides 3900 mm	86.29	105.49	4.56	117.96	nr	**223.45**
Sum of two sides 4000 mm	89.14	108.97	4.56	117.96	nr	**226.93**
Offset						
Sum of two sides 2500 mm	147.50	180.32	3.48	90.02	nr	**270.34**
Sum of two sides 2600 mm	156.95	191.87	3.49	90.28	nr	**282.15**
Sum of two sides 2700 mm	156.76	191.64	3.50	90.54	nr	**282.18**
Sum of two sides 2800 mm	175.82	214.94	4.33	112.00	nr	**326.94**
Sum of two sides 2900 mm	168.56	206.06	4.74	122.62	nr	**328.68**
Sum of two sides 3000 mm	157.42	192.45	5.31	137.36	nr	**329.81**
Sum of two sides 3100 mm	157.44	192.48	5.34	138.13	nr	**330.61**
Sum of two sides 3200 mm	157.40	192.42	5.35	138.40	nr	**330.82**
Sum of two sides 3300 mm	157.93	193.07	5.35	138.40	nr	**331.47**
Sum of two sides 3400 mm	158.13	193.31	5.35	138.40	nr	**331.71**
Sum of two sides 3500 mm	158.56	193.84	5.35	138.40	nr	**332.24**

U: VENTILATION/AIR CONDITIONING SYSTEMS

Item	Net Price £	Material £	Labour hours	Labour £	Unit	**Total rate £**
U10: DUCTWORK: RECTANGULAR – CLASS C – cont						
Extra over fittings – cont						
Offset – cont						
Sum of two sides 3600 mm	158.96	194.33	5.35	138.40	nr	**332.73**
Sum of two sides 3700 mm	159.26	194.70	5.35	138.40	nr	**333.10**
Sum of two sides 3800 mm	164.73	201.39	5.35	138.40	nr	**339.79**
Sum of two sides 3900 mm	170.21	208.08	5.35	138.40	nr	**346.48**
Sum of two sides 4000 mm	175.71	214.81	5.35	138.40	nr	**353.21**
Square to round						
Sum of two sides 2500 mm	75.66	92.49	4.79	123.91	nr	**216.40**
Sum of two sides 2600 mm	82.69	101.08	4.95	128.05	nr	**229.13**
Sum of two sides 2700 mm	89.73	109.70	4.95	128.05	nr	**237.75**
Sum of two sides 2800 mm	96.78	118.32	8.49	219.63	nr	**337.95**
Sum of two sides 2900 mm	103.83	126.94	8.88	229.71	nr	**356.65**
Sum of two sides 3000 mm	110.86	135.53	9.02	233.33	nr	**368.86**
Sum of two sides 3100 mm	115.90	141.69	9.02	233.33	nr	**375.02**
Sum of two sides 3200 mm	120.33	147.10	9.09	235.14	nr	**382.24**
Sum of two sides 3300 mm	121.65	148.72	9.09	235.14	nr	**383.86**
Sum of two sides 3400 mm	124.76	152.52	9.09	235.14	nr	**387.66**
Sum of two sides 3500 mm	129.22	157.97	9.09	235.14	nr	**393.11**
Sum of two sides 3600 mm	130.05	158.99	9.09	235.14	nr	**394.13**
Sum of two sides 3700 mm	133.37	163.05	9.09	235.14	nr	**398.19**
Sum of two sides 3800 mm	137.81	168.48	9.09	235.14	nr	**403.62**
Sum of two sides 3900 mm	142.26	173.92	9.09	235.14	nr	**409.06**
Sum of two sides 4000 mm	146.68	179.31	9.09	235.14	nr	**414.45**
90° radius bend						
Sum of two sides 2500 mm	200.13	244.66	4.35	112.53	nr	**357.19**
Sum of two sides 2600 mm	205.88	251.69	4.53	117.17	nr	**368.86**
Sum of two sides 2700 mm	223.25	272.93	4.53	117.17	nr	**390.10**
Sum of two sides 2800 mm	216.01	264.07	7.13	184.44	nr	**448.51**
Sum of two sides 2900 mm	233.07	284.93	7.17	185.48	nr	**470.41**
Sum of two sides 3000 mm	249.83	305.42	7.26	187.80	nr	**493.22**
Sum of two sides 3100 mm	262.50	320.91	7.26	187.80	nr	**508.71**
Sum of two sides 3200 mm	274.05	335.03	7.31	189.10	nr	**524.13**
Sum of two sides 3300 mm	277.87	339.69	7.31	189.10	nr	**528.79**
Sum of two sides 3400 mm	285.91	349.53	7.31	189.10	nr	**538.63**
Sum of two sides 3500 mm	289.43	353.83	7.31	189.10	nr	**542.93**
Sum of two sides 3600 mm	297.47	363.66	7.31	189.10	nr	**552.76**
Sum of two sides 3700 mm	318.34	389.18	7.31	189.10	nr	**578.28**
Sum of two sides 3800 mm	329.90	403.31	7.31	189.10	nr	**592.41**
Sum of two sides 3900 mm	338.10	413.33	7.31	189.10	nr	**602.43**
Sum of two sides 4000 mm	341.45	417.43	7.31	189.10	nr	**606.53**

U: VENTILATION/AIR CONDITIONING SYSTEMS

Item	Net Price £	Material £	Labour hours	Labour £	Unit	**Total rate £**
45° bend						
Sum of two sides 2500 mm	163.85	200.30	8.30	214.71	nr	**415.01**
Sum of two sides 2600 mm	169.98	207.80	8.56	221.43	nr	**429.23**
Sum of two sides 2700 mm	182.11	222.63	8.62	222.98	nr	**445.61**
Sum of two sides 2800 mm	181.56	221.95	9.09	235.14	nr	**457.09**
Sum of two sides 2900 mm	193.53	236.59	9.09	235.14	nr	**471.73**
Sum of two sides 3000 mm	205.27	250.94	9.62	248.86	nr	**499.80**
Sum of two sides 3100 mm	214.31	261.99	9.62	248.86	nr	**510.85**
Sum of two sides 3200 mm	222.57	272.10	9.62	248.86	nr	**520.96**
Sum of two sides 3300 mm	231.08	282.50	9.62	248.86	nr	**531.36**
Sum of two sides 3400 mm	231.56	283.09	9.62	248.86	nr	**531.95**
Sum of two sides 3500 mm	239.35	292.60	9.62	248.86	nr	**541.46**
Sum of two sides 3600 mm	239.83	293.19	9.62	248.86	nr	**542.05**
Sum of two sides 3700 mm	259.71	317.50	9.62	248.86	nr	**566.36**
Sum of two sides 3800 mm	267.97	327.59	9.62	248.86	nr	**576.45**
Sum of two sides 3900 mm	276.24	337.71	9.62	248.86	nr	**586.57**
Sum of two sides 4000 mm	277.05	338.69	9.62	248.86	nr	**587.55**
90° mitre bend						
Sum of two sides 2500 mm	214.17	261.83	8.30	214.71	nr	**476.54**
Sum of two sides 2600 mm	224.45	274.39	8.56	221.43	nr	**495.82**
Sum of two sides 2700 mm	244.07	298.38	8.62	222.98	nr	**521.36**
Sum of two sides 2800 mm	232.93	284.75	15.20	393.19	nr	**677.94**
Sum of two sides 2900 mm	252.27	308.40	15.20	393.19	nr	**701.59**
Sum of two sides 3000 mm	271.74	332.21	15.20	393.19	nr	**725.40**
Sum of two sides 3100 mm	287.07	350.94	16.04	414.94	nr	**765.88**
Sum of two sides 3200 mm	301.34	368.39	16.04	414.94	nr	**783.33**
Sum of two sides 3300 mm	315.47	385.66	16.04	414.94	nr	**800.60**
Sum of two sides 3400 mm	329.74	403.11	16.04	414.94	nr	**818.05**
Sum of two sides 3500 mm	331.15	404.83	16.04	414.94	nr	**819.77**
Sum of two sides 3600 mm	331.30	405.02	16.04	414.94	nr	**819.96**
Sum of two sides 3700 mm	343.73	420.21	16.04	414.94	nr	**835.15**
Sum of two sides 3800 mm	358.00	437.66	16.04	414.94	nr	**852.60**
Sum of two sides 3900 mm	372.27	455.11	16.04	414.94	nr	**870.05**
Sum of two sides 4000 mm	403.65	493.46	16.04	414.94	nr	**908.40**
Branch						
Sum of two sides 2500 mm	161.84	197.85	2.88	74.50	nr	**272.35**
Sum of two sides 2600 mm	170.54	208.48	2.88	74.50	nr	**282.98**
Sum of two sides 2700 mm	179.36	219.27	2.88	74.50	nr	**293.77**
Sum of two sides 2800 mm	187.96	229.78	3.94	101.92	nr	**331.70**
Sum of two sides 2900 mm	196.77	240.56	3.94	101.92	nr	**342.48**
Sum of two sides 3000 mm	205.50	251.23	4.83	124.94	nr	**376.17**
Sum of two sides 3100 mm	212.50	259.79	4.83	124.94	nr	**384.73**
Sum of two sides 3200 mm	218.91	267.61	4.83	124.94	nr	**392.55**
Sum of two sides 3300 mm	225.41	275.57	4.83	124.94	nr	**400.51**
Sum of two sides 3400 mm	231.83	283.41	4.83	124.94	nr	**408.35**
Sum of two sides 3500 mm	238.52	291.59	4.83	124.94	nr	**416.53**
Sum of two sides 3600 mm	244.94	299.44	4.83	124.94	nr	**424.38**
Sum of two sides 3700 mm	254.28	310.85	4.83	124.94	nr	**435.79**
Sum of two sides 3800 mm	260.69	318.69	4.83	124.94	nr	**443.63**
Sum of two sides 3900 mm	267.10	326.53	4.83	124.94	nr	**451.47**
Sum of two sides 4000 mm	273.60	334.48	4.83	124.94	nr	**459.42**

U: VENTILATION/AIR CONDITIONING SYSTEMS

Item	Net Price £	Material £	Labour hours	Labour £	Unit	Total rate £
U10: DUCTWORK: RECTANGULAR – CLASS C – cont						
Extra over fittings – cont						
Grille neck						
Sum of two sides 2500 mm	165.29	202.07	3.06	79.16	nr	**281.23**
Sum of two sides 2600 mm	174.80	213.69	3.08	79.67	nr	**293.36**
Sum of two sides 2700 mm	184.31	225.32	3.08	79.67	nr	**304.99**
Sum of two sides 2800 mm	193.82	236.94	4.13	106.83	nr	**343.77**
Sum of two sides 2900 mm	203.33	248.57	4.13	106.83	nr	**355.40**
Sum of two sides 3000 mm	212.83	260.19	5.02	129.85	nr	**390.04**
Sum of two sides 3100 mm	220.41	269.45	5.02	129.85	nr	**399.30**
Sum of two sides 3200 mm	227.33	277.91	5.02	129.85	nr	**407.76**
Sum of two sides 3300 mm	234.22	286.34	5.02	129.85	nr	**416.19**
Sum of two sides 3400 mm	241.13	294.79	5.02	129.85	nr	**424.64**
Sum of two sides 3500 mm	248.03	303.22	5.02	129.85	nr	**433.07**
Sum of two sides 3600 mm	254.94	311.67	5.02	129.85	nr	**441.52**
Sum of two sides 3700 mm	261.85	320.11	5.02	129.85	nr	**449.96**
Sum of two sides 3800 mm	268.76	328.56	5.02	129.85	nr	**458.41**
Sum of two sides 3900 mm	275.67	337.01	5.02	129.85	nr	**466.86**
Sum of two sides 4000 mm	282.56	345.43	5.02	129.85	nr	**475.28**
Y30 – DUCTWORK ANCILLARIES: ACCESS DOORS						
Refer to ancillaries in U10: DUCTWORK: RECTANGULAR: CLASS B for details of access doors						

U: VENTILATION/AIR CONDITIONING SYSTEMS

Item	Net Price £	Material £	Labour hours	Labour £	Unit	**Total rate £**
U10: DUCTWORK: VOLUME/FIRE DAMPERS						
Y30 – DUCTWORK ANCILLARIES: VOLUME CONTROL AND FIRE DAMPERS						
Volume control damper; opposed blade; galvanized steel casing; aluminium aerofoil blades; manually operated						
Rectangular						
Sum of two sides 200 mm	18.77	22.95	1.60	41.40	nr	**64.35**
Sum of two sides 300 mm	20.08	24.54	1.60	41.40	nr	**65.94**
Sum of two sides 400 mm	21.96	26.84	1.60	41.40	nr	**68.24**
Sum of two sides 500 mm	24.03	29.38	1.60	41.40	nr	**70.78**
Sum of two sides 600 mm	26.56	32.47	1.70	44.00	nr	**76.47**
Sum of two sides 700 mm	29.19	35.69	2.10	54.34	nr	**90.03**
Sum of two sides 800 mm	31.90	39.00	2.15	55.61	nr	**94.61**
Sum of two sides 900 mm	35.00	42.79	2.30	59.50	nr	**102.29**
Sum of two sides 1000 mm	38.01	46.46	2.40	62.08	nr	**108.54**
Sum of two sides 1100 mm	41.40	50.61	2.60	67.26	nr	**117.87**
Sum of two sides 1200 mm	46.83	57.25	2.80	72.43	nr	**129.68**
Sum of two sides 1300 mm	50.49	61.73	3.10	80.20	nr	**141.93**
Sum of two sides 1400 mm	54.44	66.55	3.25	84.07	nr	**150.62**
Sum of two sides 1500 mm	58.93	72.04	3.40	87.95	nr	**159.99**
Sum of two sides 1600 mm	63.35	77.45	3.45	89.24	nr	**166.69**
Sum of two sides 1700 mm	67.57	82.61	3.60	93.13	nr	**175.74**
Sum of two sides 1800 mm	72.63	88.79	3.90	100.88	nr	**189.67**
Sum of two sides 1900 mm	77.33	94.53	4.20	108.70	nr	**203.23**
Sum of two sides 2000 mm	82.96	101.42	4.33	112.00	nr	**213.42**
Circular						
100 mm dia.	25.06	30.63	0.80	20.69	nr	**51.32**
150 mm dia.	28.62	34.99	0.80	20.69	nr	**55.68**
160 mm dia.	29.84	36.48	0.90	23.28	nr	**59.76**
200 mm dia.	32.37	39.57	1.05	27.17	nr	**66.74**
250 mm dia.	36.04	44.06	1.20	31.05	nr	**75.11**
399 mm dia.	39.89	48.77	1.35	34.93	nr	**83.70**
315 mm dia.	41.67	50.94	1.35	34.97	nr	**85.91**
350 mm dia.	43.83	53.58	1.65	42.69	nr	**96.27**
400 mm dia.	47.86	58.51	1.90	49.18	nr	**107.69**
450 mm dia.	51.71	63.21	2.10	54.34	nr	**117.55**
500 mm dia.	56.32	68.86	2.95	76.31	nr	**145.17**
550 mm dia.	60.82	74.36	2.94	76.06	nr	**150.42**
600 mm dia.	65.88	80.54	2.94	76.06	nr	**156.60**
650 mm dia.	71.14	86.97	4.55	117.70	nr	**204.67**
700 mm dia.	76.87	93.98	5.20	134.51	nr	**228.49**
750 mm dia.	82.96	101.42	5.20	134.51	nr	**235.93**
800 mm dia.	89.26	109.12	5.80	150.04	nr	**259.16**
850 mm dia.	95.82	117.14	5.80	150.04	nr	**267.18**
900 mm dia.	102.67	125.52	6.40	165.55	nr	**291.07**
950 mm dia.	109.71	134.12	6.40	165.55	nr	**299.67**
1000 mm dia.	116.85	142.85	7.00	181.07	nr	**323.92**

U: VENTILATION/AIR CONDITIONING SYSTEMS

Item	Net Price £	Material £	Labour hours	Labour £	Unit	Total rate £
U10: DUCTWORK: VOLUME/FIRE DAMPERS – cont						
Volume control damper – cont						
Flat oval						
345 × 102 mm	42.33	51.75	1.20	31.04	nr	**82.79**
508 × 102 mm	47.68	58.29	1.60	41.40	nr	**99.69**
559 × 152 mm	53.87	65.86	1.90	49.18	nr	**115.04**
531 × 203 mm	58.85	71.95	1.90	49.18	nr	**121.13**
851 × 203 mm	62.23	76.08	4.55	117.70	nr	**193.78**
582 × 254 mm	65.33	79.87	2.10	54.34	nr	**134.21**
823 × 254 mm	71.52	87.44	4.10	106.07	nr	**193.51**
632 × 305 mm	71.98	87.99	2.95	76.31	nr	**164.30**
765 × 356 mm	80.53	98.45	4.55	117.70	nr	**216.15**
737 × 406 mm	85.78	104.86	4.55	117.70	nr	**222.56**
818 × 406 mm	87.57	107.06	5.20	134.51	nr	**241.57**
978 × 406 mm	94.70	115.77	5.50	142.27	nr	**258.04**
709 × 457 mm	89.26	109.12	4.50	116.53	nr	**225.65**
678 × 508 mm	94.42	115.43	4.55	117.70	nr	**233.13**
919 × 508 mm	105.96	129.53	6.00	155.21	nr	**284.74**
Fire damper; galvanized steel casing; stainless steel folding shutter; fusible link with manual reset; BS 476 4 hour fire-rated						
Rectangular						
Sum of two sides 200 mm	34.39	42.04	1.60	41.40	nr	**83.44**
Sum of two sides 300 mm	35.48	43.37	1.60	41.40	nr	**84.77**
Sum of two sides 400 mm	37.00	45.24	1.60	41.40	nr	**86.64**
Sum of two sides 500 mm	40.48	49.49	1.60	41.40	nr	**90.89**
Sum of two sides 600 mm	44.35	54.22	1.70	44.00	nr	**98.22**
Sum of two sides 700 mm	48.31	59.06	2.10	54.34	nr	**113.40**
Sum of two sides 800 mm	52.45	64.12	2.15	55.62	nr	**119.74**
Sum of two sides 900 mm	56.59	69.19	2.30	59.60	nr	**128.79**
Sum of two sides 1000 mm	60.82	74.36	2.40	62.18	nr	**136.54**
Sum of two sides 1100 mm	64.60	78.97	2.60	67.36	nr	**146.33**
Sum of two sides 1200 mm	69.09	84.46	2.80	72.46	nr	**156.92**
Sum of two sides 1300 mm	73.70	90.09	3.10	80.20	nr	**170.29**
Sum of two sides 1400 mm	78.66	96.16	3.25	84.07	nr	**180.23**
Sum of two sides 1500 mm	82.98	101.44	3.40	87.98	nr	**189.42**
Sum of two sides 1600 mm	87.75	107.27	3.45	89.24	nr	**196.51**
Sum of two sides 1700 mm	92.53	113.11	3.60	93.13	nr	**206.24**
Sum of two sides 1800 mm	97.58	119.29	3.90	100.88	nr	**220.17**
Sum of two sides 1900 mm	102.36	125.13	4.20	108.70	nr	**233.83**
Sum of two sides 2000 mm	106.86	130.63	4.33	112.00	nr	**242.63**
Sum of two sides 2100 mm	113.72	139.02	4.43	114.60	nr	**253.62**
Sum of two sides 2200 mm	120.31	147.08	4.55	117.70	nr	**264.78**

U: VENTILATION/AIR CONDITIONING SYSTEMS

Item	Net Price £	Material £	Labour hours	Labour £	Unit	**Total rate £**
Circular						
100 mm dia.	41.89	51.21	0.80	20.69	nr	**71.90**
150 mm dia.	44.08	44.08	0.80	16.93	nr	**61.01**
160 mm dia.	44.35	54.22	0.90	23.28	nr	**77.50**
200 mm dia.	46.54	56.90	1.05	27.16	nr	**84.06**
250 mm dia.	51.87	63.41	1.20	31.05	nr	**94.46**
300 mm dia.	57.65	70.47	1.35	34.93	nr	**105.40**
315 mm dia.	60.15	73.53	1.35	34.97	nr	**108.50**
350 mm dia.	63.80	77.99	1.65	42.69	nr	**120.68**
400 mm dia.	70.49	86.17	1.90	49.18	nr	**135.35**
450 mm dia.	77.09	94.25	2.10	54.34	nr	**148.59**
500 mm dia.	84.33	103.09	2.95	76.31	nr	**179.40**
550 mm dia.	91.57	111.94	2.94	76.06	nr	**188.00**
600 mm dia.	99.26	121.34	4.55	117.70	nr	**239.04**
650 mm dia.	107.41	131.30	4.55	117.70	nr	**249.00**
700 mm dia.	115.75	141.50	5.20	134.51	nr	**276.01**
750 mm dia.	124.53	152.23	5.20	134.51	nr	**286.74**
800 mm dia.	133.49	163.19	5.80	150.04	nr	**313.23**
850 mm dia.	142.93	174.73	5.80	150.04	nr	**324.77**
900 mm dia.	152.63	186.59	6.40	165.55	nr	**352.14**
950 mm dia.	162.59	198.77	6.40	165.55	nr	**364.32**
1000 mm dia.	173.03	211.53	7.00	181.07	nr	**392.60**
Flat oval						
345 × 102 mm	57.61	70.43	1.20	31.05	nr	**101.48**
427 × 102 mm	62.56	76.48	1.35	34.97	nr	**111.45**
508 × 102 mm	64.67	79.06	1.60	41.40	nr	**120.46**
559 × 152 mm	70.51	86.20	1.90	49.18	nr	**135.38**
531 × 203 mm	74.16	90.66	1.90	49.18	nr	**139.84**
851 × 203 mm	93.20	113.94	4.55	117.70	nr	**231.64**
582 × 254 mm	98.34	120.22	2.10	54.34	nr	**174.56**
632 × 305 mm	103.20	126.16	2.95	76.31	nr	**202.47**
765 × 356 mm	118.58	144.96	4.55	117.70	nr	**262.66**
737 × 406 mm	123.70	151.22	4.55	117.70	nr	**268.92**
818 × 406 mm	131.30	160.52	5.20	134.51	nr	**295.03**
978 × 406 mm	142.75	174.51	5.50	142.27	nr	**316.78**
709 × 457 mm	124.63	152.36	4.50	116.53	nr	**268.89**
678 × 508 mm	129.66	158.51	4.55	117.70	nr	**276.21**

U: VENTILATION/AIR CONDITIONING SYSTEMS

Item	Net Price £	Material £	Labour hours	Labour £	Unit	Total rate £
U10: DUCTWORK: VOLUME/FIRE DAMPERS – cont						
Smoke/fire damper; galvanized steel casing; stainless steel folding shutter; fusible link and 24 V d.c. electromagnetic shutter release mechanism; spring operated; BS 476 4 hour fire-rating						
Rectangular						
Sum of two sides 200 mm	263.40	322.01	1.60	41.40	nr	**363.41**
Sum of two sides 300 mm	264.16	322.94	1.60	41.40	nr	**364.34**
Sum of two sides 400 mm	264.92	323.86	1.60	41.40	nr	**365.26**
Sum of two sides 500 mm	271.27	331.63	1.60	41.40	nr	**373.03**
Sum of two sides 600 mm	277.91	339.75	1.70	44.00	nr	**383.75**
Sum of two sides 700 mm	284.84	348.22	2.10	54.34	nr	**402.56**
Sum of two sides 800 mm	291.95	356.92	2.15	55.62	nr	**412.54**
Sum of two sides 900 mm	299.35	365.96	2.30	59.60	nr	**425.56**
Sum of two sides 1000 mm	307.02	375.33	2.40	62.18	nr	**437.51**
Sum of two sides 1100 mm	314.90	384.96	2.60	67.36	nr	**452.32**
Sum of two sides 1200 mm	323.05	394.93	2.80	72.46	nr	**467.39**
Sum of two sides 1300 mm	331.49	405.25	3.10	80.20	nr	**485.45**
Sum of two sides 1400 mm	340.12	415.80	3.25	84.07	nr	**499.87**
Sum of two sides 1500 mm	349.02	426.68	3.40	87.98	nr	**514.66**
Sum of two sides 1600 mm	358.22	437.93	3.45	89.24	nr	**527.17**
Sum of two sides 1700 mm	367.61	449.41	3.60	93.13	nr	**542.54**
Sum of two sides 1800 mm	377.28	461.22	3.90	100.88	nr	**562.10**
Sum of two sides 1900 mm	387.13	473.26	4.20	108.70	nr	**581.96**
Sum of two sides 2000 mm	397.27	485.67	4.33	112.00	nr	**597.67**
Circular						
100 mm dia.	128.49	157.08	0.80	20.69	nr	**177.77**
150 mm dia.	128.49	157.08	0.90	23.28	nr	**180.36**
200 mm dia.	128.49	157.08	1.05	27.16	nr	**184.24**
250 mm dia.	135.46	165.60	1.20	31.05	nr	**196.65**
300 mm dia.	142.72	174.47	1.35	34.97	nr	**209.44**
350 mm dia.	162.43	198.57	1.65	42.69	nr	**241.26**
400 mm dia.	178.62	218.36	1.90	49.18	nr	**267.54**
450 mm dia.	186.27	227.71	2.10	54.34	nr	**282.05**
500 mm dia.	202.86	247.99	2.95	76.31	nr	**324.30**
550 mm dia.	220.00	220.00	2.95	62.42	nr	**282.42**
600 mm dia.	230.37	281.63	4.55	117.70	nr	**399.33**
650 mm dia.	249.26	249.26	4.55	96.28	nr	**345.54**
700 mm dia.	269.81	329.85	5.20	134.51	nr	**464.36**
750 mm dia.	282.58	282.58	5.20	110.03	nr	**392.61**
800 mm dia.	307.17	375.52	5.80	150.04	nr	**525.56**
850 mm dia.	329.31	329.31	5.80	122.73	nr	**452.04**
900 mm dia.	341.71	417.75	6.40	165.55	nr	**583.30**
950 mm dia.	364.66	364.66	6.40	135.42	nr	**500.08**
1000 mm dia.	381.12	465.92	7.00	181.07	nr	**646.99**

U: VENTILATION/AIR CONDITIONING SYSTEMS

Item	Net Price £	Material £	Labour hours	Labour £	Unit	**Total rate £**
Flat oval						
531 × 203 mm	227.95	278.67	1.90	49.18	nr	**327.85**
851 × 203 mm	254.46	311.08	4.55	117.70	nr	**428.78**
582 × 254 mm	248.13	303.35	2.10	54.34	nr	**357.69**
632 × 305 mm	264.52	323.38	2.95	76.31	nr	**399.69**
765 × 356 mm	285.37	348.87	4.55	117.70	nr	**466.57**
737 × 406 mm	298.25	364.61	4.55	117.70	nr	**482.31**
818 × 406 mm	302.43	369.72	5.20	134.51	nr	**504.23**
978 × 406 mm	319.47	390.56	5.50	142.27	nr	**532.83**
709 × 457 mm	306.80	375.06	4.50	116.53	nr	**491.59**
678 × 508 mm	319.39	390.45	4.55	117.70	nr	**508.15**

U: VENTILATION/AIR CONDITIONING SYSTEMS

Item	Net Price £	Material £	Labour hours	Labour £	Unit	Total rate £
U10: GRILLES/DIFFUSERS/LOUVRES						
Y46 – GRILLES/DIFFUSERS/LOUVRES						
Supply grilles; single deflection; extruded aluminium alloy frame and adjustable horizontal vanes; silver grey polyester powder coated; screw fixed						
Rectangular; for duct, ceiling and sidewall applications						
150 × 100 mm	10.41	11.76	0.60	14.34	nr	**26.10**
150 × 150 mm	12.73	14.38	0.60	14.34	nr	**28.72**
200 × 150 mm	15.13	17.09	0.65	15.52	nr	**32.61**
200 × 200 mm	13.09	14.78	0.72	17.21	nr	**31.99**
300 × 100 mm	11.24	12.70	0.72	17.21	nr	**29.91**
300 × 150 mm	13.03	14.72	0.80	19.13	nr	**33.85**
300 × 200 mm	14.81	16.73	0.88	21.03	nr	**37.76**
300 × 300 mm	18.38	20.76	1.04	24.86	nr	**45.62**
400 × 100 mm	12.58	14.21	0.88	21.03	nr	**35.24**
400 × 150 mm	14.55	16.43	0.94	22.46	nr	**38.89**
400 × 200 mm	16.53	18.67	1.04	24.86	nr	**43.53**
400 × 300 mm	20.48	23.13	1.12	26.76	nr	**49.89**
600 × 200 mm	25.21	28.47	1.26	30.11	nr	**58.58**
600 × 300 mm	40.78	46.05	1.40	33.46	nr	**79.51**
600 × 400 mm	48.58	54.87	1.61	38.47	nr	**93.34**
600 × 500 mm	56.39	63.69	1.76	42.05	nr	**105.74**
600 × 600 mm	64.20	72.51	2.17	51.86	nr	**124.37**
800 × 300 mm	47.74	53.92	1.76	42.05	nr	**95.97**
800 × 400 mm	56.83	64.19	2.17	51.86	nr	**116.05**
800 × 600 mm	75.01	84.71	3.00	71.70	nr	**156.41**
1000 × 300 mm	54.71	61.79	2.60	62.14	nr	**123.93**
1000 × 400 mm	65.08	73.50	3.00	71.70	nr	**145.20**
1000 × 600 mm	85.83	96.94	3.80	90.81	nr	**187.75**
1000 × 800 mm	106.57	120.36	3.80	90.81	nr	**211.17**
1200 × 600 mm	96.64	109.14	4.61	110.17	nr	**219.31**
1200 × 800 mm	119.95	135.47	4.61	110.17	nr	**245.64**
1200 × 1000 mm	143.26	161.79	4.61	110.17	nr	**271.96**
Rectangular; for duct, ceiling and sidewall applications; including opposed blade damper volume regulator						
150 × 100 mm	19.40	21.91	0.72	17.23	nr	**39.14**
150 × 150 mm	22.48	25.39	0.72	17.23	nr	**42.62**
200 × 150 mm	24.13	27.25	0.83	19.86	nr	**47.11**
200 × 200 mm	26.08	29.46	0.90	21.51	nr	**50.97**
300 × 100 mm	27.71	31.29	0.90	21.51	nr	**52.80**
300 × 150 mm	29.54	33.36	0.98	23.42	nr	**56.78**
300 × 200 mm	31.82	35.93	1.06	25.35	nr	**61.28**
300 × 300 mm	36.37	41.08	1.20	28.68	nr	**69.76**
400 × 100 mm	36.01	40.67	1.06	25.35	nr	**66.02**
400 × 150 mm	38.02	42.94	1.13	27.00	nr	**69.94**

U: VENTILATION/AIR CONDITIONING SYSTEMS

Item	Net Price £	Material £	Labour hours	Labour £	Unit	Total rate £
400 × 200 mm	38.20	43.15	1.20	28.68	nr	**71.83**
400 × 300 mm	43.41	49.03	1.34	32.03	nr	**81.06**
600 × 200 mm	52.90	59.74	1.50	35.88	nr	**95.62**
600 × 300 mm	70.27	79.37	1.66	39.70	nr	**119.07**
600 × 400 mm	79.89	90.23	1.80	43.06	nr	**133.29**
600 × 500 mm	90.47	102.17	2.00	47.80	nr	**149.97**
600 × 600 mm	100.09	113.04	2.60	62.23	nr	**175.27**
800 × 300 mm	96.38	108.85	2.00	47.80	nr	**156.65**
800 × 400 mm	108.07	122.05	2.60	62.23	nr	**184.28**
800 × 600 mm	132.76	149.94	3.61	86.27	nr	**236.21**
1000 × 300 mm	112.15	126.66	3.00	71.76	nr	**198.42**
1000 × 400 mm	125.65	141.91	3.61	86.27	nr	**228.18**
1000 × 600 mm	154.29	174.26	4.61	110.13	nr	**284.39**
1000 × 800 mm	228.19	257.71	4.61	110.13	nr	**367.84**
1200 × 600 mm	171.40	193.58	5.62	134.26	nr	**327.84**
1200 × 800 mm	251.50	284.04	6.10	145.78	nr	**429.82**
1200 × 1000 mm	285.97	322.97	6.50	155.34	nr	**478.31**
Supply grilles; double deflection; extruded aluminium alloy frame and adjustable horizontal and vertical vanes; white polyester powder coated; screw fixed						
Rectangular; for duct, ceiling and sidewall applications						
150 × 100 mm	10.41	11.76	0.88	21.03	nr	**32.79**
150 × 150 mm	13.43	15.17	0.88	21.03	nr	**36.20**
200 × 150 mm	15.13	17.09	1.08	25.81	nr	**42.90**
200 × 200 mm	16.86	19.04	1.25	29.88	nr	**48.92**
300 × 100 mm	16.42	18.54	1.25	29.88	nr	**48.42**
300 × 150 mm	18.54	20.94	1.50	35.84	nr	**56.78**
300 × 200 mm	20.67	23.35	1.75	41.82	nr	**65.17**
300 × 300 mm	24.93	28.15	2.15	51.38	nr	**79.53**
400 × 100 mm	19.42	21.93	1.75	41.82	nr	**63.75**
400 × 150 mm	21.95	24.79	1.95	46.60	nr	**71.39**
400 × 200 mm	24.49	27.66	2.15	51.38	nr	**79.04**
400 × 300 mm	29.56	33.38	2.55	60.95	nr	**94.33**
600 × 200 mm	54.60	61.67	3.01	71.93	nr	**133.60**
600 × 300 mm	65.98	74.52	3.36	80.30	nr	**154.82**
600 × 400 mm	77.36	87.37	3.80	90.81	nr	**178.18**
600 × 500 mm	88.74	100.23	4.20	100.37	nr	**200.60**
600 × 600 mm	100.12	113.08	4.51	107.77	nr	**220.85**
800 × 300 mm	81.71	92.28	4.20	100.37	nr	**192.65**
800 × 400 mm	95.85	108.25	4.51	107.77	nr	**216.02**
800 × 600 mm	124.13	140.19	5.10	121.88	nr	**262.07**
1000 × 300 mm	97.44	110.05	4.80	114.71	nr	**224.76**
1000 × 400 mm	114.34	129.13	5.10	121.88	nr	**251.01**
1000 × 600 mm	148.14	167.31	5.72	136.70	nr	**304.01**
1000 × 800 mm	181.95	205.49	6.33	151.28	nr	**356.77**
1200 × 600 mm	172.15	194.42	5.72	136.70	nr	**331.12**
1200 × 800 mm	211.48	238.84	6.33	151.28	nr	**390.12**
1200 × 1000 mm	250.80	283.25	6.33	151.28	nr	**434.53**

U: VENTILATION/AIR CONDITIONING SYSTEMS

Item	Net Price £	Material £	Labour hours	Labour £	Unit	**Total rate £**
U10: GRILLES/DIFFUSERS/LOUVRES – cont						
Supply grilles – cont						
Rectangular; for duct, ceiling and sidewall applications; including opposed blade damper volume regulator						
150 × 100 mm	21.23	23.98	1.00	23.90	nr	**47.88**
150 × 150 mm	25.18	28.44	1.00	23.90	nr	**52.34**
200 × 150 mm	27.76	31.35	1.26	30.11	nr	**61.46**
200 × 200 mm	29.84	33.70	1.43	34.18	nr	**67.88**
300 × 100 mm	32.89	37.15	1.43	34.18	nr	**71.33**
300 × 150 mm	35.06	39.60	1.68	40.15	nr	**79.75**
300 × 200 mm	37.67	42.54	1.93	46.12	nr	**88.66**
300 × 300 mm	42.93	48.48	2.31	55.20	nr	**103.68**
400 × 100 mm	42.85	48.40	1.93	46.12	nr	**94.52**
400 × 150 mm	45.42	51.30	2.14	51.14	nr	**102.44**
400 × 200 mm	46.15	52.12	2.31	55.20	nr	**107.32**
400 × 300 mm	52.49	59.28	2.77	66.20	nr	**125.48**
600 × 200 mm	82.29	92.94	3.25	77.67	nr	**170.61**
600 × 300 mm	95.48	107.83	3.62	86.52	nr	**194.35**
600 × 400 mm	108.68	122.75	3.99	95.36	nr	**218.11**
600 × 500 mm	122.82	138.71	4.44	106.11	nr	**244.82**
600 × 600 mm	136.01	153.61	4.94	118.05	nr	**271.66**
800 × 300 mm	130.35	147.22	4.44	106.11	nr	**253.33**
800 × 400 mm	147.08	166.11	4.94	118.05	nr	**284.16**
800 × 600 mm	181.87	205.41	5.71	136.46	nr	**341.87**
1000 × 300 mm	154.88	174.92	5.20	124.27	nr	**299.19**
1000 × 400 mm	174.91	197.54	5.71	136.46	nr	**334.00**
1000 × 600 mm	216.60	244.63	6.53	156.05	nr	**400.68**
1000 × 800 mm	303.57	342.85	6.53	156.05	nr	**498.90**
1200 × 600 mm	246.90	278.85	7.34	175.41	nr	**454.26**
1200 × 800 mm	343.03	387.42	8.80	210.30	nr	**597.72**
1200 × 1000 mm	393.51	444.42	8.80	210.30	nr	**654.72**
Floor grille suitable for mounting in raised access floors; heavy duty; extruded alumiinium; standard mill finish; complete with opposed blade volume control damper						
Diffuser						
600 mm × 600 mm	128.49	145.12	0.70	16.73	nr	**161.85**
Extra for nylon coated black finish	19.66	22.21	–	–	nr	**22.21**

U: VENTILATION/AIR CONDITIONING SYSTEMS

Item	Net Price £	Material £	Labour hours	Labour £	Unit	**Total rate £**
Exhaust grilles; aluminium						
0° fixed blade core						
150 × 150 mm	13.30	15.02	0.60	14.39	nr	**29.41**
200 × 200 mm	15.65	17.68	0.72	17.23	nr	**34.91**
250 × 250 mm	18.22	20.57	0.80	19.13	nr	**39.70**
300 × 300 mm	21.01	23.73	1.00	23.90	nr	**47.63**
350 × 350 mm	24.01	27.12	1.20	28.67	nr	**55.79**
0° fixed blade core; including opposed blade damper volume regulator						
150 × 150 mm	25.05	28.29	0.62	14.83	nr	**43.12**
200 × 200 mm	28.64	32.34	0.72	17.23	nr	**49.57**
250 × 250 mm	32.59	36.81	0.80	19.13	nr	**55.94**
300 × 300 mm	39.00	44.04	1.00	23.90	nr	**67.94**
350 × 350 mm	43.66	49.31	1.20	28.67	nr	**77.98**
45° fixed blade core						
150 × 150 mm	13.30	15.02	0.62	14.83	nr	**29.85**
200 × 200 mm	15.65	17.68	0.72	17.23	nr	**34.91**
250 × 250 mm	18.22	20.57	0.80	19.13	nr	**39.70**
300 × 300 mm	21.01	23.73	1.00	23.90	nr	**47.63**
350 × 350 mm	24.01	27.12	1.20	28.67	nr	**55.79**
45° fixed blade core; including opposed blade damper volume regulator						
150 × 150 mm	25.05	28.29	0.62	14.83	nr	**43.12**
200 × 200 mm	28.64	32.34	0.72	17.23	nr	**49.57**
250 × 250 mm	32.59	36.81	0.80	19.13	nr	**55.94**
300 × 300 mm	39.00	44.04	1.00	23.90	nr	**67.94**
350 × 350 mm	43.66	49.31	1.20	28.67	nr	**77.98**
Eggcrate core						
150 × 150 mm	8.93	10.09	0.62	14.83	nr	**24.92**
200 × 200 mm	11.03	12.45	1.00	23.90	nr	**36.35**
250 × 250 mm	13.43	15.17	0.80	19.13	nr	**34.30**
300 × 300 mm	17.10	19.31	1.00	23.90	nr	**43.21**
350 × 350 mm	19.11	21.58	1.20	28.67	nr	**50.25**
Eggcrate core; including opposed blade damper volume regulator						
150 × 150 mm	20.67	23.35	0.62	14.83	nr	**38.18**
200 × 200 mm	24.01	27.12	0.72	17.23	nr	**44.35**
250 × 250 mm	27.79	31.39	0.80	19.13	nr	**50.52**
300 × 300 mm	35.10	39.64	1.00	23.90	nr	**63.54**
350 × 350 mm	38.76	43.78	1.20	28.67	nr	**72.45**

U: VENTILATION/AIR CONDITIONING SYSTEMS

Item	Net Price £	Material £	Labour hours	Labour £	Unit	Total rate £
U10: GRILLES/DIFFUSERS/LOUVRES – cont						
Exhaust grilles – cont						
Mesh/perforated plate core						
150 × 150 mm	9.82	11.09	0.62	14.82	nr	**25.91**
200 × 200 mm	12.14	13.71	0.72	17.21	nr	**30.92**
250 × 250 mm	14.78	16.70	0.80	19.13	nr	**35.83**
300 × 300 mm	18.82	21.26	1.00	23.90	nr	**45.16**
350 × 350 mm	21.04	23.77	1.20	28.67	nr	**52.44**
Mesh/perforated plate core; including opposed blade damper volume regulator						
150 × 150 mm	21.57	24.36	0.62	14.82	nr	**39.18**
200 × 200 mm	25.12	28.37	0.72	17.21	nr	**45.58**
250 × 250 mm	29.15	32.93	0.80	19.13	nr	**52.06**
300 × 300 mm	36.82	41.59	0.80	19.13	nr	**60.72**
350 × 350 mm	40.68	45.94	1.20	28.67	nr	**74.61**
Plastic air diffusion system						
Eggcrate grilles						
150 × 150 mm	4.00	4.52	0.62	15.33	nr	**19.85**
200 × 200 mm	7.00	7.91	0.72	17.80	nr	**25.71**
250 × 250 mm	8.00	9.04	0.80	19.77	nr	**28.81**
300 × 300 mm	8.00	9.04	1.00	24.71	nr	**33.75**
Single deflection grilles						
150 × 150 mm	6.30	7.12	0.62	15.33	nr	**22.45**
200 × 200 mm	7.80	8.81	0.72	17.80	nr	**26.61**
250 × 250 mm	8.60	9.71	0.80	19.77	nr	**29.48**
300 × 300 mm	10.80	12.20	1.00	24.71	nr	**36.91**
Double deflection grilles						
150 × 150 mm	7.85	8.86	0.62	15.33	nr	**24.19**
200 × 200 mm	10.42	11.77	0.72	17.80	nr	**29.57**
250 × 250 mm	11.90	13.44	0.80	19.77	nr	**33.21**
300 × 300 mm	16.80	18.97	1.00	24.71	nr	**43.68**
Door transfer grilles						
150 × 150 mm	11.80	13.33	0.62	15.33	nr	**28.66**
200 × 200 mm	13.70	15.47	0.72	17.80	nr	**33.27**
250 × 250 mm	14.20	16.03	0.80	19.77	nr	**35.80**
300 × 300 mm	15.40	17.39	1.00	24.71	nr	**42.10**
Opposed blade dampers						
150 × 150 mm	7.15	8.08	0.62	15.33	nr	**23.41**
200 × 200 mm	8.40	9.49	0.72	17.80	nr	**27.29**
250 × 250 mm	12.60	14.23	0.80	19.77	nr	**34.00**
300 × 300 mm	13.70	15.47	1.00	24.71	nr	**40.18**

U: VENTILATION/AIR CONDITIONING SYSTEMS

Item	Net Price £	Material £	Labour hours	Labour £	Unit	**Total rate £**
Ceiling mounted diffusers; circular aluminium multi-cone diffuser						
Circular; for ceiling mounting						
141 mm dia. neck	45.61	51.51	0.80	19.13	nr	**70.64**
197 mm dia. neck	66.77	75.41	1.10	26.30	nr	**101.71**
309 mm dia. neck	90.45	102.15	1.40	33.48	nr	**135.63**
365 mm dia. neck	77.38	87.39	1.50	35.88	nr	**123.27**
457 mm dia. neck	122.00	137.78	2.00	47.80	nr	**185.58**
Circular; for ceiling mounting; including louvre damper volume control						
141 mm dia. neck	68.17	76.99	1.00	23.90	nr	**100.89**
197 mm dia. neck	89.34	100.90	1.20	28.68	nr	**129.58**
309 mm dia. neck	117.87	133.12	1.60	38.24	nr	**171.36**
365 mm dia. neck	106.86	120.69	1.90	45.43	nr	**166.12**
457 mm dia. neck	156.69	176.97	2.40	57.44	nr	**234.41**
Ceiling mounted diffusers; rectangular aluminium multi-cone diffuser; four way flow						
Rectangular; for ceiling mounting						
150 × 150 mm neck	18.76	21.19	1.80	43.06	nr	**64.25**
300 × 150 mm neck	32.02	36.16	2.30	54.97	nr	**91.13**
300 × 300 mm neck	28.60	32.30	2.80	66.91	nr	**99.21**
450 × 150 mm neck	40.79	46.06	2.80	66.91	nr	**112.97**
450 × 300 mm neck	39.93	45.09	3.20	76.47	nr	**121.56**
450 × 450 mm neck	39.88	45.04	3.40	81.28	nr	**126.32**
600 × 150 mm neck	49.97	56.44	3.20	76.47	nr	**132.91**
600 × 300 mm neck	49.70	56.13	3.50	83.64	nr	**139.77**
600 × 600 mm neck	55.46	62.64	4.00	95.59	nr	**158.23**
Rectangular; for ceiling mounting; including opposed blade damper volume regulator						
150 × 150 mm neck	24.10	27.21	1.80	43.06	nr	**70.27**
300 × 150 mm neck	48.53	54.80	2.30	54.97	nr	**109.77**
300 × 300 mm neck	35.95	40.61	2.80	66.94	nr	**107.55**
450 × 150 mm neck	65.07	73.49	2.80	66.91	nr	**140.40**
450 × 300 mm neck	63.88	72.14	3.30	78.87	nr	**151.01**
450 × 450 mm neck	50.54	57.08	3.51	83.85	nr	**140.93**
600 × 150 mm neck	79.79	90.12	3.30	78.87	nr	**168.99**
600 × 300 mm neck	79.20	89.44	4.00	95.59	nr	**185.03**
600 × 600 mm neck	81.64	92.21	5.62	134.26	nr	**226.47**

U: VENTILATION/AIR CONDITIONING SYSTEMS

Item	Net Price £	Material £	Labour hours	Labour £	Unit	Total rate £
U10: GRILLES/DIFFUSERS/LOUVRES – cont						
Slot diffusers; continuous aluminium slot diffuser with flanged frame (1500 mm sections)						
Diffuser						
1 slot	30.85	34.84	3.76	89.84	m	**124.68**
2 slot	37.56	42.42	3.76	89.82	m	**132.24**
3 slot	46.22	52.20	3.76	89.82	m	**142.02**
4 slot	52.02	58.75	4.50	107.65	m	**166.40**
6 slot	68.54	77.41	4.50	107.65	m	**185.06**
Diffuser; including equalizing deflector						
1 slot	63.86	72.12	5.26	125.78	m	**197.90**
2 slot	79.24	89.49	5.26	125.78	m	**215.27**
3 slot	96.91	109.45	5.26	125.78	m	**235.23**
4 slot	108.81	122.89	6.33	151.25	m	**274.14**
6 slot	143.15	161.67	6.33	151.25	m	**312.92**
Extra over for ends						
1 slot	3.44	3.89	1.00	23.90	nr	**27.79**
2 slot	3.63	4.10	1.00	23.90	nr	**28.00**
3 slot	3.85	4.35	1.00	23.90	nr	**28.25**
4 slot	4.04	4.56	1.30	31.07	nr	**35.63**
6 slot	4.61	5.20	1.40	33.46	nr	**38.66**
Plenum boxes; 1.0 m long; circular spigot; including cord operated flap damper						
1 slot	43.97	49.66	2.75	65.83	nr	**115.49**
2 slot	45.12	50.96	2.75	65.83	nr	**116.79**
3 slot	45.70	51.61	2.75	65.83	nr	**117.44**
4 slot	49.86	56.31	3.51	83.85	nr	**140.16**
6 slot	52.30	59.07	3.51	83.85	nr	**142.92**
Plenum boxes; 2.0 m long; circular spigot; including cord operated flap damper						
1 slot	48.69	54.99	3.26	77.84	nr	**132.83**
2 slot	51.04	57.64	3.26	77.84	nr	**135.48**
3 slot	51.76	58.46	3.26	77.84	nr	**136.30**
4 slot	57.65	65.11	3.76	89.84	nr	**154.95**
6 slot	61.47	69.42	3.76	89.84	nr	**159.26**

U: VENTILATION/AIR CONDITIONING SYSTEMS

Item	Net Price £	Material £	Labour hours	Labour £	Unit	**Total rate £**
Perforated diffusers; rectangular face aluminium perforated diffuser; quick release face plate; for integration with rectangular ceiling tiles						
Circular spigot; rectangular diffuser						
150 mm dia. spigot; 300 × 300 diffuser	44.94	50.76	1.00	23.90	nr	**74.66**
300 mm dia. spigot; 600 × 600 diffuser	67.06	75.74	1.40	33.48	nr	**109.22**
Circular spigot; rectangular diffuser; including louvre damper volume regulator						
150 mm dia. spigot; 300 × 300 diffuser	67.51	76.25	1.00	23.90	nr	**100.15**
300 mm dia. spigot; 600 × 600 diffuser	94.47	106.69	1.60	38.24	nr	**144.93**
Rectangular spigot; rectangular diffuser						
150 × 150 mm dia. spigot; 300 × 300 mm diffuser	44.94	50.76	1.00	24.71	nr	**75.47**
300 × 150 mm dia. spigot; 600 × 300 mm diffuser	60.49	68.32	1.20	29.66	nr	**97.98**
300 × 300 mm dia. spigot; 600 × 600 mm diffuser	67.06	75.74	1.40	34.60	nr	**110.34**
600 × 300 mm dia. spigot; 1200 × 600 mm diffuser	70.04	79.10	1.60	39.54	nr	**118.64**
Rectangular spigot; rectangular diffuser; including opposed blade damper volume regulator						
150 × 150 mm dia. spigot; 300 × 300 mm diffuser	67.51	76.25	1.20	29.66	nr	**105.91**
300 × 150 mm dia. spigot; 600 × 300 mm diffuser	89.98	101.62	1.40	34.62	nr	**136.24**
300 × 300 mm dia. spigot; 600 × 600 mm diffuser	94.47	106.69	1.60	39.54	nr	**146.23**
600 × 300 mm dia. spigot; 1200 × 600 mm diffuser	110.34	124.61	1.80	44.47	nr	**169.08**
Floor swirl diffuser; manual adjustment of air discharge direction; complete with damper and dirt trap						
Plastic diffuser						
150 dia.	24.90	28.12	0.50	11.94	nr	**40.06**
200 dia.	24.90	28.12	0.50	11.94	nr	**40.06**
Aluminium diffuser						
150 dia.	26.66	30.11	0.50	11.94	nr	**42.05**
200 dia.	38.97	44.01	0.50	11.94	nr	**55.95**

U: VENTILATION/AIR CONDITIONING SYSTEMS

Item	Net Price £	Material £	Labour hours	Labour £	Unit	Total rate £
U10: GRILLES/DIFFUSERS/LOUVRES – cont						
Plastic air diffusion system						
Cellular diffusers						
300 × 300 mm	11.48	12.96	2.80	69.22	nr	**82.18**
600 × 600 mm	26.57	30.01	4.00	98.85	nr	**128.86**
Multi-cone diffusers						
300 × 300 mm	9.27	10.47	2.80	69.22	nr	**79.69**
450 × 450 mm	15.24	17.21	3.40	84.05	nr	**101.26**
500 × 500 mm	16.12	18.21	3.80	93.95	nr	**112.16**
600 × 600 mm	22.30	25.18	4.00	98.85	nr	**124.03**
625 × 625 mm	38.37	43.34	4.26	105.16	nr	**148.50**
Opposed blade dampers						
300 × 300 mm	12.36	13.96	1.20	29.67	nr	**43.63**
450 × 450 mm	13.60	15.36	1.50	37.10	nr	**52.46**
600 × 600 mm	28.12	31.76	2.60	64.35	nr	**96.11**
Plenum boxes						
300 mm	7.83	8.84	2.80	69.22	nr	**78.06**
450 mm	12.15	13.72	3.40	84.05	nr	**97.77**
600 mm	19.47	21.99	4.00	98.85	nr	**120.84**
Plenum spigot reducer						
600 mm	15.45	17.45	1.00	24.71	nr	**42.16**
Blanking kits for cellular diffusers						
300 mm	5.46	6.17	0.88	21.76	nr	**27.93**
600 mm	6.49	7.33	1.10	27.18	nr	**34.51**
Blanking kits for multi-cone diffusers						
300 mm	2.88	3.25	0.88	21.76	nr	**25.01**
450 mm	8.65	9.77	0.90	22.24	nr	**32.01**
600 mm	10.09	11.39	1.10	27.18	nr	**38.57**
Acoustic louvres; opening mounted; 300 mm deep steel louvres with blades packed with acoustic infill; 12 mm galvanized mesh birdscreen; screw fixing in opening						
Louvre units; self finished galvanized steel						
900 high × 600 wide	112.62	127.19	3.00	74.13	nr	**201.32**
900 high × 900 wide	139.73	157.81	3.00	74.13	nr	**231.94**
900 high × 1200 wide	165.42	186.82	3.34	82.54	nr	**269.36**

U: VENTILATION/AIR CONDITIONING SYSTEMS

Item	Net Price £	Material £	Labour hours	Labour £	Unit	**Total rate £**
900 high × 1500 wide	215.89	243.82	3.34	82.54	nr	**326.36**
900 high × 1800 wide	242.07	273.39	3.34	82.54	nr	**355.93**
900 high × 2100 wide	268.23	302.94	3.34	82.54	nr	**385.48**
900 high × 2400 wide	293.92	331.95	3.68	90.94	nr	**422.89**
900 high × 2700 wide	332.25	375.24	3.68	90.94	nr	**466.18**
900 high × 3000 wide	355.61	401.63	3.68	90.94	nr	**492.57**
1200 high × 600 wide	147.67	166.78	3.00	74.13	nr	**240.91**
1200 high × 900 wide	181.78	205.31	3.34	82.54	nr	**287.85**
1200 high × 1200 wide	215.42	243.29	3.34	82.54	nr	**325.83**
1200 high × 1500 wide	285.52	322.46	3.34	82.54	nr	**405.00**
1200 high × 1800 wide	319.16	360.46	3.68	90.94	nr	**451.40**
1200 high × 2100 wide	353.74	399.51	3.68	90.94	nr	**490.45**
1200 high × 2400 wide	387.39	437.52	3.68	90.94	nr	**528.46**
1500 high × 600 wide	183.18	206.89	3.00	74.13	nr	**281.02**
1500 high × 900 wide	224.78	253.87	3.34	82.54	nr	**336.41**
1500 high × 1200 wide	266.36	300.83	3.34	82.54	nr	**383.37**
1500 high × 1500 wide	355.15	401.10	3.68	90.94	nr	**492.04**
1500 high × 1800 wide	396.74	448.08	3.68	90.94	nr	**539.02**
1500 high × 2100 wide	438.80	495.58	4.00	98.85	nr	**594.43**
1800 high × 600 wide	217.76	245.93	3.34	82.54	nr	**328.47**
1800 high × 900 wide	267.29	301.88	3.34	82.54	nr	**384.42**
1800 high × 1200 wide	317.29	358.35	3.68	90.94	nr	**449.29**
1800 high × 1500 wide	425.24	480.27	3.68	90.94	nr	**571.21**
Louvre units; polyester powder coated steel						
900 high × 600 wide	163.56	184.72	3.00	74.13	nr	**258.85**
900 high × 900 wide	215.42	243.29	3.00	74.13	nr	**317.42**
900 high × 1200 wide	266.36	300.83	3.34	82.54	nr	**383.37**
900 high × 1500 wide	342.06	386.33	3.34	82.54	nr	**468.87**
900 high × 1800 wide	393.46	444.37	3.34	82.54	nr	**526.91**
900 high × 2100 wide	445.34	502.96	3.34	82.54	nr	**585.50**
900 high × 2400 wide	496.27	560.48	3.68	90.94	nr	**651.42**
900 high × 2700 wide	559.35	631.73	3.68	90.94	nr	**722.67**
900 high × 3000 wide	607.96	686.62	3.68	90.94	nr	**777.56**
1200 high × 600 wide	214.96	242.77	3.00	74.13	nr	**316.90**
1200 high × 900 wide	282.72	319.30	3.34	82.54	nr	**401.84**
1200 high × 1200 wide	350.47	395.82	3.34	82.54	nr	**478.36**
1200 high × 1500 wide	453.74	512.45	3.34	82.54	nr	**594.99**
1200 high × 1800 wide	521.04	588.46	3.68	90.94	nr	**679.40**
1200 high × 2100 wide	589.26	665.51	3.68	90.94	nr	**756.45**
1200 high × 2400 wide	656.55	741.50	3.68	90.94	nr	**832.44**
1500 high × 600 wide	267.29	301.88	3.00	74.13	nr	**376.01**
1500 high × 900 wide	350.94	396.35	3.34	82.54	nr	**478.89**
1500 high × 1200 wide	434.59	490.82	3.34	82.54	nr	**573.36**
1500 high × 1500 wide	565.43	638.59	3.68	90.94	nr	**729.53**
1500 high × 1800 wide	649.55	733.59	3.68	90.94	nr	**824.53**
1500 high × 2100 wide	733.20	828.08	4.00	98.85	nr	**926.93**
1800 high × 600 wide	318.68	359.92	3.34	82.54	nr	**442.46**
1800 high × 900 wide	418.69	472.86	3.34	82.54	nr	**555.40**
1800 high × 1200 wide	519.17	586.35	3.68	90.94	nr	**677.29**
1800 high × 1500 wide	677.58	765.26	3.68	90.94	nr	**856.20**

U: VENTILATION/AIR CONDITIONING SYSTEMS

Item	Net Price £	Material £	Labour hours	Labour £	Unit	Total rate £
U10: GRILLES/DIFFUSERS/LOUVRES – cont						
Weather louvres; opening mounted; 300 mm deep galvanized steel louvres; screw fixing in position						
Louvre units; including 12 mm galvanized mesh birdscreen						
900 × 600 mm	103.18	116.54	2.25	55.66	nr	**172.20**
900 × 900 mm	149.08	168.37	2.25	55.66	nr	**224.03**
900 × 1200 mm	179.38	202.59	2.50	61.78	nr	**264.37**
900 × 1500 mm	218.30	246.54	2.50	61.78	nr	**308.32**
900 × 1800 mm	257.23	290.52	2.50	61.78	nr	**352.30**
900 × 2100 mm	322.43	364.15	2.50	61.78	nr	**425.93**
900 × 2400 mm	346.37	391.19	2.76	68.26	nr	**459.45**
900 × 2700 mm	391.04	441.64	2.76	68.26	nr	**509.90**
900 × 3000 mm	427.05	482.31	2.76	68.26	nr	**550.57**
1200 × 600 mm	139.74	157.82	2.25	55.66	nr	**213.48**
1200 × 900 mm	195.26	220.52	2.50	61.78	nr	**282.30**
1200 × 1200 mm	251.21	283.71	2.50	61.78	nr	**345.49**
1200 × 1500 mm	293.23	331.17	2.50	61.78	nr	**392.95**
1200 × 1800 mm	383.49	433.11	2.76	68.26	nr	**501.37**
1200 × 2100 mm	436.13	492.56	2.76	68.26	nr	**560.82**
1200 × 2400 mm	455.46	514.40	2.76	68.26	nr	**582.66**
1500 × 600 mm	165.16	186.53	2.25	55.66	nr	**242.19**
1500 × 900 mm	227.59	257.04	2.50	61.78	nr	**318.82**
1500 × 1200 mm	299.79	338.58	2.50	61.78	nr	**400.36**
1500 × 1500 mm	351.48	396.96	2.76	68.26	nr	**465.22**
1500 × 1800 mm	414.44	468.07	2.76	68.26	nr	**536.33**
1500 × 2100 mm	518.34	585.41	3.00	74.22	nr	**659.63**
1800 × 600 mm	184.75	208.65	2.50	61.78	nr	**270.43**
1800 × 900 mm	255.40	288.45	2.50	61.78	nr	**350.23**
1800 × 1200 mm	334.90	378.24	2.76	68.26	nr	**446.50**
1800 × 1500 mm	396.15	447.41	3.00	74.22	nr	**521.63**

U: VENTILATION/AIR CONDITIONING SYSTEMS

Item	Net Price £	Material £	Labour hours	Labour £	Unit	**Total rate £**
U10: PLANT/EQUIPMENT						
Y41 – FANS						
Axial flow fan; including ancillaries, anti-vibration mountings, mounting feet, matching flanges, flexible connectors and clips; 415 V, 3 phase, 50 Hz motor; includes fixing in position; electrical work elsewhere						
Aerofoil blade fan unit; short duct case						
315 mm dia.; 0.47 m³/s duty; 147 Pa	577.80	652.57	4.50	111.20	nr	**763.77**
500 mm dia.; 1.89 m³/s duty; 500 Pa	981.19	1108.15	5.00	123.55	nr	**1231.70**
560 mm dia.; 2.36 m³/s duty; 147 Pa	773.61	873.71	5.50	135.92	nr	**1009.63**
710 mm dia.; 5.67 m³/s duty; 245 Pa	1196.26	1351.05	6.00	148.27	nr	**1499.32**
Aerofoil blade fan unit; long duct case						
315 mm dia.; 0.47 m³/s duty; 147 Pa	664.47	750.44	4.50	111.20	nr	**861.64**
500 mm dia.; 1.89 m³/s duty; 500 Pa	1003.66	1133.53	5.00	123.55	nr	**1257.08**
560 mm dia.; 2.36 m³/s duty; 147 Pa	927.73	1047.77	5.50	135.92	nr	**1183.69**
710 mm dia.; 5.67 m³/s duty; 245 Pa	1265.81	1429.60	6.00	148.27	nr	**1577.87**
Aerofoil blade fan unit; two stage parallel fan arrangement; long duct case						
315 mm; 0.47 m³/s @ 500 Pa	1169.51	1320.84	4.50	111.20	nr	**1432.04**
355 mm; 0.83 m³/s @ 147 Pa	1219.80	1377.63	4.75	117.38	nr	**1495.01**
710 mm; 3.77 m³/s @ 431 Pa	2144.28	2421.74	6.00	148.27	nr	**2570.01**
710 mm; 6.61 m³/s @ 500 Pa	2672.86	3018.71	6.00	148.27	nr	**3166.98**
Axial flow fan; suitable for operation at 300°C for 90 minutes; including ancillaries, anti-vibration mountings, mounting feet, matching flanges, flexible connectors and clips; 415 V, 3 phase, 50 Hz motor; includes fixing in position; price includes air operated damper; electrical work elsewhere						
450 mm; 2.0 m³/s @ 300 Pa	674.10	761.33	5.00	123.55	nr	**884.88**
630 mm; 4.6 m³/s @ 200 Pa	2255.56	2547.42	5.50	135.92	nr	**2683.34**
900 mm; 9.0 m³/s @ 300 Pa	2142.14	2419.32	6.50	160.62	nr	**2579.94**
1000 mm; 15.0 m³/s @ 400 Pa	5613.22	6339.54	7.50	185.33	nr	**6524.87**

U: VENTILATION/AIR CONDITIONING SYSTEMS

Item	Net Price £	Material £	Labour hours	Labour £	Unit	**Total rate £**
U10: PLANT/EQUIPMENT – cont						
Y41 – FANS – cont						
Bifurcated fan; suitable for temperature up to 200°C with motor protection to IP55; including ancillaries, anti-vibration mountings, mounting feet, matching flanges, flexible connectors and clips; 415 V, 3 phase, 50 Hz motor; includes fixing in position; electrical work elsewhere						
300 mm; 0.50 m³/s @ 100 Pa	1156.88	1306.58	4.50	111.20	nr	**1417.78**
400 mm; 1.97 m³/s @ 200 Pa	2018.02	2279.14	5.00	123.55	nr	**2402.69**
800 mm; 3.86 m³/s @ 200 Pa	2655.74	2999.38	5.50	135.92	nr	**3135.30**
1000 mm; 6.10 m³/s @ 400 Pa	3912.67	4418.95	6.50	160.62	nr	**4579.57**
Duct mounted in line fan with backward curved centrifugal impellor; including ancillaries, matching flanges, flexible connectors and clips; 415 V, 3 phase, 50 Hz motor; includes fixing in position; electrical work elsewhere						
0.5 m³/s @ 200 Pa	1627.47	1838.06	4.50	111.20	nr	**1949.26**
1.0 m³/s @ 300 Pa	1808.30	2042.28	5.00	123.55	nr	**2165.83**
3.0 m³/s @ 500 Pa	2982.09	3367.96	5.50	135.92	nr	**3503.88**
5.0 m³/s @ 750 Pa	4934.84	5573.38	6.50	160.62	nr	**5734.00**
7.0 m³/s @ 1000 Pa	6203.86	7006.60	7.00	172.98	nr	**7179.58**
Twin fan extract unit; belt driven; located internally; complete with anti-vibration mounts and non return shutter; including ancillaries, matching flanges, flexible connectors and clips; 3 phase, 50 Hz motor; includes fixing in position; electrical work elsewhere						
0.25 m³/s @ 150 Pa	2574.42	2907.54	4.50	111.20	nr	**3018.74**
Extra for external unit	755.42	853.17	–	–	nr	**853.17**
1.00 m³/s @ 200 Pa	3287.04	3712.37	5.00	123.55	nr	**3835.92**
Extra for external unit	860.28	971.60	–	–	nr	**971.60**
2.00 m³/s @ 250 Pa	4300.33	4856.77	6.50	160.62	nr	**5017.39**
Extra for external unit	965.14	1090.02	–	–	nr	**1090.02**
Extra for auto changeover panel including BMS contacts	1225.15	1383.68	2.50	61.78	nr	**1445.46**

U: VENTILATION/AIR CONDITIONING SYSTEMS

Item	Net Price £	Material £	Labour hours	Labour £	Unit	**Total rate £**
Roof mounted extract fan; including ancillaries, fibreglass cowling, fitted shutters and bird guard; 415 V, 3 phase, 50 Hz motor; includes fixing in position; electrical work elsewhere						
Flat roof installation, fixed to curb						
315 mm; 1380 rpm	606.66	685.15	4.50	111.20	nr	**796.35**
400 mm; 900 rpm	1453.06	1641.08	5.50	135.92	nr	**1777.00**
800 mm; 700 rpm	2360.42	2665.85	7.00	172.98	nr	**2838.83**
760 mm; 710 rpm	4680.18	5285.77	8.00	197.69	nr	**5483.46**
Pitched roof installation; including purlin mounting box						
315 mm; 1380 rpm	826.37	933.30	4.50	111.20	nr	**1044.50**
400 mm; 900 rpm	1684.18	1902.11	5.50	135.92	nr	**2038.03**
800 mm; 700 rpm	2763.81	3121.43	7.00	172.98	nr	**3294.41**
760 mm; 710 rpm	5093.20	5752.24	8.00	197.69	nr	**5949.93**
Centrifugal fan; single speed for internal domestic kitchens/utility rooms; fitted with standard overload protection; complete with housing; includes placing in position; electrical work elsewhere						
Window mounted						
245 m^3/hr	187.95	212.27	0.50	12.35	nr	**224.62**
500 m^3/hr	391.65	884.66	0.50	12.35	nr	**897.01**
Wall mounted						
245 m^3/hr	245.70	503.17	0.83	20.59	nr	**523.76**
500 m^3/hr	449.40	1405.61	0.83	20.59	nr	**1426.20**
Centrifugal fan; various speeds, simultaneous ventilation from separate areas fitted with standard overload protection; complete with housing; includes placing in position; ducting and electrical work elsewhere						
Fan unit						
147–300 m^3/hr	93.45	105.54	1.00	24.71	nr	**130.25**
175–411 m^3/hr	116.55	131.63	1.00	24.71	nr	**156.34**

U: VENTILATION/AIR CONDITIONING SYSTEMS

Item	Net Price £	Material £	Labour hours	Labour £	Unit	Total rate £
U10: PLANT/EQUIPMENT – cont						
Y41 – FANS – cont						
Toilet extract units; centrifugal fan; various speeds for internal domestic bathrooms/ W.Cs, complete with housing; includes placing in position; electrical work elsewhere						
Fan unit; fixed to wall; including shutter						
Single speed 85 m^3/hr	124.95	141.12	0.75	18.53	nr	**159.65**
Two speed 60–85 m^3/hr	152.25	171.95	0.83	20.59	nr	**192.54**
Humidity controlled; autospeed; fixed to wall; including shutter						
30–60–85 m^3/hr	199.97	225.85	1.00	24.71	nr	**250.56**
Y42 – AIR FILTRATION						
High efficiency duct mounted filters; 99.997% H13 (EU13); tested to BS 3928						
Standard; 1700 m^3/hr air volume; continuous rating up to 80°C; sealed wood case, aluminium spacers, neoprene gaskets; water repellant filter media; includes placing in position						
610 × 610 × 292 mm	249.18	281.42	1.00	23.90	nr	**305.32**
Side withdrawl frame	94.39	106.60	2.50	59.74	nr	**166.34**
High capacity; 3400 m^3/hr air volume; continuous rating up to 80°C; anti-corrosion coated mild steel frame, polyurethane sealant and neoprene gaskets; water repellant filter media; includes placing in position						
610 × 610 × 292 mm	270.93	305.99	1.00	23.90	nr	**329.89**
Side withdrawl frame	94.39	106.60	2.50	59.74	nr	**166.34**
Bag Filters; 40/60% F5 (EU5); tested to BSEN 779						
Duct mounted bag filter; continuous rating up to 60°C; rigid filter assembly; sealed into one piece coated mild steel header with sealed pocket separators; includes placing in position						
6 pocket, 592 × 592 × 25 mm header; pockets 380 mm long; 1690 m^3/hr	17.61	19.89	1.00	23.90	nr	**43.79**
Side withdrawl frame	16.34	18.45	2.00	47.80	nr	**66.25**
6 pocket, 592 × 592 × 25 mm header; pockets 500 mm long; 2550 m^3/hr	17.88	20.20	1.50	35.84	nr	**56.04**

U: VENTILATION/AIR CONDITIONING SYSTEMS

Item	Net Price £	Material £	Labour hours	Labour £	Unit	**Total rate £**
Side withdrawl frame	16.34	18.45	2.50	59.74	nr	**78.19**
6 pocket, 592 × 592 × 25 mm header; pockets 600 mm long; 3380 m³/hr	21.12	23.86	1.50	35.84	nr	**59.70**
Side withdrawl frame	16.34	18.45	3.00	71.70	nr	**90.15**
Bag Filters; 80/90% F7, (EU7); tested to BSEN 779						
Duct mounted bag filter; continuous rating up to 60°C; rigid filter assembly; sealed into one piece coated mild steel header with sealed pocket separators; includes placing in position						
6 pocket, 592 × 592 × 25 mm header; pockets 500 mm long; 1688 m³/hr	17.88	20.20	1.00	23.90	nr	**44.10**
Side withdrawl frame	16.34	18.45	2.00	47.80	nr	**66.25**
6 pocket, 592 × 592 × 25 mm header; pockets 635 mm long; 2047 m³/hr	21.12	23.86	1.50	35.84	nr	**59.70**
Side withdrawl frame	16.34	18.45	2.50	59.74	nr	**78.19**
6 pocket, 592 × 592 × 25 mm header; pockets 762 mm long; 2729 m³/hr	24.54	27.71	1.50	35.84	nr	**63.55**
Side withdrawl frame	16.34	18.45	3.00	71.70	nr	**90.15**
Grease filters, washable; minimum 65%						
Double sided extract unit; lightweight stainless steel construction; demountable composite filter media of woven metal mat and expanded metal mesh supports; for mounting on hood and extract systems (hood not included); includes placing in position						
500 × 686 × 565 mm, 4080 m³/hr	366.49	413.92	2.00	47.80	nr	**461.72**
1000 × 686 × 565 mm, 8160 m³/hr;	558.89	631.21	3.00	71.70	nr	**702.91**
1500 × 686 × 565 mm, 12240 m³/hr;	766.59	865.79	3.50	83.64	nr	**949.43**
Panel filters; 82% G3 (EU3); tested to BS EN779						
Modular duct mounted filter panels; continuous rating up to 100°C; graduated density media; rigid cardboard frame; includes placing in position						
596 × 596 × 47 mm, 2360 m³/hr	4.43	5.01	1.00	23.90	nr	**28.91**
Side withdrawl frame	31.79	35.90	2.50	59.74	nr	**95.64**
596 × 287 × 47 mm, 1140 m³/hr	3.06	3.46	1.00	23.90	nr	**27.36**
Side withdrawl frame	31.79	35.90	2.50	59.74	nr	**95.64**

U: VENTILATION/AIR CONDITIONING SYSTEMS

Item	Net Price £	Material £	Labour hours	Labour £	Unit	**Total rate £**
U10: PLANT/EQUIPMENT – cont						
Y42 – AIR FILTRATION – cont						
Panel filters; 90% G4 (EU4); tested to BS EN779						
Modular duct mounted filter panels; continuous rating up to 100°C; pleated media with wire support; rigid cardboard frame; includes placing in position						
596 × 596 × 47 mm, 2560 m³/hr	6.71	7.58	1.00	23.90	nr	**31.48**
side withdrawl frame	40.09	45.28	3.00	71.70	nr	**116.98**
596 × 287 × 47 mm, 1230 m³/hr	4.63	5.23	1.00	23.90	nr	**29.13**
Side withdrawl frame	31.79	35.90	3.00	71.70	nr	**107.60**
Carbon filters; standard duty disposable carbon filters; steel frame with bonded carbon panels; for fixing to ductwork; including placing in position						
12 panels						
597 × 597 × 298 mm, 1460 m³/hr	312.34	352.76	0.33	7.88	nr	**360.64**
597 × 597 × 451 mm, 2200 m³/hr	352.11	397.67	0.33	7.88	nr	**405.55**
597 × 597 × 597 mm, 2930 m³/hr	392.67	443.48	0.33	7.88	nr	**451.36**
8 panels						
451 × 451 × 298 mm, 740 m³/hr	235.48	265.95	0.29	6.94	nr	**272.89**
451 × 451 × 451 mm, 1105 m³/hr	260.62	294.34	0.29	6.94	nr	**301.28**
451 × 451 × 597 mm, 1460 m³/hr	284.62	321.45	0.29	6.94	nr	**328.39**
6 panels						
298 × 298 × 298 mm, 365 m³/hr	166.85	188.43	0.25	5.98	nr	**194.41**
298 × 298 × 451 mm, 550 m³/hr	178.64	201.76	0.25	5.98	nr	**207.74**
298 × 298 × 597 mm, 780 m³/hr	189.92	214.50	0.25	5.98	nr	**220.48**

U: VENTILATION/AIR CONDITIONING SYSTEMS

Item	Net Price £	Material £	Labour hours	Labour £	Unit	**Total rate £**
U10: SILENCERS/ACOUSTIC TREATMENT						
Y45 – SILENCERS/ACOUSTIC TREATMENT						
Attenuators; DW144 galvanized construction c/w splitters; self securing; fitted to ductwork						
To suit rectangular ducts; unit length 600 mm						
100 × 100 mm	93.53	105.63	0.75	17.92	nr	**123.55**
150 × 150 mm	97.63	110.26	0.75	17.92	nr	**128.18**
200 × 200 mm	101.73	114.89	0.75	17.92	nr	**132.81**
300 × 300 mm	113.81	128.54	0.75	17.92	nr	**146.46**
400 × 400 mm	127.75	144.28	1.00	23.90	nr	**168.18**
500 × 500 mm	149.10	168.39	1.25	29.88	nr	**198.27**
600 × 300 mm	143.63	162.21	1.25	29.88	nr	**192.09**
600 × 600 mm	201.02	227.03	1.25	29.88	nr	**256.91**
700 × 300 mm	152.28	171.98	1.50	35.84	nr	**207.82**
700 × 700 mm	226.76	256.10	1.50	35.84	nr	**291.94**
800 × 300 mm	157.74	178.15	2.00	47.80	nr	**225.95**
800 × 800 mm	261.61	295.46	2.00	47.80	nr	**343.26**
1000 × 1000 mm	334.88	378.22	3.00	71.70	nr	**449.92**
To suit rectangular ducts; unit length 1200 mm						
200 × 200 mm	112.17	126.68	1.00	23.90	nr	**150.58**
300 × 300 mm	138.68	156.62	1.00	23.90	nr	**180.52**
400 × 400 mm	166.74	188.31	1.33	31.78	nr	**220.09**
500 × 500 mm	201.26	227.31	1.66	39.68	nr	**266.99**
600 × 300 mm	191.91	216.74	1.66	39.68	nr	**256.42**
600 × 600 mm	280.77	317.10	1.66	39.68	nr	**356.78**
700 × 300 mm	205.71	232.33	2.00	47.80	nr	**280.13**
700 × 700 mm	321.30	362.88	2.00	47.80	nr	**410.68**
800 × 300 mm	215.06	242.89	2.66	63.58	nr	**306.47**
800 × 800 mm	370.74	418.71	2.66	63.58	nr	**482.29**
1000 × 1000 mm	501.87	566.81	4.00	95.59	nr	**662.40**
1300 × 1300 mm	861.48	972.95	8.00	191.19	nr	**1164.14**
1500 × 1500 mm	986.06	1113.65	8.00	191.19	nr	**1304.84**
1800 × 1800 mm	1349.17	1523.75	10.66	254.76	nr	**1778.51**
2000 × 2000 mm	1513.92	1709.82	13.33	318.56	nr	**2028.38**
To suit rectangular ducts; unit length 1800 mm						
200 × 200 mm	141.57	159.88	1.00	23.90	nr	**183.78**
300 × 300 mm	188.13	212.48	1.00	23.90	nr	**236.38**
400 × 400 mm	224.87	253.97	1.33	31.78	nr	**285.75**
500 × 500 mm	262.06	295.97	1.66	39.68	nr	**335.65**
600 × 300 mm	274.09	309.56	1.66	39.68	nr	**349.24**
600 × 600 mm	382.99	432.55	1.66	39.68	nr	**472.23**
700 × 300 mm	290.56	328.15	2.00	47.80	nr	**375.95**
700 × 700 mm	457.60	516.81	2.00	47.80	nr	**564.61**
800 × 300 mm	307.26	347.01	2.66	63.58	nr	**410.59**
800 × 800 mm	532.43	601.32	2.66	63.58	nr	**664.90**
1000 × 1000 mm	715.66	808.26	4.00	95.59	nr	**903.85**

U: VENTILATION/AIR CONDITIONING SYSTEMS

Item	Net Price £	Material £	Labour hours	Labour £	Unit	Total rate £
U10: SILENCERS/ACOUSTIC TREATMENT – cont						
To suit rectangular ducts – cont						
1300 × 1300 mm	1195.56	1350.26	8.00	191.19	nr	**1541.45**
1500 × 1500 mm	1354.88	1530.19	8.00	191.19	nr	**1721.38**
1800 × 1800 mm	1881.00	2124.39	13.33	318.56	nr	**2442.95**
2000 × 2000 mm	2129.97	2405.58	10.66	254.76	nr	**2660.34**
2300 × 2300 mm	2662.42	3006.92	16.00	382.37	nr	**3389.29**
2500 × 2500 mm	3713.74	4194.28	18.66	445.94	nr	**4640.22**
To suit rectangular ducts; unit length 2400 mm						
500 × 500 mm	338.89	382.74	2.08	49.79	nr	**432.53**
600 × 300 mm	323.54	365.40	2.08	49.79	nr	**415.19**
600 × 600 mm	481.87	544.22	2.08	49.79	nr	**594.01**
700 × 300 mm	350.70	396.08	2.50	59.74	nr	**455.82**
700 × 700 mm	569.18	642.82	2.50	59.74	nr	**702.56**
800 × 300 mm	370.74	418.71	3.33	79.58	nr	**498.29**
800 × 800 mm	658.70	743.94	3.33	79.58	nr	**823.52**
1000 × 1000 mm	888.71	1003.70	5.00	119.49	nr	**1123.19**
1300 × 1300 mm	1439.43	1625.69	10.00	238.98	nr	**1864.67**
1500 × 1500 mm	1642.84	1855.41	10.00	238.98	nr	**2094.39**
1800 × 1800 mm	2344.03	2647.34	13.33	318.52	nr	**2965.86**
2000 × 2000 mm	2630.39	2970.75	16.66	398.15	nr	**3368.90**
2300 × 2300 mm	3254.39	3675.49	20.00	477.96	nr	**4153.45**
2500 × 2500 mm	4569.63	5160.91	23.32	557.40	nr	**5718.31**
To suit circular ducts; unit length 600 mm						
100 mm dia.	81.23	91.74	0.75	17.92	nr	**109.66**
200 mm dia.	95.57	107.94	0.75	17.92	nr	**125.86**
250 mm dia.	108.33	122.35	0.75	17.92	nr	**140.27**
315 mm dia.	126.10	142.41	1.00	23.90	nr	**166.31**
355 mm dia.	145.92	164.80	1.00	23.90	nr	**188.70**
400 mm dia.	156.85	177.14	1.00	23.90	nr	**201.04**
450 mm dia.	189.41	213.91	1.00	23.90	nr	**237.81**
500 mm dia.	208.55	235.54	1.26	30.06	nr	**265.60**
630 mm dia.	245.88	277.70	1.50	35.84	nr	**313.54**
710 mm dia.	267.75	302.40	1.50	35.84	nr	**338.24**
800 mm dia.	297.82	336.36	2.00	47.80	nr	**384.16**
1000 mm dia.	399.78	451.51	3.00	71.70	nr	**523.21**
To suit circular ducts; unit length 1200 mm						
100 mm dia.	119.49	134.96	1.00	23.90	nr	**158.86**
200 mm dia.	131.10	148.06	1.00	23.90	nr	**171.96**
250 mm dia.	147.28	166.34	1.00	23.90	nr	**190.24**
315 mm dia.	168.46	190.26	1.33	31.78	nr	**222.04**
355 mm dia.	201.95	228.08	1.33	31.78	nr	**259.86**
400 mm dia.	214.24	241.96	1.33	31.78	nr	**273.74**
450 mm dia.	248.17	280.29	1.33	31.78	nr	**312.07**

U: VENTILATION/AIR CONDITIONING SYSTEMS

Item	Net Price £	Material £	Labour hours	Labour £	Unit	**Total rate £**
500 mm dia.	254.33	287.24	1.66	39.68	nr	**326.92**
630 mm dia.	336.77	380.35	2.00	47.80	nr	**428.15**
710 mm dia.	372.29	420.46	2.00	47.80	nr	**468.26**
800 mm dia.	371.61	419.70	2.66	63.58	nr	**483.28**
1000 mm dia.	498.86	563.41	4.00	95.59	nr	**659.00**
1250 mm dia.	939.74	1061.34	8.00	191.19	nr	**1252.53**
1400 mm dia.	1021.72	1153.93	8.00	191.19	nr	**1345.12**
1600 mm dia.	1132.71	1279.27	10.66	254.76	nr	**1534.03**
To suit circular ducts; unit length 1800 mm						
200 mm dia.	183.72	207.49	1.25	29.88	nr	**237.37**
250 mm dia.	205.58	232.18	1.25	29.88	nr	**262.06**
315 mm dia.	244.31	275.92	1.66	39.68	nr	**315.60**
355 mm dia.	288.04	325.31	1.66	39.68	nr	**364.99**
400 mm dia.	298.05	336.62	1.66	39.68	nr	**376.30**
450 mm dia.	304.44	343.83	1.66	39.68	nr	**383.51**
500 mm dia.	334.68	377.98	2.08	49.70	nr	**427.68**
630 mm dia.	370.23	418.14	2.50	59.74	nr	**477.88**
710 mm dia.	433.10	489.14	2.50	59.74	nr	**548.88**
800 mm dia.	448.14	506.12	3.33	79.58	nr	**585.70**
1000 mm dia.	1170.00	1321.39	6.66	159.16	nr	**1480.55**
1250 mm dia.	1189.13	1342.99	6.66	159.16	nr	**1502.15**
1400 mm dia.	1313.79	1483.78	5.00	119.49	nr	**1603.27**
1600 mm dia.	1314.58	1484.68	9.90	236.59	nr	**1721.27**

U: VENTILATION/AIR CONDITIONING SYSTEMS

Item	Net Price £	Material £	Labour hours	Labour £	Unit	Total rate £
U10: THERMAL INSULATION						
Y50 – THERMAL INSULATION						
Concealed ductwork						
Flexible wrap; 20 kg–45 kg Bright Class O aluminium foil faced; Bright Class O foil taped joints; 62 mm metal pins and washers; aluminium bands						
40 mm thick insulation	11.40	12.87	0.40	9.88	m²	**22.75**
Semi-rigid slab; 45 kg Bright Class O aluminium foil faced mineral fibre; Bright Class O foil taped joints; 62 mm metal pins and washers; aluminium bands						
40 mm thick insulation	14.92	16.85	0.65	16.07	m²	**32.92**
Plantroom ductwork						
Semi-rigid slab; 45 kg Bright Class O aluminium foil faced mineral fibre; Bright Class O foil taped joints; 62 mm metal pins and washers; 22 swg plain/embossed aluminium cladding; pop rivited						
50 mm thick insulation	37.62	42.48	1.50	37.07	m²	**79.55**
External ductwork						
Semi-rigid slab; 45 kg Bright Class O aluminium foil faced mineral fibre; Bright Class O foil taped joints; 62 mm metal pins and washers; 0.8 mm polyisobutylene sheeting; welded joints						
50 mm thick insulation	27.45	31.00	1.25	30.89	m²	**61.89**

U: VENTILATION/AIR CONDITIONING SYSTEMS

Item	Net Price £	Material £	Labour hours	Labour £	Unit	**Total rate £**
U14: DUCTWORK: FIRE RATED						
Y30 – DUCTLINES						
The relevant BS requires that the fire-rating of ductwork meets 3 criteria; stability (hours), integrity (hours) and insulation (hours). The least of the 3 periods defines the fire-rating. The BS does however allow stability and integrity to be considered in isolation. Rates are therefore provided for both types of system.						
Care should to be taken when using the rates within this section to ensure thta the requiremements for stability, integrity and insulation are known and the appropriate rates are used						
High density single layer mineral wool fire-rated ductwork slab, in accordance with BS476, Part 24 (ISO 6944: 1985), ducts 'Type A' and 'Type B'; 165 kg class O foil faced mineral fibre; 100 mm wide bright class O foil taped joints; welded pins; includes protection to all supports.						
½ hour stability, integrity and insulation						
25 mm thick, vertical and horizontal ductwork	40.95	50.06	1.25	33.43	m²	**83.49**
1 hour stability, integrity and insulation						
30 mm thick, vertical ductwork	48.07	58.76	1.50	40.12	m²	**98.88**
40 mm thick, horizontal ductwork	56.53	69.11	1.50	40.12	m²	**109.23**
1½ hour stability, integrity and insulation						
50 mm thick, vertical ductwork	67.54	82.56	1.75	46.81	m²	**129.37**
70 mm thick, horizontal ductwork	84.49	103.29	1.75	46.81	m²	**150.10**
2 hour stability, integrity and insulation						
70 mm, vertical ductwork	87.05	106.42	2.00	53.50	m²	**159.92**
90 mm horizontal ductwork	103.98	127.11	2.00	53.50	m²	**180.61**
Kitchen extract, 1 hour stability, integrity and insulation						
90 mm, vertical and horizontal	103.98	127.11	2.00	53.50	m²	**180.61**

U: VENTILATION/AIR CONDITIONING SYSTEMS

Item	Net Price £	Material £	Labour hours	Labour £	Unit	Total rate £
U14: DUCTWORK: FIRE RATED – cont						
Galvanized sheet metal rectangular section ductwork to BS476 Part 24 (ISO 6944:1985), ducts 'Type A' and 'Type B'; provides 2 hours stability and 2 hours integrity at 1100°C (no rating for insulation); including all necessary stiffeners, joints and supports in the running length						
Ductwork up to 600 mm longest side						
Sum of two sides 200 mm	65.70	80.32	2.91	75.28	m	**155.60**
Sum of two sides 300 mm	71.32	87.19	2.99	77.35	m	**164.54**
Sum of two sides 400 mm	76.94	94.06	3.17	82.01	m	**176.07**
Sum of two sides 500 mm	82.57	100.94	3.37	87.18	m	**188.12**
Sum of two sides 600 mm	88.18	107.80	3.54	91.58	m	**199.38**
Sum of two sides 700 mm	93.81	114.68	3.72	96.23	m	**210.91**
Sum of two sides 800 mm	99.44	121.57	3.90	100.88	m	**222.45**
Sum of two sides 900 mm	105.05	128.42	5.04	130.38	m	**258.80**
Sum of two sides 1000 mm	110.68	135.31	5.58	144.34	m	**279.65**
Sum of two sides 1100 mm	116.29	142.16	5.84	151.07	m	**293.23**
Sum of two sides 1200 mm	121.92	149.05	6.11	158.05	m	**307.10**
Extra over fittings; ductwork up to 600 mm longest side						
End cap						
Sum of two sides 200 mm	18.65	22.80	0.81	20.95	nr	**43.75**
Sum of two sides 300 mm	19.78	24.18	0.84	21.72	nr	**45.90**
Sum of two sides 400 mm	20.92	25.57	0.87	22.51	nr	**48.08**
Sum of two sides 500 mm	22.05	26.96	0.90	23.28	nr	**50.24**
Sum of two sides 600 mm	23.18	28.34	0.93	24.06	nr	**52.40**
Sum of two sides 700 mm	24.32	29.74	0.96	24.83	nr	**54.57**
Sum of two sides 800 mm	25.48	31.15	0.98	25.36	nr	**56.51**
Sum of two sides 900 mm	26.61	32.53	1.17	30.27	nr	**62.80**
Sum of two sides 1000 mm	27.74	33.91	1.22	31.57	nr	**65.48**
Sum of two sides 1100 mm	28.88	35.31	1.25	32.33	nr	**67.64**
Sum of two sides 1200 mm	30.01	36.69	1.28	33.10	nr	**69.79**
Reducer						
Sum of two sides 200 mm	73.98	90.44	2.23	57.69	nr	**148.13**
Sum of two sides 300 mm	77.38	94.60	2.37	61.31	nr	**155.91**
Sum of two sides 400 mm	80.77	98.75	2.51	64.93	nr	**163.68**
Sum of two sides 500 mm	84.18	102.91	2.65	68.55	nr	**171.46**
Sum of two sides 600 mm	87.57	107.06	2.79	72.17	nr	**179.23**
Sum of two sides 700 mm	90.97	111.21	2.87	74.24	nr	**185.45**
Sum of two sides 800 mm	94.37	115.37	3.01	77.86	nr	**193.23**
Sum of two sides 900 mm	97.77	119.52	3.06	79.16	nr	**198.68**
Sum of two sides 1000 mm	101.16	123.67	3.17	82.01	nr	**205.68**
Sum of two sides 1100 mm	104.54	127.80	3.22	83.31	nr	**211.11**
Sum of two sides 1200 mm	107.96	131.98	3.28	84.84	nr	**216.82**

U: VENTILATION/AIR CONDITIONING SYSTEMS

Item	Net Price £	Material £	Labour hours	Labour £	Unit	**Total rate £**
Offset						
Sum of two sides 200 mm	178.56	218.29	2.95	76.31	nr	**294.60**
Sum of two sides 300 mm	182.98	223.70	3.11	80.45	nr	**304.15**
Sum of two sides 400 mm	187.41	229.11	3.26	84.33	nr	**313.44**
Sum of two sides 500 mm	191.85	234.54	3.42	88.48	nr	**323.02**
Sum of two sides 600 mm	196.27	239.94	3.57	92.35	nr	**332.29**
Sum of two sides 700 mm	205.14	250.78	3.23	83.55	nr	**334.33**
Sum of two sides 800 mm	200.71	245.37	3.73	96.49	nr	**341.86**
Sum of two sides 900 mm	209.58	256.21	3.45	89.24	nr	**345.45**
Sum of two sides 1000 mm	214.00	261.62	3.67	94.94	nr	**356.56**
Sum of two sides 1100 mm	218.43	267.03	3.78	97.78	nr	**364.81**
Sum of two sides 1200 mm	222.88	272.48	3.89	100.62	nr	**373.10**
90° radius bend						
Sum of two sides 200 mm	72.39	88.50	2.06	53.29	nr	**141.79**
Sum of two sides 300 mm	77.89	95.22	2.21	57.17	nr	**152.39**
Sum of two sides 400 mm	83.35	101.90	2.36	61.05	nr	**162.95**
Sum of two sides 500 mm	88.86	108.63	2.51	64.93	nr	**173.56**
Sum of two sides 600 mm	94.34	115.33	2.66	68.81	nr	**184.14**
Sum of two sides 700 mm	99.81	122.01	2.81	72.69	nr	**194.70**
Sum of two sides 800 mm	105.30	128.73	2.97	76.84	nr	**205.57**
Sum of two sides 900 mm	110.77	135.41	2.99	77.35	nr	**212.76**
Sum of two sides 1000 mm	116.28	142.15	3.04	78.64	nr	**220.79**
Sum of two sides 1100 mm	121.74	148.82	3.07	79.41	nr	**228.23**
Sum of two sides 1200 mm	127.22	155.53	3.09	79.93	nr	**235.46**
45° radius bend						
Sum of two sides 200 mm	89.26	109.12	1.52	39.32	nr	**148.44**
Sum of two sides 300 mm	91.49	111.85	1.59	41.13	nr	**152.98**
Sum of two sides 400 mm	93.72	114.58	1.66	42.95	nr	**157.53**
Sum of two sides 500 mm	95.93	117.28	1.73	44.75	nr	**162.03**
Sum of two sides 600 mm	98.14	119.97	1.80	46.57	nr	**166.54**
Sum of two sides 700 mm	100.37	122.70	1.87	48.38	nr	**171.08**
Sum of two sides 800 mm	102.57	125.39	1.93	49.92	nr	**175.31**
Sum of two sides 900 mm	104.80	128.12	1.99	51.48	nr	**179.60**
Sum of two sides 1000 mm	107.01	130.82	2.05	53.03	nr	**183.85**
Sum of two sides 1100 mm	109.22	133.52	2.11	54.59	nr	**188.11**
Sum of two sides 1200 mm	111.43	136.23	2.17	56.14	nr	**192.37**
90° mitre bend						
Sum of two sides 200 mm	75.05	91.75	2.47	63.90	nr	**155.65**
Sum of two sides 300 mm	86.99	106.35	2.65	68.55	nr	**174.90**
Sum of two sides 400 mm	98.92	120.93	2.84	73.46	nr	**194.39**
Sum of two sides 500 mm	110.87	135.54	3.02	78.12	nr	**213.66**
Sum of two sides 600 mm	122.79	150.11	3.20	82.77	nr	**232.88**
Sum of two sides 700 mm	134.73	164.71	3.38	87.44	nr	**252.15**
Sum of two sides 800 mm	146.67	179.30	3.56	92.09	nr	**271.39**
Sum of two sides 900 mm	158.61	193.90	3.59	92.86	nr	**286.76**
Sum of two sides 1000 mm	170.57	208.52	3.66	94.69	nr	**303.21**
Sum of two sides 1100 mm	182.48	223.09	3.69	95.45	nr	**318.54**
Sum of two sides 1200 mm	194.45	237.71	3.72	96.23	nr	**333.94**

U: VENTILATION/AIR CONDITIONING SYSTEMS

Item	Net Price £	Material £	Labour hours	Labour £	Unit	Total rate £
U14: DUCTWORK: FIRE RATED – cont						
Extra over fittings – cont						
Branch (Side-on Shoe)						
Sum of two sides 200 mm	27.86	34.06	0.98	25.36	nr	**59.42**
Sum of two sides 300 mm	28.02	34.26	1.06	27.42	nr	**61.68**
Sum of two sides 400 mm	28.17	34.43	1.14	29.49	nr	**63.92**
Sum of two sides 500 mm	28.33	34.63	1.22	31.57	nr	**66.20**
Sum of two sides 600 mm	28.47	34.80	1.30	33.63	nr	**68.43**
Sum of two sides 700 mm	28.63	35.00	1.38	35.70	nr	**70.70**
Sum of two sides 800 mm	28.99	35.44	1.37	35.44	nr	**70.88**
Sum of two sides 900 mm	28.76	35.16	1.46	37.76	nr	**72.92**
Sum of two sides 1000 mm	29.06	35.52	1.46	37.76	nr	**73.28**
Sum of two sides 1100 mm	29.22	35.72	1.50	38.80	nr	**74.52**
Sum of two sides 1200 mm	29.37	35.91	1.55	40.10	nr	**76.01**
Ductwork 601 to 800 mm longest side						
Sum of two sides 900 mm	105.03	128.40	5.04	130.38	m	**258.78**
Sum of two sides 1000 mm	110.66	135.28	5.58	144.34	m	**279.62**
Sum of two sides 1100 mm	116.29	142.16	5.84	151.07	m	**293.23**
Sum of two sides 1200 mm	121.92	149.05	6.11	158.05	m	**307.10**
Sum of two sides 1300 mm	127.56	155.94	6.38	165.04	m	**320.98**
Sum of two sides 1400 mm	133.19	162.82	6.65	172.02	m	**334.84**
Sum of two sides 1500 mm	138.83	169.72	6.91	178.76	m	**348.48**
Sum of two sides 1600 mm	144.46	176.60	7.18	185.74	m	**362.34**
Extra over fittings; ductwork 601 to 800 mm longest side						
End cap						
Sum of two sides 900 mm	24.73	30.23	1.17	30.27	nr	**60.50**
Sum of two sides 1000 mm	26.49	32.38	1.22	31.57	nr	**63.95**
Sum of two sides 1100 mm	28.27	34.56	1.25	32.33	nr	**66.89**
Sum of two sides 1200 mm	30.03	36.71	1.28	33.10	nr	**69.81**
Sum of two sides 1300 mm	31.79	38.86	1.31	33.89	nr	**72.75**
Sum of two sides 1400 mm	33.56	41.02	1.34	34.66	nr	**75.68**
Sum of two sides 1500 mm	35.33	43.19	1.36	35.18	nr	**78.37**
Sum of two sides 1600 mm	37.10	45.35	1.39	35.96	nr	**81.31**
Reducer						
Sum of two sides 900 mm	92.53	113.11	3.06	79.16	nr	**192.27**
Sum of two sides 1000 mm	97.68	119.42	3.17	82.01	nr	**201.43**
Sum of two sides 1100 mm	102.82	125.70	3.22	83.31	nr	**209.01**
Sum of two sides 1200 mm	107.97	131.99	3.28	84.84	nr	**216.83**
Sum of two sides 1300 mm	113.13	138.30	3.33	86.14	nr	**224.44**
Sum of two sides 1400 mm	118.28	144.60	3.38	87.44	nr	**232.04**
Sum of two sides 1500 mm	123.42	150.88	3.44	88.98	nr	**239.86**
Sum of two sides 1600 mm	128.57	157.18	3.49	90.28	nr	**247.46**

U: VENTILATION/AIR CONDITIONING SYSTEMS

Item	Net Price £	Material £	Labour hours	Labour £	Unit	**Total rate £**
Offset						
Sum of two sides 900 mm	196.79	240.58	3.45	89.24	nr	**329.82**
Sum of two sides 1000 mm	206.16	252.03	3.67	94.94	nr	**346.97**
Sum of two sides 1100 mm	215.53	263.48	3.78	97.78	nr	**361.26**
Sum of two sides 1200 mm	224.90	274.94	3.89	100.62	nr	**375.56**
Sum of two sides 1300 mm	234.26	286.38	4.00	103.47	nr	**389.85**
Sum of two sides 1400 mm	243.64	297.86	4.11	106.32	nr	**404.18**
Sum of two sides 1500 mm	253.00	309.30	4.23	109.43	nr	**418.73**
Sum of two sides 1600 mm	262.67	321.11	4.34	112.26	nr	**433.37**
90° radius bend						
Sum of two sides 900 mm	98.40	120.29	2.99	77.35	nr	**197.64**
Sum of two sides 1000 mm	108.06	132.11	3.04	78.64	nr	**210.75**
Sum of two sides 1100 mm	117.70	143.89	3.07	79.41	nr	**223.30**
Sum of two sides 1200 mm	127.37	155.72	3.09	79.93	nr	**235.65**
Sum of two sides 1300 mm	137.02	167.51	3.12	80.71	nr	**248.22**
Sum of two sides 1400 mm	146.67	179.30	3.15	81.48	nr	**260.78**
Sum of two sides 1500 mm	156.34	191.13	3.17	82.01	nr	**273.14**
Sum of two sides 1600 mm	166.01	202.95	3.20	82.77	nr	**285.72**
45° bend						
Sum of two sides 900 mm	97.70	119.44	1.74	45.01	nr	**164.45**
Sum of two sides 1000 mm	102.29	125.05	1.85	47.86	nr	**172.91**
Sum of two sides 1100 mm	106.88	130.67	1.91	49.41	nr	**180.08**
Sum of two sides 1200 mm	111.45	136.25	1.96	50.70	nr	**186.95**
Sum of two sides 1300 mm	116.02	141.83	2.02	52.25	nr	**194.08**
Sum of two sides 1400 mm	120.61	147.44	2.07	53.55	nr	**200.99**
Sum of two sides 1500 mm	125.19	153.05	2.13	55.09	nr	**208.14**
Sum of two sides 1600 mm	129.77	158.64	2.18	56.39	nr	**215.03**
90° mitre bend						
Sum of two sides 900 mm	147.31	180.09	3.59	92.86	nr	**272.95**
Sum of two sides 1000 mm	163.02	199.29	3.66	94.69	nr	**293.98**
Sum of two sides 1100 mm	178.72	218.48	3.69	95.45	nr	**313.93**
Sum of two sides 1200 mm	194.45	237.71	3.72	96.23	nr	**333.94**
Sum of two sides 1300 mm	210.13	256.88	3.75	97.01	nr	**353.89**
Sum of two sides 1400 mm	225.84	276.09	3.78	97.78	nr	**373.87**
Sum of two sides 1500 mm	241.54	295.28	3.81	98.56	nr	**393.84**
Sum of two sides 1600 mm	257.25	314.49	3.84	99.33	nr	**413.82**
Branch (Side-on Shoe)						
Sum of two sides 900 mm	27.34	33.42	1.37	35.44	nr	**68.86**
Sum of two sides 1000 mm	28.02	34.26	1.46	37.76	nr	**72.02**
Sum of two sides 1100 mm	28.70	35.08	1.50	38.80	nr	**73.88**
Sum of two sides 1200 mm	29.39	35.93	1.55	40.10	nr	**76.03**
Sum of two sides 1300 mm	30.07	36.76	1.59	41.13	nr	**77.89**
Sum of two sides 1400 mm	29.55	36.12	1.68	43.46	nr	**79.58**
Sum of two sides 1500 mm	30.77	37.62	1.63	42.17	nr	**79.79**
Sum of two sides 1600 mm	32.14	39.29	1.72	44.50	nr	**83.79**

U: VENTILATION/AIR CONDITIONING SYSTEMS

Item	Net Price £	Material £	Labour hours	Labour £	Unit	Total rate £
U14: DUCTWORK: FIRE RATED – cont						
Ductwork 801 to 1000 mm longest side						
Sum of two sides 1100 mm	116.29	142.16	5.84	151.07	m	**293.23**
Sum of two sides 1200 mm	121.92	149.05	6.11	158.05	m	**307.10**
Sum of two sides 1300 mm	127.55	155.93	6.38	165.04	m	**320.97**
Sum of two sides 1400 mm	133.17	162.80	6.65	172.02	m	**334.82**
Sum of two sides 1500 mm	138.80	169.68	6.91	178.76	m	**348.44**
Sum of two sides 1600 mm	144.43	176.56	7.18	185.74	m	**362.30**
Sum of two sides 1700 mm	150.07	183.46	7.45	192.71	m	**376.17**
Sum of two sides 1800 mm	155.70	190.34	7.71	199.44	m	**389.78**
Sum of two sides 1900 mm	161.33	197.22	7.98	206.43	m	**403.65**
Sum of two sides 2000 mm	166.96	204.10	8.25	213.42	m	**417.52**
Extra over fittings; ductwork 801 to 1000 mm longest side						
End cap						
Sum of two sides 1100 mm	28.27	34.56	1.25	32.33	nr	**66.89**
Sum of two sides 1200 mm	30.03	36.71	1.28	33.10	nr	**69.81**
Sum of two sides 1300 mm	31.79	38.86	1.31	33.89	nr	**72.75**
Sum of two sides 1400 mm	33.56	41.02	1.34	34.66	nr	**75.68**
Sum of two sides 1500 mm	35.33	43.19	1.36	35.18	nr	**78.37**
Sum of two sides 1600 mm	37.08	45.33	1.39	35.96	nr	**81.29**
Sum of two sides 1700 mm	38.85	47.49	1.42	36.74	nr	**84.23**
Sum of two sides 1800 mm	40.61	49.65	1.45	37.50	nr	**87.15**
Sum of two sides 1900 mm	42.37	51.80	1.48	38.29	nr	**90.09**
Sum of two sides 2000 mm	44.13	53.95	1.51	39.06	nr	**93.01**
Reducer						
Sum of two sides 1100 mm	102.82	125.70	3.22	83.31	nr	**209.01**
Sum of two sides 1200 mm	107.99	132.02	3.28	84.84	nr	**216.86**
Sum of two sides 1300 mm	113.17	138.35	3.33	86.14	nr	**224.49**
Sum of two sides 1400 mm	118.35	144.68	3.38	87.44	nr	**232.12**
Sum of two sides 1500 mm	123.51	150.99	3.44	88.98	nr	**239.97**
Sum of two sides 1600 mm	128.72	157.36	3.49	90.28	nr	**247.64**
Sum of two sides 1700 mm	133.87	163.66	3.54	91.58	nr	**255.24**
Sum of two sides 1800 mm	139.06	170.00	3.60	93.13	nr	**263.13**
Sum of two sides 1900 mm	144.24	176.34	3.65	94.41	nr	**270.75**
Sum of two sides 2000 mm	151.99	185.81	3.70	95.71	nr	**281.52**
Offset						
Sum of two sides 1100 mm	213.84	261.42	3.78	97.78	nr	**359.20**
Sum of two sides 1200 mm	222.88	272.48	3.89	100.62	nr	**373.10**
Sum of two sides 1300 mm	231.91	283.51	4.00	103.47	nr	**386.98**
Sum of two sides 1400 mm	240.96	294.58	4.11	106.32	nr	**400.90**
Sum of two sides 1500 mm	249.98	305.60	4.23	109.43	nr	**415.03**
Sum of two sides 1600 mm	259.04	316.67	4.34	112.26	nr	**428.93**
Sum of two sides 1700 mm	268.07	327.72	4.45	115.11	nr	**442.83**

U: VENTILATION/AIR CONDITIONING SYSTEMS

Item	Net Price £	Material £	Labour hours	Labour £	Unit	**Total rate £**
Sum of two sides 1800 mm	277.12	338.78	4.56	117.96	nr	**456.74**
Sum of two sides 1900 mm	286.14	349.81	4.67	120.81	nr	**470.62**
Sum of two sides 2000 mm	299.70	366.38	5.53	143.05	nr	**509.43**
90° radius bend						
Sum of two sides 1100 mm	117.52	143.66	3.07	79.41	nr	**223.07**
Sum of two sides 1200 mm	127.21	155.52	3.09	79.93	nr	**235.45**
Sum of two sides 1300 mm	136.89	167.34	3.12	80.71	nr	**248.05**
Sum of two sides 1400 mm	146.57	179.18	3.15	81.48	nr	**260.66**
Sum of two sides 1500 mm	156.25	191.01	3.17	82.01	nr	**273.02**
Sum of two sides 1600 mm	165.92	202.84	3.20	82.77	nr	**285.61**
Sum of two sides 1700 mm	175.60	214.67	3.22	83.31	nr	**297.98**
Sum of two sides 1800 mm	185.28	226.51	3.25	84.07	nr	**310.58**
Sum of two sides 1900 mm	194.97	238.35	3.28	84.84	nr	**323.19**
Sum of two sides 2000 mm	204.35	249.82	3.30	85.37	nr	**335.19**
45° bend						
Sum of two sides 1100 mm	107.57	131.50	1.91	49.41	nr	**180.91**
Sum of two sides 1200 mm	112.07	137.01	1.96	50.70	nr	**187.71**
Sum of two sides 1300 mm	116.57	142.51	2.02	52.25	nr	**194.76**
Sum of two sides 1400 mm	121.07	148.01	2.07	53.55	nr	**201.56**
Sum of two sides 1500 mm	125.57	153.51	2.13	55.09	nr	**208.60**
Sum of two sides 1600 mm	130.07	159.01	2.18	56.39	nr	**215.40**
Sum of two sides 1700 mm	134.57	164.51	2.24	57.95	nr	**222.46**
Sum of two sides 1800 mm	139.07	170.01	2.30	59.50	nr	**229.51**
Sum of two sides 1900 mm	143.55	175.49	2.35	60.79	nr	**236.28**
Sum of two sides 2000 mm	148.07	181.01	2.76	71.39	nr	**252.40**
90° mitre bend						
Sum of two sides 1100 mm	178.72	218.48	3.69	95.45	nr	**313.93**
Sum of two sides 1200 mm	194.45	237.71	3.72	96.23	nr	**333.94**
Sum of two sides 1300 mm	210.13	256.88	3.75	97.01	nr	**353.89**
Sum of two sides 1400 mm	225.84	276.09	3.78	97.78	nr	**373.87**
Sum of two sides 1500 mm	241.56	295.31	3.81	98.56	nr	**393.87**
Sum of two sides 1600 mm	257.25	314.49	3.84	99.33	nr	**413.82**
Sum of two sides 1700 mm	272.96	333.70	3.87	100.12	nr	**433.82**
Sum of two sides 1800 mm	288.67	352.90	3.90	100.88	nr	**453.78**
Sum of two sides 1900 mm	304.37	372.10	3.93	101.66	nr	**473.76**
Sum of two sides 2000 mm	320.08	391.30	3.96	102.43	nr	**493.73**
Branch (Side-on Shoe)						
Sum of two sides 1100 mm	28.68	35.06	1.50	38.80	nr	**73.86**
Sum of two sides 1200 mm	29.34	35.86	1.55	40.10	nr	**75.96**
Sum of two sides 1300 mm	30.06	36.75	1.59	41.13	nr	**77.88**
Sum of two sides 1400 mm	30.75	37.59	1.63	42.17	nr	**79.76**
Sum of two sides 1500 mm	31.45	38.45	1.68	43.46	nr	**81.91**
Sum of two sides 1600 mm	32.14	39.29	1.72	44.50	nr	**83.79**
Sum of two sides 1700 mm	32.83	40.14	1.77	45.78	nr	**85.92**
Sum of two sides 1800 mm	33.53	40.99	1.81	46.82	nr	**87.81**
Sum of two sides 1900 mm	34.23	41.85	1.86	48.12	nr	**89.97**
Sum of two sides 2000 mm	35.25	43.10	1.90	49.14	nr	**92.24**

U: VENTILATION/AIR CONDITIONING SYSTEMS

Item	Net Price £	Material £	Labour hours	Labour £	Unit	**Total rate £**
U14: DUCTWORK: FIRE RATED – cont						
Ductwork 1001 to 1250 mm longest side						
Sum of two sides 1300 mm	144.56	176.73	6.38	165.04	m	**341.77**
Sum of two sides 1400 mm	150.80	184.35	6.65	172.02	m	**356.37**
Sum of two sides 1500 mm	157.04	191.98	6.91	178.76	m	**370.74**
Sum of two sides 1600 mm	161.72	197.71	6.91	178.76	m	**376.47**
Sum of two sides 1700 mm	166.38	203.40	7.45	192.71	m	**396.11**
Sum of two sides 1800 mm	172.59	211.00	7.71	199.44	m	**410.44**
Sum of two sides 1900 mm	178.81	218.60	7.98	206.43	m	**425.03**
Sum of two sides 2000 mm	185.05	226.23	8.25	213.42	m	**439.65**
Sum of two sides 2100 mm	191.27	233.83	9.62	248.86	m	**482.69**
Sum of two sides 2200 mm	195.93	239.53	9.62	248.86	m	**488.39**
Sum of two sides 2300 mm	200.61	245.24	10.07	260.49	m	**505.73**
Sum of two sides 2400 mm	206.83	252.85	10.47	270.85	m	**523.70**
Sum of two sides 2500 mm	213.05	260.46	10.87	281.19	m	**541.65**
Extra over fittings; ductwork 1001 to 1250 mm longest side						
End cap						
Sum of two sides 1300 mm	31.81	38.89	1.31	33.89	nr	**72.78**
Sum of two sides 1400 mm	33.56	41.02	1.34	34.66	nr	**75.68**
Sum of two sides 1500 mm	35.30	43.16	1.36	35.18	nr	**78.34**
Sum of two sides 1600 mm	37.08	45.33	1.39	35.96	nr	**81.29**
Sum of two sides 1700 mm	38.85	47.49	1.42	36.74	nr	**84.23**
Sum of two sides 1800 mm	40.59	49.63	1.45	37.50	nr	**87.13**
Sum of two sides 1900 mm	42.37	51.80	1.48	38.29	nr	**90.09**
Sum of two sides 2000 mm	44.13	53.95	1.51	39.06	nr	**93.01**
Sum of two sides 2100 mm	45.90	56.11	2.66	68.81	nr	**124.92**
Sum of two sides 2200 mm	47.67	58.28	2.80	72.43	nr	**130.71**
Sum of two sides 2300 mm	49.44	60.44	2.95	76.31	nr	**136.75**
Sum of two sides 2400 mm	51.19	62.58	3.10	80.20	nr	**142.78**
Sum of two sides 2500 mm	52.96	64.75	3.24	83.81	nr	**148.56**
Reducer						
Sum of two sides 1300 mm	113.13	138.30	3.33	86.14	nr	**224.44**
Sum of two sides 1400 mm	118.22	144.53	3.38	87.44	nr	**231.97**
Sum of two sides 1500 mm	123.31	150.75	3.44	88.98	nr	**239.73**
Sum of two sides 1600 mm	128.44	157.02	3.49	90.28	nr	**247.30**
Sum of two sides 1700 mm	133.56	163.28	3.54	91.58	nr	**254.86**
Sum of two sides 1800 mm	138.68	169.54	3.60	93.13	nr	**262.67**
Sum of two sides 1900 mm	143.79	175.78	3.65	94.41	nr	**270.19**
Sum of two sides 2000 mm	148.92	182.05	3.70	95.71	nr	**277.76**
Sum of two sides 2100 mm	154.01	188.27	3.75	97.01	nr	**285.28**
Sum of two sides 2200 mm	159.12	194.53	3.80	98.30	nr	**292.83**
Sum of two sides 2300 mm	164.24	200.79	3.85	99.60	nr	**300.39**
Sum of two sides 2400 mm	169.36	207.04	3.90	100.88	nr	**307.92**
Sum of two sides 2500 mm	174.49	213.31	3.95	102.18	nr	**315.49**

U: VENTILATION/AIR CONDITIONING SYSTEMS

Item	Net Price £	Material £	Labour hours	Labour £	Unit	**Total rate £**
Offset						
Sum of two sides 1300 mm	231.14	282.57	4.00	103.47	nr	**386.04**
Sum of two sides 1400 mm	240.32	293.80	4.11	106.32	nr	**400.12**
Sum of two sides 1500 mm	249.48	304.99	4.23	109.43	nr	**414.42**
Sum of two sides 1600 mm	258.64	316.19	4.34	112.26	nr	**428.45**
Sum of two sides 1700 mm	267.80	327.39	4.45	115.11	nr	**442.50**
Sum of two sides 1800 mm	276.97	338.60	4.56	117.96	nr	**456.56**
Sum of two sides 1900 mm	286.13	349.80	4.67	120.81	nr	**470.61**
Sum of two sides 2000 mm	295.29	360.99	5.53	143.05	nr	**504.04**
Sum of two sides 2100 mm	304.45	372.19	5.76	149.00	nr	**521.19**
Sum of two sides 2200 mm	313.47	383.22	5.99	154.95	nr	**538.17**
Sum of two sides 2300 mm	322.78	394.60	6.22	160.91	nr	**555.51**
Sum of two sides 2400 mm	331.93	405.79	6.45	166.85	nr	**572.64**
Sum of two sides 2500 mm	341.10	417.00	6.68	172.80	nr	**589.80**
90° radius bend						
Sum of two sides 1300 mm	136.71	167.13	3.12	80.71	nr	**247.84**
Sum of two sides 1400 mm	152.12	185.97	3.15	81.48	nr	**267.45**
Sum of two sides 1500 mm	156.13	190.87	3.17	82.01	nr	**272.88**
Sum of two sides 1600 mm	165.83	202.72	3.20	82.77	nr	**285.49**
Sum of two sides 1700 mm	175.54	214.60	3.22	83.31	nr	**297.91**
Sum of two sides 1800 mm	185.24	226.46	3.25	84.07	nr	**310.53**
Sum of two sides 1900 mm	194.95	238.33	3.28	84.84	nr	**323.17**
Sum of two sides 2000 mm	204.66	250.20	3.30	85.37	nr	**335.57**
Sum of two sides 2100 mm	214.36	262.05	3.32	85.88	nr	**347.93**
Sum of two sides 2200 mm	224.07	273.93	3.34	86.40	nr	**360.33**
Sum of two sides 2300 mm	233.78	285.79	3.36	86.92	nr	**372.71**
Sum of two sides 2400 mm	243.50	297.68	3.38	87.44	nr	**385.12**
Sum of two sides 2500 mm	253.19	309.53	3.40	87.95	nr	**397.48**
45° bend						
Sum of two sides 1300 mm	110.74	135.38	2.02	52.25	nr	**187.63**
Sum of two sides 1400 mm	115.71	141.46	2.07	53.55	nr	**195.01**
Sum of two sides 1500 mm	120.70	147.56	2.13	55.09	nr	**202.65**
Sum of two sides 1600 mm	125.69	153.66	2.18	56.39	nr	**210.05**
Sum of two sides 1700 mm	130.67	159.74	2.24	57.95	nr	**217.69**
Sum of two sides 1800 mm	135.65	165.83	2.30	59.50	nr	**225.33**
Sum of two sides 1900 mm	140.63	171.92	2.35	60.79	nr	**232.71**
Sum of two sides 2000 mm	145.63	178.04	2.76	71.39	nr	**249.43**
Sum of two sides 2100 mm	150.62	184.13	2.89	74.76	nr	**258.89**
Sum of two sides 2200 mm	155.61	190.23	3.01	77.86	nr	**268.09**
Sum of two sides 2300 mm	160.59	196.32	3.13	80.97	nr	**277.29**
Sum of two sides 2400 mm	165.58	202.43	3.26	84.33	nr	**286.76**
Sum of two sides 2500 mm	170.57	208.52	3.37	87.18	nr	**295.70**
90° mitre bend						
Sum of two sides 1300 mm	209.65	256.29	3.75	97.01	nr	**353.30**
Sum of two sides 1400 mm	225.22	275.33	3.78	97.78	nr	**373.11**
Sum of two sides 1500 mm	241.39	295.11	3.81	98.56	nr	**393.67**

U: VENTILATION/AIR CONDITIONING SYSTEMS

Item	Net Price £	Material £	Labour hours	Labour £	Unit	**Total rate £**
U14: DUCTWORK: FIRE RATED – cont						
Extra over fittings – cont						
90° mitre bend – cont						
Sum of two sides 1600 mm	257.23	314.47	3.84	99.33	nr	**413.80**
Sum of two sides 1700 mm	273.08	333.84	3.87	100.12	nr	**433.96**
Sum of two sides 1800 mm	288.91	353.20	3.90	100.88	nr	**454.08**
Sum of two sides 1900 mm	304.77	372.58	3.93	101.66	nr	**474.24**
Sum of two sides 2000 mm	320.63	391.98	3.96	102.43	nr	**494.41**
Sum of two sides 2100 mm	336.47	411.34	3.99	103.22	nr	**514.56**
Sum of two sides 2200 mm	352.33	430.73	4.02	103.99	nr	**534.72**
Sum of two sides 2300 mm	368.18	450.10	4.05	104.77	nr	**554.87**
Sum of two sides 2400 mm	384.01	469.45	4.08	105.54	nr	**574.99**
Sum of two sides 2500 mm	399.85	488.82	4.11	106.32	nr	**595.14**
Branch (Side-on Shoe)						
Sum of two sides 1300 mm	30.12	36.82	1.59	41.13	nr	**77.95**
Sum of two sides 1400 mm	30.80	37.66	1.63	42.17	nr	**79.83**
Sum of two sides 1500 mm	31.48	38.48	1.68	43.46	nr	**81.94**
Sum of two sides 1600 mm	32.17	39.33	1.72	44.50	nr	**83.83**
Sum of two sides 1700 mm	32.86	40.17	1.77	45.78	nr	**85.95**
Sum of two sides 1800 mm	33.56	41.02	1.81	46.82	nr	**87.84**
Sum of two sides 1900 mm	34.24	41.86	1.86	48.12	nr	**89.98**
Sum of two sides 2000 mm	34.94	42.71	1.90	49.14	nr	**91.85**
Sum of two sides 2100 mm	35.63	43.56	2.58	66.73	nr	**110.29**
Sum of two sides 2200 mm	36.31	44.39	2.61	67.52	nr	**111.91**
Sum of two sides 2300 mm	37.01	45.25	2.64	68.29	nr	**113.54**
Sum of two sides 2400 mm	37.70	46.09	2.88	74.50	nr	**120.59**
Sum of two sides 2500 mm	38.39	46.94	2.91	75.28	nr	**122.22**
Ductwork 1251 to 2000 mm longest side						
Sum of two sides 1700 mm	158.52	193.79	7.45	192.71	m	**386.50**
Sum of two sides 1800 mm	168.70	206.24	7.71	199.44	m	**405.68**
Sum of two sides 1900 mm	178.85	218.65	7.98	206.43	m	**425.08**
Sum of two sides 2000 mm	189.01	231.07	8.25	213.42	m	**444.49**
Sum of two sides 2100 mm	199.17	243.48	9.62	248.86	m	**492.34**
Sum of two sides 2200 mm	209.33	255.90	9.66	249.89	m	**505.79**
Sum of two sides 2300 mm	219.47	268.30	10.07	260.49	m	**528.79**
Sum of two sides 2400 mm	229.63	280.73	10.47	270.85	m	**551.58**
Sum of two sides 2500 mm	239.78	293.13	10.87	281.19	m	**574.32**
Sum of two sides 2600 mm	249.96	305.58	11.27	291.53	m	**597.11**
Sum of two sides 2700 mm	260.10	317.97	11.67	301.88	m	**619.85**
Sum of two sides 2800 mm	270.27	330.40	12.08	312.48	m	**642.88**
Sum of two sides 2900 mm	280.41	342.80	12.48	322.84	m	**665.64**
Sum of two sides 3000 mm	290.57	355.23	12.88	333.18	m	**688.41**
Sum of two sides 3100 mm	300.73	367.65	13.26	343.01	m	**710.66**
Sum of two sides 3200 mm	310.88	380.05	13.69	354.13	m	**734.18**
Sum of two sides 3300 mm	330.79	404.39	14.09	364.48	m	**768.87**
Sum of two sides 3400 mm	341.24	417.17	14.49	374.84	m	**792.01**
Sum of two sides 3500 mm	351.70	429.95	14.89	385.18	m	**815.13**

U: VENTILATION/AIR CONDITIONING SYSTEMS

Item	Net Price £	Material £	Labour hours	Labour £	Unit	Total rate £
Sum of two sides 3600 mm	362.15	442.74	15.29	395.53	m	**838.27**
Sum of two sides 3700 mm	372.63	455.54	15.69	405.87	m	**861.41**
Sum of two sides 3800 mm	383.08	468.32	16.09	416.21	m	**884.53**
Sum of two sides 3900 mm	393.56	481.13	16.49	426.57	m	**907.70**
Sum of two sides 4000 mm	404.01	493.91	16.89	436.92	m	**930.83**
Extra over fittings; ductwork 1251 to 2000 mm longest side						
End cap						
Sum of two sides 1700 mm	57.71	70.56	1.42	36.74	nr	**107.30**
Sum of two sides 1800 mm	61.42	75.09	1.45	37.50	nr	**112.59**
Sum of two sides 1900 mm	65.13	79.62	1.48	38.29	nr	**117.91**
Sum of two sides 2000 mm	68.84	84.16	1.51	39.06	nr	**123.22**
Sum of two sides 2100 mm	72.56	88.70	2.66	68.81	nr	**157.51**
Sum of two sides 2200 mm	76.27	93.24	2.80	72.43	nr	**165.67**
Sum of two sides 2300 mm	79.97	97.77	2.95	76.31	nr	**174.08**
Sum of two sides 2400 mm	83.68	102.30	3.10	80.20	nr	**182.50**
Sum of two sides 2500 mm	87.39	106.83	3.24	83.81	nr	**190.64**
Sum of two sides 2600 mm	91.10	111.37	3.39	87.69	nr	**199.06**
Sum of two sides 2700 mm	94.80	115.90	3.54	91.58	nr	**207.48**
Sum of two sides 2800 mm	98.51	120.43	3.68	95.19	nr	**215.62**
Sum of two sides 2900 mm	102.26	125.01	3.83	99.08	nr	**224.09**
Sum of two sides 3000 mm	105.95	129.52	3.98	102.96	nr	**232.48**
Sum of two sides 3100 mm	109.67	134.08	4.12	106.57	nr	**240.65**
Sum of two sides 3200 mm	113.38	138.61	4.27	110.46	nr	**249.07**
Sum of two sides 3300 mm	120.63	147.48	4.42	114.34	nr	**261.82**
Sum of two sides 3400 mm	124.45	152.14	4.57	118.21	nr	**270.35**
Sum of two sides 3500 mm	127.60	155.99	4.72	122.11	nr	**278.10**
Sum of two sides 3600 mm	130.17	159.13	4.87	125.98	nr	**285.11**
Sum of two sides 3700 mm	134.00	163.81	5.02	129.85	nr	**293.66**
Sum of two sides 3800 mm	131.50	160.76	5.17	133.75	nr	**294.51**
Sum of two sides 3900 mm	143.56	175.50	5.32	137.62	nr	**313.12**
Sum of two sides 4000 mm	147.38	180.18	5.47	141.50	nr	**321.68**
Reducer						
Sum of two sides 1700 mm	124.62	152.35	3.54	91.58	nr	**243.93**
Sum of two sides 1800 mm	139.80	170.91	3.60	93.13	nr	**264.04**
Sum of two sides 1900 mm	154.96	189.44	3.65	94.41	nr	**283.85**
Sum of two sides 2000 mm	185.31	226.54	2.92	75.54	nr	**302.08**
Sum of two sides 2100 mm	170.15	208.01	3.70	95.71	nr	**303.72**
Sum of two sides 2200 mm	200.49	245.10	3.10	80.20	nr	**325.30**
Sum of two sides 2300 mm	215.67	263.66	3.29	85.11	nr	**348.77**
Sum of two sides 2400 mm	230.84	282.20	3.48	90.02	nr	**372.22**
Sum of two sides 2500 mm	246.01	300.75	3.66	94.69	nr	**395.44**
Sum of two sides 2600 mm	261.20	319.32	3.85	99.60	nr	**418.92**
Sum of two sides 2700 mm	276.36	337.85	4.03	104.25	nr	**442.10**
Sum of two sides 2800 mm	291.54	356.41	4.22	109.17	nr	**465.58**
Sum of two sides 2900 mm	306.71	374.95	4.40	113.81	nr	**488.76**
Sum of two sides 3000 mm	329.45	402.75	4.59	118.73	nr	**521.48**
Sum of two sides 3100 mm	329.45	402.75	4.78	123.64	nr	**526.39**

U: VENTILATION/AIR CONDITIONING SYSTEMS

Item	Net Price £	Material £	Labour hours	Labour £	Unit	Total rate £
U14: DUCTWORK: FIRE RATED – cont						
Extra over fittings – cont						
Reducer – cont						
Sum of two sides 3200 mm	344.64	421.32	4.96	128.31	nr	**549.63**
Sum of two sides 3300 mm	378.52	462.75	5.15	133.22	nr	**595.97**
Sum of two sides 3400 mm	394.16	481.86	5.34	138.13	nr	**619.99**
Sum of two sides 3500 mm	409.80	500.98	5.53	143.05	nr	**644.03**
Sum of two sides 3600 mm	425.42	520.08	5.72	147.97	nr	**668.05**
Sum of two sides 3700 mm	441.07	539.22	5.91	152.88	nr	**692.10**
Sum of two sides 3800 mm	456.69	558.31	6.10	157.81	nr	**716.12**
Sum of two sides 3900 mm	472.33	577.43	6.29	162.72	nr	**740.15**
Sum of two sides 4000 mm	487.96	596.53	6.48	167.63	nr	**764.16**
Offset						
Sum of two sides 1700 mm	124.62	152.35	4.45	115.11	nr	**267.46**
Sum of two sides 1800 mm	134.90	164.91	4.56	117.96	nr	**282.87**
Sum of two sides 1900 mm	145.17	177.47	4.67	120.81	nr	**298.28**
Sum of two sides 2000 mm	155.41	189.99	5.53	143.05	nr	**333.04**
Sum of two sides 2100 mm	165.68	202.55	5.76	149.00	nr	**351.55**
Sum of two sides 2200 mm	175.96	215.12	5.99	154.95	nr	**370.07**
Sum of two sides 2300 mm	186.23	227.67	6.22	160.91	nr	**388.58**
Sum of two sides 2400 mm	196.48	240.20	6.45	166.85	nr	**407.05**
Sum of two sides 2500 mm	206.76	252.76	6.68	172.80	nr	**425.56**
Sum of two sides 2600 mm	217.03	265.32	6.91	178.76	nr	**444.08**
Sum of two sides 2700 mm	227.31	277.88	7.14	184.70	nr	**462.58**
Sum of two sides 2800 mm	237.59	290.46	7.37	190.65	nr	**481.11**
Sum of two sides 2900 mm	247.83	302.97	7.60	196.61	nr	**499.58**
Sum of two sides 3000 mm	258.11	315.54	7.83	202.55	nr	**518.09**
Sum of two sides 3100 mm	268.38	328.10	8.06	208.50	nr	**536.60**
Sum of two sides 3200 mm	278.66	340.66	8.29	214.46	nr	**555.12**
Sum of two sides 3300 mm	297.66	363.89	8.52	220.39	nr	**584.28**
Sum of two sides 3400 mm	308.23	376.82	8.75	226.34	nr	**603.16**
Sum of two sides 3500 mm	318.82	389.76	8.98	232.30	nr	**622.06**
Sum of two sides 3600 mm	329.41	402.71	9.21	238.24	nr	**640.95**
Sum of two sides 3700 mm	339.98	415.63	9.44	244.19	nr	**659.82**
Sum of two sides 3800 mm	350.55	428.55	9.67	250.15	nr	**678.70**
Sum of two sides 3900 mm	361.14	441.50	9.90	256.09	nr	**697.59**
Sum of two sides 4000 mm	371.73	454.45	10.13	262.04	nr	**716.49**
90° radius bend						
Sum of two sides 1700 mm	173.43	212.02	3.22	83.31	nr	**295.33**
Sum of two sides 1800 mm	189.10	231.17	3.25	84.07	nr	**315.24**
Sum of two sides 1900 mm	199.58	243.99	3.32	85.88	nr	**329.87**
Sum of two sides 2000 mm	204.78	250.34	3.28	84.84	nr	**335.18**
Sum of two sides 2100 mm	220.46	269.51	3.30	85.37	nr	**354.88**
Sum of two sides 2200 mm	252.24	308.36	3.32	85.88	nr	**394.24**
Sum of two sides 2300 mm	267.48	326.99	3.36	86.92	nr	**413.91**

U: VENTILATION/AIR CONDITIONING SYSTEMS

Item	Net Price £	Material £	Labour hours	Labour £	Unit	**Total rate £**
Sum of two sides 2400 mm	283.13	346.13	3.38	87.44	nr	**433.57**
Sum of two sides 2500 mm	298.83	365.32	3.40	87.95	nr	**453.27**
Sum of two sides 2600 mm	314.51	384.49	3.42	88.48	nr	**472.97**
Sum of two sides 2700 mm	330.19	403.66	3.44	88.98	nr	**492.64**
Sum of two sides 2800 mm	345.86	422.82	3.46	89.50	nr	**512.32**
Sum of two sides 2900 mm	361.51	441.95	3.48	90.02	nr	**531.97**
Sum of two sides 3000 mm	377.19	461.12	3.50	90.54	nr	**551.66**
Sum of two sides 3100 mm	392.86	480.28	3.52	91.05	nr	**571.33**
Sum of two sides 3200 mm	395.04	482.94	3.54	91.58	nr	**574.52**
Sum of two sides 3300 mm	408.20	499.02	3.56	92.09	nr	**591.11**
Sum of two sides 3400 mm	393.00	480.44	3.58	92.61	nr	**573.05**
Sum of two sides 3500 mm	422.44	516.43	3.60	93.13	nr	**609.56**
Sum of two sides 3600 mm	436.96	534.19	3.62	93.65	nr	**627.84**
Sum of two sides 3700 mm	451.49	551.95	3.64	94.16	nr	**646.11**
Sum of two sides 3800 mm	466.03	569.72	3.66	94.69	nr	**664.41**
Sum of two sides 3900 mm	480.57	587.50	3.68	95.19	nr	**682.69**
Sum of two sides 4000 mm	495.10	605.26	3.70	95.71	nr	**700.97**
45° bend						
Sum of two sides 1700 mm	128.57	157.18	2.24	57.95	nr	**215.13**
Sum of two sides 1800 mm	142.78	174.55	2.30	59.50	nr	**234.05**
Sum of two sides 1900 mm	156.96	191.89	2.35	60.79	nr	**252.68**
Sum of two sides 2000 mm	171.19	209.29	2.76	71.39	nr	**280.68**
Sum of two sides 2100 mm	220.46	269.51	3.01	77.86	nr	**347.37**
Sum of two sides 2200 mm	213.81	261.38	3.13	80.97	nr	**342.35**
Sum of two sides 2300 mm	236.13	288.67	2.89	74.76	nr	**363.43**
Sum of two sides 2400 mm	227.99	278.72	3.26	84.33	nr	**363.05**
Sum of two sides 2500 mm	242.20	296.09	3.37	87.18	nr	**383.27**
Sum of two sides 2600 mm	256.39	313.44	3.49	90.28	nr	**403.72**
Sum of two sides 2700 mm	270.61	330.83	3.62	93.65	nr	**424.48**
Sum of two sides 2800 mm	284.81	348.18	3.74	96.75	nr	**444.93**
Sum of two sides 2900 mm	299.00	365.53	3.86	99.86	nr	**465.39**
Sum of two sides 3000 mm	313.20	382.89	3.98	102.96	nr	**485.85**
Sum of two sides 3100 mm	327.43	400.29	4.10	106.07	nr	**506.36**
Sum of two sides 3200 mm	341.61	417.62	4.22	109.17	nr	**526.79**
Sum of two sides 3300 mm	366.60	448.17	4.34	112.26	nr	**560.43**
Sum of two sides 3400 mm	381.23	466.05	4.46	115.37	nr	**581.42**
Sum of two sides 3500 mm	395.87	483.96	4.58	118.47	nr	**602.43**
Sum of two sides 3600 mm	410.51	501.85	4.70	121.58	nr	**623.43**
Sum of two sides 3700 mm	425.13	519.73	4.82	124.68	nr	**644.41**
Sum of two sides 3800 mm	439.77	537.62	4.94	127.79	nr	**665.41**
Sum of two sides 3900 mm	454.38	555.48	5.06	130.89	nr	**686.37**
Sum of two sides 4000 mm	469.04	573.40	5.18	134.00	nr	**707.40**
90° mitre bend						
Sum of two sides 1700 mm	259.76	317.56	3.87	100.12	nr	**417.68**
Sum of two sides 1800 mm	304.65	372.44	3.90	100.88	nr	**473.32**
Sum of two sides 1900 mm	349.55	427.33	3.93	101.66	nr	**528.99**
Sum of two sides 2000 mm	394.42	482.18	3.96	102.43	nr	**584.61**
Sum of two sides 2100 mm	439.32	537.07	3.82	98.82	nr	**635.89**
Sum of two sides 2200 mm	484.21	591.95	3.83	99.08	nr	**691.03**
Sum of two sides 2300 mm	529.11	646.84	3.83	99.08	nr	**745.92**

U: VENTILATION/AIR CONDITIONING SYSTEMS

Item	Net Price £	Material £	Labour hours	Labour £	Unit	Total rate £
U14: DUCTWORK: FIRE RATED – cont						
Extra over fittings – cont						
90° mitre bend – cont						
Sum of two sides 2400 mm	574.00	701.72	3.84	99.33	nr	**801.05**
Sum of two sides 2500 mm	618.89	756.60	3.85	99.60	nr	**856.20**
Sum of two sides 2600 mm	663.75	811.43	3.85	99.60	nr	**911.03**
Sum of two sides 2700 mm	708.66	866.34	3.86	99.86	nr	**966.20**
Sum of two sides 2800 mm	753.55	921.22	3.86	99.86	nr	**1021.08**
Sum of two sides 2900 mm	798.44	976.10	3.87	100.12	nr	**1076.22**
Sum of two sides 3000 mm	843.33	1030.98	3.87	100.12	nr	**1131.10**
Sum of two sides 3100 mm	888.22	1085.86	3.88	100.36	nr	**1186.22**
Sum of two sides 3200 mm	933.10	1140.73	3.88	100.36	nr	**1241.09**
Sum of two sides 3300 mm	1007.64	1231.85	3.89	100.62	nr	**1332.47**
Sum of two sides 3400 mm	1053.89	1288.39	3.90	100.88	nr	**1389.27**
Sum of two sides 3500 mm	1100.15	1344.94	3.91	101.15	nr	**1446.09**
Sum of two sides 3600 mm	1146.40	1401.48	3.92	101.40	nr	**1502.88**
Sum of two sides 3700 mm	1192.65	1458.03	3.93	101.66	nr	**1559.69**
Sum of two sides 3800 mm	1238.89	1514.55	3.94	101.92	nr	**1616.47**
Sum of two sides 3900 mm	1285.15	1571.10	3.95	102.18	nr	**1673.28**
Sum of two sides 4000 mm	1331.41	1627.66	3.96	102.43	nr	**1730.09**
Branch (Side-on Shoe)						
Sum of two sides 1700 mm	29.45	36.00	1.77	45.78	nr	**81.78**
Sum of two sides 1800 mm	32.18	39.34	1.81	46.82	nr	**86.16**
Sum of two sides 1900 mm	34.93	42.70	1.86	48.12	nr	**90.82**
Sum of two sides 2000 mm	37.67	46.05	1.90	49.14	nr	**95.19**
Sum of two sides 2100 mm	40.41	49.40	2.58	66.73	nr	**116.13**
Sum of two sides 2200 mm	43.14	52.74	2.61	67.52	nr	**120.26**
Sum of two sides 2300 mm	45.89	56.10	2.65	68.55	nr	**124.65**
Sum of two sides 2400 mm	48.62	59.44	2.68	69.33	nr	**128.77**
Sum of two sides 2500 mm	51.37	62.80	2.71	70.10	nr	**132.90**
Sum of two sides 2600 mm	54.11	66.15	2.75	71.14	nr	**137.29**
Sum of two sides 2700 mm	56.83	69.47	2.78	71.90	nr	**141.37**
Sum of two sides 2800 mm	59.58	72.84	2.81	72.69	nr	**145.53**
Sum of two sides 2900 mm	62.33	76.20	2.84	73.46	nr	**149.66**
Sum of two sides 3000 mm	65.06	79.54	2.87	74.24	nr	**153.78**
Sum of two sides 3100 mm	67.79	82.87	2.90	75.01	nr	**157.88**
Sum of two sides 3200 mm	70.54	86.23	2.93	75.80	nr	**162.03**
Sum of two sides 3300 mm	75.49	92.29	2.93	75.80	nr	**168.09**
Sum of two sides 3400 mm	78.32	95.75	3.00	77.60	nr	**173.35**
Sum of two sides 3500 mm	81.15	99.21	3.03	78.37	nr	**177.58**
Sum of two sides 3600 mm	83.95	102.63	3.06	79.16	nr	**181.79**
Sum of two sides 3700 mm	86.78	106.09	3.09	79.93	nr	**186.02**
Sum of two sides 3800 mm	89.61	109.55	3.12	80.71	nr	**190.26**
Sum of two sides 3900 mm	92.44	113.01	3.15	81.48	nr	**194.49**
Sum of two sides 4000 mm	95.24	116.43	3.18	82.27	nr	**198.70**

U: VENTILATION/AIR CONDITIONING SYSTEMS

Item	Net Price £	Material £	Labour hours	Labour £	Unit	**Total rate £**
Rectangular section ductwork to BS476 Part 24 (ISO 6944:1985) (Duraduct LT), ducts 'Type A' and 'Type B'; manufactured from 6 mm thick laminate fire board consisting of steel circular hole punched facings pressed to a fibre cement core; provides upto 4 hours stability, 4 hours integrity and 32 minutes insulation; including all necessary stiffeners, joints and supports in the running length						
Ductwork up to 600 mm longest side						
Sum of two sides 200 mm	221.29	270.53	4.00	103.47	m	**374.00**
Sum of two sides 400 mm	226.69	277.13	4.00	103.47	m	**380.60**
Sum of two sides 600 mm	236.89	289.59	4.00	103.47	m	**393.06**
Sum of two sides 800 mm	312.60	382.15	5.50	142.27	m	**524.42**
Sum of two sides 1000 mm	421.14	514.85	5.50	142.27	m	**657.12**
Sum of two sides 1200 mm	435.30	532.16	6.00	155.21	m	**687.37**
Extra over fittings; ductwork up to 600 mm longest side						
End cap						
Sum of two sides 200 mm	52.52	64.21	0.81	20.95	m	**85.16**
Sum of two sides 400 mm	52.52	64.21	0.87	22.51	m	**86.72**
Sum of two sides 600 mm	52.52	64.21	0.93	24.06	m	**88.27**
Sum of two sides 800 mm	84.83	103.70	0.98	25.36	m	**129.06**
Sum of two sides 1000 mm	84.83	103.70	1.22	31.57	m	**135.27**
Sum of two sides 1200 mm	129.25	158.01	1.28	33.10	m	**191.11**
Reducer						
Sum of two sides 200 mm	46.44	56.77	2.23	57.69	m	**114.46**
Sum of two sides 400 mm	46.44	56.77	2.51	64.93	m	**121.70**
Sum of two sides 600 mm	46.44	56.77	2.79	72.17	m	**128.94**
Sum of two sides 800 mm	90.88	111.11	3.01	77.86	m	**188.97**
Sum of two sides 1000 mm	90.88	111.11	3.17	82.01	m	**193.12**
Sum of two sides 1200 mm	115.12	140.74	3.28	84.84	m	**225.58**
Offset						
Sum of two sides 200 mm	46.44	56.77	2.95	76.31	m	**133.08**
Sum of two sides 400 mm	46.44	56.77	3.26	84.33	m	**141.10**
Sum of two sides 600 mm	46.44	56.77	3.57	92.35	m	**149.12**
Sum of two sides 800 mm	90.88	111.11	3.23	83.55	m	**194.66**
Sum of two sides 1000 mm	90.88	111.11	3.67	94.94	m	**206.05**
Sum of two sides 1200 mm	115.12	140.74	3.89	100.62	m	**241.36**
90° radius bend						
Sum of two sides 200 mm	155.51	190.11	2.06	53.29	m	**243.40**
Sum of two sides 400 mm	155.51	190.11	2.36	61.05	m	**251.16**
Sum of two sides 600 mm	155.51	190.11	2.66	68.81	m	**258.92**

U: VENTILATION/AIR CONDITIONING SYSTEMS

Item	Net Price £	Material £	Labour hours	Labour £	Unit	Total rate £
U14: DUCTWORK: FIRE RATED – cont						
Extra over fittings – cont						
90° radius bend – cont						
Sum of two sides 800 mm	181.77	222.21	2.97	76.84	m	**299.05**
Sum of two sides 1000 mm	181.77	222.21	3.04	78.64	m	**300.85**
Sum of two sides 1200 mm	206.01	251.85	3.09	79.93	m	**331.78**
45° radius bend						
Sum of two sides 200 mm	155.51	190.11	1.52	39.32	m	**229.43**
Sum of two sides 400 mm	155.51	190.11	1.66	42.95	m	**233.06**
Sum of two sides 600 mm	155.51	190.11	1.80	46.57	m	**236.68**
Sum of two sides 800 mm	181.77	222.21	1.93	49.92	m	**272.13**
Sum of two sides 1000 mm	181.77	222.21	2.05	53.03	m	**275.24**
Sum of two sides 1200 mm	206.01	251.85	2.17	56.14	m	**307.99**
90° mitre bend						
Sum of two sides 200 mm	155.51	190.11	2.47	63.90	m	**254.01**
Sum of two sides 400 mm	155.51	190.11	2.84	73.46	m	**263.57**
Sum of two sides 600 mm	155.51	190.11	3.20	82.77	m	**272.88**
Sum of two sides 800 mm	181.77	222.21	3.56	92.09	m	**314.30**
Sum of two sides 1000 mm	181.77	222.21	3.66	94.69	m	**316.90**
Sum of two sides 1200 mm	206.01	251.85	3.72	96.23	m	**348.08**
Branch						
Sum of two sides 200 mm	56.55	69.13	0.98	25.36	m	**94.49**
Sum of two sides 400 mm	56.55	69.13	1.14	29.49	m	**98.62**
Sum of two sides 600 mm	56.55	69.13	1.30	33.63	m	**102.76**
Sum of two sides 800 mm	70.68	86.41	1.46	37.76	m	**124.17**
Sum of two sides 1000 mm	77.07	94.21	1.46	37.76	m	**131.97**
Sum of two sides 1200 mm	94.91	116.03	1.55	40.10	m	**156.13**
Ductwork 601 to 1000 mm longest side						
Sum of two sides 1000 mm	318.21	389.01	6.00	155.21	m	**544.22**
Sum of two sides 1100 mm	443.10	541.70	6.00	155.21	m	**696.91**
Sum of two sides 1300 mm	450.50	550.74	6.00	155.21	m	**705.95**
Sum of two sides 1500 mm	461.01	563.59	6.00	155.21	m	**718.80**
Sum of two sides 1700 mm	557.16	681.13	6.00	155.21	m	**836.34**
Sum of two sides 1900 mm	638.49	780.55	6.00	155.21	m	**935.76**
Extra over fittings; ductwork 601 to 1000 mm longest side						
End cap						
Sum of two sides 1000 mm	86.35	105.56	1.25	32.33	m	**137.89**
Sum of two sides 1100 mm	131.57	160.84	1.25	32.33	m	**193.17**
Sum of two sides 1300 mm	131.57	160.84	1.31	33.89	m	**194.73**
Sum of two sides 1500 mm	131.57	160.84	1.36	35.18	m	**196.02**
Sum of two sides 1700 mm	232.31	284.00	1.42	36.74	m	**320.74**
Sum of two sides 1900 mm	232.31	284.00	1.48	38.29	m	**322.29**

U: VENTILATION/AIR CONDITIONING SYSTEMS

Item	Net Price £	Material £	Labour hours	Labour £	Unit	**Total rate £**
Reducer						
Sum of two sides 1000 mm	92.51	113.09	3.22	83.31	m	**196.40**
Sum of two sides 1100 mm	117.19	143.26	3.22	83.31	m	**226.57**
Sum of two sides 1300 mm	117.19	143.26	3.33	86.14	m	**229.40**
Sum of two sides 1500 mm	117.19	143.26	3.44	88.98	m	**232.24**
Sum of two sides 1700 mm	143.92	175.94	3.54	91.58	m	**267.52**
Sum of two sides 1900 mm	143.92	175.94	3.65	94.41	m	**270.35**
Offset						
Sum of two sides 1000 mm	92.51	113.09	3.78	97.78	m	**210.87**
Sum of two sides 1100 mm	117.19	143.26	3.78	97.78	m	**241.04**
Sum of two sides 1300 mm	117.19	143.26	4.00	103.47	m	**246.73**
Sum of two sides 1500 mm	117.19	143.26	4.23	109.43	m	**252.69**
Sum of two sides 1700 mm	143.92	175.94	4.45	115.11	m	**291.05**
Sum of two sides 1900 mm	143.92	175.94	4.67	120.81	m	**296.75**
90° radius bend						
Sum of two sides 1000 mm	185.03	226.20	3.78	97.78	m	**323.98**
Sum of two sides 1100 mm	209.70	256.36	3.07	79.41	m	**335.77**
Sum of two sides 1300 mm	209.70	256.36	3.12	80.71	m	**337.07**
Sum of two sides 1500 mm	209.70	256.36	3.17	82.01	m	**338.37**
Sum of two sides 1700 mm	275.49	336.79	3.22	83.31	m	**420.10**
Sum of two sides 1900 mm	275.49	336.79	3.28	84.84	m	**421.63**
45° bend						
Sum of two sides 1000 mm	209.70	256.36	1.91	49.41	m	**305.77**
Sum of two sides 1100 mm	209.70	256.36	2.00	51.74	m	**308.10**
Sum of two sides 1300 mm	209.70	256.36	2.02	52.25	m	**308.61**
Sum of two sides 1500 mm	209.70	256.36	2.13	55.09	m	**311.45**
Sum of two sides 1700 mm	275.49	336.79	2.24	57.95	m	**394.74**
Sum of two sides 1900 mm	275.49	336.79	2.35	60.79	m	**397.58**
90° mitre bend						
Sum of two sides 1000 mm	185.03	226.20	3.78	97.78	m	**323.98**
Sum of two sides 1100 mm	209.70	256.36	3.69	95.45	m	**351.81**
Sum of two sides 1300 mm	209.70	256.36	3.75	97.01	m	**353.37**
Sum of two sides 1500 mm	209.70	256.36	3.81	98.56	m	**354.92**
Sum of two sides 1700 mm	275.49	336.79	3.87	100.12	m	**436.91**
Sum of two sides 1900 mm	275.49	336.79	3.93	101.66	m	**438.45**
Branch						
Sum of two sides 1000 mm	74.81	91.45	2.38	61.56	m	**153.01**
Sum of two sides 1100 mm	96.61	118.11	1.50	38.80	m	**156.91**
Sum of two sides 1300 mm	96.61	118.11	1.59	41.13	m	**159.24**
Sum of two sides 1500 mm	96.61	118.11	1.68	43.46	m	**161.57**
Sum of two sides 1700 mm	123.35	150.79	1.77	45.78	m	**196.57**
Sum of two sides 1900 mm	123.35	150.79	1.86	48.12	m	**198.91**

U: VENTILATION/AIR CONDITIONING SYSTEMS

Item	Net Price £	Material £	Labour hours	Labour £	Unit	Total rate £
U14: DUCTWORK: FIRE RATED – cont						
Ductwork 1001 to 1250 mm longest side						
Sum of two sides 1300 mm	443.10	541.70	6.00	155.21	m	**696.91**
Sum of two sides 1500 mm	470.97	575.76	6.00	155.21	m	**730.97**
Sum of two sides 1700 mm	557.16	681.13	6.00	155.21	m	**836.34**
Sum of two sides 1900 mm	566.44	692.48	6.00	155.21	m	**847.69**
Sum of two sides 2100 mm	742.29	907.45	6.50	168.15	m	**1075.60**
Sum of two sides 2300 mm	782.62	956.76	6.50	168.15	m	**1124.91**
Sum of two sides 2500 mm	931.68	1138.98	6.50	168.15	m	**1307.13**
Extra over fittings; ductwork 1001 to 1250 mm longest side						
End cap						
Sum of two sides 1300 mm	131.57	160.84	1.31	33.89	m	**194.73**
Sum of two sides 1500 mm	131.57	160.84	1.36	35.18	m	**196.02**
Sum of two sides 1700 mm	232.31	284.00	1.42	36.74	m	**320.74**
Sum of two sides 1900 mm	232.31	284.00	1.48	38.29	m	**322.29**
Sum of two sides 2100 mm	316.60	387.04	2.66	68.81	m	**455.85**
Sum of two sides 2300 mm	316.60	387.04	2.95	76.31	m	**463.35**
Sum of two sides 2500 mm	316.60	387.04	3.24	83.81	m	**470.85**
Reducer						
Sum of two sides 1300 mm	117.19	143.26	3.33	86.14	m	**229.40**
Sum of two sides 1500 mm	117.19	143.26	3.44	88.98	m	**232.24**
Sum of two sides 1700 mm	143.92	175.94	3.54	91.58	m	**267.52**
Sum of two sides 1900 mm	143.92	175.94	3.65	94.41	m	**270.35**
Sum of two sides 2100 mm	170.62	208.59	3.75	97.01	m	**305.60**
Sum of two sides 2300 mm	170.62	208.59	3.85	99.60	m	**308.19**
Sum of two sides 2500 mm	170.62	208.59	3.95	102.18	m	**310.77**
Offset						
Sum of two sides 1300 mm	117.19	143.26	4.00	103.47	m	**246.73**
Sum of two sides 1500 mm	117.19	143.26	4.23	109.43	m	**252.69**
Sum of two sides 1700 mm	143.92	175.94	4.45	115.11	m	**291.05**
Sum of two sides 1900 mm	143.92	175.94	4.67	120.81	m	**296.75**
Sum of two sides 2100 mm	170.62	208.59	5.76	149.00	m	**357.59**
Sum of two sides 2300 mm	170.62	208.59	6.22	160.91	m	**369.50**
Sum of two sides 2500 mm	170.62	208.59	6.68	172.80	m	**381.39**
90° radius bend						
Sum of two sides 1300 mm	209.70	256.36	3.12	80.71	m	**337.07**
Sum of two sides 1500 mm	209.70	256.36	3.17	82.01	m	**338.37**
Sum of two sides 1700 mm	275.49	336.79	3.22	83.31	m	**420.10**
Sum of two sides 1900 mm	275.49	336.79	3.28	84.84	m	**421.63**
Sum of two sides 2100 mm	281.65	344.32	3.32	85.88	m	**430.20**
Sum of two sides 2300 mm	281.65	344.32	3.36	86.92	m	**431.24**
Sum of two sides 2500 mm	281.65	344.32	3.40	87.95	m	**432.27**

U: VENTILATION/AIR CONDITIONING SYSTEMS

Item	Net Price £	Material £	Labour hours	Labour £	Unit	**Total rate £**
45° bend						
Sum of two sides 1300 mm	209.70	256.36	2.02	52.25	m	**308.61**
Sum of two sides 1500 mm	209.70	256.36	2.13	55.09	m	**311.45**
Sum of two sides 1700 mm	275.49	336.79	2.24	57.95	m	**394.74**
Sum of two sides 1900 mm	275.49	336.79	2.35	60.79	m	**397.58**
Sum of two sides 2100 mm	281.65	344.32	2.89	74.76	m	**419.08**
Sum of two sides 2300 mm	281.65	344.32	3.13	80.97	m	**425.29**
Sum of two sides 2500 mm	281.65	344.32	3.37	87.18	m	**431.50**
90° mitre bend						
Sum of two sides 1300 mm	209.70	256.36	3.75	97.01	m	**353.37**
Sum of two sides 1500 mm	209.70	256.36	3.81	98.56	m	**354.92**
Sum of two sides 1700 mm	275.49	336.79	3.87	100.12	m	**436.91**
Sum of two sides 1900 mm	275.49	336.79	3.93	101.66	m	**438.45**
Sum of two sides 2100 mm	281.65	344.32	3.99	103.22	m	**447.54**
Sum of two sides 2300 mm	281.65	344.32	4.05	104.77	m	**449.09**
Sum of two sides 2500 mm	281.65	344.32	4.11	106.32	m	**450.64**
Branch						
Sum of two sides 1300 mm	96.61	118.11	1.59	41.13	m	**159.24**
Sum of two sides 1500 mm	96.61	118.11	1.68	43.46	m	**161.57**
Sum of two sides 1700 mm	123.35	150.79	1.77	45.78	m	**196.57**
Sum of two sides 1900 mm	123.35	150.79	1.86	48.12	m	**198.91**
Sum of two sides 2100 mm	129.50	158.31	2.58	66.73	m	**225.04**
Sum of two sides 2300 mm	129.50	158.31	2.64	68.29	m	**226.60**
Sum of two sides 2500 mm	129.50	158.31	2.91	75.28	m	**233.59**
Ductwork 1251 to 2000 mm longest side						
Sum of two sides 1800 mm	557.16	681.13	6.00	155.21	m	**836.34**
Sum of two sides 2000 mm	650.00	794.63	6.00	155.21	m	**949.84**
Sum of two sides 2200 mm	742.29	907.45	6.50	168.15	m	**1075.60**
Sum of two sides 2400 mm	834.75	1020.48	6.50	168.15	m	**1188.63**
Sum of two sides 2600 mm	879.42	1075.09	6.66	172.29	m	**1247.38**
Sum of two sides 2800 mm	964.64	1179.28	6.66	172.29	m	**1351.57**
Sum of two sides 3000 mm	1049.33	1282.81	6.66	172.29	m	**1455.10**
Sum of two sides 3200 mm	1057.17	1292.40	9.00	232.81	m	**1525.21**
Sum of two sides 3400 mm	1162.89	1421.64	9.00	232.81	m	**1654.45**
Sum of two sides 3600 mm	1183.88	1447.30	11.70	302.66	m	**1749.96**
Sum of two sides 3800 mm	1289.22	1576.09	11.70	302.66	m	**1878.75**
Sum of two sides 4000 mm	1491.44	1823.30	11.70	302.66	m	**2125.96**
Extra over fittings; ductwork 1251 to 2000 mm longest sides						
End cap						
Sum of two sides 1800 mm	232.31	284.00	1.45	37.50	m	**321.50**
Sum of two sides 2000 mm	232.31	284.00	1.51	39.06	m	**323.06**
Sum of two sides 2200 mm	316.60	387.04	2.80	72.43	m	**459.47**
Sum of two sides 2400 mm	316.60	387.04	3.10	80.20	m	**467.24**
Sum of two sides 2600 mm	433.76	530.27	3.39	87.69	m	**617.96**
Sum of two sides 2800 mm	433.76	530.27	3.68	95.19	m	**625.46**

U: VENTILATION/AIR CONDITIONING SYSTEMS

Item	Net Price £	Material £	Labour hours	Labour £	Unit	Total rate £
U14: DUCTWORK: FIRE RATED – cont						
Extra over fittings – cont						
End cap – cont						
Sum of two sides 3000 mm	433.76	530.27	3.98	102.96	m	**633.23**
Sum of two sides 3200 mm	585.90	716.26	4.27	110.46	m	**826.72**
Sum of two sides 3400 mm	585.90	716.26	4.57	118.21	m	**834.47**
Sum of two sides 3600 mm	721.56	882.11	4.87	125.98	m	**1008.09**
Sum of two sides 3800 mm	721.56	882.11	5.17	133.75	m	**1015.86**
Sum of two sides 4000 mm	721.56	882.11	5.47	141.50	m	**1023.61**
Reducer						
Sum of two sides 1800 mm	143.92	175.94	3.60	93.13	m	**269.07**
Sum of two sides 2000 mm	143.92	175.94	3.70	95.71	m	**271.65**
Sum of two sides 2200 mm	170.62	208.59	3.10	80.20	m	**288.79**
Sum of two sides 2400 mm	170.62	208.59	3.48	90.02	m	**298.61**
Sum of two sides 2600 mm	236.41	289.02	3.85	99.60	m	**388.62**
Sum of two sides 2800 mm	236.41	289.02	4.22	109.17	m	**398.19**
Sum of two sides 3000 mm	236.41	289.02	4.59	118.73	m	**407.75**
Sum of two sides 3200 mm	263.14	321.69	4.96	128.31	m	**450.00**
Sum of two sides 3400 mm	263.14	321.69	5.34	138.13	m	**459.82**
Sum of two sides 3600 mm	287.81	351.85	5.72	147.97	m	**499.82**
Sum of two sides 3800 mm	287.81	351.85	6.10	157.81	m	**509.66**
Sum of two sides 4000 mm	287.81	351.85	6.48	167.63	m	**519.48**
Offset						
Sum of two sides 1800 mm	143.92	175.94	4.56	117.96	m	**293.90**
Sum of two sides 2000 mm	143.92	175.94	5.53	143.05	m	**318.99**
Sum of two sides 2200 mm	170.62	208.59	5.99	154.95	m	**363.54**
Sum of two sides 2400 mm	170.62	208.59	6.45	166.85	m	**375.44**
Sum of two sides 2600 mm	236.41	289.02	6.91	178.76	m	**467.78**
Sum of two sides 2800 mm	236.41	289.02	7.37	190.65	m	**479.67**
Sum of two sides 3000 mm	236.41	289.02	7.83	202.55	m	**491.57**
Sum of two sides 3200 mm	263.14	321.69	8.29	214.46	m	**536.15**
Sum of two sides 3400 mm	263.14	321.69	8.75	226.34	m	**548.03**
Sum of two sides 3600 mm	287.81	351.85	9.21	238.24	m	**590.09**
Sum of two sides 3800 mm	287.81	351.85	9.67	250.15	m	**602.00**
Sum of two sides 4000 mm	287.81	351.85	10.13	262.04	m	**613.89**
90° radius bend						
Sum of two sides 1800 mm	275.49	336.79	3.25	84.07	m	**420.86**
Sum of two sides 2000 mm	275.49	336.79	3.30	85.37	m	**422.16**
Sum of two sides 2200 mm	281.65	344.32	3.34	86.40	m	**430.72**
Sum of two sides 2400 mm	281.65	344.32	3.38	87.44	m	**431.76**
Sum of two sides 2600 mm	431.72	527.78	3.42	88.48	m	**616.26**
Sum of two sides 2800 mm	431.72	527.78	3.46	89.50	m	**617.28**
Sum of two sides 3000 mm	431.72	527.78	3.50	90.54	m	**618.32**

U: VENTILATION/AIR CONDITIONING SYSTEMS

Item	Net Price £	Material £	Labour hours	Labour £	Unit	**Total rate £**
Sum of two sides 3200 mm	518.06	633.34	3.54	91.58	m	**724.92**
Sum of two sides 3400 mm	518.06	633.34	3.58	92.61	m	**725.95**
Sum of two sides 3600 mm	606.44	741.37	3.62	93.65	m	**835.02**
Sum of two sides 3800 mm	606.44	741.37	3.66	94.69	m	**836.06**
Sum of two sides 4000 mm	606.44	741.37	3.70	95.71	m	**837.08**
45° bend						
Sum of two sides 1800 mm	275.49	336.79	2.30	59.50	m	**396.29**
Sum of two sides 2000 mm	275.49	336.79	2.76	71.39	m	**408.18**
Sum of two sides 2200 mm	281.65	344.32	3.01	77.86	m	**422.18**
Sum of two sides 2400 mm	281.65	344.32	3.26	84.33	m	**428.65**
Sum of two sides 2600 mm	431.72	527.78	3.49	90.28	m	**618.06**
Sum of two sides 2800 mm	431.72	527.78	3.74	96.75	m	**624.53**
Sum of two sides 3000 mm	431.72	527.78	3.98	102.96	m	**630.74**
Sum of two sides 3200 mm	518.06	633.34	4.22	109.17	m	**742.51**
Sum of two sides 3400 mm	518.06	633.34	4.46	115.37	m	**748.71**
Sum of two sides 3600 mm	606.44	741.37	4.70	121.58	m	**862.95**
Sum of two sides 3800 mm	606.44	741.37	4.94	127.79	m	**869.16**
Sum of two sides 4000 mm	606.44	741.37	5.18	134.00	m	**875.37**
90° mitre band						
Sum of two sides 1800 mm	275.49	336.79	3.90	100.88	m	**437.67**
Sum of two sides 2000 mm	275.49	336.79	3.96	102.43	m	**439.22**
Sum of two sides 2200 mm	281.65	344.32	3.83	99.08	m	**443.40**
Sum of two sides 2400 mm	374.49	457.81	3.84	99.33	m	**557.14**
Sum of two sides 2600 mm	431.72	527.78	3.85	99.60	m	**627.38**
Sum of two sides 2800 mm	431.72	527.78	3.86	99.86	m	**627.64**
Sum of two sides 3000 mm	431.72	527.78	3.87	100.12	m	**627.90**
Sum of two sides 3200 mm	518.06	633.34	3.88	100.36	m	**733.70**
Sum of two sides 3400 mm	518.06	633.34	3.90	100.88	m	**734.22**
Sum of two sides 3600 mm	606.44	741.37	3.92	101.40	m	**842.77**
Sum of two sides 3800 mm	606.44	741.37	3.94	101.92	m	**843.29**
Sum of two sides 4000 mm	606.44	741.37	3.96	102.43	m	**843.80**
Branch						
Sum of two sides 1800 mm	123.35	150.79	1.81	46.82	m	**197.61**
Sum of two sides 2000 mm	123.35	150.79	1.90	49.14	m	**199.93**
Sum of two sides 2200 mm	129.50	158.31	2.61	67.52	m	**225.83**
Sum of two sides 2400 mm	129.50	158.31	2.68	69.33	m	**227.64**
Sum of two sides 2600 mm	154.18	188.49	2.75	71.14	m	**259.63**
Sum of two sides 2800 mm	154.18	188.49	2.81	72.69	m	**261.18**
Sum of two sides 3000 mm	154.18	188.49	2.87	74.24	m	**262.73**
Sum of two sides 3200 mm	160.36	196.04	2.93	75.80	m	**271.84**
Sum of two sides 3400 mm	160.36	196.04	3.00	77.60	m	**273.64**
Sum of two sides 3600 mm	205.58	251.32	3.06	79.16	m	**330.48**
Sum of two sides 3800 mm	205.58	251.32	3.12	80.71	m	**332.03**
Sum of two sides 4000 mm	205.58	251.32	3.18	82.27	m	**333.59**

U: VENTILATION/AIR CONDITIONING SYSTEMS

Item	Net Price £	Material £	Labour hours	Labour £	Unit	Total rate £
U30: LOW VELOCITY AIR CONDITIONING						
Y40 – AIR HANDLING UNITS						
Supply air handling unit; inlet with motorized damper, LTHW frost coil (at −5°C to +5°C), panel filter (EU4), bag filter (EU6), cooling coil (at 28°C db/20°C wb to 12°C db/11.5°C wb), LTHW heating coil (at 5°C to 21°C), supply fan, outlet plenum; includes access sections; all units located internally; Includes placing in position and fitting of sections together; electrical work elsewhere.						
Volume, external pressure						
2 m³/s at 350 Pa	5821.95	6575.28	40.00	988.45	nr	**7563.73**
2 m³/s at 700 Pa	6219.92	7024.75	40.00	988.45	nr	**8013.20**
5 m³/s at 350 Pa	9816.98	11087.25	65.00	1606.22	nr	**12693.47**
5 m³/s at 700 Pa	10160.39	11475.09	65.00	1606.22	nr	**13081.31**
8 m³/s at 350 Pa	14703.74	16606.33	77.00	1902.76	nr	**18509.09**
8 m/³s at 700 Pa	14932.95	16865.20	77.00	1902.76	nr	**18767.96**
10 m/³ at 350 Pa	16191.65	18286.76	100.00	2471.11	nr	**20757.87**
10 m³/s at 700 Pa	16964.68	19159.82	100.00	2471.11	nr	**21630.93**
13 m³/s at 350 Pa	20773.44	23461.42	108.00	2668.81	nr	**26130.23**
13 m³/s at 700 Pa	21415.03	24186.03	108.00	2668.81	nr	**26854.84**
15 m³/s at 350 Pa	23790.17	26868.50	120.00	2965.34	nr	**29833.84**
15 m³/s at 700 Pa	24223.03	27357.37	120.00	2965.34	nr	**30322.71**
18 m³/s at 350 Pa	26995.73	30488.84	133.00	3286.58	nr	**33775.42**
18 m³/s at 700 Pa	27491.99	31049.32	133.00	3286.58	nr	**34335.90**
20 m³/s at 350 Pa	31283.46	35331.38	142.00	3508.98	nr	**38840.36**
20 m³/s at 700 Pa	31733.35	35839.49	142.00	3508.98	nr	**39348.47**
Extra for inlet and discharge attenuators at 900 mm long						
2 m³/s at 350 Pa	1798.77	2031.52	5.00	123.55	nr	**2155.07**
5 m³/s at 350 Pa	3775.18	4263.67	10.00	247.12	nr	**4510.79**
10 m³/s at 700 Pa	5371.27	6066.29	13.00	321.25	nr	**6387.54**
15 m³/s at 700 Pa	8144.90	9198.81	16.00	395.38	nr	**9594.19**
20 m³/s at 700 Pa	10511.06	11871.14	20.00	494.22	nr	**12365.36**
Extra for locating units externally						
2 m³/s at 350 Pa	1332.80	1505.25	–	–	nr	**1505.25**
5 m³/s at 350 Pa	1691.06	1909.88	–	–	nr	**1909.88**
10 m³/s at 700 Pa	2705.37	3055.43	–	–	nr	**3055.43**
15 m³/s at 700 Pa	3909.65	4415.54	–	–	nr	**4415.54**
20 m³/s at 700 Pa	6776.69	7653.56	–	–	nr	**7653.56**

U: VENTILATION/AIR CONDITIONING SYSTEMS

Item	Net Price £	Material £	Labour hours	Labour £	Unit	**Total rate £**
Modular air handling unit with supply and extract sections. Supply side; inlet with motorized damper, LTHW frost coil (at −5°C to 5°C), panel filter (EU4), bag filter (EU6), cooling coil at 28°Cdb/20°Cwb to 12°Cdb/ 11.5°C wb), LTHW heating coil (at 5°C to 21°C), supply fan, outlet plenum. Extract side; inlet with motorized damper, extract fan; includes access sections; placing in position and fitting of sections together; electrical work elsewhere						
2 m^3/s at 350 Pa	8133.05	9185.43	50.00	1235.56	nr	**10420.99**
2 m^3/s at 700 Pa	8526.58	9629.88	50.00	1235.56	nr	**10865.44**
5 m^3/s at 350 Pa	14016.90	15830.61	86.00	2125.16	nr	**17955.77**
5 m^3/s at 700 Pa	14280.00	16127.76	86.00	2125.16	nr	**18252.92**
8 m^3/s at 350 Pa	19464.48	21983.09	105.00	2594.68	nr	**24577.77**
8 m/3 s at 700 Pa	19931.72	22510.79	105.00	2594.68	nr	**25105.47**
10 m/3 at 350 Pa	22475.67	25383.91	120.00	2965.34	nr	**28349.25**
10 m/s at 700 Pa	22959.87	25930.77	120.00	2965.34	nr	**28896.11**
13 m^3/s at 350 Pa	28922.86	32665.34	130.00	3212.45	nr	**35877.79**
13 m^3/s at 700 Pa	29830.04	33689.90	130.00	3212.45	nr	**36902.35**
15 m^3/s at 350 Pa	32319.27	36501.22	145.00	3583.12	nr	**40084.34**
15 m^3/s at 700 Pa	33848.24	38228.03	145.00	3583.12	nr	**41811.15**
18 m^3/s at 350 Pa	36944.82	41725.29	160.00	3953.79	nr	**45679.08**
18 m^3/s at 700 Pa	37809.79	42702.19	160.00	3953.79	nr	**46655.98**
20 m^3/s at 350 Pa	40379.19	45604.06	175.00	4324.45	nr	**49928.51**
20 m^3/s at 700 Pa	41782.94	47189.44	175.00	4324.45	nr	**51513.89**
Extra for inlet and discharge attenuators at 900 mm long						
2 m^3/s at 350 Pa	3279.23	3703.55	8.00	197.69	nr	**3901.24**
5 m^3/s at 350 Pa	6475.55	7313.45	10.00	247.12	nr	**7560.57**
10 m^3/s at 700 Pa	9265.98	10464.96	13.00	321.25	nr	**10786.21**
15 m^3/s at 700 Pa	15849.87	17900.77	16.00	395.38	nr	**18296.15**
20 m^3/s at 700 Pa	18502.18	20896.27	20.00	494.22	nr	**21390.49**
Extra for locating units externally						
2 m^3/s at 350 Pa	2624.88	2964.53	–	–	nr	**2964.53**
5 m^3/s at 350 Pa	3727.32	4209.62	–	–	nr	**4209.62**
10 m^3/s at 700 Pa	5320.50	6008.94	–	–	nr	**6008.94**
15 m^3/s at 700 Pa	10102.22	11409.39	–	–	nr	**11409.39**
20 m^3/s at 700 Pa	15530.19	17539.72	–	–	nr	**17539.72**
Extra for humidifier, self generating type						
2 m^3/s at 350 Pa (10 kg/hr)	2287.40	2583.37	5.00	123.55	nr	**2706.92**
5 m^3/s at 350 Pa (18 kg/hr)	2830.71	3196.99	5.00	123.55	nr	**3320.54**
10 m^3/s at 700 Pa (30 kg/hr)	3221.20	3638.00	6.00	148.27	nr	**3786.27**
15 m^3/s at 700 Pa (60 kg/hr)	5965.46	6737.37	8.00	197.69	nr	**6935.06**
20 m^3/s at 700 Pa (90 kg/hr)	8969.04	10129.59	10.00	247.12	nr	**10376.71**

U: VENTILATION/AIR CONDITIONING SYSTEMS

Item	Net Price £	Material £	Labour hours	Labour £	Unit	Total rate £
U30: LOW VELOCITY AIR CONDITIONING – cont						
Extra for mixing box						
$2\,m^3/s$ at 350 Pa	1098.36	1240.48	4.00	98.85	nr	**1339.33**
$5\,m^3/s$ at 350 Pa	1548.19	1748.51	4.00	98.85	nr	**1847.36**
$10\,m^3/s$ at 700 Pa	2169.02	2449.68	5.00	123.55	nr	**2573.23**
$15\,m^3/s$ at 700 Pa	2894.84	3269.42	6.00	148.27	nr	**3417.69**
$20\,m^3/s$ at 700 Pa	4413.10	4984.13	6.00	148.27	nr	**5132.40**
Extra for runaround coil, including pump and associated pipework; typical outputs in brackets (based on minimal distance between the supply and extract units)						
$2\,m^3/s$ at 350 Pa (26 kW)	3288.15	3713.62	30.00	741.34	nr	**4454.96**
$5\,m^3/s$ at 350 Pa (37 kW)	5829.97	6584.33	30.00	741.34	nr	**7325.67**
$10\,m^3/s$ at 700 Pa (85 kW)	10093.69	11399.76	40.00	988.45	nr	**12388.21**
$15\,m^3/s$ at 700 Pa (151 kW)	14202.18	16039.87	50.00	1235.56	nr	**17275.43**
$20\,m^3/s$ at 700 Pa (158 kW)	19359.60	21864.64	60.00	1482.67	nr	**23347.31**
Extra for thermal wheel (typical outputs in brackets)						
$2\,m^3/s$ at 350 Pa (37 kW)	6801.10	7681.13	12.00	296.53	nr	**7977.66**
$5\,m^3/s$ at 350 Pa (65 kW)	9768.03	11031.96	12.00	296.53	nr	**11328.49**
$10\,m^3/s$ at 700 Pa (127 kW)	13026.39	14711.94	15.00	370.67	nr	**15082.61**
$15\,m^3/s$ at 700 Pa (160 kW)	24405.92	27563.92	17.00	420.09	nr	**27984.01**
$20\,m^3/s$ at 700 Pa (262 kW)	28508.95	32197.86	19.00	469.52	nr	**32667.38**
Extra for plate heat exchanger, including additional filtration in extract leg (typical outputs in brackets)						
$2\,m^3/s$ at 350 Pa (25 kW)	4310.42	4868.16	12.00	296.53	nr	**5164.69**
$5\,m^3/s$ at 350 Pa (51 kW)	8426.31	9516.63	12.00	296.53	nr	**9813.16**
$10\,m^3/s$ at 700 Pa (98 kW)	12363.74	13963.55	15.00	370.67	nr	**14334.22**
$15\,m^3/s$ at 700 Pa (160 kW)	21351.16	24113.89	17.00	420.09	nr	**24533.98**
$20\,m^3/s$ at 700 Pa (190 kW)	27573.22	31141.06	19.00	469.52	nr	**31610.58**
Extra for electric heating in lieu of LTHW						
$2\,m^3/s$ at 350 Pa	1493.94	1687.25	–	–	nr	**1687.25**
$5\,m^3/s$ at 350 Pa	2799.06	3161.25	–	–	nr	**3161.25**
$10\,m^3/s$ at 700 Pa	3779.16	4268.16	–	–	nr	**4268.16**
$15\,m^3/s$ at 700 Pa	3962.80	4475.57	–	–	nr	**4475.57**
$20\,m^3/s$ at 700 Pa	4676.68	5281.82	–	–	nr	**5281.82**

U: VENTILATION/AIR CONDITIONING SYSTEMS

Item	Net Price £	Material £	Labour hours	Labour £	Unit	**Total rate £**
U31: VAV AIR CONDITIONING						
VAV TERMINAL BOXES						
VAV terminal box; integral acoustic silencer; factory installed and prewired control components (excluding electronic controller); selected at 200 Pa at entry to unit; includes fixing in position; electrical work elsewhere						
80 l/s – 110 l/s	448.80	506.87	2.00	47.80	nr	**554.67**
Extra for secondary silencer	109.14	123.27	0.50	11.94	nr	**135.21**
Extra for 2 row LTHW heating coil	44.27	50.00	–	–	nr	**50.00**
150 l/s – 190 l/s	472.26	533.37	2.00	47.80	nr	**581.17**
Extra for secondary silencer	119.85	135.35	0.50	11.94	nr	**147.29**
Extra for 2 row LTHW heating coil	53.09	59.96	–	–	nr	**59.96**
250 l/s – 310 l/s	526.83	595.00	2.00	47.80	nr	**642.80**
Extra for secondary silencer	158.61	179.13	0.50	11.94	nr	**191.07**
Extra for 2 row LTHW heating coil	29.66	33.50	–	–	nr	**33.50**
420 l/s – 520 l/s	570.69	644.54	2.00	47.80	nr	**692.34**
Extra for secondary silencer	180.54	203.90	0.50	11.94	nr	**215.84**
Extra for 2 row LTHW heating coil	72.29	81.64	–	–	nr	**81.64**
650 l/s – 790 l/s	691.05	780.47	2.00	47.80	nr	**828.27**
Extra for secondary silencer	231.03	260.93	0.50	11.94	nr	**272.87**
Extra for 2 row LTHW heating coil	94.40	106.61	–	–	nr	**106.61**
1130 l/s – 1370 l/s	802.74	906.61	2.00	47.80	nr	**954.41**
Extra for secondary silencer	322.83	364.60	0.50	11.94	nr	**376.54**
Extra for 2 row LTHW heating coil	120.22	135.77	–	–	nr	**135.77**
Extra for electric heater & thyristor controls, 3 kw/lph (per box)	350.92	396.33	–	–	nr	**396.33**
Extra for fitting free issue box controller	–	–	2.13	50.90	nr	**50.90**
Fan assisted VAV terminal box; factory installed and prewired control components (excluding electronic controller); selected at 40 Pa external static pressure; includes fixing in position, electrical work elsewhere						
100 l/s – 175 l/s	993.48	1122.03	3.00	71.70	nr	**1193.73**
Extra for secondary silencer	118.32	133.63	0.50	11.94	nr	**145.57**
Extra for 1 row LTHW heating coil	62.67	70.78	–	–	nr	**70.78**
170 l/s – 360 l/s	1083.75	1223.98	3.00	71.70	nr	**1295.68**
Extra for secondary silencer	149.94	169.34	0.50	11.94	nr	**181.28**
Extra for 1 row LTHW heating coil	71.49	80.74	–	–	nr	**80.74**
300 l/s – 640 l/s	1242.36	1403.11	3.00	71.70	nr	**1474.81**
Extra for secondary silencer	229.50	259.19	0.50	11.94	nr	**271.13**
Extra for 1 row LTHW heating coil	79.63	89.93	–	–	nr	**89.93**
620 l/s – 850 l/s	1242.36	1403.11	3.00	71.70	nr	**1474.81**
Extra for secondary silencer	229.50	259.19	0.50	11.94	nr	**271.13**
Extra for 1 row LTHW heating coil	79.63	89.93	–	–	nr	**89.93**
Extra for electric heater plus thyristor controls, 3 kw/lph (per box)	350.92	396.33	–	–	nr	**396.33**
Extra for fitting free issue controller	–	–	2.13	50.90	nr	**50.90**

U: VENTILATION/AIR CONDITIONING SYSTEMS

Item	Net Price £	Material £	Labour hours	Labour £	Unit	**Total rate £**
U41: FAN COIL AIR CONDITIONING						
FAN COIL UNITS						
All selections based on summer return air condition of 23°C @ 50% RH, CHW @ 6°/12°C, LTHW @ 82°/71°C (where applicable), medium speed, external resistance of 30 Pa.						
All selections are based on heating and cooling units. For waterside control units there is no significant reduction in cost between 4 pipe heating and cooling and 2 pipe cooling only units (excluding controls). For airside control units, there is a marginal reduction (less than 5%) between 4 pipe heating and cooling units and 2 pipe cooling only units (excluding controls).						
Ceiling void mounted horizontal waterside control fan coil unit; cooling coil; LTHW heating coil; multi-tapped speed transformer; fine wire mesh filter; includes fixing in position; electrical work elsewhere						
Total cooling load, heating load						
2800 W, 1000 W	294.04	332.09	4.00	98.85	nr	**430.94**
4000 W, 1700 W	341.86	386.09	4.00	98.85	nr	**484.94**
4500 W, 1900 W	377.85	426.74	4.00	98.85	nr	**525.59**
6000 W, 2600 W	442.66	499.94	4.00	98.85	nr	**598.79**
Ceiling void mounted horizontal waterside control fan coil unit; cooling coil; electric heating coil; multi-tapped speed transformer; fine wire mesh filter; includes fixing in position; electrical work elsewhere + thyristor and 2 No. HTCOs						
Total cooling load, heating load						
2800 W, 1500 W.	431.74	487.60	4.00	98.85	nr	**586.45**
4000 W, 2000 W	479.56	541.61	4.00	98.85	nr	**640.46**
4500 W, 2000 W	515.55	582.26	4.00	98.85	nr	**681.11**
6000 W, 3000 W	580.36	655.45	4.00	98.85	nr	**754.30**

U: VENTILATION/AIR CONDITIONING SYSTEMS

Item	Net Price £	Material £	Labour hours	Labour £	Unit	**Total rate £**
Ceiling void mounted horizontal airside control fan coil unit; cooling coil; LHTW heating coil; multi-tapped speed transformer; fine wire mesh filter, damper actuator & fixing kit; includes fixing in position; electrical work elsewhere						
Total cooling load, heating load						
2600 W, 2200 W	600.10	677.75	4.00	98.85	nr	**776.60**
3600 W, 3200 W	643.06	726.27	4.00	98.85	nr	**825.12**
4000 W, 3600 W	708.42	800.09	4.00	98.85	nr	**898.94**
5400 W, 5000 W	775.53	875.88	4.00	98.85	nr	**974.73**
Ceiling void mounted horizontal airside control fan coil unit; cooling coil ; electric heating coil; multi-tapped speed transformer; fine wire mesh filter, damper actuator & fixing kit; includes fixing in position; electrical work elsewhere + thyristor and 2 No. HTCOs						
Total cooling load, heating load						
2600 W, 1500 W	710.26	802.17	4.00	98.85	nr	**901.02**
3600 W, 2000 W	753.22	850.68	4.00	98.85	nr	**949.53**
4000 W, 2000 W	818.58	924.50	4.00	98.85	nr	**1023.35**
5400 W, 3000 W	885.69	1000.29	4.00	98.85	nr	**1099.14**
Ceiling void mounted slimline horizontal waterside control fan coil unit, 170 mm deep; cooilng coil; LTHW heating coil; multi-tapped speed transformer; fine wire mesh filter; includes fixing in position; electrical work elsewhere						
Total cooling load, heating load						
1100 W, 1500 W	478.02	539.88	3.50	86.49	nr	**626.37**
3200 W, 3700 W	585.30	661.03	4.00	98.85	nr	**759.88**
Ceiling void mounted slimline horizontal waterside control fan coil unit, 170 mm deep; cooilng coil; electric heating coil; multi-tapped speed transformer; fine wire mesh filter; includes fixing in position; electrical work elsewhere						
Total cooling load, heating load						
1100 W, 1000 W	586.67	662.58	3.50	86.49	nr	**749.07**
3400 W, 2000 W	693.94	783.74	4.00	98.85	nr	**882.59**

U: VENTILATION/AIR CONDITIONING SYSTEMS

Item	Net Price £	Material £	Labour hours	Labour £	Unit	Total rate £
U41: FAN COIL AIR CONDITIONING – cont						
All selections are based on heating and cooling units – cont						
Ceiling void mounted slimline horizontal airside control fan coil unit, 170 mm deep; cooling coil; LTHW heating coil; multi-tapped speed transformer; fine wire mesh filter, damper actuator & fixing kit; includes fixing in position; electrical work elsewhere						
Total cooling load, heating load						
1000 W, 1600 W	517.27	584.21	3.50	86.49	nr	**670.70**
3000 W, 3300 W	805.94	910.23	4.00	98.85	nr	**1009.08**
4500 W, 4500 W	1033.12	1166.80	4.00	98.85	nr	**1265.65**
Ceiling void mounted slimline horizontal airside control fan coil unit, 170 mm deep; cooilng coil; electric heating coil; multi-tapped speed transformer; fine wire mesh filter, damper actuator & fixing kit; includes fixing in position; electrical work elsewhere						
Total cooling load, heating load						
1000 W, 1000 W	761.99	860.58	3.50	86.49	nr	**947.07**
3000 W, 2000 W	1003.74	1133.62	4.00	98.85	nr	**1232.47**
4500 W, 3000 W	1274.83	1439.79	4.00	98.85	nr	**1538.64**
Low level perimeter waterside control fan coil unit; cooling coil; LTHW heating coil; multi-tapped speed transformer; fine wire mesh filter; includes fixing in position; electrical work elsewhere						
Total cooling load, heating load						
1700 W, 1400 W	366.34	413.74	3.50	86.49	nr	**500.23**
Extra over for standard cabinet	194.89	220.11	1.00	24.71	nr	**244.82**
2200 W, 1900 W	424.95	479.94	3.50	86.49	nr	**566.43**
Extra over for standard cabinet	237.39	268.11	1.00	24.71	nr	**292.82**
2600 W, 2200 W	468.91	529.58	3.50	86.49	nr	**616.07**
Extra over for standard cabinet	243.25	274.73	1.00	24.71	nr	**299.44**
3900 W, 3200 W	615.45	695.09	3.50	86.49	nr	**781.58**
Extra over for standard cabinet	265.23	299.55	1.00	24.71	nr	**324.26**
4600 W, 3900 W	703.36	794.38	3.50	86.49	nr	**880.87**
Extra over for standard cabinet	348.76	393.88	1.00	24.71	nr	**418.59**

U: VENTILATION/AIR CONDITIONING SYSTEMS

Item	Net Price £	Material £	Labour hours	Labour £	Unit	**Total rate £**
Low level perimeter waterside control fan coil unit; cooling coil; electric heating coil; multi-tapped speed transformer; fine wire mesh filter; includes fixing in position; electrical work elsewhere						
Total cooling load, heating load						
1700 W, 1500 W	476.25	537.88	3.50	86.49	nr	**624.37**
Extra over for standard cabinet	194.89	220.11	1.00	24.71	nr	**244.82**
2200 W, 2000 W	498.23	562.70	3.50	86.49	nr	**649.19**
Extra over for standard cabinet	237.39	268.11	1.00	24.71	nr	**292.82**
2600 W, 2000 W	622.78	703.36	3.50	86.49	nr	**789.85**
Extra over for standard cabinet	243.25	274.73	1.00	24.71	nr	**299.44**
3800 W, 3000 W	740.01	835.77	3.50	86.49	nr	**922.26**
Extra over for standard cabinet	265.23	299.55	1.00	24.71	nr	**324.26**
4600 W, 4000 W	901.20	1017.81	3.50	86.49	nr	**1104.30**
Extra over for standard cabinet	348.76	393.88	1.00	24.71	nr	**418.59**
Low level perimeter airside control fan coil unit; cooling coil; LTHW heating coil; multi-tapped speed transformer; fine wire mesh filter; damper actuator & fixing kit; includes fixing in position; electrical work elsewhere						
Total cooling load, heating load						
1200 W, 1400 W	473.31	534.55	3.50	86.49	nr	**621.04**
Extra over for standard cabinet	194.89	220.11	1.00	24.71	nr	**244.82**
1800 W, 2000 W	529.00	597.45	3.50	86.49	nr	**683.94**
Extra over for standard cabinet	237.39	268.11	1.00	24.71	nr	**292.82**
2200 W, 2400 W	570.03	643.79	3.50	86.49	nr	**730.28**
Extra over for standard cabinet	243.25	274.73	1.00	24.71	nr	**299.44**
3200 W, 3600 W	740.01	835.77	3.50	86.49	nr	**922.26**
Extra over for standard cabinet	265.23	299.55	1.00	24.71	nr	**324.26**
Low level perimeter airside control fan coil unit; cooling coil; electric heating coil; multi-tapped speed transformer; fine wire mesh filter; damper actuator & fixing kit; includes fixing in position; electrical work elsewhere						
Total cooling load, heating load						
1250 W, 1500 W	644.76	728.19	3.50	86.49	nr	**814.68**
Extra over for standard cabinet	194.89	220.11	1.00	24.71	nr	**244.82**
1900 W, 2000 W	696.06	786.12	3.50	86.49	nr	**872.61**
Extra over for standard cabinet	237.39	268.11	1.00	24.71	nr	**292.82**
2300 W, 2000 W	795.69	898.65	3.50	86.49	nr	**985.14**
Extra over for standard cabinet	243.25	274.73	1.00	24.71	nr	**299.44**
3300 W, 3000 W	908.52	1026.08	3.50	86.49	nr	**1112.57**
Extra over for standard cabinet	265.23	299.55	1.00	24.71	nr	**324.26**

U: VENTILATION/AIR CONDITIONING SYSTEMS

Item	Net Price £	Material £	Labour hours	Labour £	Unit	Total rate £
U41: FAN COIL AIR CONDITIONING – cont						
Typical DDC control pack for airside control units; controller, return air sensor (damper and damper actuator included in above rates)						
Heating and cooling, per unit	156.06	176.26	2.00	49.42	nr	**225.68**
Cooling only, per unit	183.60	207.36	2.00	49.42	nr	**256.78**
Typical DDC control pack for waterside control units; controller, return air sensor, four port valves and actuators						
Heating and cooling, per unit	137.70	155.52	2.00	49.42	nr	**204.94**
Cooling only, per unit	119.34	134.78	2.00	49.42	nr	**184.20**
Note: Care needs to be taken when using these controls prices, as they can vary significantly, depending on the equipment manufacturer specified and the degree of control required						

U: VENTILATION/AIR CONDITIONING SYSTEMS

Item	Net Price £	Material £	Labour hours	Labour £	Unit	Total rate £
U70: AIR CURTAINS						
The selection of air curtains requires consideration of the particular conditions involved; climatic conditions, wind influence, construction and position all influence selection; consultation with a specialist manufacturer is therefore advisable.						
Commercial grade air curtains; recessed or exposed units with rigid sheet steel casing; aluminium grilles; high quality motor/centrifugal fan assembly; includes fixing in position; electrical work elsewhere						
Ambient temperature; 240 V single phase supply; mounting height 2.40 m						
1000 × 590 × 270 mm	2447.00	2763.63	12.05	297.72	nr	**3061.35**
1500 × 590 × 270 mm	3132.00	3537.27	12.05	297.72	nr	**3834.99**
2000 × 590 × 270 mm	3789.00	4279.28	12.05	297.77	nr	**4577.05**
2500 × 590 × 270 mm	4212.00	4757.01	13.00	321.25	nr	**5078.26**
Ambient temperature; 240 V single phase supply; mounting height 2.80 m						
1000 × 590 × 270 mm	2827.00	3192.80	16.13	398.56	nr	**3591.36**
1500 × 590 × 270 mm	3609.00	4075.99	16.13	398.56	nr	**4474.55**
2000 × 590 × 270 mm	4499.00	5081.15	16.13	398.56	nr	**5479.71**
2500 × 590 × 270 mm	5202.00	5875.11	17.10	422.57	nr	**6297.68**
Ambient temperature 240 V single phase supply; mounting height 3.30 m						
1000 × 774 × 370 mm	3693.00	4170.85	17.24	426.05	nr	**4596.90**
1500 × 774 × 370 mm	4994.00	5640.20	17.24	426.05	nr	**6066.25**
2000 × 774 × 370 mm	6177.00	6976.27	17.24	426.05	nr	**7402.32**
2500 × 774 × 370 mm	7808.00	8818.32	18.30	452.21	nr	**9270.53**
Ambient temperature; 240 V single phase supply; mounting height 4.00 m						
1000 × 774 × 370 mm	4079.00	4606.80	19.10	471.98	nr	**5078.78**
1500 × 774 × 370 mm	5351.00	6043.39	19.10	471.98	nr	**6515.37**
2000 × 774 × 370 mm	6662.00	7524.03	19.10	471.98	nr	**7996.01**
2500 × 774 × 370 mm	7808.00	8818.32	19.90	491.75	nr	**9310.07**
Water heated; 240 V single phase supply; mounting height 2.40 m						
1000 × 590 × 270 mm; 2.30–9.40 kW output	2879.00	3251.53	12.05	297.72	nr	**3549.25**
1500 × 590 × 270 mm; 3.50–14.20 kW output	3685.00	4161.82	12.05	297.72	nr	**4459.54**
2000 × 590 × 270 mm; 4.70–19.00 kW output	4458.00	5034.84	12.05	297.72	nr	**5332.56**
2500 × 590 × 270 mm; 5.90–23.70 kW output	4956.00	5597.28	13.00	321.25	nr	**5918.53**

U: VENTILATION/AIR CONDITIONING SYSTEMS

Item	Net Price £	Material £	Labour hours	Labour £	Unit	Total rate £
U70: AIR CURTAINS – cont						
Commercial grade air curtains – cont						
Water heated; 240 V single phase supply; mounting height 2.80 m						
1000 × 590 × 270 mm; 3.30–11.90 kW output	3327.00	3757.50	16.13	398.56	nr	**4156.06**
1500 × 590 × 270 mm; 5.00–17.90 kW output	4247.00	4796.54	16.13	398.56	nr	**5195.10**
2000 × 590 × 270 mm; 6.70–23.90 kW output	5294.00	5979.02	16.13	398.56	nr	**6377.58**
2500 × 590 × 270 mm; 8.30–29.80 kW output	6121.00	6913.03	17.10	422.57	nr	**7335.60**
Water heated; 240 V single phase supply; mounting height 3.30 m						
1000 × 774 × 370 mm; 6.10–21.80 kW output	4345.00	4907.22	17.24	426.05	nr	**5333.27**
1500 × 774 × 370 mm; 9.20–32.80 kW output	5876.00	6636.32	17.24	426.05	nr	**7062.37**
2000 × 774 × 370 mm; 12.30–43.70 kW output	7268.00	8208.44	17.24	426.05	nr	**8634.49**
2500 × 774 × 370 mm; 15.30–54.60 kW output	8603.00	9716.18	18.30	452.21	nr	**10168.39**
Water heated; 240 V single phase supply; mounting height 4.00 m						
1000 × 774 × 370 mm; 7.20–24.20 kW output	4799.00	5419.96	19.10	471.98	nr	**5891.94**
1500 × 774 × 370 mm; 10.90–36.30 kW output	6296.00	7110.68	19.10	471.98	nr	**7582.66**
2000 × 774 × 370 mm; 14.50–48.40 kW output	7838.00	8852.20	19.10	471.98	nr	**9324.18**
2500 × 774 × 370 mm; 18.10–60.60 kW output	9186.00	10374.62	19.90	491.75	nr	**10866.37**
Electrically heated; 415 V three phase supply; mounting height 2.40 m						
1000 × 590 × 270 mm; 2.30–9.40 kW output	3502.00	3955.14	12.05	297.72	nr	**4252.86**
1500 × 590 × 270 mm; 3.50–14.20 kW output	4357.00	4920.78	12.05	297.72	nr	**5218.50**
2000 × 590 × 270 mm; 4.70–19.00 kW output	5193.00	5864.95	12.05	329.16	nr	**6194.11**
2500 × 590 × 270 mm; 5.90–23.70 kW output	5904.00	6667.94	13.00	355.16	nr	**7023.10**

U: VENTILATION/AIR CONDITIONING SYSTEMS

Item	Net Price £	Material £	Labour hours	Labour £	Unit	Total rate £
Electrically heated; 415 V three phase supply; mounting height 2.80 m						
1000 × 590 × 270 mm; 3.30–11.90 kW output	4123.00	4656.49	16.13	398.56	nr	**5055.05**
1500 × 590 × 270 mm; 5.00–17.90 kW output	5070.00	5726.04	16.13	398.56	nr	**6124.60**
2000 × 590 × 270 mm; 6.70–23.90 kW output	6402.00	7230.38	16.13	398.56	nr	**7628.94**
2500 × 590 × 270 mm; 8.30–29.80 kW output	7155.00	8080.82	17.10	422.57	nr	**8503.39**
Electrically heated; 415 V three phase supply; mounting height 3.30 m						
1000 × 774 × 370 mm; 6.10–21.80 kW output	6265.00	7075.66	17.24	426.02	nr	**7501.68**
1500 × 774 × 370 mm; 9.20–32.80 kW output	8394.00	9480.15	17.24	426.02	nr	**9906.17**
2000 × 774 × 370 mm; 12.30–43.70 kW output	10131.00	11441.90	17.24	426.02	nr	**11867.92**
2500 × 774 × 370 mm; 15.30–54.60 kW output	11993.00	13544.83	18.30	452.21	nr	**13997.04**
Electrically heated; 415 V three phase supply; mounting height 4.00 m						
1000 × 774 × 370 mm; 7.20–24.20 kW output	6914.00	7808.64	19.10	471.98	nr	**8280.62**
1500 × 774 × 370 mm; 10.90–36.30 kW output	8970.00	10130.67	19.10	471.98	nr	**10602.65**
2000 × 774 × 370 mm; 14.50–48.40 kW output	10849.00	12252.81	19.10	471.98	nr	**12724.79**
2500 × 774 × 370 mm; 18.10–60.60 kW output	12851.00	14513.86	19.90	491.75	nr	**15005.61**
Industrial grade air curtains; recessed or exposed units with rigid sheet steel casing; aluminium grilles; high quality motor/centrifugal fan assembly; includes fixing in position; electrical work elsewhere						
Ambient temperature; 415 V three phase supply; including wiring between multiple units; horizontally or vertically mounted; opening maximum 6.00 m						
1500 × 585 × 853 mm; 1.6 A supply	2170.00	2450.78	17.24	426.05	nr	**2876.83**
2000 × 585 × 853 mm; 2.1 A supply	2645.00	2987.25	17.24	426.05	nr	**3413.30**

U: VENTILATION/AIR CONDITIONING SYSTEMS

Item	Net Price £	Material £	Labour hours	Labour £	Unit	Total rate £
U70: AIR CURTAINS – cont						
Industrial grade air curtains – cont						
Water heated; 415 V three phase supply; including wiring between multiple units; horizontally or vertically mounted; opening maximum 6.00 m						
1500 × 585 × 956 mm; 1.6 A supply; 34.80 kW output	2511.00	2835.91	17.24	426.05	nr	**3261.96**
2000 × 585 × 956 mm; 2.1 A supply; 50.70 kW output	3030.00	3422.07	17.24	426.05	nr	**3848.12**
Water heated; 415 V three phase supply; including wiring between multiple units; vertically mounted in single bank for openings maximum 6.00 m wide or opposing twin banks for openings maximum 10.00 m wide						
1500 × 585 mm; 1.6 A supply; 41.1 kW output	2511.00	2835.91	17.24	426.05	nr	**3261.96**
2000 × 585 mm; 2.1 A supply; 57.7 kW output	3030.00	3422.07	17.24	426.05	nr	**3848.12**
Remote mounted electronic controller unit; 415 V three phase supply; excluding wiring to units						
Five speed, 7 A	481.00	543.24	5.00	123.55	nr	**666.79**

Electrical Installations

Material Costs/Measured Work Prices

DIRECTIONS

The following explanations are given for each of the column headings and letter codes.

Unit	Prices for each unit are given as singular (i.e. 1 metre, 1 nr) unless stated otherwise.
Net price	Industry tender prices, plus nominal allowance for fixings, waste and applicable trade discounts.
Material cost	Net price plus percentage allowance for overheads, profit and preliminaries.
Labour norms	In man-hours for each operation.
Labour cost	Labour constant multiplied by the appropriate all-in man-hour cost based on gang rate (See also relevant Rates of Wages Section) plus percentage allowance for overheads, profit and preliminaries.
Measured work price (total rate)	Material cost plus Labour cost.

MATERIAL COSTS

The Material Costs given are based at Second Quarter 2013 but exclude any charges in respect of VAT. The average rate of copper during this quarter is US$6,800/UK£4,400 per tonne. Users of the book are advised to register on the SPON's website www.pricebooks.co.uk/updates to receive the free quarterly updates – alerts will then be provided by email as changes arise.

MEASURED WORK PRICES

These prices are intended to apply to new work in the London area. The prices are for reasonable quantities of work and the user should make suitable adjustments if the quantities are especially small or especially large. Adjustments may also be required for locality (e.g. outside London – refer to cost indices in approximate estimating section for details of adjustment factors) and for the market conditions (e.g. volume of work secured or being tendered) at the time of use.

ELECTRICAL INSTALLATIONS

The labour rate has been based on average gang rates per man hour effective from 7 January 2013 including allowances for all other emoluments and expenses. Future changes will be published in the free SPON'S quarterly update by registering on their website. To this rate, has been added 3% and 7.5% to cover site and head office overheads and preliminary items together with a further 2% for profit, resulting in an inclusive rate of £27.70 per man hour. The rate has been calculated on a working year of 2025 hours; a detailed build-up of the rate is given at the end of these directions.

DIRECTIONS

In calculating the 'Measured Work Prices' the following assumptions have been made:

(a) That the work is carried out as a subcontract under the Standard Form of Building Contract.
(b) That, unless otherwise stated, the work is being carried out in open areas at a height which would not require more than simple scaffolding.
(c) That the building in which the work is being carried out is no more than six storey's high.

Where these assumptions are not valid, as for example where work is carried out in ducts and similar confined spaces or in multi-storey structures when additional time is needed to get to and from upper floors, then an appropriate adjustment must be made to the prices. Such adjustment will normally be to the labour element only.

DIRECTIONS

LABOUR RATE – ELECTRICAL

The annual cost of a notional eleven man gang of 11 men

		TECHNICIAN 1 NR	APPROVED ELECTRICIANS 4 NR	ELECTRICIANS 4 NR	LABOURERS 2 NR	SUB-TOTALS
Hourly Rate from 7 January 2013		**18.37**	**16.31**	**15.05**	**12.07**	
Working hours per annum per man		1,687.50	1,687.50	1,687.50	1,687.50	
x Hourly rate × nr of men = £ per annum		**30,999.38**	**110,092.50**	**101,587.50**	**40,736.25**	**283,415.63**
Overtime Rate		27.56	24.47	22.58	18.11	
Overtime hours per annum per man		337.50	337.50	337.50	337.50	
x Hourly rate × nr of men = £ per annum		**9,299.81**	**33,027.75**	**30,476.25**	**12,220.88**	**85,024.69**
Total		**40,299.19**	**143,120.25**	**132,063.75**	**52,957.13**	**368,440.31**
Incentive schemes (insert percentage)	**0.00%**	**0.00**	**0.00**	**0.00**	**0.00**	**0.00**
Daily Travel Allowance (15–20 miles each way) effective from 07/01/13		5.11	5.11	5.11	5.11	
Days per annum per man		225.00	225.00	225.00	225.00	
x nr of men = £ per annum		**1,149.75**	**4,599.00**	**4,599.00**	**2,299.50**	**12,647.25**
Daily Travel Allowance (15–20 miles each way) effective from 07/01/13		3.50	3.50	3.50	3.50	
Days per annum per man		225.00	225.00	225.00	225.00	
x nr of men = £ per annum		**787.50**	**3,150.00**	**3,150.00**	**1,575.00**	**8,662.50**
JIB Pension Scheme @ 2.5%		**1,185.98**	**4,232.11**	**3,919.07**	**1,589.35**	**10,926.52**
JIB combined benefits scheme (nr of weeks per man)		52.00	52.00	52.00	52.00	
Benefit Credit effective from 29 September 2011		62.03	56.29	52.61	44.36	
x nr of men = £ per annum		**3,225.56**	**11,708.32**	**10,942.88**	**4,613.44**	**30,490.20**
Holiday Top-up Funding		**53.17**	**47.39**	**44.03**	**35.62**	
x nr of men @ 7.5 hrs per day = £ per annum		**2,764.60**	**9,856.95**	**9,157.26**	**3,704.06**	**25,482.87**
National Insurance Contributions:						
Annual gross pay (subject to NI) each		48,226.60	150,869.25	139,812.75	56,831.63	
% of NI Contributions		13.80	13.80	13.80	13.80	
£ Contributions/annum		**5,515.60**	**16,261.26**	**14,735.47**	**5,563.42**	**42,075.75**

	SUBTOTAL		**498,725.39**
	TRAINING (INCLUDING ANY TRADE REGISTRATIONS) – SAY	**1.00%**	**4,987.25**
	SEVERANCE PAY AND SUNDRY COSTS – SAY	**1.50%**	**7,555.69**
	EMPLOYER'S LIABILITY AND THIRD PARTY INSURANCE – SAY	**2.00%**	**10,225.37**
	ANNUAL COST OF NOTIONAL GANG		**521,493.70**
MEN ACTUALLY WORKING = 10.5	THEREFORE ANNUAL COST PER PRODUCTIVE MAN		**49,666.07**
AVERAGE NR OF HOURS WORKED PER MAN = 2025	THEREFORE ALL IN MAN HOUR		**24.53**
	PRELIMINARY ITEMS – SAY	**7.50%**	**1.84**
	SITE AND HEAD OFFICE OVERHEADS – SAY	**3.00%**	**0.79**
	PROFIT – SAY	**2.00%**	**0.54**
	THEREFORE INCLUSIVE MAN HOUR		**27.70**

DIRECTIONS

Notes:

(1) Hourly wage rates are those effective from 7 January 2013.
(2) The following assumptions have been made in the above calculations: -
 (a) Hourly rates are based on London rate and job reporting own transport.
 (b) The working week of 37.5 hours is made up of 7.5 hours Monday to Friday.
 (c) Five days in the year are lost through sickness or similar reason.
 (d) A working year of 2025 hours.
(3) The incentive scheme addition of 0% is intended to reflect bonus schemes typically in use.
(4) National insurance contributions are those effective from 6 April 2013. The National Insurance concession will be withdrawn for the construction industry on 30 October 2012 therefore NI will be paid for 52 weeks.
(5) Weekly JIB Combined Benefit Credit Scheme are those effective from 7 January 2013.
(6) Overtime is paid after 37.5 hours.

V: ELECTRICAL SUPPLY/POWER/LIGHTING

Item	Net Price £	Material £	Labour hours	Labour £	Unit	Total rate £
V10: ELECTRICAL GENERATION PLANT						
STANDBY GENERATORS						
Standby diesel generating sets; supply and installation; fixing to base; all supports and fixings; all necessary connections to equipment						
Three phase, 400 volt, four wire 50 Hz packaged standby diesel generating set, complete with radio and television suppressors, daily service fuel tank and associated piping, 4 metres of exhaust pipe and primary exhaust silencer, control panel, mains failure relay, starting battery with charger, all internal wiring, interconnections, earthing and labels. Rated for standby duty; including UK delivery, installation and commissioning.						
60 kVA	16163.24	18254.69	100.00	2770.40	nr	**21025.09**
100 kVA	22659.21	25591.20	100.00	2770.40	nr	**28361.60**
150 kVA	25914.47	29267.68	100.00	2770.40	nr	**32038.08**
315 kVA	46722.26	52767.88	120.00	3324.49	nr	**56092.37**
500 kVA	55071.15	62197.08	120.00	3324.49	nr	**65521.57**
750 kVA	93717.06	105843.57	140.00	3878.57	nr	**109722.14**
1000 kVA	124558.00	140675.19	140.00	3878.57	nr	**144553.76**
1500 kVA	146592.73	165561.10	170.00	4709.69	nr	**170270.79**
2000 kVA	225807.61	255025.99	170.00	4709.69	nr	**259735.68**
2500 kVA	309161.13	349165.03	210.00	5817.86	nr	**354982.89**
Extra for residential silencer; peformance 75 dBA at 1 m; including connection to exhaust pipe						
60 kVA	867.81	980.10	10.00	277.04	nr	**1257.14**
100 kVA	1036.95	1171.12	10.00	277.04	nr	**1448.16**
150 kVA	1189.17	1343.04	10.00	277.04	nr	**1620.08**
315 kVA	2313.29	2612.62	15.00	415.56	nr	**3028.18**
500 kVA	2958.61	3341.44	15.00	415.56	nr	**3757.00**
750 kVA	4407.97	4978.33	20.00	554.08	nr	**5532.41**
1000 kVA	5936.73	6704.91	20.00	554.08	nr	**7258.99**
1500 kVA	9457.38	10681.11	20.00	554.08	nr	**11235.19**
2000 kVA	10243.23	11568.66	30.00	831.13	nr	**12399.79**
2500 kVA	11712.12	13227.60	30.00	831.13	nr	**14058.73**

V: ELECTRICAL SUPPLY/POWER/LIGHTING

Item	Net Price £	Material £	Labour hours	Labour £	Unit	Total rate £
V10: ELECTRICAL GENERATION PLANT – cont						
Synchronization panel for paralleling generators – not generators to mains; including interconnecting cables; commissioning and testing; fixing to backgrounds						
2 × 60 kVA	7228.66	8164.01	80.00	2216.33	nr	**10380.34**
2 × 100 kVA	7771.22	8776.77	80.00	2216.33	nr	**10993.10**
2 × 150 kVA	9776.14	11041.12	80.00	2216.33	nr	**13257.45**
2 × 315 kVA	16619.70	18770.20	80.00	2216.33	nr	**20986.53**
2 × 500 kVA	23994.10	27098.82	80.00	2216.33	nr	**29315.15**
2 × 750 kVA	27076.30	30579.83	80.00	2216.33	nr	**32796.16**
2 × 1000 kVA	29181.41	32957.34	120.00	3324.49	nr	**36281.83**
2 × 1500 kVA	34545.68	39015.71	120.00	3324.49	nr	**42340.20**
2 × 2000 kVA	37430.12	42273.39	120.00	3324.49	nr	**45597.88**
2 × 2500 kVA	43637.47	49283.94	120.00	3324.49	nr	**52608.43**
Prefabricated drop-over acoustic housing; performance 85 dBA at 1 m over the range from 60 kVA to 315 kVA, 75 dBA from 500 kVA to 2500 kVA.						
100 kVA	3230.51	3648.52	7.00	193.93	nr	**3842.45**
150 kVA	5245.86	5924.65	15.00	415.56	nr	**6340.21**
315 kVA	11192.98	12641.30	25.00	692.60	nr	**13333.90**
500 kVA	18486.71	20878.80	40.00	1108.16	nr	**21986.96**
750 kVA	29133.28	32902.98	40.00	1108.16	nr	**34011.14**
1000 kVA	36857.65	41626.84	40.00	1108.16	nr	**42735.00**
1500 kVA	36857.65	41626.84	40.00	1108.16	nr	**42735.00**
2000 kVA	73110.30	82570.41	60.00	1662.24	nr	**84232.65**
2500 kVA	87877.30	99248.18	70.00	1939.29	nr	**101187.47**
COMBINED HEAT AND POWER (CHP) UNITS						
Gas fired engine; acoustic enclosure complete with exhaust fan and attenuators; exhaust gas attenuation; includes 6 m long pipe connections; dry air cooler for secondary water circuit to reject excess heat; controls and panel; commissioning						
Electrical output; Heat output						
82 kW; 132 kW	92344.96	104293.94	–	–	nr	**104293.94**
100 kW, 148 kW	98362.43	111090.04	–	–	nr	**111090.04**
118 kW, 181 kW	103222.69	116579.19	–	–	nr	**116579.19**

V: ELECTRICAL SUPPLY/POWER/LIGHTING

Item	Net Price £	Material £	Labour hours	Labour £	Unit	Total rate £
130 kW, 201 kW	112442.13	126991.58	–	–	nr	**126991.58**
140 kW, 207 kW	119655.00	135137.76	–	–	nr	**135137.76**
150 kW, 208 kW	120349.32	135921.92	–	–	nr	**135921.92**
160 kW, 216 kW	126019.62	142325.92	–	–	nr	**142325.92**
198 kW, 233 kW	148816.56	168072.68	–	–	nr	**168072.68**
210 kW, 319 kW	157438.30	177810.03	–	–	nr	**177810.03**
237 kW, 359 kW	163628.79	184801.54	–	–	nr	**184801.54**
307 kW, 435 kW	221894.62	250606.68	–	–	nr	**250606.68**
380 kW, 500 kW	256899.51	290141.02	–	–	nr	**290141.02**
490 kW, 679 kW	353641.85	399401.34	–	–	nr	**399401.34**
501 kW, 518 kW	326331.81	368557.52	–	–	nr	**368557.52**
600 kW, 873 kW	388532.14	438806.25	–	–	nr	**438806.25**
725 kW, 1019 kW	525371.07	593351.46	–	–	nr	**593351.46**
975 kW, 1293 kW	652374.88	736788.93	–	–	nr	**736788.93**
1160 kW, 1442 kW	738586.65	834156.07	–	–	nr	**834156.07**
1379 kW, 1475 kW	881732.91	995824.74	–	–	nr	**995824.74**
1566 kW, 1647 kW	964530.37	1089335.78	–	–	nr	**1089335.78**
1600 kW, 1625 kW	962215.96	1086721.89	–	–	nr	**1086721.89**
1760 kW, 1821 kW	1056643.89	1193368.33	–	–	nr	**1193368.33**
2430 kW	1372350.00	1549925.23	–	–	nr	**1549925.23**
2931 kW	1659000.00	1873666.31	–	–	nr	**1873666.31**
3500 kW	1785000.00	2015970.07	–	–	nr	**2015970.07**
3900 kW	2100000.00	2371729.50	–	–	nr	**2371729.50**

Note: The costs detailed are based on a specialist subcontract package, as part of the M&E contract works, and include installation

V: ELECTRICAL SUPPLY/POWER/LIGHTING

Item	Net Price £	Material £	Labour hours	Labour £	Unit	Total rate £
V11: MV SUPPLY						
Y61 – MV CABLES						
Cable; 6350/11000 volts, 3 core, XLPE; stranded copper conductors; steel wire armoured; LSOH to BS 7835						
Laid in trench/duct including marker tape (cable tiles measured elsewhere)						
95 mm²	41.23	46.56	0.23	6.36	m	**52.92**
120 mm²	45.36	51.22	0.23	6.36	m	**57.58**
150 mm²	54.28	61.30	0.25	6.93	m	**68.23**
185 mm²	58.92	66.54	0.25	6.93	m	**73.47**
240 mm²	68.62	77.50	0.27	7.48	m	**84.98**
300 mm²	78.34	88.47	0.29	8.03	m	**96.50**
Cable tiles; single width; laid in trench above cables on prepared sand bed (cost of excavation excluded); reinforced concrete covers; concave/convex ends						
914 × 152 × 63/38 mm	12.01	13.57	0.11	3.05	m	**16.62**
914 × 229 × 63/38 mm	14.14	15.97	0.11	3.05	m	**19.02**
914 × 305 × 63/38 mm	15.74	17.78	0.11	3.05	m	**20.83**
Clipped direct to backgrounds including cleats						
95 mm²	52.29	59.06	0.47	13.03	m	**72.09**
120 mm²	56.43	63.73	0.50	13.86	m	**77.59**
150 mm²	65.02	73.43	0.53	14.68	m	**88.11**
185 mm²	71.25	80.47	0.55	15.24	m	**95.71**
240 mm²	82.39	93.05	0.60	16.63	m	**109.68**
300 mm²	93.64	105.75	0.68	18.84	m	**124.59**
Terminations for above cables, including heat-shrink kit and glanding off						
95 mm²	515.71	582.44	4.75	131.60	m	**714.04**
120 mm²	512.91	579.28	5.30	146.83	m	**726.11**
150 mm²	551.71	623.10	6.00	166.23	m	**789.33**
185 mm²	553.04	624.60	6.90	191.16	m	**815.76**
240 mm²	577.84	652.61	7.43	205.85	m	**858.46**
300 mm²	623.12	703.75	8.75	242.41	m	**946.16**

V: ELECTRICAL SUPPLY/POWER/LIGHTING

Item	Net Price £	Material £	Labour hours	Labour £	Unit	Total rate £
Cable; 6350/11000 volts, 3 core, paper insulated; lead sheathed; steel wire armoured; stranded copper conductors; to BS 6480						
Laid in trench/duct including marker tape. (cable tiles measured elsewhere)						
95 mm²	44.90	50.71	0.22	6.10	m	**56.81**
120 mm²	48.46	54.73	0.22	6.10	m	**60.83**
150 mm²	59.28	66.95	0.24	6.65	m	**73.60**
185 mm²	61.87	69.88	0.26	7.20	m	**77.08**
240 mm²	71.78	81.07	0.34	9.41	m	**90.48**
Cable tiles; single width; laid in trench above cables on prepared sand bed (cost of excavation excluded); reinforced concrete covers; concave/convex ends						
914 × 152 × 63/38 mm	12.01	13.57	0.11	3.05	m	**16.62**
914 × 229 × 63/38 mm	14.14	15.97	0.11	3.05	m	**19.02**
914 × 305 × 63/38 mm	15.74	17.78	0.11	3.05	m	**20.83**
Clipped direct to backgrounds including cleats						
95 mm²	57.38	64.80	0.55	15.24	m	**80.04**
120 mm²	60.92	68.80	0.63	17.45	m	**86.25**
150 mm²	71.38	80.62	0.66	18.29	m	**98.91**
185 mm²	75.60	85.38	0.72	19.94	m	**105.32**
240 mm²	87.14	98.42	0.82	22.72	m	**121.14**
Terminations for above cables, including compound joint and glanding off						
95 mm²	564.46	637.50	4.75	131.60	m	**769.10**
120 mm²	574.65	649.01	5.30	146.83	m	**795.84**
150 mm²	600.25	677.92	6.10	168.99	m	**846.91**
185 mm²	601.53	679.36	6.90	191.16	m	**870.52**
240 mm²	632.62	714.48	7.43	205.85	m	**920.33**
300 mm²	670.87	757.68	8.75	242.41	m	**1000.09**

V: ELECTRICAL SUPPLY/POWER/LIGHTING

Item	Net Price £	Material £	Labour hours	Labour £	Unit	Total rate £
V11: MV SUPPLY – cont						
Y70 – MV SWITCHGEAR AND TRANSFORMERS						
MV circuit breakers; installed on prepared foundations including all supports, fixings and inter panel connections where relevant. Excludes main and multicore cabling and heat shrink cable termination kits.						
Three phase 11 kV, 630 amp, Air or SF6 insulated, with fixed pattern vacuum or SF6 circuit breaker panels; hand charged spring closing operation; prospective fault level up to 25 kA for 3 seconds. Feeders include ammeter with selector switch, 3 pole IDMT, overcurrent and earth fault relays with necessary current relays with necessary current transformers; incomers include 3 phase VT, voltmeter and phase selector switch; Includes IDMT overcurrent and earth fault relays/CTs.						
Single panel with cable chamber	23027.56	26007.22	31.70	878.22	nr	**26885.44**
Three panel with one incomer and two feeders; with cable chambers	68997.46	77925.39	67.83	1879.17	nr	**79804.56**
Five panel with two incoming, two feeders and a bus section; with cable chambers	114975.66	129847.07	99.17	2747.29	nr	**132594.36**
Ring Main Unit (RMU)						
Three phase 11 kV, 630 amp, SF6 insulated RMU with vacuum or SF6 200 amp circuit breaker tee-off, Includes IDMT overcurrent and earth fault relays/CTs for tee-off and cable boxes.						
3 way ring main unit	13753.60	15533.30	67.83	1879.17	nr	**17412.47**
Extra for						
Remote actuator to ring switches (per switch)	2106.39	2378.95	–	–	nr	**2378.95**
Remote tripping of circuit breaker	1053.21	1189.49	–	–	nr	**1189.49**
3 – Phase Neon indicators (per circuit)	123.86	139.88	–	–	nr	**139.88**
Pressure gauge with alarm contacts (for SF6 only)	271.75	306.92	–	–	nr	**306.92**

V: ELECTRICAL SUPPLY/POWER/LIGHTING

Item	Net Price £	Material £	Labour hours	Labour £	Unit	Total rate £
Tripping batteries						
Battery chargers; switchgear tripping and closing; double wound transformer and earth screen; including fixing to background, commissioning and testing						
Valve regulated lead acid battery						
30 volt; 19 Ah; 3 A	2501.15	2824.79	6.50	180.07	nr	**3004.86**
30 volt; 29 Ah; 3 A	4007.03	4525.52	8.50	235.48	nr	**4761.00**
110 volt; 19 Ah; 3 A	4087.57	4616.48	6.50	180.07	nr	**4796.55**
110 volt; 29 Ah; 3 A	4460.02	5037.13	8.50	235.48	nr	**5272.61**
110 volt; 38 Ah; 3 A	4765.94	5382.63	10.00	277.04	nr	**5659.67**
Step down transformers; 11/0.415 kV, Dyn 11, 50 Hz. Complete with lifting lugs, mounting skids, provisions for wheels, undrilled gland plates to air-filled cable boxes, off load tapping facility, including UK delivery						
Oil-filled in free breathing ventilated steel tank						
500 kVA	9828.00	11099.69	30.00	831.13	nr	**11930.82**
800 kVA	11192.54	12640.80	30.00	831.13	nr	**13471.93**
1000 kVA	12685.29	14326.71	30.00	831.13	nr	**15157.84**
1250 kVA	15414.33	17408.87	35.00	969.64	nr	**18378.51**
1500 kVA	18266.10	20629.64	35.00	969.64	nr	**21599.28**
2000 kVA	23724.17	26792.75	40.00	1108.11	nr	**27900.86**
MIDEL – filled in gasket-sealed steel tank						
500 kVA	13017.33	14701.71	30.00	831.13	nr	**15532.84**
800 kVA	14836.70	16756.50	30.00	831.13	nr	**17587.63**
1000 kVA	16892.54	19078.35	30.00	831.13	nr	**19909.48**
1250 kVA	20417.53	23059.46	35.00	969.64	nr	**24029.10**
1500 kVA	24292.73	27436.09	35.00	969.64	nr	**28405.73**
2000 kVA	31456.43	35526.73	40.00	1108.16	nr	**36634.89**
Extra for,						
Fluid temperature indicator with 2 N/O contacts.	365.17	412.42	2.00	55.41	nr	**467.83**
Winding temperature indicator with 2 N/O contacts.	894.67	1010.43	2.00	55.41	nr	**1065.84**
Dehydrating breather	91.29	103.10	2.00	55.41	nr	**158.51**
Plain rollers	273.88	309.31	2.00	55.41	nr	**364.72**
Pressure relief device with 1 N/O contact	547.75	618.63	2.00	55.41	nr	**674.04**

V: ELECTRICAL SUPPLY/POWER/LIGHTING

Item	Net Price £	Material £	Labour hours	Labour £	Unit	Total rate £
V11: MV SUPPLY – cont						
Y70 – MV SWITCHGEAR AND TRANSFORMERS – cont						
Step down transformers; 11/0.415 kV, Dyn 11, 50 Hz. Complete with lifting lugs, mounting skids, provisions for wheels, undrilled gland plates to air-filled cable boxes, off load tapping facility, including UK delivery						
Cast Resin type in ventilated steel encloure, AN – Air Natural including winding temperture indicator with 2 N/O contacts						
500 kVA	13927.00	15729.08	40.00	1108.16	nr	**16837.24**
800 kVA	16201.19	18297.55	40.00	1108.16	nr	**19405.71**
1000 kVA	19598.83	22134.82	40.00	1108.16	nr	**23242.98**
1250 kVA	21213.50	23958.42	45.00	1246.68	nr	**25205.10**
1600 kVA	24065.32	27179.26	45.00	1246.68	nr	**28425.94**
2000 kVA	27249.21	30775.12	50.00	1385.20	nr	**32160.32**
Cast Resin type in ventilated steel enclosure with temperature controlled fans to achieve 40% increase to AN/AF rating. Includes winding temperature indicator with 2 N/O contacts						
500/700 kVA	15291.50	17270.14	42.00	1163.58	nr	**18433.72**
800/1120 kVA	17793.13	20095.47	42.00	1163.58	nr	**21259.05**
1000/1400 kVA	22014.92	24863.54	42.00	1163.58	nr	**26027.12**
1250/1750 kVA	22924.58	25890.90	47.00	1302.09	nr	**27192.99**
1600/2240 kVA	25899.18	29250.41	47.00	1302.09	nr	**30552.50**
2000/2800 kVA	29537.88	33359.94	52.00	1440.61	nr	**34800.55**

V: ELECTRICAL SUPPLY/POWER/LIGHTING

Item	Net Price £	Material £	Labour hours	Labour £	Unit	Total rate £
V20: LV DISTRIBUTION						
Y60 – CONDUIT AND CABLE TRUNKING						
Heavy gauge, screwed drawn steel; surface fixed on saddles to backgrounds, with standard pattern boxes and fittings including all fixings and supports (forming holes, conduit entry, draw wires etc. and components for earth continuity are included)						
Black enamelled						
20 mm dia.	3.48	3.93	0.49	13.58	m	**17.51**
25 mm dia.	5.05	5.70	0.56	15.51	m	**21.21**
32 mm dia.	10.33	11.67	0.64	17.73	m	**29.40**
38 mm dia.	12.29	13.88	0.73	20.23	m	**34.11**
50 mm dia.	19.50	22.02	1.04	28.82	m	**50.84**
Galvanized						
20 mm dia.	3.28	3.70	0.49	13.58	m	**17.28**
25 mm dia.	4.48	5.06	0.56	15.51	m	**20.57**
32 mm dia.	9.57	10.81	0.64	17.73	m	**28.54**
38 mm dia.	10.78	12.18	0.73	20.23	m	**32.41**
50 mm dia.	24.50	27.67	1.04	28.82	m	**56.49**
High impact PVC; surface fixed on saddles to backgrounds; with standard pattern boxes and fittings; including all fixings and supports						
Light gauge						
16 mm dia.	0.55	0.62	0.27	7.48	m	**8.10**
20 mm dia.	0.62	0.70	0.28	7.76	m	**8.46**
25 mm dia.	0.85	0.96	0.33	9.14	m	**10.10**
32 mm dia.	1.35	1.52	0.38	10.53	m	**12.05**
38 mm dia.	1.79	2.02	0.44	12.19	m	**14.21**
50 mm dia.	2.96	3.35	0.48	13.29	m	**16.64**
Heavy gauge						
16 mm dia.	0.65	0.73	0.27	7.48	m	**8.21**
20 mm dia.	0.73	0.83	0.28	7.76	m	**8.59**
25 mm dia.	1.00	1.13	0.33	9.14	m	**10.27**
32 mm dia.	1.59	1.80	0.38	10.53	m	**12.33**
38 mm dia.	2.10	2.38	0.44	12.19	m	**14.57**
50 mm dia.	3.48	3.93	0.48	13.29	m	**17.22**

V: ELECTRICAL SUPPLY/POWER/LIGHTING

Item	Net Price £	Material £	Labour hours	Labour £	Unit	Total rate £
V20: LV DISTRIBUTION – cont						
Flexible conduits; including adaptors and locknuts (for connections to equipment)						
Metallic, PVC covered conduit; not exceeding 1 m long; including zinc plated mild steel adaptors, lock nuts and earth conductor						
16 mm dia.	8.60	9.71	0.42	11.63	nr	**21.34**
20 mm dia.	8.09	9.14	0.46	12.74	nr	**21.88**
25 mm dia.	12.75	14.40	0.43	11.91	nr	**26.31**
32 mm dia.	19.75	22.31	0.51	14.13	nr	**36.44**
38 mm dia.	24.95	28.18	0.56	15.51	nr	**43.69**
50 mm dia.	87.06	98.33	0.82	22.72	nr	**121.05**
PVC conduit; not exceeding 1 m long; including nylon adaptors, lock nuts						
16 mm dia.	4.80	5.42	0.46	12.74	nr	**18.16**
20 mm dia.	4.80	5.42	0.48	13.29	nr	**18.71**
25 mm dia.	6.02	6.80	0.50	13.86	nr	**20.66**
32 mm.dia.	8.73	9.86	0.58	16.08	nr	**25.94**
PVC adaptable boxes; fixed to backgrounds; including all supports and fixings (cutting and connecting conduit to boxes is included)						
Square pattern						
75 × 75 × 53 mm	2.01	2.27	0.69	19.13	nr	**21.40**
100 × 100 × 75 mm	3.40	3.84	0.71	19.68	nr	**23.52**
150 × 150 × 75 mm	4.36	4.93	0.80	22.15	nr	**27.08**
Terminal strips to be fixed in metal or polythene adaptable boxes)						
20 amp high density polythene						
2 way	1.18	1.34	0.23	6.36	nr	**7.70**
3 way	1.30	1.47	0.23	6.36	nr	**7.83**
4 way	1.38	1.56	0.23	6.36	nr	**7.92**
5 way	1.45	1.64	0.23	6.36	nr	**8.00**
6 way	1.68	1.90	0.25	6.93	nr	**8.83**
7 way	1.72	1.94	0.25	6.93	nr	**8.87**
8 way	1.90	2.14	0.29	8.03	nr	**10.17**
9 way	2.01	2.27	0.30	8.31	nr	**10.58**
10 way	2.32	2.62	0.34	9.41	nr	**12.03**
11 way	2.36	2.66	0.34	9.41	nr	**12.07**
12 way	2.44	2.75	0.34	9.41	nr	**12.16**
13 way	2.82	3.18	0.37	10.25	nr	**13.43**
14 way	3.14	3.55	0.37	10.25	nr	**13.80**
15 way	3.22	3.64	0.39	10.81	nr	**14.45**
16 way	3.30	3.72	0.45	12.46	nr	**16.18**
18 way	3.54	4.00	0.45	12.46	nr	**16.46**

V: ELECTRICAL SUPPLY/POWER/LIGHTING

Item	Net Price £	Material £	Labour hours	Labour £	Unit	Total rate £
TRUNKING						
Galvanized steel trunking; fixed to backgrounds; jointed with standard connectors (including plates for air gap between trunking and background); earth continuity straps included						
Single compartment						
50 × 50 mm	5.74	6.49	0.39	10.81	m	**17.30**
75 × 50 mm	8.75	9.88	0.44	12.19	m	**22.07**
75 × 75 mm	9.36	10.57	0.47	13.03	m	**23.60**
100 × 50 mm	10.88	12.29	0.50	13.86	m	**26.15**
100 × 75 mm	11.75	13.27	0.57	15.79	m	**29.06**
100 × 100 mm	11.92	13.46	0.62	17.18	m	**30.64**
150 × 50 mm	15.84	17.89	0.78	21.60	m	**39.49**
150 × 100 mm	19.19	21.68	0.78	21.60	m	**43.28**
150 × 150 mm	20.50	23.15	0.86	23.83	m	**46.98**
225 × 75 mm	26.07	29.45	0.88	24.39	m	**53.84**
225 × 150 mm	32.58	36.79	0.84	23.28	m	**60.07**
225 × 225 mm	38.77	43.79	0.99	27.42	m	**71.21**
300 × 75 mm	26.80	30.26	0.96	26.60	m	**56.86**
300 × 100 mm	30.86	34.85	0.99	27.42	m	**62.27**
300 × 150 mm	23.61	26.66	0.99	27.42	m	**54.08**
300 × 225 mm	20.09	22.68	1.09	30.20	m	**52.88**
300 × 300 mm	51.06	57.67	1.16	32.13	m	**89.80**
Double compartment						
100 × 50 mm	7.69	8.68	0.54	14.96	m	**23.64**
100 × 75 mm	8.28	9.35	0.62	17.18	m	**26.53**
100 × 100 mm	8.60	9.71	0.66	18.29	m	**28.00**
150 × 50 mm	11.67	13.18	0.70	19.39	m	**32.57**
150 × 100 mm	14.47	16.34	0.83	22.99	m	**39.33**
150 × 150 mm	15.67	17.70	0.92	25.49	m	**43.19**
Triple compartment						
150 × 50 mm	13.44	15.18	0.79	21.89	m	**37.07**
150 × 100 mm	13.80	15.59	0.78	21.60	m	**37.19**
150 × 150 mm	16.90	19.08	1.01	27.99	m	**47.07**
Galvanized steel trunking fittings; cutting and jointing trunking to fittings is included						
Stop end						
50 × 50 mm	1.84	2.08	0.19	5.26	nr	**7.34**
75 × 50 mm	1.95	2.20	0.20	5.55	nr	**7.75**
75 × 75 mm	1.95	2.20	0.21	5.81	nr	**8.01**
100 × 50 mm	2.02	2.28	0.31	8.59	nr	**10.87**
100 × 75 mm	2.19	2.47	0.27	7.48	nr	**9.95**
100 × 100 mm	2.14	2.42	0.27	7.48	nr	**9.90**
150 × 50 mm	2.39	2.70	0.28	7.76	nr	**10.46**

V: ELECTRICAL SUPPLY/POWER/LIGHTING

Item	Net Price £	Material £	Labour hours	Labour £	Unit	Total rate £
V20: LV DISTRIBUTION – cont						
TRUNKING – cont						
Galvanized steel trunking fittings – cont						
Stop end – cont						
150 × 100 mm	3.36	3.79	0.30	8.31	nr	**12.10**
150 × 150 mm	2.68	3.03	0.32	8.86	nr	**11.89**
225 × 75 mm	2.78	3.14	0.35	9.70	nr	**12.84**
225 × 150 mm	3.04	3.44	0.37	10.25	nr	**13.69**
225 × 225 mm	3.30	3.72	0.38	10.53	nr	**14.25**
300 × 75 mm	4.82	5.45	0.42	11.63	nr	**17.08**
300 × 100 mm	4.02	4.54	0.42	11.63	nr	**16.17**
300 × 150 mm	4.82	5.45	0.43	11.91	nr	**17.36**
300 × 225 mm	5.79	6.54	0.45	12.46	nr	**19.00**
300 × 300 mm	6.05	6.83	0.48	13.29	nr	**20.12**
Flanged connector						
50 × 50 mm	1.95	2.20	0.19	5.26	nr	**7.46**
75 × 50 mm	2.99	3.38	0.20	5.55	nr	**8.93**
75 × 75 mm	2.42	2.73	0.21	5.81	nr	**8.54**
100 × 50 mm	3.11	3.51	0.26	7.20	nr	**10.71**
100 × 75 mm	3.21	3.62	0.27	7.48	nr	**11.10**
100 × 100 mm	3.11	3.51	0.27	7.48	nr	**10.99**
150 × 50 mm	2.61	2.95	0.28	7.76	nr	**10.71**
150 × 100 mm	3.36	3.79	0.30	8.31	nr	**12.10**
150 × 150 mm	3.40	3.84	0.32	8.86	nr	**12.70**
225 × 75 mm	2.78	3.14	0.35	9.70	nr	**12.84**
225 × 150 mm	3.30	3.72	0.37	10.25	nr	**13.97**
225 × 225 mm	4.77	5.39	0.38	10.53	nr	**15.92**
300 × 75 mm	4.02	4.54	0.42	11.63	nr	**16.17**
300 × 100 mm	4.82	5.45	0.42	11.63	nr	**17.08**
300 × 150 mm	5.30	5.99	0.43	11.91	nr	**17.90**
300 × 225 mm	5.83	6.59	0.45	12.46	nr	**19.05**
300 × 300 mm	6.05	6.83	0.48	13.29	nr	**20.12**
Bends 90°; single compartment						
50 × 50 mm	8.11	9.16	0.42	11.63	nr	**20.79**
75 × 50 mm	10.26	11.59	0.45	12.46	nr	**24.05**
75 × 75 mm	9.82	11.09	0.48	13.29	nr	**24.38**
100 × 50 mm	10.49	11.85	0.53	14.68	nr	**26.53**
100 × 75 mm	10.76	12.15	0.56	15.51	nr	**27.66**
100 × 100 mm	10.76	12.15	0.58	16.08	nr	**28.23**
150 × 50 mm	13.30	15.02	0.64	17.73	nr	**32.75**
150 × 100 mm	16.97	19.17	0.91	25.20	nr	**44.37**
150 × 150 mm	16.19	18.29	0.89	24.65	nr	**42.94**
225 × 75 mm	24.51	27.68	0.76	21.05	nr	**48.73**
225 × 150 mm	31.43	35.50	0.82	22.72	nr	**58.22**

V: ELECTRICAL SUPPLY/POWER/LIGHTING

Item	Net Price £	Material £	Labour hours	Labour £	Unit	Total rate £
225 × 225 mm	36.19	40.87	0.83	22.99	nr	**63.86**
300 × 75 mm	33.76	38.13	0.85	23.55	nr	**61.68**
300 × 100 mm	34.16	38.58	0.90	24.94	nr	**63.52**
300 × 150 mm	37.58	42.44	0.96	26.60	nr	**69.04**
300 × 225 mm	41.33	46.68	0.98	27.15	nr	**73.83**
300 × 300 mm	42.14	47.59	1.06	29.37	nr	**76.96**
Bends 90°; double compartment						
100 × 50 mm	9.76	11.03	0.53	14.68	nr	**25.71**
100 × 75 mm	10.27	11.60	0.56	15.51	nr	**27.11**
100 × 100 mm	10.81	12.21	0.58	16.08	nr	**28.29**
150 × 50 mm	10.73	12.12	0.65	18.00	nr	**30.12**
150 × 100 mm	11.30	12.76	0.69	19.13	nr	**31.89**
150 × 150 mm	11.89	13.43	0.73	20.23	nr	**33.66**
Bends 90°; triple compartment						
150 × 50 mm	14.10	15.92	0.68	18.84	nr	**34.76**
150 × 100 mm	19.71	22.26	0.73	20.23	nr	**42.49**
150 × 150 mm	20.75	23.44	0.77	21.34	nr	**44.78**
Tees; single compartment						
50 × 50 mm	7.37	8.32	0.56	15.51	nr	**23.83**
75 × 50 mm	11.50	12.98	0.57	15.79	nr	**28.77**
75 × 75 mm	11.14	12.58	0.60	16.63	nr	**29.21**
100 × 50 mm	13.46	15.20	0.65	18.00	nr	**33.20**
100 × 75 mm	13.54	15.29	0.72	19.94	nr	**35.23**
100 × 100 mm	12.99	14.67	0.71	19.68	nr	**34.35**
150 × 50 mm	15.40	17.39	0.82	22.72	nr	**40.11**
150 × 100 mm	19.13	21.60	0.84	23.28	nr	**44.88**
150 × 150 mm	18.55	20.95	0.91	25.20	nr	**46.15**
225 × 75 mm	33.01	37.28	0.94	26.04	nr	**63.32**
225 × 150 mm	44.59	50.36	1.01	27.99	nr	**78.35**
225 × 225 mm	51.03	57.63	1.02	28.25	nr	**85.88**
300 × 75 mm	47.65	53.82	1.07	29.65	nr	**83.47**
300 × 100 mm	49.37	55.75	1.07	29.65	nr	**85.40**
300 × 150 mm	53.20	60.08	1.14	31.58	nr	**91.66**
300 × 225 mm	57.60	65.04	1.19	32.97	nr	**98.01**
300 × 300 mm	60.92	68.80	1.26	34.91	nr	**103.71**
Tees; double compartment						
100 × 50 mm	9.48	10.71	0.65	18.00	nr	**28.71**
100 × 75 mm	11.98	13.53	0.71	19.68	nr	**33.21**
100 × 100 mm	11.73	13.25	0.72	19.94	nr	**33.19**
150 × 50 mm	12.45	14.07	0.82	22.72	nr	**36.79**
150 × 100 mm	13.04	14.73	0.85	23.55	nr	**38.28**
150 × 150 mm	14.63	16.52	0.91	25.20	nr	**41.72**

V: ELECTRICAL SUPPLY/POWER/LIGHTING

Item	Net Price £	Material £	Labour hours	Labour £	Unit	Total rate £
V20: LV DISTRIBUTION – cont						
TRUNKING – cont						
Galvanized steel trunking fittings – cont						
Tees; triple compartment						
150 × 50 mm	14.24	16.09	0.87	24.10	nr	**40.19**
150 × 100 mm	15.66	17.69	0.89	24.65	nr	**42.34**
150 × 150 mm	21.08	23.81	0.96	26.60	nr	**50.41**
Crossovers; single compartment						
50 × 50 mm	12.52	14.14	0.65	18.00	nr	**32.14**
75 × 50 mm	16.97	19.17	0.66	18.29	nr	**37.46**
75 × 75 mm	17.36	19.60	0.69	19.13	nr	**38.73**
100 × 50 mm	21.44	24.21	0.74	20.50	nr	**44.71**
100 × 75 mm	15.81	17.86	0.80	22.15	nr	**40.01**
100 × 100 mm	21.61	24.41	0.81	22.44	nr	**46.85**
150 × 50 mm	25.59	28.90	0.91	25.20	nr	**54.10**
150 × 100 mm	25.97	29.34	0.94	26.04	nr	**55.38**
150 × 150 mm	24.88	28.10	0.99	27.42	nr	**55.52**
225 × 75 mm	45.57	51.47	1.01	27.99	nr	**79.46**
225 × 150 mm	49.74	56.17	1.08	29.92	nr	**86.09**
225 × 225 mm	67.67	76.43	1.09	30.20	nr	**106.63**
300 × 75 mm	63.72	71.96	1.14	31.58	nr	**103.54**
300 × 100 mm	65.63	74.12	1.16	32.13	nr	**106.25**
300 × 150 mm	70.87	80.04	1.19	32.97	nr	**113.01**
300 × 225 mm	74.98	84.69	1.21	33.52	nr	**118.21**
300 × 300 mm	79.93	90.27	1.29	35.73	nr	**126.00**
Crossovers; double compartment						
100 × 50 mm	12.58	14.21	0.74	20.50	nr	**34.71**
100 × 75 mm	13.20	14.91	0.80	22.15	nr	**37.06**
100 × 100 mm	13.49	15.24	0.81	22.44	nr	**37.68**
150 × 50 mm	14.91	16.84	0.86	23.83	nr	**40.67**
150 × 100 mm	14.95	16.88	0.94	26.04	nr	**42.92**
150 × 150 mm	22.74	25.68	1.00	27.70	nr	**53.38**
Crossovers; triple compartment						
150 × 50 mm	16.31	18.42	0.97	26.87	nr	**45.29**
150 × 100 mm	17.94	20.26	0.99	27.42	nr	**47.68**
150 × 150 mm	24.99	28.22	1.06	29.37	nr	**57.59**

V: ELECTRICAL SUPPLY/POWER/LIGHTING

Item	Net Price £	Material £	Labour hours	Labour £	Unit	**Total rate £**
Galvanized steel flush floor trunking; fixed to backgrounds; supports and fixings; standard coupling joints; earth continuity straps included						
Triple compartment						
350 × 60 mm	42.93	48.48	1.32	36.57	m	**85.05**
Four compartment						
350 × 60 mm	44.70	50.48	1.32	36.57	m	**87.05**
Galvanized steel flush floor trunking; fittings (cutting and jointing trunking to fittings is included)						
Stop end; triple compartment						
350 × 60 mm	4.50	5.08	0.53	14.68	nr	**19.76**
Stop end; four compartment						
350 × 60 mm	10.44	11.79	0.53	14.68	nr	**26.47**
Rising bend; standard; triple compartment						
350 × 60 mm	37.14	41.94	1.30	36.02	nr	**77.96**
Rising bend; standard; four compartment						
350 × 60 mm	38.91	43.94	1.30	36.02	nr	**79.96**
Rising bend; skirting; triple compartment						
350 × 60 mm	72.85	82.27	1.33	36.84	nr	**119.11**
Rising bend; skirting; four compartment						
350 × 60 mm	82.69	93.39	1.33	36.84	nr	**130.23**
Junction box; triple compartment						
350 × 60 mm	48.42	54.68	1.16	32.13	nr	**86.81**
Junction box; four compartment						
350 × 60 mm	50.36	56.88	1.16	32.13	nr	**89.01**
Body coupler (pair)						
3 and 4 compartment	2.50	2.83	0.16	4.43	nr	**7.26**
Service outlet module comprising flat lid with flanged carpet trim; twin 13 A outlet and drilled plate for mounting 2 telephone outlets; one blank plate; triple compartment						
3 compartment	60.06	67.83	0.47	13.03	nr	**80.86**

V: ELECTRICAL SUPPLY/POWER/LIGHTING

Item	Net Price £	Material £	Labour hours	Labour £	Unit	Total rate £
V20: LV DISTRIBUTION – cont						
TRUNKING – cont						
Galvanized steel flush floor trunking – cont						
Service outlet module comprising flat lid with flanged carpet trim; twin 13 A outlet and drilled plate for mounting 2 telephone outlets; two blank plates; four compartment						
4 compartment	66.90	75.56	0.47	13.03	nr	**88.59**
Single compartment PVC trunking; grey finish; clip on lid; fixed to backgrounds; including supports and fixings (standard coupling joints)						
50 × 50 mm	10.34	11.68	0.27	7.48	m	**19.16**
75 × 50 mm	11.21	12.66	0.28	7.76	m	**20.42**
75 × 75 mm	12.70	14.34	0.29	8.03	m	**22.37**
100 × 50 mm	14.38	16.24	0.34	9.41	m	**25.65**
100 × 75 mm	15.77	17.81	0.37	10.25	m	**28.06**
100 × 100 mm	16.66	18.82	0.37	10.25	m	**29.07**
150 × 50 mm	14.46	16.33	0.41	11.36	m	**27.69**
150 × 75 mm	25.76	29.09	0.44	12.19	m	**41.28**
150 × 100 mm	31.01	35.03	0.44	12.19	m	**47.22**
150 × 150 mm	31.73	35.83	0.48	13.29	m	**49.12**
Single compartment PVC trunking; fittings (cutting and jointing trunking to fittings is included)						
Crossover						
50 × 50 mm	16.94	19.14	0.29	8.03	nr	**27.17**
75 × 50 mm	18.78	21.21	0.30	8.31	nr	**29.52**
75 × 75 mm	20.38	23.02	0.31	8.59	nr	**31.61**
100 × 50 mm	27.18	30.70	0.35	9.70	nr	**40.40**
100 × 75 mm	34.27	38.71	0.36	9.98	nr	**48.69**
100 × 100 mm	30.31	34.23	0.40	11.08	nr	**45.31**
150 × 75 mm	38.83	43.85	0.45	12.46	nr	**56.31**
150 × 100 mm	46.59	52.62	0.46	12.74	nr	**65.36**
150 × 150 mm	72.31	81.67	0.47	13.03	nr	**94.70**
Stop end						
50 × 50 mm	0.71	0.81	0.12	3.33	nr	**4.14**
75 × 50 mm	1.32	1.49	0.12	3.33	nr	**4.82**
75 × 75 mm	1.01	1.14	0.13	3.60	nr	**4.74**
100 × 50 mm	1.80	2.03	0.16	4.43	nr	**6.46**
100 × 75 mm	2.82	3.18	0.16	4.43	nr	**7.61**
100 × 100 mm	2.83	3.19	0.18	4.99	nr	**8.18**
150 × 75 mm	7.37	8.32	0.20	5.55	nr	**13.87**
150 × 100 mm	9.13	10.31	0.21	5.81	nr	**16.12**
150 × 150 mm	9.31	10.52	0.22	6.10	nr	**16.62**

V: ELECTRICAL SUPPLY/POWER/LIGHTING

Item	Net Price £	Material £	Labour hours	Labour £	Unit	**Total rate £**
Flanged coupling						
50 × 50 mm	4.25	4.80	0.32	8.86	nr	**13.66**
75 × 50 mm	4.85	5.48	0.33	9.14	nr	**14.62**
75 × 75 mm	5.82	6.57	0.34	9.41	nr	**15.98**
100 × 50 mm	6.56	7.41	0.44	12.19	nr	**19.60**
100 × 75 mm	7.43	8.39	0.45	12.46	nr	**20.85**
100 × 100 mm	7.91	8.94	0.46	12.74	nr	**21.68**
150 × 75 mm	8.38	9.47	0.57	15.79	nr	**25.26**
150 × 100 mm	8.83	9.98	0.57	15.79	nr	**25.77**
150 × 150 mm	9.26	10.45	0.59	16.34	nr	**26.79**
Internal coupling						
50 × 50 mm	1.50	1.69	0.07	1.94	nr	**3.63**
75 × 50 mm	1.77	2.00	0.07	1.94	nr	**3.94**
75 × 75 mm	1.78	2.01	0.07	1.94	nr	**3.95**
100 × 50 mm	2.38	2.69	0.08	2.21	nr	**4.90**
100 × 75 mm	2.68	3.03	0.08	2.21	nr	**5.24**
100 × 100 mm	2.95	3.34	0.08	2.21	nr	**5.55**
External coupling						
50 × 50 mm	1.64	1.86	0.09	2.50	nr	**4.36**
75 × 50 mm	1.96	2.21	0.09	2.50	nr	**4.71**
75 × 75 mm	1.94	2.19	0.09	2.50	nr	**4.69**
100 × 50 mm	2.61	2.95	0.10	2.76	nr	**5.71**
100 × 75 mm	2.97	3.36	0.10	2.76	nr	**6.12**
100 × 100 mm	3.24	3.66	0.10	2.76	nr	**6.42**
150 × 75 mm	3.71	4.19	0.11	3.05	nr	**7.24**
150 × 100 mm	3.86	4.36	0.11	3.05	nr	**7.41**
150 × 150 mm	4.00	4.52	0.11	3.05	nr	**7.57**
Angle; flat cover						
50 × 50 mm	6.01	6.78	0.18	4.99	nr	**11.77**
75 × 50 mm	8.01	9.05	0.19	5.26	nr	**14.31**
75 × 75 mm	9.22	10.41	0.20	5.55	nr	**15.96**
100 × 50 mm	13.76	15.54	0.23	6.36	nr	**21.90**
100 × 75 mm	20.96	23.67	0.26	7.20	nr	**30.87**
100 × 100 mm	19.12	21.59	0.26	7.20	nr	**28.79**
150 × 75 mm	26.06	29.43	0.30	8.31	nr	**37.74**
150 × 100 mm	30.99	35.00	0.33	9.14	nr	**44.14**
150 × 150 mm	46.97	53.05	0.34	9.41	nr	**62.46**
Angle; internal or external cover						
50 × 50 mm	7.31	8.25	0.18	4.99	nr	**13.24**
75 × 50 mm	10.08	11.38	0.19	5.26	nr	**16.64**
75 × 75 mm	12.95	14.63	0.20	5.55	nr	**20.18**
100 × 50 mm	13.95	15.76	0.23	6.36	nr	**22.12**
100 × 75 mm	22.43	25.34	0.26	7.20	nr	**32.54**
100 × 100 mm	22.53	25.45	0.26	7.20	nr	**32.65**
150 × 75 mm	27.54	31.10	0.30	8.31	nr	**39.41**
150 × 100 mm	32.46	36.66	0.33	9.14	nr	**45.80**
150 × 150 mm	45.60	51.50	0.34	9.41	nr	**60.91**

V: ELECTRICAL SUPPLY/POWER/LIGHTING

Item	Net Price £	Material £	Labour hours	Labour £	Unit	Total rate £
V20: LV DISTRIBUTION – cont						
TRUNKING – cont						
Single compartment PVC trunking – cont						
Tee; flat cover						
50 × 50 mm	5.03	5.68	0.24	6.65	nr	**12.33**
75 × 50 mm	7.68	8.67	0.25	6.93	nr	**15.60**
75 × 75 mm	7.89	8.91	0.26	7.20	nr	**16.11**
100 × 50 mm	17.05	19.26	0.32	8.86	nr	**28.12**
100 × 75 mm	18.11	20.45	0.33	9.14	nr	**29.59**
100 × 100 mm	23.63	26.68	0.34	9.41	nr	**36.09**
150 × 75 mm	31.25	35.29	0.41	11.36	nr	**46.65**
150 × 100 mm	40.09	45.28	0.42	11.63	nr	**56.91**
150 × 150 mm	54.53	61.59	0.44	12.19	nr	**73.78**
Tee; internal or external cover						
50 × 50 mm	14.26	16.11	0.24	6.65	nr	**22.76**
75 × 50 mm	15.67	17.70	0.25	6.93	nr	**24.63**
75 × 75 mm	17.42	19.68	0.26	7.20	nr	**26.88**
100 × 50 mm	22.36	25.26	0.32	8.86	nr	**34.12**
100 × 75 mm	25.55	28.86	0.33	9.14	nr	**38.00**
100 × 100 mm	28.67	32.37	0.34	9.41	nr	**41.78**
150 × 75 mm	37.14	41.94	0.41	11.36	nr	**53.30**
150 × 100 mm	44.75	50.54	0.42	11.63	nr	**62.17**
150 × 150 mm	59.75	67.48	0.44	12.19	nr	**79.67**
Division strip (1.8 m long)						
50 mm	7.74	8.74	0.07	1.94	nr	**10.68**
75 mm	9.90	11.18	0.07	1.94	nr	**13.12**
100 mm	12.59	14.22	0.08	2.21	nr	**16.43**
PVC miniature trunking; white finish; fixed to backgrounds; including supports and fixing; standard coupling joints						
Single compartment						
16 × 16 mm	1.39	1.57	0.20	5.55	m	**7.12**
25 × 16 mm	1.71	1.93	0.21	5.81	m	**7.74**
38 × 16 mm	2.13	2.41	0.24	6.65	m	**9.06**
38 × 25 mm	2.54	2.87	0.25	6.93	m	**9.80**
Compartmented						
38 × 16 mm	2.49	2.82	0.24	6.65	m	**9.47**
38 × 25 mm	2.97	3.36	0.25	6.93	m	**10.29**

V: ELECTRICAL SUPPLY/POWER/LIGHTING

Item	Net Price £	Material £	Labour hours	Labour £	Unit	Total rate £
PVC miniature trunking fittings; single compartment; white finish; cutting and jointing trunking to fittings is included						
Coupling						
16 × 16 mm	0.43	0.49	0.10	2.76	nr	**3.25**
25 × 16 mm	0.43	0.49	0.12	3.18	nr	**3.67**
38 × 16 mm	0.43	0.49	0.12	3.33	nr	**3.82**
38 × 25 mm	1.02	1.15	0.14	3.88	nr	**5.03**
Stop end						
16 × 16 mm	0.43	0.49	0.12	3.33	nr	**3.82**
25 × 16 mm	0.43	0.49	0.13	3.52	nr	**4.01**
38 × 16 mm	0.43	0.49	0.15	4.15	nr	**4.64**
38 × 25 mm	1.02	1.15	0.17	4.71	nr	**5.86**
Bend; flat, internal or external						
16 × 16 mm	0.43	0.49	0.18	4.99	nr	**5.48**
25 × 16 mm	0.43	0.49	0.20	5.49	nr	**5.98**
38 × 16 mm	0.43	0.49	0.21	5.81	nr	**6.30**
38 × 25 mm	1.02	1.15	0.23	6.36	nr	**7.51**
Tee						
16 × 16 mm	0.73	0.83	0.19	5.26	nr	**6.09**
25 × 16 mm	0.73	0.83	0.23	6.36	nr	**7.19**
38 × 16 mm	0.73	0.83	0.26	7.20	nr	**8.03**
38 × 25 mm	1.01	1.14	0.29	8.03	nr	**9.17**
PVC bench trunking; white or grey finish; fixed to backgrounds; including supports and fixings; standard coupling joints						
Trunking						
90 × 90 mm	23.42	26.45	0.33	9.14	m	**35.59**
PVC bench trunking fittings; white or grey finish; cutting and jointing trunking to fittings is included						
Stop end						
90 × 90 mm	4.96	5.60	0.09	2.50	nr	**8.10**
Coupling						
90 × 90 mm	3.07	3.47	0.09	2.50	nr	**5.97**
Internal or external bend						
90 × 90 mm	16.84	19.02	0.28	7.76	nr	**26.78**
Socket plate						
90 × 90 mm – 1 gang	0.95	1.07	0.10	2.76	nr	**3.83**
90 × 90 mm – 2 gang	1.14	1.29	0.10	2.76	nr	**4.05**

V: ELECTRICAL SUPPLY/POWER/LIGHTING

Item	Net Price £	Material £	Labour hours	Labour £	Unit	Total rate £
V20: LV DISTRIBUTION – cont						
TRUNKING – cont						
PVC underfloor trunking; single compartment; fitted in floor screed; standard coupling joints						
Trunking						
60 × 25 mm	11.49	12.97	0.22	6.10	m	**19.07**
90 × 35 mm	16.55	18.69	0.27	7.48	m	**26.17**
PVC underfloor trunking fittings; single compartment; fitted in floor screed; (cutting and jointing trunking to fittings is included)						
Jointing sleeve						
60 × 25 mm	0.76	0.86	0.08	2.21	nr	**3.07**
90 × 35 mm	1.27	1.44	0.10	2.76	nr	**4.20**
Duct connector						
90 × 35 mm	0.59	0.66	0.17	4.71	nr	**5.37**
Socket reducer						
90 × 35 mm	1.06	1.19	0.12	3.33	nr	**4.52**
Vertical access box; 2 compartment						
Shallow	59.42	67.11	0.37	10.25	nr	**77.36**
Duct bend; vertical						
60 × 25 mm	11.59	13.09	0.27	7.48	nr	**20.57**
90 × 35 mm	13.08	14.77	0.35	9.70	nr	**24.47**
Duct bend; horizontal						
60 × 25 mm	13.67	15.44	0.30	8.31	nr	**23.75**
90 × 35 mm	13.87	15.67	0.37	10.25	nr	**25.92**
Zinc coated steel underfloor ducting; fixed to backgrounds; standard coupling joints; earth continuity straps; (Including supports and fixing, packing shims where required)						
Double compartment						
150 × 25 mm	9.35	10.56	0.57	15.79	m	**26.35**
Triple compartment						
225 × 25 mm	16.57	18.72	0.93	25.77	m	**44.49**

V: ELECTRICAL SUPPLY/POWER/LIGHTING

Item	Net Price £	Material £	Labour hours	Labour £	Unit	**Total rate £**
Zinc coated steel underfloor ducting fittings; (cutting and jointing to fittings is included)						
Stop end; double compartment						
150 × 25 mm	2.94	3.33	0.31	8.59	nr	**11.92**
Stop end; triple compartment						
225 × 25 mm	3.36	3.79	0.37	10.25	nr	**14.04**
Rising bend; double compartment; standard trunking						
150 × 25 mm	18.93	21.38	0.71	19.68	nr	**41.06**
Rising bend; triple compartment; standard trunking						
225 × 25 mm	34.02	38.42	0.85	23.55	nr	**61.97**
Rising bend; double compartment; to skirting						
150 × 25 mm	42.14	47.59	0.90	24.94	nr	**72.53**
Rising bend; triple compartment; to skirting						
225 × 25 mm	49.69	56.12	0.95	26.32	nr	**82.44**
horizontal bend; double compartment						
150 × 25 mm	28.78	32.51	0.64	17.73	nr	**50.24**
Horizontal bend; triple compartment						
225 × 25 mm	33.20	37.50	0.77	21.34	nr	**58.84**
Junction or service outlet boxes; terminal; double compartment						
150 mm	29.78	33.63	0.91	25.20	nr	**58.83**
Junction or service outlet boxes; terminal; triple compartment						
225 mm	34.04	38.44	1.11	30.75	nr	**69.19**
Junction or service outlet boxes; through or angle; double compartment						
150 mm	39.55	44.67	0.97	26.87	nr	**71.54**
Junction or service outlet boxes; through or angle; triple compartment						
225 mm	43.93	49.61	1.17	32.42	nr	**82.03**
Junction or service outlet boxes; tee; double compartment						
150 mm	39.55	44.67	1.02	28.25	nr	**72.92**

V: ELECTRICAL SUPPLY/POWER/LIGHTING

Item	Net Price £	Material £	Labour hours	Labour £	Unit	Total rate £
V20: LV DISTRIBUTION – cont						
TRUNKING – cont						
Zinc coated steel underfloor ducting fittings – cont						
Junction or service outlet boxes; tee; triple compartment						
225 mm	43.93	49.61	1.22	33.80	nr	**83.41**
Junction or service outlet boxes; cross; double compartment						
up to 150 mm	39.55	44.67	1.03	28.54	nr	**73.21**
Junction or service outlet boxes; cross;triple compartment						
225 mm	43.93	49.61	1.23	34.08	nr	**83.69**
Plates for junction/inspection boxes; double and triple compartment						
Blank plate	8.64	9.76	0.92	25.49	nr	**35.25**
Conduit entry plate	10.84	12.24	0.86	23.83	nr	**36.07**
Trunking entry plate	10.84	12.24	0.86	23.83	nr	**36.07**
Service outlet box comprising flat lid with flanged carpet trim; twin 13 A outlet and drilled plate for mounting 2 telephone outlets and terminal blocks; terminal outlet box; double compartment						
150 × 25 mm trunking	61.68	69.67	1.68	46.54	nr	**116.21**
Service outlet box comprising flat lid with flanged carpet trim; twin 13 A outlet and drilled plate for mounting 2 telephone outlets and terminal blocks; terminal outlet box; triple compartment						
225 × 25 mm trunking	68.65	77.53	1.93	53.47	nr	**131.00**
PVC skirting/dado modular trunking; white (cutting and jointing trunking to fittings and backplates for fixing to walls is included)						
Main carrier/backplate						
50 × 170 mm	18.36	20.74	0.22	6.10	m	**26.84**
62 × 190 mm	20.38	23.01	0.22	6.10	m	**29.11**
Extension carrier/backplate						
50 × 42 mm	11.16	12.61	0.58	16.08	m	**28.69**

V: ELECTRICAL SUPPLY/POWER/LIGHTING

Item	Net Price £	Material £	Labour hours	Labour £	Unit	**Total rate £**
Carrier/backplate						
Including cover seal	6.51	7.35	0.53	14.68	m	**22.03**
Chamfered covers for fixing to backplates						
50 × 42 mm	3.78	4.27	0.33	9.14	m	**13.41**
Square covers for fixing to backplates						
50 × 42 mm	7.59	8.57	0.33	9.14	m	**17.71**
Plain covers for fixing to backplates						
85 mm	3.78	4.27	0.34	9.41	m	**13.68**
Retainers-clip to backplates to hold cables						
For chamfered covers	1.05	1.18	0.07	1.94	m	**3.12**
For square-recessed covers	0.90	1.02	0.07	1.94	m	**2.96**
For plain covers	3.49	3.94	0.07	1.94	m	**5.88**
Prepackaged corner assemblies						
Internal; for 170 × 50 Assy	7.33	8.28	0.51	14.13	nr	**22.41**
Internal; for 190 × 62 Assy	8.15	8.15	0.51	12.51	nr	**20.66**
Internal; for 215 × 50 Assy	9.12	10.30	0.53	14.68	nr	**24.98**
Internal; for 254 × 50 Assy	10.86	12.26	0.53	14.68	nr	**26.94**
External; for 170 × 50 Assy	7.33	8.28	0.56	15.51	nr	**23.79**
External; for 190 × 62 Assy	8.15	8.15	0.56	13.74	nr	**21.89**
External; for 215 × 50 Assy	9.12	10.30	0.58	16.08	nr	**26.38**
External; for 254 × 50 Assy	10.86	12.26	0.58	16.08	nr	**28.34**
Clip on end caps						
170 × 50 Assy	4.39	4.96	0.11	3.05	nr	**8.01**
215 × 50 Assy	5.20	5.88	0.11	3.05	nr	**8.93**
254 × 50 Assy	6.10	6.88	0.11	3.05	nr	**9.93**
190 × 62 Assy	6.14	6.94	0.11	3.05	nr	**9.99**
Outlet box						
1 gang; in horizontal trunking; clip in	3.92	4.43	0.34	9.41	nr	**13.84**
2 gang; in horizontal trunking; clip in	4.89	5.52	0.34	9.41	nr	**14.93**
1 gang; in vertical trunking; clip in	3.92	4.43	0.34	9.41	nr	**13.84**
Sheet steel adaptable boxes; with plain or knockout sides; fixed to backgrounds; including supports and fixings (cutting and connecting conduit to boxes is included)						
Square pattern – black						
75 × 75 × 37 mm	2.10	2.38	0.69	19.13	nr	**21.51**
75 × 75 × 50 mm	2.00	2.25	0.69	19.13	nr	**21.38**
75 × 75 × 75 mm	2.29	2.59	0.69	19.13	nr	**21.72**

V: ELECTRICAL SUPPLY/POWER/LIGHTING

Item	Net Price £	Material £	Labour hours	Labour £	Unit	Total rate £
V20: LV DISTRIBUTION – cont						
TRUNKING – cont						
Sheet steel adaptable boxes – cont						
Square pattern – black – cont						
100 × 100 × 50 mm	2.10	2.38	0.71	19.68	nr	**22.06**
150 × 150 × 50 mm	3.19	3.60	0.79	21.89	nr	**25.49**
150 × 150 × 75 mm	3.75	4.23	0.80	22.15	nr	**26.38**
150 × 150 × 100 mm	6.24	7.05	0.80	22.15	nr	**29.20**
200 × 200 × 50 mm	5.00	5.65	0.80	22.15	nr	**27.80**
225 × 225 × 50 mm	6.39	7.22	0.93	25.77	nr	**32.99**
225 × 225 × 100 mm	8.48	9.58	0.94	26.04	nr	**35.62**
300 × 300 × 100 mm	9.10	10.28	0.99	27.42	nr	**37.70**
Square pattern – galvanized						
75 × 75 × 37 mm	1.93	2.18	0.69	19.13	nr	**21.31**
75 × 75 × 50 mm	2.49	2.82	0.70	19.39	nr	**22.21**
75 × 75 × 75 mm	2.25	2.54	0.69	19.13	nr	**21.67**
100 × 100 × 50 mm	2.71	3.06	0.71	19.68	nr	**22.74**
150 × 150 × 50 mm	3.27	3.69	0.84	23.28	nr	**26.97**
150 × 150 × 75 mm	4.24	4.78	0.80	22.15	nr	**26.93**
150 × 150 × 100 mm	5.16	5.82	0.80	22.15	nr	**27.97**
225 × 225 × 50 mm	5.98	6.75	0.93	25.77	nr	**32.52**
225 × 225 × 100 mm	6.29	7.10	0.94	26.04	nr	**33.14**
300 × 300 × 100 mm	8.76	–	–	–	nr	**–**
Rectangular pattern – black						
100 × 75 × 50 mm	3.36	3.79	0.69	19.13	nr	**22.92**
150 × 75 × 50 mm	3.52	3.98	0.70	19.39	nr	**23.37**
150 × 75 × 75 mm	3.82	4.31	0.71	19.68	nr	**23.99**
150 × 100 × 75 mm	8.50	9.60	0.71	19.68	nr	**29.28**
225 × 75 × 50 mm	7.35	8.30	0.78	21.60	nr	**29.90**
225 × 150 × 75 mm	11.74	13.26	0.81	22.44	nr	**35.70**
225 × 150 × 100 mm	22.04	24.89	0.81	22.44	nr	**47.33**
300 × 150 × 50 mm	22.04	24.89	0.93	25.77	nr	**50.66**
300 × 150 × 75 mm	22.04	24.89	0.94	26.04	nr	**50.93**
300 × 150 × 100 mm	22.04	24.89	0.96	26.60	nr	**51.49**
Rectangular pattern – galvanized						
100 × 75 × 50 mm	3.19	3.60	0.69	19.13	nr	**22.73**
150 × 75 × 50 mm	3.34	3.77	0.70	19.39	nr	**23.16**
150 × 75 × 75 mm	3.63	4.10	0.71	19.68	nr	**23.78**
150 × 100 × 75 mm	8.07	9.12	0.71	19.68	nr	**28.80**
225 × 75 × 50 mm	6.99	7.89	0.89	24.65	nr	**32.54**
225 × 150 × 75 mm	11.15	12.60	0.81	22.44	nr	**35.04**
225 × 150 × 100 mm	20.94	23.65	0.81	22.44	nr	**46.09**
300 × 150 × 50 mm	20.94	23.65	0.93	25.77	nr	**49.42**
300 × 150 × 75 mm	20.94	23.65	0.94	26.04	nr	**49.69**
300 × 150 × 100 mm	20.94	23.65	0.96	26.60	nr	**50.25**

V: ELECTRICAL SUPPLY/POWER/LIGHTING

Item	Net Price £	Material £	Labour hours	Labour £	Unit	**Total rate £**
Y61 – LV CABLES AND WIRING						
ARMOURED CABLE						
Cable; XLPE insulated; PVC sheathed; copper stranded conductors to BS 5467; laid in trench/duct including marker tape; (cable tiles measured elsewhere)						
600/1000 volt grade; single core (aluminium wire armour)						
25 mm²	0.48	0.54	0.15	4.15	m	**4.69**
35 mm²	0.54	0.61	0.15	4.15	m	**4.76**
50 mm²	0.79	0.89	0.17	4.71	m	**5.60**
70 mm²	5.64	6.36	0.18	4.99	m	**11.35**
95 mm²	7.36	8.31	0.20	5.55	m	**13.86**
120 mm²	8.86	10.01	0.22	6.10	m	**16.11**
150 mm²	10.72	12.11	0.24	6.65	m	**18.76**
185 mm²	13.05	14.74	0.26	7.20	m	**21.94**
240 mm²	16.79	18.96	0.30	8.31	m	**27.27**
300 mm²	20.74	23.42	0.31	8.59	m	**32.01**
400 mm²	26.28	29.68	0.38	10.53	m	**40.21**
500 mm²	45.45	51.33	0.44	12.19	m	**63.52**
630 mm²	55.43	62.60	0.52	14.41	m	**77.01**
800 mm²	34.37	38.82	0.62	17.18	m	**56.00**
1000 mm²	39.80	44.95	0.65	18.00	m	**62.95**
600/1000 volt grade; two core (galvanized steel wire armour)						
1.5 mm²	0.86	0.97	0.06	1.66	m	**2.63**
2.5 mm²	1.07	1.20	0.06	1.66	m	**2.86**
4 mm²	1.22	1.38	0.08	2.21	m	**3.59**
6 mm²	1.78	2.01	0.08	2.21	m	**4.22**
10 mm²	2.31	2.61	0.10	2.76	m	**5.37**
16 mm²	3.37	3.80	0.10	2.76	m	**6.56**
25 mm²	4.81	5.44	0.15	4.15	m	**9.59**
35 mm²	6.08	6.86	0.15	4.15	m	**11.01**
50 mm²	7.84	8.85	0.17	4.71	m	**13.56**
70 mm²	10.76	12.15	0.18	4.99	m	**17.14**
95 mm²	14.99	16.93	0.20	5.55	m	**22.48**
120 mm²	18.23	20.59	0.22	6.10	m	**26.69**
150 mm²	22.47	25.38	0.24	6.65	m	**32.03**
185 mm²	28.69	32.41	0.26	7.20	m	**39.61**
240 mm²	37.00	41.79	0.30	8.31	m	**50.10**
300 mm²	45.68	51.59	0.31	8.59	m	**60.18**
400 mm²	53.20	60.09	0.35	9.70	m	**69.79**

V: ELECTRICAL SUPPLY/POWER/LIGHTING

Item	Net Price £	Material £	Labour hours	Labour £	Unit	**Total rate £**
V20: LV DISTRIBUTION – cont						
ARMOURED CABLE – cont						
600/1000 volt grade; three core (galvanized steel wire armour)						
1.5 mm²	0.85	0.96	0.07	1.94	m	**2.90**
2.5 mm²	1.16	1.31	0.07	1.94	m	**3.25**
4 mm²	1.53	1.72	0.09	2.50	m	**4.22**
6 mm²	1.94	2.19	0.10	2.76	m	**4.95**
10 mm²	3.01	3.40	0.11	3.05	m	**6.45**
16 mm²	4.35	4.92	0.11	3.05	m	**7.97**
25 mm²	6.22	7.03	0.16	4.43	m	**11.46**
35 mm²	8.22	9.28	0.16	4.43	m	**13.71**
50 mm²	10.72	12.11	0.19	5.26	m	**17.37**
70 mm²	15.19	17.16	0.21	5.81	m	**22.97**
95 mm²	20.95	23.66	0.23	6.36	m	**30.02**
120 mm²	25.89	29.24	0.24	6.65	m	**35.89**
150 mm²	31.99	36.13	0.27	7.48	m	**43.61**
185 mm²	39.61	44.74	0.30	8.31	m	**53.05**
240 mm²	51.41	58.06	0.33	9.14	m	**67.20**
300 mm²	63.30	71.49	0.35	9.70	m	**81.19**
400 mm²	73.94	83.51	0.41	11.36	m	**94.87**
600/1000 volt grade; four core (galvanized steel wire armour)						
1.5 mm²	0.98	1.11	0.08	2.21	m	**3.32**
2.5 mm²	1.34	1.51	0.09	2.50	m	**4.01**
4 mm²	1.80	2.03	0.10	2.76	m	**4.79**
6 mm²	2.53	2.86	0.10	2.76	m	**5.62**
10 mm²	3.78	4.27	0.12	3.33	m	**7.60**
16 mm²	5.50	6.21	0.12	3.33	m	**9.54**
25 mm²	8.33	9.40	0.18	4.99	m	**14.39**
35 mm²	11.12	12.56	0.19	5.26	m	**17.82**
50 mm²	14.80	16.72	0.21	5.81	m	**22.53**
70 mm²	19.88	22.45	0.23	6.36	m	**28.81**
95 mm²	26.69	30.14	0.26	7.20	m	**37.34**
120 mm²	33.55	37.89	0.28	7.76	m	**45.65**
150 mm²	41.01	46.32	0.32	8.86	m	**55.18**
185 mm²	51.25	57.89	0.35	9.70	m	**67.59**
240 mm²	66.05	74.59	0.36	9.98	m	**84.57**
300 mm²	82.13	92.76	0.40	11.08	m	**103.84**
400 mm²	107.51	121.42	0.45	12.46	m	**133.88**
600/1000 volt grade; seven core (galvanized steel wire armour)						
1.5 mm²	1.47	1.66	0.10	2.76	m	**4.42**
2.5 mm²	2.05	2.32	0.10	2.76	m	**5.08**
4 mm²	2.44	2.75	0.11	3.05	m	**5.80**

V: ELECTRICAL SUPPLY/POWER/LIGHTING

Item	Net Price £	Material £	Labour hours	Labour £	Unit	Total rate £
600/1000 volt grade; twelve core (galvanized steel wire armour)						
1.5 mm²	2.38	2.69	0.11	3.05	m	**5.74**
2.5 mm²	3.36	3.79	0.11	3.05	m	**6.84**
600/1000 volt grade; nineteen core (galvanized steel wire armour)						
1.5 mm²	3.43	3.88	0.13	3.60	m	**7.48**
2.5 mm²	5.13	5.79	0.14	3.88	m	**9.67**
600/1000 volt grade; twenty seven core (galvanized steel wire armour)						
1.5 mm²	4.88	5.51	0.14	3.88	m	**9.39**
2.5 mm²	7.10	8.02	0.16	4.43	m	**12.45**
600/1000 volt grade; thirty seven core (galvanized steel wire armour)						
1.5 mm²	6.28	7.09	0.15	4.15	m	**11.24**
2.5 mm²	9.25	10.44	0.17	4.71	m	**15.15**
Cable; XLPE insulated; PVC sheathed copper stranded conductors to BS 5467; clipped direct to backgrounds including cleat						
600/1000 volt grade; single core (aluminium wire armour)						
25 mm²	2.21	2.50	0.35	9.70	m	**12.20**
35 mm²	2.41	2.72	0.36	9.98	m	**12.70**
50 mm²	2.66	3.01	0.37	10.25	m	**13.26**
70 mm²	6.06	6.84	0.39	10.81	m	**17.65**
95 mm²	7.98	9.02	0.42	11.63	m	**20.65**
120 mm²	9.56	10.80	0.47	13.03	m	**23.83**
150 mm²	11.42	12.89	0.51	14.13	m	**27.02**
185 mm²	13.75	15.52	0.59	16.34	m	**31.86**
240 mm²	18.37	20.75	0.68	18.84	m	**39.59**
300 mm²	22.32	25.20	0.74	20.50	m	**45.70**
400 mm²	29.17	32.95	0.88	24.39	m	**57.34**
500 mm²	36.90	41.68	0.88	24.39	m	**66.07**
630 mm²	49.02	55.37	1.05	29.09	m	**84.46**
800 mm²	41.68	47.07	1.33	36.84	m	**83.91**
1000 mm²	46.61	52.64	1.40	38.78	m	**91.42**
600/1000 volt grade; two core (galvanized steel wire armour)						
1.5 mm²	0.90	1.02	0.20	5.55	m	**6.57**
2.5 mm²	1.08	1.22	0.20	5.55	m	**6.77**
4.0 mm²	1.41	1.59	0.21	5.81	m	**7.40**
6.0 mm²	1.94	2.19	0.22	6.10	m	**8.29**
10.0 mm²	2.87	3.24	0.24	6.65	m	**9.89**
16.0 mm²	3.93	4.44	0.25	6.93	m	**11.37**
25 mm²	5.28	5.97	0.35	9.70	m	**15.67**
35 mm²	6.67	7.54	0.36	9.98	m	**17.52**

V: ELECTRICAL SUPPLY/POWER/LIGHTING

Item	Net Price £	Material £	Labour hours	Labour £	Unit	Total rate £
V20: LV DISTRIBUTION – cont						
ARMOURED CABLE – cont						
Cable – cont						
600/1000 volt grade – cont						
50 mm²	8.37	9.46	0.37	10.25	m	**19.71**
70 mm²	11.41	12.88	0.39	10.81	m	**23.69**
95 mm²	14.95	16.88	0.42	11.63	m	**28.51**
120 mm²	19.00	21.46	0.47	13.03	m	**34.49**
150 mm²	22.96	25.93	0.51	14.13	m	**40.06**
185 mm²	28.70	32.42	0.59	16.34	m	**48.76**
240 mm²	36.52	41.25	0.68	18.84	m	**60.09**
300 mm²	34.12	38.54	0.74	20.50	m	**59.04**
400 mm²	42.41	47.90	0.88	24.39	m	**72.29**
600/1000 volt grade; three core (galvanized steel wire armour)						
1.5 mm²	1.03	1.16	0.20	5.55	m	**6.71**
2.5 mm²	1.34	1.51	0.21	5.81	m	**7.32**
4.0 mm²	1.75	1.98	0.22	6.10	m	**8.08**
6.0 mm²	2.36	2.66	0.22	6.10	m	**8.76**
10.0 mm²	3.63	4.10	0.25	6.93	m	**11.03**
16.0 mm²	4.97	5.61	0.26	7.20	m	**12.81**
25 mm²	6.92	7.81	0.37	10.25	m	**18.06**
35 mm²	8.92	10.08	0.39	10.81	m	**20.89**
50 mm²	11.52	13.02	0.40	11.08	m	**24.10**
70 mm²	15.52	17.52	0.42	11.63	m	**29.15**
95 mm²	21.54	24.33	0.45	12.46	m	**36.79**
120 mm²	26.48	29.91	0.52	14.41	m	**44.32**
150 mm²	32.72	36.95	0.55	15.24	m	**52.19**
185 mm²	40.34	45.56	0.63	17.45	m	**63.01**
240 mm²	52.94	59.79	0.71	19.68	m	**79.47**
300 mm²	64.83	73.22	0.78	21.60	m	**94.82**
400 mm²	83.34	94.13	0.87	24.10	m	**118.23**
600/1000 volt grade; four core (galvanized steel wire armour)						
1.5 mm²	1.16	1.31	0.21	5.81	m	**7.12**
2.5 mm²	1.56	1.76	0.22	6.10	m	**7.86**
4.0 mm²	2.22	2.51	0.22	6.10	m	**8.61**
6.0 mm²	3.15	3.56	0.23	6.36	m	**9.92**
10.0 mm²	4.40	4.97	0.26	7.20	m	**12.17**
16.0 mm²	6.20	7.00	0.26	7.20	m	**14.20**
25 mm²	9.03	10.20	0.39	10.81	m	**21.01**
35 mm²	11.92	13.46	0.40	11.08	m	**24.54**
50 mm²	15.60	17.62	0.41	11.36	m	**28.98**
70 mm²	20.47	23.12	0.45	12.46	m	**35.58**
95 mm²	27.42	30.97	0.50	13.86	m	**44.83**
120 mm²	34.28	38.72	0.54	14.96	m	**53.68**
150 mm²	46.56	52.58	0.60	16.63	m	**69.21**

V: ELECTRICAL SUPPLY/POWER/LIGHTING

Item	Net Price £	Material £	Labour hours	Labour £	Unit	**Total rate £**
185 mm²	57.52	64.96	0.67	18.56	m	**83.52**
240 mm²	74.40	84.03	0.75	20.78	m	**104.81**
300 mm²	92.47	104.44	0.83	22.99	m	**127.43**
400 mm²	98.83	111.62	0.91	25.20	m	**136.82**
600/1000 volt grade; seven core (galvanized steel wire armour)						
1.5 mm²	1.69	1.91	0.20	5.55	m	**7.46**
2.5 mm²	2.27	2.56	0.20	5.55	m	**8.11**
4.0 mm²	4.12	4.65	0.23	6.36	m	**11.01**
600/1000 volt grade; twelve core (galvanized steel wire armour)						
1.5 mm²	2.80	7.73	0.23	6.36	m	**14.09**
2.5 mm²	3.98	4.50	0.24	6.65	m	**11.15**
600/1000 volt grade; nineteen core (galvanized steel wire armour)						
1.5 mm²	4.05	4.57	0.26	7.20	m	**11.77**
2.5 mm²	5.83	6.59	0.28	7.76	m	**14.35**
600/1000 volt grade; twenty seven core (galvanized steel wire armour)						
1.5 mm²	5.58	6.30	0.29	8.03	m	**14.33**
2.5 mm²	7.90	8.93	0.30	8.31	m	**17.24**
600/1000 volt grade; thirty seven core (galvanized steel wire armour)						
1.5 mm²	7.08	8.00	0.32	8.86	m	**16.86**
2.5 mm²	10.05	11.35	0.33	9.14	m	**20.49**
Cable termination; brass weatherproof gland with inner and outer seal, shroud, brass locknut and earth ring (including drilling and cutting mild steel gland plate)						
600/1000 volt grade; single core (aluminium wire armour)						
25 mm²	8.24	9.30	1.70	47.09	nr	**56.39**
35 mm²	8.24	9.30	1.79	49.59	nr	**58.89**
50 mm²	9.02	10.19	2.06	57.07	nr	**67.26**
70 mm²	15.43	17.42	2.12	58.73	nr	**76.15**
95 mm²	15.48	17.48	2.39	66.22	nr	**83.70**
120 mm²	15.52	17.52	2.47	68.43	nr	**85.95**
150 mm²	20.06	22.65	2.73	75.63	nr	**98.28**
185 mm²	20.32	22.95	3.05	84.50	nr	**107.45**
240 mm²	25.92	29.27	3.45	95.58	nr	**124.85**
300 mm²	26.28	29.68	3.84	106.39	nr	**136.07**
400 mm²	33.42	37.74	4.21	116.64	nr	**154.38**
500 mm²	35.09	39.63	5.70	157.92	m	**197.55**
630 mm²	47.40	53.53	6.20	171.77	m	**225.30**
800 mm²	80.85	91.31	7.50	207.77	m	**299.08**
1000 mm²	97.57	110.19	10.00	277.04	m	**387.23**

V: ELECTRICAL SUPPLY/POWER/LIGHTING

Item	Net Price £	Material £	Labour hours	Labour £	Unit	Total rate £
V20: LV DISTRIBUTION – cont						
ARMOURED CABLE – cont						
Cable termination – cont						
600/1000 volt grade; two core (galvanized steel wire armour)						
1.5 mm²	5.10	5.76	0.58	16.08	nr	**21.84**
2.5 mm²	5.10	5.76	0.58	16.10	nr	**21.86**
4 mm²	5.10	5.76	0.58	16.08	nr	**21.84**
6 mm²	5.12	5.78	0.67	18.56	nr	**24.34**
10 mm²	5.76	6.51	1.00	27.70	nr	**34.21**
16 mm²	5.94	6.71	1.11	30.75	nr	**37.46**
25 mm²	6.76	7.64	1.70	47.09	nr	**54.73**
35 mm²	6.69	7.56	1.79	49.59	nr	**57.15**
50 mm²	11.86	13.39	2.06	57.07	nr	**70.46**
70 mm²	11.96	13.50	2.12	58.73	nr	**72.23**
95 mm²	14.33	16.19	2.39	66.22	nr	**82.41**
120 mm²	14.39	16.25	2.47	68.43	nr	**84.68**
150 mm²	21.03	23.76	2.73	75.63	nr	**99.39**
185 mm²	21.53	24.32	3.05	84.50	nr	**108.82**
240 mm²	23.11	26.10	3.45	95.58	nr	**121.68**
300 mm²	37.94	42.85	3.84	106.39	nr	**149.24**
400 mm²	60.35	68.16	4.21	116.64	nr	**184.80**
600/1000 volt grade; three core (galvanized steel wire armour)						
1.5 mm²	5.23	5.91	0.61	17.03	nr	**22.94**
2.5 mm²	5.23	5.91	0.62	17.18	nr	**23.09**
4 mm²	5.10	5.76	0.62	17.18	nr	**22.94**
6 mm²	5.90	6.66	0.71	19.68	nr	**26.34**
10 mm²	6.78	7.66	1.06	29.37	nr	**37.03**
16 mm²	7.05	7.97	1.19	32.97	nr	**40.94**
25 mm²	12.02	13.58	1.81	50.14	nr	**63.72**
35 mm²	12.05	13.61	1.99	55.12	nr	**68.73**
50 mm²	12.08	13.65	2.23	61.78	nr	**75.43**
70 mm²	14.48	16.35	2.40	66.48	nr	**82.83**
95 mm²	14.63	16.52	2.63	72.86	nr	**89.38**
120 mm²	14.75	16.66	2.83	78.41	nr	**95.07**
150 mm²	22.30	25.18	3.22	89.21	nr	**114.39**
185 mm²	23.08	26.07	3.44	95.30	nr	**121.37**
240 mm²	39.54	44.66	3.83	106.11	nr	**150.77**
300 mm²	40.62	45.88	4.28	118.57	nr	**164.45**
400 mm²	53.46	60.37	5.00	138.52	nr	**198.89**

V: ELECTRICAL SUPPLY/POWER/LIGHTING

Item	Net Price £	Material £	Labour hours	Labour £	Unit	**Total rate £**
600/1000 volt grade; four core (galvanized steel wire armour)						
1.5 mm²	5.36	6.05	0.67	18.56	nr	**24.61**
2.5 mm²	5.36	6.05	0.67	18.56	nr	**24.61**
4 mm²	5.36	6.05	0.71	19.68	nr	**25.73**
6 mm²	6.92	7.81	0.76	21.05	nr	**28.86**
10 mm²	6.92	7.81	1.14	31.58	nr	**39.39**
16 mm²	7.28	8.22	1.29	35.73	nr	**43.95**
25 mm²	12.22	13.80	1.99	55.12	nr	**68.92**
35 mm²	12.26	13.84	2.16	59.83	nr	**73.67**
50 mm²	14.55	16.43	2.49	68.98	nr	**85.41**
70 mm²	14.75	16.66	2.65	73.41	nr	**90.07**
95 mm²	19.75	22.31	2.98	82.56	nr	**104.87**
120 mm²	19.87	22.44	3.15	87.27	nr	**109.71**
150 mm²	23.63	26.68	3.50	96.97	nr	**123.65**
185 mm²	38.72	43.73	3.72	103.06	nr	**146.79**
240 mm²	41.88	47.30	4.33	119.95	nr	**167.25**
300 mm²	60.57	68.41	4.86	134.65	nr	**203.06**
400 mm²	56.43	63.73	5.46	151.26	nr	**214.99**
600/1000 volt grade; seven core (galvanized steel wire armour)						
1.5 mm²	5.75	6.50	0.81	22.44	nr	**28.94**
2.5 mm²	5.75	6.50	0.85	23.55	nr	**30.05**
4 mm²	9.20	10.39	0.93	25.77	nr	**36.16**
600/1000 volt grade; twelve core (galvanized steel wire armour)						
1.5 mm²	7.04	7.96	1.14	31.58	nr	**39.54**
2.5 mm²	7.92	8.95	1.13	31.30	nr	**40.25**
600/1000 volt grade; nineteen core (galvanized steel wire armour)						
1.5 mm²	8.83	9.98	1.54	42.67	nr	**52.65**
2.5 mm²	8.83	9.98	1.54	42.67	nr	**52.65**
600/1000 volt grade; twenty seven core (galvanized steel wire armour)						
1.5 mm²	9.87	11.15	1.94	53.74	nr	**64.89**
2.5 mm²	14.93	16.86	2.31	63.99	nr	**80.85**
600/1000 volt grade; thirty seven core (galvanized steel wire armour)						
1.5 mm²	16.23	18.33	2.53	70.09	nr	**88.42**
2.5 mm²	16.23	18.33	2.87	79.51	nr	**97.84**

V: ELECTRICAL SUPPLY/POWER/LIGHTING

Item	Net Price £	Material £	Labour hours	Labour £	Unit	Total rate £
V20: LV DISTRIBUTION – cont						
ARMOURED CABLE – cont						
Cable; XLPE insulated; LSOH sheathed (LSF) ; copper stranded conductors to BS 6724; laid in trench/duct including marker tape (cable tiles measured elsewhere)						
600/1000 volt grade; single core (aluminium wire armour)						
50 mm²	4.41	4.98	0.17	4.71	m	**9.69**
70 mm²	6.06	6.84	0.18	4.99	m	**11.83**
95 mm²	7.98	9.02	0.20	5.55	m	**14.57**
120 mm²	9.56	10.80	0.22	6.10	m	**16.90**
150 mm²	11.42	12.89	0.24	6.65	m	**19.54**
185 mm²	13.75	15.52	0.26	7.20	m	**22.72**
240 mm²	18.37	20.75	0.30	8.31	m	**29.06**
300 mm²	22.32	25.20	0.31	8.59	m	**33.79**
400 mm²	29.17	32.95	0.35	9.70	m	**42.65**
500 mm²	36.90	41.68	0.44	12.19	m	**53.87**
630 mm²	49.02	55.37	0.52	14.41	m	**69.78**
800 mm²	61.65	69.63	0.62	17.18	m	**86.81**
1000 mm²	78.68	88.86	0.65	18.00	m	**106.86**
600/1000 volt grade; two core (galvanized steel wire armour)						
1.5 mm²	0.90	1.02	0.06	1.66	m	**2.68**
2.5 mm²	1.08	1.22	0.06	1.66	m	**2.88**
4 mm²	1.41	1.59	0.08	2.21	m	**3.80**
6 mm²	1.94	2.19	0.08	2.21	m	**4.40**
10 mm²	2.87	3.24	0.10	2.76	m	**6.00**
16 mm²	3.93	4.44	0.10	2.76	m	**7.20**
25 mm²	5.28	5.97	0.15	4.15	m	**10.12**
35 mm²	6.67	7.54	0.17	4.71	m	**12.25**
50 mm²	8.37	9.46	0.15	4.15	m	**13.61**
70 mm²	11.41	12.88	0.18	4.99	m	**17.87**
95 mm²	14.95	16.88	0.20	5.55	m	**22.43**
120 mm²	19.00	21.46	0.22	6.10	m	**27.56**
150 mm²	22.96	25.93	0.24	6.65	m	**32.58**
185 mm²	28.70	32.42	0.26	7.20	m	**39.62**
240 mm²	36.52	41.25	0.30	8.31	m	**49.56**
300 mm²	34.12	38.54	0.31	8.59	m	**47.13**
400 mm²	52.69	59.51	0.35	9.70	m	**69.21**

V: ELECTRICAL SUPPLY/POWER/LIGHTING

Item	Net Price £	Material £	Labour hours	Labour £	Unit	**Total rate £**
600/1000 volt grade; three core (galvanized steel wire armour)						
1.5 mm²	1.03	1.16	0.07	1.94	m	**3.10**
2.5 mm²	1.34	1.51	0.07	1.94	m	**3.45**
4 mm²	1.75	1.98	0.09	2.50	m	**4.48**
6 mm²	2.36	2.66	0.10	2.76	m	**5.42**
10 mm²	3.63	4.10	0.11	3.05	m	**7.15**
16 mm²	4.97	5.61	0.11	3.05	m	**8.66**
25 mm²	6.92	7.81	0.16	4.43	m	**12.24**
35 mm²	8.92	10.08	0.16	4.43	m	**14.51**
50 mm²	11.52	13.02	0.19	5.26	m	**18.28**
70 mm²	15.52	17.52	0.21	5.81	m	**23.33**
95 mm²	21.54	24.33	0.23	6.36	m	**30.69**
120 mm²	26.48	29.91	0.24	6.65	m	**36.56**
150 mm²	32.72	36.95	0.27	7.48	m	**44.43**
185 mm²	40.34	45.56	0.30	8.31	m	**53.87**
240 mm²	52.94	59.79	0.33	9.14	m	**68.93**
300 mm²	64.83	73.22	0.35	9.70	m	**82.92**
400 mm²	91.19	102.99	0.41	11.36	m	**114.35**
600/1000 volt grade; four core (galvanized steel wire armour)						
1.5 mm²	1.16	1.31	0.08	2.21	m	**3.52**
2.5 mm²	1.56	1.76	0.09	2.50	m	**4.26**
4 mm²	2.22	2.51	0.10	2.76	m	**5.27**
6 mm²	3.15	3.56	0.10	2.76	m	**6.32**
10 mm²	4.40	4.97	0.12	3.33	m	**8.30**
16 mm²	6.20	7.00	0.12	3.33	m	**10.33**
25 mm²	9.03	10.20	0.18	4.99	m	**15.19**
35 mm²	11.92	13.46	0.19	5.26	m	**18.72**
50 mm²	15.60	17.62	0.21	5.81	m	**23.43**
70 mm²	20.47	23.12	0.23	6.36	m	**29.48**
95 mm²	27.42	30.97	0.26	7.20	m	**38.17**
120 mm²	34.28	38.72	0.28	7.76	m	**46.48**
150 mm²	46.56	52.58	0.32	8.86	m	**61.44**
185 mm²	57.52	64.96	0.35	9.70	m	**74.66**
240 mm²	74.40	84.03	0.36	9.98	m	**94.01**
300 mm²	92.47	104.44	0.40	11.08	m	**115.52**
400 mm²	123.47	139.45	0.45	12.46	m	**151.91**
600/1000 volt grade; seven core (galvanized steel wire armour)						
1.5 mm²	1.69	1.91	0.10	2.76	m	**4.67**
2.5 mm²	2.27	2.56	0.10	2.76	m	**5.32**
4 mm²	4.19	4.73	0.11	3.05	m	**7.78**

V: ELECTRICAL SUPPLY/POWER/LIGHTING

Item	Net Price £	Material £	Labour hours	Labour £	Unit	Total rate £
V20: LV DISTRIBUTION – cont						
ARMOURED CABLE – cont						
Cable – cont						
600/1000 volt grade; twelve core (galvanized steel wire armour)						
1.5 mm²	2.80	3.16	0.11	3.05	m	**6.21**
2.5 mm²	3.98	4.50	0.11	3.05	m	**7.55**
600/1000 volt grade; nineteen core (galvanized steel wire armour)						
1.5 mm²	4.05	4.57	0.13	3.60	m	**8.17**
2.5 mm²	5.83	6.59	0.14	3.88	m	**10.47**
600/1000 volt grade; twenty seven core (galvanized steel wire armour)						
1.5 mm²	5.58	6.30	0.14	3.88	m	**10.18**
2.5 mm²	7.90	8.93	0.16	4.43	m	**13.36**
600/1000 volt grade; thirty seven core (galvanized steel wire armour)						
1.5 mm²	7.08	8.00	0.15	4.15	m	**12.15**
2.5 mm²	10.05	11.35	0.17	4.71	m	**16.06**
Cable; XLPE insulated; LSOH sheathed (LSF) copper stranded conductors to BS 6724; clipped direct to backgrounds including cleat						
600/1000 volt grade; single core (aluminium wire armour)						
50 mm²	3.50	3.96	0.37	10.25	m	**14.21**
70 mm²	5.31	6.00	0.39	10.81	m	**16.81**
95 mm²	7.14	8.07	0.42	11.63	m	**19.70**
120 mm²	8.50	9.60	0.47	13.03	m	**22.63**
150 mm²	10.51	11.87	0.51	14.13	m	**26.00**
185 mm²	12.62	14.25	0.59	16.34	m	**30.59**
240 mm²	16.28	18.39	0.68	18.84	m	**37.23**
300 mm²	19.83	22.40	0.74	20.50	m	**42.90**
400 mm²	25.77	29.10	0.81	22.44	m	**51.54**
500 mm²	33.07	37.35	0.88	24.39	m	**61.74**
630 mm²	41.45	46.82	1.05	29.09	m	**75.91**
800 mm²	53.42	60.33	1.33	36.84	m	**97.17**
1000 mm²	61.86	69.86	1.40	38.78	m	**108.64**

V: ELECTRICAL SUPPLY/POWER/LIGHTING

Item	Net Price £	Material £	Labour hours	Labour £	Unit	**Total rate £**
600/1000 volt grade; two core (galvanized steel wire armour)						
1.5 mm²	0.76	0.86	0.20	5.55	m	**6.41**
2.5 mm²	0.87	0.98	0.20	5.55	m	**6.53**
4.0 mm²	1.13	1.28	0.21	5.81	m	**7.09**
6.0 mm²	1.38	1.56	0.22	6.10	m	**7.66**
10.0 mm²	1.85	2.09	0.24	6.65	m	**8.74**
16.0 mm²	2.78	3.14	0.25	6.93	m	**10.07**
25 mm²	2.78	3.14	0.35	9.70	m	**12.84**
35 mm²	5.62	6.34	0.36	9.98	m	**16.32**
50 mm²	6.95	7.85	0.37	10.25	m	**18.10**
70 mm²	10.34	11.68	0.39	10.81	m	**22.49**
95 mm²	13.89	15.69	0.42	11.63	m	**27.32**
120 mm²	17.02	19.23	0.47	13.03	m	**32.26**
150 mm²	21.11	23.84	0.51	14.13	m	**37.97**
185 mm²	26.90	30.39	0.59	16.34	m	**46.73**
240 mm²	34.43	38.88	0.68	18.84	m	**57.72**
300 mm²	46.25	52.23	0.74	20.50	m	**72.73**
400 mm²	61.18	69.09	0.81	22.44	m	**91.53**
600/1000 volt grade; three core (galvanized steel wire armour)						
1.5 mm²	0.83	0.94	0.20	5.55	m	**6.49**
2.5 mm²	1.06	1.19	0.21	5.81	m	**7.00**
4.0 mm²	1.30	1.47	0.22	6.10	m	**7.57**
6.0 mm²	1.68	1.90	0.22	6.10	m	**8.00**
10.0 mm²	2.58	2.92	0.25	6.93	m	**9.85**
16.0 mm²	3.59	4.06	0.26	7.20	m	**11.26**
25 mm²	4.85	5.48	0.37	10.25	m	**15.73**
35 mm²	6.76	7.64	0.39	10.81	m	**18.45**
50 mm²	9.64	10.88	0.40	11.08	m	**21.96**
70 mm²	14.31	16.16	0.42	11.63	m	**27.79**
95 mm²	19.70	22.25	0.45	12.46	m	**34.71**
120 mm²	24.50	27.67	0.52	14.41	m	**42.08**
150 mm²	30.44	34.37	0.55	15.24	m	**49.61**
185 mm²	37.50	42.35	0.63	17.45	m	**59.80**
240 mm²	47.99	54.20	0.71	19.68	m	**73.88**
300 mm²	60.81	68.68	0.78	21.60	m	**90.28**
400 mm²	72.22	81.57	0.87	24.10	m	**105.67**
600/1000 volt grade; four core (galvanized steel wire armour)						
1.5 mm²	0.98	1.11	0.21	5.81	m	**6.92**
2.5 mm²	1.23	1.39	0.22	6.10	m	**7.49**
4.0 mm²	1.52	1.71	0.22	6.10	m	**7.81**
6.0 mm²	2.23	2.52	0.23	6.36	m	**8.88**
10.0 mm²	3.13	3.54	0.26	7.20	m	**10.74**
16.0 mm²	4.52	5.10	0.26	7.20	m	**12.30**
25 mm²	6.52	7.36	0.39	10.81	m	**18.17**

V: ELECTRICAL SUPPLY/POWER/LIGHTING

Item	Net Price £	Material £	Labour hours	Labour £	Unit	Total rate £
V20: LV DISTRIBUTION – cont						
ARMOURED CABLE – cont						
Cable – cont						
600/1000 volt grade – cont						
35 mm²	8.50	9.60	0.40	11.08	m	**20.68**
50 mm²	12.31	13.90	0.41	11.36	m	**25.26**
70 mm²	18.99	21.45	0.45	12.46	m	**33.91**
95 mm²	25.44	28.73	0.50	13.86	m	**42.59**
120 mm²	31.95	36.09	0.54	14.96	m	**51.05**
150 mm²	38.67	43.68	0.60	16.63	m	**60.31**
185 mm²	47.46	53.60	0.67	18.56	m	**72.16**
240 mm²	61.35	69.29	0.75	20.78	m	**90.07**
300 mm²	76.97	86.93	0.83	22.99	m	**109.92**
400 mm²	80.02	90.37	0.91	25.20	m	**115.57**
600/1000 volt grade; seven core (galvanized steel wire armour)						
1.5 mm²	1.40	1.58	0.20	5.55	m	**7.13**
2.5 mm²	1.96	2.21	0.20	5.55	m	**7.76**
4.0 mm²	3.55	4.01	0.23	6.36	m	**10.37**
600/1000 volt grade; twelve core (galvanized steel wire armour)						
1.5 mm²	2.23	2.52	0.23	6.36	m	**8.88**
2.5 mm²	3.35	3.78	0.24	6.65	m	**10.43**
600/1000 volt grade; nineteen core (galvanized steel wire armour)						
1.5 mm²	3.37	3.80	0.26	7.20	m	**11.00**
2.5 mm²	5.06	5.71	0.28	7.76	m	**13.47**
600/1000 volt grade; twenty seven core (galvanized steel wire armour)						
1.5 mm²	4.63	5.23	0.29	8.03	m	**13.26**
2.5 mm²	6.46	7.29	0.30	8.31	m	**15.60**
600/1000 volt grade; thirty seven core (galvanized steel wire armour)						
1.5 mm²	5.77	6.52	0.32	8.86	m	**15.38**
2.5 mm²	8.18	9.24	0.33	9.14	m	**18.38**

V: ELECTRICAL SUPPLY/POWER/LIGHTING

Item	Net Price £	Material £	Labour hours	Labour £	Unit	Total rate £
Cable termination; brass weatherproof gland with inner and outer seal, shroud, brass locknut and earth ring (including drilling and cutting mild steel gland plate)						
600/1000 volt grade; single core (aluminium wire armour)						
25 mm²	3.22	3.64	1.70	47.09	nr	**50.73**
35 mm²	3.51	3.97	1.79	49.59	nr	**53.56**
50 mm²	3.88	4.39	2.06	57.07	nr	**61.46**
70 mm²	5.55	6.27	2.12	58.73	nr	**65.00**
95 mm²	7.43	8.39	2.39	66.22	nr	**74.61**
120 mm²	8.79	9.92	2.47	68.43	nr	**78.35**
150 mm²	10.81	12.21	2.73	75.63	nr	**87.84**
185 mm²	12.88	14.55	3.05	84.50	nr	**99.05**
240 mm²	16.87	19.05	3.45	95.58	nr	**114.63**
300 mm²	20.40	23.04	3.84	106.39	nr	**129.43**
400 mm²	29.77	33.62	4.21	116.64	nr	**150.26**
500 mm²	37.52	42.37	5.70	157.92	m	**200.29**
630 mm²	48.79	55.10	6.20	171.77	m	**226.87**
800 mm²	60.67	68.52	7.50	207.77	m	**276.29**
1000 mm²	67.85	76.63	10.00	277.04	m	**353.67**
600/1000 volt grade; two core (galvanized steel wire armour)						
1.5 mm²	0.95	1.07	0.52	14.32	nr	**15.39**
2.5 mm²	1.09	1.23	0.58	16.08	nr	**17.31**
4 mm²	1.38	1.56	0.58	16.08	nr	**17.64**
6 mm²	1.61	1.82	0.67	18.56	nr	**20.38**
10 mm²	2.08	2.35	1.00	27.70	nr	**30.05**
16 mm²	3.01	3.40	1.11	30.75	nr	**34.15**
25 mm²	2.62	2.96	1.70	47.09	nr	**50.05**
35 mm²	5.87	6.63	1.79	49.59	nr	**56.22**
50 mm²	7.22	8.15	2.06	57.07	nr	**65.22**
70 mm²	10.72	12.11	2.12	58.73	nr	**70.84**
95 mm²	13.98	15.79	2.39	66.22	nr	**82.01**
120 mm²	17.18	19.40	2.47	68.43	nr	**87.83**
150 mm²	21.28	24.03	2.73	75.63	nr	**99.66**
185 mm²	27.09	30.60	3.05	84.50	nr	**115.10**
240 mm²	33.71	38.08	3.45	95.58	nr	**133.66**
300 mm²	47.21	53.32	3.84	106.39	nr	**159.71**
400 mm²	58.77	66.37	4.21	116.64	nr	**183.01**
600/1000 volt grade; three core (galvanized steel wire armour)						
1.5 mm²	1.02	1.15	0.62	17.18	nr	**18.33**
2.5 mm²	1.21	1.37	0.62	17.18	nr	**18.55**
4 mm²	1.54	1.74	0.62	17.18	nr	**18.92**
6 mm²	1.92	2.17	0.71	19.68	nr	**21.85**
10 mm²	2.79	3.15	1.06	29.37	nr	**32.52**
16 mm²	3.84	4.33	1.19	32.97	nr	**37.30**
25 mm²	5.38	6.08	1.81	50.14	nr	**56.22**

V: ELECTRICAL SUPPLY/POWER/LIGHTING

Item	Net Price £	Material £	Labour hours	Labour £	Unit	Total rate £
V20: LV DISTRIBUTION – cont						
ARMOURED CABLE – cont						
Cable termination – cont						
600/1000 volt grade – cont						
35 mm²	6.99	7.89	1.99	55.12	nr	**63.01**
50 mm²	9.97	11.26	2.23	61.78	nr	**73.04**
70 mm²	14.46	16.33	2.40	66.48	nr	**82.81**
95 mm²	19.86	22.43	2.63	72.86	nr	**95.29**
120 mm²	24.66	27.85	2.83	78.41	nr	**106.26**
150 mm²	30.69	34.66	3.22	89.21	nr	**123.87**
185 mm²	37.73	42.62	3.44	95.30	nr	**137.92**
240 mm²	50.11	56.59	3.83	106.11	nr	**162.70**
300 mm²	63.08	71.25	4.28	118.57	nr	**189.82**
400 mm²	74.88	84.57	5.00	138.52	nr	**223.09**
600/1000 volt grade; four core (galvanized steel wire armour)						
1.5 mm²	1.17	1.33	0.67	18.56	nr	**19.89**
2.5 mm²	1.45	1.64	0.67	18.56	nr	**20.20**
4 mm²	1.77	2.00	0.71	19.68	nr	**21.68**
6 mm²	2.47	2.78	0.76	21.05	nr	**23.83**
10 mm²	3.38	3.81	1.14	31.58	nr	**35.39**
16 mm²	4.76	5.38	1.29	35.73	nr	**41.11**
25 mm²	6.77	7.65	1.99	55.12	nr	**62.77**
35 mm²	8.82	9.97	2.16	59.83	nr	**69.80**
50 mm²	12.65	14.29	2.49	68.98	nr	**83.27**
70 mm²	19.15	21.62	2.65	73.41	nr	**95.03**
95 mm²	25.66	28.98	2.98	82.56	nr	**111.54**
120 mm²	32.15	36.31	3.15	87.27	nr	**123.58**
150 mm²	40.58	45.83	3.50	96.97	nr	**142.80**
185 mm²	49.59	56.01	3.72	103.06	nr	**159.07**
240 mm²	64.20	72.51	4.33	119.95	nr	**192.46**
300 mm²	80.56	90.98	4.86	134.65	nr	**225.63**
400 mm²	83.57	94.38	5.46	151.26	nr	**245.64**
600/1000 volt grade; seven core (galvanized steel wire armour)						
1.5 mm²	1.64	1.86	0.81	22.44	nr	**24.30**
2.5 mm²	2.24	2.53	0.85	23.55	nr	**26.08**
4 mm²	4.05	4.57	0.93	25.77	nr	**30.34**
600/1000 volt grade; twelve core (galvanized steel wire armour)						
1.5 mm²	2.38	2.69	1.14	31.58	nr	**34.27**
2.5 mm²	3.58	4.04	1.13	31.30	nr	**35.34**

V: ELECTRICAL SUPPLY/POWER/LIGHTING

Item	Net Price £	Material £	Labour hours	Labour £	Unit	**Total rate £**
600/1000 volt grade; nineteen core (galvanized steel wire armour)						
1.5 mm²	3.60	4.07	1.54	42.67	nr	**46.74**
2.5 mm²	5.30	5.99	1.54	42.67	nr	**48.66**
600/1000 volt grade; twenty seven core (galvanized steel wire armour)						
1.5 mm²	4.86	5.49	1.94	53.74	nr	**59.23**
2.5 mm²	6.77	7.65	2.31	63.99	nr	**71.64**
600/1000 volt grade; thirty seven core (galvanized steel wire armour)						
1.5 mm²	6.08	6.86	2.53	70.09	nr	**76.95**
2.5 mm²	8.49	9.59	2.87	79.51	nr	**89.10**
UN-ARMOURED CABLE						
Cable: XLPE insulated; PVC sheathed 90c copper to CMA Code 6181e; for internal wiring; clipped to backgrounds; (supports and fixings included)						
300/500 volt grade; single core						
6.0 mm²	0.53	0.60	0.09	2.50	m	**3.10**
10 mm²	0.76	0.86	0.10	2.76	m	**3.62**
16 mm²	1.17	1.33	0.12	3.33	m	**4.66**
Cable; LSF insulated to CMA Code 6491B; non-sheathed copper; laid/drawn in trunking/conduit						
450/750 volt grade; single core						
1.5 mm²	0.14	0.17	0.03	0.84	m	**1.01**
2.5 mm²	0.18	0.21	0.03	0.84	m	**1.05**
4.0 mm²	0.31	0.35	0.03	0.84	m	**1.19**
6.0 mm²	0.47	0.53	0.04	1.11	m	**1.64**
10.0 mm²	0.78	0.88	0.04	1.11	m	**1.99**
16.0 mm²	1.25	1.41	0.05	1.39	m	**2.80**
25.0 mm²	1.82	2.06	0.06	1.66	m	**3.72**
35.0 mm²	2.51	2.84	0.06	1.66	m	**4.50**
50.0 mm²	3.35	3.78	0.07	1.94	m	**5.72**
70.0 mm²	4.77	5.39	0.08	2.21	m	**7.60**
95.0 mm²	6.64	7.50	0.08	2.21	m	**9.71**
120.0 mm²	8.66	9.78	0.10	2.76	m	**12.54**
150.0 mm²	9.79	11.06	0.13	3.60	m	**14.66**
185.0 mm²	12.98	12.98	0.13	3.19	m	**16.17**
240.0 mm²	16.61	16.61	0.13	3.19	m	**19.80**
300.0 mm²	20.91	20.91	0.13	3.19	m	**24.10**

V: ELECTRICAL SUPPLY/POWER/LIGHTING

Item	Net Price £	Material £	Labour hours	Labour £	Unit	Total rate £
V20: LV DISTRIBUTION – cont						
UN-ARMOURED CABLE – cont						
Cable; twin & earth to CMA code 6242Y; clipped to backgrounds						
300/500 volt grade; PVC/PVC						
1.5 mm² 2C+E	0.29	0.33	0.01	0.29	m	**0.62**
1.5 mm² 3C+E	0.36	0.41	0.02	0.55	m	**0.96**
2.5 mm² 2C+E	0.54	0.61	0.02	0.55	m	**1.16**
4.0 mm² 2C+E	0.82	0.93	0.02	0.55	m	**1.48**
6.0 mm² 2C+E	1.18	1.34	0.02	0.55	m	**1.89**
10.0 mm² 2C+E	1.97	2.22	0.03	0.84	m	**3.06**
16.0 mm² 2C+E	2.95	3.34	0.03	0.84	m	**4.18**
300/500 volt grade; LSF/LSF						
1.5 mm² 2C+E	0.40	0.45	0.01	0.29	m	**0.74**
1.5 mm² 3C+E	0.57	0.64	0.02	0.55	m	**1.19**
2.5 mm² 2C+E	0.71	0.81	0.02	0.55	m	**1.36**
4.0 mm² 2C+E	0.84	0.95	0.02	0.55	m	**1.50**
6.0 mm² 2C+E	1.23	1.39	0.02	0.55	m	**1.94**
10.0 mm² 2C+E	2.82	3.18	0.03	0.84	m	**4.02**
16.0 mm² 2C+E	4.51	5.09	0.03	0.84	m	**5.93**
EARTH CABLE						
Cable; LSF insulated to CMA Code 6491B; non-sheathed copper; laid/drawn in trunking/conduit						
450/750 volt grade; single core						
1.5 mm²	0.14	0.17	0.03	0.84	m	**1.01**
2.5 mm²	0.18	0.21	0.03	0.84	m	**1.05**
4.0 mm²	0.31	0.35	0.03	0.84	m	**1.19**
6.0 mm²	0.47	0.53	0.04	1.11	m	**1.64**
10.0 mm²	0.78	0.88	0.04	1.11	m	**1.99**
16.0 mm²	1.25	1.41	0.05	1.39	m	**2.80**
25.0 mm²	1.82	2.06	0.06	1.66	m	**3.72**
35.0 mm²	2.51	2.84	0.06	1.66	m	**4.50**
50.0 mm²	3.35	3.78	0.07	1.94	m	**5.72**
70.0 mm²	4.77	5.39	0.08	2.21	m	**7.60**
95.0 mm²	6.64	7.50	0.08	2.21	m	**9.71**
120.0 mm²	8.66	9.78	0.10	2.76	m	**12.54**
150.0 mm²	9.79	11.06	0.13	3.60	m	**14.66**
185.0 mm²	12.98	14.66	0.16	4.43	m	**19.09**
240.0 mm²	16.61	18.76	0.20	5.55	m	**24.31**

V: ELECTRICAL SUPPLY/POWER/LIGHTING

Item	Net Price £	Material £	Labour hours	Labour £	Unit	Total rate £
FLEXIBLE CABLE						
Flexible cord; PVC insulated; PVC sheathed; copper stranded to CMA Code 218*Y (laid loose)						
300 volt grade; two core						
0.50 mm²	0.13	0.15	0.07	1.94	m	**2.09**
0.75 mm²	0.16	0.19	0.07	1.94	m	**2.13**
300 volt grade; three core						
0.50 mm²	0.18	0.21	0.07	1.94	m	**2.15**
0.75 mm²	0.26	0.30	0.07	1.94	m	**2.24**
1.0 mm²	0.27	0.31	0.07	1.94	m	**2.25**
1.5 mm²	0.43	0.49	0.07	1.94	m	**2.43**
2.5 mm²	0.57	0.64	0.08	2.21	m	**2.85**
Flexible cord; PVC insulated; PVC sheathed; copper stranded to CMA Code 318Y (laid loose)						
300/500 volt grade; two core						
0.75 mm²	0.18	0.21	0.07	1.94	m	**2.15**
1.0 mm²	0.22	0.25	0.07	1.94	m	**2.19**
1.5 mm²	0.30	0.34	0.07	1.94	m	**2.28**
2.5 mm²	0.51	0.57	0.07	1.94	m	**2.51**
300/500 volt grade; three core						
0.75 mm²	0.24	0.28	0.07	1.94	m	**2.22**
1.0 mm²	0.30	0.34	0.07	1.94	m	**2.28**
1.5 mm²	0.43	0.49	0.07	1.94	m	**2.43**
2.5 mm²	0.68	0.77	0.08	2.21	m	**2.98**
300/500 volt grade; four core						
0.75 mm²	0.32	0.36	0.08	2.21	m	**2.57**
1.0 mm²	0.40	0.45	0.08	2.21	m	**2.66**
1.5 mm²	0.57	0.64	0.08	2.21	m	**2.85**
2.5 mm²	0.93	1.05	0.09	2.50	m	**3.55**
Flexible cord; PVC insulated; PVC sheathed for use in high temperature zones; copper stranded to CMA Code 309Y (laid loose)						
300/500 volt grade; two core						
0.50 mm²	0.19	0.22	0.07	1.94	m	**2.16**
0.75 mm²	0.23	0.26	0.07	1.94	m	**2.20**
1.0 mm²	0.27	0.31	0.07	1.94	m	**2.25**
1.5 mm²	0.38	0.43	0.07	1.94	m	**2.37**
2.5 mm²	0.55	0.62	0.07	1.94	m	**2.56**

V: ELECTRICAL SUPPLY/POWER/LIGHTING

Item	Net Price £	Material £	Labour hours	Labour £	Unit	Total rate £
V20: LV DISTRIBUTION – cont						
FLEXIBLE CABLE – cont						
Flexible cord – cont						
300/500 volt grade; three core						
0.50 mm²	0.27	0.31	0.07	1.94	m	**2.25**
0.75 mm²	0.32	0.36	0.07	1.94	m	**2.30**
1.0 mm²	0.37	0.42	0.07	1.94	m	**2.36**
1.5 mm²	0.50	0.56	0.07	1.94	m	**2.50**
2.5 mm²	0.70	0.80	0.07	1.94	m	**2.74**
Flexible cord; rubber insulated; rubber sheathed; copper stranded to CMA code 318 (laid loose)						
300/500 volt grade; two core						
0.50 mm²	0.32	0.36	0.07	1.94	m	**2.30**
0.75 mm²	0.40	0.45	0.07	1.94	m	**2.39**
1.0 mm²	0.48	0.54	0.07	1.94	m	**2.48**
1.5 mm²	0.61	0.69	0.07	1.94	m	**2.63**
2.5 mm²	0.90	1.02	0.07	1.94	m	**2.96**
300/500 volt grade; three core						
0.50 mm²	0.40	0.45	0.07	1.94	m	**2.39**
0.75 mm²	0.48	0.54	0.07	1.94	m	**2.48**
1.0 mm²	0.56	0.63	0.07	1.94	m	**2.57**
1.5 mm²	0.78	0.88	0.07	1.94	m	**2.82**
2.5 mm²	1.15	1.30	0.07	1.94	m	**3.24**
300/500 volt grade; four core						
0.50 mm²	0.45	0.51	0.08	2.21	m	**2.72**
0.75 mm²	0.54	0.61	0.08	2.21	m	**2.82**
1.0 mm²	0.70	0.80	0.08	2.21	m	**3.01**
1.5 mm²	1.08	1.22	0.08	2.21	m	**3.43**
2.5 mm²	1.51	1.70	0.08	2.21	m	**3.91**
Flexible cord; rubber insulated; rubber sheathed; for 90C operation; copper stranded to CMA Code 318 (laid loose)						
450/750 volt grade; two core						
0.50 mm²	0.28	0.32	0.07	1.94	m	**2.26**
0.75 mm²	0.44	0.50	0.07	1.94	m	**2.44**
1.0 mm²	0.44	0.50	0.07	1.94	m	**2.44**
1.5 mm²	0.52	0.59	0.07	1.94	m	**2.53**
2.5 mm²	0.71	0.81	0.07	1.94	m	**2.75**

V: ELECTRICAL SUPPLY/POWER/LIGHTING

Item	Net Price £	Material £	Labour hours	Labour £	Unit	Total rate £
450/750 volt grade; three core						
0.50 mm²	0.45	0.51	0.07	1.94	m	**2.45**
0.75 mm²	0.51	0.57	0.07	1.94	m	**2.51**
1.0 mm²	0.56	0.63	0.07	1.94	m	**2.57**
1.5 mm²	0.78	0.88	0.07	1.94	m	**2.82**
2.5 mm²	1.15	1.30	0.07	1.94	m	**3.24**
450/750 volt grade; four core						
1.0 mm²	0.70	0.80	0.08	2.21	m	**3.01**
1.5 mm²	1.08	1.22	0.08	2.21	m	**3.43**
2.5 mm²	1.51	1.70	0.08	2.21	m	**3.91**
Heavy flexible cable; rubber insulated; rubber sheathed; copper stranded to CMA Code 638P (laid loose)						
450/750 volt grade; two core						
1.0 mm²	0.44	0.50	0.08	2.21	m	**2.71**
1.5 mm²	0.52	0.59	0.08	2.21	m	**2.80**
2.5 mm²	0.71	0.81	0.08	2.21	m	**3.02**
450/750 volt grade; three core						
1.0 mm²	0.52	0.59	0.08	2.21	m	**2.80**
1.5 mm²	0.61	0.69	0.08	2.21	m	**2.90**
2.5 mm²	0.70	0.80	0.08	2.21	m	**3.01**
450/750 volt grade; four core						
1.0 mm²	0.71	0.81	0.08	2.21	m	**3.02**
1.5 mm²	0.77	0.87	0.08	2.21	m	**3.08**
2.5 mm²	1.07	1.20	0.08	2.21	m	**3.41**
FIRE RATED CABLE						
Cable, mineral insulated; copper sheathed with copper conductors; fixed with clips to backgrounds. BASEC approval to BS 6207 Part 1 1995; complies with BS 6387 Category CWZ						
Light duty 500 volt grade; bare						
2L 1.0	4.09	4.62	0.23	6.36	m	**10.98**
2L 1.5	4.67	5.27	0.23	6.36	m	**11.63**
2L 2.5	5.62	6.34	0.25	6.93	m	**13.27**
2L 4.0	7.94	8.97	0.25	6.93	m	**15.90**
3L 1.0	4.83	5.46	0.24	6.65	m	**12.11**
3L 1.5	5.74	6.49	0.25	6.93	m	**13.42**
3L 2.5	8.20	9.26	0.25	6.93	m	**16.19**
4L 1.0	5.52	6.23	0.25	6.93	m	**13.16**
4L 1.5	6.19	6.99	0.25	6.93	m	**13.92**
4L 2.5	9.72	10.98	0.26	7.20	m	**18.18**
7L 1.5	9.79	11.06	0.28	7.76	m	**18.82**
7L 2.5	12.43	14.04	0.27	7.48	m	**21.52**

V: ELECTRICAL SUPPLY/POWER/LIGHTING

Item	Net Price £	Material £	Labour hours	Labour £	Unit	Total rate £
V20: LV DISTRIBUTION – cont						
FIRE RATED CABLE – cont						
Cable, mineral insulated – cont						
Light duty 500 volt grade; LSF sheathed						
2L 1.0	3.19	3.60	0.23	6.36	m	**9.96**
2L 1.5	3.51	3.97	0.23	6.36	m	**10.33**
2L 2.5	4.23	4.77	0.25	6.93	m	**11.70**
2L 4.0	5.92	6.68	0.25	6.93	m	**13.61**
3L 1.0	3.75	4.23	0.24	6.65	m	**10.88**
3L 1.5	4.41	4.98	0.25	6.93	m	**11.91**
3L 2.5	6.14	6.94	0.25	6.93	m	**13.87**
4L 1.0	4.28	4.83	0.25	6.93	m	**11.76**
4L 1.5	5.05	5.70	0.25	6.93	m	**12.63**
4L 2.5	7.18	8.11	0.26	7.20	m	**15.31**
7L 1.5	6.25	7.06	0.28	7.76	m	**14.82**
7L 2.5	9.68	10.93	0.27	7.48	m	**18.41**
Heavy duty 750 volt grade; bare						
1H 10	8.22	9.28	0.25	6.93	m	**16.21**
1H 16	11.13	12.57	0.26	7.20	m	**19.77**
1H 25	15.47	17.47	0.27	7.48	m	**24.95**
1H 35	22.32	25.20	0.32	8.86	m	**34.06**
1H 50	24.44	27.60	0.35	9.70	m	**37.30**
1H 70	32.61	36.83	0.38	10.53	m	**47.36**
1H 95	48.36	54.62	0.41	11.36	m	**65.98**
1H 120	50.88	57.47	0.46	12.74	m	**70.21**
1H 150	62.40	70.47	0.50	13.86	m	**84.33**
1H 185	76.10	85.95	0.56	15.51	m	**101.46**
1H 240	99.00	111.81	0.69	19.13	m	**130.94**
2H 1.5	7.58	8.56	0.25	6.93	m	**15.49**
2H 2.5	9.15	10.33	0.26	7.20	m	**17.53**
2H 4.0	11.21	12.66	0.26	7.20	m	**19.86**
2H 6.0	14.41	16.28	0.29	8.03	m	**24.31**
2H 10.0	18.42	20.81	0.34	9.41	m	**30.22**
2H 16.0	26.35	29.76	0.40	11.08	m	**40.84**
2H 25.0	36.33	41.03	0.44	12.19	m	**53.22**
3H 1.5	8.42	9.51	0.25	6.93	m	**16.44**
3H 2.5	10.44	11.79	0.25	6.93	m	**18.72**
3H 4.0	12.72	14.36	0.27	7.48	m	**21.84**
3H 6.0	15.71	17.74	0.30	8.31	m	**26.05**
3H 10.0	23.29	26.31	0.35	9.70	m	**36.01**
3H 16.0	31.37	35.42	0.41	11.36	m	**46.78**
3H 25.0	46.02	51.98	0.47	13.03	m	**65.01**
4H 1.5	10.24	11.57	0.24	6.65	m	**18.22**
4H 2.5	12.41	14.01	0.26	7.20	m	**21.21**
4H 4.0	15.32	17.30	0.29	8.03	m	**25.33**
4H 6.0	19.38	21.89	0.31	8.59	m	**30.48**
4H 10.0	27.53	31.09	0.37	10.25	m	**41.34**
4H 16.0	39.45	44.55	0.44	12.19	m	**56.74**

V: ELECTRICAL SUPPLY/POWER/LIGHTING

Item	Net Price £	Material £	Labour hours	Labour £	Unit	Total rate £
4H 25.0	56.32	63.61	0.52	14.41	m	**78.02**
7H 1.5	13.80	15.59	0.30	8.31	m	**23.90**
7H 2.5	18.39	20.77	0.32	8.86	m	**29.63**
12H 2.5	31.37	35.42	0.39	10.81	m	**46.23**
19H 1.5	45.99	51.94	0.42	11.63	m	**63.57**
Heavy duty 750 volt grade; LSF sheathed						
1H 10	6.53	7.37	0.25	6.93	m	**14.30**
1H 16	8.61	9.72	0.26	7.20	m	**16.92**
1H 25	11.87	13.40	0.27	7.48	m	**20.88**
1H 35	16.43	18.55	0.32	8.86	m	**27.41**
1H 50	19.60	22.13	0.35	9.70	m	**31.83**
1H 70	24.95	28.18	0.38	10.53	m	**38.71**
1H 95	35.74	40.36	0.41	11.36	m	**51.72**
1H 120	39.39	44.48	0.46	12.74	m	**57.22**
1H 150	47.92	54.12	0.50	13.86	m	**67.98**
1H 185	58.17	65.70	0.56	15.51	m	**81.21**
1H 240	75.38	85.13	0.68	18.84	m	**103.97**
2H 1.5	5.87	6.63	0.25	6.93	m	**13.56**
2H 2.5	7.01	7.92	0.26	7.20	m	**15.12**
2H 4.0	8.69	9.81	0.26	7.20	m	**17.01**
2H 6.0	11.14	12.58	0.29	8.03	m	**20.61**
2H 10.0	14.75	16.66	0.34	9.41	m	**26.07**
2H 16.0	20.30	22.93	0.40	11.08	m	**34.01**
2H 25.0	28.27	31.93	0.44	12.19	m	**44.12**
3H 1.5	6.40	7.23	0.25	6.93	m	**14.16**
3H 2.5	8.02	9.06	0.25	6.93	m	**15.99**
3H 4.0	9.88	11.16	0.27	7.48	m	**18.64**
3H 6.0	12.68	14.32	0.30	8.31	m	**22.63**
3H 10.0	17.63	19.91	0.35	9.70	m	**29.61**
3H 16.0	24.73	27.93	0.41	11.36	m	**39.29**
3H 25.0	35.67	40.29	0.47	13.03	m	**53.32**
4H 1.5	7.89	8.91	0.24	6.65	m	**15.56**
4H 2.5	9.61	10.85	0.26	7.20	m	**18.05**
4H 4.0	12.21	13.79	0.29	8.03	m	**21.82**
4H 6.0	15.56	17.57	0.31	8.59	m	**26.16**
4H 10.0	21.44	24.21	0.37	10.25	m	**34.46**
4H 16.0	30.69	34.66	0.44	12.19	m	**46.85**
4H 25.0	43.52	49.15	0.52	14.41	m	**63.56**
7H 1.5	10.63	12.01	0.30	8.31	m	**20.32**
7H 2.5	14.56	16.44	0.32	8.86	m	**25.30**
12H 2.5	24.48	27.65	0.39	10.81	m	**38.46**
19H 1.5	35.61	40.22	0.42	11.63	m	**51.85**

V: ELECTRICAL SUPPLY/POWER/LIGHTING

Item	Net Price £	Material £	Labour hours	Labour £	Unit	Total rate £
V20: LV DISTRIBUTION – cont						
FIRE RATED CABLE – cont						
Cable terminations for MI cable; polymeric one piece moulding; containing grey sealing compound; testing; phase marking and connection						
Light duty 500 volt grade; brass gland; polymeric one moulding containing grey sealing compound; coloured conductor sleeving; earth tag; plastic gland shroud						
2L 1.5	10.38	11.72	0.27	7.48	m	**19.20**
2L 2.5	10.38	12.96	0.27	7.48	m	**20.44**
3L 1.5	10.41	12.58	0.27	7.48	m	**20.06**
4L 1.5	10.41	12.58	0.27	7.48	m	**20.06**
Cable terminations; for MI copper sheathed cable. Certified for installation in potentially explosive atmospheres; testing; phase marking and connection; BS 6207 Part 2 1995						
Light duty 500 volt grade; brass gland; brass pot with earth tail; pot closure; sealing compound; conductor sleving; plastic gland shroud; identification markers						
2L 1.0	10.38	11.72	0.39	10.81	nr	**22.53**
2L 1.5	10.38	11.72	0.41	11.36	nr	**23.08**
2L 2.5	10.38	11.72	0.44	12.09	nr	**23.81**
2L 4.0	10.38	11.72	0.46	12.74	nr	**24.46**
3L 1.0	10.41	11.76	0.43	11.91	nr	**23.67**
3L 1.5	10.41	11.76	0.44	12.12	nr	**23.88**
3L 2.5	10.41	11.76	0.44	12.19	nr	**23.95**
4L 1.0	10.41	11.76	0.47	13.03	nr	**24.79**
4L 1.5	10.41	11.76	0.47	13.15	nr	**24.91**
4L 2.5	10.41	11.76	0.50	13.86	nr	**25.62**
7L 1.0	10.41	11.76	0.69	19.13	nr	**30.89**
7L 1.5	25.23	28.50	0.70	19.39	nr	**47.89**
7L 2.5	25.23	28.50	0.74	20.50	nr	**49.00**
Heavy duty 750 volt grade; brass gland; brass pot with earth tail; pot closure; sealing compound; conductor sleeving; plastic gland shroud; identification markers						
1H 10	10.44	11.79	0.37	10.25	nr	**22.04**
1H 16	10.44	11.79	0.39	10.81	nr	**22.60**
1H 25	10.44	11.79	0.56	15.51	nr	**27.30**

V: ELECTRICAL SUPPLY/POWER/LIGHTING

Item	Net Price £	Material £	Labour hours	Labour £	Unit	Total rate £
1H 35	10.44	11.79	0.57	15.79	nr	**27.58**
1H 50	25.31	28.58	0.60	16.63	nr	**45.21**
1H 70	24.82	28.03	0.67	18.56	nr	**46.59**
1H 95	24.82	28.03	0.75	20.78	nr	**48.81**
1H 120	40.20	45.40	0.94	26.04	nr	**71.44**
1H 150	40.20	45.40	0.99	27.42	nr	**72.82**
1H 185	40.20	45.40	1.26	34.91	nr	**80.31**
1H 240	63.42	71.62	1.37	37.95	nr	**109.57**
2H 1.5	11.06	12.50	0.42	11.63	nr	**24.13**
2H 2.5	11.06	12.50	0.44	12.28	nr	**24.78**
2H 4	11.06	12.50	0.47	13.03	nr	**25.53**
2H 6	11.06	12.50	0.54	14.96	nr	**27.46**
2H 10	26.82	30.29	0.58	16.08	nr	**46.37**
2H 16	26.82	30.29	0.69	19.13	nr	**49.42**
2H 25	43.45	49.07	0.77	21.34	nr	**70.41**
3H 1.5	11.07	12.51	0.44	12.19	nr	**24.70**
3H 2.5	11.07	12.51	0.47	13.10	nr	**25.61**
3H 4	11.07	12.51	0.57	15.79	nr	**28.30**
3H 6	26.92	30.41	0.61	16.89	nr	**47.30**
3H 10	26.92	30.41	0.65	18.00	nr	**48.41**
3H 16	26.92	30.41	0.78	21.60	nr	**52.01**
3H 25	68.59	77.47	0.85	23.55	nr	**101.02**
4H 1.5	11.07	12.51	0.52	14.41	nr	**26.92**
4H 2.5	11.07	12.51	0.53	14.68	nr	**27.19**
4H 4	26.92	30.41	0.60	16.63	nr	**47.04**
4H 6	26.92	30.41	0.65	18.00	nr	**48.41**
4H 10	26.92	30.41	0.69	19.13	nr	**49.54**
4H 16	43.45	49.07	0.88	24.39	nr	**73.46**
4H 25	70.60	79.73	0.93	25.77	nr	**105.50**
7H 1.5	26.82	30.29	0.71	19.68	nr	**49.97**
7H 2.5	26.82	30.29	0.74	20.50	nr	**50.79**
12H 1.5	43.45	49.07	0.85	23.55	nr	**72.62**
12H 2.5	43.45	49.07	1.00	27.70	nr	**76.77**
19H 2.5	68.59	77.47	1.11	30.75	nr	**108.22**
Cable; FP100; LOSH insulated; non sheathed fire-resistant to LPCB Approved to BS 6387 Catergory CWZ; in conduit or trunking including terminations						
450/750 volt grade; single core						
1.0 mm²	0.45	0.51	0.13	3.60	m	**4.11**
1.5 mm²	0.22	0.25	0.13	3.60	m	**3.85**
2.5 mm²	0.33	0.38	0.13	3.60	m	**3.98**
4.0 mm²	0.54	0.61	0.14	3.79	m	**4.40**
6.0 mm²	1.29	1.46	0.13	3.60	m	**5.06**
10 mm²	2.19	2.47	0.16	4.43	m	**6.90**
16 mm²	3.39	3.83	0.16	4.43	m	**8.26**

V: ELECTRICAL SUPPLY/POWER/LIGHTING

Item	Net Price £	Material £	Labour hours	Labour £	Unit	Total rate £
V20: LV DISTRIBUTION – cont						
FIRE RATED CABLE – cont						
Cable; FP200; Insudite insulated; LSOH sheathed screened fire-resistant BASEC Approved to BS 7629; fixed with clips to backgrounds						
300/500 volt grade; two core						
1.0 mm²	2.68	3.03	0.20	5.55	m	**8.58**
1.5 mm²	2.22	2.51	0.20	5.55	m	**8.06**
2.5 mm²	3.46	3.91	0.20	5.55	m	**9.46**
4.0 mm²	4.34	4.91	0.23	6.36	m	**11.27**
300/500 volt grade; three core						
1.0 mm²	2.95	3.34	0.23	6.36	m	**9.70**
1.5 mm²	3.52	3.98	0.23	6.36	m	**10.34**
2.5 mm²	4.24	4.78	0.23	6.36	m	**11.14**
4.0 mm²	6.98	7.88	0.25	6.93	m	**14.81**
300/500 volt grade; four core						
1.0 mm²	3.45	3.90	0.25	6.93	m	**10.83**
1.5 mm²	4.16	4.70	0.25	6.93	m	**11.63**
2.5 mm²	5.76	6.51	0.25	6.93	m	**13.44**
4.0 mm	8.72	9.85	0.28	7.76	m	**17.61**
Terminations; including glanding-off, connection to equipment						
Two core						
1.0 mm²	0.64	0.72	0.35	9.70	nr	**10.42**
1.5 mm²	0.64	0.72	0.35	9.70	nr	**10.42**
2.5 mm²	0.64	0.72	0.35	9.70	nr	**10.42**
4.0 mm²	0.64	0.72	0.35	9.70	nr	**10.42**
Three core						
1.0 mm²	0.64	0.72	0.35	9.70	nr	**10.42**
1.5 mm²	0.64	0.72	0.35	9.70	nr	**10.42**
2.5 mm²	0.64	0.72	0.35	9.79	nr	**10.51**
4.0 mm²	0.64	0.72	0.35	9.70	nr	**10.42**
Four core						
1.0 mm²	0.64	0.72	0.35	9.70	nr	**10.42**
1.5 mm²	0.64	0.72	0.35	9.70	nr	**10.42**
2.5 mm²	0.64	0.72	0.35	9.70	nr	**10.42**
4.0 mm²	0.64	0.72	0.35	9.70	nr	**10.42**

V: ELECTRICAL SUPPLY/POWER/LIGHTING

Item	Net Price £	Material £	Labour hours	Labour £	Unit	Total rate £
Cable; FP400; polymeric insulated; LSOH sheathed fire-resistant; armoured; with copper stranded copper conductors; BASEC Approved to BS 7846; fixed with clips to backgrounds						
600/1000 volt grade; two core						
1.5 mm²	3.04	3.44	0.20	5.55	m	**8.99**
2.5 mm²	3.52	3.98	0.20	5.55	m	**9.53**
4.0 mm²	3.90	4.41	0.21	5.81	m	**10.22**
6.0 mm²	4.80	5.42	0.22	6.10	m	**11.52**
10 mm²	5.18	5.85	0.24	6.65	m	**12.50**
16 mm²	7.88	8.90	0.25	6.93	m	**15.83**
25 mm²	10.90	12.31	0.35	9.70	m	**22.01**
600/1000 volt grade; three core						
1.5 mm²	3.37	3.80	0.20	5.55	m	**9.35**
2.5 mm²	3.93	4.44	0.21	5.81	m	**10.25**
4.0 mm²	4.63	5.23	0.22	6.10	m	**11.33**
6.0 mm²	4.91	5.55	0.22	6.10	m	**11.65**
10 mm²	5.97	6.74	0.25	6.93	m	**13.67**
16 mm²	9.76	11.03	0.26	7.20	m	**18.23**
25 mm²	12.18	13.76	0.37	10.25	m	**24.01**
600/1000 volt grade; four core						
1.5 mm²	4.49	4.49	0.21	5.81	m	**10.30**
2.5 mm²	5.24	5.92	0.22	6.10	m	**12.02**
4.0 mm²	6.17	6.97	0.22	6.10	m	**13.07**
6.0 mm²	6.53	7.38	0.23	6.37	m	**13.75**
10 mm²	7.96	8.99	0.26	7.20	m	**16.19**
16 mm²	13.01	14.70	0.26	7.20	m	**21.90**
25 mm²	16.24	18.34	0.39	10.80	m	**29.14**
Terminations; including glanding-off, connection to equipment						
Two core						
1.5 mm²	1.95	2.20	0.58	16.08	nr	**18.28**
2.5 mm²	2.06	2.33	0.58	16.08	nr	**18.41**
4.0 mm²	2.26	2.55	0.61	16.98	nr	**19.53**
6.0 mm²	2.49	2.82	0.67	18.56	nr	**21.38**
10 mm²	2.74	3.09	1.00	27.70	nr	**30.79**
16 mm²	3.01	3.40	1.11	30.75	nr	**34.15**
25 mm²	3.31	3.73	1.70	47.09	nr	**50.82**
Three core						
1.5 mm²	2.93	3.30	0.62	17.18	nr	**20.48**
2.5 mm²	3.08	3.48	0.62	17.18	nr	**20.66**
4.0 mm²	3.39	3.83	0.66	18.40	nr	**22.23**
6.0 mm²	3.73	4.21	0.71	19.68	nr	**23.89**
10 mm²	4.10	4.63	1.06	29.37	nr	**34.00**
16 mm²	4.51	5.09	1.19	32.97	nr	**38.06**
25 mm²	4.96	5.60	1.81	50.14	nr	**55.74**

V: ELECTRICAL SUPPLY/POWER/LIGHTING

Item	Net Price £	Material £	Labour hours	Labour £	Unit	**Total rate £**
V20: LV DISTRIBUTION – cont						
FIRE RATED CABLE – cont						
Terminations – cont						
Four core						
1.5 mm²	3.90	4.41	0.67	18.56	nr	**22.97**
2.5 mm²	4.11	4.64	0.69	18.97	nr	**23.61**
4.0 mm²	4.32	4.88	0.71	19.68	nr	**24.56**
6.0 mm²	4.52	5.10	0.76	21.05	nr	**26.15**
10 mm²	4.65	5.25	1.14	31.58	nr	**36.83**
16 mm²	5.02	5.67	1.29	35.73	nr	**41.40**
25 mm²	5.59	6.31	1.35	37.40	nr	**43.71**
Cable; Firetuff fire-resistant to BS 6387; fixed with clips to backgrounds						
Two core						
1.5 mm²	2.28	2.57	0.20	5.55	m	**8.12**
2.5 mm²	2.75	3.10	0.20	5.55	m	**8.65**
4.0 mm²	3.57	4.03	0.21	5.81	m	**9.84**
Three core						
1.5 mm²	2.78	3.14	0.20	5.55	m	**8.69**
2.5 mm²	3.14	3.55	0.21	5.81	m	**9.36**
4.0 mm²	4.54	5.13	0.22	6.10	m	**11.23**
Four core						
1.5 mm²	3.11	3.51	0.21	5.81	m	**9.32**
2.5 mm²	3.81	4.30	0.22	6.10	m	**10.40**
4.0 mm²	5.19	5.87	0.22	6.10	m	**11.97**
MODULAR WIRING						
Modular wiring systems; including commissioning						
Master distribution box; steel; fixed to backgrounds; 6 Port						
4.0 mm 18 core armoured home run cable	156.49	176.74	0.90	24.94	nr	**201.68**
4.0 mm 24 core armoured cable home run cable	156.49	176.74	0.95	26.32	nr	**203.06**
4.0 mm 18 core armoured home run cable & data cable	166.26	187.77	0.95	26.32	nr	**214.09**
6.0 mm 18 core armoured home run cable	156.49	176.74	1.00	27.70	nr	**204.44**
6.0 mm 24 core armoured home run cable	156.49	176.74	1.10	30.47	nr	**207.21**
6.0 mm 18 core armoured home run cable & data cable	166.26	187.77	1.10	30.47	nr	**218.24**

V: ELECTRICAL SUPPLY/POWER/LIGHTING

Item	Net Price £	Material £	Labour hours	Labour £	Unit	**Total rate £**
Master distribution box; steel; fixed to backgrounds; 9 Port						
4.0 mm 27 core armoured home run cable	195.59	220.90	1.30	36.02	nr	**256.92**
4.0 mm 27 core armoured home run cable & data cable	195.59	220.90	1.45	40.17	nr	**261.07**
6.0 mm 27 core armoured home run cable	205.38	231.96	1.45	40.17	nr	**272.13**
6.0 mm 27 core armoured home run cable & data cable	205.38	231.96	1.55	42.94	nr	**274.90**
Metal clad cable; BSEN 60439 Part 2 1993; BASEC approved						
4.0 mm 18 core	13.66	15.43	0.30	8.31	m	**23.74**
4.0 mm 24 core	21.02	23.74	0.32	8.86	m	**32.60**
4.0 mm 27 core	21.18	23.92	0.35	9.70	m	**33.62**
6.0 mm 18 core	18.19	20.54	0.32	8.86	m	**29.40**
6.0 mm 27 core	25.34	28.62	0.35	9.70	m	**38.32**
Metal clad data cable						
Single twisted pair	2.66	3.01	0.18	4.99	m	**8.00**
Twin twisted pair	4.29	4.85	0.18	4.99	m	**9.84**
Distribution cables; armoured; BSEN 60439 Part 2 1993; BASEC approved						
3 wire; 6.1 m long	50.45	56.98	0.92	25.49	nr	**82.47**
4 wire; 6.1 m long	56.90	64.26	0.96	26.60	nr	**90.86**
Extender cables; armoured; BSEN 60439 Part 2 1993; BASEC approved						
3 wire						
0.9 m long	31.66	35.76	0.13	3.60	nr	**39.36**
1.5 m long	34.86	39.37	0.23	6.36	nr	**45.73**
2.1 m long	38.06	42.98	0.31	8.59	nr	**51.57**
2.7 m long	41.26	46.60	0.40	11.08	nr	**57.68**
3.4 m long	44.98	50.80	0.51	14.13	nr	**64.93**
4.6 m long	51.38	58.03	0.69	19.13	nr	**77.16**
6.1 m long	59.37	67.05	0.92	25.49	nr	**92.54**
7.6 m long	67.36	76.07	1.14	31.58	nr	**107.65**
9.1 m long	75.35	85.10	1.37	37.95	nr	**123.05**
10.7 m long	83.88	94.74	1.61	44.60	nr	**139.34**
4 wire						
0.9 m long	33.57	37.91	0.14	3.88	nr	**41.79**
1.5 m long	37.36	42.20	0.24	6.65	nr	**48.85**
2.1 m long	41.14	46.46	0.32	8.86	nr	**55.32**
2.7 m long	44.93	50.75	0.43	11.91	nr	**62.66**
3.4 m long	49.34	55.72	0.51	14.13	nr	**69.85**
4.6 m long	56.92	64.28	0.67	18.56	nr	**82.84**
6.1 m long	66.60	75.21	0.92	25.49	nr	**100.70**
7.6 m long	75.85	85.66	1.22	33.80	nr	**119.46**
9.1 m long	85.31	96.35	1.46	40.44	nr	**136.79**
10.7 m long	95.42	107.76	1.71	47.38	nr	**155.14**

V: ELECTRICAL SUPPLY/POWER/LIGHTING

Item	Net Price £	Material £	Labour hours	Labour £	Unit	Total rate £
V20: LV DISTRIBUTION – cont						
MODULAR WIRING – cont						
Extender cables – cont						
3 wire; including twisted pair						
0.9 m long	42.99	48.55	0.13	3.60	nr	**52.15**
1.5 m long	48.30	54.55	0.23	6.36	nr	**60.91**
2.1 m long	53.60	60.54	0.31	8.59	nr	**69.13**
2.7 m long	58.91	66.53	0.40	11.08	nr	**77.61**
3.4 m long	65.10	73.52	0.51	14.13	nr	**87.65**
4.6 m long	75.71	85.51	0.69	19.13	nr	**104.64**
6.1 m long	88.98	100.49	0.92	25.49	nr	**125.98**
7.6 m long	102.23	115.45	1.14	31.58	nr	**147.03**
9.1 m long	115.50	130.45	1.37	37.95	nr	**168.40**
10.7 m long	129.64	146.41	1.61	44.60	nr	**191.01**
Extender whip ended cables; armoured; BSEN 60439 Part 2 1993; BASEC approved						
3 wire; 3.0 m long	33.03	37.30	0.30	8.31	nr	**45.61**
4 wire; 3.0 m long	36.62	41.36	0.30	8.31	nr	**49.67**
T connectors						
3 wire						
6 Pole 20a Tee	37.15	37.15	0.10	2.45	nr	**39.60**
Snap fix	18.93	21.38	0.10	2.76	nr	**24.14**
0.3 metre flexible cable	21.96	24.81	0.10	2.76	nr	**27.57**
0.3 metre armoured cable	24.79	28.00	0.15	4.15	nr	**32.15**
0.3 metre armoured cable with twisted pair	31.60	35.69	0.15	4.15	nr	**39.84**
4 wire						
Snap fix	22.86	25.82	0.10	2.76	nr	**28.58**
0.3 m flexible cable	19.47	21.99	0.10	2.76	nr	**24.75**
0.3 m armoured cable	25.70	29.03	0.18	4.99	nr	**34.02**
Splitters						
5 wire	18.75	21.18	0.20	5.55	nr	**26.73**
5 wire converter	24.17	27.30	0.20	5.55	nr	**32.85**
Switch modules						
3 wire; 6.1 m long armoured cable	51.29	57.93	0.75	20.78	nr	**78.71**
4 wire; 6.1 m long armoured cable	57.72	65.19	0.80	22.15	nr	**87.34**
Distribution cables; unarmoured; IEC 998 DIN/VDE 0628						
3 wire; 6.1 m long	19.50	22.02	0.70	19.39	nr	**41.41**
4 wire; 6.1 m long	24.39	27.55	0.75	20.78	nr	**48.33**

V: ELECTRICAL SUPPLY/POWER/LIGHTING

Item	Net Price £	Material £	Labour hours	Labour £	Unit	**Total rate £**
Extender cables; unarmoured; IEC 998 DIN/VDE 0628						
3 wire						
0.9 m long	11.91	13.45	0.07	1.94	nr	**15.39**
1.5 m long	13.04	14.73	0.12	3.33	nr	**18.06**
2.1 m long	14.17	16.00	0.17	4.71	nr	**20.71**
2.7 m long	18.77	21.20	0.27	7.48	nr	**28.68**
3.4 m long	16.63	18.78	0.22	6.10	nr	**24.88**
4.6 m long	18.90	21.35	0.37	10.25	nr	**31.60**
6.1 m long	21.73	24.54	0.49	13.58	nr	**38.12**
7.6 m long	24.56	27.73	0.61	16.89	nr	**44.62**
9.1 m long	27.40	30.95	0.73	20.23	nr	**51.18**
10.7 m long	30.42	34.35	0.86	23.83	nr	**58.18**
4 wire						
0.9 m long	14.34	16.20	0.08	2.21	nr	**18.41**
1.5 m long	15.82	17.87	0.14	3.88	nr	**21.75**
2.1 m long	17.30	19.54	0.19	5.26	nr	**24.80**
2.7 m long	18.77	21.20	0.24	6.65	nr	**27.85**
3.4 m long	20.48	23.13	0.31	8.59	nr	**31.72**
4.6 m long	23.43	26.46	0.41	11.36	nr	**37.82**
6.1 m long	27.12	30.63	0.55	15.24	nr	**45.87**
7.6 m long	30.80	34.78	0.68	18.84	nr	**53.62**
9.1 m long	34.48	38.94	0.82	22.72	nr	**61.66**
10.7 m long	38.41	43.38	0.96	26.60	nr	**69.98**
5 wire						
0.9 m long	21.75	24.56	0.09	2.50	nr	**27.06**
1.5 m long	23.66	26.72	0.15	4.15	nr	**30.87**
2.1 m long	25.57	28.88	0.21	5.81	nr	**34.69**
2.7 m long	27.47	31.03	0.27	7.48	nr	**38.51**
3.4 m long	29.69	33.53	0.34	9.41	nr	**42.94**
4.6 m long	33.50	37.83	0.46	12.74	nr	**50.57**
6.1 m long	38.26	43.21	0.61	16.89	nr	**60.10**
7.6 m long	43.02	48.58	0.76	21.05	nr	**69.63**
9.1 m long	47.78	53.96	0.91	25.20	nr	**79.16**
10.7 m long	52.87	59.71	1.07	29.65	nr	**89.36**
Extender whip ended cables; armoured; IEC 998 DIN/VDE 0628						
3 wire; 2.5 mm; 3.0 m long	12.49	14.11	0.30	8.31	nr	**22.42**
4 wire; 2.5 mm; 3.0 m long	14.67	16.56	0.30	8.31	nr	**24.87**
T connectors						
3 wire						
5 pin; direct fix	11.53	13.03	0.10	2.76	nr	**15.79**
5 pin; 1.5 mm flexible cable; 0.3 m long	16.15	18.24	0.15	4.15	nr	**22.39**

V: ELECTRICAL SUPPLY/POWER/LIGHTING

Item	Net Price £	Material £	Labour hours	Labour £	Unit	Total rate £
V20: LV DISTRIBUTION – cont						
MODULAR WIRING – cont						
T connectors – cont						
4 wire						
5 pin; direct fix	14.69	16.60	0.20	5.55	nr	**22.15**
5 pin; 1.5 mm flexible cable; 0.3 m long	18.30	20.67	0.20	5.55	nr	**26.22**
5 wire						
5 pin; direct fix	16.73	18.89	0.20	5.55	nr	**24.44**
Splitters						
3 way; 5 pin	9.64	10.88	0.25	6.93	nr	**17.81**
Switch modules						
3 wire	27.47	31.03	0.20	5.55	nr	**36.58**
4 wire	28.64	32.34	0.22	6.10	nr	**38.44**
Y62 – BUSBAR TRUNKING						
MAINS BUSBAR						
Low impedance busbar trunking; fixed to backgrounds including supports, fixings and connections/jointing to equipment						
Straight copper busbar						
1000 amp TP&N	466.83	527.24	3.41	94.47	m	**621.71**
1350 amp TP&N	573.80	648.05	3.58	99.18	m	**747.23**
2000 amp TP&N	719.69	812.82	5.00	138.52	m	**951.34**
2500 amp TP&N	1087.33	1228.03	5.90	163.45	m	**1391.48**
Extra for fittings mains bus bar						
IP54 protection						
1000 amp TP&N	31.84	35.95	2.16	59.83	m	**95.78**
1350 amp TP&N	34.21	38.64	2.61	72.31	m	**110.95**
2000 amp TP&N	47.46	53.60	3.51	97.24	m	**150.84**
2500 amp TP&N	56.35	63.64	3.96	109.71	m	**173.35**
End cover						
1000 amp TP&N	36.61	41.35	0.56	15.51	nr	**56.86**
1350 amp TP&N	38.19	43.14	0.56	15.51	nr	**58.65**
2000 amp TP&N	61.28	69.21	0.66	18.29	nr	**87.50**
2500 amp TP&N	62.06	70.09	0.66	18.29	nr	**88.38**

V: ELECTRICAL SUPPLY/POWER/LIGHTING

Item	Net Price £	Material £	Labour hours	Labour £	Unit	Total rate £
Edge elbow						
1000 amp TP&N	524.43	592.29	2.01	55.69	nr	**647.98**
1350 amp TP&N	590.20	666.57	2.01	55.69	nr	**722.26**
2000 amp TP&N	915.41	1033.86	2.40	66.48	nr	**1100.34**
2500 amp TP&N	1202.32	1357.90	2.40	66.48	nr	**1424.38**
Flat elbow						
1000 amp TP&N	454.99	513.87	2.01	55.69	nr	**569.56**
1350 amp TP&N	493.35	557.19	2.01	55.69	nr	**612.88**
2000 amp TP&N	699.82	790.38	2.40	66.48	nr	**856.86**
2500 amp TP&N	871.58	984.36	2.40	66.48	nr	**1050.84**
Offset						
1000 amp TP&N	913.62	1031.84	3.00	83.11	nr	**1114.95**
1350 amp TP&N	1123.73	1269.14	3.00	83.11	nr	**1352.25**
2000 amp TP&N	1770.59	1999.70	3.50	96.97	nr	**2096.67**
2500 amp TP&N	2015.41	2276.19	3.50	96.97	nr	**2373.16**
Edge Z unit						
1000 amp TP&N	1368.59	1545.68	3.00	83.11	nr	**1628.79**
1350 amp TP&N	1752.31	1979.05	3.00	83.11	nr	**2062.16**
2000 amp TP&N	2625.73	2965.49	3.50	96.97	nr	**3062.46**
2500 amp TP&N	2998.48	3386.47	3.50	96.97	nr	**3483.44**
Flat Z unit						
1000 amp TP&N	1479.72	1671.19	3.00	83.11	nr	**1754.30**
1350 amp TP&N	1480.06	1671.58	3.00	83.11	nr	**1754.69**
2000 amp TP&N	2203.63	2488.77	3.50	96.97	nr	**2585.74**
2500 amp TP&N	2649.47	2992.30	3.50	96.97	nr	**3089.27**
Edge tee						
1000 amp TP&N	1368.59	1545.68	2.20	60.96	nr	**1606.64**
1350 amp TP&N	1752.31	1979.05	2.20	60.96	nr	**2040.01**
2000 amp TP&N	2625.73	2965.49	2.60	72.03	nr	**3037.52**
2500 amp TP&N	2999.37	3387.47	2.60	72.03	nr	**3459.50**
Tap off; TP&N integral contactor/breaker						
18 amp	222.61	251.41	0.82	22.72	nr	**274.13**
Tap off; TP&N fusable with on-load switch; excludes fuses						
32 amp	629.84	711.34	0.82	22.72	nr	**734.06**
63 amp	643.55	726.82	0.88	24.39	nr	**751.21**
100 amp	787.03	888.87	1.18	32.69	nr	**921.56**
160 amp	894.64	1010.40	1.41	39.07	nr	**1049.47**
250 amp	1155.23	1304.71	1.76	48.76	nr	**1353.47**
315 amp	1364.12	1540.63	2.06	57.07	nr	**1597.70**

V: ELECTRICAL SUPPLY/POWER/LIGHTING

Item	Net Price £	Material £	Labour hours	Labour £	Unit	Total rate £
V20: LV DISTRIBUTION – cont						
MAINS BUSBAR – cont						
Extra for fittings mains bus bar – cont						
Tap off; TP&N MCCB						
63 amp	796.53	899.60	0.88	24.39	nr	**923.99**
125 amp	954.78	1078.32	1.18	32.69	nr	**1111.01**
160 amp	1042.34	1177.21	1.41	39.07	nr	**1216.28**
250 amp	1340.91	1514.41	1.76	48.76	nr	**1563.17**
400 amp	1709.10	1930.25	2.06	57.07	nr	**1987.32**
RISING MAINS BUSBAR						
Rising mains busbar; insulated supports, earth continuity bar; including couplers; fixed to backgrounds						
Straight aluminium bar						
200 amp TP&N	162.74	183.79	2.13	59.01	m	**242.80**
315 amp TP&N	182.31	205.90	2.15	59.57	m	**265.47**
400 amp TP&N	211.15	238.48	2.15	59.57	m	**298.05**
630 amp TP&N	260.59	294.31	2.47	68.43	m	**362.74**
800 amp TP&N	391.40	442.05	2.88	79.79	m	**521.84**
Extra for fittings rising busbar						
End feed unit						
200 amp TP&N	325.48	367.60	2.57	71.20	nr	**438.80**
315 amp TP&N	326.51	368.76	2.76	76.46	nr	**445.22**
400 amp TP&N	364.62	411.80	2.76	76.46	nr	**488.26**
630 amp TP&N	365.65	412.97	3.64	100.85	nr	**513.82**
800 amp TP&N	405.82	458.33	4.54	125.78	nr	**584.11**
Top feeder unit						
200 amp TP&N	325.48	367.60	2.57	71.20	nr	**438.80**
315 amp TP&N	326.51	368.76	2.76	76.46	nr	**445.22**
400 amp TP&N	364.62	411.80	2.76	76.46	nr	**488.26**
630 amp TP&N	365.65	412.97	3.64	100.85	nr	**513.82**
800 amp TP&N	405.82	458.33	4.54	125.78	nr	**584.11**
End cap						
200 amp TP&N	27.81	31.41	0.18	4.99	nr	**36.40**
315 amp TP&N	27.81	31.41	0.27	7.48	nr	**38.89**
400 amp TP&N	30.90	34.89	0.27	7.48	nr	**42.37**
630 amp TP&N	30.90	34.89	0.41	11.36	nr	**46.25**
800 amp TP&N	89.61	101.20	0.41	11.36	nr	**112.56**

V: ELECTRICAL SUPPLY/POWER/LIGHTING

Item	Net Price £	Material £	Labour hours	Labour £	Unit	**Total rate £**
Edge elbow						
200 amp TP&N	38.11	43.04	0.55	15.24	nr	**58.28**
315 amp TP&N	39.14	44.21	0.94	26.04	nr	**70.25**
400 amp TP&N	291.49	329.21	0.94	26.04	nr	**355.25**
630 amp TP&N	292.52	330.37	1.45	40.17	nr	**370.54**
800 amp TP&N	277.07	312.93	1.45	40.17	nr	**353.10**
Flat elbow						
200 amp TP&N	125.66	141.92	0.55	15.24	nr	**157.16**
315 amp TP&N	125.66	141.92	0.94	26.04	nr	**167.96**
400 amp TP&N	169.95	191.94	0.94	26.04	nr	**217.98**
630 amp TP&N	170.98	193.11	1.45	40.17	nr	**233.28**
800 amp TP&N	235.87	266.39	1.45	40.17	nr	**306.56**
Edge tee						
200 amp TP&N	176.13	198.92	0.61	16.89	nr	**215.81**
315 amp TP&N	177.16	200.08	1.02	28.25	nr	**228.33**
400 amp TP&N	248.23	280.35	1.02	28.25	nr	**308.60**
630 amp TP&N	248.23	280.35	1.57	43.49	nr	**323.84**
800 amp TP&N	405.82	458.33	1.57	43.49	nr	**501.82**
Flat tee						
200 amp TP&N	226.60	255.92	0.61	16.89	nr	**272.81**
315 amp TP&N	177.16	200.08	1.02	28.25	nr	**228.33**
400 amp TP&N	356.38	402.49	1.02	28.25	nr	**430.74**
630 amp TP&N	357.41	403.65	1.57	43.49	nr	**447.14**
800 amp TP&N	499.55	564.19	1.57	43.49	nr	**607.68**
Tap off units						
TP&N fusable with on-load switch; excludes fuses						
32 amp	165.83	187.29	0.82	22.72	nr	**210.01**
63 amp	219.39	247.78	0.88	24.39	nr	**272.17**
100 amp	294.58	332.69	1.18	32.69	nr	**365.38**
250 amp	441.87	499.05	1.41	39.07	nr	**538.12**
400 amp	643.75	727.05	2.06	57.07	nr	**784.12**
TP&N MCCB						
32 amp	168.92	190.78	0.82	22.72	nr	**213.50**
63 amp	233.81	264.07	0.88	24.39	nr	**288.46**
100 amp	374.92	423.43	1.18	32.69	nr	**456.12**
250 amp	635.51	717.74	1.41	39.07	nr	**756.81**
400 amp	1116.52	1261.00	2.06	57.07	nr	**1318.07**

V: ELECTRICAL SUPPLY/POWER/LIGHTING

Item	Net Price £	Material £	Labour hours	Labour £	Unit	Total rate £
V20: LV DISTRIBUTION – cont						
LIGHTING BUSBAR						
Prewired busbar, plug-in trunking for lighting; galvanized sheet steel housing (PE); tin-plated copper conductors with tap-off units at 1 m intervals						
Straight lengths – 25 amp						
2 pole & PE	27.54	31.10	0.16	4.43	m	**35.53**
4 pole & PE	30.60	34.56	0.16	4.43	m	**38.99**
Straight lengths – 40 amp						
2 pole & PE	27.54	31.10	0.16	4.43	m	**35.53**
4 pole & PE	36.72	41.47	0.16	4.43	m	**45.90**
Components for prewired busbars, plug-in trunking for lighting						
Plug-in tap off units						
10 amp 4 pole & PE; 3 m of cable	23.46	26.50	0.10	2.76	nr	**29.26**
16 amp 4 pole & PE; 3 m of cable	25.50	28.79	0.10	2.76	nr	**31.55**
16 amp with phase selection, 2P & PE; no cable	19.38	21.89	0.10	2.76	nr	**24.65**
Trunking components						
End feed unit & cover; 4P & PE	29.58	33.40	0.23	6.36	nr	**39.76**
Centre feed unit	147.90	167.04	0.29	8.03	nr	**175.07**
Right hand, intermediate terminal box feed unit	31.62	35.71	0.23	6.36	nr	**42.07**
End cover (for R/hand feed)	8.16	9.22	0.06	1.66	nr	**10.88**
Flexible elbow unit	70.38	79.49	0.12	3.33	nr	**82.82**
Fixing bracket – universal	5.10	5.76	0.10	2.76	nr	**8.52**
Suspension bracket – flat	4.08	4.61	0.10	2.76	nr	**7.37**
UNDERFLOOR BUSBAR						
Prewired busbar, plug-in trunking for underfloor power distribution; galvanized sheet steel housing (PE); copper conductors with tap-off units at 300 mm intervals						
Straight lengths – 63 amp						
2 pole & PE	17.34	19.58	0.28	7.76	m	**27.34**
3 pole & PE; Clean Earth System	22.44	25.35	0.28	7.76	m	**33.11**

V: ELECTRICAL SUPPLY/POWER/LIGHTING

Item	Net Price £	Material £	Labour hours	Labour £	Unit	Total rate £
Components for prewired busbars, plug-in trunking for underfloor power distribution						
Plug-in tap off units						
32 amp 2P & PE; 3 m metal flexible prewired conduit	28.56	32.25	0.25	6.93	nr	**39.18**
32 amp 3P & PE; clean earth; 3 m metal flexible prewired conduit	34.68	39.17	0.28	7.76	nr	**46.93**
Trunking components						
End feed unit & cover; 2P & PE	30.60	34.56	0.35	9.70	nr	**44.26**
End feed unit & cover; 3P & PE; clean earth	33.66	38.02	0.38	10.53	nr	**48.55**
End cover; 2P & PE	9.18	10.36	0.11	3.05	nr	**13.41**
End cover; 3P & PE	10.20	11.52	0.11	3.05	nr	**14.57**
Flexible interlink/corner; 2P&PE; 1 m long	53.04	59.90	0.34	9.41	nr	**69.31**
Flexible interlink/corner; 3P&PE; 1 m long	60.18	67.96	0.35	9.70	nr	**77.66**
Flexible interlink/corner; 2P&PE; 2 m long	65.28	73.73	0.37	10.25	nr	**83.98**
Flexible interlink/corner; 3P&PE; 2 m long	71.40	80.64	0.37	10.25	nr	**90.89**
Y63 – CABLE SUPPORTS						
LADDER RACK						
Light duty galvanized steel ladder rack; fixed to backgrounds; including supports, fixings and brackets; earth continuity straps						
Straight lengths						
150 mm wide ladder	20.09	22.68	0.69	19.13	m	**41.81**
300 mm wide ladder	21.89	24.72	0.88	24.39	m	**49.11**
450 mm wide ladder	15.49	17.49	1.26	34.91	m	**52.40**
600 mm wide ladder	25.01	28.24	1.51	41.83	m	**70.07**
Extra over (cutting and jointing racking to fittings is included)						
Inside riser bend						
150 mm wide ladder	73.87	83.43	0.33	9.14	nr	**92.57**
300 mm wide ladder	73.87	83.43	0.56	15.51	nr	**98.94**
450 mm wide ladder	71.59	80.86	0.85	23.55	nr	**104.41**
600 mm wide ladder	103.43	116.81	0.99	27.42	nr	**144.23**
Outside riser bend						
300 mm wide ladder	73.87	83.43	0.43	11.91	nr	**95.34**
450 mm wide ladder	53.25	60.14	0.73	20.23	nr	**80.37**
600 mm wide ladder	103.43	116.81	0.86	23.83	nr	**140.64**

V: ELECTRICAL SUPPLY/POWER/LIGHTING

Item	Net Price £	Material £	Labour hours	Labour £	Unit	Total rate £
V20: LV DISTRIBUTION – cont						
LADDER RACK – cont						
Extra over (cutting and jointing racking to fittings is included) – cont						
Equal tee						
300 mm wide ladder	62.23	70.28	0.62	17.18	nr	**87.46**
450 mm wide ladder	85.34	96.38	1.09	30.20	nr	**126.58**
600 mm wide ladder	73.77	83.31	1.12	31.03	nr	**114.34**
Unequal tee						
300 mm wide ladder	62.23	70.28	0.57	15.79	nr	**86.07**
450 mm wide ladder	81.07	91.56	1.17	32.42	nr	**123.98**
600 mm wide ladder	68.38	77.22	1.17	32.42	nr	**109.64**
4 way crossovers						
300 mm wide ladder	159.83	180.51	0.72	19.94	nr	**200.45**
450 mm wide ladder	127.65	144.17	1.13	31.30	nr	**175.47**
600 mm wide ladder	179.60	202.84	1.29	35.73	nr	**238.57**
Heavy duty galvanized steel ladder rack; fixed to backgrounds; including supports, fixings and brackets; earth continuity straps						
Straight lengths						
150 mm wide ladder	26.04	29.41	0.68	18.84	m	**48.25**
300 mm wide ladder	28.29	31.95	0.79	21.89	m	**53.84**
450 mm wide ladder	30.43	34.36	1.07	29.65	m	**64.01**
600 mm wide ladder	33.14	37.42	1.24	34.35	m	**71.77**
750 mm wide ladder	38.80	43.82	1.49	41.28	m	**85.10**
900 mm wide ladder	42.80	48.34	1.67	46.27	m	**94.61**
Extra over (cutting and jointing racking to fittings is included)						
Flat bend						
150 mm wide ladder	61.63	69.60	0.34	9.41	nr	**79.01**
300 mm wide ladder	65.92	74.45	0.39	10.81	nr	**85.26**
450 mm wide ladder	75.07	84.78	0.43	11.91	nr	**96.69**
600 mm wide ladder	85.43	96.48	0.61	16.89	nr	**113.37**
750 mm wide ladder	98.19	110.89	0.82	22.72	nr	**133.61**
900 mm wide ladder	108.43	122.41	0.97	26.87	nr	**149.29**
Inside riser bend						
150 mm wide ladder	88.40	99.84	0.27	7.48	nr	**107.32**
300 mm wide ladder	89.83	101.45	0.45	12.46	nr	**113.91**
450 mm wide ladder	97.33	109.93	0.65	18.00	nr	**127.93**

V: ELECTRICAL SUPPLY/POWER/LIGHTING

Item	Net Price £	Material £	Labour hours	Labour £	Unit	Total rate £
600 mm wide ladder	104.93	118.50	0.81	22.44	nr	**140.94**
750 mm wide ladder	109.65	123.84	0.92	25.49	nr	**149.33**
900 mm wide ladder	122.45	138.29	1.06	29.37	nr	**167.66**
Outside riser bend						
150 mm wide ladder	88.40	99.84	0.27	7.48	nr	**107.32**
300 mm wide ladder	89.93	101.56	0.33	9.14	nr	**110.70**
450 mm wide ladder	97.33	109.93	0.61	16.89	nr	**126.82**
600 mm wide ladder	104.93	118.50	0.76	21.05	nr	**139.55**
750 mm wide ladder	109.65	123.84	0.94	26.04	nr	**149.88**
900 mm wide ladder	122.45	138.29	1.05	29.09	nr	**167.38**
Equal tee						
150 mm wide ladder	94.50	106.73	0.37	10.25	nr	**116.98**
300 mm wide ladder	111.58	126.02	0.57	15.79	nr	**141.81**
450 mm wide ladder	122.15	137.96	0.83	22.99	nr	**160.95**
600 mm wide ladder	137.14	154.89	0.92	25.49	nr	**180.38**
750 mm wide ladder	177.47	200.43	1.13	31.30	nr	**231.73**
900 mm wide ladder	195.58	220.81	1.20	33.25	nr	**254.06**
Unequal tee						
300 mm wide ladder	106.51	120.29	0.57	15.79	nr	**136.08**
450 mm wide ladder	103.84	117.28	1.17	32.42	nr	**149.70**
600 mm wide ladder	112.51	127.07	1.17	32.42	nr	**159.49**
750 mm wide ladder	120.20	135.75	1.25	34.63	nr	**170.38**
900 mm wide ladder	154.92	174.97	1.33	36.84	nr	**211.81**
4 way crossovers						
150 mm wide ladder	152.52	172.26	0.50	13.86	nr	**186.12**
300 mm wide ladder	161.65	182.57	0.67	18.56	nr	**201.13**
450 mm wide ladder	207.04	233.83	0.92	25.49	nr	**259.32**
600 mm wide ladder	222.30	251.06	1.07	29.65	nr	**280.71**
750 mm wide ladder	234.20	264.51	1.25	34.63	nr	**299.14**
900 mm wide ladder	257.62	290.85	1.36	37.68	nr	**328.53**
Extra heavy duty galvanized steel ladder rack; fixed to backgrounds; including supports, fixings and brackets; earth continuity straps						
Straight lengths						
150 mm wide ladder	35.06	39.60	0.63	17.45	m	**57.05**
300 mm wide ladder	36.79	41.55	0.70	19.39	m	**60.94**
450 mm wide ladder	38.61	43.60	0.83	22.99	m	**66.59**
600 mm wide ladder	39.13	44.20	0.89	24.65	m	**68.85**
750 mm wide ladder	39.24	44.32	1.22	33.80	m	**78.12**
900 mm wide ladder	40.95	46.25	1.44	39.89	m	**86.14**

V: ELECTRICAL SUPPLY/POWER/LIGHTING

Item	Net Price £	Material £	Labour hours	Labour £	Unit	Total rate £
V20: LV DISTRIBUTION – cont						
LADDER RACK – cont						
Extra over (cutting and jointing racking to fittings is included)						
Flat bend						
150 mm wide ladder	76.12	85.97	0.36	9.98	nr	**95.95**
300 mm wide ladder	79.21	89.46	0.39	10.81	nr	**100.27**
450 mm wide ladder	85.53	96.59	0.43	11.91	nr	**108.50**
600 mm wide ladder	96.75	109.27	0.61	16.89	nr	**126.16**
750 mm wide ladder	104.90	118.47	0.82	22.72	nr	**141.19**
900 mm wide ladder	113.60	128.30	0.97	26.87	nr	**155.17**
Inside riser bend						
150 mm wide ladder	90.96	102.73	0.36	9.98	nr	**112.71**
300 mm wide ladder	92.44	104.40	0.39	10.81	nr	**115.21**
450 mm wide ladder	97.71	110.35	0.43	11.91	nr	**122.26**
600 mm wide ladder	107.77	121.72	0.61	16.89	nr	**138.61**
750 mm wide ladder	110.57	124.88	0.82	22.72	nr	**147.60**
900 mm wide ladder	119.22	134.65	0.97	26.87	nr	**161.52**
Outside riser bend						
150 mm wide ladder	91.21	103.01	0.36	9.98	nr	**112.99**
300 mm wide ladder	92.44	104.40	0.39	10.81	nr	**115.21**
450 mm wide ladder	97.71	110.35	0.41	11.36	nr	**121.71**
600 mm wide ladder	107.77	121.72	0.57	15.79	nr	**137.51**
750 mm wide ladder	110.57	124.88	0.82	22.72	nr	**147.60**
900 mm wide ladder	119.22	134.65	0.93	25.77	nr	**160.42**
Equal tee						
150 mm wide ladder	109.94	124.16	0.37	10.25	nr	**134.41**
300 mm wide ladder	121.38	137.09	0.57	15.79	nr	**152.88**
450 mm wide ladder	128.56	145.20	0.83	22.99	nr	**168.19**
600 mm wide ladder	144.89	163.64	0.92	25.49	nr	**189.13**
750 mm wide ladder	175.14	197.80	1.13	31.30	nr	**229.10**
900 mm wide ladder	175.43	198.12	1.20	33.25	nr	**231.37**
Unequal tee						
150 mm wide ladder	113.19	127.84	0.37	10.25	nr	**138.09**
300 mm wide ladder	124.66	140.79	0.57	15.79	nr	**156.58**
450 mm wide ladder	131.85	148.91	1.17	32.42	nr	**181.33**
600 mm wide ladder	148.34	167.54	1.17	32.42	nr	**199.96**
750 mm wide ladder	175.14	593.41	1.25	34.63	nr	**628.04**
900 mm wide ladder	182.34	205.94	1.33	36.84	nr	**242.78**

V: ELECTRICAL SUPPLY/POWER/LIGHTING

Item	Net Price £	Material £	Labour hours	Labour £	Unit	Total rate £
4 way crossovers						
150 mm wide ladder	152.87	172.66	0.50	13.86	nr	**186.52**
300 mm wide ladder	158.87	179.43	0.67	18.56	nr	**197.99**
450 mm wide ladder	185.85	209.90	0.92	25.49	nr	**235.39**
600 mm wide ladder	206.61	233.35	1.07	29.65	nr	**263.00**
750 mm wide ladder	214.77	242.56	1.25	34.63	nr	**277.19**
900 mm wide ladder	224.42	253.46	1.36	37.68	nr	**291.14**
CABLE TRAY						
Galvanized steel cable tray to BS 729; including standard coupling joints, fixings and earth continuity straps (supports and hangers are excluded)						
Light duty tray						
Straight lengths						
50 mm wide	2.73	3.08	0.19	5.26	m	**8.34**
75 mm wide	3.44	3.89	0.23	6.36	m	**10.25**
100 mm wide	4.25	4.80	0.31	8.59	m	**13.39**
150 mm wide	5.57	6.29	0.33	9.14	m	**15.43**
225 mm wide	10.67	12.05	0.39	10.81	m	**22.86**
300 mm wide	15.02	16.96	0.49	13.58	m	**30.54**
450 mm wide	10.31	11.65	0.60	16.63	m	**28.28**
600 mm wide	21.29	24.04	0.79	21.89	m	**45.93**
750 mm wide	27.04	30.54	1.04	28.82	m	**59.36**
900 mm wide	33.69	38.05	1.26	34.91	m	**72.96**
Extra over (cutting and jointing tray to fittings is included)						
Straight reducer						
75 mm wide	11.06	12.50	0.22	6.10	nr	**18.60**
100 mm wide	11.28	12.74	0.25	6.93	nr	**19.67**
150 mm wide	14.75	16.66	0.27	7.48	nr	**24.14**
225 mm wide	19.36	21.87	0.34	9.41	nr	**31.28**
300 mm wide	24.65	27.84	0.39	10.81	nr	**38.65**
450 mm wide	19.30	21.80	0.49	13.58	nr	**35.38**
600 mm wide	31.17	35.20	0.54	14.96	nr	**50.16**
750 mm wide	40.94	46.24	0.61	16.89	nr	**63.13**
900 mm wide	47.56	53.71	0.69	19.13	nr	**72.84**
Flat bend; 90°						
50 mm wide	6.59	7.45	0.19	5.26	nr	**12.71**
75 mm wide	6.86	7.75	0.24	6.65	nr	**14.40**
100 mm wide	7.14	8.07	0.28	7.76	nr	**15.83**
150 mm wide	8.38	9.47	0.30	8.31	nr	**17.78**
225 mm wide	10.92	12.33	0.36	9.98	nr	**22.31**
300 mm wide	16.03	18.11	0.44	12.19	nr	**30.30**
450 mm wide	13.84	15.63	0.57	15.79	nr	**31.42**

V: ELECTRICAL SUPPLY/POWER/LIGHTING

Item	Net Price £	Material £	Labour hours	Labour £	Unit	Total rate £
V20: LV DISTRIBUTION – cont						
CABLE TRAY – cont						
Extra over (cutting and jointing tray to fittings is included) – cont						
Flat bend – cont						
600 mm wide	25.73	29.06	0.69	19.13	nr	**48.19**
750 mm wide	36.44	41.16	0.81	22.44	nr	**63.60**
900 mm wide	53.59	60.53	0.94	26.04	nr	**86.57**
Adjustable riser						
50 mm wide	13.34	15.07	0.26	7.20	nr	**22.27**
75 mm wide	14.43	16.30	0.29	8.03	nr	**24.33**
100 mm wide	13.97	15.78	0.32	8.86	nr	**24.64**
150 mm wide	16.17	18.26	0.36	9.98	nr	**28.24**
225 mm wide	21.30	24.05	0.44	12.19	nr	**36.24**
300 mm wide	22.83	25.79	0.52	14.41	nr	**40.20**
450 mm wide	19.72	22.27	0.66	18.29	nr	**40.56**
600 mm wide	30.48	34.43	0.79	21.89	nr	**56.32**
750 mm wide	39.23	44.31	1.03	28.54	nr	**72.85**
900 mm wide	46.16	52.13	1.10	30.47	nr	**82.60**
Inside riser; 90°						
50 mm wide	4.19	4.73	0.28	7.76	nr	**12.49**
75 mm wide	4.38	4.95	0.31	8.59	nr	**13.54**
100 mm wide	4.81	10.38	0.33	9.14	nr	**19.52**
150 mm wide	5.72	6.46	0.37	10.25	nr	**16.71**
225 mm wide	7.03	7.94	0.44	12.19	nr	**20.13**
300 mm wide	7.44	8.40	0.53	14.68	nr	**23.08**
450 mm wide	11.90	13.44	0.67	18.56	nr	**32.00**
600 mm wide	30.41	34.34	0.79	21.89	nr	**56.23**
750 mm wide	37.51	42.36	0.95	26.32	nr	**68.68**
900 mm wide	44.56	50.33	1.11	30.75	nr	**81.08**
Outside riser; 90°						
50 mm wide	4.19	4.73	0.28	7.76	nr	**12.49**
75 mm wide	4.38	4.95	0.31	8.59	nr	**13.54**
100 mm wide	4.81	5.44	0.33	9.14	nr	**14.58**
150 mm wide	5.72	6.46	0.37	10.25	nr	**16.71**
225 mm wide	7.03	7.94	0.44	12.19	nr	**20.13**
300 mm wide	7.44	8.40	0.53	14.68	nr	**23.08**
450 mm wide	11.90	13.44	0.67	18.56	nr	**32.00**
600 mm wide	16.91	19.09	0.79	21.89	nr	**40.98**
750 mm wide	22.96	25.93	0.95	26.32	nr	**52.25**
900 mm wide	22.79	25.73	1.11	30.75	nr	**56.48**

V: ELECTRICAL SUPPLY/POWER/LIGHTING

Item	Net Price £	Material £	Labour hours	Labour £	Unit	Total rate £
Equal tee						
50 mm wide	9.91	11.19	0.30	8.31	nr	**19.50**
75 mm wide	10.32	11.66	0.31	8.59	nr	**20.25**
100 mm wide	10.65	12.03	0.35	9.70	nr	**21.73**
150 mm wide	12.95	14.63	0.36	9.98	nr	**24.61**
225 mm wide	17.24	19.47	0.44	12.19	nr	**31.66**
300 mm wide	24.04	27.15	0.54	14.96	nr	**42.11**
450 mm wide	20.23	22.85	0.71	19.68	nr	**42.53**
600 mm wide	33.95	38.34	0.92	25.49	nr	**63.83**
750 mm wide	50.89	57.48	1.19	32.97	nr	**90.45**
900 mm wide	72.38	81.74	1.44	39.89	nr	**121.63**
Unequal tee						
75 mm wide	6.46	7.29	0.38	10.53	nr	**17.82**
100 mm wide	6.46	7.29	0.39	10.81	nr	**18.10**
150 mm wide	6.65	7.51	0.43	11.91	nr	**19.42**
225 mm wide	7.72	8.72	0.50	13.86	nr	**22.58**
300 mm wide	9.72	10.98	0.63	17.45	nr	**28.43**
450 mm wide	12.90	14.57	0.80	22.15	nr	**36.72**
600 mm wide	20.23	22.85	1.02	28.25	nr	**51.10**
750 mm wide	57.79	65.27	1.12	31.03	nr	**96.30**
900 mm wide	76.17	86.03	1.35	37.40	nr	**123.43**
4 way crossovers						
50 mm wide	14.72	16.63	0.38	10.53	nr	**27.16**
75 mm wide	14.99	16.93	0.40	11.08	nr	**28.01**
100 mm wide	15.78	17.82	0.40	11.08	nr	**28.90**
150 mm wide	18.69	21.10	0.44	12.19	nr	**33.29**
225 mm wide	25.18	28.44	0.53	14.68	nr	**43.12**
300 mm wide	34.66	39.15	0.64	17.73	nr	**56.88**
450 mm wide	30.14	34.04	0.84	23.28	nr	**57.32**
600 mm wide	33.95	38.34	1.03	28.54	nr	**66.88**
750 mm wide	50.89	57.48	1.13	31.30	nr	**88.78**
900 mm wide	72.38	81.74	1.36	37.68	nr	**119.42**
Medium duty tray with return flange						
Straight lengths						
50 mm wide	6.96	6.96	0.33	8.09	m	**15.05**
75 mm wide	8.12	9.17	0.33	9.14	m	**18.31**
100 mm wide	8.68	9.80	0.35	9.70	m	**19.50**
150 mm wide	10.61	11.98	0.39	10.81	m	**22.79**
225 mm wide	12.84	14.50	0.45	12.46	m	**26.96**
300 mm wide	17.77	20.07	0.57	15.79	m	**35.86**
450 mm wide	27.61	31.18	0.69	19.13	m	**50.31**
600 mm wide	38.55	43.53	0.91	25.20	m	**68.73**

V: ELECTRICAL SUPPLY/POWER/LIGHTING

Item	Net Price £	Material £	Labour hours	Labour £	Unit	Total rate £
V20: LV DISTRIBUTION – cont						
CABLE TRAY – cont						
Extra over (cutting and jointing tray to fittings is included)						
Straight reducer						
100 mm wide	16.98	19.18	0.25	6.93	nr	**26.11**
150 mm wide	18.55	20.95	0.27	7.48	nr	**28.43**
225 mm wide	21.38	24.14	0.34	9.41	nr	**33.55**
300 mm wide	24.75	27.95	0.39	10.81	nr	**38.76**
450 mm wide	31.98	36.12	0.49	13.58	nr	**49.70**
600 mm wide	39.62	44.75	0.54	14.96	nr	**59.71**
Flat bend; 90°						
75 mm wide	22.64	25.57	0.24	6.65	nr	**32.22**
100 mm wide	25.20	28.46	0.28	7.76	nr	**36.22**
150 mm wide	26.74	30.20	0.30	8.31	nr	**38.51**
225 mm wide	30.87	34.86	0.36	9.98	nr	**44.84**
300 mm wide	38.07	42.99	0.44	12.19	nr	**55.18**
450 mm wide	57.19	64.59	0.57	15.79	nr	**80.38**
600 mm wide	69.15	78.10	0.69	19.13	nr	**97.23**
Adjustable bend						
75 mm wide	25.89	29.24	0.29	8.03	nr	**37.27**
100 mm wide	28.17	31.81	0.32	8.86	nr	**40.67**
150 mm wide	32.21	36.37	0.36	9.98	nr	**46.35**
225 mm wide	36.11	40.78	0.44	12.19	nr	**52.97**
300 mm wide	40.80	46.06	0.52	14.41	nr	**60.47**
Adjustable riser						
75 mm wide	23.09	26.08	0.29	8.03	nr	**34.11**
100 mm wide	23.48	26.52	0.32	8.86	nr	**35.38**
150 mm wide	25.87	29.21	0.36	9.98	nr	**39.19**
225 mm wide	27.44	30.99	0.44	12.19	nr	**43.18**
300 mm wide	29.26	33.05	0.52	14.41	nr	**47.46**
450 mm wide	38.37	43.34	0.66	18.29	nr	**61.63**
600 mm wide	47.07	53.16	0.79	21.89	nr	**75.05**
Inside riser; 90°						
75 mm wide	13.87	15.67	0.31	8.59	nr	**24.26**
100 mm wide	14.00	15.81	0.33	9.14	nr	**24.95**
150 mm wide	15.95	18.01	0.37	10.25	nr	**28.26**
225 mm wide	19.49	22.01	0.44	12.19	nr	**34.20**
300 mm wide	23.67	26.73	0.53	14.68	nr	**41.41**
450 mm wide	34.17	38.59	0.67	18.56	nr	**57.15**
600 mm wide	54.84	61.93	0.79	21.89	nr	**83.82**

V: ELECTRICAL SUPPLY/POWER/LIGHTING

Item	Net Price £	Material £	Labour hours	Labour £	Unit	**Total rate £**
Outside riser; 90°						
75 mm wide	13.87	15.67	0.31	8.59	nr	**24.26**
100 mm wide	14.00	15.81	0.33	9.14	nr	**24.95**
150 mm wide	15.95	18.01	0.37	10.25	nr	**28.26**
225 mm wide	19.49	22.01	0.44	12.19	nr	**34.20**
300 mm wide	23.67	26.73	0.53	14.68	nr	**41.41**
450 mm wide	34.17	38.59	0.67	18.56	nr	**57.15**
600 mm wide	54.84	61.93	0.79	21.89	nr	**83.82**
Equal tee						
75 mm wide	30.83	34.82	0.31	8.59	nr	**43.41**
100 mm wide	32.87	37.13	0.35	9.70	nr	**46.83**
150 mm wide	35.39	39.97	0.36	9.98	nr	**49.95**
225 mm wide	38.46	43.43	0.74	20.50	nr	**63.93**
300 mm wide	46.68	52.72	0.54	14.96	nr	**67.68**
450 mm wide	61.49	69.44	0.71	19.68	nr	**89.12**
600 mm wide	88.34	99.77	0.92	25.49	nr	**125.26**
Unequal tee						
100 mm wide	32.87	37.13	0.39	10.81	nr	**47.94**
150 mm wide	32.87	37.13	0.43	11.91	nr	**49.04**
225 mm wide	38.46	43.43	0.50	13.86	nr	**57.29**
300 mm wide	46.68	52.72	0.63	17.45	nr	**70.17**
450 mm wide	61.49	69.44	0.80	22.15	nr	**91.59**
600 mm wide	88.34	99.77	1.02	28.25	nr	**128.02**
4 way crossovers						
75 mm wide	43.72	49.38	0.40	11.08	nr	**60.46**
100 mm wide	47.12	53.21	0.40	11.08	nr	**64.29**
150 mm wide	40.95	46.25	0.44	12.19	nr	**58.44**
225 mm wide	59.21	66.87	0.53	14.68	nr	**81.55**
300 mm wide	68.89	77.81	0.64	17.73	nr	**95.54**
450 mm wide	87.68	99.02	0.84	23.28	nr	**122.30**
600 mm wide	127.82	144.36	1.03	28.54	nr	**172.90**
Heavy duty tray with return flange						
Straight lengths						
75 mm	10.91	12.32	0.34	9.41	m	**21.73**
100 mm	11.62	13.13	0.36	9.98	m	**23.11**
150 mm	13.44	15.18	0.40	11.08	m	**26.26**
225 mm	15.06	17.01	0.46	12.74	m	**29.75**
300 mm	18.37	20.75	0.58	16.08	m	**36.83**
450 mm	27.76	31.35	0.70	19.39	m	**50.74**
600 mm	33.66	38.02	0.92	25.49	m	**63.51**
750 mm	42.45	47.94	1.01	27.99	m	**75.93**
900 mm	46.96	53.04	1.14	31.58	m	**84.62**

V: ELECTRICAL SUPPLY/POWER/LIGHTING

Item	Net Price £	Material £	Labour hours	Labour £	Unit	Total rate £
V20: LV DISTRIBUTION – cont						
CABLE TRAY – cont						
Extra over (cutting and jointing tray to fittings is included)						
Straight reducer						
100 mm wide	26.63	30.08	0.25	6.93	nr	**37.01**
150 mm wide	27.65	31.23	0.27	7.48	nr	**38.71**
225 mm wide	31.40	35.47	0.34	9.41	nr	**44.88**
300 mm wide	35.00	39.53	0.39	10.81	nr	**50.34**
450 mm wide	50.06	56.54	0.49	13.58	nr	**70.12**
600 mm wide	55.66	62.86	0.54	14.96	nr	**77.82**
750 mm wide	70.60	79.73	0.60	16.63	nr	**96.36**
900 mm wide	77.43	87.44	0.66	18.29	nr	**105.73**
Flat bend; 90°						
75 mm wide	29.79	33.64	0.24	6.65	nr	**40.29**
100 mm wide	33.87	38.25	0.28	7.76	nr	**46.01**
150 mm wide	36.09	40.76	0.30	8.31	nr	**49.07**
225 mm wide	40.82	46.10	0.36	9.98	nr	**56.08**
300 mm wide	46.51	52.53	0.44	12.19	nr	**64.72**
450 mm wide	68.19	77.01	0.57	15.79	nr	**92.80**
600 mm wide	91.14	102.93	0.69	19.13	nr	**122.06**
750 mm wide	122.14	137.94	0.83	22.99	nr	**160.93**
900 mm wide	138.50	156.42	1.01	27.99	nr	**184.41**
Adjustable bend						
75 mm wide	30.98	34.99	0.29	8.03	nr	**43.02**
100 mm wide	34.26	38.69	0.32	8.86	nr	**47.55**
150 mm wide	36.19	40.87	0.36	9.98	nr	**50.85**
225 mm wide	39.83	44.98	0.44	12.19	nr	**57.17**
300 mm wide	47.25	53.37	0.52	14.41	nr	**67.78**
Adjustable riser						
75 mm wide	27.30	30.83	0.29	8.03	nr	**38.86**
100 mm wide	26.38	29.79	0.32	8.86	nr	**38.65**
150 mm wide	31.43	35.50	0.36	9.98	nr	**45.48**
225 mm wide	33.66	38.02	0.44	12.19	nr	**50.21**
300 mm wide	36.46	41.18	0.52	14.41	nr	**55.59**
450 mm wide	44.37	50.11	0.66	18.29	nr	**68.40**
600 mm wide	52.86	59.70	0.79	21.89	nr	**81.59**
750 mm wide	64.01	72.30	1.03	28.54	nr	**100.84**
900 mm wide	74.33	83.95	1.10	30.47	nr	**114.42**
Inside riser; 90°						
75 mm wide	22.83	25.79	0.31	8.59	nr	**34.38**
100 mm wide	23.13	26.12	0.33	9.14	nr	**35.26**
150 mm wide	25.06	28.30	0.37	10.25	nr	**38.55**

V: ELECTRICAL SUPPLY/POWER/LIGHTING

Item	Net Price £	Material £	Labour hours	Labour £	Unit	**Total rate £**
225 mm wide	26.19	29.58	0.44	12.19	nr	**41.77**
300 mm wide	27.02	30.52	0.53	14.68	nr	**45.20**
450 mm wide	46.35	52.35	0.67	18.56	nr	**70.91**
600 mm wide	56.87	64.23	0.79	21.89	nr	**86.12**
750 mm wide	70.38	79.49	0.95	26.32	nr	**105.81**
900 mm wide	82.98	93.72	1.11	30.75	nr	**124.47**
Outside riser; 90°						
75 mm wide	22.83	25.79	0.31	8.59	nr	**34.38**
100 mm wide	23.13	26.12	0.33	9.14	nr	**35.26**
150 mm wide	25.06	28.30	0.37	10.25	nr	**38.55**
225 mm wide	26.19	29.58	0.44	12.19	nr	**41.77**
300 mm wide	27.02	30.52	0.53	14.68	nr	**45.20**
450 mm wide	46.35	52.35	0.67	18.56	nr	**70.91**
600 mm wide	56.87	64.23	0.79	21.89	nr	**86.12**
750 mm wide	70.38	79.49	0.95	26.32	nr	**105.81**
900 mm wide	82.98	93.72	1.11	30.75	nr	**124.47**
Equal tee						
75 mm wide	39.77	44.92	0.31	8.59	nr	**53.51**
100 mm wide	44.46	50.21	0.35	9.70	nr	**59.91**
150 mm wide	48.69	54.99	0.36	9.98	nr	**64.97**
225 mm wide	56.03	63.28	0.44	12.19	nr	**75.47**
300 mm wide	60.58	68.42	0.54	14.96	nr	**83.38**
450 mm wide	87.96	99.34	0.71	19.68	nr	**119.02**
600 mm wide	116.14	131.17	0.92	25.49	nr	**156.66**
750 mm wide	153.17	172.99	1.19	32.97	nr	**205.96**
900 mm wide	174.92	197.55	1.45	40.17	nr	**237.72**
Unequal tee						
75 mm wide	39.77	44.92	0.38	10.53	nr	**55.45**
100 mm wide	44.42	50.16	0.39	10.81	nr	**60.97**
150 mm wide	48.69	54.99	0.43	11.91	nr	**66.90**
225 mm wide	56.03	63.28	0.50	13.86	nr	**77.14**
300 mm wide	60.58	68.42	0.63	17.45	nr	**85.87**
450 mm wide	87.96	99.34	0.80	22.15	nr	**121.49**
600 mm wide	116.14	131.17	1.02	28.25	nr	**159.42**
750 mm wide	153.17	172.99	1.12	31.03	nr	**204.02**
900 mm wide	185.28	209.25	1.35	37.40	nr	**246.65**
4 way crossovers						
75 mm wide	40.36	45.58	0.40	11.08	nr	**56.66**
100 mm wide	57.85	65.33	0.40	11.08	nr	**76.41**
150 mm wide	57.85	65.33	0.44	12.19	nr	**77.52**
225 mm wide	80.44	90.85	0.53	14.68	nr	**105.53**
300 mm wide	88.70	100.17	0.64	17.73	nr	**117.90**
450 mm wide	127.58	144.09	0.84	23.28	nr	**167.37**
600 mm wide	173.40	195.84	1.03	28.54	nr	**224.38**
750 mm wide	211.50	238.86	1.13	31.30	nr	**270.16**
900 mm wide	255.12	288.13	1.36	37.68	nr	**325.81**

V: ELECTRICAL SUPPLY/POWER/LIGHTING

Item	Net Price £	Material £	Labour hours	Labour £	Unit	Total rate £
V20: LV DISTRIBUTION – cont						
CABLE TRAY – cont						
GRP cable tray including standard coupling joints and fixings (supports and hangers excluded)						
Tray						
100 mm wide	29.96	33.83	0.34	9.41	m	**43.24**
200 mm wide	38.36	43.32	0.39	10.81	m	**54.13**
400 mm wide	60.16	67.94	0.53	14.68	m	**82.62**
Cover						
100 mm wide	16.88	19.06	0.10	2.76	m	**21.82**
200 mm wide	22.36	25.26	0.11	3.05	m	**28.31**
400 mm wide	38.18	43.12	0.14	3.88	m	**47.00**
Extra for (cutting and jointing to fittings included)						
Reducer						
200 mm wide	79.62	89.92	0.23	6.36	nr	**96.28**
400 mm wide	104.02	117.48	0.30	8.31	nr	**125.79**
Reducer cover						
200 mm wide	49.83	56.27	0.25	6.93	nr	**63.20**
400 mm wide	72.83	82.25	0.28	7.76	nr	**90.01**
Bend						
100 mm wide	66.58	75.19	0.34	9.41	nr	**84.60**
200 mm wide	77.28	87.28	0.40	11.08	nr	**98.36**
400 mm wide	99.32	112.17	0.32	8.86	nr	**121.03**
Bend cover						
100 mm wide	33.08	37.36	0.10	2.76	nr	**40.12**
200 mm wide	44.18	49.90	0.10	2.76	nr	**52.66**
400 mm wide	61.05	68.95	0.13	3.60	nr	**72.55**
Tee						
100 mm wide	84.74	95.71	0.37	10.25	nr	**105.96**
200 mm wide	93.42	105.51	0.43	11.91	nr	**117.42**
400 mm wide	115.89	130.89	0.56	15.51	nr	**146.40**
Tee cover						
100 mm wide	42.69	48.22	0.27	7.48	nr	**55.70**
200 mm wide	51.39	58.04	0.31	8.59	nr	**66.63**
400 mm wide	72.48	81.85	0.37	10.25	nr	**92.10**

V: ELECTRICAL SUPPLY/POWER/LIGHTING

Item	Net Price £	Material £	Labour hours	Labour £	Unit	**Total rate £**
BASKET TRAY						
Mild steel cable basket; zinc plated including standard coupling joints, fixings and earth continuity straps (supports and hangers are excluded)						
Basket 54 mm deep						
100 mm wide	4.40	4.97	0.22	6.10	m	**11.07**
150 mm wide	4.90	5.54	0.25	6.93	m	**12.47**
200 mm wide	5.36	6.05	0.28	7.76	m	**13.81**
300 mm wide	6.29	7.10	0.34	9.41	m	**16.51**
450 mm wide	7.62	8.61	0.44	12.19	m	**20.80**
600 mm wide	9.32	10.53	0.70	19.39	m	**29.92**
Extra for (cutting and jointing to fittings is included)						
Reducer						
150 mm wide	14.38	16.24	0.25	6.93	nr	**23.17**
200 mm wide	16.87	19.05	0.28	7.76	nr	**26.81**
300 mm wide	17.15	19.37	0.38	10.53	nr	**29.90**
450 mm wide	18.82	21.26	0.48	13.29	nr	**34.55**
600 mm wide	23.45	26.49	0.48	13.29	nr	**39.78**
Bend						
100 mm wide	13.73	15.50	0.23	6.36	nr	**21.86**
150 mm wide	16.09	18.18	0.26	7.20	nr	**25.38**
200 mm wide	16.37	18.49	0.30	8.31	nr	**26.80**
300 mm wide	17.92	20.24	0.35	9.70	nr	**29.94**
450 mm wide	22.32	25.20	0.50	13.86	nr	**39.06**
600 mm wide	26.26	29.66	0.58	16.08	nr	**45.74**
Tee						
100 mm wide	17.26	19.49	0.28	7.76	nr	**27.25**
150 mm wide	18.31	20.68	0.30	8.31	nr	**28.99**
200 mm wide	18.69	21.10	0.33	9.14	nr	**30.24**
300 mm wide	23.51	26.55	0.39	10.81	nr	**37.36**
450 mm wide	30.79	34.77	0.56	15.51	nr	**50.28**
600 mm wide	31.84	35.95	0.65	18.00	nr	**53.95**
Cross over						
100 mm wide	23.55	26.60	0.40	11.08	nr	**37.68**
150 mm wide	23.95	27.05	0.42	11.63	nr	**38.68**
200 mm wide	26.01	29.38	0.46	12.74	nr	**42.12**
300 mm wide	29.55	33.37	0.51	14.13	nr	**47.50**
450 mm wide	34.64	39.13	0.74	20.50	nr	**59.63**
600 mm wide	35.77	40.40	0.82	22.72	nr	**63.12**

V: ELECTRICAL SUPPLY/POWER/LIGHTING

Item	Net Price £	Material £	Labour hours	Labour £	Unit	Total rate £
V20: LV DISTRIBUTION – cont						
BASKET TRAY – cont						
Mild steel cable basket; expoxy coated including standard coupling joints, fixings and earth continuity straps (supports and hangers are excluded)						
Basket 54 mm deep						
100 mm wide	9.84	11.12	0.22	6.10	m	**17.22**
150 mm wide	11.13	12.57	0.25	6.93	m	**19.50**
200 mm wide	12.39	13.99	0.28	7.76	m	**21.75**
300 mm wide	14.07	15.89	0.34	9.41	m	**25.30**
450 mm wide	17.37	19.61	0.44	12.19	m	**31.80**
600 mm wide	19.78	22.34	0.70	19.39	m	**41.73**
Extra for (cutting and jointing to fittings is included)						
Reducer						
150 mm wide	20.58	23.25	0.28	7.76	nr	**31.01**
200 mm wide	23.06	26.04	0.28	7.76	nr	**33.80**
300 mm wide	24.60	27.78	0.38	10.53	nr	**38.31**
450 mm wide	28.76	32.48	0.48	13.29	nr	**45.77**
600 mm wide	35.84	40.47	0.48	13.29	nr	**53.76**
Bend						
100 mm wide	19.95	22.53	0.23	6.36	nr	**28.89**
150 mm wide	22.30	25.18	0.26	7.20	nr	**32.38**
200 mm wide	22.58	25.50	0.30	8.31	nr	**33.81**
300 mm wide	25.40	28.68	0.35	9.70	nr	**38.38**
450 mm wide	32.23	36.40	0.50	13.86	nr	**50.26**
600 mm wide	38.70	43.71	0.58	16.08	nr	**59.79**
Tee						
100 mm wide	23.46	26.50	0.28	7.76	nr	**34.26**
150 mm wide	24.53	27.70	0.30	8.31	nr	**36.01**
200 mm wide	24.89	28.11	0.33	9.14	nr	**37.25**
300 mm wide	30.95	34.96	0.39	10.81	nr	**45.77**
450 mm wide	39.45	44.55	0.56	15.51	nr	**60.06**
600 mm wide	44.24	49.96	0.65	18.00	nr	**67.96**
Cross over						
100 mm wide	29.75	33.60	0.40	11.08	nr	**44.68**
150 mm wide	30.09	33.99	0.42	11.63	nr	**45.62**
200 mm wide	32.23	36.40	0.46	12.74	nr	**49.14**
300 mm wide	37.00	41.79	0.51	14.13	nr	**55.92**
450 mm wide	44.56	50.33	0.74	20.50	nr	**70.83**
600 mm wide	48.15	54.38	0.82	22.72	nr	**77.10**

V: ELECTRICAL SUPPLY/POWER/LIGHTING

Item	Net Price £	Material £	Labour hours	Labour £	Unit	**Total rate £**
Y71 – LV SWITCHGEAR & DISTRIBUTION BOARDS						
LV switchboard components, factory-assembled modular construction to IP41; form 4, type 5; 2400 mm high, with front and rear access; top cable entry/exit; includes delivery, offloading, positioning and commissioning (hence separate labour costs are not detailed below); excludes cabling and cable terminations						
Air circuit breakers (ACBs) to BSEN 60947–2, withdrawable type, fitted with adjustable instantaneous and overload protection. Includes enclosure and copper links, assembled into LV switchboard.						
ACB–100 kA fault rated						
4 pole, 6300 A (1600 mm wide)	31888.85	36015.11	–	–	nr	**36015.11**
4 pole, 5000 A (1600 mm wide)	24097.08	27215.12	–	–	nr	**27215.12**
4 pole, 4000 A (1600 mm wide)	20215.44	22831.22	–	–	nr	**22831.22**
4 pole, 3200 A (1600 mm wide)	13647.31	15413.20	–	–	nr	**15413.20**
4 pole, 2500 A (1600 mm wide)	10606.54	11978.97	–	–	nr	**11978.97**
4 pole, 2000 A (1600 mm wide)	8268.20	9338.06	–	–	nr	**9338.06**
4 pole, 1600 A (1600 mm wide)	6769.70	7645.66	–	–	nr	**7645.66**
4 pole, 1250 A (1600 mm wide)	5991.95	6767.28	–	–	nr	**6767.28**
4 pole, 1000 A (1600 mm wide)	5918.65	6684.50	–	–	nr	**6684.50**
4 pole, 800 A (1600 mm wide)	5818.89	6571.83	–	–	nr	**6571.83**
3 pole, 6300 A (1600 mm wide)	28894.91	32633.77	–	–	nr	**32633.77**
3 pole, 5000 A (1600 mm wide)	20263.29	22885.26	–	–	nr	**22885.26**
3 pole, 4000 A (1600 mm wide)	16593.40	18740.50	–	–	nr	**18740.50**
3 pole, 3200 A (1600 mm wide)	11394.47	12868.86	–	–	nr	**12868.86**
3 pole, 2500 A (1600 mm wide)	9122.30	10302.68	–	–	nr	**10302.68**
3 pole, 2000 A (1600 mm wide)	6869.46	7758.33	–	–	nr	**7758.33**
3 pole, 1600 A (1600 mm wide)	5584.75	6307.38	–	–	nr	**6307.38**
3 pole, 1250 A (1600 mm wide)	5191.80	5863.59	–	–	nr	**5863.59**
3 pole, 1000 A (1600 mm wide)	5080.84	5738.28	–	–	nr	**5738.28**
3 pole, 800 A (1600 mm wide)	4986.16	5631.35	–	–	nr	**5631.35**

V: ELECTRICAL SUPPLY/POWER/LIGHTING

Item	Net Price £	Material £	Labour hours	Labour £	Unit	**Total rate £**
V20: LV DISTRIBUTION – cont						
Y71 – LV SWITCHGEAR & DISTRIBUTION BOARDS – cont						
LV switchboard components, factory-assembled modular construction to IP41 – cont						
ACB–65 kA fault rated						
4 pole, 4000 A (1600 mm wide)	17739.67	20035.09	–	–	nr	**20035.09**
4 pole, 3200 A (1600 mm wide)	12000.18	13552.94	–	–	nr	**13552.94**
4 pole, 2500 A (1600 mm wide)	9433.81	10654.50	–	–	nr	**10654.50**
4 pole, 2000 A (1600 mm wide)	7502.66	8473.47	–	–	nr	**8473.47**
4 pole, 1600 A (1600 mm wide)	6190.46	6991.48	–	–	nr	**6991.48**
4 pole, 1250 A (1600 mm wide)	5526.72	6241.85	–	–	nr	**6241.85**
4 pole, 1000 A (1600 mm wide)	5421.87	6123.44	–	–	nr	**6123.44**
4 pole, 800 A (1600 mm wide)	5324.14	6013.05	–	–	nr	**6013.05**
3 pole, 4000 A (1600 mm wide)	15027.72	16972.23	–	–	nr	**16972.23**
3 pole, 3200 A (1600 mm wide)	10191.20	11509.89	–	–	nr	**11509.89**
3 pole, 2500 A (1600 mm wide)	7925.13	8950.60	–	–	nr	**8950.60**
3 pole, 2000 A (1600 mm wide)	6299.38	7114.49	–	–	nr	**7114.49**
3 pole, 1600 A (1600 mm wide)	5153.12	5819.91	–	–	nr	**5819.91**
3 pole, 1250 A (1600 mm wide)	4796.82	5417.51	–	–	nr	**5417.51**
3 pole, 1000 A (1600 mm wide)	4688.91	5295.64	–	–	nr	**5295.64**
3 pole, 800 A (1600 mm wide)	4593.22	5187.56	–	–	nr	**5187.56**
Extra for						
Cable box (one per ACB for form 4, types 6 & 7)	297.26	335.72	–	–	nr	**335.72**
Opening Coil	88.57	100.03	–	–	nr	**100.03**
Closing Coil	89.58	101.17	–	–	nr	**101.17**
Undervoltage Release	174.08	196.61	–	–	nr	**196.61**
Motor Operator	539.54	609.36	–	–	nr	**609.36**
Mechnical Interlock (per ACB)	506.96	572.56	–	–	nr	**572.56**
ACB Fortress/Castell Adaptor Kit (one per ACB)	126.23	142.57	–	–	nr	**142.57**
Fortress/Castell ACB Lock (one per ACB)	253.48	286.28	–	–	nr	**286.28**
Fortress/Castell Key	63.12	71.29	–	–	nr	**71.29**

V: ELECTRICAL SUPPLY/POWER/LIGHTING

Item	Net Price £	Material £	Labour hours	Labour £	Unit	Total rate £
Moulded case circuit breakers (MCCBs) to BS EN 60947–2; plug-in type, fitted with electronic trip unit. Includes metalwork section and copper links, assembled into LV switchboard						
MCCB – 150 kA fault rated						
4 pole, 630 A (800 mm wide, 600 mm high)	3018.99	3409.64	–	–	nr	**3409.64**
4 pole, 400 A (800 mm wide, 400 mm high)	2159.49	2438.92	–	–	nr	**2438.92**
4 pole, 250 A (800 mm wide, 400 mm high)	1930.56	2180.36	–	–	nr	**2180.36**
4 pole, 160 A (800 mm wide, 300 mm high)	1312.25	1482.05	–	–	nr	**1482.05**
4 pole, 100 A (800 mm wide, 200 mm high)	1092.52	1233.88	–	–	nr	**1233.88**
3 pole, 630 A (800 mm wide, 600 mm high)	2693.99	3042.58	–	–	nr	**3042.58**
3 pole, 400 A (800 mm wide, 400 mm high)	2057.29	2323.49	–	–	nr	**2323.49**
3 pole, 250 A (800 mm wide, 400 mm high)	1657.68	1872.18	–	–	nr	**1872.18**
3 pole, 160 A (800 mm wide, 300 mm high)	1213.11	1370.08	–	–	nr	**1370.08**
3 pole, 100 A (800 mm wide, 200 mm high)	969.88	1095.38	–	–	nr	**1095.38**
MCCB – 70 kA fault rated						
4 pole, 630 A (800 mm wide, 600 mm high)	2585.66	2920.23	–	–	nr	**2920.23**
4 pole, 400 A (800 mm wide, 400 mm high)	1777.26	2007.23	–	–	nr	**2007.23**
4 pole, 250 A (800 mm wide, 400 mm high)	1527.89	1725.60	–	–	nr	**1725.60**
4 pole, 160 A (800 mm wide, 300 mm high)	1146.68	1295.05	–	–	nr	**1295.05**
4 pole, 100 A (800 mm wide, 200 mm high)	864.61	976.49	–	–	nr	**976.49**
3 pole, 630 A (800 mm wide, 600 mm high)	2263.73	2556.65	–	–	nr	**2556.65**
3 pole, 400 A (800 mm wide, 400 mm high)	1534.02	1732.51	–	–	nr	**1732.51**
3 pole, 250 A (800 mm wide, 400 mm high)	1354.15	1529.37	–	–	nr	**1529.37**
3 pole, 160 A (800 mm wide, 300 mm high)	999.52	1128.85	–	–	nr	**1128.85**
3 pole, 100 A (800 mm wide, 200 mm high)	735.84	831.06	–	–	nr	**831.06**
MCCB – 45 kA fault rated						
4 pole, 630 A (800 mm wide, 600 mm high)	2507.99	2832.51	–	–	nr	**2832.51**
4 pole, 400 A (800 mm wide, 400 mm high)	1723.09	1946.05	–	–	nr	**1946.05**
3 pole, 630 A (800 mm wide, 600 mm high)	2167.66	2448.14	–	–	nr	**2448.14**
3 pole, 400 A (800 mm wide, 400 mm high)	1476.79	1667.88	–	–	nr	**1667.88**
MCCB – 36 kA fault rated						
4 pole, 250 A (800 mm wide, 400 mm high)	1462.48	1651.72	–	–	nr	**1651.72**
4 pole, 160 A (800 mm wide, 300 mm high)	1107.85	1251.20	–	–	nr	**1251.20**
3 pole, 250 A (800 mm wide, 400 mm high)	830.89	938.40	–	–	nr	**938.40**
3 pole, 160 A (800 mm wide, 300 mm high)	731.75	826.43	–	–	nr	**826.43**

V: ELECTRICAL SUPPLY/POWER/LIGHTING

Item	Net Price £	Material £	Labour hours	Labour £	Unit	Total rate £
V20: LV DISTRIBUTION – cont						
Y71 – LV SWITCHGEAR & DISTRIBUTION BOARDS – cont						
Moulded case circuit breakers (MCCBs) to BS EN 60947–2 – cont						
Extra for						
Cable box (one per MCCB for form 4, types 6 & 7)	125.71	141.97	–	–	nr	**141.97**
Shunt trip (for ratings 100 A to 630 A)	43.95	49.63	–	–	nr	**49.63**
Undervoltage release (for ratings 100 A to 630 A)	63.36	71.56	–	–	nr	**71.56**
Motor operator for 630 A MCCB	600.94	678.70	–	–	nr	**678.70**
Motor operator for 400 A MCCB	600.94	678.70	–	–	nr	**678.70**
Motor operator for 250 A MCCB	498.74	563.27	–	–	nr	**563.27**
Motor operator for 160 A/100 A MCCB	318.86	360.12	–	–	nr	**360.12**
Door handle for 630/400 A MCCB	78.69	88.87	–	–	nr	**88.87**
Door handle for 250/160/100 A MCCB	62.34	70.41	–	–	nr	**70.41**
MCCB earth fault protection	474.21	535.57	–	–	nr	**535.57**
LV SWITCHBOARD BUSBAR						
Copper busbar assembled into LV switchboard, ASTA type tested to appropriate fault level. Busbar Length may be estimated by adding the widths of the ACB sections to the width of the MCCB sections. ACBs up to 2000 A rating may be stacked two high; larger ratings are one per section. To determine the number of MCCB sections, add together all the MCCB heights and divide by 1800 mm, rounding up as necessary						
6000 A (6 × 10 mm × 100 mm)	2467.54	2786.82	–	–	nr	**2786.82**
5000 A (4 × 10 mm × 100 mm)	2028.16	2290.59	–	–	nr	**2290.59**
4000 A (4 × 10 mm × 100 mm)	2028.16	2290.59	–	–	nr	**2290.59**
3200 A (3 × 10 mm × 100 mm)	1408.70	1590.98	–	–	nr	**1590.98**
2500 A (2 × 10 mm × 100 mm)	1188.50	1342.29	–	–	nr	**1342.29**
2000 A (2 × 10 mm × 80 mm)	873.62	986.67	–	–	nr	**986.67**
1600 A (2 × 10 mm × 50 mm)	633.86	715.88	–	–	nr	**715.88**
1250 A (2 × 10 mm × 40 mm)	497.01	561.32	–	–	nr	**561.32**
1000 A (2 × 10 mm × 30 mm)	497.01	561.32	–	–	nr	**561.32**
800 A (2 × 10 mm × 20 mm)	386.90	436.97	–	–	nr	**436.97**
630 A (2 × 10 mm × 20 mm)	386.90	436.97	–	–	nr	**436.97**
400 A (2 × 10 mm × 10 mm)	331.34	374.22	–	–	nr	**374.22**

V: ELECTRICAL SUPPLY/POWER/LIGHTING

Item	Net Price £	Material £	Labour hours	Labour £	Unit	Total rate £
Automatic power factor correction (PFC); floor standing steel enclosure to IP 42, complete with microprocessor based relay and status indication; includes delivery, offloading, positioning and commissioning; excludes cabling and cable terminations						
Standard PFC (no de-tuning)						
100 kVAr	4838.36	5464.42	–	–	nr	**5464.42**
200 kVAr	6766.70	7642.28	–	–	nr	**7642.28**
400 kVAr	11551.55	13046.26	–	–	nr	**13046.26**
600 kVAr	15822.93	17870.34	–	–	nr	**17870.34**
PFC with de-tuning reactors						
100 kVAr	7871.85	8890.43	–	–	nr	**8890.43**
200 kVAr	11001.04	12424.52	–	–	nr	**12424.52**
400 kVAr	18960.35	21413.73	–	–	nr	**21413.73**
600 kVAr	27883.84	31491.87	–	–	nr	**31491.87**
AUTOMATIC TRANSFER SWITCHES						
Automatic transfer switches; steel enclosure; solenoid operating; programmable controller, keypad and LCD display; fixed to backgrounds; including commissioning and testing						
Panel Mounting type 4 pole M6 s; non BMS connection						
40 amp	1608.95	1817.14	2.40	66.48	nr	**1883.62**
63 amp	1707.49	1928.43	2.50	69.27	nr	**1997.70**
100 amp	1772.31	2001.64	2.60	72.03	nr	**2073.67**
125 amp	1820.47	2056.03	2.70	74.80	nr	**2130.83**
160 amp	2139.10	2415.89	2.80	77.57	nr	**2493.46**
Panel Mounting type 4 pole M6e; BMS connection						
40 amp	2220.24	2507.53	2.60	72.03	nr	**2579.56**
63 amp	2351.64	2655.93	2.70	74.80	nr	**2730.73**
100 amp	2438.05	2753.52	2.80	77.57	nr	**2831.09**
125 amp	2501.74	2825.45	2.90	80.35	nr	**2905.80**
160 amp	2852.13	3221.18	3.00	83.11	nr	**3304.29**
Enclosed type 3 pole or 4 pole						
63 amp	3232.52	3650.79	2.60	72.03	nr	**3722.82**
100 amp	3344.86	3777.67	2.90	80.35	nr	**3858.02**
125 amp	3295.59	3722.02	2.90	80.35	nr	**3802.37**

V: ELECTRICAL SUPPLY/POWER/LIGHTING

Item	Net Price £	Material £	Labour hours	Labour £	Unit	Total rate £
V20: LV DISTRIBUTION – cont						
AUTOMATIC TRANSFER SWITCHES – cont						
Automatic transfer switches – cont						
Enclosed type 3 pole or 4 pole – cont						
160 amp	3452.52	3899.26	2.60	72.03	nr	**3971.29**
250 amp	4296.27	4852.18	3.30	91.42	nr	**4943.60**
400 amp	6755.96	7630.15	4.30	119.13	nr	**7749.28**
630 amp	9095.82	10272.78	4.84	134.09	nr	**10406.87**
800 amp	12554.63	14179.13	5.12	141.84	nr	**14320.97**
1000 amp	13917.53	15718.39	5.50	152.37	nr	**15870.76**
1250 amp	17917.53	20235.97	6.00	166.23	nr	**20402.20**
1600 amp	29940.68	33814.86	6.20	171.77	nr	**33986.63**
2000 amp	30715.16	34689.55	6.90	191.16	nr	**34880.71**
3000 amp	33859.54	38240.80	6.90	191.16	nr	**38431.96**
4000 amp	41445.82	46808.70	6.90	191.16	nr	**46999.86**
Enclosed type 3 pole or 4 pole; with single bypass						
40 amp	7654.64	8644.72	2.60	72.03	nr	**8716.75**
63 amp	7803.25	8812.95	2.60	72.03	nr	**8884.98**
100 amp	8010.41	9046.92	2.60	72.03	nr	**9118.95**
125 amp	8424.74	9514.86	2.90	80.35	nr	**9595.21**
160 amp	9115.30	10294.78	2.90	80.35	nr	**10375.13**
250 amp	11863.64	13398.74	3.30	91.42	nr	**13490.16**
400 amp	12465.67	14078.66	4.30	119.13	nr	**14197.79**
630 amp	17098.53	19311.00	4.84	134.09	nr	**19445.09**
800 amp	18369.02	20745.88	5.12	141.84	nr	**20887.72**
1000 amp	25967.78	29327.88	5.50	152.37	nr	**29480.25**
1250 amp	27522.96	31084.30	6.00	166.23	nr	**31250.53**
1600 amp	28263.12	31920.23	6.20	171.77	nr	**32092.00**
Enclosed type 3 pole or 4 pole; with dual bypass						
40 amp	10022.68	11319.05	2.60	72.03	nr	**11391.08**
63 amp	10404.33	11750.59	2.60	72.03	nr	**11822.62**
100 amp	10680.55	12062.56	2.60	72.03	nr	**12134.59**
125 amp	11232.99	12686.48	2.90	80.35	nr	**12766.83**
160 amp	12153.72	13726.35	2.90	80.35	nr	**13806.70**
250 amp	15818.19	17864.98	3.30	91.42	nr	**17956.40**
400 amp	16620.90	18771.56	4.30	119.13	nr	**18890.69**
630 amp	22798.03	25747.98	4.84	134.09	nr	**25882.07**
800 amp	24492.03	27661.18	5.12	141.84	nr	**27803.02**
1000 amp	34623.71	39103.84	5.50	152.37	nr	**39256.21**
1250 amp	36697.28	41445.72	6.00	166.23	nr	**41611.95**
1600 amp	37684.16	42560.31	6.20	171.77	nr	**42732.08**

V: ELECTRICAL SUPPLY/POWER/LIGHTING

Item	Net Price £	Material £	Labour hours	Labour £	Unit	**Total rate £**
BREAKERS/FUSES						
MCCB panelboards; IP4X construction, 50kA busbars and fully-rated neutral; fitted with doorlock, removable glandplate; form 3b Type2; BSEN 60439–1; including fixing to backgrounds						
Panelboards cubicle with MCCB incomer						
Up to 250 A						
4 way TPN	887.72	1002.59	1.00	27.70	nr	**1030.29**
Extra over for integral incomer metering	909.20	1026.84	1.50	41.56	nr	**1068.40**
Up to 630 A						
6 way TPN	1596.89	1803.52	2.00	55.41	nr	**1858.93**
12 way TPN	1803.53	2036.90	2.50	69.27	nr	**2106.17**
18 way TPN	2099.43	2371.08	3.00	83.11	nr	**2454.19**
Extra over for integral incomer metering	1074.51	1213.54	1.50	41.56	nr	**1255.10**
Up to 800 A						
6 way TPN	2732.57	3086.15	2.00	55.41	nr	**3141.56**
12 way TPN	3278.10	3702.27	2.50	69.27	nr	**3771.54**
18 Way TPN	3489.69	3941.24	3.00	83.11	nr	**4024.35**
Extra over for integral incomer metering	1074.51	1213.54	1.50	41.56	nr	**1255.10**
Up to 1200 A						
20 way TPN	7311.67	8257.77	3.50	96.97	nr	**8354.74**
Up to 1600 A						
28 way TPN	9468.95	10694.18	3.50	96.97	nr	**10791.15**
Up to 2000 A						
28 way TPN	10307.08	11640.76	4.00	110.81	nr	**11751.57**
Feeder MCCBs						
Single pole						
32 A	74.98	84.68	0.75	20.78	nr	**105.46**
63 A	76.69	86.62	0.75	20.78	nr	**107.40**
100 A	78.39	88.54	0.75	20.78	nr	**109.32**
160 A	83.51	94.32	1.00	27.70	nr	**122.02**
Double pole						
32 A	112.48	127.03	0.75	20.78	nr	**147.81**
63 A	114.17	128.94	0.75	20.78	nr	**149.72**
100 A	167.01	188.62	0.75	20.78	nr	**209.40**
160 A	207.91	234.81	1.00	27.70	nr	**262.51**

V: ELECTRICAL SUPPLY/POWER/LIGHTING

Item	Net Price £	Material £	Labour hours	Labour £	Unit	Total rate £
V20: LV DISTRIBUTION – cont						
BREAKERS/FUSES – cont						
Feeder MCCBs – cont						
Triple pole						
32 A	149.98	169.39	0.75	20.78	nr	**190.17**
63 A	153.37	173.22	0.75	20.78	nr	**194.00**
100 A	199.39	225.19	0.75	20.78	nr	**245.97**
160 A	257.34	290.64	1.00	27.70	nr	**318.34**
250 A	386.86	436.92	1.00	27.70	nr	**464.62**
400 A	523.19	590.89	1.25	34.63	nr	**625.52**
630 A	858.94	970.08	1.50	41.56	nr	**1011.64**
MCB distribution boards; IP3X external protection enclosure; removable earth and neutral bars and DIN rail; 125/250 amp incomers; including fixing to backgrounds						
SP&N						
6 way	72.59	81.99	2.00	55.41	nr	**137.40**
8 way	86.09	97.23	2.50	69.27	nr	**166.50**
12 way	98.80	111.59	3.00	83.11	nr	**194.70**
16 way	117.45	132.65	4.00	110.81	nr	**243.46**
24 way	247.40	279.41	5.00	138.52	nr	**417.93**
TP&N						
4 way	503.03	568.12	3.00	83.11	nr	**651.23**
6 way	520.71	588.09	3.50	96.97	nr	**685.06**
8 way	545.20	615.74	4.00	110.81	nr	**726.55**
12 way	581.33	656.55	4.00	110.81	nr	**767.36**
16 way	665.19	751.26	5.00	138.52	nr	**889.78**
24 way	837.71	946.10	6.40	177.31	nr	**1123.41**
Miniature circuit breakers for distribution boards; BS EN 60 898; DIN rail mounting; including connecting to circuit						
SP&N; including connecting of wiring						
6 amp	8.67	9.79	0.10	2.76	nr	**12.55**
10–40 amp	9.02	10.19	0.10	2.76	nr	**12.95**
50–63 amp	9.44	10.66	0.14	3.88	nr	**14.54**
TP&N; including connecting of wiring						
6 amp	36.76	41.51	0.30	8.31	nr	**49.82**
10–40 amp	38.21	43.16	0.45	12.46	nr	**55.62**
50–63 amp	40.03	45.21	0.45	12.46	nr	**57.67**

V: ELECTRICAL SUPPLY/POWER/LIGHTING

Item	Net Price £	Material £	Labour hours	Labour £	Unit	**Total rate £**
Resdiual current circuit breakers for distribution boards; DIN rail mounting; including connecting to circuit						
SP&N						
10 mA						
6 amp	56.92	64.28	0.21	5.81	nr	**70.09**
10–32 amp	55.85	63.08	0.26	7.20	nr	**70.28**
45 amp	56.76	64.11	0.26	7.20	nr	**71.31**
30 mA						
6 amp	56.92	64.28	0.21	5.81	nr	**70.09**
10–40 amp	55.85	63.08	0.21	5.81	nr	**68.89**
50–63 amp	57.66	65.12	0.26	7.20	nr	**72.32**
100 mA						
6 amp	105.26	118.88	0.21	5.81	nr	**124.69**
10–40 amp	105.26	118.88	0.23	6.46	nr	**125.34**
50–63 amp	105.26	118.88	0.26	7.20	nr	**126.08**
HRC fused distribution boards; IP4X external protection enclosure; including earth and neutral bars; fixing to backgrounds						
SP&N						
20 amp incomer						
4 way	137.13	154.88	1.00	27.70	nr	**182.58**
6 way	165.55	186.98	1.20	33.25	nr	**220.23**
8 way	194.06	219.17	1.40	38.78	nr	**257.95**
12 way	251.14	283.63	1.80	49.87	nr	**333.50**
32 amp incomer						
4 way	165.11	186.48	1.00	27.70	nr	**214.18**
6 way	217.11	245.21	1.20	33.25	nr	**278.46**
8 way	255.44	288.50	1.40	38.78	nr	**327.28**
12 way	328.88	371.43	1.80	49.87	nr	**421.30**
TP&N						
20 amp incomer						
4 way	258.72	292.20	1.50	41.56	nr	**333.76**
6 way	327.22	369.56	2.10	58.17	nr	**427.73**
8 way	386.89	436.95	2.70	74.80	nr	**511.75**
12 way	542.79	613.02	3.90	108.05	nr	**721.07**

V: ELECTRICAL SUPPLY/POWER/LIGHTING

Item	Net Price £	Material £	Labour hours	Labour £	Unit	Total rate £
V20: LV DISTRIBUTION – cont						
BREAKERS/FUSES – cont						
HRC fused distribution boards – cont						
32 amp incomer						
4 way	309.52	349.57	1.50	41.56	nr	**391.13**
6 way	415.54	469.31	2.10	58.17	nr	**527.48**
8 way	507.61	573.29	2.70	74.80	nr	**648.09**
12 way	704.04	795.14	3.90	108.05	nr	**903.19**
63 amp incomer						
4 way	657.93	743.06	2.17	60.12	nr	**803.18**
6 way	843.79	952.98	2.83	78.41	nr	**1031.39**
8 way	1016.13	1147.61	2.57	71.20	nr	**1218.81**
100 amp incomer						
4 way	1040.52	1175.16	2.40	66.48	nr	**1241.64**
6 way	1360.04	1536.02	2.73	75.63	nr	**1611.65**
8 way	1662.99	1878.18	3.87	107.21	nr	**1985.39**
200 amp incomer						
4 way	2577.25	2910.73	5.36	148.49	nr	**3059.22**
6 way	3406.38	3847.14	6.17	170.93	nr	**4018.07**
HRC fuse; includes fixing to fuse holder						
2–30 amp	2.80	3.16	0.10	2.76	nr	**5.92**
35–63 amp	6.06	6.84	0.12	3.33	nr	**10.17**
80 amp	8.95	10.11	0.15	4.15	nr	**14.26**
100 amp	10.76	12.15	0.15	4.15	nr	**16.30**
125 amp	16.26	18.36	0.15	4.15	nr	**22.51**
160 amp	17.06	19.27	0.15	4.15	nr	**23.42**
200 amp	17.67	19.96	0.15	4.15	nr	**24.11**
Consumer units; fixed to backgrounds; including supports, fixings, connections/ jointing to equipment.						
Switched and insulated; moulded plastic case, 63 amp 230 volt SP&N; earth and neutral bars; 30 mA RCCB protection; fitted MCBs						
2 way	107.27	121.15	1.67	46.27	nr	**167.42**
4 way	120.37	135.95	1.59	44.04	nr	**179.99**
6 way	130.85	147.78	2.50	69.27	nr	**217.05**
8 way	141.19	159.46	3.00	83.11	nr	**242.57**
12 way	164.02	185.24	4.00	110.81	nr	**296.05**
16 way	198.19	223.84	5.50	152.37	nr	**376.21**

V: ELECTRICAL SUPPLY/POWER/LIGHTING

Item	Net Price £	Material £	Labour hours	Labour £	Unit	Total rate £
Switched and insulated; moulded plastic case, 100 amp 230 volt SP&N; earth and neutral bars; 30 mA RCCB protection; fitted MCBs						
2 way	107.27	121.15	1.67	46.27	nr	**167.42**
4 way	120.37	135.95	1.59	44.04	nr	**179.99**
6 way	130.85	147.78	2.50	69.27	nr	**217.05**
8 way	141.19	159.46	3.00	83.11	nr	**242.57**
12 way	164.02	185.24	4.00	110.81	nr	**296.05**
16 way	198.19	223.84	5.50	152.37	nr	**376.21**
Extra for						
Residual current device; double pole; 230 volt/30 mA tripping current						
16 amp	58.65	66.24	0.22	6.10	nr	**72.34**
30 amp	59.58	67.29	0.22	6.10	nr	**73.39**
40 amp	60.52	68.35	0.22	6.10	nr	**74.45**
63 amp	74.92	84.62	0.22	6.10	nr	**90.72**
80 amp	83.33	94.12	0.22	6.10	nr	**100.22**
100 amp	102.57	115.84	0.25	6.93	nr	**122.77**
Residual current device; double pole; 230 volt/100 mA tripping current						
63 amp	68.49	77.36	0.22	6.10	nr	**83.46**
80 amp	79.23	89.48	0.22	6.10	nr	**95.58**
100 amp	102.59	115.86	0.25	6.93	nr	**122.79**
Heavy duty fuse switches; with HRC fuses BS 5419; short circuit rating 65kA, 500 volt; including retractable operating switches						
SP&N						
63 amp	264.97	299.26	1.30	36.02	nr	**335.28**
100 amp	387.32	437.44	1.95	54.02	nr	**491.46**
TP&N						
63 amp	333.82	377.01	1.83	50.69	nr	**427.70**
100 amp	469.08	529.78	2.48	68.70	nr	**598.48**
200 amp	723.59	817.22	3.13	86.71	nr	**903.93**
300 amp	1257.08	1419.74	4.45	123.29	nr	**1543.03**
400 amp	1380.04	1558.61	4.45	123.29	nr	**1681.90**
600 amp	2083.09	2352.63	5.72	158.47	nr	**2511.10**
800 amp	3254.02	3675.07	7.88	218.31	nr	**3893.38**

V: ELECTRICAL SUPPLY/POWER/LIGHTING

Item	Net Price £	Material £	Labour hours	Labour £	Unit	Total rate £
V20: LV DISTRIBUTION – cont						
BREAKERS/FUSES – cont						
Switch disconnectors to BSEN 60947–3; in sheet steel case; IP41 with door interlock fixed to backgrounds						
Double pole						
20 amp	51.93	58.65	1.02	28.25	nr	**86.90**
32 amp	62.59	70.69	1.02	28.25	nr	**98.94**
63 amp	228.41	257.97	1.21	33.52	nr	**291.49**
100 amp	209.58	236.70	1.86	51.53	nr	**288.23**
TP&N						
20 amp	65.19	73.62	1.29	35.73	nr	**109.35**
32 amp	75.85	85.66	1.83	50.69	nr	**136.35**
63 amp	257.42	290.73	2.48	68.70	nr	**359.43**
100 amp	259.54	293.13	2.48	68.70	nr	**361.83**
125 amp	270.71	305.73	2.48	68.70	nr	**374.43**
160 amp	622.87	703.46	2.48	68.70	nr	**772.16**
Enclosed switch disconnector to BSEN 60947–3; enclosure minimum IP55 rating; complete with earth connection bar; fixed to backgrounds.						
TP						
20 amp	51.93	58.65	1.02	28.25	nr	**86.90**
32 amp	62.59	70.69	1.02	28.25	nr	**98.94**
63 amp	228.41	257.97	1.21	33.52	nr	**291.49**
TP&N						
20 amp	65.19	73.62	1.29	35.73	nr	**109.35**
32 amp	75.85	85.66	1.83	50.69	nr	**136.35**
63 amp	257.42	290.73	2.48	68.70	nr	**359.43**
Busbar chambers; fixed to background including all supports, fixings, connections/jointing to equipment.						
Sheet steel case enclosing 4 pole 550 volt copper bars, detachable metal end plates						
600 mm long						
200 amp	473.08	534.30	2.62	72.58	nr	**606.88**
300 amp	607.91	686.57	3.03	83.95	nr	**770.52**
500 amp	1041.64	1176.43	4.48	124.11	nr	**1300.54**

V: ELECTRICAL SUPPLY/POWER/LIGHTING

Item	Net Price £	Material £	Labour hours	Labour £	Unit	Total rate £
900 mm long						
200 amp	681.42	769.59	3.04	84.22	nr	**853.81**
300 amp	803.09	907.00	3.59	99.45	nr	**1006.45**
500 amp	1187.91	1341.62	4.42	122.45	nr	**1464.07**
1350 mm long						
200 amp	930.76	1051.19	3.38	93.64	nr	**1144.83**
300 amp	1095.50	1237.25	3.94	109.16	nr	**1346.41**
500 amp	1753.58	1980.48	4.82	133.53	nr	**2114.01**
Contactor relays; pressed steel enclosure; fixed to backgrounds including supports, fixings, connections/jointing to equipment						
Relays						
6 amp, 415/240 volt, 4 pole N/O	54.79	61.88	0.52	14.41	nr	**76.29**
6 amp, 415/240 volt, 8 pole N/O	67.00	75.67	0.85	23.55	nr	**99.22**
Push button stations; heavy gauge pressed steel enclosure; polycarbonate cover; IP65; fixed to backgrounds including supports, fixings,connections/ joining to equipment						
Standard units						
One button (start or stop)	69.33	78.31	0.39	10.81	nr	**89.12**
Two button (start or stop)	73.60	83.12	0.47	13.03	nr	**96.15**
Three button (forward-reverse-stop)	104.30	117.80	0.57	15.79	nr	**133.59**
Weatherproof junction boxes; enclosures with rail mounted terminal blocks; side hung door to receive padlock; fixed to backgrounds, including all supports and fixings (Suitable for cable up to 2.5 mm²; including glandplates and gaskets.)						
Sheet steel with zinc spray finish enclosure						
Overall size 229 × 152; suitable to receive 3 × 20(A) glands per gland plate	70.22	79.30	1.43	39.62	nr	**118.92**
Overall size 306 × 306; suitable to receive 14 × 20(A) glands per gland plate	94.21	106.40	2.17	60.12	nr	**166.52**
Overall size 458 × 382; suitable to receive 18 × 20(A) glands per gland plate	137.17	154.92	3.51	97.24	nr	**252.16**
Overall size 762 x508; suitable to receive 26 × 20(A) glands per gland plate	145.08	163.85	4.85	134.36	nr	**298.21**
Overall size 914 × 610; suitable to receive 45 × 20(A) glands per gland plate	160.36	181.11	7.01	194.21	nr	**375.32**

V: ELECTRICAL SUPPLY/POWER/LIGHTING

Item	Net Price £	Material £	Labour hours	Labour £	Unit	Total rate £
V20: LV DISTRIBUTION – cont						
BREAKERS/FUSES – cont						
Weatherproof junction boxes; enclosures with rail mounted terminal blocks; screw fixed lid; fixed to backgrounds, including all supports and fixings (suitable for cable up to 2.5 mm²; including glandplates and gaskets)						
Glassfibre reinforced polycarbonate enclosure						
Overall size 190 × 190 × 130	95.91	108.32	1.43	39.62	nr	**147.94**
Overall size 190 × 190 × 180	140.42	158.59	1.53	42.39	nr	**200.98**
Overall size 280 × 190 × 130	158.60	179.12	2.17	60.12	nr	**239.24**
Overall size 280 × 190 × 180	177.68	200.67	2.37	65.67	nr	**266.34**
Overall size 380 × 190 × 130	198.24	223.89	3.33	92.25	nr	**316.14**
Overall size 380 × 190 × 180	212.91	240.45	3.30	91.42	nr	**331.87**
Overall size 380 × 280 × 130	227.60	257.05	4.66	129.10	nr	**386.15**
Overall size 380 × 280 × 180	245.22	276.95	5.36	148.49	nr	**425.44**
Overall size 560 × 280 × 130	295.15	333.34	7.01	194.21	nr	**527.55**
Overall size 560 × 380 × 180	303.97	343.30	7.67	212.50	nr	**555.80**

V: ELECTRICAL SUPPLY/POWER/LIGHTING

Item	Net Price £	Material £	Labour hours	Labour £	Unit	Total rate £
V21: GENERAL LIGHTING						
Y73 – LUMINAIRES (GENERAL)						
LUMINAIRES						
Fluorescent luminaires; surface fixed to backgrounds						
Batten type; surface mounted						
600 mm Single – 18 W	8.85	10.00	0.58	16.08	nr	**26.08**
600 mm Twin – 18 W	15.68	17.71	0.59	16.34	nr	**34.05**
1200 mm Single – 36 W	11.72	13.24	0.76	21.05	nr	**34.29**
1200 mm Twin – 36 W	13.26	14.97	0.84	23.28	nr	**38.25**
1500 mm Single – 58 W	22.44	25.35	0.77	21.34	nr	**46.69**
1500 mm Twin – 58 W	26.46	29.89	0.85	23.55	nr	**53.44**
1800 mm Single – 70 W	15.99	18.05	1.05	29.09	nr	**47.14**
1800 mm Twin – 70 W	29.25	33.04	1.06	29.37	nr	**62.41**
2400 mm Single – 100 W	21.89	24.72	1.25	34.63	nr	**59.35**
2400 mm Twin – 100 W	38.31	43.27	1.27	35.18	nr	**78.45**
Surface mounted, opal diffuser						
600 mm Twin – 18 W	24.33	27.48	0.62	17.18	nr	**44.66**
1200 mm Single – 36 W	20.78	23.47	0.79	21.89	nr	**45.36**
1200 mm Twin – 36 W	32.36	36.55	0.80	22.15	nr	**58.70**
1500 mm Single – 58 W	23.89	26.98	0.88	24.39	nr	**51.37**
1500 mm Twin – 58 W	37.84	42.74	0.90	24.94	nr	**67.68**
1800 mm Single – 70 W	29.80	33.66	1.09	30.20	nr	**63.86**
1800 mm Twin – 70 W	40.91	46.21	1.10	30.47	nr	**76.68**
2400 mm Single – 100 W	39.15	44.22	1.30	36.02	nr	**80.24**
2400 mm Twin – 100 W	56.53	63.84	1.31	36.29	nr	**100.13**
Surface mounted linear fluorescent; T8 lamp; high frequency control gear; low brightness; 65° cut-off; including wedge style louvre						
1200 mm, 1 × 36 W	56.05	63.30	1.09	30.20	nr	**93.50**
1200 mm 2 × 36 W	59.94	67.70	1.09	30.20	nr	**97.90**
Extra for emergency pack	50.56	57.10	0.25	6.93	nr	**64.03**
1500 mm, 1 × 58 W	65.33	73.79	0.90	24.94	nr	**98.73**
1500 mm 2 × 58 W	69.82	78.86	0.90	24.94	nr	**103.80**
Extra for emergency pack	50.46	56.99	0.25	6.93	nr	**63.92**
1800 mm, 1 × 70 W	98.37	111.10	0.90	24.94	nr	**136.04**
1800 mm, 2 × 70 W	112.93	127.54	0.90	24.94	nr	**152.48**
Extra for emergency pack	83.53	94.34	0.25	6.93	nr	**101.27**

V: ELECTRICAL SUPPLY/POWER/LIGHTING

Item	Net Price £	Material £	Labour hours	Labour £	Unit	Total rate £
V21: GENERAL LIGHTING – cont						
Fluorescent luminaires – cont						
Modular recessed linear fluorescent; high frequency control gear; low brightness; 65° cut off; including wedge style louvre; fitted to exposed T grid ceiling						
600 × 600 mm, 3 × 18 W T8	44.08	49.79	0.84	23.28	nr	**73.07**
600 × 600 mm, 4 × 18 W T8	45.55	51.45	0.87	24.10	nr	**75.55**
Extra for emergency pack	50.97	57.57	0.25	6.93	nr	**64.50**
300 × 1200 mm, 2 × 36 W T8	61.52	69.48	0.87	24.10	nr	**93.58**
Extra for emergency pack	64.48	72.83	0.25	6.93	nr	**79.76**
600 × 1200 mm, 3 × 36 W T8	67.12	75.81	0.89	24.65	nr	**100.46**
600 × 1200 mm, 4 × 36 W T8	77.21	87.20	0.91	25.20	nr	**112.40**
Extra for emergency pack	59.72	67.44	0.25	6.93	nr	**74.37**
600 × 600, 3 × 14 W T5	62.62	70.73	0.84	23.28	nr	**94.01**
600 × 600, 4 × 14 W T5	65.16	73.59	0.87	24.10	nr	**97.69**
Extra for emergency pack	74.96	84.66	0.25	6.93	nr	**91.59**
Modular recessed; T8 lamp; high frequency control gear; cross-blade louvre; fitted to exposed T grid ceiling						
600 × 600 mm, 3 × 18 W	48.13	54.36	0.84	23.28	nr	**77.64**
600 × 600 mm, 4 × 18 W	65.13	73.56	0.87	24.10	nr	**97.66**
Extra for emergency pack	57.94	65.43	0.25	6.93	nr	**72.36**
Modular recessed compact fluorescent; TCL lamp; high frequency control gear; low brightness; 65° cut-off; including wedge style louvre; fitted to exposed T grid ceiling						
300 × 300 mm, 2 × 18 W	97.48	110.09	0.75	20.78	nr	**130.87**
Extra for emergency pack	90.01	101.65	0.25	6.93	nr	**108.58**
500 × 500 mm, 2 × 36 W	91.16	102.96	0.82	22.72	nr	**125.68**
600 × 600 mm, 2 × 36 W	95.38	107.72	0.82	22.72	nr	**130.44**
600 × 600 mm, 2 × 40 W	55.01	62.13	0.82	22.72	nr	**84.85**
Extra for emergency pack	53.41	60.32	0.25	6.93	nr	**67.25**
Ceiling recessed asymetric compact fluorescent downlighter; high frequency control gear; TCD lamp in 200 mm dia. luminaire; for wall-washing application						
1 × 18 W	154.38	174.36	0.75	20.78	nr	**195.14**
1 × 26 W	154.38	174.36	0.75	20.78	nr	**195.14**
2 × 18 W	171.98	194.23	0.75	20.78	nr	**215.01**
2 × 26 W	171.98	194.23	0.75	20.78	nr	**215.01**
Ceiling recessed asymetric compact fluorescesnt downlights; high frequency control gear; linear 200 mm × 600 mm luminaire with low glare louvre; for wall washing applications						
1 × 55 W TCL	62.36	70.43	0.75	20.78	nr	**91.21**

V: ELECTRICAL SUPPLY/POWER/LIGHTING

Item	Net Price £	Material £	Labour hours	Labour £	Unit	Total rate £
Wall mounted compact fluorescent uplighter; high frequency control gear; TCL lamp in 300 mm × 600 mm luminaire						
2 × 36 W	242.01	273.33	0.84	23.28	nr	**296.61**
2 × 40 W	260.27	293.94	0.84	23.28	nr	**317.22**
2 × 55 W	364.66	411.85	0.84	23.28	nr	**435.13**
Suspended linear fluorescent; T5 lamp; high frequency control gear; low brightness; 65° cut-off; 30% uplight, 70% downlight; including wedge style louvre						
1 × 49 W	165.95	187.42	0.75	20.78	nr	**208.20**
Extra for emergency pack	91.67	103.53	0.25	6.93	nr	**110.46**
Semi-recessed 'architectural' linear fluorescent; T5 lamp; high frequency control gear; low brightness, delivers direct, ceiling and graduated wall washing illumination						
600 × 600 mm, 2 × 24 W	135.96	153.55	0.87	24.10	nr	**177.65**
600 × 600 mm, 4 × 14 W	146.31	165.24	0.87	24.10	nr	**189.34**
500 × 500 mm, 2 × 24 W	133.69	150.99	0.87	24.10	nr	**175.09**
Extra for emergency pack	65.10	73.52	0.25	6.93	nr	**80.45**
Downlighter, recessed; low voltage; mirror reflector with white/chrome bezel; dimmable transformer; for dichroic lamps						
85 mm dia. × 20/50 W	15.74	17.78	0.66	18.29	nr	**36.07**
118 mm dia. × 50 W	20.81	23.50	0.66	18.29	nr	**41.79**
165 mm dia. × 100 W	95.05	107.34	0.66	18.29	nr	**125.63**
High/Low bay luminaires						
Compact discharge; aluminium reflector						
150 W	58.10	65.62	1.50	41.56	nr	**107.18**
250 W	60.16	67.94	1.50	41.56	nr	**109.50**
400 W	61.67	69.65	1.50	41.56	nr	**111.21**
Sealed discharge; aluminium reflector						
150 W	179.52	202.75	1.50	41.56	nr	**244.31**
250 W	193.76	218.83	1.50	41.56	nr	**260.39**
400 W	253.34	286.12	1.50	41.56	nr	**327.68**
Corrosion resistant GRP body; gasket sealed; acrylic diffuser						
600 mm Single – 18 W	30.03	33.92	0.49	13.58	nr	**47.50**
600 mm Twin – 18 W	39.10	44.16	0.49	13.58	nr	**57.74**
1200 mm Single – 36 W	33.81	38.19	0.64	17.73	nr	**55.92**
1200 mm Twin – 36 W	43.34	48.95	0.64	17.73	nr	**66.68**
1500 mm Single – 58 W	37.68	42.55	0.72	19.94	nr	**62.49**
1500 mm Twin – 58 W	46.84	52.90	0.72	19.94	nr	**72.84**
1800 mm Single – 70 W	55.76	62.97	0.94	26.04	nr	**89.01**
1800 mm Twin – 70 W	69.21	78.16	0.94	26.04	nr	**104.20**

V: ELECTRICAL SUPPLY/POWER/LIGHTING

Item	Net Price £	Material £	Labour hours	Labour £	Unit	Total rate £
V21: GENERAL LIGHTING – cont						
High/Low bay luminaires – cont						
Flameproof to IIA/IIB,I.P. 64; Aluminium Body; BS 229 and 899						
600 mm Single – 18 W	336.73	380.30	1.04	28.82	nr	**409.12**
600 mm Twin – 18 W	419.47	473.75	1.04	28.82	nr	**502.57**
1200 mm Single – 36 W	368.84	416.57	1.31	36.29	nr	**452.86**
1200 mm Twin – 36 W	456.02	515.03	1.18	32.69	nr	**547.72**
1500 mm Single – 58 W	394.88	445.97	1.64	45.43	nr	**491.40**
1500 mm Twin – 58 W	476.82	538.52	1.64	45.43	nr	**583.95**
1800 mm Single – 70 W	432.89	488.91	1.97	54.57	nr	**543.48**
1800 mm Twin – 70 W	499.40	564.02	1.97	54.57	nr	**618.59**
External lighting						
Ground mounted 50 W	370.87	418.86	2.25	62.33	nr	**481.19**
Ceiling mounted 50 W	166.62	188.18	2.25	62.33	nr	**250.51**
Bulkhead; aluminium body and polycarbonate bowl; vandal-resistant; IP65						
60 W	32.64	36.86	0.75	20.78	nr	**57.64**
Extra for						
Emergency version	87.73	99.08	0.25	6.93	nr	**106.01**
2D 2 pin 16 W	27.57	31.14	0.66	18.29	nr	**49.43**
2D 2 pin 28 W	55.05	62.17	0.66	18.29	nr	**80.46**
Extra for						
Emergency version	64.87	73.27	0.25	6.93	nr	**80.20**
Photocell	19.35	21.86	0.75	20.78	nr	**42.64**
1500 mm high circular bollard; polycarbonate visor; vandal-resistant; IP54						
50 W	204.34	230.79	1.75	48.48	nr	**279.27**
70 W	207.61	234.48	1.75	48.48	nr	**282.96**
80 W	248.46	280.61	1.75	48.48	nr	**329.09**
Floodlight; enclosed high performance dischargelight; integeral control gear; reflector; toughened glass; IP65						
70 W	90.52	102.23	1.25	34.63	nr	**136.86**
100 W	94.73	106.99	1.25	34.63	nr	**141.62**
150 W	100.44	113.43	1.25	34.63	nr	**148.06**
250 W	175.82	198.57	1.25	34.63	nr	**233.20**
400 W	182.13	205.69	1.25	34.63	nr	**240.32**
Extra for						
Photocell	19.33	21.83	0.75	20.78	nr	**42.61**

V: ELECTRICAL SUPPLY/POWER/LIGHTING

Item	Net Price £	Material £	Labour hours	Labour £	Unit	**Total rate £**
Lighting track						
Single circuit; 25 A 2 P&E steel trunking; low voltage with copper conductors; including couplers and supports; fixed to backgrounds						
Straight track	12.98	14.66	0.50	13.86	m	**28.52**
Live end feed unit complete with end stop	22.36	25.26	0.33	9.14	nr	**34.40**
Flexible couplers 0.5 m	47.50	53.64	0.33	9.14	nr	**62.78**
Tap off complete with 0.8 m of cable	7.47	8.44	0.25	6.93	nr	**15.37**
Three circuit; 25 A 2 P&E steel trunking; low voltage with copper conductors; including couplers and supports incorporating integral twisted pair comms bus bracket; fixed to backgrounds						
Straight track	21.15	23.89	0.75	20.78	m	**44.67**
Live end feed unit complete with end stop	38.09	43.02	0.50	13.86	nr	**56.88**
Flexible couplers 0.5 m	54.26	61.28	0.45	12.46	nr	**73.74**
Tap off complete with 0.8 m of cable	9.80	11.07	0.30	8.31	nr	**19.38**
Y74 – LIGHTING ACCESSORIES						
SWITCHES						
6 amp metal clad surface mounted switch, gridswitch; one way						
1 gang	7.32	8.27	0.43	11.91	nr	**20.18**
2 gang	10.05	11.35	0.55	15.24	nr	**26.59**
3 gang	15.93	17.99	0.77	21.34	nr	**39.33**
4 gang	18.86	21.30	0.88	24.39	nr	**45.69**
6 gang	32.40	36.59	1.10	30.47	nr	**67.06**
8 gang	37.94	42.85	1.28	35.47	nr	**78.32**
10 gang	57.24	64.65	1.67	46.27	nr	**110.92**
Extra for						
10 amp – Two way switch	2.61	2.95	0.03	0.84	nr	**3.79**
20 amp – Two way switch	3.49	3.94	0.04	1.11	nr	**5.05**
20 amp – Intermediate	6.55	7.39	0.08	2.21	nr	**9.60**
20 amp – One way SP switch	2.69	3.04	0.08	2.21	nr	**5.25**
Steel blank plate; 1 gang	1.81	2.04	0.07	1.94	nr	**3.98**
Steel blank plate; 2 gang	3.06	3.46	0.08	2.21	nr	**5.67**

V: ELECTRICAL SUPPLY/POWER/LIGHTING

Item	Net Price £	Material £	Labour hours	Labour £	Unit	Total rate £
V21: GENERAL LIGHTING – cont						
SWITCHES – cont						
6 amp modular type switch; galvanized steel box, bronze or satin chrome coverplate; metalclad switches; flush mounting; one way						
1 gang	18.00	20.33	0.43	11.91	nr	**32.24**
2 gang	24.67	27.87	0.55	15.24	nr	**43.11**
3 gang	35.65	40.26	0.77	21.34	nr	**61.60**
4 gang	42.43	47.92	0.88	24.39	nr	**72.31**
6 gang	71.54	80.79	1.18	32.69	nr	**113.48**
8 gang	85.35	96.39	1.63	45.16	nr	**141.55**
9 gang	107.05	120.90	1.83	50.69	nr	**171.59**
12 gang	127.50	143.99	2.29	63.43	nr	**207.42**
6 amp modular type swtich; galvanized steel box; bronze or satin chrome coverplate; flush mounting; two way						
1 gang	18.72	21.14	0.43	11.91	nr	**33.05**
2 gang	25.66	28.98	0.55	15.24	nr	**44.22**
3 gang	37.07	41.87	0.77	21.34	nr	**63.21**
4 gang	44.13	49.84	0.88	24.39	nr	**74.23**
6 gang	74.40	84.03	1.18	32.69	nr	**116.72**
8 gang	88.76	100.25	1.63	45.16	nr	**145.41**
9 gang	111.34	125.75	1.83	50.69	nr	**176.44**
12 gang	132.60	149.76	2.22	61.51	nr	**211.27**
Plate switches; 10 amp flush mounted, white plastic fronted; 16 mm metal box; fitted brass earth terminal						
1 gang 1 way, single pole	6.88	7.77	0.28	7.76	nr	**15.53**
1 gang 2 way, single pole	7.70	8.70	0.33	9.14	nr	**17.84**
2 gang 2 way, single pole	9.62	10.86	0.44	12.19	nr	**23.05**
3 gang 2 way, single pole	11.28	12.74	0.56	15.51	nr	**28.25**
1 gang Intermediate	15.23	17.20	0.43	11.91	nr	**29.11**
1 gang 1 Way, double pole	13.16	14.86	0.33	9.14	nr	**24.00**
1 gang single pole with bell symbol	10.53	11.89	0.23	6.36	nr	**18.25**
1 gang single pole marked ‘PRESS’	8.42	9.51	0.23	6.36	nr	**15.87**
Time delay switch, suppressed	46.62	52.63	0.49	13.58	nr	**66.21**
Plate switches; 6 amp flush mounted white plastic fronted; 25 mm metal box; fitted brass earth terminal						
4 gang 2 way, single pole	15.56	17.57	0.42	11.63	nr	**29.20**
6 gang 2 way, single pole	24.88	28.10	0.47	13.03	nr	**41.13**

V: ELECTRICAL SUPPLY/POWER/LIGHTING

Item	Net Price £	Material £	Labour hours	Labour £	Unit	**Total rate £**
Architrave plate switches; 6 amp flush mounted, white plastic fronted; 27 mm metal box; brass earth terminal						
1 gang 2 way, single pole	2.82	3.18	0.30	8.31	nr	**11.49**
2 gang 2 way, single pole	5.58	6.30	0.36	9.98	nr	**16.28**
Ceiling switches, white moulded plastic, pull cord; standard unit						
6 amp, 1 way, single pole	4.04	4.56	0.32	8.86	nr	**13.42**
6 amp, 2 way, single pole	4.86	5.49	0.34	9.41	nr	**14.90**
16 amp, 1 way, double pole	7.27	8.21	0.37	10.25	nr	**18.46**
45 amp, 1 way, double pole with neon indicator	11.29	12.75	0.47	13.03	nr	**25.78**
10 amp splash-proof moulded switch with plain, threaded or PVC entry						
1 gang, 2 way single pole	17.34	19.58	0.34	9.41	nr	**28.99**
2 gang, 1 way single pole	19.59	22.12	0.36	9.98	nr	**32.10**
2 gang, 2 way single pole	24.82	28.03	0.40	11.08	nr	**39.11**
6 amp watertight switch; metalclad; BS 3676; ingress protected to IP65 surface mounted						
1 gang, 2 way; terminal entry	16.73	18.89	0.41	11.36	nr	**30.25**
1 gang, 2 way; through entry	16.73	18.89	0.42	11.63	nr	**30.52**
2 gang, 2 way; terminal entry	49.33	55.71	0.54	14.96	nr	**70.67**
2 gang, 2 way; through entry	49.33	55.71	0.53	14.68	nr	**70.39**
2 way replacement switch	13.29	15.01	0.10	2.76	nr	**17.77**
15 amp watertight switch; metalclad; BS 3676; ingress protected to IP65; surface mounted						
1 gang 2 way, terminal entry	22.40	25.30	0.42	11.63	nr	**36.93**
1 gang 2 way, through entry	22.40	25.30	0.43	11.91	nr	**37.21**
2 gang 2 way, terminal entry	52.59	59.39	0.55	15.24	nr	**74.63**
2 gang 2 way, through entry	52.59	59.39	0.54	14.96	nr	**74.35**
Intermediate interior only	13.29	15.01	0.11	3.05	nr	**18.06**
2 way interior only	13.29	15.01	0.11	3.05	nr	**18.06**
Double pole interior only	13.29	15.01	0.11	3.05	nr	**18.06**
Electrical accessories; fixed to backgrounds (Including fixings)						
Dimmer switches; rotary action; for individual lights; moulded plastic case; metal backbox; flush mounted						
1 gang, 1 way; 250 W	13.95	15.76	0.28	7.76	nr	**23.52**
1 gang, 1 way; 400 W	18.36	20.74	0.28	7.76	nr	**28.50**

V: ELECTRICAL SUPPLY/POWER/LIGHTING

Item	Net Price £	Material £	Labour hours	Labour £	Unit	Total rate £
V21: GENERAL LIGHTING – cont						
SWITCHES – cont						
Electrical accessories – cont						
Dimmer switches; push on/off action; for individual lights; moulded plastic case; metal backbox; flush mounted						
1 gang, 2 way; 250 W	21.31	24.07	0.34	9.41	nr	**33.48**
3 gang, 2 way; 250 W	31.40	35.47	0.48	13.29	nr	**48.76**
4 gang, 2 way; 250 W	41.32	46.67	0.57	15.79	nr	**62.46**
Dimmer switches; rotary action; metal cald; metal backbox; BS 5518 and BS 800; flush mounted						
1 gang, 1 way; 400 W	38.70	43.71	0.33	9.14	nr	**52.85**
Ceiling roses						
Ceiling rose: white moulded plastic; flush fixed to conduit box						
Plug in type; ceiling socket with 2 terminals, loop-in and ceiling plug with 3 terminals and cover	7.43	8.39	0.34	9.41	nr	**17.80**
BC lampholder; white moulded plastic; heat resistent PVC insulated and sheathed cable; flush fixed						
2 core; 0.75 mm²	2.35	2.65	0.33	9.14	nr	**11.79**
Batten holder: white moulded plastic; 3 terminals; BS 5042; fixed to conduit						
Straight pattern; 2 terminals with loop-in and earth	3.44	3.89	0.29	8.03	nr	**11.92**
Angled pattern; looped in terminal	3.44	3.89	0.29	8.03	nr	**11.92**
LIGHTING CONTROLS						
Lighting control system; including software. Typical component parts indicated. System requirements dependent on final lighting design (commissioning not included)						
Y61 – CABLES						
E-Bus Cable; Twin twisted bus; LSF sheathed; aluminium conductors	1.00	1.13	0.08	2.21	m	**3.34**
T-Bus Cable; ELV 4 core 7/0.2; LSF sheathed; alumimium screened; copper conductor	1.28	1.45	0.15	4.15	m	**5.60**

V: ELECTRICAL SUPPLY/POWER/LIGHTING

Item	Net Price £	Material £	Labour hours	Labour £	Unit	**Total rate £**
Standard head end equipment						
SPC.1000 PC & Graphics head end including Gateway includes SPM300 software	4900.80	5534.94	12.00	332.45	nr	**5867.39**
Additional head end equipment (for emergency monitoring & BMS interfacing)						
SPC.2000 Integration Control Processor (ATS Processor)	4084.00	4612.45	4.00	110.81	nr	**4723.26**
SPECS DALI Emergency monitoring software	2042.00	2306.22	10.00	277.04	nr	**2583.26**
SPECS SPN.410 IP Data server software suite – BacNet Interface	2042.00	2306.22	4.00	110.81	nr	**2417.03**
Astronomical Timeclock	510.50	576.55	2.00	55.41	nr	**631.96**
Rack mounted head end equipment – replaces standard head end equipment						
SPU.1000–1U SPC.1000 with SPG.3 Gateway 1U 19' rack case	4900.80	5534.94	1.00	27.70	nr	**5562.64**
Area controllers						
SPX.310+ Area controller – standard	816.80	922.49	4.00	110.81	nr	**1033.30**
SPX.320+ Area controller with RF interface – mid range	918.90	1037.80	4.00	110.81	nr	**1148.61**
SPX.330+ Area controller with RF interface and IP enabled – advanced	1021.00	1153.11	4.00	110.81	nr	**1263.92**
SPV.92 Lighting control module; plug in; 9 output, 9 channel – switching						
Base and lid assembly	141.92	160.28	2.05	56.79	nr	**217.07**
SPV.92 Lighting control module; plug in; 9 output, 9 channel – dimming (DALI, DSI, 1–10V)						
Base and lid assembly	173.57	196.03	2.05	56.79	nr	**252.82**
SPH.12+ Lighting control module; hard wired; 4 circuit switching						
Base and lid assembly	153.15	172.97	1.85	51.26	nr	**224.23**
SPH.27r Lighting control module; hard wired; 2 circuit 40 ballast drive DALI (with relays)						
Base and lid assembly	219.52	247.92	1.85	51.26	nr	**299.18**
SPH.27 Lighting control module; hard wired; 2 circuit 40 ballast drive DALI (without relays)						
Base and lid assembly	209.31	236.40	1.85	51.26	nr	**287.66**

V: ELECTRICAL SUPPLY/POWER/LIGHTING

Item	Net Price £	Material £	Labour hours	Labour £	Unit	Total rate £
V21: GENERAL LIGHTING – cont						
Y61 – CABLES – cont						
SPV.56 Compact lighting control module; 3 output 18 ballast drive; dimmable (DALI)						
Base and lid assembly	193.99	219.10	1.85	51.26	nr	**270.36**
SPB.82 Blind control module; 8 outputs						
Base and lid assembly	153.15	172.97	1.85	51.26	nr	**224.23**
SPF.6-T AV interface; RS232 interface	132.73	149.91	1.85	51.26	nr	**201.17**
SPO.20 DALI to DMX converter	204.20	230.62	1.80	49.87	nr	**280.49**
Dual bus presence detectors (DALI & E-Bus)						
SPU.7-S Dual-bus presence detector and infra-red sensor	51.05	57.66	0.60	16.63	nr	**74.29**
SPU.6-S Dual-bus universal sensor	56.16	63.42	0.60	16.63	nr	**80.05**
Scene switch plate; anodized aluminium finish						
SPK.9–4+4 Eight button intelligent switch plate	102.10	115.31	2.00	55.41	nr	**170.72**
SPK.9–4-S Four button intelligent switch plate	61.26	69.19	1.60	44.33	nr	**113.52**
SPK.9–2-S Two button intelligent switch plate	51.05	57.66	1.20	33.25	nr	**90.91**
SPF.4d Compact DALI interface for 4 gang retractive switch plate	25.53	28.84	1.20	33.25	nr	**62.09**
External Photocell						
SPL.20 External Photocell 20,000 Lux range	122.52	138.37	1.50	41.56	nr	**179.93**
Sub-addressing for DALI strings						
Sub-addressing charge per DALI point	3.57	4.03	3.00	83.11	nr	**87.14**
Commissioning						
Not included – worked out based on complexity of project						

V: ELECTRICAL SUPPLY/POWER/LIGHTING

Item	Net Price £	Material £	Labour hours	Labour £	Unit	Total rate £
V22: GENERAL LV POWER						
Y74 – ACCESSORIES						
OUTLETS						
Socket outlet: unswitched; 13 amp metal clad; BS 1363; galvanized steel box and coverplate with white plastic inserts; fixed surface mounted						
1 gang	5.95	6.72	0.41	11.36	nr	**18.08**
2 gang	10.96	12.38	0.41	11.36	nr	**23.74**
Socket outlet: switched; 13 amp metal clad; BS 1363; galvanized steel box and coverplate with white plastic inserts; fixed surface mounted						
1 gang	10.92	12.33	0.43	11.91	nr	**24.24**
2 gang	20.21	22.83	0.45	12.46	nr	**35.29**
Socket outlet: switched with neon indicator; 13 amp metal clad; BS 1363; galvanized steel box and coverplate withwhite plastic inserts; fixed surface mounted						
1 gang	12.94	14.62	0.43	11.91	nr	**26.53**
2 gang	23.53	26.57	0.45	12.46	nr	**39.03**
Socket outlet: unswitched; 13 amp; BS 1363; white moulded plastic box and coverplate; fixed surface mounted						
1 gang	4.47	5.05	0.41	11.36	nr	**16.41**
2 gang	8.83	9.98	0.41	11.36	nr	**21.34**
Socket outlet; switched; 13 amp; BS 1363; white moulded plastic box and coverplate; fixed surface mounted						
1 gang	5.40	6.10	0.43	11.91	nr	**18.01**
2 gang	8.67	9.79	0.45	12.46	nr	**22.25**
Socket outlet: switched with neon indicator; 13 amp; BS 1363; white moulded plastic box and coverplate; fixed surface mounted						
1 gang	11.05	12.48	0.43	11.91	nr	**24.39**
2 gang	14.89	16.82	0.45	12.46	nr	**29.28**
Socket outlet: switched; 13 amp; BS 1363; galvanized steel box, white moulded coverplate; flush fitted						
1 gang	5.40	6.10	0.43	11.91	nr	**18.01**
2 gang	10.45	11.80	0.45	12.46	nr	**24.26**

V: ELECTRICAL SUPPLY/POWER/LIGHTING

Item	Net Price £	Material £	Labour hours	Labour £	Unit	Total rate £
V22: GENERAL LV POWER – cont						
OUTLETS – cont						
Socket outlet: switched with neon indicator; 13 amp; BS 1363; galvanized steel box, white moulded coverplate; flush fixed						
1 gang	11.05	24.95	0.43	11.91	nr	**36.86**
2 gang	19.10	43.15	0.45	12.46	nr	**55.61**
Socket outlet: switched; 13 amp; BS 1363; galvanized steel box, satin chrome coverplate; BS 4662; flush fixed						
1 gang	18.02	20.35	0.43	11.91	nr	**32.26**
2 gang	25.57	28.88	0.45	12.46	nr	**41.34**
Socket outlet: switched with neon indicator; 13 amp; BS 1363; steel backbox, satin chrome coverplate; BS 4662; flush fixed						
1 gang	14.17	16.00	0.43	11.91	nr	**27.91**
2 gang	25.57	28.88	0.45	12.46	nr	**41.34**
RCD protected socket outlets, 13 amp, to BS 1363; galvanized steel box, white moulded cover plate; flush fitted						
2 gang, 10 mA tripping (active control)	65.87	74.39	0.45	12.46	nr	**86.85**
2 gang, 30 mA tripping (active control)	58.68	66.27	0.45	12.46	nr	**78.73**
2 gang, 30 mA tripping (passive control)	58.68	66.27	0.45	12.46	nr	**78.73**
Filtered socket outlets, 13 amp, to BS 1363, with separate 'clean earth' terminal; galvanized steel box, white moulded cover plate; flush fitted						
2 gang (spike protected)	49.60	56.02	0.50	13.86	nr	**69.88**
2 gang (spike and RFI protected)	62.78	70.90	0.55	15.24	nr	**86.14**
Replacement filter cassette	17.39	19.65	0.15	4.15	nr	**23.80**
Non-standard socket outlets, 13 amp, to BS 1363, with separate 'clean earth' terminal; for plugs with T-shaped earth pin; galvanized steel box, white moulded cover plate; flush fitted						
1 gang	9.21	10.40	0.43	11.91	nr	**22.31**
2 gang	16.56	18.71	0.43	11.91	nr	**30.62**
2 gang coloured RED	23.01	25.99	0.43	11.91	nr	**37.90**

V: ELECTRICAL SUPPLY/POWER/LIGHTING

Item	Net Price £	Material £	Labour hours	Labour £	Unit	**Total rate £**
Weatherproof socket outlet: 40 amp; switched; single gang; RCD protected; water and dust protected to I.P.66; surface mounted						
40 A 30 mA tripping current protecting 1 socket	74.95	84.65	0.52	14.41	nr	**99.06**
40 A 30 mA tripping current protecting 2 sockets	79.94	90.28	0.64	17.73	nr	**108.01**
Plug for weatherproof socket outlet: protected to I.P.66						
13 amp plug	3.28	3.70	0.21	5.81	nr	**9.51**
Floor service outlet box; comprising flat lid with flanged carpet trim; twin 13 A switched socket outlets; punched plate for mounting 2 telephone outlets; one blank plate; triple compartment						
3 compartment	33.74	38.11	0.88	24.39	nr	**62.50**
Floor service outlet box; comprising flat lid with flanged carpet trim; 2 twin 13 A switched socket outlets; punched plate for mounting 1 telephone outlet; one blank plate; triple compartment						
3 compartment	34.99	39.51	0.88	24.38	nr	**63.89**
Floor service outlet box; comprising flat lid with flanged carpet trim; twin 13 A switched socket outlets; punched plate for mounting 2 telephone outlets; two blank plates; four compartment						
4 compartment	46.01	51.96	0.88	24.39	nr	**76.35**
Floor service outlet box; comprising flat lid with flanged carpet trim; single 13 A unswitched socket outlet; single compartment; circular						
1 compartment	49.08	55.43	0.79	21.89	nr	**77.32**
Floor service grommet, comprising flat lid with flanged carpet trim; circular						
Floor grommet	22.20	25.07	0.49	13.58	nr	**38.65**

V: ELECTRICAL SUPPLY/POWER/LIGHTING

Item	Net Price £	Material £	Labour hours	Labour £	Unit	**Total rate £**
V22: GENERAL LV POWER – cont						
POWER POSTS/POLES/PILLARS						
Power post						
Power post; aluminium painted body; PVC-U cover; 5 nr outlets	265.00	299.29	4.00	110.81	nr	**410.10**
Power pole						
Power pole; 3.6 m high; aluminium painted body; PVC-U cover; 6 nr outlets	339.80	383.76	4.00	110.81	nr	**494.57**
Extra for						
Power pole extension bar; 900 mm long	37.14	41.94	1.50	41.56	nr	**83.50**
Vertical multi compartment pillar; PVC-U; BS 4678 Part4 EN60529; excludes accessories						
Single						
630 mm long	118.36	133.67	2.00	55.41	nr	**189.08**
3000 mm long	341.08	385.21	2.00	55.41	nr	**440.62**
Double						
630 mm long	118.36	133.67	3.00	83.11	nr	**216.78**
3000 mm long	362.01	408.86	3.00	83.11	nr	**491.97**
CONNECTION UNITS						
Connection units: moulded pattern; BS 5733; moulded plastic box; white coverplate; knockout for flex outlet; surface mounted – standard fused						
DP Switched	6.89	7.78	0.49	13.58	nr	**21.36**
Unswitched	6.32	7.14	0.49	13.58	nr	**20.72**
DP Switched with neon indicator	8.74	9.87	0.49	13.58	nr	**23.45**
Connection units: moulded pattern; BS 5733; galvanized steel box; white coverplate; knockout for flex outlet; surface mounted						
DP Switched	8.68	17.57	0.49	13.58	nr	**31.15**
DP Unswitched	8.12	9.17	0.49	13.58	nr	**22.75**
DP Switched with neon indicator	10.54	11.90	0.49	13.58	nr	**25.48**
Connection units: galvanized pressed steel pattern; galvanized steel box; satin chrome or satin brass finish; white moulded plastic inserts; flush mounted – standard fused						
DP Switched	12.70	14.34	0.49	13.58	nr	**27.92**
Unswitched	11.95	13.49	0.49	13.58	nr	**27.07**
DP Switched with neon indicator	17.14	19.36	0.49	13.58	nr	**32.94**

V: ELECTRICAL SUPPLY/POWER/LIGHTING

Item	Net Price £	Material £	Labour hours	Labour £	Unit	**Total rate £**
Connection units: galvanized steel box; satin chrome or satin brass finish; white moulded plastic inserts; flex outlet; flush mounted – standard fused						
Switched	12.19	13.77	0.49	13.58	nr	**27.35**
Unswitched	11.57	13.07	0.49	13.58	nr	**26.65**
Switched with neon indicator	15.68	17.71	0.49	13.58	nr	**31.29**
SHAVER SOCKETS						
Shaver unit: self setting overload device; 200/250 voltage supply; white moulded plastic faceplate; unswitched						
Surface type with moulded plastic box	22.91	25.88	0.55	15.24	nr	**41.12**
Flush type with galvanized steel box	23.97	27.06	0.57	15.79	nr	**42.85**
Shaver unit: dual voltage supply unit; white moulded plastic faceplate; unswitched						
Surface type with moulded plastic box	27.62	31.19	0.62	17.18	nr	**48.37**
Flush type with galvanized steel box	28.72	32.44	0.64	17.73	nr	**50.17**
COOKER CONTROL UNITS						
Cooker control unit: BS 4177; 45 amp D.P. main switch; 13 amp switched socket outlet; metal coverplate; plastic inserts; neon indicators						
Surface mounted with mounting box	32.44	36.64	0.61	16.89	nr	**53.53**
Flush mounted with galvanized steel box	31.00	35.01	0.61	16.89	nr	**51.90**
Cooker control unit: BS 4177; 45 amp D.P. main switch; 13 amp switched socket outlet; moulded plastic box and coverplate; surface mounted						
Standard	22.50	25.41	0.61	16.89	nr	**42.30**
With neon indicators	26.41	29.82	0.61	16.89	nr	**46.71**
CONTROL COMPONENTS						
Connector unit: moulded white plastic cover and block; galvanized steel back box; to immersion heaters						
3 Kw up to 915 mm long; fitted to thermostat	24.12	27.24	0.75	20.78	nr	**48.02**

V: ELECTRICAL SUPPLY/POWER/LIGHTING

Item	Net Price £	Material £	Labour hours	Labour £	Unit	Total rate £
V22: GENERAL LV POWER – cont						
CONTROL COMPONENTS – cont						
Water heater switch: 20 amp; switched with neon indicator						
DP Switched with neon indicator	11.74	28.09	0.45	12.46	nr	**40.55**
SWITCH DISCONNECTORS						
Switch disconnectors; moulded plastic enclosure; fixed to backgrounds						
3 pole; IP54; Grey						
16 amp	21.81	24.63	0.80	22.15	nr	**46.78**
25 amp	25.81	29.15	0.80	22.15	nr	**51.30**
40 amp	42.03	47.47	0.80	22.15	nr	**69.62**
63 amp	65.43	73.90	1.00	27.70	nr	**101.60**
80 amp	113.44	128.12	1.25	34.63	nr	**162.75**
6 pole; IP54; Grey						
25 amp	36.27	40.96	1.00	27.70	nr	**68.66**
63 amp	61.26	69.19	1.25	34.63	nr	**103.82**
80 amp	116.10	131.12	1.80	49.87	nr	**180.99**
3 pole; IP54; Yellow						
16 amp	23.97	27.07	0.80	22.15	nr	**49.22**
25 amp	28.34	32.01	0.80	22.15	nr	**54.16**
40 amp	45.97	51.92	0.80	22.15	nr	**74.07**
63 amp	71.51	80.76	1.00	27.70	nr	**108.46**
6 pole; IP54; Yellow						
25 amp	36.27	40.96	1.00	27.70	nr	**68.66**
INDUSTRIAL SOCKETS/PLUGS						
Plugs; splash-proof; 100–130 volts, 50–60 Hz; IP 44 (Yellow)						
2 pole and earth						
16 amp	3.81	4.30	0.55	15.24	nr	**19.54**
32 amp	10.29	11.62	0.60	16.63	nr	**28.25**
3 pole and earth						
16 amp	11.27	12.73	0.65	18.00	nr	**30.73**
32 amp	13.95	15.76	0.72	19.94	nr	**35.70**

V: ELECTRICAL SUPPLY/POWER/LIGHTING

Item	Net Price £	Material £	Labour hours	Labour £	Unit	**Total rate £**
3 pole; neutral and earth						
16 amp	11.34	12.81	0.72	19.94	nr	**32.75**
32 amp	16.51	18.65	0.78	21.60	nr	**40.25**
Connectors; splash-proof; 100–130 volts, 50–60 Hz; IP 44 (Yellow)						
2 pole and earth						
16 amp	5.20	5.88	0.42	11.63	nr	**17.51**
32 amp	9.00	10.17	0.50	13.86	nr	**24.03**
3 pole and earth						
16 amp	9.77	11.04	0.48	13.29	nr	**24.33**
32 amp	14.56	16.44	0.58	16.08	nr	**32.52**
3 pole; neutral and earth						
16 amp	13.67	15.44	0.52	14.41	nr	**29.85**
32 amp	19.18	21.66	0.73	20.23	nr	**41.89**
Angled sockets; surface mounted; splash-proof; 100–130 volts, 50–60 Hz; IP 44 (Yellow)						
2 pole and earth						
16 amp	4.39	4.96	0.55	15.24	nr	**20.20**
32 amp	10.98	12.40	0.60	16.63	nr	**29.03**
3 pole and earth						
16 amp	10.47	11.82	0.65	18.00	nr	**29.82**
32 amp	19.66	22.20	0.72	19.94	nr	**42.14**
3 pole; neutral and earth						
16 amp	14.52	16.39	0.72	19.94	nr	**36.33**
32 amp	19.02	21.47	0.78	21.60	nr	**43.07**
Plugs; watertight; 100–130 volts, 50–60 Hz; IP67 (Yellow)						
2 pole and earth						
16 amp	8.40	9.49	0.55	15.24	nr	**24.73**
32 amp	14.62	16.51	0.60	16.63	nr	**33.14**
63 amp	40.11	45.30	0.75	20.78	nr	**66.08**

V: ELECTRICAL SUPPLY/POWER/LIGHTING

Item	Net Price £	Material £	Labour hours	Labour £	Unit	Total rate £
V22: GENERAL LV POWER – cont						
INDUSTRIAL SOCKETS/PLUGS – cont						
Connectors; watertight; 100–130 volts, 50–60 Hz; IP 67 (Yellow)						
2 pole and earth						
16 amp	17.14	19.36	0.42	11.63	nr	**30.99**
32 amp	29.06	32.82	0.50	13.86	nr	**46.68**
63 amp	70.33	79.43	0.67	18.56	nr	**97.99**
Angled sockets; surface mounted; Watertight; 100–130 volts, 50–60 Hz; IP 67 (Yellow)						
2 pole and earth						
16 amp	14.17	16.00	0.55	15.24	nr	**31.24**
32 amp	27.71	31.29	0.60	16.63	nr	**47.92**
Plugs; splash-proof; 200–250 volts, 50–60 Hz; IP 44 (Blue)						
2 pole and earth						
16 amp	2.02	2.28	0.55	15.24	nr	**17.52**
32 amp	7.26	8.20	0.60	16.63	nr	**24.83**
63 amp	34.98	39.50	0.75	20.78	nr	**60.28**
3 pole and earth						
16 amp	8.04	9.08	0.65	18.00	nr	**27.08**
32 amp	10.80	12.20	0.72	19.94	nr	**32.14**
63 amp	35.11	39.66	0.83	22.99	nr	**62.65**
3 pole; neutral and earth						
16 amp	8.57	9.68	0.72	19.94	nr	**29.62**
32 amp	13.00	14.68	0.78	21.60	nr	**36.28**
Connectors; splash-proof; 200–250 volts, 50–60 Hz; IP 44 (Blue)						
2 pole and earth						
16 amp	5.00	5.65	0.42	11.63	nr	**17.28**
32 amp	12.41	14.01	0.50	13.86	nr	**27.87**
63 amp	43.83	49.50	0.67	18.56	nr	**68.06**
3 pole and earth						
16 amp	13.22	14.93	0.48	13.29	nr	**28.22**
32 amp	17.48	19.74	0.58	16.08	nr	**35.82**
63 amp	36.07	40.74	0.75	20.78	nr	**61.52**

V: ELECTRICAL SUPPLY/POWER/LIGHTING

Item	Net Price £	Material £	Labour hours	Labour £	Unit	Total rate £
3 pole; neutral and earth						
16 amp	15.37	17.36	0.52	14.41	nr	**31.77**
32 amp	52.66	59.48	0.73	20.23	nr	**79.71**
Angled sockets; surface mounted; splash-proof; 200–250 volts, 50–60 Hz; IP 44 (Blue)						
2 pole and earth						
16 amp	6.44	7.27	0.55	15.24	nr	**22.51**
32 amp	11.15	12.59	0.60	16.63	nr	**29.22**
63 amp	48.99	55.16	0.75	20.78	nr	**75.94**
3 pole and earth						
16 amp	12.76	14.41	0.65	18.00	nr	**32.41**
32 amp	22.10	24.96	0.72	19.94	nr	**44.90**
63 amp	48.68	54.98	0.83	22.99	nr	**77.97**
3 pole; neutral and earth						
16 amp	11.12	12.56	0.72	19.94	nr	**32.50**
32 amp	19.45	21.97	0.78	21.60	nr	**43.57**
Plugs; watertight; 200–250 volts, 50–60 Hz; IP67 (Blue)						
2 pole and earth						
16 amp	8.40	10.00	0.41	11.36	nr	**21.36**
32 amp	14.61	16.49	0.50	13.86	nr	**30.35**
63 amp	40.11	45.28	0.66	18.29	nr	**63.84**
125 amp	102.83	116.10	0.86	23.83	nr	**139.93**
Connectors; watertight; 200–250 volts, 50–60 Hz; IP 67 (Blue)						
2 pole and earth						
16 amp	8.40	9.48	0.42	11.63	nr	**22.72**
32 amp	14.61	16.49	0.50	13.86	nr	**33.12**
63 amp	56.63	63.95	0.67	18.56	nr	**82.51**
125 amp	145.61	164.39	0.87	24.10	nr	**192.09**
Angled sockets; surface mounted; watertight; 200–250 volts, 50–60 Hz; IP 67 (Blue)						
2 pole and earth						
16 amp	14.17	16.00	0.55	15.24	nr	**31.24**
32 amp	27.70	31.28	0.60	16.63	nr	**47.91**
125 amp	197.81	223.41	1.00	27.70	nr	**251.11**

V: ELECTRICAL SUPPLY/POWER/LIGHTING

Item	Net Price £	Material £	Labour hours	Labour £	Unit	Total rate £
V32: UNINTERRUPTIBLE POWER SUPPLY						
Uninterruptible power supply; sheet steel enclosure; self contained battery pack; including installation, testing and commissioning						
Single phase input and output; 5 year battery life; standard 13 A socket outlet connection						
1.0 kVA (10 minute supply)	966.00	1090.99	0.30	8.31	nr	**1099.30**
1.0 kVA (30 minute supply)	1394.00	1574.38	0.50	13.86	nr	**1588.24**
2.0 kVA (10 minute supply)	1350.00	1524.69	0.50	13.86	nr	**1538.55**
2.0 kVA (60 minute supply)	3055.00	3450.30	0.50	13.86	nr	**3464.16**
3.0 kVA (10 minute supply)	1697.00	1916.58	0.50	13.86	nr	**1930.44**
3.0 kVA (40 minute supply)	3008.00	3397.22	1.00	27.70	nr	**3424.92**
5.0 kVA (30 minute supply)	3817.00	4310.90	1.00	27.70	nr	**4338.60**
8.0 kVA (10 minute supply)	3665.00	4139.23	2.00	55.41	nr	**4194.64**
8.0 kVA (30 minute supply)	6407.00	7236.03	2.00	55.41	nr	**7291.44**
Uninterruptible power supply; including final connections and testing and commissioning						
Medium size static; single phase input and output; 10 year battery life; in cubicle						
10.0 kVA (10 minute supply)	3586.00	4050.01	10.00	277.04	nr	**4327.05**
10.0 kVA (30 minute supply)	6260.00	7070.01	15.00	415.56	nr	**7485.57**
15.0 kVA (10 minute supply)	6836.00	7720.54	10.00	277.04	nr	**7997.58**
15.0 kVA (30 minute supply)	9449.00	10671.66	15.00	415.56	nr	**11087.22**
20.0 kVA (10 minute supply)	6836.00	7720.54	10.00	277.04	nr	**7997.58**
20.0 kVA (30 minute supply)	12163.00	13736.83	15.00	415.56	nr	**14152.39**
Medium size static; three phase input and output; 10 year battery life; in cubicle						
10.0 kVA (10 minute supply)	6083.00	6870.11	10.00	277.04	nr	**7147.15**
10.0 kVA (30 minute supply)	7822.00	8834.13	15.00	415.56	nr	**9249.69**
15.0 kVA (10 minute supply)	6945.00	7843.65	15.00	415.56	nr	**8259.21**
15.0 kVA (30 minute supply)	8810.00	9949.97	20.00	554.08	nr	**10504.05**
20.0 kVA (10 minute supply)	7011.00	7918.19	20.00	554.08	nr	**8472.27**
20.0 kVA (30 minute supply)	10529.00	11891.40	25.00	692.60	nr	**12584.00**
30.0 kVA (10 minute supply)	12281.00	13870.10	25.00	692.60	nr	**14562.70**
30.0 kVA (30 minute supply)	14602.00	16491.42	30.00	831.13	nr	**17322.55**
Large size static; three phase input and output; 10 year battery life; in cubicle						
40 kVA (10 minute supply)	12611.00	14242.80	30.00	831.13	nr	**15073.93**
40 kVA (30 minute supply)	18095.00	20436.40	30.00	831.13	nr	**21267.53**
60 kVA (10 minute supply)	16618.00	18768.29	35.00	969.64	nr	**19737.93**

V: ELECTRICAL SUPPLY/POWER/LIGHTING

Item	Net Price £	Material £	Labour hours	Labour £	Unit	Total rate £
60 kVA (30 minute supply)	19369.00	21875.26	35.00	969.64	nr	**22844.90**
100 kVA (10 minute supply)	24467.00	27632.91	40.00	1108.16	nr	**28741.07**
200 kVA (10 minute supply)	39494.00	44604.32	40.00	1108.16	nr	**45712.48**
300 kVA (10 minute supply)	61843.00	69845.17	40.00	1108.16	nr	**70953.33**
400 kVA (10 minute supply)	72562.00	81951.16	50.00	1385.20	nr	**83336.36**
500 kVA (10 minute supply)	85157.00	96175.89	60.00	1662.24	nr	**97838.13**
600 kVA (10 minute supply)	114566.00	129390.72	70.00	1939.29	nr	**131330.01**
800 kVA (10 minute supply)	143079.00	161592.70	80.00	2216.33	nr	**163809.03**
Integral diesel rotary; 400 V three phase input and output; no break supply; including ventilation and acoustic attenuation oil day tank and interconnecting pipework						
500 kVA	430000.00	485617.92	120.00	3324.33	nr	**488942.25**
800 kVA	480000.00	542085.12	140.00	3878.40	nr	**545963.52**
1000 kVA	520000.00	587258.88	140.00	3878.40	nr	**591137.28**
1670 kVA	720000.00	813127.68	170.00	4709.47	nr	**817837.15**
2000 kVA	820000.00	926062.08	200.00	5540.56	nr	**931602.64**
2500 kVA	950000.00	1072876.80	230.00	6371.64	nr	**1079248.44**

V: ELECTRICAL SUPPLY/POWER/LIGHTING

Item	Net Price £	Material £	Labour hours	Labour £	Unit	**Total rate £**
V40: EMERGENCY LIGHTING						
Y73 – LUMINAIRES						
24 volt/50 volt/110 volt fluorescent slave luminaires						
For use with DC central battery systems						
Indoor, 8 W	37.62	42.48	0.80	22.15	nr	**64.63**
Indoor, exit sign box	46.86	52.93	0.80	22.15	nr	**75.08**
Outdoor, 8 W weatherproof	41.92	47.35	0.80	22.15	nr	**69.50**
Conversion module AC/DC	46.86	52.93	0.25	6.93	nr	**59.86**
Self contained; polycarbonate base and diffuser; LED charging light to European sign directive; 3 hour standby						
Non maintained						
Indoor, 8 W	41.53	46.90	1.00	27.70	nr	**74.60**
Outdoor, 8 W weatherproof, vandal-resistant IP65	59.63	67.35	1.00	27.70	nr	**95.05**
Maintained						
Indoor, 8 W	29.73	33.58	1.00	27.70	nr	**61.28**
Outdoor, 8 W weatherproof, vandal-resistant IP65	87.96	99.34	1.00	27.70	nr	**127.04**
Exit signage						
Exit sign; gold effect, pendular including brackets						
Non maintained, 8 W	112.94	127.55	1.00	27.70	nr	**155.25**
Maintained, 8 W	121.57	137.30	1.00	27.70	nr	**165.00**
Modification kit						
Module & battery for 58 W fluorescent modification from mains fitting to emergency; 3 hour standby	38.93	43.97	0.50	13.86	nr	**57.83**
Extra for remote box (when fitting is too small for modification)	19.33	21.83	0.50	13.86	nr	**35.69**
12 volt low voltage lighting; non maintained; 3 hour standby						
2 × 20 W lamp load	134.14	151.50	1.20	33.25	nr	**184.75**
1 × 50 W lamp load	146.43	165.37	1.00	27.70	nr	**193.07**
Maintained 3 hour standby						
2 × 20 W lamp load	126.98	143.41	1.20	33.25	nr	**176.66**
1 × 50 W lamp load	139.26	157.28	1.00	27.70	nr	**184.98**

V: ELECTRICAL SUPPLY/POWER/LIGHTING

Item	Net Price £	Material £	Labour hours	Labour £	Unit	**Total rate £**
DC central battery systems BS5266 compliant 24/50/110 volt						
DC supply to luminaires on mains failure; metal cubicle with battery charger, changeover device and battery as integral unit; ICEL 1001 compliant; 10 year design life valve regulated lead acid battery; 24 hour recharge; LCD display & LED indication; ICEL alarm pack; Includes on-site commissioning on 110 volt systems only						
24 volt, wall mounted						
300 W maintained, 1 hour	2139.00	2415.78	4.00	110.81	nr	**2526.59**
635 W maintained, 3 hour	2595.00	2930.78	6.00	166.23	nr	**3097.01**
470 W non maintained, 1 hour	1907.00	2153.76	4.00	110.81	nr	**2264.57**
780 W non maintained, 3 hour	2253.00	2544.52	6.00	166.23	nr	**2710.75**
50 volt						
935 W maintained, 3 hour	2049.00	2314.14	8.00	221.64	nr	**2535.78**
1965 W maintained, 3 hour	2852.00	3221.04	8.00	221.64	nr	**3442.68**
1311 W non maintained, 3 hour	2792.00	3153.27	8.00	221.64	nr	**3374.91**
2510 W non maintained 3, hour	4264.00	4815.74	8.00	221.64	nr	**5037.38**
110 volt						
1603 W maintained, 3 hour	4019.00	4539.04	8.00	221.64	nr	**4760.68**
4446 W maintained, 3 hour	4590.00	5183.93	10.00	277.04	nr	**5460.97**
2492 W non maintained, 3 hour	5099.00	5758.79	10.00	277.04	nr	**6035.83**
5429 W non maintained, 3 hour	6960.00	7860.59	12.00	332.45	nr	**8193.04**
DC central battery systems; BS EN 50171 compliant; 24/50/110 volt						
Central power systems						
DC supply to luminaires on mains failure; metal cubicle with battery charger, changeover device and battery as integral unit; 10 year design life valve regulated lead acid battery; 12 hour recharge to 80% of specified duty; low volts discount; LCD display & LED indication; includes on-site commissioning for CPS systems only; battery sized for 'end of life' @ 20°C test pushbutton						
24 volt, floor standing						
400 W non maintained, 1 hour	2009.00	2268.96	4.00	110.81	nr	**2379.77**
600 W maintained, 3 hour	2583.00	2917.23	6.00	166.23	nr	**3083.46**

V: ELECTRICAL SUPPLY/POWER/LIGHTING

Item	Net Price £	Material £	Labour hours	Labour £	Unit	Total rate £
V40: EMERGENCY LIGHTING – cont						
Central power systems – cont						
50 volt						
2133 W non maintained, 3 hour	4516.00	5100.35	8.00	221.64	nr	**5321.99**
1900 W maintained, 3 hour	4316.00	4874.47	8.00	221.64	nr	**5096.11**
110 volt						
2200 W non maintained, 3 hour	4177.00	4717.48	8.00	221.64	nr	**4939.12**
4000 W maintained, 3 hour	4662.00	5265.24	12.00	332.45	nr	**5597.69**
Low power systems						
DC supply to luminaires on mains failure; metal cubicle with battery charger, changeover device and battery as integral unit; 5 Year design life valve regulated lead acid battery; low volts discount; LED display & LED indication; battery sized for 'end of life' @ 20°C test pushbutton						
24 volt, floor standing						
300 W non maintained, 1 hour	2203.00	2488.06	4.00	110.81	nr	**2598.87**
600 W maintained, 3 hour	2439.00	2754.59	6.00	166.23	nr	**2920.82**
AC static inverter system; BS5266 compliant; one hour standby						
Central system supplying AC power on mains failure to mains luminaires; ICEL 1001 compliant metal cubicle(s) with changeover device, battery charger, battery & static inverter; 10 year design LIfe valve regulated lead acid battery; 24 hour recharge; LED indication and LCD display; pure sinewave output						
One hour						
750 VA, 600 W single phase I/P & O/P	2603.00	2939.81	6.00	166.23	nr	**3106.04**
3 KVA, 2.55 KW single phase I/P & O/P	4270.00	4822.52	8.00	221.64	nr	**5044.16**
5 KVA, 4.25 KW single phase I/P & O/P	5639.00	6368.66	10.00	277.04	nr	**6645.70**
8 KVA, 6.80 KW single phase I/P & O/P	6643.00	7502.57	12.00	332.45	nr	**7835.02**
10 KVA, 8.5 KW single phase I/P & O/P	8576.00	9685.70	14.00	387.86	nr	**10073.56**
13 KVA, 11.05 KW single phase I/P & O/P	11473.00	12957.55	16.00	443.26	nr	**13400.81**
15 KVA, 12.75 KW single phase I/P & O/P	11891.00	13429.64	30.00	831.13	nr	**14260.77**

V: ELECTRICAL SUPPLY/POWER/LIGHTING

Item	Net Price £	Material £	Labour hours	Labour £	Unit	**Total rate £**
20 KVA, 17.0 KW 3 phase I/P & single phase O/P	16423.00	18548.06	40.00	1108.16	nr	**19656.22**
30 KVA, 25.5 KW 3 phase I/P & O/P	22194.00	25065.80	60.00	1662.24	nr	**26728.04**
40 KVA, 34.0 KW 3 phase I/P & O/P	27669.00	31249.23	80.00	2216.33	nr	**33465.56**
50 KVA, 42.5 KW 3 phase I/P & O/P	36087.00	40756.48	90.00	2493.37	nr	**43249.85**
65 KVA, 55.25 KW 3 phase I/P & O/P	44269.00	49997.19	100.00	2770.40	nr	**52767.59**
90 KVA, 68.85 KW 3 phase I/P & O/P	58500.00	66069.61	120.00	3324.49	nr	**69394.10**
120 KVA, 102 KW 3 phase I/P & O/P	65300.00	73749.50	150.00	4155.61	nr	**77905.11**
Three hour						
750 VA, 600 W single phase I/P & O/P	3046.00	3440.13	6.00	166.23	nr	**3606.36**
3 KVA, 2.55 KW single phase I/P & O/P	5126.00	5789.28	8.00	221.64	nr	**6010.92**
5 KVA, 4.25 KW single phase I/P & O/P	8126.00	9177.46	10.00	277.04	nr	**9454.50**
8 KVA, 6.80 KW single phase I/P & O/P	10110.00	11418.19	12.00	332.45	nr	**11750.64**
10 KVA, 8.5 KW single phase I/P & O/P	13336.00	15061.62	14.00	387.86	nr	**15449.48**
13 KVA, 11.05 KW single phase I/P & O/P	16444.00	18571.77	16.00	443.26	nr	**19015.03**
15 KVA, 12.75 KW single phase I/P & O/P	16714.00	18876.71	30.00	831.13	nr	**19707.84**
20 KVA, 17.0 KW 3 phase I/P & single phase O/P	24204.00	27335.88	40.00	1108.16	nr	**28444.04**
30 KVA, 25.5 KW 3 phase I/P & O/P	36371.00	41077.23	60.00	1662.24	nr	**42739.47**
40 KVA, 34.0 KW 3 phase I/P & O/P	43484.00	49110.61	80.00	2216.33	nr	**51326.94**
50 KVA, 42.5 KW 3 phase I/P & O/P	58650.00	66239.01	90.00	2493.37	nr	**68732.38**
65 KVA, 55.25 KW 3 phase I/P & O/P	70914.00	80089.92	100.00	2770.40	nr	**82860.32**
90 KVA, 68.85 KW 3 phase I/P & O/P	83927.00	94786.73	120.00	3324.49	nr	**98111.22**
120 KVA, 102 KW 3 phase I/P & O/P	103442.00	116826.87	150.00	4155.61	nr	**120982.48**
AC static inverter system; BS EN 50171 compliant; one hour standby; Low power system (typically wall mounted)						
Central system supplying AC power on mains failure to mains luminaires; metal cubicle(s) with changeover device, battery charger, battery & static inverter; 5 year design life valve regulated lead acid battery; LED indication and LCD display; 12 hour recharge to 80% duty; inverter rated for 120% of load for 100% of duty; battery sized for 'end of life' @ 20°C test pushbutton						
One hour						
300 VA, 240 W single phase I/P & O/P	1164.00	1314.62	3.00	83.11	nr	**1397.73**
600 VA, 480 W single phase I/P & O/P	1403.00	1584.54	4.00	110.81	nr	**1695.35**
750 VA, 600 W single phase I/P & O/P	2589.00	2924.00	6.00	166.23	nr	**3090.23**
Three hour						
150 VA, 120 W single phase I/P & O/P	1283.00	1449.01	3.00	83.11	nr	**1532.12**
450 VA, 360 W single phase I/P & O/P	1522.00	1718.93	4.00	110.81	nr	**1829.74**
750 VA, 600 W single phase I/P & O/P	3030.00	3422.07	6.00	166.23	nr	**3588.30**

V: ELECTRICAL SUPPLY/POWER/LIGHTING

Item	Net Price £	Material £	Labour hours	Labour £	Unit	Total rate £
V40: EMERGENCY LIGHTING – cont						
AC static inverter system central power system; CPS BS EN 50171 compliant; one hour standby						
Central system supplying AC power on mains failure to mains luminaires; metal cubicle(s) with changeover device, battery charger, battery & static inverter; LED indication and LCD display; pure sinewave output; 10 year design life valve regulated lead acid battery; 12 hour recharge to 80% duty specified; unverter rated for 120% of load for 100% of duty; battery sized for 'end of life' @ 20°C test push button; includes on-site commissioning						
One hour						
750 VA, 600 W single phase I/P & O/P	2897.00	3271.85	6.00	166.23	nr	**3438.08**
3 KVA, 2.55 KW single phase I/P & O/P	4616.00	5213.29	8.00	221.64	nr	**5434.93**
5 KVA, 4.25 KW single phase I/P & O/P	5977.00	6750.39	10.00	277.04	nr	**7027.43**
8 KVA, 6.80 KW single phase I/P & O/P	6976.00	7878.66	12.00	332.45	nr	**8211.11**
10 KVA, 8.5 KW single phase I/P & O/P	8958.00	10117.13	14.00	387.86	nr	**10504.99**
13 KVA, 11.05 KW single phase I/P & O/P	11840.00	13372.04	16.00	443.26	nr	**13815.30**
15 KVA, 12.75 KW single phase I/P & O/P	12255.00	13840.74	30.00	831.13	nr	**14671.87**
20 KVA, 17.0 KW 3 phase I/P & single phase O/P	16882.00	19066.44	40.00	1108.16	nr	**20174.60**
30 KVA, 25.5 KW 3 phase I/P & O/P	22741.00	25683.57	60.00	1662.24	nr	**27345.81**
40 KVA, 34.0 KW 3 phase I/P & O/P	28186.00	31833.13	80.00	2216.33	nr	**34049.46**
50 KVA, 42.5 KW 3 phase I/P & O/P	36708.00	41457.83	90.00	2493.37	nr	**43951.20**
65 KVA, 55.25 KW 3 phase I/P & O/P	45339.00	51205.64	100.00	2770.40	nr	**53976.04**
90 KVA, 68.85 KW 3 phase I/P & O/P	59494.00	67192.22	120.00	3324.49	nr	**70516.71**
120 KVA, 102 KW 3 phase I/P & O/P	66258.00	74831.45	150.00	4155.61	nr	**78987.06**
Three hour						
750 VA, 600 W single phase I/P & O/P	3352.00	3785.73	6.00	166.23	nr	**3951.96**
3 KVA, 2.55 KW single phase I/P & O/P	5467.00	6174.41	8.00	221.64	nr	**6396.05**
5 KVA, 4.25 KW single phase I/P & O/P	8471.00	9567.10	10.00	277.04	nr	**9844.14**
8 KVA, 6.80 KW single phase I/P & O/P	10425.00	11773.94	12.00	332.45	nr	**12106.39**
10 KVA, 8.5 KW single phase I/P & O/P	13694.00	15465.93	14.00	387.86	nr	**15853.79**
13 KVA, 11.05 KW single phase I/P & O/P	16785.00	18956.89	16.00	443.26	nr	**19400.15**
15 KVA, 12.75 KW single phase I/P & O/P	17052.00	19258.45	30.00	831.13	nr	**20089.58**
20 KVA, 17.0 KW 3 phase I/P & single phase O/P	24849.00	28064.34	40.00	1108.16	nr	**29172.50**
30 KVA, 25.5 KW 3 phase I/P & O/P	36841.00	41608.04	60.00	1662.24	nr	**43270.28**
40 KVA, 34.0 KW 3 phase I/P & O/P	43917.00	49599.64	80.00	2216.33	nr	**51815.97**
50 KVA, 42.5 KW 3 phase I/P & O/P	59150.00	66803.72	90.00	2493.37	nr	**69297.09**
65 KVA, 55.25 KW 3 phase I/P & O/P	71842.00	81137.99	100.00	2770.40	nr	**83908.39**
90 KVA, 68.85 KW 3 phase I/P & O/P	89316.00	100873.04	120.00	3324.49	nr	**104197.53**
120 KVA, 102 KW 3 phase I/P & O/P	104194.00	117676.19	150.00	4155.61	nr	**121831.80**

W: COMMUNICATIONS/SECURITY/CONTROL

Item	Net Price £	Material £	Labour hours	Labour £	Unit	Total rate £
W10: TELECOMMUNICATIONS						
Y61 – CABLES						
Multipair internal telephone cable; BS 6746; loose laid on tray/basket						
0.5 mm dia. conductor LSZH insulated and sheathed multipair cables; BT specification CW 1308						
3 pair	0.17	0.20	0.03	0.84	m	**1.04**
4 pair	0.20	0.23	0.03	0.84	m	**1.07**
6 pair	0.29	0.33	0.03	0.84	m	**1.17**
10 pair	0.51	0.57	0.05	1.39	m	**1.96**
15 pair	0.77	0.87	0.06	1.66	m	**2.53**
20 pair + 1 wire	1.09	1.23	0.06	1.66	m	**2.89**
25 pair	1.38	1.56	0.08	2.21	m	**3.77**
40 pair + earth	1.95	2.20	0.08	2.21	m	**4.41**
50 pair + earth	2.35	2.65	0.10	2.76	m	**5.41**
80 pair + earth	3.15	3.56	0.10	2.76	m	**6.32**
100 pair + earth	3.79	4.28	0.12	3.33	m	**7.61**
Multipair internal telephone cable; BS 6746; installed in conduit/trunking						
0.5 mm dia. conductor LSZH insulated and sheathed multipair cables; BT specification CW 1308						
3 pair	0.17	0.20	0.05	1.39	m	**1.59**
4 pair	0.20	0.23	0.06	1.66	m	**1.89**
6 pair	0.29	0.33	0.06	1.66	m	**1.99**
10 pair	0.51	0.57	0.07	1.94	m	**2.51**
15 pair	0.77	0.87	0.07	1.94	m	**2.81**
20 pair + 1 wire	1.09	1.23	0.09	2.50	m	**3.73**
25 pair	1.38	1.56	0.10	2.76	m	**4.32**
40 pair + earth	1.95	2.20	0.12	3.33	m	**5.53**
50 pair + earth	2.35	2.65	0.14	3.88	m	**6.53**
80 pair + earth	3.15	3.56	0.14	3.88	m	**7.44**
100 pair + earth	3.79	4.28	0.15	4.15	m	**8.43**
Low speed data; unshielded twisted pair; solid copper conductors; LSOH sheath; nominal impedance 100 Ohm; Category 3 to ISO IS 1801/EIA/TIA 568B and EN50173/ 50174 standards to current revisions						
Installed in riser						
25 pair 24 AWG	1.52	1.71	0.03	0.84	m	**2.55**
50 pair 24 AWG	3.03	3.42	0.06	1.66	m	**5.08**
100 pair 24 AWG	5.70	6.44	0.10	2.76	m	**9.20**

W: COMMUNICATIONS/SECURITY/CONTROL

Item	Net Price £	Material £	Labour hours	Labour £	Unit	Total rate £
W10: TELECOMMUNICATIONS – cont						
Multipair internal telephone cable – cont						
Installed below floor						
25 pair 24 AWG	1.52	1.71	0.02	0.55	m	**2.26**
50 pair 24 AWG	3.03	3.42	0.05	1.39	m	**4.81**
100 pair 24 AWG	5.70	6.44	0.08	2.21	m	**8.65**
Y74 – ACCESSORIES						
Telephone outlet: moulded plastic plate with box; fitted and connected; flush or surface mounted						
Single master outlet	6.47	7.30	0.35	9.70	nr	**17.00**
Single secondary outlet	4.78	5.40	0.35	9.70	nr	**15.10**
Telephone outlet: bronze or satin chromeplate; with box; fitted and connected; flush or surface mounted						
Single master outlet	10.53	11.89	0.35	9.70	nr	**21.59**
Single secondary outlet	11.51	12.99	0.35	9.70	nr	**22.69**
Frames and box connections						
Provision and Installation of a Dual Vertical Krone 108 A Voice Distribution Frame that can accommodate a total of 138 × Krone 237 A Strips	218.90	247.23	1.50	41.56	nr	**288.79**
Label Frame (Traffolyte style)	1.38	1.56	0.27	7.48	nr	**9.04**
Provision and Installation of a Box Connection 301 A Voice Termination Unit that can accommodate a total of 10 × Krone 237 A Strips	18.98	21.44	0.25	6.93	nr	**28.37**
Label Frame (Traffolyte style)	0.45	0.51	0.08	2.31	nr	**2.82**
Provision and Installation of a Box Connection 201 Voice Termination Unit that can accommodate 20 pairs	10.10	11.40	0.17	4.60	nr	**16.00**
Label Frame (Traffolyte style)	0.45	0.51	0.08	2.31	nr	**2.82**

W: COMMUNICATIONS/SECURITY/CONTROL

Item	Net Price £	Material £	Labour hours	Labour £	Unit	**Total rate £**
Terminate, test and label voice multicore system						
Patch panels						
Voice; 19" wide fully loaded, finished in black including termination and forming of cables (assuming 2 pairs per port)						
25 port – RJ45 UTP – Krone	47.50	53.64	2.60	72.03	nr	**125.67**
50 port – RJ45 UTP – Krone	64.75	73.12	4.65	128.82	nr	**201.94**
900 pair fully loaded Systimax style of frame including forming and termination of 9 × 100 pair cables	535.00	604.23	25.00	692.60	nr	**1296.83**
Installation and termination of Krone Strip (10 pair block – 237 A) including designation label strip	4.25	4.80	0.50	13.86	nr	**18.66**
Patch panel and Outlet labelling per port (Traffolyte style)	0.16	0.19	0.02	0.55	nr	**0.74**
Provision and installation of a voice jumper, for cross termination on Krone Termination Strips	0.05	0.07	0.06	1.66	nr	**1.73**
CW1308 / Cat 3 cable circuit test per pair	–	–	0.04	1.11	nr	**1.11**

W: COMMUNICATIONS/SECURITY/CONTROL

Item	Net Price £	Material £	Labour hours	Labour £	Unit	**Total rate £**
W20: RADIO/TELEVISION						
RADIO						
Y61 – CABLES						
Radio Frequency Cable; BS 2316 ; PVC sheathed; laid loose						
7/0.41 mm tinned copper inner conductor; solid polyethylene dielectric insulation; bare copper wire braid; PVC sheath; 75 ohm impedance						
Cable	0.98	1.11	0.05	1.39	m	**2.50**
Twin 1/0.58 mm copper covered steel solid core wire conductor; solid polyethylene dielectric insulation; barecopper wire braid; PVC sheath; 75 ohm impedance						
Cable	0.22	0.25	0.05	1.39	m	**1.64**
TELEVISION						
Y61 – CABLES						
Television aerial cable; coaxial; PVC sheathed; fixed to backgrounds						
General purpose TV aerial downlead; copper stranded inner conductor; cellular polythene insulation; copper braid outer conductor; 75 ohm impedance						
7/0.25 mm	0.21	0.24	0.06	1.66	m	**1.90**
Low loss TV aerial downlead; solid copper inner conductor; cellular polythene insulation; copper braid outer; conductor; 75 ohm impedance						
1/1.12 mm	0.22	0.25	0.06	1.66	m	**1.91**
Low loss air spaced; solid copper inner conductor; air spaced polythene insulation; copper braid outer conductor; 75 ohm impedance						
1/1.00 mm	0.29	0.33	0.06	1.66	m	**1.99**
Satelite aerial downlead; solid copper inner conductor; air spaced polythene insulation; copper tape and braid outer conductor; 75 ohm impedance						
1/1.00 mm	0.26	0.30	0.06	1.66	m	**1.96**

W: COMMUNICATIONS/SECURITY/CONTROL

Item	Net Price £	Material £	Labour hours	Labour £	Unit	**Total rate £**
Satelite TV coaxial; solid copper inner conductor; semi air spaced polyethylene dielectric insulation; plain annealed copper foil and copper braid screen in outer conductor; PVC sheath; 75 ohm impedance						
1/1.25 mm	0.43	0.49	0.08	2.21	m	**2.70**
Satelite TV coaxial; solid copper inner conductor; air spaced polyethylene dielectric insulation; plain annealed copper foil and copper braid screen in outer conductor; PVC sheath; 75 ohm impedance						
1/1.67 mm	0.70	0.80	0.09	2.50	m	**3.30**
Video cable; PVC flame retardant sheath; laid loose						
7/0.1 mm silver coated copper covered annealed steel wire conductor; polyethylene dielectric insulation with tin coated copper wire braid; 75 ohm impedance						
Cable	0.17	0.20	0.05	1.39	m	**1.59**
Y74 – ACCESSORIES						
TV co-axial socket outlet: moulded plastic box; flush or surface mounted						
One way Direct Connection	7.45	8.42	0.35	9.70	nr	**18.12**
Two way Direct Connection	10.39	11.73	0.35	9.70	nr	**21.43**
One way Isolated UHF/VHF	13.12	14.82	0.35	9.70	nr	**24.52**
Two way Isolated UHF/VHF	17.88	20.20	0.35	9.70	nr	**29.90**

W: COMMUNICATIONS/SECURITY/CONTROL

Item	Net Price £	Material £	Labour hours	Labour £	Unit	Total rate £
W23: CLOCKS						
Clock timing systems; master and slave units; fixed to background; excluding supports and fixings						
Quartz master clock with solid state digital readout for parallel loop operation; one minute, half minute and one second pulse; maximum of 160 clocks						
Over two loops only	642.60	725.75	4.40	121.90	nr	**847.65**
Power supplies for above, giving 24 hours power reserve						
2 6 amp hour batteries	187.43	211.68	3.00	83.11	nr	**294.79**
2 15 amp hour batteries	214.20	241.91	5.00	138.52	nr	**380.43**
Radio receiver to accept BBC Rugby Transmitter MSF signal						
To synchronize time of above Quartz master clock	139.23	157.24	7.04	195.11	nr	**352.35**
Wall clocks for slave (impulse) systems; 24 V DC, white dial with black numerals fitted with Axis Polycarbonate disc; BS 467.7 Class O						
305 mm dia. 1 minute impulse	56.27	63.55	1.37	37.95	nr	**101.50**
305 mm dia. 1/2 minute impulse	56.27	63.55	1.37	37.95	nr	**101.50**
227 mm dia. 1 second impulse	80.33	90.73	1.37	37.95	nr	**128.68**
305 mm dia. 1 second impulse	80.33	90.73	1.37	37.95	nr	**128.68**
Quartz battery movement; BS 467.7 Class O; white dial with black numerals and sweep second hand; fitted with Axis Polycarbonate disc; stove enamel case						
305 mm dia.	29.46	33.27	0.77	21.34	nr	**54.61**
Internal wall mounted electric clock; white dial with black numerals; 240v, 50 Hz						
305 mm dia.	37.48	42.33	1.00	27.70	nr	**70.03**
458 mm dia.	160.65	181.44	1.00	27.70	nr	**209.14**
Matching clock; BS 467.7 Class O; 240 V AC, 50/60 Hz mains supply; 12 hour duration; dial with 1–12; IP 66; Axis Polycarbonate disc; spun metal movement cover; semi flush mount on 6 point fixing bezel						
227 mm dia.	214.20	241.91	0.62	17.18	nr	**259.09**

W: COMMUNICATIONS/SECURITY/CONTROL

Item	Net Price £	Material £	Labour hours	Labour £	Unit	**Total rate £**
Digital clocks; 240 V, 50 hz supply; with/without synchronization from masterclock;12/24 hour display; stand alone operation; 50 mm digits						
Flush – hours/minutes/seconds or minutes/seconds/10th seconds	214.20	241.91	0.57	15.79	nr	**257.70**
Surface – hours/minutes/seconds or minutes/seconds/10th seconds	187.43	453.59	0.57	15.79	nr	**469.38**
Flush – hours/minutes or minutes/seconds	182.07	641.65	0.57	15.79	nr	**657.44**
Surface – hours/minutes or minutes/seconds	155.29	405.78	0.57	15.79	nr	**421.57**

W: COMMUNICATIONS/SECURITY/CONTROL

Item	Net Price £	Material £	Labour hours	Labour £	Unit	**Total rate £**
W30: DATA TRANSMISSION						
Cabinets						
Floor standing; suitable for 19" patch panels with glass lockable doors, metal rear doors, side panels, vertical cable management, 2 × 4 way PDUs, 4 way fan, earth bonding kit; installed on raised floor						
600 wide × 800 deep – 18U	655.20	739.98	3.00	83.11	nr	**823.09**
600 wide × 800 deep – 24U	682.50	770.81	3.00	83.11	nr	**853.92**
600 wide × 800 deep – 33U	733.00	827.84	4.00	110.81	nr	**938.65**
600 wide × 800 deep – 42U	786.25	887.99	4.00	110.81	nr	**998.80**
600 wide × 800 deep – 47U	830.00	937.40	4.00	110.81	nr	**1048.21**
800 wide × 800 deep – 42U	890.00	1005.16	4.00	110.81	nr	**1115.97**
800 wide × 800 deep – 47U	928.25	1048.36	4.00	110.81	nr	**1159.17**
Label cabinet	2.10	2.38	0.25	6.93	nr	**9.31**
Wall mounted; suitable for 19" patch panels with glass lockable doors, side panels, vertical cable management, 2 × 4 way PDUs, 4 way fan, earth bonding kit; fixed to wall						
19 wide × 500 deep – 9U	437.00	493.55	3.00	83.11	nr	**576.66**
19 wide × 500 deep – 12U	447.75	505.69	3.00	83.11	nr	**588.80**
19 wide × 500 deep – 15U	464.00	524.04	3.00	83.11	nr	**607.15**
19 wide × 500 deep – 18U	546.00	616.65	3.00	83.11	nr	**699.76**
19 wide × 500 deep – 21U	568.00	641.50	3.00	83.11	nr	**724.61**
Label cabinet	2.10	2.38	0.25	6.93	nr	**9.31**
Frames						
Floor standing; suitable for 19" patch panels with supports, vertical cable management, earth bonding kit; installed on raised floor						
19 wide × 500 deep – 25U	491.50	555.09	2.50	69.27	nr	**624.36**
19 wide × 500 deep – 39U	600.50	678.20	2.50	69.27	nr	**747.47**
19 wide × 500 deep – 42U	655.25	740.04	2.50	69.27	nr	**809.31**
19 wide × 500 deep – 47U	710.00	801.87	2.50	69.27	nr	**871.14**
Label frame	2.10	2.38	0.25	6.93	nr	**9.31**

W: COMMUNICATIONS/SECURITY/CONTROL

Item	Net Price £	Material £	Labour hours	Labour £	Unit	Total rate £
Copper data cabling						
Unshielded twisted pair; solid copper conductors; PVC insulation; nominal impedance 100 Ohm; Cat 5e to ISO 11801, EIA/TIA 568B and EN 50173/50174 standards to the current revisions						
4 pair 24 AWG; nominal outside dia. 5.6 mm; installed above ceiling	0.18	0.21	0.02	0.55	m	**0.76**
4 pair 24 AWG; nominal outside dia. 5.6 mm; installed in riser	0.18	0.21	0.02	0.55	m	**0.76**
4 pair 24 AWG; nominal outside dia. 5.6 mm; installed below floor	0.18	0.21	0.01	0.29	m	**0.50**
4 pair 24 AWG; nominal outside dia. 5.6 mm; installed in trunking	0.18	0.21	0.02	0.55	m	**0.76**
Unshielded twisted pair; solid copper conductors; LSOH sheathed; nominal impedance 100 Ohm; Cat 5e to ISO 11801, EIA/TIA 568B and EN 50173/50174 standards to the current revisions						
4 pair 24 AWG; nominal outside dia. 5.6 mm; installed above ceiling	0.21	0.24	0.02	0.55	m	**0.79**
4 pair 24 AWG; nominal outside dia. 5.6 mm; installed in riser	0.21	0.24	0.02	0.55	m	**0.79**
4 pair 24 AWG; nominal outside dia. 5.6 mm; installed below floor	0.21	0.24	0.01	0.29	m	**0.53**
4 pair 24 AWG; nominal outside dia. 5.6 mm; installed in trunking	0.21	0.24	0.02	0.55	m	**0.79**
Unshielded twisted pair; solid copper conductors; PVC insulation; nominal impedance 100 Ohm; Cat 6 to ISO 11801, EIA/TIA 568B and EN 50173/50174 standards to the current revisions						
4 pair 24 AWG; nominal outside dia. 5.6 mm; installed above ceiling	0.25	0.29	0.02	0.42	m	**0.71**
4 pair 24 AWG; nominal outside dia. 5.6 mm; installed in riser	0.25	0.29	0.02	0.42	m	**0.71**
4 pair 24 AWG; nominal outside dia. 5.6 mm; installed below floor	0.25	0.29	0.01	0.23	m	**0.52**
4 pair 24 AWG; nominal outside dia. 5.6 mm; installed in trunking	0.25	0.29	0.02	0.55	m	**0.84**

W: COMMUNICATIONS/SECURITY/CONTROL

Item	Net Price £	Material £	Labour hours	Labour £	Unit	Total rate £
W30: DATA TRANSMISSION – cont						
Copper data cabling – cont						
Unshielded twisted pair; solid copper conductors; LSOH sheathed; nominal impedance 100 Ohm; Cat 6 to ISO 11801, EIA/TIA 568B and EN 50173/50174 standards to the current revisions						
4 pair 24 AWG; nominal outside dia. 5.6 mm; installed above ceiling	0.26	0.30	0.02	0.61	m	**0.91**
4 pair 24 AWG; nominal outside dia. 5.6 mm; installed in riser	0.26	0.30	0.02	0.61	m	**0.91**
4 pair 24 AWG; nominal outside dia. 5.6 mm; installed below floor	0.26	0.30	0.01	0.31	m	**0.61**
4 pair 24 AWG; nominal outside dia. 5.6 mm; installed in trunking	0.26	0.30	0.02	0.61	m	**0.91**
Patch panels						
Category 5e; 19" wide fully loaded, finished in black including termination and forming of cables						
24 port – RJ45 UTP – Krone / 110	69.00	77.93	4.75	131.60	nr	**209.53**
48 port – RJ45 UTP – Krone / 110	129.50	146.26	9.35	259.04	nr	**405.30**
Patch panel labelling per port	0.22	0.25	0.02	0.55	nr	**0.80**
Category 6; 19" wide fully loaded, finished in black including termination and forming of cables						
24 port – RJ45 UTP – Krone / 110	132.00	149.08	5.00	138.52	nr	**287.60**
48 port – RJ45 UTP – Krone / 110	240.50	271.62	9.80	271.49	nr	**543.11**
Patch panel labelling per port	0.22	0.25	0.02	0.55	nr	**0.80**
Work station						
Category 5e RJ45 data outlet plate and multiway outlet boxes for wall, ceiling and below floor installations including label to ISO 11801 standards						
Wall mounted; fully loaded						
One gang LSOH PVC plate	6.38	7.20	0.15	4.15	nr	**11.35**
Two gang LSOH PVC plate	9.13	10.31	0.20	5.55	nr	**15.86**
Four gang LSOH PVC plate	15.62	17.65	0.40	11.08	nr	**28.73**
One gang satin brass plate	14.50	16.38	0.25	6.93	nr	**23.31**
Two gang satin brass plate	16.35	18.46	0.33	9.14	nr	**27.60**

W: COMMUNICATIONS/SECURITY/CONTROL

Item	Net Price £	Material £	Labour hours	Labour £	Unit	Total rate £
Ceiling mounted; fully loaded						
One gang metal clad plate	7.70	8.70	0.33	9.14	nr	**17.84**
Two gang metal clad plate	10.45	11.80	0.45	12.46	nr	**24.26**
Below floor; fully loaded						
Four way outlet box, 5 m length 20 mm flexible conduit with glands and starin relief bracket	26.02	29.39	0.80	22.15	nr	**51.54**
Six way outlet box, 5 m length 25 mm flexible conduit with glands and strain relief bracket	33.64	38.00	1.20	33.25	nr	**71.25**
Eight way outlet box, 5 m length 25 mm flexible conduit with glands and strain relief bracket	42.35	47.83	1.40	38.78	nr	**86.61**
Installation of outlet boxes to desks	–	–	0.40	11.08	nr	**11.08**
Category 6 RJ45 data outlet plate and multiway outlet boxes for wall, ceiling and below floor installations including label to ISO 11801 standards						
Wall mounted; fully loaded						
One gang LSOH PVC plate	8.14	9.19	0.17	4.57	nr	**13.76**
Two gang LSOH PVC plate	11.94	13.48	0.22	6.10	nr	**19.58**
Four gang LSOH PVC plate	21.56	24.35	0.44	12.19	nr	**36.54**
One gang satin brass plate	13.93	15.73	0.28	7.62	nr	**23.35**
Two gang satin brass plate	17.43	19.69	0.36	10.05	nr	**29.74**
Ceiling mounted; fully loaded						
One gang metal clad plate	9.24	10.43	0.36	10.05	nr	**20.48**
Two gang metal clad plate	13.04	14.73	0.50	13.71	nr	**28.44**
Below floor; fully loaded						
Four way outlet box, 5 m length 20 mm flexible conduit with glands and starin relief bracket	31.89	36.02	0.88	24.39	nr	**60.41**
Six way outlet box, 5 m length 25 mm flexible conduit with glands and strain relief bracket	42.42	47.91	1.32	36.57	nr	**84.48**
Eight way outlet box, 5 m length 25 mm flexible conduit with glands and strain relief bracket	54.03	61.02	1.54	42.67	nr	**103.69**
Installation of outlet boxes to desks	–	–	0.40	11.08	nr	**11.08**
Category 5e cable test	–	–	0.10	2.76	–	**2.76**
Category 6 cable test	–	–	0.11	3.05	nr	**3.05**

W: COMMUNICATIONS/SECURITY/CONTROL

Item	Net Price £	Material £	Labour hours	Labour £	Unit	Total rate £
W30: DATA TRANSMISSION – cont						
Copper patch leads						
Category 5e; straight through booted RJ45 UTP – RJ45 UTP						
Patch lead 1 m length	1.60	1.81	0.09	2.50	nr	**4.31**
Patch lead 3 m length	1.90	2.14	0.09	2.50	nr	**4.64**
Patch lead 5 m length	2.70	3.05	0.10	2.76	nr	**5.81**
Patch lead 7 m length	3.95	4.46	0.10	2.76	nr	**7.22**
Category 6; straight through booted RJ45 UTP – RJ45 UTP						
Patch lead 1 m length	4.25	4.80	0.09	2.50	nr	**7.30**
Patch lead 3 m length	6.40	7.23	0.09	2.50	nr	**9.73**
Patch lead 5 m length	7.50	8.47	0.10	2.76	nr	**11.23**
Patch lead 7 m length	9.50	10.73	0.10	2.76	nr	**13.49**
Note 1: With an intelligent Patching System, add a 25% uplift to the Cat5e and Cat 6 System price. This would be dependent on the System and IMS solution chosen, i.e. iPatch, RIT or iTRACs						
Note 2: For a Cat 6 augmented solution (10Gbps capable), add 25% to a Cat 6 System price.						
Fibre infrastructure						
Fibre optic cable, tight buffered, internal/external application, single mode, LSOH sheathed						
4 core fibre optic cable	1.27	1.44	0.10	2.76	m	**4.20**
8 core fibre optic cable	2.11	2.39	0.10	2.76	m	**5.15**
12 core fibre optic cable	2.62	2.96	0.10	2.76	m	**5.72**
16 core fibre optic cable	3.18	3.59	0.10	2.76	m	**6.35**
24 core fibre optic cable	3.84	4.33	0.10	2.76	m	**7.09**
Fibre optic cable OM1 and OM2, tight buffered, internal/external application, 62.5/125 multimode fibre, LSOH sheathed						
4 core fibre optic cable	1.36	1.54	0.10	2.76	m	**4.30**
8 core fibre optic cable	2.25	2.54	0.10	2.76	m	**5.30**
12 core fibre optic cable	2.86	3.23	0.10	2.76	m	**5.99**
16 core fibre optic cable	3.56	4.02	0.10	2.76	m	**6.78**
24 core fibre optic cable	4.12	4.65	0.10	2.76	m	**7.41**

W: COMMUNICATIONS/SECURITY/CONTROL

Item	Net Price £	Material £	Labour hours	Labour £	Unit	**Total rate £**
Fibre optic cable OM3, tight buffered, internal only application, 50/125 multimode fibre, LSOH sheathed						
4 core fibre optic cable	1.68	1.90	0.10	2.76	nr	**4.66**
8 core fibre optic cable	2.62	2.96	0.10	2.76	nr	**5.72**
12 core fibre optic cable	3.80	4.29	0.10	2.76	nr	**7.05**
24 core fibre cable	7.26	8.20	0.10	2.76	nr	**10.96**
Fibre optic single and multimode connectors and couplers including termination						
ST singlemode booted connector	6.39	7.22	0.25	6.93	nr	**14.15**
ST multimode booted connector	3.09	3.49	0.25	6.93	nr	**10.42**
SC simplex singlemode booted connector	9.06	10.23	0.25	6.93	nr	**17.16**
SC simplex multimode booted connector	3.30	3.72	0.25	6.93	nr	**10.65**
SC duplex multimode booted connector	6.59	7.45	0.25	6.93	nr	**14.38**
ST – SC duplex adaptor	16.02	18.09	0.01	0.29	nr	**18.38**
ST inline bulkhead coupler	3.43	3.88	0.01	0.29	nr	**4.17**
SC duplex coupler	6.41	7.24	0.01	0.29	nr	**7.53**
MTRJ small form factor duplex connector	6.69	7.56	0.25	6.93	nr	**14.49**
LC simplex multimode booted connector	–	–	0.25	6.93	nr	**6.93**
Singlemode core test per core	–	–	0.20	5.55	nr	**5.55**
Multimode core test per core	–	–	0.20	5.55	nr	**5.55**
Fibre; 19' wide fully loaded, labled, alluminium alloy c/w couplers, fibre management and glands (excludes termination of fibre cores)						
8 way ST; fixed drawer	68.78	77.68	0.50	13.86	nr	**91.54**
16 way ST; fixed drawer	103.16	116.50	0.50	13.86	nr	**130.36**
24 way ST; fixed drawer	143.32	161.86	0.50	13.86	nr	**175.72**
8 way ST; sliding drawer	87.50	98.82	0.50	13.86	nr	**112.68**
16 way ST; sliding drawer	120.25	135.81	0.50	13.86	nr	**149.67**
24 way ST; sliding drawer	158.25	178.72	0.50	13.86	nr	**192.58**
8 way (4 duplex) SC; fixed drawer	82.00	92.61	0.50	13.86	nr	**106.47**
16 way (8 duplex) SC; fixed drawer	120.00	135.53	0.50	13.86	nr	**149.39**
24 way (12 duplex) SC; fixed drawer	147.50	166.59	0.50	13.86	nr	**180.45**
8 way (4 duplex) SC; sliding drawer	103.75	117.18	0.50	13.86	nr	**131.04**
16 way (8 duplex) SC; sliding drawer	142.00	160.37	0.50	13.86	nr	**174.23**
24 way (12 duplex) SC; sliding drawer	169.25	191.15	0.50	13.86	nr	**205.01**
8 way (4 duplex) MTRJ; fixed drawer	93.00	105.03	0.50	13.86	nr	**118.89**
16 way (8 duplex) MTRJ; fixed drawer	142.00	160.37	0.50	13.86	nr	**174.23**
24 way (12 duplex) MTRJ; fixed drawer	169.25	191.15	0.50	13.86	nr	**205.01**
8 way (4 duplex) FC/PC; fixed drawer	97.50	110.12	0.50	13.86	nr	**123.98**
16 way (8 duplex) FC/PC; fixed drawer	149.00	168.28	0.50	13.86	nr	**182.14**
24 way (12 duplex) FC/PC; fixed drawer	177.75	200.75	0.50	13.86	nr	**214.61**
Patch panel label per way	0.22	0.25	0.02	0.55	nr	**0.80**

W: COMMUNICATIONS/SECURITY/CONTROL

Item	Net Price £	Material £	Labour hours	Labour £	Unit	Total rate £
W30: DATA TRANSMISSION – cont						
Fibre patch leads						
Single mode						
Duplex OS1 LC – LC						
Fibre patch lead 1 m length	21.09	23.82	0.08	2.21	nr	**26.03**
Fibre patch lead 3 m length	19.73	22.29	0.08	2.21	nr	**24.50**
Fibre patch lead 5 m length	20.99	23.70	0.10	2.76	nr	**26.46**
Multimode						
Duplex 50/125 OM3 MTRJ – MTRJ						
Fibre patch lead 1 m length	18.51	20.91	0.08	2.21	nr	**23.12**
Fibre patch lead 3 m length	20.12	22.73	0.08	2.21	nr	**24.94**
Fibre patch lead 5 m length	21.14	23.88	0.10	2.76	nr	**26.64**
Duplex 50/125 OM3 ST – ST						
Fibre patch lead 1 m length	23.31	26.33	0.08	2.21	nr	**28.54**
Fibre patch lead 3 m length	25.36	28.64	0.08	2.21	nr	**30.85**
Fibre patch lead 5 m length	26.64	30.09	0.10	2.76	nr.	**32.85**
Duplex 50/125 OM3 SC – SC						
Fibre patch lead 1 m length	20.72	23.40	0.08	2.21	nr	**25.61**
Firbe patch lead 3 m length	22.54	25.46	0.08	2.21	nr	**27.67**
Fibre patch lead 5 m length	23.67	26.73	0.10	2.76	nr	**29.49**
Duplex 50/125 OM3 LC – LC						
Fibre patch lead 1 m length	22.19	25.06	0.08	2.21	nr	**27.27**
Fibre patch lead 3 m length	24.15	27.27	0.08	2.21	nr	**29.48**
Fibre patch lead 5 m length	25.37	28.65	0.10	2.76	nr	**31.41**

W: COMMUNICATIONS/SECURITY/CONTROL

Item	Net Price £	Material £	Labour hours	Labour £	Unit	**Total rate £**
W40: ACCESS CONTROL						
ACCESS CONTROL EQUIPMENT						
Equipment to control the movement of personnel into defined spaces; includes fixing to backgrounds, termination of power and data cables; excludes cable containment and cable installation						
Access control						
Magnetic swipe reader	160.00	180.70	1.50	41.56	nr	**222.26**
Proximity reader	110.00	124.24	1.50	41.56	nr	**165.80**
Exit button	20.00	22.59	1.50	41.56	nr	**64.15**
Exit PIR	132.00	149.08	2.00	55.41	nr	**204.49**
Emergency break glass double pole	30.00	33.88	1.00	27.70	nr	**61.58**
Alarm contact flush	5.00	5.65	1.00	27.70	nr	**33.35**
Alarm contact surface	5.00	5.65	1.00	27.70	nr	**33.35**
Reader controller 16 door	4290.00	4845.10	4.00	110.81	nr	**4955.91**
Reader controller 8 door	2847.00	3215.39	4.00	110.81	nr	**3326.20**
Reader controller 2 door	997.00	1126.01	4.00	110.81	nr	**1236.82**
Reader interface	404.35	456.67	1.50	41.56	nr	**498.23**
Lock power supply 12 volt 3 amp	241.00	272.19	2.00	55.41	nr	**327.60**
Lock power supply 24 volt 3 amp	421.00	475.47	2.00	55.41	nr	**530.88**
Rechargeable battery 12 volt 7 AHr	12.00	13.56	0.25	6.93	nr	**20.49**
Lock equipment						
Single slimline magnetic lock, monitored	104.00	117.45	2.00	55.41	nr	**172.86**
Single slimline magnetic lock, unmonitored	90.00	101.64	1.75	48.48	nr	**150.12**
Double slimline magnetic lock, monitored	206.00	232.65	4.00	110.81	nr	**343.46**
Double slimline magnetic lock, unmonitored	177.00	199.90	4.50	124.67	nr	**324.57**
Standard single magnetic lock, monitored	132.00	149.08	1.25	34.63	nr	**183.71**
Standard single magnetic lock, unmonitored	106.00	119.72	1.00	27.70	nr	**147.42**
Standard single magnetic lock, double monitored	177.00	199.90	1.50	41.56	nr	**241.46**
Standard double magnetic lock, monitored	264.00	298.16	1.50	41.56	nr	**339.72**
Standard double magnetic lock, unmonitored	207.00	233.78	1.25	34.63	nr	**268.41**
12 V electric release fail safe, monitored	138.00	155.86	1.25	34.63	nr	**190.49**
12 V electric release fail secure, monitored	138.00	155.86	1.00	27.70	nr	**183.56**
Solenoid bolt	261.00	294.77	1.50	41.56	nr	**336.33**
Electric mortice lock	98.00	110.68	1.00	27.70	nr	**138.38**

W: COMMUNICATIONS/SECURITY/CONTROL

Item	Net Price £	Material £	Labour hours	Labour £	Unit	Total rate £
W41: SECURITY DETECTION & ALARM						
SECURITY DETECTION & ALARM EQUIPMENT						
Detection and alarm systems for the protection of property and persons; includes fixing of equipment to backgrounds and termination of power and data cabling; excludes cable containment and cable installation						
Detection, alarm equipment						
Alarm contact flush	5.00	5.65	1.00	27.70	nr	**33.35**
Alarm contact surface	5.00	5.65	1.00	27.70	nr	**33.35**
Roller shutter contact	16.00	18.07	1.00	27.70	nr	**45.77**
Personal attack button	17.00	19.20	1.00	27.70	nr	**46.90**
Acoustic break glass detectors	23.00	25.98	2.00	55.41	nr	**81.39**
Vibration detectors	30.00	33.88	2.00	55.41	nr	**89.29**
12 m PIR detector	35.00	39.53	1.50	41.56	nr	**81.09**
15 m dual detector	58.00	65.50	1.50	41.56	nr	**107.06**
8 zone alarm panel	308.00	347.85	3.00	83.11	nr	**430.96**
8–24 zone end station	302.66	341.82	4.00	110.81	nr	**452.63**
Remote keypad	102.00	115.20	2.00	55.41	nr	**170.61**
8 zone expansion	78.00	88.10	1.50	41.56	nr	**129.66**
Final exit set button	8.31	9.38	1.00	27.70	nr	**37.08**
Self contained external sounder	75.00	84.70	2.00	55.41	nr	**140.11**
Internal loudspeaker	15.00	16.94	2.00	55.41	nr	**72.35**
Rechargeable battery 12 volt 7 AHr (ampere hours)	12.00	13.56	0.25	6.93	nr	**20.49**
Surveillance equipment						
Vandal-resistant camera, colour	300.00	338.81	3.00	83.11	nr	**421.92**
External camera, colour	490.00	553.40	5.00	138.52	nr	**691.92**
Auto dome external, colour	1400.00	1581.15	5.00	138.52	nr	**1719.67**
Auto dome external, colour/monochrome	1400.00	1581.15	5.00	138.52	nr	**1719.67**
Auto dome internal, colour	1000.00	1129.39	4.00	110.81	nr	**1240.20**
Auto dome internal colour/monochrome	1000.00	1129.39	4.00	110.81	nr	**1240.20**
Mini internal domes	490.00	553.40	3.00	83.11	nr	**636.51**
Camera switcher, 32 inputs	1063.41	1201.01	4.00	110.81	nr	**1311.82**
Full function keyboard	200.00	225.88	1.00	27.70	nr	**253.58**
16 CH multiplexors, duplex	589.36	665.62	2.00	55.41	nr	**721.03**
16 way DVR, 250 GB harddrive	1500.00	1694.10	3.00	83.11	nr	**1777.21**
Additional 250 GB harddrive	200.00	221.46	3.00	81.49	nr	**302.95**
16 way DVR, 500 GB harddrive	4450.00	6719.90	3.00	83.11	nr	**6803.01**
16 way DVR, 750 GB harddrive	4450.00	15077.43	1.00	27.70	nr	**15105.13**
10" colour monitor	265.06	299.36	2.00	55.41	nr	**354.77**
15" colour monitor	311.87	352.23	2.00	55.41	nr	**407.64**
17" colour monitor, high resolution	250.00	282.35	2.00	55.41	nr	**337.76**
21" colour monitor, high resolution	300.00	338.81	2.00	55.41	nr	**394.22**

W: COMMUNICATIONS/SECURITY/CONTROL

Item	Net Price £	Material £	Labour hours	Labour £	Unit	**Total rate £**
IP CCTV						
Internal fixed dome IP camera h.264	565.00	565.00	–	–	nr	**565.00**
Internal fixed dome IP dome camera 1MP	707.00	707.00	–	–	nr	**707.00**
External fixed dome IP	773.00	773.00	–	–	nr	**773.00**
External PTZ IP	3465.00	3465.00	–	–	nr	**3465.00**
24 Port PoE switch	2805.00	2805.00	–	–	nr	**2805.00**
16 Base channel network video recorder with 4TB storage	5000.00	5000.00	–	–	nr	**5000.00**
IP camera channel licence	127.00	127.00	–	–	nr	**127.00**

W: COMMUNICATIONS/SECURITY/CONTROL

Item	Net Price £	Material £	Labour hours	Labour £	Unit	Total rate £
W50: FIRE DETECTION AND ALARM						
STANDARD FIRE DETECTION						
CONTROL PANEL						
Zone control panel; 2 × 12 volt batteries/ charge up to 48 hours standby; mild steel case; flush or surface mounting						
1 zone	161.31	182.18	3.00	83.11	nr	**265.29**
2 zone	217.13	245.23	3.51	97.21	nr	**342.44**
4 zone	255.65	288.73	4.00	110.81	nr	**399.54**
8 zone	379.28	428.36	5.00	138.52	nr	**566.88**
12 zone	457.09	516.23	6.00	166.23	nr	**682.46**
16 zone	770.71	870.44	6.00	166.23	nr	**1036.67**
24 zone	1037.17	1171.38	6.00	166.23	nr	**1337.61**
Repeater panels						
8 zone	283.20	319.84	4.00	110.81	nr	**430.65**
EQUIPMENT						
Manual call point units: plastic covered						
Surface mounted						
Call point	9.17	10.35	0.50	13.86	nr	**24.21**
Call point; Weatherproof	84.74	95.71	0.80	22.15	nr	**117.86**
Flush mounted						
Call point	9.17	10.35	0.56	15.51	nr	**25.86**
Call point; Weatherproof	84.74	95.71	0.86	23.84	nr	**119.55**
Detectors						
Smoke, ionization type with mounting base	31.76	35.87	0.75	20.78	nr	**56.65**
Smoke, optical type with mounting base	31.76	35.87	0.75	20.78	nr	**56.65**
Fixed temperature heat detector with mounting base (60°C)	25.56	28.87	0.75	20.78	nr	**49.65**
Rate of Rise heat detector with mounting base (90°C)	23.95	27.05	0.75	20.78	nr	**47.83**
Duct detector including optical smoke detector and base	203.04	229.32	2.00	55.41	nr	**284.73**
Remote smoke detector LED indicator with base	8.18	9.24	0.50	13.86	nr	**23.10**
Sounders						
6" bell, conduit box	17.68	19.97	0.75	20.78	nr	**40.75**
6" bell, conduit box; weatherproof	22.97	25.94	0.75	20.78	nr	**46.72**
Siren; 230 V	62.81	70.94	1.25	34.63	nr	**105.57**
Magnetic door holder; 230 V ; surface fixed	59.24	66.90	1.50	41.60	nr	**108.50**

W: COMMUNICATIONS/SECURITY/CONTROL

Item	Net Price £	Material £	Labour hours	Labour £	Unit	**Total rate £**
ADDRESSABLE FIRE DETECTION						
CONTROL PANEL						
Analogue addressable panel; BS EN54 Part 2 and 4 1998; incorporating 120 addresses per loop (maximum 1–2 km length); sounders wired on loop; sealed lead acid integral battery standby providing 48 hour standby; 24 volt DC; mild steel case; surface fixed						
1 loop; 4 × 12 volt batteries	1250.34	1412.13	6.00	166.23	nr	**1578.36**
Extra for 1 loop panel						
Loop card	296.70	335.09	1.00	27.70	nr	**362.79**
Repeater panel	868.04	980.36	6.00	166.23	nr	**1146.59**
Network nodes	1485.41	1677.61	6.00	166.23	nr	**1843.84**
Interface unit; for other systems						
Mains powered	444.82	502.38	1.50	41.56	nr	**543.94**
Loop powered	195.26	220.52	1.00	27.70	nr	**248.22**
Single channel I/O	101.46	114.59	1.00	27.70	nr	**142.29**
Zone module	126.20	142.52	1.50	41.56	nr	**184.08**
4 loop; 4 × 12 volt batteries; 24 hour standby; 30 minute alarm	2096.16	2367.39	8.00	221.64	nr	**2589.03**
Extra for 4 loop panel						
Loop card	439.55	496.42	1.00	27.70	nr	**524.12**
Repeater panel	1244.70	1405.75	6.00	166.23	nr	**1571.98**
Mimic panel	3055.90	3451.32	5.00	138.52	nr	**3589.84**
Network nodes	1484.25	1676.31	6.00	166.23	nr	**1842.54**
Interface unit; for other systems						
Mains powered	444.82	502.38	1.50	41.56	nr	**543.94**
Loop powered	195.26	220.52	1.00	27.70	nr	**248.22**
Single channel I/O	101.46	114.59	1.00	27.70	nr	**142.29**
Zone module	126.20	142.52	1.50	41.56	nr	**184.08**
Line modules	23.27	26.29	1.00	27.70	nr	**53.99**
8 loop; 4 × 12 volt batteries; 24 hour standby; 30 minute alarm	4273.13	4826.05	12.00	332.45	nr	**5158.50**
Extra for 8 loop panel						
Loop card	439.55	496.42	1.00	27.70	nr	**524.12**
Repeater panel	1244.70	1405.75	6.00	166.23	nr	**1571.98**
Mimic panel	3055.90	3451.32	5.00	138.52	nr	**3589.84**
Network nodes	1484.25	1676.31	6.00	166.23	nr	**1842.54**
Interface unit; for other systems						
Mains powered	444.82	502.38	1.50	41.56	nr	**543.94**
Loop powered	195.28	220.54	1.00	27.70	nr	**248.24**
Single channel I/O	42.58	48.09	1.00	27.70	nr	**75.79**
Zone module	126.20	142.52	1.50	41.56	nr	**184.08**
Line modules	23.27	26.29	1.00	27.70	nr	**53.99**

W: COMMUNICATIONS/SECURITY/CONTROL

Item	Net Price £	Material £	Labour hours	Labour £	Unit	Total rate £
W50: FIRE DETECTION AND ALARM – cont						
EQUIPMENT						
Manual call point						
Surface mounted						
Call point	58.61	66.20	1.00	27.70	nr	**93.90**
Call point; weatherproof	214.75	242.54	1.25	34.63	nr	**277.17**
Flush mounted						
Call point	58.61	66.20	1.00	27.70	nr	**93.90**
Call point; weatherproof	214.75	242.54	1.25	34.63	nr	**277.17**
Detectors						
Smoke, ionization type with mounting base	59.69	67.41	0.75	20.78	nr	**88.19**
Smoke, optical type with mounting base	59.08	66.73	0.75	20.78	nr	**87.51**
Fixed temperature heat detector with mounting base (60°C)	59.39	67.08	0.75	20.78	nr	**87.86**
Rate of Rise heat detector with mounting base (90°C)	59.39	67.08	0.75	20.78	nr	**87.86**
Duct Detector including optical smoke detector and addressable base	378.43	427.40	2.00	55.41	nr	**482.81**
Beam smoke detector with transmitter and receiver unit	583.41	658.90	2.00	55.41	nr	**714.31**
Zone short circuit isolator	37.52	42.37	0.75	20.78	nr	**63.15**
Plant interface unit	27.64	31.21	0.50	13.86	nr	**45.07**
Sounders						
Xenon flasher, 24 volt, conduit box	63.01	71.17	0.50	13.86	nr	**85.03**
Xenon flasher, 24 volt, conduit box; weatherproof	93.80	105.94	0.50	13.86	nr	**119.80**
6" bell, conduit box	17.68	97.70	0.75	20.78	nr	**118.48**
6" bell, conduit box; weatherproof	51.13	57.74	0.75	20.78	nr	**78.52**
Siren; 24 V polarized	63.81	72.06	1.00	27.70	nr	**99.76**
Siren; 240 V	62.81	70.94	1.25	34.63	nr	**105.57**
Magnetic door holder; 240 V ; surface fixed	59.24	66.90	1.50	41.60	nr	**108.50**

W: COMMUNICATIONS/SECURITY/CONTROL

Item	Net Price £	Material £	Labour hours	Labour £	Unit	**Total rate £**
W51: EARTHING AND BONDING						
EARTH BAR						
Earth bar; polymer insulators and base mounting; including connections						
Non disconnect link						
6 way	163.40	184.54	0.81	22.44	nr	**206.98**
8 way	180.93	204.34	0.81	22.44	nr	**226.78**
10 way	210.01	237.18	0.81	22.44	nr	**259.62**
Disconnect link						
6 way	184.07	207.89	1.01	27.99	nr	**235.88**
8 way	203.51	229.85	1.01	27.99	nr	**257.84**
10 way	225.69	254.90	1.01	27.99	nr	**282.89**
Soild earth bar; including connections						
150 × 50 × 6 mm	49.79	56.23	1.01	27.99	nr	**84.22**
Extra for earthing						
Disconnecting link						
300 × 50 × 6 mm	59.27	66.94	1.16	32.13	nr	**99.07**
500 × 50 × 6 mm	65.20	73.63	1.16	32.13	nr	**105.76**
Crimp lugs; including screws and connections to cable						
25 mm	0.46	0.52	0.31	8.59	nr	**9.11**
35 mm	0.63	0.71	0.31	8.59	nr	**9.30**
50 mm	0.73	0.83	0.32	8.86	nr	**9.69**
70 mm	1.30	1.47	0.32	8.86	nr	**10.33**
95 mm	2.17	2.45	0.46	12.74	nr	**15.19**
120 mm	1.54	1.74	1.25	34.63	nr	**36.37**
Earth clamps; connection to pipework						
15 mm to 32 mm dia.	1.33	1.50	0.15	4.15	nr	**5.65**
32 mm to 50 mm dia.	1.66	1.88	0.18	4.99	nr	**6.87**
50 mm to 75 mm dia.	1.95	2.20	0.20	5.55	nr	**7.75**

W: COMMUNICATIONS/SECURITY/CONTROL

Item	Net Price £	Material £	Labour hours	Labour £	Unit	Total rate £
W52: LIGHTNING PROTECTION						
CONDUCTOR TAPE						
PVC sheathed copper tape						
25 × 3 mm	20.45	23.09	0.30	8.31	m	**31.40**
25 × 6 mm	38.67	43.68	0.30	8.31	m	**51.99**
50 × 6 mm	79.04	89.27	0.30	8.31	m	**97.58**
PVC sheathed copper solid circular conductor						
8 mm	11.93	13.47	0.50	13.86	m	**27.33**
Bare copper tape						
20 × 3 mm	16.78	18.95	0.30	8.31	m	**27.26**
25 × 3 mm	19.23	21.72	0.30	8.31	m	**30.03**
25 × 6 mm	35.24	39.80	0.40	11.08	m	**50.88**
50 × 6 mm	68.17	76.99	0.50	13.86	m	**90.85**
Bare copper solid circular conductor						
8 mm	10.85	12.25	0.50	13.86	m	**26.11**
Tape fixings; flat; metalic						
PVC sheathed copper						
25 × 3 mm	20.45	23.09	0.33	9.14	nr	**32.23**
25 × 6 mm	38.67	43.68	0.33	9.14	nr	**52.82**
50 × 6 mm	79.04	89.27	0.33	9.14	nr	**98.41**
8 mm	7.27	8.21	0.50	13.86	nr	**22.07**
Bare copper						
20 × 3 mm	17.62	19.90	0.30	8.31	nr	**28.21**
25 × 3 mm	19.23	21.72	0.30	8.31	nr	**30.03**
25 × 6 mm	35.24	39.80	0.40	11.08	nr	**50.88**
50 × 6 mm	68.17	76.99	0.50	13.86	nr	**90.85**
8 mm	11.39	12.86	0.50	13.86	nr	**26.72**
Tape fixings; flat; non-metalic; PVC sheathed copper						
25 × 3 mm	2.38	2.69	0.30	8.31	nr	**11.00**
Tape fixings; flat; non-metalic; Bare copper						
20 × 3 mm	0.94	1.06	0.30	8.31	nr	**9.37**
25 × 3 mm	0.94	1.06	0.30	8.31	nr	**9.37**
50 × 6 mm	5.28	5.97	0.30	8.31	nr	**14.28**
Puddle flanges; copper						
600 mm long	108.96	123.06	0.93	25.77	nr	**148.83**

W: COMMUNICATIONS/SECURITY/CONTROL

Item	Net Price £	Material £	Labour hours	Labour £	Unit	Total rate £
AIR RODS						
Pointed air rod fixed to structure; copper						
10 mm dia.						
500 mm long	18.14	20.49	1.00	27.70	nr	**48.19**
1000 mm long	30.58	34.54	1.50	41.56	nr	**76.10**
Extra for						
Air terminal base	22.64	25.57	0.35	9.70	nr	**35.27**
Strike Pad	24.05	27.16	0.35	9.70	nr	**36.86**
16 mm dia.						
500 mm long	28.75	32.47	0.91	25.20	nr	**57.67**
1000 mm long	52.52	59.31	1.75	48.48	nr	**107.79**
2000 mm long	95.91	108.32	2.50	69.27	nr	**177.59**
Extra for						
Multiple point	48.30	54.55	0.35	9.70	nr	**64.25**
Air terminal base	26.54	29.98	0.35	9.70	nr	**39.68**
Ridge saddle	49.83	56.27	0.35	9.70	nr	**65.97**
Side mounting bracket	49.59	56.01	0.50	13.86	nr	**69.87**
Rod to tape coupling	20.01	22.60	0.50	13.86	nr	**36.46**
Strike Pad	24.05	27.16	0.35	9.70	nr	**36.86**
AIR TERMINALS						
16 mm dia.						
500 mm long	28.75	32.47	0.65	18.00	nr	**50.47**
1000 mm long	52.52	59.31	0.78	21.60	nr	**80.91**
2000 mm long	95.91	108.32	1.50	41.56	nr	**149.88**
Extra for						
Multiple point	48.30	54.55	0.35	9.70	nr	**64.25**
Flat saddle	49.59	56.01	0.35	9.70	nr	**65.71**
Side bracket	24.78	27.99	0.50	13.86	nr	**41.85**
Rod to cable coupling	20.58	23.25	0.50	13.86	nr	**37.11**
BONDS AND CLAMPS						
Bond to flat surface; copper						
26 mm	4.31	4.87	0.45	12.46	nr	**17.33**
8 mm dia.	17.78	20.08	0.33	9.14	nr	**29.22**
Pipe bond						
26 mm	4.31	4.87	0.45	12.46	nr	**17.33**
8 mm dia.	44.19	49.91	0.33	9.14	nr	**59.05**

W: COMMUNICATIONS/SECURITY/CONTROL

Item	Net Price £	Material £	Labour hours	Labour £	Unit	Total rate £
W52: LIGHTNING PROTECTION – cont						
BONDS AND CLAMPS – cont						
Rod to tape clamp						
26 mm	7.97	9.00	0.45	12.46	nr	**21.46**
Square clamp; copper						
25 × 3 mm	9.06	10.23	0.33	9.14	nr	**19.37**
50 × 6 mm	46.35	52.35	0.50	13.86	nr	**66.21**
8 mm dia.	9.68	10.93	0.33	9.14	nr	**20.07**
Test clamp; copper						
26 × 8 mm; oblong	13.49	15.24	0.50	13.86	nr	**29.10**
26 × 8 mm; plate type	41.07	46.38	0.50	13.86	nr	**60.24**
26 × 8 mm; screw down	36.91	41.69	0.50	13.86	nr	**55.55**
Cast in earth points						
2 hole	30.00	33.88	0.75	20.78	nr	**54.66**
4 hole	44.52	50.28	1.00	27.70	nr	**77.98**
Extra for cast in earth points						
Cover plate; 25 × 3 mm	33.67	38.03	0.25	6.93	nr	**44.96**
Cover plate; 8 mm	33.67	38.03	0.25	6.93	nr	**44.96**
Rebar clamp; 8 mm	23.19	26.19	0.25	6.93	nr	**33.12**
Static earth receptacle	157.76	178.17	0.50	13.86	nr	**192.03**
Copper braided bonds						
25 × 3 mm						
200 mm hole centres	15.57	17.58	0.33	9.14	nr	**26.72**
400 mm holes centres	23.87	26.96	0.40	11.08	nr	**38.04**
U bolt clamps						
16 mm	10.29	11.62	0.33	9.14	nr	**20.76**
20 mm	11.66	13.17	0.33	9.14	nr	**22.31**
25 mm	14.29	16.14	0.33	9.14	nr	**25.28**
EARTH PITS/MATS						
Earth inspection pit; hand to others for fixing						
Concrete	46.26	52.24	1.00	27.70	nr	**79.94**
Polypropylene	45.21	51.06	1.00	27.70	nr	**78.76**
Extra for						
5 hole copper earth bar; concrete pit	44.85	101.31	0.35	9.70	nr	**111.01**
5 hole earth bar; polypropylene	37.58	42.44	0.35	9.70	nr	**52.14**

W: COMMUNICATIONS/SECURITY/CONTROL

Item	Net Price £	Material £	Labour hours	Labour £	Unit	**Total rate £**
Waterproof electrode seal						
Single flange	281.95	318.43	0.93	25.77	nr	**344.20**
Double flange	474.42	535.81	0.93	25.77	nr	**561.58**
Earth electrode mat; laid in ground and connected						
Copper tape lattice						
600 × 600 × 3 mm	127.25	143.72	0.93	25.77	nr	**169.49**
900 × 900 × 3 mm	228.20	257.72	0.93	25.77	nr	**283.49**
Copper tape plate						
600 × 600 × 1.5 mm	110.25	124.51	0.93	25.77	nr	**150.28**
600 × 600 × 3 mm	218.62	246.91	0.93	25.77	nr	**272.68**
900 × 900 × 1.5 mm	245.13	276.85	0.93	25.77	nr	**302.62**
900 × 900 × 3 mm	475.41	536.93	0.93	25.77	nr	**562.70**
EARTH RODS						
Solid cored copper earth electrodes driven into ground and connected						
15 mm dia.						
1200 mm long	44.78	50.57	0.93	25.77	nr	**76.34**
Extra for						
Coupling	1.35	1.52	0.06	1.66	nr	**3.18**
Driving stud	1.66	1.88	0.06	1.66	nr	**3.54**
Spike	1.54	1.74	0.06	1.66	nr	**3.40**
Rod Clamp; flat tape	7.97	9.00	0.25	6.93	nr	**15.93**
Rod Clamp; solid conductor	3.57	4.03	0.25	6.93	nr	**10.96**
20 mm dia.						
1200 mm long	86.87	98.11	0.98	27.15	nr	**125.26**
Extra for						
Coupling	1.35	1.52	0.06	1.66	nr	**3.18**
Driving stud	2.73	3.08	0.06	1.66	nr	**4.74**
Spike	2.51	2.84	0.06	1.66	nr	**4.50**
Rod Clamp; flat tape	7.97	9.00	0.25	6.93	nr	**15.93**
Rod Clamp; solid conductor	3.95	4.46	0.25	6.93	nr	**11.39**
Stainless steel earth electrodes driven into ground and connected						
16 mm dia.						
1200 mm long	57.21	64.62	0.93	25.77	nr	**90.39**

W: COMMUNICATIONS/SECURITY/CONTROL

Item	Net Price £	Material £	Labour hours	Labour £	Unit	Total rate £
W52: LIGHTNING PROTECTION – cont						
EARTH RODS – cont						
Extra for						
Coupling	1.75	1.98	0.06	1.66	nr	**3.64**
Driving head	1.66	1.88	0.06	1.66	nr	**3.54**
Spike	1.54	1.74	0.06	1.66	nr	**3.40**
Rod Clamp; flat tape	7.97	9.00	0.25	6.93	nr	**15.93**
Rod Clamp; solid conductor	3.57	4.03	0.25	6.93	nr	**10.96**
SURGE PROTECTION						
Single Phase; including connection to equipment						
90–150 V	327.00	369.31	5.00	138.52	nr	**507.83**
200–280 V	310.00	350.12	5.00	138.52	nr	**488.64**
Three Phase; including connection to equipment						
156–260 V	660.00	745.40	10.00	277.04	nr	**1022.44**
346–484 V	615.00	694.58	10.00	277.04	nr	**971.62**
349–484 V; remote display	684.00	772.51	10.00	277.04	nr	**1049.55**
346–484 V; 60 kA	1136.00	1283.00	10.00	277.04	nr	**1560.04**
346–484 V; 120 kA	2173.00	2454.17	10.00	277.04	nr	**2731.21**

W: COMMUNICATIONS/SECURITY/CONTROL

Item	Net Price £	Material £	Labour hours	Labour £	Unit	**Total rate £**
W60: CENTRAL CONTROL/BUILDING MANAGEMENT						
EQUIPMENT						
Switches/sensors; includes fixing in position; electrical work elsewhere.						
Note: These are normally free issued to the mechanical contractor for fitting. The labour times applied assume the installation has been prepared for the fitting of the component						
Pressure devices						
Liquid differential pressure sensor	143.80	162.40	0.50	13.86	nr	**176.26**
Liquid differential pressure switch	78.80	88.99	0.50	13.86	nr	**102.85**
Air differential pressure transmitter	104.00	117.45	0.50	13.86	nr	**131.31**
Air differential pressure switch	14.22	16.07	0.50	13.86	nr	**29.93**
Liquid level switch	71.76	81.05	0.50	13.86	nr	**94.91**
Static pressure sensor	81.25	91.76	0.50	13.86	nr	**105.62**
High pressure switch	71.76	81.05	0.50	13.86	nr	**94.91**
Low pressure switch	71.76	81.05	0.50	13.86	nr	**94.91**
Water pressure switch	109.98	124.22	0.50	13.86	nr	**138.08**
Duct averaging temperature sensor	47.84	54.03	1.00	27.70	nr	**81.73**
Temperature devices						
Return air sensor (fan coils)	4.68	5.28	1.00	27.70	nr	**32.98**
Frost thermostat	30.68	34.65	0.50	13.86	nr	**48.51**
Immersion thermostat	49.66	56.09	0.50	13.86	nr	**69.95**
Temperature high limit	49.66	56.09	0.50	13.86	nr	**69.95**
Temperature sensor with averaging element	47.84	54.03	0.50	13.86	nr	**67.89**
Immersion temperature sensor	20.32	22.95	0.50	13.86	nr	**36.81**
Space temperature sensor	5.89	6.65	1.00	27.70	nr	**34.35**
Combined space temperature & humidity sensor	81.25	91.76	1.00	27.70	nr	**119.46**
Outside air temperature sensor	11.38	12.85	2.00	55.41	nr	**68.26**
Outside air temperature & humidity sensor	266.50	300.98	2.00	55.41	nr	**356.39**
Duct humidity sensor	104.81	118.37	0.50	13.86	nr	**132.23**
Space humidity sensor	81.25	91.76	1.00	27.70	nr	**119.46**
Immersion water flow sensor	20.32	22.95	0.50	13.86	nr	**36.81**
Rain sensor	190.13	214.73	2.00	55.41	nr	**270.14**
Wind speed and direction sensor	601.25	679.04	2.00	55.41	nr	**734.45**

W: COMMUNICATIONS/SECURITY/CONTROL

Item	Net Price £	Material £	Labour hours	Labour £	Unit	Total rate £
W60: CENTRAL CONTROL/BUILDING MANAGEMENT – cont						
Controllers; includes fixing in position; electrical work elsewhere						
Zone						
Fan coil controller	153.56	173.43	2.00	55.41	nr	**228.84**
VAV controller	200.81	226.80	2.00	55.41	nr	**282.21**
Plant						
Controller, 96 I/O points (exact configuration is dependent upon the number of I/O boards added)	2818.56	3183.27	0.50	13.86	nr	**3197.13**
Controller, 48 I/O points (exact configuration is dependent upon the number of I/O boards added)	2353.00	2657.47	0.50	13.86	nr	**2671.33**
Controller, 32 I/O points (exact configuration is dependent upon the number of I/O boards added)	1954.00	2206.84	0.50	13.86	nr	**2220.70**
Additional Digital Input Boards (12 DI)	465.56	525.80	0.20	5.55	nr	**531.35**
Additional Digital Output Boards (6 DO)	398.13	449.65	0.20	5.55	nr	**455.20**
Additional analogue Input Boards (8 AI)	398.13	449.65	0.20	5.55	nr	**455.20**
Additional analogue Output Boards (8 AO)	398.13	449.65	0.20	5.55	nr	**455.20**
Outstation Enclosure (fitted in riser with space allowance for controller and network device)	398.13	449.65	5.00	138.52	nr	**588.17**
Damper actuator; electrical work elsewhere						
Damper actuator 0–10v	84.50	95.43	–	–	nr	**95.43**
Damper actuator with auxiliary switches	56.06	63.31	–	–	nr	**63.31**
Frequency inverters: not mounted within MCC; includes fixing in position; electrical work elsewhere						
2.2 kW	643.00	726.20	2.00	55.41	nr	**781.61**
3 kW	711.00	802.99	2.00	55.41	nr	**858.40**
7.5 kW	1052.00	1188.13	2.00	55.41	nr	**1243.54**
11 kW	1286.00	1452.40	2.00	55.41	nr	**1507.81**
15 kW	1554.00	1755.08	2.00	55.41	nr	**1810.49**
18.5 kW	1816.00	2050.99	2.50	69.27	nr	**2120.26**
20 kW	2050.00	2315.26	2.50	69.27	nr	**2384.53**
30 kW	2484.00	2805.42	2.50	69.27	nr	**2874.69**
55 kW	4486.00	5066.46	3.00	83.11	nr	**5149.57**
Miscellaneous; includes fixing in position; electrical work elsewhere						
1 kW thyristor	57.25	64.66	2.00	55.41	nr	**120.07**
10 kW thyristor	110.25	124.51	2.00	55.41	nr	**179.92**

W: COMMUNICATIONS/SECURITY/CONTROL

Item	Net Price £	Material £	Labour hours	Labour £	Unit	**Total rate £**
Front end and networking; electrical work elsewhere						
PC/monitor	850.00	959.98	2.00	55.41	nr	**1015.39**
Dot matrix printer	250.00	282.35	2.00	55.41	nr	**337.76**
PC Software	–	–	–	–	nr	–
Network server software	2044.00	2308.48	–	–	nr	**2308.48**
Router (allows connection to a network)	103.00	116.33	–	–	nr	**116.33**

PART 4

Rates of Wages

Mechanical Installations

Rates of Wages

HEATING, VENTILATING, AIR CONDITIONING, PIPING AND DOMESTIC ENGINEERING INDUSTRY

For full details of the wage agreement and the Heating Ventilating Air Conditioning Piping and Domestic Engineering Industry's National Working Rule Agreement, contact:

Building & Engineering Services Association
ESCA House
34 Palace Court
Bayswater
London W2 4JG

Tel: 020 7313 4900
www.b-es.org

WAGE RATES, ALLOWANCES AND OTHER PROVISIONS

Hourly rates of wages

All districts of the United Kingdom

Main Grades	*From 1 April 2013 p/hr*
Foreman	15.23
Senior Craftsman (+2nd welding skill)	13.11
Senior Craftsman	12.59
Craftsman (+2nd welding skill)	12.07
Craftsman	11.56
Operative	10.46
Adult Trainee	8.82
Mate (18 and over)	8.82
Mate (16–17)	4.09
Modern Apprentices	
Junior	5.72
Intermediate	8.12
Senior	10.46

Note: Ductwork Erection Operatives are entitled to the same rates and allowances as the parallel Fitter grades shown.

HEATING, VENTILATING, AIR CONDITIONING, PIPING AND DOMESTIC ENGINEERING INDUSTRY

Trainee *Rates of Pay*

Junior Ductwork Trainees (Probationary)

Age at entry	*From 1 April 2013 p/hr*
17	5.19
18	5.19
19	5.19
20	5.19

Junior Ductwork Erectors (Year of Training)

	From 1 April 2013		
Age at entry	*1 yr p/h*	*2 yr p/h*	*3 yr p/h*
17	6.46	8.04	9.12
18	6.46	8.04	9.12
19	6.46	8.04	9.12
20	6.46	8.04	9.12

Responsibility Allowance (Craftsmen)	*From 1 April 2013 p/hr*
Second welding skill or supervisory responsibility (one unit)	0.52
Second welding skill and supervisory responsibility (two units)	0.52

Responsibility Allowance (Senior Craftsmen)	*From 1 April 2013 p/hr*
Second welding skill	0.52
Supervising responsibility	0.52
Second welding skill and supervisory responsibility	0.52

Daily travelling allowance – Scale 2

C: Craftsmen including Installers

M&A: Mates, Apprentices and Adult Trainees

Direct distance from centre to job in miles

		From 1 April 2013	
Over	*Not exceeding*	*C p/hr*	*p/hr M&A*
15	20	2.53	2.17
20	30	6.51	5.63
30	40	9.36	8.10
40	50	12.33	10.55

HEATING, VENTILATING, AIR CONDITIONING, PIPING AND DOMESTIC ENGINEERING INDUSTRY

Daily travelling allowance – Scale 1

C: Craftsmen including Installers

M&A: Mates, Apprentices and Adult Trainees

Direct distance from centre to job in miles

		From 1 April 2013	
Over	*Not exceeding*	*C p/hr*	*M&A p/hr*
0	15	6.92	6.92
15	20	9.45	9.09
20	30	13.43	12.55
30	40	16.28	15.02
40	50	19.24	17.48

Weekly Holiday Credit and Welfare Contributions

	From 1 April 2013						
	£	£	£	£	£	£	£
	a	b	c	d	e	f	g
Weekly Holiday Credit	73.77	68.50	65.98	63.46	60.95	58.48	55.96
Combined Weekly/Welfare Holiday Credit and Contribution	81.95	76.78	74.16	71.64	69.13	66.66	64.14

	From 1 April 2013			
	£	£	£	£
	h	i	j	k
Weekly Holiday Credit	50.68	42.71	39.38	27.74
Combined Weekly/Welfare Holiday Credit and Contribution	58.82	50.89	47.56	35.92

HEATING, VENTILATING, AIR CONDITIONING, PIPING AND DOMESTIC ENGINEERING INDUSTRY

The grades of H&V Operatives entitled to the different rates of Weekly Holiday Credit and Welfare Contribution are as follows:

a Foreman	*b* Senior Craftsman (RAS & RAW)	*c* Senior Craftsman (RAS)	*d* Senior Craftsman (RAW)
e Senior Craftsman Craftsman (+2 RA)	*f* Craftsman (+ 1RA)	*g* Craftsman	*h* Installer Senior Modern Apprentice
i Adult Trainee Mate (over 18)	*j* Intermediate Modern Apprentice	*k* Junior Modern Apprentice	*l* No grade allocated to this Credit Value Category

Daily abnormal conditions money	*From 1 April 2013*
Per day	3.10

Lodging allowance	*From 1 April 2013*
Per night	35.00

Explanatory Notes

1. Working Hours

   The normal working week (Monday to Friday) shall be 38 hours.

2. Overtime

   Time worked in excess of 38 hours during the normal working week shall be paid at time and a half until 12 hours have been worked since the actual starting time. Thereafter double time shall be paid until normal starting time the following morning. Weekend overtime shall be paid at time and a half for the first 5 hours worked on a Saturday and at double time thereafter until normal starting time on Monday morning.

PLUMBING MECHANICAL ENGINEERING SERVICES INDUSTRY

The Joint Industry Board for Plumbing Mechanical Engineering Services has agreed a two year wage agreement for 2011–2012 with effect from 3 January 2011. Current rates of pay for 2012 will be effective from 2 January 2012.

For full details of this wage agreement and the JIB PMES National Working Rules, contact:

The Joint Industry Board for Plumbing Mechanical Engineering Services in England and Wales

JIB-PMES
PO Box 267
PE19 9DN

Telephone: 01480 476925
Email: info@jib-pmes.org.uk

WAGE RATES, ALLOWANCES AND OTHER PROVISIONS

EFFECTIVE FROM **2 January 2012**

Basic Rates of Hourly Pay

Applicable in England and Wales

	Hourly rate £
Operatives	
Technical plumber and gas service technician	14.99
Advanced plumber and gas service engineer	13.50
Trained plumber and gas service fitter	11.58
Apprentices	
4th year of training with NVQ level 3	11.20
4th year of training with NVQ level 2	10.15
4th year of training	8.94
3rd year of training with NVQ level 2	8.83
3rd year of training	7.26
2nd year of training	6.43
1st year of training	5.61
Adult Trainees	
3rd 6 months of employment	10.09
2nd 6 months of employment	9.69
1st 6 months of employment	9.03

PLUMBING MECHANICAL ENGINEERING SERVICES INDUSTRY

Major Projects Agreement

Where a job is designated as being a Major Project then the following Major Project Performance Payment hourly rate supplement shall be payable:

Employee Category

	National Payment £	*London* Payment* £
Technical Plumber and Gas Service Technician	2.20	3.57
Advanced Plumber and Gas Service Engineer	2.20	3.57
Trained Plumber and Gas Service Fitter	2.20	3.57
All 4th year apprentices	1.76	2.86
All 3rd year apprentices	1.32	2.68
2nd year apprentice	1.21	1.96
1st year apprentice	0.88	1.43
All adult trainees	1.76	2.86

* The London Payment Supplement applies only to designated Major Projects that are within the M25 London orbital motorway and are effective from 1 February 2007.

* National payment hourly rates are unchanged for 2007 and will continue to be at the rates shown in Promulgation 138 A issued 14 October 2003.

Allowances

Daily travel time allowance plus return fares

All daily travel allowances are to be paid at the daily rate as follows:

Over	*Not exceeding*	*All Operatives*	*3rd & 4th Year Apprentices*	*1 st & 2nd Year Apprentices*
20	30	£4.24	£2.74	£1.70
30	40	£9.89	£6.35	£4.07
40	50	£11.30	£6.75	£4.24

PLUMBING MECHANICAL ENGINEERING SERVICES INDUSTRY

Responsibility/Incentive Pay Allowance

As from Monday 2 January 2012, Employers may, in consultation with the employees concerned, enhance the basic graded rates of pay by the payment of an additional amount, as per the bands shown below, where it is agreed that their work involves extra responsibility, productivity or flexibility.

Band 1	an additional rate of £ 0.26 per hour
Band 2	an additional rate of £ 0.46 per hour
Band 3	an additional rate of £ 0.68 per hour
Band 4	an additional rate of £ 0.90 per hour

This allowance forms part of an operative's basic rate of pay and shall be used to calculate premium payments.

Mileage allowance	£0.40 per mile
Lodging allowance	£28.35 per night
Subsistence Allowance (London Only)	£4.87 per night

Plumbers welding supplement

Possession of Gas or Arc Certificate	£0.29 per hour
Possession of Gas and Arc Certificate	£0.48 per hour

Weekly Holiday Credit Contributions (38th Issue Stamps Option)

	Public & Annual Gross Value, £	*Combined Holiday Credit*, £*
Technical Plumber and Gas Service Technician	62.50	62.90
Advance Plumber and Gas Service Engineer	56.20	56.60
Trained Plumber and Gas Service Fitter	48.10	48.50
Adult Trainee	45.55	48.50
Apprentice in last year of training	36.30	35.60
Apprentice 3rd year	26.15	25.55
Apprentice 2nd year	23.10	22.60
Apprentice 1st year	20.15	20.30
Working Principal	26.00	26.40
Ancillary Employee	31.90	32.30

* Public and Annual gross value is the Combined Holiday credit value less JIB administration value.

PLUMBING MECHANICAL ENGINEERING SERVICES INDUSTRY

Explanatory Notes

1. Working Hours

   The normal working week (Monday to Friday) shall be 37½ hours, with 45 hours to be worked in the same period before overtime rates become applicable.

2. Overtime

   Overtime shall be paid at time and a half up to 8.00 pm (Monday to Friday) and up to 1.00 pm (Saturday). Overtime worked after these times shall be paid at double time.

3. Major Projects Agreement

   Under the Major Projects Agreement the normal working week shall be 38 hours (Monday to Friday) with overtime rates payable for all hours worked in excess of 38 hours in accordance with 2 above. However, it should be noted that the hourly rate supplement shall be paid for each hour worked but does <u>not</u> attract premium time enhancement.

4. Pension

   In addition to their hourly rates of pay, plumbing employees are entitled to inclusion within the Industry Pension Scheme (or one providing equivalent benefits). The current levels of industry scheme contributions are 6½% (employers) and 3¼% (employees).

5. Weekly Holiday Credit Contributions

   There has been a fundamental change in the way the JIB-PMES Holiday Pay Schemes shall apply for the 38th issue, in that 60 credits will be paid over a period of 53 weeks.

   For full details please refer to the 38th issue of the Holiday Credit Values as published by the JIB for Plumbing Mechanical Engineering Services in England and Wales.

Electrical Installations

Rates of Wages

ELECTRICAL CONTRACTING INDUSTRY

For full details of this wage agreement and the Joint Industry Board for the Electrical Contracting Industry's National Working Rules, contact:

Joint Industry Board
PO Box 127
Swanley
Kent
BR8 9BH

Tel: 0333 321 230
www.jib.org.uk

WAGES (Graded Operatives)

Rates

Since 7 January 2002 two different wage rates have applied to JIB Graded Operatives working on site, depending on whether the Employer transports them to site or whether they provide their own transport. The two categories are:

Job Employed (Transport Provided)

Payable to an Operative who is transported to and from the job by his Employer. The Operative shall also be entitled to payment for Travel Time, when travelling in his own time, as detailed in the appropriate scale.

Job Employed (Own Transport)

Payable to an Operative who travels by his own means to and from the job. The Operative shall be entitled to payment for Travel Allowance and also Travel Time, when travelling in his own time, as detailed in the appropriate scale.

ELECTRICAL CONTRACTING INDUSTRY

The JIB rates of wages are set out below:

From and including 7 January 2013, the JIB hourly rates of wages shall be as set out below:

(i) **National Standard Rate:**

Grade	*Transport Provided*	*Own Transport*
Technician (or equivalent specialist grade)	£15.61	£16.40
Approved Electrician (or equivalent specialist grade)	£13.79	£14.57
Electrician (or equivalent specialist grade)	£12.64	£13.43
Senior Graded Electrical Trainee	£12.01	£12.76
Electrical Improver	£11.37	£12.10
Labourer	£10.04	£10.78
Adult Trainee	£10.04	£10.78

(ii) **London Rate:**

Grade	*Transport Provided*	*Own Transport*
Technician (or equivalent specialist grade)	£17.49	£18.37
Approved Electrician (or equivalent specialist grade)	£15.45	£16.31
Electrician (or equivalent specialist grade)	£14.15	£15.05
Senior Graded Electrical Trainee	£13.44	£14.30
Electrical Improver	£12.73	£13.55
Labourer	£11.25	£12.07
Adult Trainee	£11.25	£12.07

From and including 7 January 2013, the JIB hourly rates for Job Employed apprentices shall be:

(i) **National Standard Rates**

	Transport Provided	*Own Transport*
Stage 1	£4.42	£5.17
Stage 2	£6.51	£7.27
Stage 3	£9.42	£10.20
Stage 4	£9.97	£10.75

(ii) **London Rate**

	Transport Provided	*Own Transport*
Stage 1	£4.94	£5.79
Stage 2	£7.29	£8.14
Stage 3	£10.55	£11.43
Stage 4	£11.17	£12.04

ELECTRICAL CONTRACTING INDUSTRY

Travelling time and travel allowances

From and including 7 January 2013

Operatives required to start/finish at the normal starting and finishing time on jobs which are 15 miles and over from the shop – in a straight line – receive payment for Travelling Time and where transport is not provided by the Employer, Travel Allowance, as follows:

	Distance	*Total Daily Travel Allowance*	*Total Daily Travelling Time*
(a)	**National Standard Rate**		
	Up to 15 miles	Nil	Nil
	Over 15 & up to 20 miles each way	£3.48	£4.75
	Over 20 & up to 25 miles each way	£4.63	£6.02
	Over 25 & up to 35 miles each way	£6.10	£7.36
	Over 35 & up to 55 miles each way	£9.71	£9.71
	Over 55 & up to 75 miles each way	£11.89	£11.89

For each additional 10 mile band over 75 miles, additional payment of £2.09 for Daily Travel Allowance and £2.09 for Daily Travelling Time will be made.

	Distance	*Total Daily Travel Allowance*	*Total Daily Travelling Time*
(b)	**London Rate**		
	Up to 15 miles	Nil	Nil
	Over 15 & up to 20 miles each way	£3.50	£5.11
	Over 20 & up to 25 miles each way	£4.66	£6.69
	Over 25 & up to 35 miles each way	£6.14	£8.03
	Over 35 & up to 55 miles each way	£9.79	£10.84
	Over 55 & up to 75 miles each way	£11.97	£12.69

For each additional 10 mile band over 75 miles, additional payments of £2.11 for Daily Travel Allowance and £2.11 for Daily Travelling Time will be made.

ELECTRICAL CONTRACTING INDUSTRY

Travelling time and travel allowances

Operatives required to start/finish at the normal starting and finishing time on jobs which are 15 miles and over from the shop – in a straight line – receive payment for Travelling Time and where transport is not provided by the Employer, Travel Allowance, as follows:

From and including 7 January 2013

	Distance	*Total Daily Travel Allowance*	*Total Daily Travelling Time*
(a)	**National Standard Rate**		
	Up to 15 miles	Nil	Nil
	Over 15 & up to 20 miles each way	£3.55	£2.85
	Over 20 & up to 25 miles each way	£4.72	£4.27
	Over 25 & up to 35 miles each way	£6.22	£5.70
	Over 35 & up to 55 miles each way	£9.90	£7.13
	Over 55 & up to 75 miles each way	£12.13	£8.56

For each additional 10 mile band over 75 miles, additional payments of £2.13 for Daily Travel Allowance and £2.13 for Daily Travelling Time will be made.

	Distance	*Total Daily Travel Allowance*	*Total Daily Travelling Time*
(b)	**London Rate**		
	Up to 15 miles	Nil	Nil
	Over 15 & up to 20 miles each way	£3.57	£3.17
	Over 20 & up to 25 miles each way	£4.75	£4.75
	Over 25 & up to 35 miles each way	£6.26	£6.32
	Over 35 & up to 55 miles each way	£9.99	£7.88
	Over 55 & up to 75 miles each way	£12.21	£9.50

For each additional 10 mile band over 75 miles, additional payments of £2.15 for Daily Travel Allowance and £2.15 for Daily Travelling Time will be made.

Lodging Allowances

£35.55 from and including 7 January 2013

Lodgings weekend retention fee, maximum reimbursement

£35.55 from and including 7 January 2013

Annual Holiday Lodging Allowance Retention

Maximum £11.81 per night (£82.67 per week) from and including 7 January 2013

ELECTRICAL CONTRACTING INDUSTRY

Responsibility money

From and including 30 March 1998 the minimum payment increased to 10p per hour and the maximum to £1.00 per hour (no change)

From and including 4 January 1992 responsibility payments are enhanced by overtime and shift premiums where appropriate (no change)

Combined JIB Benefits Stamp Value (from week commencing 24 September 2010)

JIB Grade	*Weekly JIB combined credit value* £	*Holiday Value* £
Technician	£62.30	£46.49
Approved Electrician	£56.62	£41.01
Electrician	£53.01	£37.63
Senior Graded Electrical Trainee and Electrical Improver	£49.04	£33.86
Labourer & Adult Trainee	£44.88	£29.89

Explanatory Notes

1. Working Hours

   The normal working week (Monday to Friday) shall be 37½ hours, with 38 hours to be worked in the same period before overtime rates become applicable.

2. Overtime

   Overtime shall be paid at time and a half for all weekday overtime. Saturday overtime shall be paid at time and a half for the first 6 hours, or up to 3.00 pm (whichever comes first). Thereafter double time shall be paid until normal starting time on Monday.

PART 5

Daywork

page
Heating and Ventilating Industry 731
Electrical Industry 734
Building Industry Plant Hire Costs 737

When work is carried out in connection with a contract that cannot be valued in any other way, it is usual to assess the value on a cost basis with suitable allowances to cover overheads and profit. The basis of costing is a matter for agreement between the parties concerned but definitions of prime cost for the Heating and Ventilating and Electrical Industries have been published jointly by the Royal Institution of Chartered Surveyors and the appropriate bodies of the industries concerned, for those who wish to use them.

These, together with a schedule of basic plant hire charges are reproduced on the following pages, with the kind permission of the Royal Institution of Chartered Surveyors, who own the copyright.

HEATING AND VENTILATING INDUSTRY

DEFINITION OF PRIME COST OF DAYWORK CARRIED OUT UNDER A HEATING, VENTILATING, AIR CONDITIONING, REFRIGERATION, PIPEWORK AND/OR DOMESTIC ENGINEERING CONTRACT (JULY 1980 EDITION)

This Definition of Prime Cost is published by the Royal Institution of Chartered Surveyors and the Heating and Ventilating Contractors Association for convenience, and for use by people who choose to use it. Members of the Heating and Ventilating Contractors Association are not in any way debarred from defining Prime Cost and rendering accounts for work carried out on that basis in any way they choose. Building owners are advised to reach agreement with contractors on the Definition of Prime Cost to be used prior to entering into a contract or subcontract.

SECTION 1: APPLICATION

1.1 This Definition provides a basis for the valuation of daywork executed under such heating, ventilating, air conditioning, refrigeration, pipework and or domestic engineering contracts as provide for its use.
1.2 It is not applicable in any other circumstances, such as jobbing or other work carried out as a separate or main contract nor in the case of daywork executed after a date of practical completion.
1.3 The terms 'contract' and 'contractor' herein shall be read as 'subcontract' and 'subcontractor' as applicable.

SECTION 2: COMPOSITION OF TOTAL CHARGES

2.1 The Prime Cost of daywork comprises the sum of the following costs:
 (a) Labour as defined in Section 3.
 (b) Materials and goods as defined in Section 4.
 (c) Plant as defined in Section 5.
2.2 Incidental costs, overheads and profit as defined in Section 6, as provided in the contract and expressed therein as percentage adjustments, are applicable to each of 2.1 (a)–(c).

SECTION 3: LABOUR

3.1 The standard wage rates, emoluments and expenses referred to below and the standard working hours referred to in 3.2 are those laid down for the time being in the rules or decisions or agreements of the Joint Conciliation Committee of the Heating, Ventilating and Domestic Engineering Industry applicable to the works (or those of such other body as may be appropriate) and to the grade of operative concerned at the time when and the area where the daywork is executed.
3.2 Hourly base rates for labour are computed by dividing the annual prime cost of labour, based upon the standard working hours and as defined in 3.4, by the number of standard working hours per annum. See example.
3.3 The hourly rates computed in accordance with 3.2 shall be applied in respect of the time spent by operatives directly engaged on daywork, including those operating mechanical plant and transport and erecting and dismantling other plant (unless otherwise expressly provided in the contract) and handling and distributing the materials and goods used in the daywork.
3.4 The annual prime cost of labour comprises the following:
 (a) Standard weekly earnings (i.e. the standard working week as determined at the appropriate rate for the operative concerned).
 (b) Any supplemental payments.
 (c) Any guaranteed minimum payments (unless included in Section 6.1(a)–(p)).
 (d) Merit money.
 (e) Differentials or extra payments in respect of skill, responsibility, discomfort, inconvenience or risk (excluding those in respect of supervisory responsibility – see 3.5)
 (f) Payments in respect of public holidays.
 (g) Any amounts which may become payable by the contractor to or in respect of operatives arising from the rules etc. referred to in 3.1 which are not provided for in 3.4 (a)–(f) nor in Section 6.1 (a)–(p).
 (h) Employers contributions to the WELPLAN, the HVACR Welfare and Holiday Scheme or payments in lieu thereof.
 (i) Employers National Insurance contributions as applicable to 3.4 (a)–(h).
 (j) Any contribution, levy or tax imposed by Statute, payable by the contractor in his capacity as an employer.

HEATING AND VENTILATING INDUSTRY

3.5 Differentials or extra payments in respect of supervisory responsibility are excluded from the annual prime cost (see Section 6). The time of principals, staff, foremen, chargehands and the like when working manually is admissible under this Section at the rates for the appropriate grades.

SECTION 4: MATERIALS AND GOODS

4.1 The prime cost of materials and goods obtained specifically for the daywork is the invoice cost after deducting all trade discounts and any portion of cash discounts in excess of 5%.
4.2 The prime cost of all other materials and goods used in the daywork is based upon the current market prices plus any appropriate handling charges.
4.3 The prime cost referred to in 4.1 and 4.2 includes the cost of delivery to site.
4.4 Any Value Added Tax which is treated, or is capable of being treated, as input tax (as defined by the Finance Act 1972, or any re-enactment or amendment thereof or substitution therefore) by the contractor is excluded.

SECTION 5: PLANT

5.1 Unless otherwise stated in the contract, the prime cost of plant comprises the cost of the following:
 (a) Use or hire of mechanically-operated plant and transport for the time employed on and/or provided or retained for the daywork;
 (b) Use of non-mechanical plant (excluding non-mechanical hand tools) for the time employed on and/or provided or retained for the daywork;
 (c) Transport to and from the site and erection and dismantling where applicable.
5.2 The use of non-mechanical hand tools and of erected scaffolding, staging, trestles or the like is excluded (see Section 6), unless specifically retained for the daywork.

SECTION 6: INCIDENTAL COSTS, OVERHEADS AND PROFIT

6.1 The percentage adjustments provided in the contract which are applicable to each of the totals of Sections 3, 4 and 5 comprise the following:
 (a) Head office charges.
 (b) Site staff including site supervision.
 (c) The additional cost of overtime (other than that referred to in 6.2).
 (d) Time lost due to inclement weather.
 (e) The additional cost of bonuses and all other incentive payments in excess of any included in 3.4.
 (f) Apprentices' study time.
 (g) Fares and travelling allowances.
 (h) Country, lodging and periodic allowances.
 (i) Sick pay or insurances in respect thereof, other than as included in 3.4.
 (j) Third party and employers' liability insurance.
 (k) Liability in respect of redundancy payments to employees.
 (l) Employer's National Insurance contributions not included in 3.4.
 (m) Use and maintenance of non-mechanical hand tools.
 (n) Use of erected scaffolding, staging, trestles or the like (but see 5.2).
 (o) Use of tarpaulins, protective clothing, artificial lighting, safety and welfare facilities, storage and the like that may be available on site.
 (p) Any variation to basic rates required by the contractor in cases where the contract provides for the use of a specified schedule of basic plant charges (to the extent that no other provision is made for such variation – see 5.1).
 (q) In the case of a subcontract which provides that the subcontractor shall allow a cash discount, such provision as is necessary for the allowance of the prescribed rate of discount.
 (r) All other liabilities and obligations whatsoever not specifically referred to in this Section nor chargeable under any other Section.
 (s) Profit.
6.2 The additional cost of overtime where specifically ordered by the Architect/Supervising Officer shall only be chargeable in the terms of a prior written agreement between the parties.

HEATING AND VENTILATING INDUSTRY

MECHANICAL INSTALLATIONS

Calculation of Hourly Base Rate of Labour for Typical Main Grades applicable from 4 October 2010, refer to notes within Section Three – Rates of Wages.

	Foreman	Senior Craftsman (+ 2nd Welding Skill)	Senior Craftsman	Craftsman	Installer	Mate over 18
Hourly Rate from 4 October 2010	15.00	12.91	12.40	11.38	10.31	8.69
Annual standard earnings excluding all holidays, 45.6 weeks × 38 hours	25,992.00	22,370.45	21,486.72	19,719.26	17,865.17	15,058.03
Employers national insurance contributions from 6 April 2011	2,633.04	2,133.27	2,011.31	1,767.40	1,511.54	1,124.15
Weekly holiday credit and welfare contributions (52 weeks) from 6 October 2010	4,217.20	3,681.08	3,550.56	3,291.08	3,016.52	2,602.08
Annual prime cost of labour	32,654.20	28,024.59	26,894.87	24,635.44	22,265.26	18,676.75
Hourly base rate	18.84	16.17	15.52	14.22	12.85	10.78

Notes:

(1) Annual industry holiday (4.6 weeks × 38 hours) and public holidays (1.8 weeks × 38 hours) are paid through weekly holiday credit and welfare stamp scheme.
(2) Where applicable, Merit money and other variables (e.g. daily abnormal conditions money), which attract Employer's National Insurance contribution, should be included.
(3) Contractors in Northern Ireland should add the appropriate amount of CITB Levy to the annual prime cost of labour prior to calculating the hourly base rate.
(4) Hourly rate based on 1,732.80 hours per annum and calculated as follows

52 Weeks @ 38 hrs/wk	=		1,976.00
Less			
Public Holiday = 9/5 = 1.8 weeks @ 38 hrs/wk	=	68.40	
Annual holidays = 4.6 weeks @ 38 hrs/wk	=	174.80	243.20
Hours	=		1,732.80

(5) For calculation of Holiday Credits and ENI refer to detailed labour rate evaluation.
(6) National Insurance contributions are those effective from 6 April 2012.
(7) Weekly holiday credit/welfare stamp values are those effective from 4 October 2010, still curent in October 2011.
(8) Hourly rates of wages are those effective from 4 October 2010, still curent in October 2011.

ELECTRICAL INDUSTRY

DEFINITION OF PRIME COST OF DAYWORK CARRIED OUT UNDER AN ELECTRICAL CONTRACT (MARCH 1981 EDITION)

This Definition of Prime Cost is published by The Royal Institution of Chartered Surveyors and The Electrical Contractors' Associations for convenience and for use by people who choose to use it. Members of The Electrical Contractors' Association are not in any way debarred from defining Prime Cost and rendering accounts for work carried out on that basis in any way they choose. Building owners are advised to reach agreement with contractors on the Definition of Prime Cost to be used prior to entering into a contract or subcontract.

SECTION 1: APPLICATION

1.1 This Definition provides a basis for the valuation of daywork executed under such electrical contracts as provide for its use.
1.2 It is not applicable in any other circumstances, such as jobbing, or other work carried out as a separate or main contract, nor in the case of daywork executed after the date of practical completion.
1.3 The terms 'contract' and 'contractor' herein shall be read as 'subcontract' and 'subcontractor' as the context may require.

SECTION 2: COMPOSITION OF TOTAL CHARGES

2.1 The Prime Cost of daywork comprises the sum of the following costs:
  (a) Labour as defined in Section 3.
  (b) Materials and goods as defined in Section 4.
  (c) Plant as defined in Section 5.
2.2 Incidental costs, overheads and profit as defined in Section 6, as provided in the contract and expressed therein as percentage adjustments, are applicable to each of 2.1 (a)–(c).

SECTION 3: LABOUR

3.1 The standard wage rates, emoluments and expenses referred to below and the standard working hours referred to in 3.2 are those laid down for the time being in the rules and determinations or decisions of the Joint Industry Board or the Scottish Joint Industry Board for the Electrical Contracting Industry (or those of such other body as may be appropriate) applicable to the works and relating to the grade of operative concerned at the time when and in the area where daywork is executed.
3.2 Hourly base rates for labour are computed by dividing the annual prime cost of labour, based upon the standard working hours and as defined in 3.4 by the number of standard working hours per annum. See examples.
3.3 The hourly rates computed in accordance with 3.2 shall be applied in respect of the time spent by operatives directly engaged on daywork, including those operating mechanical plant and transport and erecting and dismantling other plant (unless otherwise expressly provided in the contract) and handling and distributing the materials and goods used in the daywork.
3.4 The annual prime cost of labour comprises the following:
  (a) Standard weekly earnings (i.e. the standard working week as determined at the appropriate rate for the operative concerned).
  (b) Payments in respect of public holidays.
  (c) Any amounts which may become payable by the Contractor to or in respect of operatives arising from operation of the rules etc. referred to in 3.1 which are not provided for in 3.4 (a) and (b) nor in Section 6.
  (d) Employer's National Insurance Contributions as applicable to 3.4 (a)–(c).
  (e) Employer's contributions to the Joint Industry Board Combined Benefits Scheme or Scottish Joint Industry Board Holiday and Welfare Stamp Scheme, and holiday payments made to apprentices in compliance with the Joint Industry Board National Working Rules and Industrial Determinations as an employer.
  (f) Any contribution, levy or tax imposed by Statute, payable by the Contractor in his capacity as an employer.
3.5 Differentials or extra payments in respect of supervisory responsibility are excluded from the annual prime cost (see Section 6). The time of principals and similar categories, when working manually, is admissible under this Section at the rates for the appropriate grades.

ELECTRICAL INDUSTRY

SECTION 4: MATERIALS AND GOODS

4.1 The prime cost of materials and goods obtained specifically for the daywork is the invoice cost after deducting all trade discounts and any portion of cash discounts in excess of 5%.

4.2 The prime cost of all other materials and goods used in the daywork is based upon the current market prices plus any appropriate handling charges.

4.3 The prime cost referred to in 4.1 and 4.2 includes the cost of delivery to site.

4.4 Any Value Added Tax which is treated, or is capable of being treated, as input tax (as defined by the Finance Act 1972, or any re-enactment or amendment thereof or substitution therefore) by the Contractor is excluded.

SECTION 5: PLANT

5.1 Unless otherwise stated in the contract, the prime cost of plant comprises the cost of the following:
- (a) Use or hire of mechanically-operated plant and transport for the time employed on and/or provided or retained for the daywork;
- (b) Use of non-mechanical plant (excluding non-mechanical hand tools) for the time employed on and/or provided or retained for the daywork;
- (c) Transport to and from the site and erection and dismantling where applicable.

5.2 The use of non-mechanical hand tools and of erected scaffolding, staging, trestles or the likes is excluded (see Section 6), unless specifically retained for daywork.

5.3 Note: Where hired or other plant is operated by the Electrical Contractor's operatives, such time is to be included under Section 3 unless otherwise provided in the contract.

SECTION 6: INCIDENTAL COSTS, OVERHEADS AND PROFIT

6.1 The percentage adjustments provided in the contract which are applicable to each of the totals of Sections 3, 4 and 5, compromise the following:
- (a) Head Office charges.
- (b) Site staff including site supervision.
- (c) The additional cost of overtime (other than that referred to in 6.2).
- (d) Time lost due to inclement weather.
- (e) The additional cost of bonuses and other incentive payments.
- (f) Apprentices' study time.
- (g) Travelling time and fares.
- (h) Country and lodging allowances.
- (i) Sick pay or insurance in lieu thereof, in respect of apprentices.
- (j) Third party and employers' liability insurance.
- (k) Liability in respect of redundancy payments to employees.
- (l) Employers' National Insurance Contributions not included in 3.4.
- (m) Use and maintenance of non-mechanical hand tools.
- (n) Use of erected scaffolding, staging, trestles or the like (but see 5.2.).
- (o) Use of tarpaulins, protective clothing, artificial lighting, safety and welfare facilities, storage and the like that may be available on site.
- (p) Any variation to basic rates required by the Contractor in cases where the contract provides for the use of a specified schedule of basic plant charges (to the extent that no other provision is made for such variation – see 5.1).
- (q) All other liabilities and obligations whatsoever not specifically referred to in this Section nor chargeable under any other Section.
- (r) Profit.
- (s) In the case of a subcontract which provides that the subcontractor shall allow a cash discount, such provision as is necessary for the allowance of the prescribed rate of discount.

6.2 The additional cost of overtime where specifically ordered by the Architect/Supervising Officer shall only be chargeable in the terms of a prior written agreement between the parties.

ELECTRICAL INDUSTRY

ELECTRICAL INSTALLATIONS

Calculation of Hourly Base Rate of Labour for Typical Main Grades applicable from 4 January 2010.

	Technician	Approved Electrician	Electrician	Labourer
Hourly Rate from 4 January 2010 (London Rates)	18.10	16.07	14.83	11.89
Annual standard earnings excluding all holidays, 45.8 weeks × 37.5 hours	31,086.75	27,600.23	25,470.53	20,421.08
Employers national insurance contributions from 6 April 2011	3,336.12	2,854.98	2,561.08	1,864.25
JIB Combined benefits from 29 September 2009	3,234.40	2,944.24	2,756.62	2,333.76
Holiday top up funding	1,675.85	1,494.04	1,393.46	1,132.67
Annual prime cost of labour	39,333.12	34,893.48	32,181.58	25,751.75
Hourly base rate	22.90	20.32	18.74	14.99

Notes:

(1) Annual industry holiday (4.4 weeks × 37.5 hours) and public holidays (1.8 weeks × 37.5 hours)
(2) It should be noted that all labour costs incurred by the Contractor in his capacity as an Employer, other than those contained in the hourly rate above, must be taken into account under Section 6.
(3) Public Holidays are paid through weekly holiday credit and welfare stamp scheme.
(4) Contractors in Northern Ireland should add the appropriate amount of CITB Levy to the annual prime cost of labour prior to calculating the hourly base rate.
(5) Hourly rate based on 1,717.5 hours per annum and calculated as follows:

52 Weeks @ 37.5 hrs/wk	=		1,950.00
Less			
Public Holiday = 9/5 = 1.8 weeks @ 37.5 hrs/wk	=	67.50	
Annual holidays = 4.4 weeks @ 37.5 hrs/wk	=	165.00	232.50
Hours	=		1,717.50

(6) For calculation of holiday credits and ENI refer to detailed labour rate evaluation.
(7) Hourly wage rates are those effective from 4 January 2010, still current in January 2012.
(8) National Insurance contributions are those effective from 6 April 2012.
(9) JIB Combined Benefits Values are those effective from 28 September 2011, still current in September 2011.

BUILDING INDUSTRY PLANT HIRE COSTS

SCHEDULE OF BASIC PLANT CHARGES

This Schedule is published by the Royal Institution of Chartered Surveyors and is for use in connection with Dayworks under a Building Contract.

EXPLANATORY NOTES

1 The rates in the Schedule are intended to apply solely to daywork carried out under and incidental to a Building Contract. They are NOT intended to apply to:
   (i) Jobbing or any other work carried out as a main or separate contract; or
   (ii) Work carried out after the date of commencement of the Defects Liability Period.

2 The rates apply to plant and machinery already on site, whether hired or owned by the Contractor.

3 The rates, unless otherwise stated, include the cost of fuel and power of every description, lubricating oils, grease, maintenance, sharpening of tools, replacement of spare parts, all consumable stores and for licences and insurances applicable to items of plant.

4 The rates, unless otherwise stated, do not include the costs of drivers and attendants (unless otherwise stated).

5 The rates in the Schedule are base costs and may be subject to an overall adjustment for price movement, overheads and profit, quoted by the Contractor prior to the placing of the Contract.

6 The rates should be applied to the time during which the plant is actually engaged in daywork.

7 Whether or not plant is chargeable on daywork depends on the daywork agreement in use and the inclusion of an item of plant in this schedule does not necessarily indicate that item is chargeable.

8 Rates for plant not included in the Schedule or which is not already on site and is specifically provided or hired for daywork shall be settled at prices which are reasonably related to the rates in the Schedule having regard to any overall adjustment quoted by the Contractor in the Conditions of Contract.

Note: All rates in the schedule were calculated during the first quarter of 2011.

BUILDING INDUSTRY PLANT HIRE COSTS

MECHANICAL PLANT AND TOOLS			
Item of plant	**Size/Rating**	**Unit**	**Rate per hour (£)**
PUMPS			
Mobile Pumps			
Including pump hoses, values and strainers, etc.			
Diaphragm	50 mm dia.	Each	1.17
Diaphragm	76 mm dia.	Each	1.89
Diaphragm	102 mm dia.	Each	3.54
Submersible	50 mm dia.	Each	0.76
Submersible	76 mm dia.	Each	0.86
Submersible	102 mm dia.	Each	1.03
Induced Flow	50 mm dia.	Each	0.77
Induced Flow	76 mm dia.	Each	1.67
Centrifugal, self priming	25 mm dia.	Each	1.30
Centrifugal, self priming	50 mm dia.	Each	1.92
Centrifugal, self priming	75 mm dia.	Each	2.74
Centrifugal, self priming	102 mm dia.	Each	3.35
Centrifugal, self priming	152 mm dia.	Each	4.27
SCAFFOLDING, SHORING, FENCING			
Complete Scaffolding			
Mobile working towers, single width	2.0 m × 0.72 m base × 7.45 m high	Each	3.36
Mobile working towers, single width	2.0 m × 0.72 m base × 8.84 m high	Each	3.79
Mobile working towers, double width	2.0 m × 1.35 m × 7.45 m high	Each	3.79
Mobile working towers, double width	2.0 m × 1.35 m × 15.8 m high	Each	7.13
Chimney scaffold, single unit		Each	1.92
Chimney scaffold, twin unit		Each	3.59
Push along access platform	1.63–3.1 m	Each	5.00
Push along access platform	1.80 m × 0.70 m	Each	1.79
Trestles			
Trestle, adjustable	Any height	Pair	0.41
Trestle, painters	1.8 m high	Pair	0.31
Trestle, painters	2.4 m high	Pair	0.36
Shoring, Planking and Strutting			
'Acrow' adjustable prop	Sizes up to 4.9 m (open)	Each	0.06
'Strong boy' support attachment		Each	0.22
Adjustable trench strut	Sizes up to 1.67 m (open)	Each	0.16

BUILDING INDUSTRY PLANT HIRE COSTS

MECHANICAL PLANT AND TOOLS			
Item of plant	**Size/Rating**	**Unit**	**Rate per hour (£)**
SCAFFOLDING, SHORING, FENCING – cont			
Shoring, Planking and Strutting – cont			
Trench sheet		Metre	0.03
Backhoe trench box	Base unit	Each	1.23
Backhoe trench box	Top unit	Each	0.87
Temporary Fencing			
Including block and coupler			
Site fencing steel grid panel	3.5 m × 2.0 m	Each	0.05
Anti-climb site steel grid fence panel	3.5 m × 2.0 m	Each	0.08
Solid panel Heras	2.0 m × 2.0 m	Each	0.09
Pedestrian gate		Each	0.36
Roadway gate		Each	0.60
LIFTING APPLIANCES AND CONVEYORS			
Cranes			
Mobile Cranes			
Rates are inclusive of drivers			
Lorry mounted, telescopic jib			
Two wheel drive	5 tonnes	Each	19.00
Two wheel drive	8 tonnes	Each	42.00
Two wheel drive	10 tonnes	Each	50.00
Two wheel drive	12 tonnes	Each	77.00
Two wheel drive	20 tonnes	Each	89.69
Four wheel drive	18 tonnes	Each	46.51
Four wheel drive	25 tonnes	Each	35.90
Four wheel drive	30 tonnes	Each	38.46
Four wheel drive	45 tonnes	Each	46.15
Four wheel drive	50 tonnes	Each	53.85
Four wheel drive	60 tonnes	Each	61.54
Four wheel drive	70 tonnes	Each	71.79

BUILDING INDUSTRY PLANT HIRE COSTS

MECHANICAL PLANT AND TOOLS

Item of plant	Size/Rating		Unit	Rate per hour (£)
LIFTING APPLIANCES AND CONVEYORS – cont				
Cranes – cont				
Static tower crane				
Rates inclusive of driver				
Note: Capacity equals maximum lift in tonnes times maximum radius at which it can be lifted				
	Capacity (Metre/tonnes)	Height under hook above ground (m)		
	Up to	Up to		
Tower crane	30	22	Each	22.23
Tower crane	40	22	Each	26.62
Tower crane	40	30	Each	33.33
Tower crane	50	22	Each	29.16
Tower crane	60	22	Each	35.90
Tower crane	60	36	Each	35.90
Tower crane	70	22	Each	41.03
Tower crane	80	22	Each	39.12
Tower crane	90	42	Each	37.18
Tower crane	110	36	Each	47.62
Tower crane	140	36	Each	55.77
Tower crane	170	36	Each	64.11
Tower crane	200	36	Each	71.95
Tower crane	250	36	Each	84.77
Tower crane with luffing jig	30	25	Each	22.23
Tower crane with luffing jig	40	30	Each	26.62
Tower crane with luffing jig	50	30	Each	29.16
Tower crane with luffing jig	60	36	Each	41.03
Tower crane with luffing jig	65	30	Each	33.13
Tower crane with luffing jig	80	22	Each	48.72
Tower crane with luffing jig	100	45	Each	48.72
Tower crane with luffing jig	125	30	Each	53.85
Tower crane with luffing jig	160	50	Each	53.85
Tower crane with luffing jig	200	50	Each	74.36
Tower crane with luffing jig	300	60	Each	100.00

BUILDING INDUSTRY PLANT HIRE COSTS

MECHANICAL PLANT AND TOOLS			
Item of plant	**Size/Rating**	**Unit**	**Rate per hour (£)**
LIFTING APPLIANCES AND CONVEYORS – cont			
Crane Equipment			
Muck tipping skip	Up to 200 litres	Each	0.67
Muck tipping skip	500 litres	Each	0.82
Muck tipping skip	750 litres	Each	1.08
Muck tipping skip	1000 litres	Each	1.28
Muck tipping skip	1500 litres	Each	1.41
Muck tipping skip	2000 litres	Each	1.67
Mortar skip	250 litres, plastic	Each	0.41
Mortar skip	350 litres steel	Each	0.77
Boat skip	250 litres	Each	0.92
Boat skip	500 litres	Each	1.08
Boat skip	750 litres	Each	1.23
Boat skip	1000 litres	Each	1.38
Boat skip	1500 litres	Each	1.64
Boat skip	2000 litres	Each	1.90
Boat skip	3000 litres	Each	2.82
Boat skip	4000 litres	Each	3.23
Master flow skip	250 litres	Each	0.77
Master flow skip	500 litres	Each	1.03
Master flow skip	750 litres	Each	1.28
Master flow skip	1000 litres	Each	1.44
Master flow skip	1500 litres	Each	1.69
Master flow skip	2000 litres	Each	1.85
Grand master flow skip	500 litres	Each	1.28
Grand master flow skip	750 litres	Each	1.64
Grand master flow skip	1000 litres	Each	1.69
Grand master flow skip	1500 litres	Each	1.95
Grand master flow skip	2000 litres	Each	2.21
Cone flow skip	500 litres	Each	1.33
Cone flow skip	1000 litres	Each	1.69
Geared rollover skip	500 litres	Each	1.28
Geared rollover skip	750 litres	Each	1.64
Geared rollover skip	1000 litres	Each	1.69
Geared rollover skip	1500 litres	Each	1.95

BUILDING INDUSTRY PLANT HIRE COSTS

MECHANICAL PLANT AND TOOLS			
Item of plant	**Size/Rating**	**Unit**	**Rate per hour (£)**
LIFTING APPLIANCES AND CONVEYORS – cont			
Crane Equipment – cont			
Geared rollover skip	2000 litres	Each	2.21
Multi skip, rope operated	200 mm outlet size, 500 litres	Each	1.49
Multi skip, rope operated	200 mm outlet size, 750 litres	Each	1.64
Multi skip, rope operated	200 mm outlet size, 1000 litres	Each	1.74
Multi skip, rope operated	200 mm outlet size, 1500 litres	Each	2.00
Multi skip, rope operated	200 mm outlet size, 2000 litres	Each	2.26
Multi skip, man riding	200 mm outlet size, 1000 litres	Each	2.00
Multi skip	4 point lifting frame	Each	0.90
Multi skip	Chain brothers	Set	0.87
Crane Accessories			
Multi-purpose crane forks	1.5 and 2 tonnes S.W.L.	Each	1.13
Self levelling crane forks		Each	1.28
Man cage	1 man, 230 kg S.W.L.	Each	1.90
Man cage	2 man, 500 kg S.W.L.	Each	1.95
Man cage	4 man, 750 kg S.W.L.	Each	2.15
Man cage	8 man, 1000 kg S.W.L.	Each	3.33
Stretcher cage	500 kg, S.W.L.	Each	2.69
Goods carrying cage	1500 kg, S.W.L.	Each	1.33
Goods carrying cage	3000 kg, S.W.L.	Each	1.85
Builders' skip lifting cradle	12 tonnes, S.W.L.	Each	2.31
Board/pallet fork	1600 kg, S.W.L.	Each	1.90
Gas bottle carrier	500 kg, S.W.L.	Each	0.92
Hoists			
Scaffold hoist	200 kg	Each	2.46
Rack and pinion (goods only)	500 kg	Each	4.56
Rack and pinion (goods only)	1100 kg	Each	5.90
Rack and pinion (goods and passenger)	8 person, 80 kg	Each	7.44
Rack and pinion (goods and passenger)	14 person, 1400 kg	Each	8.72
Wheelbarrow chain sling		Each	1.67
Conveyors			
Belt conveyors			
Conveyor	8 m long × 450 mm wide	Each	5.90
Miniveyor, control box and loading hopper	3 m unit	Each	4.49

BUILDING INDUSTRY PLANT HIRE COSTS

MECHANICAL PLANT AND TOOLS				
Item of plant	**Size/Rating**		**Unit**	**Rate per hour (£)**
LIFTING APPLIANCES AND CONVEYORS – cont				
Conveyors – cont				
Other Conveying Equipment				
Wheelbarrow			Each	0.62
Hydraulic superlift			Each	4.56
Pavac slab lifter (tile hoist)			Each	4.49
High lift pallet truck			Each	3.08
Lifting Trucks	Payload	Maximum Lift		
Fork lift, two wheel drive	1100 kg	up to 3.0 m	Each	5.64
Fork lift, two wheel drive	2540 kg	up to 3.7 m	Each	5.64
Fork lift, four wheel drive	1524 kg	up to 6.0 m	Each	5.64
Fork lift, four wheel drive	2600 kg	up to 5.4 m	Each	7.44
Fork life, four wheel drive	4000 kg	up to 17 m	Each	10.77
Lifting Platforms				
Hydraulic platform (Cherry picker)	9 m		Each	4.62
Hydraulic platform (Cherry picker)	12 m		Each	7.56
Hydraulic platform (Cherry picker)	15 m		Each	10.13
Hydraulic platform (Cherry picker)	17 m		Each	15.63
Hydraulic platform (Cherry picker)	20 m		Each	18.13
Hydraulic platform (Cherry picker)	25.6 m		Each	32.38
Scissor lift	7.6 m, electric		Each	3.85
Scissor lift	7.8 m, electric		Each	5.13
Scissor lift	9.7 m, electric		Each	4.23
Scissor lift	10 m, diesel		Each	6.41
Telescopic handler	7 m, 2 tonnes		Each	5.13
Telescopic handler	13 m, 3 tonnes		Each	7.18
Lifting and Jacking Gear				
Pipe winch including gantry	1 tonne		Set	1.92
Pipe winch including gantry	3 tonnes		Set	3.21
Chain block	1 tonne		Each	0.35
Chain block	2 tonnes		Each	0.58
Chain block	5 tonnes		Each	1.14
Pull lift (Tirfor winch)	1 tonne		Each	0.64
Pull lift (Tirfor winch)	1.6 tonnes		Each	0.90
Pull lift (Tirfor winch)	3.2 tonnes		Each	1.15
Brother or chain slings, two legs	not exceeding 3.1 tonnes		Set	0.21

BUILDING INDUSTRY PLANT HIRE COSTS

MECHANICAL PLANT AND TOOLS			
Item of plant	**Size/Rating**	**Unit**	**Rate per hour (£)**
LIFTING APPLIANCES AND CONVEYORS – cont			
Lifting and Jacking Gear – cont			
Brother or chain slings, two legs	not exceeding 4.25 tonnes	Set	0.31
Brother or chain slings, four legs	not exceeding 11.2 tonnes	Set	1.09
CONSTRUCTION VEHICLES			
Lorries			
Plated lorries (Rates are inclusive of driver)			
Platform lorry	7.5 tonnes	Each	16.21
Platform lorry	17 tonnes	Each	22.90
Platform lorry	24 tonnes	Each	30.68
Extra for lorry with crane attachment	up to 2.5 tonnes	Each	3.25
Extra for lorry with crane attachment	up to 5 tonnes	Each	6.00
Extra for lorry with crane attachment	up to 7.5 tonnes	Each	9.10
Tipper Lorries			
(Rates are inclusive of driver)			
Tipper lorry	up to 11 tonnes	Each	15.78
Tipper lorry	up to 17 tonnes	Each	23.95
Tipper lorry	up to 25 tonnes	Each	31.35
Tipper lorry	up to 31 tonnes	Each	37.79
Dumpers			
Site use only (excl. tax, insurance and extra cost of DERV etc. when operating on highway)	Makers Capacity		
Two wheel drive	1 tonnes	Each	1.71
Four wheel drive	2 tonnes	Each	2.43
Four wheel drive	3 tonnes	Each	2.44
Four wheel drive	5 tonnes	Each	3.08
Four wheel drive	6 tonnes	Each	3.85
Four wheel drive	9 tonnes	Each	5.65
Tracked	0.5 tonnes	Each	3.33
Tracked	1.5 tonnes	Each	4.23
Tracked	3.0 tonnes	Each	8.33
Tracked	6.0 tonnes	Each	16.03

BUILDING INDUSTRY PLANT HIRE COSTS

MECHANICAL PLANT AND TOOLS			
Item of plant	**Size/Rating**	**Unit**	**Rate per hour (£)**
CONSTRUCTION VEHICLES – cont			
Dumpers – cont			
Dumper Trucks (*Rates are inclusive of drivers*)			
Dumper truck	up to 15 tonnes	Each	28.56
Dumper truck	up to 17 tonnes	Each	32.82
Dumper truck	up to 23 tonnes	Each	54.64
Dumper truck	up to 30 tonnes	Each	63.50
Dumper truck	up to 35 tonnes	Each	73.02
Dumper truck	up to 40 tonnes	Each	87.84
Dumper truck	up to 50 tonnes	Each	133.44
Tractors			
Agricultural Type			
Wheeled, rubber-clad tyred	up to 40 kW	Each	8.63
Wheeled, rubber-clad tyred	up to 90 kW	Each	25.31
Wheeled, rubber-clad tyred	up to 140 kW	Each	36.49
Crawler Tractors			
With bull or angle dozer	up to 70 kW	Each	29.38
With bull or angle dozer	up to 85 kW	Each	38.63
With bull or angle dozer	up to 100 kW	Each	52.59
With bull or angle dozer	up to 115 kW	Each	55.85
With bull or angle dozer	up to 135 kW	Each	60.43
With bull or angle dozer	up to 185 kW	Each	76.44
With bull or angle dozer	up to 200 kW	Each	96.43
With bull or angle dozer	up to 250 kW	Each	117.68
With bull or angle dozer	up to 350 kW	Each	160.03
With bull or angle dozer	up to 450 kW	Each	219.86
With loading shovel	0.8 m^3	Each	26.92
With loading shovel	1.0 m^3	Each	32.59
With loading shovel	1.2 m^3	Each	37.53
With loading shovel	1.4 m^3	Each	42.89
With loading shovel	1.8 m^3	Each	52.22
With loading shovel	2.0 m^3	Each	57.22
With loading shovel	2.1 m^3	Each	60.12
With loading shovel	3.5 m^3	Each	87.26

BUILDING INDUSTRY PLANT HIRE COSTS

MECHANICAL PLANT AND TOOLS			
Item of plant	**Size/Rating**	**Unit**	**Rate per hour (£)**
CONSTRUCTION VEHICLES – cont			
Tractors – cont			
Light Vans			
VW Caddivan or the like		Each	5.26
VW Transport transit or the like	1.0 tonnes	Each	6.03
Luton Box Van or the like	1.8 tonnes	Each	9.87
Water/Fuel Storage			
Mobile water container	110 litres	Each	0.62
Water bowser	1100 litres	Each	0.72
Water bowser	3000 litres	Each	0.87
Mobile fuel container	110 litres	Each	0.62
Fuel bowser	1100 litres	Each	1.23
Fuel bowser	3000 litres	Each	1.87
EXCAVATIONS AND LOADERS			
Excavators			
Wheeled, hydraulic	up to 11 tonnes	Each	25.86
Wheeled, hydraulic	up to 14 tonnes	Each	30.82
Wheeled, hydraulic	up to 16 tonnes	Each	34.50
Wheeled, hydraulic	up to 21 tonnes	Each	39.10
Wheeled, hydraulic	up to 25 tonnes	Each	43.81
Wheeled, hydraulic	up to 30 tonnes	Each	55.30
Crawler, hydraulic	up to 11 tonnes	Each	25.86
Crawler, hydraulic	up to 14 tonnes	Each	30.82
Crawler, hydraulic	up to 17 tonnes	Each	34.50
Crawler, hydraulic	up to 23 tonnes	Each	39.10
Crawler, hydraulic	up to 30 tonnes	Each	43.81
Crawler, hydraulic	up to 35 tonnes	Each	55.30
Crawler, hydraulic	up to 38 tonnes	Each	71.73
Crawler, hydraulic	up to 55 tonnes	Each	95.63
Mini excavator	1000/1500 kg	Each	4.87
Mini excavator	2150/2400 kg	Each	6.67
Mini excavator	2700/3500 kg	Each	7.31
Mini excavator	3500/4500 kg	Each	8.21

BUILDING INDUSTRY PLANT HIRE COSTS

MECHANICAL PLANT AND TOOLS			
Item of plant	**Size/Rating**	**Unit**	**Rate per hour (£)**
EXCAVATIONS AND LOADERS – cont			
Excavators – cont			
Mini excavator	4500/6000 kg	Each	9.23
Mini excavator	7000 kg	Each	14.10
Micro excavator	725 mm wide	Each	5.13
Loaders			
Shovel loader	0.4 m³	Each	7.69
Shovel loader	1.57 m³	Each	8.97
Shovel loader, four wheel drive	1.7 m³	Each	4.83
Shovel loader, four wheel drive	2.3 m³	Each	4.38
Shovel loader, four wheel drive	3.3 m³	Each	5.06
Skid steer loader wheeled	300/400 kg payload	Each	7.31
Skid steer loader wheeled	625 kg payload	Each	7.67
Tracked skip loader	650 kg	Each	4.42
Excavator Loaders			
Wheeled tractor type with black-hoe excavator			
Four wheel drive			
Four wheel drive, 2 wheel steer	6 tonnes	Each	6.41
Four wheel drive, 2 wheel steer	8 tonnes	Each	8.59
Attachments			
Breakers for excavator		Each	8.72
Breakers for mini excavator		Each	1.75
Breakers for back-hoe excavator/loader		Each	5.13
COMPACTION EQUIPMENT			
Rollers			
Vibrating roller	368–420 kg	Each	1.43
Single roller	533 kg	Each	1.94
Single roller	750 kg	Each	3.43
Twin roller	up to 650 kg	Each	6.03
Twin roller	up to 950 kg	Each	6.62
Twin roller with seat end steering wheel	up to 1400 kg	Each	7.68

BUILDING INDUSTRY PLANT HIRE COSTS

MECHANICAL PLANT AND TOOLS			
Item of plant	**Size/Rating**	**Unit**	**Rate per hour (£)**
COMPACTION EQUIPMENT – cont			
Rollers – cont			
Twin roller with seat end steering wheel	up to 2500 kg	Each	10.61
Pavement roller	3–4 tonnes dead weight	Each	6.00
Pavement roller	4–6 tonnes	Each	6.86
Pavement roller	6–10 tonnes	Each	7.17
Pavement roller	10–13 tonnes	Each	19.86
Rammers			
Tamper rammer 2 stroke–petrol	225–275 mm	Each	1.52
Soil Compactors			
Plate compactor	75–400 mm	Each	1.53
Plate compactor rubber pad	375–1400 mm	Each	1.53
Plate compactor reversible plate–petrol	400 mm	Each	2.44
CONCRETE EQUIPEMENT			
Concrete/Mortar Mixers			
Open drum without hopper	0.09/0.06 m³	Each	0.61
Open drum without hopper	0.12/0.09 m³	Each	1.22
Open drum without hopper	0.15/0.10 m³	Each	0.72
Concrete/Mortar Transport Equipment			
Concrete pump incl. hose, valve and couplers			
Lorry mounted concrete pump	24 m max. distance	Each	50.00
Lorry mounted concrete pump	34 m max. distance	Each	66.00
Lorry mounted concrete pump	42 m max. distance	Each	91.50
Concrete Equipment			
Vibrator, poker, petrol type	up to 75 mm dia.	Each	0.69
Air vibrator (*excluding compressor and hose*)	up to 75 mm dia.	Each	0.64
Extra poker heads	25/36/60 mm dia.	Each	0.76
Vibrating screed unit with beam	5.00 m	Each	2.48
Vibrating screed unit with adjustable beam	3.00–5.00 m	Each	3.54
Power float	725–900 mm	Each	2.56

BUILDING INDUSTRY PLANT HIRE COSTS

MECHANICAL PLANT AND TOOLS			
Item of plant	**Size/Rating**	**Unit**	**Rate per hour (£)**
CONCRETE EQUIPEMENT – cont			
Concrete Equipment – cont			
Power float finishing pan		Each	0.62
Floor grinder	660 × 1016 mm, 110 V electric	Each	4.31
Floor plane	450 × 1100 mm	Each	4.31
TESTING EQUIPMENT			
Pipe Testing Equipment			
Pressure testing pump, electric		Set	2.19
Pressure test pump		Set	0.80
SITE ACCOMMODATION AND TEMPORARY SERVICES			
Heating equipment			
Space heater – propane	80,000 Btu/hr	Each	1.03
Space heater – propane/electric	125,000 Btu/hr	Each	2.09
Space heater – propane/electric	250,000 Btu/hr	Each	2.33
Space heater – propane	125,000 Btu/hr	Each	1.54
Space heater – propane	260,000 Btu/hr	Each	1.88
Cabinet heater		Each	0.82
Cabinet heater, catalytic		Each	0.57
Electric halogen heater		Each	1.27
Ceramic heater	3 kW	Each	0.99
Fan heater	3 kW	Each	0.66
Cooling fan		Each	1.92
Mobile cooling unit, small		Each	3.60
Mobile cooling unit, large		Each	4.98
Air conditioning unit		Each	2.81
Site Lighting and Equipment			
Tripod floodlight	500 W	Each	0.48
Tripod floodlight	1000 W	Each	0.62
Towable floodlight	4 × 100 W	Each	3.85
Hand held floodlight	500 W	Each	0.51
Rechargeable light		Each	0.41
Inspection light		Each	0.37
Plasterer's light		Each	0.65

BUILDING INDUSTRY PLANT HIRE COSTS

MECHANICAL PLANT AND TOOLS			
Item of plant	**Size/Rating**	**Unit**	**Rate per hour (£)**
SITE ACCOMMODATION AND TEMPORARY SERVICES – cont			
Site Lighting and Equipment – cont			
Lighting mast		Each	2.87
Festoon light string	25 m	Each	0.55
Site Electrical Equipment			
Extension leads	240 V/14 m	Each	0.26
Extension leads	110 V/14 m	Each	0.36
Cable reel	25 m 110 V/240 V	Each	0.46
Cable reel	50 m 110 V/240 V	Each	0.88
4 way junction box	110 V	Each	0.56
Power Generating Units			
Generator – petrol	2 kVA	Each	1.23
Generator – silenced petrol	2 kVA	Each	2.87
Generator – petrol	3 kVA	Each	1.47
Generator – diesel	5 kVA	Each	2.44
Generator – silenced diesel	10 kVA	Each	1.90
Generator – silenced diesel	15 kVA	Each	2.26
Generator – silenced diesel	30 kVA	Each	3.33
Generator – silenced diesel	50 kVA	Each	4.10
Generator – silenced diesel	75 kVA	Each	4.62
Generator – silenced diesel	100 kVA	Each	5.64
Generator – silenced diesel	150 kVA	Each	7.18
Generator – silenced diesel	200 kVA	Each	9.74
Generator – silenced diesel	250 kVA	Each	11.28
Generator – silenced diesel	350 kVA	Each	14.36
Generator – silenced diesel	500 kVA	Each	15.38
Tail adaptor	240 V	Each	0.10
Transformers			
Transformer	3 kVA	Each	0.32
Transformer	5 kVA	Each	1.23
Transformer	7.5 kVA	Each	0.59
Transformer	10 kVA	Each	2.00

BUILDING INDUSTRY PLANT HIRE COSTS

MECHANICAL PLANT AND TOOLS			
Item of plant	**Size/Rating**	**Unit**	**Rate per hour (£)**
SITE ACCOMMODATION AND TEMPORARY SERVICES – cont			
Rubbish Collection and Disposal Equipment			
Rubbish Chutes			
Standard plastic module	1 m section	Each	0.15
Steel liner insert		Each	0.30
Steel top hopper		Each	0.22
Plastic side entry hopper		Each	0.22
Plastic side entry hopper liner		Each	0.22
Dust Extraction Plant			
Dust extraction unit, light duty		Each	2.97
Dust extraction unit, heavy duty		Each	2.97
SITE EQUIPMENT – Welding Equipment			
Arc-(Electric) Complete With Leads			
Welder generator – petrol	200 amp	Each	3.53
Welder generator – diesel	300/350 amp	Each	3.78
Welder generator – diesel	4000 amp	Each	7.92
Extra welding lead sets		Each	0.69
Gas-Oxy Welder			
Welding and cutting set (including oxygen and acetylene, excluding underwater equipment and thermic boring)			
Small		Each	2.24
Large		Each	3.75
Lead burning gun		Each	0.50
Mig welder		Each	1.38
Fume extractor		Each	2.46
Road Works Equipment			
Traffic lights, mains/generator	2-way	Set	10.94
Traffic lights, mains/generator	3-way	Set	11.56
Traffic lights, mains/generator	4-way	Set	12.19
Flashing light		Each	0.10
Road safety cone	450 mm	Each	0.08
Safety cone	750 mm	Each	0.10
Safety barrier plank	1.25 m	Each	0.13
Safety barrier plank	2 m	Each	0.15
Safety barrier plank post		Each	0.13

BUILDING INDUSTRY PLANT HIRE COSTS

MECHANICAL PLANT AND TOOLS			
Item of plant	**Size/Rating**	**Unit**	**Rate per hour (£)**
SITE ACCOMMODATION AND TEMPORARY SERVICES – cont			
Road Works Equipment – cont			
Safety barrier plank post base		Each	0.10
Safety four gate barrier	1 m each gate	Set	0.77
Guard barrier	2 m	Each	0.19
Road sign	750 mm	Each	0.23
Road sign	900 mm	Each	0.31
Road sign	1200 mm	Each	0.42
Speed ramp/cable protection	500 mm section	Each	0.14
Hose ramp open top	3 m section	Each	0.07
DPC Equipment			
Damp-proofing injection machine		Each	2.56
Cleaning Equipment			
Vacuum cleaner (industrial wet) single motor		Each	1.08
Vacuum cleaner (industrial wet) twin motor	30 litre capacity	Each	1.79
Vacuum cleaner (industrial wet) twin motor	70 litre capacity	Each	2.21
Steam cleaner	Diesel/electric 1 phase	Each	3.33
Steam cleaner	Diesel/electric 3 phase	Each	3.85
Pressure washer, light duty electric	1450 PSI	Each	0.72
Pressure washer, heavy duty, diesel	2500 PSI	Each	1.33
Pressure washer, heavy duty, diesel	4000 PSI	Each	2.18
Cold pressure washer, electric		Each	2.39
Hot pressure washer, petrol		Each	4.19
Hot pressure washer, electric		Each	5.13
Cold pressure washer, petrol		Each	2.92
Sandblast attachment to last washer		Each	1.23
Drain cleaning attachment to last washer		Each	1.03
Surface Preparation Equipment			
Rotavator	5 h.p.	Each	2.46
Rotavator	9 h.p.	Each	5.00
Scabbler, up to three heads		Each	1.53
Scabbler, pole		Each	2.68
Scrabbler, multi-headed floor		Each	3.89

BUILDING INDUSTRY PLANT HIRE COSTS

MECHANICAL PLANT AND TOOLS			
Item of plant	**Size/Rating**	**Unit**	**Rate per hour (£)**
SITE ACCOMMODATION AND TEMPORARY SERVICES – cont			
Surface Preparation Equipment – cont			
Floor preparation machine		Each	1.05
Compressors and Equipment			
Portable Compressors			
Compressor – electric	4 cfm	Each	1.36
Compressor – electric	8 cfm lightweight	Each	1.31
Compressor – electric	8 cfm	Each	1.36
Compressor – electric	14 cfm	Each	1.56
Compressor – petrol	24 cfm	Each	2.15
Compressor – electric	25 cfm	Each	2.10
Compressor – electric	30 cfm	Each	2.36
Compressor – diesel	100 cfm	Each	2.56
Compressor – diesel	250 cfm	Each	5.54
Compressor – diesel	400 cfm	Each	8.72
Mobile Compressors			
Lorry mounted compressor			
(machine plus lorry only)	up to 3 m^3	Each	41.47
(machine plus lorry only)	up to 5 m^3	Each	48.94
Tractor mounted compressor			
(machine plus rubber tyred tractor)	Up to 4 m^3	Each	21.03
Accessories (Pneumatic Tools) *(with and including up to 15 m of air hose)*			
Demolition pick, medium duty		Each	0.90
Demolition pick, heavy duty		Each	1.03
Breakers (with six steels) light	up to 150 kg	Each	1.19
Breakers (with six steels) medium	295 kg	Each	1.24
Breakers (with six steels) heavy	386 kg	Each	1.44
Rock drill (for use with compressor) hand held		Each	1.18
Additional hoses	15 m	Each	0.09
Breakers			
Demolition hammer drill, heavy duty, electric		Each	1.54
Road breaker, electric		Each	2.41
Road breaker, 2 stroke, petrol		Each	4.06
Hydraulic breaker unit, light duty, petrol		Each	3.06
Hydraulic breaker unit, heavy duty, petrol		Each	3.46

BUILDING INDUSTRY PLANT HIRE COSTS

MECHANICAL PLANT AND TOOLS			
Item of plant	**Size/Rating**	**Unit**	**Rate per hour (£)**
SITE ACCOMMODATION AND TEMPORARY SERVICES – cont			
Breakers – cont			
Hydraulic breaker unit, heavy duty, diesel		Each	4.62
Quarrying and Tooling Equipment			
Block and stone splitter, hydraulic	600 mm × 600 mm	Each	1.90
Block and stone splitter, manual		Each	1.64
Steel Reinforcement Equipment			
Bar bending machine – manual	up to 13 mm dia. rods	Each	1.03
Bar bending machine – manual	up to 20 mm dia. rods	Each	1.41
Bar shearing machine – electric	up to 38 mm dia. rods	Each	3.08
Bar shearing machine – electric	up to 40 mm dia. rods	Each	4.62
Bar cropper machine – electric	up to 13 mm dia. rods	Each	2.05
Bar cropper machine – electric	up to 20 mm dia. rods	Each	2.56
Bar cropper machine – electric	up to 40 mm dia. rods	Each	4.62
Bar cropper machine – 3 phase	up to 40 mm dia. rods	Each	4.62
Dehumidifiers			
110/240 V Water	68 litres extraction per 24 hours	Each	2.46
110/240 V Water	90 litres extraction per 24 hours	Each	3.38
SMALL TOOLS			
Saws			
Masonry bench saw	350–500 mm dia.	Each	1.13
Floor saw	125 mm max. cut	Each	1.15
Floor saw	150 mm max. cut	Each	3.83
Floor saw, reversible	350 mm max. cut	Each	3.32
Wall saw, electric		Each	2.05
Chop/cut off saw, electric	350 mm dia.	Each	1.79
Circular saw, electric	230 mm dia.	Each	0.72
Tyrannosaw		Each	1.74
Reciprocating saw		Each	0.79
Door trimmer		Each	1.17
Stone saw	300 mm	Each	1.44
Chainsaw, petrol	500 mm	Each	3.92
Full chainsaw safety kit		Each	0.41
Worktop jig		Each	1.08

BUILDING INDUSTRY PLANT HIRE COSTS

MECHANICAL PLANT AND TOOLS			
Item of plant	**Size/Rating**	**Unit**	**Rate per hour (£)**
SMALL TOOLS – cont			
Saws – cont			
Pipework Equipment			
Pipe bender	15–22 mm	Each	0.92
Pipe bender, hydraulic	50 mm	Each	1.76
Pipe bender, electric	50–150 mm dia.	Each	2.19
Pipe cutter, hydraulic		Each	0.46
Tripod pipe vice		Set	0.75
Ratchet threader	12–32 mm	Each	0.93
Pipe threading machine, electric	12–75 mm	Each	3.07
Pipe threading machine, electric	12–100 mm	Each	4.93
Impact wrench, electric		Each	1.33
Hand-held Drills and Equipment			
Impact or hammer drill	up to 25 mm dia.	Each	1.03
Impact or hammer drill	35 mm dia.	Each	1.29
Dry diamond core cutter		Each	0.99
Angle head drill		Each	0.90
Stirrer, mixer drill		Each	1.13
Paint, Insulation Application Equipment			
Airless spray unit		Each	4.13
Portaspray unit		Each	1.16
HPVL turbine spray unit		Each	2.23
Compressor and spray gun		Each	1.91
Other Handtools			
Staple gun		Each	0.96
Air nail gun	110 V	Each	1.01
Cartridge hammer		Each	1.08
Tongue and groove nailer complete with mallet		Each	1.59
Diamond wall chasing machine		Each	2.63
Masonry chain saw	300 mm	Each	5.49
Floor grinder		Each	3.99
Floor plane		Each	1.79
Diamond concrete planer		Each	1.93

BUILDING INDUSTRY PLANT HIRE COSTS

MECHANICAL PLANT AND TOOLS			
Item of plant	**Size/Rating**	**Unit**	**Rate per hour (£)**
SMALL TOOLS – cont			
Other Handtools – cont			
Autofeed screwdriver, electric		Each	1.38
Laminate trimmer		Each	0.91
Biscuit jointer		Each	1.49
Random orbital sander		Each	0.97
Floor sander		Each	1.54
Palm, delta, flap or belt sander		Each	0.75
Disc cutter, electric	300 mm	Each	1.49
Disc cutter, 2 stroke petrol	300 mm	Each	1.24
Dust suppressor for petrol disc cutter		Each	0.51
Cutter cart for petrol disc cutter		Each	1.21
Grinder, angle or cutter	up to 225 mm	Each	0.50
Grinder, angle or cutter	300 mm	Each	1.41
Mortar raking tool attachment		Each	0.19
Floor/polisher scrubber	325 mm	Each	1.76
Floor tile stripper		Each	2.44
Wallpaper stripper, electric		Each	0.81
Hot air paint stripper		Each	0.50
Electric diamond tile cutter	all sizes	Each	2.42
Hand tile cutter		Each	0.82
Electric needle gun		Each	1.29
Needle chipping gun		Each	1.85
Pedestrian floor sweeper	250 mm dia.	Each	0.82
Pedestrian floor sweeper	Petrol	Each	2.20
Diamond tile saw		Each	1.84
Blow lamp equipment and glass		Set	0.50

PART 6

Tables and Memoranda

This part of the book contains the following sections:

	page
Conversion Tables	759
Geometry	763
Formulae	767
Typical Thermal Conductivity of Building Materials	768
Earthwork	769
Concrete Work	775
Reinforcement	780
Formwork	782
Masonry	783
Timber	786
Roofing	791
Glazing	795
Metal	796
Kerbs, Paving, etc.	814
Seeding/Turfing and Planting	821
Fencing and Gates	824
Drainage	829
Electrical Supply/Power/Lighting Systems	836
Rail Tracks	837
Fractions, Decimals and Millimetre Equivalents	840
Imperial Standard Wire Gauge (SWG)	841
Pipes, Water, Storage, Insulation	842

CONVERSION TABLES

CONVERSION TABLES

Length	Unit	Conversion factors			
Millimetre	mm	1 in	= 25.4 mm	1 mm	= 0.0394 in
Centimetre	cm	1 in	= 2.54 cm	1 cm	= 0.3937 in
Metre	m	1 ft	= 0.3048 m	1 m	= 3.2808 ft
		1 yd	= 0.9144 m		= 1.0936 yd
Kilometre	km	1 mile	= 1.6093 km	1 km	= 0.6214 mile

Note:	1 cm	= 10 mm	1 ft	= 12 in	
	1 m	= 1 000 mm	1 yd	= 3 ft	
	1 km	= 1 000 m	1 mile	= 1 760 yd	

Area	Unit	Conversion factors			
Square Millimetre	mm^2	1 in^2	= 645.2 mm^2	1 mm^2	= 0.0016 in^2
Square Centimetre	cm^2	1 in^2	= 6.4516 cm^2	1 cm^2	= 1.1550 in^2
Square Metre	m^2	1 ft^2	= 0.0929 m^2	1 m^2	= 10.764 ft^2
		1 yd^2	= 0.8361 m^2	1 m^2	= 1.1960 yd^2
Square Kilometre	km^2	1 $mile^2$	= 2.590 km^2	1 km^2	= 0.3861 $mile^2$

Note:	1 cm^2	= 100 mm^2	1 ft^2	= 144 in^2
	1 m^2	= 10 000 cm^2	1 yd^2	= 9 ft^2
	1 km^2	= 100 hectares	1 acre	= 4 840 yd^2
			1 $mile^2$	= 640 acres

Volume	Unit	Conversion factors			
Cubic Centimetre	cm^3	1 cm^3	= 0.0610 in^3	1 in^3	= 16.387 cm^3
Cubic Decimetre	dm^3	1 dm^3	= 0.0353 ft^3	1 ft^3	= 28.329 dm^3
Cubic Metre	m^3	1 m^3	= 35.3147 ft^3	1 ft^3	= 0.0283 m^3
		1 m^3	= 1.3080 yd^3	1 yd^3	= 0.7646 m^3
Litre	l	1 l	= 1.76 pint	1 pint	= 0.5683 l
			= 2.113 US pt		= 0.4733 US l

Note:	1 dm^3	= 1 000 cm^3	1 ft^3	= 1 728 in^3	1 pint	= 20 fl oz
	1 m^3	= 1 000 dm^3	1 yd^3	= 27 ft^3	1 gal	= 8 pints
	1 l	= 1 dm^3				

Neither the Centimetre nor Decimetre are SI units, and as such their use, particularly that of the Decimetre, is not widespread outside educational circles.

Mass	Unit	Conversion factors			
Milligram	mg	1 mg	= 0.0154 grain	1 grain	= 64.935 mg
Gram	g	1 g	= 0.0353 oz	1 oz	= 28.35 g
Kilogram	kg	1 kg	= 2.2046 lb	1 lb	= 0.4536 kg
Tonne	t	1 t	= 0.9842 ton	1 ton	= 1.016 t

Note:	1 g	= 1000 mg	1 oz	= 437.5 grains	1 cwt	= 112 lb
	1 kg	= 1000 g	1 lb	= 16 oz	1 ton	= 20 cwt
	1 t	= 1000 kg	1 stone	= 14 lb		

Force	Unit	Conversion factors			
Newton	N	1 lbf	= 4.448 N	1 kgf	= 9.807 N
Kilonewton	kN	1 lbf	= 0.004448 kN	1 ton f	= 9.964 kN
Meganewton	MN	100 tonf	= 0.9964 MN		

CONVERSION TABLES

Pressure and stress	Unit	Conversion factors	
Kilonewton per square metre	kN/m^2	1 lbf/in^2	= 6.895 kN/m^2
		1 bar	= 100 kN/m^2
Meganewton per square metre	MN/m^2	1 $tonf/ft^2$	= 107.3 kN/m^2 = 0.1073 MN/m^2
		1 kgf/cm^2	= 98.07 kN/m^2
		1 lbf/ft^2	= 0.04788 kN/m^2

Coefficient of consolidation (Cv) or swelling	Unit	Conversion factors	
Square metre per year	m^2/year	1 cm^2/s	= 3 154 m^2/year
		1 ft^2/year	= 0.0929 m^2/year

Coefficient of permeability	Unit	Conversion factors	
Metre per second	m/s	1 cm/s	= 0.01 m/s
Metre per year	m/year	1 ft/year	= 0.3048 m/year
			= 0.9651 × $(10)^8$ m/s

Temperature	Unit	Conversion factors	
Degree Celsius	°C	°C = 5/9 × (°F − 32)	°F = (9 × °C)/5 + 32

CONVERSION TABLES

SPEED CONVERSION

km/h	m/min	mph	fpm
1	16.7	0.6	54.7
2	33.3	1.2	109.4
3	50.0	1.9	164.0
4	66.7	2.5	218.7
5	83.3	3.1	273.4
6	100.0	3.7	328.1
7	116.7	4.3	382.8
8	133.3	5.0	437.4
9	150.0	5.6	492.1
10	166.7	6.2	546.8
11	183.3	6.8	601.5
12	200.0	7.5	656.2
13	216.7	8.1	710.8
14	233.3	8.7	765.5
15	250.0	9.3	820.2
16	266.7	9.9	874.9
17	283.3	10.6	929.6
18	300.0	11.2	984.3
19	316.7	11.8	1038.9
20	333.3	12.4	1093.6
21	350.0	13.0	1148.3
22	366.7	13.7	1203.0
23	383.3	14.3	1257.7
24	400.0	14.9	1312.3
25	416.7	15.5	1367.0
26	433.3	16.2	1421.7
27	450.0	16.8	1476.4
28	466.7	17.4	1531.1
29	483.3	18.0	1585.7
30	500.0	18.6	1640.4
31	516.7	19.3	1695.1
32	533.3	19.9	1749.8
33	550.0	20.5	1804.5
34	566.7	21.1	1859.1
35	583.3	21.7	1913.8
36	600.0	22.4	1968.5
37	616.7	23.0	2023.2
38	633.3	23.6	2077.9
39	650.0	24.2	2132.5
40	666.7	24.9	2187.2

CONVERSION TABLES

km/h	m/min	mph	fpm
41	683.3	25.5	2241.9
42	700.0	26.1	2296.6
43	716.7	26.7	2351.3
44	733.3	27.3	2405.9
45	750.0	28.0	2460.6
46	766.7	28.6	2515.3
47	783.3	29.2	2570.0
48	800.0	29.8	2624.7
49	816.7	30.4	2679.4
50	833.3	31.1	2734.0

GEOMETRY

GEOMETRY

Two dimensional figures

Figure	Diagram of figure	Surface area	Perimeter
Square	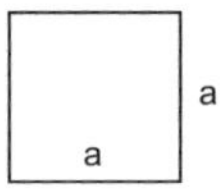	a^2	4a
Rectangle	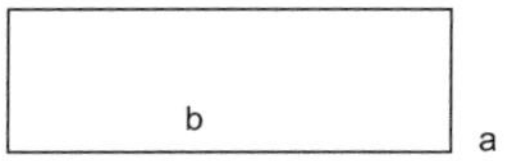	ab	$2(a+b)$
Triangle	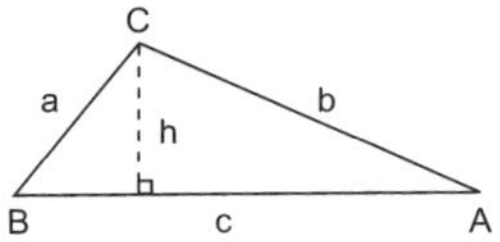	$\frac{1}{2}ch$	$a+b+c$
Circle	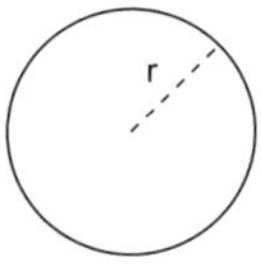	πr^2 $\frac{1}{4}\pi d^2$ where $2r = d$	$2\pi r$ πd
Parallelogram	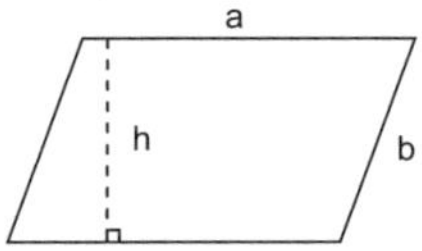	ah	$2(a+b)$
Trapezium	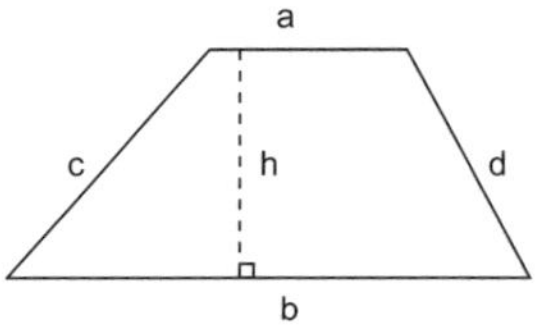	$\frac{1}{2}h(a+b)$	$a+b+c+d$
Ellipse	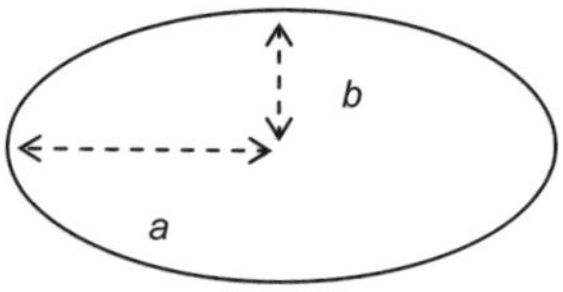	Approximately πab	$\pi(a+b)$
Hexagon	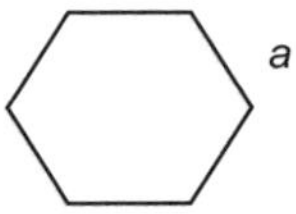	$2.6 \times a^2$	

GEOMETRY

Figure	Diagram of figure	Surface area	Perimeter
Octagon		$4.83 \times a^2$	6a
Sector of a circle	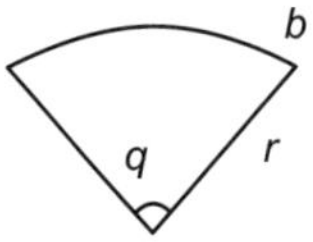	$\frac{1}{2}rb$ or $\frac{q}{360}\pi r^2$ note $b =$ angle $\frac{q}{360} \times \pi 2r$	
Segment of a circle	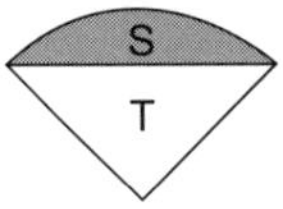	S – T where S = area of sector, T = area of triangle	
Bellmouth	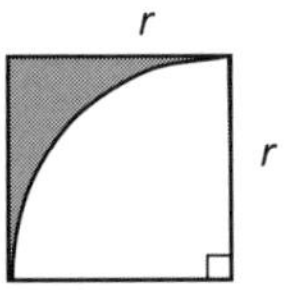	$\frac{3}{14} \times r^2$	

GEOMETRY

Three dimensional figures

Figure	**Diagram of figure**	**Surface area**	**Volume**
Cube	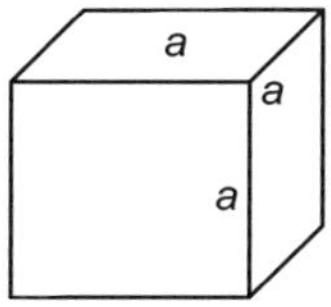	$6a^2$	a^3
Cuboid/ rectangular block	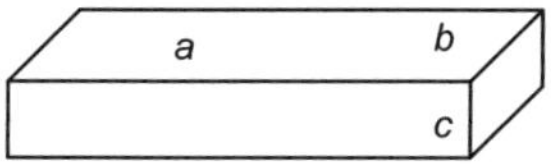	$2(ab + ac + bc)$	abc
Prism/ triangular block	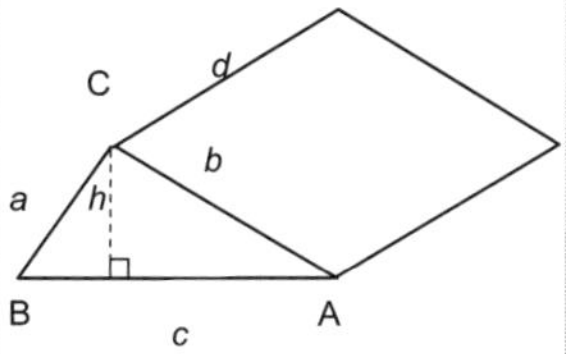	$bd + hc + dc + ad$	$\frac{1}{2}hcd$
Cylinder	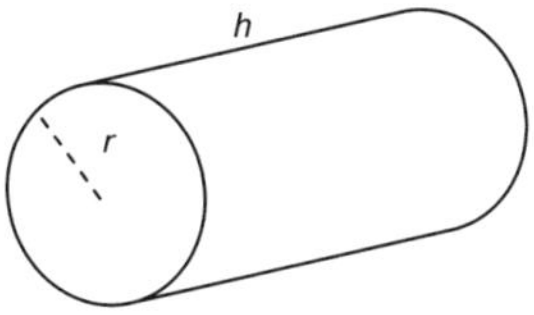	$2\pi r^2 + 2\pi h$	$\pi r^2 h$ $\frac{1}{4}\pi d^2 h$
Sphere	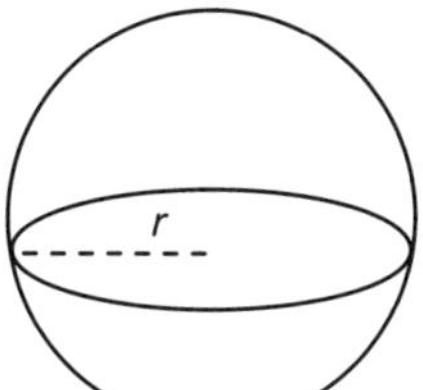	$4\pi r^2$	$\frac{4}{3}\pi r^3$
Segment of sphere	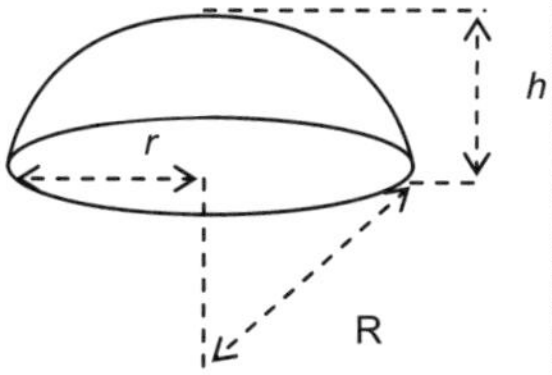	$2\pi Rh$	$\frac{1}{6}\pi h(3r^2 + h^2)$ $\frac{1}{3}\pi h^2(3R - H)$
Pyramid	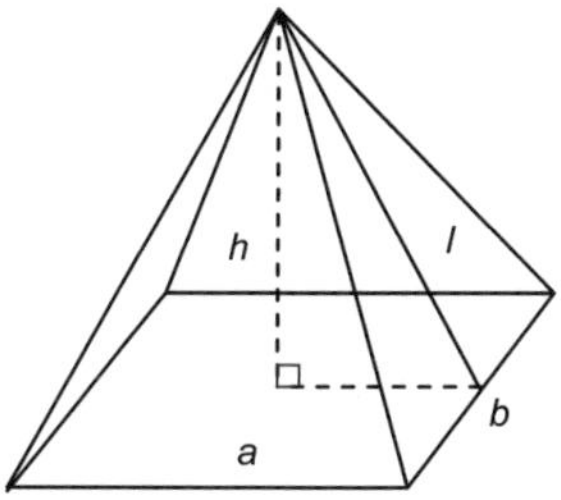	$(a + b)l + ab$	$\frac{1}{3}abh$

GEOMETRY

Figure	Diagram of figure	Surface area	Volume
Frustum of a pyramid	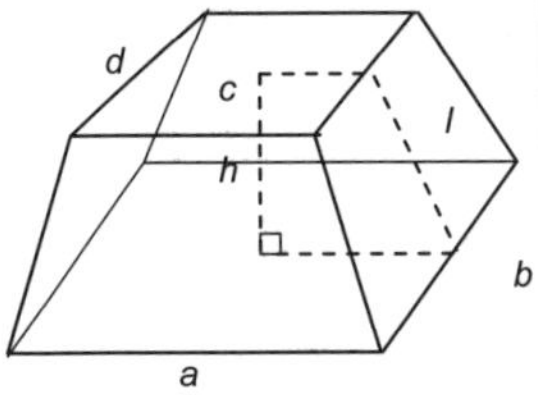	$l(a+b+c+d)+\sqrt{(ab+cd)}$ [rectangular figure only]	$\frac{h}{3}(ab+cd+\sqrt{abcd})$
Cone	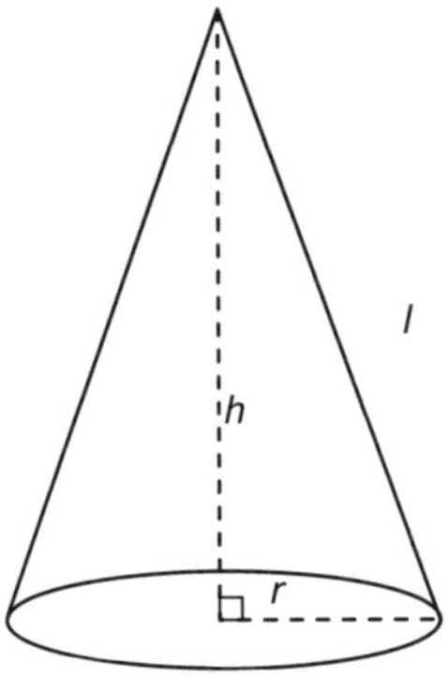	πrl (excluding base) $\pi rl+\pi r^2$ (including base)	$\frac{1}{3}\pi r^2h$ $\frac{1}{12}\pi d^2h$
Frustum of a cone	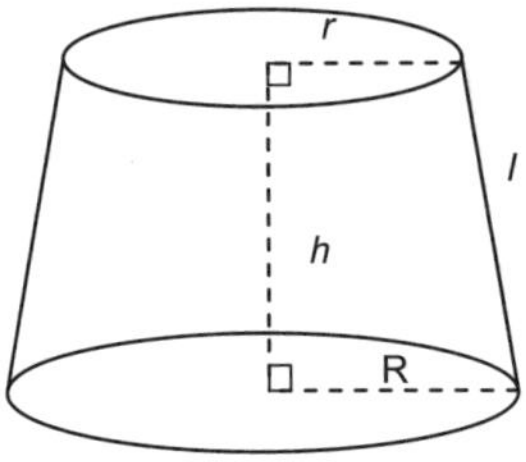	$\pi r^2+\pi R^2+\pi l(R+r)$	$\frac{1}{3}\pi(R^2+Rr+r^2)$

FORMULAE

Formulae

Formula	Description
Pythagoras' Theorem	$A^2 = B^2 + C^2$ where A is the hypotenuse of a right-angled triangle and B and C are the two adjacent sides
Simpson's Rule	The Area is divided into an even number of strips of equal width, and therefore has an odd number of ordinates at the division points area = $\frac{S(A + 2B + 4C)}{3}$ where S = common interval (strip width) A = sum of first and last ordinates B = sum of remaining odd ordinates C = sum of the even ordinates The Volume can be calculated by the same formula, but by substituting the area of each coordinate rather than its length
Trapezoidal Rule	A given trench is divided into two equal sections, giving three ordinates, the first, the middle and the last volume = $\frac{S \times (A + B + 2C)}{2}$ where S = width of the strips A = area of the first section B = area of the last section C = area of the rest of the sections
Prismoidal Rule	A given trench is divided into two equal sections, giving three ordinates, the first, the middle and the last volume = $\frac{L \times (A + 4B + C)}{6}$ where L = total length of trench A = area of the first section B = area of the middle section C = area of the last section

TYPICAL THERMAL CONDUCTIVITY OF BUILDING MATERIALS

TYPICAL THERMAL CONDUCTIVITY OF BUILDING MATERIALS

(Always check manufacturer's details – variation will occur depending on product and nature of materials)

	Thermal conductivity (W/mK)		Thermal conductivity (W/mK)
Acoustic plasterboard	0.25	Oriented strand board	0.13
Aerated concrete slab (500 kg/m^3)	0.16	Outer leaf brick	0.77
Aluminium	237	Plasterboard	0.22
Asphalt (1700 kg/m^3)	0.5	Plaster dense (1300 kg/m^3)	0.5
Bitumen-impregnated fibreboard	0.05	Plaster lightweight (600 kg/m^3)	0.16
Blocks (standard grade 600 kg/m^3)	0.15	Plywood (950 kg/m^3)	0.16
Blocks (solar grade 460 kg/m^3)	0.11	Prefabricated timber wall panels (check manufacturer)	0.12
Brickwork (outer leaf 1700 kg/m^3)	0.84	Screed (1200 kg/m^3)	0.41
Brickwork (inner leaf 1700 kg/m^3)	0.62	Stone chippings (1800 kg/m^3)	0.96
Dense aggregate concrete block 1800 kg/m^3 (exposed)	1.21	Tile hanging (1900 kg/m^3)	0.84
Dense aggregate concrete block 1800 kg/m^3 (protected)	1.13	Timber (650 kg/m^3)	0.14
Calcium silicate board (600 kg/m^3)	0.17	Timber flooring (650 kg/m^3)	0.14
Concrete general	1.28	Timber rafters	0.13
Concrete (heavyweight 2300 kg/m^3)	1.63	Timber roof or floor joists	0.13
Concrete (dense 2100 kg/m^3 typical floor)	1.4	Roof tile (1900 kg/m^3)	0.84
Concrete (dense 2000 kg/m^3 typical floor)	1.13	Timber blocks (650 kg/m^3)	0.14
Concrete (medium 1400 kg/m^3)	0.51	Cellular glass	0.045
Concrete (lightweight 1200 kg/m^3)	0.38	Expanded polystyrene	0.034
Concrete (lightweight 600 kg/m^3)	0.19	Expanded polystyrene slab (25 kg/m^3)	0.035
Concrete slab (aerated 500 kg/m^3)	0.16	Extruded polystyrene	0.035
Copper	390	Glass mineral wool	0.04
External render sand/cement finish	1	Mineral quilt (12 kg/m^3)	0.04
External render (1300 kg/m^3)	0.5	Mineral wool slab (25 kg/m^3)	0.035
Felt – Bitumen layers (1700 kg/m^3)	0.5	Phenolic foam	0.022
Fibreboard (300 kg/m^3)	0.06	Polyisocyanurate	0.025
Glass	0.93	Polyurethane	0.025
Marble	3	Rigid polyurethane	0.025
Metal tray used in wriggly tin concrete floors (7800 kg/m^3)	50	Rock mineral wool	0.038
Mortar (1750 kg/m^3)	0.8		

EARTHWORK

EARTHWORK

Weights of Typical Materials Handled by Excavators

The weight of the material is that of the state in its natural bed and includes moisture
Adjustments should be made to allow for loose or compacted states

Material	Mass (kg/m^3)	Mass (lb/cu yd)
Ashes, dry	610	1028
Ashes, wet	810	1365
Basalt, broken	1954	3293
Basalt, solid	2933	4943
Bauxite, crushed	1281	2159
Borax, fine	849	1431
Caliche	1440	2427
Cement, clinker	1415	2385
Chalk, fine	1221	2058
Chalk, solid	2406	4055
Cinders, coal, ash	641	1080
Cinders, furnace	913	1538
Clay, compacted	1746	2942
Clay, dry	1073	1808
Clay, wet	1602	2700
Coal, anthracite, solid	1506	2538
Coal, bituminous	1351	2277
Coke	610	1028
Dolomite, lumpy	1522	2565
Dolomite, solid	2886	4864
Earth, dense	2002	3374
Earth, dry, loam	1249	2105
Earth, Fullers, raw	673	1134
Earth, moist	1442	2430
Earth, wet	1602	2700
Felsite	2495	4205
Fieldspar, solid	2613	4404
Fluorite	3093	5213
Gabbro	3093	5213
Gneiss	2696	4544
Granite	2690	4534
Gravel, dry ¼ to 2 inch	1682	2835
Gravel, dry, loose	1522	2565
Gravel, wet ¼ to 2 inch	2002	3374
Gypsum, broken	1450	2444
Gypsum, solid	2787	4697
Hardcore (consolidated)	1928	3249
Lignite, dry	801	1350
Limestone, broken	1554	2619
Limestone, solid	2596	4375
Magnesite, magnesium ore	2993	5044
Marble	2679	4515
Marl, wet	2216	3735
Mica, broken	1602	2700
Mica, solid	2883	4859
Peat, dry	400	674
Peat, moist	700	1179
Peat, wet	1121	1889

EARTHWORK

Material	Mass (kg/m^3)	Mass (lb/cu yd)
Potash	1281	2159
Pumice, stone	640	1078
Quarry waste	1438	2423
Quartz sand	1201	2024
Quartz, solid	2584	4355
Rhyolite	2400	4045
Sand and gravel, dry	1650	2781
Sand and gravel, wet	2020	3404
Sand, dry	1602	2700
Sand, wet	1831	3086
Sandstone, solid	2412	4065
Shale, solid	2637	4444
Slag, broken	2114	3563
Slag, furnace granulated	961	1619
Slate, broken	1370	2309
Slate, solid	2667	4495
Snow, compacted	481	810
Snow, freshly fallen	160	269
Taconite	2803	4724
Trachyte	2400	4045
Trap rock, solid	2791	4704
Turf	400	674
Water	1000	1685

Transport Capacities

Type of vehicle	Capacity of vehicle	
	Payload	Heaped capacity
Wheelbarrow	150	0.10
1 tonne dumper	1250	1.00
2.5 tonne dumper	4000	2.50
Articulated dump truck (Volvo A20 6 × 4)	18500	11.00
Articulated dump truck (Volvo A35 6 × 6)	32000	19.00
Large capacity rear dumper (Euclid R35)	35000	22.00
Large capacity rear dumper (Euclid R85)	85000	50.00

EARTHWORK

Machine Volumes for Excavating and Filling

Machine type	Cycles per minute	Volume per minute (m^3)
1.5 tonne excavator	1	0.04
	2	0.08
	3	0.12
3 tonne excavator	1	0.13
	2	0.26
	3	0.39
5 tonne excavator	1	0.28
	2	0.56
	3	0.84
7 tonne excavator	1	0.28
	2	0.56
	3	0.84
21 tonne excavator	1	1.21
	2	2.42
	3	3.63
Backhoe loader JCB3CX excavator	1	0.28
Rear bucket capacity 0.28 m^3	2	0.56
	3	0.84
Backhoe loader JCB3CX loading	1	1.00
Front bucket capacity 1.00 m^3	2	2.00

Machine Volumes for Excavating and Filling

Machine type	Loads per hour	Volume per hour (m^3)
1 tonne high tip skip loader	5	2.43
Volume 0.485 m^3	7	3.40
	10	4.85
3 tonne dumper	4	7.60
Max volume 2.40 m^3	5	9.50
Available volume 1.9 m^3	7	13.30
	10	19.00
6 tonne dumper	4	15.08
Max volume 3.40 m^3	5	18.85
Available volume 3.77 m^3	7	26.39
	10	37.70

EARTHWORK

Bulkage of Soils (after excavation)

Type of soil	Approximate bulking of 1 m^3 after excavation
Vegetable soil and loam	25–30%
Soft clay	30–40%
Stiff clay	10–20%
Gravel	20–25%
Sand	40–50%
Chalk	40–50%
Rock, weathered	30–40%
Rock, unweathered	50–60%

Shrinkage of Materials (on being deposited)

Type of soil	Approximate bulking of 1 m^3 after excavation
Clay	10%
Gravel	8%
Gravel and sand	9%
Loam and light sandy soils	12%
Loose vegetable soils	15%

Voids in Material Used as Subbases or Beddings

Material	m^3 of voids/m^3
Alluvium	0.37
River grit	0.29
Quarry sand	0.24
Shingle	0.37
Gravel	0.39
Broken stone	0.45
Broken bricks	0.42

Angles of Repose

Type of soil		Degrees
Clay	– dry	30
	– damp, well drained	45
	– wet	15–20
Earth	– dry	30
	– damp	45
Gravel	– moist	48
Sand	– dry or moist	35
	– wet	25
Loam		40

EARTHWORK

Slopes and Angles

Ratio of base to height	Angle in degrees
5:1	11
4:1	14
3:1	18
2:1	27
1½:1	34
1:1	45
1:1½	56
1:2	63
1:3	72
1:4	76
1:5	79

Grades (in Degrees and Percents)

Degrees	Percent	Degrees	Percent
1	1.8	24	44.5
2	3.5	25	46.6
3	5.2	26	48.8
4	7.0	27	51.0
5	8.8	28	53.2
6	10.5	29	55.4
7	12.3	30	57.7
8	14.0	31	60.0
9	15.8	32	62.5
10	17.6	33	64.9
11	19.4	34	67.4
12	21.3	35	70.0
13	23.1	36	72.7
14	24.9	37	75.4
15	26.8	38	78.1
16	28.7	39	81.0
17	30.6	40	83.9
18	32.5	41	86.9
19	34.4	42	90.0
20	36.4	43	93.3
21	38.4	44	96.6
22	40.4	45	100.0
23	42.4		

EARTHWORK

Bearing Powers

Ground conditions		Bearing power		
		kg/m²	lb/in²	Metric t/m²
Rock,	broken	483	70	50
	solid	2415	350	240
Clay,	dry or hard	380	55	40
	medium dry	190	27	20
	soft or wet	100	14	10
Gravel,	cemented	760	110	80
Sand,	compacted	380	55	40
	clean dry	190	27	20
Swamp and alluvial soils		48	7	5

Earthwork Support

Maximum depth of excavation in various soils without the use of earthwork support

Ground conditions	Feet (ft)	Metres (m)
Compact soil	12	3.66
Drained loam	6	1.83
Dry sand	1	0.3
Gravelly earth	2	0.61
Ordinary earth	3	0.91
Stiff clay	10	3.05

It is important to note that the above table should only be used as a guide. Each case must be taken on its merits and, as the limited distances given above are approached, careful watch must be kept for the slightest signs of caving in.

CONCRETE WORK

Weights of Concrete and Concrete Elements

Type of material		kg/m³	lb/cu ft
Ordinary concrete (dense aggregates)			
Non-reinforced plain or mass concrete			
Nominal weight		2305	144
Aggregate	– limestone	2162 to 2407	135 to 150
	– gravel	2244 to 2407	140 to 150
	– broken brick	2000 (av)	125 (av)
	– other crushed stone	2326 to 2489	145 to 155
Reinforced concrete			
Nominal weight		2407	150
Reinforcement	– 1%	2305 to 2468	144 to 154
	– 2%	2356 to 2519	147 to 157
	– 4%	2448 to 2703	153 to 163
Special concretes			
Heavy concrete			
Aggregates	– barytes, magnetite	3210 (min)	200 (min)
	– steel shot, punchings	5280	330
Lean mixes			
Dry-lean (gravel aggregate)		2244	140
Soil-cement (normal mix)		1601	100

CONCRETE WORK

Type of material		kg/m² per mm thick	lb/sq ft per inch thick
Ordinary concrete (dense aggregates)			
Solid slabs (floors, walls etc.)			
Thickness:	75 mm or 3 in	184	37.5
	100 mm or 4 in	245	50
	150 mm or 6 in	378	75
	250 mm or 10 in	612	125
	300 mm or 12 in	734	150
Ribbed slabs			
Thickness:	125 mm or 5 in	204	42
	150 mm or 6 in	219	45
	225 mm or 9 in	281	57
	300 mm or 12 in	342	70
Special concretes			
Finishes etc.			
	Rendering, screed etc. Granolithic, terrazzo	1928 to 2401	10 to 12.5
	Glass-block (hollow) concrete	1734 (approx)	9 (approx)
Prestressed concrete		Weights as for reinforced concrete (upper limits)	
Air-entrained concrete		Weights as for plain or reinforced concrete	

CONCRETE WORK

Average Weight of Aggregates

Materials	Voids %	Weight kg/m^3
Sand	39	1660
Gravel 10–20 mm	45	1440
Gravel 35–75 mm	42	1555
Crushed stone	50	1330
Crushed granite (over 15 mm)	50	1345
(n.e. 15 mm)	47	1440
'All-in' ballast	32	1800–2000

Material	kg/m^3	lb/cu yd
Vermiculite (aggregate)	64–80	108–135
All-in aggregate	1999	125

Applications and Mix Design

Site mixed concrete

Recommended mix	Class of work suitable for	Cement (kg)	Sand (kg)	Coarse aggregate (kg)	Nr 25 kg bags cement per m^3 of combined aggregate
1:3:6	Roughest type of mass concrete such as footings, road haunching over 300 mm thick	208	905	1509	8.30
1:2.5:5	Mass concrete of better class than 1:3:6 such as bases for machinery, walls below ground etc.	249	881	1474	10.00
1:2:4	Most ordinary uses of concrete, such as mass walls above ground, road slabs etc. and general reinforced concrete work	304	889	1431	12.20
1:1.5:3	Watertight floors, pavements and walls, tanks, pits, steps, paths, surface of 2 course roads, reinforced concrete where extra strength is required	371	801	1336	14.90
1:1:2	Works of thin section such as fence posts and small precast work	511	720	1206	20.40

CONCRETE WORK

Ready mixed concrete

Application	Designated concrete	Standardized prescribed concrete	Recommended consistence (nominal slump class)
Foundations			
Mass concrete fill or blinding	GEN 1	ST2	S3
Strip footings	GEN 1	ST2	S3
Mass concrete foundations			
Single storey buildings	GEN 1	ST2	S3
Double storey buildings	GEN 3	ST4	S3
Trench fill foundations			
Single storey buildings	GEN 1	ST2	S4
Double storey buildings	GEN 3	ST4	S4
General applications			
Kerb bedding and haunching	GEN 0	ST1	S1
Drainage works – immediate support	GEN 1	ST2	S1
Other drainage works	GEN 1	ST2	S3
Oversite below suspended slabs	GEN 1	ST2	S3
Floors			
Garage and house floors with no embedded steel	GEN 3	ST4	S2
Wearing surface: Light foot and trolley traffic	RC30	ST4	S2
Wearing surface: General industrial	RC40	N/A	S2
Wearing surface: Heavy industrial	RC50	N/A	S2
Paving			
House drives, domestic parking and external parking	PAV 1	N/A	S2
Heavy-duty external paving	PAV 2	N/A	S2

CONCRETE WORK

Prescribed Mixes for Ordinary Structural Concrete

Weights of cement and total dry aggregates in kg to produce approximately one cubic metre of fully compacted concrete together with the percentages by weight of fine aggregate in total dry aggregates

Conc. grade	Nominal max size of aggregate (mm)	40		20		14		10	
	Workability	Med.	High	Med.	High	Med.	High	Med.	High
	Limits to slump that may be expected (mm)	50–100	100–150	25–75	75–125	10–50	50–100	10–25	25–50
7	Cement (kg)	180	200	210	230	–	–	–	–
	Total aggregate (kg)	1950	1850	1900	1800	–	–	–	–
	Fine aggregate (%)	30–45	30–45	35–50	35–50	–	–	–	–
10	Cement (kg)	210	230	240	260	–	–	–	–
	Total aggregate (kg)	1900	1850	1850	1800	–	–	–	–
	Fine aggregate (%)	30–45	30–45	35–50	35–50	–	–	–	–
15	Cement (kg)	250	270	280	310	–	–	–	–
	Total aggregate (kg)	1850	1800	1800	1750	–	–	–	–
	Fine aggregate (%)	30–45	30–45	35–50	35–50	–	–	–	–
20	Cement (kg)	300	320	320	350	340	380	360	410
	Total aggregate (kg)	1850	1750	1800	1750	1750	1700	1750	1650
	Sand								
	Zone 1 (%)	35	40	40	45	45	50	50	55
	Zone 2 (%)	30	35	35	40	40	45	45	50
	Zone 3 (%)	30	30	30	35	35	40	40	45
25	Cement (kg)	340	360	360	390	380	420	400	450
	Total aggregate (kg)	1800	1750	1750	1700	1700	1650	1700	1600
	Sand								
	Zone 1 (%)	35	40	40	45	45	50	50	55
	Zone 2 (%)	30	35	35	40	40	45	45	50
	Zone 3 (%)	30	30	30	35	35	40	40	45
30	Cement (kg)	370	390	400	430	430	470	460	510
	Total aggregate (kg)	1750	1700	1700	1650	1700	1600	1650	1550
	Sand								
	Zone 1 (%)	35	40	40	45	45	50	50	55
	Zone 2 (%)	30	35	35	40	40	45	45	50
	Zone 3 (%)	30	30	30	35	35	40	40	45

REINFORCEMENT

REINFORCEMENT

Weights of Bar Reinforcement

Nominal sizes (mm)	Cross-sectional area (mm^2)	Mass (kg/m)	Length of bar (m/tonne)
6	28.27	0.222	4505
8	50.27	0.395	2534
10	78.54	0.617	1622
12	113.10	0.888	1126
16	201.06	1.578	634
20	314.16	2.466	405
25	490.87	3.853	260
32	804.25	6.313	158
40	1265.64	9.865	101
50	1963.50	15.413	65

Weights of Bars (at specific spacings)

Weights of metric bars in kilogrammes per square metre

Size	Spacing of bars in millimetres									
(mm)	75	100	125	150	175	200	225	250	275	300
6	2.96	2.220	1.776	1.480	1.27	1.110	0.99	0.89	0.81	0.74
8	5.26	3.95	3.16	2.63	2.26	1.97	1.75	1.58	1.44	1.32
10	8.22	6.17	4.93	4.11	3.52	3.08	2.74	2.47	2.24	2.06
12	11.84	8.88	7.10	5.92	5.07	4.44	3.95	3.55	3.23	2.96
16	21.04	15.78	12.63	10.52	9.02	7.89	7.02	6.31	5.74	5.26
20	32.88	24.66	19.73	16.44	14.09	12.33	10.96	9.87	8.97	8.22
25	51.38	38.53	30.83	25.69	22.02	19.27	17.13	15.41	14.01	12.84
32	84.18	63.13	50.51	42.09	36.08	31.57	28.06	25.25	22.96	21.04
40	131.53	98.65	78.92	65.76	56.37	49.32	43.84	39.46	35.87	32.88
50	205.51	154.13	123.31	102.76	88.08	77.07	68.50	61.65	56.05	51.38

Basic weight of steelwork taken as 7850 kg/m^3
Basic weight of bar reinforcement per metre run = 0.00785 kg/mm^2
The value of π has been taken as 3.141592654

REINFORCEMENT

Fabric Reinforcement

Preferred range of designated fabric types and stock sheet sizes

Fabric reference	Longitudinal wires			Cross wires			
	Nominal wire size (mm)	Pitch (mm)	Area (mm^2/m)	Nominal wire size (mm)	Pitch (mm)	Area (mm^2/m)	Mass (kg/m^2)
Square mesh							
A393	10	200	393	10	200	393	6.16
A252	8	200	252	8	200	252	3.95
A193	7	200	193	7	200	193	3.02
A142	6	200	142	6	200	142	2.22
A98	5	200	98	5	200	98	1.54
Structural mesh							
B1131	12	100	1131	8	200	252	10.90
B785	10	100	785	8	200	252	8.14
B503	8	100	503	8	200	252	5.93
B385	7	100	385	7	200	193	4.53
B283	6	100	283	7	200	193	3.73
B196	5	100	196	7	200	193	3.05
Long mesh							
C785	10	100	785	6	400	70.8	6.72
C636	9	100	636	6	400	70.8	5.55
C503	8	100	503	5	400	49.0	4.34
C385	7	100	385	5	400	49.0	3.41
C283	6	100	283	5	400	49.0	2.61
Wrapping mesh							
D98	5	200	98	5	200	98	1.54
D49	2.5	100	49	2.5	100	49	0.77

Stock sheet size 4.8 m × 2.4 m, Area 11.52 m^2

Average weight kg/m^3 of steelwork reinforcement in concrete for various building elements

Substructure	kg/m^3 concrete	Substructure	kg/m^3 concrete
Pile caps	110–150	Plate slab	150–220
Tie beams	130–170	Cant slab	145–210
Ground beams	230–330	Ribbed floors	130–200
Bases	125–180	Topping to block floor	30–40
Footings	100–150	Columns	210–310
Retaining walls	150–210	Beams	250–350
Raft	60–70	Stairs	130–170
Slabs – one way	120–200	Walls – normal	40–100
Slabs – two way	110–220	Walls – wind	70–125

Note: For exposed elements add the following %:
Walls 50%, Beams 100%, Columns 15%

FORMWORK

FORMWORK

Formwork Stripping Times – Normal Curing Periods

Conditions under which concrete is maturing	Minimum periods of protection for different types of cement					
	Number of days (where the average surface temperature of the concrete exceeds 10°C during the whole period)			Equivalent maturity (degree hours) calculated as the age of the concrete in hours multiplied by the number of degrees Celsius by which the average surface temperature of the concrete exceeds 10°C		
	Other	SRPC	OPC or RHPC	Other	SRPC	OPC or RHPC
1. Hot weather or drying winds	7	4	3	3500	2000	1500
2. Conditions not covered by 1	4	3	2	2000	1500	1000

KEY
OPC – Ordinary Portland Cement
RHPC – Rapid-hardening Portland Cement
SRPC – Sulphate-resisting Portland Cement

Minimum Period before Striking Formwork

	Minimum period before striking		
	Surface temperature of concrete		
	16°C	17°C	t°C (0–25)
Vertical formwork to columns, walls and large beams	12 hours	18 hours	300 hours / t+10
Soffit formwork to slabs	4 days	6 days	100 days / t+10
Props to slabs	10 days	15 days	250 days / t+10
Soffit formwork to beams	9 days	14 days	230 days / t+10
Props to beams	14 days	21 days	360 days / t+10

MASONRY

MASONRY

Number of Bricks Required for Various Types of Work per m² of Walling

Description	Brick size	
	215 × 102.5 × 50 mm	215 × 102.5 × 65 mm
Half brick thick		
Stretcher bond	74	59
English bond	108	86
English garden wall bond	90	72
Flemish bond	96	79
Flemish garden wall bond	83	66
One brick thick and cavity wall of two half brick skins		
Stretcher bond	148	119

Quantities of Bricks and Mortar required per m² of Walling

	Unit	No of bricks required	Mortar required (cubic metres)		
Standard bricks			No frogs	Single frogs	Double frogs
Brick size 215 × 102.5 × 50 mm					
half brick wall (103 mm)	m^2	72	0.022	0.027	0.032
2 × half brick cavity wall (270 mm)	m^2	144	0.044	0.054	0.064
one brick wall (215 mm)	m^2	144	0.052	0.064	0.076
one and a half brick wall (322 mm)	m^2	216	0.073	0.091	0.108
Mass brickwork	m^3	576	0.347	0.413	0.480
Brick size 215 × 102.5 × 65 mm					
half brick wall (103 mm)	m^2	58	0.019	0.022	0.026
2 × half brick cavity wall (270 mm)	m^2	116	0.038	0.045	0.055
one brick wall (215 mm)	m^2	116	0.046	0.055	0.064
one and a half brick wall (322 mm)	m^2	174	0.063	0.074	0.088
Mass brickwork	m^3	464	0.307	0.360	0.413
Metric modular bricks			Perforated		
Brick size 200 × 100 × 75 mm					
90 mm thick	m^2	67	0.016	0.019	
190 mm thick	m^2	133	0.042	0.048	
290 mm thick	m^2	200	0.068	0.078	
Brick size 200 × 100 × 100 mm					
90 mm thick	m^2	50	0.013	0.016	
190 mm thick	m^2	100	0.036	0.041	
290 mm thick	m^2	150	0.059	0.067	
Brick size 300 × 100 × 75 mm					
90 mm thick	m^2	33	–	0.015	
Brick size 300 × 100 × 100 mm					
90 mm thick	m^2	44	0.015	0.018	

Note: Assuming 10 mm thick joints

MASONRY

Mortar Required per m² Blockwork (9.88 blocks/m²)

Wall thickness	75	90	100	125	140	190	215
Mortar m^3/m^2	0.005	0.006	0.007	0.008	0.009	0.013	0.014

Mortar group	Cement: lime: sand	Masonry cement: sand	Cement: sand with plasticizer
1	1:0–0.25:3		
2	1:0.5:4–4.5	1:2.5-3.5	1:3–4
3	1:1:5–6	1:4–5	1:5–6
4	1:2:8–9	1:5.5–6.5	1:7–8
5	1:3:10–12	1:6.5–7	1:8

Group 1: strong inflexible mortar
Group 5: weak but flexible

All mixes within a group are of approximately similar strength
Frost resistance increases with the use of plasticizers
Cement: lime: sand mixes give the strongest bond and greatest resistance to rain penetration
Masonry cement equals ordinary Portland cement plus a fine neutral mineral filler and an air entraining agent

Calcium Silicate Bricks

Type	Strength	Location
Class 2 crushing strength	14.0 N/mm²	not suitable for walls
Class 3	20.5 N/mm²	walls above DPC
Class 4	27.5 N/mm²	cappings and copings
Class 5	34.5 N/mm²	retaining walls
Class 6	41.5 N/mm²	walls below ground
Class 7	48.5 N/mm²	walls below ground

The Class 7 calcium silicate bricks are therefore equal in strength to Class B bricks
Calcium silicate bricks are not suitable for DPCs

Durability of bricks

FL	Frost resistant with low salt content
FN	Frost resistant with normal salt content
ML	Moderately frost resistant with low salt content
MN	Moderately frost resistant with normal salt content

MASONRY

Brickwork Dimensions

No. of horizontal bricks	Dimensions (mm)	No. of vertical courses	Height of vertical courses (mm)
½	112.5	1	75
1	225.0	2	150
1½	337.5	3	225
2	450.0	4	300
2½	562.5	5	375
3	675.0	6	450
3½	787.5	7	525
4	900.0	8	600
4½	1012.5	9	675
5	1125.0	10	750
5½	1237.5	11	825
6	1350.0	12	900
6½	1462.5	13	975
7	1575.0	14	1050
7½	1687.5	15	1125
8	1800.0	16	1200
8½	1912.5	17	1275
9	2025.0	18	1350
9½	2137.5	19	1425
10	2250.0	20	1500
20	4500.0	24	1575
40	9000.0	28	2100
50	11250.0	32	2400
60	13500.0	36	2700
75	16875.0	40	3000

TIMBER

TIMBER

Weights of Timber

Material	**kg/m^3**	**lb/cu ft**
General	806 (avg)	50 (avg)
Douglas fir	479	30
Yellow pine, spruce	479	30
Pitch pine	673	42
Larch, elm	561	35
Oak (English)	724 to 959	45 to 60
Teak	643 to 877	40 to 55
Jarrah	959	60
Greenheart	1040 to 1204	65 to 75
Quebracho	1285	80
Material	**kg/m^2 per mm thickness**	**lb/sq ft per inch thickness**
Wooden boarding and blocks		
Softwood	0.48	2.5
Hardwood	0.76	4
Hardboard	1.06	5.5
Chipboard	0.76	4
Plywood	0.62	3.25
Blockboard	0.48	2.5
Fibreboard	0.29	1.5
Wood-wool	0.58	3
Plasterboard	0.96	5
Weather boarding	0.35	1.8

TIMBER

Conversion Tables (for timber only)

Inches	Millimetres	Feet	Metres
1	25	1	0.300
2	50	2	0.600
3	75	3	0.900
4	100	4	1.200
5	125	5	1.500
6	150	6	1.800
7	175	7	2.100
8	200	8	2.400
9	225	9	2.700
10	250	10	3.000
11	275	11	3.300
12	300	12	3.600
13	325	13	3.900
14	350	14	4.200
15	375	15	4.500
16	400	16	4.800
17	425	17	5.100
18	450	18	5.400
19	475	19	5.700
20	500	20	6.000
21	525	21	6.300
22	550	22	6.600
23	575	23	6.900
24	600	24	7.200

Planed Softwood

The finished end section size of planed timber is usually 3/16" less than the original size from which it is produced. This however varies slightly depending upon availability of material and origin of the species used.

Standards (timber) to cubic metres and cubic metres to standards (timber)

Cubic metres	Cubic metres standards	Standards
4.672	1	0.214
9.344	2	0.428
14.017	3	0.642
18.689	4	0.856
23.361	5	1.070
28.033	6	1.284
32.706	7	1.498
37.378	8	1.712
42.050	9	1.926
46.722	10	2.140
93.445	20	4.281
140.167	30	6.421
186.890	40	8.561
233.612	50	10.702
280.335	60	12.842
327.057	70	14.982
373.779	80	17.122

TIMBER

1 cu metre = 35.3148 cu ft = 0.21403 std

1 cu ft = 0.028317 cu metres

1 std = 4.67227 cu metres

Basic sizes of sawn softwood available (cross-sectional areas)

Thickness (mm)	Width (mm)								
	75	100	125	150	175	200	225	250	300
16	X	X	X	X					
19	X	X	X	X					
22	X	X	X	X					
25	X	X	X	X	X	X	X	X	X
32	X	X	X	X	X	X	X	X	X
36	X	X	X	X					
38	X	X	X	X	X	X	X		
44	X	X	X	X	X	X	X	X	X
47*	X	X	X	X	X	X	X	X	X
50	X	X	X	X	X	X	X	X	X
63	X	X	X	X	X	X	X		
75	X	X	X	X	X	X	X	X	
100		X		X		X		X	X
150				X		X			X
200						X			
250								X	
300									X

* This range of widths for 47 mm thickness will usually be found to be available in construction quality only

Note: The smaller sizes below 100 mm thick and 250 mm width are normally but not exclusively of European origin. Sizes beyond this are usually of North and South American origin

Basic lengths of sawn softwood available (metres)

1.80	2.10	3.00	4.20	5.10	6.00	7.20
	2.40	3.30	4.50	5.40	6.30	
	2.70	3.60	4.80	5.70	6.60	
		3.90			6.90	

Note: Lengths of 6.00 m and over will generally only be available from North American species and may have to be recut from larger sizes

TIMBER

Reductions from basic size to finished size by planning of two opposed faces

Purpose		Reductions from basic sizes for timber		
	15–35 mm	36–100 mm	101–150 mm	over 150 mm
a) Constructional timber	3 mm	3 mm	5 mm	6 mm
b) Matching interlocking boards	4 mm	4 mm	6 mm	6 mm
c) Wood trim not specified in BS 584	5 mm	7 mm	7 mm	9 mm
d) Joinery and cabinet work	7 mm	9 mm	11 mm	13 mm

Note: The reduction of width or depth is overall the extreme size and is exclusive of any reduction of the face by the machining of a tongue or lap joints

Maximum Spans for Various Roof Trusses

Maximum permissible spans for rafters for Fink trussed rafters

Basic size (mm)	Actual size (mm)	Pitch (degrees)								
		15 (m)	17.5 (m)	20 (m)	22.5 (m)	25 (m)	27.5 (m)	30 (m)	32.5 (m)	35 (m)
38 × 75	35 × 72	6.03	6.16	6.29	6.41	6.51	6.60	6.70	6.80	6.90
38 × 100	35 × 97	7.48	7.67	7.83	7.97	8.10	8.22	8.34	8.47	8.61
38 × 125	35 × 120	8.80	9.00	9.20	9.37	9.54	9.68	9.82	9.98	10.16
44 × 75	41 × 72	6.45	6.59	6.71	6.83	6.93	7.03	7.14	7.24	7.35
44 × 100	41 × 97	8.05	8.23	8.40	8.55	8.68	8.81	8.93	9.09	9.22
44 × 125	41 × 120	9.38	9.60	9.81	9.99	10.15	10.31	10.45	10.64	10.81
50 × 75	47 × 72	6.87	7.01	7.13	7.25	7.35	7.45	7.53	7.67	7.78
50 × 100	47 × 97	8.62	8.80	8.97	9.12	9.25	9.38	9.50	9.66	9.80
50 × 125	47 × 120	10.01	10.24	10.44	10.62	10.77	10.94	11.00	11.00	11.00

TIMBER

Sizes of Internal and External Doorsets

Description	Internal size (mm)	Permissible deviation	External size (mm)	Permissible deviation
Coordinating dimension: height of door leaf height sets	2100		2100	
Coordinating dimension: height of ceiling height set	2300 2350 2400 2700 3000		2300 2350 2400 2700 3000	
Coordinating dimension: width of all doorsets S = Single leaf set D = Double leaf set	600 S 700 S 800 S&D 900 S&D 1000 S&D 1200 D 1500 D 1800 D 2100 D		900 S 1000 S 1200 D 1800 D 2100 D	
Work size: height of door leaf height set	2090	± 2.0	2095	± 2.0
Work size: height of ceiling height set	2285 2335 2385 2685 2985	± 2.0	2295 2345 2395 2695 2995	± 2.0
Work size: width of all doorsets S = Single leaf set D = Double leaf set	590 S 690 S 790 S&D 890 S&D 990 S&D 1190 D 1490 D 1790 D 2090 D	± 2.0	895 S 995 S 1195 D 1495 D 1795 D 2095 D	± 2.0
Width of door leaf in single leaf sets F = Flush leaf P = Panel leaf	526 F 626 F 726 F&P 826 F&P 926 F&P	± 1.5	806 F&P 906 F&P	± 1.5
Width of door leaf in double leaf sets F = Flush leaf P = Panel leaf	362 F 412 F 426 F 562 F&P 712 F&P 826 F&P 1012 F&P	± 1.5	552 F&P 702 F&P 852 F&P 1002 F&P	± 1.5
Door leaf height for all doorsets	2040	± 1.5	1994	± 1.5

ROOFING

ROOFING

Total Roof Loadings for Various Types of Tiles/Slates

		Roof load (slope) kg/m^2		
		Slate/Tile	Roofing underlay and battens[2]	Total dead load kg/m
Asbestos cement slate (600 × 300)		21.50	3.14	24.64
Clay tile	interlocking	67.00	5.50	72.50
	plain	43.50	2.87	46.37
Concrete tile	interlocking	47.20	2.69	49.89
	plain	78.20	5.50	83.70
Natural slate (18" × 10")		35.40	3.40	38.80
		Roof load (plan) kg/m^2		
Asbestos cement slate (600 × 300)		28.45	76.50	104.95
Clay tile	interlocking	53.54	76.50	130.04
	plain	83.71	76.50	60.21
Concrete tile	interlocking	57.60	76.50	134.10
	plain	96.64	76.50	173.14

ROOFING

Tiling Data

Product		Lap (mm)	Gauge of battens	No. slates per m^2	Battens (m/m^2)	Weight as laid (kg/m^2)
CEMENT SLATES						
Eternit slates	600 × 300 mm	100	250	13.4	4.00	19.50
(Duracem)		90	255	13.1	3.92	19.20
		80	260	12.9	3.85	19.00
		70	265	12.7	3.77	18.60
	600 × 350 mm	100	250	11.5	4.00	19.50
		90	255	11.2	3.92	19.20
	500 × 250 mm	100	200	20.0	5.00	20.00
		90	205	19.5	4.88	19.50
		80	210	19.1	4.76	19.00
		70	215	18.6	4.65	18.60
	400 × 200 mm	90	155	32.3	6.45	20.80
		80	160	31.3	6.25	20.20
		70	165	30.3	6.06	19.60
CONCRETE TILES/SLATES						
Redland Roofing						
Stonewold slate	430 × 380 mm	75	355	8.2	2.82	51.20
Double Roman tile	418 × 330 mm	75	355	8.2	2.91	45.50
Grovebury pantile	418 × 332 mm	75	343	9.7	2.91	47.90
Norfolk pantile	381 × 227 mm	75	306	16.3	3.26	44.01
		100	281	17.8	3.56	48.06
Renown interlocking tile	418 × 330 mm	75	343	9.7	2.91	46.40
'49' tile	381 × 227 mm	75	306	16.3	3.26	44.80
		100	281	17.8	3.56	48.95
Plain, vertical tiling	265 × 165 mm	35	115	52.7	8.70	62.20
Marley Roofing						
Bold roll tile	420 × 330 mm	75	344	9.7	2.90	47.00
		100	–	10.5	3.20	51.00
Modern roof tile	420 × 330 mm	75	338	10.2	3.00	54.00
		100	–	11.0	3.20	58.00
Ludlow major	420 × 330 mm	75	338	10.2	3.00	45.00
		100	–	11.0	3.20	49.00
Ludlow plus	387 × 229 mm	75	305	16.1	3.30	47.00
		100	–	17.5	3.60	51.00
Mendip tile	420 × 330 mm	75	338	10.2	3.00	47.00
		100	–	11.0	3.20	51.00
Wessex	413 × 330 mm	75	338	10.2	3.00	54.00
		100	–	11.0	3.20	58.00
Plain tile	267 × 165 mm	65	100	60.0	10.00	76.00
		75	95	64.0	10.50	81.00
		85	90	68.0	11.30	86.00
Plain vertical tiles (feature)	267 × 165 mm	35	110	53.0	8.70	67.00
		34	115	56.0	9.10	71.00

ROOFING

Slate Nails, Quantity per Kilogram

Length	Type			
	Plain wire	Galvanized wire	Copper nail	Zinc nail
28.5 mm	325	305	325	415
34.4 mm	286	256	254	292
50.8 mm	242	224	194	200

Metal Sheet Coverings

Thicknesses and weights of sheet metal coverings								
Lead to BS 1178								
BS Code No	3	4	5	6	7	8		
Colour code	Green	Blue	Red	Black	White	Orange		
Thickness (mm)	1.25	1.80	2.24	2.50	3.15	3.55		
Density kg/m^2	14.18	20.41	25.40	30.05	35.72	40.26		
Copper to BS 2870								
Thickness (mm)		0.60	0.70					
Bay width								
Roll (mm)		500	650					
Seam (mm)		525	600					
Standard width to form bay	600	750						
Normal length of sheet	1.80	1.80						
Zinc to BS 849								
Zinc Gauge (Nr)	9	10	11	12	13	14	15	16
Thickness (mm)	0.43	0.48	0.56	0.64	0.71	0.79	0.91	1.04
Density (kg/m^2)	3.1	3.2	3.8	4.3	4.8	5.3	6.2	7.0
Aluminium to BS 4868								
Thickness (mm)	0.5	0.6	0.7	0.8	0.9	1.0	1.2	
Density (kg/m^2)	12.8	15.4	17.9	20.5	23.0	25.6	30.7	

ROOFING

Type of felt	Nominal mass per unit area (kg/10 m)	Nominal mass per unit area of fibre base (g/m^2)	Nominal length of roll (m)
Class 1			
1B fine granule surfaced bitumen	14	220	10 or 20
	18	330	10 or 20
	25	470	10
1E mineral surfaced bitumen	38	470	10
1F reinforced bitumen	15	160 (fibre) 110 (hessian)	15
1F reinforced bitumen, aluminium faced	13	160 (fibre) 110 (hessian)	15
Class 2			
2B fine granule surfaced bitumen asbestos	18	500	10 or 20
2E mineral surfaced bitumen asbestos	38	600	10
Class 3			
3B fine granule surfaced bitumen glass fibre	18	60	20
3E mineral surfaced bitumen glass fibre	28	60	10
3E venting base layer bitumen glass fibre	32	60*	10
3H venting base layer bitumen glass fibre	17	60*	20

* Excluding effect of perforations

GLAZING

GLAZING

Nominal thickness (mm)	Tolerance on thickness (mm)	Approximate weight (kg/m²)	Normal maximum size (mm)
Float and polished plate glass			
3	+ 0.2	7.50	2140 × 1220
4	+ 0.2	10.00	2760 × 1220
5	+ 0.2	12.50	3180 × 2100
6	+ 0.2	15.00	4600 × 3180
10	+ 0.3	25.00)	6000 × 3300
12	+ 0.3	30.00)	
15	+ 0.5	37.50	3050 × 3000
19	+ 1.0	47.50)	3000 × 2900
25	+ 1.0	63.50)	
Clear sheet glass			
2 *	+ 0.2	5.00	1920 × 1220
3	+ 0.3	7.50	2130 × 1320
4	+ 0.3	10.00	2760 × 1220
5 *	+ 0.3	12.50)	2130 × 2400
6 *	+ 0.3	15.00)	
Cast glass			
3	+ 0.4 − 0.2	6.00)	2140 × 1280
4	+ 0.5	7.50)	
5	+ 0.5	9.50	2140 × 1320
6	+ 0.5	11.50)	3700 × 1280
10	+ 0.8	21.50)	
Wired glass			
(Cast wired glass)			
6	+ 0.3 − 0.7	–)	3700 × 1840
7	+ 0.7	–)	
(Polished wire glass)			
6	+ 1.0	–	330 × 1830

* The 5 mm and 6 mm thickness are known as *thick drawn sheet*. Although 2 mm sheet glass is available it is not recommended for general glazing purposes

METAL

METAL

Weights of Metals

Material	kg/m^3	lb/cu ft
Metals, steel construction, etc.		
Iron		
– cast	7207	450
– wrought	7687	480
– ore – general	2407	150
– (crushed) Swedish	3682	230
Steel	7854	490
Copper		
– cast	8731	545
– wrought	8945	558
Brass	8497	530
Bronze	8945	558
Aluminium	2774	173
Lead	11322	707
Zinc (rolled)	7140	446
	g/mm^2 per metre	**lb/sq ft per foot**
Steel bars	7.85	3.4
Structural steelwork	Net weight of member @ 7854 kg/m^3	
riveted	+ 10% for cleats, rivets, bolts, etc.	
welded	+ 1.25% to 2.5% for welds, etc.	
Rolled sections		
beams	+ 2.5%	
stanchions	+ 5% (extra for caps and bases)	
Plate		
web girders	+ 10% for rivets or welds, stiffeners, etc.	
	kg/m	**lb/ft**
Steel stairs: industrial type		
1 m or 3 ft wide	84	56
Steel tubes		
50 mm or 2 in bore	5 to 6	3 to 4
Gas piping		
20 mm or ¾ in	2	1¼

METAL

Universal Beams BS 4: Part 1: 2005

Designation	Mass (kg/m)	Depth of section (mm)	Width of section (mm)	Thickness		Surface area (m^2/m)
				Web (mm)	Flange (mm)	
1016 × 305 × 487	487.0	1036.1	308.5	30.0	54.1	3.20
1016 × 305 × 438	438.0	1025.9	305.4	26.9	49.0	3.17
1016 × 305 × 393	393.0	1016.0	303.0	24.4	43.9	3.15
1016 × 305 × 349	349.0	1008.1	302.0	21.1	40.0	3.13
1016 × 305 × 314	314.0	1000.0	300.0	19.1	35.9	3.11
1016 × 305 × 272	272.0	990.1	300.0	16.5	31.0	3.10
1016 × 305 × 249	249.0	980.2	300.0	16.5	26.0	3.08
1016 × 305 × 222	222.0	970.3	300.0	16.0	21.1	3.06
914 × 419 × 388	388.0	921.0	420.5	21.4	36.6	3.44
914 × 419 × 343	343.3	911.8	418.5	19.4	32.0	3.42
914 × 305 × 289	289.1	926.6	307.7	19.5	32.0	3.01
914 × 305 × 253	253.4	918.4	305.5	17.3	27.9	2.99
914 × 305 × 224	224.2	910.4	304.1	15.9	23.9	2.97
914 × 305 × 201	200.9	903.0	303.3	15.1	20.2	2.96
838 × 292 × 226	226.5	850.9	293.8	16.1	26.8	2.81
838 × 292 × 194	193.8	840.7	292.4	14.7	21.7	2.79
838 × 292 × 176	175.9	834.9	291.7	14.0	18.8	2.78
762 × 267 × 197	196.8	769.8	268.0	15.6	25.4	2.55
762 × 267 × 173	173.0	762.2	266.7	14.3	21.6	2.53
762 × 267 × 147	146.9	754.0	265.2	12.8	17.5	2.51
762 × 267 × 134	133.9	750.0	264.4	12.0	15.5	2.51
686 × 254 × 170	170.2	692.9	255.8	14.5	23.7	2.35
686 × 254 × 152	152.4	687.5	254.5	13.2	21.0	2.34
686 × 254 × 140	140.1	383.5	253.7	12.4	19.0	2.33
686 × 254 × 125	125.2	677.9	253.0	11.7	16.2	2.32
610 × 305 × 238	238.1	635.8	311.4	18.4	31.4	2.45
610 × 305 × 179	179.0	620.2	307.1	14.1	23.6	2.41
610 × 305 × 149	149.1	612.4	304.8	11.8	19.7	2.39
610 × 229 × 140	139.9	617.2	230.2	13.1	22.1	2.11
610 × 229 × 125	125.1	612.2	229.0	11.9	19.6	2.09
610 × 229 × 113	113.0	607.6	228.2	11.1	17.3	2.08
610 × 229 × 101	101.2	602.6	227.6	10.5	14.8	2.07
533 × 210 × 122	122.0	544.5	211.9	12.7	21.3	1.89
533 × 210 × 109	109.0	539.5	210.8	11.6	18.8	1.88
533 × 210 × 101	101.0	536.7	210.0	10.8	17.4	1.87
533 × 210 × 92	92.1	533.1	209.3	10.1	15.6	1.86
533 × 210 × 82	82.2	528.3	208.8	9.6	13.2	1.85
457 × 191 × 98	98.3	467.2	192.8	11.4	19.6	1.67
457 × 191 × 89	89.3	463.4	191.9	10.5	17.7	1.66
457 × 191 × 82	82.0	460.0	191.3	9.9	16.0	1.65
457 × 191 × 74	74.3	457.0	190.4	9.0	14.5	1.64
457 × 191 × 67	67.1	453.4	189.9	8.5	12.7	1.63
457 × 152 × 82	82.1	465.8	155.3	10.5	18.9	1.51
457 × 152 × 74	74.2	462.0	154.4	9.6	17.0	1.50
457 × 152 × 67	67.2	458.0	153.8	9.0	15.0	1.50
457 × 152 × 60	59.8	454.6	152.9	8.1	13.3	1.50
457 × 152 × 52	52.3	449.8	152.4	7.6	10.9	1.48
406 × 178 × 74	74.2	412.8	179.5	9.5	16.0	1.51
406 × 178 × 67	67.1	409.4	178.8	8.8	14.3	1.50
406 × 178 × 60	60.1	406.4	177.9	7.9	12.8	1.49

METAL

Designation	Mass (kg/m)	Depth of section (mm)	Width of section (mm)	Thickness		Surface area (m^2/m)
				Web (mm)	Flange (mm)	
406 × 178 × 50	54.1	402.6	177.7	7.7	10.9	1.48
406 × 140 × 46	46.0	403.2	142.2	6.8	11.2	1.34
406 × 140 × 39	39.0	398.0	141.8	6.4	8.6	1.33
356 × 171 × 67	67.1	363.4	173.2	9.1	15.7	1.38
356 × 171 × 57	57.0	358.0	172.2	8.1	13.0	1.37
356 × 171 × 51	51.0	355.0	171.5	7.4	11.5	1.36
356 × 171 × 45	45.0	351.4	171.1	7.0	9.7	1.36
356 × 127 × 39	39.1	353.4	126.0	6.6	10.7	1.18
356 × 127 × 33	33.1	349.0	125.4	6.0	8.5	1.17
305 × 165 × 54	54.0	310.4	166.9	7.9	13.7	1.26
305 × 165 × 46	46.1	306.6	165.7	6.7	11.8	1.25
305 × 165 × 40	40.3	303.4	165.0	6.0	10.2	1.24
305 × 127 × 48	48.1	311.0	125.3	9.0	14.0	1.09
305 × 127 × 42	41.9	307.2	124.3	8.0	12.1	1.08
305 × 127 × 37	37.0	304.4	123.3	7.1	10.7	1.07
305 × 102 × 33	32.8	312.7	102.4	6.6	10.8	1.01
305 × 102 × 28	28.2	308.7	101.8	6.0	8.8	1.00
305 × 102 × 25	24.8	305.1	101.6	5.8	7.0	0.992
254 × 146 × 43	43.0	259.6	147.3	7.2	12.7	1.08
254 × 146 × 37	37.0	256.0	146.4	6.3	10.9	1.07
254 × 146 × 31	31.1	251.4	146.1	6.0	8.6	1.06
254 × 102 × 28	28.3	260.4	102.2	6.3	10.0	0.904
254 × 102 × 25	25.2	257.2	101.9	6.0	8.4	0.897
254 × 102 × 22	22.0	254.0	101.6	5.7	6.8	0.890
203 × 133 × 30	30.0	206.8	133.9	6.4	9.6	0.923
203 × 133 × 25	25.1	203.2	133.2	5.7	7.8	0.915
203 × 102 × 23	23.1	203.2	101.8	5.4	9.3	0.790
178 × 102 × 19	19.0	177.8	101.2	4.8	7.9	0.738
152 × 89 × 16	16.0	152.4	88.7	4.5	7.7	0.638
127 × 76 × 13	13.0	127.0	76.0	4.0	7.6	0.537

METAL

Universal Columns BS 4: Part 1: 2005

Designation	Mass (kg/m)	Depth of section (mm)	Width of section (mm)	Thickness		Surface area (m^2/m)
				Web (mm)	Flange (mm)	
356 × 406 × 634	633.9	474.7	424.0	47.6	77.0	2.52
356 × 406 × 551	551.0	455.6	418.5	42.1	67.5	2.47
356 × 406 × 467	467.0	436.6	412.2	35.8	58.0	2.42
356 × 406 × 393	393.0	419.0	407.0	30.6	49.2	2.38
356 × 406 × 340	339.9	406.4	403.0	26.6	42.9	2.35
356 × 406 × 287	287.1	393.6	399.0	22.6	36.5	2.31
356 × 406 × 235	235.1	381.0	384.8	18.4	30.2	2.28
356 × 368 × 202	201.9	374.6	374.7	16.5	27.0	2.19
356 × 368 × 177	177.0	368.2	372.6	14.4	23.8	2.17
356 × 368 × 153	152.9	362.0	370.5	12.3	20.7	2.16
356 × 368 × 129	129.0	355.6	368.6	10.4	17.5	2.14
305 × 305 × 283	282.9	365.3	322.2	26.8	44.1	1.94
305 × 305 × 240	240.0	352.5	318.4	23.0	37.7	1.91
305 × 305 × 198	198.1	339.9	314.5	19.1	31.4	1.87
305 × 305 × 158	158.1	327.1	311.2	15.8	25.0	1.84
305 × 305 × 137	136.9	320.5	309.2	13.8	21.7	1.82
305 × 305 × 118	117.9	314.5	307.4	12.0	18.7	1.81
305 × 305 × 97	96.9	307.9	305.3	9.9	15.4	1.79
254 × 254 × 167	167.1	289.1	265.2	19.2	31.7	1.58
254 × 254 × 132	132.0	276.3	261.3	15.3	25.3	1.55
254 × 254 × 107	107.1	266.7	258.8	12.8	20.5	1.52
254 × 254 × 89	88.9	260.3	256.3	10.3	17.3	1.50
254 × 254 × 73	73.1	254.1	254.6	8.6	14.2	1.49
203 × 203 × 86	86.1	222.2	209.1	12.7	20.5	1.24
203 × 203 × 71	71.0	215.8	206.4	10.0	17.3	1.22
203 × 203 × 60	60.0	209.6	205.8	9.4	14.2	1.21
203 × 203 × 52	52.0	206.2	204.3	7.9	12.5	1.20
203 × 203 × 46	46.1	203.2	203.6	7.2	11.0	1.19
152 × 152 × 37	37.0	161.8	154.4	8.0	11.5	0.912
152 × 152 × 30	30.0	157.6	152.9	6.5	9.4	0.901
152 × 152 × 23	23.0	152.4	152.2	5.8	6.8	0.889

METAL

Joists BS 4: Part 1: 2005 (retained for reference, Corus have ceased manufacture in UK)

Designation	Mass (kg/m)	Depth of section (mm)	Width of section (mm)	Thickness		Surface area (m^2/m)
				Web (mm)	Flange (mm)	
254 × 203 × 82	82.0	254.0	203.2	10.2	19.9	1.210
203 × 152 × 52	52.3	203.2	152.4	8.9	16.5	0.932
152 × 127 × 37	37.3	152.4	127.0	10.4	13.2	0.737
127 × 114 × 29	29.3	127.0	114.3	10.2	11.5	0.646
127 × 114 × 27	26.9	127.0	114.3	7.4	11.4	0.650
102 × 102 × 23	23.0	101.6	101.6	9.5	10.3	0.549
102 × 44 × 7	7.5	101.6	44.5	4.3	6.1	0.350
89 × 89 × 19	19.5	88.9	88.9	9.5	9.9	0.476
76 × 76 × 13	12.8	76.2	76.2	5.1	8.4	0.411

Parallel Flange Channels

Designation	Mass (kg/m)	Depth of section (mm)	Width of section (mm)	Thickness		Surface area (m^2/m)
				Web (mm)	Flange (mm)	
430 × 100 × 64	64.4	430	100	11.0	19.0	1.23
380 × 100 × 54	54.0	380	100	9.5	17.5	1.13
300 × 100 × 46	45.5	300	100	9.0	16.5	0.969
300 × 90 × 41	41.4	300	90	9.0	15.5	0.932
260 × 90 × 35	34.8	260	90	8.0	14.0	0.854
260 × 75 × 28	27.6	260	75	7.0	12.0	0.79
230 × 90 × 32	32.2	230	90	7.5	14.0	0.795
230 × 75 × 26	25.7	230	75	6.5	12.5	0.737
200 × 90 × 30	29.7	200	90	7.0	14.0	0.736
200 × 75 × 23	23.4	200	75	6.0	12.5	0.678
180 × 90 × 26	26.1	180	90	6.5	12.5	0.697
180 × 75 × 20	20.3	180	75	6.0	10.5	0.638
150 × 90 × 24	23.9	150	90	6.5	12.0	0.637
150 × 75 × 18	17.9	150	75	5.5	10.0	0.579
125 × 65 × 15	14.8	125	65	5.5	9.5	0.489
100 × 50 × 10	10.2	100	50	5.0	8.5	0.382

METAL

Equal Angles BS EN 10056-1

Designation	Mass (kg/m)	Surface area (m²/m)
200 × 200 × 24	71.1	0.790
200 × 200 × 20	59.9	0.790
200 × 200 × 18	54.2	0.790
200 × 200 × 16	48.5	0.790
150 × 150 × 18	40.1	0.59
150 × 150 × 15	33.8	0.59
150 × 150 × 12	27.3	0.59
150 × 150 × 10	23.0	0.59
120 × 120 × 15	26.6	0.47
120 × 120 × 12	21.6	0.47
120 × 120 × 10	18.2	0.47
120 × 120 × 8	14.7	0.47
100 × 100 × 15	21.9	0.39
100 × 100 × 12	17.8	0.39
100 × 100 × 10	15.0	0.39
100 × 100 × 8	12.2	0.39
90 × 90 × 12	15.9	0.35
90 × 90 × 10	13.4	0.35
90 × 90 × 8	10.9	0.35
90 × 90 × 7	9.61	0.35
90 × 90 × 6	8.30	0.35

Unequal Angles BS EN 10056-1

Designation	Mass (kg/m)	Surface area (m²/m)
200 × 150 × 18	47.1	0.69
200 × 150 × 15	39.6	0.69
200 × 150 × 12	32.0	0.69
200 × 100 × 15	33.7	0.59
200 × 100 × 12	27.3	0.59
200 × 100 × 10	23.0	0.59
150 × 90 × 15	26.6	0.47
150 × 90 × 12	21.6	0.47
150 × 90 × 10	18.2	0.47
150 × 75 × 15	24.8	0.44
150 × 75 × 12	20.2	0.44
150 × 75 × 10	17.0	0.44
125 × 75 × 12	17.8	0.40
125 × 75 × 10	15.0	0.40
125 × 75 × 8	12.2	0.40
100 × 75 × 12	15.4	0.34
100 × 75 × 10	13.0	0.34
100 × 75 × 8	10.6	0.34
100 × 65 × 10	12.3	0.32
100 × 65 × 8	9.94	0.32
100 × 65 × 7	8.77	0.32

METAL

Structural Tees Split from Universal Beams BS 4: Part 1: 2005

Designation	Mass (kg/m)	Surface area (m²/m)
305 × 305 × 90	89.5	1.22
305 × 305 × 75	74.6	1.22
254 × 343 × 63	62.6	1.19
229 × 305 × 70	69.9	1.07
229 × 305 × 63	62.5	1.07
229 × 305 × 57	56.5	1.07
229 × 305 × 51	50.6	1.07
210 × 267 × 61	61.0	0.95
210 × 267 × 55	54.5	0.95
210 × 267 × 51	50.5	0.95
210 × 267 × 46	46.1	0.95
210 × 267 × 41	41.1	0.95
191 × 229 × 49	49.2	0.84
191 × 229 × 45	44.6	0.84
191 × 229 × 41	41.0	0.84
191 × 229 × 37	37.1	0.84
191 × 229 × 34	33.6	0.84
152 × 229 × 41	41.0	0.76
152 × 229 × 37	37.1	0.76
152 × 229 × 34	33.6	0.76
152 × 229 × 30	29.9	0.76
152 × 229 × 26	26.2	0.76

Universal Bearing Piles BS 4: Part 1: 2005

Designation	Mass (kg/m)	Depth of Section (mm)	Width of Section (mm)	Thickness	
				Web (mm)	Flange (mm)
356 × 368 × 174	173.9	361.4	378.5	20.3	20.4
356 × 368 × 152	152.0	356.4	376.0	17.8	17.9
356 × 368 × 133	133.0	352.0	373.8	15.6	15.7
356 × 368 × 109	108.9	346.4	371.0	12.8	12.9
305 × 305 × 223	222.9	337.9	325.7	30.3	30.4
305 × 305 × 186	186.0	328.3	320.9	25.5	25.6
305 × 305 × 149	149.1	318.5	316.0	20.6	20.7
305 × 305 × 126	126.1	312.3	312.9	17.5	17.6
305 × 305 × 110	110.0	307.9	310.7	15.3	15.4
305 × 305 × 95	94.9	303.7	308.7	13.3	13.3
305 × 305 × 88	88.0	301.7	307.8	12.4	12.3
305 × 305 × 79	78.9	299.3	306.4	11.0	11.1
254 × 254 × 85	85.1	254.3	260.4	14.4	14.3
254 × 254 × 71	71.0	249.7	258.0	12.0	12.0
254 × 254 × 63	63.0	247.1	256.6	10.6	10.7
203 × 203 × 54	53.9	204.0	207.7	11.3	11.4
203 × 203 × 45	44.9	200.2	205.9	9.5	9.5

METAL

Hot Formed Square Hollow Sections EN 10210 S275J2H & S355J2H

Size (mm)	Wall thickness (mm)	Mass (kg/m)	Superficial area (m^2/m)
40 × 40	2.5	2.89	0.154
	3.0	3.41	0.152
	3.2	3.61	0.152
	3.6	4.01	0.151
	4.0	4.39	0.150
	5.0	5.28	0.147
50 × 50	2.5	3.68	0.194
	3.0	4.35	0.192
	3.2	4.62	0.192
	3.6	5.14	0.191
	4.0	5.64	0.190
	5.0	6.85	0.187
	6.0	7.99	0.185
	6.3	8.31	0.184
60 × 60	3.0	5.29	0.232
	3.2	5.62	0.232
	3.6	6.27	0.231
	4.0	6.90	0.230
	5.0	8.42	0.227
	6.0	9.87	0.225
	6.3	10.30	0.224
	8.0	12.50	0.219
70 × 70	3.0	6.24	0.272
	3.2	6.63	0.272
	3.6	7.40	0.271
	4.0	8.15	0.270
	5.0	9.99	0.267
	6.0	11.80	0.265
	6.3	12.30	0.264
	8.0	15.00	0.259
80 × 80	3.2	7.63	0.312
	3.6	8.53	0.311
	4.0	9.41	0.310
	5.0	11.60	0.307
	6.0	13.60	0.305
	6.3	14.20	0.304
	8.0	17.50	0.299
90 × 90	3.6	9.66	0.351
	4.0	10.70	0.350
	5.0	13.10	0.347
	6.0	15.50	0.345
	6.3	16.20	0.344
	8.0	20.10	0.339
100 × 100	3.6	10.80	0.391
	4.0	11.90	0.390
	5.0	14.70	0.387
	6.0	17.40	0.385
	6.3	18.20	0.384
	8.0	22.60	0.379
	10.0	27.40	0.374

METAL

Size (mm)	Wall thickness (mm)	Mass (kg/m)	Superficial area (m²/m)
120 × 120	4.0	14.40	0.470
	5.0	17.80	0.467
	6.0	21.20	0.465
	6.3	22.20	0.464
	8.0	27.60	0.459
	10.0	33.70	0.454
	12.0	39.50	0.449
	12.5	40.90	0.448
140 × 140	5.0	21.00	0.547
	6.0	24.90	0.545
	6.3	26.10	0.544
	8.0	32.60	0.539
	10.0	40.00	0.534
	12.0	47.00	0.529
	12.5	48.70	0.528
150 × 150	5.0	22.60	0.587
	6.0	26.80	0.585
	6.3	28.10	0.584
	8.0	35.10	0.579
	10.0	43.10	0.574
	12.0	50.80	0.569
	12.5	52.70	0.568
Hot formed from seamless hollow	16.0	65.2	0.559
160 × 160	5.0	24.10	0.627
	6.0	28.70	0.625
	6.3	30.10	0.624
	8.0	37.60	0.619
	10.0	46.30	0.614
	12.0	54.60	0.609
	12.5	56.60	0.608
	16.0	70.20	0.599
180 × 180	5.0	27.30	0.707
	6.0	32.50	0.705
	6.3	34.00	0.704
	8.0	42.70	0.699
	10.0	52.50	0.694
	12.0	62.10	0.689
	12.5	64.40	0.688
	16.0	80.20	0.679
200 × 200	5.0	30.40	0.787
	6.0	36.20	0.785
	6.3	38.00	0.784
	8.0	47.70	0.779
	10.0	58.80	0.774
	12.0	69.60	0.769
	12.5	72.30	0.768
	16.0	90.30	0.759
250 × 250	5.0	38.30	0.987
	6.0	45.70	0.985
	6.3	47.90	0.984
	8.0	60.30	0.979
	10.0	74.50	0.974
	12.0	88.50	0.969
	12.5	91.90	0.968
	16.0	115.00	0.959

METAL

Size (mm)	Wall thickness (mm)	Mass (kg/m)	Superficial area (m^2/m)
300 × 300	6.0	55.10	1.18
	6.3	57.80	1.18
	8.0	72.80	1.18
	10.0	90.20	1.17
	12.0	107.00	1.17
	12.5	112.00	1.17
	16.0	141.00	1.16
350 × 350	8.0	85.40	1.38
	10.0	106.00	1.37
	12.0	126.00	1.37
	12.5	131.00	1.37
	16.0	166.00	1.36
400 × 400	8.0	97.90	1.58
	10.0	122.00	1.57
	12.0	145.00	1.57
	12.5	151.00	1.57
	16.0	191.00	1.56
(Grade S355J2H only)	20.00*	235.00	1.55

Note: * SAW process

METAL

Hot Formed Square Hollow Sections JUMBO RHS: JIS G3136

Size (mm)	Wall thickness (mm)	Mass (kg/m)	Superficial area (m^2/m)
350 × 350	19.0	190.00	1.33
	22.0	217.00	1.32
	25.0	242.00	1.31
400 × 400	22.0	251.00	1.52
	25.0	282.00	1.51
450 × 450	12.0	162.00	1.76
	16.0	213.00	1.75
	19.0	250.00	1.73
	22.0	286.00	1.72
	25.0	321.00	1.71
	28.0 *	355.00	1.70
	32.0 *	399.00	1.69
500 × 500	12.0	181.00	1.96
	16.0	238.00	1.95
	19.0	280.00	1.93
	22.0	320.00	1.92
	25.0	360.00	1.91
	28.0 *	399.00	1.90
	32.0 *	450.00	1.89
	36.0 *	498.00	1.88
550 × 550	16.0	263.00	2.15
	19.0	309.00	2.13
	22.0	355.00	2.12
	25.0	399.00	2.11
	28.0 *	443.00	2.10
	32.0 *	500.00	2.09
	36.0 *	555.00	2.08
	40.0 *	608.00	2.06
600 × 600	25.0 *	439.00	2.31
	28.0 *	487.00	2.30
	32.0 *	550.00	2.29
	36.0 *	611.00	2.28
	40.0 *	671.00	2.26
700 × 700	25.0 *	517.00	2.71
	28.0 *	575.00	2.70
	32.0 *	651.00	2.69
	36.0 *	724.00	2.68
	40.0 *	797.00	2.68

Note: * SAW process

METAL

Hot Formed Rectangular Hollow Sections: EN10210 S275J2h & S355J2H

Size (mm)	Wall thickness (mm)	Mass (kg/m)	Superficial area (m^2/m)
50 × 30	2.5	2.89	0.154
	3.0	3.41	0.152
	3.2	3.61	0.152
	3.6	4.01	0.151
	4.0	4.39	0.150
	5.0	5.28	0.147
60 × 40	2.5	3.68	0.194
	3.0	4.35	0.192
	3.2	4.62	0.192
	3.6	5.14	0.191
	4.0	5.64	0.190
	5.0	6.85	0.187
	6.0	7.99	0.185
	6.3	8.31	0.184
80 × 40	3.0	5.29	0.232
	3.2	5.62	0.232
	3.6	6.27	0.231
	4.0	6.90	0.230
	5.0	8.42	0.227
	6.0	9.87	0.225
	6.3	10.30	0.224
	8.0	12.50	0.219
76.2 × 50.8	3.0	5.62	0.246
	3.2	5.97	0.246
	3.6	6.66	0.245
	4.0	7.34	0.244
	5.0	8.97	0.241
	6.0	10.50	0.239
	6.3	11.00	0.238
	8.0	13.40	0.233
90 × 50	3.0	6.24	0.272
	3.2	6.63	0.272
	3.6	7.40	0.271
	4.0	8.15	0.270
	5.0	9.99	0.267
	6.0	11.80	0.265
	6.3	12.30	0.264
	8.0	15.00	0.259
100 × 50	3.0	6.71	0.292
	3.2	7.13	0.292
	3.6	7.96	0.291
	4.0	8.78	0.290
	5.0	10.80	0.287
	6.0	12.70	0.285
	6.3	13.30	0.284
	8.0	16.30	0.279

METAL

Size (mm)	Wall thickness (mm)	Mass (kg/m)	Superficial area (m^2/m)
100 × 60	3.0	7.18	0.312
	3.2	7.63	0.312
	3.6	8.53	0.311
	4.0	9.41	0.310
	5.0	11.60	0.307
	6.0	13.60	0.305
	6.3	14.20	0.304
	8.0	17.50	0.299
120 × 60	3.6	9.70	0.351
	4.0	10.70	0.350
	5.0	13.10	0.347
	6.0	15.50	0.345
	6.3	16.20	0.344
	8.0	20.10	0.339
120 × 80	3.6	10.80	0.391
	4.0	11.90	0.390
	5.0	14.70	0.387
	6.0	17.40	0.385
	6.3	18.20	0.384
	8.0	22.60	0.379
	10.0	27.40	0.374
150 × 100	4.0	15.10	0.490
	5.0	18.60	0.487
	6.0	22.10	0.485
	6.3	23.10	0.484
	8.0	28.90	0.479
	10.0	35.30	0.474
	12.0	41.40	0.469
	12.5	42.80	0.468
160 × 80	4.0	14.40	0.470
	5.0	17.80	0.467
	6.0	21.20	0.465
	6.3	22.20	0.464
	8.0	27.60	0.459
	10.0	33.70	0.454
	12.0	39.50	0.449
	12.5	40.90	0.448
200 × 100	5.0	22.60	0.587
	6.0	26.80	0.585
	6.3	28.10	0.584
	8.0	35.10	0.579
	10.0	43.10	0.574
	12.0	50.80	0.569
	12.5	52.70	0.568
	16.0	65.20	0.559
250 × 150	5.0	30.40	0.787
	6.0	36.20	0.785
	6.3	38.00	0.784
	8.0	47.70	0.779
	10.0	58.80	0.774
	12.0	69.60	0.769
	12.5	72.30	0.768
	16.0	90.30	0.759

METAL

Size (mm)	Wall thickness (mm)	Mass (kg/m)	Superficial area (m^2/m)
300 × 200	5.0	38.30	0.987
	6.0	45.70	0.985
	6.3	47.90	0.984
	8.0	60.30	0.979
	10.0	74.50	0.974
	12.0	88.50	0.969
	12.5	91.90	0.968
	16.0	115.00	0.959
400 × 200	6.0	55.10	1.18
	6.3	57.80	1.18
	8.0	72.80	1.18
	10.0	90.20	1.17
	12.0	107.00	1.17
	12.5	112.00	1.17
	16.0	141.00	1.16
450 × 250	8.0	85.40	1.38
	10.0	106.00	1.37
	12.0	126.00	1.37
	12.5	131.00	1.37
	16.0	166.00	1.36
500 × 300	8.0	98.00	1.58
	10.0	122.00	1.57
	12.0	145.00	1.57
	12.5	151.00	1.57
	16.0	191.00	1.56
	20.0	235.00	1.55

METAL

Hot Formed Circular Hollow Sections EN 10210 S275J2H & S355J2H

Outside diameter (mm)	Wall thickness (mm)	Mass (kg/m)	Superficial area (m^2/m)
21.3	3.2	1.43	0.067
26.9	3.2	1.87	0.085
33.7	3.0	2.27	0.106
	3.2	2.41	0.106
	3.6	2.67	0.106
	4.0	2.93	0.106
42.4	3.0	2.91	0.133
	3.2	3.09	0.133
	3.6	3.44	0.133
	4.0	3.79	0.133
48.3	2.5	2.82	0.152
	3.0	3.35	0.152
	3.2	3.56	0.152
	3.6	3.97	0.152
	4.0	4.37	0.152
	5.0	5.34	0.152
60.3	2.5	3.56	0.189
	3.0	4.24	0.189
	3.2	4.51	0.189
	3.6	5.03	0.189
	4.0	5.55	0.189
	5.0	6.82	0.189
76.1	2.5	4.54	0.239
	3.0	5.41	0.239
	3.2	5.75	0.239
	3.6	6.44	0.239
	4.0	7.11	0.239
	5.0	8.77	0.239
	6.0	10.40	0.239
	6.3	10.80	0.239
88.9	2.5	5.33	0.279
	3.0	6.36	0.279
	3.2	6.76	0.27
	3.6	7.57	0.279
	4.0	8.38	0.279
	5.0	10.30	0.279
	6.0	12.30	0.279
	6.3	12.80	0.279
114.3	3.0	8.23	0.359
	3.2	8.77	0.359
	3.6	9.83	0.359
	4.0	10.09	0.359
	5.0	13.50	0.359
	6.0	16.00	0.359
	6.3	16.80	0.359

METAL

Outside diameter (mm)	Wall thickness (mm)	Mass (kg/m)	Superficial area (m^2/m)
139.7	3.2	10.80	0.439
	3.6	12.10	0.439
	4.0	13.40	0.439
	5.0	16.60	0.439
	6.0	19.80	0.439
	6.3	20.70	0.439
	8.0	26.00	0.439
	10.0	32.00	0.439
168.3	3.2	13.00	0.529
	3.6	14.60	0.529
	4.0	16.20	0.529
	5.0	20.10	0.529
	6.0	24.00	0.529
	6.3	25.20	0.529
	8.0	31.60	0.529
	10.0	39.00	0.529
	12.0	46.30	0.529
	12.5	48.00	0.529
193.7	5.0	23.30	0.609
	6.0	27.80	0.609
	6.3	29.10	0.609
	8.0	36.60	0.609
	10.0	45.30	0.609
	12.0	53.80	0.609
	12.5	55.90	0.609
219.1	5.0	26.40	0.688
	6.0	31.50	0.688
	6.3	33.10	0.688
	8.0	41.60	0.688
	10.0	51.60	0.688
	12.0	61.30	0.688
	12.5	63.70	0.688
	16.0	80.10	0.688
244.5	5.0	29.50	0.768
	6.0	35.30	0.768
	6.3	37.00	0.768
	8.0	46.70	0.768
	10.0	57.80	0.768
	12.0	68.80	0.768
	12.5	71.50	0.768
	16.0	90.20	0.768
273.0	5.0	33.00	0.858
	6.0	39.50	0.858
	6.3	41.40	0.858
	8.0	52.30	0.858
	10.0	64.90	0.858
	12.0	77.20	0.858
	12.5	80.30	0.858
	16.0	101.00	0.858

METAL

Outside diameter (mm)	Wall thickness (mm)	Mass (kg/m)	Superficial area (m²/m)
323.9	5.0	39.30	1.02
	6.0	47.00	1.02
	6.3	49.30	1.02
	8.0	62.30	1.02
	10.0	77.40	1.02
	12.0	92.30	1.02
	12.5	96.00	1.02
	16.0	121.00	1.02
355.6	6.3	54.30	1.12
	8.0	68.60	1.12
	10.0	85.30	1.12
	12.0	102.00	1.12
	12.5	106.00	1.12
	16.0	134.00	1.12
406.4	6.3	62.20	1.28
	8.0	79.60	1.28
	10.0	97.80	1.28
	12.0	117.00	1.28
	12.5	121.00	1.28
	16.0	154.00	1.28
457.0	6.3	70.00	1.44
	8.0	88.60	1.44
	10.0	110.00	1.44
	12.0	132.00	1.44
	12.5	137.00	1.44
	16.0	174.00	1.44
508.0	6.3	77.90	1.60
	8.0	98.60	1.60
	10.0	123.00	1.60
	12.0	147.00	1.60
	12.5	153.00	1.60
	16.0	194.00	1.60

METAL

Spacing of Holes in Angles

Nominal leg length (mm)	Spacing of holes						Maximum diameter of bolt or rivet		
	A	B	C	D	E	F	A	B and C	D, E and F
200		75	75	55	55	55		30	20
150		55	55					20	
125		45	60					20	
120									
100	55						24		
90	50						24		
80	45						20		
75	45						20		
70	40						20		
65	35						20		
60	35						16		
50	28						12		
45	25								
40	23								
30	20								
25	15								

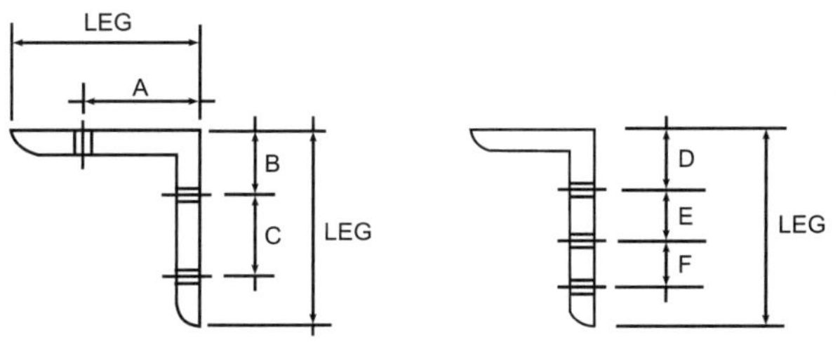

KERBS, PAVING, ETC.

KERBS/EDGINGS/CHANNELS

Precast Concrete Kerbs to BS 7263

Straight kerb units: length from 450 to 915 mm

150 mm high × 125 mm thick		
bullnosed	type BN	
half battered	type HB3	
255 mm high × 125 mm thick		
45° splayed	type SP	
half battered	type HB2	
305 mm high × 150 mm thick		
half battered	type HB1	
Quadrant kerb units		
150 mm high × 305 and 455 mm radius to match	type BN	type QBN
150 mm high × 305 and 455 mm radius to match	type HB2, HB3	type QHB
150 mm high × 305 and 455 mm radius to match	type SP	type QSP
255 mm high × 305 and 455 mm radius to match	type BN	type QBN
255 mm high × 305 and 455 mm radius to match	type HB2, HB3	type QHB
225 mm high × 305 and 455 mm radius to match	type SP	type QSP
Angle kerb units		
305 × 305 × 225 mm high × 125 mm thick		
bullnosed external angle	type XA	
splayed external angle to match type SP	type XA	
bullnosed internal angle	type IA	
splayed internal angle to match type SP	type IA	
Channels		
255 mm wide × 125 mm high flat	type CS1	
150 mm wide × 125 mm high flat type	CS2	
255 mm wide × 125 mm high dished	type CD	

KERBS, PAVING, ETC.

Transition kerb units			
from kerb type SP to HB	left handed	type TL	
	right handed	type TR	
from kerb type BN to HB	left handed	type DL1	
	right handed	type DR1	
from kerb type BN to SP	left handed	type DL2	
	right handed	type DR2	

Number of kerbs required per quarter circle (780 mm kerb lengths)

Radius (m)	Number in quarter circle
12	24
10	20
8	16
6	12
5	10
4	8
3	6
2	4
1	2

Precast Concrete Edgings

Round top type ER	Flat top type EF	Bullnosed top type EBN
150 × 50 mm	150 × 50 mm	150 × 50 mm
200 × 50 mm	200 × 50 mm	200 × 50 mm
250 × 50 mm	250 × 50 mm	250 × 50 mm

KERBS, PAVING, ETC.

BASES

Cement Bound Material for Bases and Subbases

CBM1:	very carefully graded aggregate from 37.5–75 mm, with a 7-day strength of 4.5 N/mm^2
CBM2:	same range of aggregate as CBM1 but with more tolerance in each size of aggregate with a 7-day strength of 7.0 N/mm^2
CBM3:	crushed natural aggregate or blast furnace slag, graded from 37.5 mm–150 mm for 40 mm aggregate, and from 20–75 mm for 20 mm aggregate, with a 7-day strength of 10 N/mm^2
CBM4:	crushed natural aggregate or blast furnace slag, graded from 37.5 mm–150 mm for 40 mm aggregate, and from 20–75 mm for 20 mm aggregate, with a 7-day strength of 15 N/mm^2

INTERLOCKING BRICK/BLOCK ROADS/PAVINGS

Sizes of Precast Concrete Paving Blocks

Type R blocks
200 × 100 × 60 mm
200 × 100 × 65 mm
200 × 100 × 80 mm
200 × 100 × 100 mm

Type S
Any shape within a 295 mm space

Sizes of clay brick pavers
200 × 100 × 50 mm
200 × 100 × 65 mm
210 × 105 × 50 mm
210 × 105 × 65 mm
215 × 102.5 × 50 mm
215 × 102.5 × 65 mm

Type PA: 3 kN
Footpaths and pedestrian areas, private driveways, car parks, light vehicle traffic and over-run

Type PB: 7 kN
Residential roads, lorry parks, factory yards, docks, petrol station forecourts, hardstandings, bus stations

KERBS, PAVING, ETC.

PAVING AND SURFACING

Weights and Sizes of Paving and Surfacing

Description of item	Size	Quantity per tonne
Paving 50 mm thick	900 × 600 mm	15
Paving 50 mm thick	750 × 600 mm	18
Paving 50 mm thick	600 × 600 mm	23
Paving 50 mm thick	450 × 600 mm	30
Paving 38 mm thick	600 × 600 mm	30
Path edging	914 × 50 × 150 mm	60
Kerb (including radius and tapers)	125 × 254 × 914 mm	15
Kerb (including radius and tapers)	125 × 150 × 914 mm	25
Square channel	125 × 254 × 914 mm	15
Dished channel	125 × 254 × 914 mm	15
Quadrants	300 × 300 × 254 mm	19
Quadrants	450 × 450 × 254 mm	12
Quadrants	300 × 300 × 150 mm	30
Internal angles	300 × 300 × 254 mm	30
Fluted pavement channel	255 × 75 × 914 mm	25
Corner stones	300 × 300 mm	80
Corner stones	360 × 360 mm	60
Cable covers	914 × 175 mm	55
Gulley kerbs	220 × 220 × 150 mm	60
Gulley kerbs	220 × 200 × 75 mm	120

KERBS, PAVING, ETC.

Weights and Sizes of Paving and Surfacing

Material	kg/m^3	lb/cu yd
Tarmacadam	2306	3891
Macadam (waterbound)	2563	4325
Vermiculite (aggregate)	64–80	108–135
Terracotta	2114	3568
Cork – compressed	388	24
	kg/m^2	**lb/sq ft**
Clay floor tiles, 12.7 mm	27.3	5.6
Pavement lights	122	25
Damp-proof course	5	1
	kg/m^2 per mm thickness	**lb/sq ft per inch thickness**
Paving slabs (stone)	2.3	12
Granite setts	2.88	15
Asphalt	2.30	12
Rubber flooring	1.68	9
Polyvinyl chloride	1.94 (avg)	10 (avg)

Coverage (m^2) Per Cubic Metre of Materials Used as Subbases or Capping Layers

Consolidated thickness laid in (mm)	Square metre coverage		
	Gravel	Sand	Hardcore
50	15.80	16.50	–
75	10.50	11.00	–
100	7.92	8.20	7.42
125	6.34	6.60	5.90
150	5.28	5.50	4.95
175	–	–	4.23
200	–	–	3.71
225	–	–	3.30
300	–	–	2.47

KERBS, PAVING, ETC.

Approximate Rate of Spreads

Average thickness of course	Description	Approximate rate of spread			
		Open Textured		Dense, Medium & Fine Textured	
(mm)		(kg/m²)	(m²/t)	(kg/m²)	(m²/t)
35	14 mm open textured or dense wearing course	60–75	13–17	70–85	12–14
40	20 mm open textured or dense base course	70–85	12–14	80–100	10–12
45	20 mm open textured or dense base course	80–100	10–12	95–100	9–10
50	20 mm open textured or dense, or 28 mm dense base course	85–110	9–12	110–120	8–9
60	28 mm dense base course, 40 mm open textured or dense base course or 40 mm single course as base course	–	8–10	130–150	7–8
65	28 mm dense base course, 40 mm open textured or dense base course or 40 mm single course	100–135	7–10	140–160	6–7
75	40 mm single course, 40 mm open textured or dense base course, 40 mm dense roadbase	120–150	7–8	165–185	5–6
100	40 mm dense base course or roadbase	–	–	220–240	4–4.5

KERBS, PAVING, ETC.

Surface Dressing Roads: Coverage (m^2) per Tonne of Material

Size in mm	Sand	Granite chips	Gravel	Limestone chips
Sand	168	–	–	–
3	–	148	152	165
6	–	130	133	144
9	–	111	114	123
13	–	85	87	95
19	–	68	71	78

Sizes of Flags

Reference	Nominal size (mm)	Thickness (mm)
A	600 × 450	50 and 63
B	600 × 600	50 and 63
C	600 × 750	50 and 63
D	600 × 900	50 and 63
E	450 × 450	50 and 70 chamfered top surface
F	400 × 400	50 and 65 chamfered top surface
G	300 × 300	50 and 60 chamfered top surface

Sizes of Natural Stone Setts

Width (mm)		Length (mm)		Depth (mm)
100	×	100	×	100
75	×	150 to 250	×	125
75	×	150 to 250	×	150
100	×	150 to 250	×	100
100	×	150 to 250	×	150

SEEDING/TURFING AND PLANTING

SEEDING/TURFING AND PLANTING

Topsoil Quality

Topsoil grade	Properties
Premium	Natural topsoil, high fertility, loamy texture, good soil structure, suitable for intensive cultivation.
General purpose	Natural or manufactured topsoil of lesser quality than Premium, suitable for agriculture or amenity landscape, may need fertilizer or soil structure improvement.
Economy	Selected subsoil, natural mineral deposit such as river silt or greensand. The grade comprises two subgrades; 'Low clay' and 'High clay' which is more liable to compaction in handling. This grade is suitable for low-production agricultural land and amenity woodland or conservation planting areas.

Forms of Trees

Standards:	Shall be clear with substantially straight stems. Grafted and budded trees shall have no more than a slight bend at the union. Standards shall be designated as Half, Extra light, Light, Standard, Selected standard, Heavy, and Extra heavy.
Sizes of Standards	
Heavy standard	12–14 cm girth × 3.50 to 5.00 m high
Extra heavy standard	14–16 cm girth × 4.25 to 5.00 m high
Extra heavy standard	16–18 cm girth × 4.25 to 6.00 m high
Extra heavy standard	18–20 cm girth × 5.00 to 6.00 m high
Semi-mature trees:	Between 6.0 m and 12.0 m tall with a girth of 20 to 75 cm at 1.0 m above ground.
Feathered trees:	Shall have a defined upright central leader, with stem furnished with evenly spread and balanced lateral shoots down to or near the ground.
Whips:	Shall be without significant feather growth as determined by visual inspection.
Multi-stemmed trees:	Shall have two or more main stems at, near, above or below ground.

Seedlings grown from seed and not transplanted shall be specified when ordered for sale as:

1+0	one year old seedling
2+0	two year old seedling
1+1	one year seed bed, one year transplanted = two year old seedling
1+2	one year seed bed, two years transplanted = three year old seedling
2+1	two year seed bed, one year transplanted = three year old seedling
1u1	two years seed bed, undercut after one year = two year old seedling
2u2	four years seed bed, undercut after two years = four year old seedling

SEEDING/TURFING AND PLANTING

Cuttings

The age of cuttings (plants grown from shoots, stems, or roots of the mother plant) shall be specified when ordered for sale. The height of transplants and undercut seedlings/cuttings (which have been transplanted or undercut at least once) shall be stated in centimetres. The number of growing seasons before and after transplanting or undercutting shall be stated.

0 + 1	one year cutting
0 + 2	two year cutting
0 + 1 + 1	one year cutting bed, one year transplanted = two year old seedling
0 + 1 + 2	one year cutting bed, two years transplanted = three year old seedling

Grass Cutting Capacities in m^2 per hour

Speed mph	Width of cut in metres												
	0.5	**0.7**	**1.0**	**1.2**	**1.5**	**1.7**	**2.0**	**2.0**	**2.1**	**2.5**	**2.8**	**3.0**	**3.4**
1.0	724	1127	1529	1931	2334	2736	3138	3219	3380	4023	4506	4828	5472
1.5	1086	1690	2293	2897	3500	4104	4707	4828	5069	6035	6759	7242	8208
2.0	1448	2253	3058	3862	4667	5472	6276	6437	6759	8047	9012	9656	10944
2.5	1811	2816	3822	4828	5834	6840	7846	8047	8449	10058	11265	12070	13679
3.0	2173	3380	4587	5794	7001	8208	9415	9656	10139	12070	13518	14484	16415
3.5	2535	3943	5351	6759	8167	9576	10984	11265	11829	14082	15772	16898	19151
4.0	2897	4506	6115	7725	9334	10944	12553	12875	13518	16093	18025	19312	21887
4.5	3259	5069	6880	8690	10501	12311	14122	14484	15208	18105	20278	21726	24623
5.0	3621	5633	7644	9656	11668	13679	15691	16093	16898	20117	22531	24140	27359
5.5	3983	6196	8409	10622	12834	15047	17260	17703	18588	22128	24784	26554	30095
6.0	4345	6759	9173	11587	14001	16415	18829	19312	20278	24140	27037	28968	32831
6.5	4707	7322	9938	12553	15168	17783	20398	20921	21967	26152	29290	31382	35566
7.0	5069	7886	10702	13518	16335	19151	21967	22531	23657	28163	31543	33796	38302

Number of Plants per m^2: For Plants Planted on an Evenly Spaced Grid

Planting distances

mm	0.10	0.15	0.20	0.25	0.35	0.40	0.45	0.50	0.60	0.75	0.90	1.00	1.20	1.50
0.10	100.00	66.67	50.00	40.00	28.57	25.00	22.22	20.00	16.67	13.33	11.11	10.00	8.33	6.67
0.15	66.67	44.44	33.33	26.67	19.05	16.67	14.81	13.33	11.11	8.89	7.41	6.67	5.56	4.44
0.20	50.00	33.33	25.00	20.00	14.29	12.50	11.11	10.00	8.33	6.67	5.56	5.00	4.17	3.33
0.25	40.00	26.67	20.00	16.00	11.43	10.00	8.89	8.00	6.67	5.33	4.44	4.00	3.33	2.67
0.35	28.57	19.05	14.29	11.43	8.16	7.14	6.35	5.71	4.76	3.81	3.17	2.86	2.38	1.90
0.40	25.00	16.67	12.50	10.00	7.14	6.25	5.56	5.00	4.17	3.33	2.78	2.50	2.08	1.67
0.45	22.22	14.81	11.11	8.89	6.35	5.56	4.94	4.44	3.70	2.96	2.47	2.22	1.85	1.48
0.50	20.00	13.33	10.00	8.00	5.71	5.00	4.44	4.00	3.33	2.67	2.22	2.00	1.67	1.33
0.60	16.67	11.11	8.33	6.67	4.76	4.17	3.70	3.33	2.78	2.22	1.85	1.67	1.39	1.11
0.75	13.33	8.89	6.67	5.33	3.81	3.33	2.96	2.67	2.22	1.78	1.48	1.33	1.11	0.89
0.90	11.11	7.41	5.56	4.44	3.17	2.78	2.47	2.22	1.85	1.48	1.23	1.11	0.93	0.74
1.00	10.00	6.67	5.00	4.00	2.86	2.50	2.22	2.00	1.67	1.33	1.11	1.00	0.83	0.67
1.20	8.33	5.56	4.17	3.33	2.38	2.08	1.85	1.67	1.39	1.11	0.93	0.83	0.69	0.56
1.50	6.67	4.44	3.33	2.67	1.90	1.67	1.48	1.33	1.11	0.89	0.74	0.67	0.56	0.44

SEEDING/TURFING AND PLANTING

Grass Clippings Wet: Based on 3.5 m^3/tonne

Annual kg/100 m^2	Average 20 cuts kg/100 m^2	m^2/tonne	m^2/m^3
32.0	1.6	61162.1	214067.3

Nr of cuts	22	20	18	16	12	4
kg/cut	1.45	1.60	1.78	2.00	2.67	8.00
Area capacity of 3 tonne vehicle per load						
m^2	206250	187500	168750	150000	112500	37500
Load m^3	**100 m^2 units/m^3 of vehicle space**					
1	196.4	178.6	160.7	142.9	107.1	35.7
2	392.9	357.1	321.4	285.7	214.3	71.4
3	589.3	535.7	482.1	428.6	321.4	107.1
4	785.7	714.3	642.9	571.4	428.6	142.9
5	982.1	892.9	803.6	714.3	535.7	178.6

Transportation of Trees

To unload large trees a machine with the necessary lifting strength is required. The weight of the trees must therefore be known in advance. The following table gives a rough overview. The additional columns with root ball dimensions and the number of plants per trailer provide additional information for example about preparing planting holes and calculating unloading times.

Girth in cm	Rootball diameter in cm	Ball height in cm	Weight in kg	Numbers of trees per trailer
16–18	50–60	40	150	100–120
18–20	60–70	40–50	200	80–100
20–25	60–70	40–50	270	50–70
25–30	80	50–60	350	50
30–35	90–100	60–70	500	12–18
35–40	100–110	60–70	650	10–15
40–45	110–120	60–70	850	8–12
45–50	110–120	60–70	1100	5–7
50–60	130–140	60–70	1600	1–3
60–70	150–160	60–70	2500	1
70–80	180–200	70	4000	1
80–90	200–220	70–80	5500	1
90–100	230–250	80–90	7500	1
100–120	250–270	80–90	9500	1

Data supplied by Lorenz von Ehren GmbH
The information in the table is approximate; deviations depend on soil type, genus and weather

FENCING AND GATES

FENCING AND GATES

Types of Preservative

Creosote (tar oil) can be 'factory' applied	by pressure to BS 144: pts 1&2 by immersion to BS 144: pt 1 by hot and cold open tank to BS 144: pts 1&2
Copper/chromium/arsenic (CCA)	by full cell process to BS 4072 pts 1&2
Organic solvent (OS)	by double vacuum (vacvac) to BS 5707 pts 1&3 by immersion to BS 5057 pts 1&3
Pentachlorophenol (PCP)	by heavy oil double vacuum to BS 5705 pts 2&3
Boron diffusion process (treated with disodium octaborate to BWPA Manual 1986)	

Note: Boron is used on green timber at source and the timber is supplied dry

Cleft Chestnut Pale Fences

Pales	Pale spacing	Wire lines	
900 mm	75 mm	2	temporary protection
1050 mm	75 or 100 mm	2	light protective fences
1200 mm	75 mm	3	perimeter fences
1350 mm	75 mm	3	perimeter fences
1500 mm	50 mm	3	narrow perimeter fences
1800 mm	50 mm	3	light security fences

Close-Boarded Fences

Close-boarded fences 1.05 to 1.8 m high
Type BCR (recessed) or BCM (morticed) with concrete posts 140 × 115 mm tapered and Type BW with timber posts

Palisade Fences

Wooden palisade fences
Type WPC with concrete posts 140 × 115 mm tapered and Type WPW with timber posts

For both types of fence:
Height of fence 1050 mm: two rails
Height of fence 1200 mm: two rails
Height of fence 1500 mm: three rails
Height of fence 1650 mm: three rails
Height of fence 1800 mm: three rails

FENCING AND GATES

Post and Rail Fences

Wooden post and rail fences
Type MPR 11/3 morticed rails and Type SPR 11/3 nailed rails
Height to top of rail 1100 mm
Rails: three rails 87 mm, 38 mm

Type MPR 11/4 morticed rails and Type SPR 11/4 nailed rails
Height to top of rail 1100 mm
Rails: four rails 87 mm, 38 mm

Type MPR 13/4 morticed rails and Type SPR 13/4 nailed rails
Height to top of rail 1300 mm
Rail spacing 250 mm, 250 mm, and 225 mm from top
Rails: four rails 87 mm, 38 mm

Steel Posts

Rolled steel angle iron posts for chain link fencing

Posts	Fence height	Strut	Straining post
1500 × 40 × 40 × 5 mm	900 mm	1500 × 40 × 40 × 5 mm	1500 × 50 × 50 × 6 mm
1800 × 40 × 40 × 5 mm	1200 mm	1800 × 40 × 40 × 5 mm	1800 × 50 × 50 × 6 mm
2000 × 45 × 45 × 5 mm	1400 mm	2000 × 45 × 45 × 5 mm	2000 × 60 × 60 × 6 mm
2600 × 45 × 45 × 5 mm	1800 mm	2600 × 45 × 45 × 5 mm	2600 × 60 × 60 × 6 mm
3000 × 50 × 50 × 6 mm with arms	1800 mm	2600 × 45 × 45 × 5 mm	3000 × 60 × 60 × 6 mm

Concrete Posts

Concrete posts for chain link fencing

Posts and straining posts	Fence height	Strut
1570 mm 100 × 100 mm	900 mm	1500 mm × 75 × 75 mm
1870 mm 125 × 125 mm	1200 mm	1830 mm × 100 × 75 mm
2070 mm 125 × 125 mm	1400 mm	1980 mm × 100 × 75 mm
2620 mm 125 × 125 mm	1800 mm	2590 mm × 100 × 85 mm
3040 mm 125 × 125 mm	1800 mm	2590 mm × 100 × 85 mm (with arms)

FENCING AND GATES

Rolled Steel Angle Posts

Rolled steel angle posts for rectangular wire mesh (field) fencing

Posts	Fence height	Strut	Straining post
1200 × 40 × 40 × 5 mm	600 mm	1200 × 75 × 75 mm	1350 × 100 × 100 mm
1400 × 40 × 40 × 5 mm	800 mm	1400 × 75 × 75 mm	1550 × 100 × 100 mm
1500 × 40 × 40 × 5 mm	900 mm	1500 × 75 × 75 mm	1650 × 100 × 100 mm
1600 × 40 × 40 × 5 mm	1000 mm	1600 × 75 × 75 mm	1750 × 100 × 100 mm
1750 × 40 × 40 × 5 mm	1150 mm	1750 × 75 × 100 mm	1900 × 125 × 125 mm

Concrete Posts

Concrete posts for rectangular wire mesh (field) fencing

Posts	Fence height	Strut	Straining post
1270 × 100 × 100 mm	600 mm	1200 × 75 × 75 mm	1420 × 100 × 100 mm
1470 × 100 × 100 mm	800 mm	1350 × 75 × 75 mm	1620 × 100 × 100 mm
1570 × 100 × 100 mm	900 mm	1500 × 75 × 75 mm	1720 × 100 × 100 mm
1670 × 100 × 100 mm	600 mm	1650 × 75 × 75 mm	1820 × 100 × 100 mm
1820 × 125 × 125 mm	1150 mm	1830 × 75 × 100 mm	1970 × 125 × 125 mm

Cleft Chestnut Pale Fences

Timber Posts

Timber posts for wire mesh and hexagonal wire netting fences

Round timber for general fences

Posts	Fence height	Strut	Straining post
1300 × 65 mm dia.	600 mm	1200 × 80 mm dia.	1450 × 100 mm dia.
1500 × 65 mm dia.	800 mm	1400 × 80 mm dia.	1650 × 100 mm dia.
1600 × 65 mm dia.	900 mm	1500 × 80 mm dia.	1750 × 100 mm dia.
1700 × 65 mm dia.	1050 mm	1600 × 80 mm dia.	1850 × 100 mm dia.
1800 × 65 mm dia.	1150 mm	1750 × 80 mm dia.	2000 × 120 mm dia.

Squared timber for general fences

Posts	Fence height	Strut	Straining post
1300 × 75 × 75 mm	600 mm	1200 × 75 × 75 mm	1450 × 100 × 100 mm
1500 × 75 × 75 mm	800 mm	1400 × 75 × 75 mm	1650 × 100 × 100 mm
1600 × 75 × 75 mm	900 mm	1500 × 75 × 75 mm	1750 × 100 × 100 mm
1700 × 75 × 75 mm	1050 mm	1600 × 75 × 75 mm	1850 × 100 × 100 mm
1800 × 75 × 75 mm	1150 mm	1750 × 75 × 75 mm	2000 × 125 × 100 mm

FENCING AND GATES

Steel Fences to BS 1722: Part 9: 1992

	Fence height	Top/bottom rails and flat posts	Vertical bars
Light	1000 mm	40 × 10 mm 450 mm in ground	12 mm dia. at 115 mm cs
	1200 mm	40 × 10 mm 550 mm in ground	12 mm dia. at 115 mm cs
	1400 mm	40 × 10 mm 550 mm in ground	12 mm dia. at 115 mm cs
Light	1000 mm	40 × 10 mm 450 mm in ground	16 mm dia. at 120 mm cs
	1200 mm	40 × 10 mm 550 mm in ground	16 mm dia. at 120 mm cs
	1400 mm	40 × 10 mm 550 mm in ground	16 mm dia. at 120 mm cs
Medium	1200 mm	50 × 10 mm 550 mm in ground	20 mm dia. at 125 mm cs
	1400 mm	50 × 10 mm 550 mm in ground	20 mm dia. at 125 mm cs
	1600 mm	50 × 10 mm 600 mm in ground	22 mm dia. at 145 mm cs
	1800 mm	50 × 10 mm 600 mm in ground	22 mm dia. at 145 mm cs
Heavy	1600 mm	50 × 10 mm 600 mm in ground	22 mm dia. at 145 mm cs
	1800 mm	50 × 10 mm 600 mm in ground	22 mm dia. at 145 mm cs
	2000 mm	50 × 10 mm 600 mm in ground	22 mm dia. at 145 mm cs
	2200 mm	50 × 10 mm 600 mm in ground	22 mm dia. at 145 mm cs

Notes: Mild steel fences: round or square verticals; flat standards and horizontals. Tops of vertical bars may be bow-top, blunt, or pointed. Round or square bar railings

Timber Field Gates to BS 3470: 1975

Gates made to this standard are designed to open one way only
All timber gates are 1100 mm high
Width over stiles 2400, 2700, 3000, 3300, 3600 and 4200 mm
Gates over 4200 mm should be made in two leaves

Steel Field Gates to BS 3470: 1975

All steel gates are 1100 mm high	
Heavy duty:	width over stiles 2400, 3000, 3600 and 4500 mm
Light duty:	width over stiles 2400, 3000 and 3600 mm

FENCING AND GATES

Domestic Front Entrance Gates to BS 4092: Part 1: 1966

Metal gates:	Single gates are 900 mm high minimum, 900 mm, 1000 mm and 1100 mm wide

Domestic Front Entrance Gates to BS 4092: Part 2: 1966

Wooden gates:	All rails shall be tenoned into the stiles Single gates are 840 mm high minimum, 801 mm and 1020 mm wide Double gates are 840 mm high minimum, 2130, 2340 and 2640 mm wide

Timber Bridle Gates to BS 5709:1979 (Horse or Hunting Gates)

Gates open one way only	
Minimum width between posts	1525 mm
Minimum height	1100 mm

Timber Kissing Gates to BS 5709:1979

Minimum width	700 mm
Minimum height	1000 mm
Minimum distance between shutting posts	600 mm
Minimum clearance at mid-point	600 mm

Metal Kissing Gates to BS 5709:1979

Sizes are the same as those for timber kissing gates
Maximum gaps between rails 120 mm

Categories of Pedestrian Guard Rail to BS 3049:1976

Class A for normal use
Class B where vandalism is expected
Class C where crowd pressure is likely

DRAINAGE

DRAINAGE

Width Required for Trenches for Various Diameters of Pipes

Pipe diameter (mm)	Trench n.e. 1.50 m deep	Trench over 1.50 m deep
n.e. 100 mm	450 mm	600 mm
100–150 mm	500 mm	650 mm
150–225 mm	600 mm	750 mm
225–300 mm	650 mm	800 mm
300–400 mm	750 mm	900 mm
400–450 mm	900 mm	1050 mm
450–600 mm	1100 mm	1300 mm

Weights and Dimensions – Vitrified Clay Pipes

Product	Nominal diameter	Effective length	BS 65 limits of tolerance		Crushing strength	Weight	
			min	max			
	(mm)	(mm)	(mm)	(mm)	(kN/m)	(kg/pipe)	(kg/m)
Supersleve	100	1600	96	105	35.00	14.71	9.19
	150	1750	146	158	35.00	29.24	16.71
Hepsleve	225	1850	221	236	28.00	84.03	45.42
	300	2500	295	313	34.00	193.05	77.22
	150	1500	146	158	22.00	37.04	24.69
Hepseal	225	1750	221	236	28.00	85.47	48.84
	300	2500	295	313	34.00	204.08	81.63
	400	2500	394	414	44.00	357.14	142.86
	450	2500	444	464	44.00	454.55	181.63
	500	2500	494	514	48.00	555.56	222.22
	600	2500	591	615	57.00	796.23	307.69
	700	3000	689	719	67.00	1111.11	370.45
	800	3000	788	822	72.00	1351.35	450.45
Hepline	100	1600	95	107	22.00	14.71	9.19
	150	1750	145	160	22.00	29.24	16.71
	225	1850	219	239	28.00	84.03	45.42
	300	1850	292	317	34.00	142.86	77.22
Hepduct (conduit)	90	1500	–	–	28.00	12.05	8.03
	100	1600	–	–	28.00	14.71	9.19
	125	1750	–	–	28.00	20.73	11.84
	150	1750	–	–	28.00	29.24	16.71
	225	1850	–	–	28.00	84.03	45.42
	300	1850	–	–	34.00	142.86	77.22

DRAINAGE

Weights and Dimensions – Vitrified Clay Pipes

Nominal internal diameter (mm)	Nominal wall thickness (mm)	Approximate weight (kg/m)
150	25	45
225	29	71
300	32	122
375	35	162
450	38	191
600	48	317
750	54	454
900	60	616
1200	76	912
1500	89	1458
1800	102	1884
2100	127	2619

Wall thickness, weights and pipe lengths vary, depending on type of pipe required

The particulars shown above represent a selection of available diameters and are applicable to strength class 1 pipes with flexible rubber ring joints

Tubes with Ogee joints are also available

DRAINAGE

Weights and Dimensions – PVC-u Pipes

	Nominal size	Mean outside diameter (mm)		Wall thickness	Weight
		min	max	(mm)	(kg/m)
Standard pipes	82.4	82.4	82.7	3.2	1.2
	110.0	110.0	110.4	3.2	1.6
	160.0	160.0	160.6	4.1	3.0
	200.0	200.0	200.6	4.9	4.6
	250.0	250.0	250.7	6.1	7.2
Perforated pipes					
heavy grade	As above	As above	As above	As above	As above
thin wall	82.4	82.4	82.7	1.7	–
	110.0	110.0	110.4	2.2	–
	160.0	160.0	160.6	3.2	–

Width of Trenches Required for Various Diameters of Pipes

Pipe diameter (mm)	Trench n.e. 1.5 m deep (mm)	Trench over 1.5 m deep (mm)
n.e. 100	450	600
100–150	500	650
150–225	600	750
225–300	650	800
300–400	750	900
400–450	900	1050
450–600	1100	1300

DRAINAGE

DRAINAGE BELOW GROUND AND LAND DRAINAGE

Flow of Water Which Can Be Carried by Various Sizes of Pipe

Clay or concrete pipes

	Gradient of pipeline							
	1:10	1:20	1:30	1:40	1:50	1:60	1:80	1:100
Pipe size	**Flow in litres per second**							
DN 100 15.0	8.5	6.8	5.8	5.2	4.7	4.0	3.5	
DN 150 28.0	19.0	16.0	14.0	12.0	11.0	9.1	8.0	
DN 225 140.0	95.0	76.0	66.0	58.0	53.0	46.0	40.0	

Plastic pipes

	Gradient of pipeline							
	1:10	1:20	1:30	1:40	1:50	1:60	1:80	1:100
Pipe size	**Flow in litres per second**							
82.4 mm i/dia.	12.0	8.5	6.8	5.8	5.2	4.7	4.0	3.5
110 mm i/dia.	28.0	19.0	16.0	14.0	12.0	11.0	9.1	8.0
160 mm i/dia.	76.0	53.0	43.0	37.0	33.0	29.0	25.0	22.0
200 mm i/dia.	140.0	95.0	76.0	66.0	58.0	53.0	46.0	40.0

Vitrified (Perforated) Clay Pipes and Fittings to BS En 295-5 1994

Length not specified		
75 mm bore	**250 mm bore**	**600 mm bore**
100	300	700
125	350	800
150	400	1000
200	450	1200
225	500	

Precast Concrete Pipes: Prestressed Non-pressure Pipes and Fittings: Flexible Joints to BS 5911: Pt. 103: 1994

Rationalized metric nominal sizes: 450, 500	
Length:	500–1000 by 100 increments 1000–2200 by 200 increments 2200–2800 by 300 increments
Angles: length:	450–600 angles 45, 22.5,11.25° 600 or more angles 22.5, 11.25°

DRAINAGE

Precast Concrete Pipes: Unreinforced and Circular Manholes and Soakaways to BS 5911: Pt. 200: 1994

Nominal sizes:	
Shafts:	675, 900 mm
Chambers:	900, 1050, 1200, 1350, 1500, 1800, 2100, 2400, 2700, 3000 mm
Large chambers:	To have either tapered reducing rings or a flat reducing slab in order to accept the standard cover
Ring depths:	1. 300–1200 mm by 300 mm increments except for bottom slab and rings below cover slab, these are by 150 mm increments 2. 250–1000 mm by 250 mm increments except for bottom slab and rings below cover slab, these are by 125 mm increments
Access hole:	750 × 750 mm for DN 1050 chamber 1200 × 675 mm for DN 1350 chamber

Calculation of Soakaway Depth

The following formula determines the depth of concrete ring soakaway that would be required for draining given amounts of water.

$$\mathbf{h} = \frac{\mathbf{4ar}}{\mathbf{3\pi D^2}}$$

h = depth of the chamber below the invert pipe
a = the area to be drained
r = the hourly rate of rainfall (50 mm per hour)
π = pi
D = internal diameter of the soakaway

This table shows the depth of chambers in each ring size which would be required to contain the volume of water specified. These allow a recommended storage capacity of ⅓ (one third of the hourly rainfall figure).

Table Showing Required Depth of Concrete Ring Chambers in Metres

Area m^2	**50**	**100**	**150**	**200**	**300**	**400**	**500**
Ring size							
0.9	1.31	2.62	3.93	5.24	7.86	10.48	13.10
1.1	0.96	1.92	2.89	3.85	5.77	7.70	9.62
1.2	0.74	1.47	2.21	2.95	4.42	5.89	7.37
1.4	0.58	1.16	1.75	2.33	3.49	4.66	5.82
1.5	0.47	0.94	1.41	1.89	2.83	3.77	4.72
1.8	0.33	0.65	0.98	1.31	1.96	2.62	3.27
2.1	0.24	0.48	0.72	0.96	1.44	1.92	2.41
2.4	0.18	0.37	0.55	0.74	1.11	1.47	1.84
2.7	0.15	0.29	0.44	0.58	0.87	1.16	1.46
3.0	0.12	0.24	0.35	0.47	0.71	0.94	1.18

DRAINAGE

Precast Concrete Inspection Chambers and Gullies to BS 5911: Part 230: 1994

Nominal sizes:	375 diameter, 750, 900 mm deep 450 diameter, 750, 900, 1050, 1200 mm deep
Depths:	from the top for trapped or untrapped units: centre of outlet 300 mm invert (bottom) of the outlet pipe 400 mm
Depth of water seal for trapped gullies:	85 mm, rodding eye int. dia. 100 mm
Cover slab:	65 mm min

Bedding Flexible Pipes: PVC-u Or Ductile Iron

Type 1 =	100 mm fill below pipe, 300 mm above pipe: single size material
Type 2 =	100 mm fill below pipe, 300 mm above pipe: single size or graded material
Type 3 =	100 mm fill below pipe, 75 mm above pipe with concrete protective slab over
Type 4 =	100 mm fill below pipe, fill laid level with top of pipe
Type 5 =	200 mm fill below pipe, fill laid level with top of pipe
Concrete =	25 mm sand blinding to bottom of trench, pipe supported on chocks,100 mm concrete under the pipe, 150 mm concrete over the pipe

DRAINAGE

Bedding Rigid Pipes: Clay or Concrete

(for vitrified clay pipes the manufacturer should be consulted)

Class D:	Pipe laid on natural ground with cut-outs for joints, soil screened to remove stones over 40 mm and returned over pipe to 150 m min depth. Suitable for firm ground with trenches trimmed by hand.
Class N:	Pipe laid on 50 mm granular material of graded aggregate to Table 4 of BS 882, or 10 mm aggregate to Table 6 of BS 882, or as dug light soil (not clay) screened to remove stones over 10 mm. Suitable for machine dug trenches.
Class B:	As Class N, but with granular bedding extending half way up the pipe diameter.
Class F:	Pipe laid on 100 mm granular fill to BS 882 below pipe, minimum 150 mm granular fill above pipe: single size material. Suitable for machine dug trenches.
Class A:	Concrete 100 mm thick under the pipe extending half way up the pipe, backfilled with the appropriate class of fill. Used where there is only a very shallow fall to the drain. Class A bedding allows the pipes to be laid to an exact gradient.
Concrete surround:	25 mm sand blinding to bottom of trench, pipe supported on chocks, 100 mm concrete under the pipe, 150 mm concrete over the pipe. It is preferable to bed pipes under slabs or wall in granular material.

PIPED SUPPLY SYSTEMS

Identification of Service Tubes From Utility to Dwellings

Utility	Colour	Size	Depth
British Telecom	grey	54 mm od	450 mm
Electricity	black	38 mm od	450 mm
Gas	yellow	42 mm od rigid 60 mm od convoluted	450 mm
Water	may be blue	(normally untubed)	750 mm

ELECTRICAL SUPPLY/POWER/LIGHTING SYSTEMS

ELECTRICAL SUPPLY/POWER/LIGHTING SYSTEMS

Electrical Insulation Class En 60.598 BS 4533

Class 1:	luminaires comply with class 1 (I) earthed electrical requirements
Class 2:	luminaires comply with class 2 (II) double insulated electrical requirements
Class 3:	luminaires comply with class 3 (III) electrical requirements

Protection to Light Fittings

BS EN 60529:1992 Classification for degrees of protection provided by enclosures.
(IP Code – International or Ingress Protection)

1st characteristic: against ingress of solid foreign objects		
The figure	2	indicates that fingers cannot enter
	3	that a 2.5 mm diameter probe cannot enter
	4	that a 1.0 mm diameter probe cannot enter
	5	the fitting is dust proof (no dust around live parts)
	6	the fitting is dust tight (no dust entry)
2nd characteristic: ingress of water with harmful effects		
The figure	0	indicates unprotected
	1	vertically dripping water cannot enter
	2	water dripping 15° (tilt) cannot enter
	3	spraying water cannot enter
	4	splashing water cannot enter
	5	jetting water cannot enter
	6	powerful jetting water cannot enter
	7	proof against temporary immersion
	8	proof against continuous immersion
Optional additional codes:		A–D protects against access to hazardous parts
	H	High voltage apparatus
	M	fitting was in motion during water test
	S	fitting was static during water test
	W	protects against weather
Marking code arrangement:		(example) IPX5S = IP (International or Ingress Protection) X (denotes omission of first characteristic) 5 = jetting S = static during water test

RAIL TRACKS

RAIL TRACKS

	kg/m of track	lb/ft of track
Standard gauge		
Bull head rails, chairs, transverse timber (softwood) sleepers etc.	245	165
Main lines		
Flat-bottom rails, transverse prestressed concrete sleepers, etc.	418	280
Add for electric third rail	51	35
Add for crushed stone ballast	2600	1750
	kg/m²	**lb/sq ft**
Overall average weight – rails connections, sleepers, ballast, etc.	733	150
	kg/m of track	**lb/ft of track**
Bridge rails, longitudinal timber sleepers, etc.	112	75

RAIL TRACKS

Heavy Rails

British Standard Section No.	Rail height (mm)	Foot width (mm)	Head width (mm)	Min web thickness (mm)	Section weight (kg/m)
Flat Bottom Rails					
60 A	114.30	109.54	57.15	11.11	30.62
70 A	123.82	111.12	60.32	12.30	34.81
75 A	128.59	114.30	61.91	12.70	37.45
80 A	133.35	117.47	63.50	13.10	39.76
90 A	142.88	127.00	66.67	13.89	45.10
95 A	147.64	130.17	69.85	14.68	47.31
100 A	152.40	133.35	69.85	15.08	50.18
110 A	158.75	139.70	69.85	15.87	54.52
113 A	158.75	139.70	69.85	20.00	56.22
50 'O'	100.01	100.01	52.39	10.32	24.82
80 'O'	127.00	127.00	63.50	13.89	39.74
60R	114.30	109.54	57.15	11.11	29.85
75R	128.59	122.24	61.91	13.10	37.09
80R	133.35	127.00	63.50	13.49	39.72
90R	142.88	136.53	66.67	13.89	44.58
95R	147.64	141.29	68.26	14.29	47.21
100R	152.40	146.05	69.85	14.29	49.60
95N	147.64	139.70	69.85	13.89	47.27
Bull Head Rails					
95R BH	145.26	69.85	69.85	19.05	47.07

Light Rails

British Standard Section No.	Rail height (mm)	Foot width (mm)	Head width (mm)	Min web thickness (mm)	Section weight (kg/m)
Flat Bottom Rails					
20M	65.09	55.56	30.96	6.75	9.88
30M	75.41	69.85	38.10	9.13	14.79
35M	80.96	76.20	42.86	9.13	17.39
35R	85.73	82.55	44.45	8.33	17.40
40	88.11	80.57	45.64	12.3	19.89
Bridge Rails					
13	48.00	92	36.00	18.0	13.31
16	54.00	108	44.50	16.0	16.06
20	55.50	127	50.00	20.5	19.86
28	67.00	152	50.00	31.0	28.62
35	76.00	160	58.00	34.5	35.38
50	76.00	165	58.50	–	50.18
Crane Rails					
A65	75.00	175.00	65.00	38.0	43.10
A75	85.00	200.00	75.00	45.0	56.20
A100	95.00	200.00	100.00	60.0	74.30
A120	105.00	220.00	120.00	72.0	100.00
175CR	152.40	152.40	107.95	38.1	86.92

RAIL TRACKS

Fish Plates

British Standard Section No.	Overall plate length		Hole diameter	Finished weight per pair	
	4 Hole (mm)	6 Hole (mm)	(mm)	4 Hole (kg/pair)	6 Hole (kg/pair)
For British Standard Heavy Rails: Flat Bottom Rails					
60 A	406.40	609.60	20.64	9.87	14.76
70 A	406.40	609.60	22.22	11.15	16.65
75 A	406.40	–	23.81	11.82	17.73
80 A	406.40	609.60	23.81	13.15	19.72
90 A	457.20	685.80	25.40	17.49	26.23
100 A	508.00	–	pear	25.02	–
110 A (shallow)	507.00	–	27.00	30.11	54.64
113 A (heavy)	507.00	–	27.00	30.11	54.64
50 'O' (shallow)	406.40	–	–	6.68	10.14
80 'O' (shallow)	495.30	–	23.81	14.72	22.69
60R (shallow)	406.40	609.60	20.64	8.76	13.13
60R (angled)	406.40	609.60	20.64	11.27	16.90
75R (shallow)	406.40	–	23.81	10.94	16.42
75R (angled)	406.40	–	23.81	13.67	–
80R (shallow)	406.40	609.60	23.81	11.93	17.89
80R (angled)	406.40	609.60	23.81	14.90	22.33
For British Standard Heavy Rails: Bull Head Rails					
95R BH (shallow)	–	457.20	27.00	14.59	14.61
For British Standard Light Rails: Flat Bottom Rails					
30M	355.6	–	–	–	2.72
35M	355.6	–	–	–	2.83
40	355.6	–	–	3.76	–

FRACTIONS, DECIMALS AND MILLIMETRE EQUIVALENTS

FRACTIONS, DECIMALS AND MILLIMETRE EQUIVALENTS

Fractions	Decimals	(mm)		Fractions	Decimals	(mm)
1/64	0.015625	0.396875		33/64	0.515625	13.096875
1/32	0.03125	0.79375		17/32	0.53125	13.49375
3/64	0.046875	1.190625		35/64	0.546875	13.890625
1/16	0.0625	1.5875		9/16	0.5625	14.2875
5/64	0.078125	1.984375		37/64	0.578125	14.684375
3/32	0.09375	2.38125		19/32	0.59375	15.08125
7/64	0.109375	2.778125		39/64	0.609375	15.478125
1/8	0.125	3.175		5/8	0.625	15.875
9/64	0.140625	3.571875		41/64	0.640625	16.271875
5/32	0.15625	3.96875		21/32	0.65625	16.66875
11/64	0.171875	4.365625		43/64	0.671875	17.065625
3/16	0.1875	4.7625		11/16	0.6875	17.4625
13/64	0.203125	5.159375		45/64	0.703125	17.859375
7/32	0.21875	5.55625		23/32	0.71875	18.25625
15/64	0.234375	5.953125		47/64	0.734375	18.653125
1/4	0.25	6.35		3/4	0.75	19.05
17/64	0.265625	6.746875		49/64	0.765625	19.446875
9/32	0.28125	7.14375		25/32	0.78125	19.84375
19/64	0.296875	7.540625		51/64	0.796875	20.240625
5/16	0.3125	7.9375		13/16	0.8125	20.6375
21/64	0.328125	8.334375		53/64	0.828125	21.034375
11/32	0.34375	8.73125		27/32	0.84375	21.43125
23/64	0.359375	9.128125		55/64	0.859375	21.828125
3/8	0.375	9.525		7/8	0.875	22.225
25/64	0.390625	9.921875		57/64	0.890625	22.621875
13/32	0.40625	10.31875		29/32	0.90625	23.01875
27/64	0.421875	10.71563		59/64	0.921875	23.415625
7/16	0.4375	11.1125		15/16	0.9375	23.8125
29/64	0.453125	11.50938		61/64	0.953125	24.209375
15/32	0.46875	11.90625		31/32	0.96875	24.60625
31/64	0.484375	12.30313		63/64	0.984375	25.003125
1/2	0.5	12.7		1.0	1	25.4

IMPERIAL STANDARD WIRE GAUGE (SWG)

IMPERIAL STANDARD WIRE GAUGE (SWG)

SWG	Diameter	
No.	(inches)	(mm)
7/0	0.5	12.7
6/0	0.464	11.79
5/0	0.432	10.97
4/0	0.4	10.16
3/0	0.372	9.45
2/0	0.348	8.84
1/0	0.324	8.23
1	0.3	7.62
2	0.276	7.01
3	0.252	6.4
4	0.232	5.89
5	0.212	5.38
6	0.192	4.88
7	0.176	4.47
8	0.16	4.06
9	0.144	3.66
10	0.128	3.25
11	0.116	2.95
12	0.104	2.64
13	0.092	2.34
14	0.08	2.03
15	0.072	1.83
16	0.064	1.63
17	0.056	1.42
18	0.048	1.22
19	0.04	1.016
20	0.036	0.914
21	0.032	0.813
22	0.028	0.711
23	0.024	0.61
24	0.022	0.559
25	0.02	0.508
26	0.018	0.457
27	0.0164	0.417
28	0.0148	0.376
29	0.0136	0.345
30	0.0124	0.315
31	0.0116	0.295
32	0.0108	0.274
33	0.01	0.254
34	0.009	0.234
35	0.008	0.213
36	0.008	0.193
37	0.007	0.173
38	0.006	0.152
39	0.005	0.132
40	0.005	0.122
41	0.004	0.112
42	0.004	0.102
43	0.004	0.091
44	0.003	0.081
45	0.003	0.071
46	0.002	0.061
47	0.002	0.051
48	0.002	0.041
49	0.001	0.031
50	0.001	0.025

PIPES, WATER, STORAGE, INSULATION

WATER PRESSURE DUE TO HEIGHT

Imperial

Head (feet)	Pressure (lb/in^2)		Head (feet)	Pressure (lb/in^2)
1	0.43		70	30.35
5	2.17		75	32.51
10	4.34		80	34.68
15	6.5		85	36.85
20	8.67		90	39.02
25	10.84		95	41.18
30	13.01		100	43.35
35	15.17		105	45.52
40	17.34		110	47.69
45	19.51		120	52.02
50	21.68		130	56.36
55	23.84		140	60.69
60	26.01		150	65.03
65	28.18			

Metric

Head (m)	Pressure (bar)		Head (m)	Pressure (bar)
0.5	0.049		18.0	1.766
1.0	0.098		19.0	1.864
1.5	0.147		20.0	1.962
2.0	0.196		21.0	2.06
3.0	0.294		22.0	2.158
4.0	0.392		23.0	2.256
5.0	0.491		24.0	2.354
6.0	0.589		25.0	2.453
7.0	0.687		26.0	2.551
8.0	0.785		27.0	2.649
9.0	0.883		28.0	2.747
10.0	0.981		29.0	2.845
11.0	1.079		30.0	2.943
12.0	1.177		32.5	3.188
13.0	1.275		35.0	3.434
14.0	1.373		37.5	3.679
15.0	1.472		40.0	3.924
16.0	1.57		42.5	4.169
17.0	1.668		45.0	4.415

1 bar	=	14.5038 lbf/in^2
1 lbf/in^2	=	0.06895 bar
1 metre	=	3.2808 ft or 39.3701 in
1 foot	=	0.3048 metres
1 in wg	=	2.5 mbar (249.1 N/m^2)

PIPES, WATER, STORAGE, INSULATION

Dimensions and Weights of Copper Pipes to BSEN 1057, BSEN 12499, BSEN 14251

Outside diameter (mm)	Internal diameter (mm)	Weight per metre (kg)	Internal diameter (mm)	Weight per metre (kg)	Internal diameter (mm)	Weight per metre (kg)
	Formerly Table X		Formerly Table Y		Formerly Table Z	
6	4.80	0.0911	4.40	0.1170	5.00	0.0774
8	6.80	0.1246	6.40	0.1617	7.00	0.1054
10	8.80	0.1580	8.40	0.2064	9.00	0.1334
12	10.80	0.1914	10.40	0.2511	11.00	0.1612
15	13.60	0.2796	13.00	0.3923	14.00	0.2031
18	16.40	0.3852	16.00	0.4760	16.80	0.2918
22	20.22	0.5308	19.62	0.6974	20.82	0.3589
28	26.22	0.6814	25.62	0.8985	26.82	0.4594
35	32.63	1.1334	32.03	1.4085	33.63	0.6701
42	39.63	1.3675	39.03	1.6996	40.43	0.9216
54	51.63	1.7691	50.03	2.9052	52.23	1.3343
76.1	73.22	3.1287	72.22	4.1437	73.82	2.5131
108	105.12	4.4666	103.12	7.3745	105.72	3.5834
133	130.38	5.5151	–	–	130.38	5.5151
159	155.38	8.7795	–	–	156.38	6.6056

Dimensions of Stainless Steel Pipes to BS 4127

Outside diameter (mm)	Maximum outside diameter (mm)	Minimum outside diameter (mm)	Wall thickness (mm)	Working pressure (bar)
6	6.045	5.940	0.6	330
8	8.045	7.940	0.6	260
10	10.045	9.940	0.6	210
12	12.045	11.940	0.6	170
15	15.045	14.940	0.6	140
18	18.045	17.940	0.7	135
22	22.055	21.950	0.7	110
28	28.055	27.950	0.8	121
35	35.070	34.965	1.0	100
42	42.070	41.965	1.1	91
54	54.090	53.940	1.2	77

PIPES, WATER, STORAGE, INSULATION

Dimensions of Steel Pipes to BS 1387

Nominal size (mm)	Approx. outside diameter (mm)	Outside diameter				Thickness		
		Light		Medium & heavy		Light (mm)	Medium (mm)	Heavy (mm)
		Max (mm)	Min (mm)	Max (mm)	Min (mm)			
6	10.20	10.10	9.70	10.40	9.80	1.80	2.00	2.65
8	13.50	13.60	13.20	13.90	13.30	1.80	2.35	2.90
10	17.20	17.10	16.70	17.40	16.80	1.80	2.35	2.90
15	21.30	21.40	21.00	21.70	21.10	2.00	2.65	3.25
20	26.90	26.90	26.40	27.20	26.60	2.35	2.65	3.25
25	33.70	33.80	33.20	34.20	33.40	2.65	3.25	4.05
32	42.40	42.50	41.90	42.90	42.10	2.65	3.25	4.05
40	48.30	48.40	47.80	48.80	48.00	2.90	3.25	4.05
50	60.30	60.20	59.60	60.80	59.80	2.90	3.65	4.50
65	76.10	76.00	75.20	76.60	75.40	3.25	3.65	4.50
80	88.90	88.70	87.90	89.50	88.10	3.25	4.05	4.85
100	114.30	113.90	113.00	114.90	113.30	3.65	4.50	5.40
125	139.70	–	–	140.60	138.70	–	4.85	5.40
150	165.1*	–	–	166.10	164.10	–	4.85	5.40

* 165.1 mm (6.5 in) outside diameter is not generally recommended except where screwing to BS 21 is necessary
All dimensions are in accordance with ISO R65 except approximate outside diameters which are in accordance with ISO R64
Light quality is equivalent to ISO R65 Light Series II

Approximate Metres Per Tonne of Tubes to BS 1387

Nom. Size (mm)	BLACK						GALVANIZED					
	Plain/screwed ends			Screwed & socketed			Plain/screwed ends			Screwed & socketed		
	L (m)	M (m)	H (m)	L (m)	M (m)	H (m)	L (m)	M (m)	H (m)	L (m)	M (m)	H (m)
6	2765	2461	2030	2743	2443	2018	2604	2333	1948	2584	2317	1937
8	1936	1538	1300	1920	1527	1292	1826	1467	1254	1811	1458	1247
10	1483	1173	979	1471	1165	974	1400	1120	944	1386	1113	939
15	1050	817	688	1040	811	684	996	785	665	987	779	661
20	712	634	529	704	628	525	679	609	512	673	603	508
25	498	410	336	494	407	334	478	396	327	474	394	325
32	388	319	260	384	316	259	373	308	254	369	305	252
40	307	277	226	303	273	223	296	268	220	292	264	217
50	244	196	162	239	194	160	235	191	158	231	188	157
65	172	153	127	169	151	125	167	149	124	163	146	122
80	147	118	99	143	116	98	142	115	97	139	113	96
100	101	82	69	98	81	68	98	81	68	95	79	67
125	–	62	56	–	60	55	–	60	55	–	59	54
150	–	52	47	–	50	46	–	51	46	–	49	45

The figures for 'plain or screwed ends' apply also to tubes to BS 1775 of equivalent size and thickness
Key:
L – Light
M – Medium
H – Heavy

PIPES, WATER, STORAGE, INSULATION

Flange Dimension Chart to BS 4504 & BS 10

Normal Pressure Rating (PN 6) 6 Bar

Nom. Size	Flange Outside Dia.	Table 6/2 Forged Welding Neck	Table 6/3 Plate Slip on	Table 6/4 Forged Bossed Screwed	Table 6/5 Forged Bossed Slip on	Table 6/8 Plate Blank	Raised Face		Nr. Bolt Hole	Size of Bolt
							Dia.	T'ness		
15	80	12	12	12	12	12	40	2	4	M10 × 40
20	90	14	14	14	14	14	50	2	4	M10 × 45
25	100	14	14	14	14	14	60	2	4	M10 × 45
32	120	14	16	14	14	14	70	2	4	M12 × 45
40	130	14	16	14	14	14	80	3	4	M12 × 45
50	140	14	16	14	14	14	90	3	4	M12 × 45
65	160	14	16	14	14	14	110	3	4	M12 × 45
80	190	16	18	16	16	16	128	3	4	M16 × 55
100	210	16	18	16	16	16	148	3	4	M16 × 55
125	240	18	20	18	18	18	178	3	8	M16 × 60
150	265	18	20	18	18	18	202	3	8	M16 × 60
200	320	20	22	–	20	20	258	3	8	M16 × 60
250	375	22	24	–	22	22	312	3	12	M16 × 65
300	440	22	24	–	22	22	365	4	12	M20 × 70

Normal Pressure Rating (PN 16) 16 Bar

Nom. Size	Flange Outside Dia.	Table 6/2 Forged Welding Neck	Table 6/3 Plate Slip on	Table 6/4 Forged Bossed Screwed	Table 6/5 Forged Bossed Slip on	Table 6/8 Plate Blank	Raised Face		Nr. Bolt Hole	Size of Bolt
							Dia.	T'ness		
15	95	14	14	14	14	14	45	2	4	M12 × 45
20	105	16	16	16	16	16	58	2	4	M12 × 50
25	115	16	16	16	16	16	68	2	4	M12 × 50
32	140	16	16	16	16	16	78	2	4	M16 × 55
40	150	16	16	16	16	16	88	3	4	M16 × 55
50	165	18	18	18	18	18	102	3	4	M16 × 60
65	185	18	18	18	18	18	122	3	4	M16 × 60
80	200	20	20	20	20	20	138	3	8	M16 × 60
100	220	20	20	20	20	20	158	3	8	M16 × 65
125	250	22	22	22	22	22	188	3	8	M16 × 70
150	285	22	22	22	22	22	212	3	8	M20 × 70
200	340	24	24	–	24	24	268	3	12	M20 × 75
250	405	26	26	–	26	26	320	3	12	M24 × 90
300	460	28	28	–	28	28	378	4	12	M24 × 90

PIPES, WATER, STORAGE, INSULATION

Minimum Distances Between Supports/Fixings

Material	BS Nominal pipe size		Pipes – Vertical	Pipes – Horizontal on to low gradients
	(inch)	(mm)	Support distance in metres	Support distance in metres
Copper	0.50	15.00	1.90	1.30
	0.75	22.00	2.50	1.90
	1.00	28.00	2.50	1.90
	1.25	35.00	2.80	2.50
	1.50	42.00	2.80	2.50
	2.00	54.00	3.90	2.50
	2.50	67.00	3.90	2.80
	3.00	76.10	3.90	2.80
	4.00	108.00	3.90	2.80
	5.00	133.00	3.90	2.80
	6.00	159.00	3.90	2.80
muPVC	1.25	32.00	1.20	0.50
	1.50	40.00	1.20	0.50
	2.00	50.00	1.20	0.60
Polypropylene	1.25	32.00	1.20	0.50
	1.50	40.00	1.20	0.50
uPVC	–	82.40	1.20	0.50
	–	110.00	1.80	0.90
	–	160.00	1.80	1.20
Steel	0.50	15.00	2.40	1.80
	0.75	20.00	3.00	2.40
	1.00	25.00	3.00	2.40
	1.25	32.00	3.00	2.40
	1.50	40.00	3.70	2.40
	2.00	50.00	3.70	2.40
	2.50	65.00	4.60	3.00
	3.00	80.40	4.60	3.00
	4.00	100.00	4.60	3.00
	5.00	125.00	5.50	3.70
	6.00	150.00	5.50	4.50
	8.00	200.00	8.50	6.00
	10.00	250.00	9.00	6.50
	12.00	300.00	10.00	7.00
	16.00	400.00	10.00	8.25

PIPES, WATER, STORAGE, INSULATION

Litres of Water Storage Required Per Person Per Building Type

Type of building	Storage (litres)
Houses and flats (up to 4 bedrooms	120/bedroom
Houses and flats (more than 4 bedrooms)	100/bedroom
Hostels	90/bed
Hotels	200/bed
Nurses homes and medical quarters	120/bed
Offices with canteen	45/person
Offices without canteen	40/person
Restaurants	7/meal
Boarding schools	90/person
Day schools – Primary	15/person
Day schools – Secondary	20/person

Recommended Air Conditioning Design Loads

Building type	Design loading
Computer rooms	500 W/m^2 of floor area
Restaurants	150 W/m^2 of floor area
Banks (main area)	100 W/m^2 of floor area
Supermarkets	25 W/m^2 of floor area
Large office block (exterior zone)	100 W/m^2 of floor area
Large office block (interior zone)	80 W/m^2 of floor area
Small office block (interior zone)	80 W/m^2 of floor area

PIPES, WATER, STORAGE, INSULATION

Capacity and Dimensions of Galvanized Mild Steel Cisterns – BS 417

Capacity (litres)	BS type (SCM)	Dimensions		
		Length (mm)	Width (mm)	Depth (mm)
18	45	457	305	305
36	70	610	305	371
54	90	610	406	371
68	110	610	432	432
86	135	610	457	482
114	180	686	508	508
159	230	736	559	559
191	270	762	584	610
227	320	914	610	584
264	360	914	660	610
327	450/1	1220	610	610
336	450/2	965	686	686
423	570	965	762	787
491	680	1090	864	736
709	910	1070	889	889

Capacity of Cold Water Polypropylene Storage Cisterns – BS 4213

Capacity (litres)	BS type (PC)	Maximum height (mm)
18	4	310
36	8	380
68	15	430
91	20	510
114	25	530
182	40	610
227	50	660
273	60	660
318	70	660
455	100	760

PIPES, WATER, STORAGE, INSULATION

Minimum Insulation Thickness to Protect Against Freezing for Domestic Cold Water Systems (8 Hour Evaluation Period)

Pipe size (mm)	Insulation thickness (mm)					
	Condition 1			Condition 2		
	λ = 0.020	λ = 0.030	λ = 0.040	λ = 0.020	λ = 0.030	λ = 0.040
Copper pipes						
15	11	20	34	12	23	41
22	6	9	13	6	10	15
28	4	6	9	4	7	10
35	3	5	7	4	5	7
42	3	4	5	8	4	6
54	2	3	4	2	3	4
76	2	2	3	2	2	3
Steel pipes						
15	9	15	24	10	18	29
20	6	9	13	6	10	15
25	4	7	9	5	7	10
32	3	5	6	3	5	7
40	3	4	5	3	4	6
50	2	3	4	2	3	4
65	2	2	3	2	3	3

Condition 1: water temperature 7°C; ambient temperature –6°C; evaluation period 8 h; permitted ice formation 50%; normal installation, i.e. inside the building and inside the envelope of the structural insulation
Condition 2: water temperature 2°C; ambient temperature –6°C; evaluation period 8 h; permitted ice formation 50%; extreme installation, i.e. inside the building but outside the envelope of the structural insulation
λ = thermal conductivity [W/(mK)]

Insulation Thickness for Chilled And Cold Water Supplies to Prevent Condensation

On a Low Emissivity Outer Surface (0.05, i.e. Bright Reinforced Aluminium Foil) with an Ambient Temperature of +25°C and a Relative Humidity of 80%

Steel pipe size (mm)	*t* = +10			*t* = +5			*t* = 0		
	Insulation thickness (mm)			Insulation thickness (mm)			Insulation thickness (mm)		
	λ = 0.030	λ = 0.040	λ = 0.050	λ = 0.030	λ = 0.040	λ = 0.050	λ = 0.030	λ = 0.040	λ = 0.050
15	16	20	25	22	28	34	28	36	43
25	18	24	29	25	32	39	32	41	50
50	22	28	34	30	39	47	38	49	60
100	26	34	41	36	47	57	46	60	73
150	29	38	46	40	52	64	51	67	82
250	33	43	53	46	60	74	59	77	94
Flat surfaces	39	52	65	56	75	93	73	97	122

t = temperature of contents (°C)
λ = thermal conductivity at mean temperature of insulation [W/(mK)]

PIPES, WATER, STORAGE, INSULATION

Insulation Thickness for Non-domestic Heating Installations to Control Heat Loss

Steel pipe size (mm)	*t* = 75 Insulation thickness (mm)			*t* = 100 Insulation thickness (mm)			*t* = 150 Insulation thickness (mm)		
	λ = 0.030	λ = 0.040	λ = 0.050	λ = 0.030	λ = 0.040	λ = 0.050	λ = 0.030	λ = 0.040	λ = 0.050
10	18	32	55	20	36	62	23	44	77
15	19	34	56	21	38	64	26	47	80
20	21	36	57	23	40	65	28	50	83
25	23	38	58	26	43	68	31	53	85
32	24	39	59	28	45	69	33	55	87
40	25	40	60	29	47	70	35	57	88
50	27	42	61	31	49	72	37	59	90
65	29	43	62	33	51	74	40	63	92
80	30	44	62	35	52	75	42	65	94
100	31	46	63	37	54	76	45	68	96
150	33	48	64	40	57	77	50	73	100
200	35	49	65	42	59	79	53	76	103
250	36	50	66	43	61	80	55	78	105

t = hot face temperature (°C)

λ = thermal conductivity at mean temperature of insulation [W/(mK)]

Index

Above Ground Drainage 188
- brackets, cast iron 206
- pipes, PVC-U 188

ABS pipe 248
AC Static Inverter System, AC power supply to luminaires 680
Access control
- alarm contact flush 697
- emergency break glass double pole 697
- exit button 697
- exit PIR 697
- magnetic swipe reader 697
- proximity reader 697

Access Control Equipment, all-in-rates 126
Access Control System 74
Access Door, PVC-U, spill 199
Acoustic Housings (for generators) 571
- prefabricated acoustic housing 572

Active Chilled Beam System 80
Adaptable Boxes
- square pattern 579
- steel 592

Adaptor, copper pipes 222
AHU's 547
Air Circuit Breakers (ACBs)
- ACB 65 kA fault rated 643
- ACB 100kA fault rated 643

Air Cooled Condensers 423
Air Curtain Unit
- commercial grade 562
- industrial grade 563

Air Handling Units, all-in-rates 109
Airport terminal, elemental cost 128
All-In-Rates 106
- above ground drainage 106
- access control 126
- air handling units 109
- CCTV 126
- chilled water 107–108
- chilled water/LTHW 107
- data cabling 127
- ductwork 109
- electrical in connection with mechanical services 112
- escalators 125
- EWIC 112
- external lighting 112
- extract fans 109
- fire alarms 126
- generators 110
- heat rejection 108
- heat source 108
- hose reels/dry risers 125
- hot 106
- lifts and escalators 113
- lighting 112
- LTHW 107
- LV switchgear 110
- MV switchgear 110
- pipes 106
- pipework 106
- pumps 109
- ring main units 110
- small power 111
- sprinkler systems 125
- sprinklers 109
- substation (packaged) 110
- transformers 110
- UPS 110
- water installations 106

Allowances
- annual holiday allowances 726
- daily travel allowance 720
- lodging allowance 721
- lodging allowances 726
- mileage allowance 721
- subsistence allowance 721
- weekend lodging allowances 726

Analogue Addressable Panels 701
Ancillaries 269
- low temp' hot water 353
- regulators 269

Angle, Flat Cover, PVC trunking 586
Angle, Internal/External Cover, PVC trunking 587

Approximate Estimating 84
stage A feasibility cost 84
stage B feasibility cost 84
stage C elemental rates 98
Aprons, PVC-U, fittings 202
Armoured Cable
clipped direct to background 597
laid in trench 595
SWA 596
in trench 593
Attenuators, DW144 galvanized, rectangular 528
Automatic Air Admittance, valves 201
Automatic Air Vents, valves 364
Automatic Power Factor Correction, floor standing steel enclosure 647
Automatic Transfer Switches
enclosed type, 3 or 4 pole 647
with dual by-pass 648
with single by-pass 648
panel mounting type, 3 or 4 pole 647

Balcony Outlets, PVC-U 172
Basket, tray, epoxy coated 642
Basket Tray, mild steel cable basket
epoxy coated 642
zinc plated 641
Battery Chargers, MV switchgear 576
Bell mouth, gully drains 210
Bend, PVC-U 199
Bend, Flat/Internal/External, PVC miniature trunking 588
Bend, Internal/External, PVC bench trunking 589
Bends 90°
double compartment, galvanized steel 582
single compartment, galvanized steel 582
triple compartment, galvanized steel 583
Biomass Energy 29
Black Enamelled, heavy gauge, screwed drawn steel 578
Black steel pipes, low temp' hot water pipe 322
Boards and Panels 649
Body Coupler, 3 and 4 compartment, galvanized steel 584
Boilers
all-in-rates 108
atmospheric, cast iron 309
condensing 310
domestic 304
forced draft
cast iron sectional 305
steel shell 307
Borehole System 19
Bosses, PVC-U, adaptor 200
Brackets 206
fixed to background 172
for copper Pipes 219
Brackets and Supports, ductile iron, mild steel 211
Breakers/Fuses 649
Building Information Modelling (BIM) 3
Building Management Installations, elemental cost 152
Building Management Systems 7
controls 709
Busbar
chambers 654
underfloor 628
Business Parks, elemental cost 135

Cabinets
floor standing 690
wall mounted 690
Cable
600/1000 Volt grade, 2 core, galvanized steel wire armour 595, 597, 602, 605
600/1000 volt grade, 3 core, galvanized steel wire armour 595, 597, 603, 605
600/1000 volt grade, 4 core, galvanized steel wire armour 596, 598, 603, 605
600/1000 volt grade, 7 core, galvanized steel wire armour 596, 598, 603, 606
600/1000 volt grade, 12 core, galvanized steel wire armour 597, 599, 604, 606
600/1000 volt grade, 19 core, galvanized steel wire armour 597, 599, 604, 606
600/1000 volt grade, 27 core, galvanized steel wire armour 597, 599, 604, 606
600/1000 volt grade, 37 core, galvanized steel wire armour 597, 599, 604, 606
600/1000 Volt grade, single core, aluminium wire armour 595, 597, 602, 604
earth, 450/750 volt grade, single core 610
fire-rated 613
firetuff fire-resistant
2 core 620
3 core 620
4 core 620
flexible 611

heavy duty fire-rated cable
750 volt grade, bare 614
750 volt grade, LSF sheathed 615
light duty fire-rated cable
500 volt grade, bare 613
500 volt grade, LSF sheathed 614
modular 620
XLPE insulation, LSOH
clipped 604
laid in trench 602
Cable Basket 641
Cable Clip, medium voltage 572
Cable Terminations
brass weatherproof gland 599, 607
compound joint, glanding off 573
heavy duty, 750 volt grade 616
light duty, 500 volt grade 616
Cable Tray
heavy duty galvanized steel, BS729 637
light duty galvanized steel, BS729 633
medium duty galvanized steel, BS729 635
Cable, Firetuff
2 core 620
3 core 620
4 core 620
Cable, FP100–450/750 volt grade, single core 617
Cable, FP200–300/500 volt grade, 2 core 618
Cable, FP200–300/500 volt grade, 3 core 618
Cable, FP200–300/500 volt grade, 4 core 618
Cable, FP400–600/1000 volt grade, 2 core 619
Cable, FP400–600/1000 volt grade, 3 core 619
Cable, FP400–600/1000 volt grade, 4 core 619
Cable, LSF Insulated, 450/750 volt grade, single core, non-sheathed copper 609
Cable, Twin 610
300/500 volt grade, PVC/PVC 610
Cable, XLPE Insulated, 300/500 volt grade, single core, PVC sheathed 609
Cables
armoured 595
fibre optic 694
lighting controls 664
low voltage 595
MV, XLPE 573
telecommunications, laid on tray/basket 683
Cables and Wiring, armoured Cable 595
Cables Ties 573
Calorifiers
non storage steam 416
non storage water 373
storage 281
Capital Allowances 33
Car Park (Shopping Mall), elemental cost 131
Carbon Trading 55
Cast Resin Transformer, ventilated steel enclosure
air natural 577
temperature controlled fans 578
Cat A Fit-out, elemental cost, office 134
CCTV 76
CCTV Systems, all-in-rates 126
Ceiling Recessed Asymmetric Compact, fluorescent downlighter 658
Ceiling Roses
batten holder, white moulded plastic, fixed to conduit 664
BC lampholder, white moulded plastic, flush fixed 664
white moulded plastic, flush fixed 664
Central battery system 679
Channel Sections, fabricated brackets 259
Check Valves, DRZ copper alloy and bronze 273
Chilled Beam System 80
Chilled Beams 431
Chilled Ceiling System 79
Chilled Water Pipes, pipes 426
Chillers
absorption 421
air cooled 418
all-in-rates 108
dry air cooled 422
water cooled 419
CHP Combined Heat 572
Circuit Breaker, medium voltage 575
Circuit Breakers (ACBs)
ACB 65 kA fault rated 643
ACB 100kA fault rated 643
Circular, pipes, PVC-U 167
Circular Ductwork 437
Cistern, overflow 197
Class C, pipes, PVC-U 251
Class D, pipes, PVC-U 251
Class E, pipes, PVC-U 251
Cleaning 278
electromagnetic water conditioner 277
simplex configuration 277
triplex configuration 278
ultraviolet water sterilising unit 277

Clock Timing Systems
  digital clock 689
  internal wall mounted electric clock 688
  matching clock 688
  quartz battery movement 688
  quartz master clock 688
  radio receiver 688
  wall clocks for impulse systems 688
Closed Loop System 20
Cold Water, all-in-rates 106
Combined Heat 572
Commercial Research Buildings
  RIBA stage A feasibility cost 97
  RIBA stage B feasibility cost 97
Compaction Equipment, plant hire 747
Concrete Equipment, plant hire 748
Conduit 579
  black enamelled, heavy gauge, 579
  flexible 579
  flexible conduits, metallic, PVC covered conduit 580
  galvanized, heavy gauge, 579
  high Impact PVC heavy Gauge, standard pattern boxes 579
  high Impact PVC light gauge, standard pattern boxes 579
Conduit and Cable Trunking, containment 579
Connection Unit
  galvanized steel box 670
    galvanized 670
  moulded plastic box, standard fused 670
Connector Unit to immersion heater, moulded white plastic cover 671
Connectors
  copper pipes 222
  roof vent 215
Construction Vehicles, plant hire 744
Consumer Units 652
Contactor Relays, pressed steel enclosure 655
Containment
  basket 641
  cable tray GRP 640
  conduit 579
  ladder rack 629
  tray 633
Control Components
  FCU DDC control pack 560
  immersion thermostats 383
  room thermostats 382
  surface thermostats 383
  thermostatic radiator valves 383
Controllers
  damper actuators 710
  frequency inverters 710
  plant 710
  zone 710
Conveyors, plant hire 739
Cooker Control Units
  45 amp, metal coverplate 671
  45 amp, plastic box and coverplate 671
Cooling Tower, all-in-rates 108
Cooling Towers
  closed circuit type 425
  open circuit type 424
Cooling Units, VRF, room units 434
Copper Data Cabling, unshielded twisted pair, solid copper conductor 691
Copper Patch Leads, straight through booted 694
Copper Pipes
  bronze blank flanges 234
  bronze screwed flanges 235
  cold water
    brazed (silver) fittings 230
    brazed (silver) flanges 234
    bronze screw 233
    compression fittings 229
    high duty capillary 227
    press fit mechanical joints 236
  table x, cold Water 218
Cost Indices 85
Couplers, PVC-U 168
Coupling
  EPDM 211
  PVC bench trunking 589
  PVC miniature trunking 589
Crossovers
  double compartment, galvanized steel 584
  PVC trunking 585
  single compartment, galvanized steel 584
  triple compartment, galvanized steel 584
Cylinders
  copper; indirect 281
  storage 281

Dado Modular Trunking, PVC skirting 592
Daily Travelling Allowance, hourly Rate 717
Dampers
  fire 507
  smoke 509

volume control 482
Data Centre, elemental cost 150
Data Transmission 70
Data Transmission Equipment/Cabling, all-in-rates 127
Day Care Unit
RIBA stage A feasibility cost 95
RIBA stage B feasibility cost 95
Daywork
calculation of yearly rate (electrical) 736
calculation of yearly rate (mechanical) 733
electrical 735
heating and ventilating 731
hourly rate of labour table, different grades 733
mechanical 729
Dearators 367
Department Stores
RIBA stage A feasibility cost 96
RIBA stage B feasibility cost 96
Detection/Alarm Equipment 698
Detectors, smoke/fire 700
Diagram
11 kV network connection for LV intakes up to 1000 kVA 67
11 kV network connection for MV intakes 1000 kVA to 6000 kVA 68
access control system 74
active chilled beam system 80
CCTV 76
chilled beam system 80
chilled ceiling system 79
data transmission 70
digital CCTV 76
displacement ventilation system 78
fan coil unit system 77
FCU system 77
fire detection and alarm 73
heat pump 83
intruder detection system 75
lighting control system 71
passive chilled beam system 80
reverse cycle heat pump 83
standby power system 72
static UPS system 69
structured cabling 70
variable air volume system 81
variable refrigerent flow system 82
VAV system 81
VRF system 82
VRV system 82
Diffusers
ceiling mounted, circular and rectangular 516
floor swirl 518
perforated 519
plastic air system 519
slot, aluminium 517
Digital CCTV 76
Dimmer, electrical accessories 663
Dirt Separators 366
Displacement Ventilation System 78
Distribution Boards 650
LV switchgear 643
Distribution Cables
armoured 621
unarmoured 622
Distribution Centres 93
elemental cost 149
Division Strip, PVC trunking, 1.8 m long 587
Domed Outlets, PVC-U 171
Double Compartment Trunking, jointed with standard connectors, galvanized steel 580
Drains, floor, cast iron 216
Dry Air Cooler, all-in-rates 108
Dry Risers, all-in-rates 125
Duct Bend, PVC underfloor trunking
horizontal 590
vertical 589
Duct Connector, PVC underfloor trunking 590
Ductwork
access doors 452
all-in-rates 109
circular ductwork, class B 435
circular polypropylene 447
circular PVC ductwork 446
dampers
fire 508
smoke 510
volume control 507
dura duct LT, fire-rated 547
fire-rated ductwork
duct, 2 hr protection 532
duct, 4 hr protection 533
slab, mineral wool 533
flat oval ductwork 437
flexible ductwork 441
gang rates 161
labour rates 158
rectangular ductwork, class B, galvanized sheet metal DW144 447
rectangular ductwork, class C 481

Dura duct, fire-rated ductwork 547
DW144 482
  ductwork 437, 452
DW154, ductwork 447
DX Split Units, split type 435

Earth Cable, 450/750 volt grade, single core 610
Earthing 703
Edge Elbow
  mains busbar 625
  rising mains busbar 627
Edge Tee
  mains busbar 625
  rising mains busbar 627
Edge Z Unit, mains busbar 625
Educational Buildings
  RIBA stage A feasibility cost 96
  RIBA stage B feasibility cost 96
Educational Research Buildings
  RIBA stage A feasibility cost 97
  RIBA stage B feasibility cost 97
Electrical Accessories
  dimmer switch, push on/off action 663
  dimmer switch, rotary action 663
    metal clad 664
Electrical in connection with Mechanical Services, all-in-rates 112
Electrical Installations 565
  gang rates 567
  labour rates 567
  material cost 567
  measured work prices 567
Electrical Rates of Wages 723
Electrical Supply 569
Elemental Cost
  airport terminal 128
  building management installations 152
  business parks 135
  car park (shopping mall) 131
  data centre 150
  distribution centres 149
  fit-out Cat A 134
  hospitals 141
  hotels 140
  office building 133
  performing arts 137
  residential affordable 145
  residential private 147
  school 143
  shopping mall 130
  sports hall 138
  stadium 139
  supermarket 132
Elemental Rates 93, 98
Emergency Central Battery System, supply to luminaires 679
Emergency Lighting
  110 Volt fluorescent luminaire 678
  LED charging light European sign, polycarbonate base 678
Enclosed Switch Disconnectors, enclosure minimum IP55 rating 654
End Cap, rising mains busbar 626
End Cover, mains busbar 624
End Feed Unit, rising mains busbar 626
Energy Meters
  electromagnetic 434
  ultrasonic 433
Enhanced Capital Allowance (ECAs) 39
Entertainment and Recreation
  RIBA stage A feasibility cost 95
  RIBA stage B feasibility cost 95
EPDM, pipes, cast iron 210
Escalators, all-in-rates 125
EWIC, all-in-rates 112
Excavators, plant hire 746
Expansion Joints
  angular movement, stainless steel 354
  axial movement, stainless steel 353
Extender Cables
  armoured 621
  unarmoured 623
Extender Whip Ended Cables, armoured 622–623
External Coupling, PVC trunking 587
External Lighting 660
  all-in-rates 112
  bulkhead, polycarbonate/aluminium body 660
  ceiling mounted 660
  circular bollard, polycarbonate visor 660
  floodlight 660
  ground mounted 660

Fabricated Hangers and Brackets, galvanized steel 220
Factories 93
fan coil unit system 77
Fan Coil Units 557
  ceiling void mounted 558
  low level perimeter 558

Fans
all-in-rates 109
axial flow fans 522
bifurcated 523
centrifugal 525
in-line centrifugal 524
roof fans 524
toilet fans – domestic 525
twin extract 524
FCU System 77
Feasibility Cost 93
Feed-In Tariffs (FITs) 51
Fencing, plant hire 738
Fibre Optic Cable, tight buffered, LSOH sheathed 694
Fibre Optic Connectors, single 695
Fibre Patch Leads 696
Filtered Socket Outlet, 13 amp, separate clean earth terminal, galvanized steel box 668
Filters, air
bag 526
carbon, grease 527
duct mounted 526
panel 526
Fire Control Panels 700
Fire Dampers 508
Fire Detection and Alarm 73
Fire Detection and Alarm Equipment, all-in-rates 126
Fire Extinguishers, CO_2, ABC 302
Fire Hose Reels, automatic; rubber, manual; rubber 293
Fire Hydrants 303
Fire Protection Control Panels 701
Fire-Rated Cable 613
Fit-out Cat A, elemental cost, office 134
Fittings
ABS, water 249
cast iron, EPDM 212
copper pipes 221
gutter, cast iron 173
moulded, cast iron 173
polypropylene Waste, push fit 194
Flange Connections, galvanized steel 259
Flanged Coupling, PVC trunking 587
Flanges, copper 233
Flat Elbow
mains busbar 625
rising mains busbar 627
Flat Oval Ductwork 441
Flat Tee, rising mains busbar 627
Flat Z Unit, mains busbar 625
Flexible Cable
300 volt grade, 2 core 611
300 volt grade, 3 core 611
300/500 volt grade, 2 core 611
use in high temperature 611
300/500 volt grade, 3 core 611
use in high temperature 612
300/500 volt grade, 4 core 611
rubber insulated
300/500 volt grade, 2 core 612
300/500 volt grade, 3 core 612
300/500 volt grade, 4 core 612
450/750 volt grade, 2 core 612
450/750 volt grade, 3 core 613
450/750 volt grade, 4 core 613
Flexible Conduits
metallic, PVC covered conduit, not exceeding 1 m long 580
PVC conduit, not exceeding 1 m long 580
Flexible Ductwork 446
Floor Drains, cast iron 216
Floor Outlet Box
single 13 amp Unswitched socket outlet, single compartment 669
twin 13 amp switched socket outlet
1 blank plate, 3 compartment 669
2 blank plate, 4 compartment 669
Floor Service Grommet, circular 669
Floor Trunking 585
Flue Supports 315
Flue System Fittings 313
Flue Systems
domestic 311
domestic/small commercial 318
flashings 316
Fluorescent Luminaires
batten type, surface mounted 657
surface mounted, opal diffuser 657
Flush Floor Trunking
four compartment, galvanized steel 585
triple compartment, galvanized steel 585
Flush Floor Trunking Fittings 585
Frames, floor standing,, vertical cable management 690
Fuel Cells System 25
benefits, types 26
markets, cost comparison 27
typical current project cost 27

Fused Switches Heavy Duty 653

Galvanized, heavy gauge, screwed drawn steel 579
Galvanized Steel Trunking 581
Gang Rates
- ductwork 161
- electrical 569
- mechanical 159
- public health 159

Gas Booster Sets 290
Gas Fire Engine 572
Gauges, temperature and pressure 365
Generating Sets, standby diesel 110
Generators
- all-in-rates 110
- standby 571

Getting Connection
- competition 48
- costs 48
- design 47
- network reinforcement 48
- planning 47

Goods Lifts, all-in-rates 121
Grey Water Recycling 15
Grilles
- air diffusion, plastic 515
- exhaust, aluminium 514
- floor 510
- supply, rectangular 508

Ground Water Cooling 19
GRP Cable Tray 640
GRP Tanks 276
Gutter Fittings, adaptors, angles, stop end, outlets 163
Gutters, cast iron 172

Hangers and Brackets
- fabricated 295
- galvanized steel 428

Heads, rainwater 168
- flat 187

Health and Welfare Facilities
- RIBA stage A feasibility cost 94
- RIBA stage B feasibility cost 94

Heat Emitters
- fan convectors 377
- perimeter heaters 374
- radiant strip heaters 374
- radiators
  - flat panel 376
  - pressed steel 374
- trench heating 379
- under floor heating 380
- unit heaters 417

Heat Exchangers
- CHW 429
- LTHW 372

Heat Pump 83
Heat Pumps 20
Heat Rejection 423
- air cooled condensers 424
- all-in-rates 108
- cooling towers 425
- dry air liquid coolers 423

Heater Batteries, local heating units 418
Heaters for Hot Water 284
Heavy Duty Fuse Switches, with HRC fuses 653
Heavy Fittings, galvanized steel 262
Heavy Flexible Cable
- 450/750 Volt grade, 2 core 613
- 450/750 Volt grade, 3 core 613
- 450/750 Volt grade, 4 core 613

High Impact PVC Heavy Gauge, standard pattern boxes 579
High Impact PVC Light gauge, standard pattern boxes 579
High/Low Bay Luminaires
- compact discharge, aluminium reflector 659
- corrosion resistant GRP body, acrylic diffuser 659
- flame proof, aluminium body 660
- sealed discharge, aluminium reflector 659

Hoists, plant hire 742
Horizontal Bend
- double compartment, zinc coated underfloor ducting 590
- triple compartment, zinc coated underfloor ducting 591

Hose Reels
- all-in-rates 125
- automatic; rubber, manual; rubber 293

Hospitals
- elemental cost 141
- RIBA stage A feasibility cost 94
- RIBA stage B feasibility cost 94

Hot Water, all-in-rates 106
Hotels
- elemental cost 140

RIBA stage A feasibility cost 97
RIBA stage B feasibility cost 97
RIBA stage C elemental rates 101
HRC Fused Distribution Boards, IP4X external protection enclosure 651

Industrial Angled Sockets
splash-proof, 100–130 amp, surface mounted 673
splash-proof, 200–250 amp, surface mounted 675
water tight, 100–130 amp, surface mounted 674
water tight, 200–250 amp 675
Industrial Connectors
splash-proof, 100–130 amp 672
splash-proof, 200–250 amp 674
water tight, 100–130 amp 673
water tight, 200–250 amp 675
Industrial Plugs
splash-proof, 100–130 amp 672
splash-proof, 200–250 amp 674
water tight, 100–130 amp 673
water tight, 200–250 amp 675
Inlet Boxes, steel dry riser 294
Insulations
ductwork – all types/finishes 529
mineral fibre for pipes, concealed 385
pipes – closed cell 279
Internal Coupling, PVC trunking 587
Intruder Detection System 75
Inverter, AC static to luminaires 680
IP54 Protection, mains busbar 624
Isolating Switches 654
IT / Data, all-in-rates 127

JIB (Joint Industry board) 569
JIB Apprentice Training Scheme Rate of Wages
London rate, rates with transport provided 724
national standard rate, rates with transport provided 724
JIB Rates of Wages
London rate, rates with transport provided 724
national standard rate, rates with transport provided 724
Jointing Sleeve, PVC underfloor trunking 590
Junction Box
four compartment, galvanized steel 585
triple compartment, galvanized steel 585
Junction Box Plates, double/triple compartment, zinc coated underfloor ducting 591
Junction Box/ Cross
double compartment, zinc coated underfloor ducting 592
triple compartment, zinc coated underfloor ducting 592
Junction Box/ Tee
double compartment, zinc coated underfloor ducting 591
triple compartment, zinc coated underfloor ducting 592
Junction Box/ Through or Angle
double compartment, zinc coated underfloor ducting 591
triple compartment, zinc coated underfloor ducting 591
Junction/Terminal Box
double compartment, zinc coated underfloor ducting 591
triple compartment, zinc coated underfloor ducting 591

Labour Rates
ductwork 158
electrical 567
mechanical 157
Ladder Racks
extra heavy duty galvanized steel 631
heavy duty galvanized steel 630
light duty galvanized steel 629
Lamp, T8 657
Leak Detection 432
LED Lighting 43
Leisure Centre
RIBA stage A feasibility cost 95
RIBA stage B feasibility cost 95
Lifting Appliances, plant hire 739
Lifts
all-in-rates 113
good lifts 121
machine room above 113
machine room less 117
Lighting, all-in-rates 112
Lighting Busbar, prewired busbar, plug in trunking for lighting 628
Lighting Control System 71
Lighting Controls
area controllers 665
cable, aluminium sheathed 664

head end equipment 665
lighting control modules 665
presence detectors 666
scene switch plate 666
Lighting Switches 661
Lighting Track
single circuit, 25 A 2 P661
three circuit, 25 A 2 P661
Lightning Protection
air rods 705
air terminals 705
bonds and clamps 705
conductor tape 704
earth pits 706
earth rods 707
surge Protection 708
Loaders, plant hire 746
Location Factors 87
Lock Equipment 697
Lodging Allowances, nightly rate 718
Louvres
acoustic 520
weather 520
Low Impedance Busbar Trunking, straight copper busbar 624
Low Temp' Hot Water Pipes
extra for black malleable fittings 328
extra over black steel pipes 326
mech' grooved steel pipe, fittings 347
plastic (polypropylene) pipe 348
fittings 348
press fit carbon steel pipe 345
fittings 345
screwed BS 1387 steel pipe 322
fixings 323
flanges 326
welded BS 1387 steel pipe 334
fittings 338
flanges 336
welded carbon steel 341
fittings 347
flanges 344
Low Velocity Air Conditioning, air handling units
supply 554
supply and extract 554
Luminaires
batten 657
bulkhead, 60 watt 660
ceiling recessed asymmetric 658
downlighters 659
emergency 678
external 660
high/low bay 659
surface fluorescent 657
suspended fluorescent 659
wall mounted fluorescent 659
LV Switchboard Busbar, copper busbar 646
LV switchgear
all-in-rates 110
distribution boards 643

Machine room above (MRA) Lift, all-in-rates 113
Mains Busbar 624
Major Projects Agreement, employee categories, hourly rate comparisons 720
Malleable Iron Fittings, galvanized steel 263
Manifold Connector, cast iron 213
Manual Call Points 700
Master Distribution Box
6 port, steel 620
9 port, steel 621
Material Cost, mechanical 157
Materials Cost, electrical 567
MCB Distribution Boards, IP3X external protection enclosure 650
MCCB Panel Boards 649
MCCB Panelboards
feeder MCCBs 649
panelboards 649
Measured Work Prices
electrical 567
mechanical 157
Mechanical 159
Mechanical Installations
labour rates 157
material cost 157
measured work prices 157
Mechanical Rates of Wages 715
Medium Density Polyethylene
MDPE (blue) 244
compression fittings 247
electro fusion fittings 245
MDPE (yellow), gas 286
compression fittings 286
electro fusion fittings 286
Metal Clad Cable
BSEN 60439 621
data cable 621
Meters Energy
CHW 434

LTHW, ultrasonic 382
Midel Transformer, gasket-sealed steel tank 577
Miniature Circuit Breakers, For distribution boards 650
Modular Recessed, T8 lamp 658
Modular Recessed Compact Fluorescent, TCL lamp 658
Modular Recessed, Linear Fluorescent, high frequency control gear 658
Modular Wiring 620
Moulded Case Circuit Breakers
MCCB 36 kA fault rated 645
MCCB 45 kA fault rated 645
MCCB 70 kA fault rated 645
MCCB 150 kA fault rated 645
MRL (Machine Room less) Lift, all-in-rates 117
Multi Purpose Leisure Centre
RIBA stage A feasibility cost 95
RIBA stage B feasibility cost 95
MV Cables, medium voltage 573
MV Circuit Breaker, 3 phase, 11 kV 576
MV switchgear, all-in-rates 110

Nipple, capillary fittings 225
Non-Standard Socket Outlet, 13 amp, separate clean earth terminal, galvanized steel box 668

Office Building, elemental cost 133
Offices 93
RIBA stage C elemental rates 98
Offset, mains busbar 625
Ogee Gutters, PVC-U 167
Open Loop System 19
Outlet Boxes, steel dry riser 294
Outlets
rainwater 163
roof 165
small power 667
Overflow, bath 197

P-traps 210
Packaged Steam Boilers, gas/oil fired 321
Packaged substation, all-in-rates 110
Passenger Lifts, all-in-rates 113
Passive Chilled Beam System 80
Patch Panels
19” wide, finished in black 692
voice 19” wide, finished in black 685
Performing Arts, elemental cost 137
Performing Arts Centre
RIBA stage A feasibility cost 95
RIBA stage B feasibility cost 95
Pipe Clips
copper pipes (CHW) 429
galvanized steel, plastic coated steel 197
metal fixing 189
Pipe Freezing, carbon steel, copper 381
Pipe in Pipe (Gas) 288
Pipeline Fittings
steam heating system 414
steel pipes 295
Pipes
ABS
waste 192
water 248
all-in-rates 106
black steel 322
screwed chilled water 428
screwed LTHW 322
welded LTHW 334
black steel screwed, gas 288
carbon steel 341
cast iron 183
soil EPDM rubber joints 206
copper, cold water 218
copper (plastic coated) 289
galvanized steel 256
galvanized steel, heavy 256
galvanized steel, medium 256
low temp’ hot water
extra for black malleable fittings 328
extra over black steel pipes 322
mech’ grooved steel pipe 347
plastic (polypropylene) pipe 348
press fit carbon steel pipe 345
screwed BS 1387 steel pipe 322
welded BS 1387 steel pipe 334
welded carbon steel 341
MDPE
blue 244
yellow 286
MDPE (yellow), gas
compression fittings 286
electro fusion fittings 286
mechanical grooved 347
MuPVC, plastic 189
polypropylene for LTHW 348
thermally fused joints 348
prefabricated black steel 295

press fit 345
carbon steel 345
stainless steel 242
PVC-C (cold water) 254
PVC-U
dry push fit joints 167
solvent welded joints 170
sprinkler, victaulic 297
stainless Steel 238
steel welded 288
Pipes Black Steel, Sprinkler Pipes 295
PIR, access control 697
Plant Charges 737
Plant Hire Costs 737
compaction equipment 747
concrete equipment 748
construction vehicles 744
excavators 746
hoists 742
lifting appliances and conveyors 739
pumps 738
scaffolding, shoring 738
site accommodation 749
site equipment 751
small tools 757
testing equipment 749
Plastic Ductwork, circular 447
Plug for Weatherproof socket outlet, protected to IP66 669
Plumbers Welding Supplement 721
Polypropylene, pipes, push fit joints 194
Power Factor Correction 647
Power Pole, aluminium painted 670
Power Post, aluminium painted, UPVC 670
Pressure Reducing Valves (PRV), steam 414
Pressurization Units
CHW 432
LTHW 366
Public Health Installations, gang rates 159
Pumps
accelerator 372
all-in-rates 109
belt drive – CHW 368
circulator 368
close coupled 369
glandless – domestic 371
plant hire 738
pressurization – cold water 273
sump 274
twin head belt drive 369
variable speed 370
Push Button Stations, heavy gauge pressed steel, polycarbonate cover 655
PVC Bench Trunking, white/grey finish 589
PVC Fittings, PVC-u (cold water) 252
PVC Miniature Trunking
compartmented, white finish 588
single compartment, white finish 588
PVC Miniature Trunking Fittings 589
PVC Underfloor Trunking, single compartment 590
PVC-C, PVC-u (cold water) 254

Radio Frequency Cable, PVC sheathed, laid loose 686
Rain Water Harvesting 15
Rainwater Gutters 163
Rainwater Pipework 163
Rates and Wages, hourly rates of wages, electrical 723
Rates of wages
hourly rates of wages, heating, ventilation 715
hourly rates of wages, main grades, plumbing, mechanical 719
mechanical 713
plumbing 719
trainee hourly rates of pay 716
RCD Protected Socket Outlet, 13 amp, galvanized steel box 669
Recessed Downlighter, reflector with white/chrome bezel 659
Recreation
RIBA stage A feasibility cost 95
RIBA stage B feasibility cost 95
Rectangular Ductwork 452, 482
Regional Variations 87
Renewable Energy 11
Renewable Energy Options
biomass boilers 13
borehole cooling 13
building integrated photovoltaic (BIPV) 12
combined heat and power (CHP) 13
ground source heat pumps 13
solar water heating 13
wind generators 12
Renewable Obligations Certificates (ROCs) 59
Research Buildings
RIBA stage A feasibility cost 97
RIBA stage B feasibility cost 97
Residential affordable, elemental cost 145

Residential Buildings, RIBA stage C elemental rates 103
Residential Facilities
RIBA stage A feasibility cost 97
RIBA stage B feasibility cost 97
Residential Private, elemental cost 147
Residual Current Circuit Breakers, For distribution boards 651
Responsibility Allowance
hourly rates of wages, Craftsmen 716
hourly rates of wages, Senior Craftsmen 716
weekly rates of credit 727
Retail
RIBA stage A feasibility cost 96
RIBA stage B feasibility cost 96
Reverse Cycle Heat Pump 83
RIBA Stage A Feasibility Cost 93
commercial research buildings 97
day care unit 95
department stores 96
distribution centres 93
educational buildings 96
educational research buildings 97
entertainment and recreation 95
factories 93
health and welfare facilities 94
hospitals 94
hotels 97
leisure centre 95
multi purpose leisure centre 95
office buildings 93
performing arts centre 95
recreation 95
research buildings 97
residential facilities 97
retail 96
schools 96
scientific buildings 97
shopping malls 96
sports hall 95
supermarkets 96
theatre 95
warehouses 93
RIBA Stage B Feasibility Cost 93
commercial research buildings 97
day care unit 95
department stores 96
distribution centres 93
educational buildings 96
educational research buildings 97
entertainment and recreation 95
factories 93
health and welfare facilities 94
hospitals 94
hotels 97
leisure centre 95
multi purpose leisure centre 95
office buildings 93
performing arts centre 95
recreation 95
research buildings 97
residential facilities 97
retail 96
schools 96
scientific buildings 97
shopping malls 96
sports hall 95
supermarkets 96
theatre 95
warehouses 93
RIBA Stage C Elemental Rates 98
offices 98
RIBA Stage C Feasibility Cost
hotels 101
residential 103
Ring Main Unit, 3 phase, 11 kV, 630 amp 576
Ring Main Units, all-in-rates 107
Rising Bend
double compartment, zinc coated underfloor ducting 591
four compartment, galvanized steel 585
triple compartment
galvanized steel 585
zinc coated underfloor ducting 591
Rising Mains Busbar, straight aluminium bar 626
Roof Outlets, PVC-U 165
Roof Vent Connector 215
Roses, ceiling 664

Scaffolding, plant hire 738
Schematic
11 kV network connection for LV intakes up to 1000 kVA 67
11 kV network connection for MV intakes 1000 kVA to 6000 kVA 68
access control system 74
active chilled beam system 80
CCTV 76
chilled beam system 80
chilled ceiling system 79

data transmission 70
digital CCTV 76
displacement ventilation system 78
fan coil unit system 77
FCU system 77
fire detection and alarm 73
heat pump 83
intruder detection system 75
lighting control system 71
passive chilled beam system 80
reverse cycle heat pump 83
standby power system 72
static UPS system 69
structured cabling 70
variable air volume system 81
variable refrigerent flow system 82
VAV system 81
VRF system 82
VRV system 82
School, elemental cost 143
Schools
RIBA stage A feasibility cost 96
RIBA stage B feasibility cost 96
Scientific Buildings
RIBA stage A feasibility cost 97
RIBA stage B feasibility cost 97
Semi Recessed Linear Fluorescent, T5 lamp 659
Service Outlet Box/Flat Lid
double compartment 592
triple compartment 592
Service Outlet Module
3 compartment 585
4 compartment 586
Shaver Sockets
250 voltage supply, white moulded plastic 671
dual voltage supply, white moulded plastic 671
Sheet Steel Adaptable Boxes
black rectangular pattern 593
black square pattern 593
galvanized square pattern 593
Shopping Mall, elemental cost 130
Shopping Malls
RIBA stage A feasibility cost 96
RIBA stage B feasibility cost 96
Shoring, plant hire 738
Single Compartment PVC Trunking, grey finish, clip on lid 586
Single Compartment PVC Trunking Fittings 586
Single Compartment Trunking, jointed with standard connectors, galvanized steel 581
Site Accommodation, plant hire 749
Site Equipment, plant hire 751
Small Power, all-in-rates 111
Small Tools, plant hire 757
Socket Outlet
switched, 13 amp metal clad, 1 667
unswitched, 13 amp metal clad, 1 667
Socket Plate, PVC bench trunking 590
Socket Reducer, PVC underfloor trunking 590
Sockets, outlets, waste, solvent joints 170
Soil and Waste Pipes
cast iron Pipes, rubber gasket joints 206
PVC-U 202
Sounders 700
Split Systems, for heating 435
Splitters, 5 wire 622
Sports Hall
elemental cost 138
RIBA stage A feasibility cost 95
RIBA stage B feasibility cost 95
Sprinkler Alarms/Gongs 299
Sprinkler Fixings 295
Sprinkler Heads 298
Sprinkler Systems, all-in-rates 125
Sprinkler Tanks 299
Sprinkler Valves 298
Stadium, elemental cost 139
Stage A RIBA 93
Stage B RIBA 93
Stage C RIBA 98
Stainless Steel Pipes 238
capillary fittings 239
compression fittings 241
press fit fittings 242
Standby Generators 571
Standby Power System 72
Static UPS System 69
Steam Strainers 404
Steam Traps
cast iron 414
stainless steel 414
Stop End
double compartment, zinc coated underfloor ducting 591
four compartment, galvanized steel 585
galvanized steel 581
PVC bench trunking 589
PVC miniature trunking 589
PVC trunking 586
triple compartment

galvanized steel 585
zinc coated underfloor ducting 591
Stopcocks, copper 272
Storage Calorifiers, Copper 282
Strainers 361
steam 414
Structured Cabling 70
Substation (Packaged), all-in-rates 110
Supermarket, elemental cost 132
Supermarkets
RIBA stage A feasibility cost 96
RIBA stage B feasibility cost 96
Support, cast iron, pipe spacer 183
Supports and Brackets, ductile iron, mild steel 211
Surface Mounted, Linear Fluorescent, T8 lamp 657
Surveillance Equipment 698
all-in-rates 126
Suspended Linear Fluorescent, T5 lamp 659
Switch Disconnectors
moulded Plastic Enclosure 672
steel sheet case 654
Switched Modules, armoured cable 622
Switches
6 amp modular type switch 661
galvanized steel box, 1 way 662
galvanized steel box, 2 way 662
6 amp, watertight switch 663
10 amp, splash-proof moulded switch 663
15 amp, watertight switch 663
architrave plate switch, 6 amp, white plastic 663
ceiling switch, white moulded plastic 663
plate switch, 6 amp flush mounted, white plastic 662
plate switch, 10 amp flush mounted, white plastic 662
Switches/ Sensors, pressure devices 709
Switches/Sensors, temperature devices 709
Switchgear, medium voltage 576
Synchronization Panels (for generators) 572

T Connectors, snap fix 622
Tanks, overflow 197
Tanks and Cisterns
cold water storage tanks
sectional GRP 276
sectional steel 275
fuel oil storage tanks; steel, fuel storage tanks; plastic 291
moulded glass fibre cisterns 275
polypropylene cisterns 275
sprinkler tanks
sectional GRP 300
sectional steel 299
Tap Off, mains busbar 625
Tap-Off Units
lighting busbar 628
rising mains busbar 627
Tee, Flat Cover, PVC trunking 588
Tee, Internal/External Cover, PVC trunking 588
Tees
double compartment, galvanized steel 583
PVC miniature trunking 589
single compartment, galvanized steel 583
triple compartment, galvanized steel 583
Telecommunications Cables, multipair internal telephone cable
Installed in conduit/Trunking 683
laid on tray/basket 683
Telephone Outlet
bronze or satin chromeplate and box 684
moulded plastic plate and box 684
Television Aerial Cable, coaxial, PVC Sheathed 686
Terminal Strips, 20 amp high density polythene 580
Terminations
600/1000 volt grade, 2 core, galvanized steel wire armour 599, 607
600/1000 volt grade, 3 core, galvanized steel wire armour 600, 607
600/1000 volt grade, 4 core, galvanized steel wire armour 601, 608
600/1000 volt grade, 7 core, galvanized steel wire armour 601, 608
600/1000 volt grade, 12 core, galvanized steel wire armour 601, 608
600/1000 volt grade, 19 core, galvanized steel wire armour 601, 609
600/1000 volt grade, 27 core, galvanized steel wire armour 601, 609
600/1000 volt grade, 37 core, galvanized steel wire armour 601, 609
600/1000 volt grade, single core, aluminium wire armour 599, 607
glanding off
2 core 618
3 core 618–619
4 core 618, 620
medium voltage 573

Terminations,, glanding off, 2 core 619
Testing Equipment, plant hire 749
The Energy Technology List (ETL) 39
The Water Technology List (WTL) 39
Theatre
RIBA stage A feasibility cost 95
RIBA stage B feasibility cost 95
Thermal Insulation
concealed insulation fittings 385
ductwork 532
external insulation fittings 404
extra over concealed insulation fittings 385
extra over external insulation fittings 404
extra over plantroom insulation fittings 392
mineral fibre for pipes
external 404
plant room 392
pipes – closed cell 279
plant room insulation fittings 392
Thermostat, room thermostats 382
Thermostats 382
Top Feeder Unit, rising mains busbar 626
Trace Heating System 430
Transfer Switches 647
Transformer
medium voltage 576
oil filled, free breathing tank 577
Transformers
all-in-rates 110
step down 577
Traps, polypropylene 195
Travelling 725
national standard rate, daily travelling allowance 725
Tray, GRP 640
Trench heating
fan assisted type 380
natural convection type 379
Triple Compartment Trunking, jointed with standard connectors, galvanized steel 581
Tripping Batteries, battery chargers, valve regulated lead acid battery 577
Trunking 581
dado/skirting 592
double compartment, jointed with standard connectors 581
flanged connector, galvanized steel 581
flush floor, triple compartment 585
galvanized Steel 581
PVC 586
single compartment 581
stop end, galvanized steel 581
triple compartment, jointed with standard connectors 581
underfloor 590
Tubular Fittings, galvanized steel 261
TV Socket Outlet, moulded plastic box, flush/surface mounted 687
Typical Engineering Details 63
urban network underground 64
urban network underground cost information 64
urban network underground w/ reinforcement 65
urban network underground w/ reinforcement cost information 65
urban network underground w/ substation reinforcement 66
urban network underground w/ substation reinforcement cost information 66

Un-Armoured Cable 609
Underfloor Busbar, prewired busbar, plug in trunking for power 628
Underfloor Ducting
double compartment, zinc coated underfloor ducting 590
fittings, zinc coated underfloor ducting 591
triple Compartment, zinc coated underfloor ducting 590
Underfloor heating 380
Uninterruptible Power Supply
integral diesel rotary, 3 phase 677
large size static, 3 phase, 10 year battery 676
medium size static, 1 phase, 10 year battery 676
medium size static, 3 phase, 10 year battery 676
single phase, 5 year battery, standard 13 amp socket outlet 676
UPS 676
all-in-rates 110

Valve ancillaries, steam heating system 414
Valves
automatic air vent 364
ball float valves 270
ball valves 356
bronze/gunmetal inlet 294
check valves 358

check valves – DZR 271
cocks 271, 363
commissioning valves 359
control valves 362
gate valve– DZR 272
gates 270
globe valves 357
isolating 354
pressure reducing 416
radiator valves 31
regulators 362
safety and relief valves 363, 416
stopcocks 271, 364
strainers 414
Variable Air Volume System 81
Variable Refrigerent Flow System 82
VAV System 81
VAV – Variable Air Volume System, terminal units 555
Vertical Access Box, 2 Compartment, PVC underfloor trunking 590
Vertical Multi Compartment Pillar, UPVC 670
Vertical Transport, all-in-rates 113
Video Cable, PVC flame retardant sheathed, laid loose 687
VRF System 82
VRF Units, refrigerant 435
VRV System 82

Wages, hourly rates of wages 715
Wall Mounted Compact Fluorescent, TCL lamp 659
Wall Spacer, cast iron 211
Warehouses 93
Water Heater (Electric) 285
Water Heater Switch, switched with neon indicator 672
Water Installations, all-in-rates 106
Water Treatment 277
Weatherproof Junction Boxes
glassfibre reinforced polycarbonate 655
sheet steel with zinc spray finish 655
Weekly Holiday Credit 717
rates 717
Weekly Holiday Credit Contributions 721
Work Station
data outlet 692

Ebook Single-User Licence Agreement

We welcome you as a user of this Spon Price Book ebook and hope that you find it a useful and valuable tool. Please read this document carefully. **This is a legal agreement** between you (hereinafter referred to as the "Licensee") and Taylor and Francis Books Ltd. (the "Publisher"), which defines the terms under which you may use the Product. **By accessing and retrieving the access code on the label inside the front cover of this book you agree to these terms and conditions outlined herein. If you do not agree to these terms you must return the Product to your supplier intact, with the seal on the label unbroken and with the access code not accessed.**

1. Definition of the Product

The product which is the subject of this Agreement, (the "Product") consists of online and offline access to the VitalSource ebook edition of *Spon's Mechanical & Electrical Services Price Book 2014*.

2. Commencement and licence

2.1 This Agreement commences upon the breaking open of the document containing the access code by the Licensee (the "Commencement Date").

2.2 This is a licence agreement (the "Agreement") for the use of the Product by the Licensee, and not an agreement for sale.

2.3 The Publisher licenses the Licensee on a non-exclusive and non-transferable basis to use the Product on condition that the Licensee complies with this Agreement. The Licensee acknowledges that it is only permitted to use the Product in accordance with this Agreement.

3. Multiple use

For more than one user or for a wide area network or consortium, use of the Product is only permissible with the purchase from the Publisher of additional access codes..

4. Installation and Use

4.1 The Licensee may provide access to the Product for individual study in the following manner: The Licensee may install the Product on a secure local area network on a single site for use by one user.

4.2 The Licensee shall be responsible for installing the Product and for the effectiveness of such installation.

4.3 Text from the Product may be incorporated in a coursepack. Such use is only permissible with the express permission of the Publisher in writing and requires the payment of the appropriate fee as specified by the Publisher and signature of a separate licence agreement.

4.4 The Product is a free addition to the book and the Publisher is under no obligation to provide any technical support.

5. Permitted Activities

5.1 The Licensee shall be entitled to use the Product for its own internal purposes;

5.2 The Licensee acknowledges that its rights to use the Product are strictly set out in this Agreement, and all other uses (whether expressly mentioned in Clause 6 below or not) are prohibited.

6. Prohibited Activities

The following are prohibited without the express permission of the Publisher:

6.1 The commercial exploitation of any part of the Product.

6.2 The rental, loan, (free or for money or money's worth) or hire purchase of this product, save with the express consent of the Publisher.

6.3 Any activity which raises the reasonable prospect of impeding the Publisher's ability or opportunities to market the Product.

6.4 Any networking, physical or electronic distribution or dissemination of the product save as expressly permitted by this Agreement.

6.5 Any reverse engineering, decompilation, disassembly or other alteration of the Product save in accordance with applicable national laws.

6.6 The right to create any derivative product or service from the Product save as expressly provided for in this Agreement.

6.7 Any alteration, amendment, modification or deletion from the Product, whether for the purposes of error correction or otherwise.

7. General Responsibilities of the License

7.1 The Licensee will take all reasonable steps to ensure that the Product is used in accordance with the terms and conditions of this Agreement.

7.2 The Licensee acknowledges that damages may not be a sufficient remedy for the Publisher in the event of breach of this Agreement by the Licensee, and that an injunction may be appropriate.

7.3 The Licensee undertakes to keep the Product safe and to use its best endeavours to ensure that the product does not fall into the hands of third parties, whether as a result of theft or otherwise.

7.4 Where information of a confidential nature relating to the product of the business affairs of the Publisher comes into the possession of the Licensee pursuant to this Agreement (or otherwise), the Licensee agrees to use such information solely for the purposes of this Agreement, and under no circumstances to disclose any element of the information to any third party save strictly as permitted under this Agreement. For the avoidance of doubt, the Licensee's obligations under this sub-clause 7.4 shall survive the termination of this Agreement.

8. Warrant and Liability

8.1 The Publisher warrants that it has the authority to enter into this agreement and that it has secured all rights and permissions necessary to enable the Licensee to use the Product in accordance with this Agreement.

8.2 The Publisher warrants that the Product as supplied on the Commencement Date shall be free of defects in materials and workmanship, and undertakes to replace any defective Product within 28 days of notice of such defect being received provided such notice is received within 30 days of such supply. As an alternative to replacement, the Publisher agrees fully to refund the Licensee in such circumstances, if the Licensee so requests, provided that the Licensee returns this copy of *Spon's Mechanical & Electrical Services Price Book 2014* to the Publisher. The provisions of this sub-clause 8.2 do not apply where the defect results from an accident or from misuse of the product by the Licensee.

8.3 Sub-clause 8.2 sets out the sole and exclusive remedy of the Licensee in relation to defects in the Product.

8.4 The Publisher and the Licensee acknowledge that the Publisher supplies the Product on an "as is" basis. The Publisher gives no warranties:

- 8.4.1 that the Product satisfies the individual requirements of the Licensee; or
- 8.4.2 that the Product is otherwise fit for the Licensee's purpose; or
- 8.4.3 that the Product is compatible with the Licensee's hardware equipment and software operating environment.

8.5 The Publisher hereby disclaims all warranties and conditions, express or implied, which are not stated above.

8.6 Nothing in this Clause 8 limits the Publisher's liability to the Licensee in the event of death or personal injury resulting from the Publisher's negligence.

8.7 The Publisher hereby excludes liability for loss of revenue, reputation, business, profits, or for indirect or consequential losses, irrespective of whether the Publisher was advised by the Licensee of the potential of such losses.

8.8 The Licensee acknowledges the merit of independently verifying the price book data prior to taking any decisions of material significance (commercial or otherwise) based on such data. It is agreed that the Publisher shall not be liable for any losses which result from the Licensee placing reliance on the data under any circumstances.

8.9 Subject to sub-clause 8.6 above, the Publisher's liability under this Agreement shall be limited to the purchase price.

9. Intellectual Property Rights

9.1 Nothing in this Agreement affects the ownership of copyright or other intellectual property rights in the Product.

9.2 The Licensee agrees to display the Publishers' copyright notice in the manner described in the Product.

9.3 The Licensee hereby agrees to abide by copyright and similar notice requirements required by the Publisher, details of which are as follows:

"© 2014 Taylor & Francis. All rights reserved. All materials in *Spon's Mechanical & Electrical Services Price Book 2014* are copyright protected. All rights reserved. No such materials may be used, displayed, modified, adapted, distributed, transmitted, transferred, published or otherwise reproduced in any form or by any means now or hereafter developed other than strictly in accordance with the terms of the licence agreement enclosed with *Spon's Mechanical & Electrical Services Price Book 2014*. However, text and images may be printed and copied for research and private study within the preset program limitations. Please note the copyright notice above, and that any text or images printed or copied must credit the source."

9.4 This Product contains material proprietary to and copyedited by the Publisher and others. Except for the licence granted herein, all rights, title and interest in the Product, in all languages, formats and media throughout the world, including copyrights therein, are and remain the property of the Publisher or other copyright holders identified in the Product.

10. Non-assignment

This Agreement and the licence contained within it may not be assigned to any other person or entity without the written consent of the Publisher.

11. Termination and Consequences of Termination.

11.1 The Publisher shall have the right to terminate this Agreement if:

11.1.1 the Licensee is in material breach of this Agreement and fails to remedy such breach (where capable of remedy) within 14 days of a written notice from the Publisher requiring it to do so; or

11.1.2 the Licensee becomes insolvent, becomes subject to receivership, liquidation or similar external administration; or

11.1.3 the Licensee ceases to operate in business.

11.2 The Licensee shall have the right to terminate this Agreement for any reason upon two month's written notice. The Licensee shall not be entitled to any refund for payments made under this Agreement prior to termination under this sub-clause 11.2.

11.3 Termination by either of the parties is without prejudice to any other rights or remedies under the general law to which they may be entitled, or which survive such termination (including rights of the Publisher under sub-clause 7.4 above).

11.4 Upon termination of this Agreement, or expiry of its terms, the Licensee must destroy all copies and any back up copies of the product or part thereof.

12. General

12.1 *Compliance with export provisions*

The Publisher hereby agrees to comply fully with all relevant export laws and regulations of the United Kingdom to ensure that the Product is not exported, directly or indirectly, in violation of English law.

12.2 *Force majeure*

The parties accept no responsibility for breaches of this Agreement occurring as a result of circumstances beyond their control.

12.3 *No waiver*

Any failure or delay by either party to exercise or enforce any right conferred by this Agreement shall not be deemed to be a waiver of such right.

12.4 *Entire agreement*

This Agreement represents the entire agreement between the Publisher and the Licensee concerning the Product. The terms of this Agreement supersede all prior purchase orders, written terms and conditions, written or verbal representations, advertising or statements relating in any way to the Product.

12.5 *Severability*

If any provision of this Agreement is found to be invalid or unenforceable by a court of law of competent jurisdiction, such a finding shall not affect the other provisions of this Agreement and all provisions of this Agreement unaffected by such a finding shall remain in full force and effect.

12.6 *Variations*

This agreement may only be varied in writing by means of variation signed in writing by both parties.

12.7 *Notices*

All notices to be delivered to: Spon's Price Books, Taylor & Francis Books Ltd., 3 Park Square, Milton Park, Abingdon, Oxfordshire, OX14 4RN, UK.

12.8 *Governing law*

This Agreement is governed by English law and the parties hereby agree that any dispute arising under this Agreement shall be subject to the jurisdiction of the English courts.

If you have any queries about the terms of this licence, please contact:

Spon's Price Books
Taylor & Francis Books Ltd.
3 Park Square, Milton Park, Abingdon, Oxfordshire, OX14 4RN
Tel: +44 (0) 20 7017 6000
www.sponpress.com

Spon Press
Taylor & Francis Group

Multiple-user use of the Spon Press Software and eBook

For more than one user or for a wide area network or consortium, use of *Spon's Online* is only permissible with the purchase from the Publisher of a multiple-user licence and adherence to the terms and conditions of that licence.

To buy a licence and set up access to *Spon's Online* for a number of users of in your organisation please contact:

Spon's Price Books
eBooks & Online Sales
Taylor & Francis Books Ltd.
3 Park Square, Milton Park, Oxfordshire, OX14 4RN
Tel: +44 (0) 207 017 6000
www.pricebooks.co.uk

Number of users	**Licence cost**
2	£270
5	£620
10	£1200
other	Please contact Spon for details